NEAR–WALL TURBULENCE

PROCEEDINGS OF THE INTERNATIONAL CENTRE FOR HEAT AND MASS TRANSFER
N. H. AFGAN, EDITOR
J. T. ROGERS, SENIOR CONSULTING EDITOR

ICHMT Belgrade

1. Afgan and Schlünder **Heat Exchangers: Design and Theory Sourcebook**
2. Afgan and Beer **Heat Transfer in Flames**
3. Blackshear **Heat Transfer in Fires**
4. deVries and Afgan **Heat and Mass Transfer in the Biosphere**
5. Denton and Afgan **Future Energy Production Systems**
6. Hartnett **Alternative Energy Sources**
7. Spalding and Afgan **Heat Transfer and Turbulent Buoyant Convection**
8. Zarić **Thermal Effluent Disposal From Power Generation**
9. Hoogendoorn and Afgan **Energy Conservation in Heating, Cooling, and Ventilating Buildings**
10. Durst, Tsklauri, and Afgan **Two-Phase Momentum, Heat and Mass Transfer in Chemical, Process, and Energy Engineering Systems**
11. Spalding and Afgan **Heat and Mass Transfer in Metallurgical Systems**
12. Jones **Nuclear Reactor Safety Heat Transfer**
13. Bankoff and Afgan **Heat Transfer in Nuclear Reactor Safety**
14. Zarić **Structure of Turbulence in Heat and Mass Transfer**
15. Taborek, Hewitt, and Afgan **Heat Exchangers: Theory and Practice**
16. Metzger and Afgan **Heat and Mass Transfer in Rotating Machinery**
17. Afgan **Measurement Techniques in Power Engineering**
18. Soloukhin and Afgan **Measurement Techniques in Heat and Mass Transfer**
19. Abramson **Guidebook to Light Water Reactor Safety Analysis**
20. van Swaaij and Afgan **Heat and Mass Transfer in Fixed and Fluidized Beds**
21. Radovanovic **Fluidized Bed Combustion**
22. Mori, Sheindlin, and Afgan **High Temperature Heat Exchangers**
23. Sheindlin **High Temperature Equipment**
24. Bougard and Afgan **Heat and Mass Transfer in Cryoengineering and Refrigeration**
25. Afgan **Transient Phenomena in Multiphase Flow**
26. Spalding and Afgan **Heat and Mass Transfer in Gasoline and Diesel Engines**
27. Markatos **Computer Simulation for Fluid Flow, Heat and Mass Transfer, and Combustion in Reciprocating Engines**
28. Kline and Afgan **Near-Wall Turbulence: 1988 Zoran Zarić Memorial Conference**

NEAR–WALL TURBULENCE

1988 Zoran Zarić Memorial Conference

Edited by

S. J. Kline
Mechanical Engineering Department
Stanford University, California

N. H. Afgan
International Centre for Heat and Mass Transfer
Institute Boris Kidric
P.O. Box 522
11001 Belgrade, Yugoslavia

HEMISPHERE PUBLISHING CORPORATION
A member of the Taylor & Francis Group

New York   Washington   Philadelphia   London

NEAR-WALL TURBULENCE: 1988 Zoran Zarić Memorial Conference

1 2 3 4 5 6 7 8 9 0 BRBR 8 9 8 7 6 5 4 3 2 1 0 9

Library of Congress Cataloging-in-Publication Data

Zoran Zarić Memorial Conference (1988 : Dubrovnik, Croatia)
Near-wall turbulence : 1988 Zoran Zarić Memorial Conference / edited by S. J. Kline, N. H. Afgan.
p. cm.—(Proceedings of the International Centre for Heat and Mass Transfer : 27)
Includes index.

1. Fluid dynamics—Congresses. 2. Turbulence—Congresses. 3. Boundary value problems—Congresses. 4. Zarić, Zoran—Congresses. I. Kline, S. J. (Stephen Jay), date. II. Afgan, Naim. III. Zarić, Zoran. IV. Title. V. Series.
TA357.Z67 1988
620.1'064—dc20 89-11233
CIP

ISBN 0-89116-733-1

Contents

Part 3: QUASI–COHERENT STRUCTURES

Part 4: PRESSURE FLUCTUATIONS AND RELATED NEAR–WALL EFFECTS

Part 8: PRACTICAL PREDICTIONS BY TWO-EQUATION MODELS AND OTHER FAST METHODS

Part 9: INSTRUMENTATION METHODS IN WALL TURBULENCE

Introduction

As a memorial to Professor Zoran Zarić, The International Centre for Heat and Mass Transfer of Belgrade, Yugoslavia, decided to organize a conference that would provide an up-to-date summary of Professor Zarić's field—Near-Wall Turbulence. This volume is the Proceedings of that Conference held in Dubrovnik, Yugoslavia, May 16–20, 1988.

The volume presents papers that give representative coverage to the various topics of research and design interest concerning near-wall turbulence. At the suggestion of The International Centre for Heat and Mass Transfer, the parts are organized by a leading expert in the relevant area. This part editor provides an overview and invites other authors to round out the topic. The part editors are thus primarily responsible for the high quality of the papers appearing in this volume. The papers taken together, I believe, provide a good representation of the state of knowledge concerning near-wall turbulence in 1988.

Thanks are also due to the Seminar Organization Committee and to the staff of the International Centre for Heat and Mass Transfer, particularly Dr. Joel Risnić, for the vital planning and support of the Conference and the papers presented, which made the meeting both successful and enjoyable for participants.

S. J. Kline, Editor
Chairman, Seminar Organization Committee

Zoran Zarić—
Luykov Medal of ICHMT

After this presentation, as the chairman of the Executive Committee, on behalf of all the ICHMT, its President, its Scientific Council and all the scientific community, Michel Combarnous presented to Madame Danka Zarić, who was with her son and daughter, Marko and Milana, the Luykov Medal.

This highest award given by the center is awarded to Zoran Zarić for both his scientific contribution in the field of heat and mass transfer and his permanent action in favor of international cooperation.

Opening Address

It is my duty to say some words about Zoran Zarić.

I can try to do so, as one of his numerous friends, but, in this respect, I don't pretend to be the best qualified. . . because he really had a large number of very good friends all over the world. So I will do it mainly on behalf of the Société Française des Thermiciens, which I am representing in the Executive Committee of the Center.

It is really very difficult to recall personal memories, thus I intend, very briefly, to comment on two aspects of the warm personality of Zoran: his international activities which are best represented in terms of the work he did for the center, and part of his scientific activity.

Starting with the international activities of Zoran Zarić, let me first recall for you that, in his country, he quickly reached an important scientific position that was very useful in the initial development of the center. For more than 25 years he was a professor at the University of Belgrade and was a corresponding Member of the Serbian Academy of Science.

A good indication of the deep interest Zoran Zarić had in various cultures and people was his impressive ability in foreign languages. He was fluent in English, Russian, and French. He spent a lot of time in France and was sensitive to the various accents in different parts of the country, to the evolution of the language, even including "slang" words, and popular expressions.

Through all the contacts he had with various people, telling jokes in various languages, he was always sensitive to national feelings and professional hopes.

So we should not be surprised to note the very determined action of Zoran Zarić in the formulation of the idea of the center. We will recall, during our 20th anniversary this summer in Dubrovnik, how a very active group started the idea of a better cooperation between west and east, as we used to say in the 1960s and north and south as we say today.

Professor Zarić, acting as the General Secretary, with the help of a small local group, combined his effort with that of Professor Naim Afgan, acting as Scientific Secretary, during all these years to make the idea of a center a reality. Some personal aspects of the first period of the center have been nicely recalled by Brian Spalding in three compact pages on Zoran Zarić in the December 1986 issue of the *International Journal of Heat and Mass Transfer,* and it would be difficult to do better.

During the decade between 1964 and 1974, Zoran Zarić worked both in France and with the Heat Transfer laboratory at the Boris Kidrić Institute for Nuclear Energy; an institute with which he has been always very strongly connected.

This period started with two years of study in France which resulted in a thesis in 1964. The thesis was mainly experimental, on heat transfer between a fluid and the wall of a convergent-divergent tube.

Since that time the work was mainly focused on near-wall turbulence, as can be seen in the papers that have been published in the three languages mentioned before, French, English, and Russian.

Originally, the work was to use a very simple experimental technique, a unique hot-wire anemometer, to measure at some points, at various distances of the wall, fluid velocities, temperatures, and their fluctuations. Then to treat all these data by some statistical approach to put into evidence intermittent phases along the wall.

All this work has been included in a general synthesis, the thesis of Doctorat d'Etat which is the highest degree we had in universities in France which was presented in 1974.

As long as our friend Zoran Zarić stays alive in our memory, I am sure that, in addition to tremendous individual and national competitions, each of us, in some part of ourself, will try to work for international cooperation between structures and individuals as well as greater achievements of human society.

Michel Combarnous
Chairman of the ICHMT Executive Committee

Early Contributions by Zoran Zarić on Near-Wall Turbulence

1964 Etude expérimentale de la convection forcée de la chaleur dans un conduit divergent-convergent—*Thése,* Univ. Paris.

Heat transfer intensification by use of the longitudinally variable pressure gradient—Proc. III UN Conf. Atomic Energy, p. 698

1967 Turbulent heat transfer in a divergent-convergent channel—Proc. JSME Symposium, Tokyo, 161–170

1968 Turbulentnyi teploobmen v kanale s peremennym prodol'nimgradientom davleniia; *Teplo-i Massoperenos,* v. 9, Nauka i Tehnika, Minsk, 36–56

Measurements close to the wall in a turbulent boundary layer—Intern. Sum. School, Herceg Novi in "*Heat and Mass Transfer in Boundary Layers*", Pergamon Press, 1972, 555–571

1969 Distributions de probabilité des signaux provenant d'un fil chaud en écoulement turbulent—C. R. Acad. Sc. *A 269,* 986–989

Statistical interpretation of the velocity measurements in high intensity turbulence—International Seminar, Herceg Novi.

1970 Caractéristiques statistiques d'écoulement turbulent prés des parois chauffées avec gradient de pression longitudinal—Proc. 4th Intern. Heat Transfer Conf., Versailles, v. 2, FC 3.10

1972 Statisticheskii analiz pristennoi'turbulentnosti; *Teplo-i Massoperenos,* v. 9-2, Minsk

Wall turbulence studies in *Advances in Heat Transfer,* v. 8, 285–350, Academic Press

Distribution de probabilité des vitesses et des températures prés d'une paroi—C. R. Acad. Sc., *A 275,* 459–462

Analyse stastistique de la turbulence prés d'une paroi par échantillonnage conditionnel—C. R. Acad. Sc., *A 275,* 513–515

1973 Statistical analysis of wall turbulence phenomena—II IUTAM-IUGG Symp. on Turbulent Diffusion, Charlottesville

1974 Contribution à l'étude statistique de la turbulence pariétale—*Doctorat d'Etat,* Univ. de Paris VI

Seminar Organization Committee

PART ORGANIZERS:

Parviz Moin, Professor (Part I)
Department of Mechanical Engineering
Stanford University
Stanford, CA, USA,
and NASA Ames Research Center
Moffett Field, CA, USA

Peter Bradshaw, Professor (Part 2)
Department of Aeronautics
Imperial College of Science and Technology
London, UK

S. J. Kline, Professor (Part 3)
Department of Mechanical Engineering
Stanford University
Stanford, CA, USA

Dr. Helmut Eckelmann (Part 4)
Max-Planck Institute für Strömungsforschung
Göttingen, FRG

Ronald J. Adrian, Professor (Part 5)
Department of Theoretical and Applied Mechanics
University of Illinois
Urbana, IL, USA

R. Dumas (Part 5)
Institut de Mécanique Statistique de la Turbulence
Université d'Aix-Marseille II
Unité Mixte Univresité, CNRS 380033,
France

E. M. Khabakhpasheva (Part 6)
Institute of Thermophysics
Siberian Branch of the USSR Academy of Sciences
Novosibirsk, USSR

John L. Lumley, Professor (Part 7)
Sibley School of Mechanical & Aerospace Engineering
Cornell University
Ithaca, NY, USA

Kemal Hanjalić, Professor (Part 8)
Mašinski Fakultet
Sarajevo, Yugoslavia

William G. Tiederman (Part 9)
School of Mechanical Engineering
Purdue University
West Lafayette, IN, USA

Part 1
LARGE EDDY AND FULL SIMULATIONS

SUMMARY LECTURE

Similarity of Organized Structures in Turbulent Shear Flows

PARVIZ MOIN
Department of Mechanical Engineering
Stanford University
Stanford, California 94305, USA

and

NASA Ames Research Center
Moffett Field, California 94035, USA

Recent analysis of databases generated by direct numerical simulations of *homogeneous* turbulent shear flows have revealed the presence of coherent structures similar to those in turbulent boundary layers. In this note I will summarize these findings and discuss tentative conclusions on their significance.

BACKGROUND

All the studies reported here used direct numerical simulation databases with either 128^3 or $128 \times 128 \times 512$ mesh points. Spectral methods were used in the Navier-Stokes calculations that generated the databases.

Using the results of turbulent channel flow simulations, Moin & Kim (1985) obtained direct evidence for the existence of hairpin or horseshoe vortices in wall-bounded turbulent flows as proposed by Theodorsen. These conclusions were based on the probability density of the inclination angle of instantaneous vorticity vectors, two-point correlations of velocity and vorticity, and visualizations of instantaneous vorticity vectors and vortex line patterns. The horseshoe vortices were observed in the regions away from the walls. It was apparent that the vortices were formed from the roll-up of layers of transverse vorticity and that the presence of walls was not essential for their formation. In other words, random velocity fluctuations and stretching by the mean rate of strain are sufficient ingredients for generation of hairpins. It was then conjectured that hairpin vortices are a characteristic of all shear flows, and not just boundary layers.

Recently, this conjecture passed a severe test in homogeneous shear flow. Rogers & Moin (1987), using the same statistical and flow visualization techniques as in the channel flow study, showed that hairpin vortices are a common feature of homogeneous turbulent shear flow. The similarity of organized structures was manifested in favorable comparison of the two-point velocity and vorticity correlations with various directions of "probe" separation, the histograms of the vorticity inclination angles, and one-point turbulent statistics (see below).

TABLE 1. Second order statistics in turbulent channel flow at $y/\delta = 0.5$

u'/w'	u'/v'	ω'_x/ω'_y	ω'_x/ω'_z	$-\overline{uv}/u'v'$
1.38	1.63	1.06	1.06	0.48

TABLE 2. Second-order statistics in homogeneous turbulent shear flow

	u'/w'	u'/v'	ω'_x/ω'_y	ω'_x/ω'_z	$-\overline{uv}/u'v'$
RM:	1.33	1.85	1.12	1.18	0.57
TC:	1.34	1.67	-	-	0.45

In contrast to the regions far from the wall, where hairpin vortices are found, the dominant features in the region near the wall are the wall-layer streaks. These are elongated regions of high and low speed fluid alternating in the spanwise direction. The flow fields from simulations of Rogers & Moin did not reveal these streaky structures. Recognizing that the wall-layer is under the influence of strong mean shear, Lee, Kim & Moin (1987) conjectured that streaky structures might be found in homogeneous shear flow if the flow is subjected to sufficiently high shear. In particular, if the non-dimensional shear parameter, Sq^2/ϵ, is comparable to that found in the inner region of turbulent boundary layers, such structures might be found. Here, $S = \frac{\partial U}{\partial y}$ is the uniform mean shear, q^2 is the turbulent kinetic energy, and ϵ is the dissipation rate. The implicit assumption was that the same kind of similarity between wall-turbulence and homogeneous shear flow observed by Rogers & Moin can be obtained for high values of non-dimensional shear rate. Indeed, this was the case. Numerical simulations of homogeneous shear flow with (high) non-dimensional shear rates typical of those found in the wall layer were carried out. Contours of streamwise velocity fluctuations in an (x, z) plane at $y^+ \approx 10$ from the channel flow were in qualitative agreement with the velocity contours in (x, z) planes from this homogeneous flow. In both cases elongated regions of high and low speed fluid alternating in the spanwise direction were apparent. Moreover, the non-dimensional spacing of the streaks was in fair agreement between the two flows.

It is important to point out that one-point, second-order statistical correlations also show a remarkable similarity between the homogeneous shear flow and channel flow. This similarity in statistical correlations is attributed to the dominance of the same organized structures in both flows. Table 1 shows the ratios of turbulent intensities (u'/v', u'/w'), the correlation coefficient $-\overline{uv}/u'v'$, and the ratios of vorticity fluctuation intensities (ω'_x/ω'_y, ω'_x/ω'_z), for the channel flow at $y/\delta = 0.5$, typical of the outer values. The channel data is taken from Kim, Moin & Moser (1987) where the Reynolds number based on the channel half-width and centerline mean velocity is 3300. The non-dimensional shear parameter, Sq^2/ϵ, is approximately 6 in the range $0.3 < y/\delta < 0.7$. The non-dimensional shear parameter in the high Reynolds number simulation of Rogers & Moin (1987) is 8.65 at the

TABLE 3. Second-order statistics in turbulent channel flow at $y^+ = 10$

u'/w'	u'/v'	ω'_x/ω'_y	ω'_x/ω'_z	$-\overline{uv}/u'v'$
2.95	8.09	0.69	0.51	0.48

TABLE 4. Second-order statistics in high shear homogeneous flow (St=8).

u'/w'	u'/v'	ω'_x/ω'_y	ω'_x/ω'_z	$-\overline{uv}/u'v'$
2.78	8.9	1.16	.64	0.82

non-dimensional time $St = 8.0$, and in the experiments of Tavoularis & Corrsin (1981) is about 12 at dimensionless times 8.6-12.7. The same non-dimensional, second-order statistics as in Table 1 are shown in Table 2 for the (low shear rate) homogeneous shear flow data of Rogers & Moin and Tavoularis & Corrsin. Clearly, there is considerable similarity between the data in the two tables. It is also interesting to note that the scalar flux coefficients $\overline{\theta u}/\theta' u'$, and $\overline{\theta v}/\theta' v'$ are 0.6 and 0.5, respectively, in the outer layers of turbulent boundary layer experiments of Chen & Blackwelder (1978) and are 0.64 and 0.46 in simulations of Rogers, Moin & Reynolds (1986).

Turbulent statistics from the channel flow at $y^+ \approx 10$ and from the (high shear rate) homogeneous shear flow of Lee *et al.* (1987) are shown in Tables 3 and 4. At $y^+ \approx 10$, $Sq^2/\epsilon = 35$ and in the homogeneous flow, at $St = 8$ and 12, it is 52 and 61, respectively. The significant change in the statistics at high shear rate from those in Tables 1 and 2 is apparent. The direction of changes are the same in the channel and homogeneous shear flow. In particular, the streamwise component of turbulent intensity is significantly higher than the normal and spanwise components. Note that there is a better agreement between the turbulent intensities in the two flows at $St = 8$, whereas the vorticity ratios agree more favorably at $St = 12$, i.e., at higher *total* shear.

DISCUSSION

Our computations indicate that in homogeneous shear flow, when the non-dimensional shear rate, Sq^2/ϵ, is comparable to the (high) values in the inner region of the wall layer, streaky structures are observed. On the other hand at lower values of Sq^2/ϵ, hairpin-like structures characteristic of the outer regions of turbulent boundary layers dominate the flow. The similarity of organized structures is reflected in good agreement of the second-order statistics.

It may then be argued that the wall layer streaks are the consequence of high values of the local mean shear and not the presence of the solid boundary. The experiments of Uzkan & Reynolds (1967) provide further justification for this assertion. In their experiments, flow over a moving belt was considered. It was observed that when the speed of the belt matched that of the free-stream (i.e., turbulent flow over a wall with no mean shear), wall layer streaks disappeared. Except in the immediate vicinity of the wall (y^+ =1 to 2), the large anisotropy of turbulent intensities is also largely due to the high shear rate near the wall, rather than the wall suppression effect.

Homogeneous shear flow is the most basic of shear flows and the observed similarity with the much more complicated turbulent channel flow is noteworthy. It is likely that different shear flows exhibit the same kind of resemblance at comparable non-dimensional shear rates. The rate of mean shear is the key parameter providing, to a large extent, the source of vorticity for the formation of organized vortical motions and influencing their subsequent dynamics.

We do not suggest that all shear flows can be reduced to a number of homogeneous shear flows with different shear rates. Other effects such as diffusion and turbulent transport, are important in inhomogeneous flows. In addition, some flows may be specially influenced by an instability mechanism originated from the physical boundaries (e.g., Kelvin-Helmholtz instability in mixing layers). However, these extra effects may be accounted for separately. For example, in wall-bounded flows it is probable that the effects of the solid boundary can be accounted for by separate consideration of the shear-free turbulent boundary layer (Uzkan & Reynolds 1967).

The results outlined here point to a degree of universality among turbulent shear flows that has not been considered previously. This may have favorable implications for development of more widely applicable transport models of turbulence.

This work was supported in part by the United Air Force Office of Scientific Research.

REFERENCES

Chen, C. P. & Blackwelder, R. F., 1978, Large-Scale Motion in a Turbulent Boundary Layer: a Study Using Temperature Contamination. *J. Fluid Mech.* **89,** pp. 1-31.

Kim, J., Moin, P. & Moser, R. D. , 1987, Turbulence in Fully Developed Channel Flow at Low Reynolds Number. *J. Fluid Mech.* **177,** pp. 133-166.

Lee, M. J., Kim, J. & Moin, P., 1987, Turbulence Structure at High Shear Rate. *Sixth Symposium on Turbulent Shear Flows.* Toulouse, France, Sept. 7-9, 1987.

Moin, P. & Kim, J., 1985, The Structure of the Vorticity Field in Turbulent Channel Flow. Part 1. Analysis of Vorticity Fields and Statistical Correlations. *J. Fluid Mech.* **155,** pp. 441-464.

Rogers, M. M., & Moin, P., 1987, The Structure of Vorticity Field in Homogeneous Turbulent Shear Flows. *J. Fluid Mech.* **176**, pp. 33-66.

Rogers, M. M., Moin, P. & Reynolds, W. C., 1986, The Structure and Modeling of Passive Scalar Fields in Homogeneous Turbulent Shear Flow. *Report TF-25,* Department of Mechanical Engineering, Stanford Univ.

Tavoularis, S. & Corrsin, S., 1981, Experiments in Nearly Homogeneous Turbulent Shear Flow with a Uniform Mean Temperature Gradient. Part 1 . *J. Fluid Mech.* **104**, pp. 311-347.

Uzkan, T. & Reynolds, W. C., 1967 , A Shear-Free Turbulent Boundary Layer. *J. Fluid Mech.* **28**, pp. 803-821.

Near-Wall Phenomena in Transition to Turbulence

N. GILBERT and L. KLEISER
DFVLR
Institute for Theoretical Fluid Mechanics
D-3400 Göttingen, FRG

INTRODUCTION

Laminar-turbulent transition is a classical problem of fluid mechanics receiving currently a renewed interest due to recent efforts to achieve viscous drag reduction by laminar flow control, e.g. on aircraft wings. Numerous experimental and theoretical studies have been devoted to the particular case of "ribbon-induced" transition in simple wall-bounded shear flows, such as the Blasius boundary layer and the plane channel flow. Here, with small initial disturbances, transition proceeds through a sequence of regular, increasingly complex stages initiated by flow instabilities [11,19]. The first stages of the transition process are now quite well understood. Transition begins with the amplification of two-dimensional (2D) Tollmien-Schlichting (TS) waves described by the (primary) linear stability theory. If the 2D wave amplitude exceeds some critical threshold, 3D disturbances are amplified due to a secondary instability [5,16]. After the 3D disturbance amplitudes have attained the order of magnitude of the 2D wave, nonlinear interactions take place which are very difficult to analyse theoretically. In recent years, direct numerical simulations have been employed successfully by several researchers to clarify some further stages of the transition development [3,7,10,16,18,20,21]. Many of the features known from experimental investigations in boundary layer and channel flows have been reproduced. Among these are the formation of a spanwise peak-valley structure of the streamwise rms fluctuation, the appearance of high-shear layers connected with travelling Λ-vortex structures, and low-velocity spike signals of the instantaneous streamwise velocity. However, many questions have still been left unanswered. Little is known about the final stages of transition, where the origin of wall turbulence is of particular interest.

In the present work, the classical peak-valley splitting (K-type) mode of transition in plane Poiseuille flow is simulated. Our objective has been to investigate the transition development following the early spike stages as far as possible towards the turbulent regime. We demonstrate that our simulation does cover the complete transition process, though some reservations must be made concerning the simulation of the fully developed turbulent state. In the present contribution, we first give an overview of the transition development and then report some findings concerning near-wall phenomena during late transition. A more detailed documentation of the results is given in [4].

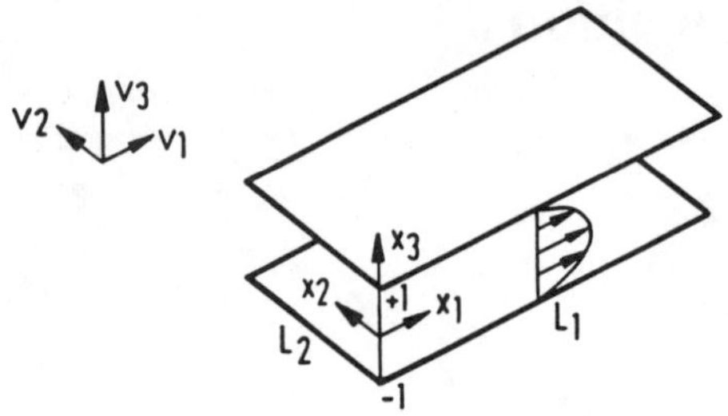

FIGURE 1. Channel flow geometry and coordinates.

NUMERICAL SIMULATION MODEL

The three-dimensional time-dependent incompressible Navier-Stokes equations are solved numerically in the spatial domain $0 \leq x_j \leq L_j$ ($j = 1,2$), $|x_3| \leq 1$ (fig. 1). No-slip conditions are applied at the walls $x_3 = \pm 1$ and periodic boundary conditions in the horizontal directions x_1 and x_2 (all variables are non-dimensionalized with channel half-width h and undisturbed laminar centerline velocity U_{CL}). In analogy with vibrating ribbon experiments, the evolution of the velocity field $\underline{v}(\underline{x}, t)$ from an initial disturbance superimposed on the basic Poiseuille flow $U(x_3) = 1\text{-}x_3^2$ is calculated. In our simulation we use the model of a streamwise and spanwise periodic flow developing in time in a frame of reference moving downstream with the phase velocity of the TS wave. Accordingly, the roles of x_1 and t are exchanged to compare experimental and numerical results in the laminar-turbulent transition phase. This model is now well established and has led to good agreement with experimental data at corresponding stages of transition development in Poiseuille flow [3,7,21] and in boundary layers [10,18,20,21].

In nonlinear stability and transition computations the treatment of the mean flow needs to be specified. In our simulations the mean mass flux is kept fixed by adjusting the mean pressure gradient at each time step. In this way, a rapid evolution on a convective time scale from the transitional stages to the developed turbulent state is made possible. If, instead, the mean pressure gradient would be prescribed, the horizontally averaged velocity $<v_1>$ could develop only on a viscous time scale [3,15], leading to prohibitively large computing times. Furthermore, a constant mass flux in our simulation does correspond to the experimental situation.

The numerical method is essentially the same as described in [7]. The spatial discretization is based on a spectral method [1] with Fourier expansions in the horizontal coordinates,

$$\underline{v}(\underline{x}, t) = \sum_{k_1} \sum_{k_2} \hat{\underline{v}}(k_1, k_2, x_3, t) \exp(ik_1\alpha_1 x_1 + ik_2\alpha_2 x_2) \qquad (1)$$

($\alpha_j = 2\pi/L_j$), and Chebyshev polynomial expansions in the normal coordinate. The sum in (1) extends over the wave number range $|k_j| < N_j/2$. In the spanwise direction the symmetry condition

$$v_j(-x_2) = v_j(x_2) \ (j=1,3), \quad v_2(-x_2) = -v_2(x_2) \qquad (2)$$

is imposed by using the appropriate cosine/sine expansions in x_2 instead of the general form (1). This reduces the computing time and storage by a factor of two (N_2 always refers to the number of grid points on the full spanwise periodicity length L_2). It may be shown that the symmetry (2) of an initial velocity field will be preserved by the Navier-Stokes equations for all $t > 0$. The transitional flows under consideration show this symmetry both in experiments and theoretical analysis. In a real turbulent flow, however, we expect that this symmetry will be broken and the use of (2) is not immediately justified.

For time integration, viscous terms are treated implicitly by the Crank-Nicolson method, and nonlinear terms explicitly by the Adams-Bashforth method. As usual, the nonlinear terms are calculated by the pseudospectral approximation. Aliasing errors in the Fourier-Chebyshev coefficients of the products are avoided by applying the 3/2 rule [1]. The pressure is calculated from a Poisson equation with the correct boundary conditions obtained from the condition of vanishing divergence at the boundary, using an influence matrix technique [7,1]. The continuity equation and boundary conditions are satisfied exactly by the discretized equations. The solution is obtained by solving sequentially a set of one-dimensional Helmholtz equations discretized by the Chebyshev Tau method. The implementation of the influence matrix method has been improved by exploiting properties of Chebyshev expansions [4]. A high efficiency of the code is obtained by vectorizing the innermost loop which runs over the (k_1, k_2) plane. The typical CPU time per grid point and time step (using one processor) is $10\mu s$ on a Cray X-MP and $18\mu s$ on a Cray 2 computer. With $N_1 \times N_2 \times N_3 = 128^3$ grid points the code needs 32 million words of core memory if no auxiliary storage is used.

RESULTS

The transition simulation starts from laminar channel flow with superimposed small initial disturbances consisting of a 2D TS wave with maximum amplitude 0.03 and a pair of oblique waves with amplitude 0.001. The physical parameters $Re_Q = 5000$, $\alpha_1 = 1.12$ and $\alpha_2 = 2.1$ are chosen to match the experimental conditions of Nishioka et al. [12,13]. Here, $Re_Q = 1.5\tilde{v}_Q h/\tilde{\nu}$ denotes the mass flux Reynolds number based on the volume-averaged velocity $\tilde{v}_Q$. The numerical resolution in the horizontal directions is gradually increased from $N_1 \times N_2 = 24 \times 32$ at $t = 0$ to 24×48, 32×64, 64×128 and finally to 128×128 at the simulation times $t = 100$, 120, 132 and 137, respectively. In the normal direction we choose $N_3 = 128$ throughout. The time step Δt varies from 0.05 initially down to 0.005 with the highest resolution during the transition phase with largest disturbance amplitudes. Up to the one-spike stage ($t \approx 128$) the present results agree with the earlier ones [3] obtained with lower resolution.

Overview of Transition Process

Based on our earlier results, and anticipating results discussed below in more detail, we distinguish four main phases of the laminar-turbulent transition process in our simulation. For later reference, these are characterized briefly in table 1. The spike stages have been identified in our simulation mainly by com-

TABLE 1. The four main phases of the transition simulation

Phase A	$(0 \leq t \leq 100)$	Initial disturbance growth Evolution of 2D Tollmien-Schlichting wave 3D disturbance growth due to secondary instability
Phase B	$(100 \leq t \leq 150)$	Formation of 3D transitional flow structures Λ-vortices, high-shear layers, spike stages t = 128 1-spike stage t = 136 2-spike stage t = 140 3-spike stage t = 144 4-spike stage t = 148 5-spike stage
Phase C	$(150 \leq t \leq 200)$	Breakdown of regular transitional structures culmination of disturbance growth, beginning of wall turbulence
Phase D	$(t \geq 200)$	Turbulent state

paring the instantaneous streamwise velocity in the "peak" plane with the corresponding experimental data by Nishioka et al. [12,13]. The spike stages extend to $t \approx 150$ in our simulation. The regular three-dimensional flow structures which have formed in the early (i.e., one and two) spike stages begin to decay in the late spike stages. The subsequent breakdown phase covers the time interval $150 \leq t \leq 200$. The transition from a highly disturbed, but in some sense still laminar, flow to a state with typical turbulent properties takes place within this breakdown phase. Violent fluctuations occur with maximum values at $t \approx 170$ by far exceeding the later, turbulent level. Towards the end of the breakdown phase the transitional flow structures have disappeared and a statistically steady, turbulent state is approached. In order to compare statistical data from phase D with known turbulent channel flow data the simulation was continued up to $t = 320$. However, the turbulent phase will be covered only briefly in the present contribution.

To get an overview of the transition development we first consider the evolution of mean flow data. Fig. 2 shows the development of the averaged centerline and wall shear velocities in terms of the Reynolds numbers $Re_{CL} = <\tilde{v}_1>_{CL} \tilde{h}/\tilde{\nu}$ and $Re_\tau = \tilde{v}_\tau \tilde{h}/\tilde{\nu}$. Here $<.>$ denotes an average over the horizontal coordinates x_1 and x_2. The wall shear velocity is calculated from the averaged velocity gradient at both channel walls. In dimensionless notation we have

$$Re_{CL} = <v_1>_{CL} Re_Q \tag{3a}$$

$$Re_\tau = v_\tau Re_Q = \sqrt{Re_Q | \partial <v_1> / \partial x_3 |_W} \,. \tag{3b}$$

In fig. 2, Re_τ and Re_{CL} remain at their laminar values up to the early spike stages. From $t = 140$ to $t = 170$ a rapid change takes place. Re_τ increases to a peak

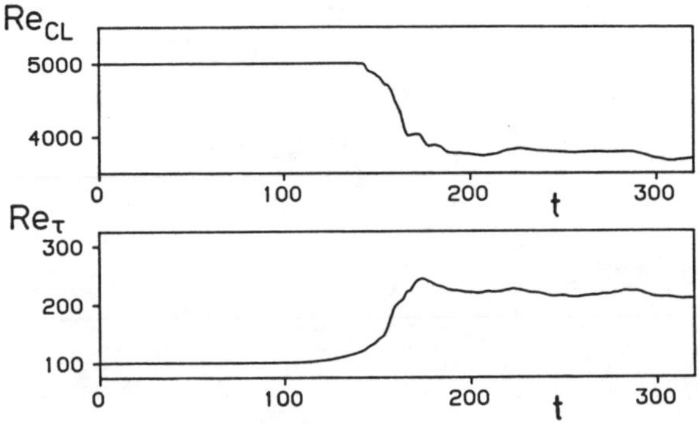

FIGURE 2. Development of $Re_{CL} = <\tilde{v}_1>_{CL}\tilde{h}/\tilde{\nu}$ and $Re_\tau = \tilde{v}_\tau \tilde{h}/\tilde{\nu}$ during the laminar-turbulent transition process.

value of 244 at t = 174 and later fluctuates around a time-mean value of ≈217. Correspondingly, Re_{CL} drops sharply from the laminar value $Re_{CL} = Re_Q = 5000$ to a time-mean value of ≈3800. These time-mean values are in excellent (Re_{CL}), or at least fair (Re_τ), agreement with experimental correlations [14,17].

Fig. 3a displays the mean velocity profiles at different times of the simulation. Up to t≈120 there is little deviation from the parabolic laminar profile. At t = 140 the mean profile has inflection points. During $140 \le t \le 170$ the shape of the profile changes rapidly. As the total mass flux is kept fixed, the increase of the wall shear stress, and thereby of the near-wall mass flux, requires the profile to become more flat in the channel interior. At t = 170 the mean velocity distribution already resembles the turbulent profile. For $t > 200$ only slight temporal variations occur. Fig. 3b shows the time-averaged mean profile $\ll v_1 \gg$ where $\ll . \gg$ denotes a time average of $<.>$ over $210 \le t \le 320$ (4.8 time units $\tilde{h}/\tilde{v}_\tau$). In addition, the experimental data of Eckelmann [2] are included which were obtained at the Reynolds number Re_{CL} = 4100 (Re_τ = 209). If both sets of

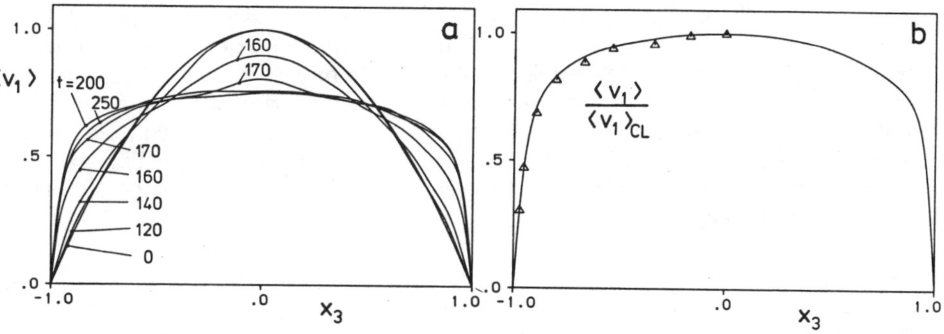

FIGURE 3. (a) Development of the mean velocity profile $<v_1>$ during transition. (b) Normalized mean turbulent profile: —— simulation, averaged over $210 \le t \le 320$; Δ Experiment Eckelmann [2].

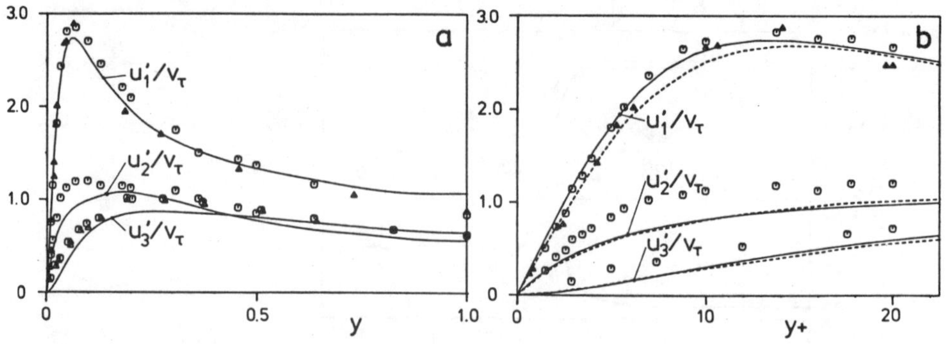

FIGURE 4. RMS fluctuations in the turbulent phase. — present results, - - - simulation Kim et al. [6], experiments Eckelmann [2] (Δ) and Kreplin and Eckelmann [9] (o).

data are normalized to unity in the channel center they are in quite close agreement.

The rms fluctuations $u_i' = \sqrt{\ll(v_i - \ll v_i \gg)^2 \gg}$ normalized with v_τ are given in fig. 4a in one channel half and in fig. 4b in the near-wall region. Agreement as well as discrepancies are found with the experimental data of Eckelmann [2] at $Re_\tau = 209$ and Kreplin & Eckelmann [9] at $Re_\tau = 194$. Differences occur with respect to the peak and centerline values of u_1'/v_τ. The normal and spanwise velocity fluctuations are higher in the experiments, especially near the wall. In the turbulent channel flow simulation of Kim et al. [6] at $Re_\tau = 180$ the same discrepancies with experimental data have been observed in u_2' and u_3' near the wall. Their results are also included in fig. 4b. Good agreement is found between the two simulations for the normal and spanwise components. The deviation of the experimental data may be caused by fundamental difficulties with hot-film measurements near walls and/or by probe interference. The gradient of u_1' from [6] and the maximum value are somewhat lower than ours. The fact that in our simulation the spanwise component u_2' falls slightly below the normal component u_3' near the channel center may be due to the symmetry condition (2) which together with periodicity forces u_2' to vanish at the edges of the spanwise integration domain $0 \leq x_2 \leq L_2/2$.

Some other statistical data, such as Reynolds stress, skewness and flatness factors, have also been evaluated in phase D. Generally, good agreement with available data is found outside the immediate vicinity of the wall, but some differences are observed close to the wall and in the interior of the channel. Two-point correlations of the velocity components become small with increasing separation near the wall only. Clearly, the periodicity lengths, which are adequate in the transitional stages, are too small in the turbulent range ($L_1^+ \approx 1220$ and $L_2^+ \approx 650$ in wall units).

In order to check whether the spatial resolution of our simulation is sufficient we have plotted the spectra of the Fourier and Chebyshev expansions. Fig. 5 shows the Fourier energy spectra, defined by

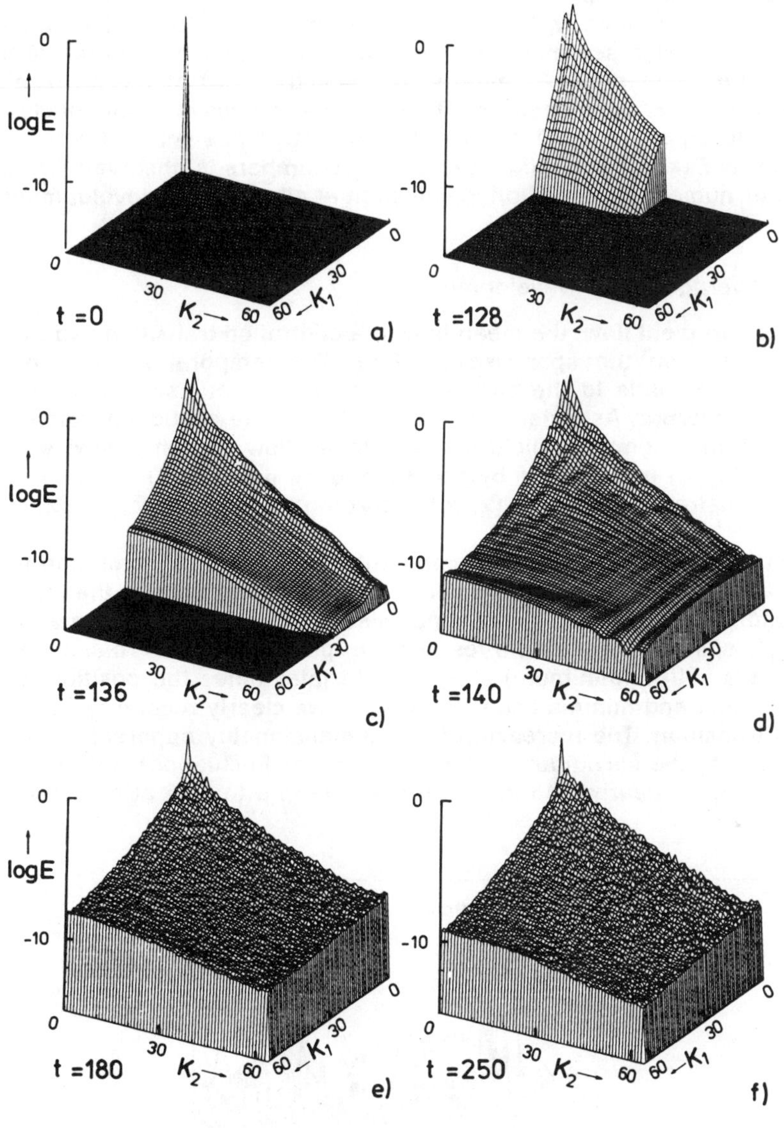

FIGURE 5. Energy $E(k_1, k_2)$ of horizontal Fourier modes.

$$E(k_1, k_2, t) = \frac{1}{2} \sum_{j=1}^{3} \int_{-1}^{1} |\hat{u}_j(k_1, k_2, x_3, t)|^2 dx_3 , \tag{4}$$

at six times of the simulation. In (4) u_j denotes the velocity disturbance defined by $\underline{u} = \underline{v}\text{-}\underline{U}$ where $\underline{U} = (1\text{-}x_3^2, 0, 0)$ is the laminar channel flow. We observe a practically monotonic decay with increasing wavenumbers at all stages. The energy at the highest wavenumbers never exceeds 10^{-8} during the simulation.

It is interesting to note that up to $t \approx 136$ the energy decreases faster in k_1 than in k_2 direction (this is independent of the cut-off in k_1 and k_2). For this reason, a higher spanwise resolution $N_2 > N_1$ has been used during the initial stages of the simulation. We also see that the energy level is higher around $t \approx 180$ than in the later, turbulent phase. Similarly, the spectra $E(k_3)$ of the horizontally averaged energy have been checked. The energy also decays continuously with k_3, e.g. down to an order of 10^{-8} at $t = 170$ and 10^{-10} at $t = 250$. No accumulation of energy $E(k_1, k_2)$ and $E(k_3)$ at large wavenumbers is observed. We conclude that our numerical resolution is sufficient at all stages of development.

Streamwise-Averaged Flow Development

In contrast to turbulent flow, the mean flow in a controlled transition experiment strongly depends on the spanwise position. The temporal average of the experiment corresponds to the streamwise average in our simulation model, denoted by an overbar. As $\underline{\overline{v}}$ depends only on two of the spatial coordinates, it is easy to obtain a complete picture of the mean flow. The streamwise mean component $\overline{v}_1(x_2, x_3)$ is distorted by the secondary flow $(\overline{v}_2, \overline{v}_3)$ which may be described by a stream-function $\psi(x_2, x_3)$ according to $\partial\psi/\partial x_3 = \overline{v}_2$, $\partial\psi/\partial x_2 = -\overline{v}_3$.

Before discussing the mean flow development we consider the evolution of the fluctuation about this mean. Fig. 6 displays the maximum (across the channel) rms fluctuation $u_1'{}_{max}(x_2, t)$ at different spanwise positions. We observe an early "peak-valley" splitting of the curves with a peak (spanwise maximum) at $x_2 = L_2/2$ and a valley (minimum) at $x_2 = 0$. At later times the position of the spanwise maxima and minima changes. In fig. 6 we clearly recognize the phases A - D of transition. The increasing three-dimensionality apparent in phases A and B is due to the secondary instability. The rms fluctuations near the peak position grow up to nearly 0.15 at $t \approx 140$ and even up to 0.25 at $t \approx 150$. At the

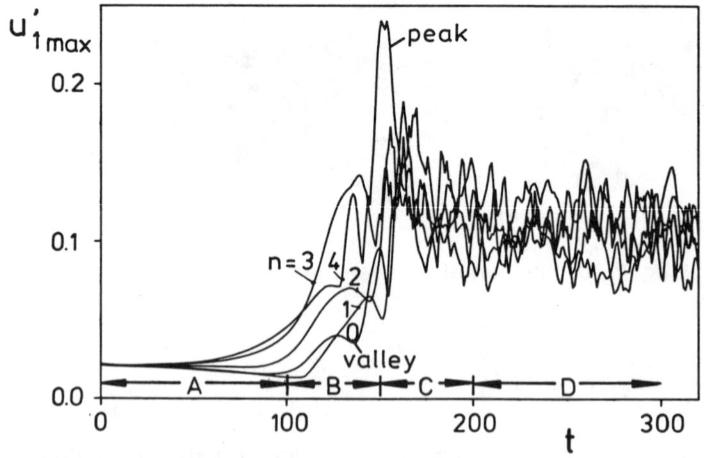

FIGURE 6. Development of maximum streamwise rms fluctuation $u_1'{}_{max}$ at spanwise positions $x_2 = nL_2/8$.

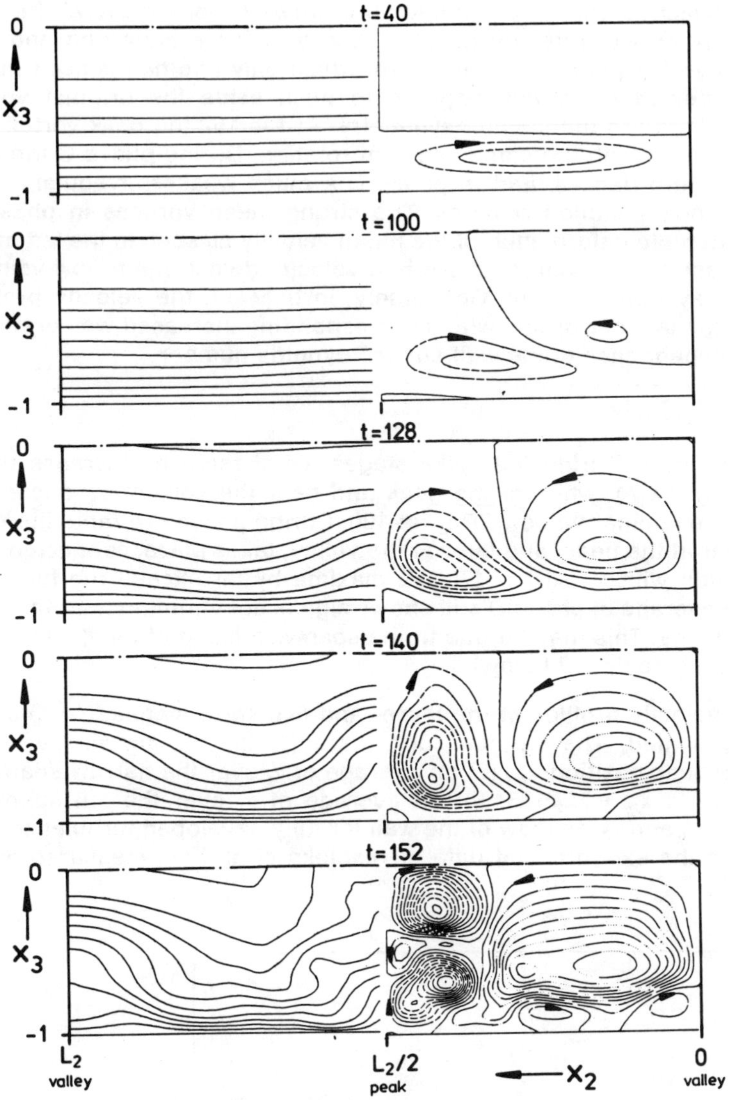

FIGURE 7. Streamwise-averaged velocity in lower channel half. Left: contours of streamwise mean velocity $\overline{v}_1$ (increment is 0.1), right: streamlines ψ = *const* of mean secondary flow ($\overline{v}_2$, $\overline{v}_3$).

valley the fluctuations are much lower in phase B, 0.06 at $t \approx 140$ and 0.10 at $t \approx 150$. In phase C the regular spanwise dependence is lost and the fluctuations settle down to the average turbulent level $u_1'{}_{max} \approx 0.12$ of phase D.

Next we consider the mean flow development. Fig. 7 shows isotachs of $\overline{v}_1$ (left half of spanwise period) and streamlines of the mean secondary flow (right half) in the lower channel half. According to (2), the distribution of $\overline{v}_1$ is mirror-symmetric and ψ is antisymmetric about the peak position $x_{2p} = L_2/2$. We observe a vortex system which is very weak initially but grows exponentially at the secondary instability growth rate. In the phases A and B we have one and two counterrotating vortex pairs, respectively (of which only one half is seen in the figure). The vortex at the valley appears to push aside the original vortex towards the peak and to induce an asymmetry. At t = 152 the peak vortex has split into two vortices with the same sense of rotation. During phase C the regular vortex structure decays, and in phase D a much weaker, irregular, time-dependent secondary motion remains. The strong mean vortices in phase B induce a considerable deformation of the mean velocity as seen in the left of fig. 7. At the peak and valley stations there is a velocity defect due to low-velocity fluid moving away from the wall. Conversely, in between the velocity profiles are fuller than the laminar profile with a corresponding increased wall gradient. The spanwise dependence of the wall-shear Reynolds number

$$\overline{Re_\tau} = \sqrt{Re_Q |\partial \bar{v}_1 / \partial x_3|_W} \tag{5}$$

is displayed in fig. 8. During the spike stages we observe an increase up to $\overline{Re_\tau} \approx 150$ near $x_2 = L_2/4$, while at the peak and near the valley $\overline{Re_\tau}$ decreases slightly below the laminar value $\sqrt{2Re_Q} = 100$. During phase C a rapid increase of the wall shear at the peak, and later at the valley, takes place, connected with a strong variation with x_2. At t = 170 the maxima by far exceed the turbulent level. The time average of phase D still shows significant variation near the peak and valley positions. This may be due to the spanwise fixing of the flow field by the symmetry (2) (see fig. 12 below).

In fig. 9 mean velocity profiles at the three- and five-spike stages (t = 140 and t = 148) are provided. The various curves are (1) the streamwise-averaged profile $\bar{v}_1$ at the peak position x_{2p}, (2) an average of $\bar{v}_1$ over the narrow spanwise interval from x_{2p} to $x_{2p} \pm 7L_2/128$, (3) an average of $\bar{v}_1$ over the full spanwise period, i.e. $<v_1>$, and (4) the law of the wall for fully developed turbulence. The symbols denote the experimental data of Nishioka et al. [12] measured at the

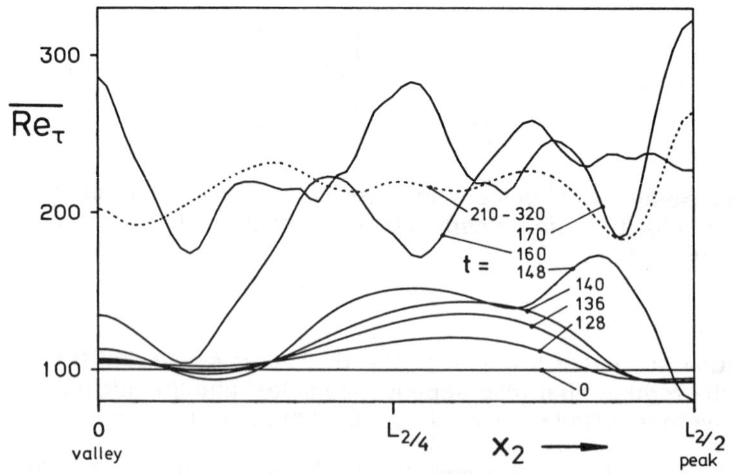

FIGURE 8. Spanwise variation of $\overline{Re_\tau}$ (x_2) based on locally streamwise-averaged wall shear at different times t in phases A - C, and time-average of phase D.

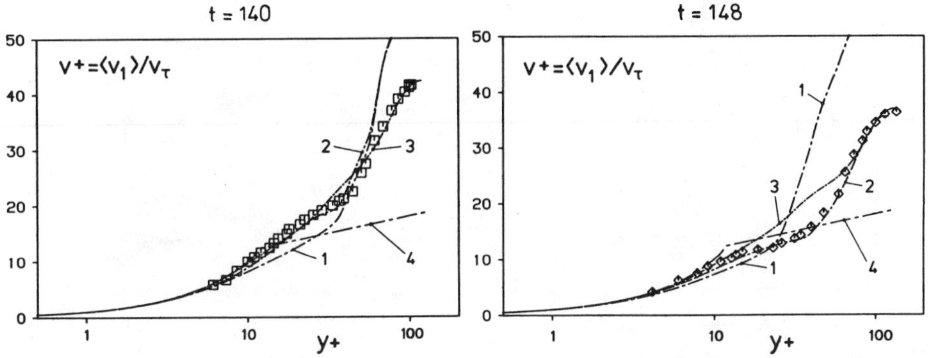

FIGURE 9. Locally and globally averaged velocity profiles (see text). Symbols: Experiment Nishioka et al. [12], 3-spike and 5-spike stage.

nominal peak position. They considerably disagree with the calculated profiles (1) at the peak. A possible explanation may be as follows. As we know from figs. 7 and 8, $\bar{v}_1$ and (at $t = 148$) $\bar{v}_\tau$ show a strong spanwise variation near x_{2p} so that a deviation from the nominal peak position, or an averaging over a small spanwise interval, would produce significant changes in the measured mean profile. A spanwise averaging could be produced by a slight spanwise meandering in time of the transitional structures, which is sometimes observed in transition experiments (the spanwise averaging zone for curves (2) corresponds to only 1.2mm ($0.16h$) in the experiment [12]). This was our motivation to include curves (2) in fig. 9, which are found to be much closer to the experimental data than curves (1). At $t = 140$ the averaged profile (2) agrees with the experiment up to $y_+ \approx 50$, whereas near the channel center the calculated profile is still more laminar than the experimental one. The globally averaged profile (3) agrees in the channel interior but does not reproduce the characteristic kink at the edge of the "buffer layer". At $t = 148$ good agreement of profile (2) with the experimental data is obtained over the full channel half. The profiles (1) and (3) show similar deviations as at the 3-spike stage.

Formation and Decay of the High-Shear Layer

In fig. 10 the instantaneous velocity profiles $v_1(x_3)$ over the downstream period in the peak plane are shown at different stages of the transition process. At $t = 100$ small deviations from the laminar, parabolic shape can be observed. At the one-spike stage ($t = 128$) an inclined shear layer appears at the height $0.4 \leq |x_3| \leq 0.65$, extending over one downstream period. During the spike stages the profiles become highly distorted in the interior of the channel, whereas in the near-wall region no strong distortions are visible. Within the breakdown phase of the transition process ($t > 150$) this behaviour changes rapidly. As observed already for the mean profiles (fig. 3a), the instantaneous profiles tend to a nearly rectangular shape. The velocity gradient at the wall increases to four times its laminar value. In the bulk of the channel ($|x_3| \leq 0.9$) the flow becomes more calm and only small localized regions with significant velocity gradients are observed.

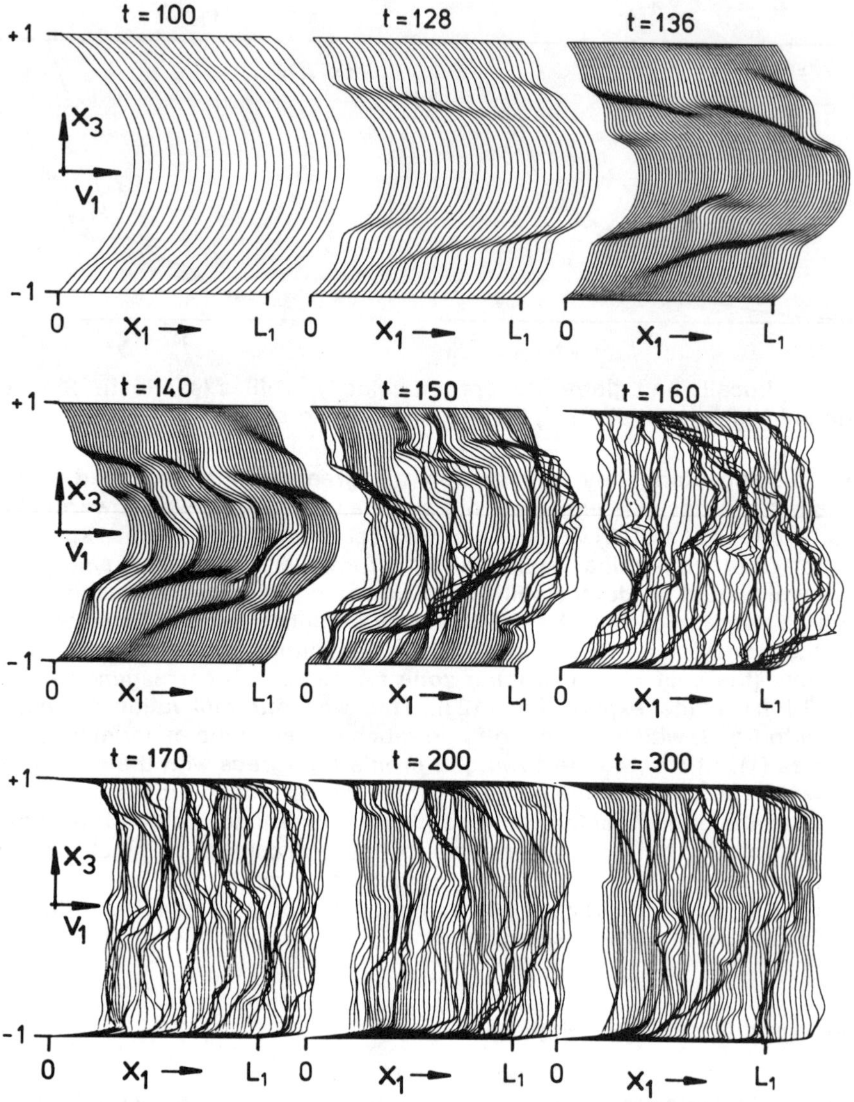

FIGURE 10. Development of instantaneous velocity profiles in peak plane.

The inflectional instantaneous velocity profiles are associated with regions of high shear (i.e., large $\partial v_1/\partial x_3$). In fig. 11 the development of the high-shear layer in the lower channel half is illustrated. Fig. 11a gives contours of $\partial v_1/\partial x_3$ at the peak plane. Two periods in downstream direction are shown. The narrow detached high-shear layer, which separates the low-velocity regions from the high-speed fluid above, elongates and is overtaking the central part of the high-shear layer from the previous period. It starts to wrinkle and has two relative maxima per period at the two-spike-stage (t = 136), and a kink at its downstream end. The local maxima of $\partial v_1/\partial x_3$ at the wall, which are due to the Toll-

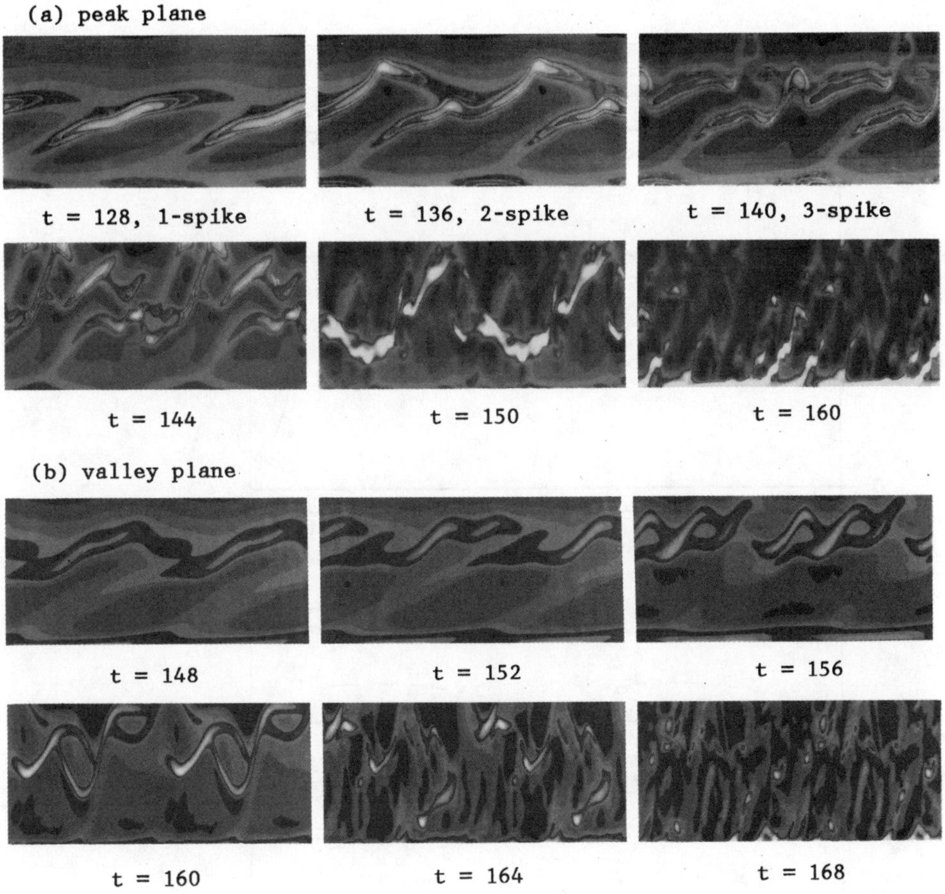

FIGURE 11. Formation and decay of the high-shear layer in lower channel half, illustrated by contours of instantaneous shear $\partial v_1/\partial x_3$. (a) peak plane, (b) valley plane.

mien-Schlichting wave, remain unchanged. At the three-spike stage the high-shear layer starts to break up. Small parts with high shear appear and lift towards the channel center. Values of $\partial v_1/\partial x_3 > 4$, twice the laminar value at the wall, are observed there. As breakdown continues, an opposite evolution takes place. The decaying parts of the high-shear layer intensify but now move towards the wall. At $t = 160$ the maximum of $\partial v_1/\partial x_3$ is obtained at the wall and strong gradients are concentrated to a thin near-wall layer $|x_3| > 0.95$. Away from the wall only sporadic patches of high-shear appear. An interesting process takes place at later times. Tongue-like structures of high shear grow from the near-wall region. They are inclined to the wall at a small angle (about 20°). A more detailed investigation of this phenomenon is in progress.

Up to now the discussion of the high-shear layer development was restricted to the peak-position. A similar process takes place at other spanwise positions. In fig. 11b the formation and decay of the high-shear-layer at the valley is illus-

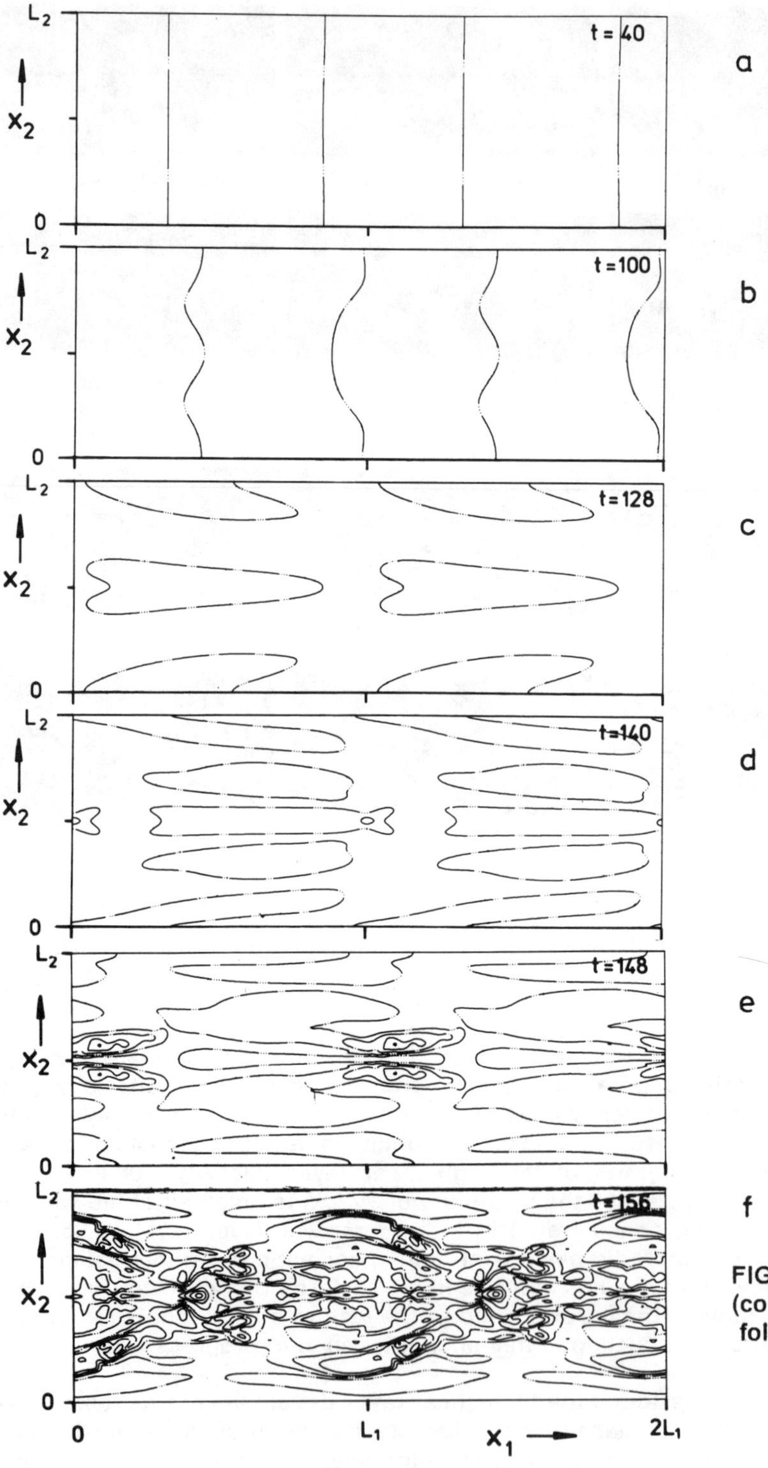

FIGURE 12.
(continued on following page)

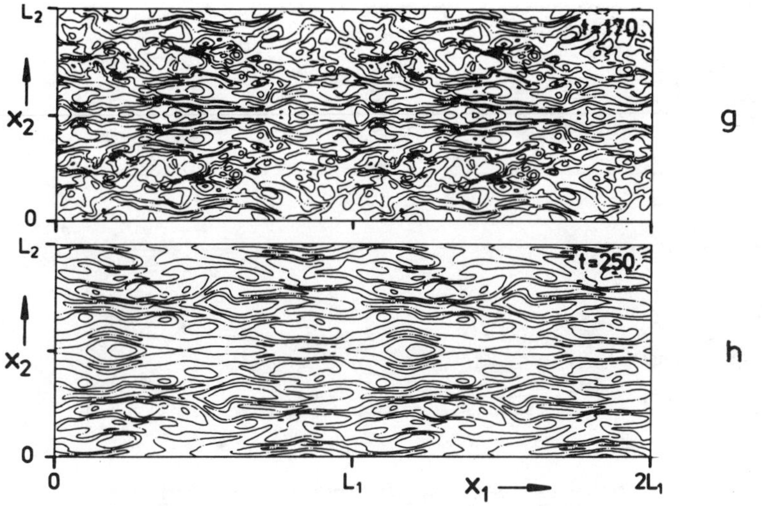

FIGURE 12. Contours of instantaneous streamwise velocity at x_3 = -0.95 (solid line at v_1 = 0.4, contour level increment Δv_1 = 0.1).

trated. Compared with the peak-position, the evolution is retarded. At $t = 148$ (corresponding to the five-spike stage at the peak) a distinct high-shear layer is observed at the height $0.15 < |x_3| < 0.45$. Similarly as at the peak, this shear layer elongates, wrinkles and finally breaks up into smaller parts moving close to the wall. The near-wall evolution of $\partial v_1/\partial x_3$ at the valley position is in contrast to the evolution at the peak. At the valley the velocity gradient near the wall is growing already while the high-shear layer forms near the channel center.

Near-Wall Phenomena of the Instantaneous Flow Field

Organized structures of high- and low-speed streaks are characteristic of near-wall turbulence. They have been observed in the viscous sublayer and buffer region and have a mean spacing in the spanwise direction of about 100 wall units [8,6]. In this section we consider the formation of such streaks during the transition process. Fig. 12 shows contour lines of the instantaneous streamwise velocity near the lower wall at $x_3 \approx -0.95$. Due to the dominant Tollmien-Schlichting wave disturbance at $t = 0$ only a streamwise variation of v_1 is observed. As the secondary instability sets in, the three-dimensional disturbances grow rapidly and at $t = 140$ streaks of faster and slower fluid appear. But the variation in v_1 does not exceed 0.1 and the streamwise extent is nearly one TS period. The breakdown of the regular transitional flow structures, which starts at the peak, is apparent from figs. 12 e and f. At $t = 170$ the large structures have completely decayed into small streaks mainly pointing in the streamwise direction. In the fully turbulent regime ($t = 250$) these low- and high-speed streaks are more pronounced than in the breakdown phase. The different scales in figs. 12 g and h indicate again that the breakdown phase of transition requires a higher resolution than the fully developed turbulence.

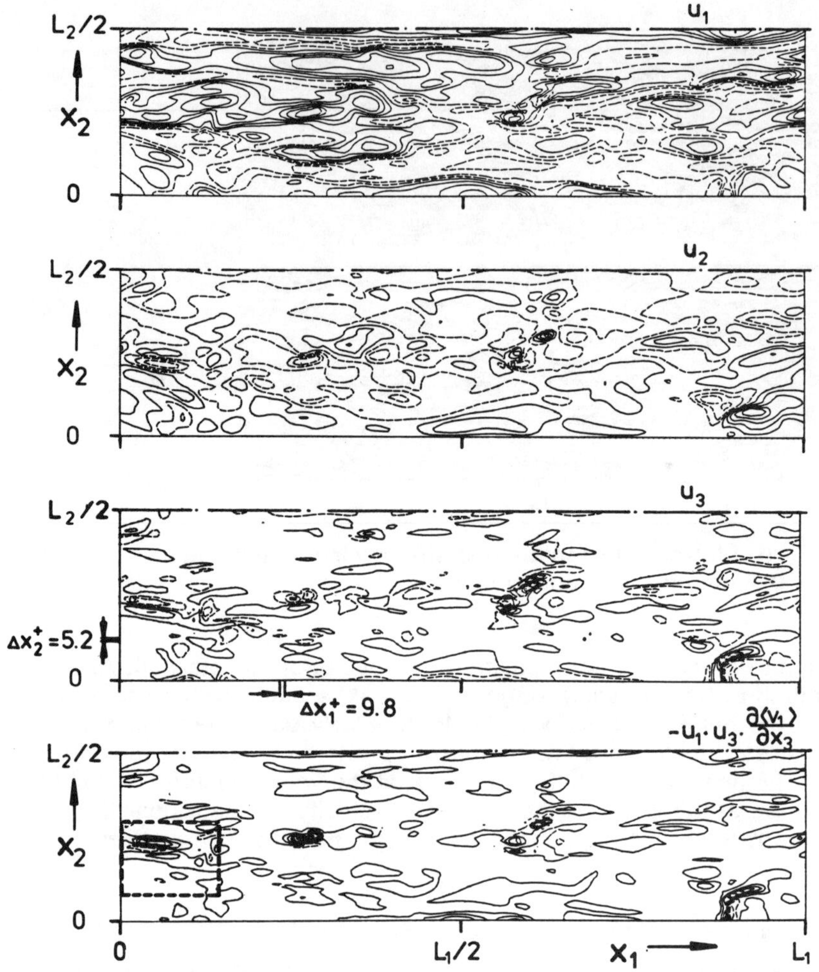

FIGURE 13. Contours of instantaneous velocity disturbance components u_j and production term $-u_1u_3\,\partial<v_1>/\partial x_3$ in plane x_3 = -0.976 (y_+ = 5.43) at t = 168 (—— positive, – – – negative values, increment 0.1).

Contour plots of all three velocity disturbance components $u_j = v_j - <v_j>$ and the turbulence production term $-u_1u_3\,\partial<v_1>/\partial x_3$ in the plane $y_+ \approx 5$ are given in fig. 13 for t = 168. In the spanwise direction only half the box is plotted because of symmetry. The horizontal extents in wall units (computed here from the averaged wall shear at the lower wall) are L_1^+ = 1253 and $L_2^+/2$ = 334. Positive values are contoured by solid lines and negative values by dashed lines. The contours of the streamwise velocity component show elongated high- and low-speed streaks alternating in the spanwise direction. All other components are dominated by small regions of intense fluctuations. As observed by other investigators in developed turbulence, the regions of large normal velocity

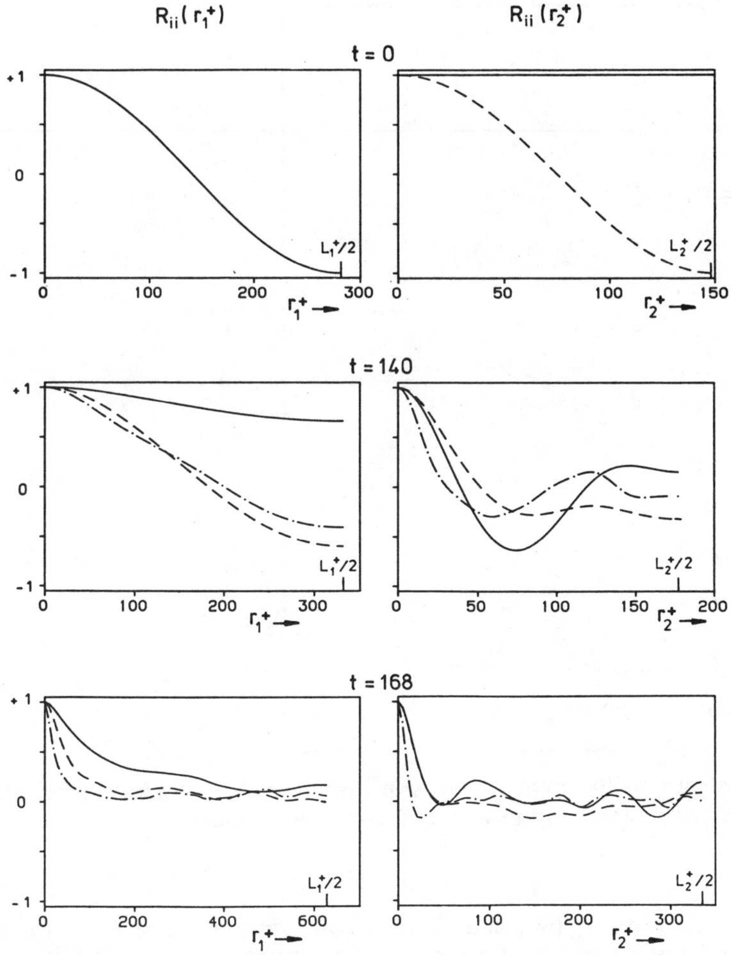

FIGURE 14. Two-point velocity correlations at 3 phases of transition ($y_+ \approx 6$).
Left: streamwise correlations $R_{ii}(r_1^+)$, Right: Spanwise correlations $R_{ii}(r_2^+)$
—— R_{11} ; – – R_{22} ; – · – R_{33}

fluctuations u_3 are composed of pairs or triplets of sections with alternating sign. The picture of the turbulence production term $-u_1 u_3 \partial < v_1 > / \partial x_3$ closely resembles that of the normal velocity fluctuations. To get quantitative statistical information about the structures of the instantaneous flow field, two-point autocorrelations in streamwise and spanwise directions have been calculated. Fig. 14 shows the correlation function for $t = 0$ (cf. fig. 12a), $t = 140$ (cf. fig. 12d) and $t = 168$ (cf. fig. 13) near the lower wall ($y^+ \approx 6$). Up to the 3-spike stage strong correlations exist in streamwise (r_1) and spanwise (r_2) direction. At $t = 168$ the correlations have small, but nonzero, values for large separations at this wall distance. As expected from fig. 13, $R_{11}(r_1^+)$ decays more slowly than R_{22} and R_{33}. In spanwise direction the correlations $R_{ii}(r_2^+)$ decay very quickly and reach

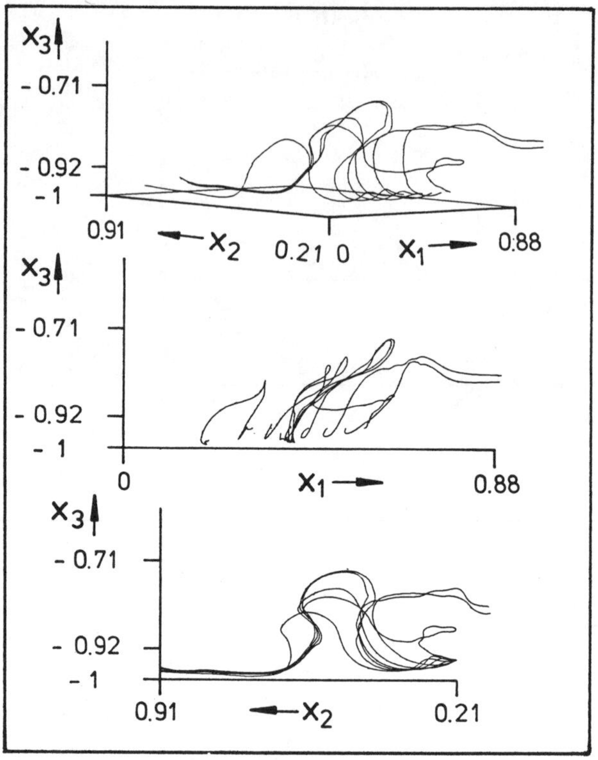

FIGURE 15. Hairpin vortex near lower wall identified by selected vortex lines (t = 168). (a)-(c) perspective, spanwise and streamwise view.

negative minima $R_{11}(50)$, $R_{22}(50)$ and $R_{33}(25)$. The separation $r_2^+ = 50$ for the minimum of $R_{11}(r_2^+)$ is a measure for the mean distance between the high- and low-speed streaks. Together with the minimum of R_{33} at $r_2^+ \approx 25$ these results support the existence of streamwise vortical structures in the wall region. The weak negative minimum of R_{22} suggests that only a few of the vortical structures consist of pairs of counter-rotating vortices. To examine these vortical structures in the transition regime, vortex lines in regions with high turbulence production have been calculated. Fig. 15 gives an example from t = 168. The vortex lines are traced from the lower side of the dotted box in fig. 13d. They show the typical hairpin-like structure. The outer part of the vortex has an Ω-shape. From the spanwise view we observe an inclination angle of the vortex-lines of about 45° at the Ω-legs and a higher angle in the tip region.

We conclude that typical phenomena of wall turbulence, such as low- and high-speed streaks and hairpin vortices, are observed already during the breakdown phase of transition.

CONCLUSIONS

The complete transition process from laminar to turbulent channel flow has been investigated by numerical integration of the three-dimensional time-dependent Navier-Stokes equations. Starting from slightly disturbed laminar Poiseuille flow, the classical peak-valley splitting (K-type) mode of "ribbon-induced" transition is simulated. The numerical resolution is increased up to $N_1 \times N_2 \times (N_3 + 1) = 128 \times 128 \times 129$ gridpoints in the streamwise (x_1), spanwise (x_2) and normal (x_3) directions. This appears to be sufficient to resolve all relevant scales of motion at the present Reynolds number Re_Q of 5000. In the spanwise direction we exploit the symmetry of the flow field which is observed in early transition and is preserved by the Navier-Stokes equations. In order to obtain a rapid evolution from the transitional stages to the fully developed turbulent stage we keep the mean mass flux fixed in our simulations. The mean flow, rms fluctuations and instantaneous velocity fields have been evaluated at the successive stages of the transition process. The temporal development of the velocity field in the simulation is in good agreement with the streamwise development observed in experiments.

In our simulation we distinguish four main phases of the transition process (see fig. 6 and table 1). During the *first phase* the 2-D Tollmien-Schlichting wave develops, and 3-D disturbances are amplified due to secondary instability. This phase ends when the 3-D disturbances attain the order of magnitude of the 2-D wave amplitude (dimensionless time $t \approx 100$ in our simulation). The *second phase* ($100 < t < 150$) is characterized by the formation of pronounced 3-D flow structures. The fluctuations grow rapidly (streamwise rms fluctuations up to 15% of the centerline velocity), whereas integral data, such as the mean wall shear stress, still remain near their laminar values. The instantaneous velocity field is dominated by travelling Λ-vortex structures. These elongate as they move downstream. In the region of the Λ-vortices large gradients of the instantaneous streamwise velocity appear. Near the "peak" plane a narrow detached high-shear layer forms with a kink at its downstream tip. During the "spike" stages this high shear layer elongates, wrinkles and finally breaks up into smaller parts. This marks the beginning of the *third phase* of the transition process, in which the transitional flow structures (on the scale of the TS wave) decay completely. The rms fluctuations become very large (locally up to 25%), by far exceeding the turbulent level. The wall shear stress and the centerline velocity attain their turbulent values. In this phase a rapid development near the wall takes place. Contour plots of the Reynolds stress near the wall indicate isolated regions with large contributions to the turbulence production. Vortex lines traced from the vicinity of those regions display hairpin-shaped vortices. Two-point autocorrelations of the velocity components indicate the formation of high-speed and low-speed streaks in the wall layer. At $t \approx 200$ the *fourth phase*, the fully developed turbulent state, is attained. Statistical results obtained in the turbulent range have been compared with experimental data and with simulation results by other investigators. Generally good agreement is found, but there are also some discrepancies. Some of these may be due to the spanwise symmetry enforced, and due to insufficient periodicity lengths in the turbulent range.

In this contribution we presented some features of the late transition process. Much more work is necessary to clarify details of the final stages of transition and their relation to wall turbulence.

REFERENCES

1. Canuto, C., Hussaini, M.Y., Quarteroni, A., and Zang, T.A., *Spectral Methods in Fluid Dynamics,* Springer-Verlag, Berlin, 1988.

2. Eckelmann, H., Experimentelle Untersuchungen in einer turbulenten Kanalströmung mit starken viskosen Wandschichten, Mitt. MPI f. Strömungsforschung und AVA Göttingen Nr. 48, 1970.

3. Gilbert, N., and Kleiser, L., Subcritical Transition to Turbulence in Channel Flow, in *Direct and Large Eddy Simulation of Turbulence*, eds. U. Schumann and R. Friedrich, pp. 1-18, Vieweg, Braunschweig, 1986.

4. Gilbert, N., DFVLR Report in preparation, 1988.

5. Herbert, Th., Secondary Instability of Boundary Layers, *Ann. Rev. Fluid Mech.*, vol. 20, pp. 487-526, 1988.

6. Kim, J., Moin, P., and Moser, R., Turbulence Statistics in Fully Developed Channel Flow at Low Reynolds Number, *J. Fluid Mech.,* vol. 177, pp. 133-166, 1987.

7. Kleiser, L., and Schumann, U., Spectral Simulations of the Laminar-Turbulent Transition Process in Plane Poiseuille Flow, in *Spectral Methods for Partial Differential Equations*, eds. R.G. Voigt et al., pp. 141-163, SIAM, Philadelphia, 1984.

8. Kline, S.J., Reynolds, W.C., Schraub, F.A., Runstadler, P.W., The Structure of Turbulent Boundary Layers, *J. Fluid Mech.,* vol. 30, pp. 741-773, 1967.

9. Kreplin, H.-P., and Eckelmann, H., Behavior of the Three Fluctuating Velocity Components in the Wall Region of a Turbulent Channel Flow, *Phys. Fluids,* vol. 22, pp. 1233-1239, 1979.

10. Laurien, E., and Kleiser, L., Numerical Simulation of Boundary-Layer Transition and Transition Control, accepted for publication in J. Fluid Mech., 1988.

11. Michel, R., Special Course on Stability and Transition of Laminar Flow, AGARD-R-709, 1984.

12. Nishioka, M., Asai, M., and Iida, S., Wall Phenomena in the Final Stage of Transition to Turbulence, in *Transition and Turbulence,* ed. R.E. Meyer, pp. 113-126, Academic Press, 1981.

13. Nishioka, M., and Asai, M., Evolution of Tollmien-Schlichting Waves into Wall Turbulence, in *Turbulence and Chaotic Phenomena in Fluids,* ed. T. Tatsumi, pp. 87-92, North-Holland, 1984.

14. Nishioka, M., and Asai, M., Some Observations of the Subcritical Transition in Plane Poiseuille Flow, *J. Fluid Mech.,* vol. 150, pp. 441-450, 1985.

15. Orszag, S.A., and Patera, A.T., Calculation of von Karman's Constant for Turbulent Channel Flow, *Phys. Rev. Lett.,* vol. 47, pp. 832-835, 1981.

16. Orszag, S.A., and Patera, A.T., Secondary Instability of Wall-Bounded Shear Flows, *J. Fluid Mech.,* vol. 128, pp. 347-385, 1983.

17. Patel, V.C., and Head, M.R., Some Observations on Skin Friction and Velocity Profiles in Fully Developed Pipe and Channel Flows, *J. Fluid Mech.*, vol. 38, pp. 181-201, 1969.

18. Spalart, P.R., and Yang, K.-S., Numerical Study of Ribbon-Induced Transition in Blasius Flow, *J. Fluid Mech.,* vol. 178, pp. 345-365, 1987.

19. Tani, J., Boundary-Layer Transition, *Ann. Rev. Fluid Mech.*, vol. 1, pp. 169-196, 1969.

20. Wray, A., and Hussaini, M.Y., Numerical Experiments in Boundary-Layer Stability, *Proc. R. Soc. Lond.,* A 392, pp. 373-389, 1984.

21. Zang, T.A., Krist, S.E., Erlebacher, G., and Hussaini, M.Y., Nonlinear Structures in the Later Stages of Transition, AIAA Paper 87-1204, 1987.

This work is supported by the Deutsche Forschungsgemeinschaft.

Bifurcations and Turbulence in Plane Poiseuille Flow

JAVIER JIMÉNEZ
School of Aeronautics
Universidad Politécnica Madrid
and IBM Scientific Centre
P. Castellana 4, 28046 Madrid, Spain

Real turbulence is known to be three dimensional, and there are indications that laminar to turbulent transition in many shear flows is mediated by three-dimensional (3D) secondary instabilities of two-dimensional (2D) nonlinear disturbances (Orszag & Patera, 1983). However, 2D disordered motion is known to be associated to the large scales of some turbulent flows, even after the small scales become fully turbulent and three-dimensional (Roshko, 1976, Papailiou & Lykoudis, 1974). This suggests that there are solutions for those flows which are two dimensional and somehow act as "attractors" for the system. These "attractors" are probably unstable, but the fact that they are observed proves that the flow is pulled towards them from, at least, a substantial portion of its phase space, even if it is then eventually repelled by the instability. If the attraction exerted by a given state is strong and its instability is relatively weak, the system will spend a substantial percentage of time in its neighbourhood, and, as a consequence, not only will we be able to observe such a state fairly often, while observing the flow, but we should also be able to describe a large part of the dynamics of the system in terms of the neighbourhood of the attractor.

A very simple model for this situation is a rotating pendulum, which has two points of equilibrium. If the energy of the pendulum is chosen correctly, it is possible to get a situation in which the pendulum expends most of its time in the neighbourhood of the upper (unstable) equilibrium point, even if makes periodic excursions away from it. The key is, not only that the unstable point has an attracting direction in phase space, but that there is a re-injection mechanism that is able to bring the system to the neighbourhood of this direction even after it has been repelled away from equilibrium along the other (unstable) direction. Dissipative systems, such as turbulent flows in viscous fluids, are more complicated, but one can think of them as in a dynamic equilibrium between dissipation, which tries to slow the system down to a "steady" state, and an instability mechanism that pumps energy into the system and plays the role of the re-injection mechanism that takes it back to the neighbourhood of whatever "attractor" describes the unsteady motion of the flow. It is clear that understanding both the nature of the equilibrium states of these flows, even when they are unstable, and of the possible re-injection processes, is essential for the understanding of the nature of the motion.

It is, of course, not obvious that the instability mechanism that is responsible for the initial laminar to turbulent transition is the same one that acts as a re-injector in a fully turbulent situation, but it is clear that any available instability should be examined as a candidate and should not be discarded *a priori*. It has been noted often that turbulent flows are governed by two very different time scales. A *convective* one, $T_i = L/U$, that governs inertial processes, and a much longer viscous one, $T_v = ReT_i$, that measures the time needed by viscosity to damp the inertial motion in the absence of especially strong gradients. Here U and L are characteristics, "external", scales of the flow, and $Re = UL/\nu$. The importance of a given process in the flow can be judged by its characteristic time scale. Processes which develop in viscous times are too weak, and are swamped by the convective effects. Only processes with time scales of the order of T_i can be expected to play important roles in fully developed turbulence.

In free shear flows, there is no mechanism for the production of strong gradients and energy is dissipated slowly, in viscous times, while it is injected at the large scales by convective processes. As a consequence, free shear flows can be analysed to a good approximation as inviscid processes. In wall turbulence, however, the presence of the wall induces strong gradients, and it is possible to have viscous processes with convective time scales. Those processes are usually nonlinear, and they are generally known as viscous-inviscid interactions. From what we have said up to now, they are prime candidates for important mechanisms in fully developed turbulence.

The best known example is unsteady separation (Didden & Ho, 1985, Walker *et al.*, 1987). There, the approach of a strong vortex to a wall if followed by what is usually described as *explosive* local growth of the boundary layer. A more recent example is the ejection instability of nonlinear Tollmien-Schlichting waves in plane Poiseuille flow observed numerically by Jiménez (1987) and described later in this paper. Both processes are two dimensional, and the resulting structures are almost surely unstable to 3D instabilities. However, their formation time scales are inertial, comparable to those of any other phenomenon in fully developed turbulent flows, and we will take here the point of view that they could compete, and maybe coexist, with other more classical 3D instability mechanisms (e.g. hairpins) in those flows.

The first part of this paper will thus be dedicated to the study of purely 2D Poiseuille flow, and to the possibility of identifying unstable 2D modes with inertial time scales, away from the complicating influence of 3D phenomena. We will do this through the numerical simulation of the initial value problem for the full Navier-Stokes equations. In the second part, we will show that structures and mechanisms similar to those identified in this way are indeed present in fully developed, three dimensional, turbulent flows, even if, as expected, they are heavily modified by 3D effects.

The Numerical Method

Poiseuille flow is established between two infinite parallel plates at $y = \pm 1$ and, at equilibrium, has a parabolic velocity profile $U(y) = 1 - y^2$. It is governed by the 2D vorticity equation

$$\omega_t + \psi_y \omega_x - \psi_x \omega_y = Re^{-1}\nabla^2\omega, \tag{1}$$

where ψ is a stream function, and

$$\omega = \nabla^2\psi \tag{2}$$

is the vorticity. Re is a Reynolds number, based in the unperturbed centre line velocity, and on the channel half width. We study the behaviour of periodic disturbances of the form

$$\omega = -2y + \sum_{k=-N}^{N} \omega_k(y,t)\, e^{ik\alpha x}, \tag{3}$$

and impose as longitudinal boundary condition that the total flux in the channel remains constant. Other boundary conditions can be chosen, and the details of the nonlinear behaviour seem to depend on the choice (Saffman, 1983).

At low Re, the parabolic profile is stable, but at $Re = 5772$ ($\alpha = 1.02$), it bifurcates into a family of periodic travelling shear waves. This bifurcation is subcritical and can be continued at finite amplitudes to lower Re, until this "lower branch" turns around near $Re = 2700$, and proceeds towards higher Re, forming an "upper" branch with larger amplitudes (Herbert, 1976). Other wavenumbers enlarge this picture into a continuous, cigar shaped, surface with upper and lower sheets, although restricted to a narrow range of wavenumbers near $\alpha = 1$. This gives some justification to the use of the monochromatic expansion in (3). The lower sheet is known to be unstable to 2D disturbances, while the upper one is, at least at moderate Re , not only stable but attracting (Zahn *et al.*, 1974). Both branches are known to be unstable to 3D perturbations (Orszag & Patera, 1983), but we will not consider such solutions here, and restrict ourselves to the bifurcations of the upper branch into other 2D solutions, as a function of Re.

For this, we integrate numerically the time evolution of the expansion (3) substituted in equation (1). The numerical procedure is described next in some detail. Briefly, it uses Tchebichev expansions in y for each of the coefficients in (3), and a Galerkin approximation in both x and y for the nonlinear terms of the equation. The time integration is a split-step finite difference scheme, and the boundary conditions are applied at the end of the last partial step using an influence matrix technique.

We define perturbation velocities with respect to the parabolic profile as $\psi_y = U(y) + u$ and $\psi_x = -v$. The resulting evolution equation for each of the Fourier components in (3) is

$$\frac{\partial \omega_k}{\partial t} + \left[ik\alpha U(y) + k^2\alpha^2/Re \right] \omega_k = -\left[u\frac{\partial \omega}{\partial x} + v\frac{\partial \omega}{\partial y} \right]_k + \frac{1}{Re}\, \frac{\partial^2 \omega_k}{\partial y^2}. \tag{4}$$

The bracket on the right hand side represents the k-th Fourier component of the convective term inside, and it is computed pseudo-spectrally with full de-aliasing by using the 3/2 rule (Orzsag, 1971). The rest of the equation is handled directly in spectral representation. For each time step, the left hand side of (4) can be integrated exactly and results in

$$\omega_k(t+\Delta t) = e^{-\Delta t S_k}\left[\omega_k(t) + \int_0^{\Delta t} e^{\xi S_k} F_k(t+\xi)\, d\xi \right] + \frac{1}{Re}\int_0^{\Delta t} e^{(\xi-\Delta t)S_k} \frac{\partial^2 \omega_k(t+\xi)}{\partial y^2}\, d\xi. \tag{5}$$

where

$$S_k(y) = ik\alpha U(y) + k^2\alpha^2/Re, \tag{6.a}$$

and

$$F_k(y,t) = \left[u\frac{\partial \omega}{\partial x} + v\frac{\partial \omega}{\partial y} \right]_k . \tag{6.b}$$

The first integral in the right hand side of (5) is computed by a second order Adams-Bashforth formula, while the second one is approximated by simple reverse Euler. The result can be written as a split-step scheme

$$\tilde{\omega}_k(t) = e^{-\Delta t S_k} \left[\omega_k(t) + \int_0^{\Delta t} e^{\xi S_k} F_k(t+\xi)\, d\xi \right] \tag{7.a}$$

$$\omega_k(t+\Delta t) - \frac{\Delta t}{Re} \frac{\partial^2 \omega_k(t+\Delta t)}{\partial y^2} = \tilde{\omega}_k(t). \tag{7.b}$$

The first step is fully explicit, while the second one is a Poisson solver, which is easily handled in the Tchebichev representation (Gottlieb & Orszag, 1980). The same is true of the solution of equation (2), which is needed to close the time step, and which results in

$$\frac{\partial^2 \psi_k(t+\Delta t)}{\partial y^2} - k^2\alpha^2 \psi_k(t+\Delta t) = \omega_k(t+\Delta t), \tag{8}$$

where the $\psi_k(y)$ are coefficients of a Fourier expansion for the stream function, similar to (3).

The boundary conditions are

$$\frac{\partial \psi_k}{\partial y}(\pm 1) = 0, \tag{9.a}$$

and

$$\psi_k(\pm 1) = 0. \tag{9.b}$$

They are applied using a Tchebichev *tau* method. Two of them can be used directly when solving equation (8), but the other two have to be applied to the vorticity, even if they are expressed in terms of the stream function. This is done using an influence matrix technique. The partial steps (7.b) and (8) are applied three times for each time step, in each case satisfying the conditions (9.a), but with a different value for $\tilde{\omega}_k$. The first application, whose result we will call ω_{k0}, ψ_{k0} , uses as a starting point the result of (7.a). The other two, which are equivalent to the *tau* approximation, are started with $\tilde{\omega}_{k1} = T_M(y)$, and $\tilde{\omega}_{k2} = T_{M-1}(y)$, which are the two highest order Tchebichev polynomials used in the expansions for the vorticity components. The final result for $t + \Delta t$ is formed as the linear combination

$$\omega_k(t+\Delta t) = \omega_{k0}(t+\Delta t) + \beta_1\omega_{k1} + \beta_2\omega_{k2},$$
$$\psi_k(t+\Delta t) = \psi_{k0}(t+\Delta t) + \beta_1\psi_{k1} + \beta_2\psi_{k2},$$

where the coefficients are chosen so that the the final stream function satisfies also the conditions (9.b). Note that the vectors ω_{k1}, ω_{k2} , and the corresponding stream functions, are constant, and only have to be computed once at the beginning of the integration. Essentially similar methods have been used by Deville *et al.* (1984) for a Stokes problem, and by Kim *et al.* (1987), for a problem similar to the one discussed here.

Two Dimensional Results

This numerical scheme is used to follow the evolution in time of various initial conditions. Initially, a random perturbation of given wavenumber is superimposed on the equilibrium parabolic flow, and left to evolve. In most cases, the wave number is set arbitrarily to $\alpha = 1.0$, although perturbations with other wavenumbers can be shown to behave similarly.

At $Re = 5000$, a small, but not infinitesimal, perturbation to the uniform flow grows, and eventually converges to a new state, corresponding to a uniform, "upper branch", travelling wavetrain, which can be considered as a new equilibrium point, in moving axes. Convergence can be measured by checking $\omega_0(y,t)$ at the wall, $y = -1$, which is real and constant in time for a uniform wavetrain, independent of the frame of reference. From this new equilibrium state, we scan Re by changing it in discrete steps, and use the converged result for one Reynolds number as the initial condition for the next one. For each Re the integration is allowed to converge for several thousand time units or until no change is apparent. At about $Re = 5500$, $\omega_0(-1,t)$ becomes unstable and sheds a limit cycle of small amplitude, and period $T \simeq 14.2$ (Fig. 1-a). As the Reynolds number increases, the amplitude of this cycle also increases, in a square root fashion which is typical of supercritical bifurcations (Fig. 2), and, at $Re = 9100$, it bifurcates again into a torus (Fig. 1-b). The ratio of the two periods of this torus stays fixed at $T_{slow}/T_{fast} = 5$, for a range of Reynolds number up to $Re \simeq 10{,}000$, at which point the torus changes into new, more complex, states, which may represent either new tori with very long periods, or perhaps mild chaos (Fig. 1-c). This regime is still being investigated, but it appears that, at $Re = 12{,}000$, the flow settles back into a new torus with $T_{slow}/T_{fast} = 4$.

The site and structure of the Hopf bifurcation which gives rise to the original limit cycle, can be studied through vorticity maps of the flow. Below $Re = 5500$, where the uniform wavetrain is stable, the wave consists of an array of large counter-rotating vortices at the core of the channel, which induce strong secondary vorticity peaks near the walls. At each wall, each wavelength contains a strong vortex, corrotating with the main flow, and a weaker, counter-rotating one, marking a local separation bubble (see Figs. 3 to 5). The larger corrotating vortex has two individual peaks. After the bifurcation, the bimodal vortex becomes unstable, interchanging vorticity between its two peaks, and shedding some of it into the core in a relatively short "burst" (Figs. 4 and 5). Both walls eject alternately, and the core flow seems to play a relatively passive role in the process.

From our point of view, it is important to understand why this bifurcation has a period which is of the order of the inertial time scale, instead of the viscous evolution times which are characteristic of the linear channel flow instabilities. In general, Tollmien-Schlichting waves have a phase velocity which is O(1), and, as a consequence, when seen in a frame of reference moving with the wave, the wall moves towards the left, while the fluid outside the boundary layer moves to the right. Assume that a vorticity inhomogeneity appears near the wall; it will be convected backwards by the flow while it diffuses outwards, until it reaches the critical layer, where it begins to be convected forwards, while still diffusing, towards its original streamwise position. As it moves away from the wall, it induces near it a secondary vortex, to satisfy the no-slip condition, and, when it reaches its original streamwise station, the induced vortex reproduces, although with an opposite sign, the initial perturbation. A second half of the cycle starts again, feeding the instability (Betchov

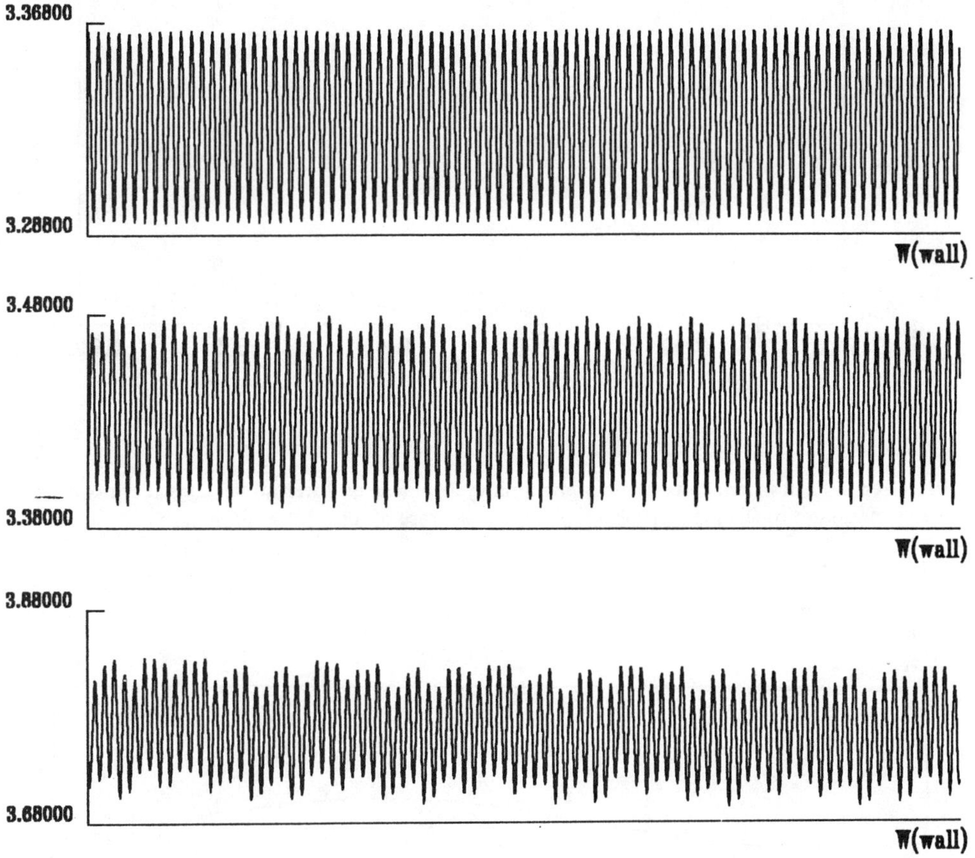

FIGURE 1. Time evolution of the space averaged wall stress for three different Reynolds numbers. Top: 9,000; middle: 9,500; Bottom: 11,000. The time span is 1000 in the three cases.

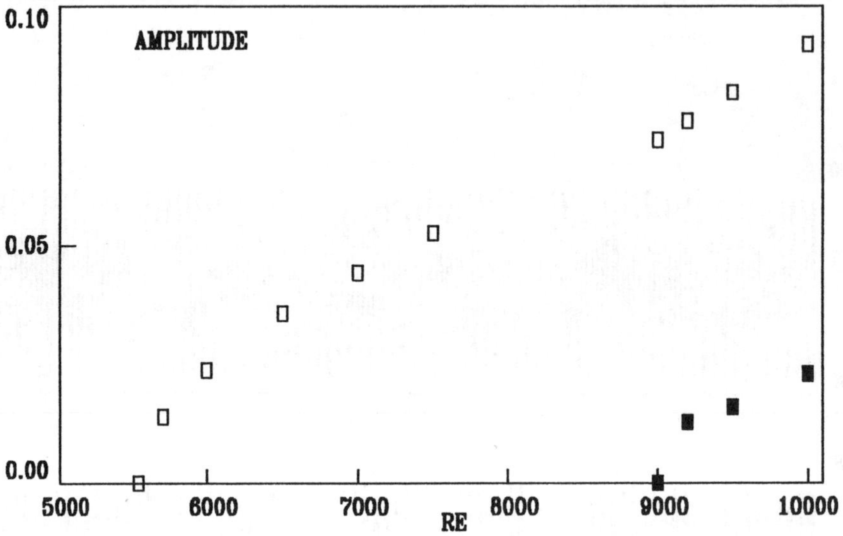

FIGURE 2. Peak to peak amplitude of the limit cycle of averaged wall stress. Closed symbols represent maximum oscillation of peak amplitudes in the case of tori.

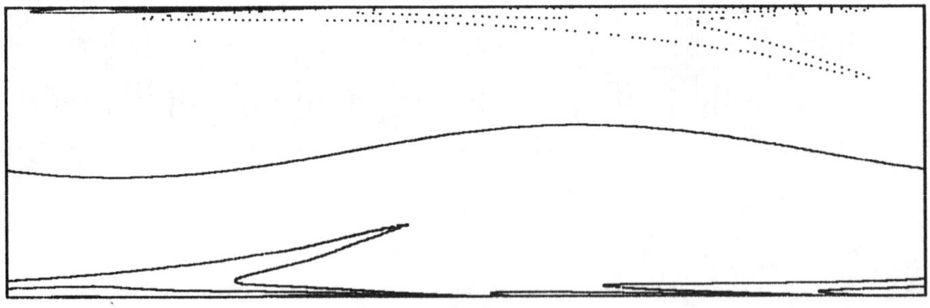

FIGURE 3. Vorticity map for one wavelength of the equilibrium wavetrain. Solid isolines are $\omega = 0, 1.5, 3, 5, 7$; Dotted lines are negative values. Flow is from left to right. $Re = 5500$.

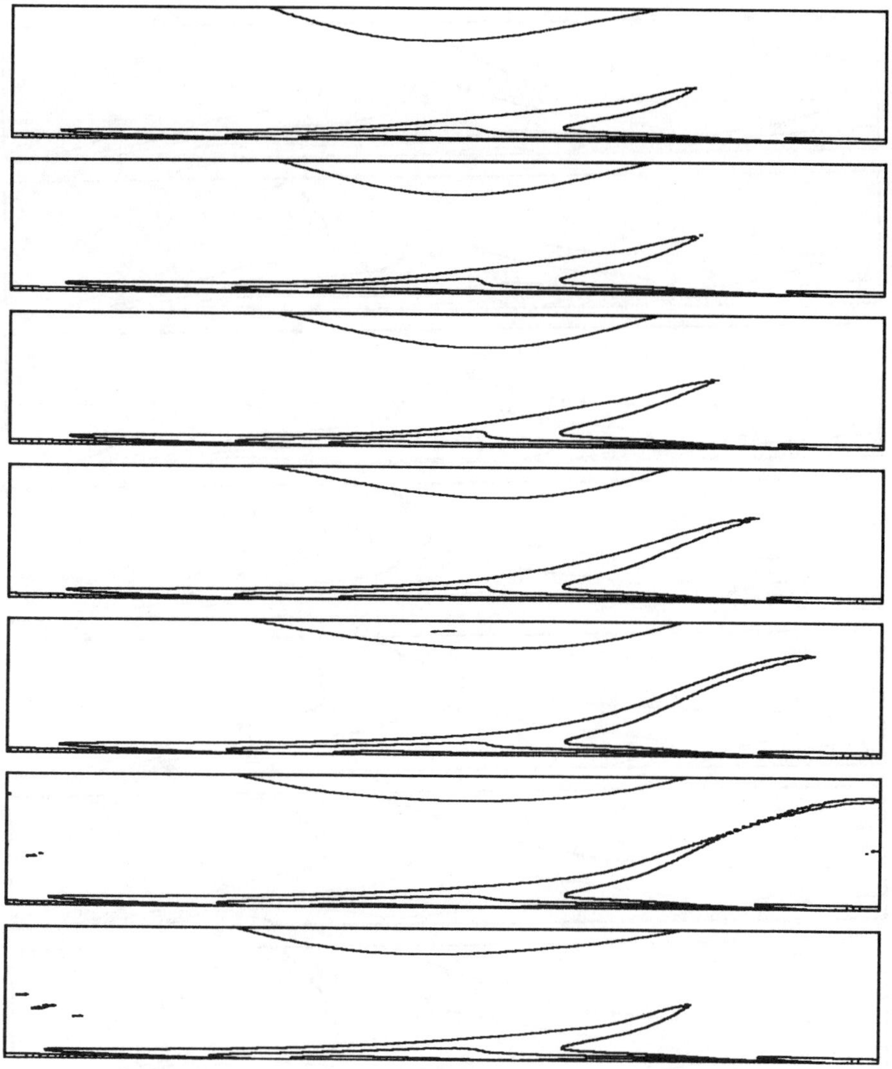

FIGURE 4. Time evolution of vorticity maps for the limit cycle at $Re = 7000$. Flow is from left to right and time runs from top to bottom, $t = 0$ to 9.6, approximately three quarter period. Only half channel is represented, and isolines are the same as in figure 3.

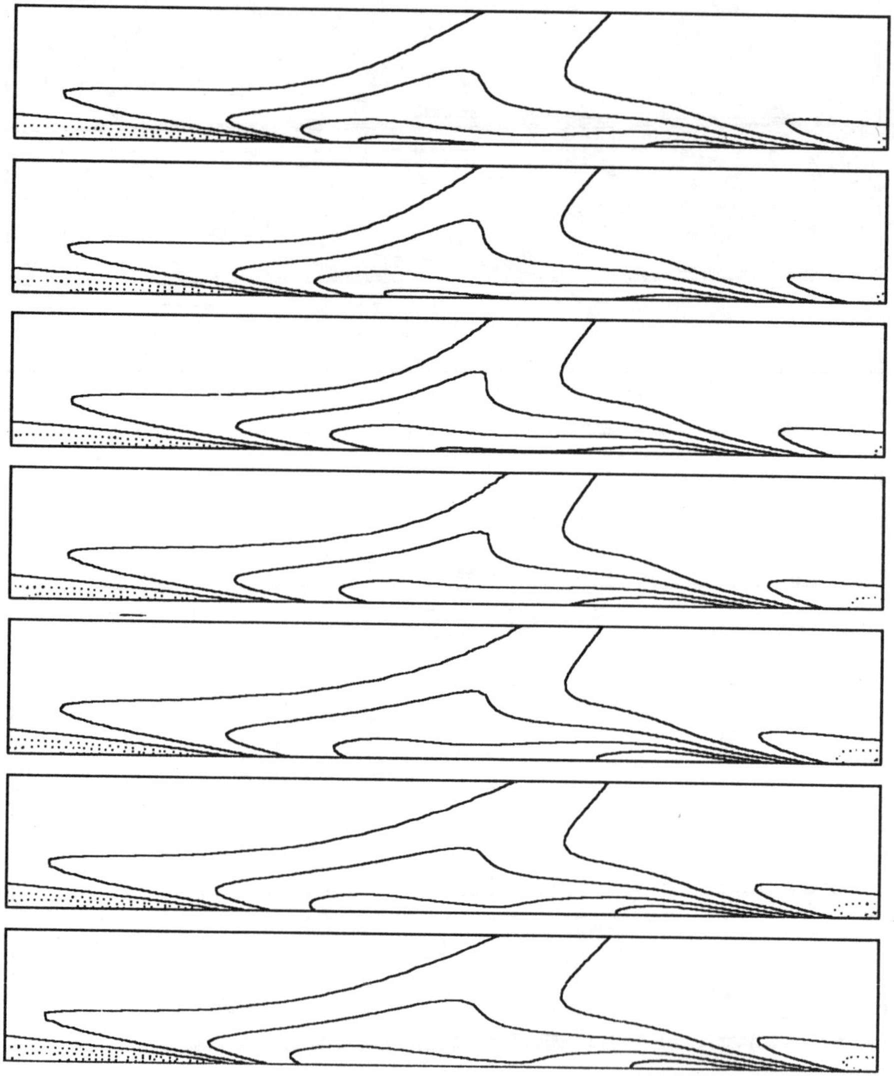

FIGURE 5. Close up of lower boundary layer for sequence in figure 4. Vertical axis spans $y = -1$ to -0.8.

& Criminale, 1967). The result is a sequence of vorticity perturbations of alternating signs, stacked one on top of the other, turning direction at about the location of the critical layer.

With infinitesimal perturbations, the only mechanism available to carry the vorticity away from the wall is viscous diffusion, and this is what determines the slow time scales characteristic of the linear instability. When the wave attains finite amplitudes, the situation is different. The alternation of vorticity near the wall is now strong enough to induce transverse velocities which are, locally, of the same order of magnitude as the free stream one. In fact, the critical layer is now substituted by a periodic array of strong saddle points, along one of whose branches the vorticity can be convected away from the wall in inertial times (Fig. 6). Now, when a perturbation appears, it is convected backwards, outwards and forward again, in times which are $O(T_i)$, and reproduces itself, through the formation of a secondary counter-vortex, with periods of that same order.

The result is also an alternation of vortex signs, but on a much faster time scale than in the linear case. Moreover, this alternation produces periodic reinforcements of the transverse velocity near the saddle point, which produce the periodic "ejections" of the limit cycle, and which can be seen as a swelling of the streamlines in some of the frames in figure 6.

As mentioned in the introduction, the generation of ejections by a strong vortex near a wall, through the formation of a secondary counter-vortex, had been reported previously in the context of forced unsteady separation (Didden & Ho, 1985, Walker *et al.*, 1987). The limit cycle behaviour that we have described here is presumably a self-reproducing instance of the same phenomenon.

It is interesting to inquire which is the relation of this instability to the classical three-dimensional secondary instability of channel flow, described, for instance in (Orzsag & Patera, 1983, Herbert, 1983). The two cases seem to be different, since, in that case, most of the perturbation is reported to be concentrated near the centre of the large elliptical eddies in the core of the channel, while, in ours, it seems to be connected with the saddle points between them. A somewhat similar situation was already known in the periodic free mixing layer, where it is relatively well understood, and it may be useful to review it here.

In that flow, there appears to be two distinct 3-D instabilities, which were first identified in (Pierrehumbert & Widnall, 1982). The first one, the *helical* mode, is a long wave phenomenon, whose zero wavenumber limit is the classical 2-D "pairing" instability of an array of point vortices (Lamb, 1945). In this case, each vortex core, when seen in the appropriate frame of reference, sits in a saddle point induced by the rest of the array, and, as soon as its position is perturbed, it tends to fly off exponentially along one of the outgoing legs of the saddle. As it does so, it is deformed and induces on itself velocities which may oppose or enhance the instability. In this particular case, they oppose it, and the motion dies for short wavelengths.

A completely different mechanism is the *translational* mode, which is a short wave phenomenon. In this case, the destabilising effects are precisely the self induced velocities, due to the core deformation, which opposed the instability in the previous case. In fact, Pierrehumbert (1986) has shown that this mechanism is quite general, an applies to many inviscid columnar vortices, whether in arrays or isolated. More recently, Landman &

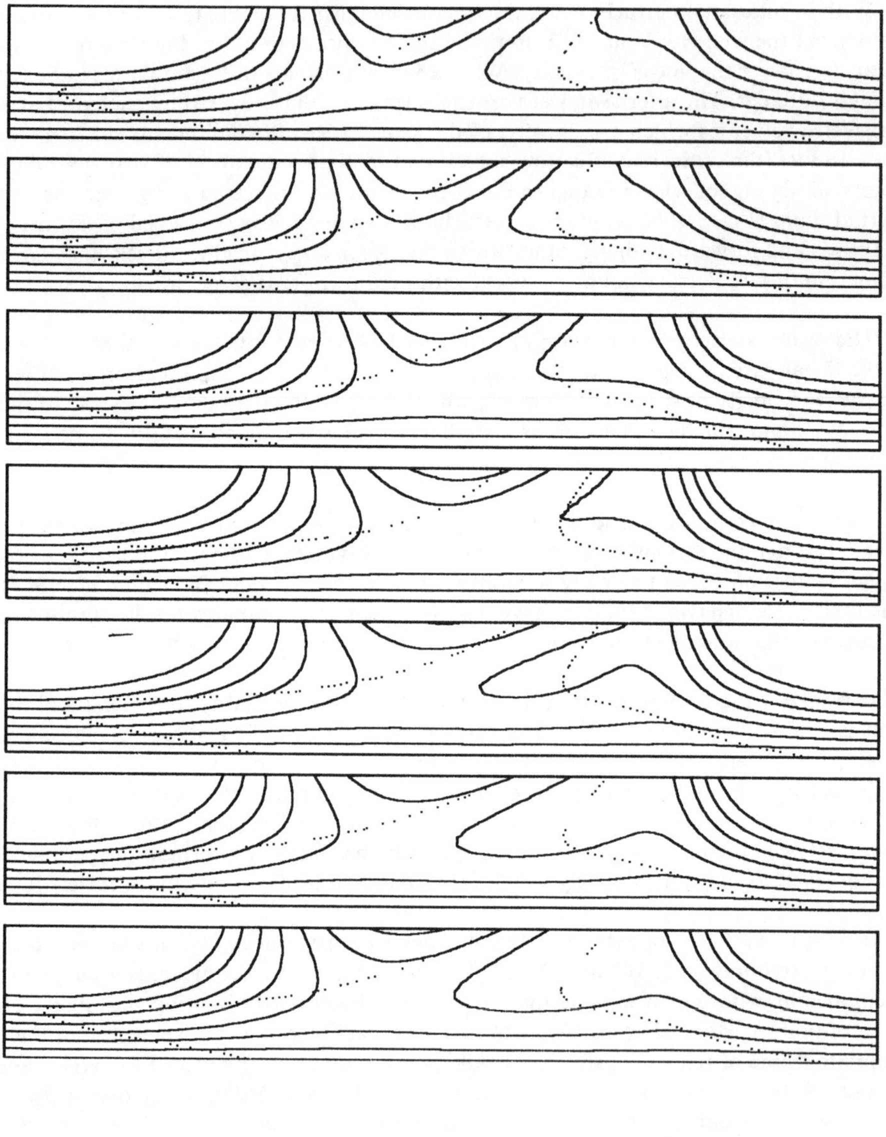

FIGURE 6. Streamlines for sequence in figure 5, on axes moving with the wave. Distance between streamlines, $\Delta\psi = .005$.

Saffman (1987) have extended this result to the viscous case, and Pugh (1987) has applied it to the 3-D instability of channel flow, giving evidence that the classical instability is indeed of this sort. As pointed before, both the simulation results and these theoretical analyses show that the disturbances associated to this mode are concentrated near the centre of the large vorticity concentrations on the core flow, away from the wall, and are, therefore, probable unrelated to the effects discussed here.

In the shear layer, the growth rates of both instabilities are of the same order, and they co-exist in physical flows. Little is known of the comparable situation in channel flow, and we believe that this is the first report of an instability that could correspond to the helical mode. Since we have shown that the relevant time scales are also inertial, it could be expected that both modes also compete, and possibly co-exist. In the next section, we show evidence that they do.

Three dimensional effects.

We review here some recent evidence that thin layers of concentrated spanwise vorticity are a common feature in the sublayer of fully turbulent channel flow (Jiménez *et al.*, 1987). Although these sheets are three dimensional, and have a limited spanwise extent, a cross section through their plane of symmetry looks remarkably similar to the equilibrium two dimensional wave trains discussed in the previous section (Fig. 7).

The observation was done on a previously computed time series of flow fields extracted from a fully resolved, three dimensional, numerical simulation of a turbulent channel (Kim *et al.*, 1987). The Reynolds number of this flow, based on the mean bulk velocity, U_m, and on the channel half width, h, is $Re = 2800$, which would be equivalent to $Re = 4200$ if based on the centre line velocity of a equivalent parabolic profile, as in the previous section. Its statistical properties are in good agreement with those of experimental flows, and we will consider this simulation here as a "natural" turbulent channel. To our knowledge, this observation of thin vorticity sheets in the sublayer of a fully developed turbulent flow is new, although there have been several previous reports of the formation of thin layers of high dU/dy in the form of "spikes" in transitional flows (Cantwell, 1981), or as part of the evolution of isolated "hairpin" vortices in the neighbourhood of a wall (Kim, 1987).

There are some important differences between these structures and those in the 2-D calculations. First, the "wavelength" is shorter, with an average longitudinal separation between consecutive features of the order of 1 to $3h$ (200-500 wall units), while the 2-D nonlinear T-S waves can only exist, as equilibrium solutions, for wavelengths in a range about π-$2\pi h$. Also, they penetrate less in the core of the channel, levelling off at a distance of $0.2h$ (35 wall units) away from the wall, while the 2-D solutions extend all the way to the channel centre. On the other hand, there is some evidence, in the natural flow, of weaker layers that do extend deeper into the core.

Also, the shear layers in the natural channel are three dimensional structures, with a spanwise extent of no more than $0.3h$, or about 55 wall units, at $y^+ = 6$ (or of $0.4h$ at the wall). They appear to be rooted at the wall in elliptical "hot" spots in which the spanwise vorticity is at least 25% higher than its average wall value, and which can be used to detect and count the protruding layers and to follow their motion. These spots move with a convection velocity of 60% of the centre line velocity, or $0.66U_m$. When they are ex-

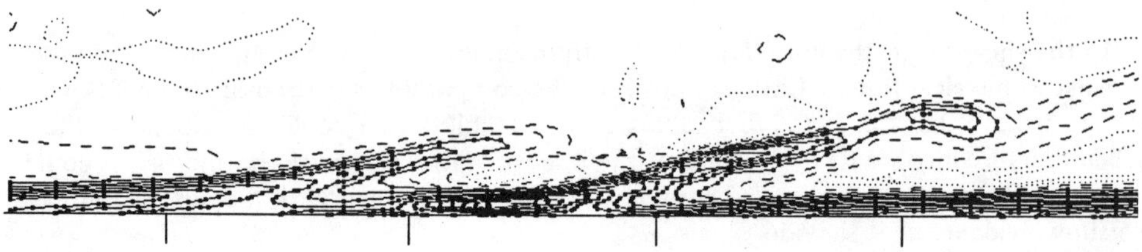

FIGURE 7. Lateral view of a the projecting z-vorticity structures in the viscous sublayer. Dotted lines correspond to $\omega_z h/U_m = -1.6$ and 0 ; dashed: 1.6 and 3.2; solid: 4.8 to 26 . Average vorticity at the wall is $\omega_z h/U_m = 11.5$. Each horizontal tick mark represents one half channel width ($x^+ = 180$) . Total vertical extent, $y^+ = 75$.

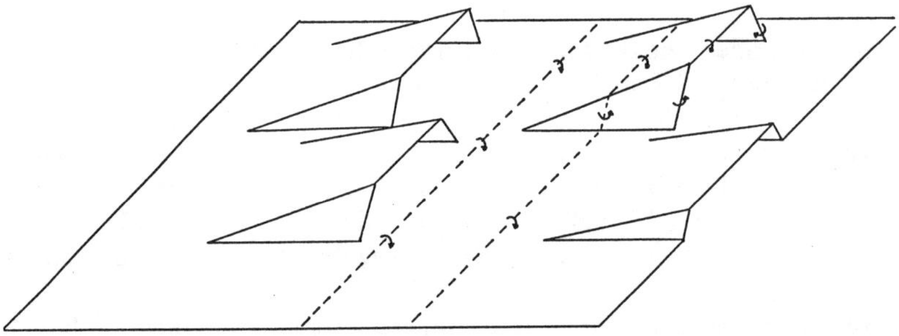

FIGURE 9. A model for the sublayer structures. The raised structures are lifted portions of wall vorticity, and are "supported" by vorticity sidewalls that are the origin of the y -vorticity streaks.

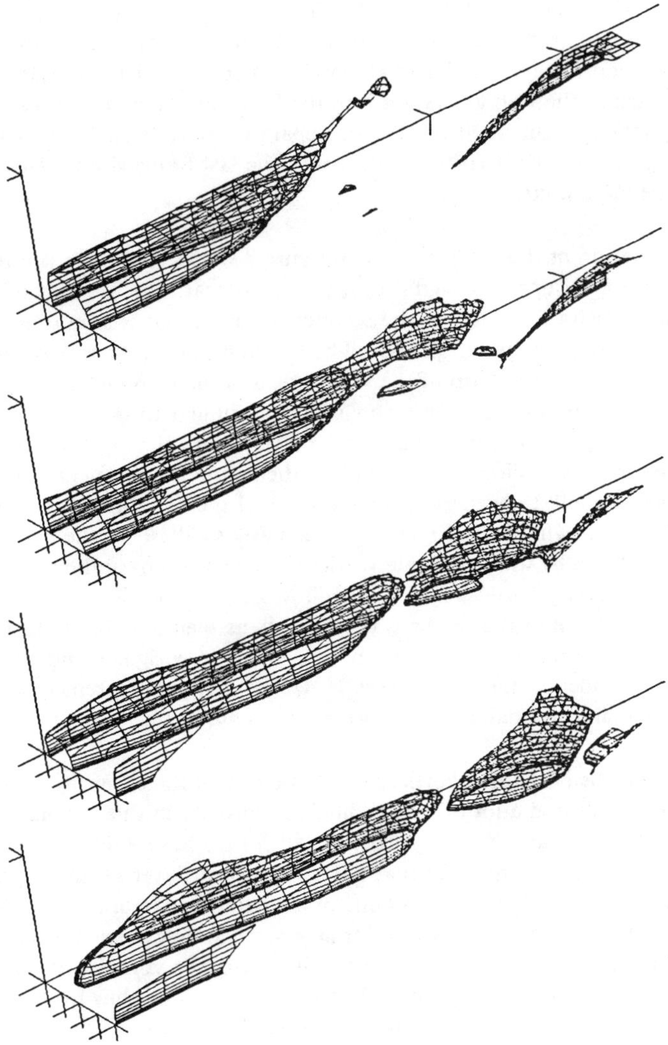

FIGURE 8. Splitting process of a structure to create a new one. See text for explanation. Time difference between frames is approximately 6 wall units, running from top to bottom.

tended into the flow in the form of three dimensional iso-surfaces of z-vorticity, they form a "forest" of leaning curving "necks" that covers much of the wall.

From our point of view, the main difference is that these are not equilibrium structures, and that they can be seen to evolve and regenerate in time, giving origin to new structures, in a process which is shown in Fig. 8. In the first frame of that time sequence, a structure has began to stretch, producing a small vorticity blob at its top end. In the next one, the blob has grown considerably, and a small patch of strong vorticity appears at the wall. Finally the vortex at the wall grows out into the boundary layer and fuses with the tip of the stretching layer. At this moment, the tip separates from its parent structure, forming what appears to be the "embryo" of a new spot. The last frame shows both spots as essentially independent units.

A closer examination of the vorticity field in figure 7 shows that there is a region of concentrated z-vorticity of opposite sign (negative), underneath the top part of the structure, and this is the vorticity that induces the secondary vortex that is seen to appear in the wall in figure 8. Later, the updraft produced by this induced vortex and by the old negative one, drives the splitting of the tip of the structure, and the reconnection to the new one. The whole regeneration mechanism is now strikingly similar to the limit cycle in the two dimensional calculations. The protruding shear layers mark the locations of saddle points, which provide the basic updrafts that maintain them, and the regeneration mechanism shares the same site and the general characteristics of the two dimensional instability. It is not clear, however, what is that maintains the basic saddles. In the two dimensional flow, they are induced by the large scale vorticity concentrations in the core flow, and only in part by the secondary vortices near the wall. Moreover, the effect of the ejection instability on the position of the saddle is quite small, as seen in figure 6. In this case, the ejections are much stronger, and the alternation of vorticity signs seems to be strong enough to induce much of the saddle flow. However, it should be remembered that the flow is now three dimensional, and that other effects are at play.

There seems to be no *hairpin* vortices in the vicinity of the shear layers described here. At least we could not find a lot of longitudinal (x) vorticity in this region. There is, however, a fair amount of transverse (y) vorticity, which may have a dynamic effect. In fact, it takes only a little reflection to see that a detached shear layer of finite spanwise extent in the vicinity of the wall should look as one of the patches in figure 9. The "walls" of these patches are just the vortex lines of the shear layers, when they come down to join the wall, which is the only source of concentrated vorticity. Also, the vorticity in this vertical walls tends to pump fluid underneath the detached shear layers, and thus helps in the maintenance of the saddle points. In fact, it is easy to see that a series of such layers will induce something like the low velocity streaks in the sublayer, and evidence to this effect is presented in (Jiménez *et al.*, 1987).

Conclusions

We have presented evidence that a two dimensional saddle-point instability mechanism is present in strictly two dimensional Poiseuille flows, and that it results in vorticity being ejected away from the wall, in a manner reminiscent of the ejections in natural turbulent channels. Then, we have shown that the sublayer of natural wall turbulent flows contains structures which have many features in common with the two dimensional phenomena, and which also seem to be responsible for at least some of the vorticity ejections.

It should be emphasised that these observations relate only to the sublayer, and that no claim is made about the outer regions of the boundary layer. Indeed we observed strong three dimensional motion in that outer part, that was not necessarily related to the sublayer phenomena. Also, we have noted that there seems to be two kinds of secondary instabilities associated to plane Poiseuille flow, and that the site of the "classical" one are the vorticity concentrations in the body of the flow, away from the wall. Therefore, that process is probably relevant to the highly three dimensional motion in the outer region, while the phenomena described here are presumably most closely related to the behaviour of the sublayer.

Acknowledgements.

The two dimensional work was supported in part by grants of computer time at the IBM-3090s of the CIEMAT Centre at Madrid and of the Thomas J. Watson IBM Research Centre at Yorktown Heights (New York), as well as by the Joint Spain-US Committee for Scientific and Technological Cooperation, under contract CCA-8510/057. The three dimensional work was carried out mostly during the 1987 Summer Program of the Centre for Turbulence Research at Stanford University/ NASA Ames. The support, and the cooperation of the Centre staff, especially of Drs. Moin, Moser and Keefe, are gratefully acknowledged.

References

Cantwell,B.J., Organized motion in turbulent flow, *Ann. Rev. Fluid Mech.*, **13**, 457-515, 1981

Deville,M., Kleiser,L. and Montigny-Rannou,F., Pressure and time treatment for Chebyshev spectral solution of a Stokes problem, *Int. J. Num. Methods Fluids*, **4**, 1149-1163, 1984

Didden,N. and Ho,C.H., Unsteady separation in a boundary layer produced by an impinging jet, *J. Fluid Mech.*, **160**, 235-256, 1985.

Gottlieb,D. and Orszag,S.A., *Numerical analysis of spectral methods*, p. 119, SIAM, Philadelphia, 1977.

Herbert,T., Periodic secondary motions in a plane channel, in *Proc. 5th. Int. Conf. on Numerical Methods in Fluid Dynamics*, edited by A.I. Van de Vooren and P.J.Zandbergen, pp. 235-240, Springer, Berlin, 1976

Herbert,T., Secondary instability of plane channel flow to subharmonic disturbances, *Phys. Fluids*, **26**, 871-874, 1983

Jiménez,J., Bifurcations and bursting in two dimensional Poiseuille flow, *Phys. Fluids*, **30**, 3644-3646, 1987.

Jiménez,J., Moin,R., Moser,R. and Keefe,L., Ejection mechanisms in the sublayer of a turbulent channel, *Proc. Summer Program Center Turb. Research*, ed. P.Moin, 1987.

Kim,J., Evolution of a vortical structure associated with the bursting event in a channel flow, in *Turbulent Shear Flows*, **5**, pp. 221-233, Springer, Berlin, 1987.

Kim,J., Moin,P. and Moser,R., Turbulence statistics in fully developed channel flow at low Reynolds number, *J.Fluid Mech.*, **177**, 133-166, 1987.

Lamb,H., *Hydrodynamics*, p. 225, Dover, N.Y., 1945.

Landman,M.J. and Saffman,P.G., On the three dimensional instability of strained vortices in a viscous fluid, *submitted to Phys. Fluids*, 1987

Orszag,S.A., Numerical simulation of incompressible flows with simple boundaries. I. Galerkin (spectral) representations. *Studies in Appl. Maths.*, **50** (4), 293-327, 1971

Orszag,S.A. and Patera,A.T., Secondary instability of wall bounded shear flows, *J. Fluid Mech.*, **128**, 347-385, 1983.

Papailiou,D.D. and Lykoudis,P.S., Turbulent vortex streets and the entrainment mechanism of the turbulent wake, *J. Fluid Mech.*, **62**, 11-31, 1974.

Pierrehumbert,R.T., Universal short-wave instability of two-dimensional eddies in an inviscid fluid, *Phys. Rev. Lett.*, **57**, 2157, 1986.

Pierrehumbert,R.T. and Widnall,S.E., The two and three dimensional instabilities of a spatially periodisc shear layer, *J.Fluid Mech.*, **114**, 59-82, 1982.

Pugh,J.D., Finite amplitude waves in plane Poiseuille flow, *Ph. D. Thesis*, Caltech, Pasadena, August 1987.

Roshko,A., Structure of turbulent shear flows: a new look, *AIAA J.*, **14**, 1349-1357, 1976

Saffman,P.G., Vortices, Stability and Turbulence, *Annals N.Y. Acad. Sci.*, **404**,12-24, 1983

Walker,J.D.A., Smith,C.R., Cerra,A.W. and Doligalski,T.L., The impact of a vortex ring on a wall, *J. Fluid Mech.*, **181**, 99-140, 1987.

Zahn,J.P., Toomre,J., Spiegel,E.A. and Gough,D.O., Nonlinear cellular motions in Poiseuille channel flow, *J. Fluid Mech.*, **64**, 319-345, 1974.

Statistical Analysis of Near-Wall Structures in Turbulent Channel Flow

ROBERT D. MOSER
NASA Ames Research Center
Moffett Field, California 94035, USA

1. INTRODUCTION

Over three decades ago, the existence of organized motions or eddies in turbulent shear flows was recognized by Theodorsen (1952), and Townsend (1956) . In the past twenty years , a great deal of insight has been gained into the characteristics of organized structures in turbulent shear flows primarily by means of flow visualization and conditional sampling techniques (see Cantwell 1981). Combined flow visualization and quantitative techniques (*e.g.*, Kline *et al.* 1967, Falco 1977) have demonstrated the significance of certain events (*e.g.*, bursting process) or structures in the turbulence production mechanism.

Unfortunately, the present body of knowledge regarding organized motions has seldom been used in turbulence theories or quantitative models of turbulence. This is due in part to the lack of a quantitative definition of organized structures and an objective means for assessing their contribution to turbulence stresses, particularly their importance in the production of turbulence.

Much of our knowledge of coherent motions is limited to those structures that can be *seen* in flow visualization experiments. Most flow visualization studies have been carried out at low Reynolds numbers, where the limited range of turbulence scales makes it easier to identify organized motions. It is desirable to have the means to extract coherent motions from fields and evaluate their contribution to turbulence statistics, regardless of how chaotic the fields are.

Lumley (1967) proposed a mathematically attractive definition of organized structures and a statistical method for their extraction from turbulent velocity fields. The method is based on the decomposition of the fluctuating velocity field into a sum of mutually orthogonal eigenfunctions of the two-point correlation tensor, weighted by random coefficients. The dominant (most energetic) eddy is defined to be the eigenfunction with the largest eigenvalue. This is the Karhunen-Loeve expansion (Loeve 1955, Papoulis 1965), which is used for directions in which turbulence is statistically inhomogeneous. An important feature of this decomposition, which is a result of its orthogonality properties, is that the contribution of the extracted eddies to second-order turbulence statistics can be determined. For directions in which the turbulence is homogeneous, the Karhunen-Loeve decomposition is *not* instructive, so a generalization of the shot noise decomposition of Rice (1944) is used.

Another technique for extracting coherent structures is to determine the best mean-square estimate of the velocity field in a region of the flow, when data or conditions are given at one or more points in the region. This is the conditional average with the given data as conditions. In linear stochastic estimation we determine the best mean-square estimate of the velocity which is *linear* in the given data (Adrian 1975). With this restriction, the stochastic estimate can be easily determined in terms of the two-point correlation tensor. One advantage of this approach is that the two-point correlation, and therefore the linear estimate, can be computed from a much larger statistical sample than would be possible for conditional averaging since the entire velocity field can contribute to the averages rather than just the points satisfying the conditions. Also, once the two-point correlation tensor is available, stochastic estimates for any desired conditions may be obtained without collecting more statistics. Linear stochastic estimation has been shown to produce a very good approximation to the conditional average (Adrian, Moin & Moser 1987, Adrian & Moin 1988, see also the paper by Adrian in this volume), and can therefore be used as an approximation to conditional averages. This also implies that the information available from a conditional average is contained in long-time averaged *unconditioned* statistics (the two-point correlation).

Both proper orthogonal decomposition and linear stochastic estimation require knowledge of the two-point correlation tensor, preferably with separations in all three spatial directions (and possibly in time). This tensor is a difficult quantity to measure experimentally because of the requirement for two probes with arbitrary spatial separation. The three-dimensional correlation tensor has been measured in boundary layers (Herzog 1987) and free shear flows (Glauser & George 1987), but such measurements usually suffer from inadequate spatial resolution of the correlation tensor, especially in the inhomogeneous spatial direction. These difficulties are avoided when results from direct numerical simulation of turbulence are used to compute the two-point correlation tensor. Numerical simulation can provide the entire velocity field as a function of space and time, which can be used to compute the two-point correlation tensor in a straightforward way (Moin & Moser 1988).

The two-point correlation used to produce the results presented here was obtained from the direct numerical simulation of turbulence in a channel (Kim, Moin & Moser 1987). The simulation database consists of instantaneous three-dimensional velocity and pressure fields collected at widely separated flow times. Calculations were performed at Reynolds number 3200 based on the centerline velocity, $U_\circ$, and channel half width, δ. The channel centerline corresponds to $y^+ = yu_\tau/\nu = 180$, where $u_\tau = \sqrt{\tau_w/\rho}$ is the wall shear velocity. The computations were carried out with $128 \times 129 \times 128$ grid points in the x, y, and z directions, respectively. The mean flow is in the x-direction and y is in the direction normal to the walls. Total averaging time was about $190\delta/U_\circ$. The physical realism of the data has been verified by detailed comparison of statistical correlations and both instantaneous and conditionally averaged flow patterns with available experimental data.

In §2, the theoretical background for proper orthogonal decomposition and stochastic estimation is briefly reviewed. Results from both techniques along with relevant discussion are presented in §3. Finally, §4 contains some concluding remarks.

2. THEORETICAL BACKGROUND

The theory of proper orthogonal decomposition as used here in the study of coherent structures has been presented in detail by Lumley (1970, 1981) and Moin & Moser (1988). A brief review of the important ideas is included here.

In one dimension, the Karhunen-Loeve decomposition (or proper orthogonal decomposition) requires solution of the following eigenvalue problem

$$\int_D R_{ij}(y,y')\phi_j(y')dy' = \lambda\phi_i(y), \tag{1}$$

where $R_{ij} =< v_i(y)v_j(y') >$ is the two-point correlation of the velocity v, $<>$ denotes ensemble average, and D is the domain over which the decomposition is performed. It can be shown that (1) does not have a unique solution; instead there are a denumerable infinity of solutions, $\phi_i^{(n)}(y)$, which can be normalized such that

$$\int_D \phi_i^{(n)}(y)\phi_i^{(m)}(y)dy = \delta_{nm}. \tag{2}$$

Orthogonality implies that structures of different order contribute independently to second order statistics. Each eigenfunction $\phi_i^{(n)}(y)$ is associated with a real positive eigenvalue $\lambda^{(n)}$ and the eigenfunctions form a complete set. That is, the random vector field v_i, can be reconstructed from the eigenfunctions

$$v_i(y) = \sum_n a_n\phi_i^{(n)}(y), \tag{3}$$

where coefficients (a_n) of different order are uncorrelated

$$< a_n a_m^* >= \begin{cases} \lambda^{(n)} & n = m, \\ 0 & n \neq m. \end{cases} \tag{4}$$

Equation (3) can be interpreted as the decomposition of the stochastic field v_i into deterministic elements (or eddies) with random coefficients. We expect more deterministic velocity fields to be more efficiently represented by the expansion in (3). An important consequence of equation (4) is that the contribution of each structure to the turbulent kinetic energy and turbulence stresses can be determined

$$< v_i(y)v_j(y) >= \sum_n \lambda^{(n)}\phi_i^{(n)}(y)\phi_j^{(n)}(y), \tag{5}$$

and

$$E = \int_D < v_iv_i > dy = \sum_n \lambda^{(n)}, \tag{6}$$

where E is the total turbulent kinetic energy in the domain. The eigenvalues $\lambda^{(n)}$ thus represent the contribution of each structure to total turbulent kinetic energy.

Since the eigenvalues are real, positive, and bounded, they can be ordered from largest to smallest. It can then be shown that the expansion in (3) converges optimally in the sense that a truncation of the sum in (3) to l terms yields the l-term approximation with maximum contribution to the turbulence kinetic energy in (6). In particular, the first term gives the best possible representation (in the above sense) of the stochastic velocity by a deterministic function. The first term $\phi_i^{(1)}$ will thus be referred to as the dominant or characteristic eddy. It is also optimum in the sense that it is the normalized deterministic function which maximizes $< \int_D v_i \phi_i \, dy >$. This analysis is easily extended to multiple dimensions.

The above analysis is valid for any vector field; in particular, the vorticity could be used instead of the velocity. In that case, R_{ij} would be replaced by Ω_{ij}, the two-point vorticity correlation, and the energy E in (6) would be replaced by the enstrophy. The resulting characteristic eddy would then be the best representation of the vorticity field. Note that Ω_{ij} can be computed from R_{ij} by differentiation.

If the stochastic process v_i is homogeneous in a spatial direction (*e.g.*, z) then it can be shown that the eigenfunctions of (1) are the complex exponentials e^{ikz}. This is not acceptable as a characteristic eddy because it is not compact in space (we expect an "eddy" to be compact), and it reflects only the homogeneity of the turbulence, not its structure. For this reason, a generalization of the "shot noise" decomposition (Rice 1944, Lumley 1981) is used in homogeneous spatial directions. In this decomposition a homogeneous stochastic process ($a(z)$, for example) is represented as

$$a(z) = \int f(z - z')g(z')\, dz' \tag{7}$$

where $f(z)$ is a deterministic function and $g(z)$ is another stochastic process. This decomposition is not unique. To select a particular decomposition of the form (7), properties of $g(z)$ and/or $f(z)$ must be specified. In the results presented here, g was required to satisfy

$$< g(z)g(z') >= \delta(z - z'), \tag{8}$$

which allows the magnitudes of the Fourier coefficients of f to be determined (Moin & Moser 1988). To determine the phases of the Fourier coefficients, three different criteria have been used: 1) the bispectrum of g is made to be as close as possible to real and positive, which is the condition analogous to (8) for the third order statistics of g; 2) the characteristic eddy is made to be compact in space, and 3) the characteristic eddy is made to have a Fourier transform which is continuous with respect to the wave numbers (see Moin & Moser (1988) for more details). The results from all three criteria are essentially the same, therefore, the results in §3 include only one of the criteria (compactness in space). Note that there are other criteria which could be used in place of (8) and the phase conditions discussed above (see Moin & Moser 1988). One of the consequences of using (8) is that the magnitudes of the velocity in the characteristic eddy will be very large. However, this magnitude is essentially arbitrary since the right-hand side of (8) could equally well be multiplied by a real positive constant.

The theory of linear stochastic estimation as applied to turbulence was worked out by Adrian (Adrian 1975, Adrian & Moin 1988), and is covered in the paper by Adrian in this volume. In this work we will seek the best mean-square estimate $\tilde{v}_i(x,y,z)$ of the velocity field $v_i(x,y,z)$, given the velocity at a point ($u_i(0,y',0)$ say), which is linear in the given velocity. Thus

$$\tilde{v}_i(x,y,z) = E_{ij}(x,y,z)u_j, \tag{9}$$

where E_{ij} is found by requiring that

$$\left\langle (v_i(x,y,z) - \tilde{v}_i(x,y,z))^2 \right\rangle = \text{minimum}, \tag{10}$$

where it is understood that the average in (10) is the average conditioned on $v_i(0,y',0) = u_i(0,y',0)$. It can be shown that (10) is satisfied when

$$R_{ij}(0,y',y',0)\tilde{v}_j(x,y,z) = R_{ij}(x,y,y',z)u_j, \tag{11}$$

which can be solved to find E_{ij} and thus $\tilde{v}_j$. This analysis can be extended to conditions at multiple points, and conditions on quantities other than velocity (Adrian, Moin & Moser 1987, Adrian & Moin 1988).

The major difficulty in stochastic estimation, or any conditional averaging technique, is the selection of conditions. Great care must be exercised in selecting conditions which are likely to occur in the flow and which are of dynamical importance. Two methods were used to select conditions in this work. 1) Conditions were chosen to have maximum contribution to Reynolds shear stress as determined from the joint-probability-density function of velocity components at a single point. 2) Conditions were chosen to be velocities observed in instantaneous velocity fields associated with instantaneous events of interest. It will be seen that significantly different structures are obtained when different conditions are used.

3. RESULTS AND DISCUSSION

The proper orthogonal decomposition technique discussed in §2 can be applied using the two-point correlation tensor with separations in one (y), two or three dimensions. Also, the technique can be used in any subdomain of the inhomogeneous flow domain. The decomposition results presented below include results in one (y), two (y-z), and three dimensions as well as several subdomains. Stochastic estimation can be applied using any number of different conditioning events. A variety of such stochastic estimates have been computed. The results presented below have been selected to illustrate some of the properties of the techniques and to facilitate comparisons of the results from these two methods with each other and with instantaneous velocity fields. Results pertaining to a particular structure which has been observed in instantaneous velocity fields are also included. Other results obtained by these techniques may be found elsewhere (Moin & Moser 1988, Adrian, Moin & Moser 1987).

TABLE 1. Contributions of the Eigenfunctions to Energy and Production

Domain	Dimensions	$E^{(1)}/E$	$E^{(2)}/E$	$E^{(3)}/E$	$E^{(1)}/E^{(2)}$	$\Pi^{(1)}/P^{(t)}$
$0 \leq y^+ \leq 40$	1	0.61	0.15	0.08	4.2	1.03
$140 \leq y^+ \leq 180$	1	0.44	0.22	0.20	2.0	1.94
$0 \leq y^+ \leq 180$	1	0.32	0.16	0.08	2.0	0.66
$0 \leq y^+ \leq 40$	2	0.64	0.17	0.07	3.8	1.05
$140 \leq y^+ \leq 180$	2	0.46	0.24	0.18	1.9	1.55
$0 \leq y^+ \leq 180$	2	0.38	0.16	0.09	2.3	0.77
$0 \leq y^+ \leq 40$	3	0.76	0.12	0.06	6.0	1.04
$140 \leq y^+ \leq 180$	3	0.57	0.22	0.12	2.5	1.53
$0 \leq y^+ \leq 180$	3	0.50	0.16	0.08	3.1	0.75

3.1. Proper Orthogonal Decomposition

The key measure of the usefulness of the proper-orthogonal decomposition is the speed of convergence of the series (6), that is, the fraction of the turbulent kinetic energy contributed by the first few eigenfunctions. In three dimensions, with the x- and z-directions homogeneous, there is a set of eigenvalues $\lambda^{(n)}$ and eigenvectors $\phi^{(n)}$ for each wave number pair (k_x, k_z). In this case we are interested in the contribution of the first eigenmodes of all the wave numbers to the kinetic energy. Therefore, in table 1, the quantity

$$E^{(n)} = \sum_{k_x, k_z} \lambda^{(n)}(k_x, k_z) \tag{12}$$

is shown for the first three characteristic eddies in one, two, and three dimensions and for three subdomains. For dimensions less than three, the sum in (12) over the wave numbers of the missing dimensions does not appear. As an indicator of the contribution of the dominant eddy to Reynolds shear stress and therefore turbulence production, the quantity $\Pi^{(1)}/P^{(t)}$ is also tabulated, where

$$\Pi^{(1)} = \sum_{k_x} \sum_{k_z} \int_{y_l}^{y_u} \lambda(k_x, k_z) \hat{\phi}_1^{(1)}(k_x, y, k_z) \hat{\phi}_2^{(1)*}(k_x, y, k_z)\, dy \tag{13}$$

and $P^{(t)}$ is the integral over the subdomain of the Reynolds shear stress.

In all cases the dominant eddy makes a significant contribution to turbulent kinetic energy (as much as 76%) and production. The contribution of the dominant eddy to turbulent kinetic energy becomes somewhat higher as the number of dimensions in which the decomposition is performed increases. Note also that the energy contribution of the dominant eddy is much greater for the near-wall domain than for the same size domain at the center of the channel. This suggests that the near-wall turbulence is more organized than that away from the wall.

As another indication of the convergence of the decomposition, partial sums of the first five terms of the sum in (5) for $i = j = 1$ are plotted in figure 1. Note that

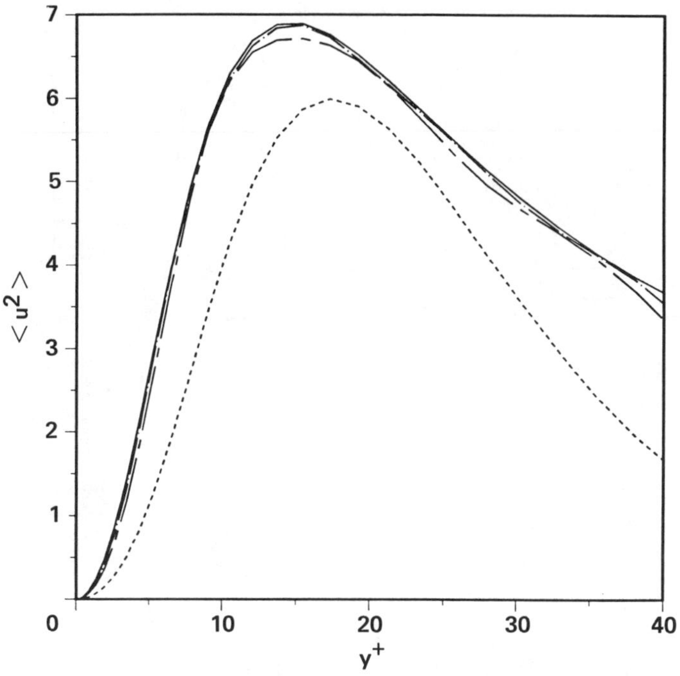

FIGURE 1. Convergence of u^2 in the near-wall domain: ——— , total; - - - - , first eigenfunction; — - — sum of first three eigenfunctions; — · — sum of first five eigenfunctions.

three to five terms are enough to provide a good representation of the variance of u. A more complete set of convergence plots are given in Moin & Moser (1988).

It is also of interest to determine the extent to which characteristic eddies can be "seen" in the instantaneous velocity fields. To this end, consider the two-dimensional decomposition of the velocity in the y-z plane in the near-wall subdomain ($y^+ < 40$). In the two-dimensional decomposition, we wish to extract the dominant eddy as viewed in planes. The relevant laboratory experiment would be to illuminate a y-z plane with a light source in a smoke-filled channel flow and to identify characteristic eddies in this plane. The characteristic eddy (figure 2) is seen to consist of a region of low streamwise velocity straddled by a pair of streamwise counter-rotating vortices. Note that the symmetry of the eddy in z is an artifact of the symmetries in the two-point correlation tensor and the phase recovery technique (see Moin & Moser 1988). To locate a characteristic eddy in an instantaneous velocity field, sample functions of the stochastic function $g(z)$ (equation 7) are computed. An example obtained at a particular x and time is shown in figure 3a. In this figure, there is a large, isolated positive excursion of g from zero at about $z = 3.1\delta$, which should correspond to a structure similar to the characteristic eddy. A vector plot of the velocity in the y-z plane centered at this location is shown in figure 3b for comparison with figure 2b. As in the characteristic eddy, there is a large

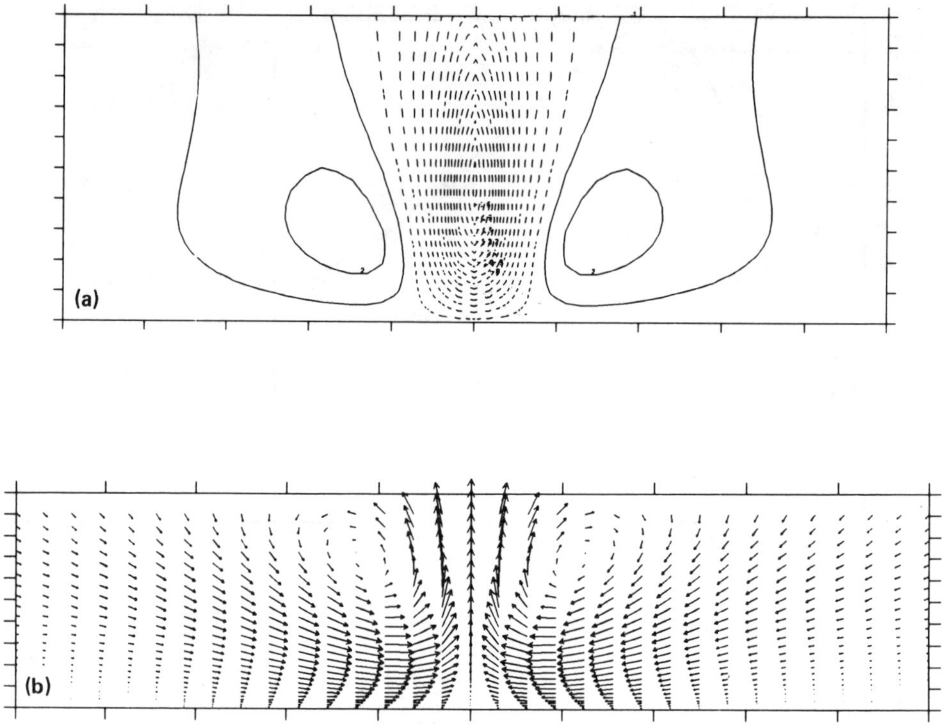

FIGURE 2. Two-dimensional (y-z) characteristic eddy in the near-wall domain ($y^+ \leq 40$), *(a)* contours of u and *(b)* velocity vectors projected into the y-z plane. Contour increment in *(a)* is 1. Negative contours are dashed. The plot domain extends *(a)* 320 and *(b)* 190 wall units in z and 40 wall units in y. In *(a)* , the vertical scale is magnified by a factor of 3.

streamwise vortex to the left of the domain, with a flow away from the wall at the center of the domain. The shape of this vortex is somewhat different from the vortices in the characteristic eddy. The major difference between the instantaneous structure and the characteristic eddy is the lack of a counterrotating partner in the instantaneous field. Thus, the major difference in this case is caused by the "artificial" symmetry imposed on the characteristic eddy.

The structure of the three-dimensional characteristic eddy (based on velocity) in the near-wall subdomain ($y^+ < 40$) is shown in figure 4. Near the wall, it consists of an elongated region of low streamwise velocity coincident with a much shorter region of positive vertical velocity. The spanwise extent of the low speed region is approximately 45 wall units. This eddy may be considered an *ejection* from the wall, and is the type of event detected by quadrant analysis (quadrant 2) (Wallace, Eckelmann & Brodkey, 1972). A pair of counter-rotating streamwise vortices

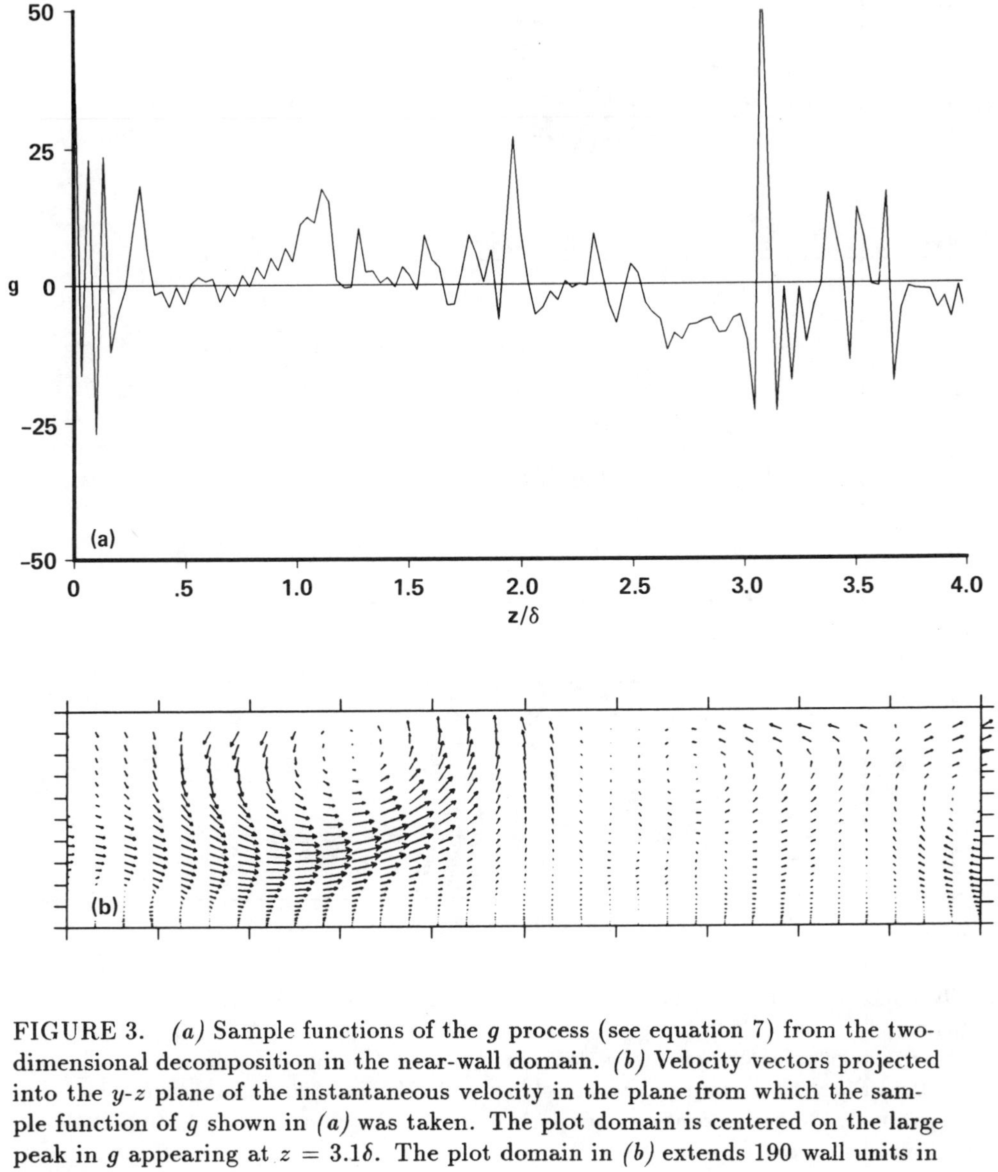

FIGURE 3. *(a)* Sample functions of the g process (see equation 7) from the two-dimensional decomposition in the near-wall domain. *(b)* Velocity vectors projected into the y-z plane of the instantaneous velocity in the plane from which the sample function of g shown in *(a)* was taken. The plot domain is centered on the large peak in g appearing at $z = 3.1\delta$. The plot domain in *(b)* extends 190 wall units in z and 40 wall units in y.

(figure 4c) straddles the low-speed region. Again the appearance of a *pair* is an artifact of the statistical symmetries. The vertical location of these vortices is dependent on the streamwise location as shown in figure 5. Also shown in figure 5 is the vertical location of the vortices in the wall-to-centerline domain ($y^+ < 180$). Note that the angle of inclination of the vortex increases with downstream location. At $x^+ = -36$ the angle is $10°$, while at $x^+ = 18$ the angle is $31°$ in the near-wall domain and as much as $60°$ in the wall-to-centerline domain. Upstream of $x^+ = -54$

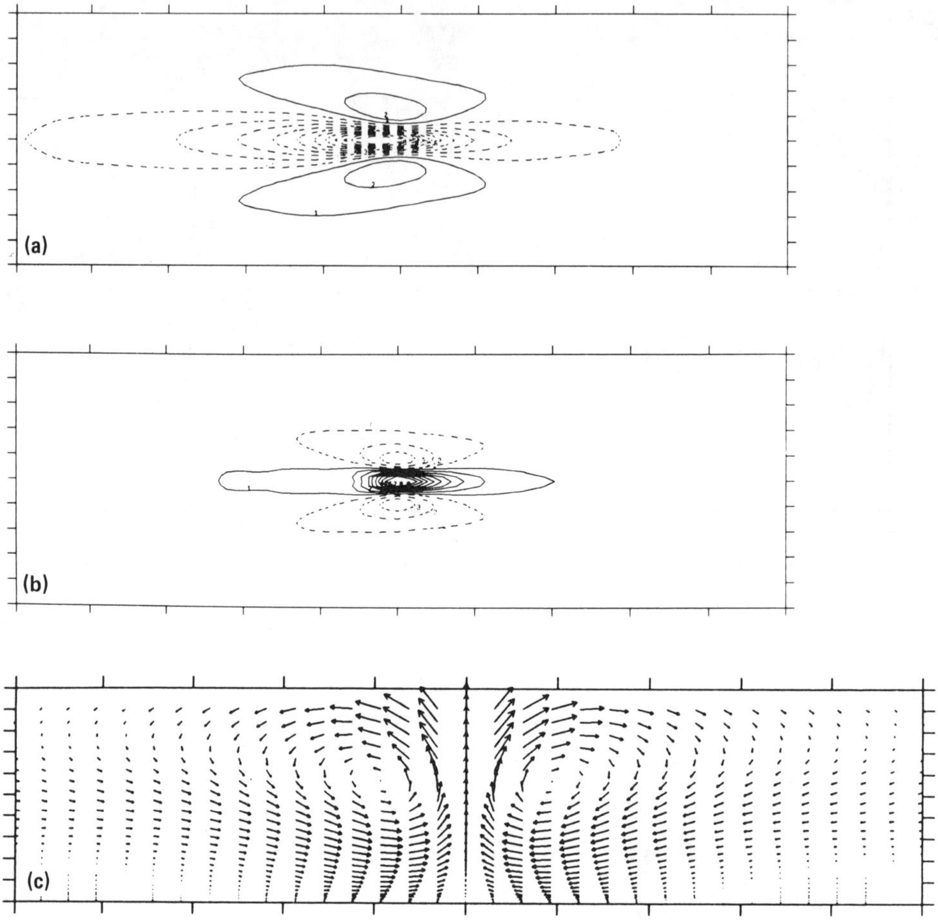

FIGURE 4. Three-dimensional velocity characteristic eddy in the near-wall domain ($y^+ \leq 40$). Contours of *(a)* u *(b)* v in the x-z plane at $y^+ = 10$, and *(c)* velocity vectors projected into y-z planes at $x = 0$. Contour increment in *(a)* 5 and *(b)* 0.6. The plot domain in *(a)* and *(b)* extends 960 and 320 wall units in x and z, respectively. Negative contours are dashed. In *(c)* the domain extends 40 and 190 wall units in y and z, respectively.

(-90) in the near-wall (wall-to-centerline) domain, the vortices are so weak that they are not discernible. Thus, in the wall region ($y^+ < 40$), the vortices have a streamwise extent of less than 100 wall units. This is consistent with the observations of streamwise vortices in instantaneous velocity fields made by Moser & Moin (1984), in which the vortices were observed to go beyond $y^+ = 40$ in a very short streamwise distance (150 wall units).

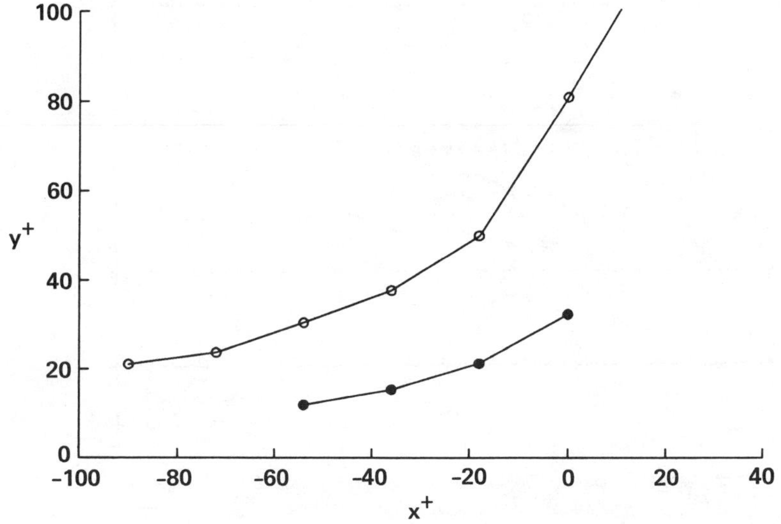

FIGURE 5. Vertical (y) location of the center of the streamwise vortices as a function of streamwise distance (x): •, near-wall domain ($y^+ \leq 40$); ○ , wall-to-centerline domain ($y^+ \leq 180$).

When the characteristic eddy is based on the decomposition of the vorticity field (eigenfunctions of the two-point vorticity-vorticity correlation), a different eddy structure is obtained. The result of the vorticity decomposition in the wall-to-centerline domain is shown in figure 6. In the x-z plane at $y^+ = 30$, the streamwise and normal velocity both have a complicated pattern. Note that near the center of the domain there is a strong streamwise gradient of both u and v. Upstream of this point, the velocities correspond to a sweep (quadrant 4) and downstream they correspond to an ejection (quadrant 2). This strong gradient in streamwise velocity is the type of event detected by the VITA technique (Blackwelder & Kaplan 1976). The region of strong streamwise gradients is a cross section of an inclined shear-layer, which coincides with the inclined region of large, negative, spanwise vorticity (ω_z) shown in figure 6*c*. This protruding region of negative ω_z is similar to that observed by Jimenez *et al* (1987); an example of the structures observed by Jimenez taken from an instantaneous velocity field is shown in figure 7.

3.2. Stochastic Estimation

Two representative examples of stochastic estimation are presented here for comparison with the decomposition results presented above. In the first case, a single event is specified at $y^+ = 19$. The velocity condition is chosen to correspond to the second quadrant maximum of $uvf(u,v)$, where $f(u,v)$ is the joint probability density function of the the velocities u and v at $y^+ = 19$ ($u = -4.3$, $v = 0.55$, $w = 0$ in wall units). This corresponds to the event which has the largest contribution to the

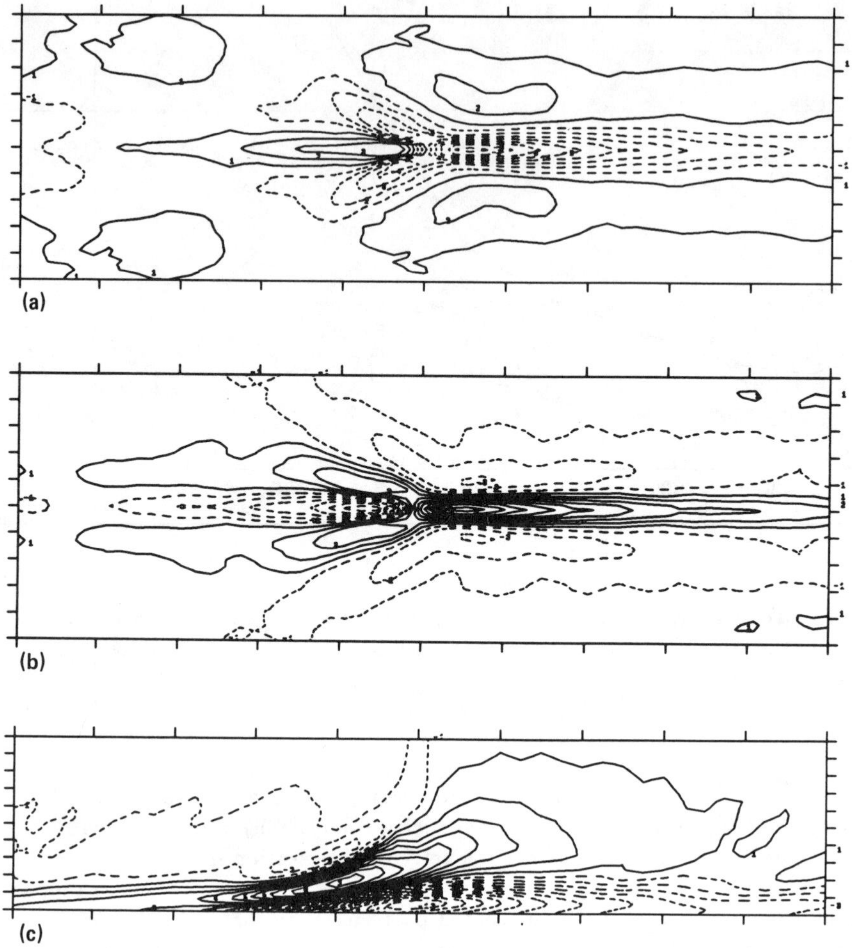

FIGURE 6. Three-dimensional vorticity characteristic eddy in the wall-to-centerline domain ($y^+ \leq 180$). Contours of *(a)* u *(b)* v in the x-z plane at $y^+ = 30$, and *(c)* contours of ω_z in the x-y plane at $z = 0$. Contour increment in *(a)* 3, *(b)* 0.7 and *(c)* 60. The plot domain in *(a)* and *(b)* extends 960, and 320 wall units in x and z respectively. In *(c)* the domain extends 700 and 76 wall units in x and y respectively, and the vertical scale is magnified by a factor of 2. In *(a)* and *(b)*, negative contours are dashed, in *(c) Positive contours are dashed.*

Reynolds shear stress. Contours of the streamwise and normal velocities in the x-z plane at $y^+ = 10$ are shown in figure 8. As in the decomposition results (figure 4) there is an elongated region of low streamwise velocity and a corresponding region of positive normal velocity. Note that in the stochastic estimation results, the region of positive v is more elongated than in the decomposition results; however the spanwise dimensions are comparable.

FIGURE 7. Contours of instantaneous ω_z in the turbulent channel (Kim, Moin & Moser 1987) in a x-y plane. Contour increment is 20, and *positive* contours are dashed. The plot domain extends 700 and 76 wall units in x and y, respectively, and the vertical scale is magnified by a factor of 2.

A two-point stochastic estimate was used to investigate the structure of the velocity field near the inclined shear layer shown in figure 7. It was observed in the instantaneous velocity field that upstream of the shear-layer there was a local maximum in the value of uv which corresponded to a sweep (quadrant 4), and downstream there was a local maximum corresponding to an ejection (quadrant 2). The velocities at these two local maxima in uv were used as conditions for stochastic estimation; the result is shown in figure 9. Note that the inclined region of negative ω_z is reproduced (figure 9*c*), though the exact shape and size are different than in figure 7. Also, in the instantaneous field below the "tip" of the shear layer, there is a region of negative ω_z, which is the base of another shear layer structure. This is not reproduced in the stochastic estimate or the proper orthogonal decomposition. Contours of u and v in the x-z plane at $y^+ = 30$ are qualitatively similar to the result of the vorticity decomposition. In particular, the sharp streamwise gradients of u and v appear in the center of the domain, and the regions of positive and negative velocity roughly correspond. However, the stochastic estimation produces a smoother, more organized, velocity field. This may be related to the method by which the phases of the characteristic eddy are determined in the decomposition.

The results of proper orthogonal decomposition indicate that the shear layers discussed above are a dominant feature in the vorticity field, but not the velocity field. Results of stochastic estimation suggest that the shear layers will occur when there is an ejection event (quadrant 2) just downstream of a sweep event (quadrant 4). Unfortunately neither technique as implemented here provides any information about how such structures are formed. In particular these results cannot determine whether the shear layer arises as a consequence of the sweep and ejection or the other way around. To help address these questions, stochastic estimation or proper orthogonal decomposition could be applied in three spatial dimensions *and* time. However, the four dimensional two-point correlation tensor required for such a computation is not currently available from experiments or simulations.

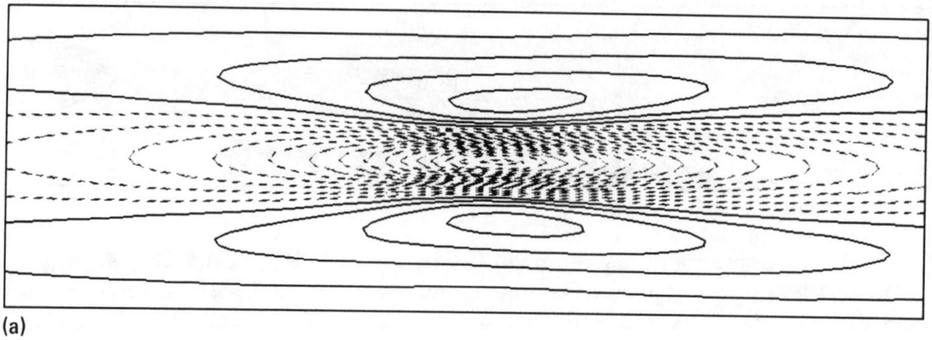

(a)

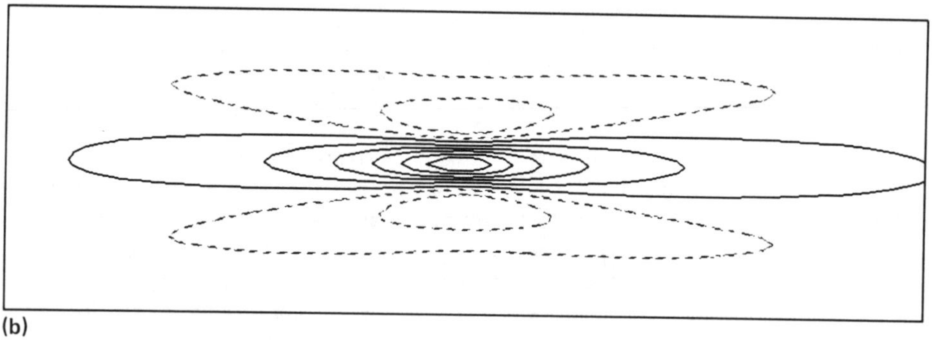

(b)

FIGURE 8. Result of stochastic estimation with a single point condition at $y^+ = 19$. Condition velocities are $u = -4.3$, $v = 0.55$ and $w = 0$. Contours of *(a)* u and *(b)* v in the x-z plane at $y^+ = 10$. Contour levels are *(a)* 0.25 and *(b)* 0.04; negative contours are broken. Plot domains extend 850 and 280 wall units in x and z, respectively.

4. CONCLUSIONS

The results presented in §3 indicate the richness of the structural information which can be extracted from the two-point correlation tensor. These results were possible because of the availability of velocity fields from the direct numerical simulation of turbulence. The use of simulation results to compute the correlation tensor avoids the difficulties associated with measuring this quantity. In particular, the use of simulation results made it possible to obtain the correlation tensor with sufficient spatial resolution to allow accurate differentiation for computing the vorticity correlation, and the resolution of sharp features such the shear layer and its corresponding region of spanwise vorticity discussed in §3.

Proper orthogonal decomposition yielded characteristic eddies which had large contributions to kinetic energy (as much as 76%), and the series in equation (5) converged rapidly. The energy contributions of the characteristic eddy in subdomains

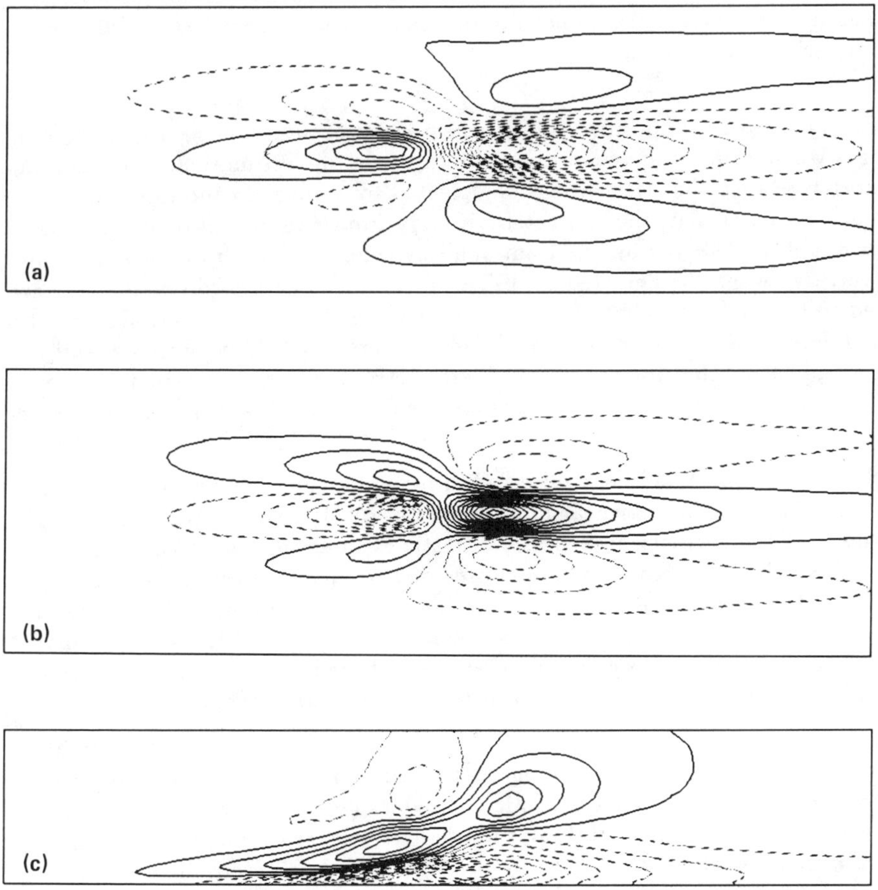

FIGURE 9. Result of two-point stochastic estimation with condition points separated 106 wall units in x and no separation in z. The upstream and downstream conditions are at $y^+ = 21$ and $y^+ = 28$ respectively. Conditions are $u = 2.48\,(-6.53)$, $v = -1.09\,(2.01)$ and $w = -.251\,(-1.17)$ for the upstream (downstream) location. Contours of *(a)* u *(b)* v in the x-z plane at $y^+ = 30$, and *(c)* contours of ω_z in the x-y plane at $z = 0$. Contour increment in *(a)* 0.5, *(b)* 0.15 and *(c)* 10. The plot domain in *(a)* and *(b)* extends 960, and 320 wall units in x and z, respectively. In *(c)* the domain extends 700 and 76 wall units in x and y respectively, and the vertical scale is magnified by a factor of 2. In *(a)* and *(b)* , negative contours are dashed, in *(c) Positive contours are dashed.*

near the wall and away from the wall indicate that the near-wall turbulence is *more* organized than that away from the wall. It was also encouraging that structures similar to the characteristic eddy were observable in instantaneous flow fields. The major differences between characteristic eddies and instantaneous flow structures were caused by the artificial symmetries imposed on the characteristic eddies by the symmetries in the statistics.

Proper orthogonal decomposition and stochastic estimation were seen to yield similar results when appropriate conditions were used for the estimation. The resulting structures were such that they would be detected by standard conditional sampling techniques. In particular, the characteristic eddy from decomposition of the velocity field would be detected by quadrant analysis, while the result of decomposition of the vorticity would be detected by VITA. Thus the structures discussed here are expected to be similar to what would be obtained by these techniques. Further, the results of decomposing the velocity field and the vorticity field were significantly different, indicating that if one finds a coherent structure which dominates the velocity fields, the curl of that velocity structure will *not* dominate the vorticity field.

An inclined shear layer had been observed in instantaneous velocity fields from the channel simulations. Similar shear layers were observed in the results of the decomposition of the vorticity as well as stochastic estimation. The decomposition results and stochastic estimation results provided complementary information in the study of the shear layer. From the proper orthogonal decomposition of the vorticity, we learned that the shear layers are dominant features of the vorticity field. From stochastic estimation we learned that these shear layers are likely to occur when there is an ejection event just downstream of a sweep event.

Finally, the results presented here suggest that caution is required when applying conditional averaging or other stochastic techniques (*e.g.*, proper orthogonal decomposition) to recover structural information. To a large extent, the structures obtained depend on the criteria used by the investigator to determine the structures. For example, the proper orthogonal decomposition may be based on the decomposition of vorticity or velocity or possibly some other quantity, the shot decomposition (equation 7) requires an externally supplied criterion to make it unique, and stochastic estimates as well as conditional averages may be based on virtually any set of conditions. It thus seems impossible to remove the prejudices of the researcher in determining coherent structures within the context of the methods discussed here. Therefore, great care must be exercised to ensure that the conditions and criteria used are realistic and suit the purposes of the investigation.

ACKNOWLEDGMENTS

The author wishes to thank Profs. P. Moin and R. J. Adrian for their numerous contributions to the work discussed here and Dr. L. R. Keefe who made useful comments on a draft of this manuscript.

REFERENCES

Adrian, R. J., 1975, On the Role of Conditional Averages in Turbulence Theory. in *Turbulence in Liquids* ed. J. L. Zakin & G. K. Patterson, Science Press, pp. 323.

Adrian, R. J., Moin, P. & Moser, R. D., 1987, Stochastic Estimation of Conditional Eddies in Turbulent Channel Flow. in *Proceedings of the 1987 Summer Program of the NASA-Stanford Center for Turbulence Research.* **CTR-S87,** pp. 7.

Adrian, R. J. & Moin, P., 1988, Stochastic Estimation of Organized Turbulent Structure: Homogeneous Shear Flow. accepted for publication *J. Fluid Mech.*

Cantwell, B. J., 1981, Organized Motion in Turbulent Flow. *Ann. Rev. Fluid Mech.*. **13,** pp. 457-515.

Blackwelder, R. F. & Kaplan, R. E., 1976, On the Wall Structure of the Turbulent Boundary Layer. *J. Fluid Mech.* **76,** p. 89.

Falco, R. E., 1977, Coherent Motions in the Outer Region of Turbulent Boundary Layers. *Phys. Fluids.* **12,** p. 426.

Glauser, M. N. & George, W. K., 1985, An Orthogonal Decomposition of the Axisymmetric Jet Mixing Layer Utilizing Cross-Wire Velocity Measurements. *Proc. Sixth Symposium on Turbulent Shear Flows, Toulouse, France.*

Herzog, S., 1986, The Large Scale Structure in the Near-Wall Region of Turbulent Pipe Flow. Ph.D. Thesis, Cornell University.

Jimenez, J., Moin, P., Moser, R. D. & Keefe, L. R., 1987, Ejection Mechanisms in the Sublayer of a Turbulent Channel. in *Proc. 1987 Summer Program of the NASA-Stanford Center for Turbulence Research.* **CTR-S87,** p. 37.

Kim, J., Moin, P. & Moser, R., 1987, Turbulence in Fully Developed Channel Flow at Low Reynolds Number. *J. Fluid Mech.* **177,** pp. 133-166.

Kline, S. J., Reynolds, W. C., Schraub, F. A. & Runstadler, P. W., 1967, The Structure of Turbulent Boundary Layers. *J. Fluid Mech.* **30,** p. 741.

Loeve, M. M., 1955, *Probability Theory.* Van Nostrand, Princeton NJ.

Lumley, J. L., 1967, The Structure of Inhomogeneous Turbulent Flows. In *Atmospheric Turbulence and Radio Wave Propagation* ed. A. M. Yaglom & V. I. Tatarsky, p. 166. NAUKA, Moscow.

Lumley, J. L., 1970, *Stochastic Tools in Turbulence.* Academic Press, New York.

Lumley, J. L., 1981, Coherent Structures in Turbulence. in *Transition and Turbulence,* ed. R. E. Meyer. Academic Press, New York.

Moin, P. & Moser, R. D., 1988, Characteristic Eddy Decomposition of Turbulence in a Channel. submitted to *J. Fluid Mech.*

Moser, R. D. & Moin, P., 1984, Direct Numerical Simulation of Curved Turbulent Channel Flow. NASA **TM-85974**.

Papoulis, A., 1965, *Probability, Random Variables, and Stochastic Processes.* McGraw-Hill.

Rice, S. O., 1944, Mathematical Analysis of Random Noise. *Bell System Tech. J.*. **23,** p. 283.

Theodorsen, T., 1952, Mechanism of Turbulence. In *Proc. 2nd Midwestern Conf. on Fluid Mech.* Ohio State University, Columbus, Ohio.

Townsend, 1956, *The Structure of Turbulent Shear Flow.* Cambridge University Press.

Wallace, J. M., Eckelmann, H. & Brodkey, R. S., 1972, The Wall Region in Turbulent Shear Flow. *J. Fluid Mech.* **54,** p. 39.

Connecting Coherent Structures and Strange Attractors

LAURENCE R. KEEFE
Center for Turbulence Research
NASA Ames Research Center
Moffett Field, California 94035, USA

and

Stanford University
Stanford, California 94305, USA

INTRODUCTION

The current belief in "coherent structures" is a synthesis of two historically persistent notions of turbulence. On the one hand has been the perception that turbulence is a stochastic process, while on the other has been the recognition that "eddies", identifiable structures within the stochastic stew, are ubiquitous in shear turbulence under all conditions. From these two views emerges the belief that turbulence contains structures, similar in character, but random in occurrence, whose dynamics strongly influence evolution of the flow.

A third view of turbulence, neither historic, nor yet persistent,(nor even widely accepted!) derives from nonlinear dynamical systems theory. This theory suggests that turbulent solutions to the Navier-Stokes equations are confined to "strange attractors", and, by implication, that turbulent phenomenology must find some expression or source in the structure of these mathematical objects. This view is advocated here, with arguments and examples to link coherent structures to some of the commonly known characteristics of strange attractors. Fundamental to this link is a geometric interpretation of conditional sampling techniques used to "educe" coherent structures, which offers an explanation for their appearance in measurements as well as their size.

Development of the link between attractors and structures first requires a brief discussion of both structures and conditional sampling techniques, as well as dynamical systems theory. Following this, a generic sampling technique is applied to the Lorenz equations to demonstrate the eduction of "coherent structures" from this model system. Attention then focusses on describing the Navier-Stokes equations as a dynamical system, after which the geometrical meaning of the VITA technique(Blackwelder & Kaplan, 1976) and Quadrant analysis(Lu & Willmarth, 1973) within this system is explained. Finally, some comments are made regarding the arbitrariness of the sampling process and the implications for questions of significance.

COHERENT STRUCTURES AND COHERENT SAMPLING

Historically, the view that turbulent flow contains structure has arisen from flow visualization studies. Two better known examples of this are the bubble pictures of Kim, Kline & Reynolds(1969) in the sublayer of a turbulent boundary layer, and the shadowgraph pictures of Brown & Roshko(1974) in the mixing layer. In both cases a recognizable sequence of events or physical structures, occurring irregularly, could be identified in the pictures. The belief that structures(or sequences) such as these were important to the flow dynamics led to the desire to extract their signature from the stochastic background for independent study. A problem in pattern recognition resulted, although the formalism and language of that field has rarely been applied(Blackwelder, 1978) to the problem of defining a conditional sampling scheme. In each case investigators first construct a "feature" space from those characteristics they believe are unique to the event or structure. In this they are guided by their sense of what important physical processes are associated with a typical event, and how best to identify such processes when they occur. Thus, choice of features includes an element of arbitrariness. In addition this choice can be affected by the kind, or amount, of data not measured. (Readers are reminded of the story of the blind men examining an elephant. Each touches something the other doesnot; each conjectures the essential nature of the elephant to be that which they touched.) The result of this attempted characterization is a detection criterion that is a function of the features. When applied to available transducer signals, it announces occurrences of an event or structure. Once an event's temporal origin has been determined by this scheme, it is aligned with transducer time histories from other events so that they can be ensemble averaged and the coherent signature extracted. Thus development and application of a conditional sampling scheme follows the sequence: choice of features that are functions of available transducer signals; development of a detection criterion based on the features; application of the criterion to determine event origins; ensemble averaging of transducer signals aligned on event origins.

In the sampling schemes used to date the transducer signals have often been hotwire signals for the velocities U_{ij}(i denotes velocity component, j indexes the physical position of the measurement in an array). From these signals M features, f_n, have been constructed. Thus

$$f_n = f_n(U_{ij}) \tag{1}$$

Typical features have been time and space derivatives of the U_{ij}, as well as powers and short term averages of these quantities. Choosing M features establishes a M-dimensional "feature" space.

The state of the flowfield at the measurement array is characterized in time by a feature vector

$$\vec{\mathbf{f}}(t) = \{f_1(U_{ij}), f_2(U_{ij}), \cdots f_M(U_{ij})\} \tag{2}$$

which sweeps out some trajectory in the feature space, and whose components are the value of the features at each instant. The detection criterion, call it G, is a function of the feature variables. In simple cases it can be written as

$$G(f_i) = 0. \tag{3}$$

Often this represents some surface in the feature space. When the feature trajectory $\vec{\mathbf{f}}(t)$ crosses the surface $G(f_i) = 0.$, an event or structure is signalled. The instant of crossing, or some time referenced to the crossing , becomes the event origin. It is used to align transducer signals or feature traces from different events for ensemble averaging. The result: a time trace "typical" of all qualitatively similar, but randomly occurring, events. Since averaging extracts a non-zero trace from a stochastic background, the average is considered to represent an underlying "coherent" structure, shorn of the incoherent fluctuations which obscured it.

DYNAMICAL SYSTEMS, STRANGE ATTRACTORS AND CONDITIONAL SAMPLING

Consider a system of *ode's*(not necessarily finite)

$$\frac{d\vec{\mathbf{X}}}{dt} = h(\mathbf{X}) \tag{4}$$

$$\vec{\mathbf{X}} = (X_1(t), X_2(t), X_3(t), \cdots) \tag{5}$$

With the initial condition

$$\vec{\mathbf{X}}_0 = (X_1(0), X_2(0), X_3(0), \cdots)$$

the solution to the system (4) can be written

$$\vec{\mathbf{X}} = \mathbf{\Phi}_{X_o}(\vec{\mathbf{X}}) = \vec{\mathbf{\Phi}}_{X_0}(t) \ , \tag{6}$$

or

$$X_i = \mathbf{\Phi}_i(t)$$

$\vec{\mathbf{X}}$ is the phase space or "state" variable. Each point on the solution trajectory $\vec{\mathbf{X}} = \vec{\mathbf{\Phi}}_{X_0}(t)$ completely characterizes the configuration of the system at an instant.

If the system of *ode's* (4) has only two members, the time asymptotic behavior of $\vec{\mathbf{X}}$ has only three possibilities: steady state, periodic motion, divergence. If the system is dimension three or greater, and is dissipative (the trace of the Jacobian matrix $[\partial h_i/\partial X_j] < 0$), then the asymptotic solution may be confined to a "strange attractor" in the phase space. Solutions confined to strange attractors are almost everywhere locally unstable, but are globally stable to perturbations which do not cross the bounds of their basins of attraction. Thus system states which are initially nearby diverge exponentially from each other for short times. Trajectories on strange attractors never repeat themselves exactly, but ultimately will pass arbitrarily close to any previous portion of the trajectory.

Consider the character of a bundle of trajectory segments which all pass through some "window" in the space of the attractor. If the window is small enough, the system states occurring on each trajectory segment at the instant of passage through the window will all be similar. The sequence of states along each segment remains correlated with those along other segments for some time after (and before) passage through the window. However, exponential divergence of nearby trajectories guarantees that this correlation time is finite. An ensemble average over trajectories passing through the window, and aligned on the window passage time, will yield a non-zero trace over finite time for any phase space component(or function of them) which has a zero mean value.

The above concepts are illustrated in Figures 1-3. The strange attractor which represents the asymptotic solution to the Lorenz system for a particular set of parameter values is pictured in Figure 1a. The X-Z plane intersects the attractor transversely, and a large rectangular window has been outlined in this plane. Each time the solution trajectory passes through the window, the next 500 points in the time series of the Y component of the solution are traced in Figure 1b. In Figure 1c the ensemble average of the 152 traces so obtained is displayed. Because the window is large the selected traces are not substantially correlated with each other. Despite the fact their Y coordinates are identically zero at window passage, the range of X and Z values at intersection means that the system states (X,Y,Z) are "far" apart. In Figure 2 the window area has been decreased by a factor of five, and the correlation between selected traces increases substantially. The ensemble average now clearly shows a non-zero value for finite time after passage. A further decrease in window size, now barely discernible in Figure 3, again reduces the number of traces selected from a given record length, but sharpens their correlation. The non-zero portion of the average increases both in duration and peak magnitude as a result.

It is easy to see that the "window" constitutes a conditional sampling criterion in the phase space of the Lorenz attractor. Here the "features" are the primitive variables (X,Y,Z) themselves, and the detection criterion, though not in the form of equation (3), is a simple function of the features, namely

$$\begin{aligned} Y &= 0. \\ a \le X &\le b \\ c \le Z &\le d \end{aligned} \qquad (7)$$

The event criterion (7) simply chooses the bundle of trajectories on the attractor which are ensemble averaged. If the detection criterion is too broad, selected trajectory segments are uncorrelated to begin with, and averaging fails to extract a structure. Tightening the selection criteria(decreasing the window size) increases the likelihood that trajectory segments are initially correlated, but exponential divergence of trajectories on the attractor guarantees that their mutual correlation time is finite. At times beyond the correlation limit, the ensemble average of the trajectories will be zero. If the size of a structure is defined as the length of non-zero trace which results from averaging, then the tighter the selection criterion, the greater the size of the educed event. Given the minimum distance, Δ, between trajectories at the event window, information theory(Shaw,1981) suggests that an upper bound on the event size or duration is given by a constant times the logarithm of $1/\Delta$. This separation between trajectories must be calculated in the

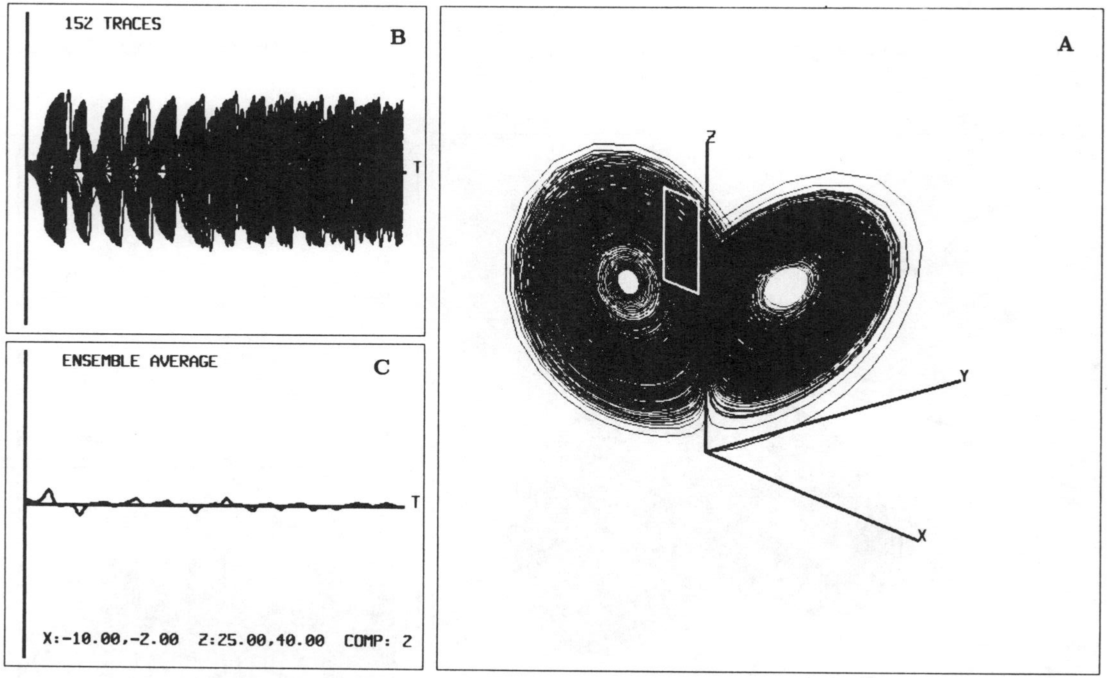

Figure 1: a) Lorenz attractor with window $-10 \leq X \leq -2$, $25 \leq Z \leq 40$. b) Aligned Y traces. c) Ensemble average of Y traces.

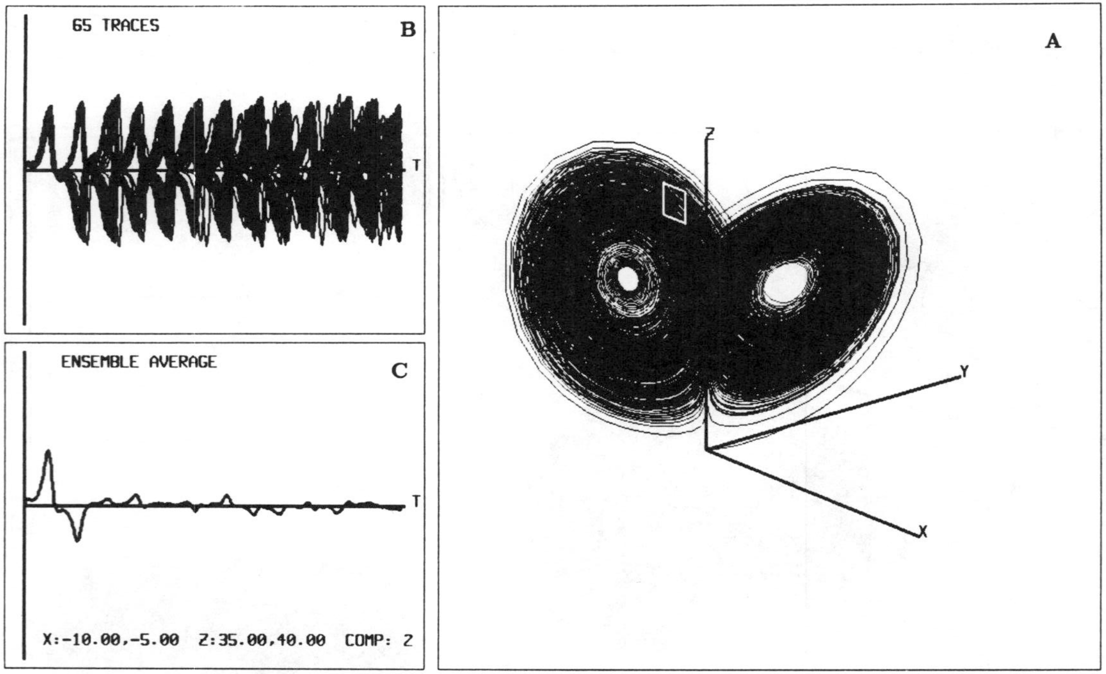

Figure 2: a) Lorenz attractor with window $-10 \leq X \leq -5$, $35 \leq Z \leq 40$. b) Aligned Y traces. c) Ensemble average of Y traces.

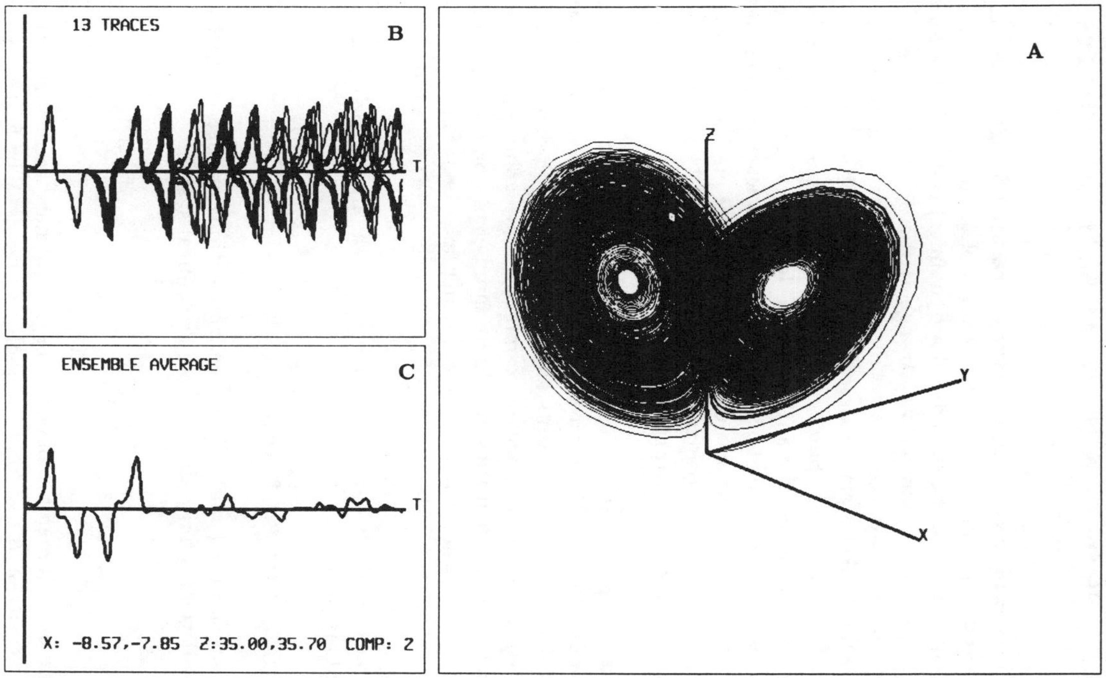

Figure 3: a) Lorenz attractor with window $-8.57 \leq X \leq -7.85$, $35 \leq Z \leq 35.7$ b) Aligned Y traces. c) Ensemble average of Y traces.

full phase space of the dynamical system, not in some reduced dimension feature space.

THE NAVIER-STOKES EQUATIONS AS A DYNAMICAL SYSTEM AND APPLICATIONS OF CONDITIONAL SAMPLING TO ITS SOLUTIONS

Constructing the link between coherent structures measured by experimentalists and their hypothesized origin in the structural details of an attractor requires that the Navier-Stokes equations can be considered a dynamical system in the form (4), and that its solutions are confined to a strange attractor. The first is shown below. For the second, there are rigorous results to prove that solutions to the two-dimensional Navier-Stokes equations are confined to a finite dimensional attractor(Constantin & Foias ,1985). Similar rigor has not yet been obtained in the three-dimensional case. However, the author has performed calculations on low Reynolds number, turbulent channel flow which indicate, at least in this important case, that solutions to the three-dimensional Navier-Stokes equations are confined to a finite (but large) dimension strange attractor(Keefe et al., 1987). The point may not be proven, but the circumstantial evidence is strong.

Formally, the Navier-Stokes equations are an infinite dimensional system because they require functional, rather than discrete, initial data. However, through the practice of disretization, both experiments and numerical simulation project the system onto a finite one. An experimenter examines a finite volume of flow, and makes measurements of flow quantities at a finite number of points. A numerical simulation operates in a computatioinal domain which may be infinite, but flow quantities are calculated at a finite number of points(the grid) within it.

Imagine a phase space vector $\vec{\mathbf{S}}$ whose components are the flow velocities u, v, w at N different points in space

$$\vec{\mathbf{S}} = (u_1, v_1, w_1, u_2, v_2, w_2, \cdots u_N, v_N, w_N) \tag{8}$$

Spatial discretization schemes(finite difference, collocation) applied to the Navier-Stokes equations will yield a finite dimensional dynamical system

$$\frac{d\vec{\mathbf{S}}}{dt} = g(\vec{\mathbf{S}}) \tag{9}$$

The right hand side of each member of equation (9) will have terms linear in the velocities due to viscosity, and two kinds of nonlinear terms. The first are due to the convective terms, and couple nearby velocities together through their space derivatives. The second kind will be due to the pressure gradient, and globally couple all components together in incompressible flow. As the number of points at which velocities are known increases towards infinity, system (9) becomes equivalent to the Navier-Stokes in the sense that each requires the same initial data, and solution of either gives the entire flow field.

Because of discretization neither experiments nor calculations have access to the full dynamical system represented by (9) as $N \to \infty$. Making a single point measurement of one velocity component records the solution to but one equation from the infinite dimensional system. It is impossible to characterize the entire quantitative state of a multidimensional dynamical system by looking at only one of its components. However, in theory(Packard et al.,1980, Takens, 1981), it is possible to recover the full qualitative behavior of a system from one of its components by use of embedding techniques. Conceptually, then, it is possible to characterize an entire flow field from one hotwire measurement, provided that one time history can be processed enough to yield sufficient numbers of independent coordinates. If they are sufficient, the qualitative dynamics of the flow on its attractor are reproduced on a different attractor in this new phase space.

With more hotwire measurements fewer additional independent coordinates need be derived for a complete description, but enough that the qualitative dynamics are reproduced. Given the choice between describing a system by ten coordinates derived from a single measurement of a primitive variable, or ten coordinates each of which is a different primitive measurement, the latter is the better choice. Both should give the same qualitative description of the dynamics, but the second also provides a quantitative description.

It is important to realize that discretization is a projection process, and that in this process quantitative information about the system under study is obscured or lost. Construction of a "feature" space for event recognition is such a discretization. It projects the full solution attractor onto a reduced dimension phase space and may make inaccessible the phase coordinates needed to discriminate between apparently similar states. In applications of conditional sampling techniques to discrete numbers of hot wire signals this loss produces the phase jitter and temporal misalignment problems which reduce coherence between events, and which may require extensive iterative processing of the signals for a structure to be educed (Hussain, 1983).

In light of the previous paragraphs, eduction of coherent structures by conditional sampling can be viewed as the following set of operations on the full solution attractor of the Navier-Stokes equations. First, discretization of the flow field collapses the full attractor onto a reduced dimension subspace. Second, a "feature" space is constructed out of components from this subspace, or functions of them. The number of features selected determines the dimension of the new space, which may actually be higher than that resulting from the discretization. This new space contains a finite size object to which the feature trajectory is confined, but this object is not an attractor in the rigorous sense. Third comes construction or specification of an event criterion in the feature space. The criterion is a subset of the feature space(curve, surface, surface patch, etc.), but may be infinite in extent. Fourth, the intersection points between the feature trajectory and the event criterion are determined. Finally, an ensemble average of primitive, or feature, traces aligned on these intersection points is obtained. Using velocity signals derived from numerical simulations of low Reynolds number channel flow (Kim, Moin & Moser, 1987) these processes are illustrated for both the VITA technique and

Quadrant analysis, two conditional sampling schemes which have played important roles in studies of near wall structure.

In Blackwelder and Kaplan's 1976 application of the VITA technique discretization of the flow field projected the full attractor onto a one dimensional subspace, namely a single hotwire measurement at $y^+ = 15$. A two dimensional feature space was constructed from this signal using the moving average operation on the raw signal and its square. Thus the feature coordinates were

$$U_{lm}(t) = \frac{1}{T} \int_{t-\frac{T}{2}}^{t+\frac{T}{2}} U(\tau)d\tau \tag{10}$$

$$U_{lm}^2(t) = \frac{1}{T} \int_{t-\frac{T}{2}}^{t+\frac{T}{2}} U^2(\tau)d\tau$$

An additional constant was derived, the meansquare value

$$U_{ms} = \lim_{T \to \infty} U_{lm}^2$$

The sampling criterion was the parabola

$$U_{lm}^2 = (U_{lm})^2 + k_1 U_{ms} \tag{11}$$

where k_1, the threshold constant, had a value 1.2. Passage of the feature trajectory from outside the parabola to inside signalled an event. The temporal origin of the event was assigned to the midpoint of the trajectory's residence within the parabola. Figures 4-6 illustrate the geometry of this scheme.

Figure 4 shows the effect of using too short an averaging time to create the feature coordinates. As the averaging time T in equation (10) approaches zero, the feature trajectory will become a plot of $U^2(t)$ against $U(t)$. This will be a parabola opening upwards with vertex at the origin. The sampling criterion (11) is always a similar shaped parabola whose vertex has been displaced upwards from the origin an amount dependent upon k_1. Since the two parabolas are the same shape, the feature trajectory can never intersect the event criterion in this case. Such a situation can be seen in Figure 4a.

Figure 5 shows the VITA technique using values of k_1 and T similar to those in Blackwelder and Kaplan. In Figure 5a the feature trajectory and event criterion are plotted. Increasing the averaging time now causes the feature trajectory to occasionally cross the parabola. Each time it does the 300 time points of $U(t)$ centered around the event origin are plotted in 5b, and the origin is marked on the time line in 5d by a vertical tick. Figure 5c shows the ensemble average of the aligned events. It should be noted that the Reynolds number of the simulation data is much lower than that of the boundary layer employed in the original measurements. The measured bursting frequency, and the amplitude of the ensemble average, are lower than found by Blackwelder, but the shape and duration of

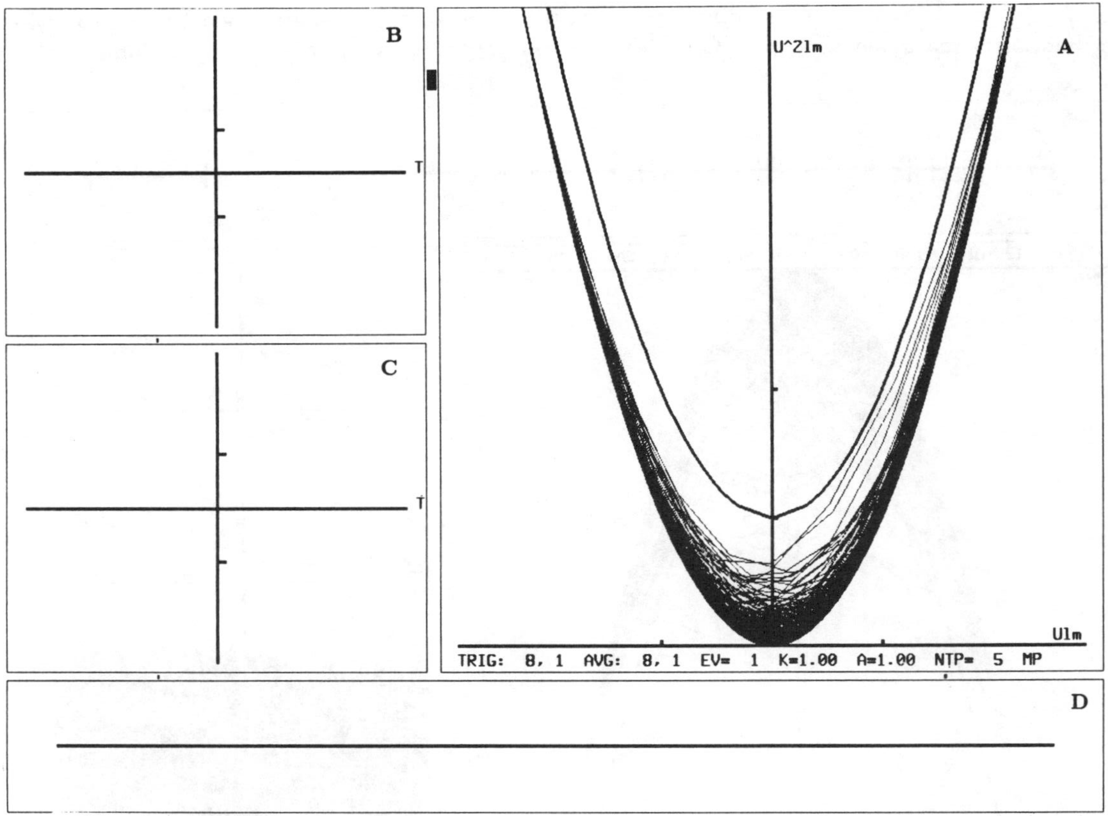

Figure 4: VITA technique, $Tu_\tau^2/\nu = 3.45$, $k_1 = 1.2$ a) Feature trajectory and parabolic event criterion. b) Aligned U' traces. c) Ensemble average of U' traces. d) Timeline of event origins.

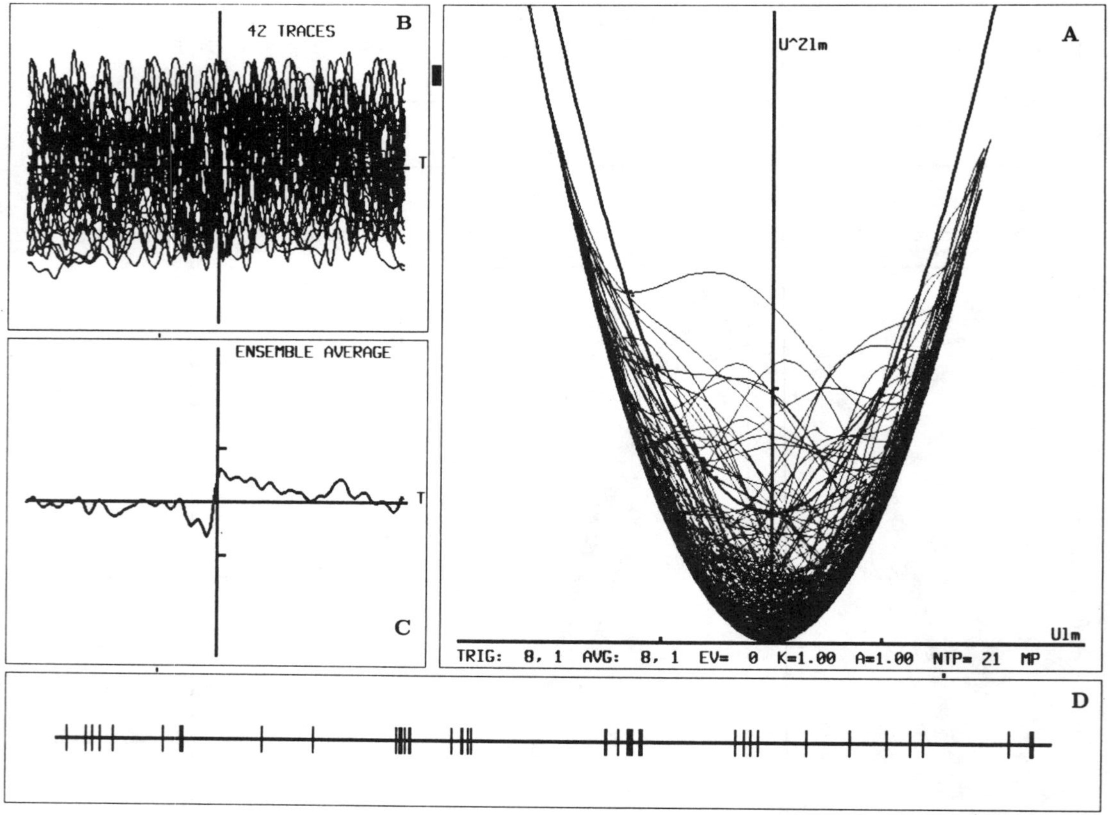

Figure 5: VITA technique, $Tu_\tau^2/\nu = 14.5$, $k_1 = 1.2$ a) Feature trajectory and parabolic event criterion. b) Aligned U' traces. c) Ensemble average of U' traces. d) Timeline of event origins.

the event are similar. Here the vertical tick marks on the average plot represent the rms value of $U(t)$. More important than these comparisons however, is the relation between the shape of the feature trajectory and the sampling criterion. With very high probability, the feature trajectory retains a parabolic character even with longer averaging times. The event criterion is essentially parallel to this "fuzzy" parabola in feature space, so the chances of intersection are low. Those intersections that do occur cover a wide range of the feature coordinates. This suggests that the events detected are not "close" to each other in the full phase space, and may decorrelate rapidly. With this feature trajectory the detector isn't sharp and doesnot necessarily pick out high probability events. However, multiplying the burst duration times the number of events suggests that some part of the process is going on 25% of the time.

In later work Blackwelder and Haritonidis(1983) sharpened the detection criterion by adding a third feature coordinate: $\partial U/\partial t$. They chose to select only those events for which $\partial U/\partial t > 0$. In this case the feature trajectory is a space curve, and the event criterion is that half of a parabolic cylinder which extends into positive $\partial U/\partial t$ space. Figure 6 shows this scheme projected back onto two dimensions. Fewer events are detected from the same time history, but their correlation is sharper, with the result that the ensemble average has a larger peak amplitude.

In an early study of the bursting process Lu & Willmarth invented the technique of Quadrant analysis. Here discretization projects the full attractor down to a 2-space consisting of U and V measurements at $y^+ = 15$. The primitive velocities are also the feature variables in this case. Two constants, the rms values of U and V, are formed. Then the event criterion is a rectangular hyperbola in hodograph(U,V) space

$$UV = -k_2 U_{rms} V_{rms} \tag{12}$$

A further distinction on events detected by this scheme was to separate the quadrant 2(bursts) events from those in quadrant 4(sweeps). Figure 7 displays this scheme applied to the same velocity fields used for the VITA technique. A threshold constant $k_2 = 4.$(within the range used by Lu and Willmarth) is used, and only quadrant 2 events have been selected. Again the 300 time points of U(t) centered on the event are plotted in 7b. Clearly the temporal behavior of the event selected here is different from that chosen by VITA. However, comparison of the time lines shows that the two schemes do choose events at roughly the same time, occasionally almost exactly. In contrast to VITA, Quadrant analysis seems to be a sharper detector, since the feature trajectory at $y^+ = 15$ is more transverse to the detector that parallel. This produces a tighter clustering of feature states at the event window. While such continues to be true for velocities measured at greater distances from the wall, it is not true for those measured at lesser distances, because the smallness of V makes the feature trajectory almost parallel to the U axis. In addition, clustering in feature space is no surety for similar behavior in the full phase space. Thus sharpness of this kind may be an entirely spurious indicator of goodness in a detector.

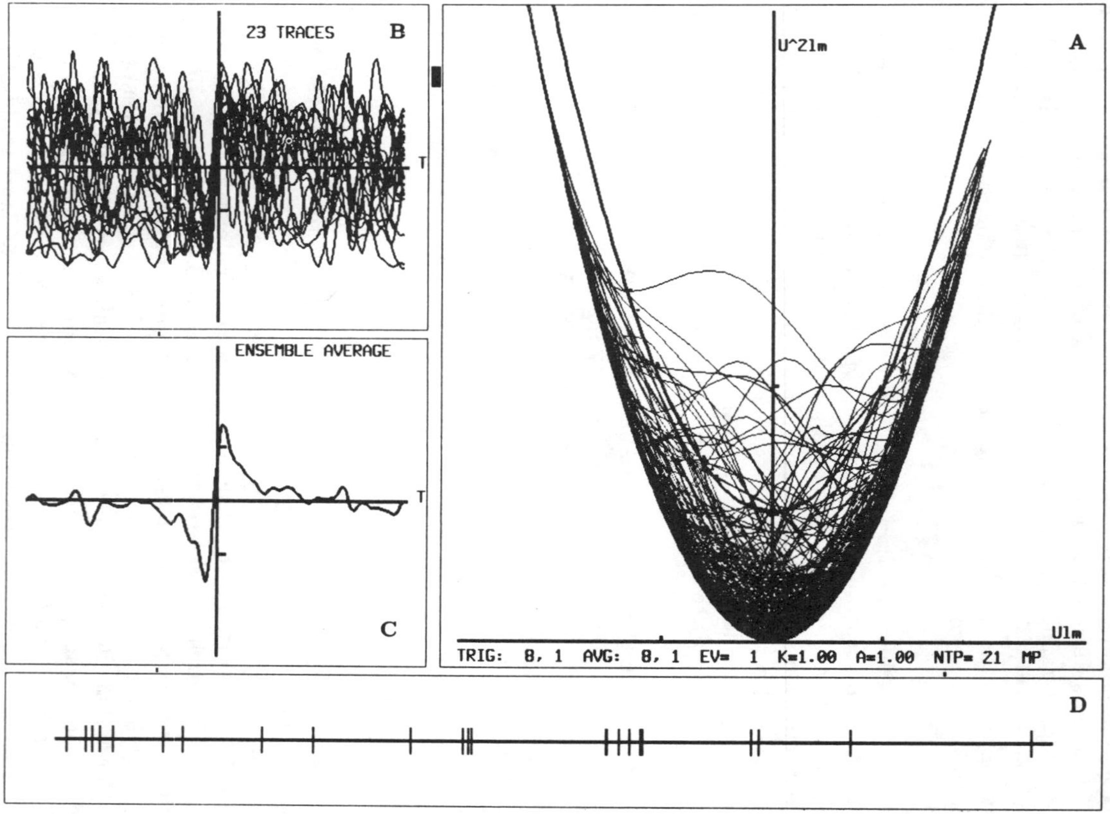

Figure 6: VITA technique, $Tu_\tau^2/\nu = 14.5, k_1 = 1.2,\ \partial U/\partial t > 0$ a) Projection of feature trajectory and parabolic cylinder event criterion. b) Aligned U' traces. c) Ensemble average of U' traces. d) Timeline of event origins.

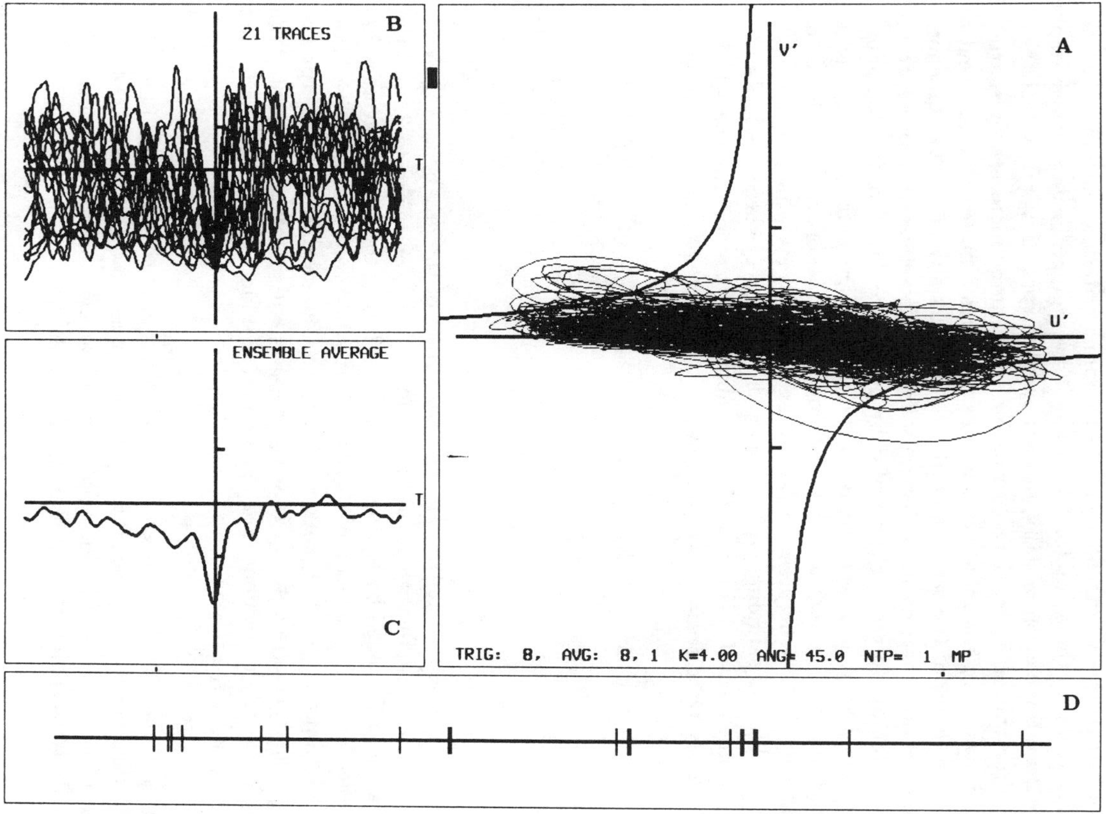

Figure 7: Quadrant analysis, $k_2 = 4$, quadrant 2 events only a) Feature trajectory and hyperbolic event criterion. b) Aligned U' traces. c) Ensemble average of U' traces. d) Timeline of event origins.

CONCLUSIONS

Discretization connects the feature space of a conditional sampling technique to subspaces of the full solution attractor of the Navier-Stokes equations. A geometric interpretation of the sampling criterion, coupled to the exponential divergence property of trajectories on strange attractors, provides an explanation for the measurement of "structure" in turbulence. The sampling criterion selects a group of nearby trajectories on the attractor for ensemble averaging. Their nearness implies a certain mutual correlation initially, but exponential spreading guarantees the correlation time will be finite. This correlation time can be increased by sharpening the criterion. An ensemble average thus produces a finite duration event or structure. The choice of the sampling criterion is essentially arbitrary. An enormous amount of flow information can be lost by the discretization process, so each investigator attempts to recover some of it by perceptive design of the sampling criterion. Carried to its logical extreme, it is clear that any sufficiently restrictive sampling criterion will extract a structure from a dynamical process confined to a strange attractor. If structures are defined as those objects which can be extracted by conditional sampling criteria, then they are everywhere one looks in turbulence. Some idea of this arbitrariness is obtained by modifying the VITA technique slightly to to see what it will detect.

Write the sampling criterion

$$U_{lm}^2 = A(U_{lm})^2 + k_1 U_{ms} \tag{13}$$

The additional constant A controls the spreading rate of the parabola. Figure 8 shows the result of sampling when the threshold k_1 has been raised to 3., but the parabola has been flattened by letting A=.1. In addition, the moving average has been shortened and only those events where the moving average is negative have been selected. Equation (13) no longer has a straightforward physical interpretation, but it is sharper as a detector in feature space, and selects more events than unmodified VITA. By coincidence the "structure" it educes is very similar to that obtained by Quadrant analysis.

Arbitrariness in the selection process begs the question of significance. There is structure in turbulence. However, it is up to each investigator to supply the physical arguments which raise their ensemble average to the status of a structure which contributes significantly to the dynamics of the flow.

ACKNOWLEDGEMENTS

The author is indebted to S. K. Lele and R. D. Moser for both stylistic and conceptual clarifications of the manuscript. This research supported, in part, by the Air Force Office of Scientific Research, under contract AFOSR-88-0056.

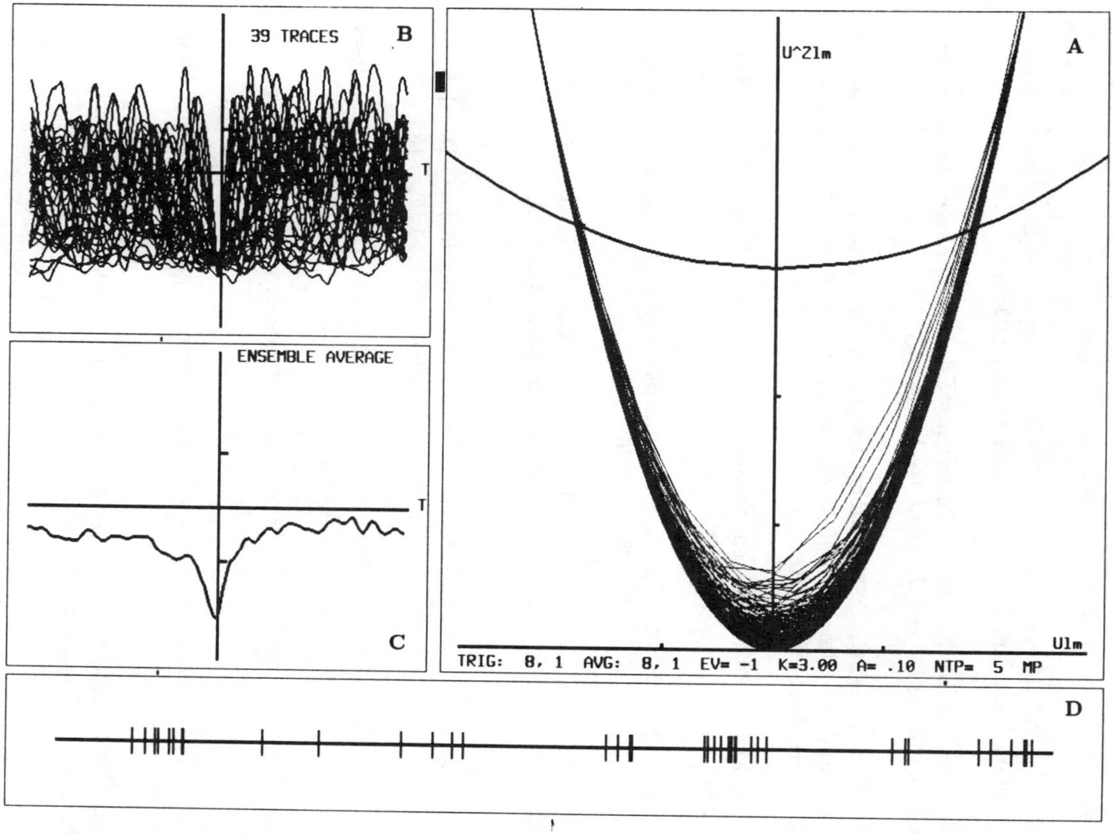

Figure 8: Modified VITA technique, $Tu_\tau^2/\nu = 3.45$, $k_1 = 3.$, $A = .1$ a) Feature trajectory and parabolic event criterion. b) Aligned U' traces. c) Ensemble average of U' traces. d) Timeline of event origins.

REFERENCES

Blackwelder, R. F. & Kaplan, R. E., 1976, On the Wall Structure of the Turbulent Boundary Layer. *J. Fluid Mech.*. **76,** p.89

Blackwelder, R. F., 1979, Eddy Detection Within Turbulent Flows. *Proceedings of the Dynamic Flow Conference, Skovlunde, Denmark*

Blackwelder, R. F. & Harotinidis, J. H., 1983, Scaling of the Bursting Frequency in Turbulent Boundary Layers. *J. Fluid Mech.*. **132,** p.87

Brown, G. L. & Roshko, A., 1974, On Density Effects and Large Structures in Turbulent Mixing Layers. *J. Fluid Mech.*. **64,** p.775

Constantin, P. & Foias, C., 1985, Global Lyapunov Exponents, Kaplan-Yorke Formulas and the Dimension of Attractors for the Navier-Stokes Equations. *Comm. Pure Appl. Math.*. **38,** p.1

Hussain, A.K.M.F., 1983, Coherent Structures-Reality and Myth. *Phys. Fluids.* **26,** p.2816

Keefe, L., Moin, P. & Kim, J., 1987, The Dimension of an Attractor in Turbulent Poiseuille Flow. *Bull. A.P.S.* **32,** p.2026

Kim, H. T., Kline, S. J. & Reynolds, W. C., 1971, The Production of Turbulence Near a Smooth Wall in a Turbulent Boundary Layer. *J. Fluid Mech.*. **50,** p.133

Kim, J., Moin, P. & Moser, R., 1987, Turbulence Statistics in Fully Developed Channel Flow at Low Reynolds Number. *J. Fluid Mech.*. **177,** p.133

Lu, S. S. & Willmarth, W. W., 1973, Measurements of the Structure of the Reynolds Stress in a Turbulent Boundary Layer. *J. Fluid Mech.*. **60,** p.481

Packard, N. H., Crutchfield, J. P., Farmer, J. D. & Shaw, R. S., 1980, Geometry From a Time Series. *Phys. Rev. Lett.*. **45,** p.712

Shaw, R. S., 1981, Strange Attractors, Chaotic Behavior, and Information Flow. *Z. Naturforsch.* . **36a,** p.80

Takens, F., 1981, Detecting Strange Attractors in Turbulence. *Lecture Notes in Mathematics, Springer-Verlag.* **898,** p.366

A Conceptual Model of the Viscous Wall Region

THOMAS J. HANRATTY
Department of Chemical Engineering
University of Illinois
Urbana, Illinois 61801, USA

INTRODUCTION

The principal theoretical problem of wall turbulence is to determine how the turbulence is created and sustained, i.e., to explain how energy is transferred from the mean flow to the turbulence. The focus of much of the work has been to relate turbulence production to turbulence structure.

Extensive laboratory and, more recently, computer studies of this problem have been carried out in this laboratory over the past twenty-five years (starting with the work of Reiss & Hanratty, 1963). A novel feature of the laboratory experiments is the use of multi-probe flush mounted electrodes to detect eddy structures. The computer experiments solve the unaveraged Navier-Stokes equations for the region $0<y^+<y_o^+$. Here, the principal question asked is what is the simplest set of boundary conditions at the edge of the viscous wall region, $y_o^+ = 30\text{-}40$, that can represent laboratory measurements. An interesting aspect of these computer experiments is that unnatural conditions can be explored in order to understand nature. This paper uses the results of these studies to support a simple conceptual model of how turbulence is produced.

A CONCEPTUAL MODEL OF THE VISCOUS WALL REGION

Turbulence measurements made in the 1950's [Hinze, 1975; Townsend, 1976] show that both the production of turbulent energy and the dissipation of turbulent energy are quite large, and of the same order, in the viscous wall region. However, the difference of these two large numbers is such that there is a net positive production which supplies energy to the outer region where the local dissipation is the same (in the log-layer) or greater (in the core) than the local production. The viscous wall region is thus the "engine" that drives wall turbulence.

We have adopted the picture that the level of Reynolds stress and the production of turbulence in the viscous wall region can be associated with eddy motion in a plane perpendicular to the flow direction which appears intermittently close to the wall and which has an average spanwise dimension of about 50 wall units. Various measurements discussed in the next section suggest that these eddies are elongated in the flow direction, having lengths which are 6-10 times their widths. In the region $y^+<15$ the wall eddies have inflows and outflows of the same strenth. They, thus, could be pictured "on average" as coupled inflows and outflows, as in Figure 1.

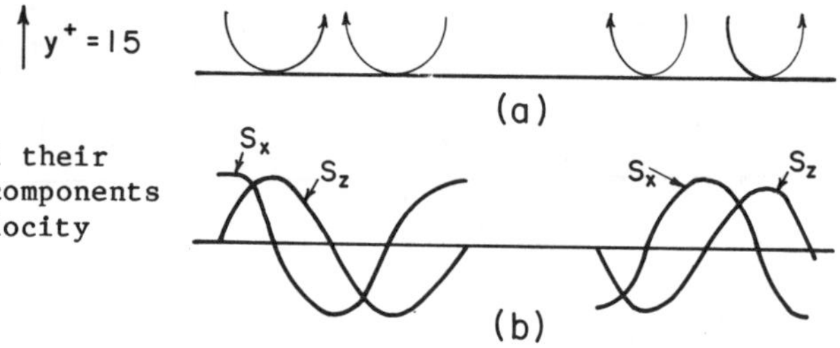

Figure 1. Wall eddies and their effect on the components of the wall velocity gradient

These eddies create streamwise velocity fluctuations, u, by bringing high momentum fluid to the wall, exchanging momentum with the wall and carrying low momentum fluid away from the wall. Since the high momentum fluid is transported by negative normal velocities, v, and low momentum fluid by positive normal velocities, these flows are associated with large values of $-uv$. They are, thus, important contributors to the Reynolds stress and to the production of turbulence. There is a temporal lag between the development of the velocity pattern associated with the eddy and the u-pattern that it causes. Thus the period over which strong Reynolds stress exists is less than the lifetime of the eddies. Furthermore, it would be expected, from this picture, that large Reynolds stresses at $y^+ = 15$ would most often would occur in pairs separated by a distance $\ell^+ = 50$.

Wall eddies with $\ell^+ \ll 50$ do not extend to the edge of the viscous wall region, y_o, and could be looked upon as separation bubbles formed by the interaction large scale eddies with the wall. However, correlation measurements and frequency-wavenumber measurements [Grant, 1958; Tritton, 1967; Nikolaides, et al, 1983; Nikolaides, 1984] suggest that large eddies with average spanwise dimensions of about 50 wall units ($\lambda^+ \approx 100$) extend through the entire viscous wall layer. They can be identified close to the wall from the spanwise variation of the spanwise and streamwise velocity components and can be related to $\lambda^+ \approx 100$ normal velocity fluctuations at $y_o^+ \approx 30$-40. They are not so easily related to the u and w velocities at y_o^+ since these components are dominated by eddies with $\lambda^+ \approx 200$-700 in the outer part of the viscous wall region.

By using special conditional sampling methods, Nikolaides, et al (1983) found that the eddy motions close to the wall (shown in Fig. 1) are, on average, associated with spanwise velocity components (having a small fraction of the total $\overline{w^2}$ energy) at $y^+ = 40$ which have the same direction earlier in time and which change to a flow in the opposite direction while the strong eddy pattern exists at the wall. Such measurements suggest that the extent of the wall eddies is about 300 wall units in the flow direction and that a closed eddy with a center below $y^+ = 40$ and an average streamwise length of about 150 wall units could exist. During the appearance of this "closed eddy" Reynolds stress production is most important. These "closed eddies" are not visualized as vortices in the classical sense. They contain streamwise vorticity in shear layers within them but they are not pictured to be a revolving mass of fluid driven by a core of high vorticity at its center.

The type eddy motions in the outer flow with which the $\lambda^+ = 100$ wall eddies are associated is also not well defined. They could be part of a structure extending well into outer flow. Another possibility is that they are caused by larger outer-flow eddies that interact with the wall. In this context they would have the appearance of "separation bubbles".

The model outlined above is really an extension of ideas presented by Townsend (1976). Townsend describes the log-layer as being dominated by attached wall eddies (elongated in the flow direction) that increase in size linearly with distance from the wall. He suggested that the observed characteristics of the flow in the constant stress region of a boundary-layer can be considered to be made up from inputs of attached eddies of a large range of sizes, for which the contributions to normal velocity and Reynolds stress are concentrated in the general neighborhood of the distance from the wall of their center. (This picture would also seem to imply a small contribution to the spanwise velocity component by an eddy close to its center.)

In the context of Townsend's model, the $\lambda^+ \approx 100$ wall patterns shown in Fig. 1 are associated with eddies which are strong contributors to the normal velocity and Reynolds stress in the viscous wall region $o<y^+<30-40$, and which are the smallest of any consequence in the log-layer. These eddies are the principal contributors to the spanwise velocity very close to the wall. However, larger attached eddies which are not important contributors to the Reynolds stress in the viscous wall region dominate the spanwise flow in the outer part of the viscous wall region.

The spanwise dimension of the wall eddies appears to be of particular importance. This is emphasized by the finding that decreases in drag caused by high molecular weight polymers or by large favorable pressure gradients are associated with an increase in λ^+ (Fortuna and Hanratty, 1972; Eckelman, et al, 1972).

As discussed above, length λ^+ is related to the distance in the z-direction between zero crossings of the v-velocity at y_o. For a signal whose amplitude and derivative are Gaussian and independent the zero crossing scale, defined as $\Lambda^{-1} = \sqrt{2}/(\pi N_o)$, (where N_o is the frequency of crossing per unit length in the z-direction) is equal to the Taylor microscale, λ_g (Rice, 1954). The probability distribution of u-velocity fluctuations shows a significant departure from Gaussian behavior at y_o. This is not the case for the v- and w-velocities. (Barlow and Johnston, 1985). Consequently, Nikolaides assumed that the spanwise wavelength of the wall eddies scales with the Taylor microscale at $y = y_o$; i.e.,

$$\lambda_g\big|_{y_o} \approx \lambda/(\sqrt{2}\,\pi). \tag{1}$$

For an isotropic turbulence the microscale is related to the dissipation of turbulent energy.

$$\varepsilon = 15\,\nu\,\overline{u^2}/\lambda_g^2 \tag{2}$$

This equation has also been used to define the Taylor microscale in anisotropic turbulence with $3\,\overline{u^2} = (\overline{u_1^2} + \overline{u_2^2} + \overline{u_3^3}) = q^2$.

The picture of the wall eddies presented focuses on an interaction (or feedback) between the viscous wall layer and the outer flow which could control the spanwise scale. From (2) one would expect that λ^+ decreases if dissipation in the outer flow increases. Numerical calculations to be presented in Section 4 suggest that this decrease would cause a net decrease in the production of turbulent energy in the viscous wall region which, in turn, would lower the energy available for dissipation in the core. This, in turn, would cause an increase in λ^+.

LABORATORY MEASUREMENTS

Velocity Gradients at the Wall

Electrochemical techniques which are the mass transfer analog of the hot film anemometer [Hanratty and Campbell, 1983; Mao and Hanratty, 1985] have been used to study flow very close to the wall without interfering with the flow. Velocity gradients at the wall in the mean-flow direction $S_x = (\partial U/\partial y)_{y=o}$ have been determined by measuring the mass transfer to electrodes flush with the wall that are circular in shape or rectangular with the long side perpendicular to the flow direction. The use of two rectangular electrodes in a \/ arrangement allows the simultaneous measurement of S_x and of the spanwise component of the velocity gradient at the wall $S_z = (\partial W/\partial y)_{y=o}$.

Values of the ratio of the root-mean-square of the fluctuations in S_x to the time average, $(\overline{s_x^2})^{1/2}/\overline{S_x}$, of 0.30-0.36 have been reported. The lower values are for situations in which the measurements are being affected by spatial averaging over the electrode surface and by the time-response of the concentration boundary-layer. The most reliable estimate of this quantity is a value of 0.35-0.36 obtained by using very small electrodes and by correcting for frequency response (Fortuna and Hanratty, 1971; Mao and Hanratty, 1985).

Measurements of $(\overline{s_z^2})^{1/2}/\overline{S_x} = 0.11$ to 0.12 have been obtained with an arrangement of rectangular electrodes having an angle of 15 degrees. The choice of this angle is a compromise. Too large a value results in spatial averaging; too small a value could result in underestimating s_z because large angular deflections from the mean flow direction are missed. This would seem to indicate that the value of 0.11 to 0.12 could be low. However, the few measurements that have been made with larger angles do not support this suggestion (Sirkar and Hanratty, 1969; Lee, 1975).

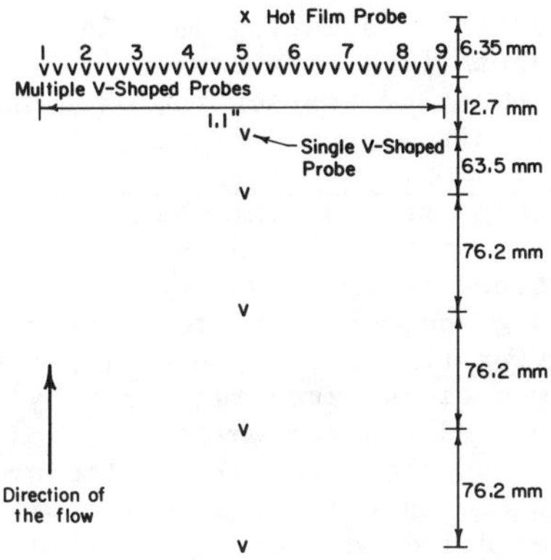

Figure 2. A schematic diagram of the arrangement of sensors

Detection of Wall Eddies

The large values of s_z obtained from these measurements and the elongated pattern for s_x fluctuations obtained from correlation measurements (Reiss and Hanratty, 1963; Mitchell and Hanratty, 1966) prompted Sirkar and Hanratty (1970) to interpret their experiments on mass transfer to a wall in terms of an eddy pattern close to the wall of the type shown in Fig. 1, which is periodic in the z-direction and not intermittent. The accuracy of this picture was explored by Lee, Eckelman and Hanratty (1974) with an array of \/ electrode pairs arranged in the spanwise direction on the wall of a 19.37 cm pipe, as shown in Fig. 2. The observed $s_z(z)$ variations differ from the idealized pattern in Fig. 1 in that there is a distribution of wavelengths, the eddies on both sides of an updraft or downdraft can be quite different in size, and a large wall eddy may contain a smaller separation bubble. Thus, Eckelman found that the Sirkar model shown in Fig. 1 is an "average behavior" and that regular pairings of eddies of the same size might not be a standard occurrence. He did, however, agree with the eddy pattern suggested by Sirkar in that downdrafts are found, on average, to have the same strength as updrafts (Eckelman, 1971). An example of measurements suggesting a pattern of eddies with sizes of $l^+ \approx 30$ and with wavelengths of $\lambda^+ \approx 60$ is shown in Fig. 3a.

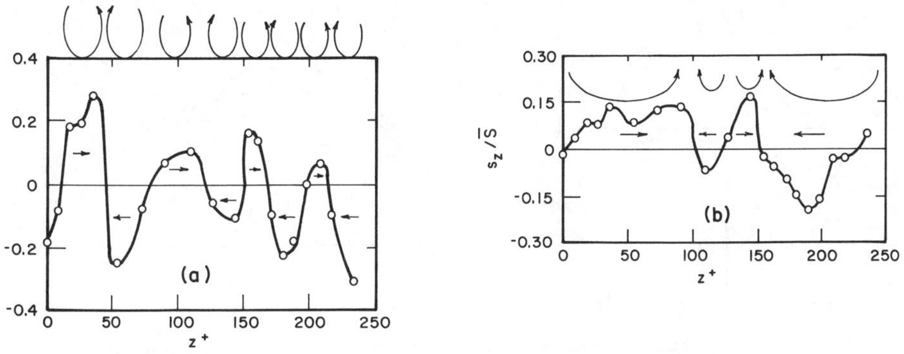

Figure 3. Samples of measured spanwise variations of the spanwise component of the velocity gradient at the wall

A pattern suggesting much larger wavelengths is shown in Fig. 3b. Eckelman (1971) interpreted the smaller eddies contained in Fig. 3b as representing two separation bubbles formed at the wall by a large sized eddy pair with $\lambda^+ \approx 220$.

Patterns such as this have been analyzed to obtain an estimate of the size of the wall eddies. An eddy length, ℓ^+, was defined as the distance between two zero crossings. It was quite clear from examining a large number of these patterns that there were numerous small eddies (having the appearance of separation bubbles) which would not reflect eddy motions extending out to y_o^+. Eckelman, therefore, decided to ignore eddies for which the amplitude of variation of $s_z(z)$ in the eddy was less than one-half of the average amplitude for all eddies. This amounted to combining eddies with $l^+<10$ with larger adjacent eddies. After doing this the size of eddy pairs was defined as the distance between two successive updrafts or two successive downdrafts. Approximately 20 percent of the l^+ were ignored in this way and eddies with $\lambda^+ = 20$-210 were detected. Eddies with $20 < \lambda^+ < 180$ accounted for 90 percent of the energy. Average values of $\lambda^+ \approx 75$ and of $l^+ \approx 40$ were obtained. It is clear that the method for neglecting $l^+ < 10$ was somewhat arbitrary. For example, it would not have discounted the two small eddies in Fig. 3b. Values of λ^+ closer to the value of 100 suggested by Kline et al (1967) would be obtained if $l^+ < 20$ were ignored.

The strong connection between the wall eddies and the creation of streamwise velocity fluctuations is demonstrated from a consideration of the simultaneous measurements of $s_x(z)$ and $s_z(z)$ reported by Lee, et al (1974). The idealized eddy pattern shown in Fig. 1a would be associated with the $s_z(z)$ variation in Fig. 1b. We picture the eddy to bring high momentum fluid to the wall in the inflow region, to exchange momentum with the wall and to transport low momentum fluid away from the wall in the outflow region. This type of process would be manifested by an $s_x(z)$ that is 90° out of phase with $s_z(z)$. (See Fig. 1b.) The correlation function of $s_x(z)$ and $s_z(z + \Delta z)$ for this idealized eddy pattern would be zero at $z^+ = 0$, would have a maximum of +1 at $\Delta z^+ = \lambda^+/4$ and would have a minimum of -1 at $\Delta z^+ = -\lambda^+/4$.

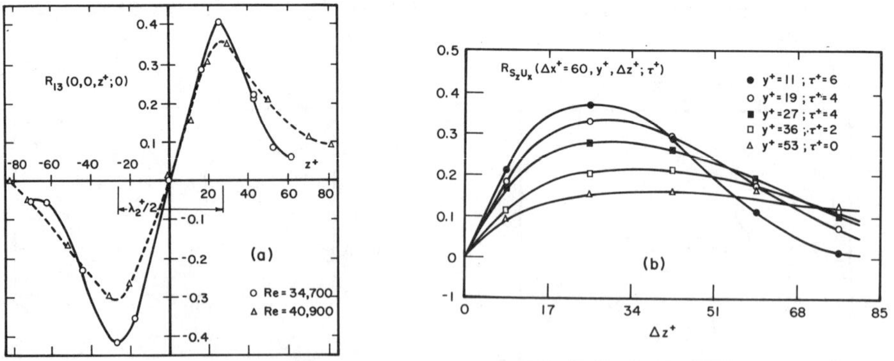

Figure 4. (a) Spatial correlation coefficient between s_x and s_z; (b) Spatial correlation coefficient between $u(y)$ and s_z

Measurements of R_{xz}, given in Fig. 4a, support this picture. They show a maximum of 0.3-0.4 at $\Delta z^+ \cong 27$ and a minimum of -0.3 to -0.4 at $\Delta z^+ \cong$

-27, indicating that eddies with $l^+ \cong 54$, are most effective in creating streamwise deficient fluid. An examination of the time variation of s_z (z) and s_x (z) profiles indicates that spatial phase relations of the type shown in Fig. 1 were triggered by the emergence of a strong s_z (z). This was followed by the appearance of a strong s_x (z) profile, as is demonstrated in measurements of R_{xz} for different amounts of time delay, τ^+, of the s_z (z). These indicate a maximum in the correlation for $\tau^+ \approx 10.5$.

The above result would seem to suggest the lifetime of an event producing a strong Reynolds stress could be shorter than the lifetime of strong s_z (z) patterns or strong s_x (z) patterns since a certain phase relation is required to produce Reynolds stress.

Lee et al (1974) attempted to develop a technique to detect wall eddies from a consideration that Reynolds stress close to the wall is given by

$$- \overline{uv}/u^{*2} = \sigma_{12}\, y^{+3} \tag{3}$$

and that $\partial s_x/\partial x$ is small compared to $\partial s_z/\partial z$. Then

$$\sigma_{12} = \frac{1}{2} s_x^+ \, \partial s_z^+/\partial z^+ \tag{4}$$

and

$$\langle\sigma_{12}\rangle = \frac{1}{\lambda^+} \int_o^{\lambda^+} \sigma_{12} dz^+, \tag{5}$$

where λ^+ is the average spanwise wavelength of the stress producing eddies. A plot of $\langle\sigma_{12}\rangle$ versus time shows an intermittent occurrence (about 59 per cent of the time) even though $s_x(t)$ and $s_z(t)$ are not intermittent. The same type behavior is exhibited in the Reynolds stress measurements of Eckelmann (1974) in a wall layer. The magnitudes of inflows (negative $\partial s_z/\partial z$ at s_z=0) and of outflows (positive $\partial s_z/\partial z$ at s_z=0) in these stress producing eddies were about equal. A quadrant analysis of σ_{12} gives results close to what had been measured by Brodkey, et al (1974) at y^+=3.4: Quadrant I = -40%; Quadrant II = 63%; Quadrant III = -20%; Quadrant IV = 101%.

Scaling of Wall Eddies

Measurements of the frequency spectra of s_x were correlated by Mitchell and Hanratty (1966) and by Hanratty (1967) by scaling with bulk parameters. However, later more accurate measurements in our laboratory agreed with the suggestion of Bakewell and Lumley that the time variation of the velocity field scales with wall parameters. Frequency spectra for s_x and for u at y^+ = 2.56, 5.91 agree with one another over a wide range of Reynolds number when scaled in this fashion. The median frequency is n^+ = 0.009; i.e., the dimensionless period in wall parameters is roughly equal to the dimensionless wavelength (Hanratty, et al 1977). Measurements of the median frequency of the s_z fluctuations are about the same as for s_x fluctuations (Sirkar and Hanratty, 1970). Estimates of the spectra of normal velocity fluctuations have been obtained by measuring the spectra of $\Delta s_z/\Delta z$ (Campbell, 1981). These indicate a median frequency which is 3-4 times that of s_x fluctuations. Measurements of the space-time correlations of s_x by Sirkar (1966) give a convection velocity of $C_x^+ \approx 15$. Measurements by Kreplin and Eckelman (1979) indicate that the convection velocity for the streamwise and spanwise velocity components are about equal in the viscous wall region. They obtained a value of $C_x^+ = 12.1$ for s_x and s_z and value of $C_x^+ = 15$ for u and w at $y^+ \approx 6$.

Different estimates of the streamwise extent of the wall eddies can be obtained from measurements of s_x and of s_z. The largest values are obtained from the ratio of the macrosales of s_x in the x- and z-directions; e.g., Hanratty, et al (1977) give an estimate of 40. A wavelength in the flow direction can be obtained from the convection velocity and the median frequency $\lambda_x^+ = C_x^+/\langle n^+\rangle = 15/0.009 = 1700$, indicating a length/width ratio of 1700/100=17.

Spanwise spatial correlation coefficients for s_x and s_z are approximately the same and scale with wall parameters. From these measurements, microscales of 12 ± 2 wall units are obtained. Streamwise spatial correlations of s_x and s_z, however, are not the same. The streamwise macroscale and microscale of s_z are roughly about one-half the values for s_x [Kreplin and Eckelmann, 1979; Lau, 1980]. A detailed study of the streamwise correlation coefficient of s_x by Finnicum and Hanratty (1985) indicates that it parameterizes with two length scales, the pipe diameter and the viscous length ν/v^*. The macroscale is given as $\Lambda_{sx}/D=35\,(Re)^{-1/2}$. If it is assumed that the macroscale is being influenced by large scale streamwise motions that are not directly related to the wall eddies then the ratio of x- and z- microscales of s_x gives an estimate of a length/width ratio of 10 for the wall eddies. A length/wide even less than this is suggested from the microscales of s_z.

Simultaneous measurements from arrays of wall electrodes, to be discussed next, give an estimated length/width of 6 for the wall eddies and of 3 for Reynolds stress producing events.

Streamwise Velocity Field Away From the Wall

Experiments aimed at relating the behavior of eddies at the wall to velocity fluctuations away from the wall are reported in papers by Hogenes and Hanratty (1982), Nikolaides, et al (1983a), Nikolaides, et al (1983b). The experiments were carried out in a 19.37 cm pipe with the arrangement of electrodes in the wall shown in Fig. 2. An array of \/ probes around the circumference of the pipe are used to detect when a strong wall eddy exists at a certain location in the pipe. An array of \/ electrodes arranged in the streamwise direction can be used to measure the streamwise extent of these eddies. Probes in the fluid located slightly downstream of the wall probes (so they wouldn't interfere with the flow) were used to relate happenings at the wall to the behavior away from the wall.

Four aspects of the wall eddies are defined. Negative or positive values of $\partial s_z/\partial z$ at the same z- location as the fluid probes (z=0) were used to indicate that a strong inflow or a strong outflow, respectively, will be sampled by the fluid probes. Maxima and minima in $s_z(z)$ indicate strong positive or negative spanwise flows and, on average, a coupled inflow and outflow at distances of $\Delta z^+ \cong \pm 25$.

In the first set of experiments a rake of probes was used to measure simultaneously the streamwise fluid velocity at a fixed Δx^+ downstream of the array of wall probes and at multiple y^+ locations. Time zero was defined by Hogenes when one of the four events detected by the wall probes reached its maximum intensity. Measurements from the fluid probes and from the wall probes were collected for $-26<T^+<26$. This was repeated for a number of non-overlapping events and an average behavior over the time interval was determined.

The strong relation of the streamwise velocity field to eddy patterns detected at the wall is best illustrated by an experiment in which the $s_z(z)$ patterns were conditionally sampled for a strong outflow followed by a strong inflow. This is detected by a change of $\partial s_z/\partial z$ from a large negative value to a large positive value at z=0.

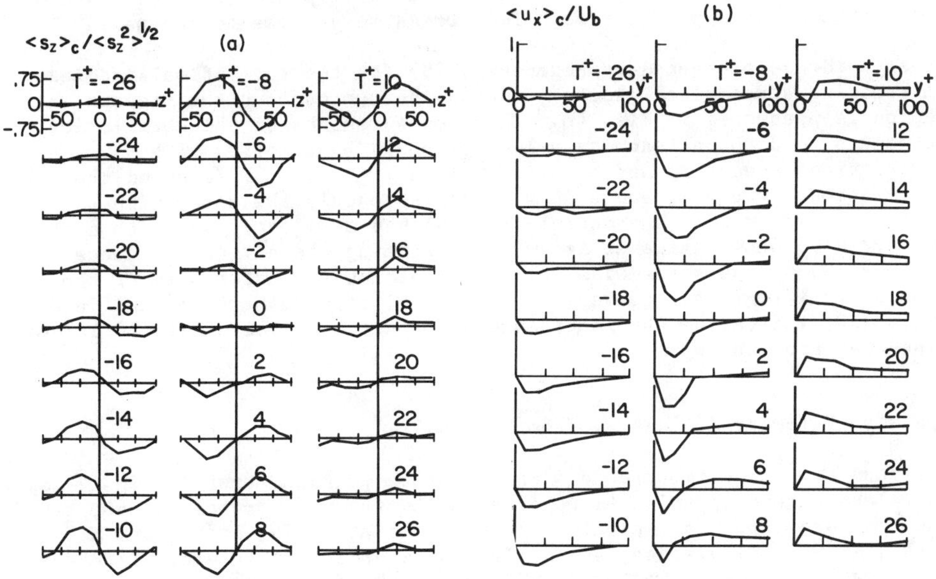

Figure 5. (a) $s_z(z)$ profiles conditionally sampled for an inflow at z=0 followed by an outflow at z=0 and (b) the average of the fluctuations in the streamwise velocity fluctuations

Six-six such events were detected with the rake located $\Delta x^+ = 90$ behind the wall probes over a dimensionless period of $T^+ = 20,000$. Conditionally averaged $s_z(z)$ profiles are shown in Fig. 5a. Note that a maximum and minimum are separated by a distance of $\Delta z^+ = 50$, indicating a spanwise wavelength of $\lambda^+ = 100$. The conditionally averaged positive and negative slopes are of approximately the same magnitude, in agreement with previous results of Lee et al (1974) that showed inflows and outflows are of roughly equal magnitude. The magnitudes of the maxima and minima of $0.75\left(\overline{s_z^2}\right)^{1/2}$ are large considering that too many samples were probably selected. If the length of an event is $\Delta T^+ \approx 50$ and if there is a 1/4 chance of having an inflow pattern at z=0, then the 66 samples assume a 2/3 chance that an inflow is followed by an outflow. A closer examination of the profiles indicates a value of 1/3 and 33 samples would have been more reasonable choices. Conditionally averaged streamwise velocity measurements are presented in Fig. 5b. Term $\langle u_c\rangle$ is the conditional average of the velocity fluctuations, $u = U - \bar{U}$. Term U_B is the bulk-average velocity. The measurements of ⟨u⟩ show a pronounced minimum close to the wall which can be associated with an outflow pattern at $z^+ = 0$. It is of interest to note how this minimum is annihilated with the change of the pattern. A wave of positive momentum is observed to move from $y^+ = 40$, $T^+ = 2$ to the wall at $T^+ = 10$. This is interpreted as being associated with an inflow disturbance propagating from the outer flow to the wall. A strong acceleration of the fluid is noted at $y^+ = 15$, so that this event probably could be detected with a measurement of the streamwise velocity at $y^+ = 15$. It has been observed by Hogenes that it is equally likely that a particular mode at the wall (outflow, inflow, positive cross flow, negative cross flow) will be followed by any one of the other modes. This means that the behavior captured in Fig. 5 would contribute about 1/6 of the Reynolds stress very close to the wall.

Other evidence presented by Hogenes (1979) for the role of wall eddies in creating streamwise velocity fluctuations throughout the viscous wall region is presented in Fig. 4b. These measurements of the correlation between s_z and the streamwise velocity should be compared with correlation between s_z and s_x presented in Fig. 4a. It is noted the correlation between s_z and u at $y^+ = 11$ is roughly the same, with a maximum at $\Delta z^+ \approx 27$. The correlations with u at $y^+ = 19$ and with u at $y^+ = 27$ also show maxima at $\Delta z^+ \approx 30$ indicating the direct influence of $\lambda^+ \approx 100$ on the streamwise velocity field at the outer part of the viscous wall region. However, the correlation curves are more diffuse for $z^+>30$, indicating the importance of larger spanwise scales at larger distances from the wall.

Spanwise Velocity Field Away from the Wall

A second type experiment done with the probe pattern shown in Fig. 2 was to relate strong spanwise flows detected by the wall probes at z=0, t=0 with measured spanwise velocity fluctuations at $y^+=20$, $\Delta x^+=25$ and at $y^+=40$, $\Delta x^+=25$ (Nikolaides, et al, 1983). A detection scheme, different from that of Hogenes, was used. The strength of an eddy was defined from the average value the absolute value of s_z over a spanwise distance of $\lambda^+=100$. Its magnitude was measured in terms of the value for a cosine wave with a wavelength of $\lambda^+ = 100$ and an amplitude equal to $\left(\overline{s_z^{+2}}\right)^{1/2} = 0.11$. Eddies stronger than this were detected roughly about 50 percent of the time. These strong eddies were identified depending on whether a positive transverse flow, a negative transverse flow, an outflow or an inflow was located at z = 0. Time zero, used in the conditional averaging, was located at the temporal midpoint of one of these identified patterns. Fig. 6a gives an example of the average of 50 positive transverse flows identified in this way. These show a cosine pattern with zero crossings approximately at $z^+ = 25$ and at $z^+ = -25$. The conditionally averaged $s_x(z)$ patterns are also given in Fig. 6a. These show a sine wave with a maximum occurring at the location the zero crossing of s(z) corresponding to a positive $\partial s_z/\partial z$ (an inflow) and a minimum occurring at the zero crossing corresponding to a negative $\partial s_z/\partial z$ (an outflow). It is clear that these conditionally sampled strong cross flows are accompanied by strong stress producing events which, on average, occur at $z^+ = \pm 25$.

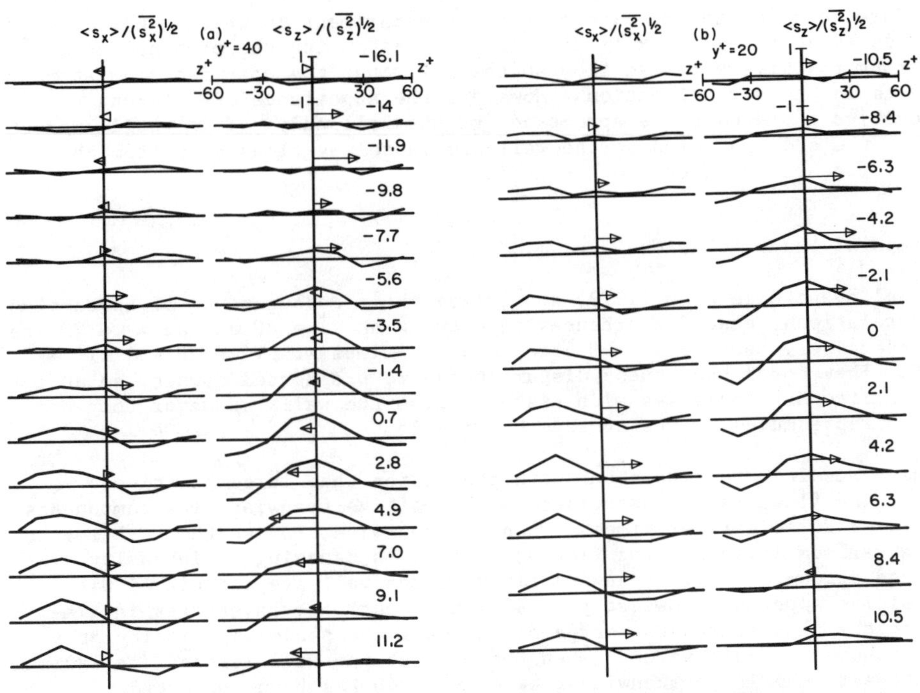

Figure 6. Conditionally sampled for positive streamwise flows at the wall. (a) The arrows indicate the streamwise and spanwise velocity components at $y^+ = 40$; (b) The arrows indicate the streamwise and spanwise velocity components at $y^+ = 20$.

It is noted that the strong eddy had an average duration of about $\Delta T^+ = 20$. If the convection velocity is taken to be $C_x^+ = 15$, this gives an estimated average length of these eddies of about 300 wall units. It is noted that the s_x profiles lag the development of the s_z profiles by $\Delta T^+ \cong 8$. The period over which both the s_x and s_z profiles are strong corresponds to $-1.4 < T^+ < 9.1$ or to a $\Delta T^+ \approx 10$. This would give an estimate of a length for the stress producing event of 150 wall units.

The maximum amplitudes of the conditionally averaged profile shown in Fig. 6a are roughly equal to $1.0\,(\overline{s_x^2})^{1/2}$ and $1.0\,(\overline{s_z^2})^{1/2}$. The arrows in Fig. 6a,b indicate the average values of u and w measured at $z^+ = 0$, $y^+ = 40$, $\Delta x^+ = 25$. An arrow extending to $z^+ = 60$ would correspond to a value of $\langle w\rangle/(\overline{w^2})^{1/2}$ or of $\langle u\rangle/(\overline{u^2})^{1/2}$ equal to + 1. The appearance of the strong positive spanwise flow at the wall is preceded by a strong positive w, which has an amplitude of about $0.5\,(\overline{w^2})^{1/2}$. The flow at $y^+ = 40$ changes direction and has a value of about $-0.5\,(\overline{w^2})^{1/2}$ at $T^+ = 4.9$. It is noted that the direction of flow at $y^+ = 40$ is opposite that at $y^+ = 0$, giving the suggestion of a "closed eddy" over the period corresponding to Reynolds stress production. This gives an estimate of the length of this "closed eddy" of about 150 wall units. This estimate appears to agree with estimates obtained from measurements of spatial correlation between w and s_z obtained with the array of wall electrodes aligned in the direction of flow as shown in Fig. 2 of Nikolaides, et al [1983].

Figure 6b shows the results of another experiment in which w was measured at $y^+ = 20$, $\Delta x^+ = 25$. It is noted during the presence of a strong positive spanwise flow at the wall that the spanwise flow at $y^+ = 20$ is in the same direction. However, the appearance of a strong positive w precedes the appearance of the wall eddy. This would suggest that the eddy patterns at the wall are caused by flows away from the wall that occurred earlier in time.

Temporal Phase Relations

Kreplin and Eckelmann (1979) show, from their measurements of space-time correlations, that disturbances in u and w at $y^+ = 40$ and at $y^+ = 20$ are respectively seen at $y^+ = 0$ after elapsed times of $\Delta T^+ = 20$ and $\Delta T^+ = 15$. They found that these disturbances are propagated downstream at a velocity that increases with distance from the wall. However they are well represented on the average by $C_x^+ = 15$.

These results are consistent with the notion that large positive spanwise flows can be associated with positive spanwise flow components at $y^+ = 20$ at earlier times. When the spanwise flow at the wall reaches its maximum intensity the flow at $y^+ = 20$ is decaying in intensity. These strong positive spanwise flows at the wall are associated with positive spanwise flows at $y^+ = 40$ that occurred even earlier in time. When the flow starts to emerge at the wall the spanwise velocity at $y^+ = 40$ assumes a small value. When the flow at the wall reaches its maximum intensity the large spanwise flow at $y^+ = 40$ has been convected downstream to $x^+ \approx 300$ and the flow at $y^+ = 40$, $x^+ = 0$ has started to change its direction to give the appearance of a closed eddy.

COMPUTER MODELS

The "Slender Body" Assumption

Information of the type presented in the previous section has been used for some time in this laboratory to develop a simple computational model for the viscous wall region, $y^+<30\text{-}40$. The goal in these efforts is not to describe exactly the time-dependent flow. Rather, the question is explored as to whether one can substitute for the real behavior a simple eddy model which represents important features of the viscous wall region such as the production of turbulence and of Reynolds stress.

The main simplification that has been made in all of this work is to assume that the flow is homogeneous in flow direction. This has been called "slender body" turbulence or a "2 1/2 D model" since it solves for three components of the velocity in a two-dimensional domain. The simplified Navier-Stokes equations are as follows for a zero pressure gradient flow:

$$\frac{\partial v}{\partial t} + \frac{\partial}{\partial y}(v^2) + \frac{\partial}{\partial z}(vw) = -\frac{\partial p}{\partial y} + \frac{1}{Re}\left(\frac{\partial^2 v}{\partial y^2} + \frac{\partial^2 v}{\partial z^2}\right) \tag{6}$$

$$\frac{\partial w}{\partial t} + \frac{\partial}{\partial y}(vw) + \frac{\partial}{\partial z}(w^2) = -\frac{\partial p}{\partial z} + \frac{1}{Re}\left(\frac{\partial^2 w}{\partial y^2} + \frac{\partial^2 w}{\partial z^2}\right) \tag{7}$$

$$\frac{\partial U}{\partial t} + \frac{\partial}{\partial y}(vU) + \frac{\partial}{\partial z}(wU) = \frac{1}{Re}\left(\frac{\partial^2 U}{\partial y^2} + \frac{\partial^2 U}{\partial z^2}\right) \tag{8}$$

$$\frac{\partial w}{\partial z} + \frac{\partial v}{\partial y} = 0 \tag{9}$$

All variables have been made dimensionless with respect to a characteristic length and a characteristic velocity. It is noted that (6) and (7) can be solved independently of (8) and that the velocity in the flow direction, U, can be obtained from (8) after v and w are calculated.

The Pseudo-Steady State Model

The first attempt to use this approach was by Fortuna [Hanratty, Chorn and Hatziavramidis, 1977; Fortuna, 1971] who explored the simple eddy model of Sirkar and Hanratty. Fortuna used a pseudo-steady state approximation whereby $\partial v/\partial t$, $\partial w/\partial t$ and $\partial U/\partial t$ are ignored. He solved (8) using

$$w = \hat{w}\, y \sin (2\pi z/\lambda) \tag{10}$$

and boundary conditions U=0, y=0 and $U=U_L$ with the additional assumption that $\nu\, \partial^2 U/\partial z^2$ can be ignored. Velocity U_L was selected so that $d\bar{U}^+/dy^+ = 1$ at $y^+ \cong 0$. Wavelength, λ, was taken as equal to $\lambda^+ = 100$. Amplitude $\hat{w}$ was assumed to be given by a Gaussian function which has r.m.s. values of the spanwise fluctuations of $\left(\overline{w^2}\right)^{1/2}/v^* = 0.0077\, y^+$ in order to approximate available turbulence measurements for $0<y^+<20$. Excellent agreement with average velocity measurements was obtained and an explanation was provided as to why increases in λ^+ are associated with drag-reduction (Hanratty, Chorn and Hatziavramidis, 1977). One of the more interesting results coming from these solutions of (8) is that eddies of the type shown in Fig. 1 can produce large streamwise velocity fluctuations. A value of $\left(\overline{s_x^2}\right)^{1/2}/\bar{S}_x = 0.44$ was obtained for $\lambda^+ = 100$. This, then, suggested a way of creating velocity deficient fluid near the wall.

In order to obtain some understanding of parameters which control the wavelength of the wall eddies, Fortuna (1971) solved (6), (7), (9) numerically with $\partial v/\partial t = \partial w/\partial t = 0$. For boundary conditions he used periodicity in the z-direction, v=w=0 at y=0 and $w=\hat{w} \sin (2\pi z/\lambda)$ at large y. The Reynolds number characterizing the problem was defined as $Re=\lambda\hat{w}/\nu$. Solutions were obtained for Re=80, 160 and 400, which correspond to $\lambda^+=43.5$, 87 and 217. The steady solution for $\lambda^+=217$ showed a separation bubble at the wall, while those for $\lambda^+=43.5$ and 87 did not. This suggested a mechanism for creating small eddies at the wall and that the size of the wall eddies could have an upper limit because of a separation phenomenon.

Single Eddy Time-Varying Model

Hatziavramidis and Hanratty (1979) followed up on the work of Fortuna by including the influence of the transient terms $\frac{\partial w}{\partial t}$, $\frac{\partial v}{\partial t}$ and $\frac{\partial U}{\partial t}$. Equations (6), (7), (9) were solved using the boundary conditions of v=w=0 at $y^+=0$ and

$$w^+ = \hat{w}_L \cos(2\pi t^+/T_B^+)\sin(2\pi z^+/\lambda^+) \qquad (11)$$

at $y_o^+ = 45$. They used $\lambda^+=100$ and $T_B^+ = 100$ to correspond to the observed wavelength and bursting frequency of the wall eddies. Term w_L^+ was selected to agree with turbulence measurements at $y_o^+=45$. These calculated values of v(y,z,t) and w(y,z,t) were used in (8) to calculate U(y,z,t) using the boundary conditions of U=0 at y=0 and $U=U_L$ at $y_o^+=45$. The streamwise patterns showed the appearance of small separation bubbles with a spanwise size of about 10 wall units.

Average behavior at a given distance from the wall was calculated by averaging in time and with respect to the spanwise coordinate. The value of U_L^+ was selected so that $\overline{U}^+=y^+$ for $y^+ \to 0$. The results are consistent with previous results of Fortuna in that $\overline{U}^+(y^+)$ agrees with experimental data and that the wall eddies produce large streamwise velocity fluctuations. This is illustrated by a calculation of a maximum in the r.m.s. value of the streamwise velocity fluctuations, $(\overline{u^2})^{1/2}$, at $y^+=12$, in agreement with laboratory measurements. The calculated ratio $(\overline{s_x^2})^{1/2}/(\overline{s_z^2})^{1/2}$ and the calculated quadrant analysis of the Reynolds stress also roughly agrees with measurements. The calculated time variation of the streamwise velocity fluctuations at $z^+=0$ showed features similar to what is displayed in Fig. 5. Space time correlations between s_x and u(y) were also calculated. In agreement with the experiments of Kreplin and Eckelman the u pattern close to the wall is calculated to lag behind the u pattern far from the wall. The calculated lag time increases with increasing y and is equal to about 20 wall units at $y^+=40$.

Two Eddy Time-Varying Model

The poor agreement with measurements of the $\overline{v^2}(y)$ and $\overline{w^2}(y)$ calculated by Hatziavramidis and Hanratty lead Chapman and Kuhn (1981) and Nikolaides (1984) to reexamine the consistency between the boundary conditions used by Hatziavramidis and Hanratty and the experimental results discussed in Section 2. Nikolaides concluded that the size of eddy patterns at the wall are probably dictated by the spanwise variation of the normal velocity fluctuations of $\lambda_1^+=100$ eddies at $y_o^+=40$ and with separation bubbles associated with larger scale eddies at $y_o^+=40$. He also concluded that the spanwise velocity fluctuations for $15<y^+<40$ are dominated by eddies with larger spanwise scales than $\lambda_1^+=100$ and that normal velocity fluctuations at $y_o^+=40$ are dominated by $\lambda_1^+=100$ eddies. He chose to represent the influence of all larger scale eddies by a single eddy with $\lambda_2^+=400$. These observations would suggest that (11) is a poor first order choice for the behavior of the stress producing eddies in the viscous wall layer. A better approximation for a single eddy model would be

$$v_1 = \hat{v}_1 \cos(2\pi z^+/\lambda_1)\cos\left(\frac{2\pi t}{T_1} + \phi_{VI}\right) \qquad \frac{\partial v_1}{\partial y^+} \approx 0,\ w_1 \approx 0 \qquad (12)$$

with $\lambda_1^+=100$ and $T_1^+=100$ and ϕ_{VI} arbitrary. In order to account for the w variation at large y^+ Nikolaides assumed the presence of a second eddy which to a first approximation behaves at $y_o^+=30-40$ as

$$w_2 = \hat{w}_2 \cos\left(\frac{2\pi t}{T_2} + \phi_{W2}\right)\sin(2\pi z/\lambda_2) \qquad v_2 \ll w_2, \qquad (13)$$

with $\lambda_2^+=400$, $T_2^+=400$. The period $T_2^+=400$ was chosen so that inertia terms and transient terms in the momentum equations are of the same order. [See Hatziavramidis and Hanratty, 1979.]

It is noted that (13) also introduces a small amount of v-component of $\lambda_2^+=400$ at $y_o^+=30\text{-}40$ since, from conservation of mass, $\partial v/\partial y + \partial w/\partial z=0$. Therefore boundary conditions (12) and (13) have to be reformulated for a two eddy model:

$$v=\hat{v}_1 \cos(2\pi z/\lambda_1)\cos(\frac{2\pi t}{T_1} + \phi_{V2}) + \hat{v}_2 \cos(2\pi z/\lambda_2)\cos(\frac{2\pi t}{T_2} + \phi_{V2}) \tag{14}$$

$$w=\hat{w}_1 \sin(2\pi z/\lambda_1)\cos(2\pi t/T_1) + \hat{w}_2 \sin(2\pi z/\lambda_2)\cos(\frac{2\pi t}{T_2} + \phi_{W2})$$

$$(\phi_{V2} - \phi_{W2}) = \pm 180 \quad , \quad 0<\phi_{V1}<180^0 \tag{15}$$

Conservation of mass requires that at y_o^+

$$-\frac{\partial v^+}{\partial y^+} = \frac{2\pi}{\lambda_1^+}\hat{w}_1 \cos(2\pi z/\lambda_1)\cos(2\pi t/T_1)$$

$$+ \frac{2\pi}{\lambda_2^+}\hat{w}_2 \cos(2\pi z/\lambda_2)\cos(\frac{2\pi t}{T_2} + \phi_{W2}) \tag{16}$$

Amplitudes $\hat{w}_2$ and $\hat{v}_1$ are much larger than $\hat{w}_1$ and $\hat{v}_2$, and chosen to agree approximately with measurements of $\overline{w^2}$ and $\overline{v^2}$ at y_o^+. Thus Nikolaides explored whether the behavior of the viscous wall region could be described by the interaction of two eddies which were behaving like (14) and (15) at y_o^+. A simplistic picture of these eddies is given in Fig. 7a,b,c. Figure 7a depicts the $y_2^+=400$ eddy. The outer edge of the viscous wall region is located in the bottom of the eddy where $v \ll w$. The velocity field below y_o^+ is always "open" in that w at y_o^+ is in the same direction as w at the wall. This requires that $(\phi_{V2} - \phi_{W2}) = \pm 180°$. Figures 7b,c depict the $\lambda_1^+ = 100$ eddy. The outer edge of the viscous wall region is close to the center of the eddy where w is small compared to v. On occasion y_o^+ will lie above the center and the eddy will be "open" below y_o^+, as shown in Fig. 7b, and on other occasions it will be "closed" in that the w velocity at y_o^+ will be in the opposite direction of w close to the wall. If $\phi_{V1}=0$ the eddy will always be "closed" so available experimental results would seem to suggest $0<\phi_{V1}<180°$. The value of ϕ_{V1}, thus, dictates the percent of the time the eddy is closed.

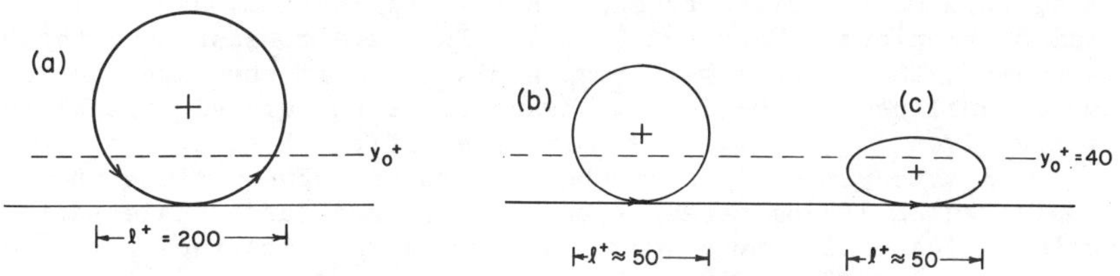

Figure 7. Pictures of the eddies: (a) $\lambda^+ = 400$ with centerline above y_o^+; (b) $\lambda^+ = 100$ with centerline above y_o^+; (c) $\lambda^+ = 100$ with centerline below y_o^+

Phase angle ϕ_{V2} (or ϕ_{W2}) seems to dictate the relative phase between the λ_1^+ and λ_2^+ eddies. No criteria exist for selecting this. It would seem that all values are equally probable if the large and small scale eddies are independent. Likewise, no clearcut criteria are available to select the amplitudes $\hat{w}_1$ and $\hat{v}_2$. It could be that the method for specifying these values as well as phase angle ϕ_{V2} and y_o^+ affect the smoothness of the velocity variation near y_o^+.

Eddy patterns of the type described by (14) and (15) would create a streamwise velocity field at y_o^+ described by the following equation:

$$U - \overline{U} = \hat{u}_1 \cos[\frac{2\pi t}{T_1} + \phi_{U1}] \cos \frac{2\pi z}{\lambda_1} + \hat{u}_2 \cos[\frac{2\pi t}{T_2} + \phi_{U2}]\cos\frac{2\pi z}{\lambda_2}$$

$$(\phi_{U1} - \phi_{V1}) \text{ close to } 180^0 \quad , \quad (\phi_{U2} - \phi_{V2}) \text{ close to } 90^0 \qquad (17)$$

According to the model explored by Nikolaides the λ_1^+ eddies are highly correlated and carry most of the Reynolds stress at y_o^+. Consequently, $(\phi_{U1} - \phi_{V1})$ should be close to 180° and $(\phi_{U2} - \phi_{V2})$, close to 90°. The magnitude $\hat{u}_1$ would be chosen to give the correct Reynolds stress at y_o^+ and the value of $\hat{u}_2$, to give a value of $\overline{u^2}$ that agrees with measurements.

In order to test this eddy model Nikolaides (1984) carried out numerical solutions of (6), (7), (8), and (9) using boundary conditions (14), (15), and (17). From the measurements of Nikolaides, et al (1983) it was estimated that the centers of the $\lambda^+ = 100$ eddies are below $y_o^+ = 40$ more than 70 percent of the time. Nikolaides, therefore, chose ϕ_{V1} = 36° to give a closed flow for 80 percent of the time. The choice of $(\phi_{U2} - \phi_{V2}) = 104.4°$ allows the λ_1^+ eddies to contribute 75 percent of the Reynolds stress at $y_o^+ = 40$. The choices of the fraction of the $\overline{u^2}$ energy (0.85) contributed by the λ_2^+ eddies and the percent of the $\overline{v^2}$ energy (0.75) contributed by the λ_1^+ eddies were determined so that the Reynolds stress and the Reynolds stress coefficient agreed with measurements at $y_o^+ = 40$. The fraction of the $\overline{w^2}$ energy (0.60) contributed by the λ_2^+ eddies was estimated from measurements given in the paper by Nikolaides, et al (1983). The calculations were somewhat insensitive to the choice of ϕ_{W2} (= 270°). Phase angles $(\phi_{V2} - \phi_{W2})$ and $(\phi_{U1} - \phi_{V1})$ were, respectively set at -180° and 180°.

Results of these calculations are presented in a thesis by Nikolaides (1984) and, to a more limited extent, in an article by Lyons, et al (1988) and by Finnicum & Hanratty (1988). They are consistent with the main features of the viscous wall region displayed in the experimental measurements outlined in Section 3. Examples of calculated streamline patterns are given in Fig. 8. The most interesting characteristic of the calculated streamlines is that they alternate between times when the viscous wall region is dominated by small scale and large scale eddies. Large scale eddies of the type shown in Fig. 8a were observed for t^+ = 30 - 80 and for t^+ = 230 - 280. These show isolated eddies originating at the wall which can persist for periods of $\Delta t^+ \approx 40$ and which resemble separation bubbles of the type seen by Fortuna in his pseudo-steady state calculations. Figure 8b shows a well developed pattern of λ^+ = 100 eddies. These started to develop at $t^+ = 90$ and lasted until t^+ = 200. A particularly strong pattern of the type shown in Fig. 8b lasted for a period $\Delta t^+ \approx$ 20-30. Patterns of these small eddies rotating in the opposite direction were also observed for t^+ = 290-400.

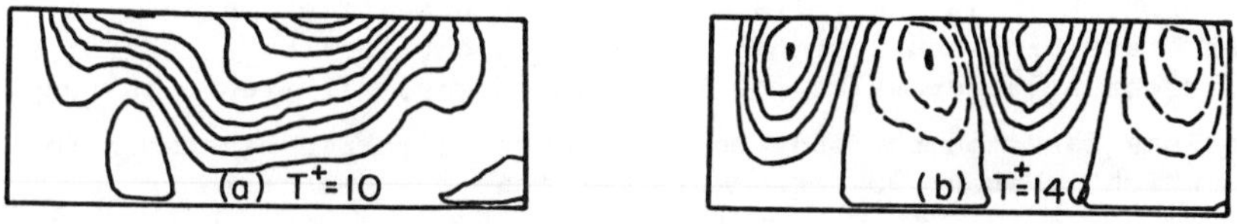

Figure 8. Streamwise contours at $t^+ = 10$ and at $t^+ = 140$

The change of the scale of the eddy pattern with distance from the wall, suggested from laboratory measurements, is also displayed in the calculated flow field. The calculations show the spanwise variation of the w-velocity component to be dominated by the $\lambda^+ = 400$ eddies for $y^+ > 10$ as evidenced by a zero crossing in the spatial correlation coefficient $R_{ww}(\Delta z)$ at $\Delta z^+ \approx 100$. Figure 9 shows, however, that at the wall it is dominated by $\lambda^+ = 100$ eddies. Here the spatial correlation $R_{s_z s_z}(\Delta z)$ has a pattern very similar to what is shown in laboratory measurements. It has a zero crossing at $\Delta z^+ \approx 35$ and a minimum at $\Delta z^+ \approx 50$. Figure 9 also shows the calculated correlation coefficient of the [Kontomaris, 1988] correlation of the normal velocity component close to the wall, $R_{vv}(y^+=0.8,\Delta z^+)$. It is of interest that the first zero crossing and minimum occur at $\Delta z^+ = 19$ and at $\Delta z^+ = 34$. This is to be compared with the calculated zero crossing and minimum for R_{vv} at $\Delta z^+ \approx 25$ and at $\Delta z^+ \approx 50$ for $y^+ \geq 20$. The difference seems to reflect the influence of "separation bubbles" at the wall with spanwise dimensions less than 50 wall units.

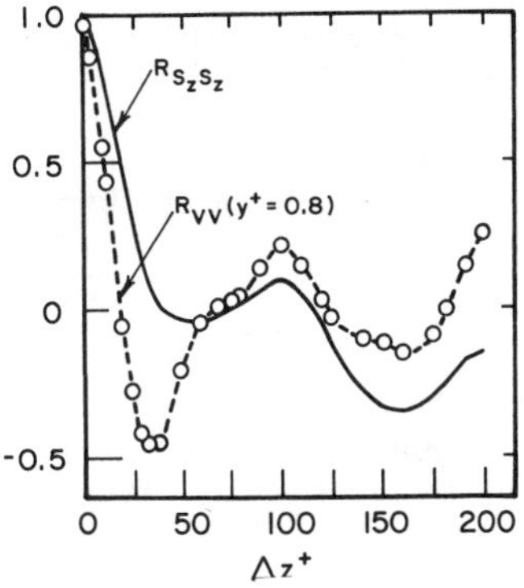

Figure 9. Spatial correlations of the spanwise velocity gradient at the wall and of the normal velocity component at $y^+ = 0.8$

The results of this computer simulation were also used to calculate the variation of various statistical properties of the velocity at different distances from the wall by averaging with respect to z and with respect to t. [See Nikolaides (1984); and Lyons, et al (1988); and Finnicum and Hanratty (1988.)] Excellent agreement between the calculated and measured profiles of the average velocity, the Reynolds stress and the Reynolds stress correlation coefficient was obtained. Approximate agreement was also obtained with measurements of $\overline{u^2}(y)$, $\overline{v^2}(y)$, $\overline{w^2}(y)$, $\overline{u^3}(y)$ and $\overline{u^4}(y)$. The calculated $\overline{v^2}$ profile is significantly lower than the measurements of Laufer at $y^+<20$, but this seems consistent to what has been found in the recent measurements of Finnicum and Hanratty (1985). The calculated $\overline{w^2}$ profile agrees quite well with Laufer's measurements but shows two slight inflections marking the transition from a region in which the w-velocity field is controlled by $\lambda^+ = 100$ wall eddies to a region in which it is controlled by $\lambda^+ = 400$ outer flow eddies. A smoother transition could probably have been obtained by selecting a smaller y_o^+ or a larger value for λ_2^+. The largest difference is that the computations based on the eddy model produced a maximum in the $\overline{u^2}(y)$ profile which is about 30 percent larger than the measurements.

Prediction of λ^+

This good agreement between calculations and experiments gives some support to the conceptual model outlined in Section 2 and prompted a calculation aimed at obtaining an understanding of how the size of the wall eddies is determined (Lyons, et al, 1988). The approach taken was simply to carry out calculations of the flow in the viscous wall region for different λ_1^+. However, it should be recognized that the flow characteristics depend not only on the scaling of λ_1 but also on the thickness of the viscous wall region, y_o, the magnitude of the velocity fluctuations, q, and the time scale of the velocity fluctuations, T. Consequently, a critical problem to be dealt with was the determination of how y_o^+ should vary with λ_1^+.

This scaling problem was dealt with by assuming that y_o lies at the inner edge of the log-layer where

$$\frac{d\overline{U}}{dy} = \frac{u^*}{\kappa y} \tag{18}$$

and where the production of turbulent energy equals the dissipation,

$$\left[-\overline{uv}\,\frac{d\overline{U}}{dy}\right]_{y=y_o} \cong [\varepsilon]_{y=y_o} \tag{19}$$

Outside the viscous wall layer an inviscid relation can be used for the dissipation

$$\varepsilon = a_1 \frac{q^3}{\ell}, \tag{20}$$

where $q^2 = \overline{u^2} + \overline{v^2} + \overline{w^2}$ and ℓ is a characteristic length of the large eddies. Equations (18), (19), and (20) reveal that $\ell = \kappa y$ and that the turbulent velocity fluctuations scale with u*. The use of (2) and (19) and the assumption that λ_1 is proportional to λ_g gives

$$\lambda_1^{+2} = C_5 y_o^+ \tag{21}$$

Therefore, in carrying out the calculations with different λ_1^+ the amplitudes in (14), (15) and (17) were kept constant and y_o^+ was changed in accordance with (21). In order that transient and convective terms are of the same order T_1^+ was changed so that $T_1^+ = \lambda_1^+$. No theoretical guidance was available for selecting λ_2^+ or for selecting the phase angles, so these parameters were kept constant for the calculations with different λ_1^+.

The most important result obtained from the calculations is the effect of λ_1^+ on the balance of turbulent kinetic energy = $q^2/2$. The energy balance terms calculated for $\lambda_1^+ = 100$ are shown in Fig. 10a. Close to the wall the energy dissipation is a maximum and the production is small. Energy is supplied to $0 < y^+ < 0.1\, y_o^+$ by a viscous diffusion of energy produced at $y^+ > 0.1\, y_o^+$. At the outer part of the viscous wall region the production of turbulent energy is approximately equal to the dissipation. The important property to be noted is that both the production, P, and the dissipation, ε, are quite large in the viscous wall region and that the flow of energy out is represented by the integral of the difference of two large numbers.

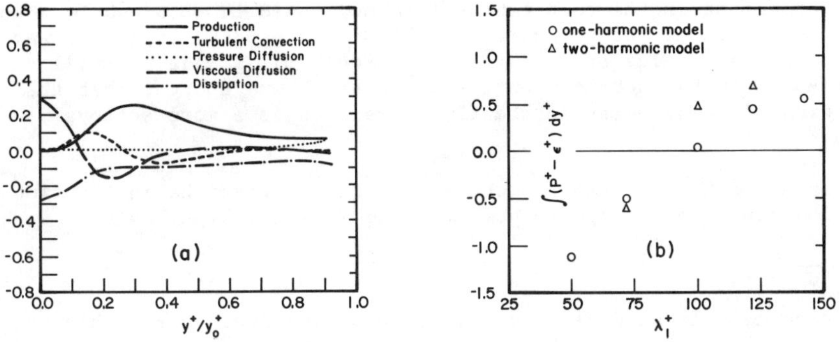

Figure 10. (a) Balance of turbulent kinetic energy for $\lambda_1^+ = 100$; (b) Net production of turbulent energy by the viscous wall region for various λ_1^+

Values of $\int_0^{y_o^+} (P - \varepsilon)dy^+$ calculated for different values of λ_1^+ are shown in Fig. 10b. It is noted that for the calculations with two harmonics this integral is positive for $\lambda_1^+ = 100$ and for $\lambda_1^+ = 122$ and is negative for $\lambda_1^+ = 72$. This calculation would suggest that $\lambda_1^+ = 72$ would give an impossible stationary situation since it requires a diffusion of turbulent energy into the viscous wall region.

Of interest is the large change in the net production of turbulent energy with an increase of λ_1^+ from 72 to 100. Laufer's (1954) pipe flow measurements indicate a dimensionless net dissipation of turbulent energy of 0.29 in the core region. Using a linear interpolation, a value of $\lambda_1^+ = 93$ is obtained as needed to satisfy the balance of mechanical energy.

Calculations using a single eddy representation of the viscous wall region are also given in Fig. 10b. These show the same qualitative behavior as the model with two harmonics and give a value of $\lambda_1^+ = 114$ as needed to provide a net production of 0.29.

Both estimates for λ_1^+ are close to the observed characteristic scale for the viscous wall region of 100 wall units. This simple eddy model has also been used to interpret the effect of large favorable pressure gradients on turbulent boundary-layers. Details regarding the results of this investigation may be found in an article by Finnicum and Hanratty (1988).

Improvement of the Simple Eddy Model

A flaw in the calculated statistical properties of the turbulence is the appearance of rapid changes near y_o^+ which are non-physical. These suggest that the outer boundary conditions are too restrictive and that this causes the existence of a viscous region of high vorticity close to

y_o. Fortunately, this error affects the calculations only over the outer 10-20 percent of the viscous wall region. Consequently, discussions of energy balances presented in this paper disregard calculations for $0.85\ y_o < y < 1.0\ y_o$.

Some improvement in the computational method could be obtained by a more judicious choice of the parameters. This could be done in such a way that the average value of $\partial v^+/\partial y^+$ at y_o^+ [Eqn. 16] is made as small as possible. This has not been pursued because it was thought that the assumption of homogeneity in the flow direction is a more serious drawback.

The present method can capture the small scale variations in the flow direction observed in experiments by using Taylor's hypothesis,

$$\frac{\partial}{\partial x} = -\frac{1}{C_x}\frac{\partial}{\partial t}, \tag{22}$$

to convert calculated derivatives in t to derivatives in x. This motivated a recent work by Kontomaris (1988b) to improve the "slender body" model by approximating derivatives in the flow direction, in the Navier Stokes equations, with (22). Numerical difficulties precluded any definite conclusions as to how the inclusion of this approximation affects calculated results.

ACKNOWLEDGEMENT

This work was supported by the Office of Naval Research under Grant N000 14-82-K0324 and by the Shell Companies Foundation.

REFERENCES

1. Barlow, R. S. and J. P. Johnston, "Structure of Turbulent Boundary Layers on a Concave Surface", Report MD-47, Department of Mechanical Engineering, Stanford University, Stanford, California, 1985.

2. Brodkey, R. S., Wallace, J. M. and H. Eckelmann, J. Fluid Mech., vol. 63, p. 209, 1974.

3. Campbell, J. A., "The Velocity-Concentration Relationship in Turbulent Mass Transfer to a Wall", Ph.D. Thesis, University of Illinois, Urbana, 1981.

4. Chapman, D. R. and G. D. Kuhn, "Two-Component Navier Stokes Computational Model of Viscous Sublayer Turbulence", AIAA Paper 81-1024 Proc. A AIAA 5th CFD Conf., Palo Alto, 1981.

5. Corrsin, S., in "Symposium on Naval Hydrogynamics" Publication 515, p. 373, National Academy of Sciences - National Research Council, Washington, D.C., 1957.

6. Eckelmann, H., "The Structure of the Viscous Sublayer and the Adjacent Wall Region in a Turbulent Channel Flow", J. Fluid Mech., vol. 65, part 3, pp. 439-459, 1974.

7. Eckelman, L. D., "The Structure of Wall Turbulence and Its Relation to Eddy Transport", Ph.D. Thesis, University of Illinois, Urbana, 1971.

8. Eckelman, L. D., Fortuna, G. and T. J. Hanratty, "Drag Reduction and the Wavelength of Flow-Oriented Eddies", Nature Physical Sciences, vol. 236, no. 67, pp. 94-96, 1972.

9. Finnicum, D. S. and T. J. Hanratty, "Effect of Favorable Pressure Gradients on Turbulent Boundary Layers", AIChE Journal, 34, pp. 529-540, 1988.

10. Finnicum, D. S. and T. J. Hanratty, "Turbulent Normal Velocity Flucutations Close to a Wall", Phys. Fluids, vol. 28, pp. 1654-1658, 1985.

11. Fortuna, G., "Effect of Drag-Reducing Polymers on Flow Near a Wall", Ph.D. Thesis, University of Illinois, Urbana, 1971.

12. Fortuna, G. and T. J. Hanratty, "Frequency Response of the Boundary Layer and Wall Transfer Probes", Int. J. Heat Mass Transfer, vol. 14, pp. 1499-1507, 1971.

13. Fortuna, G. and T. J. Hanratty, "The Influence of Drag-Reducing Polymers on Turbulence in the Viscous Sublayer", J. Fluid Mech., vol. 53, part 3, pp. 575-586, 1972.

14. Grant, H. L., "The Large Eddies of Turbulent Motion", J. Fluid Mech., vol. 4, p. 149, 1958.

15. Hanratty, T. J., "Study of Turbulence Close to a Solid Wall", The Phys. of Fluids Supplement, p. S126, 1967.

16. Hanratty, T. J. and J. A. Campbell, "Measurements of Wall Shear Stress", in Fluid Mechanics Measurements" (ed. Goldstein, R. J.), p. 559, Hemisphere Publishing, Washington, D.C., 1983.

17. Hanratty, T. J. and L. G. Chorn, "Turbulent Fluctuations in the Viscous Wall Region for Newtonian and Drag Reducing Fluids", Phys. Fluids, vol. 20, no. 10, pp. S112-S119, 1977.

18. Hatziavramidis, D. T. and T. J. Hanratty, "The Representation of the Viscous Wall Region by a Regular Eddy Pattern", J. Fluid Mech., vol. 95, part 4, pp. 655-679, 1979.

19. Hinze, J. O., "Turbulence", 2nd ed, McGraw-Hill, New York, 1975.

20. Hogenes, J. H. A., "Identification of the Dominant Flow Structure in the Viscous Wall Region of a Turbulent Flow", Ph.D. Thesis, University of Illinois, Urbana, 1979.

21. Hogenes, J. H. A. and T. J. Hanratty, "The Use of Multiple Wall Probes to Identify Coherent Flow Patterns in the Viscous Wall Region", J. Fluid Mech., vol. 124, pp. 363-390, 1982.

22. Kline, S. J., Reynolds, W. C., Schraub, F. A. and P. W. Runstadler, "The Structure of Turbulent Boundary Layers", J. Fluid Mech., vol. 30, p. 741 (1967).

23. Kline, S. J. and P. W. Runstadler, "Some Preliminary Results of Visual Studies of the Flow Model of the Wall Layers of the Turbulent Boundary Layer", J. Appl. Mech., vol. 2, p 166, 1959.

24. Kontomaris, Konstantinos, "A Turbulent Channel Flow Model Based on Taylor's Frozen-Turbulence Hypothesis", M.S. Thesis, University of Illinois, Urbana, 1988b.

25. Kontomaris, Konstantinos, private communication to appear in Ph.D. Thesis, 1988.

26. Kreplin, H. P. and H. Eckelmann, "Propagation of Perturbations in the Viscous Sublayer and Adjacent Wall Region", J. Fluid Mech., vol. 95, part 2, pp. 305-322, 1979.

27. Lau, K. K., "Study of Turbulent Structure Close to a Wall Using Conditional Sampling Techniques", Ph.D. Thesis, University of Illinois, Urbana, 1980.

28. Lee, M. K., "Turbulent Wall Eddy Structure and the Reynolds Stress Production is the Wall Region of a Pipe Flow", Ph.D. Thesis, University of Illinois, Urbana, 1975.

29. Lee, M. K., Eckelman, L. D. and T. J. Hanratty, "Identification of Turbulent Wall Eddies Through the Phase Relation of the Components of the Fluctuating Velocity Grandient", J. Fluid Mech., vol. 66, part 1, pp. 17-33, 1974.

30. Lyons, S. L., Nikolaides, C. and T. J. Hanratty, "The Size of Turbulent Eddies Close to a Wall", AIChE J., 34, pp. 938-945, 1988.

31. Mao, Z. X. and T. J. Hanratty, "The Use of Scalar Transport Probes to Measure Wall Shear Stress in a Flow with Imposed Oscillations", Experiments in Fluids, vol. 3, pp. 129-135, 1985.

32. Mitchell, J. E. and T. J. Hanratty, "A Study of Turbulence at a Wall Using an Electrochemical Wall Shear-Stress Meter", J. Fluid Mech., vol. 26, part 1, pp. 199-221, 1966.

33. Morrison, W. R. B. and R. E. Kronauer, "Structural Similarity for Fully Developed Turbulence in Smooth Tubes", J. Fluid Mech., vol. 39, p. 117, 1969.

34. Nikolaides, C., "A Study of the Coherent Structures in the Viscous Wall Region of a Turbulent Flow", Ph.D. Thesis, University of Illinois, Urbana, 1984.

35. Nikolaides, C., Lau, K. K. and T. J. Hanratty, "A Study of the Spanwise Structure of Coherent Eddies in the Viscous Wall Region", J. Fluid Mech., vol. 130, pp. 91-108, 1983.

36. Nikolaides, C., Lau, K. K., Hogenes, J. H. A. and T. J. Hanratty, "Structure of Turbulence Close to a Solid", Annals of the New York Academy of Sciences, vol. 44, pp. 374-390, 1983.

37. Reiss, L. P. and T. J. Hanratty, "An Experimental Study of the Unsteady Nature of the Viscous Stublayer", AIChE J., vol. 9, no. 2, pp. 154-160, 1963.

38. Rice, S. O., "Selected Papers on Noise and Stochastic Processes" (ed. Wax, N.), Dover Publishing, New York, 1954.

39. Sirkar, K. K., "Convection Velocity of Turbulent Velocity Fluctuations Close to the Wall", M.S. Thesis, University of Illinois, Urbana, 1966.

40. Sirkar, K. K., "Turbulence in the Immediate Vicinity of a Wall and Fully Developed Mass Transfer at High Schmidt Numbers", Ph.D. Thesis, University of Illinois, Urbana, 1969.

41. Sirkar, K. K. and T. J. Hanratty, "The Limiting Behavior of the Turbulent Transverse Velocity Component Close to a Wall", J. Fluid Mech., vol. 44, part 3, pp. 605-614, 1970a.

42. Sirkar, K. K., "Turbulence in the Immediate Vicinity of a Wall and Fully Developed Mass Transfer at High Schmidt Numbers", Ph.D. Thesis, University of Illinois, Urbana, 1969.

43. Sirkar, K. K. and T. J. Hanratty, "The Limiting Behavior of the Turbulent Transverse Velocity Component Close to a Wall", J. Fluid Mech., vol. 44, part 3, pp. 605-614, 1970.

44. Sirkar, K. K. and T. J. Hanratty, "The Use of Electrochemical Techniques to Study Turbulence Close to a Wall", in "Symposium on Turbulence Measurements in Liquids" (ed. by Patterson, G. K. and J. L. Zakin) pp. 43-50, Continuing Education Series, University of Missouri-Rolla, 1969a.

45. Sirkar, K. K. and T. J. Hanratty, "Relation of Turbulent Mass Transfer to a Wall at High Schmidt Numbers to the Velocity Field", J. Fluid Mech., vol. 44, part 3, pp. 589-603, 1970.

46. Sirkar, K. K. and T. J. Hanratty, "Limiting Behavior of the Transverse Turbulent Velocity Fluctuations Close to a Wall", Ind. Eng. Chem. Fundamentals, vol. 8, pp. 189-192, 1969b.

47. Townsend, A. A., "The Structure of Turbulent Shear Flow", 2nd ed, Cambridge Univ. Press, 1976.

48. Tritton, D. J., "Some New Correlation Measurements in a Turbulent Boundary Layer", vol. 28, p. 439, 1967.

Part 2
EFFECTS OF EXTRA STRAINS AND FORCE FIELDS

SUMMARY LECTURE

Effects of Extra Rates of Strain—Review

P. BRADSHAW
Department of Aeronautics
Imperial College of Science and Technology
Prince Consort Road
London SW7 2BY, UK

INTRODUCTION

The first session of this meeting was concerned with solutions of the full time-dependent Navier-Stokes equations. These equations are generally accepted as accurate, and it must follow that the final use of *experiments* on fluid flows is to avoid or simplify the basic problem of solving the full Navier-Stokes equations. In particular, experiments on turbulence should eventually contribute to the development of simplified prediction methods to be used by engineers or planetary scientists. Thus the final purpose of experiments is to determine the empirical coefficients in such simplified (time-averaged) prediction methods, which correspond to dimensionless structural parameters of the turbulence, such as Reynolds-stress ratios. (This definition of "structure" emphasizes the connection between basic experiments and turbulence modelling.) This does not reduce all experimental work to "code verification", and qualitative ideas from basic experiments may be as useful to modellers as quantitative data. However, the time-dependent Navier-Stokes solutions (turbulence simulations) reported at this meeting are a warning that we experimenters may have to say goodbye to "eddy-chasing" in two-dimensional ducts and flat-plate boundary layers, and carry out even our basic studies in more complex flows. One group of complex flows is the subject of the present session.

At the 1968 Stanford conference on computation of turbulent boundary layers (Kline, Morkovin, Sovran & Cockrell 1969), a wide variety of time-averaged calculation methods adequately predicted two-dimensional (2-D) incompressible turbulent boundary layers, with simple boundary conditions and with a dominant mean rate of strain $\partial U/\partial y$. This well-devised and well-timed meeting did not explicitly consider more complex flows. In fact, most of the turbulent flows that last long enough to do any good or any harm are dominated by $\partial U/\partial y$, and can be identified as boundary layers or other "species" of shear layer, usually perturbed by extra rate-of-strain components, body forces, interaction with other turbulence fields, etc. Perturbed shear layers were the focus of the 1980-81 Stanford meeting on Computation of Complex Turbulent Flows (Kline, Cantwell & Lilley 1982): despite more than a decade of intense, computer-assisted activity in turbulence research and turbulence modelling, the main conclusion of the 1980-81 meeting was that none of the existing calculation methods could predict a worthwhile range of perturbed shear layers without arbitrary adjustment of empirical constants. This conclusion still holds in 1988. Although there are several special correction formulas which force conventional shear-layer calculation methods to agree with experiments on — special — perturbations, no Reynolds-averaged calculation method covers a worthwhile range of perturbations except by linearly summing these correction formulas (which is not a reliable way to treat a non-linear phenomenon).

The epithet "Reynolds-averaged" (including time-averaged) in the previous sentence is important. Full time-dependent turbulence simulations and "large-eddy" simulations of the Navier-Stokes equations are currently being extended from 2D ducts and equilibrium boundary layers to complex flows, such as the curved duct flow of Moser & Moin (1987). However, the cost of computer time restricts full simulations to flows at low Reynolds number, and even large-eddy simulations are limited by the expense of using any non-trivial grid geometry. Therefore, simulations of geometrically-complex flows are still out of range, even as research techniques to provide information for Reynolds-averaged modelling. The present review of complex flows is written mainly from the classical viewpoint, in which laboratory experiments provide data for Reynolds-averaged models of the Reynolds stresses.

The different species of two-dimensional "simple" shear layers, with dominant mean rate of strain $\partial U/\partial y$ and dominant stress gradient $\partial(-\rho\overline{uv})/\partial y$, are the boundary layers, wakes, jets, etc. described in academic textbooks, which obey the relation $d\delta/dx << 1$ — or, more fundamentally and in tensor notation, $\partial U_i/\partial x_j << \partial U_1/\partial x_2 \equiv \partial U/\partial y$. (Unless otherwise stated we use rectangular cartesian coordinates: other systems can disguise extra strain rates.) In shear layers perturbed by extra strain rates, this order-of-magnitude inequality (<<) can be violated sufficiently strongly to change the turbulence structure, thus requiring adjustment of model coefficients. The matrix in figure 1 distinguishes the primary strain rate or "simple shear" $\partial U/\partial y$ or $\partial U_1/\partial x_2$, from the "extra" strain rates for other i, j.

$$\begin{pmatrix} \partial u_1/\partial x_1 & \partial u_1/\partial x_2 & \partial u_1/\partial x_3 \\ \partial u_2/\partial x_1 & \partial u_2/\partial x_2 & \partial u_2/\partial x_3 \\ \partial u_3/\partial x_1 & \partial u_3/\partial x_2 & \partial u_3/\partial x_3 \end{pmatrix}$$

FIGURE 1. The nine elements of the rate-of-strain tensor.

There are two types of *three*-dimensional flows which have $d\delta/dx << 1$. One comprises flows like boundary layers on swept wings (with the z axis spanwise, say), for which we have the additional inequality $d\delta/dz << 1$, or more strictly $\partial U/\partial z << \partial U/\partial y$. The other type of 3-D flow occurs in wing-body junctions, non-circular ducts, or interactions between shear layers and longitudinal vortices, in which $d\delta/dz \sim 1$ and $\partial U/\partial z \sim \partial U/\partial y$. Unfortunately, despite their importance in practice, 3-D flows were omitted from the 1980-81 Stanford meeting for administrative reasons, and also happen not to feature very strongly in the present meeting despite the continuing importance of both types of 3-D flow. (Most of us came here in swept-wing aircraft, and a large part of the body surface of a supersonic military or civil aircraft is occupied by junctions with wings, strakes, fins etc., each with its imbedded "horseshoe" vortex.)

The present position is that the best calculation methods for turbulent flows can handle most of the range of simple 2-D shear layers, as defined above, but only a restricted range of "complex" perturbed flows. As a particularly important example, the extra "crossflow" strain rate $\partial W/\partial y$, which is the essential feature of 3-D flows, seems to produce changes in structure which are not reproduced by existing calculation methods and which seem to be difficult to represent even by empirical correction formulas. It seems unlikely that we shall ever develop a universal calculation method, orders of magnitude *simpler* than a complete three-dimensional, time-dependent Navier-Stokes simulation but yet producing results of *acceptable* accuracy over *virtually* all 2-D or 3-D flows of interest. There is no logical reason why such a method should exist for any useful definitions of "simpler", "acceptable" and "virtually".

To compensate for the present lack of a universal calculation method, Tzuoo, Ferziger and Kline (1986) quantified the concept of "zonal" modelling, originally advanced by Kline at the 1980-81 Stanford meeting. Here a "zone" is an identifiable species of turbulent flow, as distinct from a perturbation of a previously-known species. The purpose of zonal modelling is to deal with the case of a shear layer of one species which encounters a sudden change of boundary conditions, such as reattachment to a solid surface, which nominally imposes an instantaneous change of species (in this case, from mixing layer to boundary layer). In fact, the changeover to a structure typical of the new species is *not* instantaneous, so that we need a model which clearly identifies the different species/zones and can provide a realistic basis for inter-zone interpolation functions (which may well be differential "lag" equations). . One aim of the present paper is to merge "zonal modelling" with the well-established (and universally-disliked) concept of "variable coefficients", in which the structural parameters of turbulent flow are allowed to be (at best) empirical functions of dimensionless perturbation parameters, representing the effects of extra rates of strain, interaction with other turbulence fields, etc.

Figure 2, which is a purely conceptual 2-D drawing of a multi-dimensional parameter space, shows the different species of shear layer as rectangular boxes. The dimensions of the space shown in figure 2 are the geometrical and other (dimensionless!) parameters which define shear-layer species and complex-flow effects. For practical purposes, the boxes are the regions in which turbulence models for simple shear layers give adequate results. In the fuzzy regions around each zone, the turbulence models will still work with the aid of empirical corrections for effects of "extra strain rate" or other complex-flow perturbations. A perturbation consisting of a complete change of species appears in figure 2 as a line in function space, joining the two special zones and possibly passing through several different kinds of perturbations — for example, reattachment involves streamline curvature and strong longitudinal acceleration. A turbulence model which can handle a wide range of species/zones/simple shear layers, combined with empirical corrections for complex-flow perturbations about those special zones, may fill most of the space of important turbulent flows.

We must not expect that the blank regions in figure 2 can be filled by *linear* algebraic perturbation formulas, based on the local ratio of some rate-of-strain perturbation to the mean shear $\partial U/\partial y$. Unfortunately, our present correlation formulas are mainly linear and may even ignore lag effects; and the blank regions outside the validity of these formulas — especially regions where two or more extra strain rates interact — are as unknown as the blank regions on a medieval map. At present our "map-making" is imperfect, and we are liable to find America when we seek India. However, *if* the correction formulas do fill the whole of parameter space *and* a computer program can be written to select the right correction formulas for an given case *then* the resulting computer program is a true universal calculation method, in the sense that it requires no extra empirical input from the user. The program may be an unlovely mess, but Occam's razor,

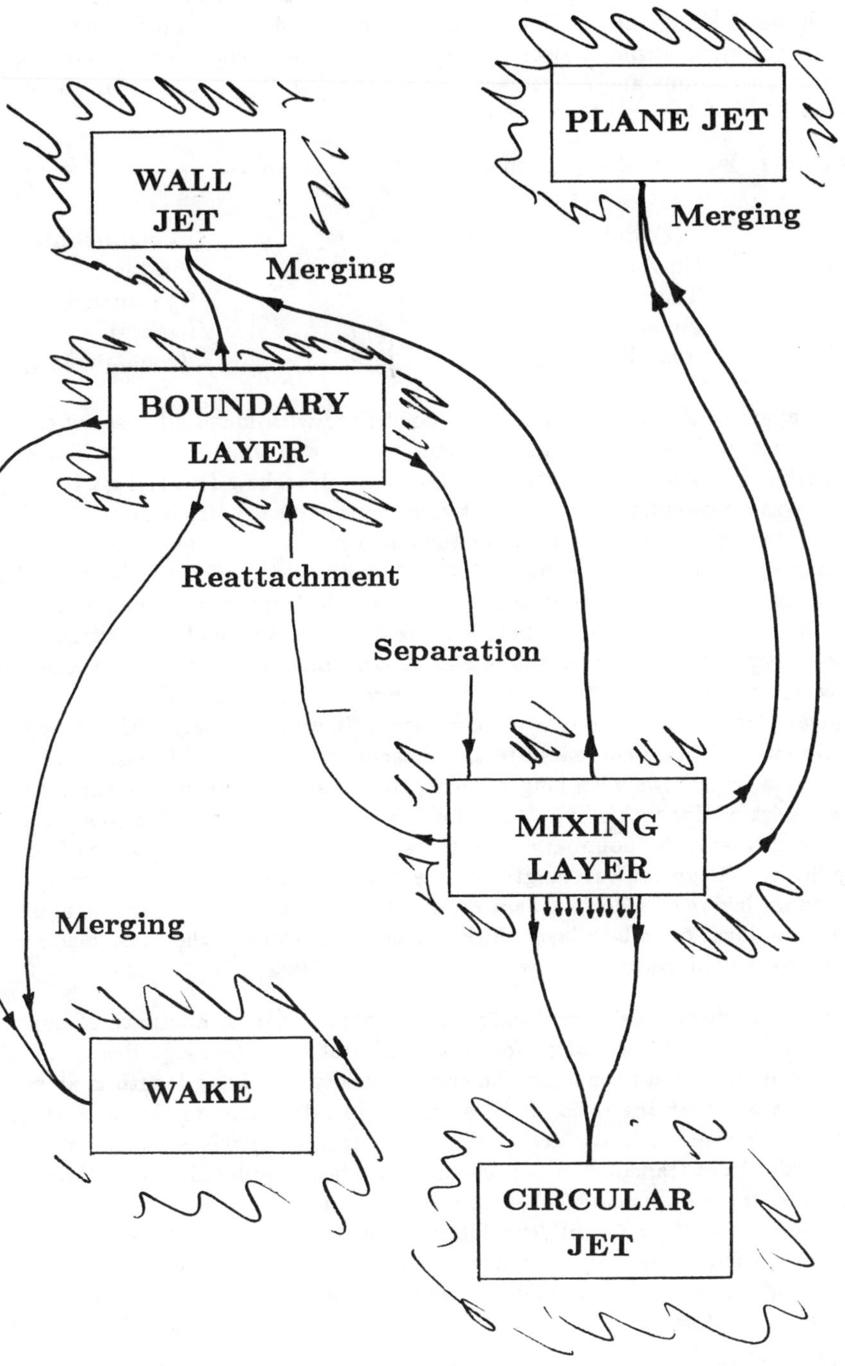

FIGURE 2. Shear-layer space.

"Entia non sunt multiplicanda praeter necessitatem", merely requires that our explanation of a phenomenon should be as simple as possible. Einstein's version is "Everything should be as simple as possible — but not too simple". Nobody says that a very simple but inconvenient explanation (the Navier-Stokes equations) can be replaced by a model that is both simple and convenient. To summarize, the descent from the unattainable universal model to real life goes as follows:-

TABLE 1. Functional dependence of model coefficients

Model	**Coefficients**
Universal	constant
Tuned for complex flow	f(parameters)
Zonal	f(zones)
Real life	f(zone, params)

The papers in this session are mainly concerned with idealized cases of complex flow, in which a well-defined turbulent "zone" — i.e. a given species of 2-D shear layer — is perturbed by well-defined external influences. Bradshaw (1971) took it for granted that there was an essential difference between "extra (mean) strain rates" $\partial U_i/\partial x_j$, and perturbations by the effect of adjacent turbulence fields (i.e. fluctuating strain rates). The finding by Hancock & Bradshaw (1987), that the extra effects of isotropic free-stream turbulence can be correlated in terms of the r.m.s. strain rate, *may* imply that interaction with another turbulence field is merely a special case of interaction with another rate-of-strain field, the relevant rate-of-strain value being an r.m.s. rather than a mean. However, perturbations by a mean (i.e. time-averaged) rate of strain can be formally equivalent to perturbations by rate-of-strain fluctuations only if the typical wavelength of the latter is large compared to the shear-layer thickness δ (quasi-steady state). Therefore, data from flows with long-wavelength free-stream turbulence should not be used in support of Hancock's hypothesis for turbulence whose wavelengths are of the order of magnitude of the boundary layer thickness. In the important case of turbomachinery flows, the typical wavelengths of the wake turbulence of a stator row, as seen by the boundary layers of a downstream rotor row, are indeed "large" in the above sense, (because the local boundary-layer thickness near the front of the rotor blade is small compared to the thickness of the wake of the stator blades).

At present, effects of different extra strain rates (mean or fluctuating) must be represented by different correlation formulas. In most cases (see e.g. Bradshaw 1975), the fractional change in properties like eddy viscosity or mixing length is an order of magnitude larger than the ratio of the extra strain rate, e say, to the primary strain rate $\partial U/\partial y$. This seems to contradict the fact that any explicit e-dependent terms in the Reynolds-stress transport equations are of no larger order than $e/(\partial U/\partial y)$ times a typical main term — consider for example the production term in the turbulent energy equation, $-\overline{uv}(\partial U/\partial y + \partial V/\partial x)$ for the case of longitudinal curvature, $e = \partial V/\partial x$. However, extra strain rates can have a large implicit effect on the pressure-strain "redistribution" terms in the Reynolds-stress transport equations, and it seems to be this that results in the final order-of-magnitude change in quantities like mixing length.

In the following sections we give a qualitative description of extra-strain effects; discuss details of empirical corrections for different levels of turbulence model; and briefly review the literature.

EFFECTS OF EXTRA MEAN STRAIN RATES: SUMMARY

The rate-of-strain tensor $\partial U_i/\partial x_j$ has six independent components (figure 1), and most of these significantly affect the behaviour of an initially two-dimensional shear layer with the mean strain rate $\partial U_1/\partial x_2$ or $\partial U/\partial y$. The dimensionless parameter describing extra rates of strain is the product of the extra rate of strain e (with dimensions time^{-1}) and a typical time scale of the basic turbulence. The time scale can be defined as the ratio of turbulent energy, $\overline{u_i^2} \equiv (\overline{u^2} + \overline{v^2} + \overline{w^2})/2$, to turbulent energy dissipation rate ϵ, and if the basic shear layer is near energy equilibrium so that energy production $-\overline{uv}\partial U/\partial y$ is closely equal to energy dissipation, this time scale can be replaced by $(\partial U/\partial y)^{-1}$, the ratio of $-\overline{uv}$ to $\overline{u_i^2}$ being roughly constant at about 0.15. Thus $e/(\partial U/\partial y)$ is an adequate parameter for many purposes. Note that since $\partial U/\partial y$ is largest near a wall, extra strain rates are likely to have the least effect here, so that changes in the law of the wall are often small, at least as far out as the start of the logarithmic region: for a warning that the effects are not negligible in general, see the paper by Gibson in this session.

The best-known and most-studied extra strain rate is streamline curvature in the plane of the original mean shear, $e = \partial V/\partial x$. Arguments relating "centrifugal force" to buoyancy force, originally due to Prandtl (1930) suggest that the mechanisms of curvature effects and buoyancy effects are analogous. Convex curvature in a boundary layer is stabilizing, and attenuates the existing large eddies, whereas concave curvature is destabilizing, and encourages the growth of a new mode of large eddies (related to Taylor-Görtler vortices). These are totally different processes, and empirical correlations for the two need not even be continuous at zero curvature - a conclusion still not fully accepted in meteorology where corresponding arguments apply to the two signs of buoyancy.

It is also well known that lateral divergence, $\partial W/\partial z$, can affect turbulence structure, nominally by the spanwise stretching of spanwise vortex lines (see Winter, Rotta and Smith 1968, who also demonstrated the significant effect of "bulk compression" in compressible flow, where the "extra strain rate" is $\partial U_i/\partial x_i$, summed over i — i.e. the sum of the three diagonal terms of the matrix in figure 1). As pointed out by Green (private communication 1970) the effect of bulk compression in the $x-y$ plane is similar to that of lateral divergence $\partial W/\partial z$: in both cases, an elementary area in the $x-y$ plane, $dxdy$, decreases so that spanwise vortex lines are reduced in cross-section.

The effect of longitudinal acceleration or vertical divergence, $\partial U/\partial x$ or $\partial V/\partial y$, is more controversial. There is certainly some evidence (Gartshore, 1965) that streamwise acceleration or normal strain rate can have a significant effect on turbulence: also, Townsend (1956) attributes the differences between jets and wakes to this extra strain rate, and it has also been used to correlate differences in structure between plane and axisymmetric jets (e.g. Pope 1978).

The extra strain rates discussed above all act in, or normal to, the plane of the original shear. The effect of "3-D" perturbations is less well documented: lateral skewing in 3-D flow, by $\partial W/\partial x$ or $\partial U/\partial z$, leads to an increase in x-component vorticity by rotation of z-component vortex lines in the $x-z$ plane. (Note that if the y-component mean vorticity is initially zero, $\partial W/\partial x \sim \partial U/\partial z$.) The measurements of van den Berg et al. (1975) and Elsenaar & Boelsma (1976), followed by Bradshaw & Pontikos (1985) and Anderson & Eaton (1987), showed that an initially two-dimensional boundary layer, subjected to lateral skewing in a three-dimensional pressure gradient, suffers a significant reduction in the absolute value of shear stress and also in the ratio of shear-stress magnitude to turbulent energy (a "structural parameter" in the above sense). These effects are not reliably predicted by simple extensions of 2D models (but see Abid, Delery & Schmitt 1985), so that we must regard $\partial W/\partial x$ as an extra rate of strain that affects turbulence structure, perhaps via the skewing of z-wise vorticity $\partial u/\partial y$ into longitudinal vorticity $\partial w/\partial y$).

A final strain rate which undoubtedly has large effects on turbulence structure is $\partial V/\partial z$ typically, an embedded longitudinal vortex with $\partial V/\partial z \sim \partial W/\partial y$. Crudely, $\partial V/\partial z$ rotates the uv-producing eddies out of the $x-y$ plane, just as $\partial W/\partial y$ does. However it is probably more logical to regard a shear layer with an embedded vortex as a new species rather than just a perturbed shear layer, simply because the perturbations of the mean flow and the Reynolds-stress tensor are so large. Embedded vortex flows were reviewed by Bradshaw (1987): see Eibeck & Eaton (1987), Westphal, Eaton & Pauley (1987), Bradshaw & Cutler (1987) for recent research results — including the effect of embedded vortices on heat transfer, which there is no space to discuss here.

TURBULENCE MODELLING FOR COMPLEX FLOWS

The paper by Lakshminarayana (1986) reviews the full range of Reynolds-averaged turbulence models for engineering use, with specific reference to complex flows: Rodi (1986 reviews a 1984 colloquium on turbulence modelling in general: a detailed treatment her is therefore unnecessary. It *is* perhaps necessary to mention that complex-flow predictions can be seriously affected by using the boundary-layer equations in flows where the boundary-layer approximation is inaccurate. As pointed out by Huang (1986), even if the turbulence model does in principle predict the stresses, and thus the *total* pressure, correctly, improper neglect, or incorrect prediction, of $\partial p/\partial y$ yields an incorrect U profile (and thus incorrect turbulence production, etc.). Mean-flow approximations can cause as much grief as approximate turbulence models!

If a "zonal" model, with empirical allowances for extra strain rates and other complex-flow effects as perturbations around a set of simple shear layers, is ever to cover a large part of the parameter space of figure 2, empirical data from experiments or time-depen ent simulations) are needed for each kind of extra strain rate or other perturbation. Such a model can — and must — be incorporated in a computer program whose empir ical input needs no attention from the operator, and is therefore indistinguishable from a "universal" method, except in one vital way:— any computer program should *indicate its own limits of validity* automatically, but by definition a universal method has n limits of validity! It is essential that a complex-flow prediction method should produce warning messages when it enters one of the blank inter-zone areas in the map of figure 2, just as medieval maps had occasional labels **"Here be Dragons"**: in both cases the message is "proceed at your own risk". Programming such warning messages should no be too difficult, because the edges of the fuzzy areas correspond to over-large values of complex-flow parameters, like the ratio of extra strain rates to the mean shear. Admittedly, in spite of the large body of experimental information about the effects of extra strain rate, there is little reliable information about likely limits of validity of empirical correlation formulas, even for the best-known example of streamline curvature - but thi is partly because the originators of correlation formulas are reluctant to specify limits of validity, which is not a proper attitude for an engineer to take.

Correlation formulas for the effects of various extra rates of strain e can be divided into "first-order" formulas, containing a factor such as $(1 + \alpha e/(\partial U/\partial y))$, where α is an empirical constant generally of order 10, and those which take a higher view.

A slight improvement on "first-order" formulas is the rate equation introduced by Bradshaw (1973), in which the effective value of the extra strain rate e is a low-pass-filtered version of the actual strain rate, reflecting the slow response of the large, stress-bearing eddies to a perturbation: the "time constant" is taken to be the time scale of the flow as a whole, (turbulent energy) / (dissipation rate), roughly the time in which the flow moves a streamwise distance of 10δ. The value of such crude allowances for flow "history" is diminished by the observation of Gillis & Johnston (1980), followed by Hoffmann, Muck & Bradshaw (1985) and Muck, Hoffmann & Bradshaw (1985), that the optimum "time constant" was significantly different for concave and for convex curvature, and for inception and recovery: this conclusion seems to apply to mild curvature, as well as to strong curvature where changes in turbulences quantities such as time scale are bound to be large.

Adams & Johnston (1985) pointed out that in the outer part of a highly-convex boundary layer, the mixing length decreased to a value proportional to the radius of curvature and independent of the boundary layer thickness (equivalent to a shear stress coefficient $-\overline{uv}/U^2 = (e/(\partial U/\partial y)^{-2}$. Perhaps an interpolation between a linear formula for small $e/(\partial U/\partial y)$ and a result like this for large $e/(\partial U/\partial y)$ may suffice for the complete range. There are analogous formulas for stably-buoyant flows.

In the most refined current examples of Reynolds-averaged turbulence modelling, the modelled equations are the full Reynolds-stress transport equations themselves, together with an equation for eddy length scale — almost always the equation for the dissipation rate of turbulent energy, ϵ. (This is modelled in a way that reproduces the behaviour of the transport equation for the rate of transfer of energy from the large eddies to the dissipating ones, and thus provides a length scale of the large, energy-containing eddies rather than the dissipating ones. The "energy-transfer-transport" equation itself is extremely complicated in inhomogeneous flows.)

Several workers (e.g. Launder, Tselepidakis & Younis 1987) have found that Reynolds-stress transport models can give fair quantitative predictions of mild curvature or rotation effects without explicit curvature correction terms in the stress equations or the dissipation equation. Corrections for extra-strain effects in Reynolds-stress transport models have been proposed by Launder, Priddin & Sharma (1977), Bardina, Ferziger & Reynolds (1983), and others. They are necessarily additional terms in the dissipation or length-scale equation, typically $c_e e\epsilon$ where c_e is of order 0.1 and of a sign that will decrease dissipation in "unstable" flow: unfortunately the sign depends on the sign of the mean shear (e.g. if $e = \partial V/\partial x$, then the flow is destabilized for $\partial U/\partial y > 0$, and conversely). Launder et al. (1987) report that the effect of Bardina's correction on predictions of a rotating channel flow was small. In any case, extra terms in a length-scale transport equation are only a modest conceptual improvement on the above-mentioned low-pass-filtered factors applied directly to the terms in the Reynolds-stress equations, because the length-scale equation is a partial differential equation which low-pass filters any modifications to its terms before passing them on to the Reynolds-stress transport equation. The improvement in prediction accuracy is not great and the difference in apparent response times between stable and unstable cases is not reproduced. Improvements are needed in the modelling of the dissipation equation for simple flows before extra-strain corrections can be usefully discussed; and although all the terms in the dissipation equation have been deduced from full turbulence simulations (Mansour, Kim & Moin 1987), the modelled dissipation equation is only very distantly related to the exact one, so that it is difficult to see how to use the information. Hong & Murthy (1986a,b) present results of their large-eddy interaction model, in which turbulent shear flows are represented as a combination of large eddies, calculated by essentially time-dependent methods, with the smaller-scale motion represented by time-averaged models. It may be that a model of this level is needed to deal with complex flows.

LITERATURE REVIEW

Bradshaw (1986) offered a selected, computer-readable reference list for complex flows and other turbulent flows, and the present paper is only a brief résumé of recent work.

The main results of recent experiments have been a better understanding of the effects of curvature/rotation and of free-stream turbulence. Other "extra strain rates" have received less attention, except for the effects of compression or dilatation in compressible flow, necessarily combined with streamline curvature in the plane of the mean shear.

Streamline Curvature and Rotation

A series of papers from Stanford (Simonich & Moffat 1982, Gillis & Johnston 1983, Jeans & Johnston 1983, Barlow & Johnston 1985a,b, Bardina, Ferziger & Rogallo 1985, Plesniak & Johnston 1987) treats a variety of low-speed curved or rotating flows, especially strongly-curved flows with shear-layer thicknesses as much as 0.1 of the radius of curvature. This work has provided a better quantitative understanding of the qualitatively different responses of a turbulent boundary layer to concave and to convex curvature: see also Hoffmann et al. (1985) and Muck et al. (1985) who studied a pair of mildly-curved boundary layers with $|\delta/R| \approx 0.01$. A series of experiments from Princeton (e.g. Hayakawa, Smits & Bogdonoff 1983; Alving & Smits 1986; Jayaram, Taylor & Smits 1987) documents the combined effects of (concave) streamline curvature and compression (flow deflection) in boundary layers or reattaching shear layers in supersonic flow. Dussauge and colleagues (e.g. Dussauge & Gaviglio 1987) have studied dilatation and compression in free shear layers. The work of Simon and collaborators (e.g. Simon & Moffat 1982, You, Simon & Kim 1986, Kim & Simon 1987) reminds us that heat transfer is as important as momentum transfer in mechanical engineering: the effect of streamline curvature on heat transfer rate seems to be generally larger than the effect on momentum transfer, so that Stanton number changes more rapidly than skin-friction coefficient.

Gibson and co-workers at Imperial College (e.g. Gibson & Younis 1983, Gibson & Verriopoulos 1984, Gibson, Verriopoulos & Vlachos 1984, Gibson 1984, Gibson & Servat-Djoo 1985) have also studied momentum and heat transfer in a variety of curved flows — mixing layers, wall jets or boundary layers. Savill (1983) and Witt & Joubert (1985) have documented curved or rotating wakes. The Melbourne workers under Joubert have also studied combinations of extra strain rates in 2-D flow (see Baskaran , Smits & Joubert 1987), and pure rotation in duct flow (Watmuff, Witt & Joubert 1985), while Kikuyama et al. (1987) studied combined curvature in, and rotation about, the $x - y$ plane. Honami and colleagues (Shizawa & Honami 1985, 1987) have undertaken basic experiments in curved flows, including detailed measurements of Taylor-Görtler "vortices" (better called "roll cells" since they do not have concentrated cores) concave surfaces. A related subject is the effect of a body force due to buoyancy, qualitatively related to centrifugal acceleration over a curved surface: see Rohr, Helland, Itsweire & Van Atta (1987). Kohama (1985, 1986) , Hall & Bennett 1987, Peerhossaini & Wesfried 1987) have studied the role of Taylor-Görtler vortices in transition in curved flows: crudely speaking, the inviscid instability modes of laminar flow may also amplify in turbulent flow if the turbulent Reynolds number based on *eddy* viscosity exceeds the Reynolds number for laminar flow instability, and Baskaran & Bradshaw (1987) have discussed the appearance, or otherwise, of Taylor-Görtler vortices in 3-D turbulent flows on concave surfaces. Mason (1985) simulated buoyancy in the Earth's boundary layer, and found longitudinal-vortex rolls whose updraft leads to condensation and the formation of cloud streets: although the basic flow was three-dimensional, the instabilities seem to be quasi-2-D convection rolls rather than crossflow instability modes.

Swirling flows are part of the curved-flow family, but are likely to have sufficiently large radial gradients to confuse simple calculation methods. Examples include the studies of flow in the boundary layer of a rotating body by Fulachier, Arzoumanian & Dumas (1982) and Driver & Hebbar (1987) — essentially a 3-D boundary layer with suddenly-applied crossflow — and the measurements in swirling flow in a diffuser by Clausen & Wood (1987). Solid-body rotation of a spatially-homogeneous flow is fairly easy to simulate in full time-dependent Navier-Stokes calculations: see Bardina, Ferziger & Rogallo (1985) who find internal waves in this highly-stabilized situation.

Bulk Compression or Dilatation

Recent work on supersonic boundary layers, in which inviscid-flow formulas couple compression with concave curvature, and expansion (dilatation) with convex curvature, has been reviewed above. Work on supersonic free shear layers is rare: perhaps the supersonic plane mixing layer should not be counted as a complex flow, since the mean dilatation $div\mathbf{U}$ is small, but it represents almost the only firm evidence of the effects of density fluctuations (represented by $div\mathbf{u}$) on turbulence (Kline et al. 1982). In internal combustion engines, relative rates of bulk compression — measured as the product of $div\mathbf{U}$ and a turbulence timescale — are comparatively small: thus, although modelling the turbulence in cylinder heads, where no dominant shear layers can be distinguished, is a severe problem, it is not aggravated by extra strain effects. In unconfined combustion (at constant pressure rather than constant volume) $div\mathbf{U}$ is large, and viscosity increases with temperature: for both reasons, turbulence intensity decreases and fine-scale motion is strongly attenuated, leading to the typical "bunch of grapes" appearance of turbulent combustion regions.

Settles, Horstmann & McKenzie (1986), Stollery, Fomison & Hussain (1986), Fisher (1986) and Tan, Tran & Bogdonoff (1987) have studied different kinds of 3-D compressible flows, but without specific conclusions about turbulence structure.

It is clear that much of the apparent change in "turbulence" structure due to shock-wave impingement is the result of shock-wave unsteadiness, at frequencies typical either of the shear-layer turbulence or of larger-scale free-stream unsteadiness: the former is real, in the sense that the energy-containing spectrum is affected, but the latter may be calculated as a quasi-steady process.

Free-stream Turbulence

The main contributions to this problem relate to the effect of nearly-isotropic turbulence, typically from square-mesh grids, on flat-plate boundary layers. The response of, say, skin-friction coefficient to free-stream turbulence is not a linear function of r.m.s. intensity. For very small intensity, too small to alter shear-layer turbulence structure, superposition principles in fact demand that c_f shall vary as $\overline{u^2}$ as shown by the measurements of Meier & Kreplin (1980): see also Blair (1983), Hancock & Bradshaw (1983) and Castro (1984). It has also been found fairly recently that response depends strongly on free-stream length scale as well as intensity. Hancock & Bradshaw (1983) showed that most data sets fell close to one line in the intensity/length-scale plane, disguising the separate effects: they offered a correlation for length scale effects, based mainly on their own results. The main effects of free-stream turbulence are felt by the outer layer, as, say, a decrease in wake parameter Π, so (i) the logarithmic "law of the wall" is at first undisturbed, (ii) low-Reynolds-number effects on the outer layer, e.g. the dependence of Π on Re_θ documented by Coles (1962), interact with, and confuse, free-stream turbulence effects (Castro 1984). Hancock & Bradshaw (1987) show that, at least over a limited range of free-stream intensities and length scales, the effects on a boundary layer can be correlated by the extra r.m.s. rate of strain, proportional to $\overline{(u^2)}^{1/2}$/(lengthscale)

Free-stream turbulence of large length scale can affect free shear layers (jets and wakes) simply by bulk lateral displacement, leading to "waving" of the shear layer and thus, on a Reynolds-averaged basis, to an increased spreading rate. This could be predicted, in principle, by an unmodified calculation method for the appropriate free shear layer, with slowly-changing time-dependent boundary conditions. Genuine effects on shear-layer turbulence — requiring modification of empirical constants in the turbulence model — will appear only if the length scale of the free-stream turbulence is of the same order as that of the shear layer. This is a particular case of a useful general principle: if a turbulence model can deal with large spatial contributions to the rate of change of Reynolds stress along a streamline, $D\overline{u_i u_j}/Dt$, it can deal with equally large temporal contributions. A recent experiment, concentrating on mean-flow measurements, is that of Pai (1985).Free-stream turbulence effects on separated or reataching flows are less straightforward: see Gartshore (1984), Hillier & Dulai (1985), Nakamura & Ohya (1985), Huot, Rey & Arbey (1986), Melbourne & Saathoff (1986) and Nakamura & Ozono (1987) on bluff-body flows. Marchman, Sumantran & Schaefer (1987) discuss effects on aerofoil stall hysteresis. Useful basic discussions of the interaction between two isotropic turbulence fields are provided by Veeravalli & Warhaft (1987) and Pope & Howarth (1987).

Other Extra Strain Rates

Lateral divergence increases turbulence intensity, nominally by stretching spanwise vortex lines; and conversely. Patel & Baek (1987) discuss the attenuation of turbulence over the rear of a body of revolution, and Cutler & Bradshaw (1987) mention the diverging boundary layer between two imbedded vortices with the "common flow' towards the surface. However one of the key studies of divergence and of multiple strain rates is still the paper by Winter, Rotta & Smith (1968), on the flow over a waisted body of revolution, which combined convergence/divergence, longitudinal curvature and dilatation/compression in amounts which varied with Mach number, and which Green et al. (1972) were able to correlate with a lag-entrainment method augmented with linear extra-strain formulas. This addition of linear formulas can at best be valid only for small extra strain rates, and other workers have found it unreliable.

CONCLUSIONS

One should not draw conclusions from an introductory survey, but it can be said that the study of "complex" flows is alive and well, and contributing to better understanding of the even-more-complex (*real?*) flows that appear in engineering.

REFERENCES

Abid, R., Delery, J. and Schmitt, R., An Examination of Turbulence Models for a Separating Three-dimensional Turbulent Boundary Layer, *4th Int. Conf. on Num. Methods in Lam. and Turb. Flow, Swansea*, 1985.

Adams, E.W. and Johnston, J.P., Effects of the Upstream Boundary Layer Thickness and State on the Structure of Reattaching Flows, *5th Symposium on Turbulent Shear Flows, Cornell*, p. 5-1, 1985.

Alving, A.E. and Smits, A.J., The Recovery of a Turbulent Boundary Layer from Longitudinal Curvature, AIAA-86-0435, 1986.

Anderson, S. and Eaton, J.K., Experimental Study of a Pressure-driven, Three-dimensional, Turbulent Boundary Layer, *AIAA J.* 25, 1086, 1987.

Bardina, J., Ferziger, J.H. and Reynolds, W.C., Improved Turbulence Models Based on Large Eddy Simulation of Homogeneous, Incompressible Turbulent Flows, Stanford Univ. Thermosciences Divn. Rept TF-19, 1983.

Bardina, J., Ferziger, J.H. and Rogallo, R.S., Effect of Rotation on Isotropic Turbulence — Computation and Modelling, *J. Fluid Mech.* 154, 321, 1985.

Barlow, R.S. and Johnston, J.P., Roll-cell Structure in a Concave Turbulent Boundary Layer, AIAA-85-0297, 1985.

Barlow, R.S. and Johnston, J.P., Velocity Spectra for Turbulent Boundary Layers on a Concave Surface, *5th Symposium on Turbulent Shear Flows, Cornell*, p. 21-13, 1985.

Baskaran, V. and Bradshaw, P., Experimental Investigation of a Three-dimensional Turbuent Shear Layer over an "Infinite" Swept Concave Surface, *6th Symposium in Turbulent Shear Flows, Toulouse*, p. 11-2, 1987.

Baskaran, V., Smits, A.J. and Joubert, P.N., A Turbulent Flow over a Curved Hill. Part 1. Growth of an Internal Boundary Layer, *J. Fluid Mech.* 182, 47, 1987.

Berg, B. v.d., Elsenaar, A., Lindhout, J.P.F. and Wesseling, P., Measurements in an Incompressible Three-dimensional Turbulent Boundary Layer under Infinite Swept-wing Conditions, and Comparison with Theory. *J. Fluid Mech.* 70, 127, 1975.

Blair, M.F., Influence of Free Stream Turbulence on Turbulent Boundary Layer Heat

Transfer and Mean Profile Development. Parts I and II, *J. Heat Transf.* 105, 33 and 41, 1983.

Bradshaw, P., Variations on a Theme of Prandtl, *AGARD Conf. Proc.* 93, 1971.

Bradshaw, P., Review — Complex Turbulent Flows, *J. Fluids Engg.* 97, 146, 1975.

Bradshaw, P., A Bibliography of Turbulent Flows, Imperial College Aero Report 86-03, 1986.

Bradshaw, P., Turbulent Secondary Flows, *Ann. Rev. Fluid Mech.* 19, 53, 1987.

Bradshaw, P. and Cutler, A.D., Three-dimensional Flows with Imbedded Longitudinal Vortices, *Perspectives in Turbulence Studies* (H.U. Meier and P. Bradshaw, Eds.), Springer, p. 382, 1987.

Bradshaw, P. and Pontikos, N.S., Measurements in the Turbulent Boundary Layer on an "Infinite" Swept Wing, *J. Fluid Mech.* 159, 105, 1985.

Castro, I.P., Effects of Free Stream Turbulence on Low Reynolds Number Boundary Layers, *J. Fluids Engg.* 106, 298, 1984.

Clausen, P.D. and Wood, D.H., Some Measurements of Swirling Flow through an Axisymetric Diffuser, *6th Symposium on Turbulent Shear Flows, Toulouse,* p. 1-3, 1987.

Dussauge, J.P. and Gaviglio, J., The Rapid Expansion of a Supersonic Turbulent Flow: Role of Bulk Dilatation, *J. Fluid Mech.* 174, 81, 1987.

Coles, D.E., The Turbulent Boundary Layer in a Compressible Fluid, RAND Rept. R-403-PR, 1962.

Driver, D.M. and Hebbar, S.K., Experimental Study of a Three-dimensional, Shear-driven, Turbulent Boundary Layer, *AIAA J.* 25, 35, 1987.

Eibeck, P.A. and Eaton, J.K., Heat Transfer Effects of a Longitudinal Vortex Embedded in a Turbulent Boundary Layer, *J. Heat Transfer* 109, 16, 1987.

Elsenaar, A. and Boelsma, S.H., Measurements of the Reynolds Stress Tensor in a Three-dimensional Turbulent Boundary Layer under Infinite Swept Conditions, Netherlands NLR TR 74095 U, 1974.

Fisher, S.A., Three-dimensional Flow Effects in a Two-dimensional Supersonic Air Intake, *J. Prop. Power* 2, 546, 1986.

Fulachier, L., Arzoumanian, E. and Dumas, R., Effect on a Developed Turbulent Boundary Layer of a Sudden Local Wall Motion, *Three-dimensional Turbulent Boundary Layers,* (H.H. Fernholz and E. Krause, Eds.), Springer, p. 188, 1982.

Gartshore, I.S., The Streamwise Development of Two-dimensional Wall Jets and Other Two-dimensional Turbulent Shear Flows., McGill Univ. Mech. Engg. Dept. PhD Thesis, 1965.

Gartshore, I.S., Some Effects of Upstream Turbulence on the Unsteady Lift Forces Imposed on Prismatic Two-dimensional Bodies, *J. Fluids Engg.* 106, 418, 1984.

Gibson, M.M., Effects of Streamline Curvature on Turbulence, *Frontiers in Fluid Mechanics* (S.H. Davis and J.L. Lumley, Eds.) Springer, p. 184, 1984.

Gibson, M.M. and Servat-Djoo, K., Measurements in the Heated Turbulent Boundary Layer on a Mildly Curved Concave Surface, *5th Symposium on Turbulent Shear Flows, Cornell*, p. 14-7, 1985.

Gibson, M.M., Verriopoulos, C.A. and Vlachos, N.S., Turbulent Boundary Layer on a Mildly Curved Convex Surface. Part 1 — Mean Flow and Turbulence Measurements, *Expts. in Fluids* 2, 17, 1984.

Gibson, M.M. and Verriopoulos, C.A., Turbulent Boundary Layer on a Mildly Curved Convex Surface. Part 2 — Temperature Field Measurements, *Expts. in Fluids* 2, 73, 1984.

Gibson, M.M. and Younis, B.A., Turbulence Measurements in a Developing Mixing Layer with Mild Destabilizing Curvature, *Expts. in Fluids* 1, 23, 1983.

Gillis, J.C. and Johnston, J.P., Turbulent Boundary Layer Flow and Structure on a Convex Wall and Its Redevelopment on a Flat Plate, *J. Fluid Mech.* 135, 123, 1983.

Green, J.E., Weeks, D.J. and Brooman, J.W.F., Prediction of Turbulent Boundary Layers and Wakes by a Lag-entrainment Method, British Aero. Res. Council R. & M. 3791, 1972.

Hall, P. and Bennett, J., Taylor-Gortler Instabilities of Tollmien-Schlichting Waves and Other Flows Governed by the Interactive Boundary-layer Equations,*J. Fluid Mech.* 171, 441, 1986.

Hancock, P.E. and Bradshaw, P., The Effect of Free Stream Turbulence on Turbulent Boundary Layers, *J. Fluids Engg.* 105, 284, 1983.

Hancock, P.E. and Bradshaw, P., The Structure of a Turbulent Boundary Layer Beneath a Turbulent Free Stream, *6th Turbulent Shear Flows Symposium, Toulouse*, paper 1-1, 1987.

Hayakawa, K., Smits, A.J. and Bogdonoff, S.M., Hot-wire Investigation of an Unseparated Shock Wave / Turbulent Boundary Layer, *AIAA J.* 22, 579, 1984.

Hillier, R. and Dulai, B.S., Pressure Fluctuations in a Turbulent Separated Flow, *5th Symposium on Turbulent Shear Flows, Cornell,* p. 5-13, 1985.

Hoffmann, P.H., Muck, K.C. and Bradshaw, P., The Effect of Concave Surface Curvature on Turbulent Boundary Layers, *J. Fluid Mech.* 161, 371, 1985.

Hong, S.K. and Murthy, S.N.B., Effective Velocity of Transport in Curved Wall Boundary Layers, *AIAA J.* 24, 361, 1986.

Hong, S.K. and Murthy, S.N.B., Pressure-strain Correlations in Curved Wall Boundary Layers, *AIAA J.* 24, 971, 1986.

Huang, P.G., The Computation of Parabolic Turbulent Flow with Second-moment Closure Models, Manchester University PhD thesis, 1986.

Huot, J.P., Rey, C. and Arbey, H., Experimental Analysis of the Pressure Field Induced on a Square Cylinder by a Turbulent Flow, *J. Fluid Mech.* 162, 283, 1986.

Jayaram, M., Taylor, M.W. and Smits, A.J., The Response of a Compressible Turbulent Boundary Layer to Short Regions of Concave Surface Curvature, *J. Fluid Mech.* 175, 343, 1987.

Jeans, A.J. and Johnston, J.P., The Effects of Concave Curvature on Turbulent Boundary Layer Structures, *Structure of Complex Turbulent Shear Flow* (R. Dumas and L. Fulachier, Eds.) Springer, p. 89, 1983.

Kim, J. and Simon, T.W., Measurements of the Turbulent Transport of Heat and Momentum in Convexly Curved Boundary Layers — Effects of Curvature, Recovery and Free-stream Turbulence, ASME 87-GT-199, 1987.

Kline, S.J., Morkovin, M.V., Sovran, G. and Cockrell, D.J. (Eds.) *Proceedings, Computation of Turbulent Boundary Layers — 1968 AFOSR-IFP-Stanford Conference,* Thermosciences Divn, Stanford Univ., 1968.

Kline, S.J., Cantwell, B.J. and Lilley, G.M., *Proceedings, 1980-81 AFOSR-HTTM-Stanford Conference on Complex Turbulent Flows - Comparison of Computation and Experiment,* Vols. 1-3, Stanford University, 1982.

Kohama, Y., Three-dimensional Boundary-layer Transition on a Concave-convex Curved Wall, *IUTAM Sympo. on Turbulence Management and Relaminarization, Bangalore,* 1986.

Lakshminarayana, B., Turbulence Modelling for Complex Shear Flows, *AIAA J.* 24, 1900, 1986.

Launder, B.E., Priddin, C.H., and Sharma, B.I. The Calculation of Turbulent Boundary Layersd on Curved and Spinning Surfaces, *J. Fluids Engg.* 99, 237, 1977.

Launder, B.E., Tselepidakis, D.P. and Younis, B.A., A Second Moment Closure Study of Rotating Channel Flow, *J. Fluid Mech.* 183, 63, 1987.

Mansour, N.N., Kim, J. and Moin, P., Reynolds Stress and Dissipation Rate Budgets in a Turbulent Channel Flow, NASA Ames Paper, 1987.

Marchman, J.F., Sumantran, V. and Schaefer, C.G., Acoustic and Turbulence Influence on Stall Hysteresis, *AIAA J.* 25, 50, 1987.

Mason, P.J., A Numerical Study of Cloud Streets in the Planetary Boundary Layer, *Boundary-Layer Met.* 32, 281, 1985.

Meier, H.U. and Kreplin, H.-P., Influence of Freestream Turbulence on Boundary-Layer Development, *AIAA. J.* 18, 11, 1980.

Melbourne, W.H. and Saathoff, P.J., Under the Reattaching Shear Layer, *Ninth Australasian Fluid Mechanics Conference, Auckland, New Zealand,* p. 43, 1986.

Moser, R.D. and Moin, P., The Effects of Curvature in Wall-bounded Turbulent Flows, *J. Fluid Mech.* 175, 479, 1987.

Muck, K.C., Hoffmann, P.H. and Bradshaw, P., The Effect of Convex Surface Curvature on Turbulent Boundary Layers, *J. Fluid Mech.* 161, 347, 1985.

Nakamura, I., Yamashita, S. and Watanabe, T., An Effect of Sudden Circumferential Strain on the Laminar Boundary Layer on a Rotating Cylinder in an Axial Flow, *Acta Mech.* 57, 69, 1985.

Nakamura, Y. and Ozono, S., The Effects of Turbulence on a Separating and Reattaching Flow, *J. Fluid Mech.* 178, 477, 1987.

Patel, V.C. and Baek, J.H., Boundary Layers in Planes of Symmetry. Part 1. Experiments in Turbulent Flow, *AIAA J.* 25, 550, 1987.

Peerhossaini, H. and Wesfreid, J.E., Experiments on Görtler Rolls, *6th Symposium on Turbulent Shear Flow, Toulouse,* paper no. 2-6, 1987.

Plesniak, M.W. and Johnston, J.P., The Effect of Stabilizing and Destabilizing Curvature on a Turbulent Mixing Layer, *2nd Int. Symposium on Transport Phenomena in Turbulent Flows, Tokyo,* 1987.

Pope, S.B., An Explanation of the Turbulent Round-jet/Plane-jet Anomaly, *AIAA J.* 16, 279, 1978.

Pope, S.B. and Haworth, D.C., The Mixing Layer between Turbulent Fields of Different Scales, *Turbulent Shear Flows 5* (F. Durst et al., Eds., Springer), p. 44, 1987.

Rodi, W., Turbulence Modelling for Incompressible Flows — a Report on the Euromech 180 Colloquium, *Physico-Chem. Hydrod.* 7, 297, 1986.

Rohr, J.J., Helland, K.N., Itsweire, E.C. and Van Atta, C.W., Turbulence in a Stably Stratified Shear Flow: a Progress Report, *Turbulent Shear Flows 5* (F. Durst et al., Eds., Springer) p. 67, 1987.

Savill, A.M., The Turbulence Structure of a Highly Curved Two-dimensional Wake, *Structure of Complex Turbulent Shear Flow* (R. Dumas and L. Fulachier, Eds.) Springer, p. 185, 1983.

Settles, G.S., Horstman, C.C. and McKenzie, T.M., Experimental and Computational Study of a Swept Compression Corner Interaction Flowfield, *AIAA J.* 24, 744, 1986.

Shizawa, T. and Honami, S., Experiments on Turbulent Boundary Layers over a Concave Surface — Response of Turbulence to Curvature., *5th Symposium on Turbulent Shear Flows, Cornell,* p. 21-1, 1985.

Shizawa, T. and Honami, S., Measurements of Longitudinal Vortices in a Turbulent Boundary Layer over a Concave Surface by the Flying Hot-wire System, Science University of Tokyo paper, 1987.

Simon, T.W. and Moffat, R.J., Turbulent Boundary Layer Heat Transfer Experiments — a Separate Effects Study on a Convexly Curved Wall, *J. Heat Transf.* 105, 835, 1983.

Simonich, J.C. and Moffat, R.J., Local Measurements of Turbulent Boundary Layer Heat Transfer on a Concave Surface, Thermosciences Divn., Stanford Univ. Rept HMT-

35, 1982.

Smits, A.J. and Joubert, P.N., Turbulent Boundary Layers on Bodies of Revolution, *J. Ship Research* 26, 135, 1982.

Stollery, J.L., Fomison, N.R. and Hussain, S., The Effects of Sweep and Bluntness on Glancing Interactions at Supersonic Speeds, *Proc. 15th ICAS Congress, London,* p. 172, 1986.

Townsend, A.A., *The Structure of Turbulent Shear Flow,* Cambridge, 1956.

Tzuoo, K.L., Ferziger, J.H. and Kline, S.J., Zonal Models of Turbulence and Their Application to Free Shear Flows, Thermosciences Divn, Stanford Univ. Rept TF-27, PB87-183562, 1986.

Veeravalli, S. and Warhaft, Z., The Interaction of Two Distinct Turbulent Velocity Scale in the Absence of Mean Shear, *Turbulent Shear Flows 5* (F. Durst et al., Eds., Springer) p. 31, 1987.

Watmuff, J.H., Witt, H.T. and Joubert, P.N., Developing Turbulent Boundary Layers with System Rotation., *J. Fluid Mech.* 157, 405, 1985.

Westphal, R.V., Eaton, J.K. and Pauley, W.R., Interaction Between a Vortex and a Turbulent Boundary Layer in a Streamwise Pressure Gradient, *Turbulent Shear Flows 5* (F. Durst et al., Eds., Springer) p. 266, 1987.

Winter, K.G., Rotta, J.C. and Smith, K.G., Studies of the Turbulent Boundary Layer on a Waisted Body of Revolution in Subsonic and Supersonic Flow., *British Aero. Res. Council* R. and M. 3633, 1968.

Witt, H.T. and Joubert, P.N., Effect of Rotation on a Turbulent Wake, *5th Symposium on Turbulent Shear Flows, Cornell,* p. 21-25, 1985.

You, S.M., Simon, T.W. and Kim, J., Boundary Layer Heat Transfer and Fluid Mechanics Measurements on a Mildly-curved Convex Wall, *Proc. 8th Int. Heat Transfer Conf.* vol. 3, p. 1089, 1986.

Free-Stream Turbulence Effects on Separated Shear Layers

I. P. CASTRO
Department of Mechanical Engineering
University of Surrey
Guilford, Surrey
GU2 5XH, UK

1. INTRODUCTION

Although separated flows of one kind or another are very common, understanding of the physical processes that are dominant in such flows is much less well developed than for attached (or 'free') shear flows. One reason for this is simply that it is relatively difficult to make measurements in highly turbulent flows so that the available database is extremely sparse. A priori, the major influences on the turbulence structure of such flows might be anticipated to include streamline curvature, upstream effects caused by highly turbulent fluid being returned at reattachment and the presence of a wall which may effect the larger scale structures developing within the outer shear layer. Which, if any, of these has a significant effect on the flow may in general depend on the particular flow in question, although one might postulate that, alternatively, just one of them (or some other) will usually be dominant. Many separated flows of engineering importance have the separated region embedded in an outer turbulent flow which will itself impose extra fluctuating strains on the separated shear layer; this again may or may not have important effects on the overall structure of the flow.

At the University of Surrey we have for some years been using pulsed wire anemometry to study a number of (geometrically) simple and nominally two-dimensional flows containing substantial regions of reverse flow. Most separated flows contain regions which superficially at least look similar to the more well-known thin shear flows. The most notable feature is perhaps the turbulent shear layer, growing in thickness from separation up to reattachment. This forms the boundary between the outer flow and the upstream moving fluid nearer the wall and has similarities with the ordinary plane mixing layer. Our intention has been, in part, to determine to what extent the characteristics of the more easily recognisable parts of such flows - like this 'separated' shear layer - differ from the corresponding thin shear flows, and to endeavour to identify the major reasons for the differences.

Quite apart from the fundamental interest in separated flows, the work has received added impetus recently because of the increasing interest in 'zonal' turbulence modelling. This approach was originally proposed by Kline (1981) and has been developed thus far largely at Stanford (see Tzuoo et al, 1986, for an example of recent progress). The modelling technique largely rests on the assumption that <u>any</u> complex turbulent flow can essentially be broken into a finite number of 'zones' each of which is identifiable as being essentially one of a finite set of simpler and well understood shear flows. It is then necessary to provide suitable patching models for the transition regions between each of these zones. For many kinds of separated flows, therefore, one might attempt to represent the flow by a suitable combination of modelled adverse pressure gradient boundary layers (upstream of separation), distorted mixing layers (between separation and reattachment), re-adjusting wall boundary layers (downstream from reattachment) and a much thinner and lower Reynolds number boundary layer developing upstream from around reattachment. Now even ignoring the possibility that the inherently unsteady nature of many separation and all reattachment processes implies that models based on the time-averaged Reynolds equations are unlikely to be very successful, effective zonal modelling for these flows will only be possible if these various zones are not too unlike the 'base' flows chosen to represent them. In terms of Bradshaw's discussion in the preceding paper, one might hope that the consecutive species changes - from boundary layer to mixing layer and back to boundary layer - which encompass the separation/reattachment sequence (see his figure 2) are recognisable as such. The mixing layer is itself one of the zones, albeit somewhat fuzzy (to use Bradshaw's terminology). If the perturbations represented by lines between the special zones (in figure 2) are too complex, they may effectively obliterate one of the zones entirely (the mixing layer, in this case) so that this 'separated-shear-layer-with-wall-influence-plus-reattachment' must become a zone in its own right. Quite apart from the modelling required to represent the transition regions between the zones, this zone would then require its own modelling procedures which would presumably have to be more sophisticated than those needed for adequate prediction of distorted mixing layers. A detailed understanding of some 'typical' separated flows would therefore be of considerable help in formulating appropriate zonal modelling techniques or, indeed, deciding whether such techniques are likely to be very useful.

Most of the work at Surrey has concentrated on just two geometries. Firstly, we have extensively studied the 'flat plate plus splitter plate' combination (figure 1). This work has involved measurements of all the Reynolds stresses, amongst other things, and has begun to yield detailed evidence of quite substantial differences between the turbulence structure of the separated shear layer and that of an ordinary plane mixing layer (Castro & Haque, 1987, hereafter denoted by CH for convenience). Secondly, we have made comprehensive measurements in a separated boundary layer on a flat plate, with separation initiated by a strong adverse pressure gradient generated by a cylinder-with-controlled-circulation mounted above the plate (Dianat & Castro, 1986,1988). This latter experiment has involved measurements of the velocity triple-correlations and so has enabled turbulent energy balances to be obtained throughout the separated region. In addition to these two flows we have also addressed the particular question of the effect of free-stream turbulence on a separated flow, since this is such a common feature of practically important flows.

The intention in this paper, bearing the above remarks in mind, is firstly to review briefly the major conclusions of our work thus far (in Section 2). In the spirit of this meeting, and particularly this session, we concentrate on the various conflicting effects of the additional rates of strain present in the flow. Secondly, we present and discuss some detailed data recently obtained in a separated flow subjected to free-stream turbulence at its outer edge (Section 3), emphasising the changes brought about by this added complication. The implications are summarised in Section 4.

2. SUMMARY OF PREVIOUS WORK

It is appropriate first to outline the experimental arrangement used for much of the work previously undertaken and for the work to be described in the following section. Figure 1 shows the general set-up. A 6mm thick flat plate spanned the shorter width of the 0.77 x 0.62m working section of the No.1 open circuit wind tunnel in the Mechanical Engineering Department. The plate height (2H) was 50mm and its edges were bevelled to provide fixed separation lines with no possibility of reattachment onto the plate itself. A central, downstream splitter plate was mounted on the axis of symmetry. The mean velocity in the empty working section was uniform to within ±½%, with a longitudinal turbulent intensity below 0.25%. Endplates ensured adequate two-dimensionality of the flow, although the influence of the vortex structures generated at the endwall-fence-splitter plate junction can never be entirely suppressed (see Jaroch, 1987, for a detailed discussion of these effects).

Instrumentation was standard and included the use of pulsed wire anemometry. Probes were designed to give an accurately cosinusoidal yaw response, so that measurements of the various Reynolds stresses throughout the separated region are believed to be at least as accurate as those which can be obtained by crossed hot wire anemometry in regions where the latter can sensibly be used. More complete descriptions of all the instrumentation and the general arrangements are contained in CH and Castro & Haque (1988).

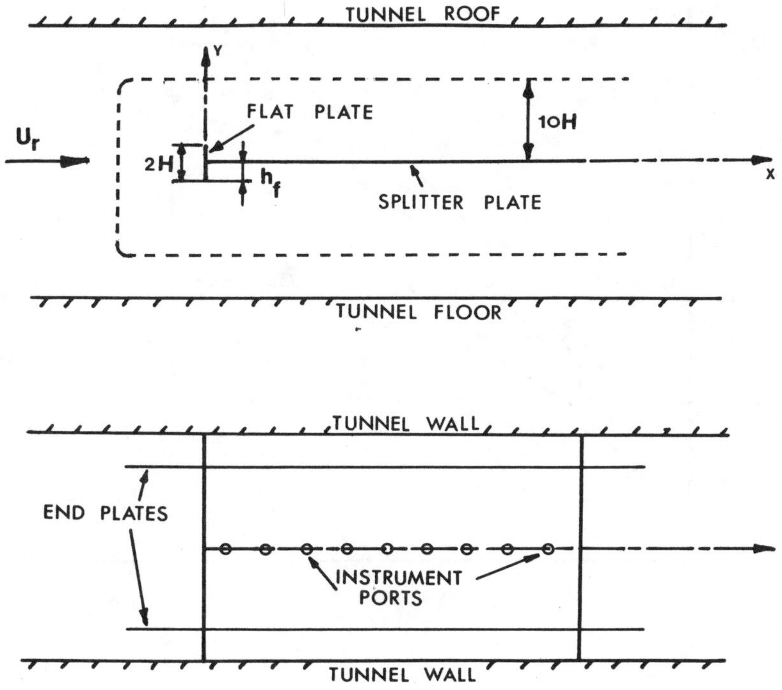

FIGURE 1. Experimental arrangement; not to scale. -----, endplate boundaries.

Now it is well known that even quite small streamline curvature often leads to large changes in turbulence structure. Even before making any measurements in this flow it is apparent that the mean flow curvature of the separated shear layer will be significant and consequently might well influence the structure of the shear layer. Figure 2 shows the variation of the mean flow curvature parameter, Λ/R, where Λ is the value of the vorticity thickness, defined in the usual way as:

$$\Lambda = 1/(\frac{d(U/\Delta U)}{dy}).$$

ΔU is the total (local) velocity difference across the shear layer ($U_{max}-U_{min}$). The curvature, R, is taken for convenience as the curvature of the separation streamline although in one sense this is a pathological choice: there is a discontinuity at $x/x_R=1$. For streamlines just below (or just above) the separation streamline the curvature would be rather higher (or lower) as reattachment is approached and there would be no discontinuity. Without the addition of free-stream turbulence (and those data will be discussed in the following section) Λ/R is not very much smaller than typical values in the curved mixing layer studied by Castro & Bradshaw (1976). Figure 3, however, shows that the changes in the turbulence structure parameter, $\overline{uv}/\overline{q^2}$, across the shear layer are quite different from those that occured in the latter study. $\overline{uv}/\overline{q^2}$ is shown plotted against η, which is defined by $\eta = -(y-y_c)/\Lambda$, where y_c is the

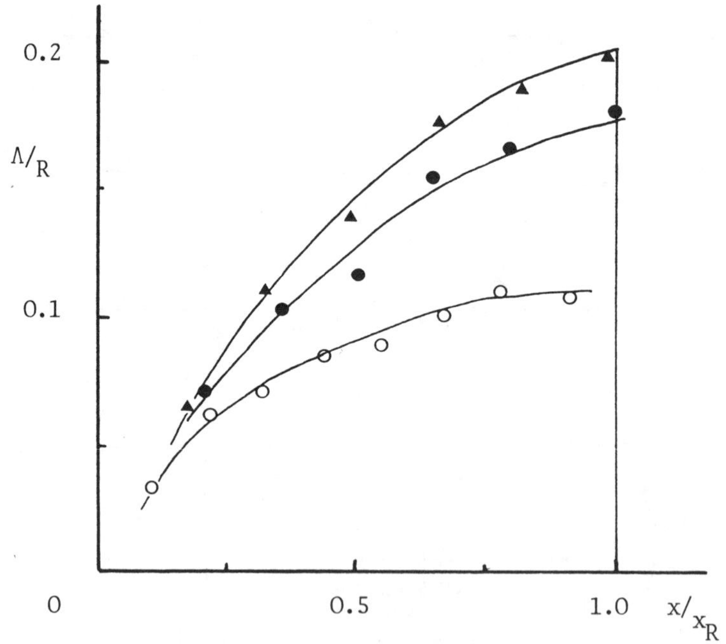

FIGURE 2. Streamline curvature parameter, Λ/R. ○, no free-stream turbulence; ●, u'/U=3.4%; ▲, 8.7% (values measured at x=0 in empty tunnel)

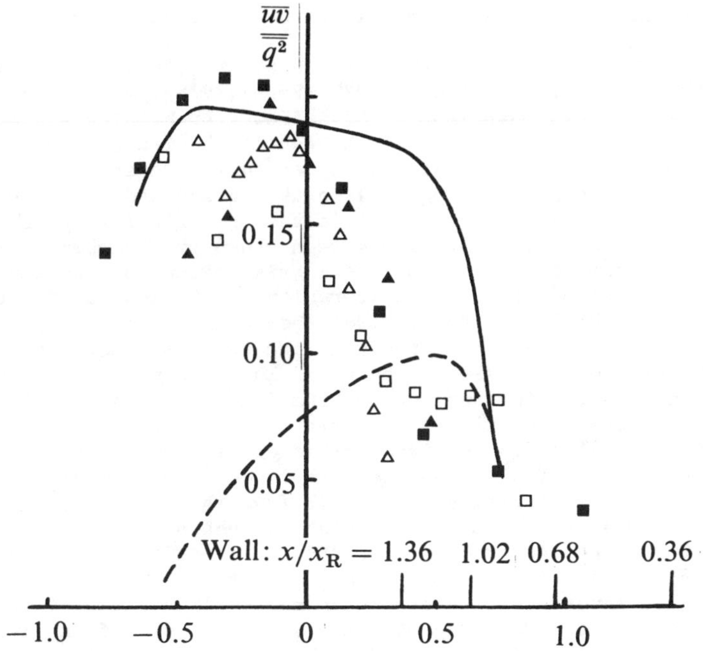

FIGURE 3. Turbulence structure function, $\overline{uv}/\overline{q^2}$. ■ , x/x_R = 0.35; □ ,0.68; ▲ ,1.02; △ ,1.36. No free stream turbulence. Solid line is the plane mixing layer variation and the dotted line is Castro & Bradshaw's result for a highly curved mixing layer, at the station where Λ/R=0.28.

height of the position at which the mean velocity is ($0.67\Delta U + U_{min}$), and U_{min} is the minimum velocity on the low speed side of the shear layer. Note that the negative sign ensures that $\eta < 0$ on the high velocity side of the flow, in common with the usual mixing layer similarity variable. The measurements shown in the figure are for various axial stations between separation and reattachment (with x_R being the distance to the mean reattachment point). It is clear that the behaviour is everywhere not only quite different from the ordinary plane mixing layer behaviour but also from the variation found by Castro & Bradshaw in the curved mixing layer.

This and other data have led to what is perhaps the most significant conclusion of the work published thus far, at least in the context of effects of extra strain rates. That is that although the streamline curvature <u>is</u> significant, the differences between the turbulence structure of this separated shear layer and a plane mixing layer <u>cannot</u> be primarily a result of this curvature. Evidence discussed in detail in CH suggests that the major changes in the turbulence structure are dominated by the re-entrainment of turbulent fluid returned upstream around reattachment. This provides a mechanism for 'positive feedback' - the presence of the wall initially acts to amplify the normal Reynolds stresses and this more energetic fluid is then fed back into the shear layer, causing even greater amplification of the turbulence energy. The impermeability condition limits the amplification of the $\overline{v^2}$ component, but the $\overline{w^2}$ stress continues rising across the whole shear layer all through the reattachment region.

Wall boundary layers (e.g. Hancock & Bradshaw,1983,1987) and two-stream mixing layers (e.g. Pui & Gartshore,1979, Patel,1978) can be significantly modified by free-stream turbulence. Mean flow growth rates and turbulence Reynolds stresses are generally all enhanced if the length scale of the free stream turbulence is of a similar order to the shear flow scales. One therefore expects a separated region like that discussed above to respond in some way to outer flow turbulence, via the latter's influence on the separated shear layer. Indeed, it has been shown by a number of authors that such turbulence normally reduces the axial extent of the separated flow region, consistent with the invigoration of the separated shear layer (see, for example, Hillier & Cherry,1981, Bearman & Morel,1983, Nakamura & Ozono,1987). This implies an increase in the mean flow curvature. However, the detailed response of the shear layer is not yet clear and this provided the motivation for our work on the flat-plate-plus-splitter flow with added-free stream turbulence. Initial experiments with a free-stream intensity of about 3% and a longitudinal integral length scale of 1.3H (at x=0) have suggested that although the turbulence structure around reattachment is rather different in detail from that in the absence of free-stream turbulence, the positive feedback mechanism around reattachment is still dominant (see Castro & Haque,1988). But this was a relatively low free-stream turbulence level - very much lower than the typical energies within the shear layer itself, so perhaps one would not expect too significant an effect. More recent measurements at significantly higher free-stream intensities have been undertaken in an attempt to answer the question of whether higher levels of added free-stream turbulence will lead to a switch in the most dominant process in the separated flow. It is these measurements that we present and discuss in the following sections.

3. MEASUREMENTS WITH FREE-STREAM TURBULENCE.

3.1 Experimental Arrangements

Both in the earlier and in the present work the free-stream turbulence was added in the normal way, using a standard bi-planar grid with a mesh size of 50mm, located at various positions upstream of the plate. Hot wire measurements showed that over the axial region spanned by the plate plus its separated wake the longitudinal turbulence intensity decayed according to:

$$(\overline{u^2}/U^2)^{-0.8} = 8.1(x'/M - 5.8),$$

where x' is measured from the grid position and M is the mesh spacing. Crossed wire measurements showed that the degree of anisotropy was within the typical ranges found by other workers for this type of grid (e.g. Comte-Bellot & Corrsin,1966). Longitudinal velocity autocorrelations led, via Taylor's hypothesis, to an integral scale given by:

$$L_x/M = 0.015(x'/M + 10.4).$$

Whilst not quite consistent with the assumption that L_x grows at a rate proportional to the growth of the dissipation length parameter,

$$L_\varepsilon = (\overline{q^2})^{3/2}/\varepsilon,$$

these expressions fitted the experimental data well over the appropriate range of x'/M. In the present work the grid was placed 12M or 16M upstream of the plate, leading to a turbulent intensity of about 8.7% or 6.1%, respectively, at the plate location with corresponding integral scales of about 0.7H or 0.8H. These figures should be compared with 3.4% and 1.3H for the earlier work described in Castro & Haque (1988). As far as the overall flow is concerned it has become clear from other published work that where separation at large angles to the approach flow occurs (90° in the present case) the free-stream length scale is not a dominant parameter provided it is sufficiently small - less than 4H according to Nakamura & Ozono (1987) (see also Bearman & Morel,1983). In all the above cases the outer turbulence has scales inevitably much larger than those in the separated shear layer just after separation, but as reattachment is approached the length scales in the latter become more comparable with those in the free stream. We believe that the dominant parameter in the above cases is the level of turbulence energy in the free stream, although further study of flows developing with the <u>same</u> L_x/H (at x=0) but different u'/U might be instructive.

It would not be possible to increase the free-stream intensity at x=0 much above the value given with the grid 12M upstream of the plate (8.7%) without reducing the level of spanwise homogeneity in the free stream flow. With the grid in that position spanwise traverses in the free stream at x=0 showed variations in u'/U which were roughly periodic with a wavelength equal to the mesh size. However, the peak to peak variations were less than ±2% and this was felt to an adequate level of homogeneity.

3.2 The Mean Flow

Figure 4 shows the variation in the distance to the mean reattachment point with free-stream intensity; the expected fall with increasing intensity is quite apparent. There is a maximum reduction in x_R/H of about 32% for the

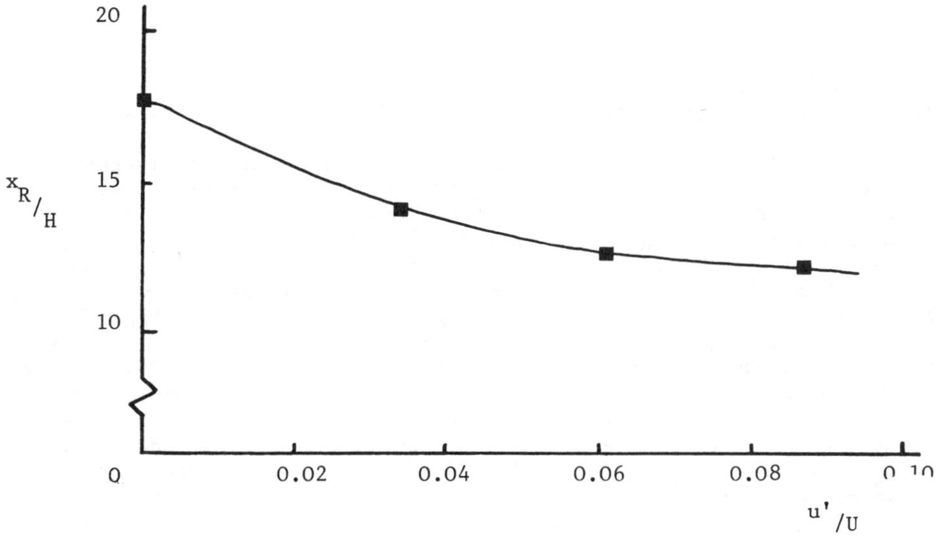

FIGURE 4. Variation of distance to reattachment with free stream intensity

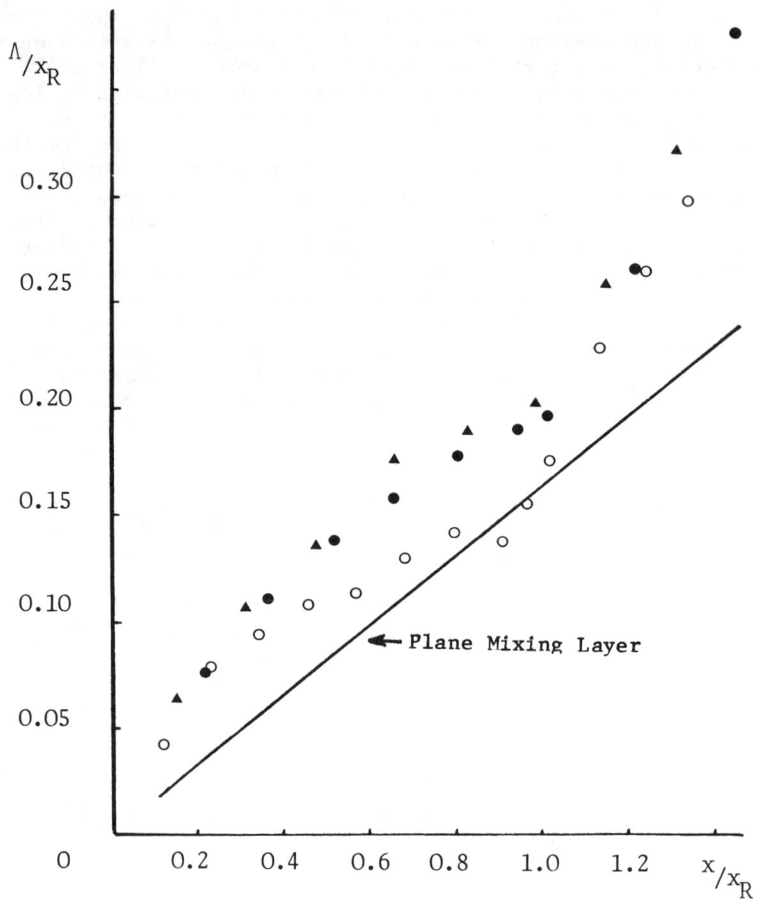

FIGURE 5. Shear layer growth. ○, u'/U=0; ●, 3.4%; ▲, 8.7%.

case of the highest free-stream intensity and the results imply that further increases in intensity would lead to only very small additional reductions in x_R. This shortening of the reverse flow region, although substantial, is not as great as that found by Hillier & Cherry (1981) in the case of a blunt flat plate. In that case x_R was reduced by some 44% at a free-stream intensity of 6.4%, compared with the present 28% at a similar intensity. The difference is presumably simply a result of the rather different geometry; the former case can be though of as a flat plate plus splitter plate but with the thickness of the latter equal to 2H. It should also be noted that Hillier & Cherry found that increasing the free-stream turbulence integral scale from 0.82H to 1.92H whilst keeping the intensity constant had an insignificant effect on the surface pressure distribution and x_R. This provides further evidence that u'/U is the important parameter in all these cases.

A shorter reverse flow region implies that the mean streamline curvature over much of the flow must be increased. This is demonstrated in figure 2, where the curvature of the separation streamline is plotted as a function of downstream distance normalised by x_R. The addition of free-stream turbulence clearly leads to significant increases in the curvature parameter, which is always very small just after separation because the shear layer thickness is small there, but further downstream it increases significantly as the free stream intensity rises. It is interesting that the rise in the effective curavature is larger for the change in intensity from (nominally) zero to 3.4% than it is for the change from 3.4 to 8.7%, which is in line with the previous observation that the flow 'saturates' so that further increases in intensity would only have relatively small effects.

Figure 5 shows the corresponding growth in shear layer thickness for the three cases of zero, 3.4% and 8.7% free-stream intensity. As pointed out in our earlier work (CH), the growth rate is initially significantly higher than it is in the ordinary plane mixing layer and is not constant. Free-stream turbulence causes an increase in the growth rate but in all cases the growth rate falls slowly as reattachment is approached. Note again the relatively small change between the 3.4% and the 8.7% free-stream intensity cases.

3.2 The Turbulence Stresses

Figure 6 shows the changes in the maximum values of the Reynolds stresses caused by the addition of free-stream turbulence. These were deduced from detailed cross-stream profiles at a number of axial stations in each case. Note firstly that the stresses are in all cases much higher than the values found in the ordinary plane mixing layer but the qualitative behaviour is hardly changed at all by the addition of free-stream turbulence - if the stresses were plotted against x/x_R a reasonable collapse would be achieved in most cases. As an example, Figure 7 shows

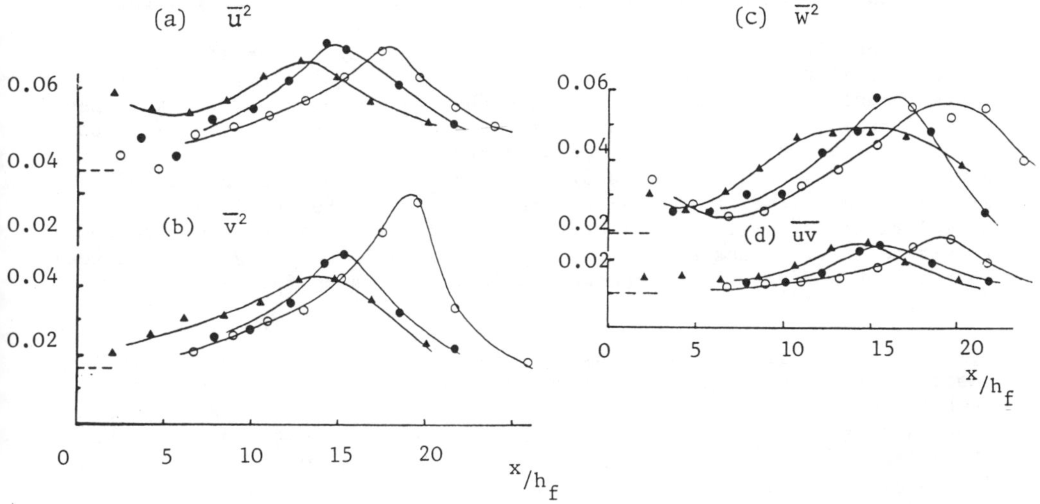

FIGURE 6. Maximum turbulence stresses along the separated shear layer, all normalised by $(\Delta U)^2$. ○ , u'/U = 0; ●,3.4%; ▲,8.7%. a) $\overline{u^2}$; b) $\overline{v^2}$; c) $\overline{w^2}$; d) $\overline{uv}$. Reattachment locations are indicated for each case. Lines added for clarity. - - - -, Plane mixing layer values

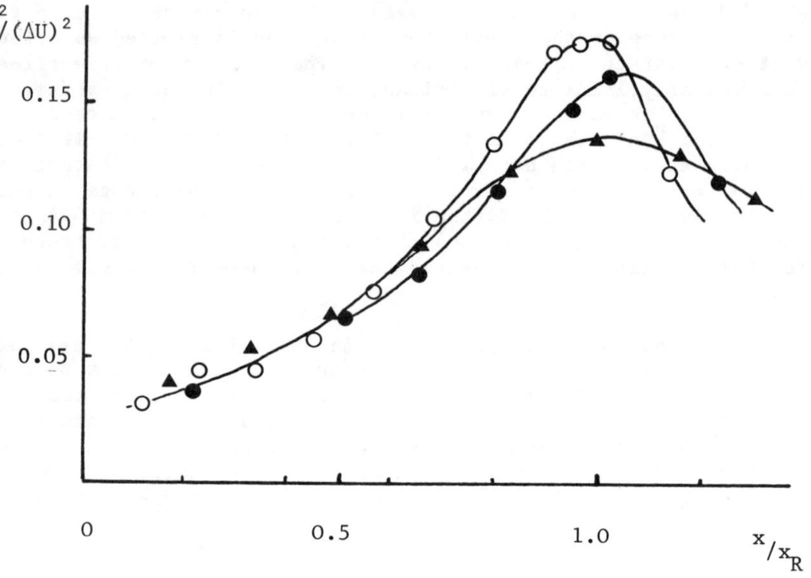

FIGURE 7 Turbulence energy, $\overline{q^2}/(\Delta U)^2$, along y/H=1.0. Legend as in figure 6

the variation of the turbulence energy as a function of distance along the reversed flow region at, in this case, a fixed height (y/H=1). Only around reattachment is there a significant effect of free-stream turbulence and nearly all of this difference is caused by the reduction in the peak levels reached by the normal component (figure 6). This latter reduction in peak $\overline{v^2}_{max}$ is, in fact, the most notable of the quantitative changes caused by free-stream turbulence and fuller discussion is deferred until later.

Note secondly that free-stream turbulence generally increases the turbulence stresses at a given axial position sufficiently far upstream of reattachment. This is consistent with the higher mean flow growth rates shown in figure 5 and is obviously in line with other published work on the effects of free-stream turbulence on standard shear flows. Now stabilising curvature has generally been shown to reduce turbulence stresses and layer growth rates so the above results imply that any effects of the considerably 'stronger' curvature (figure 2) resulting from the presence of free-stream turbulence are completely dominated by the effects of increased entrainment and/or other processes; this is discussed further in the following Section.

Figures 6 and 7 also suggest that the total turbulence energy within the shear layer around reattachment may actually be reduced by free-stream turbulence, which seems contrary to expectations. Figure 8 shows the cross-stream profiles of the turbulence stresses at reattachment. Whilst the axial stress ($\overline{u^2}$) is increased all across the flow, both the lateral ($\overline{w^2}$) and, particularly, the normal ($\overline{v^2}$) stresses are significantly reduced. Although these latter reductions are collectively greater than the increase in $\overline{u^2}$ an appropriate integral of the total turbulence energy across the flow actually shows an increase with free stream intensity. Defining the 'total' turbulence energy by

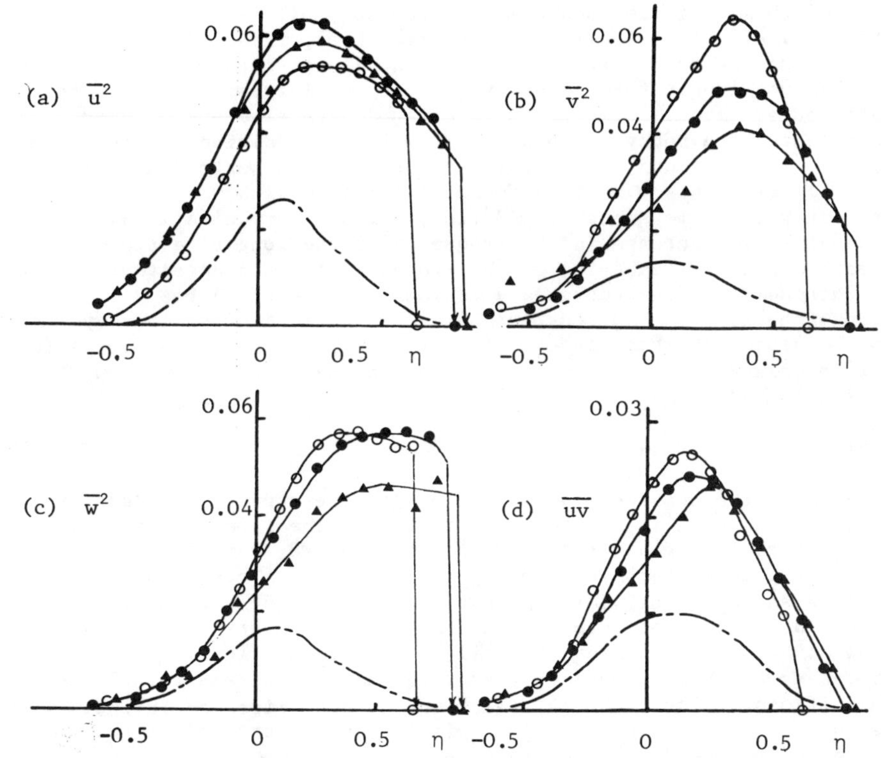

FIGURE 8. Cross-stream profiles of the turbulence stresses at x/x_R=1.0, normalised by $(\Delta U)^2$. The wall locations are shown in each case - note that the wall shear stress is zero only because x/x_R=1. a) $\overline{u^2}$; b) $\overline{v^2}$; c) $\overline{w^2}$; d) $\overline{uv}$. ——-, plane mixing layer values.

$$E_t = 1/H \int_0 (\overline{q^2}/U_r^2).dy ,$$

the data indicate that E_t rises from 0.41 in the absence of free-stream turbulence to 0.49 in the case with u'/U=8.7% so that, as expected, free-stream turbulence leads to increased production of turbulence energy within the shear flow.

It is significant that $\overline{u^2}$ increases all across the flow, and particularly at the outer edge, whereas $\overline{v^2}$ and $\overline{w^2}$ do not. This is also true for profiles compared at the same axial station (i.e. the same x/H rather than x/x_R) and Castro & Haque (1988) argued that it constitues evidence of an increase in the 'flapping' motion of the separated shear layer, which is associated with the relatively long time scale spectral peak noticable in, say, surface pressure spectra. Measurements of the frequency of this long time scale motion provided additional confirmation of this conclusion and it was noted that the time scale associated with the turbulence in the free stream ($L_\varepsilon/(\overline{q^2})^{\frac{1}{2}}$, say) is of the same order as the 'flapping' time scale so that some interaction would be expected. Other workers have noticed the presence of this long time scale motion (notably Hillier and his co-workers, e.g. Hillier & Cherry,1984, Cherry et al, 1984, Hillier & Dulai,1985) and it should be emphasised that it corresponds to motions on a time scale much longer than those typical of the largest eddies in the separated shear layer. Some of the published data also suggests an increase in the frequency of this motion with the addition of free-stream turbulence, as was found in our earlier work. It is worth mentioning here that the frequency has also been found to increase with the addition of significant three-dimensionality (Sutton, private communication). It could be argued that an increase in the frequency of the long time scale motion represents a general weakening of the process which produces that motion - if the increase is large enough any residual motion would be indistinguishable from that corresponding to the largest scale eddy motions within the separated shear layer. In any event, the precise reasons for the low frequency motion are not yet clear so that further work on this point is required.

3.3 The Turbulence Structure

Figure 9 shows the cross-stream variations of the structure parameters, $\overline{uv}/\overline{q^2}$, $\overline{v^2}/\overline{u^2}$ and $\overline{w^2}/\overline{u^2}$. In the outer region of the flow the shear layer is entraining turbulent fluid (of, far enough outboard, nominally zero shear stress) so the local turbulent energy rises without a corresponding increase in shear stress, leading to a reduction in $\overline{uv}/\overline{q^2}$. Similar effects are found in the case of zero-pressure gradient boundary layers (Hancock & Bradshaw, 1984,1987). Nearer the wall, however, the flow is dominated by the processes ocurring around reattachment and free-stream turbulence has a less significant effect on $\overline{uv}/\overline{q^2}$ (although it does rise consistently with increasing free-stream intensity). This is almost certainly a result of the large reductions in $\overline{v^2}$ and, to a lesser extent, $\overline{w^2}$.

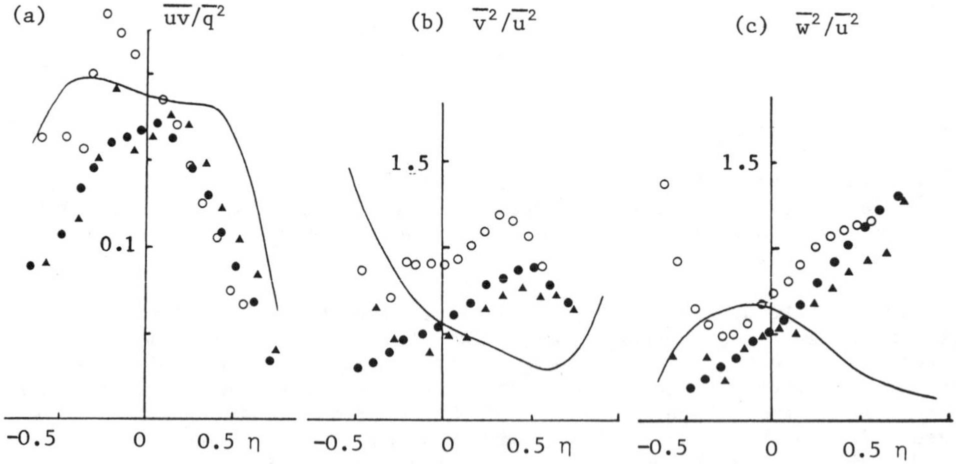

FIGURE 9. Structural parameters at x/x_R=1.0. Legend as in figure 6. a), $\overline{uv}/\overline{q^2}$; b), $\overline{v^2}/\overline{u^2}$; c), $\overline{w^2}/\overline{u^2}$. ———, Plane mixing layer.

One of the more remarkable structural differences between the ordinary plane mixing layer and the separated shear layer near reattachment (in the absence of free-stream turbulence) is the large level of $\overline{v^2}/\overline{u^2}$ in the inner part of the flow, caused by large increases in the normal component stress - see figures 6b and 8b. The rise in $\overline{w^2}$ is also proportionally greater than that in $\overline{u^2}$ as reattachment is approached and seems largely unaffected by the immediate influence of the wall, whereas $\overline{v^2}$ inevitably falls as the wall is approached (compare figures 8b and 8c). It was argued in CH that the initial response of the separated shear layer to the decreasing wall distance as reattachment is approached is to exhibit an increase in $\overline{v^2}$ - as found by Wood & Bradshaw (1982) in the case of a shear layer approaching a wall on its high velocity side. This increase must eventually feed into the other two components via the pressure-strain correlations but, in any case, is amplified by the entrainment of further high energy fluid being returned upstream. Eventually, of course, the wall must act to reduce the normal stress component and it is evident from figure 6b that beyond reattachment $\overline{v^2}$ falls more rapidly than the other components. Now with the reduced distance to reattachment caused by the addition of free-stream turbulence this amplification process starts earlier, but there is less time for it to operate. The stresses therefore rise earlier but do not reach such high levels (figure 6). This is particularly noticeable for the $\overline{v^2}$ component, as would be expected if it is this component which first responds to the presence of the wall and triggers the whole 'positive feedback' mechanism. With the addition of free-stream turbulence the large increase in $\overline{v^2}$ over the plane mixing layer value is monotonically reduced as the level of free-stream intensity rises (figure 6b); for the present case of u'/U=8.7% it is actually halved. It is this process which accounts for the fall in $\overline{v^2}/\overline{u^2}$ in the near-wall region with increasing free-stream turbulence.

Notice particularly that it cannot be argued that the large reduction in the $\overline{v^2}$ rise is a result of the significant stabilising curvature. It is certainly in the right sense (curvature tends to reduce turbulence stresses) but curvature would also lead to roughly equivalent reductions in $\overline{u^2}$; this does not occur (see figure 6a). The evidence therefore suggests again that mean flow curvature must be relatively unimportant and the major changes resulting from the addition of free-stream turbulence are those arising from the weakened effects of the reattachment/wall-influence processes.

Notice also that the changes in $\overline{uv}/\overline{q^2}$ in the inner region around reattachment, caused by increasing free-stream turbulence, are opposite to those that would be expected in the absence of wall-plus-reattachment effects. In that case one would expect $\overline{uv}/\overline{q^2}$ to fall uniformly all across the flow - although in the inner layer of a boundary layer, for example, the fall would be partly a result of changes in the 'inactive' motion. In the present case therefore, the effects of free-stream turbulence are greater than superficially apparent from the results shown in figure 9.

4. FINAL COMMENTS AND CONCLUSIONS

The major conclusion of the work described here is that free-stream turbulence does not appear to lead to a switch in the dominant processes effecting the turbulence structure of a separated flow. Although it can induce significant increases in the mean flow curvature because of the increased growth rates and consequent earlier reattachment, this curvature does not play a significant role - although it may, perhaps, act as a constraint on the changes that would occur in its absence. As in the case without free-stream turbulence, the dominant processes effecting the turbulence structure would seem to be those occuring upstream of and around reattachment. Free stream turbulence, because it leads to earlier reattachment, actually reduces the increases in the normal Reynolds stresses that occur via the feedback mechanism.

Current turbulence models, even the most sophisticated ones, are unable even to model properly the effects of free-stream turbulence on an ordinary boundary layer (e.g. Hancock & Bradshaw,1987). The complex processes around reattachment and on the upstream lower velocity edge of the separated shear layer would seem to pose problems of a higher order of difficulty. In view of the very different turbulence structure in the shear layer, with or without free-stream turbulence, from that in the ordinary mixing layer it would appear that zonal models would need to take 'the reattaching shear layer' as one of the fundamental zones, rather than consider it simply as a distortion of the plane mixing layer. Whether this would be necessary for other kinds of separated flows remains to be seen. Our complimentary work on a separating and reattaching flat plate boundary layer suggests that in that case the turbulence structure is rather different from that discussed here (see Dianat & Castro,1987 for an initial comparison).

Space has not allowed a detailed discussion of the effects of the free-stream turbulence on the large-eddy motions in the flow. This is the subject of ongoing study but, as discussed earlier, it is already clear that the lowest-frequency motions are weakened by the addition of free-stream turbulence. Whilst we have in this paper approached the whole subject on the basis of the usual time-mean parameters, a proper understanding - and, indeed, successful modelling - of flows like these may only be possible when explicit account is taken of their unsteady nature. Recent measurements of spatial velocity correlations within the separated region itself have shown that although, near reattachment, the longitudinal and lateral integral scales are very similar, further upstream they are certainly not (Nasser, private communication). Indeed, deep within the reverse flow region, the spanwise integral scales can bu up to a factor of five <u>smaller</u> than the axial scales. This is demonstrated in figure 10, which shows the axial variation of the longitudinal and spanwise integral scales (L_x and L_z, respectively) at both the outer edge of the flow and within the reverse flow region - in the case without free-stream turbulence. The outer flow variation is roughly what one would expect from the usual growth of the large-scale structures in the shear layer, but the evidence suggests that in the near wall region there is a very 'streaky' eddy structure. This could be a result of preferential stretching of hairpin-type vortices as their legs are distorted and swept upstream from the reattachment region. In any case, it seems evident that the dominant eddies in that region are not simply just some of the large eddies in the shear layer being returned upstream. A proper modelling of the flow will may have to take these large structural changes into account explicitly.

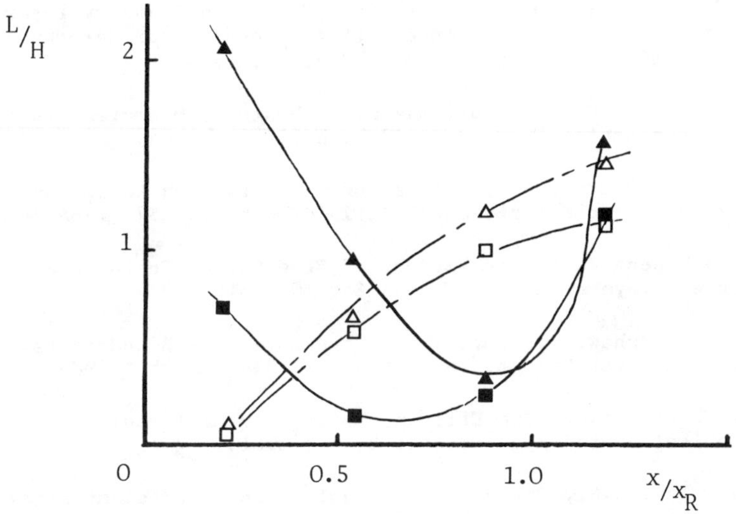

FIGURE 10. Growth of the integral scales, L_x (△ , ▲) and L_z (□ , ■). Solid symbols: y/H=0.3; open symbols: line of u'/U=10%.

5. ACKNOWLEDGEMENTS

The author is grateful for on-going funding from the Science and Engineering Research Council in support of this work and is also indebted to his colleagues Drs P.E.Hancock and M.Dianat for many stimulating discussions. He is also grateful to Dr.A.Haque for obtaining some of the measurements herein.

6. REFERENCES.

Bearman, P.W. & Morel, T., Effect of Free-Stream Turbulence on the Flow around Bluff Bodies, Prog.Aerospace Sci. 20,97-123,1983.

Bradshaw, P., Effects of Extra Strain - Review, presented at Zaric Memorial Seminar on Near-Wall Turbulence, Dubrovnik, May 1988.

Castro, I.P. & Bradshaw, P., The Turbulence Structure of a Highly Distorted Mixing Layer, J.Fluid Mech. 73,265,1976.

Castro, I.P. & Haque, A., The Structure of a Turbulent Shear Layer Bounding a Separation Region, J.Fluid Mech., 179,439, 1987

Castro, I.P. & Haque, A., The Structure of a Shear Layer Bounding a Separation Region. Part II: Effects of Free-Stream Turbulence, J.Fluid Mech. (in press) 1988.

Castro, I.P., Dianat, M. & Haque, A., Shear Layers Bounding Separated Regions, paper presented at VIth International Symposium on Turbulent Shear Flows, Toulouse, Sept.1987; to appear in 'Turbulent Shear Flows VI', ed. Durst et al, Springer-Verlag, 1988.

Dianat, M. & Castro, I.P., Measurements in a Separated Boundary Layer, paper presented at 6th Meeting of International Council for Aeronautical Sciences, London, 1986; to appear in AIAA Journal, 1988.

Cherry, N.J., Hillier, R. & Latour, M.E.M.P., Unsteady Measurements in a Separating and Reattaching Flow, J.Fluid Mech.144,13, 1984.

Compte-Bellot, G. & Corrsin, S., The Use of a Contraction to Improve the Isotropy of Grid-Generated Turbulence, J.Fluid Mech. 25,657, 1966.

Hancock, P.E. & Bradshaw, P., The Effect of Free-Stream Turbulence on Turbulent Boundary Layers, ASME J.Fluids Eng.105,284, 1983.

Hancock, P.E. & Bradshaw, P., Turbulence Structure of a Boundary Layer beneath a Turbulent Free Stream, submitted to J.Fluid Mech., 1988

Hillier, R. & Cherry, N.J., The Effects of Free-Stream Turbulence on Separation Bubbles, J.Wind Eng.Ind.Aero.8, 49, 1981.

Hillier, R. & Dulai, B.S., Pressure Fluctuations in a Turbulent Separated Flow, Proc.5th Int.Symp. on Turbulent Shear Flows, Cornell, 1985.

Jaroch, M., Eine Kritische Betrachtung der Methode Diskreter Wirbel als Modell fur eine Abgeloste Stromung mit Geschlossener Abloseblase auf der Basis Experimenteller Erkenntnisse, Ph.D. Thesis, Technical University of Berlin, 1987.

Kline, S.J., Universal or Zonal Modelling - the Way Ahead, in proceedings of the 1980/1981 AFOSR-HTTM-Stanford Conference on complex turbulent flows - comparison of computation and experiment, Vols.1-3, Stanford University, 1982.

Nakamura, Y.N. & Ozono, S., The Effects of Turbulence on a Separated and Reattaching Flow, J.Fluid Mech.178,477, 1987.

Patel, R.P., Effects of Stream Turbulence on Free Shear Flows, Aero.Quart.29,33, 1978.

Pui, N.K. & Gartshore, I.S., Measurements of Growth Rate and Structure in Plane Turbulent Mixing Layers, J.Fluid Mech.91,111, 1979.

Tzuoo, K-L., Ferziger, J.H. & Kline, S.J., Zonal Models of Turbulence and their Application to Free Shear Flows, Rep.No.TF-27, Thermosciences Division, Dept.Mech.Eng., Stanford University, 1986.

Wood, D.H. & Bradshaw, P., A Turbulent Mixing Layer Constrained by a Solid Surface. Part 1: Measurements Before Reaching the Surface, J.Fluid Mech.122,57 1983.

External Flow Turbulization Effect of the Wall Turbulence Structure

E. P. DYBAN and E. Ya. EPIK
Institute of Engineering Thermophysics
Ukrainian SSR Academy of Sciences
Kiev, USSR

ABSTRACT

Experimental data on the external flow turbulization on the wall boundary structure layer are presented. Turbulization is estimated by the effective viscosity at the external boundary of the layer. An overlayer is revealed where the boundary layer interacts with the turbulized flow and transport properties of the latter are attenuated. Based on the integral and local characteristics of the internal turbulence structure, generalized relations have been obtained for calculation of transport processes in pseudo-laminar and quasiturbulent boundary layers.

I. INTRODUCTION

During the last two decades experimental studies have shown that the external flow turbulence Tu_∞ affects substantially the values of friction and heat transfer coefficients in the wall boundary layers (see, for instance, /6/). In a number of works /II-I4, I8,23/ it has been established that the intensity of momentum and heat transfer depends also on the external flow turbulence scale. At the same time the information on the effect of Tu_∞ on the internal structure of the wall boundary layers is more limited and concerns primarily the turbulent boundary layer /4,5,9-I2,I4-I8,20,2I,23,25,30/.

The systematic study of the wall turbulence internal structure carried out by the authors for a number of years at the Institute of Engineering Thermophysics of the Ukrainian SSR Academy of Sciences makes up this deficiency to a certain extent as will be shown below.

2. TRANSPORT PROPERTIES OF TURBULIZED FLOWS

Experimental studies of turbulized flows internal structure in wind tunnels, channels, elements of heat power equipment, etc. have shown that at close of Tu_∞ one-dimensional energy spectrum of velocity pulsation components $E_i(k)=f(k)$ or $E_i(n)=f(n)$ can differ substansionally (Fig.I).

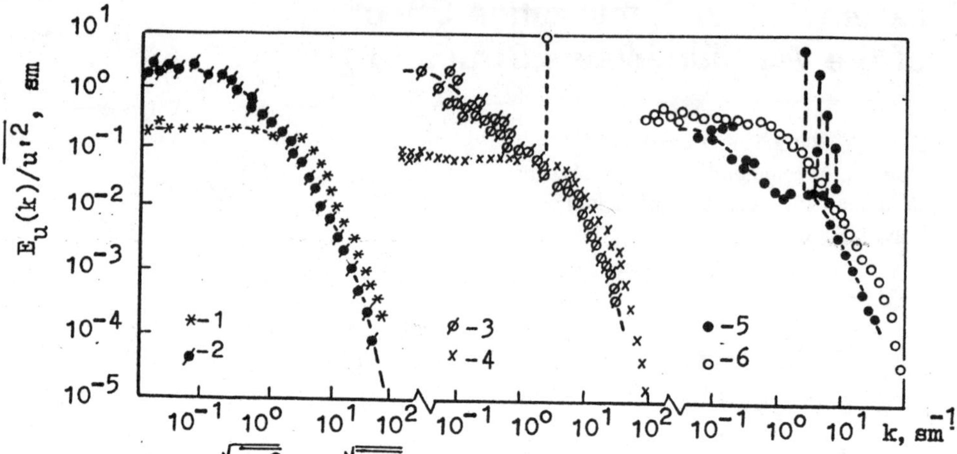

N	U ms^{-1}	$\sqrt{\overline{u'^2}}/U$ %	$\sqrt{\overline{v'^2}}/U$ %	L_{ux} sm	l_{eux} sm	λ_{ux} sm	η sm
1	30.0	6.70	6.50	0.42	0.33	0.123	0.0041
2	30.0	6.77	3.88	3.34	1.43	0.210	0.0056
3	30.6	1.38	3.49	4.40	0.40	0.324	0.0145
4	30.0	1.41	1.38	0.18	0.16	0.150	0.0097
5	23.3	9.26	9.20	0.39	0.29	0.250	0.0056
6	30.0	9.50	9.40	0.68	0.58	0.125	0.0035

FIGURE 1. Energy spectra of u'-component and turbulence scales in typical turbulized flows: 1 - past perforated plate PP-16 with contraction; 2- turbulent stabilazed flow in tube, 2y/D=0.3; 3 -before the nozzle cascade of 1-st turbine stage; 4 - past the grid M/d=2; 5 - past the rotor wheel; 6 - past perforated plate PP-12 with contraction.

In other words, at close Tu_∞ the same turbulence scales can differ from one another several times. For example, in wind tunnel at the equal Tu_∞ obtained by means of different generators, integral scales of the longitudinal velocity pulsation components can differ 7-8 times and Kolmogorov's ones - 2-2.5 times. Still larger differences can be observed between turbulized flows in real and laboratory plants: in air turbine blading and past the generator in the wind tunnel integral scales of the same pulsation component can differ almost 20 times. The flow structure in the turbine is more complex due to both the turbulence anisotropy and the superposition of the periodical velocity non-stationarity. Thus it is insufficient to characterize the external disturbed flow only by the turbulence degree which gives an idea of relative integral energy of turbulence but does not reflect its frequency distribution.

That is why in our works the disturbed external flow is called the turbulized one as the term "turbulization" as we understand it reflects two important circumstances: artificial origin of disturbations in the flow and their multiparametricity. The latter manifests itself in the presence of the effect of both the turbulence degree and its scale characteristics on transport processes.

Combination of energy and scale characteristics of the pulsation motion is known to be used in two-parametric models of turbulence for determining turbulent (often called eddy) viscosity ν_t in shear flows. If turbulence characteristics mentioned above are peculiar to turbulized flows (without transverse velocity gradient) then it can be assumed that such flows possess their own transport properties, effective viscosity $\nu_{ef\infty}$ in particular.

This gives all grounds firstly to consider the relative effective viscosity (i.e. turbulence Reynolds number $Re_{ef\infty} = \nu_{ef\infty}/\nu$) the main criterion determining the transport processes taking place in turbulized flows and secondly to use for describing their transport properties the relation of the turbulence models of Kolmogorov-Prandtl and Launder-Jones by introducing into them a correction which takes into account non-linear interaction of molecular and eddy nature effects.

This is in what an approach developed by the authors differs from that of other authors who assume that the turbulized flow transport properties do not depend on its structure, while the external turbulence effect is concentrated only within the wall boundary layer. The main disadvantage of these approaches is not only disregard of the external flow scale characteristics but also neglecting the very possibility of non-linear interaction between the external and the wall turbulences.

Assuming the values of turbulent viscosity to be independent of the method of their determination, one can show for the turbulized external flow that in the model "energy-scale" (k-L)

$$\nu_{t\infty} = C^L L_\infty \sqrt{k_\infty} = 0.55\, l_{evx\infty} \sqrt{k_\infty}. \qquad (I)$$

While determining the turbulence kinetic energy k_∞ as a sum of energies of velocity pulsation components

$$k_\infty = (\overline{u'^2_\infty} + \overline{v'^2_\infty} + \overline{w'^2_\infty})/ 2, \qquad (2)$$

the turbulence anisotropy is taken into account by choosing the characteristic scale $L_\infty = l_{evx\infty}$. The latter is the longitudinal scale of energy containing vortices of the velocity pulsation component normal to the surface, this scale being determined by the maximum one-dimensional spectral distribution $kE_v(k) = f(k)$ /5,6/. It should be mentioned that in turbulent boundary layer this scale is the analogue of Prandtl mixing length.

In isotropic or close to them turbulized flows, the scale is directly connected with the dissipative scale of u' -component

$$L^{\varepsilon}_{u\infty} = 1.5(\overline{u'^2_\infty})^{3/2} / \varepsilon_\infty ; \qquad (3)$$

$$l_{evx\infty} \cong 0.25\, L^{\varepsilon}_{u\infty},$$

which has certain advantages due to comparatively simple experimental determination.

It can be shown similarly that in the turbulence model "energy-dissipation (k-ε)

$$\nu_{t\infty} = C^{\varepsilon} k_{\infty}^2 / \varepsilon_{\infty} = 0.25(C_{\infty} \overline{v'^2})^2 / \varepsilon_{\infty} \tag{4}$$

the turbulence kinetic energy should be determined by v'-component

$$k_{\infty} = 1.5\ C_{\infty} \overline{v'^2_{\infty}}\ , \tag{5}$$

taking into account the external flow anisotropy

$$C_{\infty} = (\ 1 + \overline{u'^2_{\infty}}/\overline{v'^2_{\infty}} + \overline{w'^2_{\infty}}/\overline{v'^2_{\infty}})/3. \tag{6}$$

Dissipation of the turbulence kinetic energy can be determined both from the laws of its decay and from one-dimensional energy spectra of velocity pulsation components within the region of the "5/3-law" action

$$\varepsilon_{\infty} = -dk_{\infty}/dt \quad \text{or in common case} \quad \varepsilon = [\ E_i(k)k^{5/3}/C_i\ ]^{3/2}. \tag{7}$$

It should be remembered that for locally isotropic turbulence C_u=0.46, C_v=C_w=0.6I3.

Experimental studies of internal structure of turbulent flows have shown that C^{ε}=f(G/ε) where G is generation of turbulence kinetic energy. In turbulized flow without shear (the flows behind the grids including) when dU/dy=0 and G=0 the value C^{ε} is equal to 0.II25 rather than 0.09 as it assumed in the most works.

While determining the effective viscosity

$$\nu_{ef} = \nu_t\, f(\nu_t/\nu) + \nu\ , \tag{8}$$

the effects of non-linear interaction of molecular and molar nature can be taken into consideration by Rotta modified correction

$$f(\nu_t/\nu) = [\ 1 + 10.7/(\ \nu_t/\nu)\ ]^{-1} \tag{9}$$

which differs favourably from other similar corrections by providing the limiting transition: $f(\nu_t/\nu) \to 1$ at $\nu_t/\nu \to 0$.

3. TURBULIZED FLOW INTERNAL STRUCTURE NEAR THE STREAMLINED SURFACE

Investigation of the mechanism of interaction between the turbulized external flow with the effective viscosity peculiar to it and the wall boundary layer allowed to discover (irrespective of the layer type) the existence of the intermediate region between the turbulized flow and the layer external boundary determinated by the condition U_{δ}=0.99U_{∞}. This region formation of which is conditioned by the streamlined surface bounding the flow in the transverse direction is called here an "overlayer".

In the overlayer, attenuation of the transport properties is taking place due to which the effective viscosity at the layer external boundary $\nu_{ef\delta}$ can appear to be much less than in the external flow, i.e. $\nu_{ef\delta} \leq \nu_{ef_\infty}$.

The overlayer thickness (by analogy with the boundary layer δ) can be determined by the condition $\overline{v'^2_\delta} = 0.99\ \overline{v'^2_\infty}$ as ($\delta_v - \delta$). Decrease of transport properties in it is explained by attenuation only of v'-component (due to energy loss in the largest vortices) and of the turbulence characteristics connected with it, first of all, the scale l_{evx} . Attenuation in the overlayer $\nu_{t\infty}$ is the more intensive the higher the values of the relative scale $\bar{L} = L^\varepsilon_{u\infty}/\delta$ which specify the layer selective properties (Fig.2).

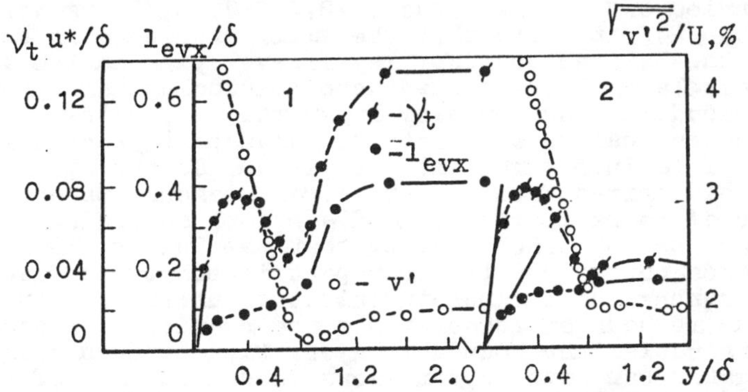

FIGURE 2. Interaction between the turbulized flow and turbulent boundary layer: 1 - $\bar{L}$ = 1.8; $Re_x = 7.6 \cdot 10^5$, Tu_∞ = 2.20%;
2 - $\bar{L}$ = 0.5; $Re_x = 4 \cdot 10^5$, Tu_∞ = 1.72%.

According to our experimental data complemented by /I8/ the following experimental dependences have been obtained for estimation of attenuation of the turbulized flow transport properties in the overlayer

$$k_v = \overline{v'^2_\delta}/\overline{v'^2_\infty} = \exp(-0.16\,\bar{L}); \tag{10}$$

$$k_\nu = \nu_{t\delta}/\nu_{t\infty} = [1 + 10(\lg\bar{L} - 0.1) + 1.5(\lg\bar{L} - 0.1)^2 + 1.25(\lg\bar{L} - 0.1)^3]^{-1} \tag{11}$$

In order to take into account the overlayer existence, i.e. decrease of effective transport coefficients in it, the approach developed presupposes to use the so called effective Reynolds number of turbulence at the layer external boundary $Re_{ef\delta} = \nu_{ef\delta}/\nu$ as the main criterion determining effect of the external flow turbulization on the wall turbulence structure and the intensity of transport processes. The turbulence scale plays two roles here: its increase provides both $\nu_{t\infty}$ growth and its attenuation in the overlayer. Combination of the two factors at the given δ determines the total effect of the turbulence scale. At the same time Tu_∞ increase almost always causes the growth of both $\nu_{ef\infty}$ and $\nu_{ef\delta}$ as in the most cases $Tu_\delta \geq Tu_\infty$ due to increase of energy of u'- and w'-pulsation components close to the external boundary of the layer inspite of energy decrease of v'-component in the overlayer. Thus in turbulized flows a substantial difference is observed in the boundary layer thicknesses determined by various parameters: the boundary layer thickness δ determined by the data average velocity is the least while that determined by the normal pulsation component δ_v is the greatest.

4. STRUCTURE OF PSEUDOLAMINAR WALL BOUNDARY LAYER

The term "wall turbulence" is most often connected with the turbulent boundary layer in which great kinetic energy of turbulence is taking place near the wall /26,28,29/. Experimental studies carried out by the authors /3,4,5-8,I6,I7/ as well as the data from /24/ indicate that the wall turbulence is also peculiar to the laminar boundary layer developing in the turbulized flow, this layer being called here "pseudolaminar". Unfortunately the pseudolaminar boundary layer was beyond the scope of the most studies dealing with both the transport processes and the problem of laminar-turbulent transition. To our opinion such situation has arized from two principal reasons. On one hand, for a number of years it was assumed a priori according to opinion /22/ supported by H.Schlichting that the Tu_∞ in flows without gradients does not affect the transport intensity and facilitates shift upstream of laminar-turbulent transition. On the other hand fine methods of measurements are required for diagnostics of the pseudolaminar boundary layer. Without such diagnostics an investigator can come to erroneous identification of this layer with the turbulent boundary one (see e.g. /I2/).

At the given $Tu_\infty > 0$ the existence of the pseudolaminar boundary layer is provided by the following conditions: the value of Re^{**} should be less than the same for the point of stability loss which at $Tu_\infty > 2\%$ coincides with the point of the beginning of laminar-turbulent transition determinated by Re_b^{**} ; the value of $Re_{ef\delta}$ should be less than the same for this latter point.

Under the influence of Tu_∞ in the laminar boundary layer, deformation of the time average velocity is taking place both near the wall and at its external boundary. The velocity profile becomes fuller, shear stresses at the wall increase and the boundary layer becomes thicker. Thus at U_∞=0.88 m/c, x_{pl} =0.34, and Tu_∞=7.79% its thickness achieves 20mm while at Tu_∞=0.67% it is I2mm. The velocity gradient at the wall grows by 45%; the form parameter $H=\delta^*/\delta^{**}$ decreases due to the growing momentum loss δ^{**} at almost constant displacement thickness δ^*.

The main peculiarity of the pseudolaminar boundary layer structure is the presence in it of turbulent pulsations penetrating through the whole its thickness (Fig.3a).

u'-component changes nonmonotonously with maximum η_{max} irrespective of Tu_∞ :

$$\eta_{max} = 3 \text{ at } Re^{**}/Re_b^{**} \leq 0.2; \qquad (I2)$$

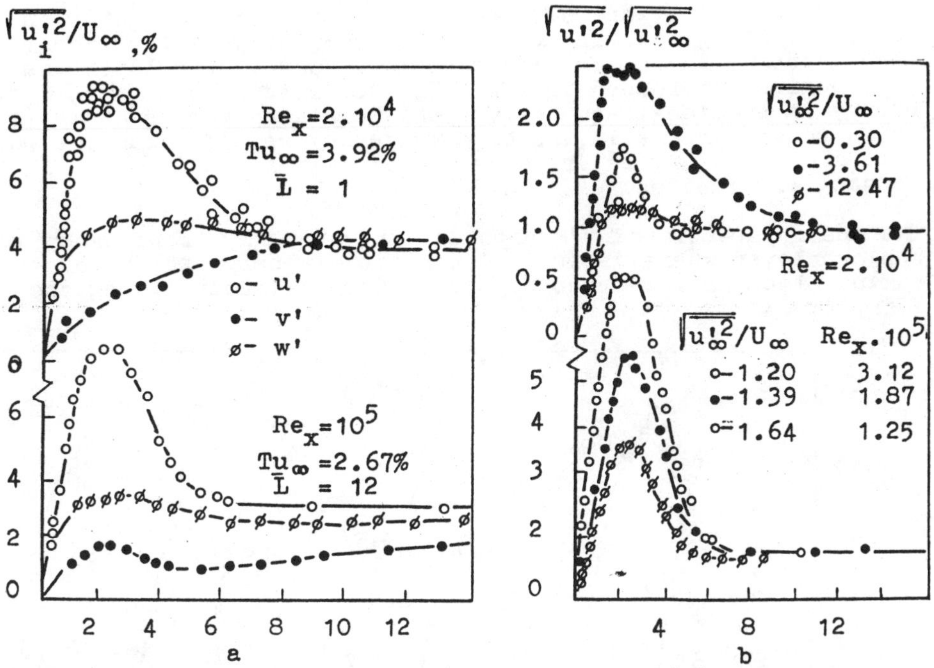

FIGURE 3. Distribution of velocity pulsations (a) and increase of longitudinal component in pseudolaminar boundary layer(b).

$$\eta_{max} = 3(1 - 6B^2 + 8B^3 - 3B^4) \text{ at } Re^{**}/Re^{**}_b > 0.2, \quad (12a)$$

where $B = (Re^{**}/Re^{**}_b - 0.2)/0.8$.

The ratio $\sqrt{\overline{u'^2_{max}}} / \sqrt{\overline{u'^2_\infty}}$ grows with Re^{**} and Re_x, i.e. at approaching the beginning of laminar-turbulent transition (Fig.3b):

$$\sqrt{\overline{u'^2_{max}}} / \sqrt{\overline{u'^2_\infty}} = 0.15(\sqrt{\overline{u'^2_\infty}} / U_\infty)^{-1} Re^{**}/Re^{**}_b \quad (13)$$

The transverse pulsation components vary in another way across the pseudolaminar boundary layer thickness. The w'-component grows slightly (up to ~15%) having its maximum at η =2.5-3. At the same time, v'-component can either decrease monotonously (at $\bar{L} \leq 1$) or be maximum as well at $L > 1$.

The profiles of all components velocity pulsation were described by empirical relations given in /6/.

The overlayer thickness for u'- and w'-components is $(0.2-0.3)\delta$ whereas for the v'-component it can be much higher and depends on the external flow turbulence scale. For instance, at $\bar{L}$ =12 it reaches 10 . Non-linear character of interaction between the pseudolaminar boundary layer and the external flow manifests itself in the overlayer which can be observed at spectral distributions of velocity pulsation components.

Spectral distributions of u'-component transform slightly in the overlayer at approaching the external boundary of the layer, i.e. they decrease in low-frequency region. Just in the layer the process begins to proceed in the opposite direction: at $\eta_\delta > \eta > \eta_{max}$ energy in the low-frequency region increases sharply at its simultaneous decrease in the high-frequency region. Thus the pseudolaminar boundary layer is a sort of filter cutting high and extending low frequencies.

The energy spectra of v'-component are strongly transformed in the overlayer where intensive decrease of energy pulsations is taking place in low-frequency and moderate-frequency regions (Fig.4).

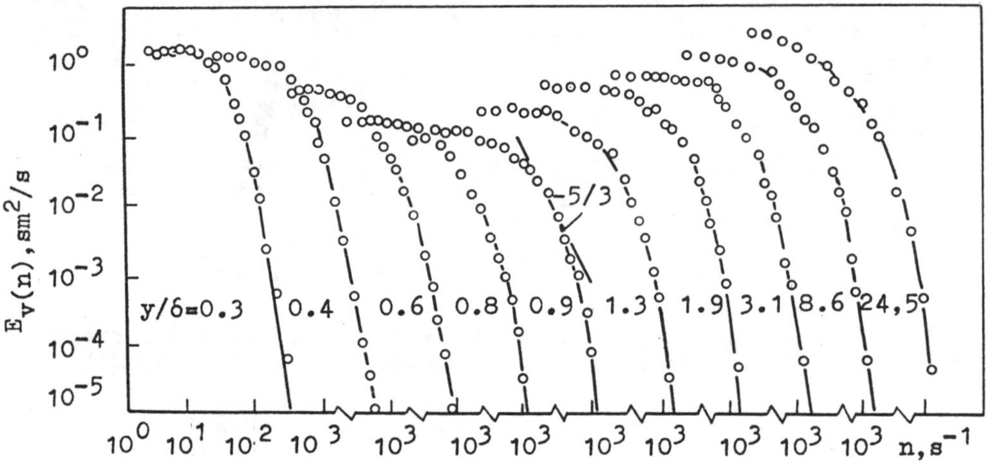

FIGURE 4. Energy spectra of v'-component in the pseudolaminar boundary layer (Re_x=7.10^4, Tu_∞=2.74%, $\bar{L}$ = 16).

In the boundary layer at $\eta_\delta > \eta > \eta_{max}$ the values of $E_v(n)$ proove to be weaker with respect to the external flow despite a certain increase of pulsation energy within low frequences which can be explained by energy transport from u'-component.

The analysis of characteristic turbulence scales in the pseudolaminar boundary layer has shown that all the scales of v'-component at $\bar{L} > I$ is subject to main changes in the overlayer. At that time the scales of u'- and w'-components (irrespective of $\bar{L}$) as well as of v'-component at $\bar{L} \leq I$ at the external boundary of the layer are close to same in the external flow and dimish only in the boundary layer. The experimental data obtained confirm that near the wall l_{eyx} = 0.4yf($Re_{ef\delta}$) in the general case; at $Re_{ef\delta} \geqslant 4$ f($Re_{ef\delta}$)=I while at $Re_{ef\delta} \leq 4$ f($Re_{ef\delta}$)=0.25$Re_{ef\delta}$.

Substantial anisotropy of turbulence within the pseudolaminar boundary layer (for example at $\bar{L}$ =I2 $\overline{u'^2}/\overline{v'^2}$=25 at η_{max} and u'^2/v'^2=3.6 at η_δ=5.6) causes nonmonotonous change of the turbulence kinetic energy and anisotropy of its dissipation. In the example cited above, $\varepsilon_u = 200\varepsilon_v = 30\,\varepsilon_w$ at η_{max}, i.e. the total dissipation of turbulence in the boundary layer is determined first of all by its longitudinal component.

Peculiarities of the pseudolaminar boundary layer described above are seen from its distribution over the effective viscosity thickness which decreases both within the boundary layer itself and in the overlayer. In large scale turbulized flows, decrease of $\nu_{t\infty}$ in the overlayer can exceed substantially decrease of Tu_∞. In some tests, $\nu_{t\infty}$ in the overlayer decreased I5 times while Tu_∞ decreased only by 7%. The experimental data obtained in the present study are generalized by the empirical dependence

$$\nu_{ef}/\nu = 1+(Re_{ef\delta}-1)\bar{y}(2-\bar{y})[1-2(1-\bar{y})^2][1-\exp(-\bar{y}/0.16)]^3, \quad (I4)$$

which is valid up to $\bar{L} \leq I6$ and $Re_{ef\delta} \leq Re_{ef\delta b}$. In the pseudolaminar boundary layer the values of ν_{ef} can be much lower than in the wall part of the turbulent boundary layer.

From the experimental data obtained for the pseudolaminar boundary layer, balances of the effective viscosity and turbulence kinetic energy were struck. The latter shows not only the turbulent energy generation in the wall part of the boundary layer but also penetration into it of a considerable energy flux from the turbulized external flow. This energy transport playes a modest but a rather sensible role in k-balance due to normal Reynolds stresses and convection. Energy transport from the u'-component to the transverse one is also observed. In the boundary layer outer part, convection plays the main role in transport, while in the wall and the middle parts it is the gradient diffusion which is of importance.

A generalized empirical dependence describing heat transfer intensification and friction in the pseudolaminar boundary layer was obtained:

$$NuC_{\gamma o}/Nu_o C_\gamma = C_f/C_{f_o} = 1+[5.5Re_{ef\delta}/(22000+Re_{ef\delta})^{1/3}](1+5.5Re^{**}/Re^{**}_b), \quad (15)$$

where C_γ is the coefficient of nonisothermity at $Tu_\infty \geq 0$. Here $Re_{ef\delta}$ takes into account not only the external flow turbulization but also non-linearity of its interaction with the boundary layer in the overlayer; Re^{**}/Re^{**}_b allows for increase of sensitivity of the boundary layer to external disturbances at approaching the beginning of the laminar-turbulent transition and contribution of normal Reynolds stresses. This dependence describes experimental data not only at the flat plate at $dU_\infty/dx=0$ but also at the curvilinear surface in the presence of the sign-alternating pressure gradient which is indicative of the physical validity and universality of the approach developed.

5. STRUCTURE OF QUASITURBULENT WALL BOUNDARY LAYER

The turbulent boundary layer developing in the turbulized flow is called here "quasiturbulent boundary layer".

Under the influence of Tu_∞ the turbulent boundary layer thickness increases and the velocity profile deforms in the same way as in the laminar flow: its fullness and gradient near the wall are growing. From presenting the velocity profile in coordinates of the "wall law" / 6,27 /

$$U/u^* = (1/k)\ln y^* + C + (\Pi/k)\omega \quad (C = 5.1) \tag{I6}$$

it is seen that the external flow turbulization at high Re** and moderate $Re_{ef\delta}$ does not affect the wall part of the boundary layer (despite u* increase) but deforms the velocity distribution in its outer part creating false impression of disappearing the wake zone. Really the wake law at the external flow turbulence growth undergoes rather weak changes which can be neglected. That is why to describe this zone, Coles formula was used. The velocity profile deformation in the boundary layer outer part is caused by sharp decrease of the profile parameter Π up to its sign alternation (Fig.5). The present study results have shown

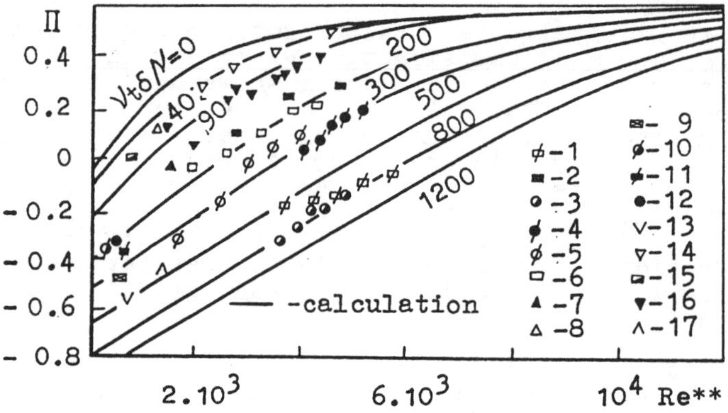

N	$\nu_{t\delta}/\nu$	Tu_∞,%	N	$\nu_{t\delta}/\nu$	Tu_∞,%
1	38-48	1.08-2.14	9	380	11.32
2	86-115	1.67-3.64	10	245	6.74
3	161-207	2.25-4.72	11	366	8.69
4	276-366	3.05-7.13	12	231	5.69
5	325-327	2.90-3.40	13	700	12.54
6	800-825	4.80-5.70	14	25	1.70
7	219-269	2.40-3.99	15	40	2.21
8	681-829	3.46-5.79	16	90	3.52
			17	344	6.85

FIGURE 5. Parameter of velocity profile Π in quasiturbulent boundary layer: 1-4 - /21/; 5,6 - /20/; 7,8 - /18/; 9-17 -/6,16,17/

that in the turbulent boundary layer at $dU_\infty/dx=0$ and $Tu_\infty > 0$ the value of Π is determined by Re ** and $Re_{ef\delta}$:

$$\Pi = 0.55 [1 - \exp(a\sqrt{Re^{**}} - bRe^{**}] \quad \text{at } \Pi \geq 0; \tag{I7}$$

$\Pi = 0.55(0.5a/\sqrt{Re^{**}_{\Pi=0}} - b)(Re^{**}_{\Pi=0} - Re^{**})$ at $\Pi \leq 0$,

where $Re^{**}_{\Pi=0} = a^2/b^2 = 100[(100+Re_{ef\delta})/(100+0.4Re_{ef\delta})]^2$; (17a)

$a = [1+3Re_{ef\delta}/(100+Re_{ef\delta})] \cdot 10^2$; $b = [1-0.6Re_{ef\delta}/(100+Re_{ef\delta})] \cdot 10^{-3}$.

The analysis of dependences (I7) and (I7a) shows that the effect of Tu_∞ on the profile parameter Π diminishes with Re** growth, stabilization of Π taking place at higher Re**, the larger $Re_{ef\delta}$.

The velocity profile deformation is the consequence of penetration of pulsations from the external flow into the boundary layer which in some cases intensify the turbulence generation.

At Tu_∞ = 1.0-I.2% increase of all the velocity pulsation components begins at y/δ=I.5. However small the values of Tu_∞ are (even at Tu_∞=0.02-0.03%) at the layer external boundary, Tu_∞ reaches 0.8-I.0%. At higher Tu_∞ in the overlayer, increase of u'- and w'-components takes place while v'-component remain almost in such as in the external flow (if $\bar{L}<I$) or decrease (at $\bar{L}>I$).

Inside the boundary layer, u'-component grows from its external boundary to the wall at least up to $\sqrt{\overline{u'^2_\infty}}/U_\infty$<I6%. The most intensive growth of u'-component is observed at $Tu_\infty \to 0$. One more distinction of u'-component distribution is its slight change near the wall. The ratio $\sqrt{\overline{u'^2_{max}}}/u^*$ = 2.6-3.2 at y^*=15 and different Tu_∞.

The overlayer thickness for u'- and w'-components at $\bar{L}$ =I.8 does not exceed 0.3δ and for v'-component it reaches 0.7δ. In this case change of $\sqrt{\overline{v'^2}}$ within the boundary layer takes place in different ways depending on $\sqrt{\overline{v'^2_\delta}}/U_\infty$: at $\sqrt{\overline{v'^2_\delta}}/U_\infty$<5.5% it grows and at $\sqrt{\overline{v'^2_\delta}}/U_\infty$>5.5% it drops towards the wall. As the values of $\sqrt{\overline{v'^2_\delta}}$ are closely connected with the external flow turbulence scale it is obvious that the latter influences on v'-component distribution within the boundary layer.

Peculiarities of distribution of pulsation components described above reflect amplitude-frequency characteristics contained in one-dimensional energy spectra. Detailed analysis of u'-component spectrum has shown, in particular, that the turbulent boundary layer amplifies the energy for all frequencies; coefficient of energy amplification grows from the external boundary towards the wall, this growth being the higher the greater the frequency of pulsations.

Decrease of v'-component energy in the overlayer is determined only by attenuation at low frequencies. In the boundary layer the range of high frequencies sharply increases and that of low frequencies is reduced at $\sqrt{\overline{v'^2_\delta}}/U_\infty$< 5.5%. Thus as to the contribution of v'-component into the total energy, the high-frequency range is more substantial than for u'-component. At the descending character of v'-component distribution at $\sqrt{\overline{v'^2_\delta}}/U_\infty$ > 5.5 % the low frequency part of the spectrum attenuates while its high-frequency part increases.

The analysis of distributions of the turbulence kinetic energy and its dissipation, $\overline{u'v'}$-correlations and coefficients connected with them, mixing length l_{mix} and its analogue l_{evx}, turbulent viscosity et. all has shown that changes in the internal structure characteristics of the quasiturbulent boundary layer are determined by ratio $\nu_{t\delta}/\nu_{to}$ where ν_{to} is the turbulent viscosity of the boundary layer outer part at $Tu_{\infty} \rightarrow 0$. The value of the latter can be determined from the known Klauser's relation.

With $\nu_{t\delta}/\nu_{to}$ increase, the mixing length in the boundary layer outer part changes substantially (Fig.6):

$$l_{mix}/\delta = l_{mix\delta}/\delta = 0.08+0.42(\nu_{t\delta}/\nu_{to})^{0.8} = 0.08+13Re_{ef\delta}^{0.8}Re_x^{-0.64}, \quad (18)$$

while in its wall part it is of the form

$$l_{mix}/\delta \doteq k(\bar{y}/\delta)f(y^*). \quad (18a)$$

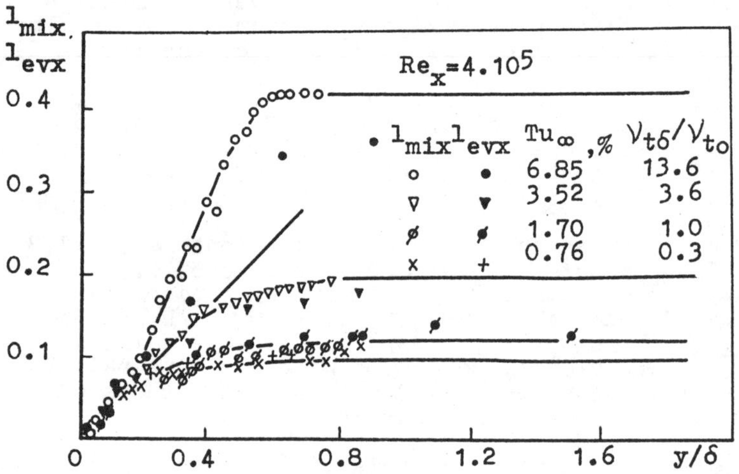

FIGURE 6. Mixing length in quasiturbulent boundary layer.

Numerical calculation results make an impression that at $Re_{ef\delta} > 250$ coefficient k slightly rises and in a first approximation

$$k = 0.4[\exp-(0.001765Re^{**} + 7.8928)](Re_{ef\delta}-250). \quad (19)$$

The turbulent viscosity in the outer part of the quasiturbulent boundary layer changes similar to the mixing length

$$\nu_t/u^*\delta = \nu_{t\delta}/u^*\delta, \quad (20)$$

and in the wall part

$$(\nu_{ef} - \nu)/u^*\delta = k(y/\delta)f(\nu_t/\nu). \quad (20a)$$

The analysis of our experimental data on intensification of heat transfer and friction in the turbulent boundary layer on the plate as well as similar data of other authors has shown that the most of measurements were carried out within the range of $Re^{**}=1500-6000$ and $Re_x=4.10^5-3.10^6$. In this case intensification of transport processes is comparatively small (i.e. does not exceed 25% at Tu_∞ up to 7% and $Re_{ef\delta}$ up to 300). That is why some authors could use dependences of the type C_f/C_{f_o} or $Nu/Nu_o=f(Tu_\infty)$ to generalize the experimental data. Thus in our previous works an empirical dependence was proposed

$$C_f/C_{f_o} = 1 + 3Tu_\infty - 6Tu_\infty^2 , \tag{21}$$

which was valid for slightly anisotropic flows at $\bar{L} < 1.3$

Empirical dependences (17), (17a) and (19) describing the effect of the external flow turbulization on the velocity profile parameter Π and k (together with equation (16)) allowed to calculate by an integral-numerical method the quasiturbulent boundary layer. These calculation results within the range $200 \leq Re^{**} \leq 2.10^5$, $6.10^4 \leq Re_x \leq 2.10^8$ and $Re_{ef\delta} \leq 1000$ are approximated by relations

$$C_{fo}=0.027[1+0.05(\lg Re^{**}-3.3)+0.1(\lg Re^{**}-3.3)^2]Re^{**-0.268}, \tag{22}$$

$$C_f/C_{f_o}= 1 + 0.0076Re_{ef\delta}(\lg Re^{**}-0.237\lg^2 Re^{**}). \tag{23}$$

Comparison of calculation and experiments made it possible to clarify the dependence of the turbulent boundary layer characteristic thicknesses and its form parameter H on $Re_{ef\delta}$ and Re_x. It appeared in particular that increase of all the characteristic thicknesses of the layer progresses with decrease of Re_x. The growth of δ (for instance, for $Re_x=10^5$ up to 2 and 5.6 times at $Re_{ef\delta}=300$ and 800) is the most remarkable whereas δ^* and δ^{**} under the same conditions increase not more than by 12-35%. Changes of formparameter H at Π =var conditioned by both the high $Re_{ef\delta}$ and the low Re^{**} (or Re_x) is occuring nonmonotonously; H changes only by 17% (from 1.414 to 1.208) within the range $6\cdot10^4 \leq Re_x \leq 10^8$ and $Re_{ef\delta} \leq 800$.

On the basis of the experimental data available, at Re_x=const the following empirical dependences are obtained which describe intensification of heat transfer and friction in turbulized flows:

$$Nu/Nu_o=1+0.05(\nu_{t\delta}/\nu_{to})^{0.5}= 1 + 1.65Re_{ef\delta}^{0.5}Re_x^{-0.4}; \tag{24}$$

$$C_f/C_{f_o}=1+0.037(\nu_{t\delta}/\nu_{to})^{0.5}=1 + 1.34Re_{ef\delta}^{0.5}Re_x^{-0.4} . \tag{24a}$$

From the analysis of dependences (24) and (24a) it follows that

$$Nu/Nu_o=(C_f/C_{f_o})f(Pr_{t\delta})=(C_f/C_{f_o})(1 + 5.10^{-3}Re_{ef\delta}^{0.4}). \tag{25}$$

For more strict solution of the problem of observing Reynolds analogy in turbulized flows, special study is necessary to establish the relation $Pr_{t\delta}=f(Re_{ef\delta})$ the presence of which is obvious in the light of the data obtained.

6. TURBULENCE STRUCTURES TAKEN INTO ACCOUNT IN CALCULATING BOUNDARY LAYERS OF TURBULIZED FLOWS

Experimental data considered above show the presence of external disturbance relaxation within the wall boundary layer which is the more pronounced in the transverse rather than in the longitudinal direction. Relaxation effects become apparent, in particular, in substantial changes of the turbulence structure in the boundary layer outer part at conservation in its inner region of the "wall laws" not only for time-average velocity and temperature profiles but also for such pulsation characteristics as the turbulence kinetic energy, its dissipation, one-point correlations, etc.

A typical feature of the wall boundary layers of turbulized flows is disagreement in them of coordinate points at which dU/dy and turbulent shear stresses $\overline{u'v'}$ equal zero. This imposes certain limitations on application for boundary layer calculation of Boussinesq hypothesis which is directly or indirectly used in the most semiempirical models of turbulence.

While developing the computational methods these circumstances can be allowed for in the following way:

- use of the equation of motion in integral methods

$$\frac{d\delta^{**}}{dx} + (n+2)\frac{\delta^{**}}{U_\infty}\frac{dU_\infty}{dx} = \frac{C_f}{2}\left(1+\frac{|\overline{u'v'}|_\delta}{u^{*2}}\right) + \frac{1}{U_\infty^2}\frac{d}{dx}\int_0^\infty (\overline{u'^2}-\overline{v'^2})dy, \qquad (26)$$

where $\overline{u'v'}_\delta$ are turbulent shear stresses at the external boundary of the boundary layer over the mean velocity;

- use of relaxation type turbulence models at solving the system of differential equations

$$|\overline{u'v'}| = \nu_t\frac{dU}{dy} + L_y\frac{d|\overline{u'v'}|}{dy} + L_x\frac{d\,\overline{u'v'}}{dx}\ , \qquad (27)$$

where L_x and L_y are longitudinal and transverse scales of turbulence respectively.

The first approach was used in /I9/ were function $|\overline{u'v'}_\delta|\ /u^{*2} = f(Tu_\infty)$ was determined from the experimental data neglecting the contribution of normal Reynolds stresses.

The second approach was applied in /I,2/ at numerical modelling of dynamic and thermal boundary layers in turbulized flows.

In using relation (27), the main difficulties are connected with determination of numerical values of scales L_x and L_y. From the experimental data available it follows that in turbulized flows the value of L_y depends on Tu_∞ the law of its decay, characteristic internal scales, local characteristics of the wall boundary layer, etc. As it follows from variant calculation of boundary layers studied experimentally, that for most turbulized flows in wind tunnels $L_x=0$ and $L_y=l_{evx}$ can be assumed. In a first approximation, the value of L_y within the boundary layer can be determined from the dependence

$$L_y = l_{evx} = l_{evx\delta}[1 - \exp(-y/\delta)^3], \qquad (28)$$

where according to the present experimental data $l_{evx\delta}$ is calculated by dependence (I8) for the quasiturbulent boundary layer.

Taking into account the relaxation effects, the results of modeling of dynamic and thermal boundary layers on the plate in turbulized flow agree well with the experimental data. This is valid not only for coefficients of friction and heat transfer (which can be obtained also at using Boussinesq hypothesis and semiempiric turbulence models) but also for some other important characteristics of the wall turbulence internal structure.

Thus as it is seen from Fig.7 calculating and experimental velocity and temperature profiles agree not only in the wall but also in the outer part of the boundary layer which is the most important. It should be emphasized that as it follows from variant calculation results, such an agreement cannot be obtained within the limits of Boussinesq hypothesis and its modifications by means of changes in distribution of effective viscosity and turbulent Prandtl number across the boundary layer thickness. Boundary layer thickness distributions of $\overline{u'v'}$-correlations are indicative of effectiveness of using the relaxation dependence (27) at investigating the quasiturbulent boundary layer. Calcula-

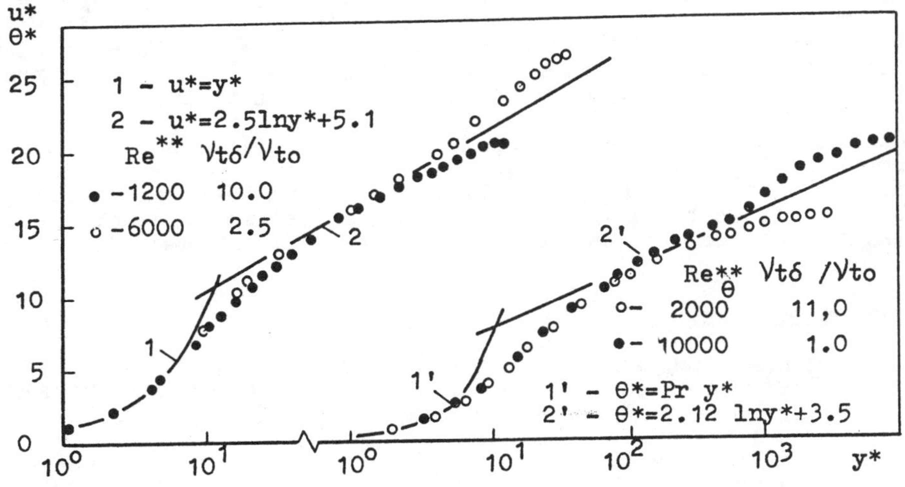

FIGURE 7. Calculation of velocity and temperature profiles in quasiturbulent boundary layer.

tion results are reflected the tendency of $|\overline{u'v'}_\delta|$ growth at increase of the external flow turbulization. The boundary layer thickness determined by $\overline{u'v'}$-correlation exceed δ ; thus at $\nu_{t\delta}/\nu_{to}= 5$ their ratio is 2.

7. CONCLUSIONS

Extensive experimental study of the external flow turbulization effect on the structure of the wall boundary layers (pseudolaminar and quasiturbulent ones) has been carried out within the range of $Re_x=6\cdot10^3-3\cdot10^6$, $Tu_\infty=0.02-15\%$, $Re_{ef\delta}=1-400$, $\bar{L}=0.5-16$.

An approach has be worked out in which the turbulized external flow is considered to be characterized by transport properties inherent to its internal structure, the effective viscosity in particular. Determination of the latter is possible on the basis of modified two-parametric models of turbulence "energy-scale" and "energy-dissipation" in view of turbulence of the external flow and the effects of interaction of molar and molecular nature.

An overlayer existence has been discovered, i.e. the region of interaction between the wall boundary layer and the turbulized flow. The external flow transport properties are attenuated in the overlayer due to decrease of the normal pulsation component in the low frequency range. For this reason the dimensionless effective viscosity $Re_{ef\delta}= \nu_{ef\delta}/\nu$ at the external boundary of the layer is considered to be the main criterian determining the transport processes taking place in the wall boundary layers of the turbulized flows. As additional criteria it is advisable to use the following ratios: for pseudolaminar boundary layer - Re^{**}/Re^{**}_b characterizing the increase in the layer sensitivity to external disturbances with approaching the laminar-turbulent transition; for quasiturbulent boundary layer - ν_{to}/ν characterizing the turbulence viscosity generation at $Tu_\infty\to0$.

On the basis of integral and local characteristics of the wall boundary layer empirical dependences are obtained which allow not only to calculate directly friction and heat transfer in turbulized flows but also to realize numerical solution of boundary layer equations in integral and differential forms.

REFERENCES

1. Дыбан Е.П.,Фридман Э.А., Релаксационная модель переноса теплоты и импульса в пристенных пограничных слоях турбулизированных потоков, Промышленная теплотехника, т.9,№4, с.16-27,1987.

2. Дыбан Е.П.,Фридман Э.А., Численное моделирование динамического пограничного слоя на пластине, обтекаемой турбулизированным потоком, Промышленная теплотехника, т.9,№5,с.6-11,1987.

3. Дыбан Е.П.,Эпик Э.Я.,Теплоперенос в ламинарном пограничном слое при повышенной турбулентности внешнего потока, Тепло-и массоперенос, т.1,ч.2,с.187-191,ИТМО АН БССР,Минск,1972.

4. Дыбан Е.П.,Эпик Э.Я., Микроструктура пограничных слоев и процессы переноса в них при повышенной степени турбулентности внешнего потока, Пристенное турбулентное течение (Труды ХУШ Сибирского теплофизического семинара), ч.П,с.30-42, ИТФ СО АН СССР, г.Новосибирск, I975.

5. Дыбан Е.П.,Эпик Э.Я., Определение длины пути смешения при турбулентном течении в трубах и пограничных слоях по измеряемым в опытах характеристиками турбулентности, Теплофизика и теплотехника, вып.34, с.I8-23, I978.

6. Дыбан Е.П., Эпик Э.Я., Тепломассообмен и гидродинамика турбулизированных потоков, с.1-296, Наукова думка , Киев, I985.

7. Дыбан Е.П., Эпик Э.Я., Супрун Т.Т., Характеристики ламинарного пограничного слоя при повышенной турбулентности внешнего потока, Теплофизика и теплотехника, вып.30,с.86-90,I976.

8. Косорыгин В.С.,Поляков Н.Ф., Супрун Т.Т., Эпик Э.Я.,Развитие возмущений в ламинарном пограничном слое пластины при повышенной турбулентности внешнего потока, Неустойчивость до-и сверхзвуковых течений, с.85-92, ИТПМ СО АН СССР,Новосибирск, I982.

9. Пядишус А.А.,Кажимекас П.-В.А.,Шланчяускас А.А., Теплоотдача пластины потоку воздуха повышенной турбулентности, Труды АН Лит.ССР, сер.Б, т.5(I08),с.9I-99, I978.

I0.Эпик Э.Я.,Румянцева Л.А., Характеристики турбулентного пограничного слоя при повышенной турбулентности внешнего потока, Теплофизика и теплотехника, вып.33,с.5I-53,I977.

II.Эпик Э.Я., Влияние масштаба турбулентности внешнего потока на теплообмен и трение в турбулентном пограничном слое, Тепломассообмен-УП, т.I,ч.2,с.I69-I73, ИТМО АН БССР, Минск,I984.

I2. Blair M.F., Influence of free-stream turbulence on turbulent boundary layer heat transfer and mean profile development, parts 1 and 2, Trans. ASME, J.Heat Transfer,vol.105,no.1, pp.33-47,1983.

13.Bradshaw P., Effect of free-stream turbulence on turbulent shear layers, I.C.Aero.Report, no 74-10,pp.1-13, 1974.

14.Castro J.P.,Effects of free-stream turbulence on low Reynolds number boundary layers, Trans. ASME, J.Fluids Eng., vol.106, no.3.pp.298-306,1984.

15.Charney G.,Comte-Bellot G.,Mathie J., Development of a turbulent boundary layer on a flat plate in an external turbulent flow, AGARD Conf.Proc.,vol.93,paper no 27,pp.1-10,1971.

16.Dyban E.P.,Epik E.Ya.,Heat transfer in a boundary layer in turbulized flow, Heat Transfer 1978, vol.2,FC(a)-4, pp.507-512, Toronto,1978.

17.Dyban E.P. Epik E.Ya.,Generalization of results of experimental and numerical studies on heat transfer in flows with in -

creased external turbulence on the bases of two-parameter models of turbulence, Heat Transfer 1982, vol.2,FC 43,pp.241-246, Munchen, 1982.

18.Hancock P.E.,Bradshaw P.,The effect of free-stream turbulence on turbulent boundary layers, Trans. ASME, J.Fluids Eng.,vol. 105, no 3, pp.284-289, 1983.

19.Horlock G.H.,Evans R.L., The momentum-integral equation for a boundary layer with turbulence or ordered unsteadiness in the free stream, Trans.ASME, J.Fluids Eng., vol.197,no1,pp.126-129, 1975.

20.Huffman G.D.,Zimmerman D.R., Bennet W.A., The effect of free-stream turbulence level on turbulent boundary layer behaviour, AGARDograph, vol.164,pp.91-115, 1972.

21. Jonas P.,Nguen tat Dat, Profily stredni rychlosti v turbulentni mezni vrstve na desce v proudu za rovinnymi mrizemi se ctvercovymo oky, Preprint Ustav Termomechaniky CSAV,no Z-603/78,pp.1-43, Praga,1978.

22. Kestin J., The effect of free-stream turbulence on heat transfer rates, Advances in Heat Transfer, vol.3,pp.1-32,Acad. Press, New-York-London,1966.

23. Simonich J.C.,Bradshaw P., Effect of free-stream turbulence on heat transfer through a turbulent boundary layer, Trans. ASME,J.Heat Transfer, vol.100,no 4,pp.671-677, 1978.

24. Smith M.C.,Kuethe A.M., Effects of turbulence on laminar skin friction and heat transfer, Physics Fluids, vol.9,no 12, pp.2337-2344, 1966.

25. Tsuji,Y.,Iida S., Influence of free-stream turbulence on mean velocity of turbulent boundary layer without pressure gradient, Trans. Jap.Soc. Aeronaut.Space Sci.,vol.15, no 29,pp.105-116, 1972.

26. Ueda H.,Hinze J.O., Fine-structure turbulence in wall region of a turbulent boundary layer, J.Fluid Mech.,vol.67,pt 1, pp.125-143,1973.

27. Yaglom A.M., Similarity laws for constant-pressure and pressure gradient turbulent wall flows, Ann.Rev.Fluid Mech.,vol.11, pp.505-540, 1979.

28. Zaric Z., Wall turbulence structure, Advances in Heat Transfer,vol.8,pp.285-350,Acad.Press,New-York-London,1972.

29. Zaric Z.,Wall turbulence structure and convection heat transfer, Int. J. Heat Mass Transfer, vol.18,no 6,pp.831-842,1975.

30. Zhubrin S.V.,Lukoshiavichius L.K.,Motulevich V.P.,Sergievsky E.D.,Heat Transfer and shear stress in boundary layer in turbulized flow, Heat Transfer 1982, vol.1, FC 56,pp.319-324, Munchen,1982.

Effects of Surface Curvature on the Law of the Wall

M. M. GIBSON
Mechanical Engineering Department
Imperial College of Science and Technology
Exhibition Road
London, SW7 2BX, UK

1. INTRODUCTION

Shear layer turbulence is highly sensitive to the effects of extra strain associated with curvature of the streamlines. In boundary layers the fractional changes in skin-friction and heat-transfer coefficients may be an order of magnitude greater than the ratio of boundary-layer thickness to the radius of curvature of the wall. They are reduced when the angular momentum of the mean flow increases in the direction of the radius of curvature, as in the boundary layer on a convex wall, and increased in concave wall flow. The present paper is concerned primarily with the effects of curvature on the turbulence in convex wall boundary layers.

The transport equations for the turbulent stresses and temperature fluxes written in curvilinear coordinates contain terms in an extra strain rate U/r which affect the rates of production. As the strain-rate ratio $S = (U/r)/(\partial U/\partial r)$ increases with distance from the wall so do the direct effects of curvature. The shear stress profiles in Figure 1, collected from various sources by Gibson et al (1984), show that these effects can be large enough to reverse the direction of the shear stress in the outer half of a convex boundary layer. The effects on the turbulence in the inner layer, $y^+ < 300$ say, are not negligible. Although S ($= Ku^+y^+ / R^+$) may be relatively small, changes in outer flow conditions are transmitted to the wall where they appear as changes in the wall shear. It has, however, been usual to assume that the law of the wall,

$$u^+ = \frac{U}{u_*} = \frac{1}{K} \log_e y^+ + C \qquad (1)$$

is unaffected, and that the skin friction coefficient may be obtained by the Clauser (1954) method of fitting (1) to the measured mean velocity profiles.

The purpose of the present study is to examine this assumption on the light of measurements made in heated convex wall layers, where the analogous process of fitting

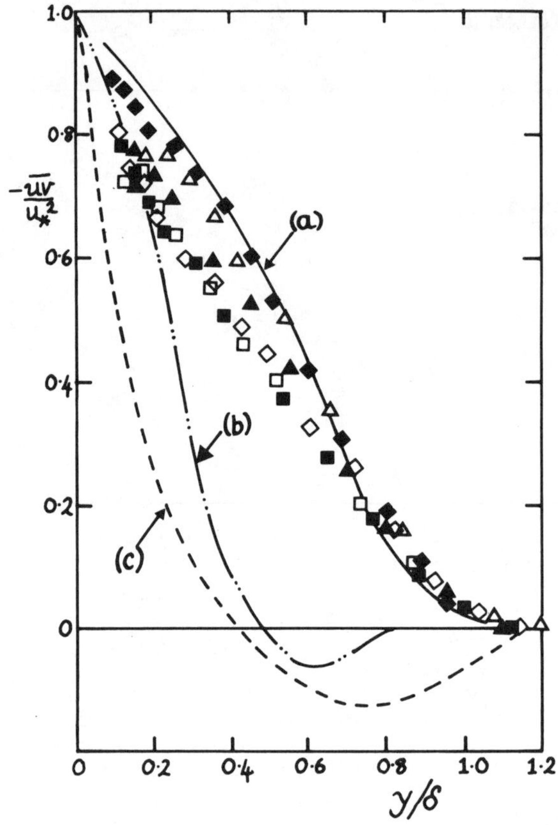

FIGURE 1. Shear stress profiles from flat and convex wall boundary layers. (a) Klebanoff (1955), flat; (b) So and Mellor (1973), convex; (c) Gillis and Johnston (1980) / Simon and Moffat (1979), convex. Data points from Gibson et al (1984), slight convex.

temperature profiles to a temperature law of the wall provides an alternative method for the skin friction, and deviations from both wall laws can also be inferred. The results are not conclusive (because the data are relatively few) but they are sufficient to raise doubts about the validity of the Clauser method for curved flows, and the use of unmodified wall functions in calculation methods.

In each of the two wind-tunnel experiments to be considered, a turbulent boundary layer developed first on a flat heated plate to a thickness δ_0, before being turned on a heated convex wall. In the flow described by Simon and Moffat (1979) (=SM) the bend radius was 45 cm, the turning angle was ninety degrees and, as shown in Figure 1, the curvature was sufficiently strong and prolonged to reverse the shear stress in outer half of the flow. At exit from the bend the distorted boundary layer recovered to flat-wall conditions on a heated plate downstream. The flow was a replica of one in which Gillis and Johnston (1980) had previously obtained velocity data. The data used in the present study are from the "baseline" experiment documented in the Stanford University report by Simon et al (1980).

Figure 2 shows the longitudinal variation of skin-friction coefficient and Stanton number in the SM flow, for which δ_0/R was 0.095. The skin-friction coefficient, c_f, was obtained by

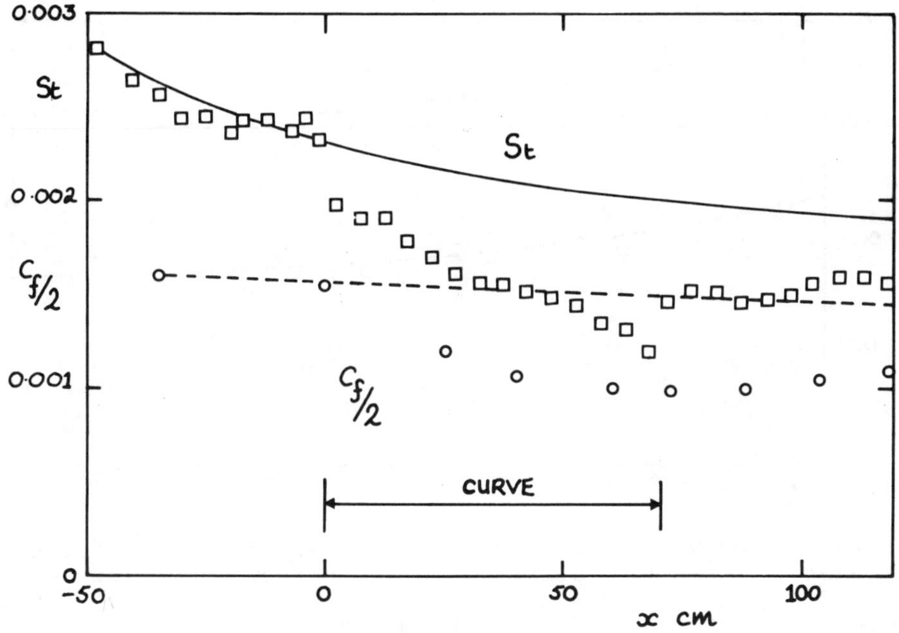

FIGURE 2. Streamwise variation of skin-friction coefficient and Stanton number in the highly curved convex wall flow of Simon and Moffat (1979).

fitting velocity profiles to (1), and the Stanton number, St, by direct measurement of the surface heat flux. At the end of the bend c_f had apparently fallen by 33% of the estimated equivalent flat-plate value. The corresponding fall in St was approximately 40%. Simon et al (1980) also note, significantly, that the mean temperature profiles in the curved and recovery flow could not be fitted to an assumed temperature law of the wall without an adjustment of the constants which is equivalent to an increase in the turbulent Prandtl number in the inner layer.

Figure 3 shows the effects of slighter convex curvature in the heated boundary layer investigated by Gibson et al (1982) (=GVN). The shear stress profiles are represented by the data points in Figure 1. In this experiment the flow was turned through 24 deg on a heated plate of 2.44m radius, with δ_0/R = 0.08. The decrease in c_f and St was proportionately much greater than in the SM experiment: to 90% and 82% of the equivalent flat-plate values respectively, when c_f was evaluated by the wall-law method. These results led the authors to the arguable conclusion that heat transfer might be in some way more sensitive than momentum transfer to the effects of curvature. Also shown in Figure 3 are the results of hot-film measurements of shear stress by Okamoto and Gibson (1986) in a replica of the GVN tunnel. These are some 4% lower than the wall-law values in the plane approach flow, but up to 11% lower on the convex wall. As a consequence of

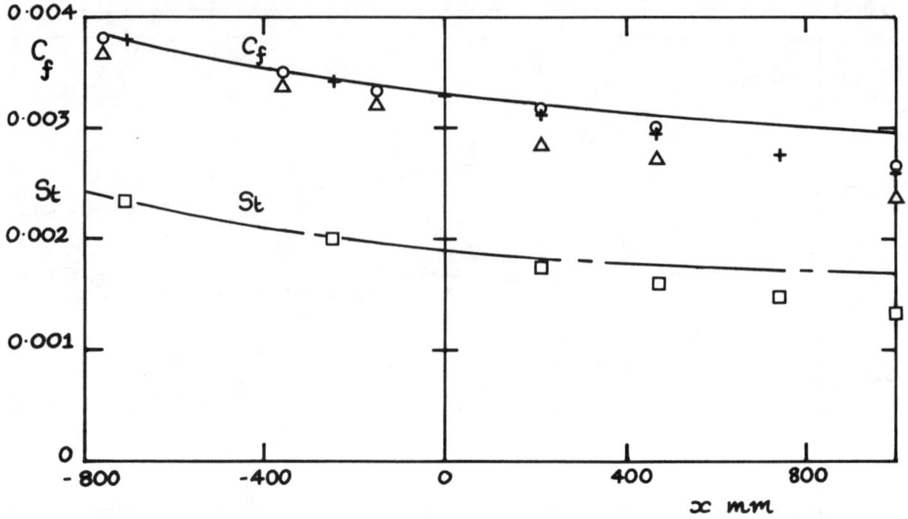

FIGURE 3. Streamwise variation of skin-friction coefficient and Stanton number in slightly curved convex wall flow. +, O, Clauser chart c_f values; △ c_f by hot-film gauge. □ Stanton number. From Okamoto and Gibson (1986).

these changes, which are also supported by the results of Okamoto's sublayer probe measurements, the reduction in c_f is the same as the reduction in St.

Typical velocity and temperature profiles from the two experiments are plotted in inner-layer coordinates in Figure 4. The velocity profiles fit the law of the wall (1) because they have been forced to do so in order to determine c_f and u_*. The temperature profiles are not, however, fitted by

$$T^+ = \frac{u_*(T - T_W)}{-q_W} = \frac{1}{K_T}\log_e y^+ + C_T \qquad (2)$$

when the same constants K_T and C_T are used for flat and curved flow. The constants used to plot (1) and (2) in Figure 4 are those used by GVN in their original paper $(K, C, K_T, C_T) = (0.41, 5.0, 0.482, 3.8)$, taken from Coles and Hirst (1968) and Yaglom (1979). The slightly different values used by SM do not materially affect the result. Of the two flat wall profiles in Figure 4, the GVN data are fitted by (2) but the SM results fall slightly below the line and show only a small logarithmic region because, at this point, the temperature boundary layer extended only to 0.55δ.

If the laws of the wall are assumed to hold, u_* or q_W can be determined by fitting measured temperature profiles to (2). When q_W is independently measured, (2) then becomes an alternative equation for u_*. Applied to the convex wall data this method produces c_f values some 12% lower than those

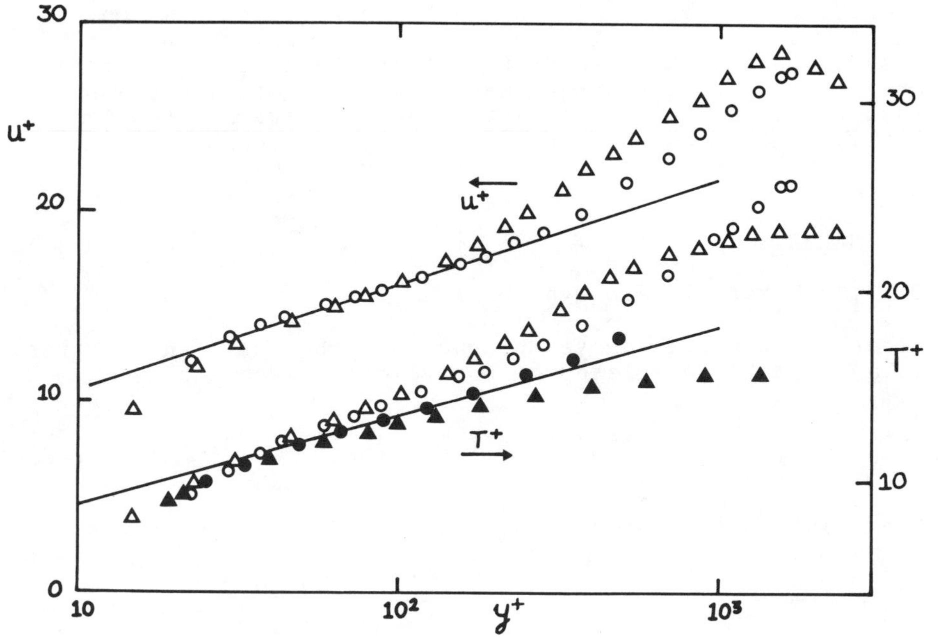

FIGURE 4. Velocity and temperature profiles in inner-layer coordinates. O GVN curved wall station 4, ● flat; △ SM curved wall station 10, ▲ flat. Calculated profiles from (1) and (2) with $(K, C, K_T, C_T) = (0.41, 5.0, 0.482, 3.8)$.

obtained from (1). A disparity of this magnitude is not likely to have been the result of measurement errors repeated in both experiments. A more probable explanation is that neither of the wall laws fit the data with the usual constants, and that the true skin-friction coefficient lies somewhere between the extremes. It will be shown that the inner-layer velocity and temperature profiles are correlated by (1) and (2) when the shear velocity is correctly chosen and when the constants C and C_T are adjusted from flat-wall values in a logical way.

The method is based on Huffman and Bradshaw's (1972) mixing-length analysis of low-Reynolds-number turbulent boundary layers. These authors adjusted the constants in the van Driest (1956) formula

$$l^+ = Ky^+[1 - \exp(-y^+/A^+)] \tag{3}$$

so as to fit the law of the wall (1) to measured velocity profiles in the inner layer. It was argued that, in the low-Reynolds-number flows considered, the outer layer and outer boundary conditions affect the inner layer primarily via the shear-stress gradient in the inner layer. The effect on the velocity distribution could be adequately accounted for by a change in A^+ in (3), or C in (1), rather than a change in K, and large negative shear-stress gradients were associated with high values of A^+

There are some similarities in the flow close to a convex wall, where the direct effects of the extra strain rate exerted in the outer layer result in a negative gradient of shear stress in the inner layer. The data of Figure 1 show that the stress gradients in curved flow are relatively small, so that changes in A^+ indicated by the Huffmann-Bradshaw correlation are also relatively small.

2. ANALYSIS.

2.1 The Mixing-Length Hypothesis.

For flow on a wall curved to radius R the total shear stress is assumed to be related to the velocity gradient by

$$\tau^+ = \tau/\tau_W = [1 + l^{+2}\, \partial u^+/\partial y^+][\partial u^+/\partial y^+ - u^+/(R^+ + y^+)] \qquad (4)$$

The solution for the dimensionless velocity gradient is

$$\frac{\partial u^+}{\partial y^+} = \frac{1 - l^{+2}\, u^+/r^+}{2l^{+2}} \left\{ \left[1 + 4l^{+2}\, \frac{\tau^+ + u^+/r^+}{1 - l^{+2}\, u^+/r^+} \right]^{0.5} - 1 \right\} \qquad (5)$$

which, when $\tau^+(y^+)$ is known, and $l^+(y^+)$ is specified from (3), may be integrated numerically to obtain the velocity distribution. For a constant pressure layer on a flat wall, $\tau^+ = 1$, R^+ is infinitely large, and the logarithmic law (1) is recovered for large y^+ as l tends to Ky. Laboratory boundary layers with moderate stress gradients have well-defined logarithmic velocity profiles that can be calculated to good accuracy with this constant-stress assumption and with l = Ky. For example, the stress gradient shown in Figure 1 for the flat-plate boundary layer of Klebanoff (1955) is $\partial\tau^+/\partial y^+ = -0.00025$ for $Re_2 = 7900$; in the GVN plane approach flow (slight favourable pressure gradient) $\partial\tau^+/\partial y^+ = -0.001$ for $Re_2 = 2750$. Compare these values with -0.003 and -0.0015 in the inner layers of the curved SM and GVN flows respectively. These figures are seen in perspective when they are compared with a pipe-flow stress gradient of -0.0032 for a Reynolds number of 10^4. For the two curved flows to be considered it appears to be permissible to assume $\tau^+ = 1$.

Just as the actual mixing-length distribution in a pipe deviates from Ky, the fall in τ with distance from a curved wall is accounted for by Bradshaw's (1973) empirical modification for curved flow

$$l = l_0/(1 + \beta S) \qquad (6)$$

where l_0 is the plane-flow value from (3) and β is a constant of order 10. The effect of this modification on the law of the wall is significant only for highly curved flows. Even in these cases there remains a region close to the wall where the deviation from (1) is small: in the SM boundary layer the difference is about 3% at $y^+ = 100$. In the first instance, at any rate, it will suffice to ignore (6) and match the measured profiles to the logarithmic part of the curve.

The total temperature flux is related to the temperature gradient by

$$q^+ = \frac{q}{q_W} = \left[\frac{1}{Pr} + l_T^+ l^+ \frac{\partial u^+}{\partial y^+}\right] \frac{\partial T^+}{\partial y^+} \tag{7}$$

where Pr is the fluid Prandtl number and l_T is the temperature mixing length given by

$$l_T^+ = K_T y^+[1 - \exp(-y^+/B^+)] \tag{8}$$

The dimensionless temperature distribution $T^+(y^+)$ is calculated by numerical integration of equation (7) with $q^+ = \tau^+ = 1$.

2.2 Values of the Constants for a Flat-Plate Boundary Layer

Studies by Pierce and Gold (1977) and de Brederode and Bradshaw (1974) show that Coles and Hirst's (1968) wall-law constants, K = 0.41, C = 5.0, imply c_f values that are about 2-3% too high, and that a better fit to flat-plate data is obtained when C is increased to about 5.2. The results of integrating (5) numerically with K = 0.41 are fitted by

$$C = 0.203A^+ \tag{9}$$

so that C = 5.2 is obtained with A^+ = 25.6. This is rounded up to the usual value of 26, to be equal to B^+, and to give C = 5.28.

Reference	K	K_T	K/K_T	B^+	C_T
Yaglom (1979)	0.41	0.482	0.85	30.4	3.8
Simon et al (1980)	0.41	0.57	0.72	36.8	5.0
Cebeci (1973) Re_2=5000	0.414	0.459	0.90	36.7	4.38
$Re_2=\infty$	0.40	0.44	0.91	35.0	4.04
Jayatillaka (1969)	0.41	0.45	0.91	27.0	3.12
Present paper	0.41	0.45	0.91	26.0	3.0

TABLE 1. Constants in the Temperature Law of the Wall from Different Sources.

Table 1 gives a selection of the values of K_T and C_T suggested by various authors for boundary layers on flat surfaces. The quality of the fit when the Yaglom (1979) constants are used in (5) is shown in Figure 4 for the GVN and SM flat-wall temperature profiles. Simon and Moffat adjusted K_T to fit their flat-wall measurements to the Kays and Crawford (1980) formula in which the ratio K/K_T (the

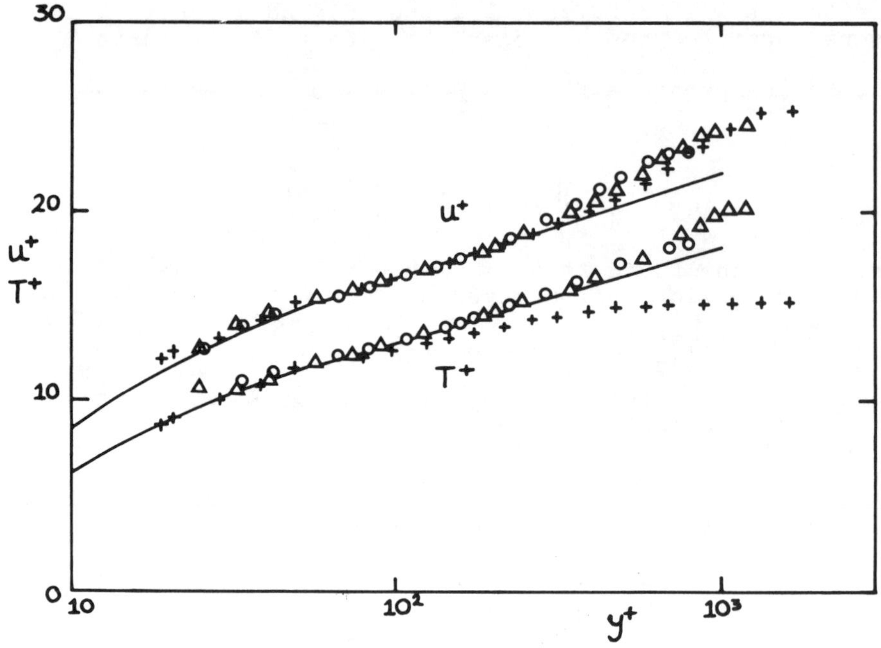

FIGURE 5. Flat wall profiles in inner-layer coordinates. O, △ GVN; + SM. Velocity and temperature wall laws are plotted for $K = 0.41$, $K_T = 0.45$, $A^+ = B^+ = 26$.

turbulent Prandtl number) appears. The value $K/K_T = 0.72$ is the lowest in the table. Cebeci expresses K, K_T, A^+ and B^+ as functions of the momentum-thickness Reynolds number, Re_2. In each of these cases the damping constant B^+ is appreciably greater than A^+. One merit of the Jayatillaka (1969) constants is that they are consistent with values of A^+ and B^+ that are nearly equal and, moreover, they produce a temperature distribution from (2) that differs by less than ±1% from the distribution calculated with the Yaglom constants (and hence with the data) in the region of greatest interest, $50 < y^+ < 300$. The Jayatillaka values in Table 1 have been obtained by setting $K_T = 0.45$ so that $K/K_T = 0.91$. C_T is then evaluated as 3.12 from

$$T^+ = (u^+ + P)K/K_T \tag{10}$$

in which P is the sublayer resistance factor expressed by Jayatillaka as

$$P = 11.52\,[(Pr.K_T/K)^{3/4} - 1] \tag{11}$$

The quality of the data fit is not perceptibly altered when $B^+ = A^+ = 26$ and, since there appears to be no reason why the damping constants should be not be equal, and it is convenient when they are, this equality will be assumed in the analysis.

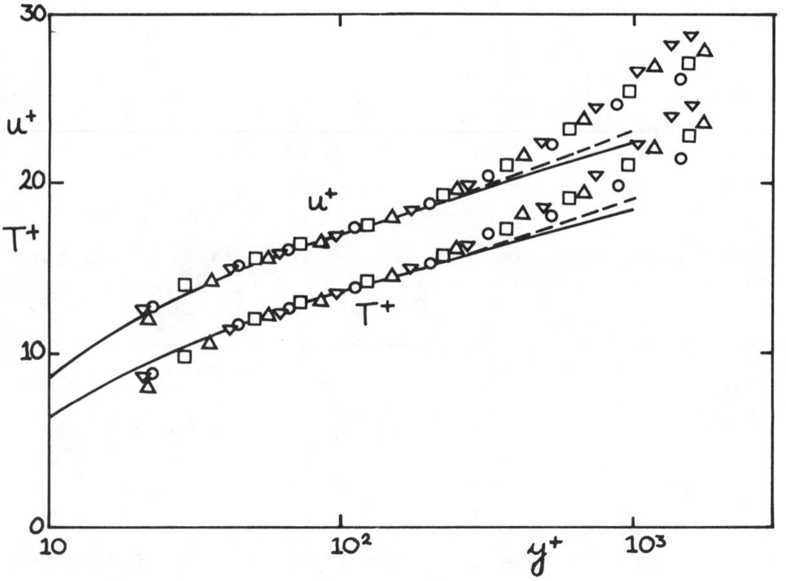

FIGURE 6. GVN curved-wall profiles at stations 1: ○ 2: □ 3: △ 4: ▽. Velocity and temperature wall laws (1) (2) are plotted for $K = 0.41$, $K_T = 0.45$, $A^+ = B^+ = 28.5$. The broken line shows the effect of the curvature modification to the mixing lengths (6), (13), when $\beta = \beta_T = 7$.

Figure 5 shows that the flat-wall profiles are fitted by (2) when $K_T = 0.45$ and $C_T = 3.0$. It is common for the measured velocities to lie above the calculated profile for $y^+ < 70$, even when a pitot-tube displacement correction is applied (Pierce and Gold 1977). The result of integrating (7) for T^+ is fitted by

$$C_T = 0.155B^+ - 1.05 \qquad (12)$$

2.3 Analysis of the Curved Wall Data

Scrutiny of the convex wall measurements suggests that the velocity and temperature profiles are logarithmic over at least some portion of the inner layer, and that departures from the flat wall distributions are accounted for mainly by changes in C and C_T rather than in K and K_T. We therefore follow the example of Huffmann and Bradshaw (1972) in retaining the flat-wall values $K = 0.41$, $K_T = 0.45$, and in adjusting C and C_T. If, additionally, it is assumed that $A^+ = B^+$, then there are two unknowns, u_* and A^+, to be found by matching the measured profiles to the wall laws calculated from the mixing-length formulae.

The curved wall profiles in the GVN flow are both fitted by the laws of the wall when the damping constants A^+ and B^+ are increased from 26 to 28.5. Thus, from (9) and (12), $C = 5.78$

Measurement Station		1	2	3	4
x/δ_0		10.6	23.6	37.5	50.6
$c_f \times 10^3$	(original)	3.12	2.96	2.76	2.59
$c_f \times 10^3$	(revised)	2.86	2.68	2.52	2.38
Okamoto	(measured)	2.82	2.72	-	2.37
$2St/c_f$	(revised)	1.22	1.19	1.19	1.15

TABLE 2. Original and Revised Skin-Friction Coefficients for the Convex Boundary Layer of Gibson et al (1982) compared with Hot-Film Measurements by Okamoto and Gibson (1986).

Measurement Station		7	8	9	10
x/δ_0		3.7	9.0	14.3	21.7
$c_f \times 10^3$	(original)	2.86	2.41	2.14	2.01
$c_f \times 10^3$	(revised)	2.66	2.24	1.95	1.80
$2St/c_f$	(revised)	1.56	1.51	1.53	1.51

TABLE 3. Original and Revised Skin-Friction Coefficients for the Convex Boundary Layer of Simon and Moffat (1979)

and C_T = 3.36. The results of this adjustment are shown in Figure 6. The revised values of c_f are compared with those of the original paper in Table 2. The streamwise distance x is measured along and from the start of the curved wall where the boundary-layer thickness is δ_0. The length over which the energy-containing eddies adjust to the new conditions is probably about 10δ (Bradshaw 1973), or roughly to the point at which the first GVN measurements were taken. The revised values of c_f are about 9% lower at all stations than the values originally estimated by GVN and they agree closely with the Okamoto and Gibson (1986) sublayer probe and hot-film gauge measurements shown in Figure 3. It is now seen that the reduction in c_f on the convex wall is practically the same as the reduction in Stanton number, where previously it had been estimated to be only about half as large.

Velocity and temperature profiles from the more highly curved SM flow show greater deviations from the logarithmic distributions. Results from the first station in Table 2 are not representative because the flow at less than four boundary-layer thicknesses from the start of the bend has not had time to adjust to the new conditions. The best fit at this station is obtained for A^+ = 27.5. For the three profile sets measured further downstream, the fit shown in Figure 7 was obtained with A^+ = 28.5, as for the GVN case. Figure 7 also shows the effect of the curvature terms in equation (5), and the much greater changes that result in this case from the use of the modified mixing length (6). The curves in Figure 7 were obtained for β = 7.0 in equation (6), and β_T = 7.0 in the equivalent expression for l_T:

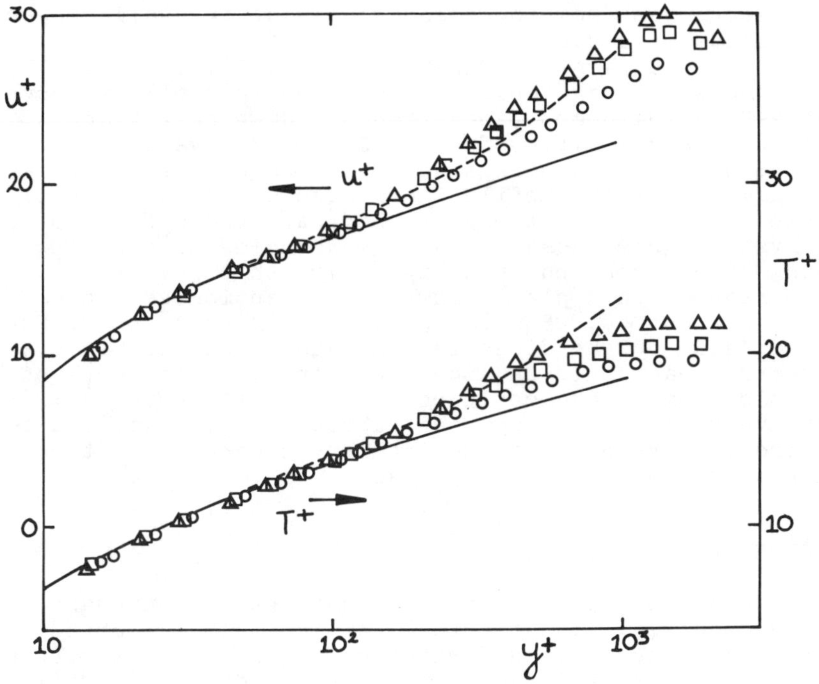

FIGURE 7. SM curved wall profiles at stations 8: ○, 9: □, 10: △. Velocity and temperature wall laws (1) (2) are plotted for K = 0.41, K_T, $A^+ = B^+ = 28.5$. The broken lines show the effect of the curvature modification to the mixing lengths (6), (13), when $\beta = \beta_T = 7$.

$$l_T = l_{TO}/(1 + \beta_T S) \tag{13}$$

The velocity and temperature distributions are correlated in the inner layer by a suitable choice of c_f, and in this case the deviations from the logarithmic distributions are equalized and accounted for by the mixing-length modifications (6) and (13). The revised values in Table 2 show that the corrections in c_f (of about 9%) are nearly the same as those of the GVN flow, although the observed effects of curvature are much greater. A minor point: for station 9 the value of St tabulated by SM appears to be about 5% too high, and it has been reduced by this amount to conform to to the general level shown in the originators graph.

Some concave-wall data from Gibson and Servat-Djoo (1985) have also been analysed. The measurements were made in the GVN wind tunnel with a working section curved to the same radius but in the opposite sense. The effect of this destabilizing type of curvature on the inner-layer profiles was a change in A^+ and B^+ of the same magnitude, but in the opposite sense, from 26 to 23 approximately, to give C = 4.67 and C_T = 2.52.

Similar departures from the standard law of the wall appear in the results of a direct numerical solution of the Navier-Stokes equations by Moser and Kim (1987). These calculations were made for turbulent flow in a mildly curved duct at a Reynolds number, based on the duct height and centreline mean velocity, of 5980. Calculated velocity distributions on the convex side of the channel fell above the standard wall-law profile, and those on the concave side fell below. But it is not easy to separate the effects of the duct curvature from those of the low Reynolds number. According to Huffman and Bradshaw (1972) this Reynolds number would correspond in a plane flow to a dimensionless stress gradient of about -0.005 and, consequently, to $A^+ = 30$. For the convex side of the directly-simulated flow the velocities were greater than could be accounted for by the effects of low Reynolds number and could be fitted by (1) with $A^+ = 35$. The concave-side velocities were substantially lower and the convex and concave profiles, which were congruent in the sublayers, started to diverge at as low a y^+ as 20.

3. CONCLUSION.

Heat and momentum transport through a turbulent boundary layer are about equally affected when the wall is curved. Velocity and temperature distributions near the wall differ from those of flat-wall flow even when the degree of wall curvature is relatively small. The differences are accounted for by changes in the additive constants in the logarithmic laws of the wall or, equivalently, in the mixing- length damping constant. The effects of wall curvature on the velocity law of the wall are qualitatively similar to those of low Reynolds number, and may be linked to the gradient of shear stress near the wall.

The changes required in the wall-law constants appear to be the same whether the wall is slightly or highly curved. This finding is consistent with the view that mild curvature of the streamlines has a disproportionately large effect on the turbulence. Thus while the general rule is that changes in the shear stress will be an order of magnitude greater than δ/R, we have observed here that although the skin-friction coefficient is reduced by 20% when $\delta/R = 0.01$, the change is only doubled by a tenfold increase in δ/R to 0.1. For slight curvature the effect on A^+ is also larger than the presumed effect on the mixing length expressed through equation (6). In this case it seems possible that, as Huffmann and Bradshaw (1973) surmised, the stress gradient plays a dominant role in transmitting outer layer conditions to the inner layer. With increased δ/R the effects expressed directly through changes in the mixing length become more apparent.

The standard wall-law method for skin friction underestimates changes when the wall is curved, and it is probable that this has occurred in most of the test cases available to turbulence modellers. The use in calculation methods of wall functions based on an unmodified wall law will produce similar underestimates.

4. ACKNOWLEDGEMENTS.

Financial support for the GVN measurements discussed in this paper was provided by the Science and Engineering Research Council through Grant GR/A76634. Additional measurements were made at Imperial College by Professor S. Okamoto during a period of study leave from the Shibaura Institute of Technology, Japan.

5. NOMENCLATURE.

A^+, B^+	damping constants in equations (3) & (8)
C, C_T	wall-law constants, equations (1) & (2)
c_f	local skin-friction coefficient
K, K_T	wall-law constants, equations (1) & (2)
l, l_T	mixing lengths
P	sublayer resistance factor, equation (11)
Pr	Prandtl number
q	turbulent temparature flux
R	radius of curvature of a surface
r	local radius of curvature $r = R + y$
Re_2	momentum thickness Reynolds number
S	strain-rate ratio $(U/r)/(\partial U/\partial r)$
St	Stanton number
T	mean temperature
U	mean velocity in x direction
u,v,w	fluctuating velocity components
u_*	friction velocity
x, y	distances measured along and normal to a wall
β, β_T	coefficients in equations (6) & (13)
δ	boundary layer thickness
τ	shear stress

Subscripts and superscripts

W	conditions at the wall
+	variables in inner-layer coordinates

6. REFERENCES.

BRADSHAW, P. 1973 Effects of streamline curvature on turbulent flow, AGARDograph No 169.

CEBECI, T. 1973 A model for eddy conductivity and turbulent Prandtl number, Trans. ASME, J. Heat Transfer 95C, 227.

CLAUSER, F.H. 1954 Turbulent boundary layers in adverse pressure gradients, J. Aero. Sci. 21, 2, 91.

COLES, D.E. and HIRST, E.A. (ed) 1968 Computation of Turbulent Boundary Layers, Proc. AFOSR-IFP-Stanford Conference.

DE BREDERODE, V. and BRADSHAW, P. 1974 A note on the empirical constants appearing in the logarithmic law for turbulent wall flows, Imperial College Aero Rep. 74-03.

GIBSON, M.M. and SERVAT-DJOO, K. 1985 Measurements in the heated turbulent boundary layer on a mildly curved concave surface, Proc. 5th Turbulent Shear Flows Symposium.

GIBSON, M.M., VERRIOPOULOS, C.A. and NAGANO, Y. 1982 Measurements in the heated boundary layer on a mildly curved convex surface, Turbulent Shear Flows 3, 80, Springer.

GIBSON, M.M., VERRIOPOULOS, C.A. and VLACHOS, N. 1984 Turbulent boundary layer on a mildly curved convex surface, 1: mean flow and turbulence measurements, Expts. Fluids 2,17.

GIBSON, M.M. and VERRIOPOULOS, C.A. 1984 Turbulent boundary layer on a mildly curved convex surface, 2: temperature field measurements, Experiments in Fluids 2, 73.

GILLIS, J.C. and JOHNSTON, J.P. 1980 Experiments on the turbulent boundary layer over convex walls and its recovery to flat wall conditions, Turbulent Shear Flows 2, 116, Springer.

HUFFMAN, G.D. and BRADSHAW, P. 1972 A note on von Karman's constant in low Reynolds number turbulent flows, J. Fluid Mech. 53, 45.

JAYATILLAKA, C.L.V. 1969 The influence of Prandtl number and surface roughness on the resistance of the laminar sublayer to momentum and heat transfer, Prog. Heat Mass Transf. 1,193.

KAYS, W.M. and CRAWFORD, M.E. 1980 Convective Heat and Mass Transfer, McGraw-Hill.

KLEBANOFF, P.S. 1955 Characteristics of turbulence in a boundary layer with zero pressure gradient, NACA Rep. 1247.

MOSER, R.D. and MOIN, P. 1987 The effects of curvature in wall-bounded turbulent flows, J. Fluid Mech. 175, 479.

OKAMOTO, S. and GIBSON, M.M. 1986 Skin-friction measurements on a convex wall, Imperial College MED Rep. FS/86/12.

PIERCE, F.J. and GOLD, D.S. 1977 Near-wall velocity measurements for wall shear inference in turbulent flows, NBS Special Publication 484.

SIMON, T.W. and MOFFAT, R.J. 1979 Heat transfer through turbulent boundary layers - the effects of introduction of and recovery from convex curvature. ASME paper 79-WA/GT-10.

SIMON, T.W., MOFFAT, R.J., JOHNSTON,J.P. AND KAYS, W.M. 1980 Turbulent boundary layer heat transfer experiments: convex curvature effects including introduction and recovery, Stanford University Mechanical Engineering Rep. HMT-32.

SO, R.M.C. and MELLOR, G.L. 1973 Experiment on convex curvature effects in turbulent boundary layers, J. Fluid Mech. 60, 43.

VAN DRIEST, E.R. 1956 On turbulent flow near a wall, J. Aero. Sci. 23, 1007.

YAGLOM, A.M. 1979 Similarity laws for constant-pressure and pressure- gradient turbulent wall flows, Ann. Rev. Fluid Mech. 11, 505.

The Influence of Free-Stream Turbulence and Surface Ribbing on the Characteristics of a Transitional Boundary Layer

V. E. KOZLOV, V. R. KUZNETSOV, B. I. MINEEV, and A. N. SEKUNDOV
Central Institute of Aviation Engines
Moscow, USSR

ABSTRACT

The influence of free-stream turbulence, the form of the plate leading edge, and the surface microstructure (riblets and waves) on the transitional boundary layer has been analyzed. Extensive experimental research has been carried out, and a theoretical model of turbulence inside a stable boundary layer has been developed.

1. INTRODUCTION

According to the classical scheme, natural transition on a flat plate at low free-stream turbulence ($u_e/U_e < 0.002$, where U_e is free-stream mean velocity, u_e is free-stream rms velocity) passes through several consecutive stages [1], that can be grouped into two main stages. During the first one, external disturbances (free-stream turbulence, acoustic waves etc.) are transformed into Tollmien-Schlichting (TS) waves with wave-number k_1, on the order of $k_1 b^* = 0.5$, where b^* is the displacement thickness, k_1 is a longitudinal wave-number. This process occurs inside the boundary layer near the leading edge of the plate. At the second stage, the amplitude of TS wave decreases as longitudinal distance increases. After the local Reynolds number exceeds its critical value the amplitude of the TS wave begins to grow. This growth continues along the plate until non-linear effects become important. These effects are the ultimate cause of boundary-layer turbulence. The second stage, i.e. the amplification of TS waves has been rather well studied theoretically. The first stage, i.e. the receptivity problem (as was defined by Morkovin) has been studied very poorly. Exact theoretical results have been obtained only when the influence of localized small surface disturbances on transformation of external acoustic waves [2] or two-dimensional potential eddy (Ryzhov, private communication) into TS waves were studied.

The transition process is probably different in the presence of free-stream turbulence with large length-scale and rather high intensity ($u_e/U_e > 0.002$). There are some peculiarities in this case, as reported in an unpublished paper of Klebanoff and in the papers [3, 4]. Low-frequency disturbances ($k_1 b^* << 1$) of longitudinal velocity component are amplified inside the boundary layer at all Reynolds numbers. As the Reynolds number exceeds the critical value given by linear stability theory the rate of amplification may rise up to one hundred, the maximum value of rms velocity is on the order of $(0.05\text{-}0.1)U_e$ which is comparable to, or exceeds, the intensity of velocity fluctuations in a developed turbulent boundary layer. Nevertheless the mean velocity profile differs little from Blasius one. The turbulence in such a boundary layer is essentially three-dimensional and highly anisotropic, i.e. its lateral length scale is much less than the longitudinal one, and the phase velocity of the disturbances is $0.5\text{-}0.6\ U_e$ [5]. As the Reynolds number grows up to transition Reynolds number (Re_{tr}), distinctly-observed separate spots of high-frequency fluctuations, resembling Emmons spots, appear in oscillograms in low-frequency background. No preferred frequency can be seen in the velocity spectrum.

Thus there are probably at least two different types of transition. This paper deals with the second type of natural transition which takes place in the presence of large-scale free-stream turbulence ($L >> b^*$, where L stands for free-stream integral length-scale). Major attention is paid to the investigation of disturbance structures inside a stable boundary layer over a large range of flow conditions. A theoretical model of this structure is also presented.

2. EXPERIMENTAL PROCEDURE

The measurements were performed in a wind tunnel with a closed test section $0.4 \times 0.5\ m^2$. The flow velocity U_e was veried from 1 m/s up to 30 m/s. The boundary layers on 200-400 mm length plexiglass plates were studied. The plates were mounted in the middle of the test section, on special supports to prevent vibration. To prevent the flow separation at the edge of the plate two schemes were chosen. In scheme A a 30-70 mm height interceptor was installed on the upper wall of the channel in the plane of the plate trailing edge. Thus a favourable angle α_0 between the plate and dividing stream-line was achieved at the edge of the plate. The longitudinal pressure gradient on the plate surface was close to zero in this scheme. The same aim was achieved in the scheme B by tilting the plate to angle of attack α_0 to the mean flow direction. In this case a negative pressure gradient appeared on the windward plate surface, i.e.

$$\pi x_1\, d\langle P\rangle/dx_1 = -\rho U_e^2\, tg(\alpha_0) \qquad (1)$$

where x_1 is a distance from the leading edge of a plate, and ρ is density. Plates with either sharp or elliptical or rounded leading edge were investigated. The curvature radius of the leading edge was varied from 0.1 mm to 10 mm. No influence either of type of leading edge or of curvature radius (r) on turbulence structure in the middle part of the plate was noticed. Probably it was because $r << x_1$ and $r << L$. Plates with riblets and waves were investigated as well as plates with smooth polished surface. Large-scale turbulence with variable length scale $L \equiv u_e^3/\varepsilon$ = 20-75 mm and intensity u_e/U_e = 0.006-0.16 was created in working section by installation of screens and grids in the receiver of the wind tunnel. Overall isotropic conditions, $u_e^2 = \langle u_{1e}^2\rangle = \langle u_{2e}^2\rangle = \langle u_{3e}^2\rangle$, were obtained. Here indices 1,2,3 stand for longitudinal, lateral and transverse directions. Isotropic interrelation

$$2E_2 = E_1 - k_1\, dE_1/dk_1 \tag{2}$$

between lateral (E_2) and longitudinal (E_1) spectra was in qualitative agreement with experimental data. However a rather big quantitative difference was noticed. The longitudinal velocity spectrum was approximated as

$$4\pi E_1(k_1) = 4 <u_e^2> L \left[1+D\left(k_1L\right)^{5/3}\right]^{-1}$$

$$\exp\left[-5.6\, k_1L\ Re_t^{-3/4}\right], \tag{3}$$

$$D = 1.35\left[1+35\, Re_t^{-3/4}\right]^{-1}$$

In the range of wave numbers $k_1 = 2\pi\, f/U_e$ from $k_1L = 0.01$ up to $k_1L\, Re_t^{-3/4} = 10$ this approximation is good enough if turbulence Reynolds number $Re_t = u_eL/\upsilon$ is more than 100, where f stands for frequency. The function $D(Re_t)$ was chosen to fulfill the normalization condition

$$\int_0^\infty E_1 dk_1 = <u_e^2>/2$$

Kolmogorov scale $\eta_t = L\, Re_t^{-3/4}$ in most experiments was more than 0.2 mm. Thus the boundary-layer thickness corresponds to the viscous part of the free-stream turbulence spectrum which is typical for various technical problems.

About 20 various regimes differing in fluctuation intensity, turbulence scale, Reynolds number, spectrum form, plate incidence angle etc., have been investigated. For example, the spectra for two typical cases are presented in Fig. 1. Case 1 corresponds to u_e/U_e = 0.07, U_e = 2 m/s, L=40 mm, case 2 corresponds to u_e/U_e = 0.0022, U_e = 2 m/s, L=75 mm. As can be seen in the first case, when turbulence intensity is large, experimental data are approximated better by relation (3). As turbulence intensity decreases, discrete frequencies appear and deviations from relation (3) increase. The velocity measurements were done by DISA hot-wire anemometer with linearizer. Time and space-time velocity correlations were measured by DISA equipment. Spectra of velocity signals were obtained by SK4-72 spectrum analyzer. The major part of measurements was done using standard 1.2 mm long hot wires. To obtain more accurate results some measurements were done with 0.7 mm long hot wires. Wall static pressure fluctuations were measured by 8501 sensor and Bruel and Kjaer microphones (the size of the sensors was 4 and 6 mm respectively). Flow visualization was performed with the help of a laser sheet from 2w power argon laser and stroboscopy. Smoke particles from a specially-designed generator were put into the flow from holes in a special pylon. In some cases the smoke-wire method was used. In this case oil-covered wire with 40 μm diameter was installed parallel to the wall inside the boundary layer. Smoke was generated by switching on a current through the wire for a short time.

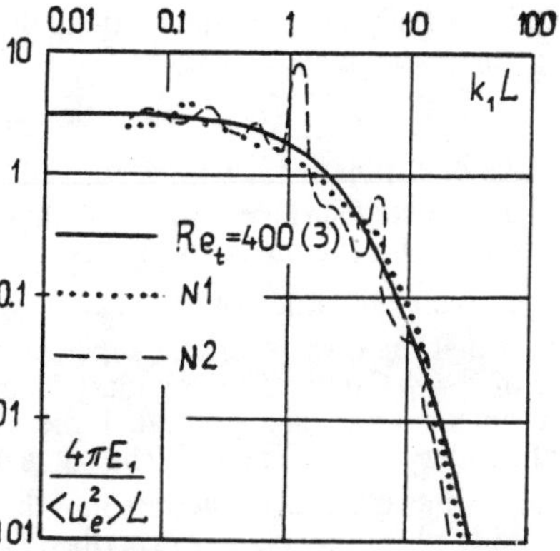

FIGURE 1

3. TRANSITION

For greater reliability the transition in the boundary layer was determined by three different methods: 1) by the appearance on the oscillograms of intermittent high-frequency spots with probability approximately equal to $\gamma = 0.1$; 2) by the surface friction measured by wire probes located at the distance 15-20 μm from the wall; 3) by longitudinal velocity spectra E_b measured at a point with Blasius coordinate $\eta = x_2(U_e/x_1\upsilon)^{1/2} = 1.6$, where $U/U_e = 0.5$. It was shown that the ratio of spectrum E_b to the free-stream spectrum E_1 defined as a function of dimensionless wave-number k_1x_1 decreases from values greater than unity when $k_1x_1 < (k_1x_1)_*$ to the values less than unity when $k_1x_1 > (k_1x_1)_*$. It was shown also that the quantity $(k_1x_1)_*$ is essentially constant in the stable boundary layer and, as is seen in Fig. 2, this quantity is approximately 10-20. Contrary to the behavior of the stable boundary layer, the quantity $(k_1x_1)_*$ increases sharply with Reynolds number growth in the transitional boundary layer. It was shown that $(k_1x_1)_* = 50\text{-}100$ when $\gamma = 0.1$, and $(k_1x_1)_* \to \infty$ when $Re = U_ex_1/\upsilon > Re_{tr}$. The use of all three methods allowed us to define transition Reynolds number Re_{tr} with satisfactory accuracy (=20%).

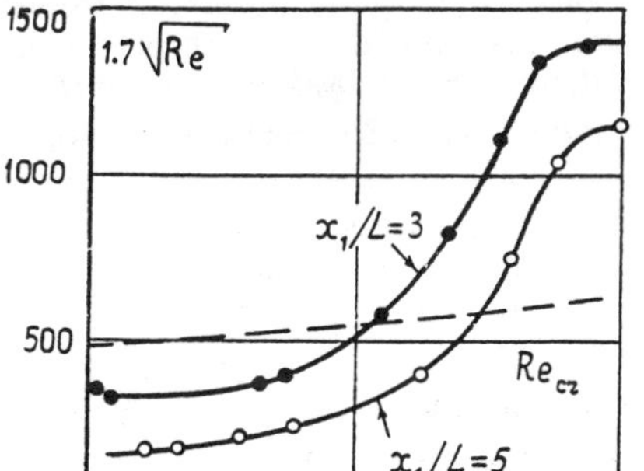

FIGURE 2 (Ordinate $\simeq U_e\,\delta^*/\upsilon$)

If the scheme B is used, i.e. a flat plate is installed along the stream line with zero mean angle of attack $\alpha_0 = 0$, the instantaneous angle of attack α' satisfies the condition $tg(\alpha') = u'_{2e}/U_e A \neq 0$ due to lateral velocity fluctuation in the free stream. At some moments the angle of attack is positive i.e. $\alpha' > \alpha_0$. At such moments, regions of separated flow appear on the leeward surface. If a quasistationary model is used, then the dimensions of these regions can be evaluated from Rice theory. Thus the dimension of the separated flow region is on the order of the Taylor microscale $\ell = 4L\ Re_t^{-1/2}$. Measurements have shown that the dimensions of separated flow regions at the edge of the plate are on the order of $\ell = 4\text{-}6$ mm. The boundary layer becomes turbulent in these regions and separate turbulent spots are generated there. Turbulent spots decay at distances $x_1 >> \ell$, if the boundary layer is stable. The spots decay slowly if the Reynolds number is high enough. Thus the spots can penetrate into the region of unstable flow and stimulate transition. A similar phenomenon was described elsewhere, for example, in the paper [6], where the dynamics of such spots was investigated.

If the negative angle of attack of the plate is increased, then the separated flow regions at the edge of the plate should appear with very small probability, depending on the probability of a very big lateral velocity fluctuation. To check this phenomenon measurements of transition Reynolds number were made at two various distances from the edge of the plate. The turbulence intensity has been held at a constant level $u_e/U_e = 0.01$ and the angle of attack has been varied in these experiments.

The results are presented in Fig. 3. The dotted line corresponds to critical Reynolds number Re_{cr} given by linear stability theory. To calculate Re_{cr} the relation (1) for pressure gradient has been used. As can be seen in Fig. 3, the transition takes place even at $Re < Re_{cr}$, if angle of attack is small enough. The transition Reynolds number depends weakly on angle of attack, if $tg\ (\alpha_0) > 4\ (u_e/U_e)$ and $Re_{tr} = 8 \times 10^5$ in this case. The dependence of Re_{tr} on α_0 can be attenuated by rounding the leading edge. However, as mentioned above, the qualitative features of the transition were preserved.

Flow separation at the leading edge of the plate was excluded in further experiments and analysis since we were interested in whether a transition can occur without specific dependence on the flow conditions at the leading edge. Then either the scheme B, where $tg(\alpha_0) > (3\text{-}4)(u_e/U_e)$, or the scheme A (with interceptor) was used. The existence of attached flow at the leading edge of the plate was checked by hot-wire probe in the latter case. In all cases studied, there were no peculiarities in spectra or oscillograms obtained near the edge of the plate. Transition begins far from the leading edge. The main features of the transition can be deduced from the Reynolds-number influence on the position and value of the maximum value of rms velocity fluctuations u_m at given x. Typical results are presented in Fig. 3 for the scheme $A(u_e/U_e = 0.016)$.

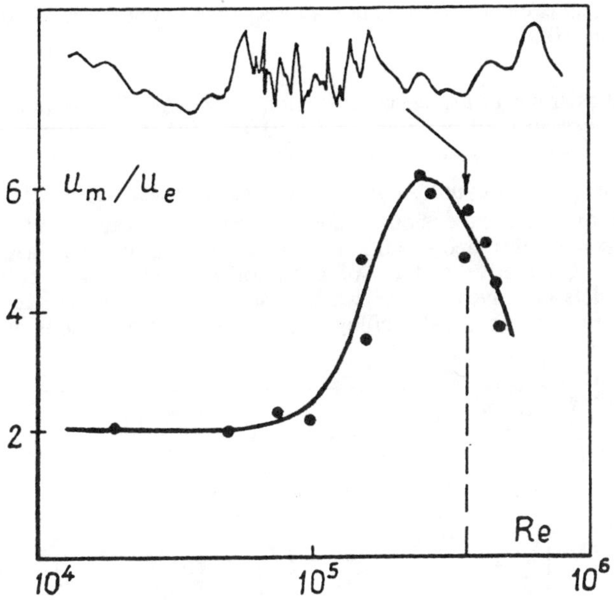

FIGURE 3

Until the latest stages of transition, the rms longitudinal velocity maximum is found to be at $\eta = 1.6$. Up to Reynolds number $Re = Re_{cr} = 8.10^4$, the amplification rate (u_m/u_e) remains approximately at a constant level. Then the amplification rate begins to grow up to 6 at $Re = 2.5 \cdot 10^5$. In this case spectra and oscillograms reveal only low-frequency fluctuations that are the same as in the steady boundary layer. As Reynolds number increases further, the amplification rate decreases and the early stages of transition begin, i.e. turbulent spots in oscillograms begin to appear. One such oscillogram is shown in the upper portion of Fig. 3. The spots are observed with low probability, $\gamma == 0.05$, in this case, and a small portion of the oscillogram containing the spot was specially selected. The typical frequency inside a spot roughly corresponds to a TS wave, i.e. it is 20-30 times higher than outside the spot. Instantaneous velocity gradient was measured by two hot wire probes. It was shown that the velocity gradient changes its sign i.e. $|du_1/dx_2| > dU/dx_2$.

As Reynolds number increases further, rms longitudinal velocity does not increase. However the form of the rms distribution changes and the position of the intensity maximum moves nearer to the wall.

4. TRANSITION IN BOUNDARY LAYER ON A SURFACE WITH REGULAR MICROSTRUCTURE

In recent years many articles have been published concerning the influence of regular surface microstructure (riblets, waves, etc.) on drag reduction [7]. The influence of such a structure on transition is unknown. We have investigated transition in the boundary layer on a plate with 200 mm length, longitudinal riblets beginning at the leading edge of a plate. The cross section of each riblet was a triangle, with height h and base width s. Typical results are shown in Fig. 4, where turbulence spectra inside a boundary layer are presented for scheme B (u_e/U_e=0.015, x_1=110 mm, U_e=16 m/s, $tg(\alpha_0)$=0.04, η=1.6). The strong damping of high-frequency disturbances inside the boundary layer on the surface with riblets is clearly seen. One may guess the boundary layer to be stable [for instance $(k_1x_1)_* = 30$].

A considerable positive influence on transition in scheme B was noticed only if (i) turbulence intensity was not too high ($u_e/U_e<0.03$), (ii) flow separation at the leading edge of the plate or mean pressure gradient were present ($tg\alpha_0 < 3u_e$ or $tg\alpha_0 > 0$). In all these cases, transition Reynolds number Re_{tr} was increased up to 20-50% through the optimal choice of riblet dimensions. The surfaces with waves with length s and amplitude h (s = 17 mm and h = 0.3 mm) were also mainly investigated. The results were the same. Qualitative features of transition on the smooth surface and on the surface with riblets or waves are the same. As on the smooth surface, low-frequency disturbances were amplified, and oscillograms of velocity fluctuations were the same as in Fig. 3.

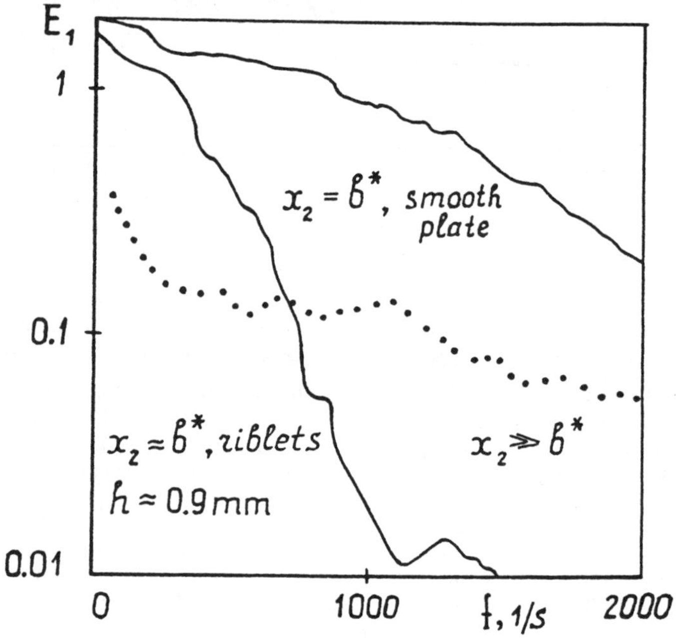

FIGURE 4

5. STABLE BOUNDARY LAYER

Consider the structure of disturbances in a stable boundary layer ($Re < 8 \cdot 10^4$) in scheme A or B ($tg\alpha_0 > 3u_e/U_e$ in the latter case). The typical data presented here were obtained in a wide range of flow conditions ($0.006 < u_e/U_e < 0.12$, $10 < L/b^* < 100$, $1 < x_1/L < 20$, $0 < -x_1\rho^{-1}U_e^{-2}d\langle P\rangle/dx_1 < 0.1$). The variation of the edge form and addition of riblets with $h=b^*$ or waves with $h = 0.3\ b^*$ did not change the results qualitatively. The distributions of average velocity $U(x_2)$, rms longitudinal (u_1) and the lateral (u_2) velocity-component fluctuations are shown in Fig. 5. All velocities are divided by the free-stream average velocity. In this case the free-stream turbulence intensity was $u_e/U_e = 0.07$. Maximum turbulence intensity inside the boundary layer was $u_m/U_e = 0.13\text{-}0.14$, i.e. much higher than in the turbulent boundary layer. Nevertheless, the average velocity is essentially the same as in a laminar boundary layer (solid line corresponds to the Blasius profile). Therefore it seems that the Reynolds stress $-\langle u_1 u_2 \rangle$ is less than viscous stress $\upsilon\, dU/dx_2$ since the strong dampling of lateral velocity disturbances is clearly seen in Fig. 5. The accuracy of lateral velocity u_2 measurements was not high enough, since the size of "X"-hot-wire probe was comparable with turbulence length scales inside a boundary layer (see Fig. 6). It seems that the levels of the lateral velocity data in Fig. 5 are too high. As seen in Fig. 5, neither rms velocity component at the edge of the boundary layer (η=5-6) equals the free-stream value $u_{1e}=u_{2e}=0.07U_e$.

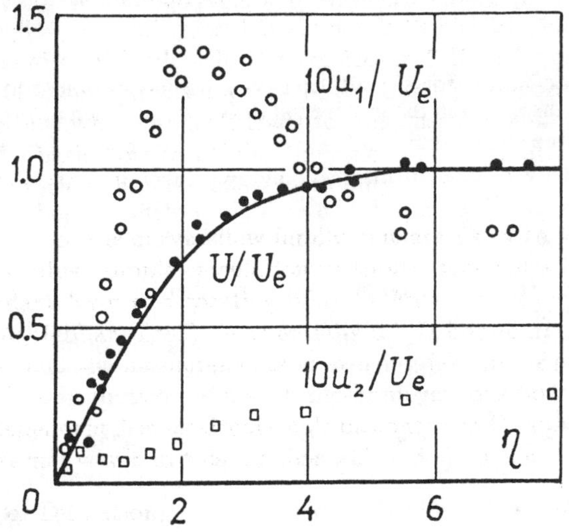

FIGURE 5

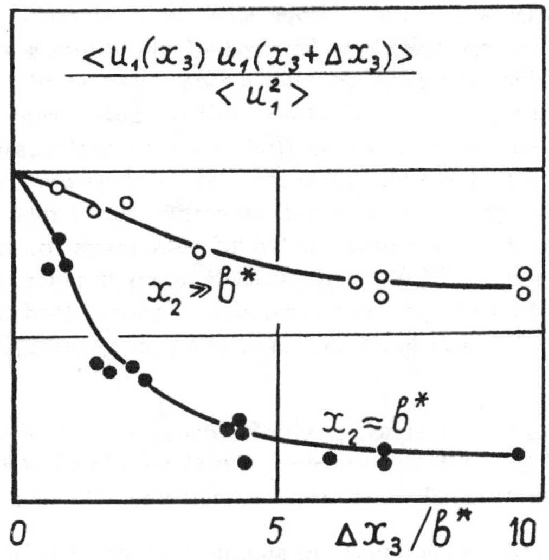

FIGURE 6

Such a phenomenon is due to the existence of the external unsheared boundary layer where only the turbulence characteristics vary [8]. It has the thickness of order L, i.e. it is much thicker than classical viscous boundary layer. The longitudinal and transversal velocity components increase, and the lateral one decreases, in the external layer through the inviscid distortion of turbulence by the presence of the wall. It was shown in [8] that the solution of linearlized Euler equations approximates experimental data fairly well. The solution has the form

$$<u_1^2>/U_e^2 = <u_3^2>/U_e^2 = 1.5\text{-}2.7(x_2/L)^{2/3}, \quad <u_2^2>/U_e^2 = 1.4(x_2/L)^{2/3} \qquad (4)$$

if $b^* < x_2 << L$, $x_1 >> L$ and there is a Kolmogorov spectrum in the free-stream turbulence.

The correlation coefficient for signals from two wires positioned at the same distance from the wall and displaced in the transversal direction is shown in Fig. 6. Measurements were done in the free stream and inside the boundary layer ($\eta = 1.6$, $x_2 = b^*$). It is clear that the integral length scale inside the boundary layer is approximately equal to the displacement thickness b^*, i.e. it decreases for an order of magnitude. Much the same picture can be seen when both wires are displaced in lateral direction.

Thus the turbulence structure is quite anomalous: there is a considerable increase of the low-frequency part of the turbulence spectrum, i.e. longitudinal scales are not decreased and lateral and transversal scales are drastically decreased. Some additional information can be extracted from the measurements of velocity differences by a probe with two wires parallel to the wall and displaced by a fixed lateral distance $\Delta x_2 = 0.6$ mm. Instantaneous velocity difference $\Delta u = u_1(x_2, t) - u_1(x_2 + \Delta x_2, t)$ was measured by hot wires. The probe with two wires was continuously driven in the x_2 direction and mean $<\Delta u>$ and rms $\Delta u' = <(\Delta u - <\Delta u>)^2>^{1/2}$ values were continuously registered on the X-Y recorder. The results are presented in Fig. 7 where the velocity fluctuation scale is stretched five times. The spectra of $\Delta u'$ signals near the wall and at the boundary layer edge are presented in Fig. 8. The amplitude of velocity difference fluctuations increases as wires move closer to the wall. Thus there is an energy transfer towards large wave-numbers k_2. At the same time, as can be seen in Fig. 8, there is an increase of energy of disturbances with small wave-numbers k_1.

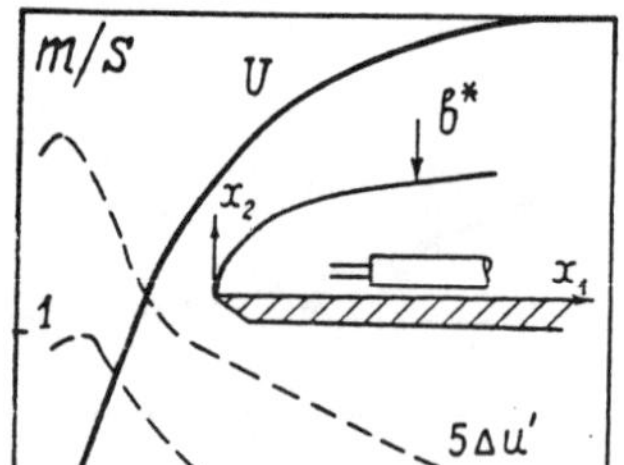

FIGURE 7

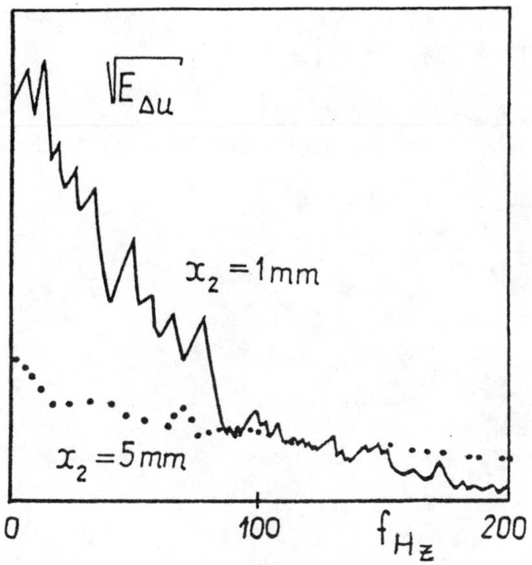

FIGURE 8

Flow visualization in three perpendicular planes was done. The picture presented in Fig. 9 corresponds to x_1-x_2 plane. The exposure time was 1/2000 s, flow velocities were U_e=2 m/s, u_e/U_e=0.06. One can see the elongated laminar-like structure with little distorted smoke threads. A single vortex flight in the free-stream can be seen also. An elongated quasi-planar spur of smoke inside the boundary layer is caused by a distortion of this vortex. Through the analysis of flow visualization in this and other planes, it becomes evident that the vortices are thin threads elongated in x_1 direction. Therefore the turbulence inside a stable boundary layer is essentially three-dimensional.

The flow visualization and hot-wire measurements also reveal the strongly anisotropic, three-dimensional character of disturbance amplification: turbulence energy is increased in those regions of wave-number space where $k_1 b^* \ll 1$, $k_2 b^* \sim k_3 b^* \sim 1$.

Space-time velocity correlations were measured by two hot wires displaced in the flow direction. The phase velocity extracted from these measurements was $u_f/U_e = 0.7 - 0.8$ at $\eta = 1.6$, i.e. slightly higher than the value reported in [5].

To confirm further theoretical considerations it was very important to investigate the role of pressure-gradient fluctuations. We have tried to measure the rms wall static pressure P_w. The accuracy of measurements was very poor because of the bad spatial resolution of the pressure probes and the rather high level of noise in wind tunnel. The upper bound of rms wall pressure was $P_w/\rho u_e U_e < 0.5$.

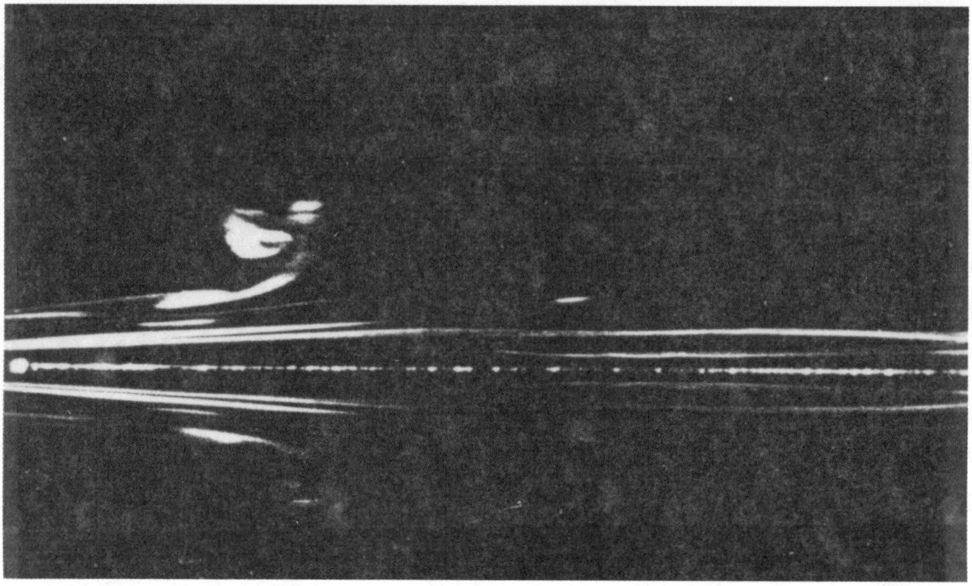

FIGURE 9

6. THEORY OF RECEPTIVITY

Experimental data and theoretical considerations reveal the following features of turbulence generation in a stable boundary layer.

1) The longitudinal fluctuations u_1 are amplified only at small wave-numbers k_1, i.e. $k_1 b^* << 1$.

2) The integral length-scale L of free-stream turbulence is much bigger than the boundary layer displacement thickness b^*(L/b^*=10-100 in most cases) and Kolmogorov length scale $\eta_t = L/Re^{3/4}$ is typically larger than b^*. Thus the energy of fluctuations with scales of order b^* is rather small. Therefore the fluctuation field inside boundary-layer is governed by the boundary -layer equations.

3) We have not noticed any nonlinear phenomena in our experiments performed at $0.006 < u_e/U_e < 0.07$. We have also shown that numerical solutions of the full and linearized unsteady two-dimensional boundary -layer equations are practically equal if $u_e/U_e < 0.1$. Thus the problem of receptivity is a linear one, for this range of disturbances.

4) Our experiments have revealed the three-dimensionality of fluctuations inside the boundary layer. For instance, the transverse integral scale is much smaller than the longitudinal one, and transverse and longitudinal velocity fluctuations are of the same order of magnitude. Moreover we have compared the measured longitudinal velocity fluctuations with numerical calculations based upon unsteady two-dimensional boundary-layer equations. Both results differ qualitatively. Thus the problem of receptivity is essentially three-dimensional.

5) As was noticed earlier, there is inviscid deformation of turbulence in the region $b^*<x_2<L$, i.e. outside the sheared part of the viscid boundary layer. This process is governed by the linearized Euler equations, if $u_e/U_e \ll 1$. Solutions of these equations were found in [8]. It was shown that transverse and longitudinal velocity fluctuations are increased and rms lateral velocity fluctuations are decreased as x_2 tends to zero [see eq. (4)]. If $x_2/L \to 0$, $x_1/L \gg 1$, then the following relations hold:

$$P_w = (\pi x_1)^{1/2} \rho U_e \int \exp\left(-|k_3| x_1 + ik_3x_3 - ik_1U_et\right)\left(k_3+ik_1\right)^{-1/2} dZ_2 \quad (5)$$

$$u_1 = \int \exp\left[ik_1(x_1-U_et) + ik_3x_3\right]\left(dZ_1 + ik_1/\kappa dZ_2\right); \; \kappa^2 = k_1^2 + k_3^2 \quad (6)$$

$$u_3 = \int \exp\left[ik_1(x_1-U_et) + ik_3x_3\right]\left(dZ_3 + ik_3/\kappa dZ_2\right); \; u_2=0 \quad (7)$$

where dZ_n is a Fourier transform of the velocity field at $x_2 = \infty$, i.e.

$$u_n = \int \exp\left[ik_1(x_1-U_et) + ik_2x_2 + ik_3x_3\right] dZ_n$$

In homogeneous isotropic turbulence we have

$$\langle dZ_n(\vec{k})\, dZ_m(\vec{k}')\rangle = E(k)\,(4\pi k^4)^{-1}\left(k^2\delta_{nm} - k_n k'_m\right)\, \delta\left(\vec{k} - \vec{k}'\right) d^3k\, d^3k' \quad (8)$$

where E is a three-dimensional energy spectral density. It is easily seen from (5) that $P_w \to 0$ as $x_1 \to \infty$. Thus the pressure gradient is negligible at large distances from the edge of the plate ($x_1/L \gg 1$).

Assuming the plate to be smooth, acoustic disturbances and pressure gradient to be negligible and that there is no flow separation at the edge of the plate, we have the linearized boundary-layer equations

$$du_1/dt + Udu_1/dx_1 + u_1dU/dx_1 + u_2\, dU/dx_2 + Vdu_1/dx_2 = \upsilon\, d^2u_1/d^2x_2 \quad (9)$$

$$du_3/dt + Udu_3/dx_1 + V\, du_3/dx_2 = \upsilon\, d^2u_3/d^2x_2 \quad (10)$$

$$du_1/dx_1 + du_2/dx_2 + du_3/dx_3 = 0 \quad (11)$$

No slip conditions ($u_n=0$) are set at the plate surface ($x_2=0$). Boundary conditions at $x_2 \to \infty$ are given by expressions (6), (7), i.e. asymptotic solution of the viscid eqs. (9)-(11) must be matched to a solution of linearized inviscid Euler equations.

Boundary conditions (6), (7) and equations (9)-(11) are valid in the region $x_1/L \gg 1$. Nevertheless, it was assumed that the asymptotic solution ($x_1/L \to \infty$) of the problem (6), (7), (9)-(11) gives a correct behavior of the real solution also for $x_1 > 0$. Then it can be shown that $x_1=0$ is a singular point of eq. (9), (10). Thus there is no need to set initial conditions.

There are three unknown functions (u_1, u_2, u_3) and four arguments (t, x_1, x_2, x_3). Moreover, boundary conditions (6), (7) are given by random functions. Therefore a numerical solution of the problem is as complicated as the solution of full Navier-Stokes equations. However, there are some simplifications. First of all, we see from Eq. (6), (7) that independent variables x_1 and t are met only in combination $x_1-U_e t$, i.e. the number of independent variables can be reduced. Let dv_n be a Fourier transform of velocity field, i.e.

$$u_n = \int \exp\left[ik_1(x_1-U_e t) + ik_3x_3\right] dv_n \qquad (12)$$

Then boundary conditions are

$$dv_n = 0, \quad x_2 = 0$$

$$dv_1 = dZ_1 + ik_1/\kappa dZ_2, \quad dv_3 = dZ_3 + ik_3/\kappa dZ_2; \quad x_2 = \infty \qquad (13)$$

Introducing new independent variables

$$x = k_1x_1, \quad \eta = x_2/(x_1\upsilon/U_e)^{1/2} \qquad (14)$$

and taking Fourier transform of Eq. (9)-(11) we have three equations for dv_n with only three independent variables: x, η and k_3/k_1.

The well-known Squire transform cannot be applied here, since the boundary-layer equations contain terms proportional to the lateral mean velocity. However the three-dimensional problem (9)-(11) can be split to two-dimensional subsystems. This can be done since there is only one unknown variable, dv_3, in Eq. (10), the Fourier transform of the continuity equation (11) depends linearly on k_3, and the boundary conditions (13) depend also linearly on k_3/κ. Now we introduce new dependent variables a_1 - a_5 such as

$$dv_1 = \left(dZ_1+ik_1/\kappa\, dZ_2\right) a_1 + ik_3/k_1 \left(dZ_3+ik_3/\kappa\, dZ_2\right) a_3 \qquad (15)$$

$$\left(U_e/\left(x_1k_1^2\upsilon\right)\right)^{1/2} dv_2 = \left(dZ_1+ik_1/\kappa\, dZ_2\right) a_2+ik_3/k_1\left(dZ_3+ik_3/\kappa\, dZ_2\right) a_4$$

$$dv_3 = \left(dZ_3+ik_3/\kappa\, dZ_2\right) a_5$$

It can be shown that the Fourier transfor of Eq. (9)-(11) is equivalent to two-dimensional subsystems

$$K(a_1)+d/d\eta\,(a_2) = 0 \qquad K(a_3)+d/d\eta\,(a_2) = -a_5$$

$$H(a_1)+a_2\,dF/d\eta = 0 \qquad H(a_3)+a_5\,dF/d\eta = dF/d\eta \int_0^{\eta} a_5\,d\eta \tag{16}$$

$G(a_5)=0$
where

$$U = U_e\,F(\eta), \quad V = (\upsilon\,U_e/x_1)^{1/2}\,f(\eta) \tag{17}$$

are longitudinal and lateral velocities in the Blasius solution and

$$H(a) = Fda/dx + (f-\eta F/2)/x\,da/d\eta - d^2a/d^2\eta\,/x + i\,(F-1)\,a - a\,dF/d\eta\,/2x \tag{18}$$

$$K(a) = ia + da/dx - \eta/(2x)da/d\eta, \quad G(a) = H(a) + (\eta a)/(2x)dF/d\eta$$

where (a) is any of the new dependent variables. Boundary conditions are

$$\begin{gathered} a_n=0, \quad n=1-5, \quad \eta=0 \\ a_1=a_2=a_5=1, \quad a_3=a_4=0, \quad \eta \rightarrow \infty \end{gathered} \tag{19}$$

Initial conditions can be deduced from the solution in the vicinity of the singularity point $x_1=0$ ($a_1\sim a_5\sim a_4\sim 1$, $a_2\sim 1/x$, $a_3\sim x$).

7. CALCULATIONS, COMPARISON WITH EXPERIMENT

The numerical calculations give us functions a_n depending on two variables x and η. Having these functions, we can calculate all components of Reynolds stress tensor. For example:

$$\langle u_1 u_1 \rangle = \iiint_{-\infty}^{\infty} E(k)/(4\pi k^4)^{-1} \left(k^2 a_1 a_1^* + k^2 k_3/k_1^2\, a_3 a_3^*\right) dk_1 dk_2\, dk_3 \tag{20}$$

Here terms antisymmetric in k_2 were not included, since integration of these terms gives zero. To compare calculations with experimental data, it is convenient to introduce a one-dimensional spectrum E_1

$$E(k) = k^3\, d\left(k^{-1}\, dE_1/dk\right)/dk \tag{21}$$

After simple but tedious transformations, we have

$$<u_1u_1> = \int_0^\infty E_1^* dk_1, \quad E_1^* = a_1a_1^* I_0 + a_3a_3^* I_1$$

$$I_0 = 2E_1 - k_1 dE_1/dk_1, \quad I_1 = 2E_1 + 8k_1^{-2} \int_{k_1}^\infty k_1 E_1 dk_1 \tag{22}$$

There is no singularity in I_i since $a_3 \sim x$, as $x = k_1x_1 \to 0$. To perform the calculations we have assumed the Eq. (3) to describe a one-dimensional spectrum. Thus from (3), (22) we can calculate $<u_1u_1>$ and E_b inside the boundary layer, where E_b is the one-dimensional spectrum E_1^* at $\eta=1.6$. Functions $<u_2u_2>$, $<u_3u_3>$ and $<u_1u_2>$ can be calculated in a similar manner. Results of the calculations together with experimental data for $<u_1u_1>^{1/2}$ are shown in Fig. 10. All results were divided by u_e. It can be shown from Eq. (3), (22) that the results of the calculations depend on three non-dimensional parameters x_1/L, $Re_t = U_eL/\upsilon$, and u_e/U_e (the last is essential only for terms depending on lateral velocity fluctuations). In Fig. 10 one has $x_{1/L} = 5$, $Re_t = 400$, $u_e/U_e = 0.07$. It should be noted that $<u_1u_1>^{1/2} = <u_3u_3>^{1/2} = 1.22\ u_e$, $<u_2u_2>^{1/2} = 0$ as $(x_2/L)^{2/3} = 0$ for inviscid solution [8]. Thus in the vicinity of the boundary-layer edge it can be seen from Fig. 10 that $<u_1u_1>^{1/2}/u_e = <u_3u_3>^{1/2}/u_e \neq 1$ and $<u_2u_2>^{1/2}/u_e$ increases as $(x_2/L)^{1/3}$. As can be seen from Fig. 10, the experimental results do confirm theoretical considerations. The increase of velocity fluctuations in the middle of the stable boundary layer is confirmed also.

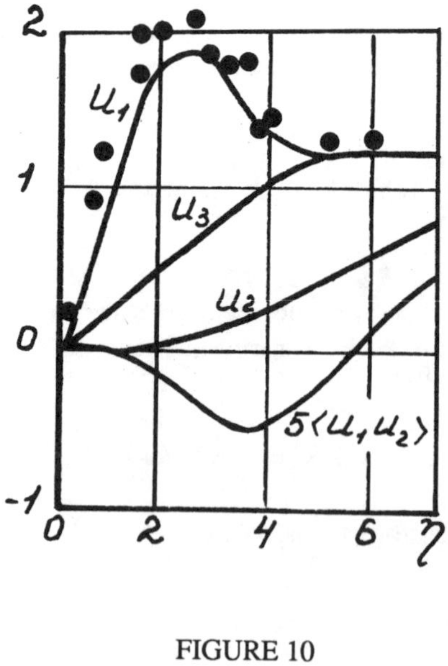

FIGURE 10

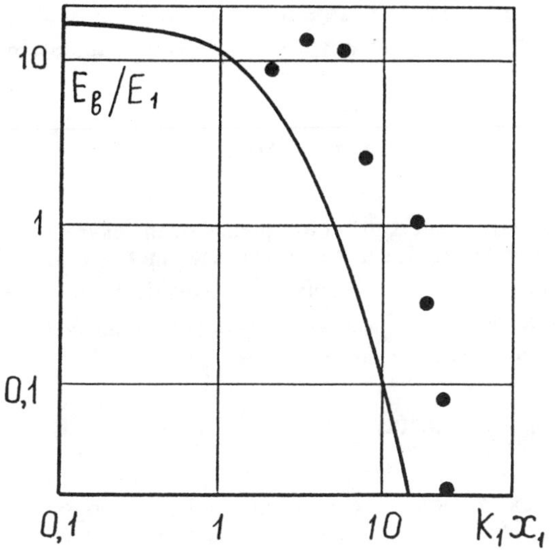

FIGURE 11

The one-dimensional spectrum of longitudinal velocity fluctuations is presented in Fig. 11 in non-dimensional form $\left(E_b/E_1 = E_1^*(k_1x_1,\ \eta=1.6)/E_1(k_1x_1,\ \eta=\infty)\right)$. The calculations are in qualitative agreement with experimental data, i.e. low-frequency fluctuations are amplified, high-frequency fluctuations are damped. However there is a considerable quantitative discrepancy. We may guess three causes of such a discrepancy: 1) neglect of pressure gradient in Eq. (9), (10); 2) anisotropy of free-stream turbulence; 3) some defects of rapid distortion theory [8], especially at the edge of the plate.

It must be kept in mind that amplification is possible only for three-dimensional disturbances and is not present in the two-dimensional case. The rate of amplification in Eq. (22) depends only on terms proportional to a_3, i.e. it depends on the solution of Eq. (10) for transversal velocity component. It can be seen also that the rate of amplification is proportional to I_1, and $I_1 = \infty$ at any k_1, if the inertial range expression $E_1 \sim k_1^{-5/3}$ is valid for $k_1 \to \infty$. Thus viscous small-scale three-dimensional fluctuations with large k_2 and k_3 wave-numbers play a dominant role. The large-scale velocity fluctuations with small k_1 wave-numbers are significant also, since $I_1 \to \infty$, as $k_1 \to 0$ (as it was mentioned earlier, there is no singularity because $a_3 \sim k_1x_1$ as $k_1x_1 \sim 0$). These considerations are supported by our flow visualizations since we have seen, inside the boundary layer, thin threads quite elongated in x_1 direction.

8. ASYMPTOTIC SOLUTION

As was mentioned above, our theory is correct only for long-wave disturbances, i.e. $(k_1x_1) << Re^{1/2}$. Since Reynolds number is usually very large, this inequality can be valid at rather large $x = k_1x_1$. There are great simplifications of Eq. (16), (17) in the limit $x \to \infty$, since derivatives in x-direction and some other terms vanish.

Thus we have a system of ordinary differential equations where x is presented only as a large parameter. Asymptotic analysis reveals a formation of a new boundary (or, better to say, critical) layer situated at a large distance from the wall, say $\eta_o >> 1$. It has a thickness of order $1/\eta_o$. The distance η_o is a very slowly increasing function of x such as

$$x = \eta_o^3 \exp\left(\eta_o^2/4\right) \tag{23}$$

The structure of this expression is the consequence of the asymptotic behavior ($\eta \to \infty$) of Blasius mean velocity profile. It can be shown that functions a_1, a_2, a_5 have jump-like behavior and functions a_3, a_4 are sharply peaked inside a critical layer. Since $\eta_o \to \infty$ as $x \to \infty$ turbulence is pushed out of the boundary layer. However, the peak value u_m of rms velocity fluctuations increases infinitely as

$$u_m/u_e = 0.4(x_1/L)^{1/2} \, Re_t^{1/8}, \quad x_1/L \to \infty, \quad Re_t \to \infty \tag{24}$$

The structure of this expression is the consequence of Kolmogorov viscous scale laws. It can be deduced analytically. The constant presented in (24) was matched numerically. The theory presented here gives the intermediate asymptotics of the exact Navier-Stokes equations. If x is large enough the boundary layer equations are not correct since $k_1 b^* > 1$. Thus we must match an asymptotic expansion of the solution of Eq. (9)-(11) ($k_1 x_1 \to \infty$) to a solution of linearized Navier-Stokes equations in the long-wave limit ($k_1 b^* \to 0$). One may guess the latter equations to be equivalent to Orr-Sommerfeld (OS) equation. Such a procedure was performed in [9].

It was shown the critical layer is in the vicinity of the wall. Thus this procedure cannot be correct here since the critical layer tends to infinity as $k_1 x_1 \to \infty$. The development of the new matching procedure is hampered because OS equation gives a correct asymptotics of linearized Navier-Stokes equations only if η = const and $Re \to \infty$. Accordingly the Eq. (9)-(11) in critical layer ($\eta \to \infty$, $Re \to \infty$) are not equivalent to the long-wave limit of OS equation: there is an extra term depending upon the rate of growth of boundary layer thickness.

9. CONCLUDING REMARKS

Our experimental data and theoretical considerations have confirmed the results of investigations [3-5] on the influence of large-scale turbulence on stable and transitional boundary layers. If acoustic disturbances are small, there is no flow separation at the leading edge of the plate: then, regardless of geometry of plates, free-stream turbulence characteristics and mean pressure gradient, the following features are present: (1) amplification of low-frequency disturbances, (2) large anisotropy of length scales inside a stable boundary layer, (3) transition to turbulence without two-dimensional TS wave generation. We have developed the theoretical model based on three-dimensional boundary-layer equations without a pressure gradient. The model preducts the main features of the structure of fluctuations inside a stable boundary layer. At present time we do not know the role of investigated disturbances in the process of transition.

ACKNOWLEDGMENTS

It was a great pleasure for us to discuss some results with N. F. Polyakov and N. S. Kosorygin. Some of the numerical calculations were performed by A. N. Gulyaev. N. N. Bychcov and O. N. Konstantinovsky have done some of the experiments.

REFERENCES

1. Kachanov Yu. S., Kozlov V. V., Levchenko V.Ya., Turbulence Generation in Boundary Layer, pp. 1-353, Nauka, Novosibirsk, 1982 (in Russian).

2. Ruban, A.I., Generation of Tollmein-Schlichting Waves by Acoustic Waves, Izv. Acad. Nauk USSR, Mech. Zhidk. i Gaza, Nt 5, pp. 44-52, 1984 (in Russian).

3. Polyakov N. Ph., Natural Transition of Laminar Boundary Layer to Turbulent Flow, ed. Levchenko V.Ya., pp. 23-67, Novosibirsk, 1979 (in Russian).

4. Kendall, J. M., Experimental Study of Disturbances Produced in a Pretransitional Laminar Boundary Layer by Weak Freestream Turbulence, AIAA Paper 85-1695, 1985.

5. Kosorygin V.S., Polyakov N.Ph., Suprun T.T., Epik E. Ya., Influence of External Turbulence on Structure of Disturbances in Laminar Boundary Layer, Wall Turbulent Flow, ed. Kutateladze S.S., pp. 79-83, Novosibirsk, 1984 (in Russian).

6. Grek G.R., Kozlov V.V., Ramazanov M.P., The Influence of High Free-stream Turbulence on Laminar-Turbulent Transition, ITPM, Rep. 8-87, 1987 (in Russian).

7. Bandyopandhyay D.R., Review Mean Flow Turbulent Boundary Layers Disturbed to Alter Skin Friction, AIAA Paper 85-1126, 1986.

8. Hunt J.C.R., Graham J.M.R., Free-stream Turbulence Near Plane Boundaries, Journal Fluid Mechanics, 1978, vol. 84, pt. 2, pp. 209-235.

9. Zhuk V.I., Rhzhov O.S., Free-stream Interaction with Boundary Layer and Stability Boundary Layer in Incompressible Fluids, Dokl. Akad. Nauk USSR, V. 253, No. 1326, 1980 (in Russian).

The Dynamic Behavior of a Shock-Wave/ Turbulent Boundary Layer Interaction

MICHAEL S. SELIG and ALEXANDER J. SMITS
Department of Mechanical and Aerospace Engineering
Princeton University
Princeton, New Jersey 08544, USA

ABSTRACT

Periodic blowing at frequencies up to 5 kHz was used to control the unsteadiness of a two-dimensional shock wave/turbulent boundary layer interaction. The pertinent incoming conditions were $M_\infty = 2.84$, $Re_\infty/\ell = 6.5 \times 10^7$, and $\delta_o = 26$ mm. The interaction was formed by a 24° compression corner, and the periodic blowing was introduced into the shock-induced separation bubble. Measurements of the fluctuating mass flux and wall pressure were made, and the unsteady flowfield was visualized through stroboscopic schlieren videography. By comparing this flow with the 24° compression corner flow without blowing, the effect of the unsteady shock could be examined. Although the control of the shock wave ranged from negligible to strong, the turbulence amplification showed no substantial changes, suggesting that the unsteady shock was not the principal cause for the increased turbulence as had been previously thought. Instead, there is strong evidence to suggest that large-scale instabilities associated with inflectional velocity profiles in the downstream boundary layer contribute strongly to the large mixing observed downstream of the interaction.

1. INTRODUCTION

Two-dimensional shock wave/turbulent boundary layer interactions have been the subject of numerous experimental and theoretical investigations in the past. Despite this attention, much remains to be understood about these complex interactions, especially for flows with boundary layer separation. Experimental studies have mostly reported the mean-flow properties; recently, however, researchers have begun to examine the more detailed time-dependent nature of these flows. The present experimental investigation adds to this body of work by examining the separated flow over a two-dimensional 24° compression corner at Mach 2.84 with special emphasis given to the effect of the unsteady shock on the boundary layer turbulence, and the possibility of controlling the unsteady shock motion by periodic blowing.

Already, much is known about this 24° compression corner flow from previous experiments involving flow visualization, mean-flow measurements, and measurements of the fluctuating wall-pressure and mass-flux [Settles, Fitzpatrick & Bogdonoff (1979), Dolling & Murphy (1983), and Selig et al. (1988)]. In the mean, the flow is known to separate at approximately $1.3\delta_o$ upstream of the corner and reattach at approximately $0.5\delta_o$ downstream, as depicted in Figure 1. Traditionally, the initial mean pressure rise, which takes place over a streamwise distance on the order of δ_o, was thought to be caused by a continuous family of compression wave (Green 1970). From the measurements of Dolling & Murphy (1983), however, it is known that this mean pressure rise is produced by the average of the undisturbed wall pressure and the higher pressure behind the unsteady separation shock which oscillates with an amplitude on the order of δ_o. With this view, a higher mean pressure indicates that the shock spends more time upstream of that station than at a station with a lower mean pressure. Shadowgrams were thought to show that the unsteady shock near the wall was composed of many subsidiary shocks. It has since been shown that this picture is probably an illusion caused by optical spanwise integration of a single leading shock front that ripples in the spanwise direction with a wavelength less than δ_o (Dussauge, Muck & Andreopoulos 1985).

From wall-pressure measurements taken at the foot of the shock, Tran (1987) and Andreopoulos & Muck (1987) showed that the average frequency of the oscillation is around 1-2 kHz for the 24° compression corner, but the distribution of frequencies beyond 2 kHz is very skewed and extends up to 10 kHz. Andreopoulos and Muck suggest that the unsteadiness may be triggered by the bursting of the incoming boundary layer, although this remains at issue (Dolling 1985; Narlo 1986).

Selig et al. (1987) found that the interaction strongly amplifies the boundary layer turbulence levels. Their results, shown in Figure 2, indicate that the maximum of the mass-flux turbulence intensity in the boundary layer is amplified by a factor of 4.8. It was suggested by Smits and Muck (1987) that the unsteady shock motion plays an important role in this amplification, and the present investigation was initiated to study the effects of the shock oscillation on the turbulence amplification more closely. Periodic blowing through a spanwise slot into the separated zone was used to control the otherwise random oscillations of the shock. By changing the position of the slot, and the strength and frequency of the blowing, the shock motion could be controlling in both frequency and amplitude, and the effect on the downstream turbulence levels could be determined. Measurements of the wall pressure fluctuations in the region of shock motion, and the mass-flux fluctuations downstream of the interaction were made. By forcing the unsteadiness at a known frequency, stroboscopic flow visualization could be used, as well as conditional sampling based on the phase of the blowing cycle.

2. EXPERIMENTAL FACILITY, APPARATUS, AND TEST CONDITIONS

The experiment was performed in the Princeton University 203 x 203 mm (8 x 8 in.), Mach 3, blowdown wind tunnel. The periodic blowing was introduced through a spanwise slot located in the nozzle wall 1.95 m downstream of the nozzle throat. The 24° compression corner model could be placed at varying distances downstream of the slot.

The periodic blowing apparatus is shown in Figure 3. The major components consist of the two-dimensional 24° moveable model, a sub-floor into which miniature pressure transducers can be mounted along the centerline of the tunnel floor, the main two-piece floor, and a slotted drum held within an enclosure that forms a sealed plenum chamber when the side walls of the test section (not shown) are joined to the main floor.

The apparatus operates like a siren in that a stream of air is clipped periodically to produce a high-frequency, pulsed jet of air that issues normal to the floor (see Fig. 4). The frequency of the blowing was controlled by the speed of the drum and could be safely set as high as 5 kHz. The amount of air injected through the injection slot into the boundary layer was controlled by varying the supply pressure in the plenum chamber and by varying the width of the slot. Since the flow mostly responded to the net amount of blowing rather than on how it was achieved, the slot width was fixed at 3.2 mm (1/8 inch), and only the plenum chamber pressure was varied for control of the blowing flow rate. The maximum amount of blowing used was 9% of the freestream mass flux, that is, the maximum value of $(\rho V)_{slot}/(\rho U)_\infty$ was 0.09, corresponding to a mass-flow rate $(\rho Vw)_{slot}/(\rho U\delta^*)_\infty$ of 0.04, where w is the slot width and δ^* is the boundary layer displacement thickness.

The 24° compression corner model was designed to be moved upstream and downstream relative to the fixed injection slot. The model was 152.4 mm wide, which left room between the sides of the model and the sidewalls of the tunnel for the sidewall boundary layer to pass by without interference. The model width was the same as that used by Settles (1975), Murphy (1983), and Or (1985). To improve the two-dimensionality of the flow, aerodynamic fences were cut out of clear, 6.4 mm thick plexiglass and screwed to the sides of the model. The edges of the fences were bevelled at 30° to reduce the disturbance caused by the presence of the fence.

Surface flow patterns by the kerosene and lampblack method can be used to investigate the flow two-dimensionality. The original surface flow patterns from Settles (1975), Murphy (1983), and Or (1985), as well as the most recent surface flow taken during this investigation (without blowing), show considerable spanwise variation in the position of the separation line.

On average, however, the separation line was located 34 mm upstream of the corner with a waviness amplitude of 2 to 6 mm. The principal aim of the present research is to outline and characterize the influence of the unsteadiness on the turbulent boundary layer. The relatively mild three-dimensionality revealed in the flow visualization does not seriously compromise this goal, although it does indicate that three dimensional effects may be an integral part of this interaction.

The incoming floor boundary layer made a natural transition to turbulence in the nozzle and developed along the approximately adiabatic wall of the tunnel. At the position of the injection slot, 1.95 m downstream of the throat, the boundary layer was typical of a zero-pressure gradient fully turbulent boundary layer and it obeyed both the law of the wall and law of the wake (Taylor 1984; and Settles, Vas & Bogdonoff 1976). Its characteristics are summarized in Table I. Over the total 50.8 mm range in the streamwise movement of the 24° model, Spina & Smits (1987) indicated a small variation in boundary layer thickness (< 5%), and the streamwise change in Reynolds number over this distance was therefore neglected.

The following data were taken: (1) mean and fluctuating wall pressure in the region near the base of the unsteady shock, (2) mean and fluctuating mass flux downstream of the shock, (3) 1.5 μsec schlieren video records to document the global flow-field unsteadiness and how it is affected by periodic blowing, and (4) conditional sampling based on the phase of the blowing cycle.

Wall static pressure flucations were measured using miniature differential pressure transducers manufactured by Kulite Semiconductor Inc., Model XCQ-062-25-D. The manufacturer quotes the natural frequency of this model transducer at 500 kHz. However, a well-damped resonance appears in the 60 to 80 kHz range which limits its reliable performance range to frequencies below this. Since in this study the transducers are used primarily to reveal the dynamic character of the low-frequency shock motion in the 500 Hz to 5 kHz range, this damped resonance at high frequencies was not of much concern.

The instantaneous mass flux was measured using the constant-temperature hot-wire technique described by Smits, Hayakawa & Muck (1983). The hot-wire was powered by a DISA 55M10 constant-temperature anemometer. To reduce the temperature sensitivity of the anemometer, so that the output was sensitive only to variations in mass flux, the probe was operated at an overheat ratio of 0.8, yielding an uncertainty in the measured $<(\rho u)'>/(\rho U)$ of -5% to 9% (Smits & Dussauge 1988). The frequency response was always in excess of 100 kHz.

For digitizing, the anemometer output signal was split into two components, a fluctuating and a mean. The mean was obtained by low-passing the signal at 10 Hz and sampled at a relatively low rate. The fluctuating part of the signal was extracted by high-passing the signal with a low-frequency cut-off of 10 Hz. This fluctuating signal was amplified to fill the range of the A/D converter, filtered at a cut-off point of 250 kHz, then sampled at 500 kHz or 1 MHz.

Every run was visualized using the schlieren videography system shown schematically in the lower part of Figure 4. The spark source of the schlieren system was strobed slightly out of phase with the periodic blowing frequency to reveal the flow response in 'slow motion'. In this way, the conventional video camera was converted into a pseudo-high-speed camera to provide 'movies' of the flow. It is important to note that the visualization was stroboscopic and not a true time-resolved 'movie'.

3. RESULTS AND DISCUSSION

The frequency, amount, and location of the blowing were the primary variables under experimental control. In selecting the physical ranges of these three variables, every effort was made to keep the strength of the perturbation at the minimum level required to produce experimentally observable trends. After considerable preliminary flow visualization using the stroboscopic video schlieren apparatus, the possible combinations of variables were reduced to a manageable number of permutations. Blowing frequencies of 1 kHz and 2 kHz were used, and the slot was positioned at 0, 25.4, and 50.8 mm upstream of the corner. The slot width was kept constant at 3.2 mm, and the blowing level was set at either 9% or approximately 2.5% of the freestream mass flux. The complete results were reported by Selig (1988). Here, only three cases with blowing will be considered. In each case, the blowing frequency was 2 kHz, the blowing level was 9%, and the only variable was the position of the slot; for Case A, the slot was located in the corner, for Case B it was 25.4 mm upstream of the corner, and for Case C it was located 50.8 mm upstream.

The mean and fluctuating wall-pressure distributions without blowing are shown in Figures 6 and 7. The agreement with the earlier measurements by Settles, Vas & Bogdonoff (1976), Dolling & Murphy (1983), and Dolling & Or (1983) is within the experimental error.

The wall pressure with blowing is shown in Figure 8. As the slot is moved upstream, the mean wall-pressure rise moves upstream as well, with little change in the shape of the distribution. The corresponding rms wall-pressure results are given in Figure 9 and show the same shift forward with little change in the region near the peak rms. As the slot is moved upstream, the shock begins to oscillate at the frequency of the blowing (see Figs. 10-13) and the mean shock location is shifted upstream. Despite these significant changes, neither the mean wall-pressure gradient nor the rms levels in the vicinity of the shock show any appreciable change. They appear as if the entire compression corner were moved upstream rather than the slot.

Energy spectra of the wall-pressure distributions were examined for each case. For the case of no blowing, the energy spectra of the wall pressure non-dimensionalized by the square of the local mean pressure are shown in Figure 14a. In this form, the spectra directly reveal the frequencies of the wall pressure that contribute most to the rms wall-pressure fluctuations. Note that the small rise in the spectra between 60 and 80 kHz is non-flow-related and comes from the damped resonance of the transducer as mentioned in Section 2. Between x = -55 and -34 mm the wall-pressure fluctuations are clearly dominated by the broad-band low-frequency oscillations of the shock. As noted earlier, the shock randomly oscillates at a mean frequency between 1.5 to 2 kHz (Dolling & Murphy 1983), although the distribution is highly skewed and extends up to a frequency of 10 kHz (Andreopoulos & Muck 1987). One might expect that these low-frequency shock oscillations would produce low-frequency pressure fluctuations which convect downstream into the separated region. On the contrary, immediately behind the shock the low-frequency fluctuations are small compared with the high-frequency fluctuations which grow in strength towards the corner. For case A, the flow visualization and wall-pressure time histories gave no obvious indication of periodic shock motion (Fig. 11). In the spectra shown in Figure 14b, however, spikes at the blowing frequency and higher harmonics are present in the shock region. These spikes do not come from acoustic noise produced by the periodic blowing; the amplitude of the fluctuation at the blowing frequency is far greater than the amplitude of an acoustic perturbation. For case B the spikes in the spectra are larger (see Fig. 14c), and even larger for case C as the oscillations become more regular. In the separated region, there is indication of a periodic fluctuation at the blowing frequency. It cannot be deduced from the spectra whether this periodicity originates from upstream traveling disturbances produced by blowing or from downstream convecting pressure fluctuations caused by the oscillating shock. The longitudinal (or streamwise) space-time correlations of the wall pressure, however, show that in fact these fluctuations are convected downstream and hence are related to the shock oscillation.

The mass-flux intensity profiles for the case of no blowing were briefly discussed in the Introduction (see Fig. 2). More detailed information can be derived from the p.d.f.s of the mass flux shown in Figure 15 (no blowing). The upstream p.d.f.s. show Gaussian-like distributions with a mean that gradually increases up through the boundary layer. In sharp contrast, the downstream p.d.f.s center around two distinct mass-flux levels. The mass flux begins to center around $0.75\rho U_\infty$ for $y < 11.7$ mm, and for $y > 18.1$ mm the mass flux centers around $2.2\rho U_\infty$ which represents the freestream value downstream of the interaction. Between y = 11.7 and 18.1 mm, the p.d.f.s. are bimodal, and here the turbulence intensities are largest. Apparently, the mass flux fluctuates between the two distinct levels, one level representative of the freestream behind the shock and another level representative of the flow very near the wall. A similar behavior was found in studying a compressible reattaching shear layer (Hayakawa, Smits & Bogdonoff 1984). Since the fluctuations between the two levels occur randomly in phase, there is no periodicity apparent in the frequency spectra.

The turbulence intensity profiles for cases A, B, and C shown in Figure 16. Contrary to expectation, there are virtually no discernible differences due to blowing. The spectra with blowing show no influence of the periodic blowing. In the boundary layer, no dominant spectral peaks are seen at the blowing frequency. Aside from slight differences, spectra with and without blowing for the same x-station are essentially the same. Furthermore, the upstream turbulence spectra are very much like the downstream, even though in the region of the shock a large amount of low-frequency energy must be associated with the shock motion. In all cases, the maxima in the spectra moved from about 10 kHz in the upstream boundary layer to about 6 kHz in the downstream case, but no peak was found with frequencies typical of the shock motion. Contrary to this result, Ardonceau et al. (1978) found for a separated 18° compression corner two broad peaks in the downstream spectra. One peak closely resembled the peak observed in the present flow - a high frequency peak above the dominant unsteady shock frequency, which was present throughout the boundary layer. A second peak appeared near the wall (for $y < 0.56\delta_o$, where $\delta_o \approx 8$ mm) at a frequency believed to be characteristic of the low-frequency fluctuations of the separated region. No such low-frequency peak was observed in the present flow. Perhaps these differences are related to the flow regimes that were studied. In the present study Re_∞/ℓ was 5.9 times larger, and $Re_{\delta o}$ was 17 times larger than that of Ardonceau. Finally, it was found that the periodic blowing had no effect on the bimodal character of the mass-flux p.d.f.

It is striking that the level and basic character of the downstream turbulence has not changed substantially in its profile or spectra, although the mean location and the behavior of the shock oscillations has been markedly changed by moving the slot. In addition, the separated region must have been affected by the blowing and by the upstream movement of the mean shock position. Yet there is little impact on the broadband turbulence intensities. The phase-averaged conditional sampling did reveal a small periodic component in downstream mass-flux fluctuations due to the periodic blowing but the periodic component contributed less than 1% to the mean-square intensities for all cases considered.

5. CONCLUSIONS

The unsteady motions of the shock can be controlled by periodically blowing into the separated zone, depending on the amount and location of the blowing. Increasing the amount of blowing and moving the blowing closer to the foot of the shock both moves the mean position of the shock upstream and increases the degree of the forced periodicity.

It is a widely held view that part of the turbulence amplification in shock wave boundary layer interactions is caused by the unsteady shock motion (see, for example, Smits & Muck 1987). The direct effects of the unsteady shock motion on the downstream turbulence, however, were observed to be small.

The p.d.f.s of the upstream turbulence are characterized by Gaussian-type distributions, while downstream the p.d.f.s. are bimodal, centering about two distinct levels, one representative of the freestream mass flux and a lower level found near the wall. Downstream, the mean velocity and mass-flux profiles are inflectional and the instability of the profile may give rise to lateral 'vortices' like those observed between the mixing layer of two unequal parallel jets (Brown & Roshko 1974). Such large-scale lateral structures can contribute strongly to the intense mixing observed downstream of the interaction.

ACKNOWLEDGEMENTS

This work was supported by the U.S. Army Research Office, under Contract DAAG29-K0255, monitored by Dr. T. L. Doligalski.

REFERENCES

Andreopoulos, J. & Muck, K. C. (1987), "Some Aspects of the Shock Wave Boundary Layer Interaction in Compression Ramp Flows," J. of Fluid Mech., Vol. 180, pp. 405-428.

Ardonceau, P. L., Lee, D. H., Roquefort, T., A. & Goethals, R. (1978), "Turbulence Behavior in a Shock Wave/Boundary Layer Interaction," AGARD-CP-271.

Ardonceau, P. L. (1984), "The Structure of Turbulence in a Supersonic Shock-Wave/Boundary Layer Interaction," AIAA J., Vol. 22, No. 9, pp. 1254-1262.

Brown, G. L. and Roshko, A. (1974), "On Density Effects and Large Scale Turbulent Mixing Layers," J. Fluid Mech., Vol. 64, pp. 775-816.

Dolling, D. S. (1985), "Unsteadiness of Shock Wave-Induced Turbulent Boundary Layer Separation - A Review, " IUTAM Symposium on Turbulent Shear Layer/Shock Wave Interaction," Palaiseau, France, September.

Dolling, D. S. & Murphy, M. T. (1983), "Unsteadiness of the Separation Shock Wave in a Supersonic Compression Ramp Flowfield," AIAA J., Vol. 21, No. 12, pp. 1628-1634.

Dolling, D. S. & Or, C. T. (1983), "Unsteadiness of the Shock Wave Structure in Attached and Separated Compression Corner Flow Fields," AIAA Paper 83-1715.

Dussauge, J. P., Muck, K. C. & Andreopoulos, J. (1985), "Properties of Wall Pressure Fluctuations in a Separated Flow over a Compression Ramp," IUTAM Symposium on Turbulent Shear Layer/Shock Wave Interaction, Palaiseau, France, September.

Fernholz, H. H., Smits, A. J., Dussauge, J. P. & Finley, P. H. (1988), "A Survey of Measurements and Measurement Techniques in Rapidly Disturbed Compressible Flows," AGARDograph (to be published).

Green, J. E. (1970), "Interaction between Shock Waves and Turbulent Boundary Layers," Prog. in Aerospace Sci., Vol. 11, pp. 235-340.

Hayakawa, K., Smits, A. J. & Bogdonoff, S. M. (1984), "Turbulence Measurements in a Compressible Reattaching Shear Layer," AIAA J., Vol. 22, No. 7, pp. 889-895.

Murphy, M. T. (1983), "An Experimental Investigation of the Separation Shock Wave Unsteadiness in a Compression Corner Flowfield," M.Sc. Thesis, Department of Mechanical and Aerospace Engineering, Princeton University.

Narlo, J. C. (1986), "Experimental Investigation of the Driving Mechanism of Separation Shock Wave Motion in Interaction Flows," M.Sc. Thesis, The University of Texas at Austin.

Or, C. T. (1985), "Unsteadiness of the Shock Wave Structure in Attached and Separated Compression Ramp Flowfields," M.Sc. Thesis, Mechanical and Aerospace Engineering Department, Princeton University.

Samimy, M., Petrie, H. L. & Addy, A. L. (1986), "A Study of Compressible Turbulent Reattaching Free Shear Layers," AIAA J., Vol. 24, No. 2.

Selig, M. S. (1988), "Unsteadiness of Shock-Wave/Turbulent Boundary Layer with Dynamic Control," M.Sc. Thesis, Department of Mechanical and Aerospace Engineering, Princeton University.

Selig, M.S., Andreopoulos, J., Muck, K. C., Dussauge, J. P. & Smits, A. J. (1988), "Simultaneous Wall-Pressure and Mass-Flux Measurements Downstream of a Shock Wave/Turbulent Boundary Layer Interaction," AIAA Paper 87-0550 (to be published in AIAA J.).

Settles, G. S. (1975), "An Experimental Study of Compressible Turbulent Boundary Layer Separation at High Reynolds Numbers," Ph. D. Thesis, Department of Mechanical and Aerospace Engineering, Princeton University.

Settles, G. S., Vas, I. E. & Bogdonoff, S. M. (1976), "Incipient Separation of a Supersonic Turbulent Boundary Layer at High Reynolds Numbers," AIAA J., Vol. 14, No. 1, pp. 50-56.

Settles, G. S., Fitzpatrick, T. J. & Bogdonoff, S. M. (1979), "Detailed Study of Attached and Separated Compression Corner Flowfield in High Reynolds Number Supersonic Flow," AIAA J., Vol. 17, No. 6, pp. 579-585.

Smits, A. J. & Dussauge, J. P. (1988), "Hot-Wire Anemometry in Supersonic Flow", Chapter 5 of Fernholz et al. (1988) op.cit.

Smits, A. J., Hayakawa, K. & Muck, K. C. (1983), "Constant Temperature Hot-Wire Anemometer Practice in Supersonic Flows, Part I - The Normal Wire," Experiments in Fluids, Springer-Verlag.

Smits, A. J. & Muck, K. C. (1987), "Experimental Study of Three Shock Wave/Turbulent Boundary Layer Interactions," J. Fluid Mech., Vol. 182, pp. 291-314.

Taylor, M. W. (1984), "A Supersonic Turbulent Boundary Layer on Concavely Curved Surfaces," M. Sc. Thesis, Mechanical & Aerospace Engineering Department, Princeton University.

Part 3
QUASI–COHERENT STRUCTURES

Quasi-Coherent Structures in the Turbulent Boundary Layer: Part I. Status Report on a Community-Wide Summary of the Data

S. J. KLINE
Department of Mechanical Engineering
Stanford University
Stanford, California 94305, USA

S. K. ROBINSON
NASA Ames Research Center
Moffett Field, California 94305, USA

ABSTRACT

This paper describes a current comprehensive study of turbulent boundary layer structure. The project involves two major efforts, and will be reported upon in the two parts of the paper.

Part I will describe the motivation, approach, and early results of a community-wide reevaluation of the knowledge collected by the research community in three decades of research on the structure of turbulent boundary layers. Part I will briefly review what is known about the major structural features, as has been learned through experimental techniques. The contributions by numerical simulations are discussed in Part II.

Part II presents preliminary results of a study of a numerically simulated turbulent boundary layer. The work has focused upon the open questions raised by the cooperative project of Part I. Most of the known structural features are confirmed in the simulation, and their spatial relationships are described.

INTRODUCTION

This paper (Part I of a pair) is organized into four sections. The first presents four distinctions which must be made in order to clarify results in the field of quasi-coherent structures in boundary layer turbulence. The second section describes the processes of the community-wide review project. The third section presents a categorization of known structural features and summarizes some of the experimentally gained knowledge about each class. The final section draws on the available data for a wide variety of boundary layer flows in order to illustrate the type of structural inferences that can be made by considering more than flat-plate, zero pressure-gradient boundary layers. This part is not intended to serve as a thorough review of the field, but rather to introduce the community-wide study and its potential payoffs. A much broader and more complete set of boundary layer structure information will be made available upon completion of the project.

PART I. COMMUNITY-WIDE REEVALUATION OF THE STATUS OF KNOWLEDGE OF QUASI-COHERENT STRUCTURES IN TURBULENT BOUNDARY LAYERS

A. Background—Four Distinctions Needed to Clarify Work on Quasi-Coherent Structures in the Turbulent Boundary Layers

The present state of knowledge concerning the structure of quasi-coherent motions in the turbulent boundary layer is in a highly mixed state. This has motivated the initiation of a community-wide effort to collect and evaluate the information gathered in 30 years of turbulence structure research. A major task of the community-wide project will be to try to distinguish what we know from matters still in doubt.

On the one hand, we know a good bit about coherent structures. They not only exist, but, indeed, are necessary for the production of Reynolds stresses. A number of types of coherent structures have been documented, and many details are known about each. These structures can be categorized into eight classes (at least), most with subclasses. Of these eight classes, there is strong evidence that at least seven of them are important in the dynamics in some portion of the boundary layer.

On the other hand, when it comes to understanding how the classes of structures relate to and/or interact with each other, we still have much to learn. As a result, we have nearly as many views of the complete structural model as there are active research groups studying the problem.

What can we do about this situation? How do we move toward better understanding of the physics and better bases for models of turbulence in boundary layers? To begin, we can suggest four distinctions, or criteria, that are needed in order to improve the practices in the field. These criteria have been recognized by others; however, a summary of them in one place seems timely in view of what will be reported in this paper.

1. <u>Data versus inferences</u>. We need to distinguish between results which are reliable data, or are direct deductive consequences from reliable data, and inferences (or hypotheses, or suggestions, or hunches) that go beyond the data and its deductive consequences. There will be some grey cases, but the distinction is of obvious first importance. For example, the existence of the low-speed/high-speed streak structure and the lifting and ejection of the low-speed streaks are well-documented data. The direct role of these quasi-coherent structural elements in the dynamics is the subject of several differing inferences. As another example, the existence of large-scale "backs" on the upstream faces of large-scale motions is also clearly documented. Again, however, there are differing inferences concerning the dynamical role played by these backs. There are many other examples.

2. <u>Possible versus actual versus significant events</u>. The authors believe it is necessary (probably essential) to distinguish between: (i) what <u>can</u> happen in a boundary layer, for example with forcing; (ii) what does happen naturally in turbulent boundary layers; (iii) what happens in turbulent boundary layers often enough and with sufficient correlation of motions so that events or structures observed have "statistical relevance" to the production of turbulence or for particular terms in model equations.

3. <u>One structure versus many</u>. Since there are a number of different important structural features in the turbulent boundary layer, there is a need to avoid overemphasizing any one structure. There is, as a part of the processes of normal science, a professional demand

that we explain, justify, and defend our own results in the crucible of public testing. There is also an all-too-human tendency to "identify" with our own brainchildren and data. However, the facts suggest strongly that more than one process is involved in the production and dissipation of turbulence, and the available laboratory techniques typically show us clearly only one structure at a time, and even that sometimes dimly. Hence, it seems timely that we expand our views to take in all the structures and not argue that any one or two structures is sufficient for full understanding. There is a need to step back and take a broader view which extracts what is valid in each set of data and then reassemble a new, improved whole. This process will demand from researchers a willingness to reevaluate long-cherished ideas, and that is very difficult for most of us. However, the authors believe the sooner this idea is accepted, the sooner we can expect to see important progress toward unification in the field of quasi-coherent structures.

4. One or two flows versus the totality of data. Work on coherent structures has been concentrated largely on the canonical flat-plate boundary layer (Fig. 1) and to a lesser extent on the canonical planar channel flow. However, a great deal of data, much of it taken for other purposes, exists for many other flow cases: curved-wall flows, flows with pressure gradient, reattached free-shear layers following separations, flows with incipient separation short of full detachment, flows with rotation, flows with body forces, flows with many kinds of unsteadiness in the free stream, and so on. Some particularly accurate and detailed data exist in the convection heat transfer literature. All of these flows provide information about what happens, and all of them act, at a minimum, as constraints on what we must be able to explain and predict in any complete view or model of turbulent boundary layers.

The four distinctions discussed above are, in concept, self-evident. However, decisions on these matters require judgments, and those judgments are not discussed in this paper; instead they are left for discussion by the research community after the results of the community-wide project have been gathered and circulated.

B. Processes of the Community-Wide Project

In late 1986 the authors of this paper began a community-wide study of the available knowledge on coherent structures in the turbulent boundary layer. The phrase "community-wide" is used to indicate that the project would involve as much of the entire international research community as possible in order to utilize all the available knowledge and to avoid parochial views that any one group, including our own, might hold. This evaluation is proceeding by two primary processes.

First, we are conducting personal interviews with as many of the highly active workers as possible. The objective of these individual meetings is to carry out extended discussions about 1) the researcher's beliefs and opinions (both published and unpublished), 2) the experimental and/or computational bases for these beliefs, 3) any revisions to the researcher's previously published conclusions, and 4) the remaining open questions in the field.

Second, we are extracting from roughly 175 papers the major truth assertions, item by item. A major portion of this extraction work is being carried out by O. Manickam and L. Portela, and an early trial study also included Prof. J. P. Johnston, P. Johnson, and M. Plesniak whose assistance is gratefully acknowledged.

The response of the research community in the work of the community-wide project has been extremely positive; almost without exception the researchers in the field have done all they could to assist in the work; often they have volunteered more than has been asked. For example, a set of special terms used in the study of quasi-coherent structures was collected in order to improve semantics and thereby reduce confusion in communications. The first draft of this nomenclature list was sent to about 70 workers worldwide; of these, 26 took the time to provide very detailed written comments. Without exception, each of these 26 sets of comments has not only raised important questions but also made useful suggestions for improvements. This iterative clarification of the bases of communications will be continued.

The interviews of active researchers have proven to be productive for several reasons. First, each scientist has been found to know things, often unpublished, which are not known to most other research groups. Second, a much greater degree of convergence in beliefs has emerged regarding the data than is suggested by the published papers. Also, open questions where there is not agreement have come into sharper focus. Third, the limited discussion time available at typical technical meetings is insufficient for a full exchange of views between workers in this complex field. Thus, the longer discussions afforded by the interview process have clarified many individual views which will add to our collective knowledge.

C. A Classification of Observed Quasi-Coherent Structures in the Turbulent Boundary Layer

The term "quasi-coherent" is used to emphasize that the structures observed are not wholly deterministic, but rather exhibit significant variation between realizations. The taxonomy which follows, like all human-made taxonomies, has a considerable degree of arbitrariness. It is believed to cover the known, observed quasi-coherent structures; however, the selection of categories, while containing some logic, could have been done in many other ways. No significance is implied by the ordering of the list.

EIGHT CATEGORIES OF QUASI-COHERENT STRUCTURES IN THE CANONICAL TURBULENT BOUNDARY LAYER AND CHANNEL

1. Low-speed streaks in the region $0 < y^+ \gtrsim 10$

2. Ejections of low-speed fluid outward from the wall

3. Sweeps of high-speed fluid toward the wall

4. Vortical structures of several proposed forms

5. Strong internal shear layers in the wall zone; $y^+ \gtrsim 80$

6. Near-wall "pockets," observed as areas clear of marked fluid in certain types of visualizations

7. Large (delta-scale) discontinuities in streamwise velocity

8. Large-scale motions in the outer layers (including, for boundary layers, bulges, superlayers and deep valleys of free-stream fluid)

In surveying what is known of these structures, the first thing that must be noted is that the dynamic processes in turbulent boundary layers are very complicated. Most phenomena in physical nature, when understood, are seen to be relatively simple; turbulent shear flows are not. Many different facts support the true complexity: (i) the boundary layer is a two-scale problem, with an inner and an outer layer having fundamentally different scaling variables; (ii) the motions and the observed structure are inherently three-dimensional, even when the mean flow is two-dimensional; (iii) there exists not one but many types of quasi-coherent structures; (iv) there exists what appear to be at least two classes of "local instabilities" which play significant roles in the production processes; (v) there is no single cause nor effect, but rather a complex flow of recurring events in the production processes; (vi) the variation between realizations of structures is large and the background noise level high; (vii) many aspects of the observed quasi-coherent structures vary parametrically with boundary conditions such as pressure gradient, wall curvature, force fields, rotation, etc.

The paper now discusses some of the knowledge about each of the classes of quasi-coherent structure that have been observed in the turbulent boundary layers. Emphasis will be placed on experimental results, while recent findings from numerical simulations will be discussed in Part II of this paper. It is reemphasized that what is given here is a sample from a much larger set of truth assertions that has been assembled during the course of the community-wide project.

1. Low-speed streaks in the region $0 < y^+ \gtrsim 10$. The existence of low-speed streaks in the sublayer of turbulent boundary layers has been documented by many laboratories. These low-speed streaks have a well-established mean spanwise spacing ($100\ \nu/u_\tau$) from the wall through the sublayer zone, $0 < y^+ < 7$. They are ubiquitously present in this zone in turbulent boundary layers, and are visible to roughly $y^+ = 40$, but not farther from the wall.

Smith and Metzler (1983) demonstrated that the mean streak spacing of $100\ \nu/u_\tau$ was invariant with Reynolds numbers to moderately high values. Smith and Metzler (1983) and also Falco (private communication) have found that the probability distribution function of the low/high-speed structure around the mean value of $100\ \nu/u_\tau$ is lognormal with a typical dense range from about 60 to 180 wall units. Schraub and Kline (1965) demonstrated that the mean spacing of the wall-low-speed streaks is the same for positive and negative pressure gradients as for the zero-pressure gradient case. Only when the flow began to relaminarize did streak spacing in wall units increase. The same study shows that the location of the streaks is random over long times; that is, not tunnel dependent. Elder (1960), and Kline and Meyer (1961) measured a mean spanwise streak spacing of $100\ \nu/u_\tau$ beneath turbulent spots in transitional boundary layers. As soon as a spot forms, and from there on downstream, this spanwise spacing is observed.

In the canonical flat-plate flow, the low-speed streaks have a streamwise velocity typically 1/2 of the local mean speed, while the adjacent high-speed streaks have a velocity of roughly 3/2 local mean speed. The wall-low-speed streaks are not in themselves vortices; they are simply regions of low-speed motion.

The wall-low-speed streaks are not straight but are observed to meander spanwise over ranges about equal to the streak width of 30-50 wall units. If one allows for the meanders, many low-speed streaks extend for surprising distances, hundreds and in some cases thousands of wall units in the streamwise direction. Wall markers typically do not stay in the near-wall region for distances this long, but wall temperature contours by Hirata and Kasagi (1979) demonstrate this feature clearly.

Gad-el-Hak (private communication), when stimulating laminar layers and in attempts to control turbulent layers using suction through two holes found the scheme effective only when the holes had a spacing within the dense range of wall-streak spacing documented by Smith and Metzler (1983) and by Falco. Chu and Falco (1986) demonstrated that when a ring vortical structure strikes a wall, three-dimensional structures produced a near-wall streaky structure with a spanwise spacing of 100 wall units. These results when taken together suggest not only an inherent three-dimensionality of the dynamics of the inner layers, but also that this three-dimensionality has a mean spanwise dimension of 100 wall units for a wide range of flows. However, the reasons for this spanwise spacing remain unknown.

2. Ejections of low-speed fluid outward from the wall. As described by Kline and co-workers, Willmarth and Lu (1972), Corino and Brodkey (1969), Smith and Metzler (1983), Tiederman and co-workers, and others, the low-speed streaks seen by marking fluid essentially at the wall, migrate slowly outward as they move downstream, and at some downstream station turn sharply and move far more rapidly away from the wall in ejection-motions, called "lifting." The lifted streaks of marked fluid then undergo rapid oscillations with a wavelike motion when viewed in lab coordinates (Kline et al., 1967; H. T. Kim et al., 1971), and with an oscillatory motion when viewed by an observer moving with the flow at a speed equal to the near velocity at $y^+ = 40$ (Corino and Brodkey, 1969). Conversations between Prof. Brodkey and the authors suggest the differences are those of frame of reference and not differences in the motions.

The oscillatory motions described by Kline and Runstadler (1959), Runstadler et al. (1963), Kline et al. (1967), Smith and Metzler (1983), Tiederman et al. (1985), and others, lead to a "breakdown" of the structure in the sense that finer scales of motion rapidly appear, suggesting locally high values of turbulence production. The entire sequence of events from low-speed streak formation through streak lifting, ejection and breakup was called "bursting" by Kline and co-workers. However, much more is now known of the structural details, and it seems best to describe these events separately rather than as a group at this time.

Runstadler et al. (1963) plotted the probability density distribution of the outward lifting and ejection motions in x-y coordinates. They found a strong peak of most common trajectory, but also a long tail at large y^+ values. These data suggest interaction between fluid elements originating essentially at the wall and fluid in the log and even the wake region for the Reynolds number studied ($Re_\theta = 1400$). (See also discussion in Section D below.)

Many workers believe the streak structure in the very-near-wall region is dynamically important. However, others believe that the streaks are passive and therefore not a significant factor in turbulence production. This disagreement may arise in part from failure to distinguish between low-speed streaks in the sublayer, $0 < y^+ < 7$, and lifted low-speed streaks which are observed primarily in the region $10 < y^+ < 40$. The near-wall streaks do appear to be passive and the lifted streaks are observed to be involved in interactions that involve much of the turbulence production in the layer. Beyond the specific disagreement, the data are clear; what is not clear is the inferences drawn from the data concerning the dynamical role of the low-speed streak structures.

3. Sweeps of high-speed fluid toward the wall. Markers of particles placed in the near-wall region show ejection motions but do not clearly show wallward sweeps of high-speed fluid. It follows from continuity that the outward ejection motions must be compensated by inward sweeps. This was in fact made explicit by Corino and Brodkey (1969), Grass (1971), and many other observers since. However, the details of the high-speed regions

near the wall associated with sweeps do not seem as well understood, partially because of the difficulty in marking sweeps in laboratory boundary layers.

A number of studies, particularly those by Willmarth and Lu (1972), Willmarth and Bogar (1977), and Brodkey, Wallace and others, have shown that sweeps do play an important role in the production of turbulence in the wall layers. More specifically, these studies have shown that ejections create the largest fraction of Reynolds shear stress for $y^+ > 12$, but nearer the wall sweeps produce the larger fraction of Reynolds shear stress; however, both ejections and sweeps are significant on each side of $y^+ = 12$.

4. Vortical structures* of several proposed forms. Theodorsen (1952) hypothesized the existence of "horseshoe-shaped vortical structures" in the wall layers and suggested that they were the "molecule of turbulence." Townsend (1956) deduced the existence of wall-attached eddies (coherent structures) in the inner region of the turbulent boundary layer from available correlation measurements.

Runstadler et al. (1963) proposed that the streak patterns, ejections and sweeps of items 1, 2, and 3 were related to an arch-like "vortical structure" which overrode the streaks. However, Runstadler et al. (1963) failed to emphasize sufficiently that this vortical structure is the ensemble averaged pattern, and need not be present completely in every realization. The distinction between ensemble averages and individual realizations is an important issue in several matters that appear later in this paper.

H. T. Kim et al. (1971) showed that turbulence production equal to all the net production in the entire boundary layer (70% of the total positive production) occurred during times that involved lifting of the wall-low-speed streaks (ejections). The observations of the bubble motions from combined-time-streak markers indicated the existence of both finite-length transverse vortices and also spiralling motions which suggested tilted vortices with primary alignment in the streamwise x-direction but with some component of outward y-direction also. However, the marker techniques do not make these spiralling motions clear; the bubble pictures obtained are more suggestive than definitive regarding vortical structures even though they are clearer in this regard than dye markers inserted near the wall. Lian (1987), using improved bubble marker methods to make vortices more visible, remarks that it is still difficult to make a complete horseshoe-like vortical structure visible with marker techniques. Pearson and Abernathy (1984) have demonstrated the existence both theoretically and experimentally of true hydrodynamic instability when vortices with streamwise orientation exist in a mean strain field, dU/dy.

Head and Bandyopadhyay (1981), using flow visualization techniques, also produced evidence of "horseshoe-shaped" vortical structures that are visible throughout much of the boundary layer on a flat plate at modest Reynolds numbers, when the flow is viewed at 45° inclination.

Falco (Falco, 1974, 1977, 1980a,b; Falco and Wiggert, 1980; Falco, 1982, 1983, 1985, 1987) has suggested that the important vortical structures are not "horseshoe-like" in shape, but are instead ring-like vortices. In recent work, Chu and Falco (1986) suggest the rings are generated from the "pinching-off" of horseshoe-like vortical structures occurring along the edge of "pockets" at the wall. Falco has also suggested that vorticity redistributions may give rise to ring-like vortical structures.

---

*The phrase "vortical structure" in this paper refers to connected vortices with more than one orientation and not to distributed vorticity.

The controversy concerning the shape of the vortical structures remains unresolved in the accumulated results from laboratory measurements. The problem seems to stem from two sources: (i) the difficulty in making vortices visible, using available laboratory techniques, particularly: (a) a complete complex vortical shape, such as a horseshoe, or (b) a number of vortices all at once; (ii) insufficient distinction between data (and its direct logical consequences) and inferences.

Several controversies have arisen concerning the details of the "horseshoe," or "arch," or "hairpin-like" structure. The important distinction between ensemble averages and realizations is again noted here. These have been understood for some time. For example, Brown and Thomas (1977) give a very lucid description of the difficulties of inferring structure from ensemble averages even when the educed signals are carefully and cleverly "enhanced." The authors suggest that these fine-grain distinctions concerning the vortical structures seem premature until more complete evidence on many individual realizations are available.

Several sets of workers have inferred from ensemble average measurements that streamwise vortices exist near the wall. Blackwelder and Eckelmann (1979) inferred this result from two-point space-time correlations of hot-wire data. Bakewell and Lumley (1967) inferred the existence of streamwise vortices from sophisticated statistical considerations (proper orthogonal decomposition). Smith and Schwartz (1983), using crossed Y-Z bubble marker wires, showed that streamwise vortices do not exist in the zone $0 < y^+ < 10$, but that strong intermittent streamwise rotation does exist centered on the zone $10 < y^+ < 30$, that is, in the buffer layer. The ensemble averages suggest streamwise vortices are continuously present in time; however, the data of Smith and Schwartz (1983) indicate the vortices are only intermittently present at a given point a low fraction of the time, pointing again to the important difference between ensemble averages and realizations.

None of the results cited in this section, given the techniques employed, would clearly distinguish between: (i) streamwise vortices with pure x-orientation, and (ii) vortices slanted away from the wall downstream, with xy-orientation, extending from the near wall farther out into the flow. Physically, however, the difference between vortices with x- and xy-orientation is important since the x-oriented vortices will not be stretched, and the xy-oriented vortices will be stretched strongly owing to the high values of dU/dy in the near-wall region.

In summary, the totality of data from laboratory experiments thus far have not made convincingly clear the spatial/temporal relationships between low-speed streaks, high-speed regions, and vortical structures near the wall for individual realizations. The data clearly indicate that all three of these structures are present, but the spatio-temporal relationships remain unclear for the following reasons: (i) laboratory techniques for showing all of the structures simultaneously are lacking; (ii) marker methods show vortex motions, particularly complex vortical structures, only dimly; (iii) the need to eliminate "jitter" requires ensemble averaging of probe data so that phase relations are lost. The incompleteness of the picture given by the laboratory data has thus left open the possibility for a variety of inferences about the vortical structures and also the spatio-temporal relations of vortical structures to the other important structural features of the boundary layer.

5. Internal shear layers in the wall zone. Corino and Brodkey (1969), H. T. Kim et al. (1971), Kreplin and Eckelman (1979), Blackwelder and Haritonidis (1983), Alfredsson

and Johansson (1984), Jimenez et al. (1988), and several others have all observed sharp internal shear layers in the inner regions of the boundary layer; that is, local regions where the shear rate significantly exceeds its local mean value. Near-wall internal shear layers exist with both dU/dy and dU/dz gradients, and they appear to occur frequently. These internal shear layers should not be confused with another type of shear layer, called "backs," which are described in item 7 below.

H. T. Kim et al. (1971), Smith (1984), and Swearingen et al. (1987) are among those who have suggested that a portion of the dynamics of the boundary layer could possibly arise from the roll-up of vortices in a Kelvin-Helmholtz-like instability. The word "like" is appended because the theory is for two-dimensional disturbances, and the observed internal shear layers are relatively short in extent normal to the velocity gradient. Kelvin-Helmholtz-like roll-ups along sharp internal shear layers have been observed in the laboratory as reported by Doligalski et al. (1980), Smith (1984), and Acarlar and Smith (1987b), both in perturbed laminar boundary layers and in fully turbulent boundary layers.

As Tiederman and others have shown, the VITA-detection technique as originally used in the near-wall region by Blackwelder and Kaplan (1976) triggers on the near-wall shear layers, which are seen in probe output as a rapid acceleration or deceleration. Bogard and Tiederman (1986, 1987) also showed that these shear layers appear upstream of a burst event. Bogard and Tiederman showed two things are important in understanding the differences between VITA-detected events and visual observations of the bursting sequence. First, it must be recognized that there is typically more than one ejection per burst (two on the average); this is seen clearly in the observations of a number of workers. Second, while the probes are more precise in giving quantitative data than visual markers, the visual recognition is more accurate in keying on specific structures. Probe detectors, therefore, need to be calibrated against the appropriate visual signatures when they are intended to reliably detect certain structures or events characterized by their visual properties. The work of Alfredsson and Johansson (1984) is also important in understanding the detection of events; however, the focus of this paper is on structures rather than techniques and further details are not discussed here.

Conventional probe techniques do not allow a clear distinction between sharp, small-scale shear layers in the near-wall region and the delta-scale "backs" discussed in item 7 below, and this has led to difficulties in interpretations of VITA-detected events. In this instance also the inability to see spatial relationships between structures with laboratory methods has inhibited full understanding of phase relations of various structures and this has led to conflicting inferences by various researchers.

6. <u>Near-wall "pockets," observed as areas clear of marked particles in certain types of visualizations</u>. Pockets (Falco, 1983) are areas in the sublayer that are clear of markers; the clear areas appear when a layer near the wall is made visible by distributed markers and observed in plan view.

Both R. Falco and co-workers and C. R. Smith and co-workers have observed pockets in a number of studies. Falco has also shown, in unpublished data, that the pockets have a mean spanwise dimension of 60 wall units and a lognormal probability distribution function. The agreement of these spanwise dimensions with the wall-low-speed streak structure, coupled with visual study of the edge of pockets and simulations of the effect of ring vortices slanting downstream onto a wall, have together led Chu and Falco (1986) to suggest that the formation of ring vortices arises from pinching off of horseshoe vortices which form at the sides and upstream and downstream ends of pockets. Smith, on the other hand, suggests that the pockets are merely artifacts of the marking technique and do not play a dynamical role. Here again we see agreement on the data but various views in what is inferred from the data.

7. Large (delta-scale) discontinuities in streamwise velocity. Brodkey et al. (1984) and also Eckelmann et al. (1977), and others, have produced a number of studies showing that "backs," or sharp jumps in U extend from the wall through the entire layer, thus having a scale of delta. Brown and Thomas (1977), and other workers, show that these backs are the upstream face of large-scale motions (see item 8) and the ensemble averages of the backs are inclined at 12 to 25 degrees to the wall. Hence they are here called "backs" rather than "fronts," as they have sometimes been named.

Some results suggest a causal relationship between backs, bulges, and high-production events near the wall. Praturi and Brodkey (1978) suggest a one-to-one correspondence, that is, high-production events are in phase with the passage of backs. However, techniques for making backs and ejections simultaneously visible in the laboratory are difficult at best. Hence the phase relation between delta-scale backs and near-wall ejections remains unclear, based on the existing laboratory data. Recall here also the distinction between what does occur and what is statistically relevant in the sense that passage of backs might set off some but not all of the high turbulence production events in the inner layers.

Praturi and Brodkey (1978) also suggest that the delta-scale backs lie at the center of the large-scale motions of item 8. However, Brown and Thomas (1977) provide correlations indicating the discontinuities in U are at the upstream side, not the center of the large-scale motions. Brown and Thomas also provide conditional velocity data, demonstrating that the vorticity and rate of shear along the backs is high.

Brown and Thomas (1977) have also shown a relatively high correlation of the passage of backs and both a low-amplitude, low-frequency oscillation of the wall shear and a high-amplitude, high-frequency oscillation of wall shear. They suggest both correlations are causal. They also suggest that the creation (or reinforcement) of the near-wall low-speed, high-speed streak structure could arise from the concavity of the streamlines near the wall under the large-scale motion and a resulting Taylor-Görtler instability giving rise to roll cells, as suggested by Coles. Brown and Thomas present an analysis showing the streamlines and rotation configuration is Taylor-Görtler unstable, but do not give calculations indicating if the instability acts for long enough times to provide the observed structures. Brown and Thomas also suggest that this interaction is the cause of the observed low-speed streak, lifting, ejection, and breakup sequence discussed above. (See, however, the discussion of inner/outer interactions in Section D below.)

8. Large-scale-motions in the outer layers (for boundary layers: with backs, superlayers, and deep crevasses of free-stream fluid). Corrsin and Kistler (1954) showed that there was a sharp interface between the turbulent fluid within a boundary layer and the outer potential flow. Murlis et al. (1982) used the term "viscous superlayer" to describe this interface as "the region at the instantaneous edge of the turbulent boundary layer in which vorticity is transferred to originally irrotational fluid by the action of viscosity." Many workers, using both probes and visual studies, have shown that for a boundary layer, the superlayer is highly contorted and occasionally extends nearly to the wall at low Re, in narrow valleys at the upstream and downstream end of large-scale motions of turbulent fluid. In a fully developed channel flow, no superlayer exists; the entire fluid volume is turbulent, and the valleys to the extent they may exist are not of potential flow but of outer turbulent fluid.

Nevertheless, the U^+ vs y^+ correlations of the inner layers is the same for the two flows. This suggests the effect of the superlayer itself on the inner layers is limited.

Kovasznay et al. (1970) showed that the large-scale motions have a slow rotation with the sense of the mean strain. The picture of the movements in the large-scale motions was amplified by Cantwell et al. (1978), who showed a quasi-stagnation point near the trailing surface of the bulge with downward flow beneath it and upward flow above. Brown and Thomas (1977), and also Falco (1977, 1980a), each using different techniques, also confirm this picture.

Many studies, including Klebanoff (1956), have shown that the outer flow contributes relatively little to the Reynolds stresses and the turbulence production. Coles, in current work on the synthetic boundary layer, presented to the Summer Study Group at the NASA Ames/Stanford Center for Turbulence Research in 1987, reached similar conclusions, specifically, that the largest scales of the boundary layer carry little Reynolds stress.

In ending this section, two summary remarks are appropriate. First, all eight classes of structure are present. Each structure has been documented by more than one independent observer and in more than one laboratory. The data provide many details and deny the possibility that any of the eight structures does not exist. However, the laboratory data have not made clear the relationships between the various structures for reasons that have been elaborated, case by case. This lacuna of the relationships among the structures in the laboratory data leaves open the possibility for various inferences concerning: (i) the spatio-temporal relations among the structures; (ii) the flow of causality among the structures; and (iii) the underlying dynamics. Second, the results given for each of the eight structures is not exhaustive but rather a sample of what will ultimately appear in the compiled results of the community-wide project.

D. The Inner/Outer Controversy

As noted in the introduction, the use of data from the wide variety of flow situations available, as contrasted with study of one or two canonical flows, makes possible the deduction of significant conclusions. This section provides one illustration of this kind of work. By discussing what can be called "the inner/outer controversy," it will be shown that examination of a variety of data can begin to close this controversy.

There has been a long-standing controversy concerning the relative importance of the inner and outer layers of the turbulent boundary layers as causal factors in the dynamics of the turbulent motions. Here, each of these views will be oversimplified in order to state them concisely and to draw the contrast between the views sharply.

Workers too numerous to mention have suggested an "outerdominant" view, in which the inner layers are "driven by the fluctuations of the outer layers." This is an older view, and often it is held more implicitly than explicitly.

Many other workers have suggested a "wall-dominant view", in which the "energetic actions including most of the turbulence production occur in the inner layers and the fluctuations so produced diffuse outward, and hence the outer layers are merely collections of tired turbulence diffused outward from events nearer the wall over the upstream surface."

A third view suggests that the two layers "interact" so that both are important in the totality of events which occur in a boundary layer. It can be argued, using negative inference, that the data for the canonical flat-plate layer deny both the outer-dominant and the inner-dominant view, and hence the interaction view must be correct, since it is the only option not denied by the data. Many researchers have reached this conclusion. However, an interaction view is a continuum running from: (i) most of the important events occur in the outer layer to (ii) most of the important events occur in the inner layer. The collection of data from the community-wide project gives a first-order answer to the question of relative importance of inner/outer layers, and illuminates some related questions.

Two types of data are particularly relevant: those for turbulent boundary layers on convex walls, and the data from boundary layers manipulated by "Large-Eddy Breakup Devices" (LEBUs). Both situations create flows in which the inner walls are largely isolated from outer fluctuations. In the LEBU data, the inserted control plate physically isolates the two layers. In the convex wall flow of Gillis et al. (1980) after 20 degrees of turn (out of 90 degrees), the outer fluctuations are essentially isotropic and follow the laws for isotropic decay of fluctuations as they move downstream, thus eliminating the Reynolds stresses in the outer flow. Still farther downstream, even the magnitude of the remaining isotropic fluctuations decays to low values owing to the stabilizing curvature of the mean streamlines of the flow. If the outer-dominant view were correct, we would have to expect that the turbulence production would stop in the inner layers in these flows, or at least continue to decay, once the outer fluctuations were removed. However, this does not occur in either case. Instead, the value of C_f and Stanton number are reduced by 10-20% depending on flow parameters, and the turbulence continues in the inner layers. Moreover, in the convex wall flow, renormalization of the wall-layer mean profiles, based on the reduced value of u_τ, produces the well-known standard inner wall function for U^+ vs y^+, suggesting that in some sense the inner layers have the usual structure.

In a convection heat transfer study, Ligrani et al. (1978) have added additional important pieces of knowledge. First, Ligrani et al. show there is a unique mean flow correlation, in agreement with the wall-wake correlation of Coles (1956, 1968) to the uncertainty level cited. Second, they demonstrate that alterations in the fluctuations of any of: the inner layers, the outer layers, or the free stream can create departure from the equilibrium between the inner and outer layers inherent in a flow obeying the wall-wake correlation in the mean motion. A large body of rough-wall data indicates that changes at the wall alone can also destroy the inner/outer equilibrium. Third, Ligrani et al. demonstrate for several cases that the departures from equilibrium between inner and outer flows vanish asymptotically downstream once the disturbances are removed or die away, and that, for modest disturbances, this approach to equilibrium occurs nearly completely in a few tens of boundary layer thicknesses.

There are many other data documenting the relaxation of a flat-plate layer from an out-of-equilibrium situation toward the canonical flow, that is, equilibrium between the inner and outer layers. Some of these data show extraordinary lengths of flow required, particularly after reattachment of a separated flow, as, for example, those by Bradshaw and Wong (1972). There are some flows, for example those of Ligrani et al. (1978), where the return to canonical equilibrium flow is at a moderate speed. However, if we take equilibrium between the inner and outer flow to be measured by agreement of the mean profile with the standard form, say the Coles wall-wake functions, there are no data seen in this study that show rapid return to an equilibrium once the flow is in a nonequilibrium state.

Data for a number of flows with unsteady free streams also provide useful evidence on the nature of free-stream boundary layer interactions. More specifically, four studies (Hussain and Reynolds, 1975; Acharya and Reynolds, 1975; Jayaraman et al., 1982; and Brereton and Reynolds, 1987) in which the free-stream unsteadiness was made as two-dimensional as feasible failed to show significant alterations of the turbulence statistics, except for alteration of the phase of the Stokes layer at the wall when passing through the bursting frequency. Since these studies encompassed a wide range of amplitudes, wave-forms, and also pressure gradients in channels and boundary layers, the results suggest that free-stream disturbances do not influence the wall structure. However, recent results by Maciejewski and Moffat, reported in this seminar, show that when three-dimensional, high-amplitude fluctuations (up to 60% of the rms) are introduced in the free stream, changes in Stanton number up to 2-5 times the canonical values are measured for various Reynolds numbers. The precise wave forms and spanwise extent of the fluctuations responsible for these alterations in Stanton number (and presumably also in C_f and in wall layer structure) are not known, but it is known they are not the largest-scale fluctuations—those of the scale of boundary layer thickness. These data serve to remind us of two important aspects of the turbulent boundary layer: (i) the preferred receptivity to three-dimensional perturbations; and (ii) the potential for direct interaction between the outer flow and the wall.

The results summarized in this section indicate several conclusions concerning the "inner/outer controversy":

a. The inner layers are capable of maintaining turbulent flow without any fluctuations from the outer layers. It appears that once the flow becomes turbulent (the streamwise location where a turbulent spot first appears), from that point downstream the inner layers contain some sequence of events capable of maintaining turbulence by themselves.

b. The level of fluctuations and turbulence production in the inner layers can be significantly affected by fluctuations in the outer layers of the boundary layer, or by some kinds of disturbances in the free stream.

c. In the canonical flat-plate layer the major proportion (roughly 70% or more) of the turbulence production in the inner layers is self-maintaining in the sense that it is apparently triggered by events occurring within the inner layer itself, and a smaller proportion (roughly 30% or less) is added on by the effect of fluctuations from the outer layer.

d. The inner and outer layers do interact, but only weakly, and hence the interaction is slow. This is evident not only in the results when outer fluctuations are removed, but also in the fact that return to equilibrium is at most moderate in speed. If the inner/outer interactions were in fact driven by pressure fluctuations, we would have to expect a rapid return to equilibrium since the propagation times would be very short across the layer. Similarly, if the outer flow were dominated directly, rather than diffusively by energetic events in the inner layer, then we would have to expect a more rapid return to equilibrium than exists.

These conclusions appear to contradict the notion that in the canonical flat-plate boundary layer, <u>all</u> highly active events in the inner layers are phase-locked to and triggered by outer-region motions. However, as in many results cited for turbulence structures, the laboratory data do not give direct evidence on the question since phase relations between the various structures are generally lacking, and in no instance have the phase relations among all the structures been reported in the experimental data.

E. Conclusions (Part I)

In the discussion of the known quasi-coherent structures and in discussing the inner/outer controversy we have seen that three factors significantly reduce the information available from laboratory data as a basis for forming a complete model of the quasi-coherent structures in the turbulent boundary layer: (i) the necessity to ensemble average probe output with resulting loss of phase information; (ii) the inability to see more than one, or two, of the various types of structures simultaneously; and (iii) the inability to see the spatial relations between the various quasi-coherent structures resulting from (i) and (ii). The repeated appearance in the discussion of these difficulties when coupled with the fact that most of the disagreements among current researchers are not about the data, but are rather about inferences that go beyond the data, suggests a basic insufficiency in the laboratory data which makes possible many differing logical inferences concerning the complete model of the quasi-coherent structures. Looking back at the results that were summarized from this point of view, what seems striking to the authors is how much information researchers have been able to infer from available data through penetrating insight and careful technique. This remark leads naturally to the topic of computer study of full numerical simulation databases, which is taken up in Part II of this paper.

ACKNOWLEDGMENTS

The community-wide project, as well as the time of S. J. Kline, O. Manickam, and L. Portela, is supported by the United States Air Force Office of Scientific Research and by the United States Office of Naval Research. The time of S. K. Robinson is supported by NASA Ames Research Center. The continuing interest and encouragement of all these agencies is gratefully acknowledged.

REFERENCES

Açarlar, M. S., and Smith, C. R., A Study of Hairpin Vortices in a Laminar Boundary Layer. Part 2: Hairpin Vortices Generated by Fluid Injection, J. Fluid Mech., vol. 175, p. 43, 1987b.

Acharya, M., and Reynolds, W. C., Measurements and Predictions of a Fully Developed Turbulent Channel Flow with Imposed Controlled Oscillations, Rept. TF-8, Thermosciences Div., Dept. Mech. Eng., Stanford Univ., 1975.

Alfredsson, P. H., and Johansson, A. V., On the Detection of Turbulence-Generating Events, J. Fluid Mech., vol. 139, p. 325, 1984.

Bakewell, H. P., Jr., and Lumley, J. L., Viscous Sublayer and Adjacent Wall Region in Turbulent Pipe Flow, Phys. Fluids, vol. 10, no. 9, p. 1880, 1967.

Blackwelder, R. F., and Kaplan, R. E., On the Wall Structure of the Turbulent Boundary Layer, J. Fluid Mech., vol. 76, part 1, p. 89, 1976.

Blackwelder, R. F., and H. Eckelmann, Streamwise Vortices Associated with the Bursting Phenomenon, J. Fluid Mech., vol. 94, part 3, p. 577, 1979.

Blackwelder, R. F., and Haritonidis, J. H., Scaling of the Bursting Frequency in Turbulent Boundary Layers, J. Fluid Mech., vol. 132, p. 87, 1983.

Bogard, D. G., and Tiederman, W. G., Burst Detection with Single-Point Velocity Measurements, J. Fluid Mech., vol. 162, p. 389, 1986.

Bogard, D. G., and Tiederman, W. G., Characteristics of Ejections in Turbulent Channel Flow, J. Fluid Mech., vol. 179, p. 1, 1987.

Bradshaw, P., and Wong, F. Y. F., The Reattachment and Relaxation of a Turbulent Shear Layer, J. Fluid Mech., vol. 52, part I, p. 113, 1972.

Brereton, G. J., and Reynolds, W. C., Experimental Study of the Fluid Mechanics of Unsteady Turbulent Boundary Layers, Rept. TF-29, Thermosciences Div., Dept. Mech. Eng., Stanford Univ., 1987.

Brodkey, R. S., Wallace, J. M., and Lewalle, J., The Delta Conferences: A Discussion of Coherent Structures in Bounded Shear Flows, Res. Found. Proj. 761061/711063 Rept., Ohio State University Research Foundation, 1984.

Brown, G. L., and Thomas, A. S. W., Large Structure in a Turbulent Boundary Layer, Phys. Fluids, vol. 20, no. 10, part 2, p. 5243, 1977.

Cantwell, B., Coles, D., and Dimotakis, P., Structure and Entrainment in the Plane of Symmetry of a Turbulent Spot, J. Fluid Mech., vol. 87, part 4, p. 641, 1978.

Chu, C. C., and Falco, R. E., A Vortex Ring/Viscous Wall Layer Interaction Model of the Turbulence Production Process Near Walls, in Proc. Tenth Symp. on Turbulence, Univ. Missouri-Rolla, 1986.

Coles, D. E., The Law of the Wake, J. Fluid Mech., vol. 1, 1956.

Coles, D. E., A Young Person's Guide to the Data, in Proc. AFOSR-IFP-Stanford Conference on Computation of Turbulent Boundary Layers, ed. D. E. Coles and E. A. Hirst, 1968.

Corino, E. R., and Brodkey, R. S., A Visual Investigation of the wall Region in Turbulent Flow, J. Fluid Mech., vol. 37, part 1, p. 1, 1969.

Corrsin, S., and Kistler, A. L., The Free-Stream Boundaries of Turbulent Flows, NACA TN-3133, 1954.

Doligalski, T. L., Smith, C. R., and Walker, J. D. A., A Production Mechanism for Turbulent Boundary Layers, in Viscous Drag Reduction, G. Hough (Ed.), Prog. Astro. Aero., vol. 72, p. 47, 1980.

Eckelmann, H., Nychas, S. G., Brodkey, R. S., and Wallace, J. M., Vorticity and Turbulence Production in Pattern Recognized Turbulent Flow Structures, Phys. Fluids, vol. 20, no. 10, part 2, p. 5225, 1977.

Elder, J. W., An Experimental Investigation of Turbulent Spots and Breakdown to Turbulence, J. Fluid Mech., p. 235, 1960.

Falco, R. E., Coherent Motions in the Outer Region of Turbulent Boundary Layers, Phys. Fluids, vol. 20, no. 10, part 2, p. S124, 1977.

Falco, R. E., The Production of Turbulence Near a Wall, AIAA Paper 80-1356, 1980a.

Falco, R. E., Combined Simultaneous Visualization/Hot-Wire Anemometry for the Study of Turbulent Flows, J. Fluids Eng., vol. 102, p. 174, 1980b.

Falco, R. E., A Synthesis and Model of Turbulence Structure in the Wall Region, Physics of Isothermal Flows, p. 43, 1982.

Falco, R. E., New Results, a Review and Synthesis of the Mechanism of Turbulence Production in Boundary Layers and its Modification, AIAA Paper 83-0377, 1983.

Gillis, J. G., Johnston, J. P., Kays, W. M., and Moffat, R. J., Turbulent Boundary Layer Heat Transfer Experiments: Convex Curvature Effects, Including Introduction and Recovery, Rept. HMT-31, Thermosciences Div., Dept. Mech. Eng., Stanford Univ., 1980.

Grass, A. J., Structural Features of Turbulent Flow over Smooth and Rough Boundaries, J. Fluid Mech., vol. 50, part 2, p. 233, 1971.

Head, M. R., and Bandyopadhyay, P., New Aspects of Turbulent Boundary Layer Structure, J. Fluid Mech., vol. 107, p. 297, 1981.

Hirata, M., and Kasagi, N., Studies of Large-Eddy Structures in Turbulent Shear Flows with the Aid of Flow-Visualization Techniques, Studies in Heat Transfer, J. P. Hartnett et al. (Eds.), Hemisphere, 1979.

Hussain, A. K. M. F., and Reynolds, W. C., Measurements in Fully Developed Turbulent Channel Flow, J. Fluids Eng., p. 568, Dec. 1975.

Jayaraman, R., Parikh, P., and Reynolds, W. C., An Experimental Study of the Dynamics of an Unsteady Turbulent Boundary Layer, Rept. TF-18, Thermosciences Div., Dept. Mech. Eng., Stanford Univ., 1982.

Jimenez, J., Moin, P., Moser, R. D., and Keefe, L. R., Ejection Mechanisms in the Sublayer of a Turbulent Channel, Proceedings of the Summer Program 1987, p. 37, Center for Turbulence Research, NASA/Stanford, 1987.

Kim, H. T., Kline, S. J., and Reynolds, W. C., The Production of Turbulence Near a Smooth Wall in a Turbulent Boundary Layer, J. Fluid Mech., vol. 50, part 1, p. 133, 1971.

Klebanoff, P. S., Characteristics of Turbulence in a Boundary Layer with Zero Pressure Gradient, NACA Report 1247, 1956.

Kline, S. J., and Runstadler, P. W., Some Preliminary Results of Visual Studies on the Flow Model of the Wall Layers of the Turbulent Boundary Layer, J. Appl. Mech., Trans. ASME, Ser. D., vol. 26, no. 2, 1959.

Kline, S. J., and Meyer, K. A., A Visual Study of the Flow Model in the Later Stages of Laminar-Turbulent Transition on a Flat Plate, Rept. MD-7, Thermosciences Div., Dept. Mech. Eng., Stanford Univ., 1961.

Kline, S. J., Reynolds, W. C., Schraub, F. A., and Runstadler, P. W., The Structure of Turbulent Boundary Layers, J. Fluid Mech., vol. 30, part 4, p. 741, 1967.

Kovasznay, L. S. G., Kibens, V., and Blackwelder, R. F., Large-Scale Motion in the Intermittent Region of a Turbulent Boundary Layer, J. Fluid Mech., vol. 41, part 2, p. 283, 1970.

Kreplin, H.-P., and Eckelmann, H., Propagation of Perturbations in the Viscous Sublayer and Adjacent Wall Region, J. Fluid Mech., vol. 95, part 2, p. 305, 1979.

Lian, Q. X., A Visual Study on Coherent Structures of the Turbulent Boundary Layer in Flow with Adverse Pressure Gradient, manuscript submitted for publication, 1987.

Ligrani, P. M., Kays, W. M., and Moffat, R. J., Report HMT-29, Thermosciences Div., Dept. Mech. Eng., Stanford University, 1978.

Murlis, J., Tsai, H. M., and Bradshaw, P., The Structure of Turbulent Boundary Layers at Low Reynolds Numbers, J. Fluid Mech., vol. 122, p. 13, 1982.

Pearson, C. F., and Abernathy, F. H., Evolution of the Flow Field Associated with a Streamwise Diffusing Vortex, J. Fluid Mech., vol. 146, p. 271, September 1984.

Praturi, A. K., and Brodkey, R. S., A Stereoscopic Visual Study of Coherent Structures in Turbulent Shear Flow, J. Fluid Mech., vol. 89, part 2, p. 251, 1978.

Runstadler, P. W., Kline, S. J., and Reynolds, W. C., An Experimental Investigation of the Flow Structures of the Turbulent Boundary Layer, Rept. MD-8, Thermosciences Div., Dept. Mech. Eng., Stanford Univ., 1963.

Schraub, F. A., and Kline, S. J., A Study of the Structure of the Turbulent Boundary Layer with and without Longitudinal Pressure Gradient, Rept. MD-12, Thermosciences Div., Dept. Mech. Eng., Stanford Univ., 1965.

Smith, C. R., A Synthesized Model of the Near-Wall Behavior in Turbulent Boundary Layers, in Proc. Eighth Symp. on Turbulence, Univ. Missouri-Rolla, 1984.

Smith, C. R., and Metzler, S. P., The Characteristics of Low-Speed Streaks in the Near-Wall Region of a Turbulent Boundary Layer, J. Fluid Mech., vol. 129, p. 27, 1983.

Smith, C. R., and Schwartz, S. P., Observation of Streamwise Rotation in the Near-Wall Region of a Turbulent Boundary Layer, Phys. Fluids, vol. 26, no. 3, p. 641, 1983.

Theodorsen, T., Mechanism of Turbulence, Proceedings 2nd Midwestern Conference on Fluid Mechanics, Ohio State Univ., Columbus, Ohio, 1952.

Tiederman, W. G., Luchik, T. S., and Bogard, D. G., Wall-Layer Structure and Drag Reduction, J. Fluid Mech., vol. 156, p. 419, 1985.

Townsend, A., The Structure of Turbulent Shear Flows, Cambridge Univ. Press, 1956.

Willmarth, W. W., and Lu, S. S., Structure of the Reynolds Stress Near the Wall, J. Fluid Mech., vol. 55, part 1, p. 65, 1972.

Willmarth, W. W., and Bogar, T. J., Survey and New Measurements of Turbulent Structure Near a Wall, Phys. Fluids, vol. 20, no. 10, part 2, p. 59, 1977.

- **Flat wall, smooth, stationary, solid, undisturbed since transition**
- **Fully turbulent boundary layer**
- **Zero pressure gradient**
- **2-D mean flow**
- **Steady outer flow with low free-stream turbulence**
- **Zero density gradient**
- **Newtonian fluid**
- **Single-phase fluid**
- **Incompressible**
- **No force fields**

FIGURE 1. The "canonical case."

Quasi-Coherent Structures in the Turbulent Boundary Layer: Part II. Verification and New Information from a Numerically Simulated Flat-Plate Layer

S. K. ROBINSON
NASA Ames Research Center
Moffett Field, California 94035, USA

S. J. KLINE
Department of Mechanical Engineering
Stanford University
Stanford, California 94305, USA

P. R. SPALART
NASA Ames Research Center
Moffett Field, California 94035, USA

INTRODUCTION

In spite of decades of experimental research, the structure of turbulent boundary layers, and the dynamical processes by which turbulence is created and maintained in boundary layers are only partially understood. The incomplete nature of our understanding has prevented the results of turbulence structure research from making any significant contributions to either predictive modeling schemes or to turbulence control methodologies.

This continuing situation has given rise to a novel approach to the problem: a community-wide re-evaluation of the knowledge. In simple terms, the aim is to develop a clear picture of what we know with certainty and to form a "strategic plan" for attacking the voids in our knowledge. The eventual goal is to lay the foundation for a class of "statistical-structural" models for turbulent boundary layers. The new and critical element of this effort is the involvement of the research community. A detailed description of the process and status of the cooperative evaluation project is given in Part I of this paper (Kline and Robinson, 1988).

The limitations of experimental methods for unravelling the puzzle of turbulence physics are well understood but are, to a large degree, inherent. Fixed probes at a small number of points in space can be weak tools for determining the characteristics of three-dimensional structures that exhibit large variations in time and space. Although many groups agree that simultaneous use of flow-visualization and probe arrays are optimum, it remains difficult to resolve Eulerian probe-based quantitative data with visual information from markers in a three-dimensional unsteady flow-field. It is to the credit of the researchers in the field that so much has been learned about the structure of turbulence through experimentation.

To complement the large volume of experimental results in the field, this companion study of the cooperative project is designed to take advantage of the direct Navier-Stokes turbulence simulation databases available at NASA Ames. Although the simulations are currently limited to simple geometries at low Reynolds numbers, there are many advantages to studying turbulence structure numerically, as discussed later in this paper, and also by Moin (1984).

Community involvement also plays a fundamental role in the study of the numerical simulation data. As the major unknowns are identified through the cooperative evaluation work, their resolution can be pursued in the simulations. The questions and suggestions of the participants in the cooperative project have helped pose the critical questions for study via the numerical simulation databases. This paper will present an overview of the recent results of this approach.

The objective of the current effort is to combine the many known structural elements and events with new insight gained from the numerical studies, to form a complete dynamical picture of flat-plate boundary layer turbulence, at least for low Reynolds numbers. This picture should include information on: (1) the spatial character of each class of structure; (2) the spatial relationships between different structures; (3) the generation and evolution of structures; (4) the statistical relevance of each class of structure; and (5) descriptions of the dominant sequences of events that are responsible for maintaining turbulence.

An additional goal is to investigate the effects of Reynolds number on boundary layer structural features, at least over the limited range of Reynolds numbers available in the simulations. A final important objective is to simulate a variety of experimental techniques in the databases to improve and possibly extend the understanding gained from experimental results.

In the remainder of this paper, a brief description of Spalart's boundary layer simulation will be given as will an outline of the strategy we are employing to study the database. A preliminary taxonomy of structures is presented, and examples of each class are illustrated and described. Finally, several samples of spatial relationships between structures will be shown. The results confirm a number of published experimental findings and also present new information on the structure of turbulent boundary layers.

Since the results of this work are voluminous and are still in progress, conclusions concerning time-evolutions of structural features will be deferred to a subsequent paper. The results presented herein are meant to serve as an introduction to the character of turbulence structure in the simulated boundary layer.

ADVANTAGES/DISADVANTAGES OF THE USE OF FULL SIMULATION DATABASES FOR THE STUDY OF TURBULENCE STRUCTURE

The central advantage of the numerical simulations over experimental investigation is that in the databases, we can collect a large sample of structural features first, and then decide what statistics are pertinent to compute. In the laboratory, the process is usually reversed. The flow can be made visible, but quantitative structural information from probes must be somehow extracted by clever statistical analysis of the data. This method is necessarily incomplete, and has proven to be seriously misleading in some cases. The averaging and filtering inherent in most statistical techniques can discard phase information, disguise true characteristics of individual realizations, create false symmetries, and contaminate ensemble averages by the inadvertent inclusion of extraneous flow elements. The current availability of numerical turbulence marks the first time in history that a sufficient sample of each structure can be inspected a priori and then quantitative statistics gathered for the significant features.

In addition to this subtle, but crucial point, the full simulation results offer a number of other more obvious, but important advantages:

1) Selected structural features can be pursued both forward and backward in time, and this process can be iteratively refined.

2) All structures of a given type that exist in the computational domain can be made simultaneously visible, allowing evaluation of case-to-case variations.

3) The spatial relationships between the several types of co-existing structures can be studied.

4) Structural features that are virtually unmeasurable in the lab can be displayed (e.g., low-pressure cores of unsteady vortices).

5) By combining time evolution with the ability to display different types of structures simultaneously, it becomes possible to examine the spatio-temporal relationships among the various elements of the dominant structures.

6) Rapid turnaround time of structural queries and the availablity of specific frozen data volumes and time-sequences allow iterative reprocessing.

The structural aspects of the direct numerical simulation databases have been studied in depth for some years by their authors (e.g., Kim, 1985; Kim and Moin, 1986; Moser and Moin, 1987; Moin and Spalart, 1988). Recently, experimentalists (including the first two authors of this paper) have begun to explore the simulations as "numerical wind tunnels" to confirm and extend experimental results, and to help design future experiments. Participation by turbulence structure experimentalists in the 1987 Summer Program at the NASA/ Stanford Center for Turbulence Research resulted in a number of significant findings from investigations of the numerical databases.

The direct Navier-Stokes simulation work is still in its infancy, and has several limitations. These include high costs in time and dollars (both for the simulation itself and for the later analyses), and restriction to low Reynolds numbers, simple geometries, and periodic boundary conditions. Nevertheless, the study of the simple flows now available will not only enlighten us about those exemplary cases, but will also aid in establishing appropriate procedures for application to the more complex flows of the future.

NUMERICAL SIMULATION SPECIFICATIONS

Spalart's direct Navier-Stokes numerical simulation of a flat-plate turbulent boundary layer (Spalart, 1988) has been chosen as the initial simulation database to explore. The code has been run to statistical equilibrium for four Reynolds numbers, $Re_\theta = 225$, 300, 670, and 1410. Most of the analysis so far has utilized the $Re_\theta = 670$ case, due to the high computational cost of the $Re_\theta = 1410$ case. At $Re_\theta = 670$, each time-step is computed on a 384×288×85 grid, comprising 9.4 million nodes. At each node, for each time-step, pressure and all three components of velocity (and thus vorticity) are available. Grid resolution is 12.8 viscous lengths in the streamwise (x) direction, and 4.3 viscous lengths in the spanwise (z) direction. Resolution in the wall-normal (y) direction varies from 0.03 to 16.0 viscous lengths, with 14 grid points between the wall and $y^+ = 10$. The grid spacing for the $Re_\theta = 670$ case gives a computational domain with streamwise, spanwise, and wall-normal dimensions of 4900, 2500, and 1100 viscous units, respectively (Figure 1). The boundary layer is approximately 300 viscous units thick at this Reynolds number. Further details concerning the computational method and the results are given in Spalart (1988).

The simulation provides access to a turbulent region approximately 25 wall-streak spacings wide and 16 boundary layer thicknesses long. The velocity and pressure field of the computation has been saved on tape for 104 time steps, each three viscous time-units apart. This database offers a massive amount of information with high resolution in both space and time. Estimates suggest the amount of data available will provide sufficient sample sizes for spatial statistics of the major structural features.

A subvolume of the total computational domain is shown in Figure 2. In the wall-normal (y) direction, the subvolume reaches from $y^+ = 3$ to $y^+ = 250$, which is approximately 80% of the mean boundary layer thickness. Over 1600 subvolumes of this size are available in the stored turbulence database. For comparison purposes, the same subvolume of data (referred to as "subvolume S") will be used repeatedly to illustrate several structural features. Many other subvolumes are being included in the overall analysis, however, so the figures presented here should be considered examples, and not necessarily representative of all realizations. It is worth noting that examples of each class of structural feature are detectable in nearly every subvolume studied so far.

Spalart (1988) has performed a number of statistical checks of the numerical boundary layer, which generally show good agreement with available experimental results. These include turbulence intensities, skewnesses, flatness factors, Reynolds stress and turbulence kinetic energy budgets, and spectra. The most significant known discrepancy is a 5% overprediction of the friction coefficient for the highest Reynolds number case. In addition, near-wall fourth-order statistics, especially for v', show a dependence on grid resolution. It is unclear how significant these may be in terms of the structural behavior of the turbulence. Results of our analyses so far indicate that the simulation is structurally indistinguishable from a physical low Reynolds number boundary layer.

ANALYSIS STRATEGY

The problem at hand is to understand the physics of turbulence creation and maintenance in boundary layers. Spalart's numerical simulation offers access to over 31 gigabytes of flow-field data. The computational and graphical power available to study the simulation results makes possible the pursuit of hundreds of "interesting" questions. Effective progress toward an understanding of turbulence production therefore demands a well thought-out strategy in order to set work priorities.

We have adopted the following six-phase strategy for the study of the simulation databases (Figure 3):

I. (a) Identify structures in the numerical boundary layer that are known to exist in laboratory boundary layers. These can be grouped into eight classes, which are delineated in the next section.

   (b) Once examples of these structures have been identified in the simulation results, obtain detailed descriptions of their spatial character by studying individual time-steps.

II. Determine the spatial relationships between the various structures by visualizing them together in individual "frozen" time steps.

III. Study the spatio-temporal relationships between structures by using all the stored time-steps and, where necessary, computing intermediate solutions at very small time increments ($\Delta t^+ = 0.3$) between the stored time-steps.

IV. Determine the statistical significance of the various structures, their apparent generation mechanisms and any interaction events. Populations, frequencies in space and time, and statistics of occurrence will be computed. Contributions to future statistical/structural models will be made at this stage by computing probability distributions of events and structures, and their contributions to specific terms in the Reynolds-averaged model equations.

V. Combine the results of stages I through V to develop a phenomenological model of the maintenance of turbulence in flat-plate boundary layers. This should include a description of the important sequences of events, and should describe the roles of all classes of experimentally observed structure. The flow of cause and effect within these sequences should be modeled, as well as any significant interactions between structural features.

VI. Simulate various experimental techniques, including multi-sensor probes and convected visual markers.

Phases I through IV involve only the kinematics and statistics of turbulence structure. As J. M. Wallace has reminded us, a picture of the dynamics does not emerge from passive observation of phenomena. Some kind of model must be proposed which links observed sequential events with causative mechanisms. Thus, only Phase V addresses the problem of dynamics in the turbulent boundary layer.

In the context of this strategy, most quantitative experimental results have addressed Phases I, II, and IV, while most flow-visualization studies could be categorized as Phase III. A number of authors have proposed structural models which include dynamical hypotheses and so belong in Phase V. Our Phase V work will involve testing the many models published by the community in the simulated turbulence, with the goal of combining the best parts of each into a single, and hopefully more complete, phenomenological model.

Phase VI of the strategy involves the simulation of various probe-based and visual experimental methods. In the numerical simulation, spatial variations can be considered separately from temporal variations. This may lead to clarification of probe-based results, which are generally Eulerian and hence treat space and time as a single, inherently combined variable.

The six-phase analysis strategy allows us to build up knowledge in a logical order, and facilitates prioritization of the large number of possible queries that might otherwise become disorderly and overwhelming. In the current study, the phases of the strategy represent an iterative, rather than a one-pass, sequential process. So far, work has been done under all six phases, but none have been completed. As noted, the current paper focuses largely upon results of Phases I and II of the research program.

TAXONOMY OF STRUCTURES

A consistent and accepted definition of "structure" (in the turbulence sense) has not yet surfaced. Thus, it may be viewed as a hopeless task to categorize the many flavors of turbulence structure to be found in the literature. However, a taxonomy, even though somewhat arbitrary, can help focus thinking and inject organization into the somewhat tangled mass of information on the subject.

Therefore, we have divided the various structures, events, flow modules, and organized motions found in the literature into eight classes. The intent is to include all types of non-random events that have been singled out for study by the researchers in the field, rather than utilize a definition that would necessarily exclude some of the data. This classification is only one of several possibilities, and the following list is not a conclusion of the current work, but rather a starting point for it.

A Classification of Boundary Layer Turbulence Structures:

(1) Wall low-speed streaks

(2) Ejections of low-speed fluid outward from the wall

(3) Sweeps of high-speed fluid inward toward the wall

(4) Vortical structures of various forms

(5) Near-wall shear layers, exhibiting strong spanwise vorticity and du'/dx

(6) Near-wall "pockets," seen as regions swept clean of near-wall marked fluid in experiments

(7) Large (delta-scale) discontinuities in the streamwise velocity, or "backs"

(8) Large (delta-scale) motions capped by bulges in the outer turbulent/potential interface

Part I of these papers (Kline and Robinson, 1988) gives descriptions and some of the known characteristics for each of these structural classes.

The present classification is meant to provide a skeleton of organization on which the study of the numerical simulations can build. Reflected in the broad coverage of the list is the objective of including all of the known structural features of turbulence in the current study, rather than just a few. Thus the categorization is comprehensive if not final.

PHASES I AND II: EXAMPLES OF THE SPATIAL CHARACTER OF EACH STRUCTURAL CLASS AND SELECTED SPATIAL RELATIONSHIPS

Sublayer Streaky Structure

Figure 5 shows an instantaneous plan view of the x-z plane at $y^+ = 2$ in the simulated turbulent boundary layer. The low-speed regions (streamwise velocity less than the local mean) are elongated and thin, in agreement with many experimental observations. (Here and elsewhere in the paper, "local mean" refers to the spatial average over the entire computational x-z plane at constant y^+, for at least five time steps.) Streamwise coherence of single low-speed streaks often exceeds 1500 viscous lengths, with widths ranging from 20 to 80 Δz^+. The streaky character of the instantaneous low-speed regions is somewhat more pronounced than shown by marked fluid in laboratory flows. This is because markers in low-speed streaks often lift away from the wall during ejections, leaving the remaining near-wall streak unmarked and invisible. However, the current results appear consistent with the wall-temperature data of Hirata and Kasagi (1979) in which continuous visualization at the wall was achieved using liquid crystals.

The high-speed regions in the sublayer (Figure 5) are considerably less elongated and somewhat wider than the low-speed regions. Streamwise lengths of high-speed regions seldom exceed 800 x^+, and widths range from 40 to 110 Δz^+. This difference in "streakiness" is to be expected, since high-speed fluid generally originates from outside the sublayer where the mean velocity gradient is lower and the flow is known to be less streak-like. Experimental flow visualizations have not shown the character of high-speed regions clearly, however, since fluid markers introduced near the wall collect in the low-speed regions. Since high-speed regions in the sublayer are less elongated than the low-speed regions, the term "high-speed streaks" is not used in the current study.

Another characteristic of low-speed streaks is shown in Figure 6, which is an instantaneous plan-view of the boundary layer at $y^+ = 15$. The colored contours mark regions of significant value of the products (-u')(+w') and (-u')(-w'). This product serves as a "streak-turning" parameter, since it highlights low-speed streaks (-u') that simultaneously exhibit spanwise motion (+ or - w'). This correlation may be termed u'w'2 and u'w'3, after the quadrants it occupies in the u'w' hodograph plane. It was noticed by one of the authors and J. Swearingen that u'w'2 and u'w'3 contours display a vaguely "cross-hatched" pattern, suggestive of possible transverse propagation mechanisms near the wall. Swearingen et al. (1987) have drawn parallels between this behavior and spanwise oscillations related to breakdown instabilities in curved-wall flows. In the simulated boundary layer, significant near-wall spanwise velocities are commonly associated with "leg" or "neck" vortices (see below). A complete explanation for the spanwise motions of the low-speed streaks awaits further investigation.

Ejections

The quadrant technique will be used to define and identify ejections, or regions of low streamwise velocity with motion outward from the wall. The quadrant method involves splitting the instantaneous product of u' and v' into four categories according to the quadrant occupied in the u'v' hodograph. Since its introduction by Wallace et al. (1972) and by Willmarth and Lu (1972), the quadrant method has been widely used to assign structural significance to motions that contribute to the boundary layer closure term, the Reynolds shear stress.

It is important to distinguish between the Reynolds shear stress $-\overline{u'v'}$, which is defined only as an average, and the instantaneous value of u'v'. To facilitate this distinction, the new term "shear product" will be used to refer to the instantaneous value of u'v' at a location in space.

Figure 4 shows the four quadrants into which the shear product can be split, and the names associated with their respective motions (Wallace et al., 1972). The contributors to the positive Reynolds shear stress are quadrants 2 and 4. The concepts of ejection and sweep imply a coherent movement of fluid, rather than the occurrence of quadrant 2 or 4 motion at a single point in space. In the current work, ejections are arbitrarily defined as instantaneous regions of quadrant 2 shear product that extend at least ten viscous lengths in any direction. Sweeps are defined similarly. Thus, every ejection is a quadrant 2 motion, but not necessarily vice versa. Note that under this definition, sweeps and ejections may occur anywhere in the boundary layer.

Figure 7 shows contour surfaces of quadrant 2 shear product in the instantaneous computational subvolume, S. For the figure, the contour value was set to detect strong quadrant 2 motions, with u'v'2 equal to $4u_\tau^2$. This and many other realizations show that strong ejections occur with two types of instantaneous spatial character: 1) thin and elongated in the buffer and lower log regions, and 2) broader and more blob-like in the outer log and wake regions. The spanwise extent of the outer-region ejections ranges up to $\Delta z^+ = 200$. Since ejections near the wall originate from low-speed streaks by definition, their character is expected to be at least somewhat elongated. Ejections are not observed to extend over the entire streamwise length of a streak, however. Only portions of the low-speed regions extend out of the sublayer.

Sweeps

Sweeps are regions in which the shear product u'v' occupies the fourth quadrant in the u'v' plane (Figure 4). Instantaneous contour surfaces of quadrant 4 shear product in the subvolume S are shown in Figure 8. The contour threshold is the same as for Figure 7; that is, the iso-surfaces correspond to quadrant 4 shear product equal to $4u_\tau^2$. At any given time, strong sweeps occupy less volume in the simulated boundary layer than equally strong ejections. This spatial result agrees with the probe-based statistics of Wallace et al. (1972), Willmarth and Lu (1972), and others, who have shown ejections to dominate the contributions to the Reynolds shear stress outside the buffer region. The spatial character of high-speed fluid with a wallward velocity component is somewhat different than that of ejections. Strong sweeps are only occasionally elongated in the streamwise direction, and tend to have spanwise dimensions of less than 100 Δz^+, regardless of distance from the wall.

Low-Speed Streaks with Ejections and Sweeps

Figure 9 is a plan-view of an x-z plane in the simulated boundary layer at $y^+ = 15$. Ejections and sweeps are shown for the same contour levels as listed above. Low-speed streaks are identified by contours of streamwise velocity $3u_\tau$ below the local mean. Strong ejections are seen to occur only on low-speed streaks, as they must by definition (both are regions of -u'). The simulated boundary layer is seen to agree with Corino and Brodkey's (1969) and Bogard and Tiederman's (1986) finding that several ejections may arise from a single low-speed streak, although the number of ejections per streak is dependent upon the contour levels chosen for -u' and u'v'2. The contour value for second quadrant u'v' in Figure 9 corresponds to a value of approximately 2.6, when normalized by the product of u'(rms) and v'(rms) at $y^+ = 15$. Strong ejections seem to originate from the central portion of streaks more often than from either end.

As expected from experimental results and from previous simulations (Moin, 1984), the contours of strong quadrant 2 and 4 motions are highly intermittent in space; neither displays the extended streamwise coherence of the low-speed regions, although quadrant 2 motions are somewhat elongated in the x-direction. The spanwise scale of sweeps is slightly larger than the low-speed streaks and the ejections, but their extent is still less than 80 z^+. Thus, sweep motions in the buffer zone are narrow and localized, rather than broad inrushes.

When strong ejections and sweeps are found in close proximity at a given time, they are often laterally associated, in a side-by-side pair. This characteristic was also noted by Moin (1984) in simulated turbulent channel flow, and is evident in Figure 9. Sweeps and ejections virtually never follow each other closely in the streamwise direction, contrary to several published structural models. However, slight skewing of the sweep/ejection pair in the x-z plane would cause a fixed probe to detect one followed closely in time by the other. Most single-point conditional sampling schemes do not allow for spanwise variations, so ensemble averages that show sweeps following ejections are to be expected, but are misleading if interpreted via Taylor's hypothesis, with the additional assumption that the motion lacks significant spanwise component.

The lateral pairing of vertical motions near the wall is more clearly shown in Figure 10, in which instantaneous contours of v' are plotted for the x-z plane at $y^+ = 15$. Figure 10 shows that outward motions (+v') nearly always occur closely beside inward vertical motions (-v'). This was also noted by Moin (1984) in channel flow simulations. This character suggests that in the buffer region, the occurrence of strong outward motion is strongly associated with an occurrence of wallward motion. The spanwise dimensions of significant wallward motions are limited to less than approximately 70 viscous lengths. The pattern in Figure 10 is highly suggestive of localized, quasi-streamwise vortical motions, which have been discussed in a number of publications (Bakewell and Lumley, 1967; H. T. Kim et al., 1971; Blackwelder and Eckelman, 1979; Smith and Schwartz, 1983). The spanwise dimensions of the $\pm$v' pairs range from 40 to 70 viscous lengths. The streamwise extent of the pairs ranges from 50 to 250 viscous lengths; however, this is not indicative of the streamwise dimensions of near-wall quasi-streamwise vortices. Since the figure depicts a plane of data at constant y, a vortex with an inclination angle to the wall would be intersected in a limited region by the plane. Note that laterally arranged triplets of alternating positive and negative v', which suggest the occurrence of vortex pairs, are relatively uncommon, but do exist. Most significant vertical motion in the buffer region appears to be associated with single vortices.

<u>Vortical Structures</u>

For the current work, vortical structures are defined as identifiable vortices, usually with segments of varying orientation in space. Under this definition, regions of significant vorticity do not necessarily qualify as vortical structures. The problem that arises from any reference to vortices is the definition of a vortex in unsteady viscous flow. In the cross-stream (y-z) plane of a zero pressure-gradient boundary layer, the time-averaged flow is nearly zero. In this plane, a working definition of a vortex is as follows: A vortex exists when instantaneous streamlines mapped onto the cross-stream plane display a roughly circular or spiral pattern; the core of this vortex will be oriented substantially streamwise. In any plane with significant mean motion, this definition becomes ambiguous due to the necessary choice of a reference frame velocity. From a reference frame moving with the core of the vortex, this same definition should hold, although it requires identifying the vortex core before defining the vortex itself. Thus, the above definition of a vortex is useful as a check of other vortex-identification methods, but not as an a priori tool in any plane other than the cross-stream (y-z) plane.

Distinction must be made, of course, between vortices and vorticity. Since regions of strong vorticity are not uniquely associated with vortices, the spatial character of the vorticity field in the turbulent boundary layer tends to be noisy and inconclusive with respect to vortical structures. This is demonstrated in Figure 11, in which iso-surfaces of the three fluctuating components of vorticity in subvolume S are shown. Clearly, understanding of vortical structures requires a technique for marking only instantaneous vortices in the turbulent flow field.

Vortex lines have been used extensively by Kim and Moin (1986) to study the structure of vortices in a numerically simulated turbulent channel flow. This method, however, requires some meaningful criteria with which to choose starting locations for the vortex line tracing. In addition, vortex lines do not reflect the local strength of vorticity, so that dense bundles that occur in regions of very low vorticity could attract undue attention. Finally, although vortex lines must move with the fluid particles (neglecting viscosity) during time evolution, vortices, which are bunched sections of vortex lines, need not. Thus the method of vortex lines can be arbitrary and inconclusive unless coupled with an independent means of identifying vortex locations.

Near-wall vortices with mostly streamwise orientation have been the subject of a number of experimental investigations (Praturi and Brodkey, 1978; Blackwelder and Eckelman, 1979; Smith and Schwartz, 1983; Swearingen and Blackwelder, 1987). By taking a cross-stream cut through the simulated boundary layer and plotting velocity vectors or streamlines, such vortical motions are easily found with the above definition of a vortex. Figure 12 shows an instantaneous cut through a low-speed streak and its neighboring high-speed region in subvolume S. Velocity vectors in the cross-plane (v'w') clearly meet the criteria of roughly circular streamlines and so a trial vortex has been identified in the data.

A method is required to trace the instantaneous, three-dimensional extent of vortical structures in space. Our proposal is to utilize the fact that vortex cores should be regions of low pressure. In Figure 13, low-pressure and cross-plane velocity vectors are plotted for the same instantaneous plane of data as shown in Figure 12. Fluctuating pressure is shown in color in Figure 13, with lighter shades indicating lower pressure. The instantaneous v'w' velocity vectors are overlaid as in Figure 12. Marking the low pressure clearly identifies the vortex core in the example. In Figure 13, the radial pressure gradient (normal to the core) is fairly large at the outer edge of the core, so the value of the pressure contour used to locate the vortex is not critical. A white contour line has been drawn in Figure 13 for values of pressure equal to $4\rho u_\tau^2$ below the local mean. (Local mean refers to the average value at each y-location over the entire computational x-z plane, for several time-steps of the simulation.) For reference, the rms of the pressure fluctuations, normalized by ρu_τ^2, varies from about 2.5 near the wall to 0.5 at the mean boundary layer edge (Spalart, 1988).

In three dimensions, contour lines become contour surfaces. The 3-D contour surface of constant low pressure that corresponds to the 2-D contour line in Figure 13 is shown for subvolume S in Figure 14. As can be seen in Figure 14, low-pressure regions tend to be elongated and occur in a variety of shapes and sizes, apparently indicating a wide variation in the three-dimensional form of vortical structures. This observation is confirmed by the study of many other subvolumes in the database. Compared to the vorticity field shown in Figure 11, the instantaneous pressure field of Figure 14 is relatively noise-free, with a variety of structural features clearly evident. (Figures 11 and 14 display the same data set.) Comparison of Figures 11 and 14 emphasizes the importance of distinguishing vortical structures from regions of distributed vorticity.

The technique of marking low-pressure regions allows application of the vortex definition advanced earlier. The velocity of a candidate low-pressure region is first determined. Then the instantaneous streamlines in a plane normal to the axis of elongation of the low-pressure zone are computed from a frame of reference moving with the low-pressure region. Roughly circular or spiral streamlines confirm the existence of a vortex.

It is not immediately clear whether a discernible low-pressure region is both a necessary and a sufficient condition for the existence of a vortex in the flow field. Many cuts normal to the axes of elongated low-pressure regions have been made. In every case so far, the velocity vectors in a cut-plane moving with the mean velocity of the low-pressure region has displayed roughly circular or spiral instantaneous streamlines. Although most low-pressure regions are elongated, there are several convoluted volumes of low pressure that are globule-shaped. So far, the analyses show that virtually all elongated low-pressure regions are vortices. The non-elongated regions are not always identifiable as vortices, and are assumed to be due to other fluid interactions, such as local accelerations. Thus, the technique of mapping elongated low-pressure regions provides a means of identifying the instantaneous spatial extent of vortical structures in the simulation.

The loop-like vortical structures shown in Figure 14 are strongly suggestive of the hairpin or horseshoe vortices discussed in many previous papers, beginning with Theodorsen (1952). To facilitate discussion of the vortical structure topology, various sections of the structures are labeled as follows: heads are vortices with mostly transverse orientation; necks are vortices that generally extend wallward and upstream from heads at an angle of approximately 45 degrees to the wall; legs are vortices with mostly streamwise orientation. Legs sometimes appear connected to and trailing from a head/neck vortical structure.

Although statistical compilations have not yet been completed, some preliminary statements may be made concerning the structural nature of the elongated low-pressure regions. First, arch- or horseshoe-like vortical structures (head/neck) combinations are common, but long leg vortices are not. The legs that do exist can be quite long in the streamwise direction, reaching up to at least $\Delta x^+ = 400$. True hairpin vortices, with two trailing legs, are very rare. Many vortical structures are strongly asymmetric. Hook-like shapes, as in Figure 14, appear relatively often. In some cases, the missing neck can be identified on an asymmetric hook-like vortical structure when the contour level of the isobaric surfaces is varied. Legs can be found without necks, and are seen both near the wall (center at $y^+ < 60$) and in the wake region. Vortical arches occur with a variety of scales, and their heads are found at y-locations ranging from the buffer region to the edge of the boundary layer. Arches are generally as tall (in the wall-normal direction) as they are wide. The spanwise extent of vortical arches ranges from 50 to 350 viscous lengths in the simulated layer. The above conclusions hold true over a range of $-p'$ contour values. In general, varying the $-p'$ contour level affects the diameter and, in some cases, the topological connectivity of the low-pressure regions, but does not significantly alter the overall form of the vortical structures marked by elongated regions of low pressure. A variety of statistics on the spatial character of vortical structures in the boundary layer will be reported in subsequent publications.

Vortices with Low-Speed Streaks, Sweeps, and Ejections

An example of the spatial relationship between low-pressure regions, low-speed regions, and quadrant 2 shear product is shown for the sample subvolume S in Figure 15. Low-speed regions are identified by iso-surfaces of streamwise velocity $3u_\tau$ below the local mean. Surface contour values for uv2 and pressure are the same as for Figures 7 and 14, respectively. Portions of at least two elongated low-speed streaks are visible near the wall, as are significant regions of locally low-speed fluid away from the wall, both within and beyond the buffer zone. Figure 15 shows low-speed fluid concentrated below and upstream of the vortical heads, and alongside the trailing vortex leg. The arch-like vortical structures in the figure straddle elongated low-speed regions, similar to the manner described by Runstadler et al. (1963) and by Smith (1984). Just upstream of the smaller of the two vortical arches is a small transverse region of low pressure, which appears to be a newly formed secondary vortex, possibly rolled up on the low-speed fluid lifted by the action the downstream vortical arch in the manner described by Acarlar and Smith (1987). The formation of this particular secondary vortex is clearly visible in time evolutions of the simulation.

The regions where low-speed fluid posesses a significant outward velocity component are marked in red in Figure 15 as quadrant 2 ejections. Most regions of lifting low-speed fluid are in close proximity to the isobaric surface.

The spatial relationship between low-pressure regions, strong ejections, and strong sweeps is shown for subvolume S in Figures 16 (oblique view) and 17 (plan view). The instantaneous spatial association of both sweeps and ejections with the vortical structures is striking.

Strong ejection regions ($u'v'_2 \geq 4u_\tau^2$ in Figure 16) apparently occur in two locations: (1) Alongside quasi-streamwise leg vortices, on the inboard (towards the head) side of the leg. This is the upward-rotating side of the vortex, so the picture is kinematically consistent. (2) Underneath and upstream of a head vortex. Ejections in location 2 are significantly farther from the wall than those in location 1. Both the head and neck seem to contribute to the outward induction of low-speed fluid from upstream and below the head. Since leg vortices are relatively rarer than head/neck combinations, the majority of the strong ejections seem to be associated with the head/neck vortical structure.

For ejections along quasi-streamwise vortices, Figure 17 clearly shows that a single vortex can be responsible for the lifting of low-speed fluid, in agreement with Pearson and Abernathy's (1984) numerical studies of vortices in viscous shear layers, and with the ensemble-averaged simulation results of Guezennec et al. (1987). This view differs from hairpin vortex models (e.g., Offen and Kline, 1975; Smith, 1984) of the boundary layer, which have fluid lifting due to the combined action of a pair of counter-rotating quasi-streamwise vortices.

Strong sweeps are observed mainly in two locations: (1) On the outboard, wallward-rotating side of the leg vortex. Here, high-speed outer fluid is induced toward the wall in a sweep motion. (2) A wallward induction of high-speed fluid by the outboard side (away from the head) of a neck vortex. Sweeps are more common in location 2, since neck vortices are more common than leg vortices in the simulated boundary layer. In either case, sweeps are found to be closely related to the existence of vortical structures.

A large percentage of the Reynolds shear stress in the boundary layer is known to be produced by highly localized strong ejections and sweeps (e.g., H T. Kim et al., 1971; Brodkey et al., 1974; Willmarth, 1975). Visualizations of the simulated boundary layer (Figure 16 is an example) suggest that the generation of Reynolds shear stress is closely tied to the existence of coherent vortical structures. The conclusion that vortical structures are a critical element of turbulence production and momentum transfer in the boundary layer is in agreement with a number of previous experimental and numerical studies.

Near-Wall Shear Layers

The existence of sloping du'/dy shear layers near the wall ($y^+ < 80$) of turbulent channels and boundary layers has been established both by flow-visualization studies (Corino and Brodkey, 1969) and by probe-based results (Kreplin and Eckelmann, 1979; Johansson et al. 1987a). The VITA technique in particular has been employed in a number of experimental studies to detect rapid changes in the streamwise velocity. Correlations of the experimental data have shown these velocity jumps to correspond to the passage of local shear layers with a steep downstream slope. Bogard and Tiederman (1987) have shown that such shear layers occur on the upstream face of an ejection from the buffer region, at the interface between the lifting low-speed fluid and the higher-speed fluid overtaking it from behind. The limited ability of the VITA technique to detect ejections implies that near-wall shear layers also occur in other locations. The origin and dynamical significance of these shear layers has remained poorly understood, however.

Recent work by Jimenez et al. (1987) in numerically simulated channel flows has shed new light on the role of near-wall shear layers. These authors have proposed a mechanism for the generation and maintenance of the shear layers which is essentially equivalent to that responsible for the instability of two-dimensional Tollmien-Schlichting waves.

Additional recent work on near-wall shear layers in numerically simulated turbulent flows has been reported by Johansson et al. (1987b). These authors found that shear layers in the near-wall region propagate with a velocity of about 10.6 u_τ, and retain their coherence over streamwise distances of 1000 viscous lengths. Johansson et al. also show that near-wall shear layers make important contributions to the conditionally averaged production of turbulent kinetic energy.

In the current study of the simulated boundary layer, the sublayer and buffer region are found to be well-populated with near-wall shear layers. Shear layers can be found in any randomly selected x-y slice through the data, and are most common below $y^+ = 80$. Several examples of near-wall shear layers in the simulation are shown in three unrelated side-views in Figure 18. The structures are nearly parallel to the wall in the sublayer and slope at an increasing but shallow (< 20°) angle farther from the wall. The spanwise dimension of most near-wall shear layers ranges from 30 to 60 viscous lengths, implying that highly three-dimensional mechanisms are involved in their formation.

Near-Wall Shear Layers with Vortices and Ejections

A new observation made in the current study is that some of the near-wall shear layers roll up into transverse vortices. Transverse roll-up of internal shear layers has previously been observed outside the wall region ($y^+ > 200$) in experiments by Nychas et al. (1973). Figures 19 and 20 show an example of near-wall shear layer roll-up. In Figure 19, the shear layer is visualized by contours of spanwise vorticity. Although the vorticity is high everywhere along the shear layer, a low-pressure region is seen to exist only at the outer tip, where a transverse vortex is indicated. The existence of a vortex at the tip of the shear layer is confirmed by the instantaneous vector field in Figure 20, which has been plotted in a reference frame moving downstream with the shear layer tip at its propagation velocity of

15 u_τ. Observation of computed time-sequences confirms that the particular transverse vortex in Figure 20 does appear after the formation of the shear layer.

Contours of quadrant 2 shear product show ejection-type contributions to the Reynolds shear stress in Figure 19. A region of strong quadrant 2 motion is observed to be associated with the transverse roll-up of the shear layer, but is not present along the remainder of the shear layer. Although the dynamic relationship between near-wall shear layer roll-up and the ejection of low-speed streak fluid needs clarification, the current findings support the suggestion of several experimental studies (Acarlar and Smith, 1987; Swearingen and Blackwelder, 1987; Johansson et al., 1987a) that shear-layer instabilities in the wall region may play a major role in the production of turbulence.

High-Pressure "Potatoes" with Low-Pressure Vortical Structures

The simulation database presents opportunities not available in the laboratory, and this allows the study of an additional class of "structure," namely high-pressure regions. Figure 21 shows the high-pressure regions in the instantaneous subvolume S of the computation, and their spatial relationship to the low-pressure regions.

High- and low-pressure regions in the simulated boundary layer exhibit very different characters, as noted by Moin and Spalart (1988). Whereas the low-pressure regions tend to be elongated, the high-pressure regions are roundish and potato-shaped. These "pressure potatoes" often occupy the space just upstream of the head portions of arch-like vortical structures, and appear to exist at a convected local stagnation point which lies at the interface of high- and low-speed fluid. The dynamic consequences of the local convecting pressure gradient set up between the low-pressure vortex core and the following high-pressure region have yet to be fully explored.

Wall-Pressure Fluctuations

Contours of instantaneous wall-pressure fluctuations are shown for the simulated boundary layer in Figure 22. The patterns are rounded rather than elongated, in agreement with experimental results (Dinkelacker et al., 1977) and previous simulations (Grötzbach and Schumann, 1979; Moin, 1984). High- and low-pressure regions appear similar in dimensions, with most regions ranging from 50 to 200 wall units in both directions. However, occasional high- and low-pressure regions with extents of over 400 viscous lengths occur. Although wall-pressure disturbances of this scale would be expected to be associated with events in the outer flow, their specific sources are unclear.

Pockets

When distributed markers are introduced into the sublayer of a turbulent boundary layer or channel, roughly circular regions devoid of marked fluid appear in the plan view. These have been named pockets (Falco, 1980a), and they give the visual impression of being a "footprint" of some outer structure that induces fluid toward the wall. Kim et al. (1987) have shown that pockets become visible in numerically simulated channel flow only with distributed fluid markers, not with simulated bubble-lines. This observation confirms a similar experimental finding of Falco. However, time evolution of the marked flow field is necessary to detect pockets with distributed markers. In the current analysis of the simulated boundary layer, we have endeavored to detect pockets in instantaneous flow fields.

If pockets are associated with wallward-moving fluid in the sublayer, the instantaneous streamlines in the x-z plane would be expected to diverge in the pocket region. Figure 23 shows instantaneous streamlines computed in the $y^+ = 2$ plane of the simulated boundary layer. Regions of abrupt spanwise divergence of the streamlines are evident in the figure. The spanwise dimensions of these regions are generally between 50 and 100 viscous lengths, in agreement with pocket dimensions reported by Falco (1983). A time evolution of simulated fluid markers is being performed by the authors to check whether the regions of streamline divergence in the numerical boundary layer are indeed pockets.

Pockets with Sweeps

Spanwise divergence of instantaneous streamlines in the sublayer is, by continuity and the streaky nature of the streamwise velocity in the sublayer, most likely to be associated with wallward velocity. Wallward fluid in the sublayer arrives from regions with relatively higher streamwise velocity, so diverging instantaneous streamlines are generally expected to be associated with uv4 motions, or sweeps. Since sweeps are observed to be commonly associated with vortices in the simulated boundary layer, pockets may be the footprint of fluid swept toward the wall by vortical structures. Preliminary studies in the simulated data suggest that this reasoning is correct for at least some of the observed pockets. This idea is nominally consistent with Falco's (1980a) theory concerning pocket formation, although the form and motion of the responsible vortex remains in question.

In Figure 23, regions of converging streamlines may be likened to "detachment" lines in a separating flow, and regions of diverging streamlines to "reattachment" lines. The figure shows that most "detachment" and "reattachment" lines lie at a significant angle to the mean (streamwise) velocity vector. Since near-wall $\pm v'$ motions have a strong association with quasi-streamwise vortical structures (legs), it appears that vortex legs are generally oriented at an angle to the mean flow in the x-z plane as well as in the x-y plane.

Pockets with Wall-Pressure Fluctuations

Figure 24 shows instantaneous streamlines at $y^+ = 2$ together with contours of instananeous wall pressure. As expected, local regions of high pressure coincide with spanwise divergence of the streamlines (wallward motion) and low pressure regions occur along lines of converging streamlines (outward motion). (See also Moin and Spalart, 1988.) Thus, pockets may be expected to be evidence of local wallward motions above regions of high wall pressure.

Bulges in the Outer Interfaces/Large-Scale Motions

Kovasnay et al. (1970) and others have shown that the instantaneous outer interface between the turbulent boundary layer and the potential free stream consists of large-scale bulges, separated by deep, narrow incursions of free-stream flow into the layer. In the simulated boundary layer, the instantaneous edge of the rotational flow can be identified by contours of low total vorticity magnitude $\left(\sqrt{\omega_x^2 + \omega_y^2 + \omega_z^2}\right)$, as shown in Figure 25. The picture is essentially the same as given by Spalart (1988), and exhibits broad outward bulges and narrow irrotational incursions. The interface bulges have dimensions of the scale of the boundary layer thickness in both the streamwise (x) and transverse (z) directions. Figure 25 resembles visualizations of smoke-filled low Reynolds number boundary layers (e.g. Falco, 1980b).

The large-scale motion beneath bulges in the simulated boundary layer show a weak rotation in the direction of the mean shear, when observed from a frame of reference moving at 0.8 U_e. This is in qualitative agreement with experimental findings for the large-scale motions (e.g., Blackwelder and Kovasznay, 1972; Falco, 1983).

Backs

Delta-scale, sloping structures with strong local values of du'/dx have been investigated by a number of groups using multi-sensor temperature and velocity probes (e.g., Brown and Thomas, 1977; Chen and Blackwelder, 1978; Subramanian et al. 1982). Space-time correlations and conditional sampling procedures have given from 12 to 30 degrees for the outer-region angle between the wall and these large structures. Because these structures are commonly associated with the upstream side of the large-scale motions discussed above (Kovasznay et al., 1970), we will refer to them as "backs." (Though some publications refer to them as "fronts.")

At this time, we are drawing a distinction between "backs" (which also have transverse vorticity) and the previously discussed near-wall shear layers. Backs are tall (delta-scale) discontinuities in u, with limited (10-40 viscous lengths) streamwise thickness, and with spacing in the streamwise direction of the order of the boundary layer thickness. Near-wall shear layers have dimensions on the scale of the near-wall layer, and exist mostly below $y^+ = 80$. It is not yet clear if the size and spacing of these shear-layer structures are distributed smoothly from small to large, or if there is a bimodal distribution, suggesting two different types of structure. Recent statistical analyses of simulated turbulence by Moin et al. (1987) have shown some evidence of a "two-layer" structural makeup of a low Reynolds number channel. Experimental ensemble averages have not generally separated the data into two structures possibly because the commonly used VITA technique will trigger on either, independent of vertical scale. Also, the scale separation between the two structures, if such a separation exists, would be most pronounced at high Reynolds numbers, where probe sensors are usually too long to independently detect the small-scale structures near the wall. Recent multi-point detection techniques (e.g., Britz and Antonia, 1987; Fernando et al., 1987) should help clarify this point.

Following Thomas and Brown (1977), Figure 27 shows instantaneous velocity vectors in an x-y plane moving at 0.8 U_e in the simulated boundary layer. A back structure is clearly evident as a large-scale, sloping shear layer with transverse vorticity. The picture is strikingly similar to Thomas and Brown's ensemble-averaged velocity field in the vicinity of backs. Figure 28 shows the instantaneous transverse vorticity associated with the back in Figure 27.

Backs with Vortical Structures

Several researchers have reported on outer-flow vortical structures that are inclined to the wall downstream at approximately 45 degrees (e.g., Head and Bandyopadhyay, 1981; Kim and Moin, 1986). Hot-wire rake studies have consistently given from 12 to 30 degrees for the slope of the backs of large-scale motions in the outer region (Brown and Thomas, 1977; Rajagopalan and Antonia, 1979; Robinson, 1982). Bandyopadhyay (1980) attempted to resolve this discrepancy by speculating that the shallow-angle backs are composed of the heads of arrays of 45 degree hairpin vortices.

In the current work, an example of the spatial relationship between a back and a hook-like vortical structure is shown in Figure 26. In this example, the back, which is sloped at about 30 degrees, is located upstream of the 45 degree vortical structure. Apparently, the low-speed fluid being left behind by the action of the vortex is crowded from behind by the faster upstream flow. The interface between the upstream fluid and the low-speed remnants of vortex induction forms the velocity discontinuity which is the back. In this scenario, the back is generated without need for the row of hairpin vortices suggested by Bandyopadhyay. Final conclusions on the relationships between backs, large-scale motions, and vortical structures will require the study of more examples, especially in the simulation at the higher Reynolds number, $Re_\theta = 1410$. The generality of the descriptions of the large scale motions in low Reynolds number simulations may be limited in view of Murlis et al.'s (1982) finding that the outer structure depends on viscosity up to $Re_\theta = 5000$.

CONCLUDING REMARKS

The current ongoing investigation of the quasi-coherent structures in a numerically simulated turbulent boundary layer has produced the following preliminary results:

1) Examples of all known classes of boundary layer structure have been identified in the simulated boundary layer.

2) High-speed regions in the sublayer are observed to be less elongated than the low-speed regions.

3) Low-pressure regions tend to be elongated, and apparently serve as a useful spatial detection method for instantaneous vortex cores.

4) Arch-like (head + neck) vortical structures with a broad range of sizes are common in the boundary layer, while quasi-streamwise vortices ("legs") with significant streamwise extent are relatively rare.

5) High-pressure regions are spheroid and rarely elongated, and typically occur just upstream and below the heads of vortical arches, at the interface between high- and low-speed regions.

6) There is a strong instantaneous spatial association between vortical structures and both ejections (uv2) and sweeps (uv4).

7) Both ejections and sweeps occur along quasi-streamwise "leg" vortices, but ejections are most common just upstream and below vortical arches. Sweeps occur most often on the outboard sides of the "necks" of vortical arches.

8) Some near-wall shear layers have been observed to roll up into transverse vortices, which are in turn associated with significant local contribution to the Reynolds shear

stress through ejection (uv2) motions. Shear layers not in the process of rolling up contribute only weakly to the Reynolds shear stress.

9) "Backs" of outer-region, large-scale motions have been identified in the simulation data. In some cases at least, the backs (at a characteristic angle of 18 to 30 degrees) are the interface between low-speed fluid trailing a single large-scale vortical structure and the following high-speed fluid.

Work on the spatio-temporal relationships between the various structures is under way, and results will be reported this year. Among the questions of primary focus are (1) the nature and importance of the interaction of the outer, large-scale motions with turbulence-producing events near the wall; and (2) the generation mechanisms and evolution histories of the common vortical structures. As noted, a variety of population and contribution statistics are being compiled to establish the statistical relevance of the observed structural features and interactions. Finally, simulated probes and visual markers are being applied to draw more direct comparisons between the simulated boundary layer and experimental observations.

The current work, along with the 1987 Center for Turbulence Research Summer Program, marks the opening of a new era in experimental fluid dynamics, in which full-simulation results are studied by experimentalists working in "numerical wind tunnels." The long-term benefits of this new activity should include 1) increased experimental emphasis on flows with more complex boundary conditions and higher Reynolds numbers, 2) development of advanced instrumentation with specialized measurement capabilities, 3) increased understanding of probe resolution, conditional averaging, and phase-jitter effects, and 4) joint experimental/numerical investigations of flows in which the two techniques are complementary.

It is hoped that the current study, as part of a community-wide assessment of the state of boundary layer structure knowledge (Part I), will accelerate progress toward a broad-based understanding of the dynamics of wall-layer turbulence.

ACKNOWLEDGMENTS

This project is jointly funded by NASA Ames Research Center, The Air Force Office of Scientific Research, and The Office of Naval Research.

All of the computational work has been done at NASA Ames. The simulation code was written and run by Philippe Spalart at Ames. The core of the graphical display software was developed by P. Buning of Ames, and by F. Merritt and G. Bancroft of Sterling Software. Support for this project has been provided by the personnel and equipment of the Ames Workstation Applications Office and the Imaging Technology Branch, and by the Experimental Fluid Dynamics Branch, of which S. K. Robinson is a member.

The time of Prof. S. J. Kline and of O. Manickham and L. Portela is supported by AFOSR and ONR.

Useful discussions of this work have been provided by Dr. J. Kim at Ames, by Professors P. Moin and B. Cantwell at Stanford, and by the many participants of the cooperative program in turbulence structure. Reviews of this paper were provided by Prof. J. Wallace at University of Maryland, by Prof. R. Brodkey at the Ohio State University, and by L. Portela of Stanford University.

REFERENCES

Acarlar, M. S., and Smith, C. R., A Study of Hairpin Vortices in a Laminar Boundary Layer. Part 1: Hairpin Vortices Generated by a Hemisphere Protuberance, J. Fluid Mech., vol. 175, p. 1, 1987.

Bakewell, H. P., Jr., and Lumley, J. L., Viscous Sublayer and Adjacent Wall Region in Turbulent Pipe Flow, Phys. Fluids, vol. 10, no. 9, p. 1880, 1967.

Bandyopadhyay, P., Large Structure with a Characteristic Upstream Interface in Turbulent Boundary Layers, Phys. Fluids, vol. 23, no. 11, p. 2326, 1980.

Blackwelder, R. F., and Kovasznay, L. S. G., Time Scales and Correlations in a Turbulent Boundary Layer, Phys. Fluids, vol. 15, no. 9, p. 1545, 1972.

Blackwelder, R. F., and H. Eckelmann, Streamwise Vortices Associated with the Bursting Phenomenon, J. Fluid Mech., vol. 94, part 3, p. 577, 1979.

Bogard, D. G., and Tiederman, W. G., Burst Detection with Single-Point Velocity Measurements, J. Fluid Mech., vol. 162, p. 389, 1986.

Bogard, D. G., and Tiederman, W. G., Characteristics of Ejections in Turbulent Channel Flow, J. Fluid Mech., vol. 179, p. 1, 1987.

Britz, D., and Antonia, R. A., A Computer Algorithm for the Identification of Temperature Fronts in a Turbulent Shear Flow, Exp. Fluids, vol. 5, p. 134, 1987.

Brodkey, R. S., Wallace, J. M., and Eckelmann, H., Some Properties of Truncated Signals in Bounded Shear Flows, J. Fluid Mech., vol. 63, part 2, p. 209, 1974.

Brown, G. L., and Thomas, A. S. W., Large Structure in a Turbulent Boundary Layer, Phys. Fluids, vol. 20, no. 10, part 2, p. 5243, 1977.

Chen, C. H. P., and Blackwelder, R. F., Large-Scale Motion in a Turbulent Boundary Layer: a Study Using Temperature Contamination, J. Fluid Mech., vol. 89, part 1, p. 1, 1978.

Corino, E. R., and Brodkey, R. S., A Visual Investigation of the Wall Region in Turbulent Flow, J. Fluid Mech., vol. 37, part 1, p. 1, 1969.

Dinkelacker, A., Hessel, M., Meier, G. E. A., and Schewe, G., Investigation of Pressure Fluctuation Beneath Turbulent Boundary Layer by Means of an Optical Method, Bericht, vol. 105, Max-Planck Institut fu;r Stro;mungsforschung, Go;ttingen, 1977.

Falco, R. E., The Production of Turbulence Near a Wall, AIAA Paper 80-1356, 1980a.

Falco, R. E., Combined Simultaneous Visualization/Hot-Wire Anemometry for the Study of Turbulent Flows, J. Fluid Mech., vol. 102, p. 174, 1980b.

Falco, R. E., New Results, a Review and Synthesis of the Mechanism of Turbulence Production in Boundary Layers and its Modification, AIAA Paper 83-0377, 1983.

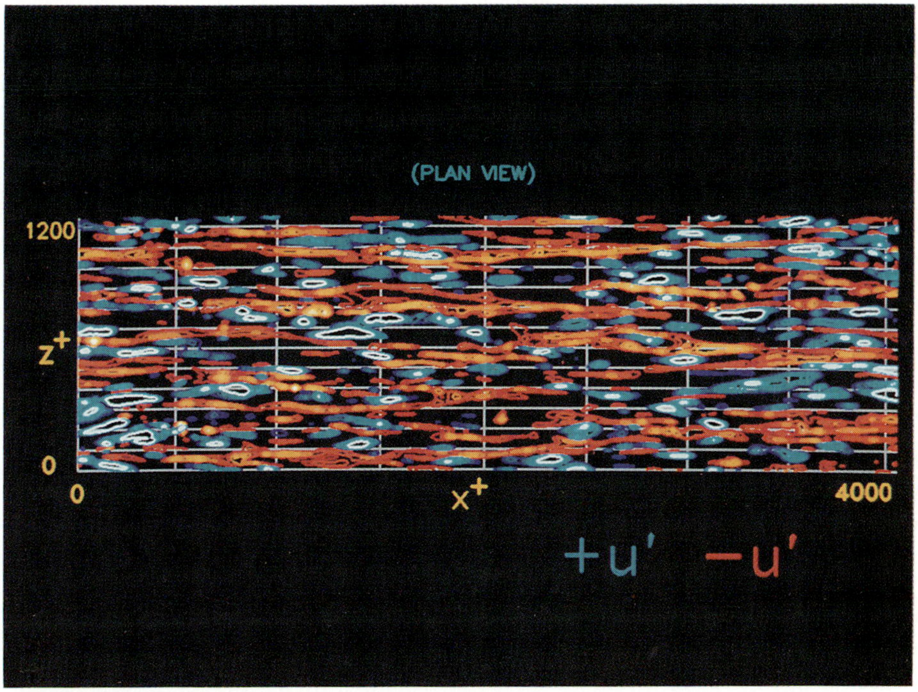
(PLAN VIEW)
1200
z+
0
0
x+
4000
+u' −u'

Fig. 5

Fig. 6

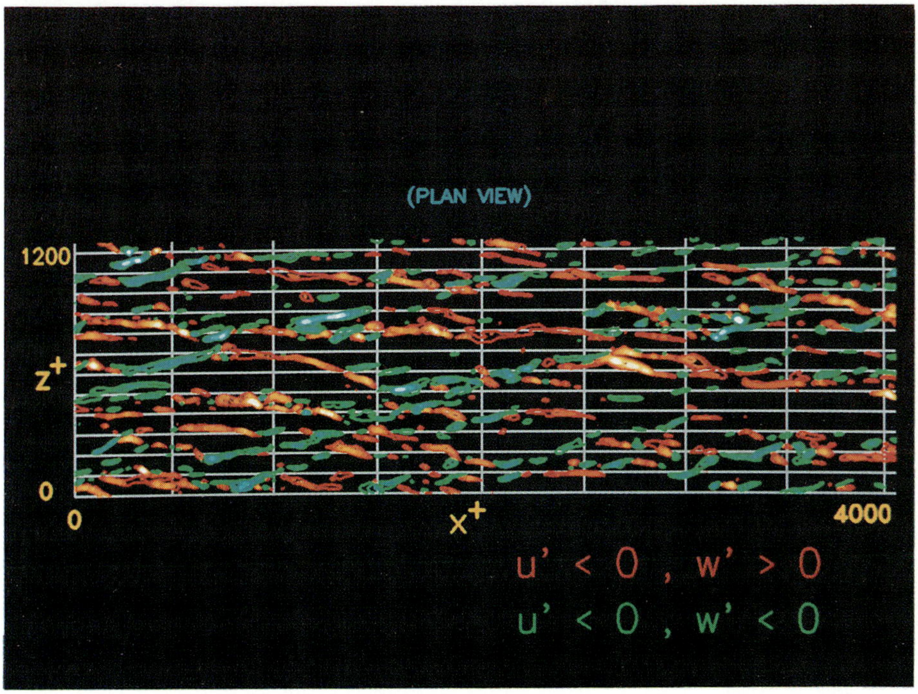
(PLAN VIEW)
1200
z+
0
0
x+
4000
u' < 0 , w' > 0
u' < 0 , w' < 0

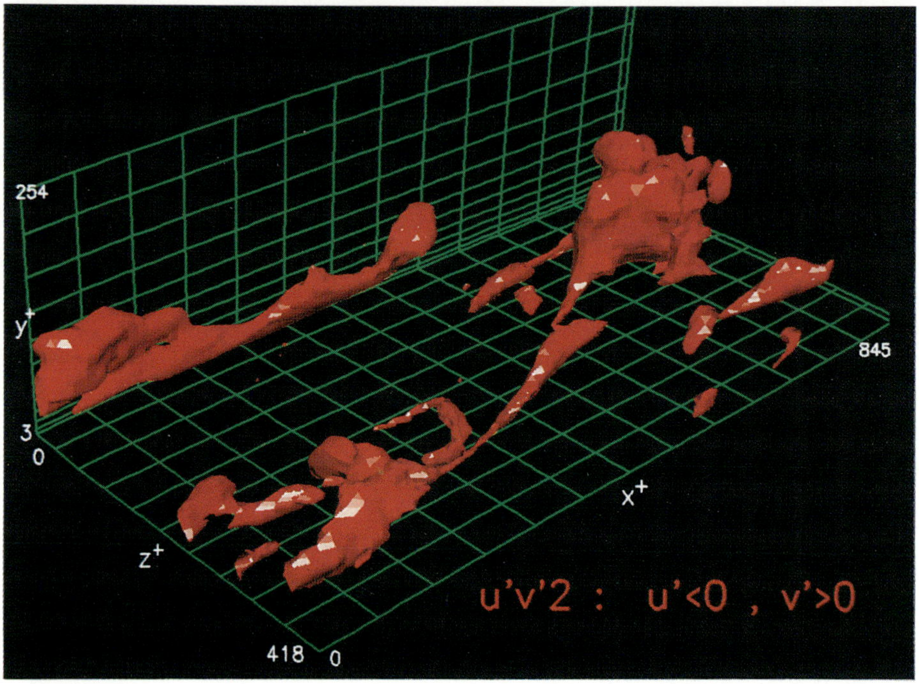

Fig. 7

Fig. 8

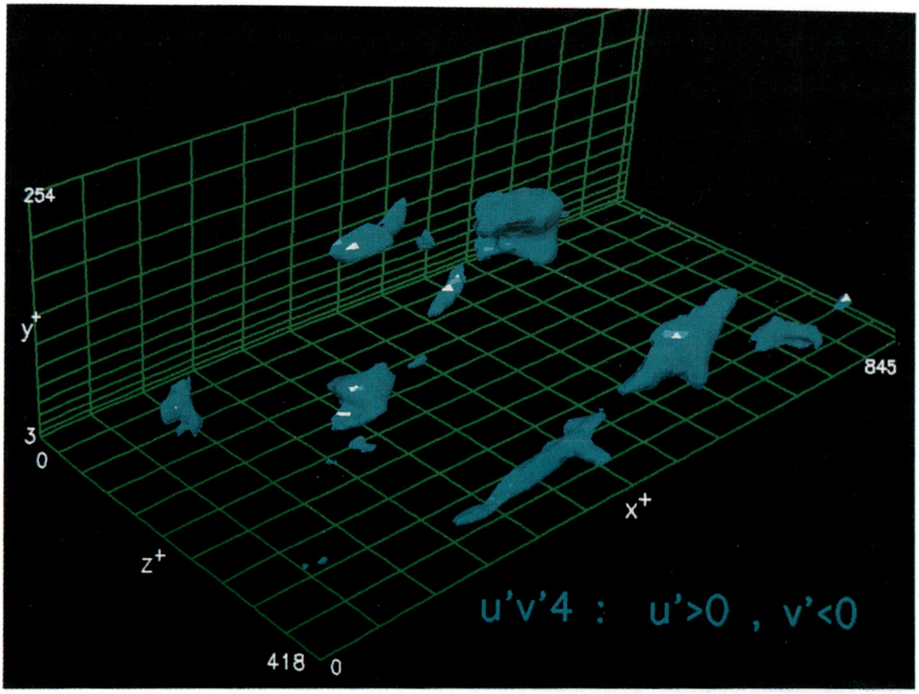

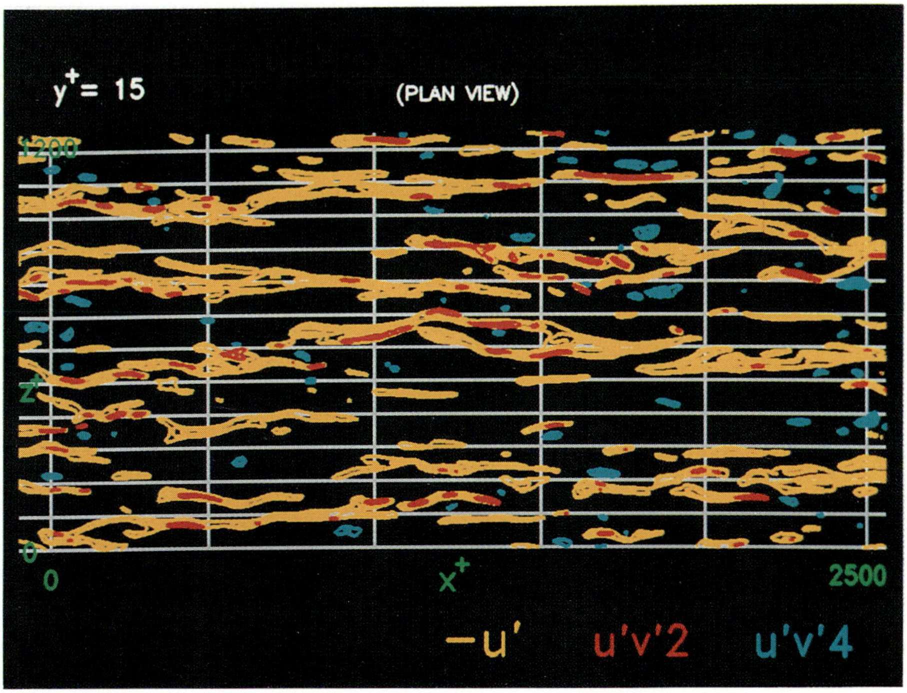
y+= 15
(PLAN VIEW)
1200
z+
0
0
x+
2500
−u'
u'v'2
u'v'4

Fig. 9

Fig. 10

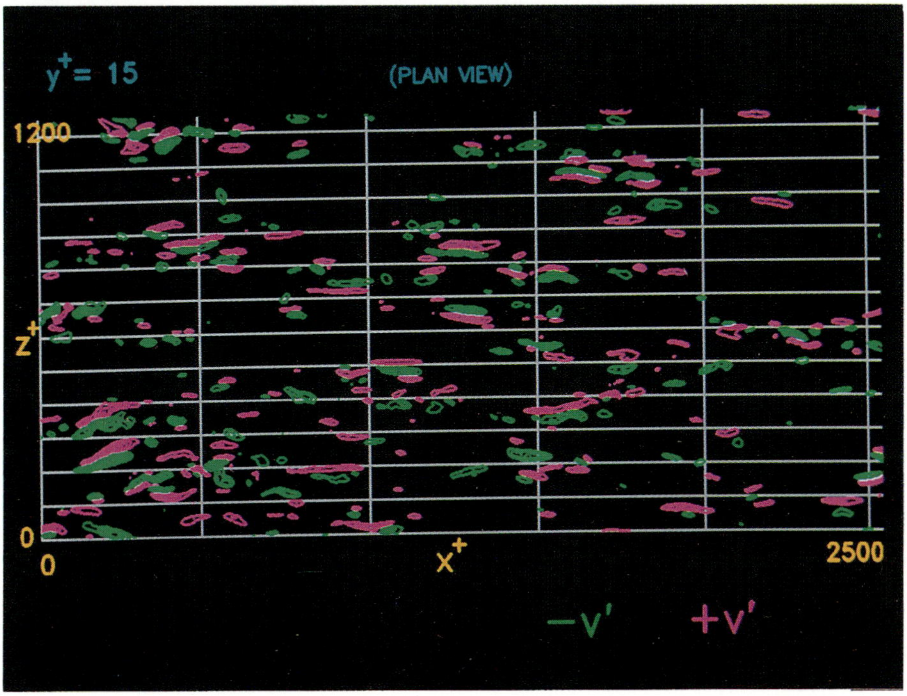
y+= 15
(PLAN VIEW)
1200
z+
0
0
x+
2500
−v'
+v'

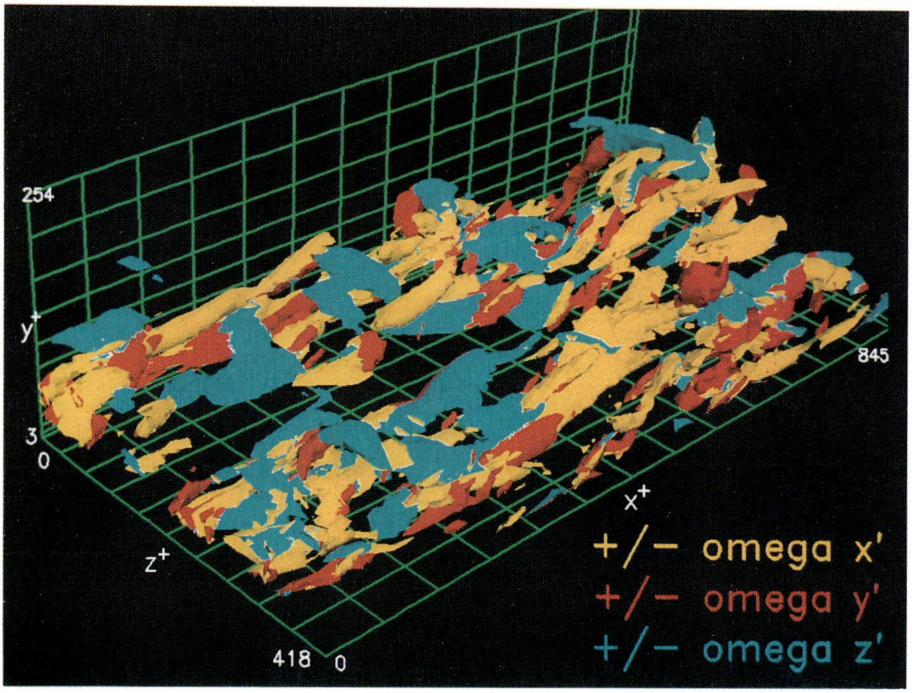
254
y+
3
0
z+
418
0
845
x+
+/− omega x'
+/− omega y'
+/− omega z'
Fig. 11

Fig. 12

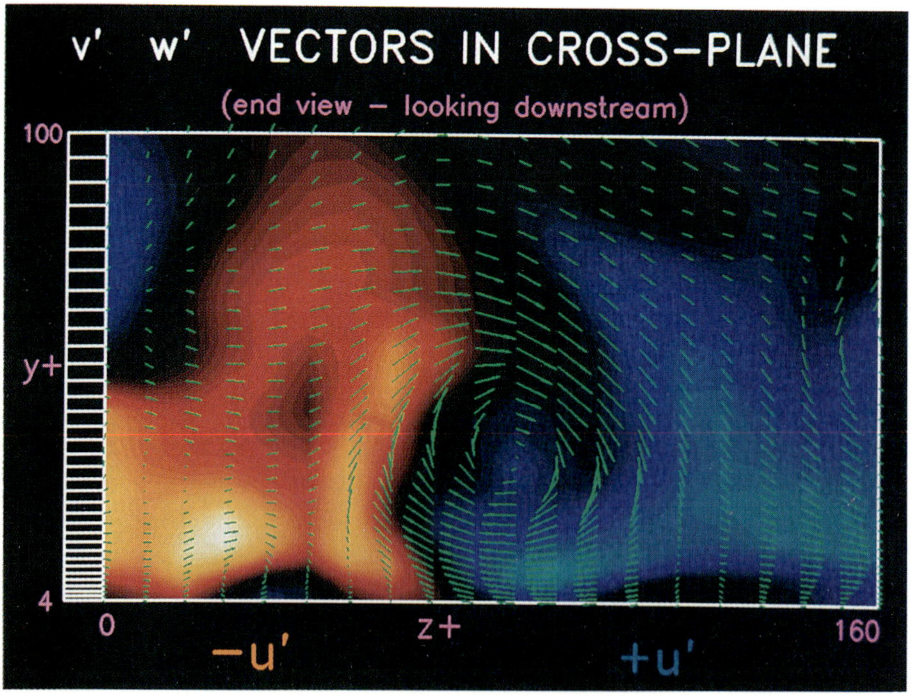
v' w' VECTORS IN CROSS-PLANE
(end view – looking downstream)
100
y+
4
0
z+
160
−u'
+u'

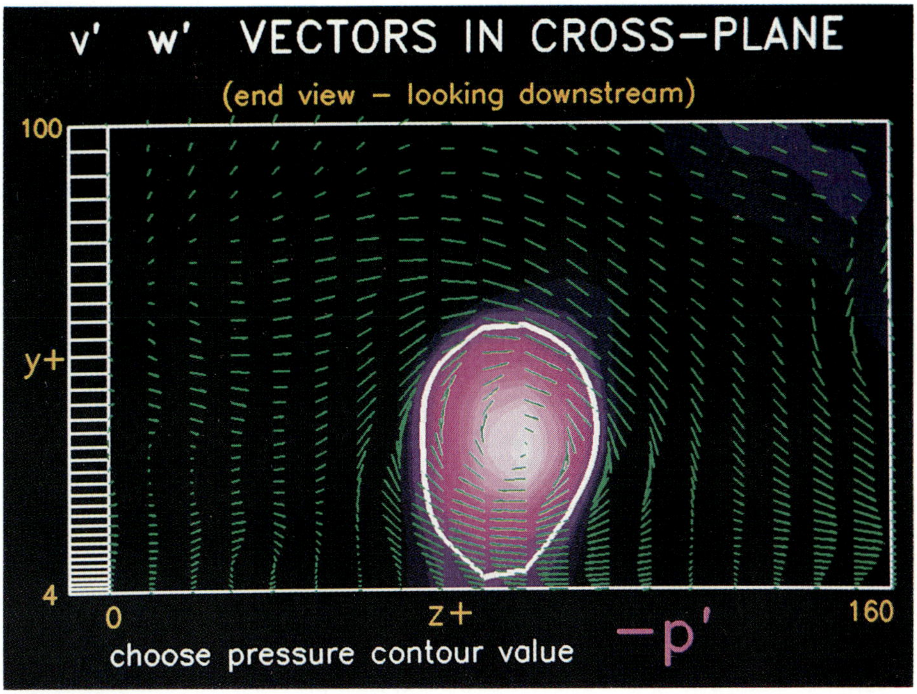
v' w' VECTORS IN CROSS-PLANE
(end view - looking downstream)
100
y+
4
0
z+
160
choose pressure contour value
−p'

Fig. 13

Fig. 14

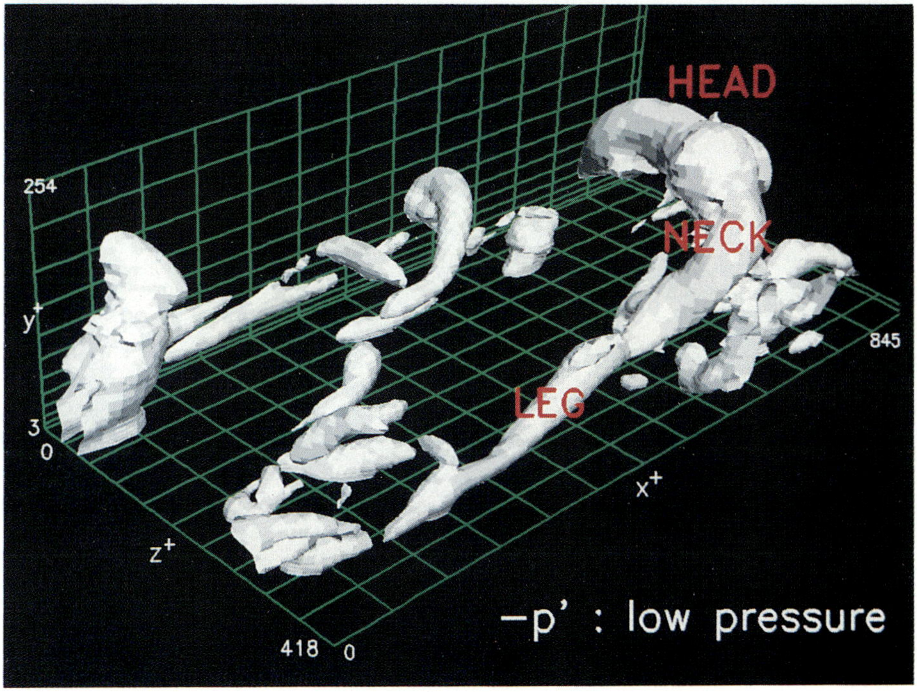
HEAD
NECK
LEG
254
y+
3
0
z+
418
0
x+
845
−p' : low pressure

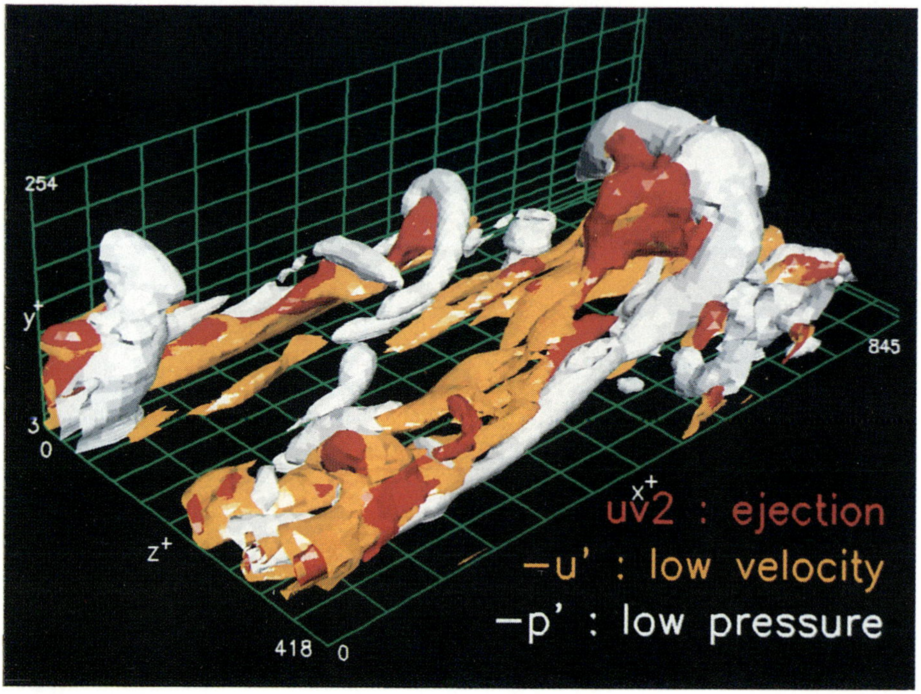
254
y+
3
0
845
x+
uv2 : ejection
-u' : low velocity
-p' : low pressure
z+
418
0

Fig. 15

Fig. 16

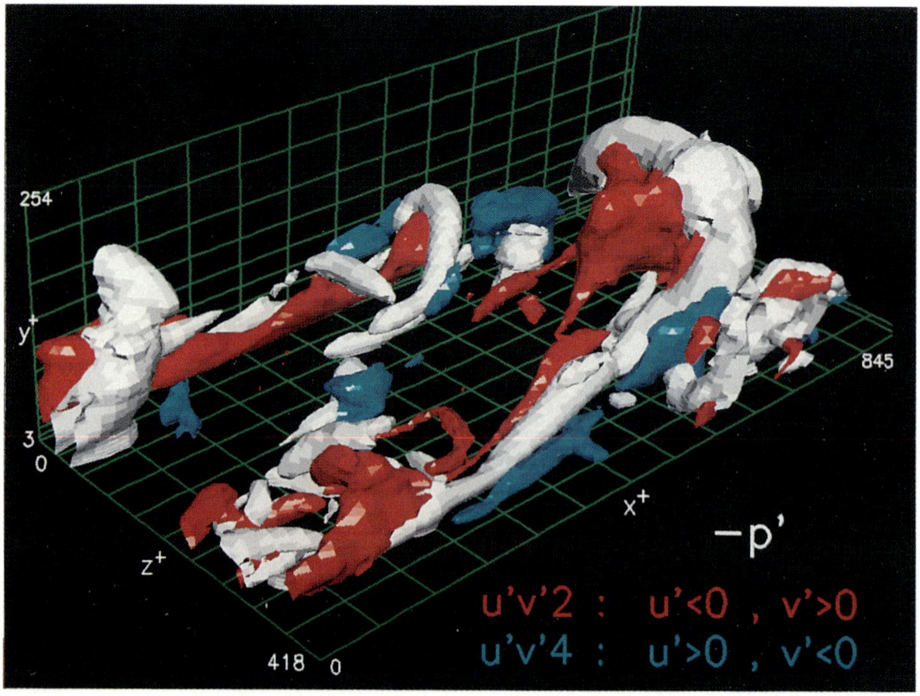
254
y+
3
0
845
x+
-p'
u'v'2 : u'<0 , v'>0
u'v'4 : u'>0 , v'<0
z+
418
0

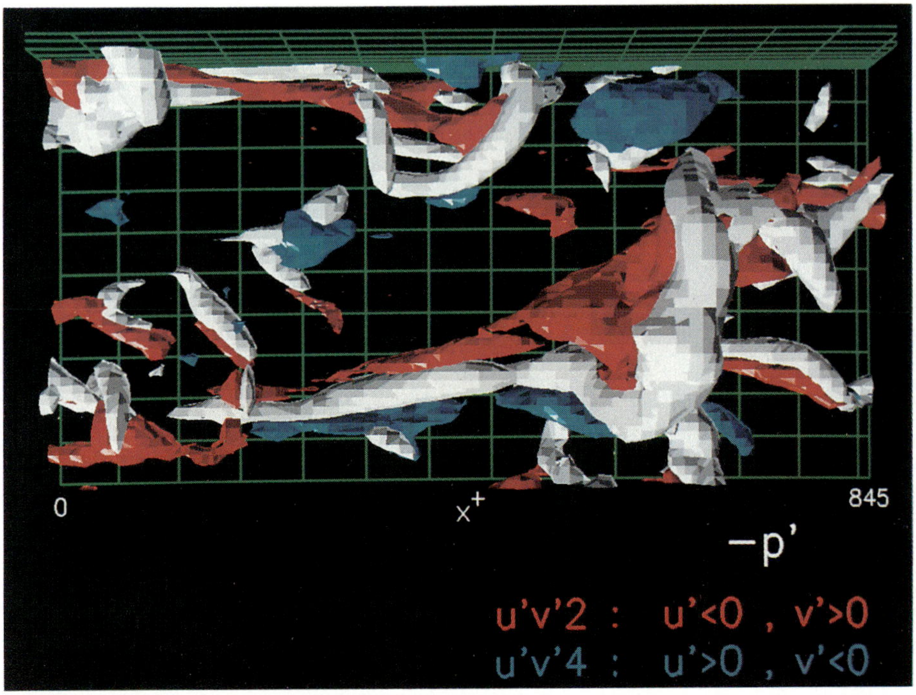
0
x+
845
−p'
u'v'2 : u'<0 , v'>0
u'v'4 : u'>0 , v'<0

Fig. 17

Fig. 18

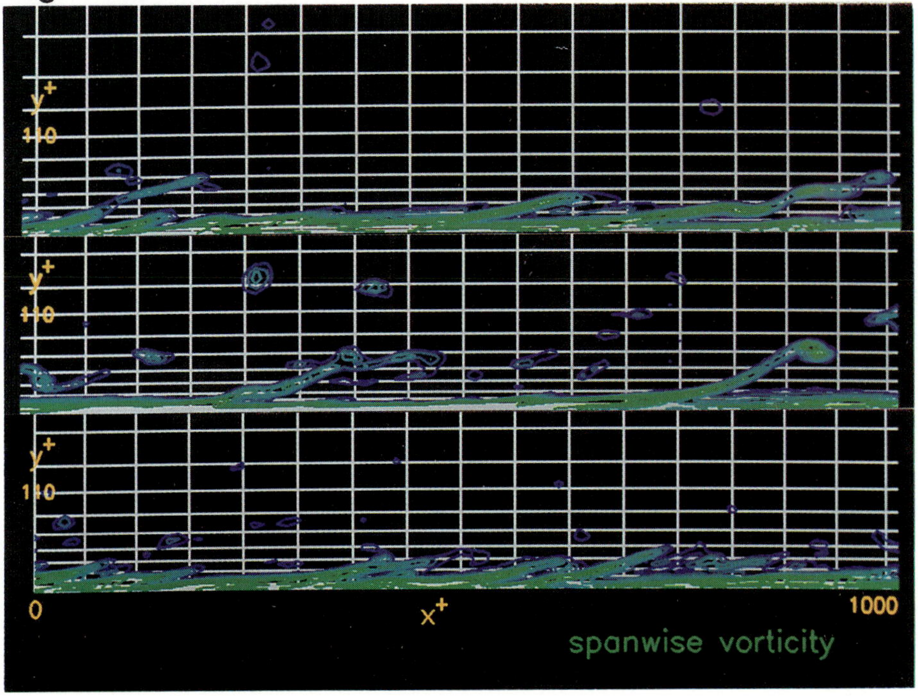
y+
110
y+
110
y+
110
0
x+
1000
spanwise vorticity

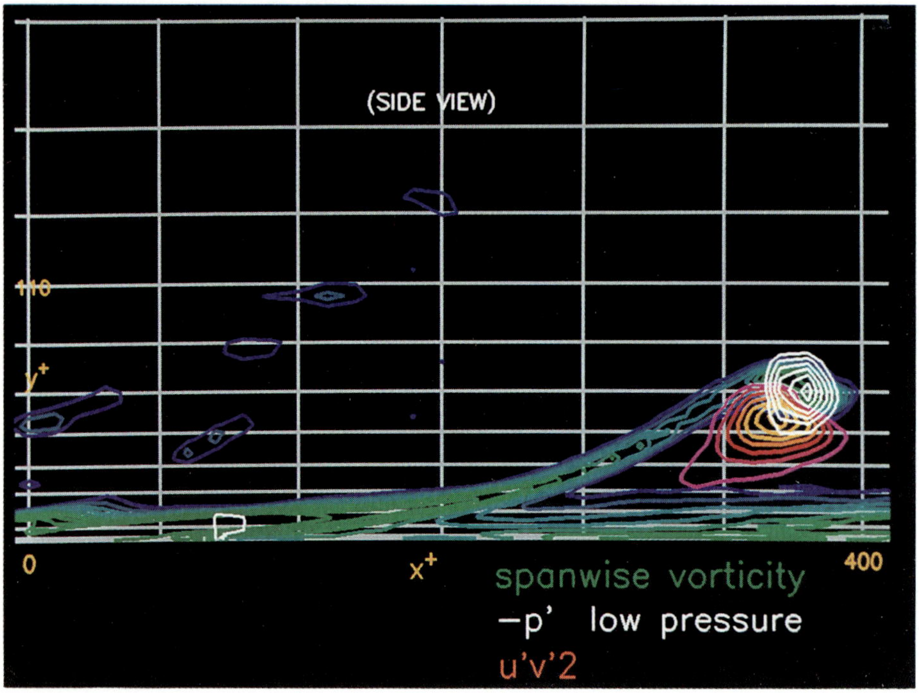

(SIDE VIEW)
110
y+
0
x+
400
spanwise vorticity
−p' low pressure
u'v'2

Fig. 19

Fig. 20

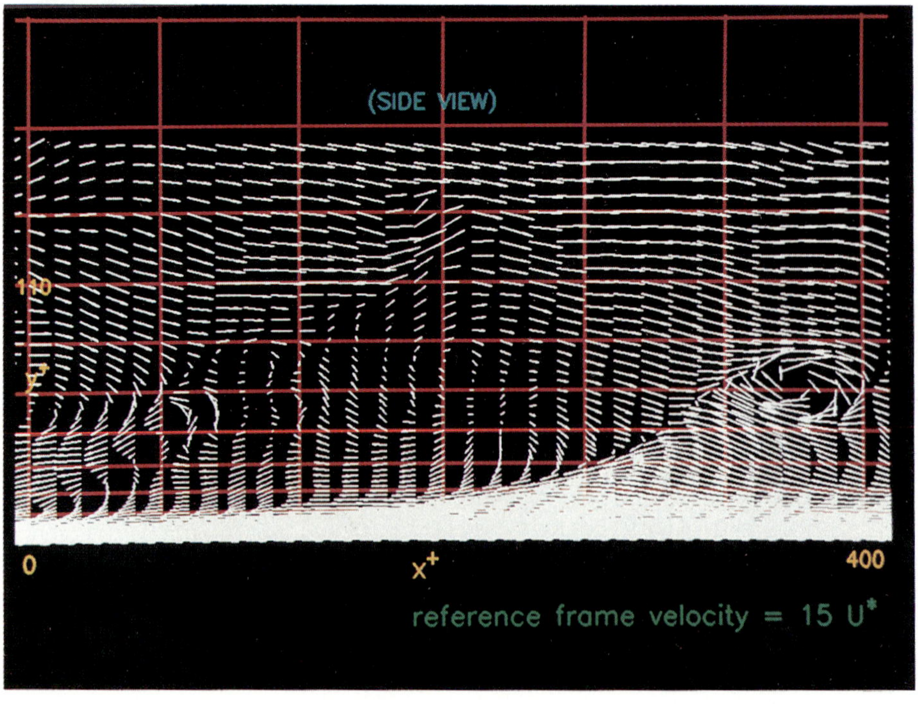

(SIDE VIEW)
110
y+
0
x+
400
reference frame velocity = 15 U*

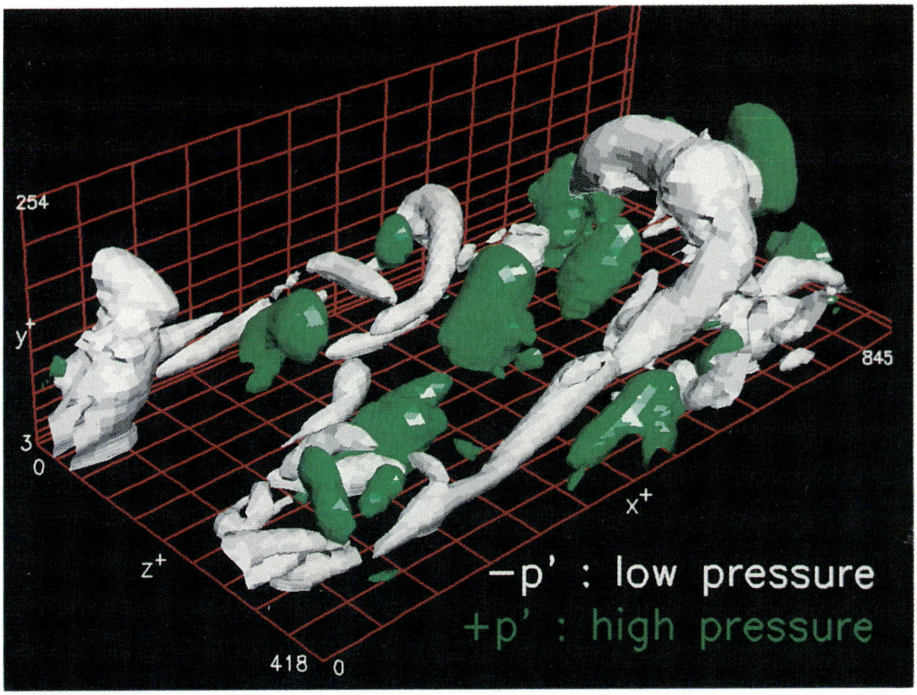
254
y+
3
0
845
x+
z+
418
0
−p' : low pressure
+p' : high pressure

Fig. 21

Fig. 22

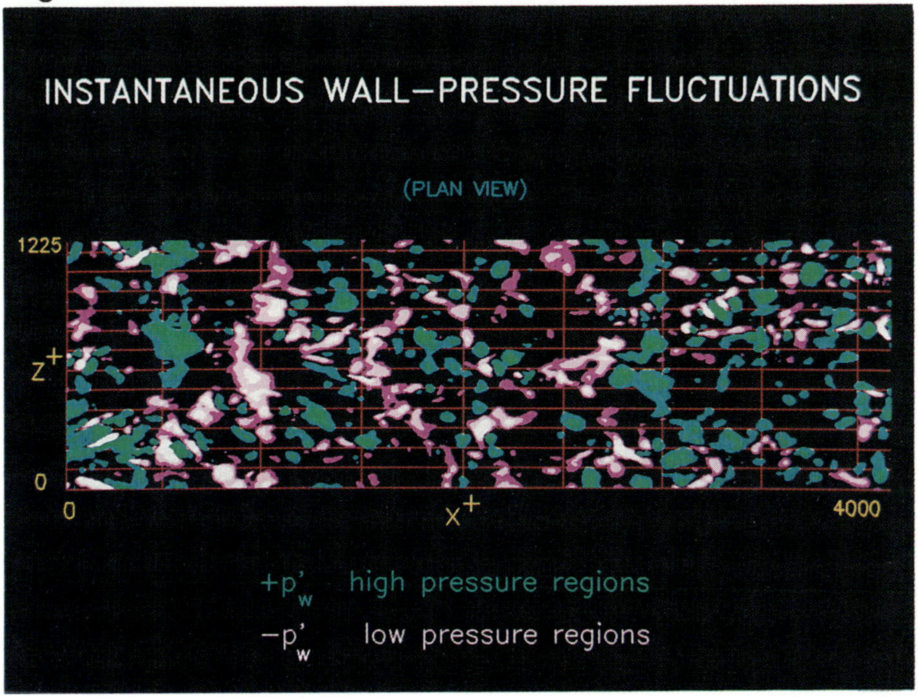
INSTANTANEOUS WALL-PRESSURE FLUCTUATIONS
(PLAN VIEW)
1225
z+
0
0
x+
4000
+p'w high pressure regions
−p'w low pressure regions

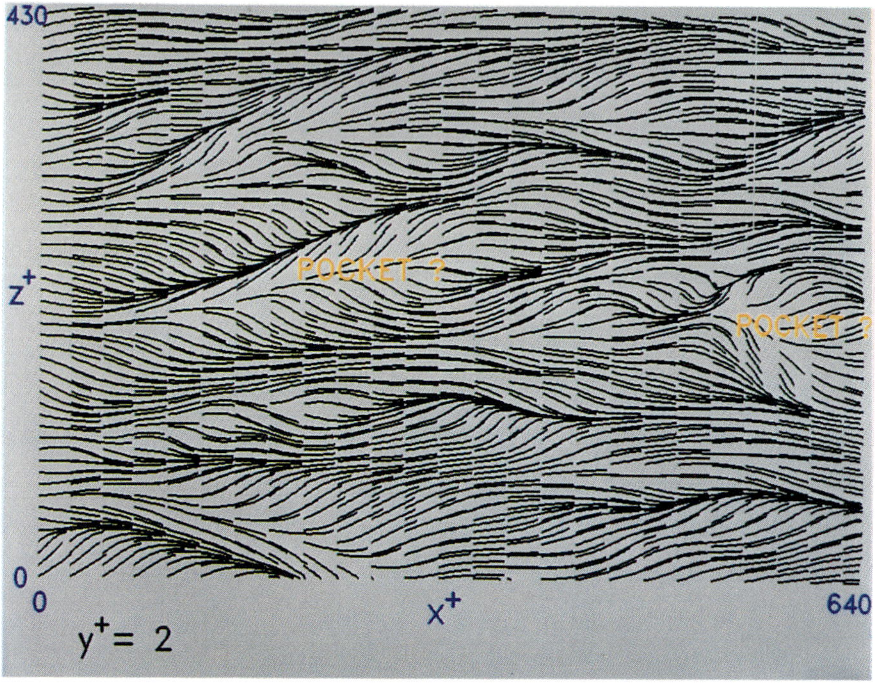

Fig. 23

Fig. 24

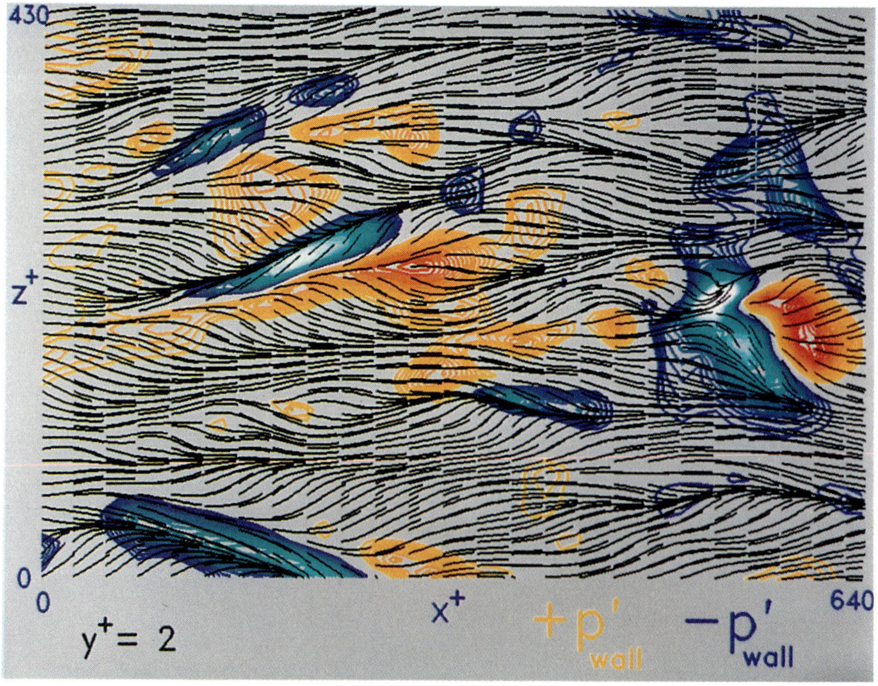

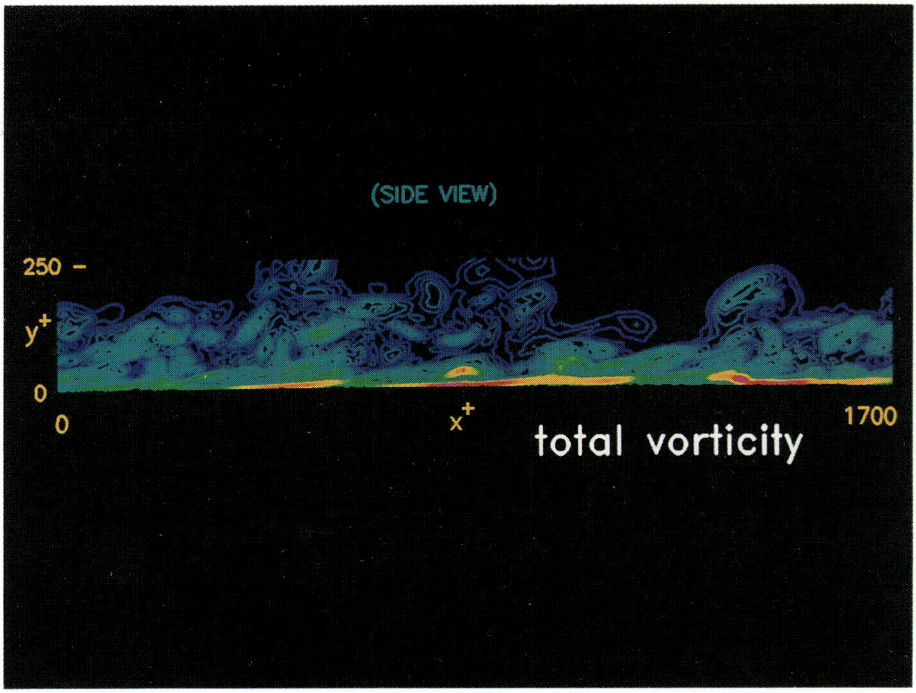
(SIDE VIEW)
250 –
y+
0
0
x+
1700
total vorticity

Fig. 25

Fig. 26

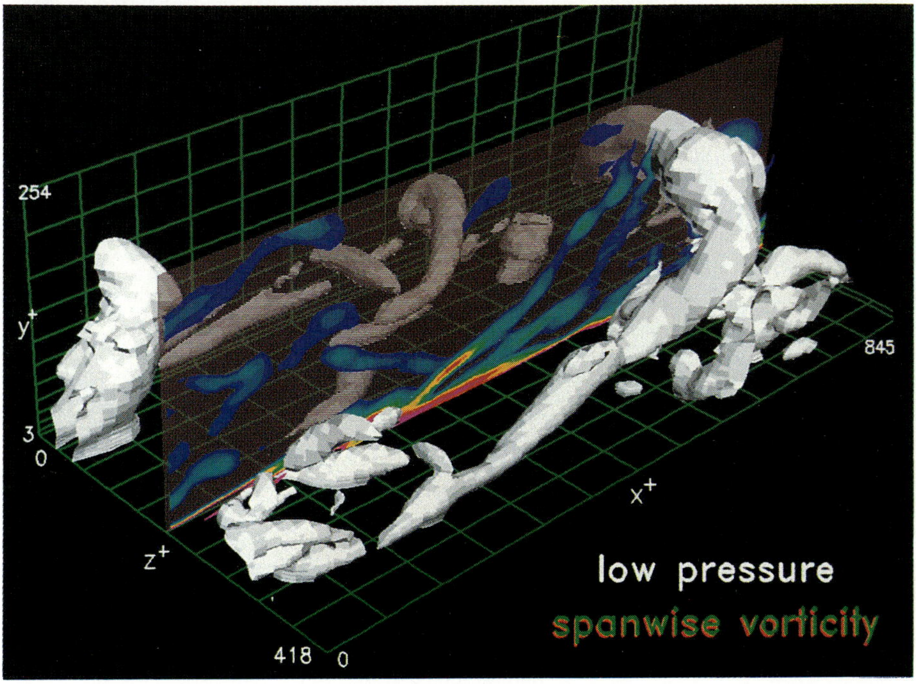
254
y+
3
0
845
x+
z+
418
0
low pressure
spanwise vorticity

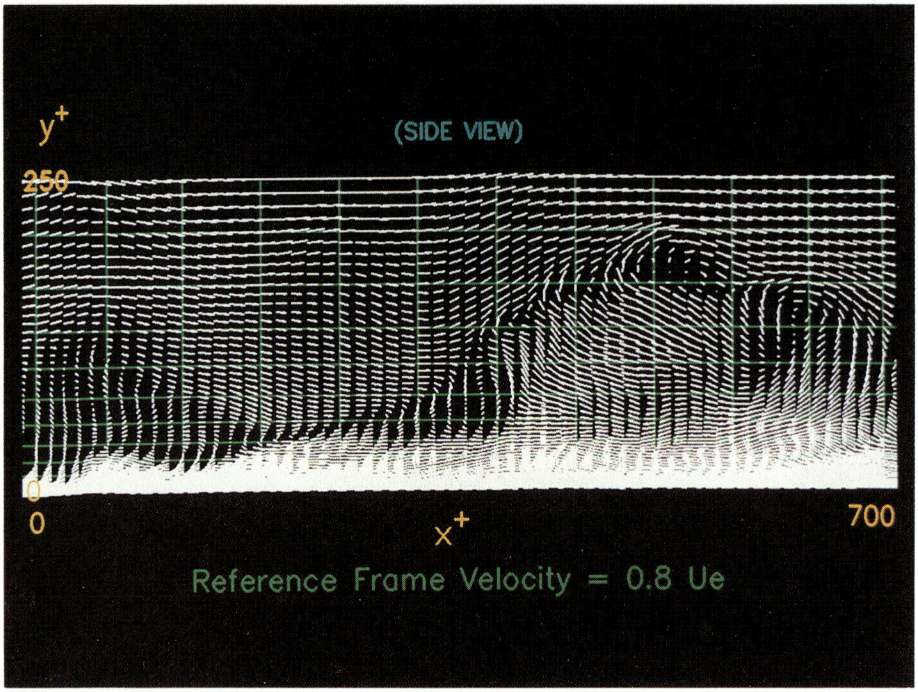
(SIDE VIEW)
y+
250
0
0
x+
700
Reference Frame Velocity = 0.8 Ue

Fig. 27

Fig. 28

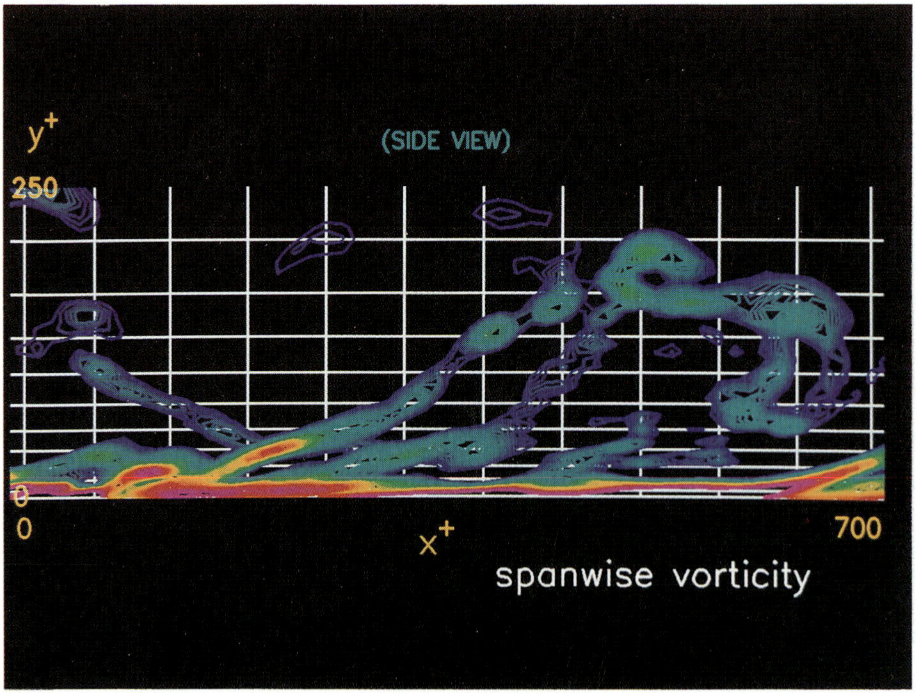
(SIDE VIEW)
y+
250
0
0
x+
700
spanwise vorticity

Fernando, E. M., Spina, E. F., Donovan, J. F., and Smits, A. J., Detection of Large-Scale Organized Motions in a Turbulent Boundary Layer, Proceedings of the 6th Symposium on Turbulent Shear Flows, Toulouse, France, September 7-9, 1987.

Grötzbach, G., and Schumann, U., Direct Numerical Simulation of Turbulent Velocity, Pressure, and Temperature Fields in Channel Flows, in Turbulent Shear Flows I, F. Durst, B. E. Launder, F. W. Schmidt, and J. H. Whitelaw (Eds.), p. 370, Springer-Verlag, Berlin, 1979.

Guezennec, Y. G., Piomelli, U., and Kim, J., Conditionally-Averaged Structures in Wall-Bounded Turbulent Flows, Proceedings of the Summer Program 1987, p. 263, Center for Turbulence Research, NASA/ Stanford, 1987.

Head, M. R., and Bandyopadhyay, P., New Aspects of Turbulent Boundary Layer Structure, J. Fluid Mech., vol. 107, p. 297, 1981.

Hirata, M., and Kasagi, N., Studies of Large-Eddy Structures in Turbulent Shear Flows with the Aid of Flow-Visualization Techniques, Studies in Heat Transfer, J. P. Hartnett et al. (Eds.), Hemisphere, 1979.

Jimenez, J., Moin, P., Moser, R. D., and Keefe, L. R., Ejection Mechanisms in the Sublayer of a Turbulent Channel, Proceedings of the Summer Program 1987, p. 37, Center for Turbulence Research, NASA/ Stanford, 1987.

Johansson, A. V., Alfredsson, P. H., and Eckelmann, H., On the Evolution of Shear-Layer Structures in Near-Wall Turbulence, in Advances in Turbulence, G. Comte-Bellot and J. Mathieu (Eds.), Springer-Verlag, 1987a.

Johansson, A. V., Alfredsson, P. H., and Kim, J., Shear-Layer Structures in Near-Wall Turbulence, Proceedings of the Summer Program 1987, p. 237, Center for Turbulence Research, NASA/Stanford, 1987b.

Kim, H. T., Kline, S. J., and Reynolds, W. C., The Production of Turbulence Near a Smooth Wall in a Turbulent Boundary Layer, J. Fluid Mech., vol. 50, part 1, p. 133, 1971.

Kim, J., Turbulence Structures Associated with the Bursting Event, Phys. Fluids, vol. 28, no. 1, p. 52, 1985.

Kim, J., Overview of Research by the Turbulence Structure Group, Proceedings of the Summer Program 1987, p. 231, Center for Turbulence Research, NASA/Stanford, 1987.

Kim, J., and Moin, P., The Structure of the Vorticity Field in Turbulent Channel Flow, Part 2. Study of Ensemble-Averaged Fields, J. Fluid Mech., vol. 162, p. 339, 1986.

Kim, J., Moin, P., and Moser, R. D., Turbulence Statistics in Fully-Developed Channel Flow at Low Reynolds Number, J. Fluid Mech., vol. 177, p. 133, 1987.

Kline, S. J., and Robinson, S. K., Quasi-Coherent Structures in the Turbulent Boundary Layer: Part I: Status Report on a Communitywide Summary of the Data, Zoran P. Zarić Memorial Seminar on Near-Wall Turbulence, Dubrovnik, Yugoslavia, May 16-20, 1988.

Kovasznay, L. S. G., Kibens, V., and Blackwelder, R. F., Large-Scale Motion in the Intermittent Region of a Turbulent Boundary Layer, J. Fluid Mech., vol. 41, part 2, p. 283, 1970.

Kreplin, H.-P., and Eckelmann, H., Propagation of Perturbations in the Viscous Sublayer and Adjacent Wall Region, J. Fluid Mech., vol. 95, part 2, p. 305, 1979.

Moin, P., Probing Turbulence via Large-Eddy Simulation, AIAA Paper 84-0174, 1984.

Moin, P., and Spalart, P. R., Contributions of Numerical Simulation Data Bases to the Physics, Modeling, and Measurement of Turbulence in Recent Advances in Turbulence, R. E. Arndt and W. K. George (Eds.), Hemisphere Publishing Corp., Washington, D.C., 1988.

Moin, P., Adrian, R. J., and Kim, J., Stochastic Estimation of Conditional Eddies in Turbulent Channel Flow. Proceedings of the 6th Symposium on Turbulent Shear Flows, Toulouse, France, September 7-9, 1987.

Moser, R. D., and Moin, P., Direct Numerical Simulation of Curved Turbulent Channel Flow, NASA TM 85974. Also, J. Fluid Mech., vol. 175, p. 479, 1987.

Murlis, J., Tsai, H. M., and Bradshaw, P., The Structure of Turbulent Boundary Layers at Low Reynolds Numbers, J. Fluid Mech., vol. 122, p. 13, 1982.

Nychas, S. G., Hershey, H. C., and Brodkey, R. S., A Visual Study of Turbulent Shear Flow, J. Fluid Mech., vol. 61, p. 513, 1973.

Offen, G. R., and Kline, S. J., A Proposed Model of the Bursting Process in Turbulent Boundary Layers, J. Fluid Mech., vol. 70, part 2, p. 209, 1975.

Pearson, C. F., and Abernathy, F.H., Evolution of the Flow Field Associated with a Streamwise Diffusing vortex, J. Fluid Mech., vol. 146, September, p. 271, 1984.

Praturi, A. K., and Brodkey, R. S., A Stereoscopic Visual Study of Coherent Structures in Turbulent Shear Flow, J. Fluid Mech., vol. 89, part 2, p. 251, 1978.

Rajagopalan, S., and Antonia, R. A., Some Properties of the Large Structure in a Fully Developed Turbulent Duct Flow, Phys. Fluids, vol. 22, no. 4, p. 614, 1979.

Robinson, S. K., An Experimental Search for Near-Wall Boundary Conditions for Large Eddy Simulation, AIAA/ASME 3rd Joint Thermophysics, Fluids, Plasma and Heat Transfer Conference, St. Louis, Missouri, June 7-11, 1982.

Runstadler, P. W., Kline, S. J., and Reynolds, W. C., An Experimental Investigation of the Flow Structures of the Turbulent Boundary Layer, Rept. MD-8, Thermosciences Div., Dept. Mech. Eng., Stanford Univ., 1963.

Smith, C. R., A Synthesized Model of the Near-Wall Behavior in Turbulent Boundary Layers, in Proc. Eighth Symp. on Turbulence, Univ. Missouri-Rolla, 1984.

Smith, C. R., and Schwartz, S. P., Observation of Streamwise Rotation in the Near-Wall Region of a Turbulent Boundary Layer, Phys. Fluids, vol. 26, no. 3, p. 641, 1983.

Spalart, P. R., Direct Simulation of a Turbulent Boundary Layer up to $Re_\theta = 1410$, J. Fluid Mech., vol. 187, p. 61, 1988.

Subramanian, C. S., Rajagopalan, S., Antonia, R. A., and Chambers, A. J., Comparison of Conditional Sampling and Averaging Techniques in a Turbulent Boundary Layer, J. Fluid Mech., vol. 123, p. 335, 1982.

Swearingen, J. D., and Blackwelder, R. F., The Growth and Breakdown of Streamwise Vortices in the Presence of a Wall, J. Fluid Mech., vol. 182, p. 255, 1987.

Swearingen, J. D., Blackwelder, R. F., and Spalart, P. R., Inflectional Instabilities in the Wall Region of Bounded Turbulent Shear Flows, Proceedings of the Summer Program 1987, p. 291, Center for Turbulence Research, NASA/Stanford, 1987.

Theodorsen, T., Mechanism of Turbulence, Proceedings 2nd Midwestern Conference on Fluid Mechanics, Ohio State Univ., Columbus, Ohio, 1952.

Thomas, A. S. W., and Brown, G. L., Large Structure in a Turbulent Boundary Layer, 6th Australasian Hydraulics and Fluid Mechanics Conference, Adelaide, Australia, December 5-9, 1977.

Wallace, J. M., Brodkey, R. S., and Eckelmann, H., Pattern-Recognized Structures in Bounded Turbulent Shear Flows, J. Fluid Mech., vol. 83, part 4, p. 673, 1977.

Willmarth, W. W., Advances in Fluid Mechanics, vol. 15, p. 158, Acad., New York, 1975.

Willmarth, W. W., and Lu, S. S., Structure of the Reynolds Stress Near the Wall, J. Fluid Mech., vol. 55, part 1, p. 65, 1972.

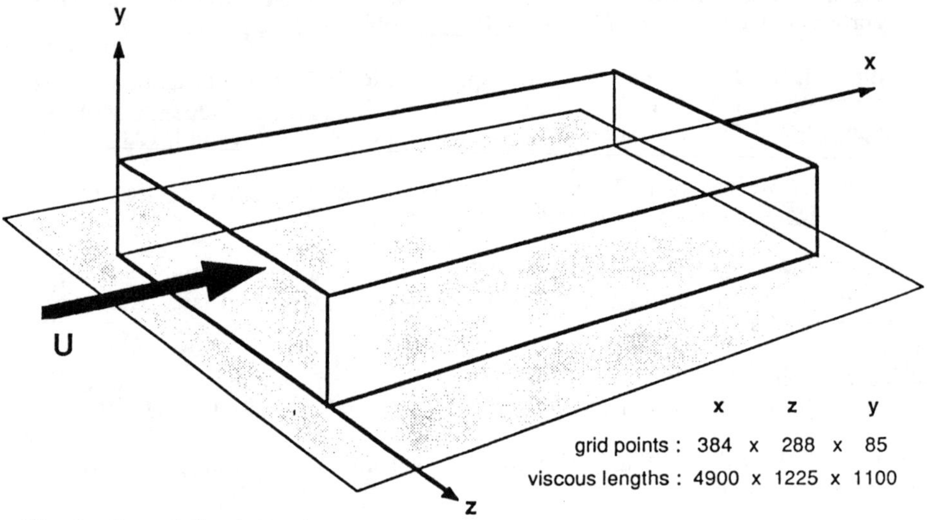

Fig. 1. Computational domain.

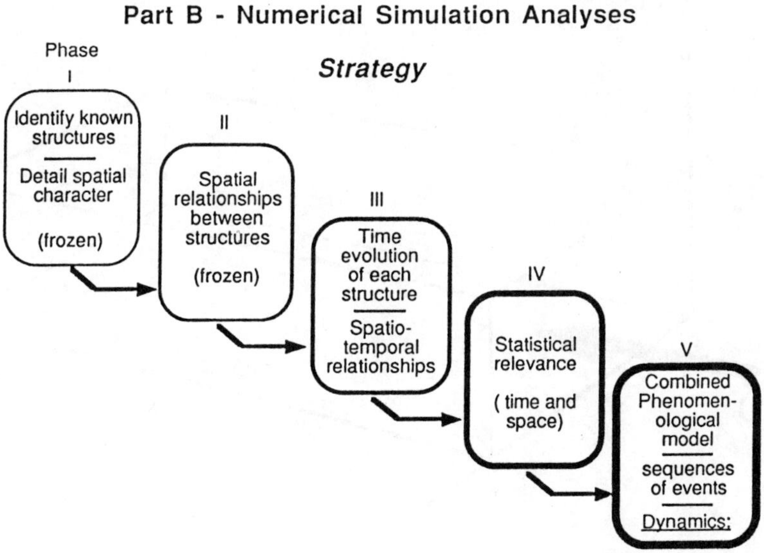

Fig. 2. Strategy for structural analysis of turbulence simulations.

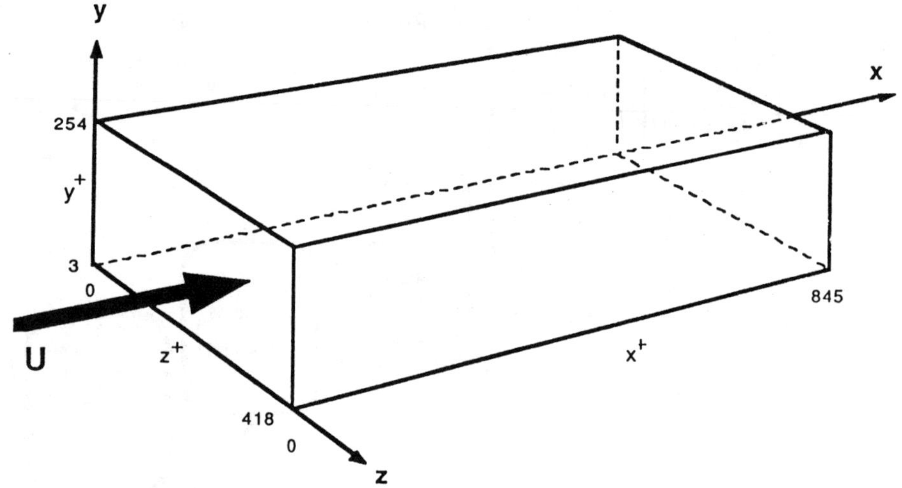

Fig. 3. Computational subvolume.

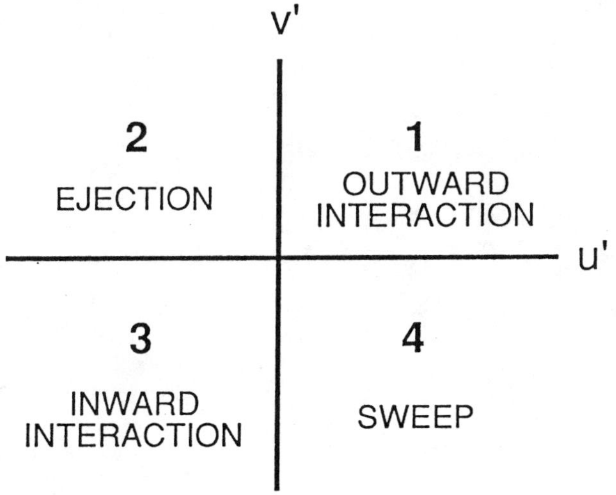

Fig. 4. Quadrants of the u' v' plane.

See figures 5 to 28 on color plates following page 236.

See next page for figure legends for the color plates.

Fig. 5. Contours of fluctuating streamwise velocity in an x-z plane at $y^+ = 2$.
Blue to white: $u'^+ = +0.7$ to $+2.0$.
Red to yellow: $u'^+ = -0.7$ to -2.0.

Fig. 6. Contours of u'w'2 and u'w'3 in an x-z plane at $y^+ = 15$.
Green to white: $u'w'2/u_\tau^2 = 2.1$ to 14.9.
Red to white: $u'w'3/u_\tau^2 = -2.1$ to -14.9.

Fig. 7. Quadrant 2 shear product in subvolume S.
Red: $u'v'2/u_\tau^2 = -4.2$.

Fig. 8. Quadrant 4 shear product in subvolume S.
Blue: $u'v'4/u_\tau^2 = -4.2$.

Fig. 9. Low-speed streaks, ejections, and sweeps in an x-z plane at $y^+ = 15$.
Yellow: $u'^+ = -3.0$.
Red: $u'v'2/u_\tau^2 \le -4.2$.
Blue: $u'v'4/u_\tau^2 \le -4.2$.

Fig. 10. Contours of $\pm v'$ in an x-z plane at $y^+ = 15$.
Magenta: $v'^+ = +1.0$.
Green: $v'^+ = -1.0$.

Fig. 11. Fluctuating vorticity in subvolume S (lowest y^+ shown is 28).
Yellow: $\omega_x'^+ = \pm 0$.
Red: $\omega_y'^+ = \pm .0$.
Blue: $\omega_z'^+ = \pm .0$.

Fig. 12. y-z plane cut through low/high-speed streak interface.
Yellow: $u'^+ = 0$ to -6.0.
Blue: $u'^+ = 0$ to $+6.0$.
Green: instantaneous v'w' vectors.

Fig. 13. y-z plane cut through low/high-speed streak interface.
Magenta to white: $p'^+ = 0$ to -12.0.
White contour line: $p'^+ = -4.2$.
Green: instantaneous v'w' vectors.

Fig. 14. Regions of low pressure in subvolume S.
White: $p'^+ = -4.2$.

Fig. 15. Regions of low pressure, low streamwise velocity, and quadrant 2 shear product in subvolume S.
White: $p'^+ = -4.2$.
Yellow: $u'^+ = -3.0$.
Red: $u'v'2/u_\tau^2 = -4.2$.

Fig. 16. Regions of low pressure, quadrant 2 shear product, and quadrant 4 shear product in subvolume S.
White: $p'^+ = -4.2$.
Red: $u'v'2/u_\tau^2 = -4.2$.
Blue: $u'v'4/u_\tau^2 = -4.2$.

Fig. 17. Top view of Figure 16.

Fig. 18. Near-wall shear layers in an instantaneous x-y plane.
Blue to green: $\omega_z^+ = 0.2$ to 1.0.

Fig. 19. Roll-up of near-wall shear layer.
Blue to green: $\omega_z^+ = +0.2$ to +1.0.
White: $p'^+ = -4.2$ to –17.0.
Magenta to yellow: $u'v'2/u_\tau^2 = -2.1$ to –12.8.

Fig. 20. Roll-up of near-wall shear layer. Instantaneous uv velocity vectors in an x-y plane moving at $15u_\tau$.

Fig. 21. Regions of high and low pressure in subvolume S.
White: $p'^+ = -4.2$.
Green: $p'^+ = +4.2$.

Fig. 22. Instantaneous wall-pressure fluctuations.
Blue to green: $p_w'^+ = +2.1$ to +21.3.
Magenta to white: $p_w'^+ = -2.1$ to –21.3.

Fig. 23. Instantaneous streamlines in an x-z plane at $y^+ = 2$.

Fig. 24. Instantaneous streamlines in an x-z plane at $y^+ = 2$, underlaid with contours of wall pressure.
Yellow to red: $p'^+ = +3.0$ to +25.0.
Blue to white: $p'^+ = -3.0$ to –25.0.

Fig. 25. Contours of total vorticity magnitude in an x-y plane.
Blue to magenta: $|\omega^+| = 0.02$ to 1.0.

Fig. 26. "Back" of large-scale motion with regions of low pressure in subvolume S.
White: $p'^+ = -4.2$.
Blue to magenta: $\omega_z = +0.2$ to +3.0.

Fig. 27. Instantaneous uv vectors in an x-y plane moving at 0.8 u_e.

Fig. 28. Instantaneous transverse vorticity in same x-y plane as for Figure 27.
Blue to magenta: $\omega_z = +0.2$ to +3.0.

The Use of a Template-Matching Technique to Identify Hairpin Vortex Flow Structures in Turbulent Boundary Layers

C. R. SMITH and L. J. LU
Department of Mechanical Engineering and Mechanics
Lehigh University
Bethlehem, Pennsylvania 18015, USA

INTRODUCTION

Over the past several decades, it has been shown that turbulence consists of a complicated process of vorticity generation, transfer, and dissipation which is generally accepted to occur in discrete events termed bursts. These bursts are widely believed to be precipitated by discrete vortex flow structures within the flow. The accepted scenario is that the "bursts" then give rise to new vortex flow "structures," which precipitate more "bursting" behavior. The form of the vortex flow structures which precipitate the bursting behavior and those which are generated by the bursting have been the subject of much debate and extensive research. It has been generally suggested by numerous researchers [1,2,3,4,5,6,7,8,9,10] that the structure of turbulent boundary layers is comprised of agglomerations of hairpin-type votices (i.e. horseshoe vortices or U-shaped loops). This premise is based primarily on qualitative flow visualization and analytical simulations. Quantitative, experimental detection of such structures within turbulent boundary layers has not been demonstrated.

Until recently, little experimental information was available on the details of hairpin-type flow structures. However, in recent studies, Smith [8] and Acarlar and Smith [9,10] provide qualitative support for the hypothesis that hairpin vortices are a characteristic structure in a turbulent boundary layer. They demonstrate that many of the organized structures and events observed in the near-wall region of turbulent boundary layers (e.g. streaks, bursts, pockets, ejections, etc.) can be synthetically generated by the stimulation of hairpin vortices in a subcritical laminar boundary layer.

The present paper, as a logical extension of the studies of Acarlar and Smith [9,10], examines a technique for quantitative identification and detection of hairpin-type vortices within a turbulent boundary layer. Employing experimentally generated hairpin votices in a subcritical laminar boundary layer, the quantitative velocity flow-field patterns or "signatures" for a hairpin vortex are evaluated using image-processing techniques in conjunction with hydrogen bubble flow visualization (see 11). These "signatures" are then employed to "search" for hairpin characteristics within the velocity field of a turbulent boundary layer which is determined using the same image-processing techniques. The search technique is a modified template matching approach which uses the spatial-temporal hairpin velocity patterns as a "template," which is then employed to two-dimensionally "search" the spatial-temporal velocity history of the turbulent boundary layer for comparable patterns using two-dimensional cross-correlation techniques.

The following section briefly describes the flow system and image-processing techniques employed to determine the velocity data for both the turbulent boundary layer and hairpin vortex patterns. The subsequent sections review the turbulent boundary layer data, the hairpin velocity patterns, and the application and results of a template matching process.

EXPERIMENTAL SYSTEM/PROCEDURE

The test facility employed in this study is the same free-surface water channel utilized in previous studies of turbulent boundary layers [11,12]. Both the turbulent boundary layer and hairpin vortex studies were done using a 2.5 m flat plate with elliptical leading edge. Hydrogen bubble flow visualization was done using a 20 μm platinum wire oriented normal to and passing through the test plate. For the present study, hydrogen bubble time-lines were generated at 30 Hz using a specially designed pulse generator which is phase-locked to the framing rate of a video viewing system. The bubble time-lines were viewed and recorded using a high-speed INSTAR video system, which incorporates synchronized strobe lights or provide 120 frames/s with an effective shutter speed of 10 μs. The present studies were done with side-view visualization of the inner 3 cm of the plate boundary layer formed on a flat plate test section [see 11]. Figure 1a is a photograph of a single frame from the turbulent boundary layer sequence employed in the present study. Further details of the flow and visualization systems my be found in [11].

To determine quantitative velocity profile data from the hydrogen bubble visualization patterns, individual video frames were converted to eight-bit, 512x512 digital intensity arrays using a Gould 8500 image processor interfaced with a DEC MicoVax. The data sequences employed for the present studies consist of extended series of frames/intensity arrays, each separated by 1/30 s in real time. Figure 1b is a digitized, image-enhanced display of figure 1a. To allow quantitative evaluation of the time-line patterns, a special bubble-line identification and smoothing algorithm was applied to the intensity arrays [see 11]; this procedure allows the location of either the leading edge or the point of maximum intensity of the bubble-lines (the maximum intensity locations are shown in figure 1c). For each bubble-line picture, the local bubble-line velocity was established for 300 points normal to the surface using time-of-flight techniques as outlined in [11,13]. The true local velocity is then determined for each point by application of a wake defect correction [11]. The result is a series of essentially instantaneous velocity profiles (note that the profiles actually represent very short time, Lagrangian averages). For detailed discussion of the velocity profile evaluation procedure, the reader is referred to [11].

RESULTS: TURBULENT BOUNDARY LAYER VELOCITY DATA

The turbulent boundary layer sequence employed in the present study was documented previously in [11], and is a flat plate boundary layer in water at 18.3 cm/s (Re_θ = 1120, u = 0.786 cm/s). The velocity data obtained by image processing consists of 1,012 velocity profiles obtained for $0<y^+<155$ and spaced $\Delta t^+ = 2$ apart. Figure 2 shows the time-averaged statistics for the first four moments of the data, compared with standard wall-layer correlations and comparable probe data [14]. In addition, power spectra and mean wall shear-stress calculated from the data correlate well with accepted standards [15].

An example of velocity and velocity-derived property profiles covering a time segment of Δt^+=40 (out of a total sequence record of Δt^+= 2024) are shown in figure 3 in conjunction with a selected set of actual visualization frames from the original video sequence. The first two profile rows show profiles of velocity (u) and velocity fluctuation (u'=u-u). The next three rows show profiles of du/dy, d^2u/dy^2, and du/dt derived using quadratic central differencing of the original u(y,t) data. Note that very near the wall a special third-order fitting routine developed by Lu [15] was employed to allow wall values to be established. The sixth row of figure 3 shows the corresponding short-time variance parameter employed in probe-based VITA burst detection techniques [14,16]. The averaging time-window for this parameter was taken as T^{+}=10. The final row in figure 3 shows the convolution of the VITA parameter and du/dt; since VITA and local acceleration (du/dt) are often employed as indicators of local turbulent "bursting" behavior, the convolution of these two properties (VITA x du/dt) can be employed to discriminate between accelerating and decelerating "bursts".

An alternative method of presentation of the cross-stream flow property behavior is the use of isocontour plots. Figure 4 is an isocontour presentation of the same data shown in the multiple profile plots of figure 3. This mode of presentation is particularly nice for the study of the time-dependent chacteristics of turbulent flow, since the closed contours illustrate local "regions" of property changes. Note how the VITA profiles and contours of figures 3 and 4 suggest that regions which would be characterized as "bursting" (beyond a threshold of, say k=1) appear to be quite localized, indicating that it would be difficult to detect the breadth of "burst-like" activity in the boundary layer using a single-sensor measurement at a fixed location.

Utilizing the present space-time data, the VITA property was used to construct ensemble averages of flow property profile behavior in the vicinity of burst detections. Using VITA>1 as a criteria for a detected burst, the flow property profiles in the vicinity of $-10<T^{+}<10$ of each detected burst were ensemble averaged over the entire turbuence sequence. The respective ensemble averages of fluctuating velocity are shown in figure 5 at selected heights above the surface for both accelerating and decelerating burst detections centered at y^{+}=15, 30, 65, and 100. Note the similarity of the ensemble averages in figure 5 to comparable probe-rake results [16]. Figure 6 is an expanded series of isocontour plots of all of the ensemble-averaged flow properties for both accelerating and decelerating bursts detected at y^{+}=15, the commonly employed detection location for single-probe VITA studies. Note the localized, closed VITA contours centered at the detection location and the descending or ascending "humps" in both the velocity and du/dy contours.

Clearly, ensemble averages represent only the lowest common denominator for a flow, averaging out many of the flow characteristics. Figure 7 is shown to illustrate an example of a typical, accelerated VITA-detected burst showing both selected segments of the visualization sequence and the cumulative property behavior in proximity to the detected burst location (centered at T^{+}=0 in the figure). Presented in isocontour form, this event displays many similarities with figure 6, such as the localized, closed VITA contours and the descending, hump-shaped velocity contours. in addition, the accompanying visual sequence appears to display representative patterns which could be construed as representing the passage of a hairpin-like flow structure. The further similarities between these types of patterns and synthesized hairpin vortex flow structures are illustrated in the following two sections

RESULTS: HAIRPIN VORTEX VELOCITY DATA

The hairpin vortices examined were generated experimentally on the same test plate employed for the turbulent boundary layer, employing hairpin vortex shedding from a hemisphere protuberance in a subcritical boundary layer, as examined previously [9]. The visualization was done using a vertical hydrogen bubble wire located 36 cm downstream from the leading edge of the plate. Flow visualization was done with a hemisphere of radius R=0.8 cm located X/R=16 and 24 upstream of the fixed wire location. At each streamwise location, visualization sequences were taken with the wire located spanwise at Z/R=0 (on the centerline of the hemisphere), 0.5, and 1. Note that the hydrogen bubble wire passed through a pin hole in the plate, thus the variation in the wire location was accomplished by translating the hemisphere relative to the fixed wire.

The hairpin vortex shedding behavior was consistent with previous results [9]. The laminar boundary layer properties without the hemisphere present were Re_L=23,000 (based on the hydrogen bubble wire location from the leading edge of the test plate) and δ=1.3 cm (laminar boundary layer thickness at $0.99U_\infty$) at the free-stream velocity of U_∞=6.5 cm/s. The hairpin vortex properties for the free-stream conditions were Re_R=560 (base on hemisphere radius) and a $f_{v.s.}$=1.67 Hz vortex shedding frequency (Strouhal number, S= $f_{v.s.}R/U_\infty$ =0.206)

Instantaneous velocity profiles, over a shedding cycle for the hairpin street, were evaluated using the same image-processing techniques employed for the turbulent boundary layer data. The velocity behavior over the shedding cycle was comprised of a series of 18 profiles. Using the initial velocity data and data reduction techniques employed previously for the turbulent boundary layer data, the profile properties of u, u', du/dy, d^2u/dy^2, du/dt, VITA, and VITA x du/dt were all established. Figure 8 and 9 show both selected visualization images and isocontour property results over a hairpin shedding cycle for the two centerline locations (X/R=16, Z/R=0, and X/R=24, Z/R=0). For results at the four off-axis loacations, the reader is referred to [15]. It should be noted that the averaging time interval for determination of the VITA property was taken as one-half the shedding cycle, which was felt to be comparable to the typical T^+=10-20 averaging times commonly used for turbulent burst detection.

In general, the characteristics of the hairpin-vortex patterns appear similar to many of the patterns observed for the turbulent boundary layer (cf. figures 4, 7 and figures 8, 9). In order to quantify this comparison process, a cross-correlation procedure employing template matching was applied to allow quantitative identification of the hairpin patterns in the turbulence data.

HAIRPIN TEMPLATE MATCHING PROCEDURE

The process of image matching has been developed and applied broadly in computer-vision, object identification approaches. One of the simplest and most efficient methods for detection of a particular feature within an image is a process called template matching [17]. The template is a sub-image that has the basic characteristics of the object (i.e. image) to be identified. A similarity measure is computed which reflects how well the image data matches the template for each possible template location. The basic process is to establish the two-dimensional cross-correlation between the sub-image template and all possible points on the reference image (i.e. a data sequence); the points of strongest correlation are then taken as matches or potential matches of the sub-image template.

Utilizing the space-time data "windows" established by image-processing techniques for the hairpin vortices, a template matching approach using regional, two-dimensional correlation techniques is applied to the space-time velocity data for the full turbulent boundary layer. Note that since the pattern we are seeking is a transient phenomena, we need to compare the profile excursions relative to the local mean (i.e. the excursions with regard to the short-time mean calculated over the time-window of comparison). This in essence is an extension of the short-time averaging process employed for determination of the VITA parameter for single-signal probe detections. Thus, in the present template matching procedure, at each time-step we correlate the velocity variation about the cycle-averaged mean for the hairpin pattern (u_h) against the velocity variation about a short-time average of the turbulence data (u_t), where the averaging time-window for both the hairpin cycle and turbulence are identical.

Since the space-time scales of the cyclic hairpin pattern relative to the turbulence space-time scales are unknown, we must parametrically vary the size of the space-time window of correlation in order to optimize the template matching process. In other words, the space and time scales of a hairpin vortex spatial-temporal velocity pattern (i.e. the "template" pattern) are shrunk or expanded relative to the turbulence spatial-temporal velocity pattern, and then a correlation is done. By systematically varying the size of the hairpin template pattern, one attempts to optimize both the space and time scales of the template pattern to yield the highest correlation value with the turbulence pattern at a particular location in time and space. The locations, in time and space, where the a template of a particular scale generates the highest correlation values are then considered the be locations where a "match" occurs (i.e. where a hairpin pattern of a particular scale is detected in the turbulence data pattern). With the additional need to optimize the template patterns obtained at different flow cross-sections for the evolving hairpin structures, the parameter variation process required to establish optimal "matches" of the hairpin patterns within the turbulence data becomes unwieldy.

One simplification that was done in the present study was to restrict the lower edge of the hairpin template pattern, regardless of the vertical scale, to be flush with the lower edge of the turbulence pattern. This restriction is invoked because hairpin vortices generally project outward from the bounding surface. Under this restriction, the variation of the vertical scale of the hairpin template will generate only one correlation coefficient at each vertical location within the turbulence pattern. Thus, the correlation of a hairpin pattern will yield a correlation coefficient which is a function of the vertical location, equivalent to the vertical scale of the hairpin window, and the temporal location of the center of the hairpin pattern [e.g. $Ru_hu_t(\tau,l)$, where τ is the temporal location of the hairpin window, and l is the height of the hairpin scale relative to the turbulence pattern]. Note that $Ru_hu_t(\tau,l)$ is normalized by the local window-averaged rms values of u_h and u_t, so that $0\leq Ru_hu_t(\tau,l) \leq 1$. For the details of the calculation procedure for $Ru_hu_t(\tau,l)$, the reader is referred to [15].

Another simplification was to restrict the examination of the hairpin template time windows used to evaluate the correlations to only T^+=12, 18, and 36. These window sizes were selected based on qualitative observation of the typical turbulent profile behavior and because of their similarity to the typical averaging times employed for VITA burst detection studies. Correspondingly, the vertical scales of the hairpin templates were varied from 100% to 10% of the height of the turbulence data (from approximately Δy^+=134 to 13).

RESULTS: HAIRPIN TEMPLATE MATCHING

Figure 10 shows a comparison of the velocity contours for hairpin vortex patterns at X/R=16, Z/R=0 and X/R=24, Z/R=0 with the velocity, velocity fluctuation, and VITA contours for the first T^+=200 of the turbulent boundary layer data, and the corresponding correlation of the hairpin patterns against the turbulence patterns. In the correlation results (figures 10f,g), only the positive correlation contours are shown, with the contour levels beginning at 0.4 and increasing in increments of 0.2 (i.e. 0.4, 0.6, and 0.8). The time-window employed for the correlations was T^+=18, the extent of which is illustrated by the width of the hairpin velocity patterns shown in figure 11a,b. For other time windows of T^+=12 or 36, the temporal extent of the hairpin patterns is increased or decreased proportionately. During the calculation of correlation values, $Ru_hu_t(\tau,l)$, the vertical scale of the hairpin template is varied systematically from l=100% to 10% of the height of the turbulence reference data, corresponding to y^+=134 to 13. Thus, the correlation contours in figure 10f,g indicate the degree to which the particular spatial size of the hairpin pattern correlates with the turbulence data at a specific temporal location. The peaks in the contours (denoted by an "H" and an accompanying numerical value) indicate the specific spatial scale and temporal locations for which the highest correlations (i.e. a match) are achieved.

An examination of the correlation results in figure 10f,g indicate that the template pattern at X/R=24 appears to correlate better than the X/R=16 template pattern, as indicated by the appearance of more closed-contour regions of higher correlation level. Note also that in locatons where the two patterns appear to yield overlapping regions of high correlation, the X/R=24 pattern seems to be more highly correlated, implying a more probable "match".

Employing the X/R=24, Z/R=0 template pattern, the effect of the width of the time-window on the degree of correlation is shown in figure 11, again in comparison with the corresponding turbulent boundary layer data. Comparison of the correlation results for the three time-windows (T^+=12, 18, and 36) indicates that the smaller the time-window, the more positive peaks appear in the isocontour plots. This illustrates that the window acts to some degree as a low-pass filter, with fewer "small-scale" patterns revealed as the time window becomes broader. However, note that the regions of highest correlation generally appear at the same vertical scale and at the same temporal location for all three time-windows, indicating a limited sensitivity to the particular time-window within the range examined.

One concern in interpreting figures 10 and 11 is what level of correlation indicates a "good" match between the template pattern and the reference (i.e turbulent boundary layer) pattern? To examine this, the six different hairpin vortex patterns were cross-correlated against each other to examine the degree of correlation of the hairpin patterns to each other. As is shown in [15], the correlation of each hairpin pattern to the other patterns, displaced either laterally or in streamwise location, results in a significant loss in correlation. Even for self-correlation of each pattern with itself, the degree of correlation drops relatively rapidly with a change in scale of the template pattern or with cyclical phase. An evalution of these self-correlation results suggests that a correlation value of 0.6 or higher indicates that the template scale is within 20% of the true pattern scale. This suggests that to demonstrate a valid "detection" of a hairpin-type pattern requires a correlation of approximately 0.6 or higher. As shown in [15], and as can be seen in figure 10, this criteria is most generally satisfied by the present hairpin patterns at X/R=24.

Figures 10 and 11 suggest that high positive correlation peaks ($R_{u_h u_t}(\tau, l)$) >0.6) are found over a range of scales and temporal locations. To illustrate this , figure 12 was compiled, which shows the spatial scale and temporal locations for all the correlation peaks of magnitude 0.6 or greater over the entire turbulence data sequence. Clearly, the X/R=24 pattern is matched much more frequently than the X/R=16 pattern. This seems to suggest that most patterns detected within the turbulence field are more characteristic of the more dispersed vorticity patterns at X/R=24 than the more concentrated vorticity patterns at X/R=16. To further illustrate the variation in the scales of the matched hairpin patterns, a distribution of the matched locations in figure 12b are shown in figure 13. This figure indicates that the distribution is skewed toward the wall, with the highest number of matches occurring between y^+=26 and 53, and the number of matches decreasing with distance from the wall . Note that the behavior of this distribution appears to be consistent with the suggested model of [6], where it is hypothesized that a turbulent boundary layer consists of a hierarchy of hairpins which undergo a process of cross-stream amalgamation and anihilation of vorticity, resulting in larger, but more disperse hairpins.

CONCLUSION

The initial use of template matching of experimentally generated hairpin-vortex patterns against a turbulent boundary layer flow pattern suggests that such hairpin patterns can be identified within a turbulent boundary layer. Although not optimized, it appears that shorter time-windows ($T^+ \approx 12$ to 18) commensurate with the time-scale normally associated with an ensemble-averaged burst patterns ($\Delta T^+ \approx 10$ to 20) provide the best matches of the hairpin patterns. In addition, an evaluation of the hairpin pattern demonstrating the best template match suggests that the predominant hairpin-type flow pattern within the turbulent boundary layer has a more dispersed vorticity distribution in the region of the hairpin head. Finally, the distribution of the identified hairpin patterns is found to be skewed toward the wall, with the scale of the largest number of identified patterns extending out from the wall to $y^+ \approx 26$ to 53.

ACKNOWLEDGEMENTS

The authors greatfully acknowledge the support of this research by the Air Force Office of Scientific Research. We would also like to thank Dr. J.D.A. Walker for many helpful discussions with regard to the research.

REFERENCES

1. Theodorsen, T., Mechanism of turbulence, *Proc. 2nd Midwestern Conf. of Fluid Mech.*, Bulletin No. 149, Ohio State Univ., 1952

2. Theodorsen, T., The structure of turbulence, *Grenzschichtforschung*, 50 Jahre, Eine Festschrift in Originalbeitragen, H. Gortler and W. Tolmien, eds., Friedr. Vieweg and Sohn, Braunschweig, p. 55, 1955

3. Hinze, J.O., *Turbulence*, 2nd ed., p.682, Mc-Graw Hill, New York, 1975.

4. Offen, G.R. and Kline, S.J., A proposed model of the bursting process in turbulent boundary layers, *J. Fluid Mech.*, vol. 70, p. 209, 1975.

5. Head, M.R. and Bandyopadhyay, P., New aspects of turbulent boundary-layer structure, *J. Fluid Mech.*, vol. 107, p. 297, 1982.

6. Perry, A.E. and Chong, M.S., On the mechanism of wall turbulence, *J. Fluid Mech.*, vol. 119, p 173, 1982.

7. Wallace, J.M., On the structure of bounded turbulent shear flow: a personal view, *Dev. in Theo. and App. Mech., XI*, T.J. Chung, G.R. Karr, eds., Univ. of Alabama, Huntsville, p. 509, 1982.

8. Smith, C.R., A synthesized model of the near-wall behavior in turbulent boundary layers, *Proc. 8th Symp. on Turb.*, J.L. Zakin, G.K. Patterson, eds., Univ. of Missouri-Rolla, p. 35, 1984.

9. Acarlar, M.S. and Smith, C.R., A study of hairpin vortices in a laminar boundary layer. part 1. hairpin vortices generated by a hemisphere protuberance, *J. Fluid Mech.*, vol. 175, p. 1, 1987.

10. Acarlar, M.S. and Smith, C.R., A study of hairpin vortices in a laminar boundary layer. part 2. hairpin vortices generated by fluid injection, *J. Fluid Mech.*, vol. 175, p. 431, 1987.

11. Lu, L.J. and Smith, C.R., Image processing of hydrogen bubble flow visualization for determination of turbulence statistics and bursting characteristics, *Exp. in Fluids*, vol. 3, p. 349, 1985.

12. Smith, C.R. and Metzler, S.P., The characteristics of low-speed streaks in the near-wall region of a turbulent boundary layer, *J. Fluid Mech.*, vol. 129, p. 27, 1983.

13. Schraub, F.A., Kline, S.J., Henry, J., Runstadler, P.W., and Little, A., Use of hydrogen bubbles for quantitative determination of time-dependent velocity fields in low-speed water flows, *ASME J. Basic Engrg.*, vol. 87, p. 429, 1965.

14. Johansson, A.V. and Alfredsson, P.H., On the structure of turbulent channel flow, *J. Fluid Mech.*, vol. 122, p. 295, 1982.

15 Lu, L.J., An evaluation of turbulent boundary layers for hairpin votex flow structures,Ph.D thesis, Dept. of ME/Mech., Lehigh University, 1988.

16 Blackwelder, R.F. and Kaplan, R.E., On the wall structure of the turbulent boundary layer, *J. Fluid Mech.*, vol. 76, p. 89, 1976.

17. Ballard, D.H. and Brown, C.M., *Computer Vision*, Prentice-Hall, New York, 1982.

Figure Notes

1. Shaded regions in figures 4 and 7 to 10 indicate negative values.

2. In figures 7 to 9, the arrows below the isocontour plots correspond to the respective picture in the time sequence shown.

3. Contour intervals :
   (a) in figures 4, 6, 7, 10, 11
   $\Delta u^+=2$, $\Delta u'^+=1$, $\Delta du^+/dy^+=0.1$, $\Delta d^2u^+/dy^{+2}=0.02$,
   $\Delta du^+/dt^+=0.4$, $\Delta VITA=0.25$, $\Delta(VITA\times du^+/dt^+)=0.4$,
   where : $u^+=u/u_\tau$, $y^+=yu_\tau/\nu$, and $t^+=tu_\tau^2/\nu$

   (b) in figures 8 to 11
   $\Delta u^*=0.1$, $\Delta u'^*=0.05$, $\Delta du^*/dy^*=0.5$, $\Delta d^2u^*/dy^{*2}=2$,
   $\Delta du^*/dt^*=0.0002$, $\Delta VITA=0.25$, $\Delta(VITA\times du^*/dt^*)=0.0002$.
   where : $u^*=u/U_\infty$, $y^*=y/R$, and $t^*=tU_\infty/R$,

   (c) in figures 10 and 11
   $\Delta Corr=0.2$; $(Corr)_{min}=0.4$ and $(Corr)_{max}=1$.

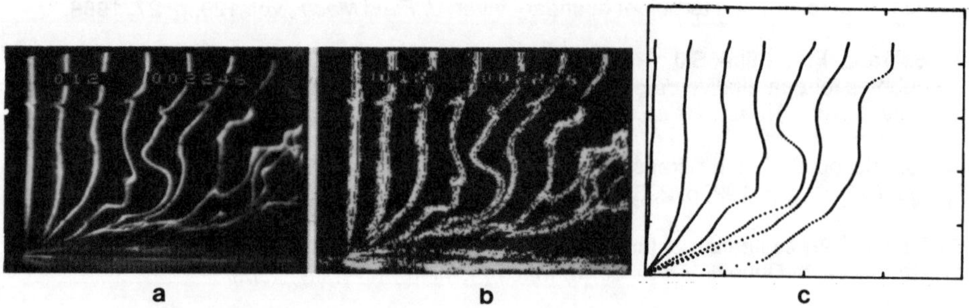

Figure 1 Computer-identified hydrogen bubble lines (a) original frame from video sequence (b) computer enhanced picture from CRT display (c) plot of identified bubble lines (Lu and Smith, 1985).

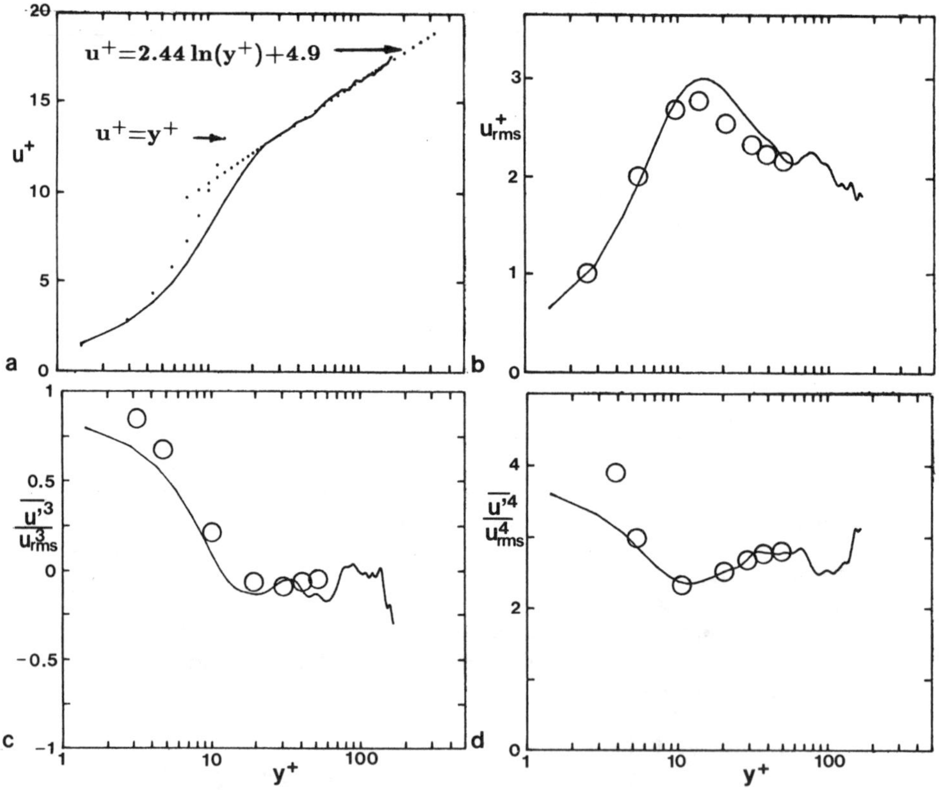

Figure 2 Time-averaged turbulence statistics established from image processed hydrogen bubble-wire data. —, present data; ..., empirical law-of-the-wall; o, data of Johansson and Alfredsson (1982). (a) mean velocity (b) turbulence intensity (c) skewness (b) flatness. (Lu and Smith, 1985)

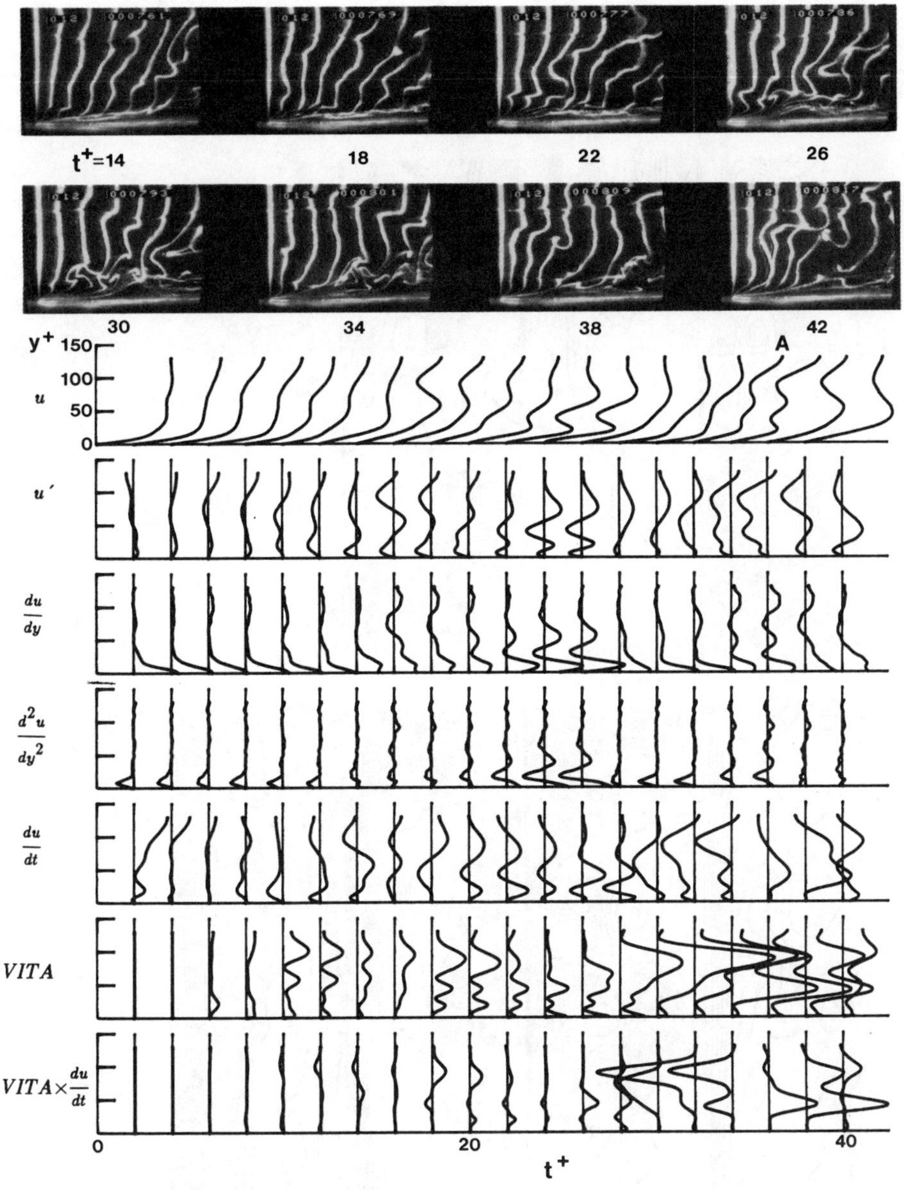

Figure 3 Comparison of visual sequence at region "A" with profiles of velocity and velocity-derived properties obtained from results for a turbulent boundary layer sequence, $0<t^+<40$. Property interval between each axis is : $\Delta u^+=8$, $\Delta u'^+=8$, $\Delta du^+/dy^+=1$, $\Delta d^2u^+/dy^{+2}=0.2$, $\Delta du^+/dt^+=2$, $\Delta VITA=1$, $\Delta(VITA\times du^+/dt^+)=2$.

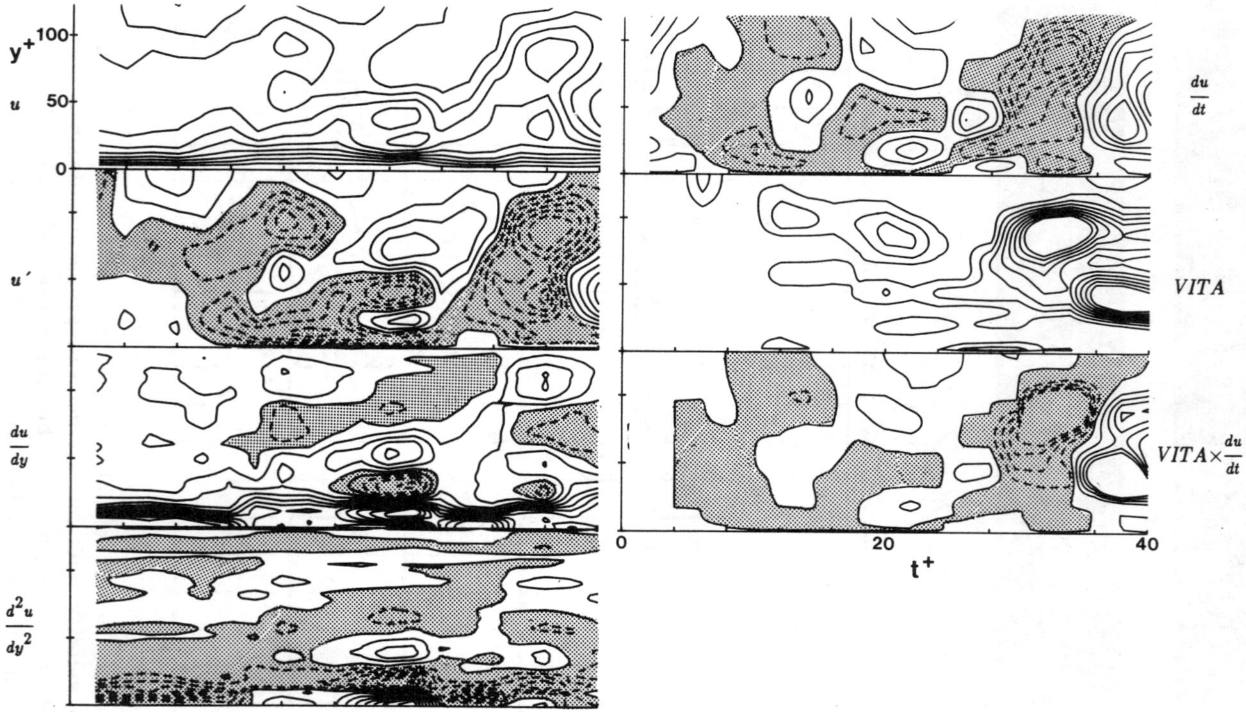

Figure 4 Contour plots of velocity and velocity-derived properties corresponding to the profiles in figure 3.

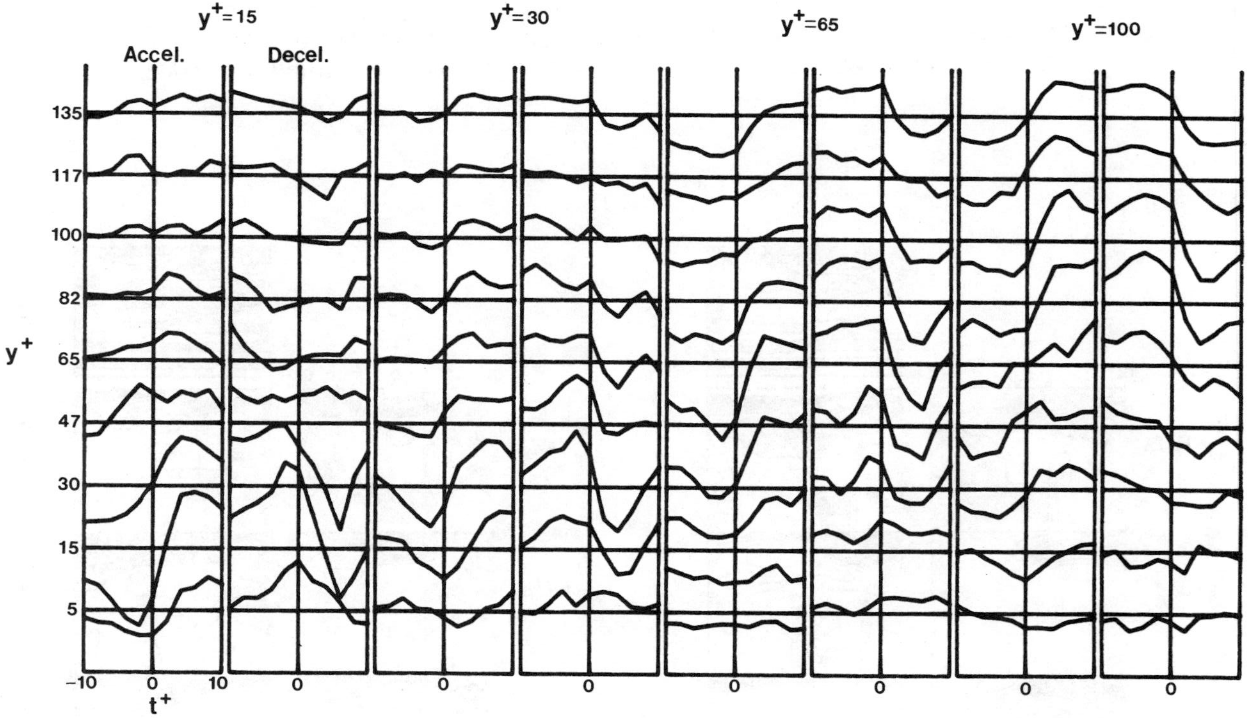

Figure 5 Conditional ensemble-averaged results for streamwise velocity behavior at selected heights above surface. VITA$\geq$1, detection level. Averages for both accelerating and decelerating detections shown ($\Delta u'^{+} = 4$).

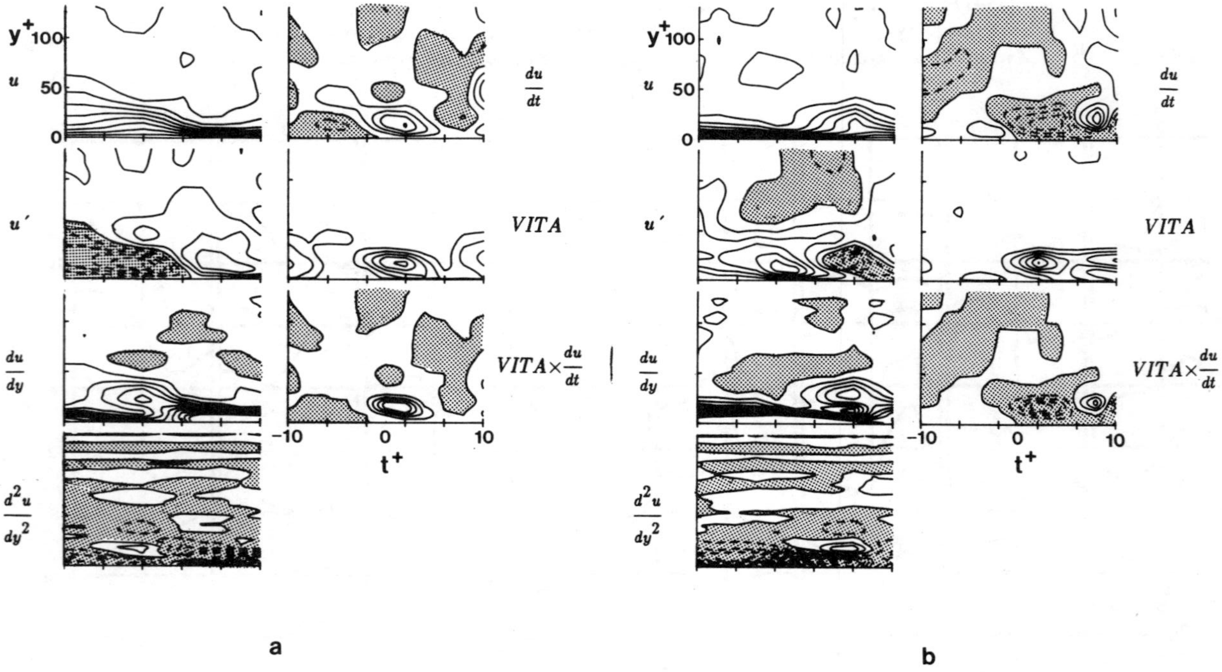

Figure 6 Conditional ensemble-averaged isocontours of velocity and velocity-derived properties for VITA$\geq$1 detection level and y^+=15 detection point. (a) du/dt>0 and (b) du/dt<0.

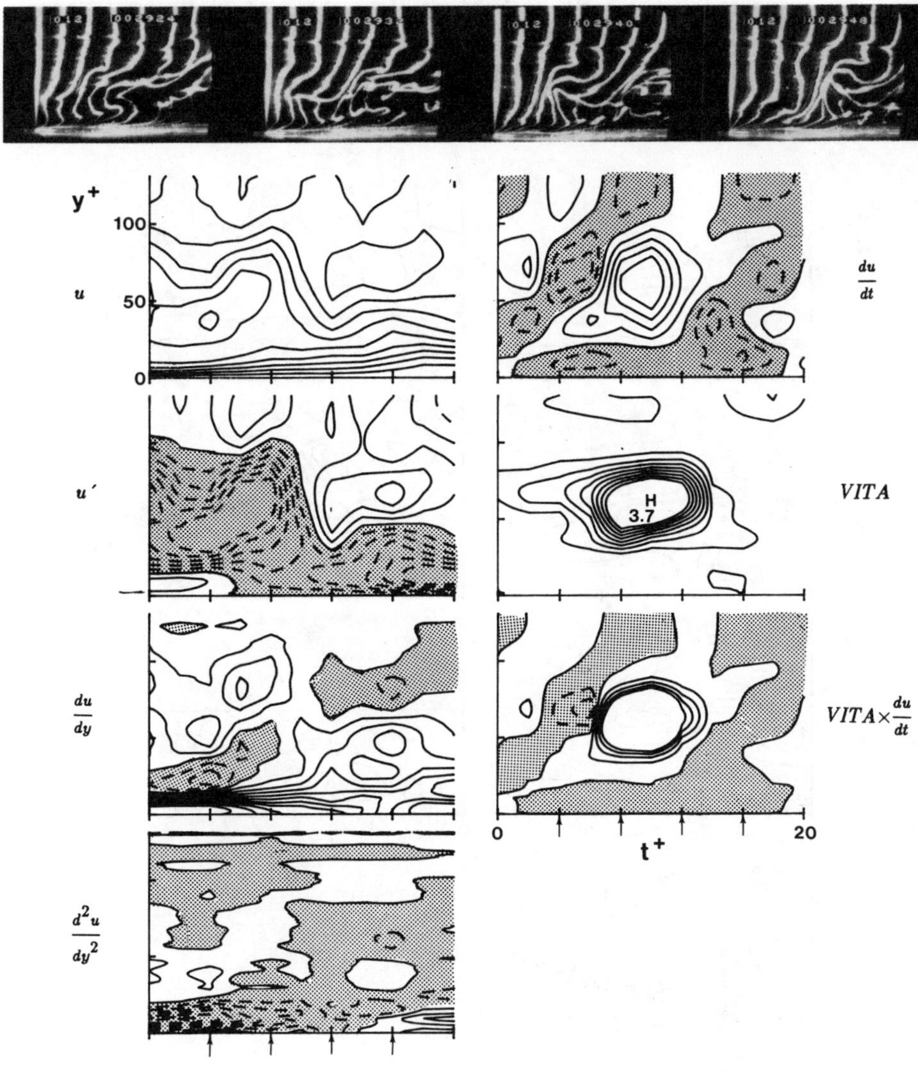

Figure 7 Single accelerated event, VITA$\geq$1. Centered at t^+=1090 in data sequence.

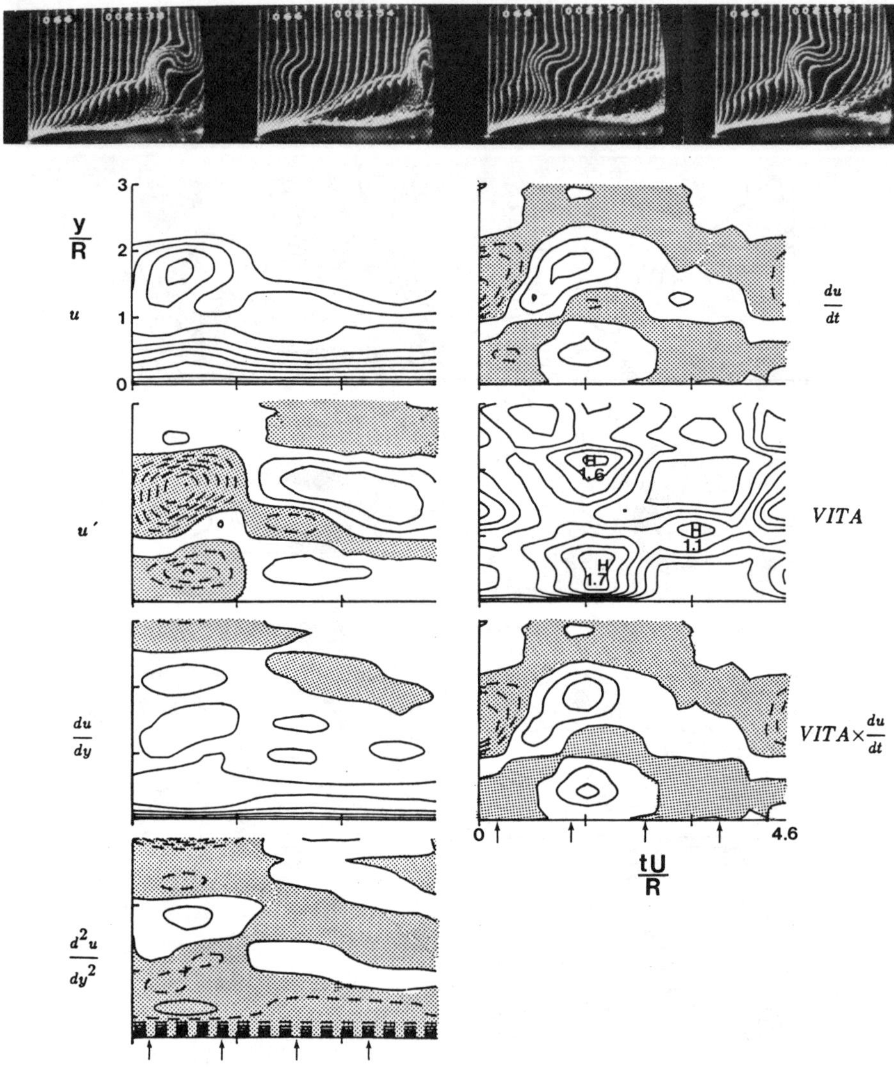

Figure 8 Contours of velocity and velocity-derived properties obtained from hairpin-vortex flow pattern at X/R=16, X/R=0 in comparison with the visualization sequence.

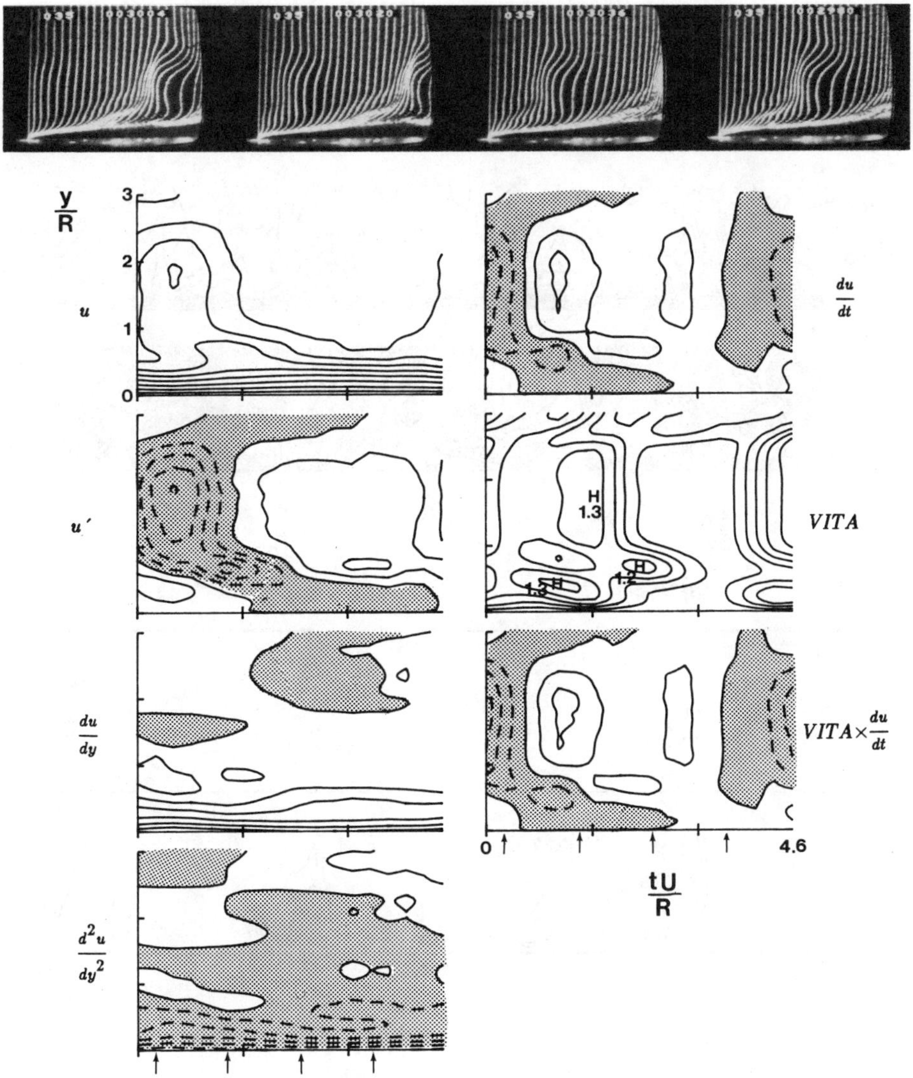

Figure 9 Contours of velocity and velocity-derived properties obtained from hairpin-vortex flow pattern at X/R=24, and X/R=0 in comparison with visualization sequence.

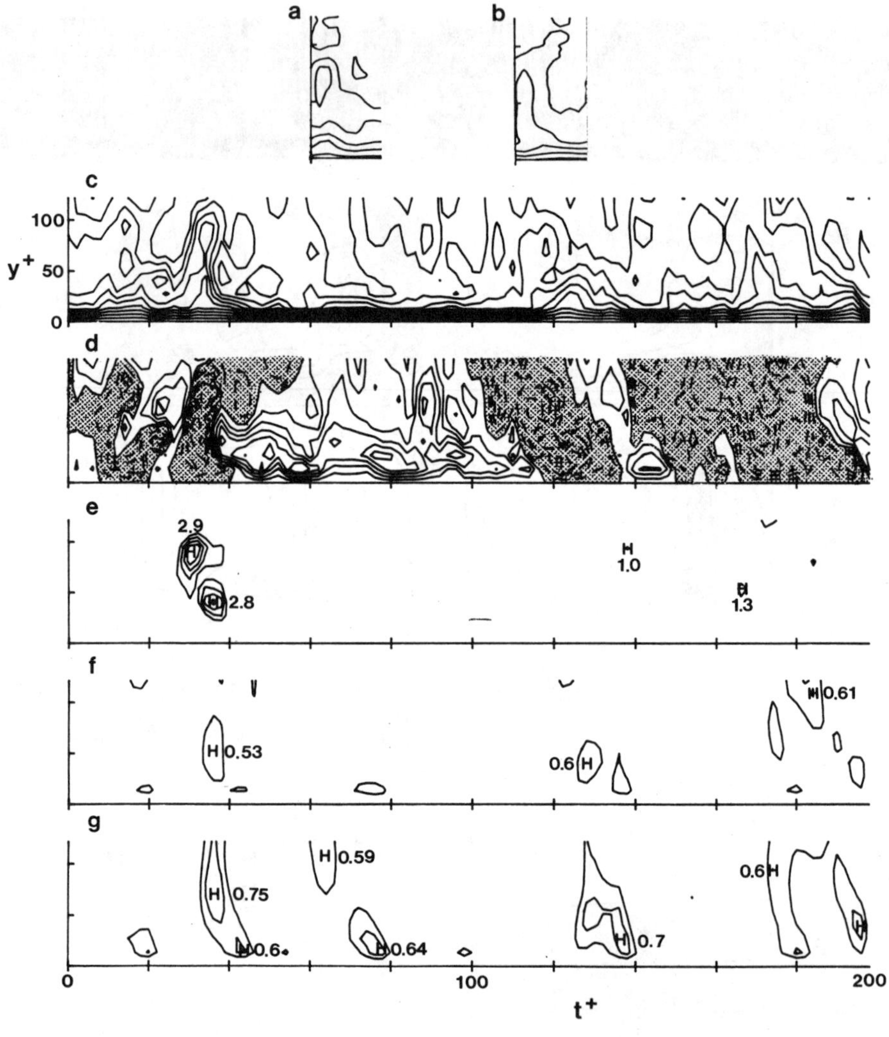

Figure 10 Contour plots of : hairpin vortex velocity at (a) X/R=16, Z/R=0, and (b) X/R=24, Z/R=0. Turbulent bounday layer: (c) velocity, (d) velocity fluctuation (about the long-time mean), and (e) VITA. Correlation results for template matches: (f)-pattern (a) against (c); (g)-pattern (b) against (c).

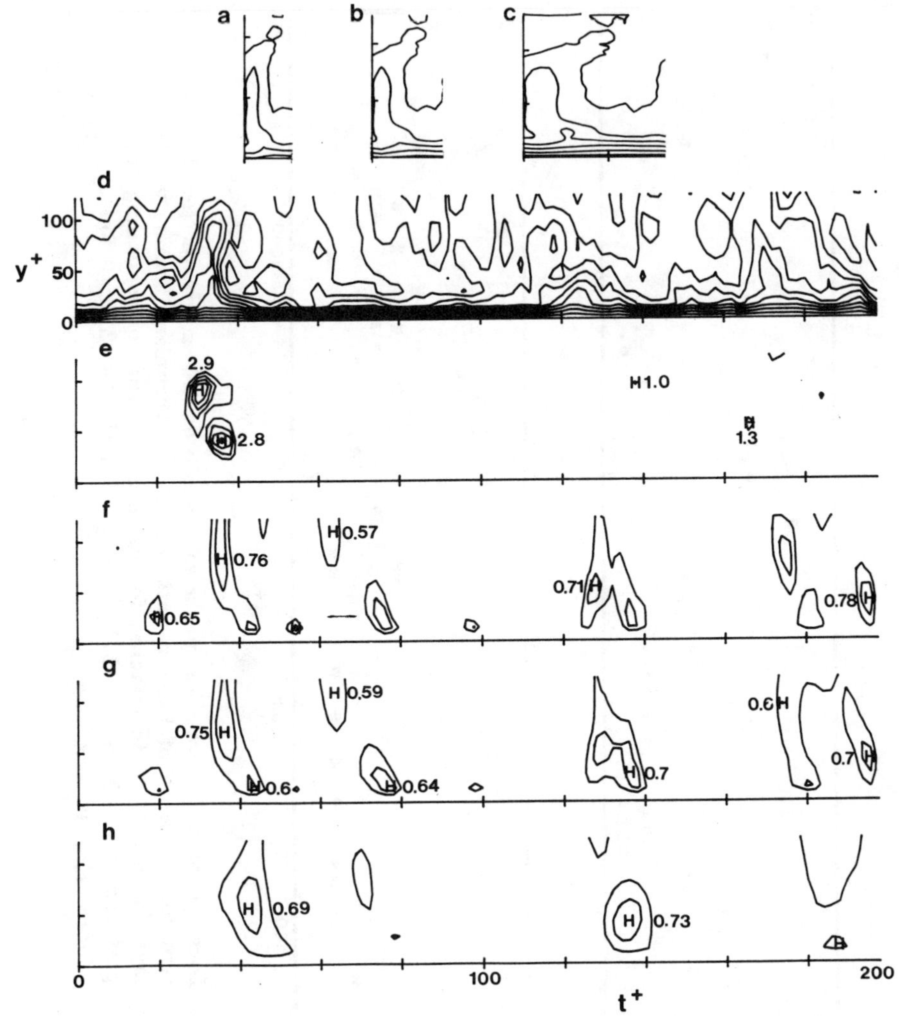

Figure 11 Contour plots of : hairpin vortex velocity at X/R=24, Z/R=0 for window sizes T^+= (a) 12, (b)18, and (c) 36; (d) turbulent bounday layer velocity, and (e) VITA. Correlation results for template matches : (f)-pattern (a) against (d); (g)-pattern (b) against (d); (h)-pattern (c) against (d).

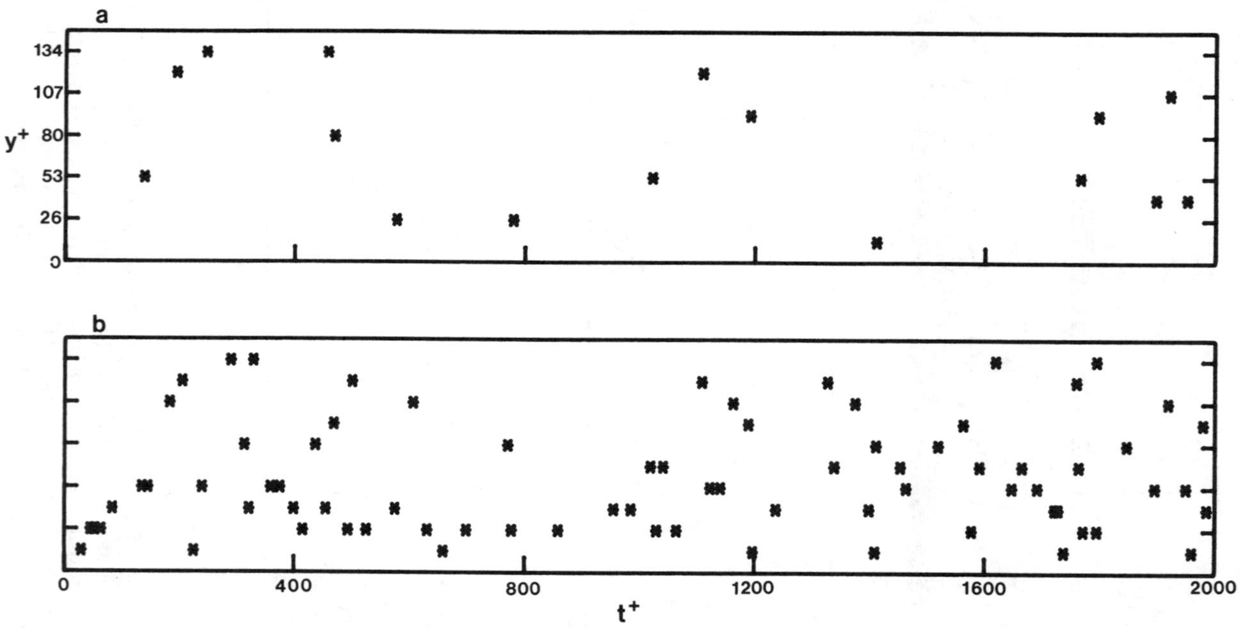

Figure 12 Location of the correlation peaks of magnitude 0.6 or greater for template matches of hairpin patterns against turbulent boundary layer data. Template matches for hairpin patterns corresponding to patterns at (a) $X/R=16$, $Z/R=0$ and (b) $X/R=24$, $Z/R=0$ for window size $T^+=18$.

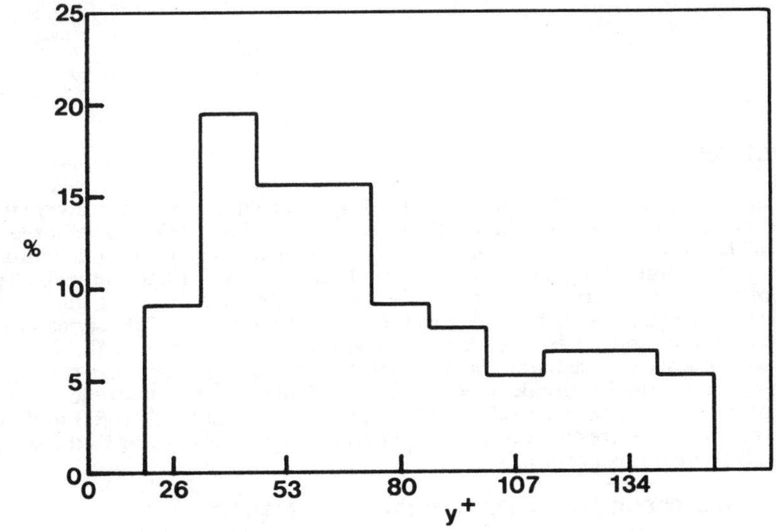

Figure 13 Probability distribution of the scales of the matched hairpin locations shown in figure 12(b). Hairpin pattern X/R=24, Z/R=0 and window size T^+=18.

The Role of Inflectional Velocity Profiles in Wall Bounded Flows

RON F. BLACKWELDER
University of Southern California
Los Angeles, California 90089-1191, USA

JERRY D. SWEARINGEN
Naval Research Laboratory
Washington, D.C. 20375-5000, USA

INTRODUCTION

Much of the research on turbulent boundary layer during the past twenty years has concentrated on the coherent eddies that control the flow field. The large scale structure of the outer flow field dominates the flow features there and contols the entrainment of irrotational fluid into the turbulent regions. Near the wall the flow field is dominated by a series of events, collectively called the bursting phenomenon, that are responsible for the turbulence production and the drag. This series of events consists of counter-rotating streamwise vortices, low speed streaks of fluid, lift up of that fluid off the wall, an oscillation of the lifted low speed streak and the rapid growth of that oscillation and its break down into turbulence. This sequence has been described in detail by Kline et al. (1967), Willmarth (1975), Cantwell (1981) and others. The main thrust of the present work is to examine the oscillation stage and suggest a mechanism by which this occurs.

STABILITY THEORY FOR INFLECTIONAL VELOCITY PROFILES

One of the earliest successes in stability theory was due to Rayleigh (1880) who showed that inflectional velocity profiles in an inviscid fluid are unstable to small disturbances. Viscosity has only a damping influence in these flow fields as discussed by Lin (1955). Michalke (1965) analysed the hyperbolic tangent profile for a mean parallel flow field given by

$$U(\xi) = \frac{U_o}{2} * [\tanh(\xi/\Delta)]$$

where U_o is the maximum velocity difference across the shear layer and 2Δ is its vorticity thickness as illustrated in figure 1. He assumed a spatially growing two dimensional disturbance in the form of a travelling wave given by

$$u'(x,\xi,t) = \hat{u}(\xi)e^{i(\alpha x-\omega t)}$$
$$v'(x,\xi,t) = \hat{v}(\xi)e^{i(\alpha x-\omega t)}$$

α is the complex wave number, $\alpha = \alpha_r + i\alpha_i$, and ω is a real frequency where x and ξ are the streamwise and cross stream coordinates, u' and v' are the respective velocity perturbation. His results yielded a broad range of growing harmonic disturbances with the most amplified wave number being $\alpha_r\Delta = 0.40$. The corresponding wavelength is $\lambda_x = 15.7\Delta$ which is long compared to the shear layer thickness.

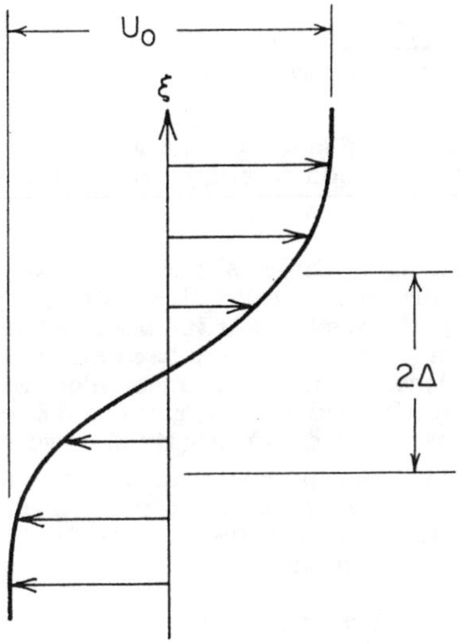

Fig. 1. Sketch of an inviscidly unstable velocity profile in the ξ direction.

More surprising however is the growth rate of these disturbances, α_i. Michalke's analysis shows that the disturbances grow like $e^{\alpha_i x}$ where the maximum growth rate was found to be $\alpha_i\Delta = -0.23$. Thus while the wavelike disturbance travels one wavelength downstream, $x\text{-}x_o = \lambda_x$, it is amplified by a factor of 36 and its energy increases more than 1000 fold! For comparison, the maximum amplification of the most unstable Tollmien-Schlichting wave is typically 1.1 (i.e. they experience a 10% increase in amplitude) during a journey of one wave length downstream because its growth rate is an order of magnitude smaller. Hence disturbances associated with inviscid inflectional velocity profiles are amplified much more than those associated with viscous instabilities. In some transitional flow fields, the initial viscous instability amplifies an initial disturbance in a manner that produces an inflectional profile as discussed by Blackwelder (1983). This opens the door to another type of instability, usually called a secondary or higher order instability. If the secondary instability is an inflectional type, the disturbances will grow much faster than the initial disturbance and will dominate the transition process.

One of the more interesting characteristics of the inviscid instability is that its amplification is strongly dependent upon the shear rate of the inflectional profile. To illustrate this, note that the maximum amplification over the distance $x_2\text{-}x_1$ is given by

$$e^{0.23\,(x_2-x_1)/\Delta} \qquad (1$$

Thus the smaller the length scale Δ, the larger the growth over the same distance. Figure 1 shows that Δ is inversely proportional to the maximum shear of the profile, i.e.

$$\left.\frac{\partial U}{\partial \xi}\right|_m = \frac{U_o}{2\Delta}$$

Eliminating Δ between these two expressions yields a growth of

$$e^{\frac{.46}{U_o}\frac{\partial U}{\partial \xi}\Big|_m (x_2-x_1)} \quad (2$$

Consequently the maximum amplification experienced by a disturbance over the distance x_2-x_1 downstream depends strongly on the mean shear at the point of inflection.

The above arguments following Michalke's analysis have been for a steady two-dimensional inflectional profile in a parallel flow. To apply these arguments in a transitional or turbulent flow, the restrictions imposed by these assumptions must be addressed. The effects of non-parallel flow have been examined by Gaster et al. (1985). They found that when the local value of the mean flow was used for the stability calculations, the theory accurately predicted the measured fluctuations even though the flow was nonlinear. Greenspan and Benney(1963) have studied the effects of an unsteady mean flow by allowing the thickness, Δ, to vary sinusoidally with time. The forcing period was much larger than the time scale of the instability. They found that the unsteadiness increased the growth rates so that the energy of the disturbance increased by 10,000 over one wavelength!

Another limitation of the above analysis is that it pertains only to unbounded flow domains. However Huerre (1983) has considered the same inflectional velocity profile confined by two side walls located at $\xi = \pm d$. He assumed the normal velocity on the walls was zero but allowed a slip condition to occur to simplify the analysis. He found that if $d > 1.2\Delta$, the linear stability characteristics were similar to the inviscid theory and the wall had little effect on the growth of the disturbances.

The two-dimensionality of the mean flow has not yet been addressed and is still an open question. However, Nishioka et al. (1980) applied the inviscid two-dimensional theory to an inflectional U(y) velocity profile in a transitional channel flow. The instantaneous velocity profile was formed by the Tollmien-Schlichting waves and was strongly three-dimensional. Experimentally they observed that a secondary instability appeared at the location of the inflection point. To examine this in more detail, they applied the two-dimensional inviscid calculation discussed above and found that it predicted the measured wavelength and eigenfunction. The measured growth factor was approximately 60% of the theoretical value. Similar results have been observed in transitional boundary layers on flat and concave plates. The reduced growth rate may be due to the spreading of energy into the third dimension which is not accounted for in the two-dimensional theory. This effect indicates that the disturbance would be amplified by a factor of 9 after travelling one wavelength downstream rather than by 36 mentioned earlier. The net result is that even though the growth rate may be reduced by the three-dimensionality, it is still very large.

GÖRTLER INSTABILITY ON A CONCAVE WALL

The Görtler instability is one member of a class of centrifugal instabilities including the Taylor, Dean, etc. instabilities. These instabilities arise as a result of an imbalance between a centrifugal force imposed on the fluid and the restoring pressure force. The Görtler instability develops within a boundary layer developing on a concave wall. The curvature of the wall imposes a centrifugal force within the fluid that acts to displace fluid particles toward the wall. The pressure force acts to restore the particles to their original position. This class of instabilites is inviscid in nature and the viscosity only acts to inhibit the motion, i.e. provides a damping force, just as in the inflectional profile instabilities. Therefore if the centrifugal force is smaller than the pressure force, the flow is stable. However if the centrifugal force sufficiently exceeds the pressure force such that the viscosity can no longer damp the motion, the flow field is unstable and small disturbances will be amplified.

Görtler(1939) first analysed this problem using temporal stability theory and found that the governing parameter for this flow was

$$\ddot{Go} = \frac{\theta U_\infty}{\nu}\sqrt{\frac{\theta}{R}} \tag{3}$$

This dimensionless number is known today as the Görtler number. If it exceeds a critical value of the order of unity, the flow is unstable and develops counter-rotating streamwise vortices lying within the boundary layer. These vortices are periodic in the spanwise direction and are characterized by a spanwise wave length λ_z. Hall (1983) has shown that for vortices with wavelengths λ_z comparable to the boundary layer thickness, the growth of the boundary layer cannot be neglected in the analysis. Consequently a unique neutral stability curve does not exist but rather depends upon the history of the boundary layer. Nevertheless, Hall found that each neutral stability curve had a minimum at $\ddot{Go} \sim 1.5$.

In a spatial analysis of the Görtler instability in a parallel flow, Smith (1955) assumed the disturbances on a concave wall grew by the factor $e^{\beta(x_2-x_1)}$ as they travelled downstream from x_1 to x_2. β is the growth rate and varies slowly in the downstream distance. As the fluid moves downstream from the leading edge, the disturbances are initially damped, i.e. $\beta < 0$. They crossed the neutral stability curve at $\beta = 0$ and are unstable, i.e. $\beta > 0$, further downstream. To compare this growth rate with the inviscid one, β must be determined. Liepmann (1945) has shown that Görtler instability transitions to turbulence when $\ddot{Go} \approx 9$. The nominal value of amplification rate slightly before transition, i.e. $\ddot{Go} \approx 7$, is $\beta\theta Re_\theta \sim 3$ according to Smith. For this value of the Görtler number, Re_θ ranges from 230 reported by Bippes (1972) to 360 from Swearingen and Blackwelder (1987). For $Re_\theta = 300$ and with the boundary layer thickness, δ, equal to 7.5θ as in a Blasius layer, the amplification becomes $e^{0.075\,(x_2-x_1)/\delta}$. Thus the streamwise vortical disturbance will double its amplitude after travelling 10δ downstream. For comparison, the most unstable Tollmien-Schlichting wave at Re_θ = 300 has a wavelength of 8δ and an amplification factor of $e^{0.06(x_2-x_1)/\delta}$. These waves are amplified approximately 80% after travelling 10δ downstream. Hence at this Reynolds number, the Görtler instability has a slightly greater amplification than Tollmien-Schlichting waves. The conclusion is that the growth rates in both boundary layer flows are at least an order of magnitude less than that of the inflectional profile instability discussed earlier

THE BREAKDOWN OF GÖRTLER VORTICES

The above linear theory only describes the initial growth of a small disturbance on a concave wall. Like most linear theories, it can say nothing about the later growth stages and breakdown of the perturbation. In a visual study of the vortices on a curved plate Bippes (1972) has shown that the perturbation velocity profiles agreed well with the computed eigenfunctions and the growth rates were comparable to the values given by the linear theory. He also found that the disturbances grew to an amplitude of a few percent of the streamwise mean velocity while maintaining their streamwise vortical structure. However, after a limited domain of linear growth, the flow field developed a strong oscillatory motion before breaking down into turbulence.

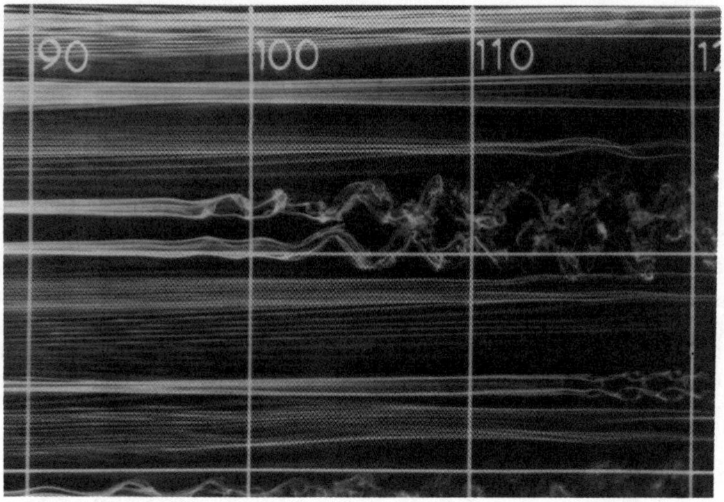

Fig. 2. Instantaneous visualizations of the Görtler instability and transition on a concave wall. The concentrated smoke marks the low speed streaks and the sinuous motion is a result of the inflectional instability.

This aspect of the motion was considered to be important in the transition process because it represented a shift from a primary to a secondary instability in the flow. Since similar observations have been made in the wall region of turbulent flows (see below), an experiment was set up to measure the velocity components in more detail to study this phenomenon. The specially designed wind tunnel used for this experiment had a curved test section with a cross sectional area of 15 x 120 cm. A boundary layer was developed on the concave wall having a radius of curvature of R = 3.2m with a free stream velocity of 5.0 m/sec. A smoke wire was used to visualize the flow field. The streamwise velocity was measured with rakes of hot-wires aligned in the spanwise and normal directions. Thus the U(z,t) and U(y,t) profiles could be measured instantaneously. The signals were digitized and stored on disc for later processing. Further details of the experimental arrangement may be found in Swearingen and Blackwelder (1987).

At a freestream velocity of 5 m/sec, the critical Görtler number occurred at approximately 10 cm downstream of the leading edge of the curved plate. When the smoke wire was placed into the lower regions of the boundary layer upstream, i.e. at $y = 0.25\delta$, and x = 20 cm, the resulting flow downstream appeared as in figure 2. By the time the smoke was convected downstream, it had also been coagulated into elongated regions as seen in the photo. Swearingen and Blackwelder(1987) showed that the regions of strong smoke concentrations seen in the figure are also regions of low streamwise velocity. They also showed that the ensuing dynamics were concentrated around and directly related to the low speed regions.

A sketch of the streamwise vortices and the low speed regions is shown in figure 3. The initial instability produces the counter-rotating streamwise vortices with two vortices per spanwise wavelength. The strength of these vortices is quite small. To illustrate this, consider a mean parallel flow U(y) with a disturbance field u such that the total velocity is

$$\begin{aligned} u(x,y,z) &= U(y) + u'(x,y,z) \\ v(x,y,z) &= v'(x,y,z)/Re \\ w(x,y,z) &= w'(x,y,z)/Re \end{aligned} \qquad (4$$

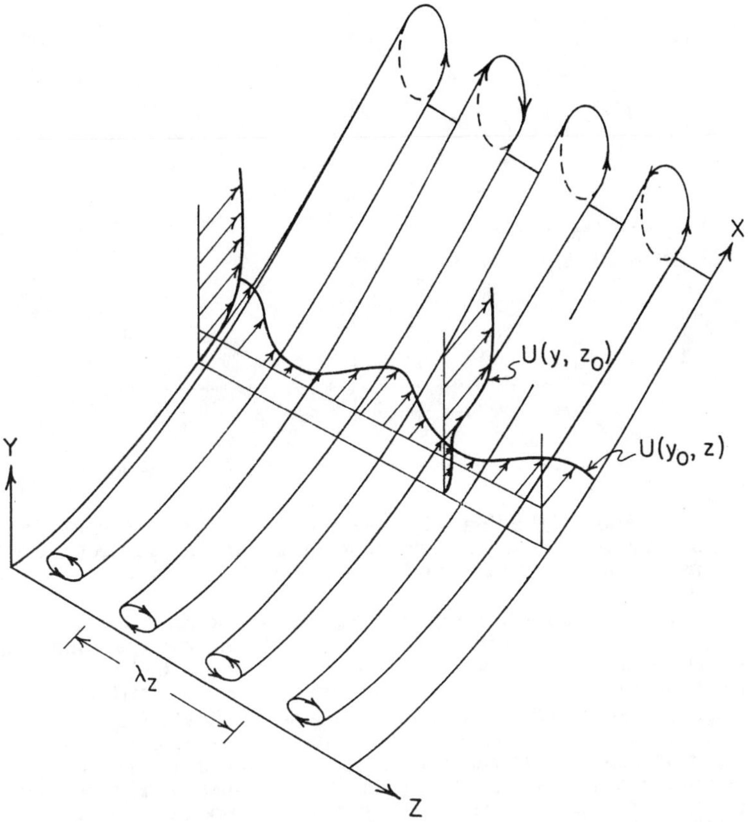

Fig. 3. The streamwise vortex system downstream of the critical Görtler number, the streamwise scale is compressed with respect to the other axes.

where Re is the Reynolds number. This scaling leaves u´, v´, and w´ of the same order as suggested by Diprima and Stuart (1972) and verified by Floryan and Saric (1982). When the vorticity components are obtained from these equations, it is easily seen that the disturbance streamwise vorticity ω_x is smaller than both ω_y and ω_z by the factor Re^{-1}. Hence even though the flow field is generally described as a laminar boundary layer with embedded growing streamwise vortices, they contain only small amounts of streamwise vorticity.

The above illustration indicates that the phase of coherent eddies is much more important than the amplitude of their velocity or vorticity fields. It also suggests that detection of coherent structures should not be based on amplitude alone without considering phase information also. In the Görtler vortex case, the vortices do generate large vertical disturbances, namely $\omega_y \approx \frac{\partial U}{\partial z}$ and $\omega_z \approx \frac{\partial U}{\partial y}$. But any detection scheme based on amplitude alone would probably fail to discern the underlying eddy structure, namely the streamwise vortices. Of course the reason the ω_x vortices are so effective is that they exist in a region of strong mean shear, $\frac{\partial U}{\partial y}$, and thus a very small ω_x rotation can produce large disturbances across the span.

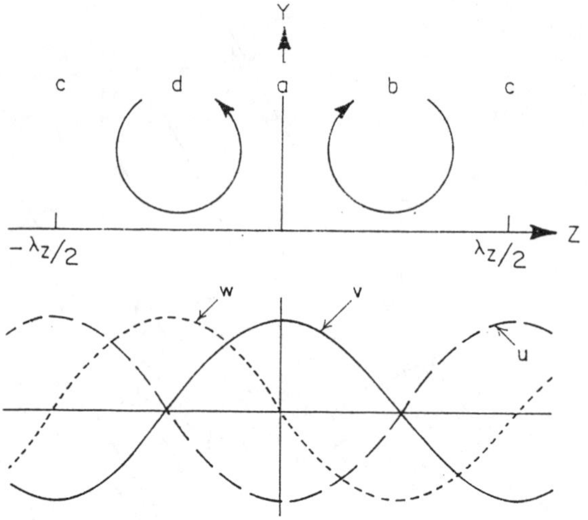

Fig. 4. Illustration of the relative phase of the disturbance velocity components at positions below the centers of the streamwise vortices. The velocity amplitudes depend on y and are not to scale.

Between the vortices in figure 3 are alternating regions of low and high speed fluid. These regions are a natural result of the counter-rotating vortices lying is a strong shear flow. The velocity field associated with the vortices has a normal velocity component with a relative maximum or minimum between the vortices at positions a and c respectively as seen in figure 4. Simultaneously the fluid lying below the centers of the vortices has a spanwise velocity component that alternates in sign with the vortices so that its motion is directed towards the region of positive normal velocity. The combined action of two vortices is to displace lower speed fluid originally near the wall into a region lying between the vortices shown at position a in figure 4. This creates the elongated low speed regions seen in figure 3. On the other hand, the higher speed fluid lying above the centerline of the vortices experiences a relative motion that forms concentrations of high speed fluid between alternate vortices at positions c in figure 4.

One of the principal results of this vortical motion and the unique aspect of the Görtler instability is that it creates inflectional velocity profiles in a flow field in which there were none before the disturbance started to grow. It is interesting to note that a spanwise inflectional profile will develop as soon as the vortices begin to grow. This is best illustrated by considering the usual velocity perturbations, i.e.

$$\begin{aligned} u'(x,y,z) &= -\hat{u}(y)\cos\alpha z\, e^{\beta x} \\ v'(x,y,z) &= \hat{v}(y)\cos\alpha z\, e^{\beta x} \\ w'(x,y,z) &= -\hat{w}(y)\sin\alpha z\, e^{\beta x} \end{aligned} \quad (5$$

As soon as this disturbance begins to grow, there are inflection points in the u(z) profile at $\alpha z = \pm(n + \frac{1}{2})\pi$, $n = 0,1,2...$ The strength of the strain rate at these points is

$$\left| \frac{\partial u}{\partial z} \right| = \alpha\hat{u}(y)\, e^{\beta x}$$

and it increases in magnitude with the growth of the instability. Note that the sign of the strain rate is not essential in the inviscid instability; only the absolute value is important. Later in the development of the flow, inflectional U(y) profiles develop as predicted and observed by several investigators. These inflectional profiles open the door to an inviscid instability as discussed above.

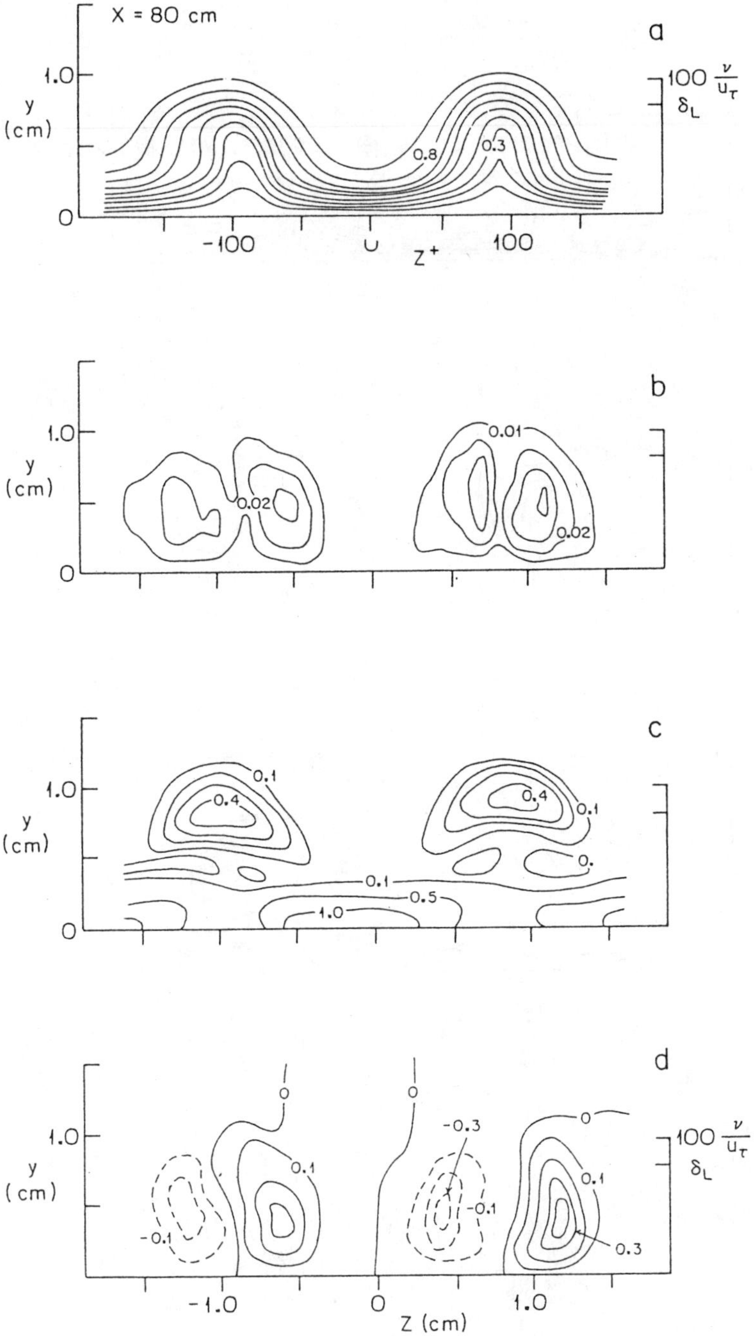

Fig. 5. From top to bottom are the iso-contours of a) the mean velocity, b) the rms velocity, c) the normal shear $\partial U/\partial y$ and d) the spanwise shear $\partial U/\partial z$. The data were taken at y = 80 cm and are normalized with the free stream velocity and the velocity gradient at the wall.

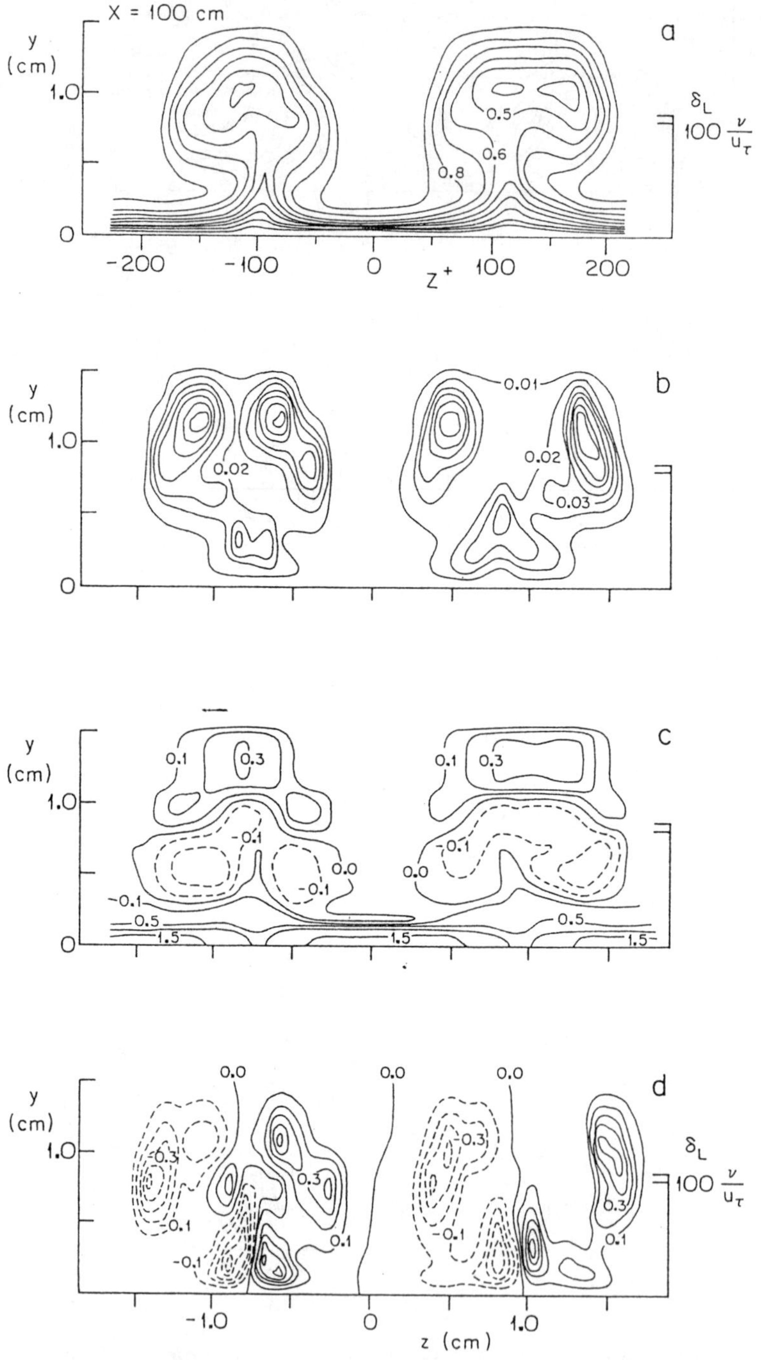

Fig. 6. Similar to figure 5 at x = 100 cm.

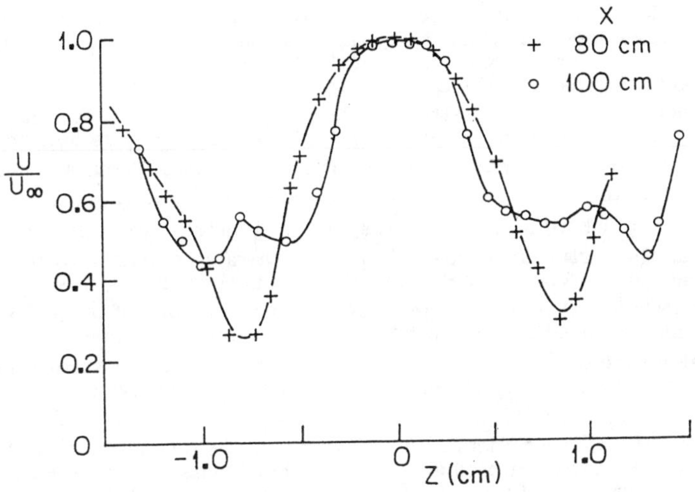

Fig. 7. Two examples of the spanwise velocity profile having inflection points and large shear.

local $\partial U/\partial z$ maxima occurs at the identical location of the maxima in the rms contours and their contour shapes are very similar (compare with figures 6b and 6d). Even the new local maximum near the wall at x=100cm in the fluctuation contours is also found in the iso-shear plot. Hence it it obvious that the velocity fluctuations are associated primarily with the spanwise shear rate, $\partial U/\partial z$.

Examples of the mean spanwise velocity profiles obtained at the y position near the maximum in the $\partial U/\partial z$ contours are shown in figure 7 for x = 80 and 100cm. The mean spanwise averaged velocity is approximately $0.7U_\infty$ in both cases and the maximum shear is typically $0.5\ \partial U/\partial y|_w$. To illustrate the stability theory predictions, the theoretical wavelength associated with these profiles should be $\lambda_x = 15.7\Delta$. From figure 7 at 100 cm the shear layer thickness is approximately 0.25cm. This yields $\lambda_x \sim 3.9$cm which compares favorably with the wavelength seen in figure 2.

Similar profiles in the normal direction are shown in figure 8 at four downstream locations. The inflection points are noted and it is easily seen that the inflections move rapidly away from the wall as the flow moves downstream. The velocity at the inflection points is typically $0.65U_\infty$.

WALL REGIONS OF BOUNDED TURBULENT SHEAR FLOWS

The wall region of bounded turbulent shear flows is dominated by the bursting process which is composed of a series of events similar to that described above for the Görtler problem. Namely there are streamwise vortices, elongated regions of low speed fluid called low speed streaks, lift-up and oscillation of these streaks and a rapid disintegration of the coherent motion followed by strong mixing.

The role of the streamwise vortices in the turbulent flow is less clear than in the Görtler case. Indeed, there is no generally accepted definition of such vortices found in the literature. Let us make the following definition:

> A vortex consists of a coherent vortical parcel of fluid such that the instantaneous streamlines in the plane perpendicular to the vortex lines are closed.

By the time the flow has reached 100cm downstream in the experiment, the velocity field was strongly nonlinear as evidenced by the variations in the measured streamwise velocity component. Figures 5 & 6 illustrate the iso-velocity contours measured over two wavelengths in the spanwise direction at two downstream locations, x = 80 and 100 cm. Two low speed regions at z ≅ -0.8 and +0.8cm are evidenced by the relative lower speed fluid lying at the elevated regions. The U(y) = 0.9 contour lies well beyond the nominal boundary layer thickness defined by $U(\delta_L) = 0.99U_\infty$. The growth of the low speed region has been quite large; the $U = 0.9U_\infty$ contour has been displaced outward from 1.25 to 1.75 δ_L where δ_L is the corresponding Blasius layer thickness at each location. Hence the disturbance has a large growth rate compared to the nominal laminar boundary layer. The displacement thickness at these locations was measured to be approximately $\delta^* = 0.7$cm compared with the Blasius value of $\delta^* = 0.28$cm. The corresponding higher speed regions at z = 0 and ±1.6cm have thinned the boundary layer considerably such that $\delta^* = 0.1$cm there. In these locations the shear is especially strong near the wall compared with the shear in the lower speed regions.

As noted earlier, the Görtler instability grows spatially downstream and has no temporal dependence. However oscillations do develop downstream as seen in figure 2 which indicate the presence of a secondary instability. The temporal fluctuations were measured and their iso-contours are given in figures 5b and 6b at the two locations. Note that the maximum at each position occurs on the sides of the low speed streak and not near the top. The maximum magnitude increases from roughly $0.03U_\infty$ to $0.08U_\infty$ bewteen 80 and 100 cm downstream. The total energy associated with the temporal oscillations is proportional to the square of the u´ values integrated over the y-z plane. Clearly this energy has increased by more than an order of magnitude.

It is also significant that the locii of the iso-contours have moved outward from the wall. The maximum at x = 80cm is at 0.5cm and the corresponding maximum at x = 100cm is at 1.15cm. The corresponding Blasius thickness, δ_L, has only increased from 0.78cm to 0.86cm. This outward movement corresponds to an average normal velocity greater than $0.03U_\infty$ which is more than an order of magnitude greater than the Blasius velocity normal to the plate.

Another important feature of figure 6b is that as the primary fluctuations have moved outward, a new local maximum has evolved in the wall region at x = 100cm. At 110cm downstream, this maximum was approximately $0.12U_\infty$ whereas the outer maximum had remained constant at $0.08U_\infty$. After 110cm, the entire layer became turbulent and the local maxima were no longer discernable.

The reason for the occurrence of two maximum rms values per wave length is apparent when the iso-shear contours are examined in figures 5cd and 6cd. The $\partial U/\partial y$ and $\partial U/\partial z$ gradient fields were calculated by fitting the averaged velocity data with a spline fit and then differentiating. They are normalized with the average gradient at the wall which was 1650 and 2300 sec^{-1} at x = 80 and 100cm respectively. The $\partial U/\partial y$ iso-shear contours are presented in figures 5c and 6c and the corresponding $\partial U/\partial z$ contours in 5d and 6d at the two streamwise locations. The normal shear $\partial U/\partial y$ has its maximum at the wall as expected, but also develops a strong local maxima above the low speed streak. This maxima moves outward as it travels downstream and is reduced slightly in magnitude. Distinct regions of negative shear are seen at x = 100cm due to the low speed fluid riding above higher velocity fluid as seen in figure 6a.

The spanwise shear rate, $\partial U/\partial z$, has a distinct minimum and maximum on both sides of the low speed streaks. Symmetry indicates the minimum and maximum must have the same magnitudes; however for these two streaks studied, the negative amplitude was always slightly less. The maximum magnitude of $\partial U/\partial z$ is comparable to that of $\partial U/\partial y$ at x = 80cm and exceeds $\partial U/\partial y$ at x = 100cm. The local maxima of the spanwise shear increases from 0.4 to $0.5u_\tau^2/\nu$ downstream in contrast to the decrease in $\partial U/\partial y$. The

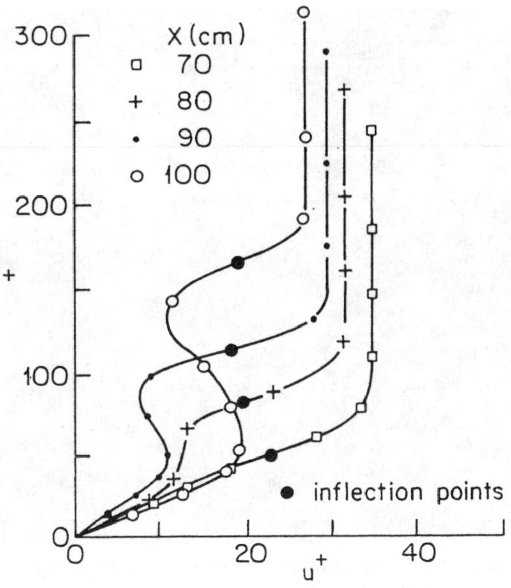

Fig. 8. Evolution of the $\partial U/\partial y$ inflectional velocity profiles in the x direction along the centerline of a low speed region.

This definition emphasizes the coherent aspects of the motion which are associated with and due to its phase and not necessarily due to its amplitude. Thus it includes the weak Görtler streamwise vortices, vortex rings, trailing wing tip vortices, etc., but excludes the mean vortex lines in a two-dimensional shear flow.

Evidence of such vortices in the wall region has been assumed by many authors, however very little quantiative data exists supporting their existance. Most of the data that are available are visual results such as those of Kline et al. (1967) and Smith and Schwartz (1983). Both of these authors have used hydrogen bubbles to visualize the flow field and have shown that often the hydrogen bubbles are coagulated into a confined region with definite rotation perpendicular to its principal axis consistent with the above definition of a vortex. Smith and Schwartz speculated that the observed streamwise rotations were due to the legs of horseshoe type vortices. Acalar and Smith (1987ab) have set up two model laminar flow fields to study the vortical nature of the motion near a solid boundary. In the first, they studied the flow over a hemisphere and the second was the flow over lower speed fluid injected through a streamwise slit in the wall. In both cases they observed the periodic formation of repeatable horseshoe vortices. The structure was studied by placing the hydrogen bubbles at different positions and photographing the vortices at different times after their formation. Their visual results are very similar to photographs obtained in the turbulent wall region and suggest that similar vortices appear there also. Kim and Moin (1986) have observed similar type structures in their numerical data by tracing vortex lines. By examining regions of strong streamwise vorticity, they concluded that the associated vortex lines rarely remained parallel to the wall for distances greater than 100 ν/u_τ. This result is consistent with the above observations by Smith and Schwartz.

The conditionally averaged data of Blackwelder and Eckelmann (1979) and of Kim (1983) also support the idea that streamwise vortices exist in the wall region. Even though their results yielded symmetrical vortices, the symmetry was a result of the technique and does not imply that symmetrical pairs will occur. Guezennec et al. (1987) have shown that the vortices rarely appear in a symmetrical manner, but one vortex is usually much larger than the other. Obviously spanwise symmetry requires that the average streamwise vorticity be zero.

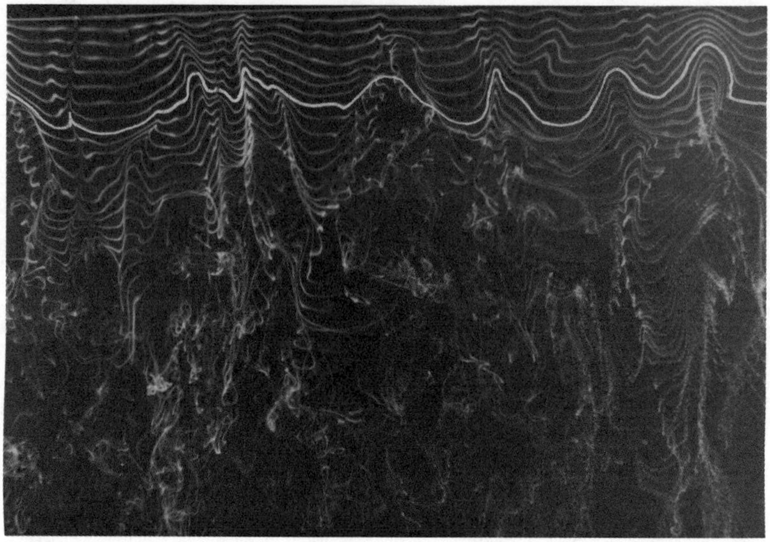

Fig. 9. Visualization of the low speed streaks and eddy structure in the wall region made visible by a hydrogen bubble wire at $y^+ = 12$.

The most ubiquitous structures of the wall layers are the low speed streak. They have a width of $20\nu/u_\tau$, an average spanwise spacing of $100\nu/u_\tau$, and have steamwise lengths up to and exceeding $1000\nu/u_\tau$. The low speed streaks are most readily observed by hydrogen bubbles as seen in figure 9. The hydrogen bubble wire is seen at the top left hand corner of the photograph and was pulsed at 40 Hz. The wire was at y+ = 12 and the mean flow velocity was 10 cm/sec. The low speed streaks are readily recognizable by the closeness of the bubble lines and the higher concentration of bubbles. The spanwise scale of the low speed regions is roughly $20\nu/u_\tau$. The spanwise scale of the higher speed fluid is much larger, typically about $60\nu/u_\tau$, so that the spanwise spacing between low speed streaks is typically $80\nu/u_\tau$. All of these dimensions are random variables and hence only average values can be used for the scales.

Since the pulsed line of bubbles is a time line, its x position downstream of the bubble wire is proportional to the Langrangian streamwise velocity component. A high-lighted time line is shown in figure 9 to illustrate the U(z) velocity profile. Neglecting the difference between the Langrangian and Eulerian velocities for the moment, one can see that the U(z) profile has a strong variation across the field; its approximate minimum is $0.2U_\infty$ and its corresponding maximum is $0.8U_\infty$. Also of importance is the fact that due to this variation, there are points of inflection in the profile on both sides of the low speed streaks. These regions are possible sites for an inviscid inflectional instability as discussed earlier. In addition, the shear rate there is found to be extremely large, i.e. of the same order of magnitude as the shear rate at the wall which is u_τ^2/ν.

This phenomenon was studied in more detail in a turbulent boundary layer at $Re_\theta = 2200$ by using multiple hot-wire probes spanning the y and z directions. Since they are sensitive to the streamwise velocity, they measure the U(y) and U(z) profiles as seen in figure 10. The data from each sensor is plotted and the solid line is a spline fit of the data. The spline fit was differentiated to determine the locations of the inflection points which are denoted. A comparison of figure 9 and 10 indicates that the shear rates are slightly higher in the visualization results probably due to the finite resolution of the hot-wire sensors. Nevertheless, it is obvious that there are many inflection points in both U(y) and U(z) in the wall region.

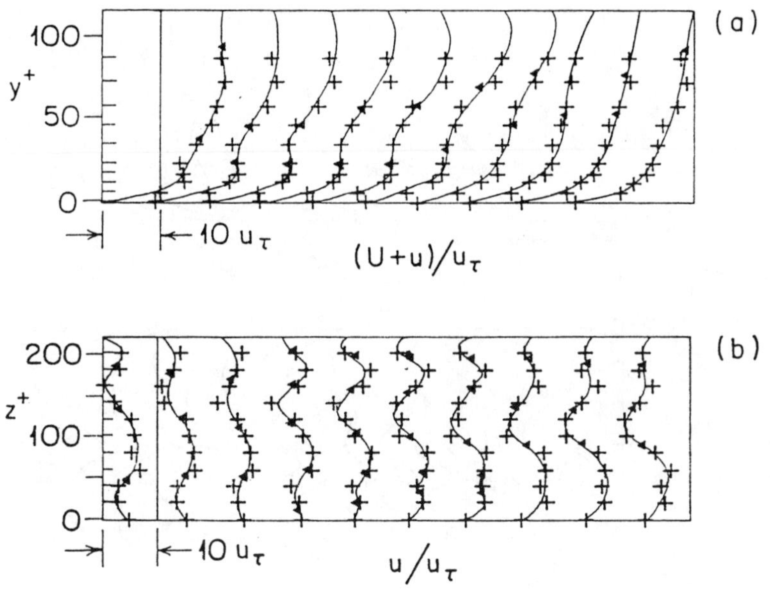

Fig. 10. Instantaneous turbulent boundary layer profiles from hot-wire sensors of the streamwise velocity component in the (a) normal and (b) span-wise directions. The data points are indicated by (+) and a spline fit yielded the lines. The points of inflection (◀) were obtained from the spline function. The time between profiles is $1.5\nu/u_\tau^2$ in (a) and $1.0\nu/u_\tau^2$ in (b).

Figure 11 gives another example of the ubiquitous nature of the inflectional profiles. These results were obtained from the numerical simulation of a turbulent channel flow at $Re_\theta = 200$. The data were searched for points of inflection which are denoted on the figure. Comparing figures 10 and 11 suggests that the lower Reynolds number channel flow has more inflection points; however, this is most probably an illusion due to different techniques used for the two data sets. It is quite evident that both the experimental and numerical data indicate that points of inflection are very evident in both the U(y) and U(z) data.

A comparison of figures 11b and 11c indicate that inflection points are much more predominate in the wall region at $y^+ = 15$ than in the logarithmic region at $y^+ = 66$. This agrees with the earlier idea that the low speed streaks are concomitant with inflectional profiles. Since the low speed streaks do not extend outward beyond $y^+ \sim 25$, one expects that fewer inflection points will be observed above that level.

In addition to the inflectional character of the velocity, several other conditions are necessary before the simple stability anaylsis can be applied. First the flow should be steady; i.e. the changes in time in the base flow must have a much longer time scale that the time scale of the instability. Smith and Metzler (1983) have measured the persistence of the low speed streaks and find that they have an average life time of $480\nu/u_\tau^2$ with some streaks persisting up to $2580\nu/u_\tau^2$. These values are much longer than the time scale of the instability which is of the order ν/u_τ^2. Thus, a quasi-steady

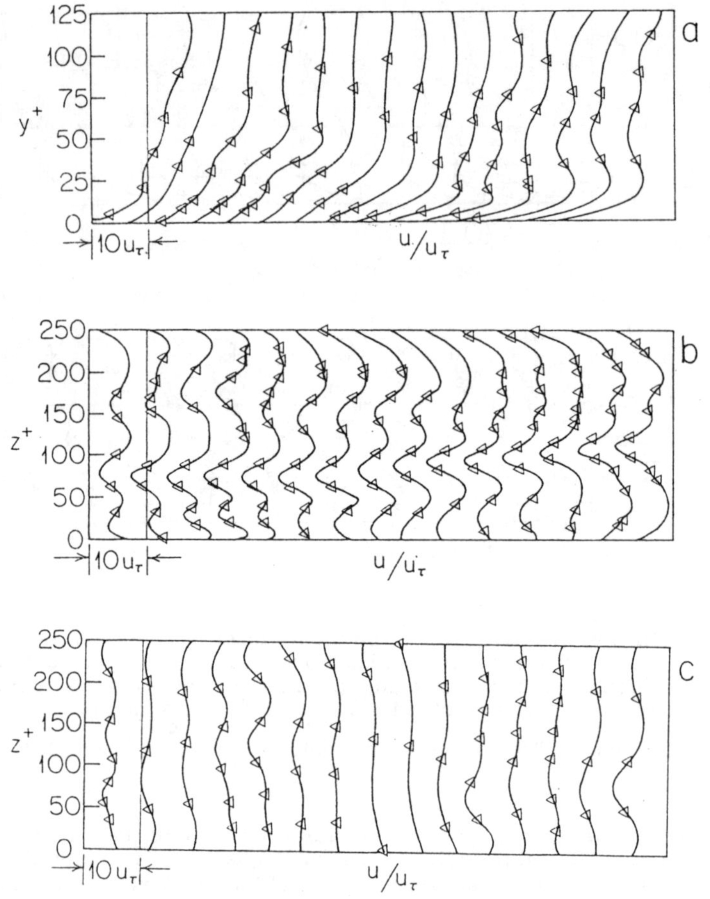

Fig. 11. Instantaneous velocity profiles in the wall region from the direct numerical simulation of a turbulent channel flow in (a) the normal direction, (b) in the spanwise direction at $y^+ = 15$, and (c) in the spanwise direction at $y^+ = 66$. The inflection points are denoted by ◁.

state exists for the inflectional instability. Secondly the site of the instability should be far removed from a solid boundary to insure its inviscid character. Huerre (1983) has shown that the characteristics of the instability are hardly altered by the presence of a wall as long as the inflectional region is removed at least 1.2Δ from the wall. Since the length scale of the inflection is $5\text{-}10\nu/u_\tau^2$, Huerre's criteria appears to be satisfied for the inflections above $y^+ \cong 10$. The third condition necessary to apply the stability criteria is that of two-dimensionality. No theory is presently available to explain how this will effect the instability, but Nishioka et al's (1980) results discussed earlier suggest that the two-dimensional theory is valid in highly three-dimensional fields.

To determine the growth rates of the incipient instability, the data were explored further to obtain the strength of the shear layers associated with the inflectional profile. This was accomplished by first locating the position of the inflection points

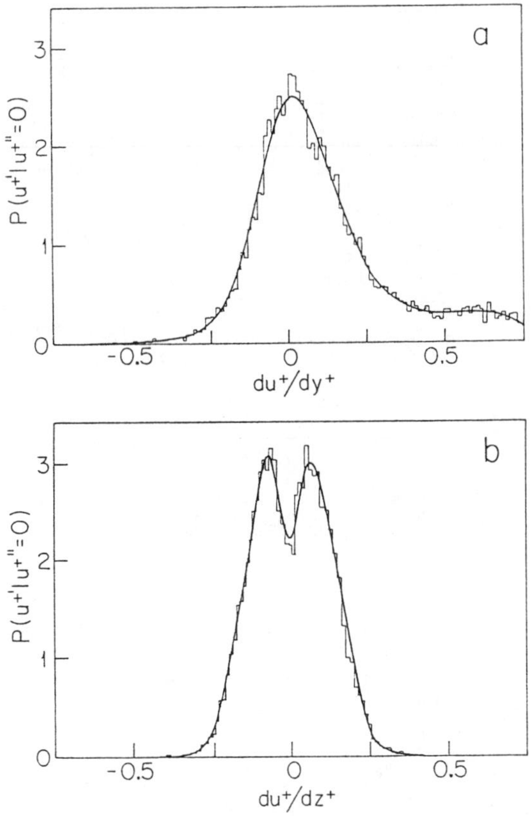

Fig. 12. Conditional probability distribution of the velocity shear at the inflection points from experimental data with Re_θ = 2500. a) velocity shear perpendicular to the wall at $10 < y^+ < 20$. b) Velocity shear parallel to the wall at $y^+ = 15$.

within a given spatial domain. All of the inflection points were included in the averaging process, although some could have been excluded using Fjørtoft's theorem. The value of the shear rate was then measured at each of these points. Thus the conditional probability of the shear rate could be obtained by sampling the shear at the points of inflection. The results are shown in figures 12 and 13 where the shear rates are normalized by the mean wall values. The first of these figures is obtained from the experimental hot-wire data base used in figure 10. The normal shear distribution in figure 12a is asymmetrical because of the mean shear in that direction. The most probable value of the shear is approximately zero and its mean value equals $\partial U/\partial y$ at that location. It is also interesting to note that the negative gradients are often observed as reported earlier by Willmarth and Sharma (1984). They also observed $\partial U/\partial y$ gradients which exceeded the mean shear at the wall. The corresponding spanwise shear in figure 12b is symmetrical as required and has dual peaks at a nondimensional shear of approximately ±0.1.

Numerical data obtained from Spalart (1987) were also analysed to compare with the experimental results. The same processing was used to obtain figure 13. The Reynolds number of this data is Re_θ = 670. The results are very similiar to that in figure 12; the main difference being that the magnitudes of the probabilities are slightly lower and the distribution is slightly wider than in the experimental case. This difference may be due to the spatial averaging imposed by the hot-wires and the coarser grid spacing in the experimental case.

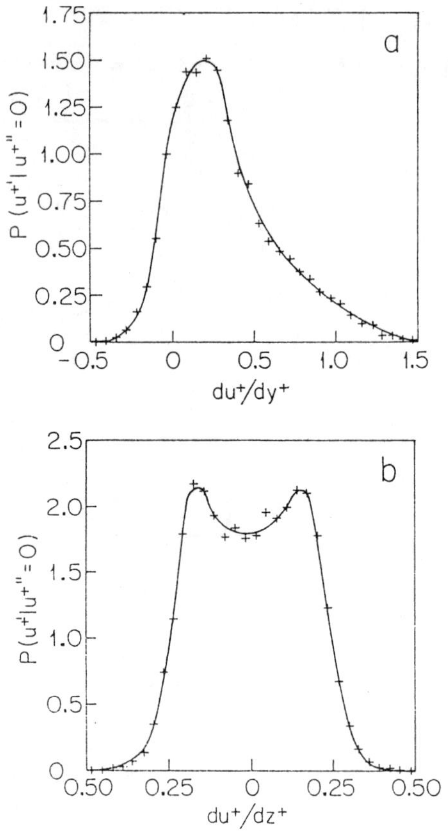

Fig. 13. Similar results as in Figure 12 but for the direct numerical simulation of a turbulent boundary layer at $Re_\theta = 670$.

If the inflectional profiles are indeed unstable, then one should be able to predict the wavelength of the amplified oscillation. Michalke's results suggest that the wavelength will be 15.7Δ. Close examination of figures 10 and 11 and other similar data indicate that Δ is roughly $10 \pm 5\nu/u_\tau$. Thus the inflectional profiles should produce disturbances with wavelengths between 75 and 225 ν/u_τ. Some distinct data exist in the literature which support this prediction. Dinkelacher et al. (1977) measured the pressure fluctuations on the wall and found wavelike patterns propogating downstream with wavelengths of $200\nu/u_\tau$ travelling at typically 12 - $15u_\tau$. Kline et al. (1967) also reported observations of wavelengths of $240\nu/u_\tau$. One reason there are few data of this type reported in the literature is that the oscillations are occurring within a random background and thus they are difficult to distinguish from the uncorrelated fluctuations. In addition, since the background provides the initial disturbance for the instability and the energy of the disturbance can grow by a factor of 1000 while travelling one wavelength downstream, an identifiable perturbation can probably only be observed for one or at most two wavelengths before it is distorted beyond the point of recognition.

DISCUSSION AND CONCLUSIONS

The primary result of this research is that inflectional velocity profiles are a prevalent and ubiquitous feature of the wall region of turbulent flow fields. The presence of low speed streaks in this region has been reported by numerous authors and is recognized as one of the clearest indications of the existence of wall layer eddy structures. The mere presence of low speed streaks is sufficient to guarantee that there will be inflectional velocity profiles in the spanwise direction because U(z) must vary from high to low values about a zero mean as seen in figures 9 through 11.

The instantaneous data also shows that there are numerous inflection points in the direction normal to the wall as well. This is somewhat surprising because of the restraining influence of the wall, but has been verified by different authors in several flows. The density of the $\partial^2U/\partial y^2 = 0$ points is comparable to the $\partial^2U/\partial z^2 = 0$ points in the region $10 < y^+ < 20$.

It is interesting to note that an "inflection point" is actually an intersection of a locus of such points. In the simplified two-dimensional $\tanh(y/\Delta)$ profile, the locus of the inflection points is a plane at $y = 0$; i.e. an inflectional plane. The resulting instability waves have crests parallel to the inflectional plane with their crests aligned along the third axis, i.e. perpendicular to both the mean flow and to the direction of the gradient,. These waves are convected along the x axis. For the spatial growth case, the wave amplitude increases in x at each point in time.

In the three-dimensional cases considered here, the locus of inflection points will be a non-planar surface. For example, a wake has a locus of inflection points which forms a cylindrical surface aligned in the mean flow direction with equal amplitude of the shear, $\partial U/\partial\xi$, at every streamwise position. In fully developed turbulent flow, the inflectional surface is much more convoluted. For example, the side of a low speed streak will always be such a surface because $\partial^2U/\partial z^2 = 0$ there. If the streak has an inflectional U(y) profile, the inflectional surface may extend from one side of the low speed streak over its top to the other side engulfing the entire low speed region. If the strength of the shear $\partial U/\partial\xi$, is constant over the entire inflectional sheet, each instability wave will have a similar growth rate. The waves would be locally parallel to the inflectional sheet and hence would have the shape of a horseshoe. The vortices generated by Acarlar and Smith (1987ab) clearly illustrate this phenomenon in a well controlled laminar flow. Similar vortices were often observed in the transitional flow of Swearingen and Blackwelder (1987).

The inflectional surface is not necessarily of infinite extent. For example, a momentary disturbance at a point in a laminar boundary layer must have an inflectional surface of finite extent because the flow at large distances from the disturbance may be devoid of inflection points. In tracing out the inflectional surfaces in data such as in figures 10 and 11, one often sees that the surface ends abruptly at some point in the flow field although it may continue in the third dimension. In the turbulent case, the inflectional surfaces are random in space and time as well.

The locii of inflection points are important because they present a mechanism by which turbulence is produced. Namely the shear associated with the inflections is generally quite strong which suggests that an inviscid instability dominates the flow field. Through this mechanism, the unstable inflectional profiles can extract energy from the instantaneous base flow and produce a rapidly growing oscillation. The wavelength of the oscillation scales with the width of the inflectional instability. Since the scale of the inflections is random, so is the scale of the growing disturbance; however, the scale of all of the observed oscillations was well within the spectral domain of the turbulence. The instability grows extremely rapidly; i.e. the two-dimensional theory implies that the oscillations increase their amplitude by a factor of 30 (and its energy by 100) as it travels one wavelength downstream. The actual growth is probably tempered by energy spreading in the third dimension as well.

A model flow field emulating the turbulent wall region was studied which consisted of Görtler streamwise vortices in a laminar boundary. This flow had the advantage that it was steady and thus provided a better test bed for studying the velocity profiles and oscillations. Similar low speed streaks and inflectional profiles were observed in the model flow. Inflectional profiles in the normal and spanwise directions yielded regular and repeatable oscillations with wavelengths consistent with the predictions of Michalke (1965). The oscillations grew rapidly in the streamwise direction and broke down into turbulence after two to four wavelengths. In this flow field, it was clear that virtually all of the turbulence was produced by the inviscid instability. Not only were the dynamics similar between the model and turbulent flows but the length and velocity scales were also comparable when normalized with the friction velocity and the kinematic viscosity, as noted by Blackwelder(1983). This scaling is common in analyzing turbulent wall flows but has not been used frequently in transitional flows. The present results suggest that this scaling provides a means of comparing these two flow regimes in more detail than heretofore and indicates that their dynamics are indeed similar.

Finally, the evidence suggests that the instability may be more successful in producing u and w fluctuations than in producing u and v oscillations. Note that the instability will always produce an oscillation in the direction of the base flow, u. In a two-dimensional case, the second velocity component produced by the instability is the component parallel to the direction of the gradient. Since an inflection point is always present on the sides of the low speed streaks, one expects these to produce oscillations of u and w. Oscillations involving v may be less prevalent because inflections do not always occur above the low speed streak. However, it appears that by the time that large growth is observed, the locus of inflection points extends upward from the sides and over the low speed streak. In this case the observed oscillations will depend upon the integrated growth rates that the disturbances have experienced which may favor the spanwise oscillation since they have existed over a longer time period. In addition, the growth of the v oscillation will be hindered at some point by the presence of the wall, thus suggesting that the spanwise oscillations will dominate and produce a larger oscillation amplitude. This is of course consistent with the observations that $w' > v'$ in the wall region of turbulent shear flows.

All of the above are devoid of Reynolds averaging and its implications. In particular, these arguments view turbulent production as an instantaneous event related to a particular mechanism, e.g. the unstable inflectional velocity profile. The reported observations show that the growing oscillations are three dimensional and produce u, v, and w fluctuations in a random manner depending upon the orientation of the inflectional profile. This observation is not inconsistent with Reynolds averaging but does provide a different description of the production process. For example in the above formulation, the turbulent energy is fed directly into the u, v, and w fluctuations. The Reynolds averaged equations show that the average turbulent energy is only fed into the u fluctuations and then is redistributed by the pressure into the other components. The difference is that the averaged equations do not allow for instantaneous variations in the base flow, i.e. the mean averaged flow, whereas the instantaneous descriptions does.

ACKNOWLEDGEMENTS

The authors wish to thank Philippe Spalart for the use of his numerical boundary layer data and the Stanford-NASA/Ames Center for Turbulence Research for the opportunity to utilize the numerical data. The turbulent channel flow data base was provided by Richard Leighton and Robert Handler of the Naval Research Laboratory.

This research was supported by the Office of Naval Research under contract number N00014-82-K-0084 and the Air Force Office of Scientific Research through contract number F49620-85-C-0080. JDS thanks the National Research Council for their support through an NRC-NRL Research Associateship.

REFERENCES

Acarlar, M. S. and Smith, C. R., 1987a, "Haripin Vortices Generated by a Hemisphere Protuberance," Journal of Fluid Mechanics, 175, 1.

Acarlar, M. S. and Smith, C. R., 1987b, "Haripin Vortices Generated by Fluid Injection," Journal of Fluid Mechanics, 175, 43.

Bippes, H., 1972, "Experimentelle Untersuchung des laminar-turbulenten Umschlags an einer parallel angeströmten konkaven Wand," Sitzungsberischte der Heidelberger Akademie der Wissenschaften Mathematisch-naturwissenshaftliche Klasse, Jahrgang 1972, 3, 103.

Blackwelder, Ron F., 1983, "Analogies between Transitional and Turbulent Boundary Layers," Phys. Fluids, 26, 2807.

Blackwelder, R. F. and Eckelmann, H., 1979, "Streamwise Vortices Associated with the Bursting Phenomenon," Journal of Fluid Mechanics, 94, 577.

Cantwell, Brian J., 1981, "Organized Motion on Turbulent Flow", Ann. Rev. Fluid Mech., 13, 457.

Diprima, R. C. and Stuart, J. T., 1972, "Non-local Effects in the Stability of Flow Between Eccentric Rotating Cylinders," Journal of Fluid Mechanics, 54, 393.

Dinkelacher, A., Hessel, M., Meier, G. E. A. and Schewe, G., 1977, "Investigation of Pressure Fluctuations beneath a Turbulent Boundary Layer by Means of an Optical Method," Phys. Fluids, 20, S216.

Floryan, J. M. and Saric, W. S., 1982, "Stability of Görtler Vortices in Boundary Layers," A.I.A.A. Journal, 20, 316.

Gaster, M., Kit, E. and Wygnanski, I., 1985, "Large Scale Structures in a Forced Turbulent Mixing Layer," Journal of Fluid Mechanics, 150, 23.

Görtler, H., 1940, " Über eine dreidimensionale Instabilität laminarer Grenschichten an Konkaven Wänden," Nachr, Wiss, Ges. Göttingen, Math-Phys Kl, Neue Folge I, 2, 1.

Greenspan, H. P. and Benney, D. J., 1963, "On Shear Layer Instability and Breakdown," Journal of Fluid Mechanics, 15, 133.

Guezennec, Y., Piomelli, U. and Kim, J., 1987, "Asymmetric Characterization of Conditionally Averaged Structures in Wall Bounded Flows" Bulletin of American Physics Society, 32, 2046.

Hall P. (1983), "The Linear Development of Görtler Vortices in growing Boundary Layers" Journal of Fluid Mechanics, 130, 41.

Hall, Philip, 1987, "The Nonlinear Development of Görtler Vortices on Growing Boundary Layers," a manuscript.

Huerre, P., 1983, "Finite Amplitude Evolution of Mixing Layers on the Presence of Solid Boundaries," J. de Mecanique Theorique et appliquee, Numero Special, p. 121.

Kim, John and Moin, Parviz, 1986, "The Structure of the Vorticity Field in Turbulent Channel Flow, Part 2," Journal of Fluid Mechanics, 162, 339.

Kim, John, 1983, "On the Stucture of Wall-bounded Vortical Flows," Phy. Fluids, 26, 2088.

Kline, S. J., Reynolds, W.C., Schraub, F. A. and Runstadler, P.W., 1967, "The Structure of Turublent Boundary Layers," Journal of Fluid Mechanics, 30, 741.

Liepmann, H. W., 1945, "Investigation of Boundary Layer Stability and Transition on Concave Walls," NACA Wartime Report W-87.

Lin C. C., 1955, "The Theory of Hydrodynamical Stability," Cambridge University Press

Michalke, A., 1965, "On Spatially Growing Disturbances in an Inviscid Shear Layer," Journal of Fluid Mechanics, 23, 521.

Nishioka, M., Asai, M. and Iida, S., 1980, "An Experimental Investigation of the Secondary Instability," Laminar-Turbulent Transition, Ed. R. Eppler and H. Fasel, Springer Verlag, Berlin, p. 37.

Rayleigh, Lord, 1880, "On the Stability or Instability of Certain Fluid Motions," Proc. London Math Soc., 11, 57-70.

Sabry, A. S. and Liu, J. T. C., 1988, "Nonlinear Development of Görtler vortices and the Generation of High Shear Layers in the Boundary Layer," Proc. of Symposium in Honor of C. C. Lin, World Scientific Pub. Co., Singapore.

Smith, C. R. and Metzler, S. P., 1983, "The Characteristics of Low Speed Streaks in the Near Wall Region of a Turbulent Boundary Layer," Journal of Fluid Mechanics, 129, 27.

Smith, A. M. O., 1955, "On the Growth of Taylor-Görtler Vortices Along Highly Concave Walls,: Quat. Appl. Math, 13, 233.

Smith, C. R. and Schwartz, S. P., 1983, "Observation of Streamwise Rotation in the Near Wall Region of a Turbulent Boundary Layer," Phy. Fluids, 26, 641.

Spalart, Philippe, 1987, data provided through the Summer Program of the Center for Turbulence Research, Stanford-NASA/Ames.

Swearingen, J. D. and Blackwelder, R. F., 1987, "The Growth and Breakdown of Streamwise Vortices in the Presence of a Wall," Journal of Fluid Mechanics, 182, 255.

Willmarth, W. W. 1975, "Structure of Turbulence on Boundary Layers," Adv. Appl Mech., 15, 159.

Willmarth, W. W. and Sharma, L. K., 1984, "Study of Turbulent Structure with Hot-wires Smaller than the Viscous Length," Journal of Fluid Mechanics, 142, 121.

Flow Visualization with Image Processing of Three-Dimensional Features of Coherent Structures in an Open-Channel Flow

T. UTAMI[1] **and R. F. BLACKWELDER**
Department of Aerospace Engineering
University of Southern California
Los Angeles, California 90089-1191, USA

T. UENO
Ujigawa Hydraulics Laboratory, DPRI
Kyoto University
Yoko-Ohji, Fushimi-Ku, Kyoto, Japan

ABSTRACT

A particle tracking method of flow visualization was introduced into an open-channel flow with $Re_d = 3900$. A rapid time series of photographs of the horizontal cross-section of the flow at various heights from the channel bed were taken. The pictures were digitized and processed by a computer and the two dimensional distribution of velocity vectors in each plane was obtained. The third velocity component was calculated by integration of the continuity equation and the resulting statistics and correlations are compared with existing data.

INTRODUCTION

Intensive research during the past twenty five years has revealed many important faces of the coherent structure of turbulent shear flows. Flow visualization methods and the conditional sampling technique have been used effectively in experimental studies and, recently, numerical simulations have yielded other detailed features of the turbulence structure. Flow visualization has been instrumental because it has provided an encompassing view of the flow structures but quantitative information by the method has been difficult to obtain. On the other hand, the transducer methods have furnished quantitative information but only at limited spatial locations. At the present time, three-dimensional quantitative data of fully developed turbulence have not been obtained experimentally. This was the goal of the present research.

To obtain the two dimensional velocity distribution in a plane by tracking particles, several image processing techniques have been proposed. These can be classified into the streakline method, the sequential images method and the combined method. In the first method (Dimotakis et.al. (1981), Ohmi (1986)), the processing is applied to a single photograph containing streaklines to identify the coordinates of both end points of each streakline. In the second method, a single photograph of tracers illuminated with sequential light pulses or sequential photographs are taken of the tracers and an image processing method is applied to identify tracers which are composing the same streak. In the combined method (Kobayashi and Saga (1986), Rignot and Spedding (1988)), photographs of sequential images of tracers and

---

1 On leave from Ujigawa Hydraulics Laboratory, DPRI, Kyoto Univ.

streaklines are taken simultaneously and the tracer images belonging to the same streakline are identified.

For processing photographs obtained sequentially, two different methods have been proposed. The local correlation method (Yano (1983)) compares the density distribution in a small area of one picture with that of the corresponding area in a second photograph taken slightly later in time. The spatial separation giving the maximum correlation coefficient is used to identify the end points of the velocity vector. The second technique is the compiled tracer image method (Yamakawa and Iwashige (1986)), in which several sequential photographs taken with small time differences are compiled into one image and the partially superimposed tracer images are used to determine the same velocity vectors.

In the first part of this paper, the method of picture processing is described in detail. In the second part, some statistical properties calculated from the data are discussed and shown compatible with those obtained by probe measurements and numerical simulation. In conclusion, the method offers a technique to experimentally obtain the entire velocity field and shows promise as a technique to study complicated three-dimensional flow fields.

EXPERIMENTAL METHOD AND CONDITION

A uniform open-channel water flow with a rectangular cross-section (40 cm wide and 3.95 cm of water depth) was visualized using polystyrene beads with an average diameter of 0.4 mm as tracers. The experimental setup is shown in figure 1. A horizontal cross-section of the flow was illuminated by a sheet of white light 3 mm thick and a pair of photographs were taken with a small time difference (typically 35 msec). The level of the light was then shifted a small distance (2.3 mm) upward

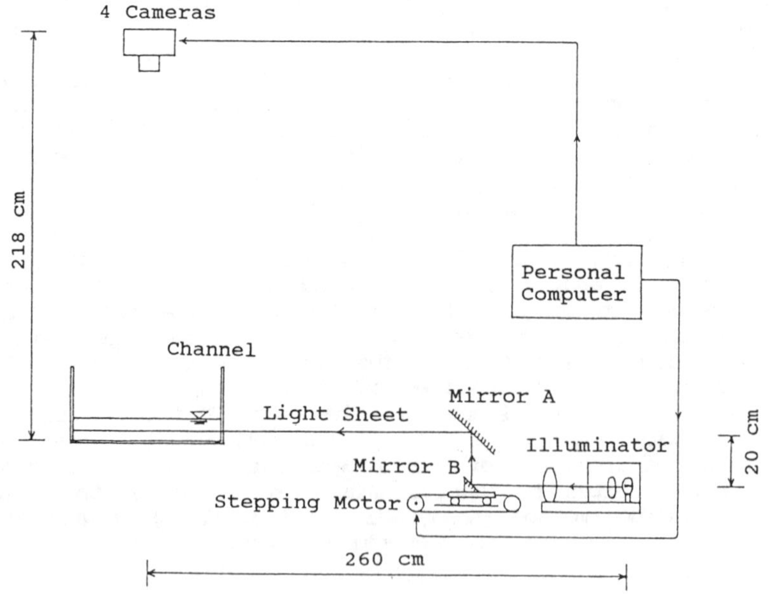

FIGURE 1. The experimental setup.

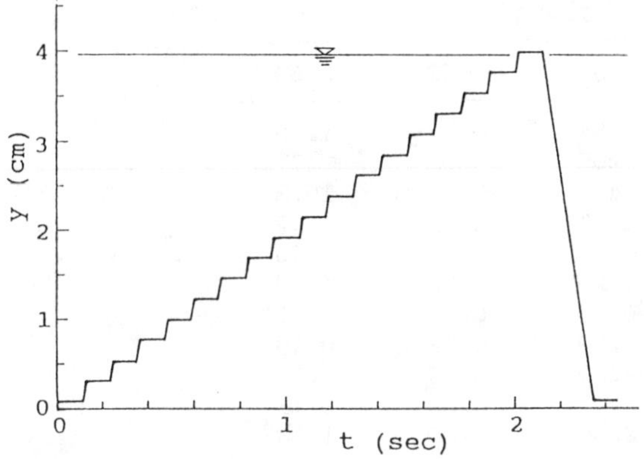

FIGURE 2. The space-time motion of the sheet light.

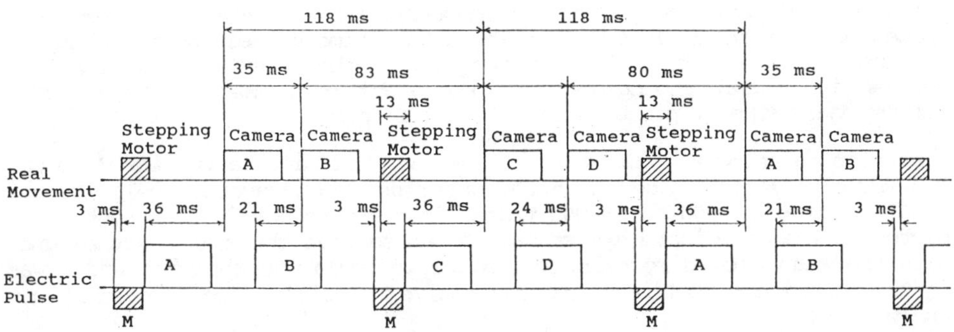

FIGURE 3. The timing relationship between the electric pulses, shutter openings and the mirror movement. The electrical signal are shown on the lower line and the physical motion of the mirror's stepping motor and the shutters is shown above. All times are in msec.

and another pair of pictures were taken in rapid succession. By repeating the procedure, 18 pairs of photographs were taken from near the channel bed (y = 1.15 mm) to the water surface (y = 39.5 mm). The space-time motion of the light sheet is shown in figure 2. The entire procedure was repeated 4 times yielding 18 x 2 x 4 = 144 photographs.

As one camera with a motor drive could take only 5 pictures per second, four Nikon F3 cameras were used as two pairs to reduce the time between the pairs of photographs. Thus 15 pictures were taken every second and the time interval between any two photographs which constitute a pair was approximately 1.3 viscous scales. The cameras were all triggered by individual electric pulses generated by the computer. The light sheet was positioned by two 45° mirror that moved parallel to each other as sketched in figure 1. The movable mirror was driven by a stepping motor synchronized by the personal computer. The timing relationship among electric pulse, shutter timing and mirror movement is shown in figure 3.

Table 1 Flow conditions in each plane.

y^+	8	21	37	53	68
y(cm)	0.115	0.31	0.54	0.77	1.00
T(ms)	0	120	235	355	471
U_m/U_τ	9.0	12.3	14.6	15.2	16.3
u'/U_τ	3.24	2.32	2.11	2.07	1.92
v'/U_τ	0.76	0.83	0.86	0.87	0.86
w'/U_τ	1.27	1.27	1.25	1.28	1.20
S_u	0.27	-0.07	-0.13	-0.12	-0.01
S_w	0.05	0.13	0.07	0.02	0.09
F_u	2.67	2.87	2.89	3.02	3.21
F_w	4.13	3.53	3.41	3.23	3.11
a	5.00	4.00	2.00	1.50	1.50

Each camera had a unique delay time between the electric pulse shown on the lower line of figure 3 and the opening of its shutter shown on the upper line of figure 3. Considerable effort was expended to correlate the timing of shutters and mirror movement. The camera's shutters were opened for approximately 17 msec. The actual time between photographs was obtained from a clock in the pictures so that the uncertainty in the time between individual photographs was known within 1 msec. The interval between two pairs of pictures was 115 - 120 msec.

The friction velocity, u_τ, was 0.57 cm/sec, the water depth was 3.95 cm. The Reynolds number based upon the water depth and average velocity was 3900. Between 20,000 to 25,000 tracer images were obtained in each picture. The coordinate system was chosen so that the x axis was aligned with the mean flow direction, y was perpendicular to the wall and z was in the spanwise direction. The respective velocities are denoted by u,v and w.

In this paper the results of analyzing the first 10 photographs (5 pairs) among 144 photographs are presented. The analysis was limited to a 35 cm x 20 cm area which was the central part of the photographs and contained 2300 - 3000 particles. The flow parameters of each plane are listed in Table 1. U_m is the mean velocity in the plane, u', v' and w' are the turbulence intensities, S_u and S_w are the skewness factors for u and w, and F_u and F_w are the corresponding flatness factors. All the u and w velocity data were obtained from the 2-dimensional vector data before interpolation; on the other hand, the normal velocity component, v, was calculated from the interpolated data. T(ms) is the time measured from the instant of the first picture and the parameter a is described in the interpolation section.

PICTURE PROCESSING

Digitizing Process

All the photographs were digitized by an Image Scanner PIS-30 (I-O Data Instrument Co.) with a resolution of 12 pixels per mm. The scale of photographs was about 1/3.7 of the flow field; thus the real resolution was 0.31 mm per pixels. The resolution is almost the same as the diameter of the tracers; however the images of the tracers occupied

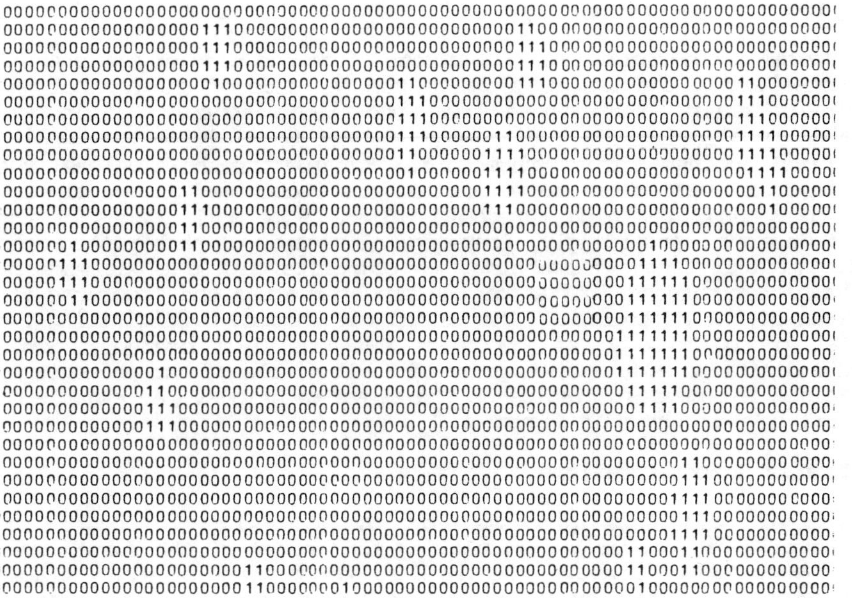

FIGURE 4. An example of the digital data at $y^+ = 53$. The spatial resolution between successive digits is 0.31 mm.

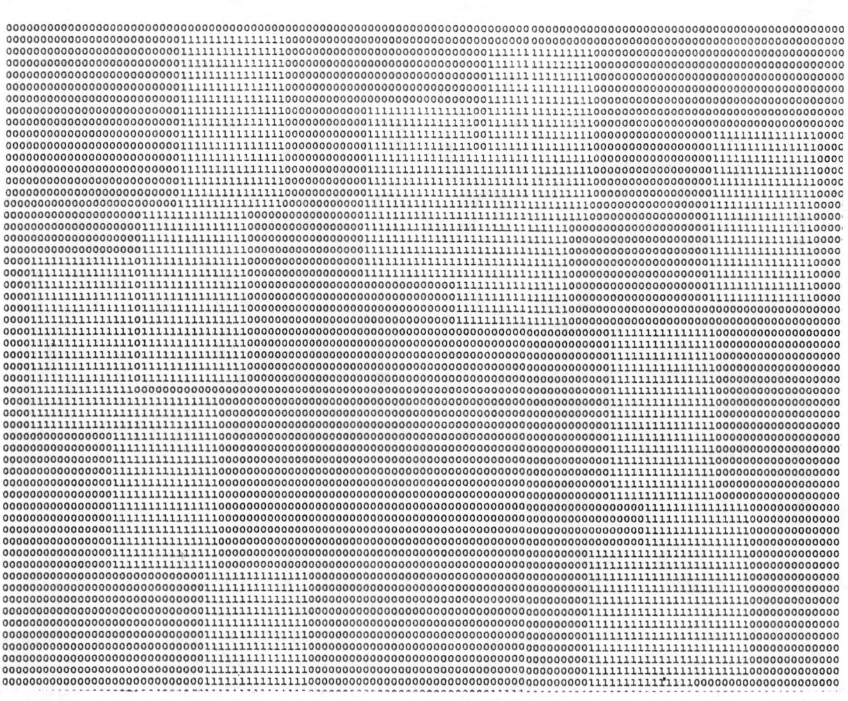

FIGURE 5. An example of the modified digital data. The area is almost the same with that of figure 4. The distance between neighboring digits is 0.2 mm.

several pixels on the average because of halation.

The horizontal cross-section was illuminated from one side of the channel only. Consequently the tracer image on the far side of the channel was comparatively darker than those near the light source. This was compensated by allowing the 1 bit image scanner to digitize the density of each pixel against a given threshold. For low threshold values, the tracers that are very close to each other are recognized as one tracer. On the other hand, for high threshold values, some dark images of the tracers fade away. Accordingly, four digital records of each photographs were made with different thresholds. A composite data set was made by using the higher threshold data for the brighter areas and lower one for darker area of the photograph. The method was checked by digitizing two sequential photographs on an Optronics C-4100hs drum scanner. This provided a digitized 2048 x 2048 array of 8 bit pixels of the photographs which yielded the same statistics as the above technique. An example of the digital data is shown in figure 4.

Calculation of Tracer Locations

The digitized data were processed by the computer to locate the coordinate of every tracer by calculating the center of the first-order moment of each tracer image. The physical coordinates of every tracer location were calculated by using six reference marks on the channel which appeared in the photographs and whose coordinates had been measured beforehand. The positional shift of each particle image introduced by refraction at the water surface was removed by calculations using geometrical optics and the relative location of the tracers, cameras and the water surface.

Calculation of Velocity Vectors by Local Correlation Analysis

Velocity vectors were calculated by comparing the tracer locations in each pair of pictures that had been processed as above. For the following analysis, the plane to be analyzed was divided into new pixels having the size of 0.2 mm. Then each tracer's image was enlarged to the same size and shape as follows. Around the centroid of each tracer, the pixels were set to 1 over a 15 x 15 matrix and the pixels in the remaining area were set to 0. An example of the digitized plane thus modified is shown in figure 5. Then an 81 x 81 area around the tracer under study in one photograph was compared with a corresponding area in the second photograph at the later time. The correlation between the two areas was obtained with a spatial shift in the x and z direction. The spatial shift which provided the maximum correlation coefficient was used to determine the trajectory of each particle during the elapsed time between the photographs.

Elimination of Irregular Vectors

Because some particles were captured in one photograph but not in the other one, the above correlation method sometimes failed to obtain a correct trajectory. These erratic vectors were eliminated by two methods. They were thrown out either if correlation was less than a prescribed threshold (typically 0.6) or if their trajectory differed from the average of the surrounding vectors by 3.5 times the local rms values. Through this procedure, 5-7 % of the vectors were eliminated. An example of the retained velocity vectors is shown in figure 6 in which the tracers in the first photograph are shown by righthand directed triangles and that in the second photograph by lefthand directed ones. The

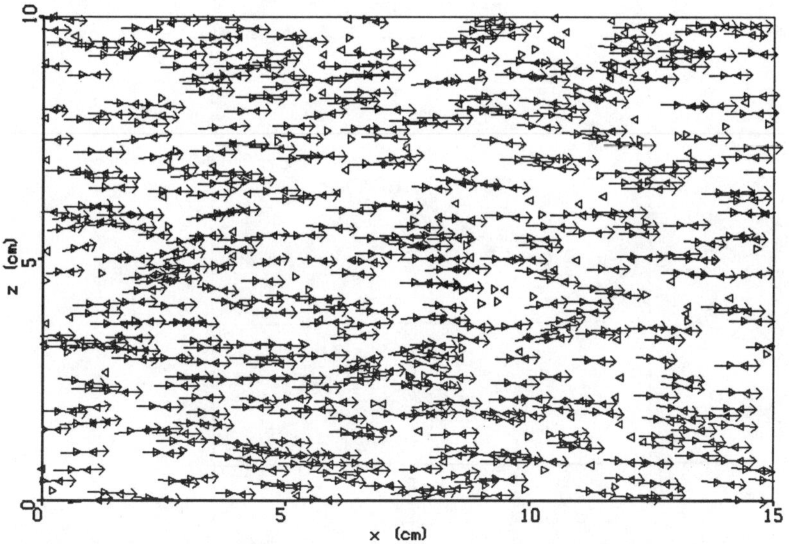

FIGURE 6. An example of the velocity vectors obtained by image processing in a portion of the $y^+ = 68$ plane. The tracers in the first photograph are denoted by ▷ and those in the second one by ◁.

particles which passed the above test are spanned by an arrow indicating their velocity. The singular particles shown in figure 6 were not utilized in the ensuing analysis.

Interpolation

To calculate the vertical velocity component, the u and w data were transformed by interpolation into a regular grid with grid intervals of 2.5 mm (17 viscous length scales). The total number of grid points utilized was 141 and 81 in the x and z direction respectively. The original tracer positions were randomly distributed with respect to the grid points; thus a mapping algorithm was necessary to transform the random data onto each grid point. Grant (1958), Compte-Bellot (1963) and Moin and Kim (1982) have shown that the R_{uu} and R_{ww} correlations are strongly elliptical in the x-z plane; thus it was decided to weight each tracer's velocity by an elliptical distance of $\sqrt{(x-x_o)^2 + a(z-z_o)^2}$ where (x,z) is the tracer's coordinate, (x_o,z_o) is the grid point, and a is the aspect ratio of the ellipse of the correlation coefficients. The value of a was determined as an function of y^+ from the data of Grant and Moin and Kim and is given in Table 1. At each grid point, all quadrants defined by $|x^+-x_o^+| < 165$ and $|z^+-z_o^+| < 60$ were searched to locate the four tracers with the shortest elliptical distance from the grid point. The u and w velocity at the grid point was calculated by the linear interpolation using the velocity of these four tracers. If less than three particles were found to satisfy the above criteria, the velocity was computed with the elliptically weighted average of the tracers.

Calculation of the vertical component of velocity

The vertical velocity component, v, was calculated with a revised method from that of Utami and Ueno (1984). By integrating the continuity

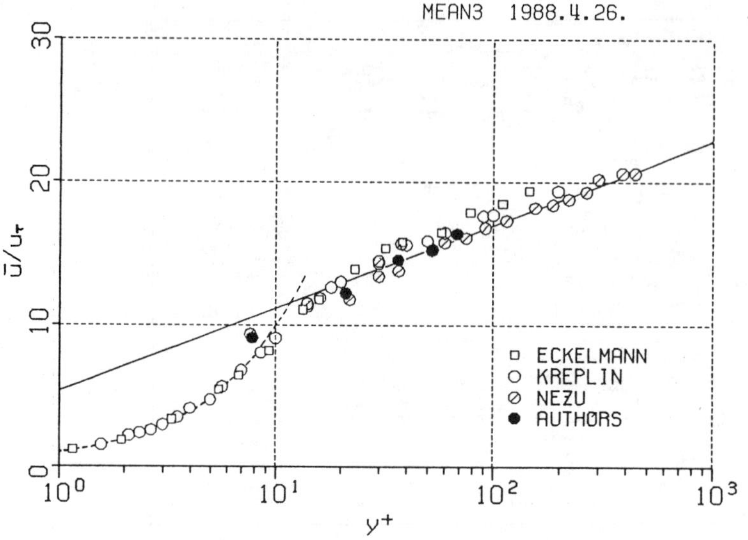

FIGURE 7. Distribution of the average longitudinal velocity.

equation, the vertical component v of velocity vector is given by the equation,

$$v_u = v_o - \{(du/dx+dw/dz)_o + (du/dx+dw/dz)_u\}\Delta y/2$$

where the o subscript denotes the value in the lower plane and the subscript u denotes those in the corresponding upper plane. However some error is always involved in calculating differential quantities from limited grid data. To avoid accumulation of these errors, the values of v_o and the two-dimensional divergence of the lower plane were smoothed. The smoothing was made by weighting and summing the values at the grid point under study and neighboring grid points. The weighting matrix given by

$$\begin{pmatrix} 0.00625 & 0.0125 & 0.0625 & 0.0125 & 0.00625 \\ 0.02500 & 0.1250 & 0.5000 & 0.1250 & 0.02500 \\ 0.00625 & 0.0125 & 0.0625 & 0.0125 & 0.00625 \end{pmatrix}$$

was used so that the rms value of v agreed better with the experimental data obtained by a velocimeter.

STATISTICAL PROPERTIES

To check the reliability of the data and obtain the characteristics of the flow, the mean statistical properties were calculated and compared with published data.

Average Velocity

The average velocity in each plane was calculated and shown in figure 7. The oil channel data of Kreplin (1976) and Eckelmann (1974) and the open-channel data of Nezu (1977) are also plotted for comparison. The present

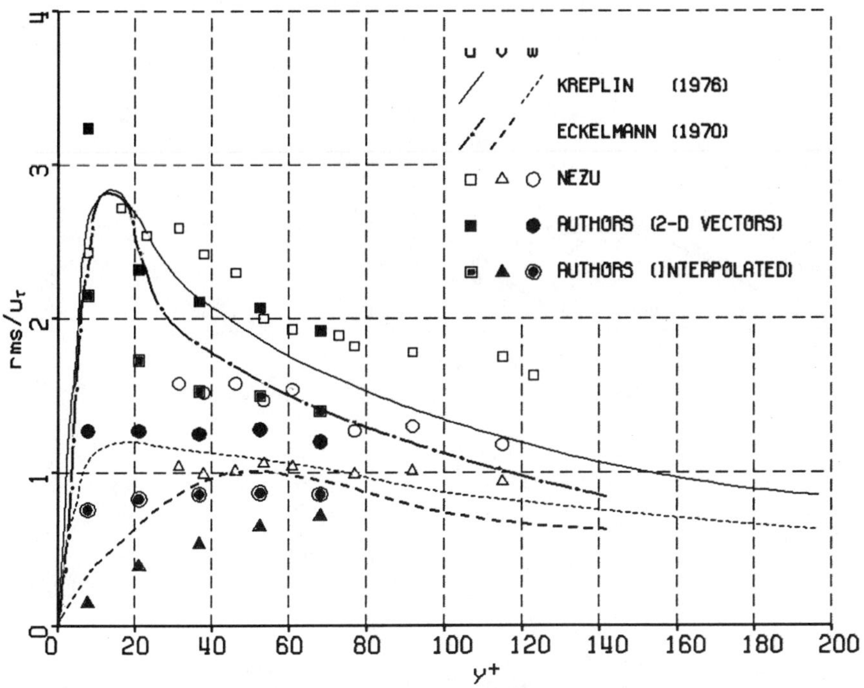

FIGURE 8. Distribution of the rms values of the turbulence velocity components.

data agrees well with the previous authors and follows the logarithmic law with k = 0.4 and C = 5.5.

Intensity of Turbulence

The rms values of the turbulence is compared with Eckelmann's (1970), Kreplin's (1976) and Nezu's data in figure 8. The u' and w' intensities were calculated from the raw two-dimensional vector data and are denoted by the solid symbols. All three rms values could be calculated from the interpolated data at the grid points. These are also shown in figure 8. Some smoothing was necessary during interpolation as approximately 3000 vectors were mapped into 11421 grid points. Thus the rms values were decreased in the process. By tracking more particles, the resolution of the technique would be improved and the difference between the rms values calculated by the two methods should decrease.

To obtain the data at the lowermost position, i.e. $y^+ = 8$, a portion of the light sheet was intersected by the wall so that its actual thickness was 2.3 mm. This thickness covered a region where u varied considerably which possibly yielded the large value of u' in figure 8.

Probability Density Distribution

The probability density distribution of u and w are shown in figure 9(a) and (b). figure 9(a) is quite similar to the Eckelmann's (1974) data and 9(b) agrees well with Kreplin's (1976) data. Table 1 shows the

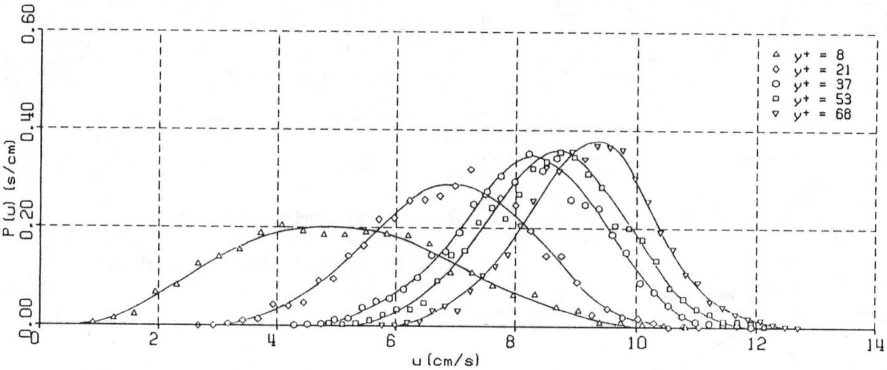

(a)

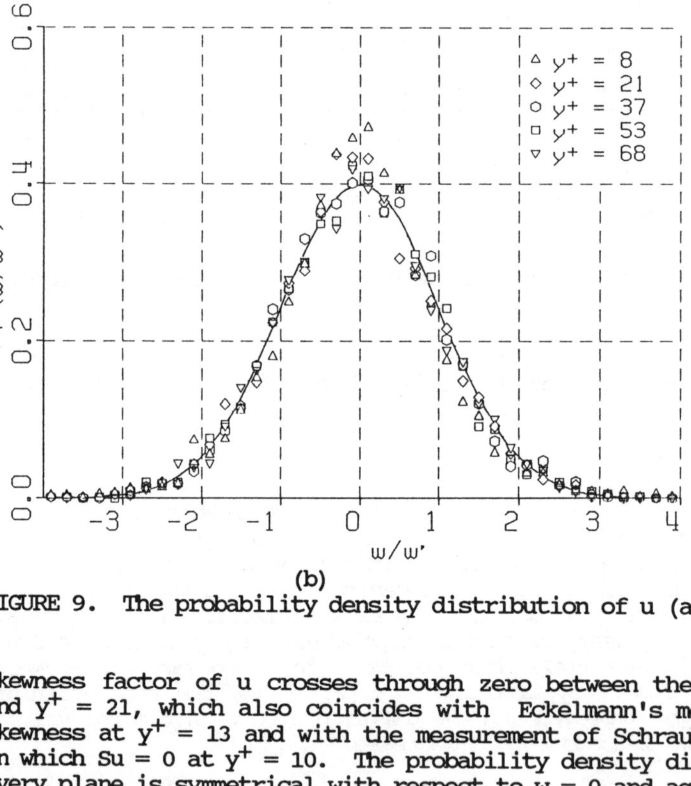

(b)

FIGURE 9. The probability density distribution of u (a) and w (b).

skewness factor of u crosses through zero between the planes of $y^+ = 8$ and $y^+ = 21$, which also coincides with Eckelmann's measurement of zero skewness at $y^+ = 13$ and with the measurement of Schraub and Kline (1965) in which Su = 0 at $y^+ = 10$. The probability density distribution of w in every plane is symmetrical with respect to w = 0 and agrees well with the Gaussian distribution shown by solid line as expected.

Correlation Coefficient of One-Dimensional Spatial Separation.

The spatial correlations were calculated as defined by

$$R_{u_i u_j}(\Delta x, \Delta y, \Delta z, y_o) = \frac{\Sigma \{ u_i(x_\alpha, y_o, z_\beta) \cdot u_j(x_\alpha+\Delta x, y_o+\Delta y, z_\beta+\Delta z) \}}{\sqrt{\Sigma \{u_i(x_\alpha, y_o, z_\beta)\}^2} \sqrt{\Sigma \{u_j(x_\alpha, y_o+\Delta y, z_\beta)\}^2}}$$

The streamwise correlations, $\Delta x \neq 0$, for the three velocity components are presented in figures 10(a)-(c) at the five y^+ positions. The correlations of Kim, Moin and Moser (1987) from the numerical simulation of channel flow at $y^+ = 5.6$ and a Reynolds number of 3300 are also shown for comparison with the data at $y^+ = 8$. The comparison is excellent for u and v although the present data has a slightly larger integral scale for w.

The integral scales were calculated by integrating the correlation coefficient from $\Delta x = 0$ to the first zero crossing. The results are given in Table 2. $R_{uu}(\Delta x)$ is the largest scale as found by previous investigators. It has typical values of 160-220 ν/u_τ and has a maximum value at about $y^+ = 50$. Clearly this must scale with the large eddy structure at higher y^+, hence the decrease at $y^+ = 68$ may be a statistical aberration. The integral scale for $R_{vv}(\Delta x)$ is the smallest of the three streamwise scales in the near wall region. It increases to a maximum of 80 ν/u_τ at $y^+ \approx 40$ and has a gradual decrease at higher elevations. On the other hand, the integral scale of $R_{ww}(\Delta x)$ shows a continuous decrease throughout the wall region from its maximum at $y^+ = 8$. The $R_{ww}(\Delta x)$ correlation is especially interesting at $y^+ > 37$ where it appears to have a periodic variation in Δx. The streamwise scale of this structure is 320 ν/u_τ which is comparative to the channel depth of 290 ν/u_τ. This type of correlation could be caused by vortices aligned predominantly with the normal axis. Such vortices may be the continuance of longitudinal vortices reported by Utami and Ueno (1987) in open-channel flow.

The spanwise correlations, $\Delta z = 0$, are shown in figures 11(a)-(c) for the three velocity components. The data at all five y^+ positions are given and the data from Kim, Moin and Moser (1987) are shown for comparison at the lowest y^+ position. The numerical simulation correlations display a more predominate negative region than the present data. This may be partially a result of Reynolds number because McLean (1989) has shown that the negative region of $R_{uu}(\Delta z)$ decreases considerably at higher Reynolds numbers. The negative region is probably also smaller in the present data because of the relatively poor resolution at the lowermost position. Nevertheless, all of the spanwise correlations do have a negative region consistent with the known presence of low speed streaks in the wall region. It is also worthy to note that all of the spanwise correlation maintain a negative region at higher elevations as well.

All three of the length scales show similar trends; namely they increase with y^+. The integral scales of $R_{uu}(\Delta z)$ and $R_{ww}(\Delta z)$ are approximately equal at all positions and the scale of $R_{vv}(\Delta z)$ is smaller by a factor of two. All three scales increase in magnitude at higher elevations.

Spatial Correlations with Two-Dimensional Spatial Separations

Some preliminary correlations with simultaneous spatial separations in both Δx and Δz have been obtained. Two of the more interesting correlations are shown in figure 12a and b which display $R_{uw}(\Delta x, \Delta z, y^+)$ at $y^+ = 8$ and 21. These contour plots have intervals of 0.05 between contours. The heavy line is the zero correlation, the unbroken lines are positive correlation values and the dashed lines are negative correlations.

The w velocity component must be statistically antisymmetrical with respect to the xy plane; hence the correlation must be antisymmetrical

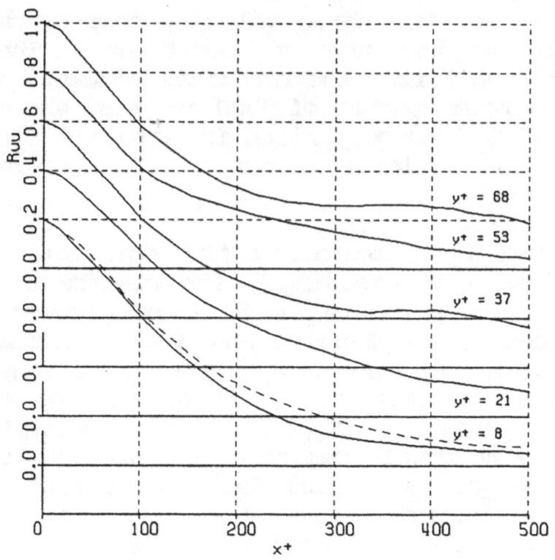

(a)

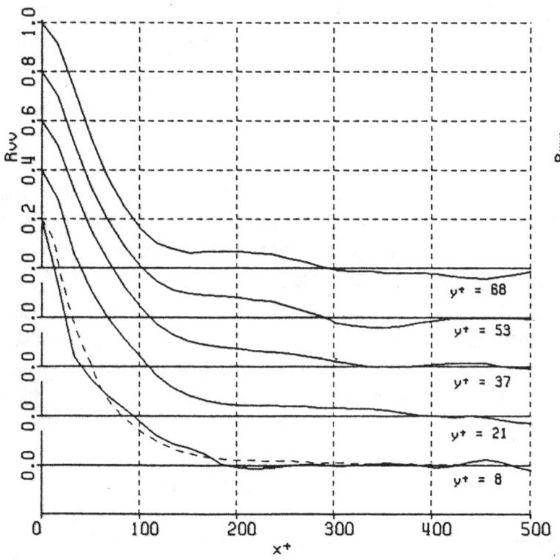

(b)

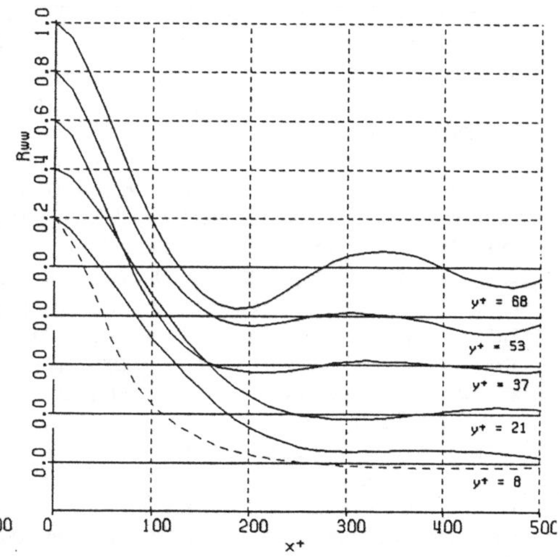

(c)

FIGURE 10. The correlation coefficients of streamwise separation for u (a), v (b) and w (c).

where the summation is over α and β. The present results were obtained within a single plane and thus $\Delta y = 0$; i.e. no cross correlations between planes are reported here. The separation distances, Δx of Δz were varied individually in the results reported in this section. Since Δx typically spanned 61 grid points, i.e. 1000 viscous scales, α was limited to 80 or less during the summation of streamwise correlations. Likewise Δz was limited to 41 grid points, i.e. approximately 700 viscous scales, and hence β could be no more than 41 during spanwise correlations. Only the results for i = j are presented in this section. The correlations were calculated for positive and negative Δx and Δz and found to be symmetrical within 4 % as required by homogeneity.

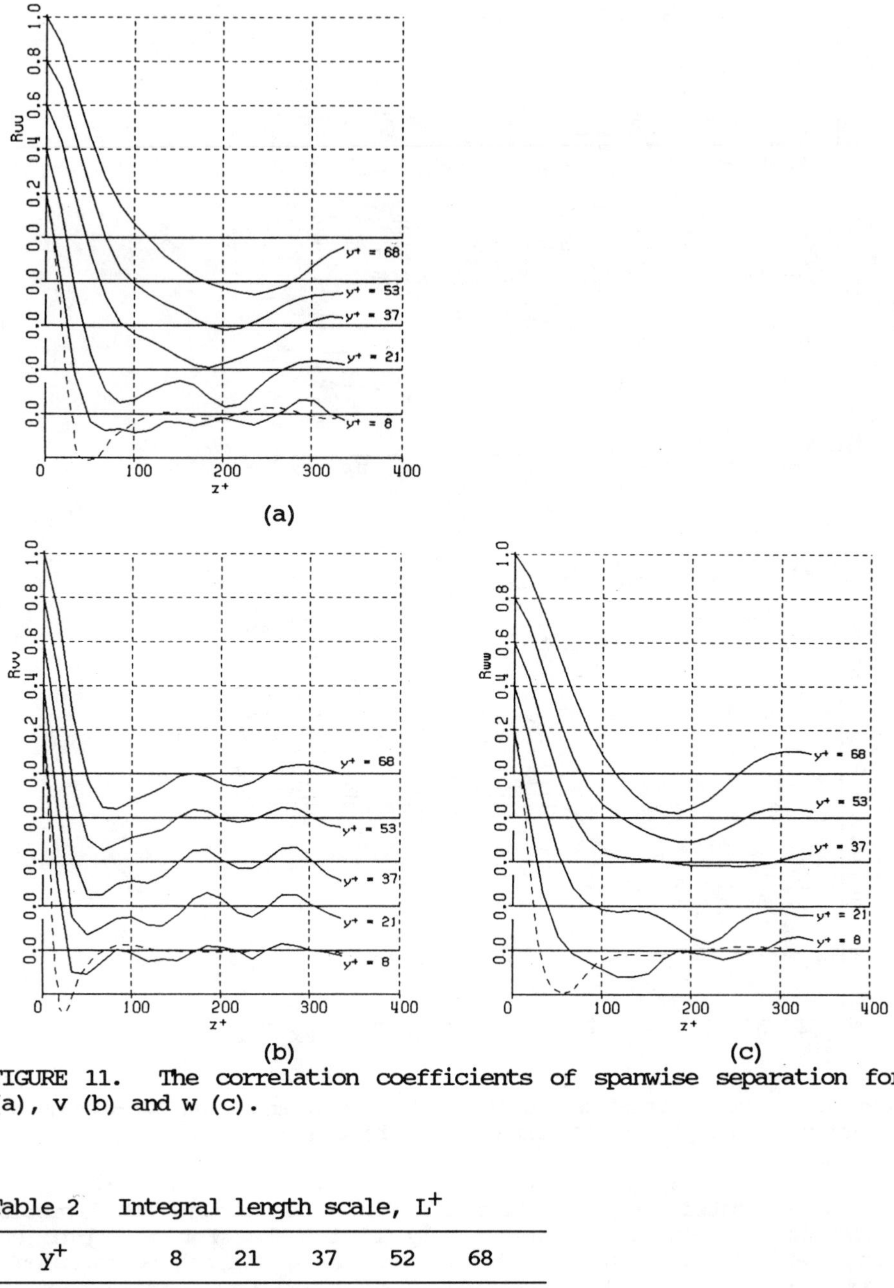

FIGURE 11. The correlation coefficients of spanwise separation for u (a), v (b) and w (c).

Table 2 Integral length scale, L^+

y^+	8	21	37	52	68
$R_{uu}(\Delta x)$	159	189	193	221	188
$R_{vv}(\Delta x)$	51	78	82	75	70
$R_{ww}(\Delta x)$	125	106	69	70	66
$R_{uu}(\Delta z)$	22	28	40	46	50
$R_{vv}(\Delta z)$	13	16	18	22	25
$R_{ww}(\Delta z)$	24	34	47	51	55

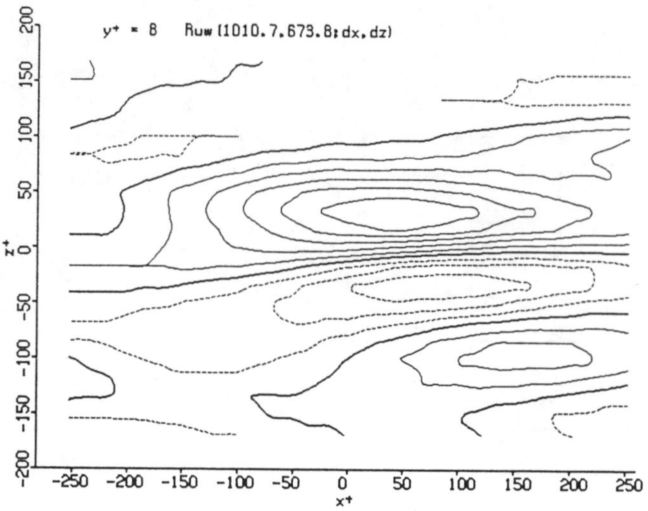

(a)

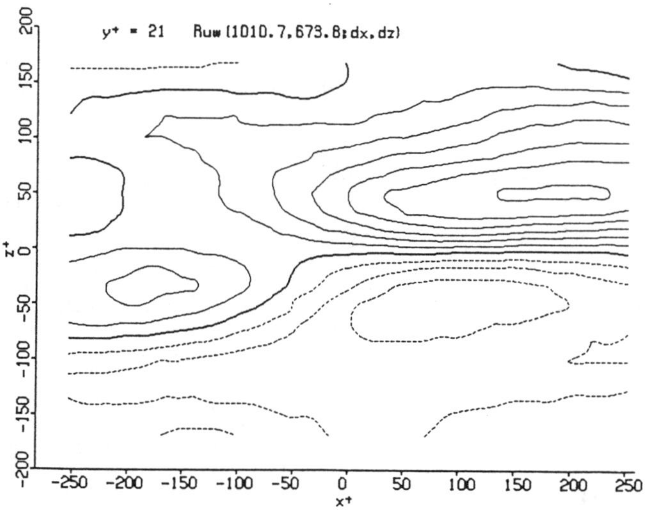

(b)

FIGURE 12. The correlation coefficients for u and w of two-dimensional separation in the $y^+ = 8$ (a) and $y^+ = 21$ (b) plane.

also as indicated by the figures. In the streamwise direction, $R_{u_iu_j}(\Delta x,\Delta z)$ is symmetrical in Δx only if $i = j$. For $i \neq j$ no such criteria exists and figure 12 shows that the $R_{uw}(\Delta x,\Delta z)$ correlation is asymmetrical with the peaks occurring at $\Delta x^+ \neq 0$. On the other hand, the $R_{uv}(\Delta x,\Delta z)$ and $R_{vw}(\Delta x,\Delta z)$ correlations were found to have their maximum negative values occurring at $\Delta x \approx 0$ for all y^+ positions.

The correlations in figure 12 imply that some aspect of the eddy structure related to u and w has a characteristic angle of approximately ±25° with respect to the streamwise direction. At the higher elevations this angle increased to 45°. The spanwise distance between the positive and negative peak increase from approximately 60 ν/u_τ at $y^+ \approx 8$ to 120 at $y^+ \approx 68$. The Δx separation distance of the peaks also increased with y^+ but the magnitude of the peaks appeared to have a constant value of 0.2 - 0.3 for all y^+.

CONCLUSIONS

The novel technique of tracking particles in several planes introduced originally by Utami and Ueno (1984) has been automated to include digitization via a computer. This has greatly reduced the amount of manual labor required and increased the potential of the method. Three component velocity data were obtained in a three dimensional region near the wall of an open channel flow. The mean statistics and correlations compare well with those of previous investigators. Hence the technique offers another method of obtaining data by automated particle tracking. The method may be especially attractive for gathering velocity data in three-dimensional flow fields where conventional techniques are often more limited.

Extensive correlations were calculated from the data in the wall region over $8 < y^+ < 70$. The correlation length scales were all much longer in the streamwise direction than in the spanwise direction consistent with the known eddy structures in this region. Somewhat surprising is that the streamwise length scale of $R_{ww}(\Delta x)$ decreased from 125 ν/u_τ at $y^+ = 8$ to 70 ν/u_τ at $y^+ \approx 30$. This suggests that this scale provides a more dramatic measure of the difference between the low speed streaks at $y^+ < 15$ to the vigorous mixing eddies at the higher elevations. The fact that the integral scale of $R_{vv}(\Delta x)$ is much less than the other two at $y^+ < 15$ agrees with the concept that the strong v motions, i.e. ejections, occur on a much shorter length scale than the low speed streaks.

The two dimensional spatial correlations of $R_{uw}(\Delta x, \Delta z)$ have shown that the correlated motion of u and w has a unique and distinct spatial pattern. It is the only correlation $R_{u_i u_j}(\Delta x, \Delta z)$ for all possible values of i and j that did not have its maximum magnitude at $\Delta x = 0$. The correlation peaks suggest the u and w velocity structure has a characteristic direction of ±25° with respect to the x axis in the xz plane. Further research is exploring the implications of this in terms of the known eddy structure.

ACKNOWLEDGEMENTS

The authors gratefully acknowledge the support of the AFOSR under contract number F49620-85-C-0080 monitored by James McMichael for that portion of the research conducted at USC.

REFERENCES

Blackwelder, R. F. and Eckelmann, H., Streamwise Vortices Associated with the Bursting Phenomenon, Jour. Fluid Mech., vol. 94, part 3, pp. 577-594, 1979.

Compte-Bellot, G., Contribution a L'etude de la Turbulence de Conduit, Doctoral Thesis, University of Grenoble, 1963.

Dimotakis, P. E., Debussy, F. D. and Koochesfahani, M. M., Particle Streak Velocity Field Measurements in a Two-Dimensional Mixing Layer, Phys. Fluids, vol. 24, no. 6, pp. 995-999, 1981.

Eckelmann, H., Experimentelle Untersuchungen in einer turbulenten Kanalströmung mit Starken Viskosen Wandschichten, Mitteilungen, Nr. 48, Max-Planck-Institute für Strömungsforschung und der Aerodynamischen Versuchsanstalt, 1970.

Eckelmann, H., The Structure of the Viscous Sublayer and the Adjacent Wall Region in a Turbulent Channel Flow, Jour. Fluid Mech., vol. 65, part 3, pp. 439-459, 1974.

Grant, H. L., The Large Eddies of Turbulent Motion, Jour. Fluid Mech., vol. 4, pp. 149-190, 1958.

Kobayashi, T. and Saga, T., How to Determine Velocity Fields from Pathline, Flow Visualization, Flow Visualization Soc. of Japan, vol. 6, no. 20, pp. 42-49, 1986 (in Japanese).

Kim, J., Moin, P. and Moser R., Turbulence Statistics in Fully Developed Channel Flow at low Reynolds Number, Jour. Fluid Mech., vol. 177, pp. 133-166, 1987.

Kreplin, H. P., Experimentelle Untersuchungen der Längsschwankungen und der Wandparallelen Querschwankungen der Geschwindigkeit in einer Turbulenten Kanalströmung, Doctoral Thesis, Georg-August-Universität zu Göttingen, 1976.

McLean, I., Boundary Layer Structure in Adverse Pressure Gradients, Doctoral Thesis, University of Southern California, 1989.

Moin, P. and Kim, J., Numerical Investigation of Turbulent Channel Flow, Jour. Fluid Mech., vol. 118, pp. 341-377, 1982.

Nezu, I., Turbulent Structure in Open-Channel Flows, Doctoral Thesis, Kyoto University, 1977.

Ohmi, K., Image Analysis of a Two-Dimensional Streakline Photograph, Flow Visualization, Flow Visualization Soc. of Japan, vol. 6, no. 20, pp. 19-27, 1986 (in Japanese).

Rignot, E. J. M. and Spedding, G. R., Performance Analysis of Automated Image Processing and Grid Interpolation Techniques for Fluid Flows, USCAE 143, University of Southern California, Dept. of Aerospace Engineering, 1988.

Schraub, F. A. and Kline, S. J., A Study of the Structure of the Turbulent Boundary Layer with and without Longitudinal Pressure Gradients, Report MD-12, Thermosciences Div., Mech. Engrg. Dept., Stanford University. 1965.

Utami, T and Ueno, T., Visualization and Picture Processing of Turbulent Flow, Experiments in Fluids, no. 2, pp.25-32, 1984.

Utami, T. and Ueno, T., Experimental Study on the Coherent Structure of Turbulent Open-Channel Flow Using Visualization and Picture Processing, Jour. Fluid Mech., vol. 174, pp. 399-440, 1987.

Yano, M., Velocity Measurement Using Correlation Concerning with Digital Tracer Image, Flow Visualization, Flow Visualization Soc. of Japan, vol. 3, no. 10, 1983, pp. 189-192 (in Japanese).

Yamakawa, M. and Iwashige, K., On-line Velocity Distribution Measuring System Applying Image Processing, Flow Visualization, Flow Visualization Soc. of Japan, vol. 6, no. 20, 1986, pp. 50-58 (in Japanese).

On the Character of Turbulence-Producing Events in Near-Wall Turbulence

CANDACE E. WARK and HASSAN M. NAGIB
Fluid Dynamics Research Center
Illinois Institute of Technology
Chicago, Illinois 60616, USA

INTRODUCTION

One of the most extensively studied turbulent flows is the turbulent boundary layer. Among the interests of current researchers in the field are the basic turbulence regenerative mechanism and attempts at control of some aspect of the turbulence. Research is largely focused on identifying dominant structures occurring in the flowfield and the nature of the interaction between two distinct parts of the turbulent boundary layer denoted as the inner region and outer regions. Fundamental understanding of this basic flowfield would be considered a major step in the subject of turbulence, in view of its importance to all wall-bounded shear flows that are ever present in engineering problems.

One distinct feature of the turbulent boundary layer visually observed by Kline et al. (1967) and Corino and Brodkey(1969) among others, is low speed streaks. A distinct characteristic of low speed streaks as described by Kim et al. (1971) is the bursting process. Briefly summarized, bursting occurs in three stages: i) Lift up of low speed streaks; ii) Growth of an oscillatory motion; iii) Breakup of the oscillatory motions followed by a more quiescent flow. Many detection techniques relying on velocity measurements have been introduced to provide quantitative information regarding the bursting process. The occurrence of low speed streaks and the insensitivity of the spanwise spacing of the streaks to Reynolds number is accepted (Smith and Metzler, 1983). Many independent investigations have found that the spanwise spacing of streaks when normalized by inner variables is roughly $\lambda_s^+ = 100$.

Burst Detection Techniques

Lu and Willmarth (1973) used both a u-level technique and a quadrant technique to detect the bursting event. The u-level technique counts a burst when the streamwise velocity falls below a certain threshold with respect to the long-time rms value. The quadrant technique sorts the u-v signal into quadrants based on the signs of the individual u and v signals, where u and v are the streamwise and normal velocity fluctuations. The Reynolds stress-producing events (negative uv events) fall into quadrants two ($u < 0$; $v > 0$) and four ($u > 0$; $v < 0$); hereafter, referred to as Q2 and Q4 events respectively. It is also helpful to note here that a Q2 event is regarded as an ejection while the Q4 event is referred to as a sweep.

A widely used technique introduced by Blackwelder and Kaplan (1976) relies on the magnitude of the short-term rms of a fluctuating signal compared to the long-term rms value for the same signal. The VITA technique (Variable Interval Time Averaging) computes a short term rms value for the streamwise velocity component (or temperature signal in some cases) for a specified length of time. If the short-term rms value is above a specified threshold with respect to the rms value for the entire time series, then a burst is counted. Later Chen and Blackwelder (1978) added a slope condition as well so that only events with an accelerating streamwise velocity component were counted. Without the slope condition it can be shown that the VITA technique mixes different types of events.

Blackwelder and Haritonidis (1983) using the VITA technique show that a sensor length correction is needed due to the spatial averaging effect of a hot-wire probe. Namely, the frequency observed will be too low due to the averaging of the velocity signal from more than one burst over the length of the probe. Using this correction they determine that the burst frequency scales with inner variables and suggest that the results of Rao et al. (1971), which conclude that outer scaling is the appropriate choice, are in error due to the spatial averaging effect. Willmarth and Sharma (1984) also using VITA with very small sensors conclude that the burst frequency does not depend on the Reynolds number when scaled with inner variables. Guezennec (1985) using the VITA technique with the spatial-averaging correction on the frequency finds that inner scaling is the correct choice. However he also introduces an analogous correction for the hold time.

Alfredsson and Johansson (1984) using the VITA technique in a turbulent channel flow, introduced a mixed scaling which collapsed their burst frequency. More recently Sreenivasan (1987) illustrated that although the production of turbulent energy -uv(dU/dy) scales with y^+, the Reynolds stress -uv does not scale with inner variables but more precisely scales with the geometric mean of the inner and outer scales, i.e. mixed scales.

As described above, there are several burst detection techniques which rely on single point velocity measurements. The results from these techniques are more quantitative than those from visual studies; however, it has been shown that the probe results can be far from accurate, depending on the particular technique. Bogard and Tiederman (1986) performed an evaluation of the effectiveness of these various techniques by comparison with the results of a visual study. They found varying degrees of effectiveness of the various techniques, and also found them to be highly dependent on the user specified parameters. The results of their study indicate that the quadrant technique was the most reliable in the sense that it had a high probability of detecting ejections and a low probability of detecting false ejections. For single sensor probes they conclude that the u-level technique was the most accurate.

Using the quadrant technique on the numerical simulation data of a flat-plate boundary layer, Kim and Spalart (1987) determine that the frequency scales with inner variables. Luchik and Tiederman (1987) also using the quadrant technique in a turbulent channel flow conclude that inner variables provide the best collapse of the bursting frequency for the range of Reynolds numbers based on channel height from 8,700 to 17,800.

The use of flat-plate manipulator blades has shown that there is a very definite link between the inner and outer regions of a turbulent boundary layer; although the exact nature of the interaction is difficult to answer at the present time. As has been noted by many investigators, alteration of the large outer structure by manipulator blades has resulted in significant drag reduction downstream of the blades. Corke et al. (1982) achieved substantial drag reduction and later Plesniak et al. (1987) confirmed this result and documented conditions leading to optimization of boundary layer manipulation. It is argued by Corke et al. (1982) that the suppression of the large outer scales by the manipulator blades had an effect on the activity in the wall region. In addition the gradual return of the large scales after manipulation (Plesniak et al., 1985) also suggests a link between the outer scales and turbulence production at the wall.

More recently Coughran and Bogard (1986) used the u-level detection technique to compare frequency of bursting in regular and manipulated turbulent boundary layers. They did find a clear effect on the ejection frequency and a lesser effect on the burst frequency. The distinction between burst and ejection frequency, as illustrated by Bogard and Tiederman (1986), is that a burst may have several ejections associated with it and care should be taken when counting bursts by dictating an appropriate minimum time delay between ejections.

Chang (1987) using temperature as a passive contaminant in regular and manipulated boundary layers showed that the entrainment process was the key parameter influencing the wall activities. Effects of the manipulator blades include decreased mixing, the reduction of the temperature and velocity fluctuations, as well as a reduction of the Taylor microscale and integral length scale. A reduced spatial extent of the iso-correlation contours was documented for the manipulated boundary layer as compared with the regular boundary layer.

EXPERIMENTAL FACILITIES AND PROCEDURE

All results were obtained in the high-speed test section of the Mark V. Morkovin closed-circuit wind tunnel at I.I.T. A flexibile ceiling in the test section allows for a zero streamwise pressure gradient. The manipulator blades used in the present study were cut from 0.127 mm thick blue tempered spring-steel shim stock. The manipulators were placed at a height (h) corresponding approximately to $h/\delta = 0.6$.

To address the question of appropriate scaling, measurements were performed over a large Reynolds number range at several vertical positions in the boundary layer. A single-sensor hot-wire was used to acquire data at heights corresponding to $y^+ = 15$, 25, 40, 55 and 110. In addition an X-wire probe was used for $y^+ = 55$, 110, 165 and 220. Three free-stream velocities with four streamwise locations gave the Reynolds number range of $3000 < Re_\theta < 10000$. A total of 409,600 points per wire were acquired at 12,000 Hz giving a record length of 34 seconds. The above measurements were performed in both regular and flat-plate manipulated boundary layers. The local skin-friction reduction was observed to be approximately 30% while the net drag reduction was approximately 10%.

DETECTION AND SCALING OF EVENTS

It is evident when comparing the implementation of the various detection schemes, that the outcome of the results will be highly dependent on the user-specified threshold parameters. However, if the various schemes are indeed detecting the bursting process accurately, then features such as the qualitative trend with Reynolds numbers should be similar between the detection schemes.

The issue regarding appropriate scaling is investigated by normalizing the frequency of bursting by inner, mixed and outer variables, over a wide range of Reynolds numbers. Many investigators examine Reynolds number dependence by keeping the streamwise measurement location fixed and varying the free-stream velocity. By varying both the free-stream velocity and streamwise measurement location, a more stringent evaluation of Reynolds number dependence is gained.

Results for all detection techniques were normalized by inner, mixed and outer variables. The length scales are ν/u_τ, $(\delta\nu/u_\tau)^{.5}$, and δ for the inner, mixed and outer scaling laws respectively. The associated velocity scales are u_τ, $(u_\tau U_\infty)^{.5}$, and U_∞ respectively. Mixed scaling, as introduced by Alfredsson and Johannsson (1984) is simply the geometric mean of inner and outer scaling.

The following results are presented to investigate the dependence of burst frequency on Reynolds number. Therefore, they provide insight into the applicability of various detection techniques and scaling laws, as well as compare the bursting frequency of a regular turbulent boundary layer with those manipulated by flat plate manipulator blades. When using the quadrant detection technique, a turbulence-producing event is detected when the product uv falls below a certain threshold i.e., $uv < k_Q u(rms)v(rms)$, where u and v are the fluctuating streamwise and normal velocity components, and k_Q is the threshold parameter. To further discriminate between sweeps and ejections, the sign of the individual signals are used; that is, Q2$(u < 0,\ v > 0)$ and Q4$(u > 0,\ v < 0)$. The VITA technique counts an event if $var(u) > k_v u(rms)$, where var(u) is the short term variance calculated over a specified hold time and again k_v is a threshold parameter. The u-level technique detects an event when $u < Lu(rms)$, where L is the threshold parameter.

Figure 1 represents the location of the various burst detection techniques on the joint probability density distribution of u and v. The probability distributions were determined using an X-wire probe located at $y^+ = 35$ in a regular boundary layer. The quadrant technique, shown in the bottom of Figure 1 results in detecting the events associated with strong turbulence production. For $k_Q = 3.0$, there is approximately twice as many Q2 events as Q4 events at $y^+ = 35$. The u-level technique is depicted in the middle of Figure 1. There is a significant contribution of non-turbulence producing events from quadrant three. The points for the VITA technique were chosen based on the minimum and maximum streamwise velocity just before and after the point of detection. For the most part, VITA picks up Q2 events followed by Q4 events; however, with no apparent threshold on the quantity uv.

The correction on the measured frequency to account for spatial-averaging effects of the sensor, as introduced by Blackwelder and Haritonidis (1983) is given by $f_c = f_m[1.0 + (l^+/l_o^+)^2]^{.5}$, where f_c is

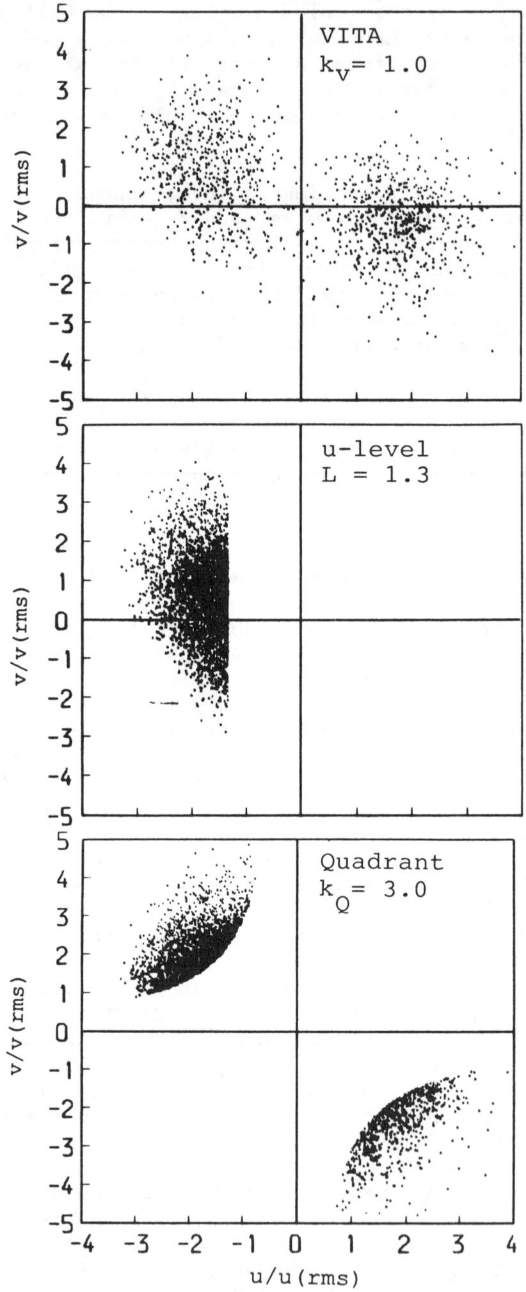

Figure 1. Joint Probability Density Distribution of u and v Associated with VITA, u-level, and Quadrant detection techniques

the corrected frequency, l^+ is the non-dimensional sensor length ($l^+ = lu_\tau/\nu$), and $l_o^+ = 20$ was determined to produce optimal results. Guezennec (1985) also investigated the sensor length averaging effect in a similar way. However, unlike Blackwelder and Haritonidis (1982) who present results of frequency vs l^+ over a range of Reynolds numbers, for only one value of the averaging time ($T_h u_\tau/\nu = 10$), Guezennec studied this effect for several values of averaging times. Using the frequency correction scheme presented above, he was not able to obtain independence of corrected non-dimensional frequency f_c^* with l^+ over a wide range of averaging times. With an analogous correction to the averaging times, he achieved excellent collapse of the data. In addition, digitization rate and record length were found to be important; namely, the sampling frequency should be sufficiently high and the record length sufficiently long to ensure proper counting of VITA detected events ($T_S^+ < 4$ and $T_R^+ > 8000$).

Figure 2 represents the effects of the three scaling laws for a regular turbulent boundary layer with the sensor length correction on the frequency. The data were taken with a single sensor probe at a distance from the wall corresponding to $y^+ = 15$. The corrected non-dimensional frequency is shown as a function of averaging time for the 15 different Reynolds numbers. To give an indication of the trend with Reynolds number, the lowest and highest Reynolds numbers are marked at the two extremes of the averaging time. Also presented on each plot is the average non-dimensional frequency over all hold times and Reynolds numbers ($\bar{f}^*$). For all three scaling laws, there is a range of averaging times for which the non-dimensional frequency is fairly constant.

Outer scaling (top of Figure 2) results in poor collapse over the entire range of averaging times. When comparing mixed scaling with inner scaling for the entire range of hold times, it can be argued that mixed scaling results in the best collapse. However, when applying the analogous correction to the hold time (Guezennec 1985), an excellent collapse over all hold times is observed for inner scaling (see Guezennec 1985 and Wark 1988). The VITA detection results are commonly investigated for a representative non-dimensional hold time (using inner scaling) of approximately 12. At this hold time, there is a better collapse of the frequency for inner scaling as compared with mixed scaling.

Bogard and Tiederman (1986) showed that several ejections could be grouped into a single burst; therefore, proper counting of bursts is done by analysing the probability distribution of time between ejections (T_E). The same method as used by Bogard and Tiederman (1986) and Luchik and Tiederman (1987) was used to determine the appropriate time delay to leave between ejections before counting another burst. Figure 3 is the result for the regular boundary layer at $y^+ = 15$ when counting bursts using a time delay of of $\tau = .3\bar{T}_E$, where $\bar{T}_E$ is the average time between ejections. The results exhibit the same qualitative trend with Reynolds number for the three scaling laws, as the frequency of ejection results (see Wark 1988). The best linear fit to the data was calculated and is shown on the plots, along with plus and minus 10% to this fit. Mixed scaling provides the best collapse of the frequency with Reynolds number as indicated by the slope of the best fit line. However, this is due to lack of sensor-length correction. When properly correcting for the sensor-length effect, inner scaling is observed to collapse the burst frequency slightly better than mixed scaling (Figure 4).

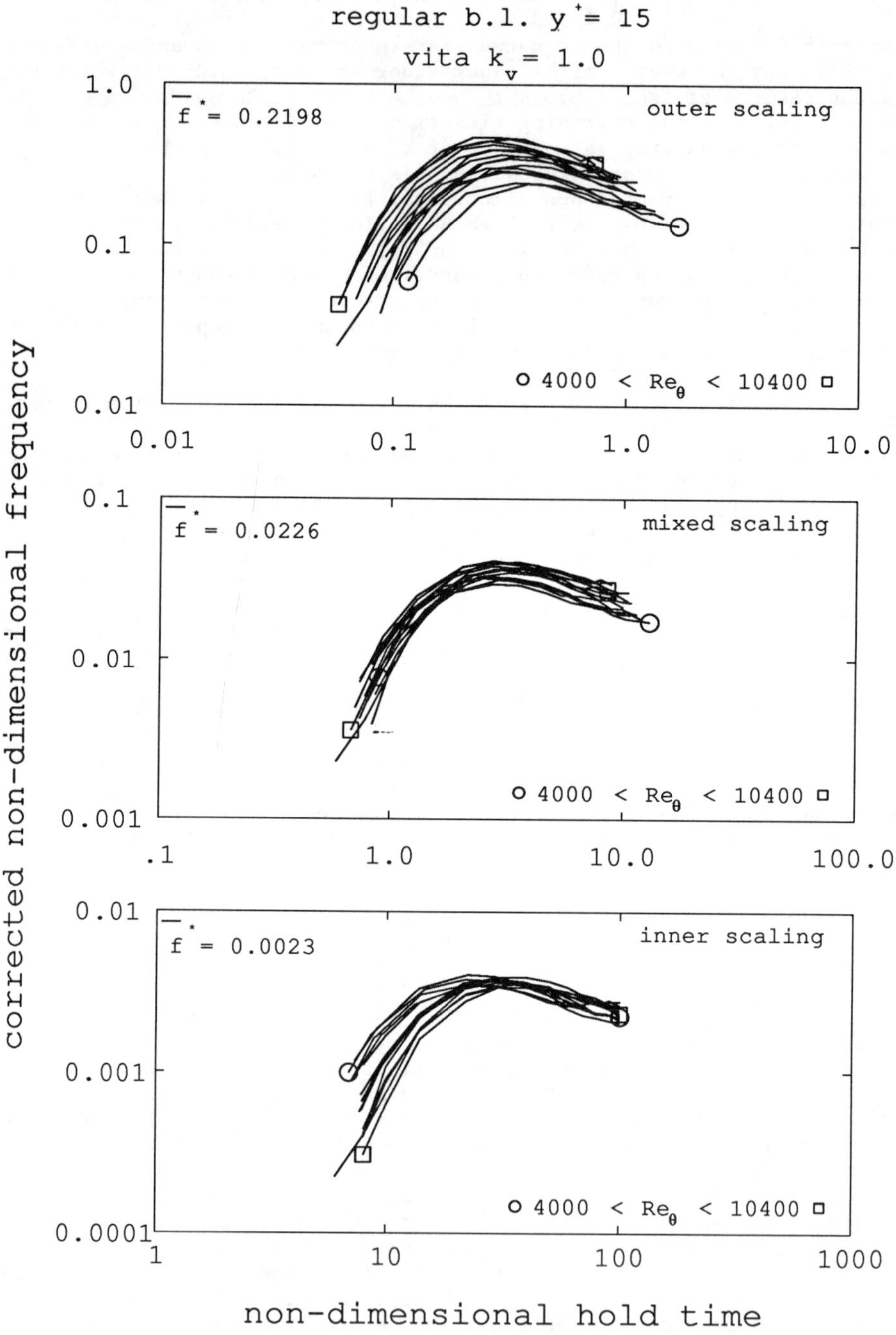

Figure 2. Corrected Frequency of Occurrence of VITA Events for all Three Scaling Laws as a Function of Hold Time for Regular Boundary Layers at $y^+ = 15$ ($4000 < Re_\theta < 10400$)

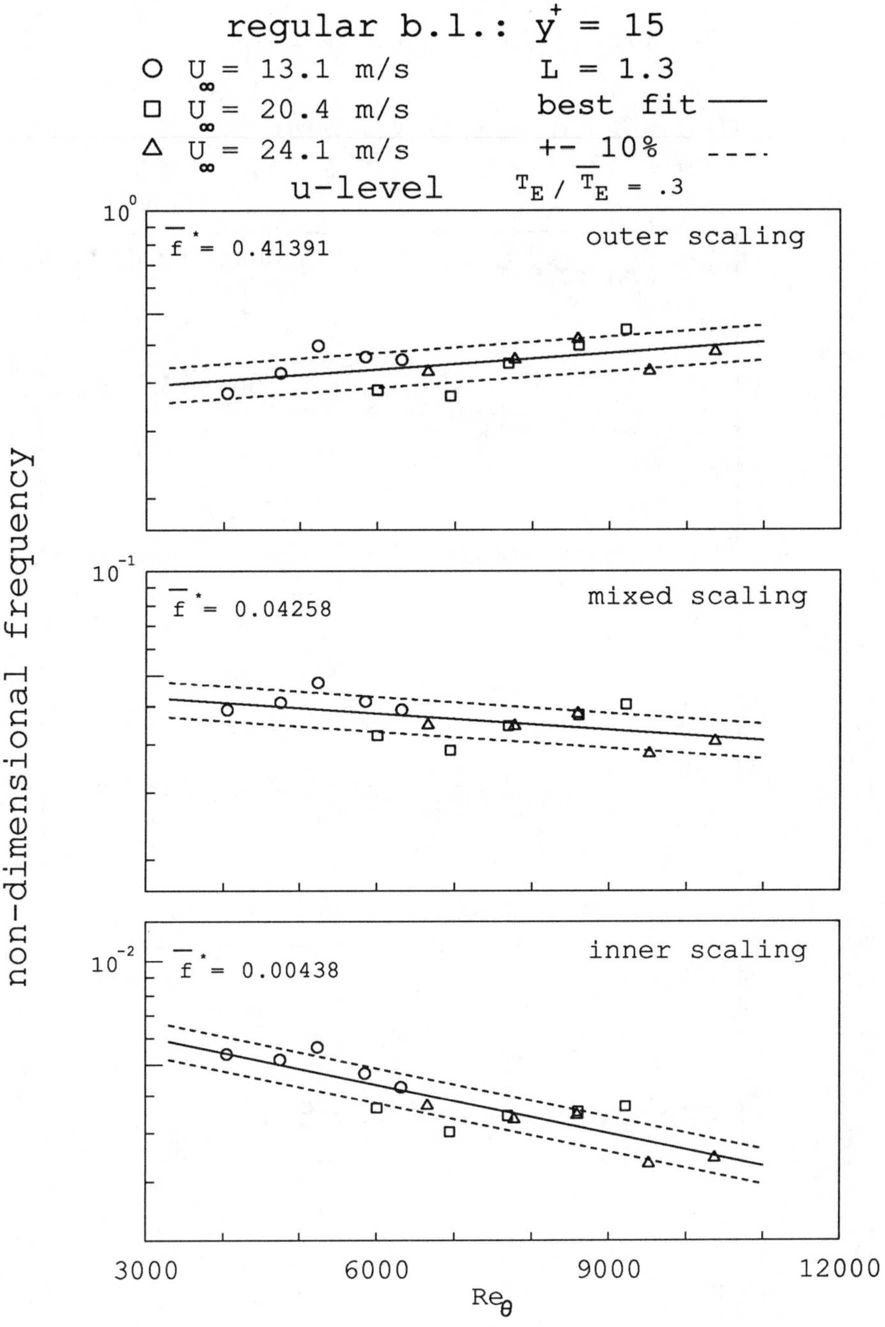

Figure 3. Non-corrected Frequency of Occurrence of u-level Bursts for all Three Scaling Laws as a Function of Reynolds Number for Regular Boundary Layers at $y^+ = 15$

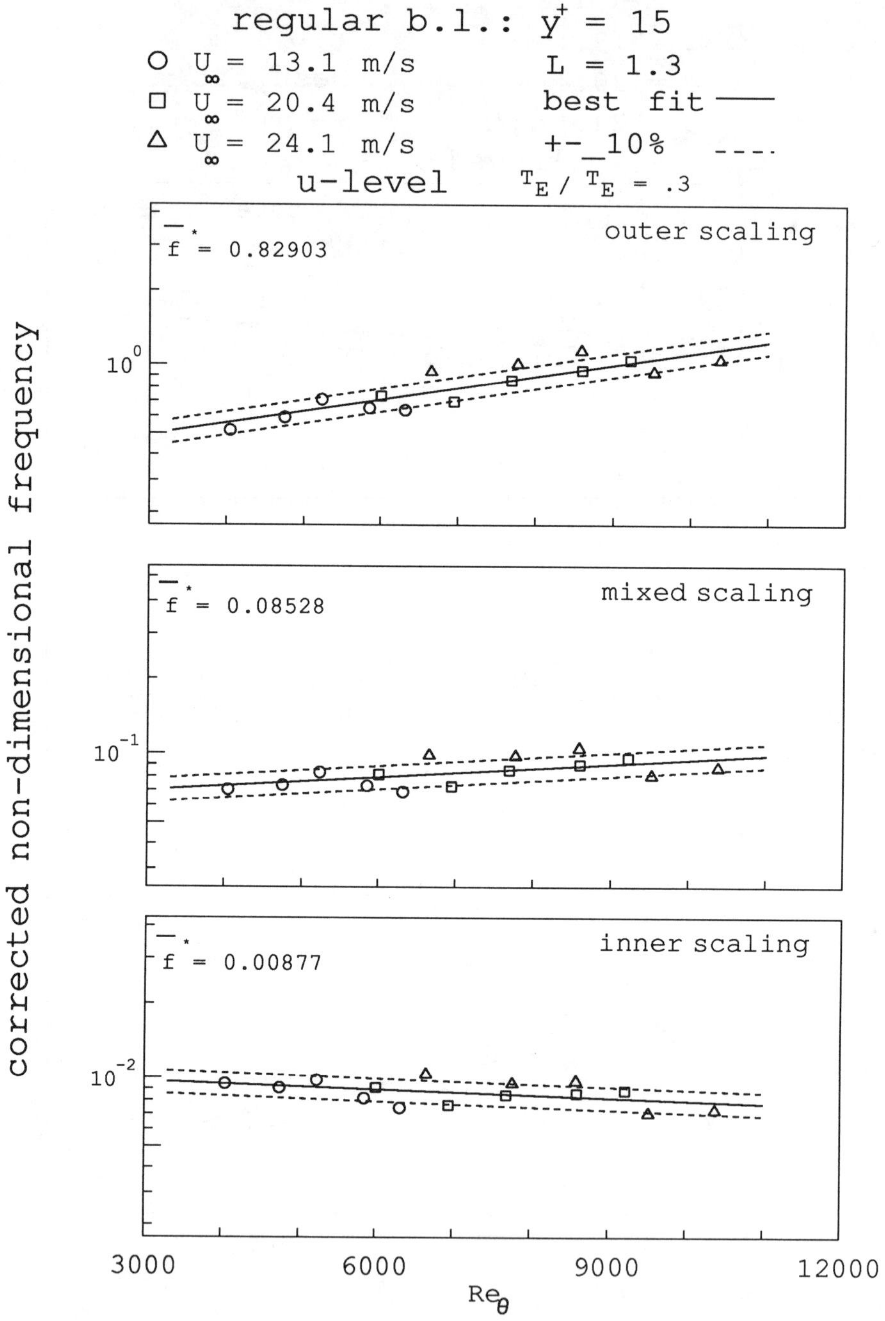

Figure 4. Corrected Frequency of Occurrence of u-level Bursts for all Three Scaling Laws as a Function of Reynolds Number for Regular Boundary Layers at $y^+ = 15$

The sensor-length correction for an X-wire probe is not as straight forward as for the single-sensor probe. The present authors are not aware of any systematic investigation to determine the spatial-averaging correction for X-wire probes. Therefore, the correction to the frequency results shown using an X-wire probe was determined in the following manner. Data were acquired at $y^+ = 55$ and 110 using both a single-sensor and an X-wire probe. The frequency of VITA and u-level events was determined for the single-sensor data using the correction scheme described earlier. The correction to the frequency using the X-wire data was determined such that the corrected frequency for the X-wire data was equal to the corrected frequency for the single-sensor data. This resulted in an l^+ value for the X-wire results which was 18% larger than the corresponding value for the single probe with an equivalent sensing length.

Inner scaling is also shown to be the appropriate choice when using VITA and quadrant techniques (Figure 5). The top plot is the frequency of ejections using the VITA technique. A slight shift towards lower frequency is observed with increasing Reynolds number. The corrected burst and "inrush" frequency results obtained using the quadrant technique are shown in the bottom two plots of Figure 5. Analogously to bursts being composed of several ejections, inrushes are groups of sweeps. These results are for a regular boundary layer at $y^+ = 55$. Again, when proper correction for sensing length is performed, inner scaling results in the best collapse of the frequency with Reynolds number. A more thorough evaluation of the height dependence is presented in the following section, along with the results for a manipulated boundary layer.

Height Dependence of Frequency and Scaling

The variation of the corrected non-dimensional frequency with height in the boundary layer using the u-level technique is shown in Figure 6 for inner and mixed scaling. The 12 curves represent the 12 Reynolds number cases. The frequency at $y^+ = 55$ and 110 is the result of averaging the values for the single sensor and X-wire probe. It is clear that inner scaling results in a better collapse of the curves as compared to mixed scaling. Also it is observed that the frequency is approximately constant for $y^+ > 55$. The trend below $y^+ = 55$ does not seem to be due to sensor-length effects since it is independent of Reynolds number over the wide range of conditions. The results for the manipulated boundary layer are shown in the bottom of Figure 6. More scatter of the data is observed as compared to the results for the regular boundary layer. The magnitudes of the frequencies are approximately the same, and again inner scaling was found to collapse the results better as compared with mixed or outer scaling (not shown).

The average ejection and bursting frequency over all 12 Reynolds numbers is shown as a function of y^+ in Figure 7 using the u-level, VITA and quadrant techniques. Also shown is the sweep and inrush frequency (Q4) as a function of height. The quadrant technique results are consistent with the u-level results; namely, the ejection frequency is constant for $y^+ > 55$. The VITA technique indicates a decrease in ejection frequency with increasing y^+. The ratio of ejection and bursting frequency for the VITA and quadrant methods to that detected by the u-level is given by the results of Figure 7. Approximately 40% of the events are detected by the Q2 scheme and 20% by the VITA technique. Also the strong sweeps

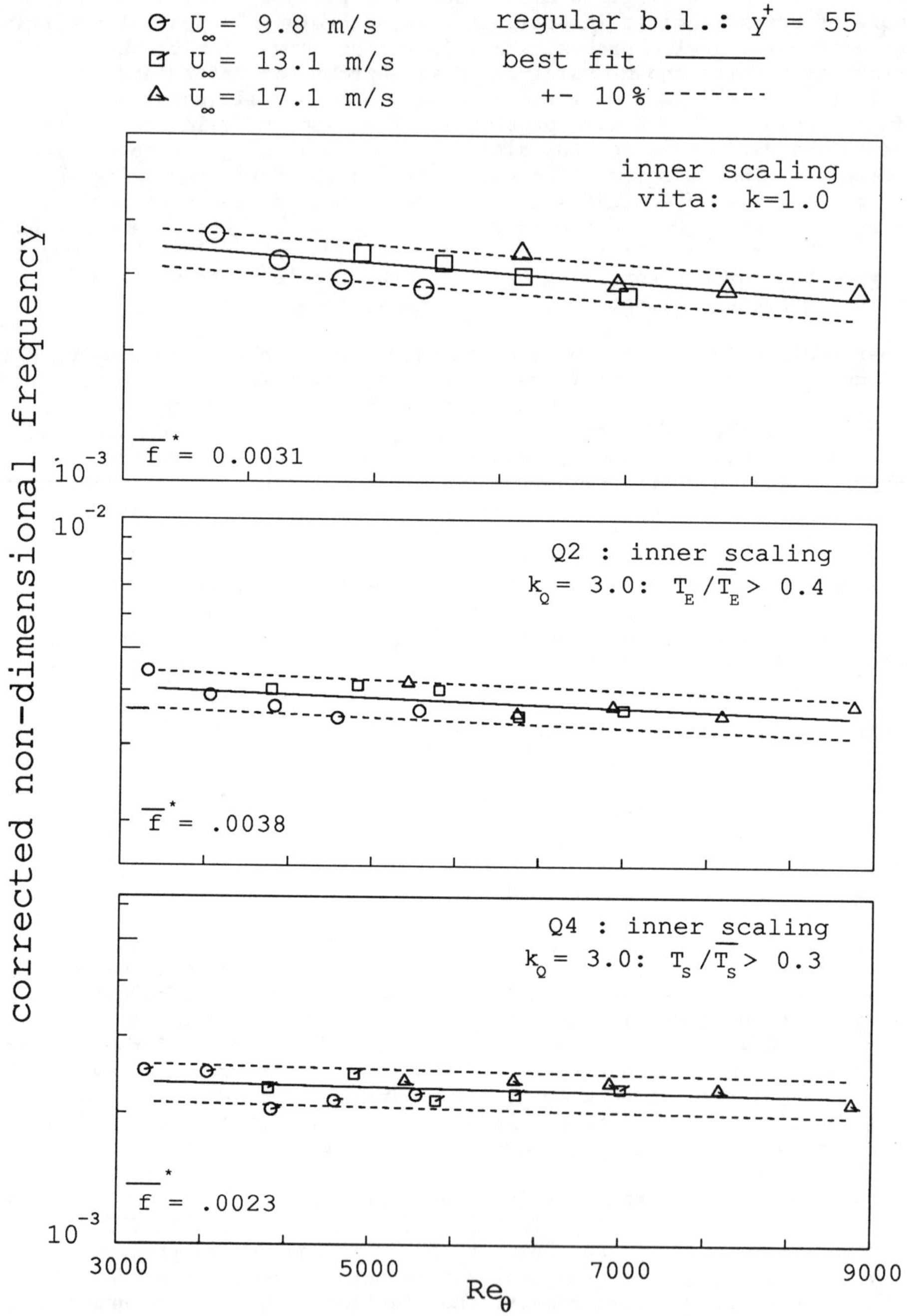

Figure 5. Corrected Frequency of Occurrence of Vita ejections and Quadrant Bursts and Inrushes for Regular Boundary Layers at $y^+ = 55$ using inner scaling

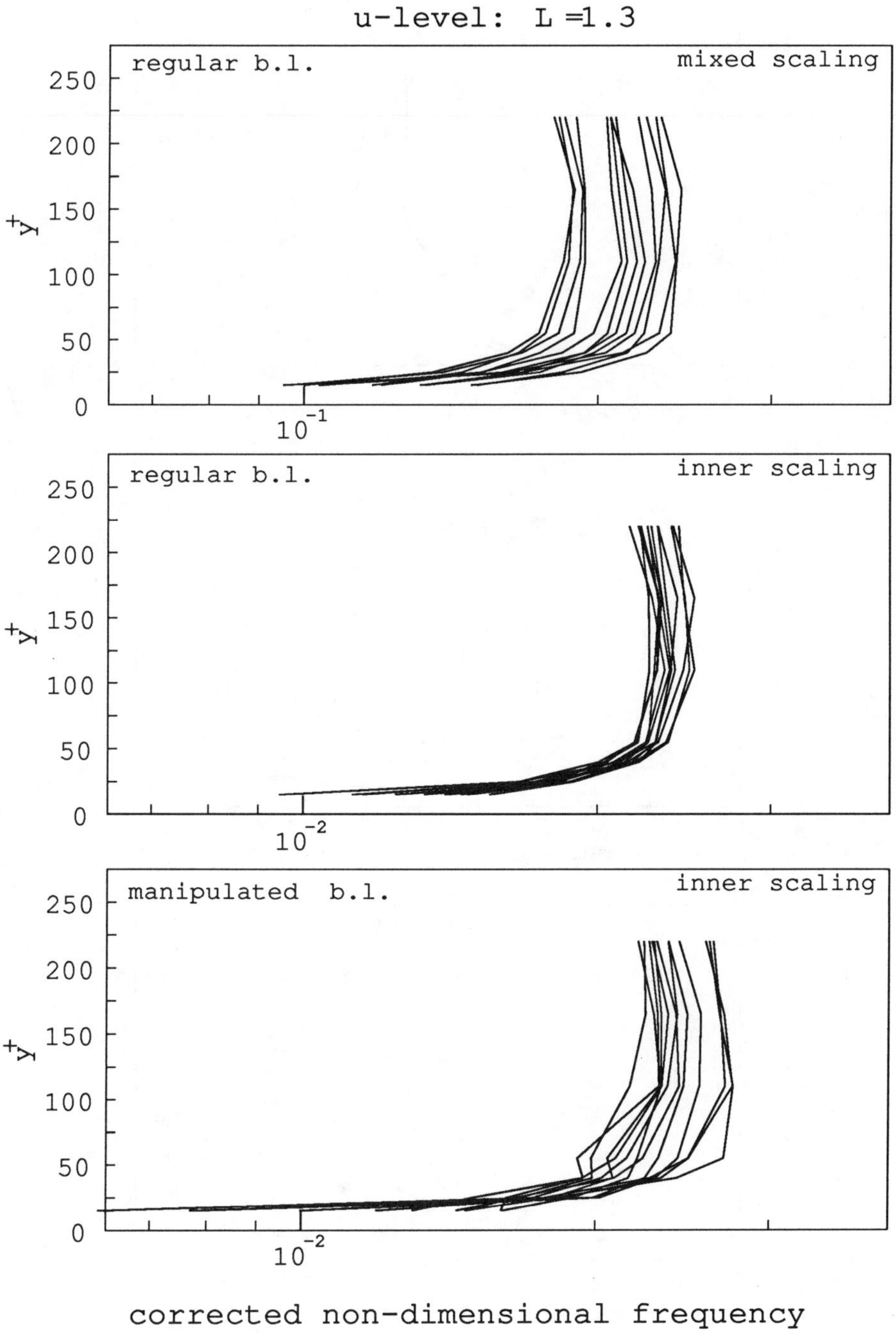

Figure 6. Variation of Corrected u-level Ejection Frequency With Height in Regular Boundary Layers Using Inner and Mixed Scaling for $3000 < Re_\theta < 10000$.

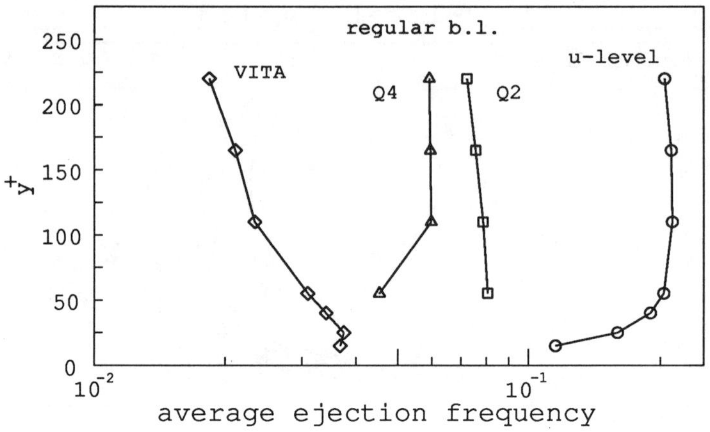

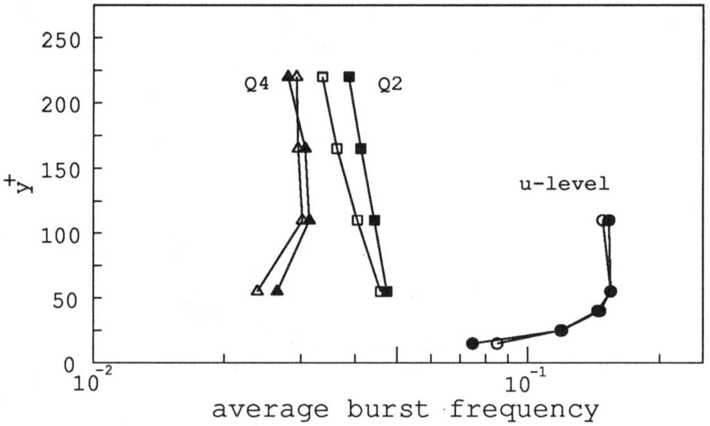

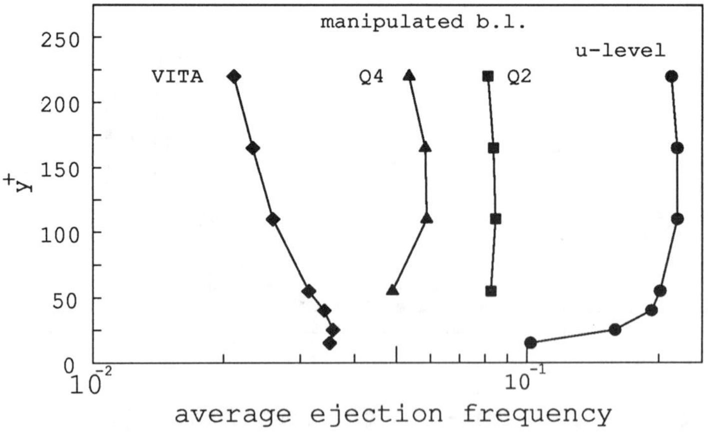

Figure 7. Ejection, Bursting, Sweep and Inrush Frequency Obtained From u-level, VITA and Quadrant Detection, as a Function of Height, Averaged Over 3000 < Re_θ < 10000, Using Mixed Scaling

(Q4) are 30% of the visually observed bursting events. The number of ejections per burst is also available from the results, and is observed to be approximately 2 for both the u-level and quadrant techniques. Another indication of these ratios is given by the results of Figure 1.

The manipulated boundary layer results (closed symbols) do not show any significant deviation from the regular boundary layer results for the burst, inrush, sweep or ejection frequencies. This is consistent with the results of Wark (1988) for $y^+ = 15$, where the results for a manipulated boundary layer provided a 3% increase in ejection frequency, accompanied with a 7% decrease in burst frequency. Also the results of Coughran and Bogard (1986) show a moderate effect on the bursting process in a manipulated boundary layer: 7% decrease in ejection frequency and a 4% decrease in burst frequency.

For all techniques, the best fit of log frequency versus Reynolds number was obtained in a similar manner as shown by Figures 3, 4 and 5. The slope of this line is an indication of the dependence on Reynolds number. Figure 8 (middle) clearly demonstrates that inner scaling results in the smallest dependence on Reynolds number for all y^+ locations investigated when using the u-level detection technique. This is also reflected in the results for the manipulated boundary layer (bottom figure). The importance of frequency correction is again depicted by the comparison between the slope using the corrected frequency (middle) with the non-corrected frequency (top) results of Figure 8. If the sensor length is not properly accounted for in the determination of the frequency, the results would incorrectly suggest that the frequency scales with mixed variables.

The dependence of this slope on height in the boundary layer is shown in Figure 9 using the VITA and quadrant techniques. For both the Q2 and VITA techniques the frequency scales better with mixed variables for $y^+ > 125$. This is somewhat surprising based on the fact that frequency is essentially constant for $y^+ > 55$. This is attributed to the high sensitivity of the frequency on the correction scheme, particularly near small values of the slope (Q2). The VITA technique has the added difficulty that it fails to detect a constant bursting frequency for $y^+ > 55$ as shown in Figure 7.

SUMMARY

The ejection and bursting frequency show a strong y^+ dependence for $y^+ < 55$ in both regular and flat-plate manipulated boundary layers. The effect of manipulators on the bursting process was observed to be small. More importantly, the conclusions regarding the effect of the manipulator blades on the bursting frequency are very dependent on the conditions for a given investigation. It has been shown that the local skin friction and net drag reduction depend strongly upon several geometrical parameters (Plesniak et al. 1985). Therefore, when studying the bursting process in this modified flow condition, one should expect the alteration of the bursting process to also be sensitive to the exact setup of the blades. For the present investigation, care was taken to ensure that net drag reduction had been achieved by the measurement location. Furthermore, the analysis of the data was performed via several detection methods which yielded equivalent results.

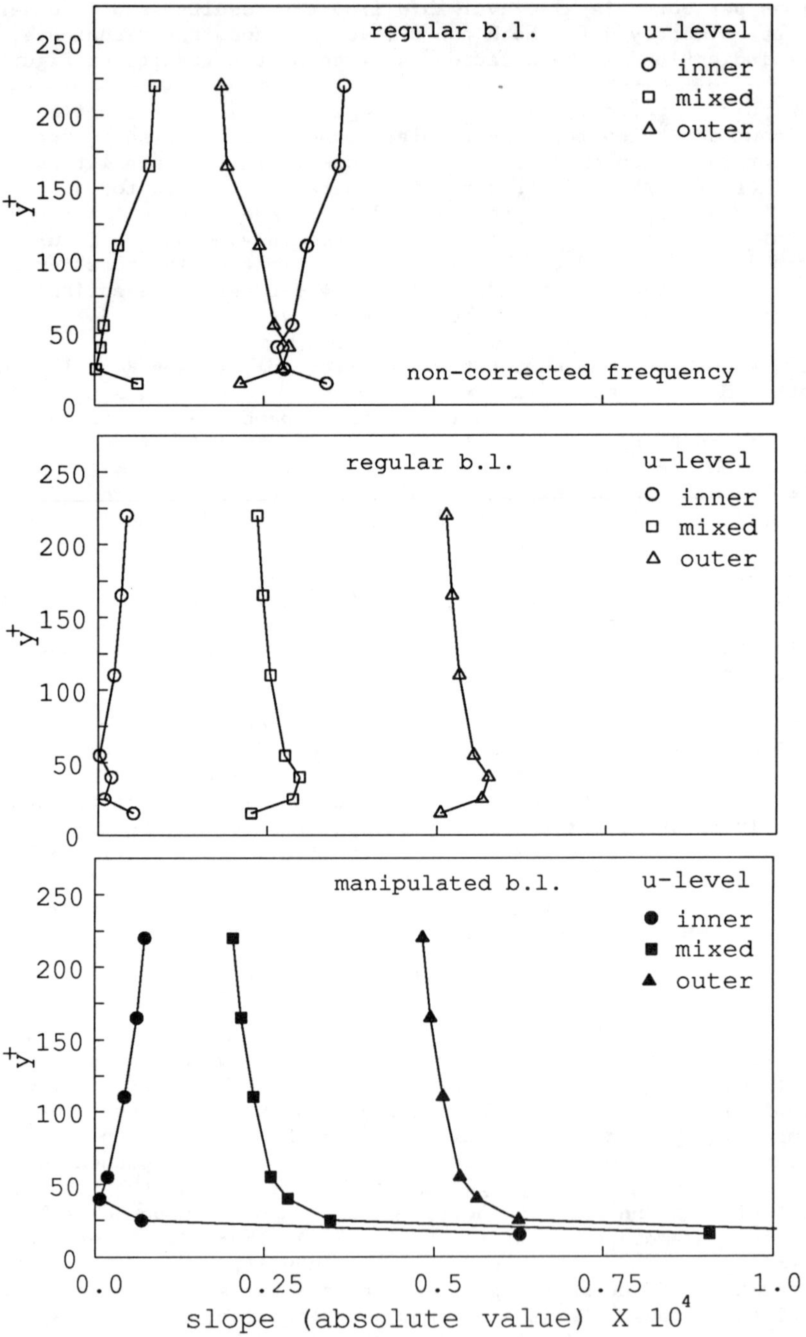

Figure 8. Variation of a Measure of Reynolds Number Independence With Height, for Corrected and Non-corrected u-level Ejection Frequencies, Using All Three Scaling Laws

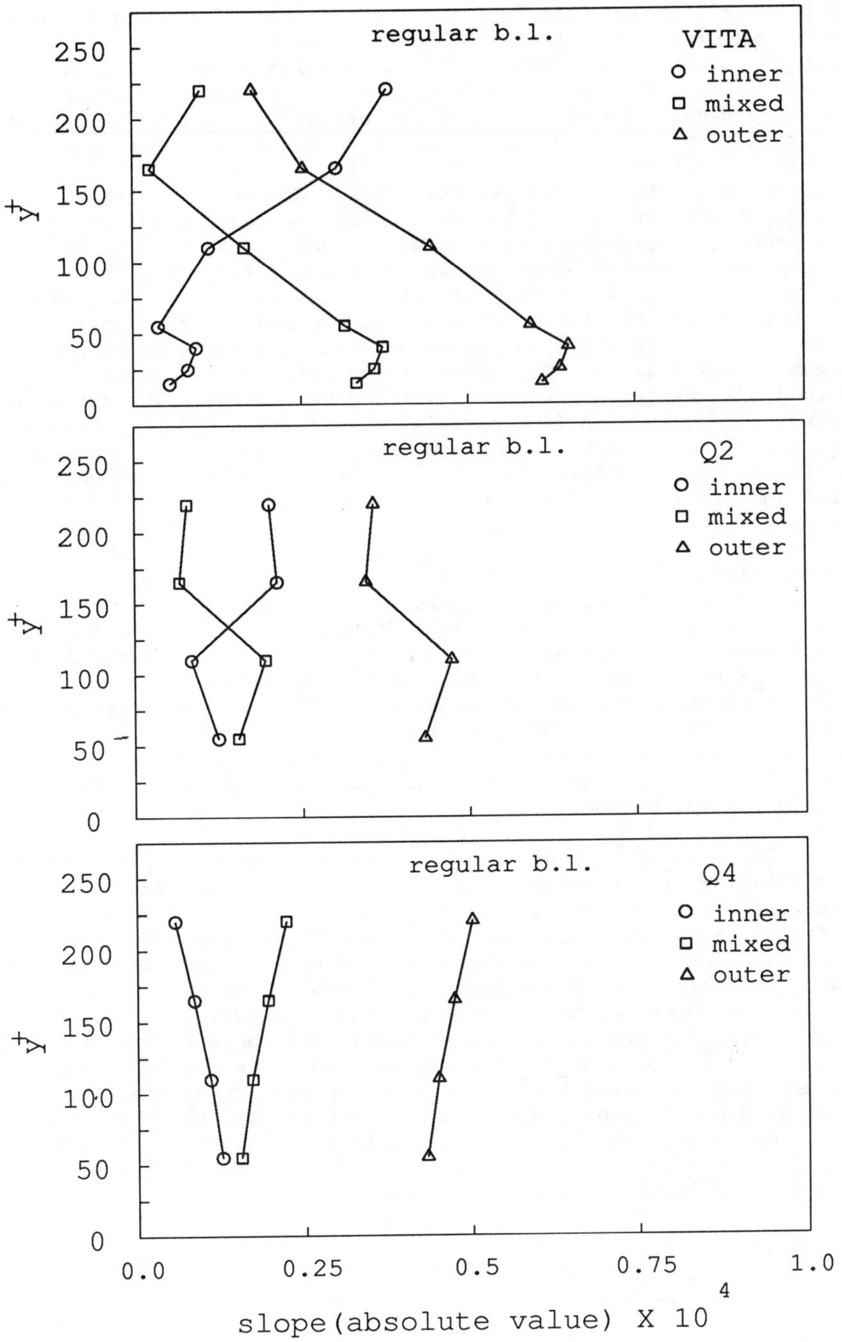

Figure 9. Variation of a Measure of Reynolds Number Independence With Height for Corrected VITA and Quadrant Frequencies, Using All Three Scaling Laws

A possible explanation for this small effect on the bursting frequency is given by Wark (1988). Briefly summarized, a hierarchy of scales was shown to be associated with the turbulence-production process. It is speculated that the manipulators placed at relatively large heights in the boundary layer act only on the largest of the events associated with the bursting process. By placement of the blades at lower y^+ values, an increased reduction in the spatial extent and frequency of events was found. This is depicted by the results of Figure 10 for a position corresponding to 3.8 boundary layer thicknesses downstream of the leading edge of the manipulator blades. Using the shear-stress detection technique introduced by Guezennec (1985), the ensemble-averaged coherent structure was documented for a regular boundary layer at $Re_\theta = 4600$. The velocity vector maps in the x-y plane result when averaging the fluctuating streamwise and normal velocity components at all measurement points associated with the detection of a large negative shear stress, T-, event at the wall (the T- event is related to a Q2 event, see Guezennec 1985 and Wark 1988). The effect of lowering two blades, placed in a parallel configuation, to $y^+ = 700$ and 440 (Figure 10b) is shown by comparison with the regular boundary layer results (Figure 10a). Further reduction is observed with blade locations of $y^+ = 560$ and 275 (Figure 10c). The most striking reduction was achieved when the blades were lowered to $y^+ = 440$ and 150 (Figure 10d).

Measurements similar to those of Figure 10 were made at several positions downstream of the manipulators. They depict a redevelopment of the structures within a relatively small distance ($< 2\delta$) downstream of the manipulators, further emphasizing the importance of the inner layer in the development of the structures. The redevelopment of the structures is given by the percentage of detected events compared with a regular boundary layer as a function of ξ' ($\xi' = 3.8$, 50%; $\xi' = 5.4$, 53%; $\xi' = 7.1$, 77%). Additional evidence that a hierarchy of sizes are associated with the turbulence-production process is given by Wark (1988) for a regular boundary layer.

In summary, it is clear that bursting frequency is very sensitive to the spatial resolution of the measuring probe; therefore, care must be exercised to minimize the sensing length and/or volume as well as to correct for their effect. Based upon results from several detection techniques, we conclude that the bursting frequency scales best with inner layer variables over a wide range of Reynolds numbers. This was found to be independent of height in the boundary layer; that is, the turbulence production process in the boundary layer depends strongly on the inner layer. The present results suggest that if a single-sensor probe is used for the detection of turbulence-producing events, the u-level technique may be more accurate as compared with the VITA technique, and caution should be exercised in interpretation if VITA is used. Also, it may be useful to investigate sweep events using a high, positive, u-level detection technique.

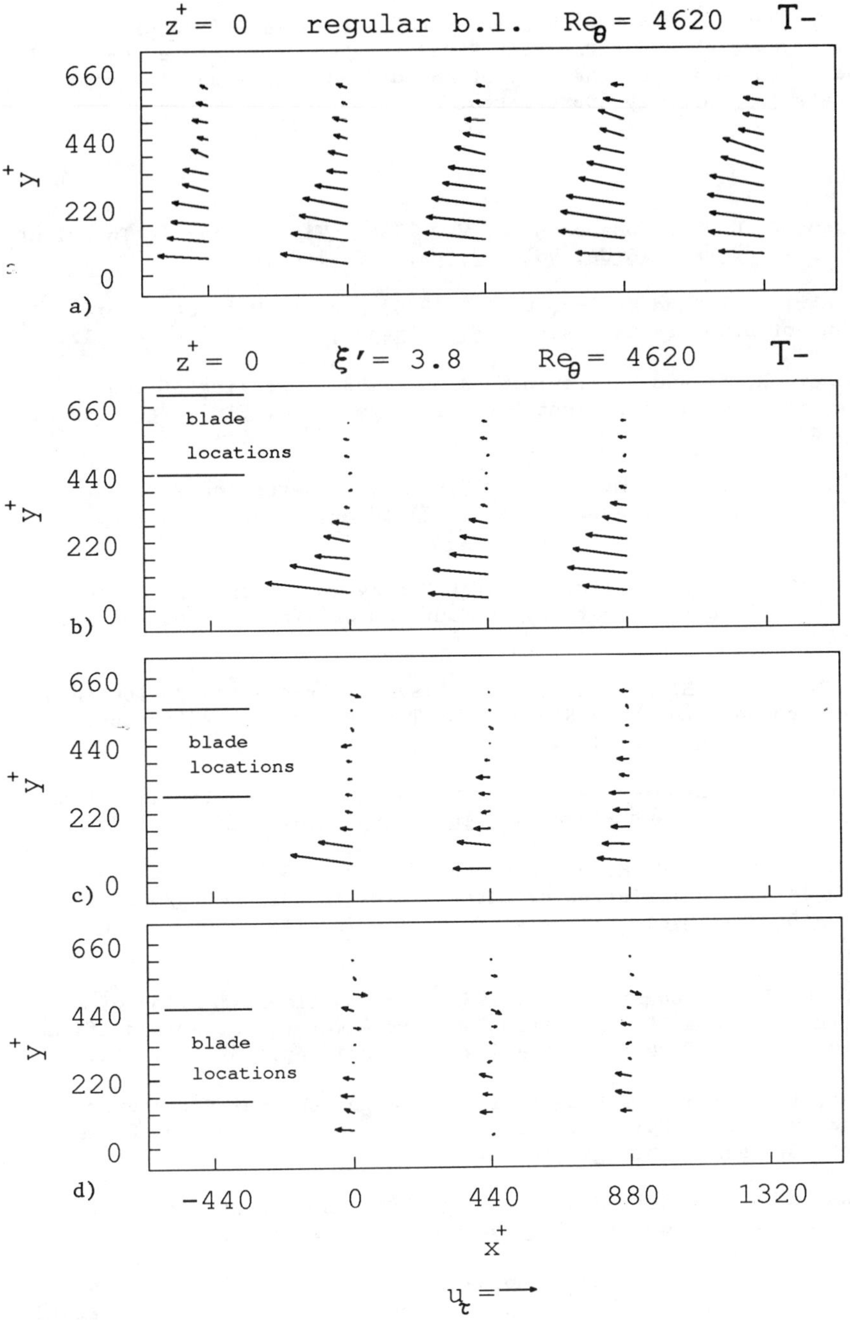

Figure 10. Velocity Vector Maps in the x-y Plane at $\xi' = 3.8$ for a Manipulated Boundary Layer with Blades Located as Indicated for the T-Event at $z^+ = 0$, Compared With a Regular Boundary Layer

ACKNOWLEDGMENT

The authors acknowledge the support by the Air Force Office of Scientific Research under Contract F49620-86-C-0133 monitored by Dr. J. McMichael. In addition, the help of Ahmed Naguib in acquiring some of the data is sincerely appreciated.

REFERENCES

Alfredsson, P. H. and Johansson, A. V. 1984. Time Scales in Turbulent Channel Flow. Phys. Fluids, Vol. 27, pp. 1974-1981.

Blackwelder, R. F. and Kaplan, R. E. 1976. On the Wall Structure of the Turbulent Boundary Layers. J. Fluid Mech., Vol. 76, pp. 89-112.

Blackwelder, R. F. and Haritonidis, J. H. 1983. Scaling of the Bursting Frequency in Turbulent Boundary Layers. J. Fluid Mech., Vol. 132, pp. 87-103.

Bogard, D. G. and Tiederman, W. G. 1986. Burst Detection with Single-Point Velocity Measurements. J. Fluid Mech., Vol. 162, pp. 389-413.

Chang, S. 1987. Modification of Large Eddies in a Turbulent Boundary Layer. Ph.D. Thesis, University of Southern California, Los Angeles, California.

Chen, C. H. P. and Blackwelder, R. F. 1978. Large Scale Motion in a Turbulent Boundary Layer: A Study Using Temperature Contamination. J. Fluid Mech., Vol. 89, pp. 1-31.

Corino, E. R. and Brodkey, R. S. 1969. A Visual Investigation of the Wall Region in Turbulent Flows. J. Fluid Mech., Vol. 37, pp. 1-30.

Corke, T. C., Nagib, H. M. and Guezennec, Y. G. 1982. A New View on Origin, Role and Manipulation of Large Scales in Turbulent Boundary Layers. NASA CR-165861. See also Ph.D. Thesis, Ill. Inst. of Tech., Chicago, 1981.

Coughran, M. T. and Bogard, D. G. 1986. An Experimental Study of the Burst Structure in a LEBU-Modified Boundary Layer. Proceedings of the Tenth Symposium on Turbulence. Rolla, Missouri, pp. 45-1 - 45-10.

Guezennec, Y. G. 1985. Documentation of Large Coherent Structures Associated with Wall Events in Turbulent Boundary Layers. Ph.D. Thesis, Ill. Inst. of Tech., Chicago, IL.

Johansson, A. V. and Alfredsson, P. H. 1982. On the Structure of Turbulent Channel Flow. J. Fluid Mech., Vol. 122, pp. 295-314.

Kim, H. T., Kline, S. J. and Reynolds, W. C. 1971. The Production of Turbulence near a Smooth Wall in a Turbulent Boundary Layer. J. Fluid Mech., Vol. 50, pp. 133-160.

Kim, J., and Spalart, P. R. 1987. Scaling of the Bursting Frequency in Turbulent Boundary Layers at Low Reynolds Numbers. Phys. of Fluids, Vol. 30, pp. 3326-3328.

Kline, S. J., Reynolds, W. C., Schraub, F. A. and Runstadler, P. W. 1967. The Structure of Turbulent Boundary Layers. J. Fluid Mech., Vol. 30, pp. 741-773.

Lu, S. S. and Willmarth, W. W. 1973. Measurements of the Structure of the Reynolds Stress in a Turbulent Boundary Layer. J. Fluid Mech., Vol. 60, pp. 481-511.

Luchik, T. S. and Tiederman, W. G. 1987. Timescale and Structure of Ejections and Bursts in Turbulent Channel Flows. J. Fluid Mech., Vol. 174, pp. 529-552.

Nagib, H. M., Wark, C. E. and Guezennec, Y. G. 1987. Documentation of Turbulence Producing Structures in Regular and Manipulated Turbulent Boundary Layers. Proceedings of the IUTAM Sumposium on Turbulence Management and Relaminarisation. Bangalore, India. Springer-Verlag.

Plesniak, M. W., Nagib, H. M. and Guezennec, Y. G. 1987. Optimized Manipulation of Turbulent Boundary Layers Aimed at Net Drag Reduction. NASA.

Rao, K. N., Narasimha, R., and Narayanan, M. A., 1971. The 'Bursting' Phenomenon in a Turbulent Boundary Layer. J. Fluid Mech., Vol. 48, pp.339-352.

Smith, C. R. and Metzler, S. P. 1983. The Characteris- tics of Low-Speed Streaks in the Near Wall Region of a Turbulent Boundary Layer. J. Fluid Mech., Vol. 129, pp. 27-46.

Sreenivasan, K. R. 1987. A Unified View of the Origin and Morphology of the Turbulent Boundary Layer Structure. Proceedings of the IUTAM Symposium on Turbulence Management and Relaminarisation. Bangalore, India. Springer-Verlag.

Wark, C. E. 1988. Experimental Investigation of Coherent Structures in Turbulent Boundary Layers. Ph.D. Thesis, Illinois Institute of Technology, Chicago Illinois.

Willmarth, W. W. and Lu, S. S. 1972. Structure of the Reynolds Stress near the Wall. J. Fluid Mech., Vol. 55, pp. 65-92.

Willmarth, W. W. and Sharma, L. K. 1984. Study of Turbulent Structure with Hot Wires Smaller Than the Viscous Length. J. Fluid Mec., Vol. 142, pp. 121-149.

Part 4
PRESSURE FLUCTUATIONS AND RELATED NEAR–WALL EFFECTS

A Review of Knowledge on Pressure Fluctuations

HELMUT ECKELMANN
Max-Planck-Institut für Strömungsforschung
Bunsenstr. 10
D 3400 Göttingen, FRG

1. INTRODUCTION

This review of the present knowledge of the understanding of pressure fluctuations in turbulent flows shall be limited to pressure fluctuations beneath turbulent flows. No reliable pressure fluctuation measurements have been made within turbulent flows so far. The problem is, that velocity fluctuations present in the flow also produce random pressure fluctuations on a probe, which would not occur if the probe were absent. It is nowadays, however, possible to calculate the pressure fluctuations by a numerical simulation of the flow. Contour plots of pressure patterns at the wall or in planes parallel to the wall have been calculated by Schumann (1975 a,b) and Moin & Kim (1982). Wall pressure fluctuations are small and extremely difficult to measure with a reasonable accuracy. In addition, these fluctuations have a wide range of scales from a few times the boundary layer thickness, pipe diameter or channel width to viscous scales.

Two excellent review articles that cover the time up to 1975 have been given by Willmarth (1975 a,b), Thus, this paper shall concentrate primarily on the time thereafter and on the character and role of pressure fluctuations due to organized or coherent structures in the flow. For a review of these structures the reader should refer to the articles by Cantwell (1981) and Hussain (1983).

Most of the experimental investigations were carried out in turbulent boundary layers (Dinkelacker, Hessel, Meier & Schewe (1977), Hessel (1978), Panton, Goldman, Lowery & Reischman (1980), Dinkelacker (1982), Thomas & Bull (1983), Schewe (1983), Kobashi, Komoda & Ichijo (1984), Kobashi & Ichijo (1986), Johansson, Her & Haritonides (1987)), or in turbulent pipe flows (Langeheineken (1981), Dinkelacker & Langeheineken (1983), Tanifuji, Ichijo, Iida & Kobashi (1986), Sieber (1987), Dinkelacker & Sieber (1987)). Numerical simulations were made both for turbulent boundary layers (Spalart (1986)) and for turbulent channel flows (Grötzbach & Schumann (1979), Moin & Kim (1982), Kim (1983)). For a review of numerical simulations of turbulent flows see Rogallo & Moin (1984).

Wall bounded turbulent flows show ordered structures in the vicinity of the wall which lead to characteristic wall pressure fluctuations. These ordered or coherent structures have been extensively studied visually by Kline, Reynolds, Schraub & Rundstadler (1967), Kim, Kline & Reynolds (1971), Offen & Kline (1974) and independently by Corino & Brodkey (1969), Nychas, Hershey & Brodkey (1973) and Praturi & Brodkey (1978). The wall region can be characterized by a sequence of events in which the various shear layers produced either by a motion of high speed fluid towards the wall (sweep) or by eruptions of low speed fluid away from the wall (ejections, lift up of low speed streaks) lead to high amplitude wall pressure fluctuations. In turbulent boundary layers large-scale motions of the outer part of the layer also lead to shear layers and hence to wall pressure fluctuations. These large scale motions have been investigated in detail by Kaplan & Laufer (1969), Kovasznay, Kibens & Blackwelder (1970), Antonia (1972) and Falco (1977). The various pressure fluctuations propagate along the wall in mean flow direction. Their different propagation velocities can be an indication for the wall distance from which the major contribution of the pressure fluctuations originate.

In an incompressible turbulent flow the Poisson equation

$$\frac{\partial^2 p}{\partial x_i^2} = -\rho q$$

with the source term

$$q = 2\,\frac{\partial \overline{U}_i}{\partial x_j}\,\frac{\partial u_j}{\partial x_i} + \frac{\partial^2}{\partial x_i\,\partial x_j}\left(u_i u_j - \overline{u_i u_j}\right) \qquad (1)$$

describes the relation between the pressure fluctuations p and the fluctuating velocity. Here $\overline{U}_i$ and u_i ($i = 1,2,3$) denote the mean and fluctuating velocity components in x_i directions respectively; ρ is the density. Basically two different types of processes are responsible for the pressure fluctuations. Firstly, mean shear velocity fluctuation interactions and secondly, interactions of the turbulent velocity fluctuations with themselves. The first term of equ. (1) is linear, while the second is quadratic in the turbulent velocity fluctuations. Since, in wall bounded flows the velocity fluctuations are small compared to the mean shear $d\overline{U}/dy$, the first term which then reads

$$2\,\frac{d\overline{U}}{dy}\,\frac{\partial v}{\partial x} \qquad (2)$$

will dominate. Here x is the coordinate in mean flow direction and y is normal to the wall. The respective velocity components are $\overline{U} + u$, v.

The fluctuating pressure at a point $\underline{x} = (x,0,z)$ on the wall (z: spanwise direction) is, if contributions from surface integrals can be neglected, given by

$$p(\underline{x}) = \frac{\rho}{\pi}\int_{y>0} \frac{d\overline{U}}{dy}\,\frac{\partial v}{\partial x}\,\frac{dV(\underline{x}_s)}{|\underline{x} - \underline{x}_s|} \;,$$

where the integration volume is the entire upper half-space containing the flow. All sources in the upper half will contribute to the pressure fluctuations at a given point. The fluctuating pressure will therefore not be highly correlated with the velocity fluctuations at any one neighboring point, and the contributions from various sources will become rapidly smaller with increasing distance from the point under consideration.

2. NUMERICAL INVESTIGATIONS

Contours of wall-pressure fluctuations for a turbulent channel flow have been calculated with the large-eddy simulation method by Schumann (1975a,b) and Grötzbach & Schumann (1979). Figure 1 shows the instantaneous pressure distribution at the wall calculated by Grötzbach & Schumann. The distances Δx, Δz in mean flow and spanwise direction respectively, which are normally nondimenionalized by inner variables, e.g. $\Delta x^+ = \Delta x u_\tau/\nu$ (u_τ: friction velocity, ν: kinematic viscosity), are estimated here from a comparison with corresponding results of Dinkelacker, Hessel, Meier & Schewe (1977) shown in the lower left corner of Figure 1. A magnification of the small figure is given in Figure 2. It represents an instantaneous wall pressure pattern which was measured in a turbulent boundary layer with an interferometry technique by Emmerling (1973) and evaluated later by Dinkelacker et al. There is a remarkable resemblance between both figures, which shows that, in agreement with the experiments of Bull (1967) and Willmarth (1975a), the large-scale pressure fluctuations are correlated at considerably greater distance in the spanwise direction than they are in the mean flow direction. The two point pressure correlations obtained from large-eddy simulation of a turbulent channel flow by Moin & Kim (1982) (see Rogallo & Moin (1984)), and which are depicted in Figure 3, also show this feature. The correlations calculated for a nondimensional wall distance $y^+ = 38$ ($y/\delta = 0.06$, δ: channel half-width) indicate that the spanwise extension of the pressure eddies is about 5 times as large as the streamwise extension. Kim (1983) pointed out that, in the simulations of Moin & Kim the scaling of the large scale structures is not captured properly because of numerical problems. Therefore the quantitative information given in Figure 3 must be used with some care.

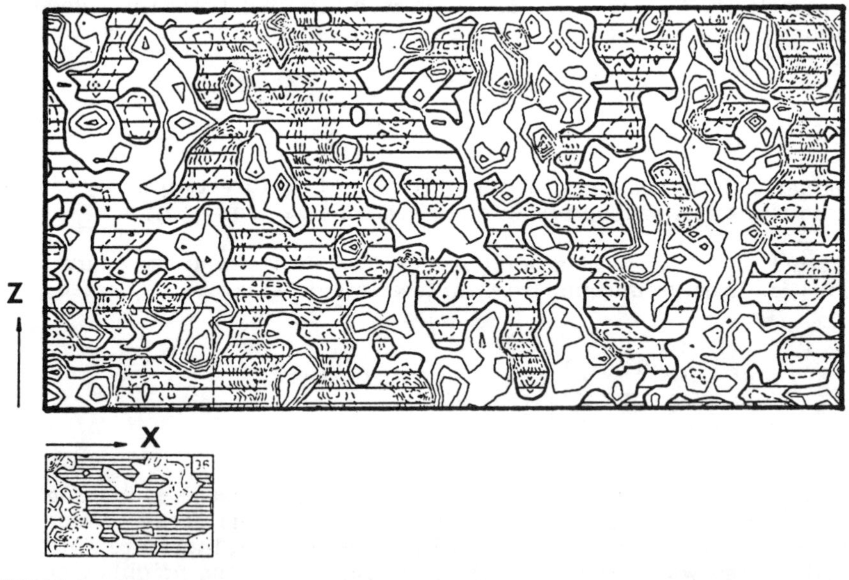

FIGURE 1. Instantaneous pressure distribution at the wall in a turbulent channel flow. Field of view $\Delta x^+ \approx 5400$, $\Delta z^+ \approx 2700$. Solid lines are isobars $p \geqslant 0$ and dashed lines $p < 0$. Low pressure regions are hatched. For comparison of the spatial dimensions Figure 2 is reproduced in the lower left corner. From Grötzbach and Schumann (1979).

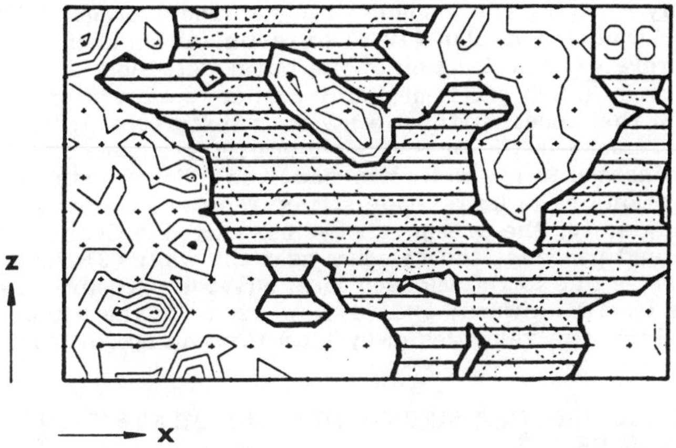

FIGURE 2. Instantaneous pressure distribution at the wall under a turbulent boundary layer. Field of observation $\Delta x^+ = 1140$, $\Delta z^+ = 681$. Lines and shading as in Figure 1. From Dinkelacker, Hessel, Meier & Schewe (1977).

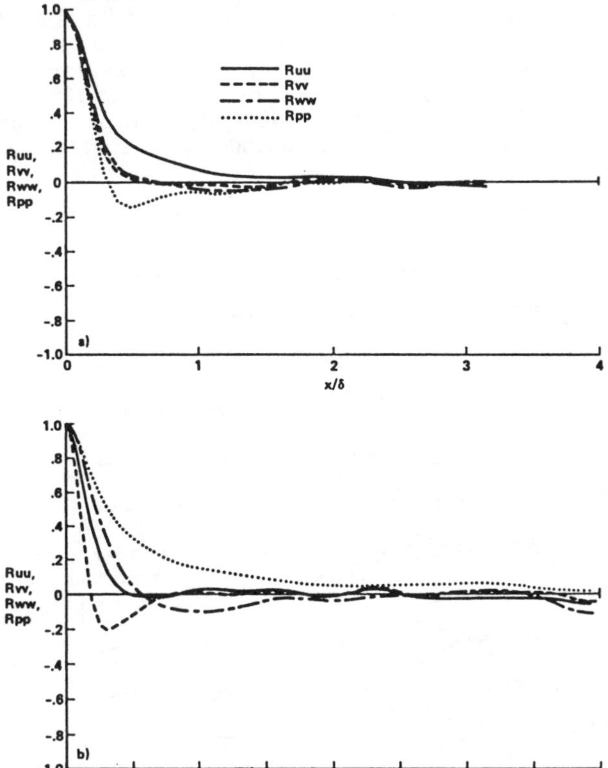

FIGURE 3. Two-point correlations of pressure and velocity near the wall ($y^+ = 38$) in a turbulent channel flow, calculated from data of Moin & Kim (1982). From Rogallo and Moin (1984).

Kim (1983) calculated pressure and streamwise velocity fluctuations during a burst event by applying the variable-integral time average (VITA) technique from Blackwelder & Kaplan (1976) to turbulent channel flow data of Moin & Kim (Figure 4). The location of the burst detection is $y^+ = 21$. The velocity signatures, which are remarkably similar to experimental results, are skewed such that those furthest away from the arrive first at the wall. In contrast to this, peaks in the pressure signatures are more or less aligned in the direction normal to the wall. Kim made the following statements: (1) the sweeping motion of the large-scale structures move towards the wall at an oblique angle; (2) the pressure wave associated with the sweep has localized peaks and precedes the velocity near the wall; (3) the flow near the wall decelerates due to the imposed local adverse pressure gradient, and the fluid is ejected upward; and (4) as a result of the lift-up of the low-momentum fluid, the velocities away from the wall are decreased.

3. EFFECT OF TRANSDUCER SIZE ON MEASUSREMENTS OF WALL PRESSURE FLUCTUATIONS

Emmerling (1973) was the first, who realized that the small-scale wall pressure fluctuations do scale with the wall parameters and have the characteristic scale ν/u_τ. Schewe (1983) carried out measurements at a Reynolds number of 13,000 with differently sized pressure transducers ($19 < d^+ = du_\tau/\nu < 333$) in Emmerling's boundary layer flow. He could show that a transducer with $d^+ = 19$ is sufficient to resolve the essential structures of the turbulent pressure fluctuations (Figure 5). The measurements of Langeheineken & Dinkelacker (1977) (see also Langeheineken (1979)) given in the figure, were obtained in fully developed pipe flow at various Reynolds numbers ($14,000 < Re = U_\infty \delta/\nu < 160,000$) with an 1/8 inch condensor microphone.

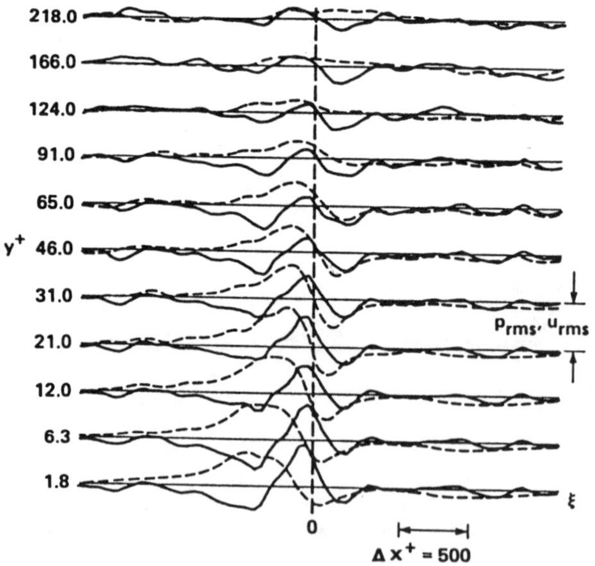

FIGURE 4. Pressure and velocity signatures during a burst event in a turbulent channel flow. ---- streamwise velocity, —— pressure. From Kim (1983).

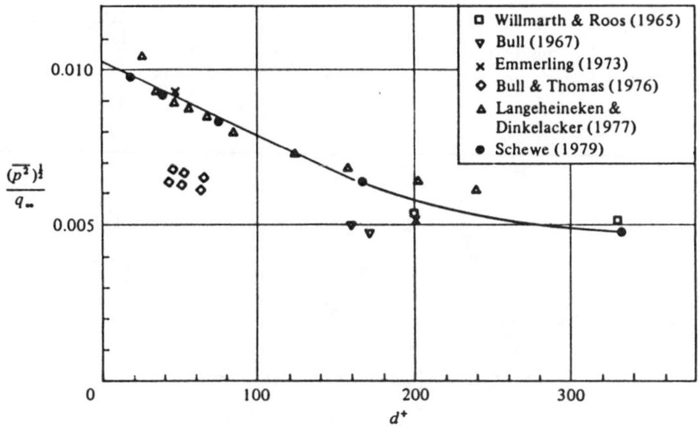

FIGURE 5. Dependence of the measured rms wall pressure fluctuations upon dimensionless pressure transducer diameter. From Schewe (1983).

Using the various pressure transducers, Schewe also measured the probability density distrituion of the pressure fluctuations (Figure 6). The pressure is normalized with the dynamic pressure q_∞ in the free stream such that the area under a curve has the value 1. When the same measurements are normalized with the rms values of the respective sensors, Figure 7 is obtained. With increasing transducer diameter d, the probability distribution approaches a Gaussian distribution. This is conceivable because the instantaneous total pressure on the transducer surface is composed more and more by independent stochastic processes. For the smallest transducer the skewness and flatness of the probability distribution reaches values of about 0.2 and 4.8 respectively which highly deviates from the values 0 and 3 belonging to a Gaussian distribution.

Over about one decade of Reynolds number, Bull & Langeheineken (1981) measured wall pressure fluctuations in a turbulent pipe flow with both a small piezoelectric transducer and a small condensor microphone. They found large discrepancies (of the order of 50 % or about 3 dB) between values of the ratio of the root mean square pressure to the dynamic pressure of the free stream p_{rms}/q_∞ measured by the two types of instruments. The common range of transducer size was $35 < d^+ < 85$. The discrepancy was not attributed to measuring errors but to genuine effects of the differences in Reynolds number which necessarily occur when two sensors of different diameter have equal values of d^+. Bull & Langeheineken conclude that the dependence of the measured ratio p_{rms}/q_∞ on d^+ is not unique.

4. EXPERIMENTAL INVESTIGATIONS IN TURBULENT BOUNDARY LAYERS

Panton, Goldman, Lowery & Reischman (1980) made wall pressure measurements under, as well as velocity measurements inside and outside of a turbulent boundary layer which developed on the fuselage of a sailplane. This flow offered a noise-free environment with a low free stream turbulence level. For the pressure velocity correlations an X-wire was used which was located slightly behind the surface of a 3/8 inch microphone. Although the rms values of the velocity fluctuations in the non-turbulent region were

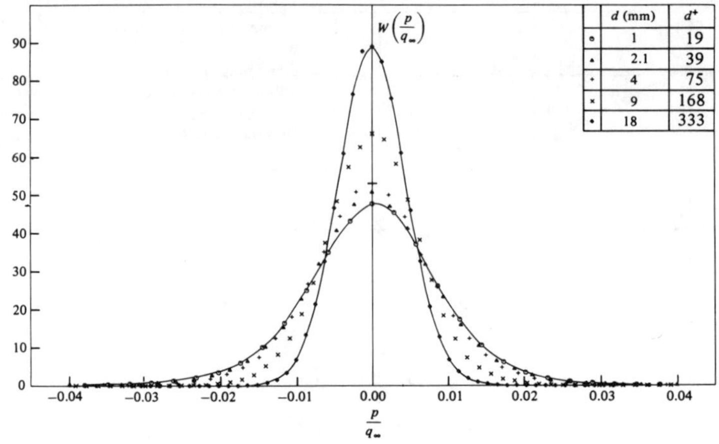

FIGURE 6. Probability density distribution measured with variously sized pressure tranducers and normalized with the dynamic pressure p_∞ in the free stream. From Schewe (1983).

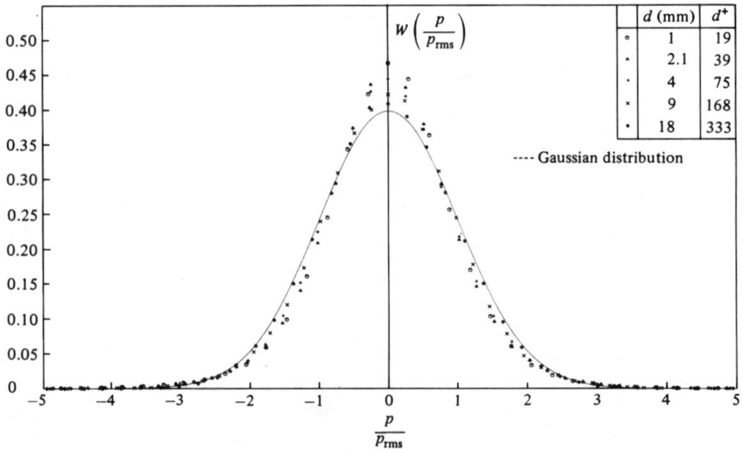

FIGURE 7. Probability density distribution from Figure 6, but normalized with corresponding rms value. From Schewe (1983).

less or equal to 0.45 %, the fluctuations maintain a strong correlation with the wall pressure (Figure 8). The R_{pu} and R_{pv} correlations are 90° out of phase and have shapes which can be deduced by considering the flow over a wavy wall. By conditional correlations taken during either turbulent or non-turbulent periods, it was found that the irrotational motions are almost completely responsible for the contribution to the wall pressure.

Kobashi, Komoda & Ichijo (1984) and Kobashi & Ichijo (1986) repeated the wall pressure velocity correlation measurements of Panton et al. To separate the pressure fluctuations produced by the boundary layer bulges from those originating from inside the boundary layer they divided the correlations into two parts. Assuming that in their case the bulges have a scale

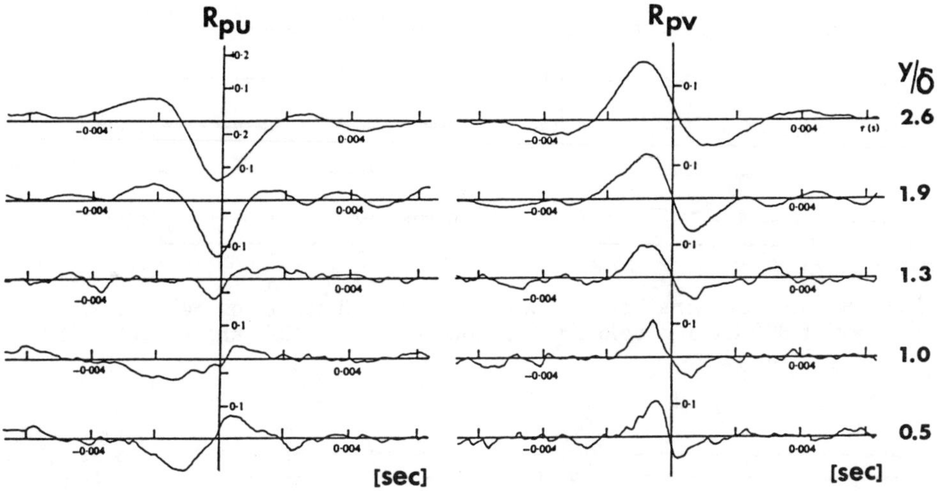

FIGURE 8. Wall pressure velocity correlations for different X-probe positions inside and outside the boundary layer. u: velocity fluctuation in mean flow direction, v: velocity fluctuation normal to the wall. From Panton, Goldman, Lowery & Reischman (1980).

of 2.25 δ = 5.6 cm and that they move with about the free stream velocity U_∞ = 17 m/s, wall pressure fluctuations of about 300 Hz can be expected. A filter cutoff frequency of 800 Hz is thus appropriate. The correlations between the wall pressure fluctuations below the cutoff frequency of 800 Hz and the streamwise and wall normal velocity component respectively are shown in Figure 9. These correlations resemble those of Panton et al. The time t = 0.004 sec in Figure 8 corresponds to $T = U_\infty t/\delta = 6$ (U_∞: free-stream velocity, δ: boundary layer thickness). The respective correlations of the high frequency part of the wall pressure fluctuations are shown in Figure 10. They are restricted to a small region near the wall ($y/\delta = 0.1$ corresponds to $y^+ \approx 120$). These measurements suggest the existence of at least two groups of pressure fluctuations: (1) large-scale structures which originate from the outer part of the boundary layer and (2) small scale structures (sweep, ejection, burst) which are limited to the wall region. The propagation velocity of the low frequency part is, according to Kobashi et al. (1984), almost U_∞ and that of the high frequency part is about $U_\infty/2$ which corresponds to about 12.5 u_τ. The interaction between both large- and small-scale structures is not clear yet.

Thomas & Bull (1983) investigated a possible relationship between the low and high frequency parts of the wall-pressure signal. They correlated the low frequency part with the rectified high frequency part, which was adjusted to have zero mean value and then smoothed by low-pass filtering. The nondimensional cutoff frequency used for both separating and smoothing was $\omega_c \delta^*/U_\infty = 0.43$ (δ^*: displacement thickness). For comparison the 800 Hz of Kobashi and co-workers corresponds to 1.2. The result of the correlation, which is shown in Figure 11, is that low- and high-frequency components of the wall pressure fluctuations are interdependent. The extreme values of the correlation coefficient are, however, rather low. From these correlations and from extensive measurements and correlations between the low-frequency and high-frequency components of both the wall

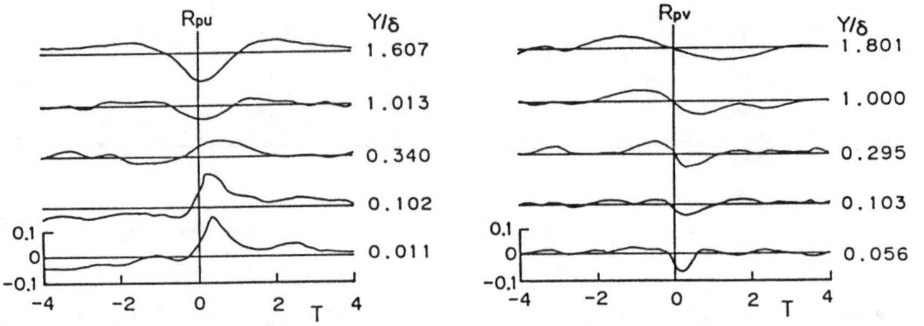

FIGURE 9. Correlations between wall pressure fluctuations below a cutoff frequency of 800 Hz and velocity fluctuations. From Kobashi & Ichijo (1986).

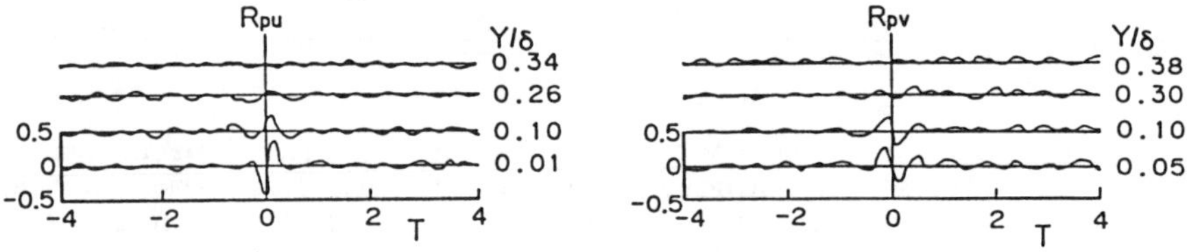

FIGURE 10. Correlations between wall pressure fluctuations above a cutoff frequency of 800 Hz and velocity fluctuations. From Kobashi & Ichijo (1986).

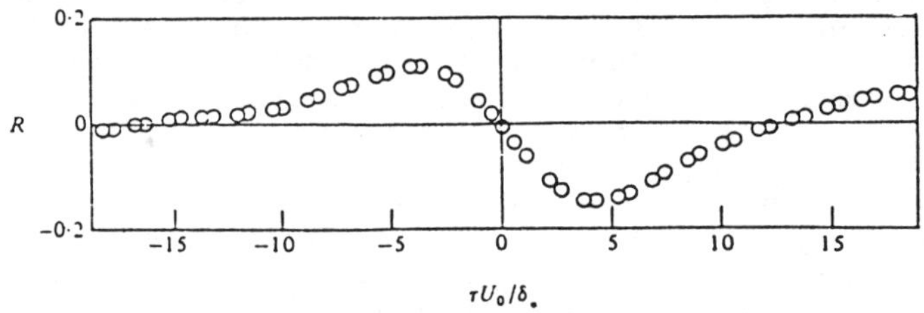

FIGURE 11. Correlation between the low-frequency component and the smoothed rectified high-frequency component of the wall pressure fluctuations. From Thomas & Bull (1983).

shear stress fluctuations and velocity fluctuations throughout the boundary layer (made by Thomas (1977) and Brown & Thomas (1977)), Thomas & Bull proposed the flow model shown in Figure 12. The essential structure is an inclined horseshoe vortex. The likely form of the characteristic wall pressure and wall shear stress patterns associated with a succession of the large-scale structures are also given in the figure. The pressure pattern is a result of the velocity step across the shear layer of the large scale structure, whose convection velocity is lower closer to the wall than it is in the outer part of the boundary layer, indicating, that the structure is slowly rotating. A similar conclusion follows also from the measurements of Kreplin & Eckelmann (1979) and Johansson, Alfredsson & Eckelmann (1987). The measured convection speed of 0.67 U_∞ or about 19 u_τ of the large scale structure, however, is an indication that the observed pressure patterns originate from shear layers located in the logarithmic region ($y^+ \approx 250$) of the boundary layer.

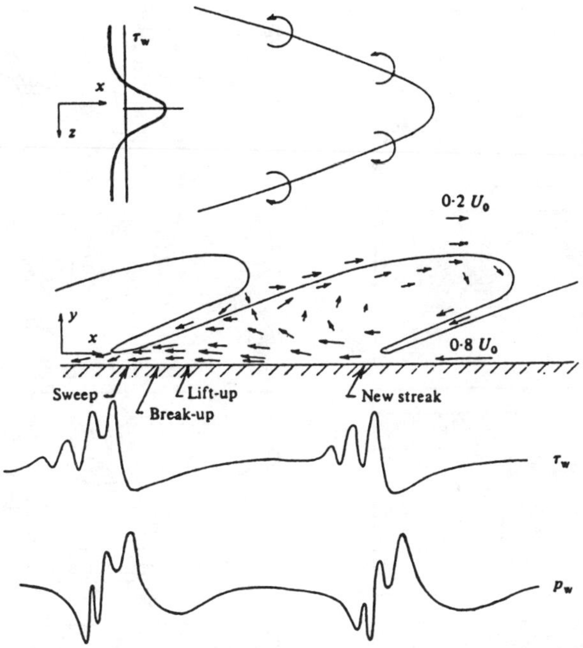

FIGURE 12. Large-scale structure, associated wall pressure and wall shear stress distribution, and location of events in the burst-sweep cycle in frame of reference moving with the large-scale structure. From Thomas & Bull (1983).

The convection speeds of large wall pressure regions have been determined by Hessel (1978) who further evaluated Emmerling's (1973) films. As an example, Figure 13 shows a sequence of consecutive wall pressure patterns. The flow is from left to right (U_∞ = 8.5 m/s) and the time between two frames 0.28 msec. The field of view of 50 mm × 30 mm corresponds to about 1.7 δ × 1 δ (1140 ν/u_τ × 681 ν/u_τ). The motion of the zero μ-bar line (enhanced in Figure 13) from left to right along the centerline of a frame is depicted in Figure 14 for 65 consecutive motion pictures. Three large scale wall pressure regions, each consisting of a high and low pressure part with an extension of about δ in mean flow direction, can be recognized. The small scale pressure regions are inbedded into the large-scale regions. For the most part, they travel with a lower speed. The probability distribution of the convection speed for the small-scale pressure regions is also given by Hessel (Figure 15). A rather high convection speed U_c = 0.67 U_∞ is most probable.

Schewe (1983), who worked in Emmerling's windtunnel with an array of four pressure transducers consecutively arranged in mean flow direction, obtained the time functions of the wall pressure shown in Figure 16. A nearly frozen pressure pattern moves along the wall. Relatively short time segments with large pressure fluctuations exist. Events that have a maximum amplitude exceeding about 5 p_{rms} (events E_1 and E_3) grow and decay faster than events with about 3 p_{rms} (like E_2), which keep their amplitude over a distance of 225 ν/u_τ or about 0.4 δ. Ensemble averages of wall pres-

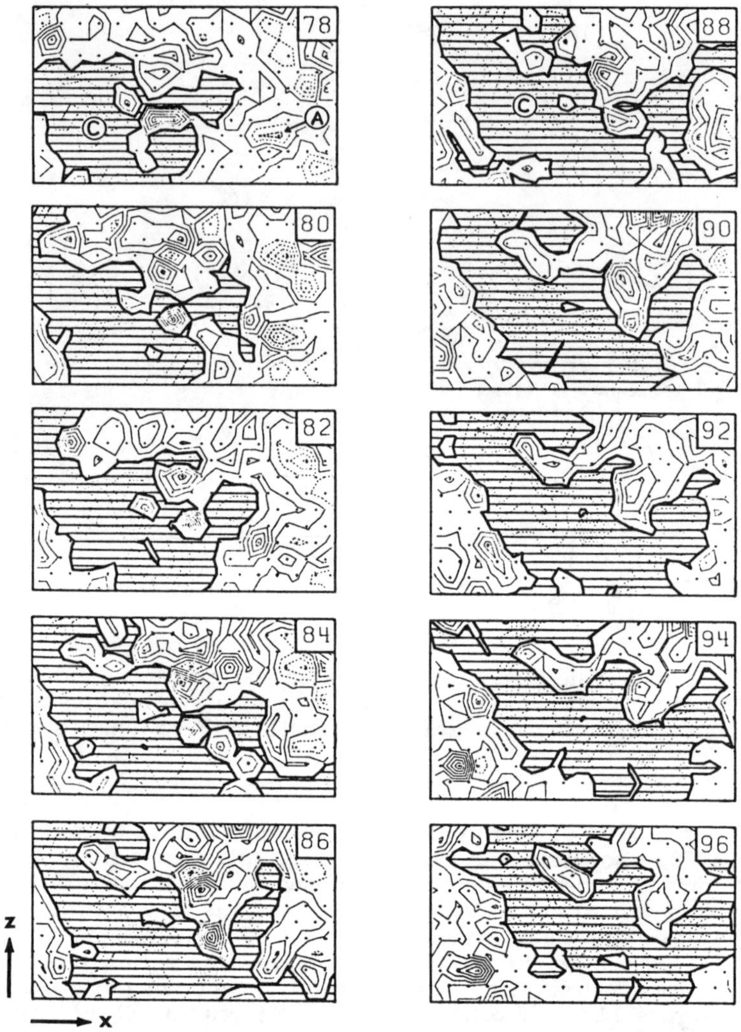

FIGURE 13. Sequence of consecutive wall pressure distributions. Solid lines are isobars $p > 0$ and dashed lines $p < 0$. Low pressure regions are hatched. From Dinkelacker, Hessel, Meier & Schewe (1977).

sure peaks exceeding a threshold of $+3\,p_{rms}$ or $-3\,p_{rms}$ are shown in Figure 17. A convection velocity $U_c = 0.53\ U_\infty = 11.9\ u_\tau$ was determined from the respective distance of the transducers and the corresponding time shift of the signals. This value is nearly the same as the convection velocity $c_x = 12.1\ u_\tau$ of the streamwise and the spanwise fluctuations of the velocity gradient at the wall as determined by Kreplin & Eckelmann (1979). The characteristic duration time of the pressure peaks exceeding $\pm 3\,p_{rms}$ indicated in Figure 17 correspond to a nondimensional T^+ of about 10.

Schewe's probability distribution of the wall pressure fluctuations for the smallest pressure transducer ($d^+ = 19$, Figure 7) shows an asymmetry such

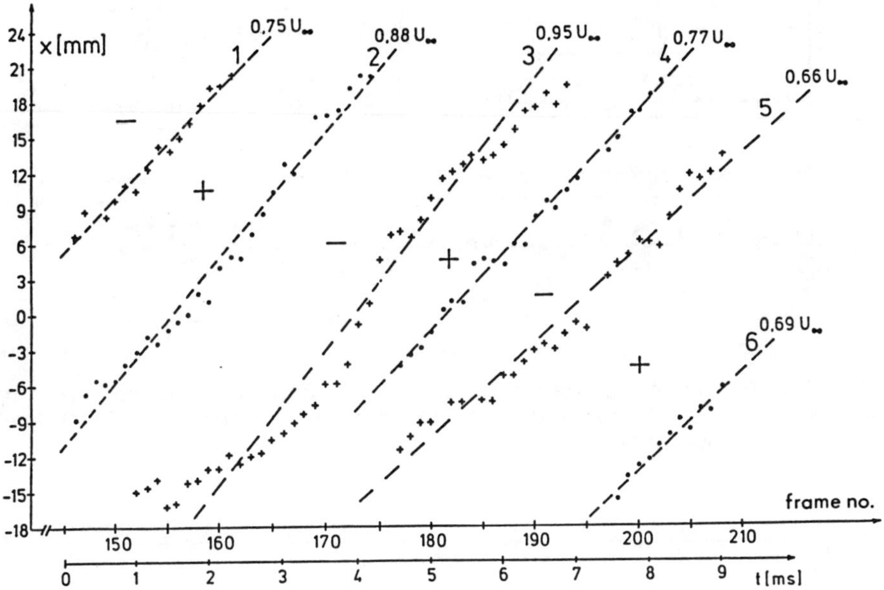

FIGURE 14. Convection speed of large-scale wall pressure areas. The front is represented by the zero μ-bar line. From Hessel (1978).

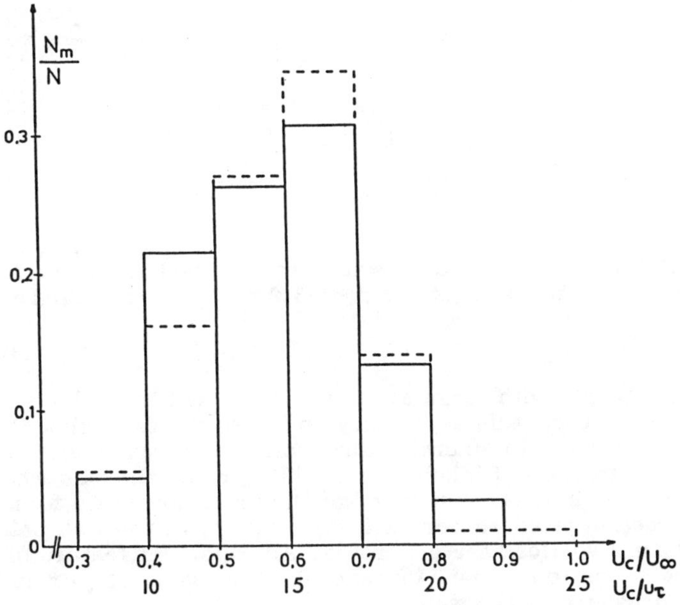

FIGURE 15. Probability distribution of the convection speed of small scale wall pressure areas. Solid line: distribution of high pressure, dashed line: distribution of low pressure, N_m: number of regions in a class, N: total number of regions. U_∞ = 8.5 m/s, u_τ = 0.34 m/s. From Hessel (1978).

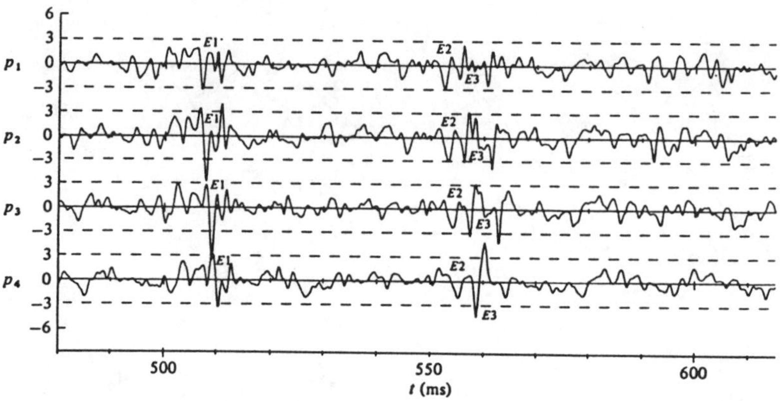

FIGURE 16. Pressure fluctuation obtained from four consecutive pressure transducers (Δx^+ = 75, 150 and 225). The signals are normalized with their p_{rms}. From Schewe (1983).

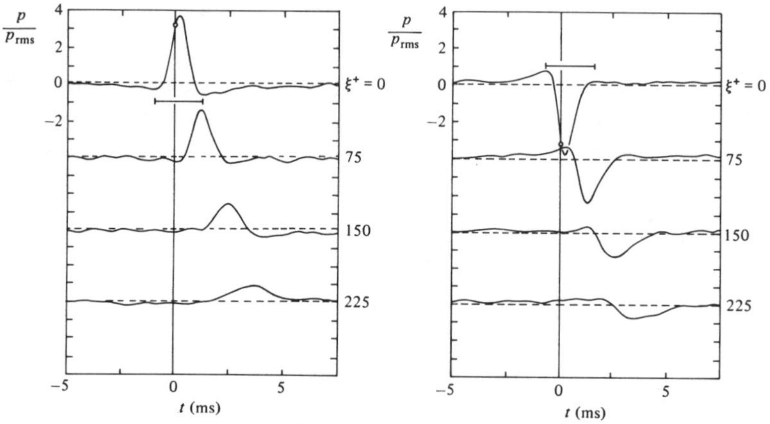

FIGURE 17. Ensemble average of wall pressure peaks. On the left for $p > 3.2\ p_{rms}$ and on the right for $p < -3.2\ p_{rms}$. o indicated triggering point. $\xi^+ = \Delta x^+$. From Schewe (1983).

that from the events with large amplitude those with negative sign occur more frequently. From this asymmetry and from the fact, that the wall pressure is correlated in spanwise direction over a larger distance than in streamwise direction, Dinkelacker (1982) derived the flow model shown in Figure 18. The idea is, that through the breakdown of a formerly in spanwise direction oriented vortex in the vicinity of the wall pairs of counter rotating vortices are generated. An important feature of these vortex pairs seems to be that, tornado-like they generate a relative small area of low pressure at the wall.

The coupling between high-amplitude wall-pressure peaks and flow structures in the near wall region was studied by Johansson, Her & Haritonidis (1987). By conditional averageing of the wall pressure and the streamwise

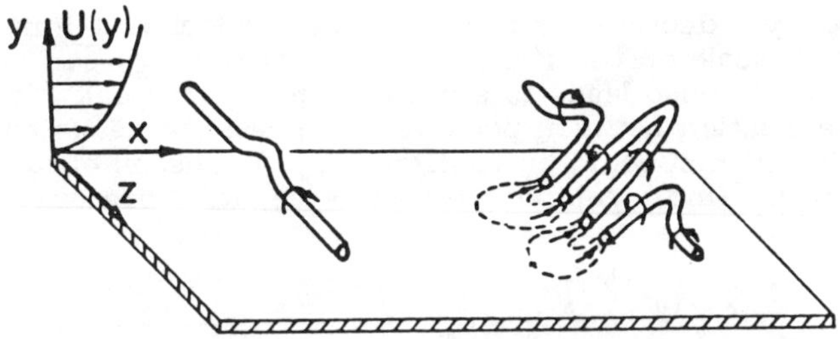

FIGURE 18. Flow model for the generation of pairs of "tornado-like" vortices in the vicinity of a wall. From Dinkelacker (1982).

velocity at various wall distances, they could show that the positive wall pressure peaks are related to shear layers in the buffer region (Figure 19). The left-hand side of the figure shows ensemble averages of the streamwise velocity when the wall pressure p is greater than three times p_{rms}. The right-hand side was obtained by detecting the shear layer at various wall distances with the VITA technique ($k = 1$, $T^+ = 24$) and by ensemble averaging the wall pressure. Also, in this case a positive pressure peak was found only for shear layers ocurring in the buffer region. Since the pressure transducer was not directly below the probe in the flow, the pressure maxima in Figure 19 do not coincide with the centers of the shear layers. The necessary shift for the pressure patterns of 5 T^+ to the right can be derived from a propagation velocity of 11 u_τ (Schewe 1983).

Due to the considerations made in the introduction which lead to equ. (2), one should expect a positive pressure to be associated with a positive value of $\partial v/\partial x$ or applying Taylor's hypothesis, with a negative value of $\partial v/\partial t$ which for the shear layer signatures presented for $y^+ = 15$ in Figure 19 is associated with a sharp increase in u. For negative wall pressure peaks a coupling to shear layers was not found, although due to Johansson et al. for the same amplitude negative wall pressure peaks occur approximately as frequently as positive ones. Instead, the negative pressure peaks can be associated to a period of high streamwise velocity occurring around the detection time.

5. EXPERIMENTAL INVESTIGATIONS IN TURBULENT PIPE FLOWS

Wall pressure fluctuations and their relation to u and v velocity fluctuations in fully developed turbulent pipe flows have been performed at the Max-Planck-Institut für Strömungsforschung in Göttingen by A. Dinkelacker and co-workers and at the Hokkaido Institute of Technology in Sapporo by Y. Kobashi and co-workers.

Langeheineken (1981) and Dinkelacker & Langeheineken (1983) could show that events of instantaneous high positive pressure at the pipe wall are accompanied by a rapid increase in the streamwise velocity (Figure 20 left). The curve on the bottom shows the conditionally averaged and normalized wall pressure for $p \geqslant 4\,p_{rms}$. The other curves are a proof for the strong coupling between positive pressure peaks and shear layers. It could be shown by Langeheineken (1981), who triggered at $y^+ = 16$ on $\partial u/\partial t > 0$ and obtained a positive pressure peak, that this relation is also bi-directional. Instanteneous high negative pressure peaks at the wall were found to be accompanied by a decrease in the streamwise velocity (Figure 20 right). Note the different u scale on the right hand side of the figure. Triggering at y^+ = 16 on $\partial u/\partial t < 0$ also leads to a negative pressure peak. For a boundary layer flow the relation between positive wall pressure peaks and shear layers in the buffer region could be confirmed by Johansson et al. (1987), however, they did not find a relation between a negative pressure peak and shear layers.

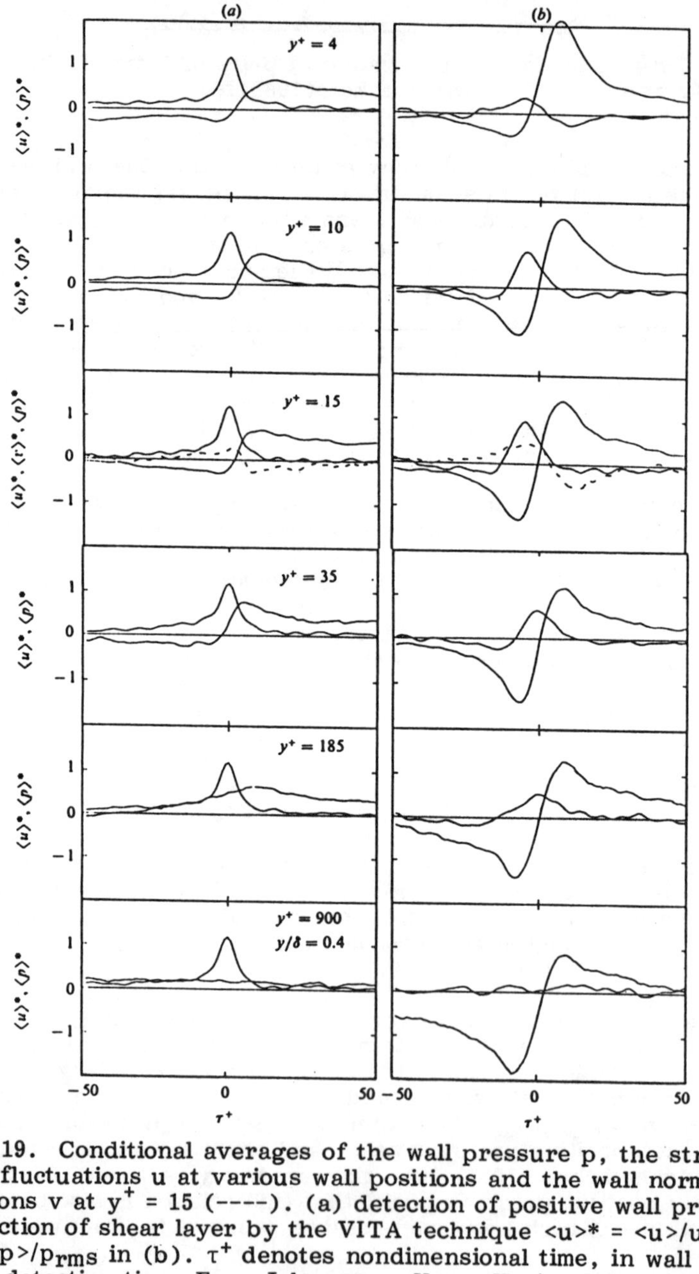

FIGURE 19. Conditional averages of the wall pressure p, the streamwise velocity fluctuations u at various wall positions and the wall normal velocity fluctuations v at y^+ = 15 (----). (a) detection of positive wall pressure, (b) detection of shear layer by the VITA technique $\langle u\rangle^* = \langle u\rangle/u_{rms}$, $\langle p\rangle^* = \langle p\rangle/p_{rms}$ in (b). τ^+ denotes nondimensional time, in wall units, relative to detection time. From Johansson, Her & Haritonidis (1987).

While Langeheineken mainly investigated the flow field in the neighbourhood of the pressure transducer, Sieber (1987) and Dinkelacker & Sieber (1987) extended their measurements to larger distances in mean flow direction. When triggering on positive wall pressure ($p > 4 p_{rms}$), they could clearly follow u signatures traveling at wall distances $0.4 < y/R < 0.6$ ($312 < y^+ < 468$) over an upstream distance of 13 pipe diameters ($-x^+ \approx 20,000!$) (<u>Figure 21 left</u>). Even in the jet leaving the pipe at $x^+ \approx 4,000$ these signa-

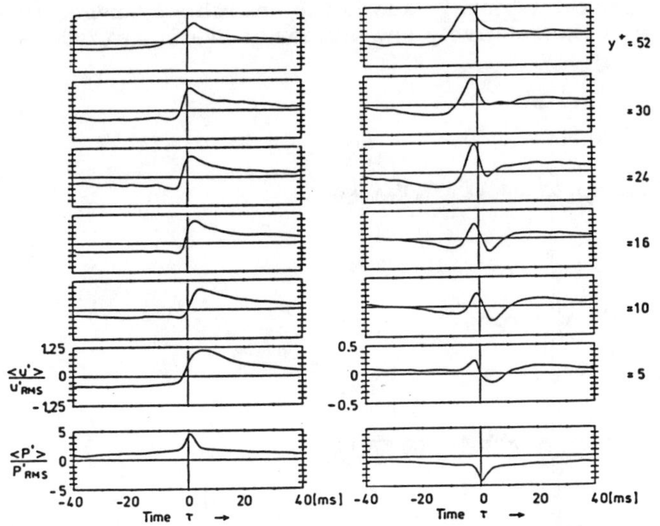

FIGURE 20. Conditionally averaged streamwise velocity fluctuations at different distances from the pipe wall for $p \geqslant 4 p_{rms}$ (left) and for $p < 4 p_{rms}$ (right). Re = 9600, $y^+ = 52$ corresponds to $y/R = 0.2$. From Langeheineken (1981).

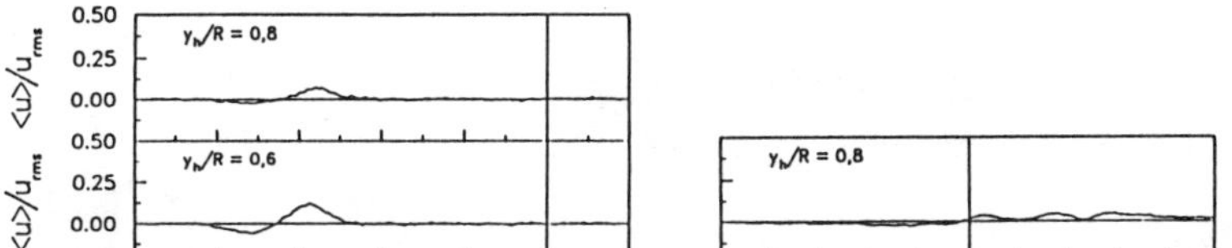

FIGURE 21. Conditionally averaged streamwise velocity fluctuations at different distances from the pipe wall. On the left, the hot wire probe is 13 pipe diameters upstream and on the right 2.6 pipe diameters downstream (in the jet leaving the pipe) of the wall pressure transducer. From Dinkelacker & Sieber (1987).

tures could be detected (Figure 21 right). Note that due to the time scales used in the last figure the pressure peak at $\tau = 0$ is so narrow that it cannot be resolved.

The relationship between the wall pressure and shear layers detected by the VITA technique at various wall distances was investigated by Tanifuji, Ichijo, Iida & Kobashi (1986). They found, like Johansson et al. (1987), that shear layers in the buffer region contribute most to the sharp positive wall pressure peaks. Figure 22 shows the ensemble averaged wall pressure, which because of the relatively low number of averages (200) is still fluctuating.

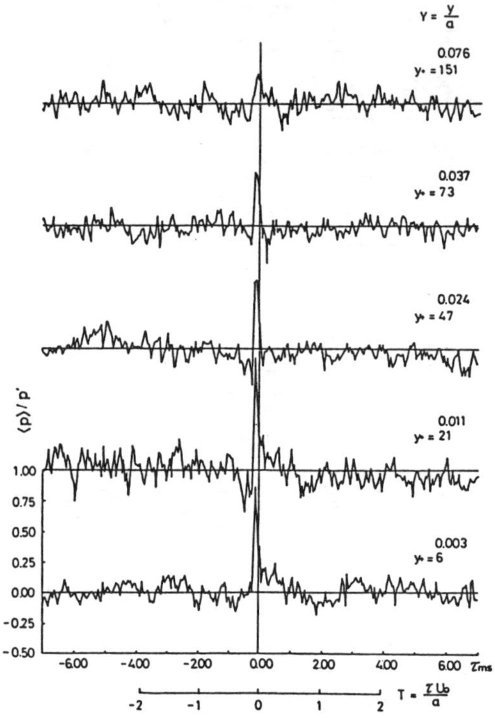

FIGURE 22. Conditional averages of the wall pressure. Detection of the shear layer by the VITA technique. From Tanifuji, Ichijo, Iida & Kobashi (1986).

6. REFERENCES

Antonia, R.A., Conditionally sampled measurements near the outer edge of a turbulent boundary layer, J. Fluid Mech. 56, pp. 1-18, 1972.

Blackwelder, R.F., and Kaplan, R.E., On the wall structure of the turbulent boundary layer, J. Fluid Mech. 76, pp. 89-112, 1976.

Brown, G.L., and Thomas, A.S.W., Large structure in a turbulent boundary layer, Phys. Fluids Suppl. 20, pp. S243-S252, 1977.

Bull, M.K., Wall pressure fluctuations associated with subsonic turbulent boundary layer flow, J. Fluid Mech. 28, pp. 719-754, 1967.

Bull, M.K., and Langeheineken, T., On the wall pressure field in turbulent pipe flow. Mitt. MPI Strömungsf. no. 73, Göttingen, 1981.

Cantwell, B.J., Organized motion in turbulent flow, Ann. Rev. Fluid Mech. 13, pp. 457-515, 1981.

Corino, E.R., and Brodkey, R.S., A visual investigation of the wall region in turbulent flow, J. Fluid Mech. 37, pp. 1-30, 1969.

Dinkelacker, A., Do tornado-like vortices play a role in turbulent mixing processes? Proc. IUTAM/ICHMT Symp. Dubrovnik, ed. Z.P. Zaric, Hemisphere Publishing Co., Washington, New York, London, pp. 59-72, 1982.

Dinkelacker, A., Hessel, M., Meier, G.E.A., and Schewe, G., Investigation of pressure fluctuations beneath a turbulent boundary layer by means of an optical method, Phys. Fluids Suppl. 20, pp. S216-S224, 1977.

Dinkelacker, A., and Langeheineken, T., Relations between wall pressure fluctuations and velocity fluctuations in turbulent flow, Proc. IUTAM Symp. Marseille, eds. R. Dumas, L. Fulachier, Springer, Berlin, Heidelberg, New York, pp. 1-9, 1983.

Dinkelacker, A., and Sieber, M., Coherent structures in turbulent pipe flow, Advances in Turbulence, eds. G. Comte-Bellot, J. Mathieu, Springer, Berlin, Heidelberg, pp. 391-398, 1987.

Emmerling, R., Die momentane Struktur des Wanddruckes einer turbulenten Grenzschichtströmung, Mitt. MPI Strömungsf. u. Aerodyn. Vers., no. 56, Göttingen, 1973.

Falco, R.E., Coherent motions in the outer region of turbulent boundary layers, Phys. Fluids Suppl. 20, pp. S124-S132, 1977.

Grötzbach, G., and Schumann, U., Direct numerical simulation of turbulent velocity-, pressure-, and temperature-fields in channel flows, eds. Durst et al., pp. 370-385, 1979.

Hessel, M., Untersuchung von Wanddruckfeldern unter turbulenten Grenzschichten, MPI Strömungsf. rep. no. 11/1978, Göttingen, 1978.

Hussain, A.K.M.F., Coherent structures - reality and myth, Phys. Fluids 26, pp. 2816-2850, 1983.

Johansson, A.V., Alfredson, P.H., and Eckelmann H., On the evolution of shear-layer structures in near-wall turbulence, Advances in Turbulence, eds. C. Comte-Bellot, J. Mathieu, Springer Berlin, Heidelberg, pp. 383-390, 1987.

Johansson, A.V., Her, J.-Y., and Haritonidis, J.H., On the generation of high-amplitude wall-pressure peaks in turbulent boundary layers and spots, J. Fluid Mech. 175, pp. 119-142, 1987.

Kaplan, R.E., and Laufer, J. The intermittently turbulent region of the boundary layer, Proc. 12th Int. Congr. Mech., pp. 236, 1969.

Kim, J., On the structure of wall-bounded turbulent flows, Phys. Fluids 26, pp. 2088-2097, 1983.

Kim, H.T., Kline S.J., and Reynolds, W.C., The production of turbulence near a smooth wall in a turbulent boundary layer, J. Fluid Mech. 50, pp. 133-160, 1971.

Kline, S.J., Reynolds, W.C., Schraub, F.A., and Rundstadler, P.W., The structure of turbulent boundary layers, J. Fluid Mech. 30, pp. 741-773, 1967.

Kobashi, Y., Komoda, H., and Ichijo, M., The wall pressure fluctuation and the turbulent structure of a boundary layer in "Turbulence and Chaotic Phenomena in Fluids" ed. T. Tatsumi , Elsevier Science Publishers B.V., pp. 461-466, 1984.

Kobashi, Y., and Ichijo, M., Wall pressure and its relation to turbulent structure of a boundary layer. Exp. in Fluids 4, pp. 49-55, 1986.

Kovasznay, L.S.G., Kibens, V., and Blackwelder, R.F., Large-scale motion in the intermittent region of a turbulent boundary layer., J. Fluid Mech 41, pp. 283-325, 1970.

Kreplin, H.-P., and Eckelmann, H., Propagation of perturbations in the viscous sublayer and adjacent wall region, J. Fluid Mech. 95, pp. 305-322, 1979.

Langeheineken, T., Experimentelle Untersuchung der Wanddruckschwankungen einer ausgebildeten turbulenten Rohrströmung, MPI Strömungsf., rep. no. 106/1979, Göttingen, 1979.

Langeheineken, T., Zusammenhänge zwischen Wanddruck- und Geschwindigkeitsschwankungen in turbulenter Rohrströmung, Mitt. MPI Strömungsf. no. 70, Göttingen, 1981.

Langeheineken, T., and Dinkelacker, A., Wanddruckschwankungen einer ausgebildeten, turbulenten Rohrströmung, DAGA'78, Fortschritte der Akustik, VDE-Verlag, Berlin, pp. 391-394, 1978.

Moin, P., and Kim, J., Numerical investigation of turbulent channel flow, J. Fluid Mech. 118, 341-377, 1982.

Nychas, S.G., Hershey, H.C., and Brodkey, R.S., A visual study of turbulent shear flow, J. Fluid Mech. 61, pp. 513-540, 1973.

Offen, G.R., and Kline, S.J., Combined dye-streak and hydrogen bubble visual observations of a turbulent boundary layer, J. Fluid Mech. 62, pp. 223-239, 1974.

Panton, R.L., Goldman, A.L., Lowery, R.L., and Reischman, M.M., Low-frequency pressure fluctuations in axisymmetric turbulent boundary layers, J. Fluid Mech. 97, pp. 299-319, 1980.

Praturi, A.K., and Brodkey, R.S., A stereoscopic visual study of coherent structures in turbulent shear flow, J. Fluid Mech. 89, pp. 251-272, 1978.

Rogallo, R.S., and Moin, P., Numerical simulation of turbulent flow, Ann. Rev. Fluid Mech. 16, pp. 99-137, 1984.

Schewe, G., On the structure and resolution of wall-pressure fluctuations associated with turbulent boundary-layer flow, J. Fluid Mech. 134, pp. 311-328, 1983.

Schumann, U., Numerical investigation of the wall pressure fluctuations in channel flows, Nucl. Eng. Design 32, pp. 37-46, 1975a.

Schumann, U., Subgrid scale model for finite difference simulations of turbulent flows in plane channels and annuli, J. Comp. Phys. 18, pp. 376-404, 1975b.

Sieber, M., Großräumige Strukturen in der turbulenten Rohrströmung, Mitt. MPI Strömungsf. no. 83, Göttingen, 1987.

Spalart, P.R., Direct simulation of a turbulent boundary layer up to $R_\theta = 1410$, NASA Techn. Mem. 89407, 1986.

Tanifuji, N., Ichijo, M., Iida, S., and Kobashi, Y., Coherent structures in a fully developed turbulent pipe flow, Bull. JSME 29, paper 250-12, pp. 1156-1162, 1986.

Thomas, A.S.W., Ph.D. thesis, Dept. Mech. Engng., University of Adelaide, 1977.

Thomas, A.S.W., and Bull, M.K., On the role of wall-pressure fluctuations in deterministic motions in the turbulent boundary layer, J. Fluid Mech. 128, pp. 283-322, 1983.

Willmarth, W.W., Structure of turbulence in boundary layers, Adv. Appl. Mech. 15, pp. 159-254, 1975a.

Willmarth, W.W., Pressure fluctuations beneath turbulent boundary layers, Ann. Rev. Fluid Mech. 7, pp. 13-38, 1975b.

Relations between Wall Pressure Fluctuations and Velocity Fluctuations in Turbulent Pipe Flow

A. DINKELACKER
Max-Planck-Institut für Strömungsforschung
Bunsenstr. 10
D 3400 Göttingen, FRG

1. INTRODUCTION

At the Max-Planck-Institut für Strömungsforschung several investigations on wall pressure fluctuations in pipes with turbulent flow have been performed [1] to [9]. Aims of these investigations were firstly, to learn more about the wall pressure fluctuations and secondly, to learn more about turbulent pipe flow by using wall pressure transducers. In this paper a survey of these investigations will be given with special reference to relations between turbulent wall pressure fluctuations and turbulent velocity fluctuations.

2. EXPERIMENTAL ARRANGEMENT

Figure 1 shows the flow facility, the main part of which is a long straight pipe with a pipe diameter $D = 2R = 5$ cm and a pipe length $L = 9.8$ m, so that $L/D = 196$. Air was used as flow medium. At the pipe inlet early transition is ensured by means of a strip of sandpaper. The experimental arrangement was so conceived as to get low disturbance levels. Fluctuations of the wall pressure and of the flow velocity were investigated in the experiments. The wall pressure fluctuations were measured with a "pinhole microphone" consisting of a hole in the pipe wall of diameter $d = 0.5$ mm and a 1/8" Brüel and Kjaer microphone. The pinhole was 110 mm upstream of the pipe outlet. The velocity fluctuations were measured with hot-wire probes. Evaluation of the measurements was mainly done using the method of "conditional signal averaging". The analogue data were fed directly into a computer, where comparatively large data sets were stored for further evaluation.

The measurements in the pipe were performed at different flow velocities. The choice of the flow velocity is a compromise: Higher velocities give a better signal to noise ratio; lower velocities give a better spatial resolution in terms of wall units (see Table 1). Cylindrical coordinates r, φ, x were used with the origin positioned so, that the wall pressure transducer had the coordinates $r = R$; $\varphi = 0$; $x = 0$, and with the x-axis directed in the mean flow direction. The suffix h denotes the position of the hot-wire. (For easier description of measurements in some cases, the distance from the pipe wall y and a coordinate at the wall in the azimuthal direction z were used.) The notation $\langle u \rangle$, $\langle v \rangle$ and $\langle p \rangle$ is used for conditional averages of the velocity fluctuations $u(t)$ in the x-direction, $v(t)$ in r-direction and of the wall pressure fluctuations $p(t)$.

Centerline velocity	U_{cl}	2.9 m/s	11.0 m/s	33.0 m/s
shear velocity	U_τ	0.147 m/s	0.47 m/s	1.26 m/s
velocity ratio	U_τ/U_{cl}	0.051	0.043	0.038
Reynolds number	$Re = U_{cl} \cdot D/\nu$	9600	36400	109000
wall time	$t_w = \nu/U_\tau^2$	0.71 ms	0.068 ms	0.010 ms
wall length	$l_w = \nu/U_\tau$	0.105 mm	0.032 mm	0.012 mm
pipe diameter in wall units	D/l_w	480	1560	4200
pinhole diameter in wall units	$d^+ = d/l_w$	4.8	15.6	42

TABLE 1. Flow Parameters (ν = kinematic viscosity)

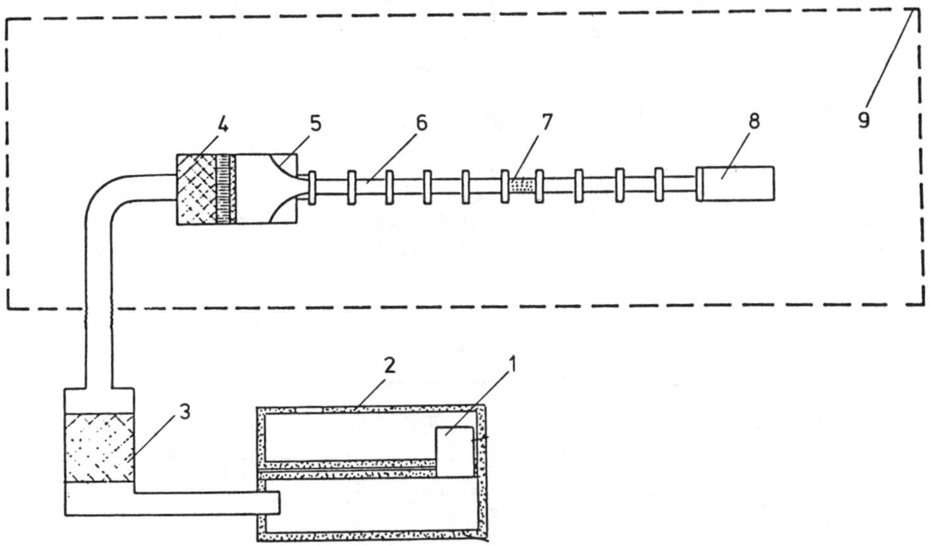

FIGURE 1. Flow facility (1 low noise fan, 2, 3, 4 sound insulation, 5 flow contraction, 6 pipe, 7 sound damping pipe section, 8 measuring section with pinhole microphone and hot-wire, 9 elastically suspended frame).

3. EXPERIMENTAL RESULTS

In this section a brief report on experimental results will be given. The data are taken from the theses of T. Langeheineken [3], M. Sieber [6] and T. Reinhardt [9]. The reader is referred to these theses for details.

Figure 2 shows a typical set of experimental results. The lowest curves in Figures 2a and 2b show the conditionally averaged functions <p> of the wall pressure fluctuations normalized with the root mean square value p_{rms}. The other curves show conditionally averaged functions <u> of velocity fluc-

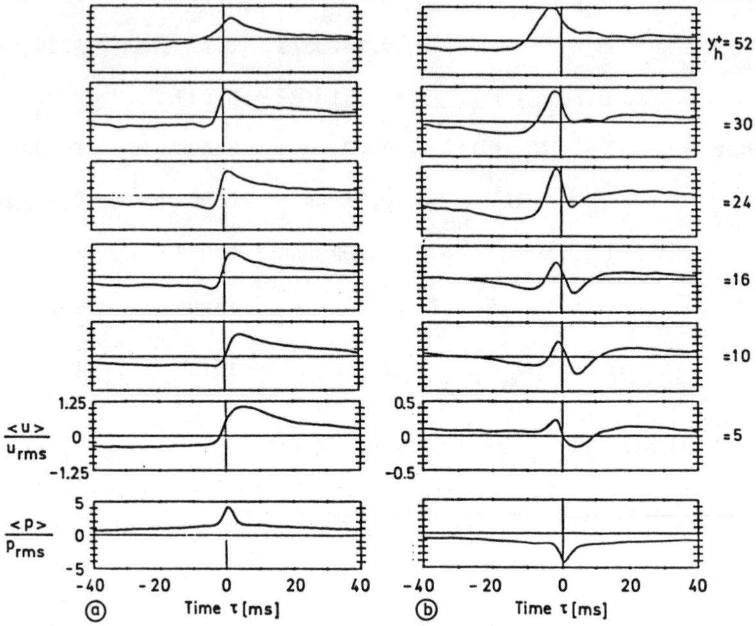

FIGURE 2. Conditionally averaged streamwise velocity fluctuations <u> (in short: u-signatures) at different distances from the pipe wall for $p > 4p_{rms}$ (left) and for $-p < -4p_{rms}$ (right) ($U_{cl} = 2.9$ m/s; $x_h = 0$, $y_h = \ldots$, $\varphi_h = 0$).

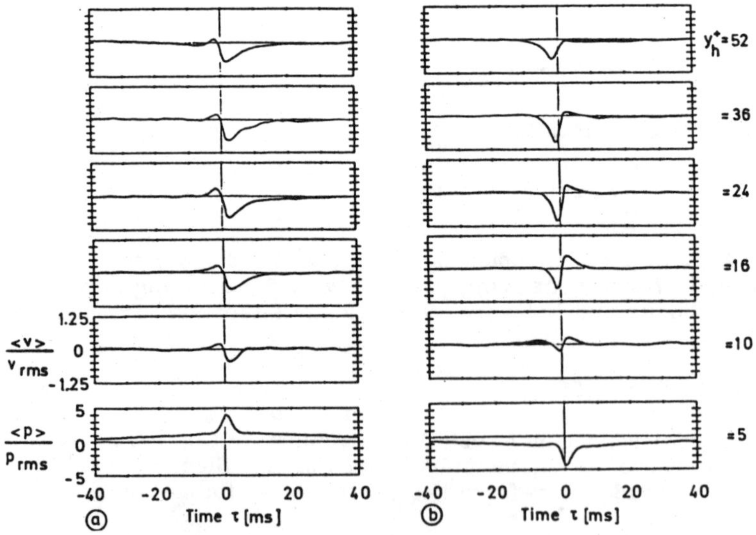

FIGURE 3. Conditionally averaged normal velocity fluctuations <v> at different distances from the pipe wall (conditions: $p > 4p_{rms}$ (left) and $p < -4p_{rms}$ (right); $U_{cl} = 2.9$ m/s; $x_h = 0$, $y_h = \ldots$, $\varphi_h = 0$).

tuations measured with the hot-wire probe at different positions in the pipe, again normalized with rms-values. We shall call these functions p- and u-signatures. They are averages over many events - usually several thousand - in which the condition "high" resp. "low" wall pressure (here $p > 4\,p_{rms}$ resp. $p < -4\,p_{rms}$) was fulfilled.

Figure 3 shows a similar set of measurements with conditionally averaged functions $\langle v \rangle$ of velocity fluctuations normal to the wall (v positive = direction away from the wall). These measurements were performed with an "x-probe". For the evaluation of the v-measurements the limited spatial resolution of such a probe (hot-wire length l = 1 mm) must be taken into account. Of course such curves depend on the choice of the condition. This influence, however, is not critical, as can be seen from Figure 4, where p-signatures and u-signatures are given for 3 different levels of the p-condition. As can be seen, even for a rather low level of the condition ($p > 1.25\,p_{rms}$) the signatures come out quite clearly.

From Figures 2 and 3 one can see especially: (I) Events with "high" wall pressure are related to local increases in $\langle u \rangle$, events with "low" wall pressure to local decreases in $\langle u \rangle$. (II) Close to the wall the u-signatures measured with the condition "high" wall pressure have much larger amplitudes than those measured with the condition "low" wall pressure (note that the ordinate scales between Figures 2a and 2b differ by a factor 2.5). Evidently the correlation between increases in u(t) and high wall pressure is larger than that between decreases in u(t) and low wall pressure. (III) At some distance from the wall the velocity signatures for both types of wall pressure conditions are very similar to one another (compare the curves for $y^+ = 52$ ind Figures 2a and 2b resp. in Figures 3a and 3b). The main difference between the corresponding curves is a shift in time τ. This time shift indicates certain relations between the occurrence of events of "high" pressure and events of "low" pressure. (IV) The u-signatures are different from zero practically for the whole time of observation (here for 80 ms or $\Delta\tau^+ = 113$). This indicates that the wall pressure events which, taken alone, are "short-time" events, are related to "long-time" processes in the turbulent flow. As one can see from Figure 2 for both sorts of wall pressure events this "long-time" process consists of a long period with $\langle u \rangle$ smaller than zero, followed by a long period with $\langle u \rangle$ larger than zero. It is remarkable that v-signatures do not show these "long-time" deviations from zero (see Figure 3).

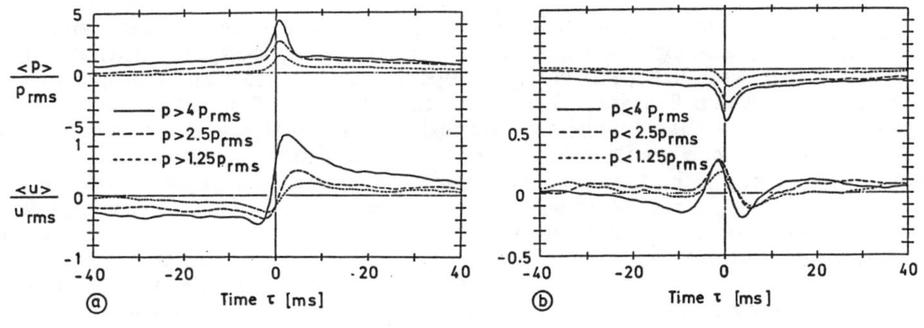

FIGURE 4. p- and u-signatures for different levels of the pressure condition (U_{cl} = 2.9 m/s; $x_h = 0$, $y_h^+ = 16$, $\varphi_h = 0$).

Figure 5 shows u-signatures with the hot-wire probe placed in different azimuthal positions. Sampling conditions were again $p > 4 p_{rms}$ resp. $p < -4 p_{rms}$. From this figure one can see that the amplitudes of the u-signatures decrease rapidly in azimuthal direction and that a systematic change occurs in the shape of the u-signatures. The shape found for the condition "high" wall pressure changes to that found for "low" wall pressure and vice versa (see the arrow marks in Figure 5).

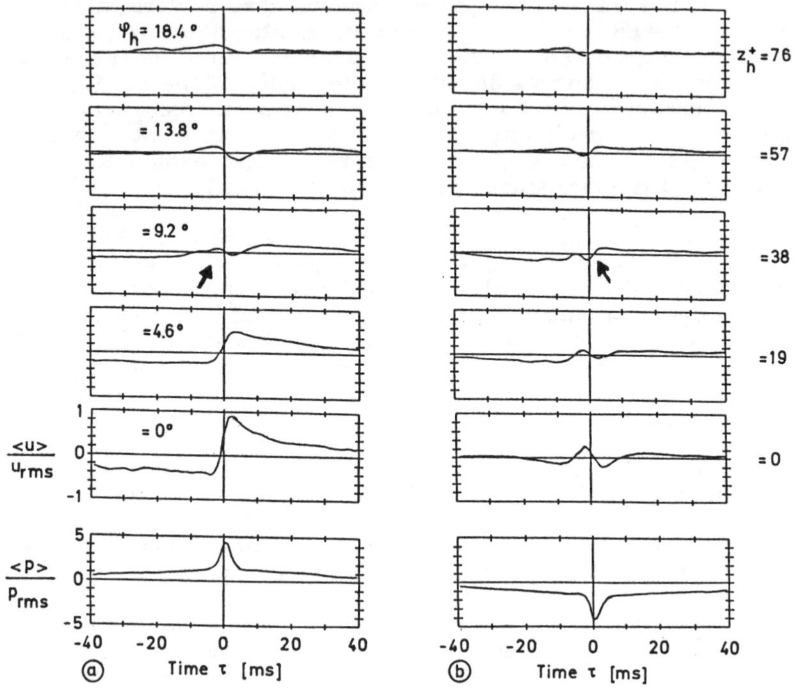

FIGURE 5. u-signatures measured with the hot-wire probe at different azimuthal positions (conditions: $p > 4 p_{rms}$ (left) and $p < -4 p_{rms}$ (right); $U_{cl} = 2.9$ m/s; $x_h = 0$, $y_h^+ = 16$, $z_h^+ = \ldots$)

Figure 6 shows measurements in which not the wall pressure was used as condition but "strong" changes in the streamwise velocity (conditions in Figure 6: $\partial u/\partial t > 8.9\ (\partial u/\partial t)_{rms}$ resp. $\partial u/\partial t < -5.5\ (\partial u/\partial t)_{rms}$; the inequality between 8.9 and -5.5 was chosen because "large" positive $\partial u/\partial t$ occur more often than "large" negative $\partial u/\partial t$). In these measurements the hot-wire probe was positioned at a constant distance from the wall ($y_h^+ = 16$) and at different azimuthal positions (an azimuthal distance of $z_h = 2$ mm corresponds to $\varphi_h = 4.6°$ or with $U_{cl} = 2.9$ m/s to $z_h^+ = 19$). From Figure 6 one can see: For the condition "large" positive $\partial u/\partial t$ the p-signatures show maxima for positions $z_h^+ = 0$ and $z_h^+ = 19$. For $z_h^+ > 19$ the function $\langle p \rangle$ is zero for all times τ. The p-signatures for the condition "large" negative $\partial u/\partial t$ are very interesting: for $z_h^+ = 0$ the p-signature shows a minimum; for $z_h^+ = 19$ it is zero for all times τ and for $z_h^+ = 38$ and $z_h^+ = 57$ maxima in $\langle p \rangle$ are found. These results show firstly, that for $z^+ = 0$ the relations between wall pressure and velocity described in the context of Figure 2 are bi-directional, and secondly, they confirm the systematic relations between p and u in spanwise direction given in Figure 5.

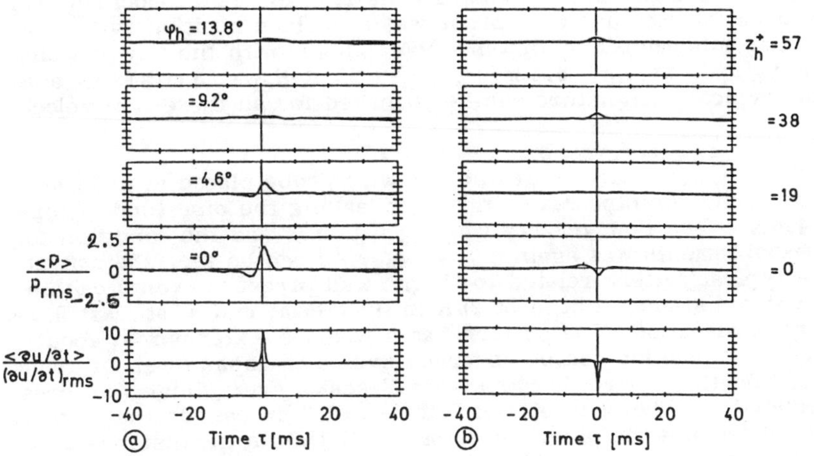

FIGURE 6. p-signatures measured with the hot-wire probe at different azimuthal positions (conditions $\partial u/\partial t > 8.9\ (\partial u/\partial t)_{rms}$ (left) and $\partial u/\partial t < -5.5\ (\partial u/\partial t)_{rms}$ (right); $U_{cl} = 2.9$ m/s; $x_h = 0$, $y_h^+ = 16$, $z_h^+ = \ldots$).

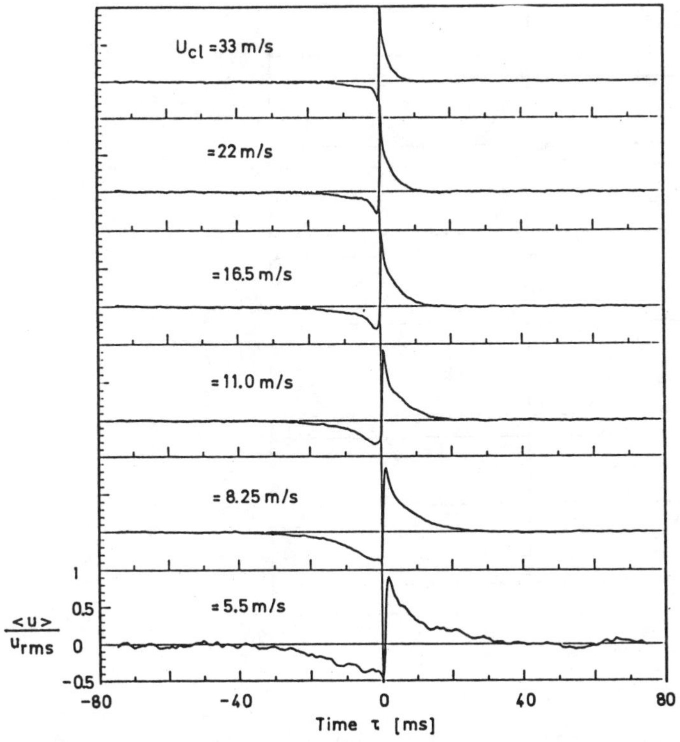

FIGURE 7. u-signatures measured with different flow velocities (condition: $p > 4p_{rms}$; $x_h = 0$, $y_h = 1$ mm, $z_h = 0$).

Figure 7 shows u-signatures measured with different flow velocities. For these measurements the hot-wire probe was at a fixed position (y_h = 1 mm), and hence the nondimensional distance y_h^+ changes with the flow velocity (here from y_h^+ = 17 at U_{cl} = 5.5 m/s to y_h^+ = 83 at U_{cl} = 33 m/s). As one can see the typical u-signature can be observed for all these flow velocities.

Figure 8 shows u-signatures with the hot-wire probe placed at different positions x_h along the pipe resp. in the jet leaving the pipe (end of pipe is at x = 4.4 R). The flow velocity was U_{cl} = 11 m/s and the condition for all these measurements was again $p > 4 p_{rms}$. As can be seen the convection of the typical pattern related to "high" wall pressure events can be followed over a distance of at least 26 R in upstream, and of at least 5.2 R in downstream direction. The pattern has a temporal extension of about 25 ms and the convection can be observed over a timespan of about 100 ms. This means that the "short-time" and "small-scale" event "high wall pressure" is related to a "long-time" and "large-scale" process in the pipe flow and in the jet leaving the pipe. (In terms of wall units the distance 26 R corresponds here to Δx^+ = 20 000 and the timespan 100 ms corresponds here to $\Delta \tau^+$ = 1500.)

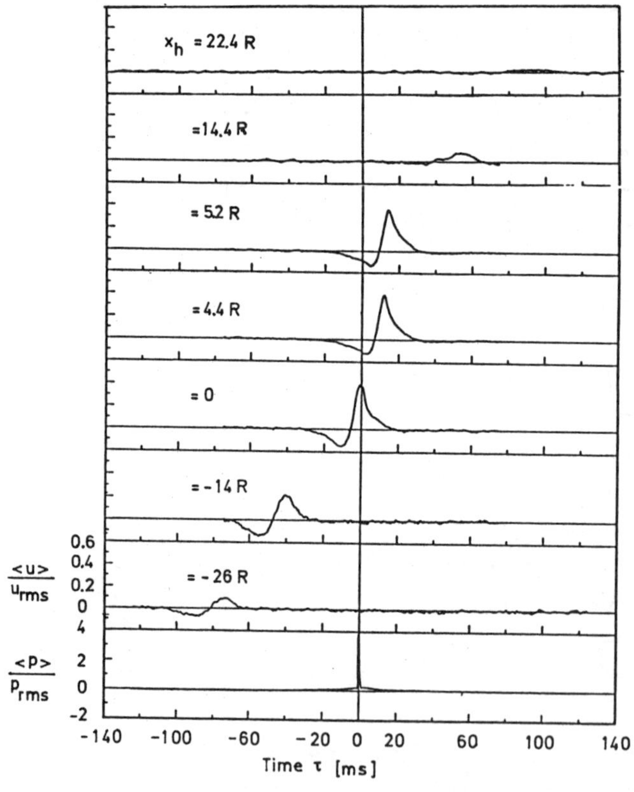

FIGURE 8. u-signatures measured with the hot-wire probe at different positions along the pipe and in the jet leaving the pipe (condition: $p > 4 p_{rms}$; U_{cl} = 11.0 m/s; x_h = ... , y_h/R = 0.4, φ_h = 0°; pipe end is located at x = 4.4 R).

Figure 9 shows more details of the x_h- and y_h-dependence of the u-signatures. The flow velocity was U_{cl} = 11 m/s, and the condition was again $p > 4p_{rms}$ (note that the ordinate scales in Figures 9c and 9d are different from those in Figures 9a and 9b). From these figures one can see the following: (I) The amplitudes of the u-signatures related to "high" wall pressure events are different from zero in the whole field which was observed here (i.e. nearly to the opposite pipe wall and over a distance along the

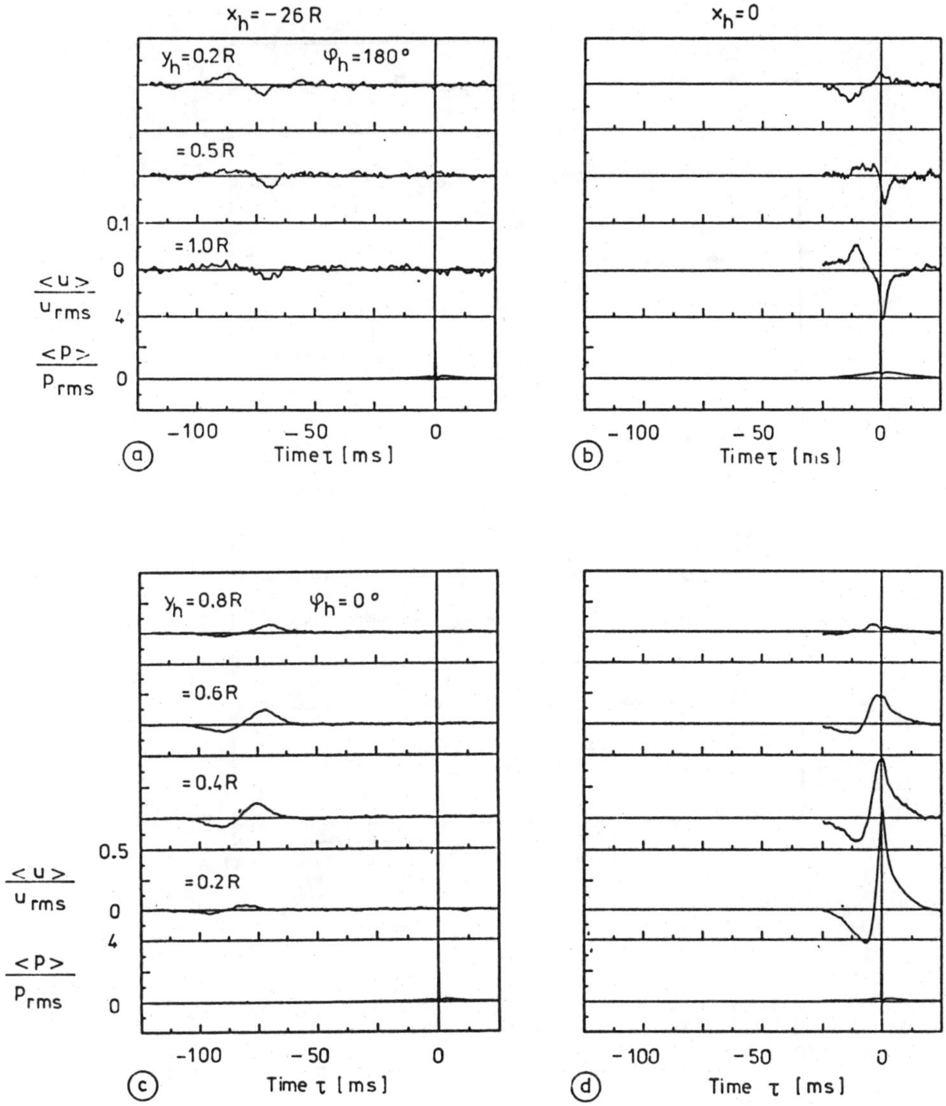

FIGURE 9. u-signatures measured with the hot-wire probe in different positions x_h and y_h (U_{cl} = 11,0 m/s; condition $p > 4p_{rms}$; x_h = -26R (left) and x_h = 0 (right), y_h = ... , φ_h = 180° (top) and φ_h = 0° (bottom)).

pipe of at least 26R). (II) The shapes of the u-signatures undergo systematic changes. (III) Most interesting is the fact that in the space close to the opposite pipe wall (at position $x_h = 0$, $y_h = 0.2R$, $\varphi_h = 180°$) the signature has a rather similar shape (with much smaller amplitude) as the signatures close to the wall pressure transducer. From Figure 10 one can see that this shape of the u-signature is maintained for azimuthal positions from at least $90° \leqslant \varphi_h \leqslant 270°$. This observation indicates a "ring-like" structure in the flow.

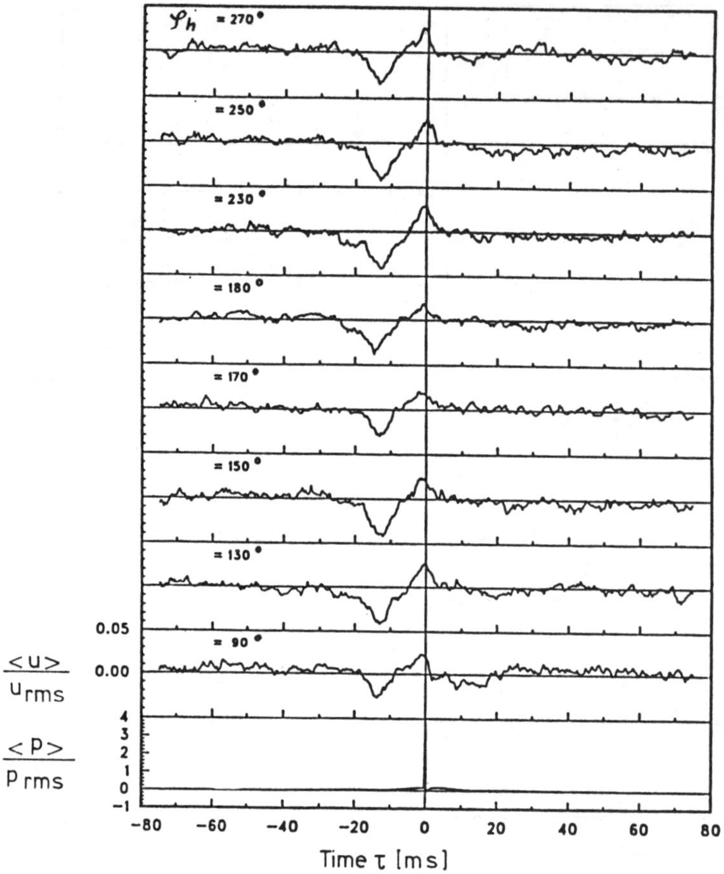

FIGURE 10. u-signatures measured with the hot-wire probe at different azimuthal positions in the half space opposite to the wall pressure transducer (condition $p > 4p_{rms}$; $U_{cl} = 11$ m/s; $x_h = 0$, $y_h = 0{,}2R$, $\varphi_h = \ldots$).

4. DISCUSSION

(1) The experiments reported in the previous section suffer from unwanted influences. Considerable effort was made to understand these influences and reduce them (for details see especially [9]).

(2) Both our measurements, as well as measurements from other investigators (Kobashi and Ichijo [10] and Johansson, Her and Haritonidis [11]), show that events of "high" wall pressure are related to local increases in <u>. Applying Taylor's hypothesis which, over short time intervalls, seems to be permissible, one can interprete this observation spatially. High wall-pressure is related to a flow process in which "low" speed fluid ($u < 0$) is followed by "high" speed fluid ($u > 0$). The high speed fluid "impinges" on the low speed fluid. The "high" wall pressure can then be understood as a sort of "stagnation pressure" transmitted to the wall from the nearby flow. (Of course the influences of other processes in the flow are superimposed, but evidently this local process dominates.) Following R. Betchov [12] we call this process "pancake making", because a ball shaped control volume moving with the flow will be deformed into an ellipsoid (a "pancake") with the minor axis directed approximately in flow direction. We regard this "pancake-making" as an important part of turbulence mechanics because in this process, for reasons of continuity, mean flow material is deviated into other flow directions. Hence mean flow energy is transformed into turbulence energy.

(3) The measurements also show that events of "low" wall pressure are related to local decreases in <u>. Using the above spatial argument again one can say that "low" speed fluid follows "high" speed fluid. Following R. Betchov we call this process "banana-making". While "pancake-making" is a process which generates a "sharp" front ("steep" increase in <u>) "banana-making" is a "soft" process ("moderate" decrese in <u>). This could explain the fact that events of "high" wall pressure give u-signatures with larger amplitudes than those found for events with "low" wall pressure. Furthermore from this consideration one can deduce that in turbulent flows "steep" increases in $\partial u/\partial t$ occur more often than "steep" decreases. In turbulent pipe flow we could confirm this conjecture. It seems that this asymmetry is a general quality of turbulent flow.

(4) "Pancake-making" detected here with the condition "high" wall pressure is evidently the same flow process as "ocurrence of bursts", detected in other investigations with the condition "high instantaneous turbulence" (VITA-technique). We will show this by comparing our results with those of Blackwelder and Kaplan [13]. Firstly the conditionally averaged u-signatures look very similar for both detection criteria (compare Figure 11b and Figure 2a). Secondly, the instantaneous timefunctions published in [13] show this characteristic "steep" increase in u for all cases in which the sampling condition "high turbulence intensity" was fulfilled (see the time functions and the marks of the VITA-detector function in Figure 11a).

(5) Spatially this process of "pancake-making" is a small scale process (the extension in z-direction is about $\Delta z^+ = 50$ and in direction normal to the wall the range with a "steep" increase of u has about the same extension). Our measurements show that this small-scale process is related to a large-scale process (characterized by <u> ≠ 0) which has an extension in x-direction in the order of several pipe radii. The convection of this large-scale process can be followed over distances of at least 26R in the pipe, and even in the jet leaving the pipe (see Figure 8). The distance 26R corresponds at $U_{cl} = 11$ m/s to $l^+ = 20\,000$.

(6) Asking what sort of "flow structures" could be behind these observations we came to the idea that "ring-like" structures might be involved, and that these rings undergo instabilities similar to those of vortex-rings. There are several reasons for this idea: (I) The mean vorticity in pipe flow has a ring structure and occurrence of local concentrations of this vorti-

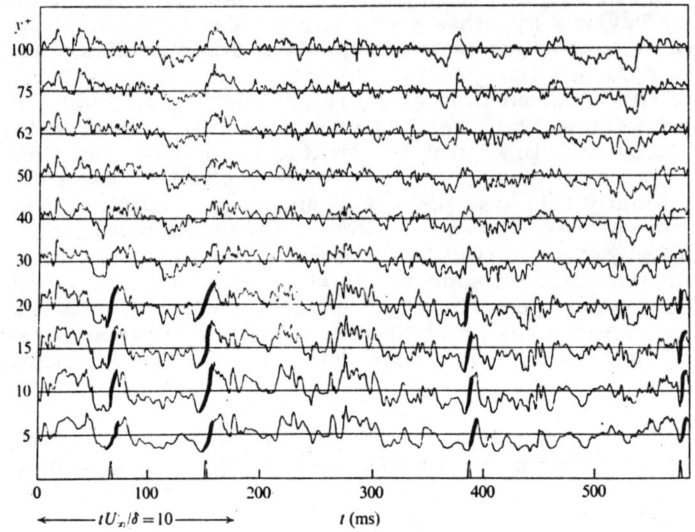

(a) FIGURE 7. Instantaneous streamwise velocities as a function of the normal direction. The detector function obtained at $y^+ = 15$ is shown between the tick marks.

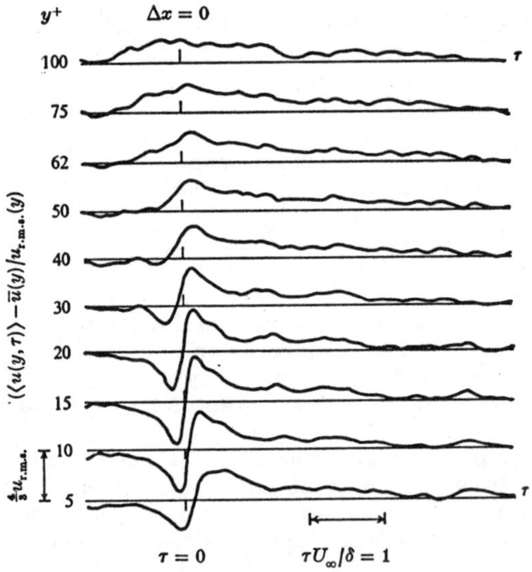

(b) FIGURE 10. Conditionally averaged streamwise velocities at ten y^+ locations. The detection criterion was applied at $y^+ = 15$. $R_\theta = 2550$.

FIGURE 11. Two figures from Blackwelder and Kaplan [13]. (To show the relation between the VITA-detector marks and the increases in u(t) these increases have been drawn thicker in the above figure.)

city can be expected. (II) Our measurements of u-signatures in the half space opposite to the wall pressure transducer indicate the existence of "ring-like" structures (see Figures 9b and 10). (III) The wavy nature of the ring instabilities could explain the existence of hairpin or horseshoe vortices in the flow which have been discussed in the context of wall pressure fluctuations by Tu and Willmarth [14] as well as by Thomas and Bull [15]. Furthermore these structures could explain the existence of low speed and high speed streaks in the flow as observed in many investigations, especially by Kline et al. [16].

(7) Carrying the speculations one step further it seems to be not unlikely that in fully developed turbulent pipe flow similar processes occur as those observed by Knapp and Roache [17] in the flow along a body of revolution for the transition process (see Figure 12).

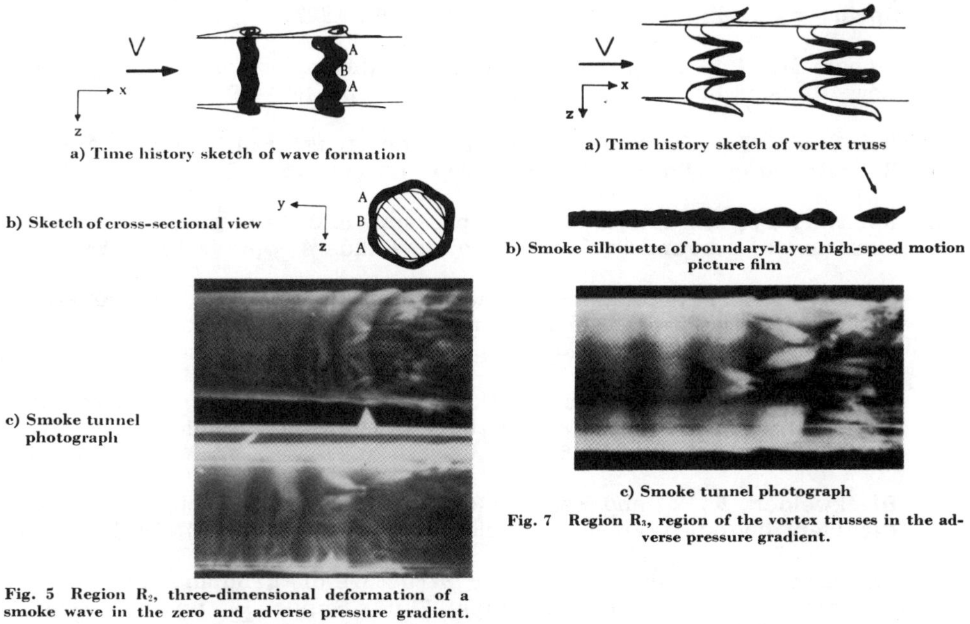

FIGURE 12. Two figures from Knapp and Roache [17] showing the time history of the transition process in boundary layer flow at a body of revolution (reprinted with permission of the American Institute of Aeronautics and Astronautics).

REFERENCES

1. Langeheineken, T., and Dinkelacker, A., Wanddruckschwankungen einer ausgebildeten, turbulenten Rohrströmung, DAGA'78, Fortschritte der Akustik, VDE-Verlag, Berlin, pp. 391-394, 1978.

2. Langeheineken, T., Experimentelle Untersuchung der Wanddruckschwankungen einer ausgebildeten turbulenten Rohrströmung, MPI Strömungsf., rep. no. 106/1979, Göttingen, 1979.

3. Langeheineken, T., Zusammenhänge zwischen Wanddruck- und Geschwindigkeitsschwankungen in turbulenter Rohrströmung, Mitt. MPI Strömungsf. no. 70, Göttingen, 1981.

4. Bull, M.K., and Langeheineken, T., On the wall pressure field in turbulent pipe flow. Mitt. MPI Strömungsf. no. 73, Göttingen, 1981.

5. Dinkelacker, A., and Langeheineken, T., Relations between wall pressure fluctuations and velocity fluctuations in turbulent flow, Proc IUTAM Symp. Marseille, eds. R. Dumas, L. Fulachier, Springer, Berlin; Heidelberg, New York, pp. 1-9, 1983.

6. Sieber, M., Großräumige Strukturen in der turbulenten Rohrströmung, Mitt. MPI Strömungsf. no. 83, Göttingen, 1987.

7. Dinkelacker, A., and Sieber, M., Coherent structures in turbulent pipe flow, Advances in Turbulence, eds. G. Comte-Bellot, J. Mathieu, Springer, Berlin, Heidelberg, pp. 391-398, 1987.

8. Sieber, M., Experiments on the attractor-dimension for turbulent pipe flow, Physics Letters A. Vol. 122, no. 9, pp. 467-470, 1987.

9. Reinhardt, T., Untersuchung kohärenter Strukturen in turbulenter Rohrströmung, MPI Strömungsf., rep. no. 19/1987, Göttingen, 1987.

10. Kobashi, Y., and Ichijo, M., Wall pressure and its relation to turbulent structure of a boundary layer. Exp. in Fluids 4, pp. 49-55, 1986.

11. Johansson, A.V., Her, J.-Y., and Haritonidis, J.H., On the generation of high-amplitude wall-pressure peaks in turbulent boundary layers and spots, J. Fluid Mech. 175, pp. 119-142, 1987.

12. Betchov, R., private communication (partly contained in "An inequality concerning the production of vorticity in isotropic turbulence, J. Fluid Mech. 1, pp. 497-504, 1956).

13. Blackwelder, R.F., and Kaplan, R.E., On the wall structure of the turbulent boundary layer, J. Fluid Mech. 76, pp. 89-112, 1976.

14. Willmarth, W.W., and Tu, Bo-Jang,Structure of turbulence in the boundary layer near the wall, Phys. Fluids Suppl. 10, pp. S134-S137, 1967.

15. Thomas, A.S.W., and Bull, M.K., On the role of wall-pressure fluctuations in deterministic motions in the turbulent boundary layer, J. Fluid Mech. 128, pp. 283-322, 1983.

16. Kline, S.J., Reynolds, W.C. Schraub, F.A., and Rundstadler, P.W., The structure of turbulent boundary layers, J. Fluid Mech. 30, pp. 741-773, 1967.

17. Knapp, C.F., and Roache, P.J., A Combined visual and hot-wire anemometer investigation of boundary-layer transition, AIAA Journal, Vol. 6, no. 1, pp. 29-36, 1968.

Relation between Wall Pressure and Turbulence Structure

YASUJIRO KOBASHI
Hokkaido Institute of Technology
Department of Mechanical Engineering

MAKOTO ICHIJO
Hokkaido University
Department of Mechanical Engineering
419-2 Teine-Maeda, Nishi-ku, Sapporo
Hokkaido, Japan 006

INTRODUCTION

Experiments and calculations were made in order to study the structure of coherent motions of a turbulent boundary layer and their relations to the wall pressure.

1) The flow outside of the boundary layer which is closely related with large scale motions inside was ensemble averaged by referring both to the flow pattern of itself and the burst signals near the wall. The results show that the large scale motions are composed of periodic and aperiodic structures and that the bursts appear only at some fixed phase of the periodic one. This fact suggests that the bursts are caused not from the instbility of the flow near the wall, but from the interaction between large scale motions and the wall.

2) Calculations of the wall pressure were made by using very simple assumptions that the pressure related with large scale motion is caused by the series of vortices in an inviscid flow and that the one which is related with burst mostly comes from the viscous terms of the Navier Stokes equation. The results agree qualitatively well with the experiments.

3) The history of a burst which was found by referring to the maximum peaks of dhe wall pressure fluctuations shows that the burst moves at the same speed as the large scale motion in its growing phase, while it decelerates to half speed in its decaying phase. Based on this interesting result, a structural model which connects the burst to the large scale motion was established.

COHERENT MOTIONS OF A TURBULENT BOUNDARY LAYER

Two types of coherent motions are known to characterize a turbulent boundary layer, the one is the burst which produces turbulent energies near the wall and the other is the large scale motion which has a vortical structure prevailing whole through the turbulent region and plays important roles in the entrainment of the fluid outside. The relation between bursts and large scale motions was examined by taking ensemble averages of the flow outside the boundary layer ($y/\delta = 1.7$, where y is the distance from the wall and δ is the boundary layer thickness) which is directly related to the large scale motions by referring both to the flow itself and to the burst near the wall [1]. The flow patterns, which are given in Fig.1 a and b together with the corresponding wall pressure patterns, show clear periodicity if they are referred

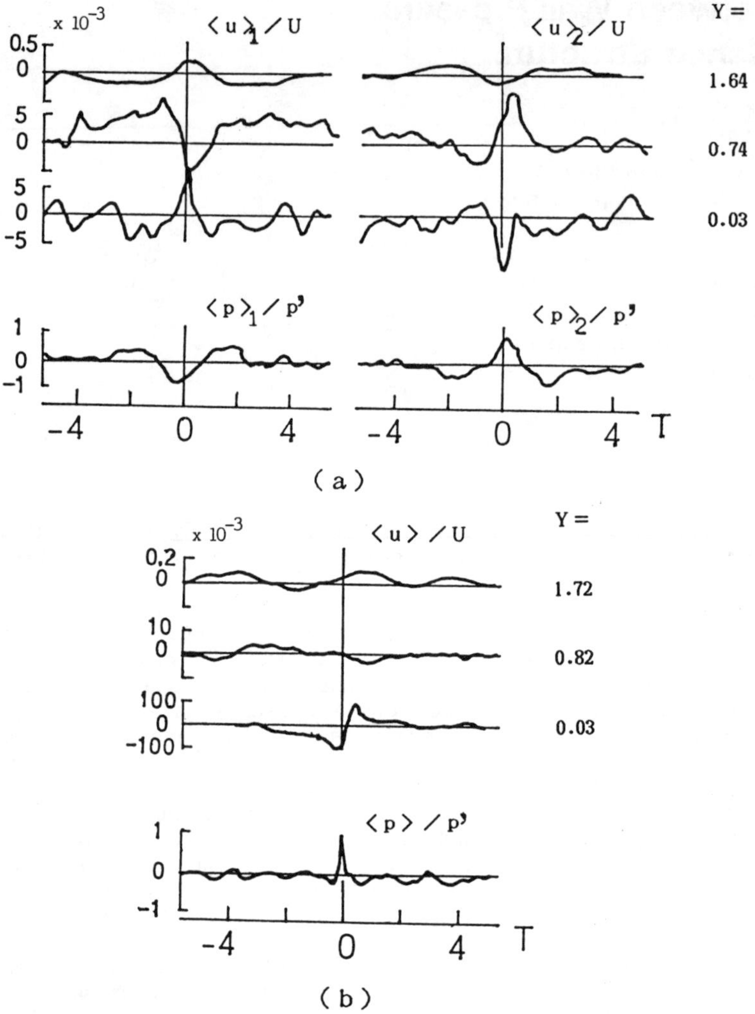

FIGURE 1. Velocity and wall pressure distributions of a boundary layer referred to large scale motion (a) and burst (b)

to the burst (expressed by $\langle u\rangle/U$ against T, where U is the main flow speed and T is the non-dimensional time defined by Ut/δ), but not when they are referred to the peaks ($\langle u\rangle_1/U$) or valleys ($\langle u\rangle_2/U$) of the flow itself. This rather funny result leads us to conclude that the large scale motions observed are composed of periodic and aperiodic structures and that the bursts appear only at some fixed phase of the periodic one. Townsend [2] has found, in this connection, that the aperiodic motions are the parts of large scale structures which have been broken down from the coherent periodic structure, which was produced probably from the instability of the turbulent shear flow. Because of this reason Townsend named the periodic structure as the "new" and the aperiodic one as the "old" structure.

Large Scale Motion and Entrainment

The periodic, or "new" coherent structure in Townsend sense, is quite 2-dimensional at its beginning, but it easily induces longitudinal paired vortices (Görtler vortices) along the separation stream surface as was mentioned by Hussain [3]. These paired vortices are stretched and intensified by the rolling action of the periodic large scale structure, and act to induce the non-turbulent fluid outside of the boundary layer inwards and carry the turbulent fluid outwards. This is the entrainment mechnism of the turbulent shear flows. The non-turbulent fluid induced inwards seems to get vorticity of the same sence as the large scale motion due to the viscosity of the super-layer at the boundary. These vortices will be brought over almost instantaneously into the large scale vortical structure, as was discovered by Kiya et al [4].

Origine and Phenomena of a Burst

The fact that the burst appears at some fixed phase of the periodic part of the large scale motions suggests that the burst is not caused by the instability of the flow near the wall as is usually understood, but it occurs due to the interaction between large scale motion and the wall. Again in this case, the Görtler votices which have been caused by the second instability of the separation stream surface play an essential role. They wind around the large scale structure and work to lift the low momentum fluid near the wall upwards in front of the structure and inrush the high momentum fluid in the back. These are the normal ejectin and sweep phenomena of a burst. There are also possibilities of the inverse burst phenomena to occur in between the normal ones.

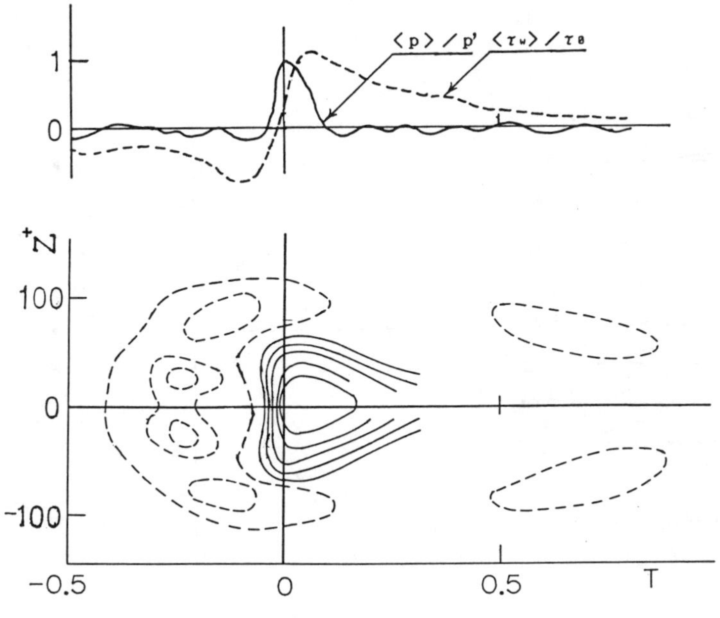

FIGURE 2. Wall pressure and shear stress patterns of a burst.

The pressure is not the direct measure of the flow fluctuations because it is influenced by the whole flow field in the sense that it is a solution of a Poison type equation. But because of its integrating effect of the flow, it is expected to give more clear informations of the coherent motions than the flow velocities themselves, which are contaminated by incoherent eddies. This is the reason why the wall pressures ensemble averaged by referring to either the flow outside of the boundary layer or the burst near the wall reviel flow structure very clearly. Fig.2 shows the wall pressure pattern when the peak of center section of a burst passes through the fixed point of the wall [5]. It has a heart-shape cross section just like the turbulent spot which is observed in the laminar to turbulence transition. It also has an isolated sharp positive peak being followed by shallow valleys in its front and back. The height of pressure peak corresponds to the energy content of the burst, and therefore to the history of its growth. Fig.3 a shows the history of a burst ensemble averaged under the condition that the peak value of pressure observed at a fixed point exceeds some threshold value which is high enough to guarantee that the burst has grown to its maximum peak. It shows how a burst grows in the upsteam and how it decays in the downstream. Fig.3 b is the convection speed of the burst which is calculated from Fig.3 a and expressed in X-T diagram, where X is the nondimensional streamwise distance defined by X= x/δ and T is the time defined by T= x/U respectively. Here x is the streamwise distance measured from the fixed point, positive in the streamwise direction, and U is the main flow speed. The figure shows that the burst moves at nearly the same speed as the main flow in the upstream and therefore in its growing region, while it decelerates suddenly to half speed in its decaying region downstream. This is a rather surprising result,

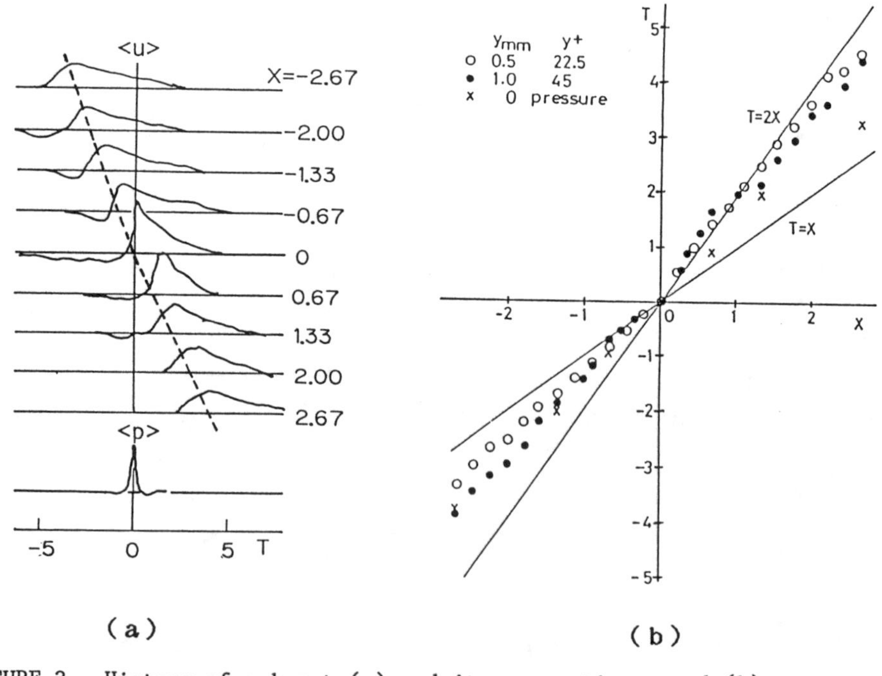

FIGURE 3. History of a burst (a) and its convection speed (b)

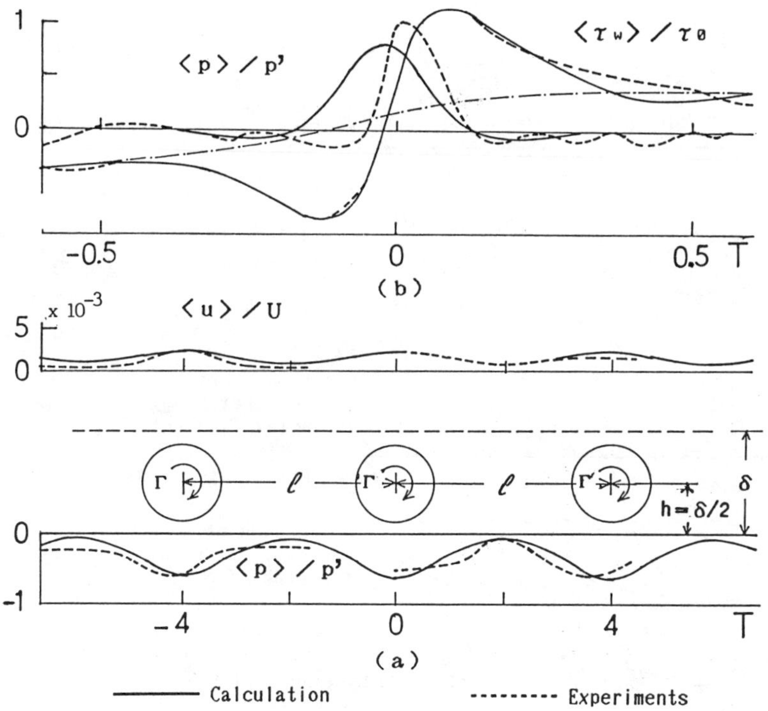

FIGURE 4. Wall pressure distributions of a boundary layer related with large scale motion (a) and burst (b)

but it turns out to be very important when the structure of the coherent motions, or the causality between them is to be considered.
The pressure and velocity of a turbulent flow are the interchangeable quantities with each other and the former is known to controle the turbulence structure by redistributing the energies between various velocity components. And because of these functions, the explicit analysis of the pressure of any turbulent flow is usually very difficult. However, because the coherent motions are not very much affected by this interchangeability, the wall presure related with them can be evaluated with rather simple assumptions. The assumptions used are that the wall pressure related with large scale motions is given by the Berboulli's equation of the series of vortices arranged regularly along the wall and that the one related with bursts mainly comes from the viscous terms of the Navier Stokes equation. The calculated results are shown in Fig.4, a and b together with the velocity outside of the boundary layer and the shear stress at the wall respectively. They agree with the experiments qualitatively well in spite of the crude assumptions.

STRUCTURE AND RELATION OF COHERENT MOTIONS OF A BOUNDARY LAYER

The existence of periodic and aperiodic, or "new" and "old" large scale motions in a boundary layer suggests that each of the large scale motions observed at a fixed point has different origine of its growth. This

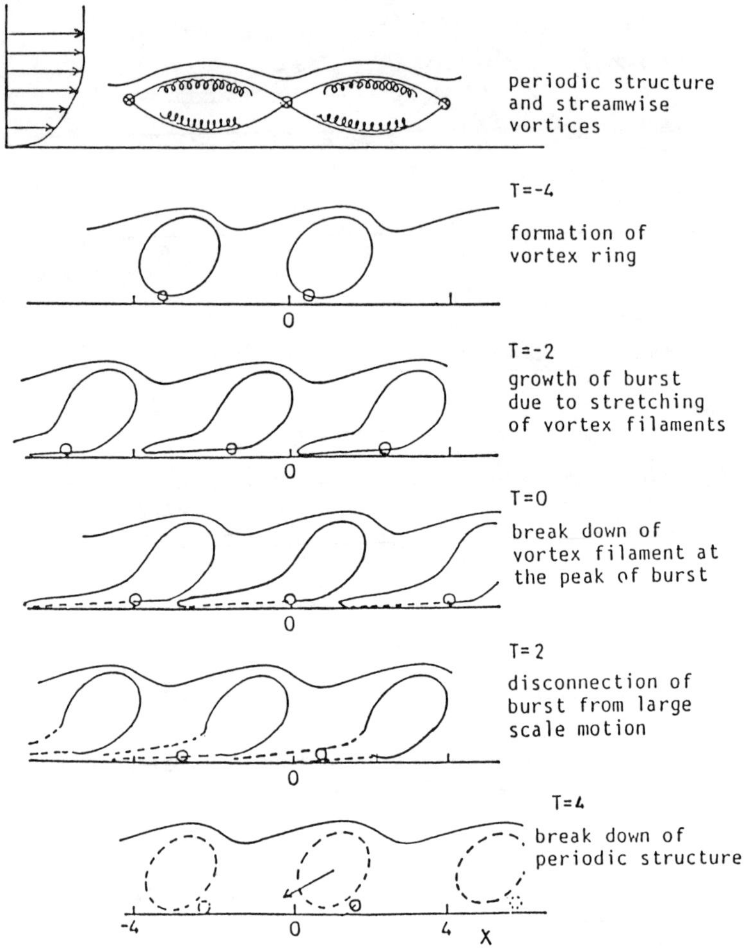

FIGURE 5. Relation between large scale motions and bursts

is also true for the case of bursts, because they are closely connected with the newly borne large scale motions.
The fact that the moving speed of a burst changes suddenly as soon as the peak of pressure reaches its maximum value (Fig.3 b) gives further informations about the relation between coherent motions. Because the maximum peak of pressure related with burst corresponds to its most energetic state, the above fact shows that the burst moves together with large scale coherent structure absorbing energies from which until it grows to the most energetic state, and thereafter, it departs from the large scale structure and starts to move at its own speed. The separation of burst from the large scale motion may either because the former has become too energetic to absorve more energy or the latter has exhausted too much to supply more energy. The burst thus departed from large scale motion starts to decay due to the viscosity near the wall. Fig.5 is the historical model of the coherent motions to relate the burst with the large scale motion.

CONCLUDING REMARKS

The large scale motions and bursts which characterize the turbulent boundary layer are closely related with each other being intermediated by Görtler vortices which are caused from the second instability of the flow along the separation surface between the neighbouring large scale motions.

The wall pressure is found to be a very effective tool for the study of coherent motions because it is less contaminated by the randomness of incoherent eddies and therefore more directly related with coherent motions than the flow velocities. And this is the reason why both ensemble averagings and correlation measurements of the flow field in relation to the wall pressure give clear informations of the cohernt motions of a boundary layer.

REFERENCES

1, Kobashi,Y., Ichijo,M., Structure of a turbulent boundary layer and its relation to wall pressure fluctuation (in Japanese), Japan J. Fluid Mech. (Nagare), vol. 1, no. 4, pp. 350-359, 1982

2. Townsend,A.A., The pattern of large eddies in a wake and in a boundary layer, J. Fluid Mech. vol. 95, pp. 515-537 (1979)

3. Hussain,A.K.M.F., Coherent structures and incoherent turbulence, Proc. Turbulence and Chaotic Phenomena in Fluids (IUTAM Sympo. Kyoto), pp. 453-460 (1984)

4. Kiya,M., Ohyama,M., Hunt,J.C.R., Vortex pairs and rings interacting with shear layer vortices, J. Fluid Mech., vol. 172, pp. 1-15 (1986)

5. Hayakawa,M., Kobashi,Y., Flow pattern related with coherent motions of a turbulent boundary layer (in Japanese), Proc. 14th Turbulence Sympo. pp. 79-83 (1986)

Velocity and Pressure Fields Associated with Near-Wall Turbulence Structures

ARNE V. JOHANSSON and P. HENRIK ALFREDSSON
Department of Mechanics
The Royal Institute of Technology
S-100 44 Stockholm, Sweden

JOHN KIM
NASA Ames Research Center
Moffet Field, California 94035, USA

ABSTRACT

Computer generated databases containing velocity and pressure fields in three-dimensional space at a sequence of time-steps, were used for the investigation of near-wall turbulence structures, their space-time evolution, and their associated pressure fields. The main body of the results were obtained from simulation data for turbulent channel flow at a Reynolds number of 180 (based on half-channel height and friction velocity) with a grid of 128* 129*128 points. The flow was followed over a total time of 141 viscous time units. Spanwise centering of the detected structures was found to be essential in order to obtain a correct magnitude of the associated Reynolds stress contribution. A positive wall-pressure peak is found immediately beneath the center of the structure. The maximum amplitude of the pressure pattern was, however, found in the buffer region at the center of the shear-layer. It was also found that these flow structures often reach a maximum strength in connection with an asymmetric spanwise motion, which motivated the construction of a conditional sampling scheme that preserved this asymmetry.

1. INTRODUCTION

Inclined internal shear-layers have been shown, by use of various conditional sampling methods, stochastic estimation etc. to be prevalent and dynamically important structures (for the process of turbulence production) in the most active part of turbulent boundary layers and other types of wall-bounded shear flows. Their creation is coupled to the process of lift-up of low speed fluid (from low-speed streaks) from the viscous sublayer. These buffer region shear-layers are highly three-dimensional and are inclined at a relatively shallow angle with respect to the wall. They are, by definition, associated with a high value of the instantaneous velocity gradient $\partial u/\partial y$[#]. The strength and inclination is affected by the shearing action of the mean velocity gradient. At maximum strength the instantaneous gradient is often close to that of the mean gradient at the wall, and the inclination around y = 15 corresponds to an angle of about 20°. This means that also the streamwise velocity gradient ($\partial u/\partial x$) across the shear-layer becomes of comparable magnitude.

A fixed observer will see a high value of the temporal derivative of the streamwise velocity and a large change in u as the shear-layer passes by. This feature is utilized in the so called VITA (variable interval time averaging) technique which has been used in various experimental investigations (see e.g. Blackwelder & Kaplan 1976 and Johansson &

[#] x, y and z will in the following be used to denote the streamwise, normal and spanwise coordinates normalized by the viscous lengthscale. Similarly, u, v and w will denote the corresponding fluctuating parts of the velocity, with the same scaling. Also other quantities will be normalized by viscous scales if nothing else is stated explicitly.

Alfredsson 1982) to detect shear-layer structures in boundary-layers and channel flows. With this detection scheme an event is considered to occur when the variance of u averaged over a 'short' time T exceeds a chosen threshold level, k, times the long time averaged variance (i.e. u_{rms}^2). The detection position is normally chosen in the buffer region. In the analysis of computer generated data-bases a spatial counterpart, VISA, is used, where the averaging length will be denoted L (see Kim 1985). In the present study, this method was extended by use of the three-dimensional spatial information to center the events in the spanwise direction.

Most of the information on the spatial scales of near-wall turbulence structures has been obtained from flow visualizations whereas measurements have provided some quantitative information on their development. Computer generated databases, obtained from numerical simulation of turbulent flows, which contain velocities and pressure fields in three-dimensional space and time, makes it possible to study these structures in more detail, especially regarding their space-time evolution.

The main body of the results presented here have been obtained from simulation data for turbulent channel flow at a Reynolds number of 180 (based on half-channel height and friction velocity) with a grid of 128*129*128 points. The grid-spacing in terms of viscous units was 17.67 in the streamwise and 5.89 in the spanwise direction.

Results concerning the three-dimensional spatial structure of internal shear-layer type structures, and their evolution in time are presented in section 2. Many of these aspects are compared with previously published (and some unpublished) data from the Göttingen oil channel (Johansson, Alfredsson & Eckelmann 1987) for a comparable Reynolds number. The agreement between numerical and experimental results concerning the spatial coherence of the structure in the symmetry (xy) plane of the structure was found to be excellent. This was also the case for the frequency of occurrence of these structures.The wall pressure patterns, as well as the pressure field in the flow, associated with the detected turbulence structures could be constructed from the simulation data, and some comparisons with previously published results are presented.Conclusions are summarized in section 3.

2. NEAR-WALL VELOCITY AND PRESSURE PATTERNS IN CHANNEL FLOW

As a first step in the detection of shear-layer structures in the buffer region the locally averaged variance (or VISA-variance) was computed for various xz-planes. For the following channel flow results the detection was applied at y=15 with a threshold level (k) of 1.0. The resulting regions of high variance, indicating the existence of strong shear-layers, gives a spatially spotty picture (figure 1). In figure 1 the variance was averaged over 11 grid points in the x-direction, corresponding to about 200 viscous length units, and about 30 regions with levels above 1.0 can be observed. In the present approach these regions were identified, i.e. their maxima were located and followed from their formation to their disintegration, for a time sequence of 141 viscous time units (t_*), each time step separated by 3 viscous time units.

As examples, the space-time history of three high-variance islands marked by A, B and C in figure 1 is shown in figure 2, where consecutive plots are separated by 12 t_*. Event C is here followed over more than 70 t_* during which it has travelled approximately 900 viscous units, which here corresponds to 5 channel half-heights. The shear-layer can, however, be identified over considerably larger distances. The mean survival time of the shear-layers, such as those indicated in figure 1, was approximately 50 t_*, but some strong shear layers could be followed throughout the entire extent of the time studied (141 t_*).

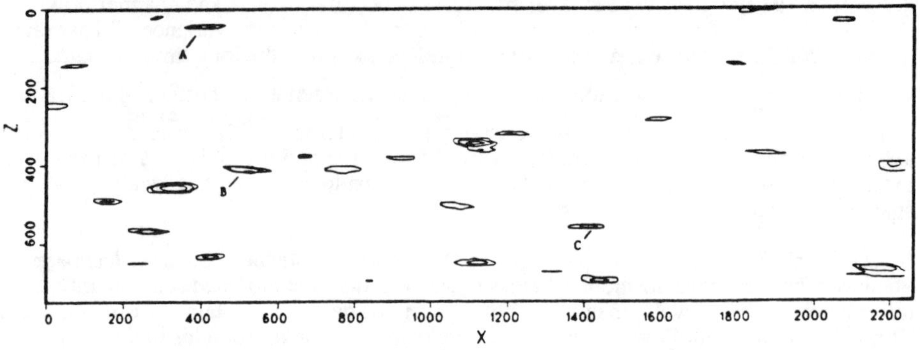

Figure 1. Contours of the streamwise velocity variance averaged over a distance of 200 viscous length units at y=15. Contours start at 1.0 with an increment of 0.5.

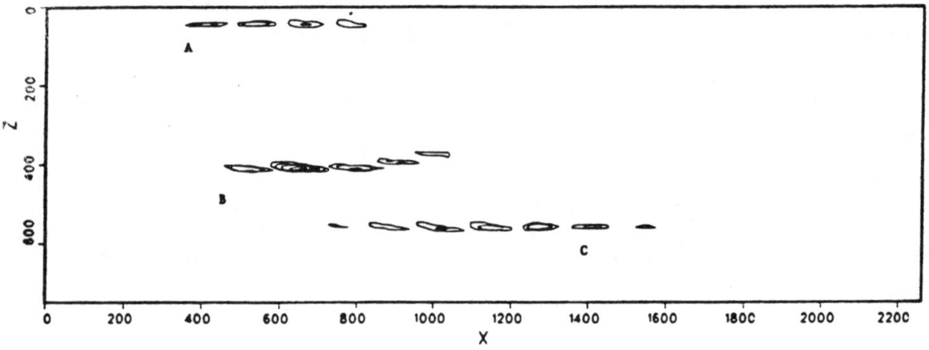

Figure 2. The space-time development of the variance associated with the events marked A, B and C in figure 1. Time separation between plots is 12 viscous time units.

If the events in figure 1, and those detected at y=15 near the opposite wall, are divided into intervals of the maximum variance amplitude, the low amplitudes will dominate, i.e. at a given time there are many events (regions of high variance) with a maximum variance just above the threshold. On the other hand, if all these events are traced in time to locate the time of maximum shear-layer strength (i.e. global variance maximum), the most probable variance maximum is instead found in the interval 1.5-2.0, and values close to 3.5 were observed. As should be expected, the strong shear-layers could be followed over longer times than the weaker ones.

The propagation velocity of the shear-layers could be determined from their space-time history, and was found to be 10.6 with a standard deviation of ±1.0. This is substantially lower than the value, 13.0, obtained from an experimental investigation (Johansson, Alfredsson & Eckelmann 1987) at about the same Reynolds number (10 % higher). In that study the VITA-detection was carried out at y=15 and a second probe was positioned at various y and x-positions downstream of the detector probe (e.g. at y=20). However, the averaging time was 10 t_*, which would correspond to a shorter averaging length than that used in the analysis of the numerical data. Also, there is an inherent difference between spatial and temporal data evaluation. It is here noteworthy that the propagation velocity of high-amplitude wall-pressure fluctuations has been found (experimentally) to be roughly 12 in wall units (see e.g. Schewe, 1983).

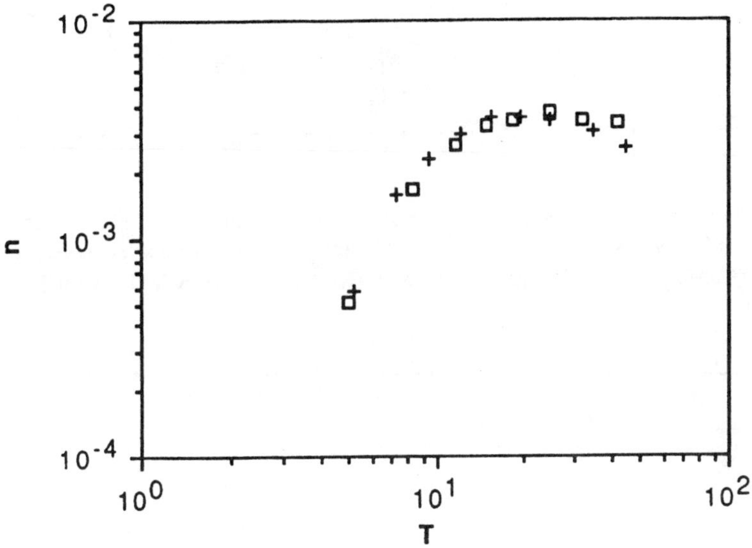

Figure 3. Number of events that would be detected per unit time with a stationary probe as function of the averaging time. Simulation (□) and measurements in the Göttingen oil channel (+) at about the same Reynolds number (k=1.0, y =15).

The significance of these flow structures for the flow dynamics in the near-wall region depends on their frequency of occurrence (or probability to occur within a given area). In order to enable comparisons with experimental results obtained with a stationary probe the average number of detections per z-position was determined and normalized by the streamwise extent of the domain (2260) divided by the propagation velocity (10.6). This means that a shear-layer, if signified by a region of high variance extending over several z-positions, gets counted several times, but gives the equivalent of the number of detections per unit time for a fixed probe. Also, the averaging length was converted to an equivalent averaging time by use of the propagation velocity. The agreement is excellent (see figure 3) between the present results obtained in this manner and the experimental results (unpublished results by the present authors and H. Eckelmann). One should bear in mind that the scaling used is immaterial since the Reynolds number is approximately the same for the two sets of data.

Johansson, Alfredsson & Eckelmann (1987) studied the spatial structure in the symmetry plane of the shear layers. Their results are shown in figure 4 where contour lines of the streamwise disturbance velocity normalized with the local rms-value are plotted in the xy-plane. These results may be compared with the corresponding results (figure 5 a) obtained from the computer generated data base for channel flow. An ensemble average of 60 events detected at both sides of the channel at one time-step are shown. The ensemble averaging here involves centering of the individual realizations in both the x (as in conventional conditional averaging) and the z-direction. Hence, boxes 600 long, 200 wide and 180 high (the entire half channel height) centered around the point of maximum variance (at y=15) are averaged. The agreement between the experimental and simulation data is seen to be good. In the following, conditional averages (denoted by < >) will be presented in absolute scale, i.e. normalized by u_τ (or other appropriate viscous scales).

Figures 5 b and c shows the conditionally averaged u and v velocities in the xy-plane. An interesting feature is that close to the wall the lifted low speed fluid is pushed back towards

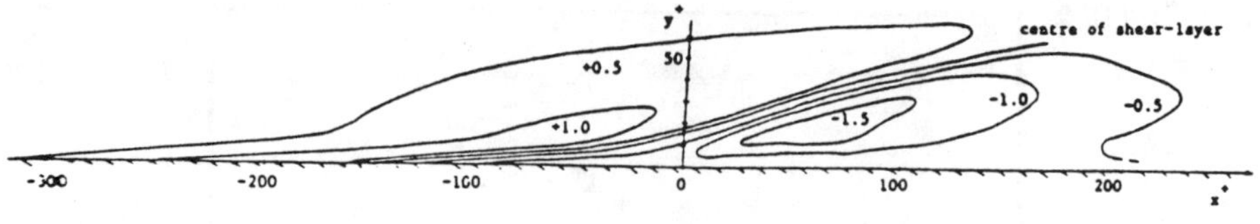

Figure 4. Conditionally averaged shear-layer structure in the near-wall region of channel flow at Re_τ=200 (Johanson, Alfredsson & Eckelmann 1987) showing streamwise velocity disturbance ($<u>/u_{rms}(y)$) in the symmetry plane. Events detected by VITA with T=10 and k=1.0.

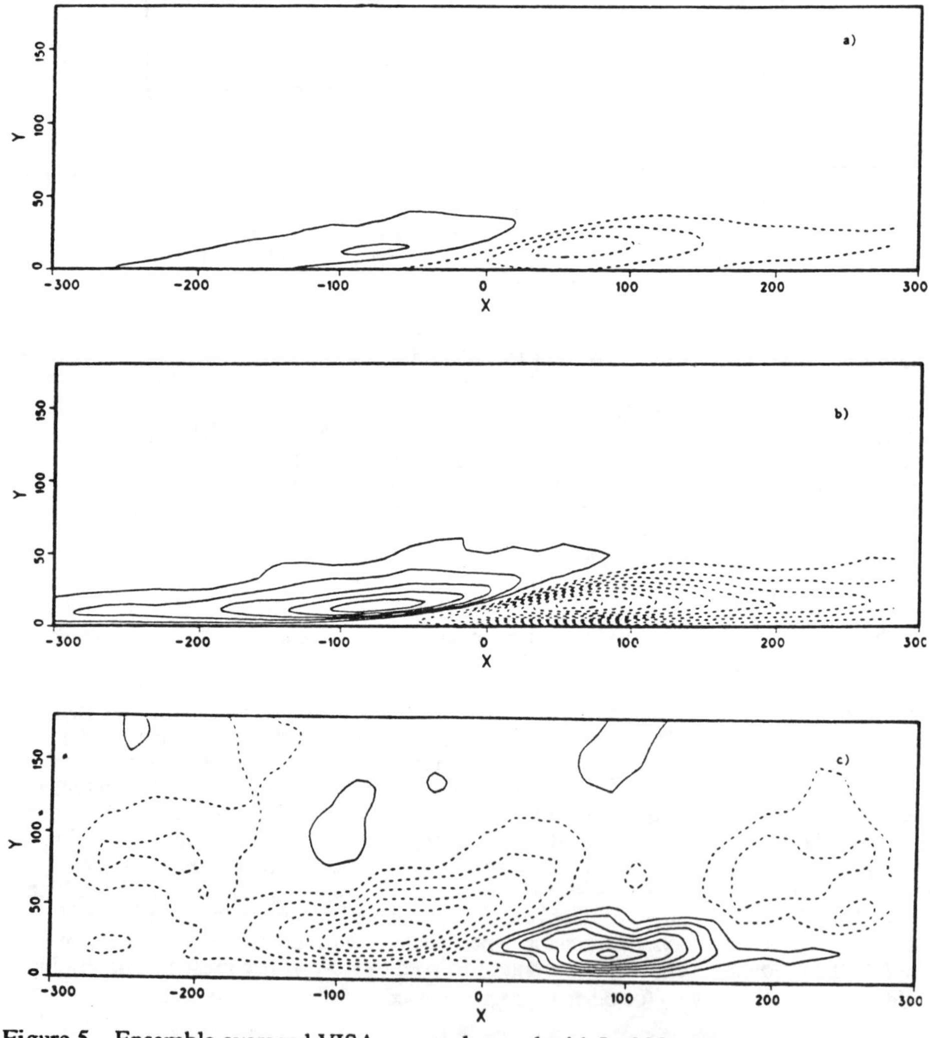

Figure 5. Ensemble averaged VISA-events detected with L=200 and k=1.0. Full drawn lines and broken lines denote positive and negative contours, respectively. a) $<u>/u_{rms}(y)$, contour increment is 0.5. b) <u>, contour increment is 0.5. c) <v>, contour increment is 0.1.

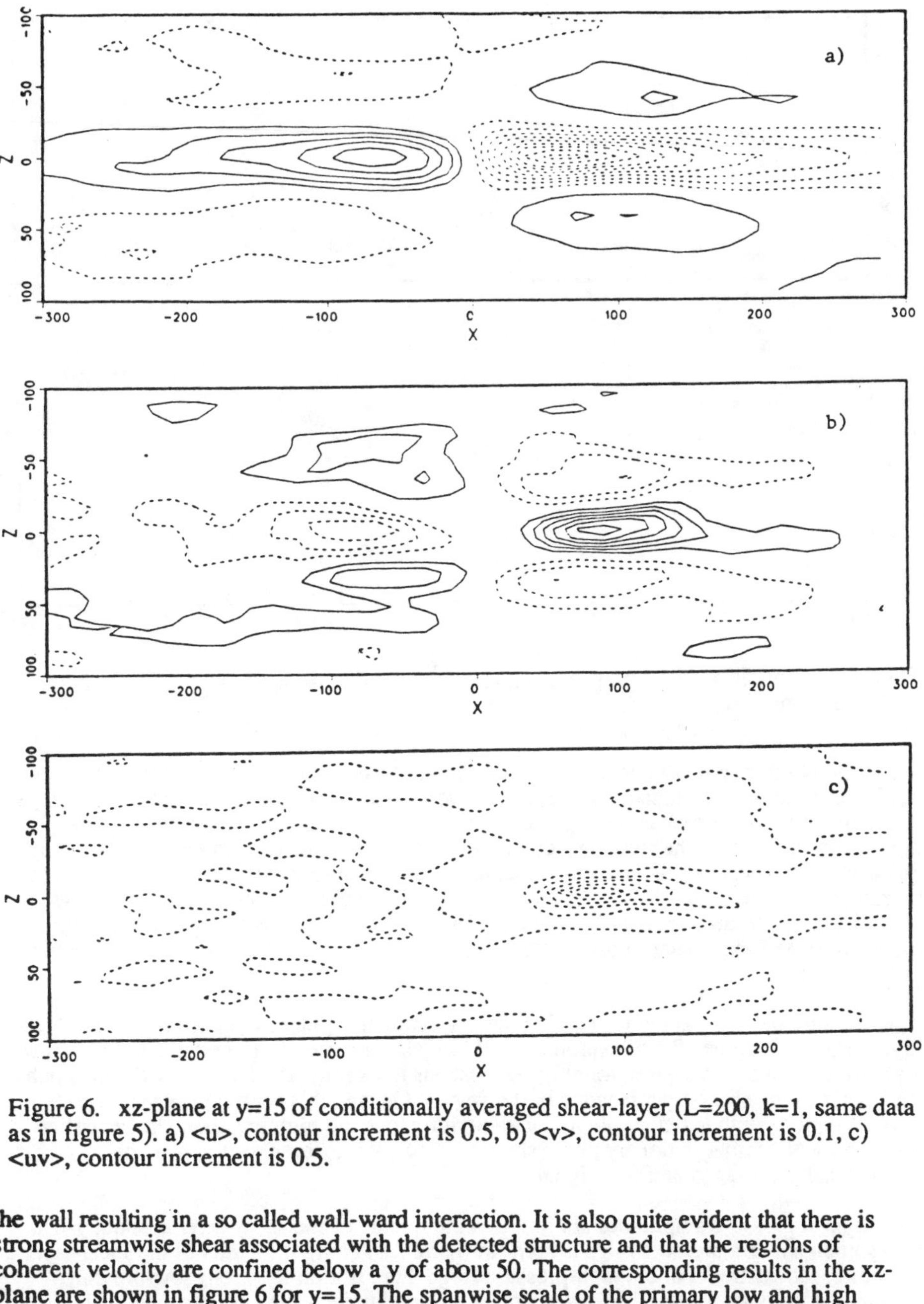

Figure 6. xz-plane at y=15 of conditionally averaged shear-layer (L=200, k=1, same data as in figure 5). a) <u>, contour increment is 0.5, b) <v>, contour increment is 0.1, c) <uv>, contour increment is 0.5.

the wall resulting in a so called wall-ward interaction. It is also quite evident that there is strong streamwise shear associated with the detected structure and that the regions of coherent velocity are confined below a y of about 50. The corresponding results in the xz-plane are shown in figure 6 for y=15. The spanwise scale of the primary low and high speed regions is seen to be about 50, which is the same as the distance between high and low-speed streaks in the viscous sublayer. The maximum deviation from the long-time mean in the low-speed region, in figure 5 b, is almost twice as large as on the high speed side of the shear-layer. Similarly the amplitude of the normal velocity is also much higher on the downstream side, which also gives a strong uv-peak in that region (figure 6 c). The uv-contribution on the sweep side is practically negligible at this y-location.

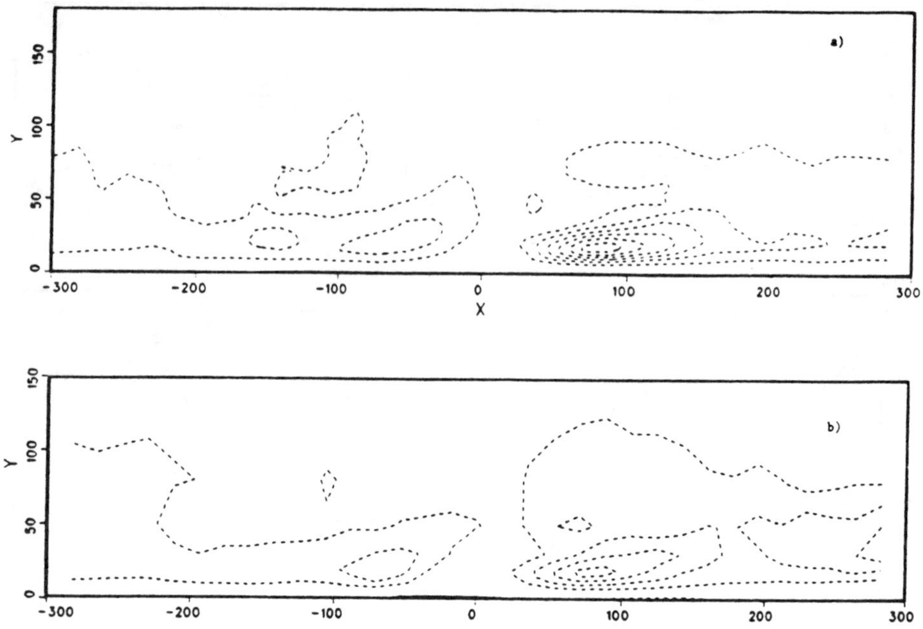

Figure 7. Conditionally averaged uv (L=200, k=1, contour increment is 0.5), a) with spanwise centering, b) without centering.

It was shown that spanwise centering of the events is essential to obtain reasonable quantitative estimates of the associated Reynolds stress contributions, as is illustrated by figures 7 a,b. In the latter no such centering was applied and the resulting maximum uv is drastically reduced on the ejection side. It should here be noted that (to the present authors' knowledge) all previous results, experimental as well as numerical, presented in the literature has been obtained without centering. The reduction that results if no spanwise centering is used can be ascribed to the resulting jitter in the spanwise separation between the uv-peak and the detector position, and the fact that the spanwise scale of the uv-peak is rather small.

The pressure patterns, at the wall and in the x-y symmetry plane, associated with the shear-layer structure (figures 8 a,b) exhibit several interesting features. The wall-pressure shows a positive peak over a region, roughly 100 viscous units long and 50 units wide, in which the center of the shear-layer is in the buffer region. Outside this central peak there are several regions with lower amplitude of alternating low and high pressure. Off-centerline measurements with a stationary probe should hence exhibit some waviness in the conditionally averaged pressure signal.

The highest pressure amplitudes are found in the middle of the shear-layer at about y=15, and somewhat of a pressure ridge is seen (in figure 8 b) along the center of the shear-layer structure. Downstream of the high-pressure ridge there is a rather strong low-pressure region at a y-distance from the wall of 25-30. Its amplitude is even greater than the high-pressure peak, but its imprint on the wall pressure is of significantly weaker amplitude.

The general characteristics, such as streamwise extent in viscous units, of the wall-pressure pattern agree closely with earlier reported experimental results obtained in pipe and boundary layer flows (see e.g. Langeheineken 1981, Dinkelacker & Langeheineken 1983, Kobashi & Ichijo 1986 and Johansson, Her & Haritonidis 1987). However, the peak amplitude in the present case is only a fraction, approximately one third, of that

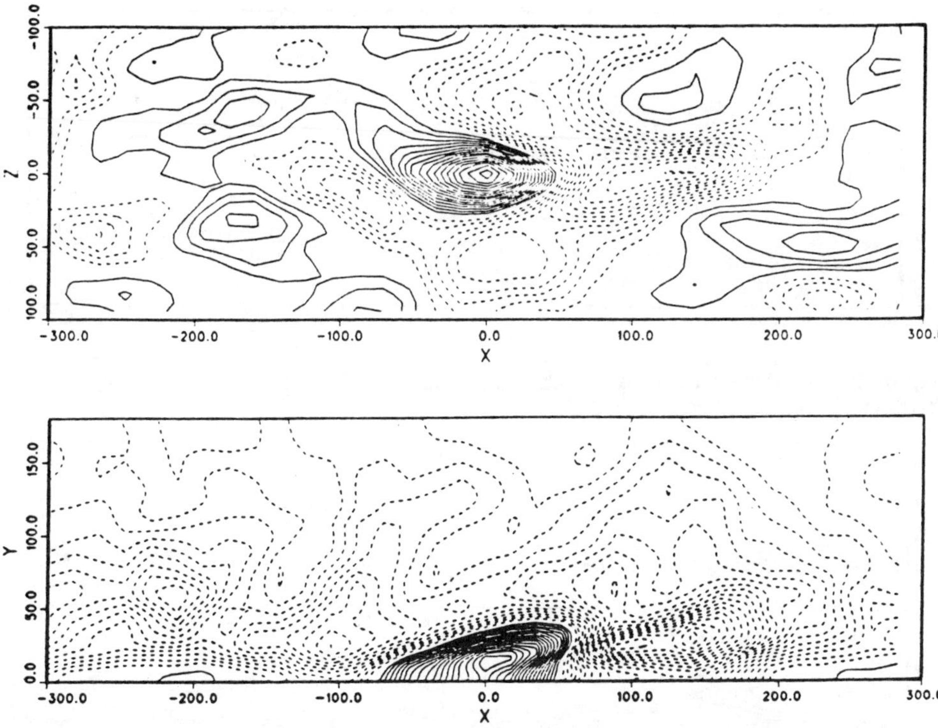

Figure 8. Pressure patterns associated with conditionally averaged shear-layer (L=200, k=1, spanwise centering applied), a) wall-pressure pattern, b) pressure pattern in x-y symmetry plane. Normalization is with τ_w and contour spacing is 0.1 (solid curves >0).

obtained experimentally in a high Reynolds number boundary-layer flow (cf figure 7 of Johansson, Her & Haritonidis). This conclusion refers to an amplitude scaled in inner variables, and it is interesting to note that the amplitudes become comparable when normalized by the rms-value. One should here also note that the rms-value of the wall-pressure normalized by τ_w ($\approx$1.43) is quite small (less than half) compared with experimental results, while the rms-value normalized by the centerline dynamic pressure is in fair agreement with experimental findings. This behaviour may signify a rather strong Reynolds number for properties that are influenced by the scale separation between the inner and outer regions of the flow.

An interesting detail is also that the rms-value of the pressure fluctuations was found to be larger in the buffer region than at the wall, reaching a maximum of about 1.8 at y=30. This behaviour is hence similar to that of the conditionally averaged pressure field, which may be taken as an indication that these structures are dominant contributors to the pressure field in the near-wall region.

In conventional conditional averaging procedures, with or without spanwise centering, symmetric patterns are the result of homogeneity of the flow in the spanwise direction. Instantaneous shear-layer structures, on the other hand, often tend to develop strong asymmetries as they propagate downstream. This can give large values of the spanwise gradient of the streamwise velocity at the center of the shear-layer. The average of the (absolute) spanwise gradient at the detection point for the events shown in figures 5 and 6 was found to be about 0.20 of the mean gradient at the wall, or 2/3 of the local mean gradient. Asymmetric structures were in fact found to be more probable than symmetric ones. In order to retain the spanwise asymmetry also in the conditional average, an asymmetry

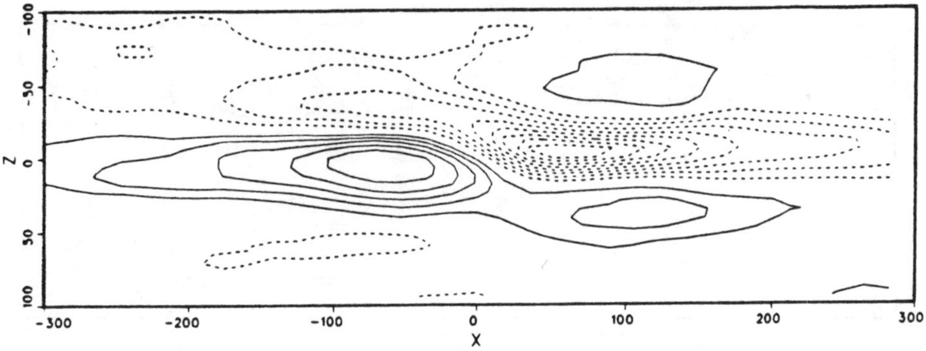

Figure 9. <u> detected with asymmetry condition modified VISA technique (L=200, k=1, y=15, contour increment =0.5).

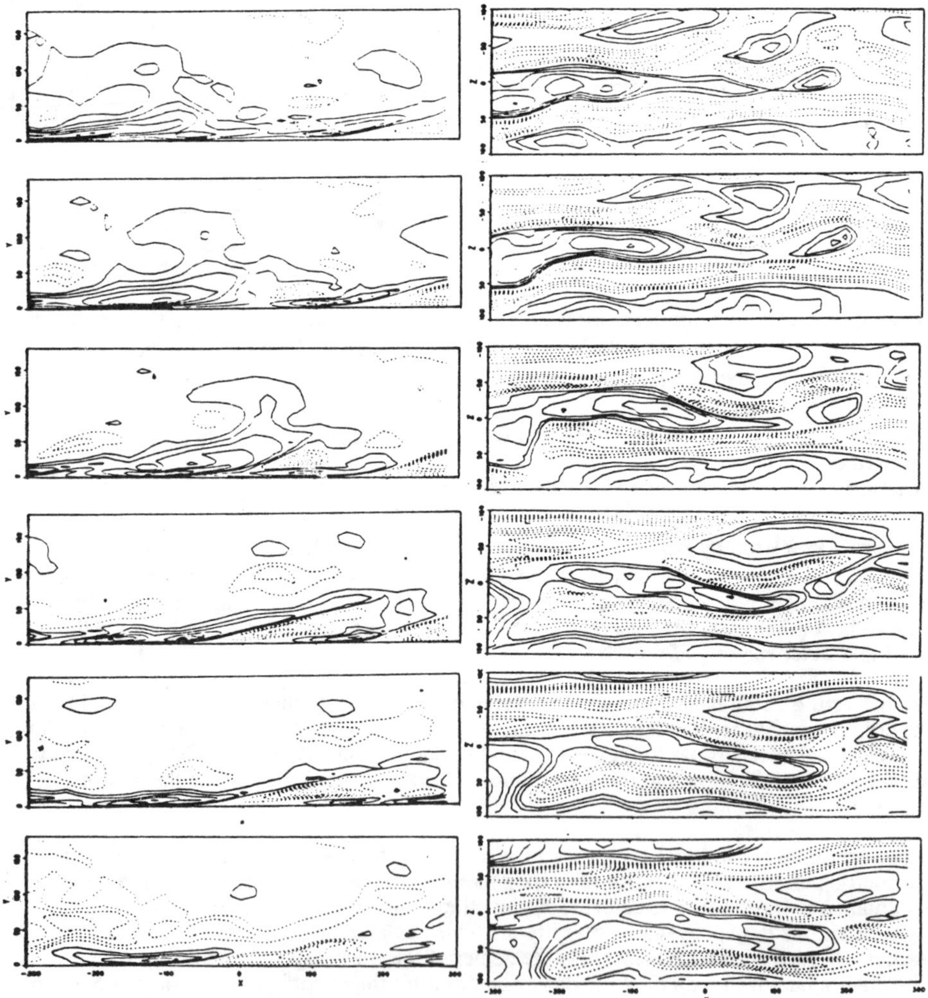

Figure 10. xy- and xz-plane contours of instantaneous u for event B in figure 1 followed in time. Time increment between plots is 12 viscous units.

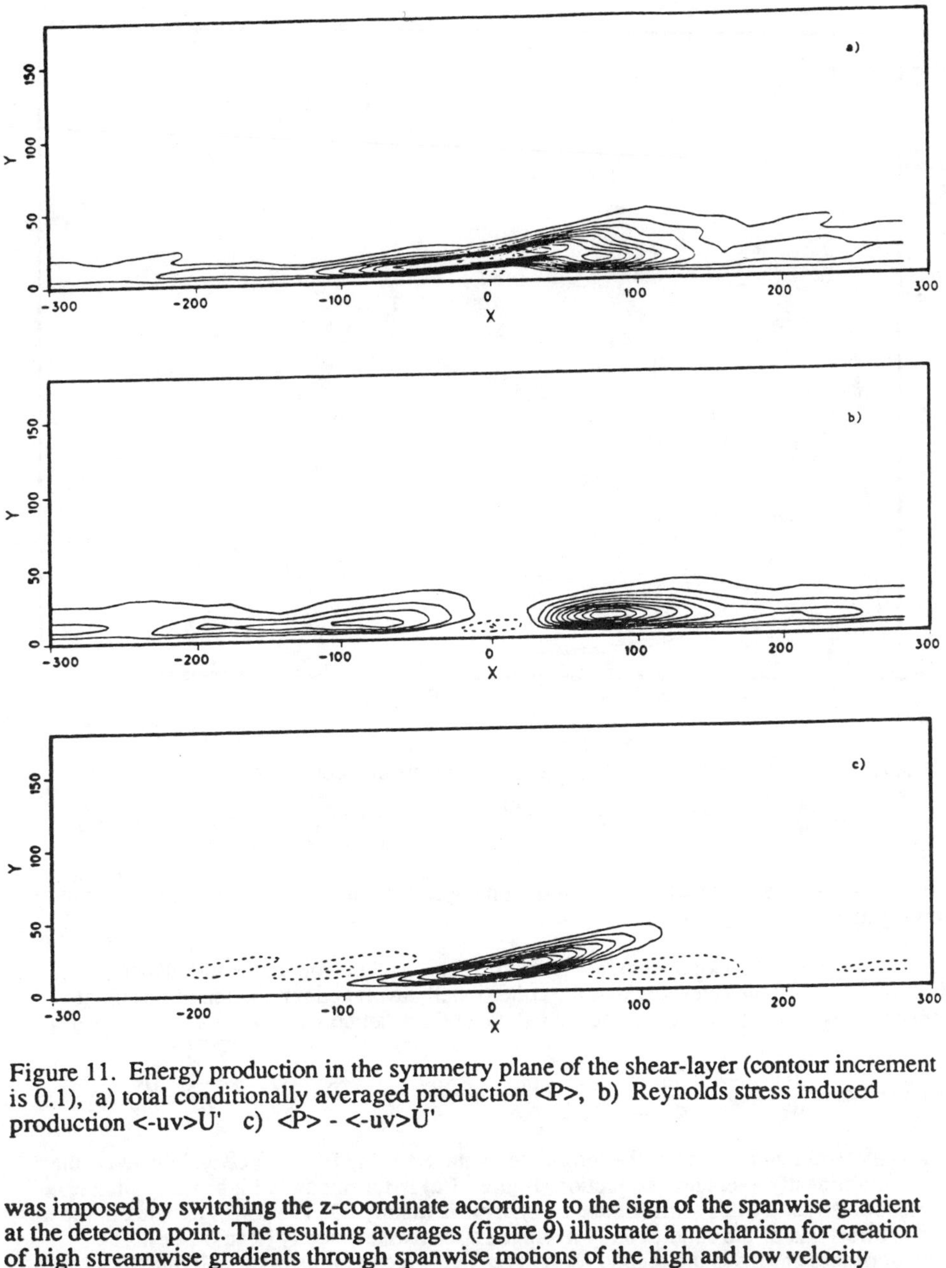

Figure 11. Energy production in the symmetry plane of the shear-layer (contour increment is 0.1), a) total conditionally averaged production <P>, b) Reynolds stress induced production <-uv>U' c) <P> - <-uv>U'

was imposed by switching the z-coordinate according to the sign of the spanwise gradient at the detection point. The resulting averages (figure 9) illustrate a mechanism for creation of high streamwise gradients through spanwise motions of the high and low velocity regions. Such streamwise gradients play a relatively important role for the generation of conditionally averaged turbulence production. It was, in this context, found that normal stress production is of comparable magnitude to the regular shear production in the conditionally averaged sense.

The often occurring asymmetric features of the shear-layers is illustrated by the space-time development of a single event in figure 10, where the frame of reference is chosen so that the center of the shear-layer is stationary. The xy-plane view clearly shows the sharpening of the shear layer due to the action of the mean shear, whereas the planar view illustrates a

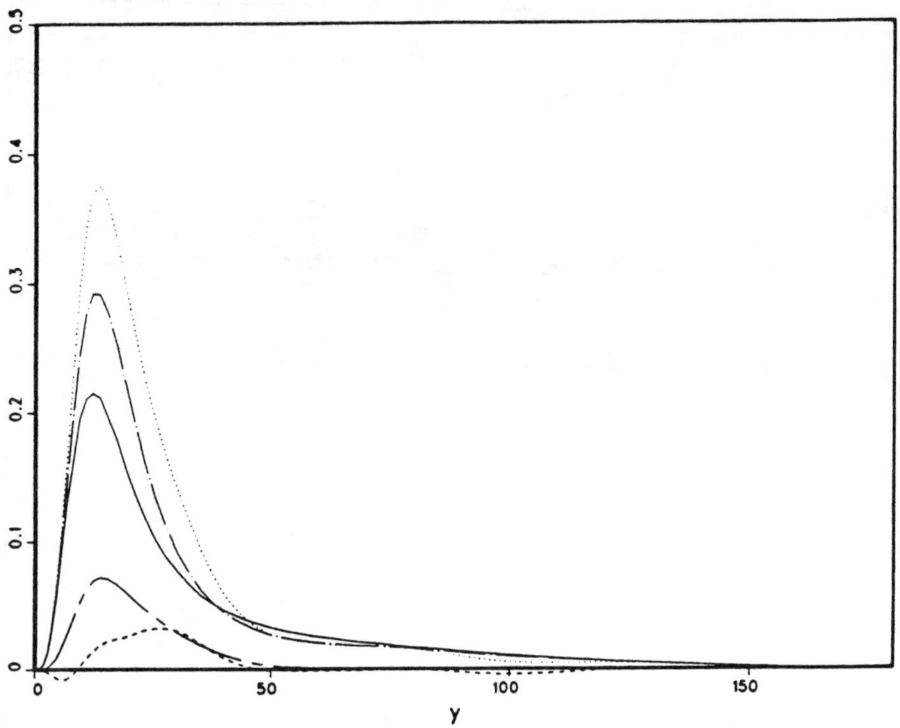

Figure 12. Energy production averaged over area around center of event.
—— : $-\overline{uv}\frac{dU}{dy}$, ··· : <P> , - - - : $-\overline{uv} <\frac{\partial u}{\partial y}>$, -·-·-· : $<-uv>\frac{dU}{dy}$, — -— -— : $-\overline{u^2} <\frac{\partial u}{\partial x}>$

meandering of the high and low-speed regions, resulting in strong streamwise and spanwise gradients.

A key issue for the importance of coherent structures is their role in the turbulence production process. One can show that the conditionally averaged production of turbulent kinetic energy for a plane two-dimensional mean flow can be written

$$<P> = -<uv>\frac{dU}{dy} - \overline{u^2} <\frac{\partial u}{\partial x}> - \overline{uv} (<\frac{\partial u}{\partial y}> + <\frac{\partial v}{\partial x}>) - \overline{v^2} <\frac{\partial v}{\partial y}> - \overline{w^2} <\frac{\partial w}{\partial z}> \quad (1)$$

The only term that remains in the long-time averaged sense is <uv>dU/dy. However, the total conditionally averaged production (figure 11 a) is substantially higher than what can be accounted for by this term (figure 11 b). This is mainly due to strong gradients in the x and y-directions of the conditionally averaged streamwise velocity (figure 11 c). The largest of these terms is the second term in eq. (1), which is illustrated in figure 12 where the production terms have been averaged over an area of 300 x 40 (streamwise and spanwise extent, respectively). Integrating the conditional production out to y=30 gives an average over this volume which is almost twice as large as the long-time mean value. However, also the dissipation of turbulent energy is large in the vicinity of the shear-layer (figure 13). It is noteworthy here that the conditionally averaged $\partial u/\partial z$ does not enter the left hand side expression of eq. (1). Hence, although there exist high values of the spanwise gradient of u at the edges of the shear-layers they do not in themselves imply contributions to the associated turbulence production.

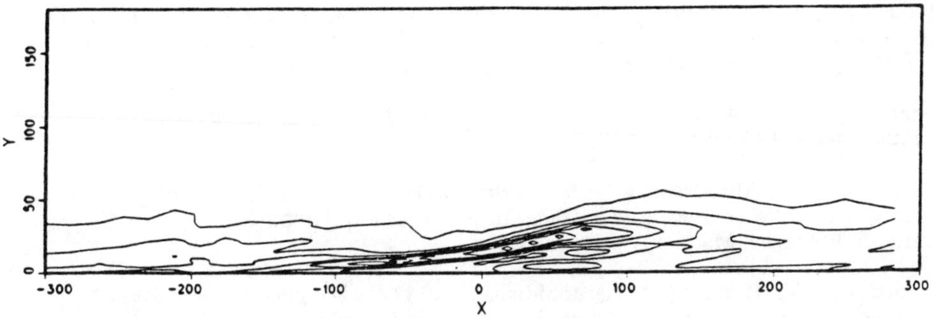

Figure 13. Dissipation of turbulent energy in the symmetry plane of the conditionally averaged structure in figure 5. Contour increment is 0.1.

3. CONCLUSIONS

* Shear-layer structures are found frequently in near-wall turbulence and often have many features in common with the conditionally averaged structure obtained with the VITA or VISA methods.

* The associated pressure pattern exhibits a high-pressure ridge along the center of the shear-layer, with a maximum at about y=15. The high pressure region extends down to the wall over a horizontal region roughly 100 long and 50 wide.

* The amplitudes of the wall-pressure pattern, and the wall-pressure rms-value, found from the numerical simulation data at a low Reynolds number compare well with experimental data at higher Reynolds numbers when normalized with centerline dynamic pressure, but is much smaller when scaled with the wall shear stress.

* Shear-layer structures were found to be important contributors to the turbulence production, and a substantial part of the conditionally averaged production was shown to be generated by strong gradients in the streamwise disturbance velocity.

* The dominant associated uv-contribution was found to be associated with an ejection type of motion on the downstream side of the shear-layer.

* Individual shear-layers retain their coherence over streamwise distances of the order of a thousand viscous length units, and they often develop a strong asymmetry (which is lost in conventional averaging procedures) as they propagate downstream, with a velocity of about 10.6.

REFERENCES

Alfredsson, P.H. & Johansson, A.V. 1984 *Time scales in turbulent channel flow* Phys. Fluids **27**, 1974.

Blackwelder, R.F. & Kaplan, R.E. 1976 *On the wall structure of the turbulent boundary layer* J. Fluid Mech. **76**, 89.

Dinkelacker, A. & Langeheineken, T. 1983 *Relations between wall-pressure fluctuations and velocity fluctuations in turbulent flow.* In Proc. Structure of Turbulent Shear Flow, IUTAM Symposium, Marseille 1982. Springer Verlag

Johansson, A.V. & Alfredsson, P.H. 1982 *On the structure of turbulent channel flow* J. Fluid Mech. **122**, 295.

Johansson, A.V., Alfredsson, P.H. & Eckelmann, H. 1987 *On the evolution of shear-layer structures in near-wall turbulence* **In Advances in Turbulence** (editors G. Comte-Bellot and J. Mathieu) Springer Verlag.

Johansson, A.V., Her, J.-Y. & Haritonidis, J.H. 1987 *On the generation of high-amplitude wall-pressure peaks in turbulent boundary layers and spots.* J. Fluid Mech. **175**, 119.

Kim, J. 1985 *Turbulence structures associated with the bursting event* Phys. Fluids **28**, 52.

Kobashi, Y. & Ichijo, M. 1986 *Wall pressure and its relation to turbulent structure of a boundary layer.* Experiments in Fluids **4**, 49.

Langeheineken, T. 1981 *Zusammenhänge zwischen Wanddruck- und Geschwindigkeitsschwankungen in turbulenter Rohrströmung.* Mitt. Max-Planck Institut für Strömungsforschung, no. 70, Göttingen.

Schewe, G. 1983 *On the structure and resolution of wall-pressure fluctuations associated with turbulent boundary layer flow.* J. Fluid Mech. **134**, 311.

Inner-Outer Structure of the Wall-Pressure Correlation Function

RONALD L. PANTON
Mechanical Engineering Department
University of Texas
Austin, Texas 78712, USA

INTRODUCTION

The two layer structure of turbulent wall layers is the central physical fact that gives a framework to our theories, models, and experiments. Any proposition about turbulent motions must be compatible with the two layer structure and how it changes with Reynolds number.

To be specific and absolutely clear about the nomenclature, the term *outer region* will refer to the log layer plus the wake; say $y^+ = 40$ to $y/\delta = 1$ and beyond. The log layer and the wake layer are put together because their main turbulent motions are inviscid. The outer scales are $u*$ and δ, with U_∞ as a reference. The term *inner region* will refer to the viscous sublayer, the buffer layer and the log layer; $y^+ = 0$ to say $y/\delta = 0.2$. The scales are $u*$ and ν. The Reynolds number $Re* = u*\delta/\nu$ is a direct measure of the size of the outer region compared to the inner region.

The major time-averaged interaction between the regions is the Reynolds stress. Researchers propose that there are sequences of motions that characterize Reynolds stress producing events. The bursting process is the most well known sequence, although the details are not completely certain. However, there are other significant patterns of activity that also contribute to the Reynolds stress.

Because pressure forces are propagated for large distances, they have frequently been proposed as a mechanism by which the outer layer can initiate, control, or modulate Reynolds stress producing events in the inner region. An opposing concept is that a stress producing motions grow as a local instability without any help from remote events. Pressure forces are undoubtedly important in both ideas; the question is whether the pressure is generated locally, or is produced as a part of a remote and statistically

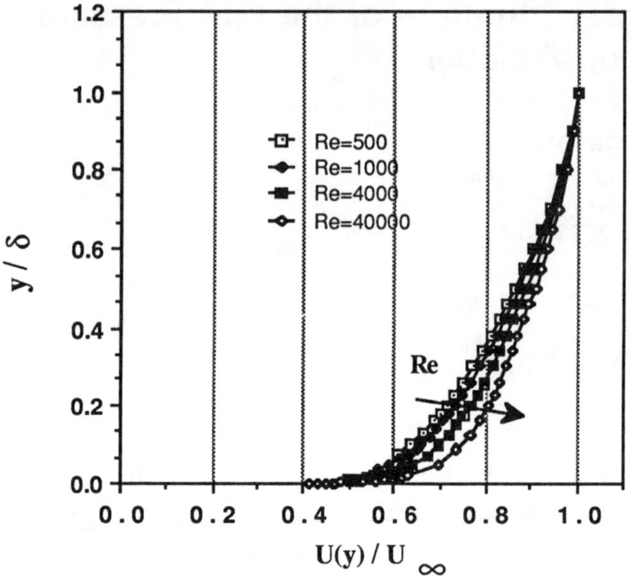

FIGURE 1. Wall layer velocity profiles as the Reynolds number is increased.

independent event in further away. In the second case the question becomes from what part of the flow does the event originate and what is its character?

The bursting process is again a good example. For many years workers have sought a wake layer pressure event that would initiate bursting. At one time the bursting frequency, or time between bursts, was thought to scale with U_∞ and δ as evidence of this. However, although it is not universally accepted, the work of Blackwelder and Haritonidis (1983) established that the "time between bursts" is relatively constant when expressed with inner scales. Luchik and Tiederman (1987), using the techniques developed by Bogard and Tiederman (1986), determined the value as $T^+ \equiv T\, u_*^2/\nu = 90$ over a large range of Reynolds number. This, as we shall see, does not mean that the local instability theory is confirmed.

For the sake of argument, assume that there is a *critical pressure pattern,* produced in the outer region, that is imposed on the inner region where it initiates a motion that produces a Reynolds stress. There may be more than one of these patterns as previously noted. The pattern has a characteristic footprint in the wall pressure. An important question is: if a characteristic pressure foot print exists, how must it change spatial scale and frequency as the Reynolds Number increases? When the Reynolds number increases, the

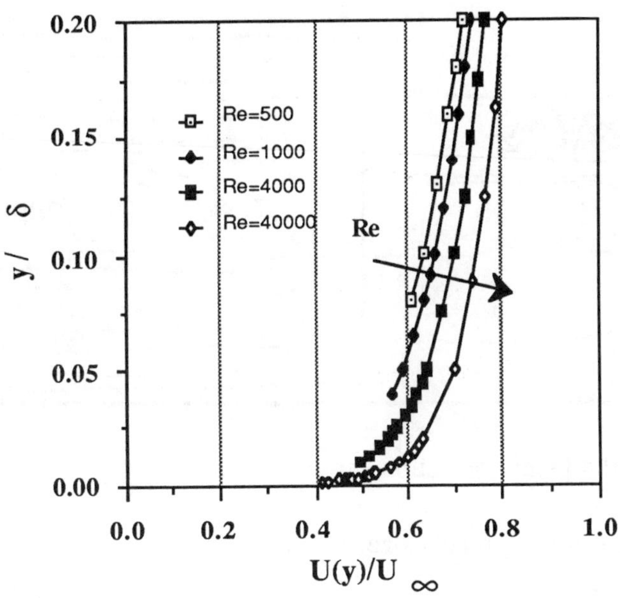

FIGURE 2 Velocity profiles in the log layer at different Reynolds numbers.

size of inner region structures decreases compared to that of the outer region. How does the characteristic pressure pattern adjust to a new Reynolds number? In order to illuminate these questions we will investigate the structure of the wall-pressure correlation function.

VELOCITY PROFILES AND SCALES AS $Re_* \Rightarrow \infty$

The overall character of the outer region is given in Figure 1, which displays velocity profiles for a zero-pressure-gradient boundary layer . The graph gives U/U_∞ as a function of y/δ computed for various Reynolds numbers, Re_*. Since the curves in Figures 1 have the same δ and U_∞, one might visualize that the Reynolds number is increased by decreasing the viscosity. As the Reynolds number increases, the outer region becomes like a large block of turbulence traveling at the free stream velocity U_∞. This means that U_∞ is a reference from which to measure the profiles, but it is not an outer region scale. It is well know that u_* is the proper scale and that Figure 1 curves collapse in the defect form $(U-U_\infty)/u_*$. The block-like flow is even more apparent when one expands the log layer as shown in Figure 2. On each curve in this graph the point nearest the wall is at $y^+ = 30$. As Re_* increases, U/U_∞ at $y+ = 30$ decreases in direct proportion to the decrease in u_*/U_∞. At

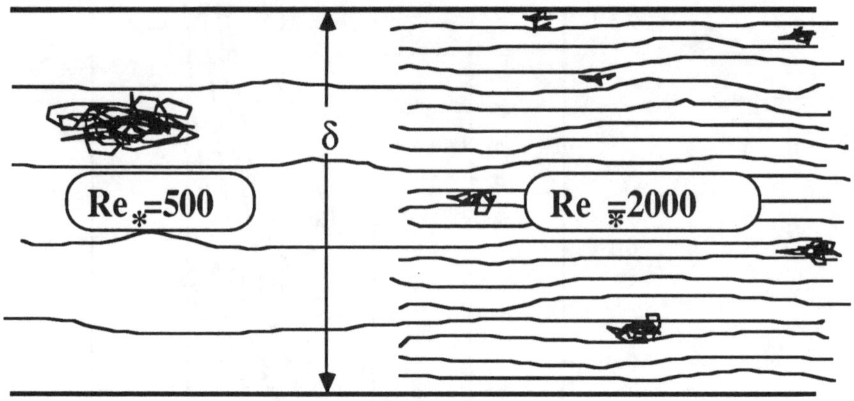

FIGURE 3. Plan view of a sublayer wall streaks.

infinite Re∗, ρu_*^2 approaches zero.

Figure 3 illustrates how length scales change with Reynolds number. This figure depicts a plan view of streaks in the inner region; first for Re∗ = 500 on the left and then for Re∗ = 2000 on the right. In each case the drawings have the same size outer scale. With regard to the critical pressure pattern, is it reasonable to assume that the inner region selects the spatial scale of the outer region motion to which it is sensitive? If so, then that scale must change drastically as the Reynolds number changes. In fact, to be compatible, the outer region motions need to scale on the inner scales u∗ and ν. Obviously, this happens only for motions in the log region.

WALL PRESSURE CORRELATIONS AND SPECTRA

In order to introduce the nomenclature, consider the wall pressure p(x, t) and its space-time correlation defined by; (the symbol < > means a time average)

$$R_{pp}(\zeta, \tau) = \langle p(x, t) \cdot p(x+\zeta, t+\tau) \rangle$$

The double Fourier transform of Rpp is;

$$\underline{S}_{pp}(k_1, \omega) = (2\pi)^{-2} \iint R_{pp}(\zeta, \tau)\ \exp(i\, k_1 \zeta)\ \exp(i\, \omega\, \tau)\, d\zeta\, d\tau$$

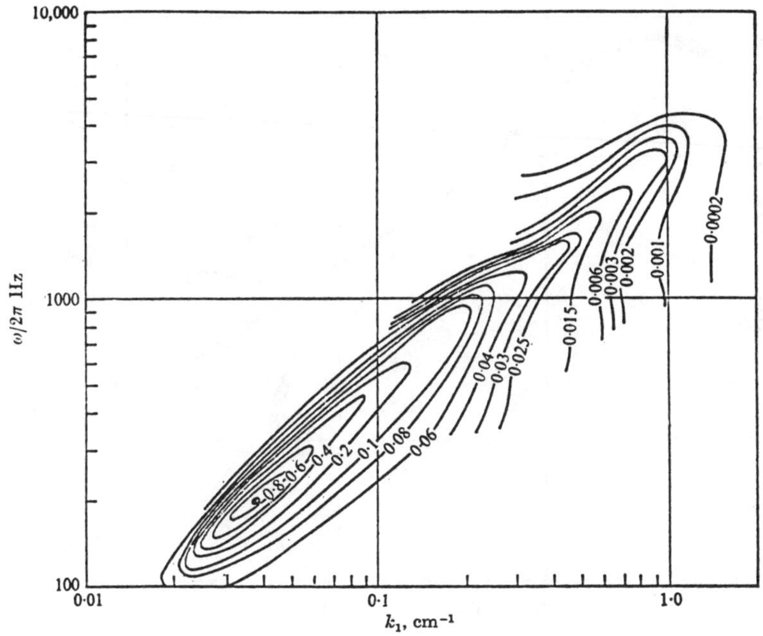

FIGURE 4. Contour plot of $\underline{S}pp(k_1, \omega)$ for a boundary layer with dp/dx = 0; from Wills(1970).

Figure 4 shows a contour plot of $\underline{S}pp$ taken from Wills'(1970) boundary layer measurements ($\underline{S}pp$ is actually complex. A field convecting in one direction has symmetry properties that allow us to only deal with the read part). Wills' data is for dp/dx = 0; similar information with a pressure gradient has been given by Bradshaw (1967). The familiar frequency spectrum $\Phi(\omega)$ is the integral of $\underline{S}pp(k_1, \omega)$ over all wavenumbers. On the other hand, the integral of $\underline{S}pp(k_1, \omega)$ over all frequencies produces the wavenumber spectrum $\phi(k_1)$.

A theoretical model of $\phi(k_1)$ from Panton and Linebarger (1974) is given in Figure 5. In constructing this model the turbulence-mean shear term is assumed to dominate, and thus the calculation requires only information about the mean velocity and fluctuating vertical velocity fields. Two very important facts are: (a) that these velocity fields have the well known inner-outer structure; and (b) that they scale on the wall shear stress (or u∗). Many pressure spectrum measurements are normalized with the dynamic

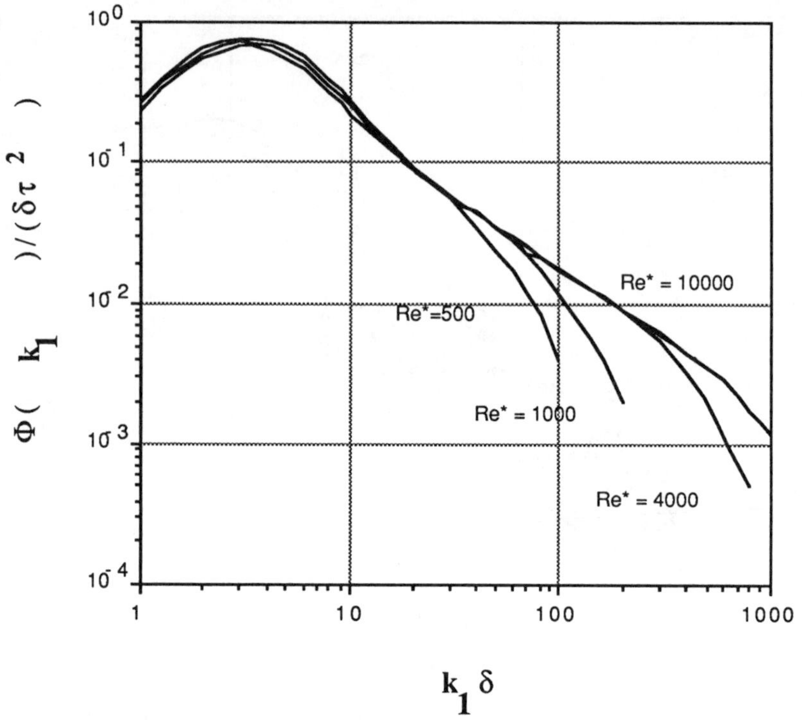

FIGURE 5. Computed $\phi(k_1)$ spectrum of wall pressure in outer variables.

pressure q instead of the wall shear. There seems to be no theoretical reason to use q, while the use of u∗ should minimize Reynolds number variations.

If one imagines that all the $\underline{S}_{pp}$ energy is concentrated in a Dirac delta function along the ridge in Figure 4, then the ridge defines a convective velocity $U_c(k_1)$. This allows the $\phi(k_1)$ spectrum to be reflected into a $\Phi(\omega)$ spectrum and compared with experimental data. The spectra have roughly the same shape because U_c is not a strong function of k_1.

Experimental wall pressure data at low frequencies, or equivalently low wavenumbers, is susceptible to contamination from vibration and acoustic noise. For this reason most experiments do not show the rise in $\Phi(\omega)$ implied by Figure 5. Note, however, that Wills' data in Figure 4, which was corrected to delete acoustic noise, shows a pronounced peak at just above 200 Hz. Previously, Hodgson (1962) took data on a sailplane wing which also showed a rising spectrum at low frequencies. In later measurements,

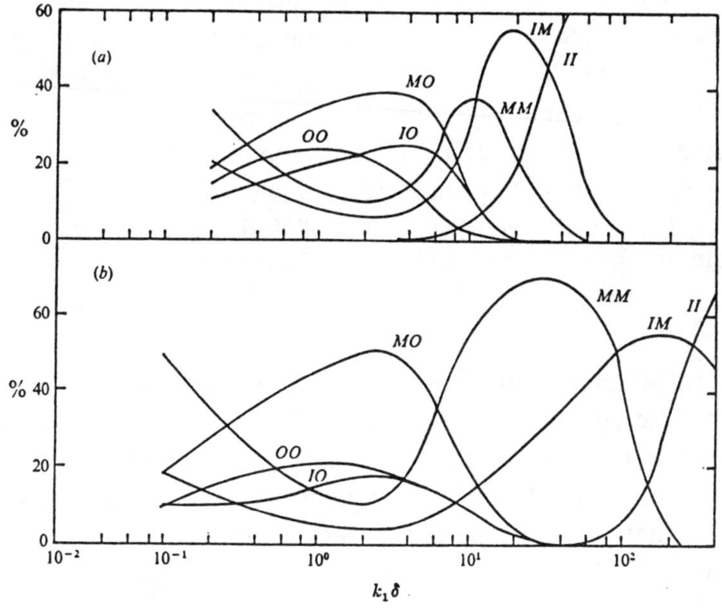

FIGURE 6. Contributions to the $\phi(k_1)$ spectrum from different regions. I = inner region $0 \leq y^+ \leq 33.2$; M = log layer $33.2 \leq y^+$, $y/\delta \leq 0.2$; O = wake layer $0.2 \leq y/\delta \leq 1.0$. (a) $Re_* = 1{,}000$, (b) $Re^* +10{,}000$

Hodgson (1971), the spectrum was flat, however, for these experiments the response of the pin-hole, tube, microphone system is thought to be a significant source of error. Subsequently, Panton et. al.(1980) obtained a rising spectrum on a sailplane fuselage where an extensive zero pressure gradient was documented. These measurements, together with those of Wills, are the best evidence that the rising spectrum caused by turbulent motions is a fact.

An overlap layer in the $\phi(k_1)$ spectrum between an outer part scaling on $k_1\delta$ and an inner part scaling on $k_1\nu/u_*$ must have the form $\phi \sim (k_1)^{-1}$. The region of -1 slope in Figure 5 begins about $k_1\delta = 10\text{-}20$. The existence of the overlap layer is a dimensional requirement; calculations using any turbulence model having an inner-outer structure will produce region with a -1 slope. In the outer variable $k_1\delta$, the high wavenumber drop off from the -1 line depends on the Reynolds number. Four curves for different Reynolds numbers are shown on Figure 5. If these curves are transformed into the inner representation, ie. $\phi(k_1 \nu / u_*) \cdot u_* / (\tau_o^2 \nu)$, they collapse into a single line at high wavenumbers. The exact nature of the drop off is suspect because the model is inaccurate in this region. It is also noteworthy that the -1 slope is retained when any arbitrary convective law is used to convert the wavenumber spectrum into a frequency spectrum, that is $\Phi(\omega) \sim \omega^{-1}$(as first proposed by Bradshaw (1967)). Another point of importance is that adverse pressure gradients have much higher peaks for the outer region but troughly the same log region.

Experimental confirmation of the overlap region in the frequency spectrum is very difficult. Microphones must have small diameters, $d^+ \equiv du_* / \nu$, and equally as important, the flow must have a high Reynolds number. The measurements of Benarrous (1979) come closest to meeting both of these requirements, and they have a slope slightly less than one. These measurements were made in an air channel with a pin hole microphone. Later on we shall see how the measurements of Wills relate to this question of an overlap region. Schewe (1983) reported measurements with a microphone diameter of only $d^+ = 19$ and he claims (a very small region) where $\Phi \sim (\omega)^{-7/3}$; over twice the theoretical overlap exponent of -1. However, the Reynolds number for these experiments was only $Re_* = 550$. From Figure 5 we see that at this low Reynolds number, the overlap region is practically gone. Hence, Schewe's unusually steep slope can be attributed to the low experimental Reynolds number.

In performing the calculations of Figure 5, it was possible to keep track of the region of the boundary layer that was causing wall pressure fluctuations at a given wavenumber. These regional contributions are displayed in Figure 6 for two Reynolds numbers. On the graph "I" corresponds to the inner region ($0 < y^+ < 33$), "M" to the overlap or log layer ($33 < y^+$, $y/\delta < 0.2$), and "O" to the wake layer ($0.2 < y/\delta < 1.$). Consistent with the existence of an overlap region, the outer layer "O" does not contribute to wavenumbers above $k_1\delta = 10 - 20$. Furthermore, the inner layer contributions "I" shift to higher wavenumbers as the Reynolds number increases.

WAVENUMBER PHASE-VELOCITY REPRESENTATION

Consider again the wavenumber-frequency spectrum $\underline{S}pp(k_1, \omega)$. Incidentally, Chase(1976) and Benarrous(1979) have constructed

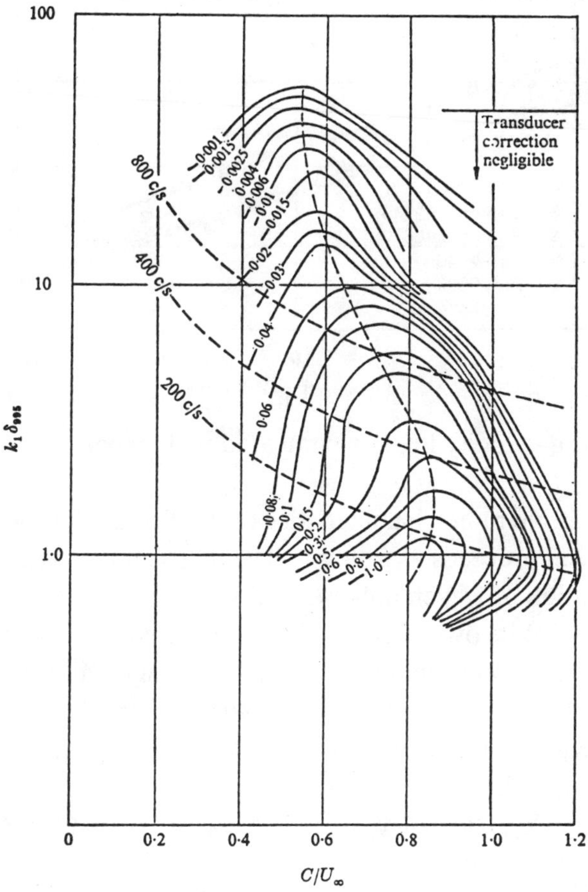

FIGURE 7. Contour plot of Spp (k_1, C) from Wills(1970)

structural vibrations calculations. For fields with convection in only the downstream direction, such as the turbulent wall pressure, Wills realized the advantages of eliminating the frequency in favor of the phase velocity, $C = \omega/k_1$. On the log-log plot of Figure 4, lines of constant phase velocity are 45 degree lines running to the right. Figure 7, taken from Wills(1970), gives the data of Figure 4 converted into a wavenumber, phase-velocity contour plot of Spp($k_1\delta$, C/U_∞). In this plot the ridge where $\partial Spp/\partial C = 0$ is defined as the overall convective velocity U_c (k_1).

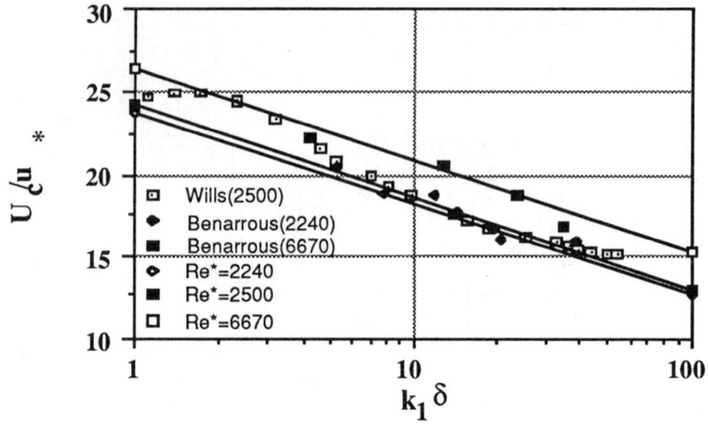

FIGURE 8. Convection velocity log law compared with experiments.

It is also possible to derive more characteristics of the $Spp(k_1, C)$ spectral function. The argument proceeds as follows. The physical fact employed is that a velocity fluctuation of wavenumber k_1 influences only a region proportional to the size of its own wavelength. A small wavenumber fluctuation in the outer region is not felt at the wall. In this light we assume that as $Re* \Rightarrow \infty$, there is a high wavenumber domain where the variables Spp, k_1, and C scale with $u*$ and ν, ie.

$$S^+ = S^+(k_1{}^+, C^+) \equiv Spp(k_1, C)\cdot u*^3/(\tau_o\nu^2)\ ; \quad k_1{}^+ \equiv k_1\nu/u*\ ; \quad C^+ \equiv C/u*$$

The fact that the high wavenumber contributions in Figure 6 come only from the inner region "I" means that the form given above is actually independent of Re*. On the other hand, the outer layers "O" and "M", according to Figure 6, dominate the contributions to the low wavenumber spectrum. For low wavenumbers, we assume that as $Re* \Rightarrow \infty$ the proper scales are δ and $u*$. Moreover, the block-like flow of the outer region displayed in Figures 1 and 2 implies that U_∞ acts as a reference that convects the wake region along. In the low wavenumber region the proper variables are assumed to be;

$$\mathbf{S} = \mathbf{S}(\mathbf{k}_1, \mathbf{C}) \equiv Spp(k_1, C)\cdot u*/(\tau_o{}^2\delta^2)\ ; \quad \mathbf{k}_1 \equiv k_1\delta\ ; \text{ and } \quad \mathbf{C} \equiv (C - U_\infty)/u*$$

Note that the two different sets of variables obey the relations;

$$S^+ = \mathbf{S}\cdot Re*^2, \ \mathbf{k}_1 = k^+ Re*, \text{ and } \mathbf{C} = C^+ - U_\infty/u*$$

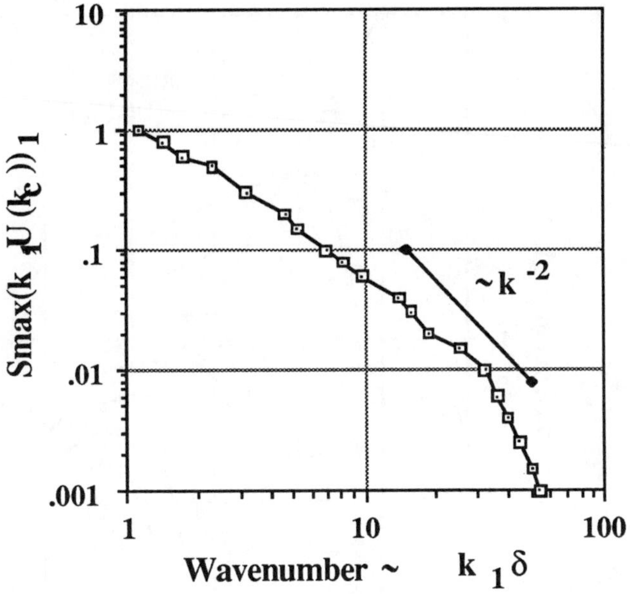

FIGURE 9. Spp along the convective ridge;data adapted from Wills(1970)

For a moment we digress to consider the convective velocity ridge $U_c(k_1)$, defined by $\partial Spp/\partial C = 0$. Arguments previously given, Panton and Linebarger (1976), showed that the convective velocity has an overlap region where the dependence is a log function of the wavenumber. The constants in this function are not arbitrary, but must be identical with the constants in the velocity profile log layer. The relations valid for wavenumbers in the overlap region may be expressed in inner variables,

$$C_c^+ \equiv U_c / u_* = (1/\kappa) \ln (k^+) + C_o$$
$$= - (1/\kappa) \ln (\mathbf{k}_1) + (1/\kappa) \ln Re_* + C_o$$

or in outer variables,

$$\mathbf{C}_c \equiv (U_c - U_\infty)/u_* = (1/\kappa) \ln (\mathbf{k}_1) - 2\,\Pi/\kappa$$

Here, κ is the von Karman constant, C_o the roughness constant , and Π the pressure gradient parameter. On Figure 8, this equation is compared with the experimental data of Wills and that of Benarrous. It appears that the overlap law covers a wide range of wavenumbers.

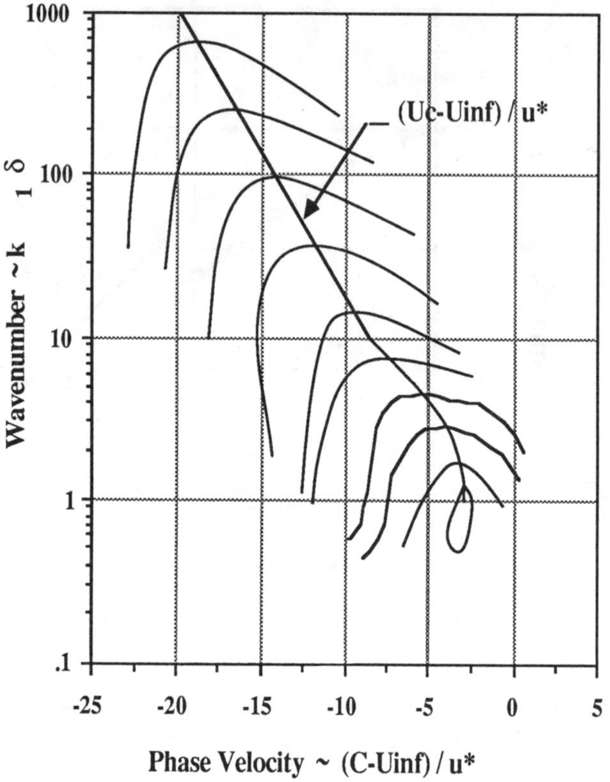

FIGURE 10. Contour plot of Spp in outer variables.

OVERLAP LAWS FOR THE CONVECTIVE RIDGE

Now we consider the question of how the convective ridge drops off as one goes to high wavenumbers. Denote the level along the ridge as S_{max};

$$S^+_{max}(k_1^+) \equiv S^+(k_1^+,\ C_c^+(k_1^+))$$

$$\mathbf{S}_{max}(\mathbf{k}_1) \equiv \mathbf{S}(\mathbf{k}_1, \mathbf{C}_c(\mathbf{k}_1))$$

Assume there is an overlap region of wavenumbers where both functions are valid and we may write;

$$S^+_{max}(k_1^+) = Re_*^2\, \mathbf{S}_{max}(\mathbf{k}_1)$$

Since the region is finite, we may differentiate to get

$$\frac{dS^+_{max}}{dk_1^+} = Re_*^2 \frac{d\mathbf{S}_{max}}{d\mathbf{k}_1} \frac{d\mathbf{k}_1}{dk_1^+}$$

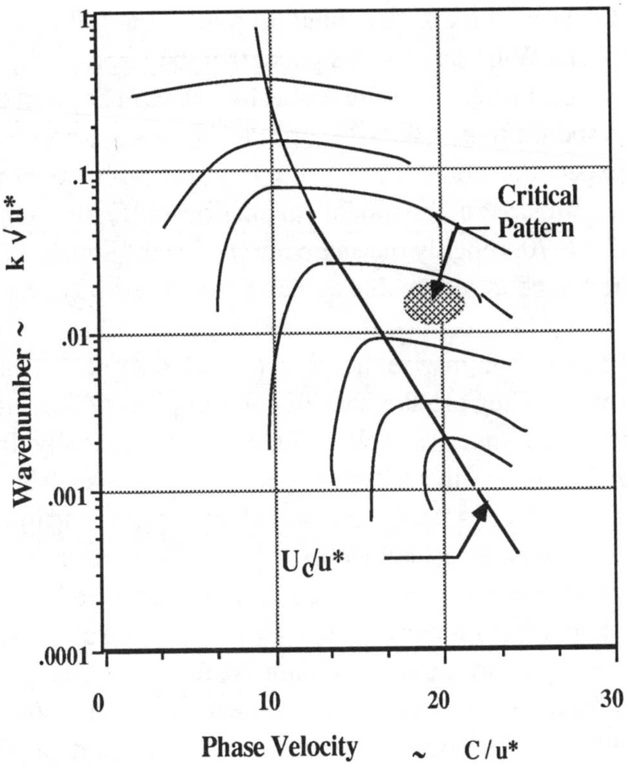

FIGURE 11. Contour plot of Spp in inner variables.

But, $d\mathbf{k}_1 / dk_1^+ = Re* = \mathbf{k}_1 / k_1^+$; thus, we can separate variables to obtain

$$(k_1^+)^3 \frac{dS^+_{max}}{dk_1^+} = (\mathbf{k}_1)^3 \frac{d\mathbf{S}_{max}}{d\mathbf{k}_1} = \text{constant}$$

Finally, integration yields the laws

$$S^+_{max} \sim (k_1^+)^{-2} \text{ and } \mathbf{S}_{max} \sim (\mathbf{k}_1)^{-2}$$

This analysis has been tested against the experimental data of Wills in Figure 9 where S_{max} is plotted as a function of $k_1\delta_{995}$. For comparison, a line of -2 slope is drawn beginning at $k_1\delta_{995} = 15$. The slope of S_{max} is definitely lower than -2 at low wavenumbers, however, extracting points from a contour plot is difficult and the data are not precise enough to chose a specific beginning of overlap region. The final drop off is caused by transducer resolution effects; Wills states in his paper that the last two points are definitely low for this reason. The highest wavenumber point on Figure 9 at $\mathbf{k}_1 = 55$ corresponds to $k_1^+ = 0.018$, or in terms of a wavelength, $\lambda^+ = 345$. One would expect this point to still be in the overlap region by a considerable amount. For comparison, the model calculations of Figure 5 begin to drop off at about $\lambda^+ = 70$, roughly the same number as the streak spacing. Figure 9 is considered to be in reasonable agreement with the theory above.

The last two figures summarize the expected behavior of S_{pp} based on the arguments above. They are sketches of contour plots of S_{pp}, first in outer variables and then in inner variables. Figure 10 is essentially the same as the lower part of Figure 4 with a different phase velocity variable. The theory is in principle for $Re_* \Rightarrow \infty$, so the plot could be a slight function of Re_*. Wills' data shows a peak just below $k_1 = 1$, but Wills was not certain of this feature because large corrections to the data were required in this region . On the other hand, there is considerable evidence noted above that such a peak exists. At low wavenumbers Figure 10 displays a convection velocity that deviates from the log law. At high wavenumbers the overlap region is established which carries over to the lower wavenumbers of Figure 11. In inner variables the character of the pressure field decay at the high wavenumbers is very uncertain. Figure 11 arbitrarily shows a broadening and relatively constant convection velocity. Measurements in this area are subuct to transducer resulution errors.

CRITICAL PRESSURE PATTERN AS $Re_* \Rightarrow \infty$

Consider, once again, the idea that the inner layer selects a critical pressure pattern as a trigger for (or as an associated part of) a certain category of Reynolds stress producing events. While it is true that an eddy is not a Fourier component, a Fourier component is certainly a major characteristic of an eddy. On Figure 11 a randomly chosen area is labeled "Critical Region" to indicate the wavenumber and phase velocity where the critical pattern contributes to S_{pp}. Of course, not all contributions to this region are necessarily from critical pressure patterns. The description of a critical pattern is necessarily incomplete; a certain spanwise pattern in k_3 may also be necessary, or a certain accompanying (either previously, simultaneously, or subsequently) Fourier component may be necessary. The important point is that the scales on this plot are inner scales and the wavenumber, phase speed and intensity level bear a constant relation to the inner region as the Reynolds number varies.

For events that are insensitive to Reynolds number it is required that the motions generating the characteristic pressure pattern occur in the log region (or lower) of the velocity profile. It is not reasonable that the motions occur in the wake region unless the Reynolds stress event shows a Reynolds number dependence.

SUMMARY

Statistical Properties of the wall pressure display overlap layers in the wavenumber variable. The layer probably begins about $k_1\delta = 15\text{-}20$ and most likely ends around $k_1\nu/u_* = 0.05$. The overlap region is negligibly small for experiments or computations at Reynolds numbers of $Re_* = 500$. Previous research found that the overlap behavior for the frequency spectrum is $\Phi(\omega) \sim \omega^{-1}$, for the wavenumber spectrum $\Phi(k_1) \sim k_1^{-1}$, and for the convective velocity $U_c \sim \ln(k_1)$. This work added the overlap relation for the value of the convective ridge as $S_{pp_{max}} \sim k_1^{-2}$.

Model calculations of wall pressure spectra that assume the turbulence mean-shear source term dominates, produce overlap layers. Although many experiments show evidence of overlap layers, no measurements extend completely through all wavenumbers in the layer. Clear, unambiguous, experimental confirmation of the overlap layers will require small transducers, high Reynolds numbers, and a clean flow situation.

It has been proposed that some Reynolds stress events rely on a pressure pattern from the outer region for initiation. If one assumes that the inner region sets the spatial scale for the critical pattern (an important assumption) they are limited to sources in the log layer in order to maintain scale relations as the Reynolds number increases. Such patterns will also have intensities, phase velocities, and frequencies that scale in inner variables.

REFERENCES

Benarrous , E., 1979, "Contribution a L'Etude des Fluctuations de Pression Parietale sous une Couche Limite Turbulente," These, L'Universite Claude Bernard- Lyon, France.

Blackwelder, R. F., and Haritonidis, J. H., 1983, "Scaling of the bursting Frequency in Turbulent Boundary Layers," J. Fluid Mechanics, **132**, p87.

Bradshaw, P., 1967, "'Inactive' Motion and Pressure Fluctions in Turbulent Boundary Layers," J. Fluid Mechanics, **30**, p241.

Chase, D. M., 1976, " The Wave Vector Frequency Spectrum of Turbulent Boundary Layer Wall Pressure," J.S.V., **70**, p29.

Hodgson, T. H.1962, "Pressure Fluctions in Shear Flow Turbulence," Ph. D. Thesis, University of London.

Hodgson, T. H.,1971, "On the Dipole Radiation from a Rigid and Plane Surface," Proc. Purdue Noise Control Conference.

Luchik, T. S., and Tiederman, W. G., 1987, "Timescale and Structure of Ejections and Bursts in Turbulent Channel Flow,"J. Fluid Mechanics, **174,** p529.

Bogard, D. G., and Tiederman, W. G.,1986, "Characteristics of Ejections in Turbulent Channel Flow," J. Fluid Mechanics, **179**, p1.

Panton, R. L., and Linebarger, J. H., 1974, "Wall pressure Spectra Calculations for Equilibrium Boundary Layers," J. Fluid Mechanics, **65**, p261.

Panton, R. L., Goldman, A. L., Lowery, R. L. and Reischman, M. M., 1980, " Low Frequency Pressure Fluctions in Axisymmetric Turbulent Boundary Layers," J. Fluid Mechanics, **97**, p299.

Schewe, G., " On the Structure and Resolution of Wall-Pressure Fluctions Associated with Turbulent Boundary Layer Flow," 1983, J. Fluid Mechanics, **134**, p311.

Wills, J. A. B., 1970, "Measurements of the Wave-number / Phase Velocity Spectrum of Wall Pressure Beneath a Turbulent Boundary Layer," J. Fluid Mechanics, **45**, p65.

Wall Pressure Peaks and Waves

JOSEPH H. HARITONIDIS, LAWRENCE S. GRESKO, and KENNETH S. BREUER
Department of Aeronautics and Astronautics
Massachusetts Institute of Technology
Cambridge, Massachusetts, USA

ABSTRACT

Large amplitude wall pressure peaks under a turbulent boundary layer are associated with near-wall events. Both isolated peaks and wave trains are primarily associated with the normal component of velocity. The isolated peaks appear to be the result of rapid normal velocity fluctuations while the wave trains are associated with a flow instability in the buffer region. In either case, the flow appears to have a strong spanwise component of velocity. The wave trains appear to be similar to waves observed in the evolution of a localized disturbance in a laminar boundary layer.

INTRODUCTION

Much of the early experimental work on wall pressure fluctuations (see the review by Willmarth, 1975) dealt with spectral and correlation measurements to determine the influence of the different regions of the boundary layer on the pressure. Of particular interest was the question of whether linear or nonlinear mechanisms were responsible for producing the measured pressures. More recent work, with the advent of higher quality and smaller pressure transducers, has concentrated on relating particular pressure signatures with the flow above the wall. Burton (1974), Dinkelacker & Langeheineken (1982), Thomas & Bull (1983), Her (1986) and Johansson, Her & Haritonidis (1987) used conditional sampling techniques to relate flow structures with wall pressure signatures. Johansson et al. (1987) used conditions on both the velocity and the pressure to assess if there was a bidirectional correspondence between pressure and flow signatures. It was found that this is not the case. In addition, high amplitude pressure peaks appeared to be linearly related to the velocity field through the turbulence-mean shear interaction term.

Emmerling (1973) and Dinkelacker, Hessel, Meir & Schewe (1977) found, through the use of a novel array of pressure transducers, that a variety of pressure patterns existed on the wall. In particular, large pressure fluctuations of the order of $5p_{rms}$ were observed as isolated peaks or parts of wave trains. Schewe (1983), using the smallest transducers to date ($d^+ = 19$, where a + superscript indicates normalization with the wall variables ν and u^*; for future reference, $l^* = \nu/u^*$ and $t^* = \nu/u^{*2}$) found very clear evidence of wave trains in a turbulent boundary layer. The wavelength, of pressure signatures with $|p| > 3p_{rms}$, was $145l^*$ and the frequency $\omega^+ = 2\pi f^+ = .52$. The average propagation velocity was found to be $11.9u^*$, indicating that the waves are most likely associated with a flow structure located in the buffer region. It is worth noting that the measured pressure amplitudes reached values of $\pm 7p_{rms}$ and that while pressure fluctuations higher (lower) than $3p_{rms}$ occur only 1% of the time, they contribute 40% to p_{rms}. Thus, these energetic pressure fluctuations must be of dynamical importance.

Her (1986) investigated the relation of wall pressure wave trains to flow structures in the buffer region. He found some evidence, through spectral analysis of pressure and velocity signals, that the wave trains are related to a flow instability. Breuer (1988) studied, numerically and experimentally, the evolution of a three-dimensional, localized, disturbance in a laminar boundary layer. His numerical (full Navier-Stokes) simulation shows the growth of a disturbance, on a developing internal shear layer, whose initial wavelength is approximately $145l^*$, in good agreement with the results of Schewe (1983) for a fully turbulent boundary layer.

The present study addresses the origin of the pressure peaks and waves in a turbulent boundary layer. Measurements of all three velocity components at $y^+ \approx 15$ and the pressure at the wall, below the measuring probe, are used.

EXPERIMENTAL ARRANGEMENT

The experiments were carried out in the Low Turbulence Wind Tunnel at MIT. Details can be found in Mangus (1984). An aluminum flat plate, 1.2m wide by 5.8m long, was used to realize the turbulent boundary layer. The flow was tripped 10cm downstream of the leading edge using Dymo (TM) tape imprinted with the letter V; the Vs pointed downstream. At x =

401cm and a free stream velocity of 11.0 m/s, the free stream turbulence level (f > 1Hz) was less than .05% and the pressure gradient, dc_p/dx, was less than ± 10^{-4} cm^{-1}.

An x-wire was used tmeasure the streamwise and normal components of velocity, u and v, and a v-wire to measure the streamwise and spanwise components u and w. The x-wire had a box size of .125mm, corresponding to $3.6l^*$, and the wires were .6μm in diameter giving an l/d ≈ 295. The resistance "overheat" ratio was 7.5%. The wires of the v-wire had the same dimensions as the x-wire except that they were now parallel to, and at the same height from, the wall with the v pointing upstream. The overall width of the v-wire arrangement was $11l^*$. The overheat ratio in this case was 20% . The wire material was 10%Pt-Rh and the frequency response of both probes was in excess of 25KHz as determined through the square wave technique. Both the x-wire and the v-wire were calibrated using an improved version of the table calibration method described in Lueptow, Breuer & Haritonidis (1988). The improvement consists of using a polar rather than a rectangular coordinate system in the E_1-E_2 plane, for u and v(or w), where E_1 and E_2 are the wire voltages. Thus the table resolution is the same at both high and low velocities. The probes were typically calibrated at 7 different velocities and 11 different angles.

A 1/8" Bruel & Kjaer, model 4138, microphone was used to measure the wall pressure fluctuations. The frequency response is flat from 30Hz to 35KHz. Care was taken to ensure that the microphone diaphragm was flush with the wall; however, a small annular gap 0.3mm wide ($8.6l^*$) and 0.05mm deep ($1.4l^*$) was present between the wall and the microphone diaphragm, due to the construction of the microphone. No effort was made to eliminate it as it is believed that it should have no effect on the measurements. The actual diaphragm diameter is 1.95mm, corresponding to $56l^*$.

All the measurements were taken at a free stream velocity of 11.0m/s, 401cm from the leading edge of, and on the centerline of, the plate. For these conditions, Re_θ = 4340, θ = 0.60cm, δ^* = 0.79cm, H = 1.32, u^* = 0.43m/s, l^* = 0.035mm and t^* = 81μs. Except for the calculation of mean flow variables, continuous data sets of 1,489,920 points were taken for two velocity components (u,v or u,w) and wall pressure, i.e. 496,640 data points per variable. The sampling rate was 80μs.

Figure 1a shows the mean velocity distribution. Figure 1b shows the higher statistics of u and v and Fig. 1c shows the higher statistics of u and w. In spite of the small size of the measuring probes, the measured values appear to be in error for $y^+ < 10$. From our measurements it appears that, for both x-wire and v-wire measurements, it is important for the overall probe size to be less than 51^*. For the x-wire in particular, it appears that the overheat ratio should be low and that the l/d ratio of the wire should be in excess of 400. Interestingly enough, a calibration of the x-wire using the cosine law provides v_{rms} values in agreement with previous measurements but the Reynolds stress is too low.

Figure 2 shows the probability density of the pressure signal. The rms value of the pressure according to Schewe (1983) should be about $.008q_\infty$; no attempt was made to calibrate the output of the microphone. The maximum positive pressure observed was $8p_{rms}$ and the minimum $-7.5p_{rms}$.

DATA ANALYSIS TECHNIQUES

The data was analyzed using conditional sampling and spectral techniques. The VITA method of Blackwelder & Kaplan (1976) was used to detect high shear layers. The VITA algorithm was implemented by taking the local, short time, variance of the signal in question and comparing this value to a predetermined threshold level ku^2_{rms}, for example, in the case of the u-signal. If the computed short time variance exceeded the threshold level, then an event had been detected. In addition to

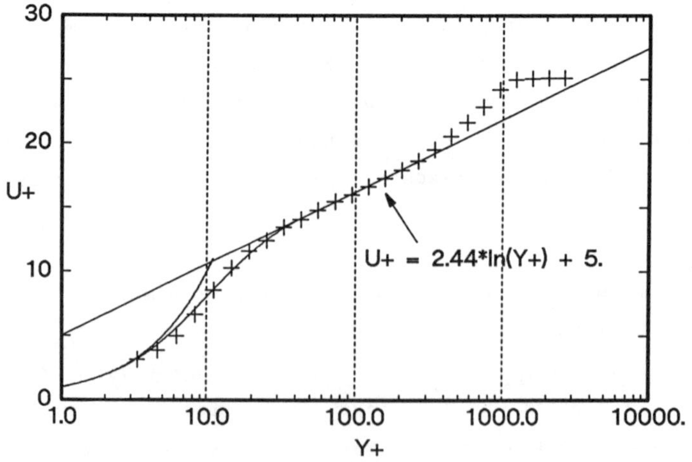

FIGURE 1a. Mean velocity distribution measured with the x-wire. U_∞ = 11.0 m/s, x = 4.01m

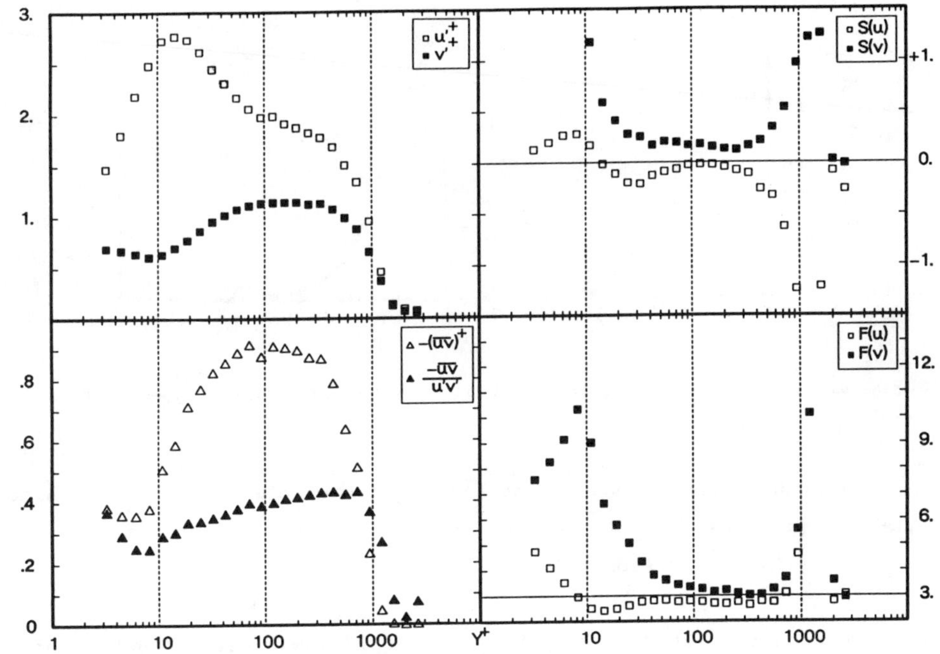

FIGURE 1b. u and v statistics.

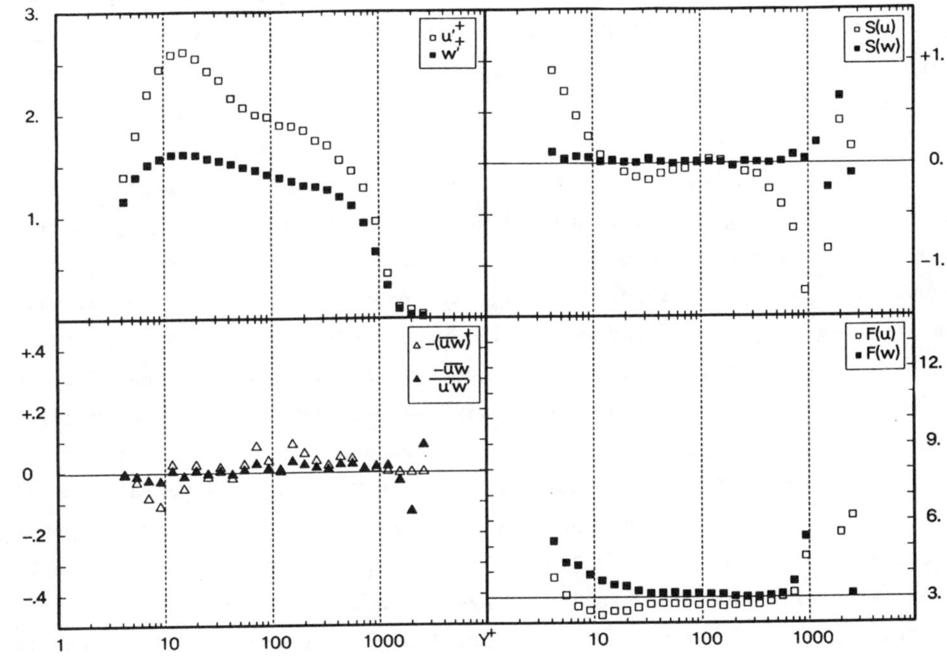

FIGURE 1c. u and w statistics.

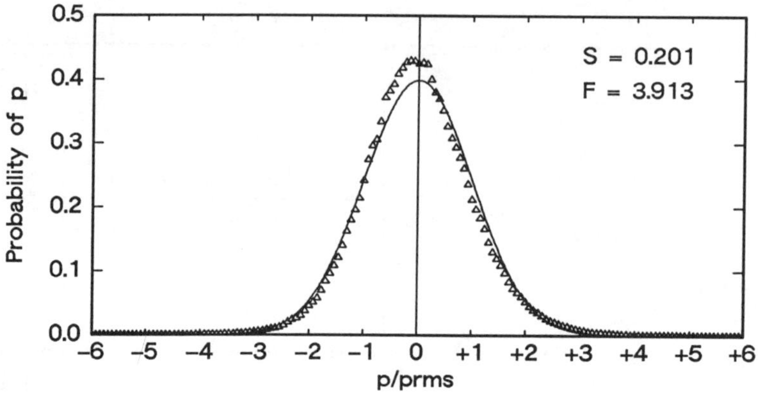

FIGURE 2. Probability density distribution of pressure. U_∞ = 11.0m/s, x = 4.01m.

this criterion, the slope of the signal at the detection time was also checked so that events could be classified as accelerating or decelerating (see Johansson & Alfredsson, 1982).

The uv-quadrant method of Lu & Willmarth (1973) was used to detect high Reynolds (shear) stress events. Again, a threshold level was set, and events were declared for excursions of the uv signal above $ku_{rms}v_{rms}$. A peak detection method (Her, 1986) was used to find the high amplitude pressure peaks. A positive event was declared whenever the pressure signal exceeded kp_{rms} and a negative peak event was declared whenever it was less than $-kp_{rms}$.

A spectral peak detection (SPD) method was developed to detect waves in a particular signal. The method was applied to signals extracted from the total record using one of the above-mentioned techniques. Figure 3a shows such a signal, obtained using the pressure peak technique, as well as the same signal windowed using a gaussian window. This procedure enhances the activity around the detection time without sacrificing frequency resolution. Normalized power spectra, $E(\omega^+)$, are then averaged and a threshold level is set at all frequencies. Figure 3b shows the power spectrum of the windowed signal in Fig. 3a, the average of all such spectra and the threshold level. (The bumps in the averaged power spectra at $\omega^+ \approx 1.0$ and 2.0 can be attributed to background noise as they are also present in the long-time power spectra of the pressure signal). If the power under <u>any</u> particular spectral peak was found to exceed the threshold level, then the center frequency of that spectral

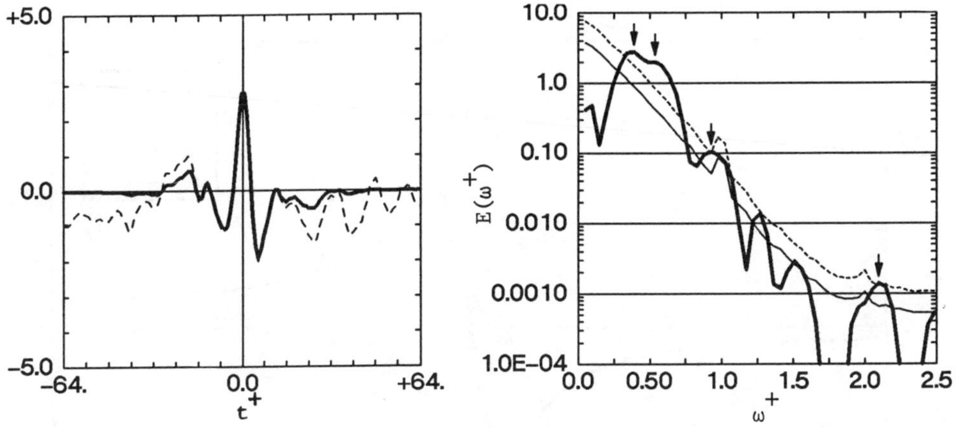

FIGURE 3. The SPD technique applied to positive pressure peak events with k = 2.0. a) ----, one of the events detected and ——, the same event after using a gaussian window. b) The heavy solid line is the normalized power spectrum of the windowed signal in a); ——, average of normalized power spectra; ----, threshold level. The detected peaks are shown by the arrows.

peak was counted as being of dynamical importance. In this example, the threshold level is the same at all frequencies. Thus a histogram of occurrences of different frequencies in the signal, weighted by the power at that frequency, vs frequency was constructed to determine the dynamically important frequencies.

VITA AND PRESSURE PEAK RESULTS

The VITA method was used on the u-signal to detect high shear layer events. Figure 4a shows the ensemble average of accelerating events and Fig. 4b shows the ensemble average of decelerating events. While the u-signals for the two cases are almost identical in shape, but inverted with respect to each other, the v-signals and the p-signals are quite different. The accelerating events produce a pressure peak coincident with the middle of the event with an associated high negative dv/dt which implies a positive dv/dx. This type of relation between dv/dx and p is strongly suggestive of a linear relation between the pressure and the velocity field as shown by Johansson et al. (1987). The decelerating events produce a high positive peak in the v-signal and a negative pressure. However, the relation between the pressure and the normal velocity component is very close to that found in the accelerating events.

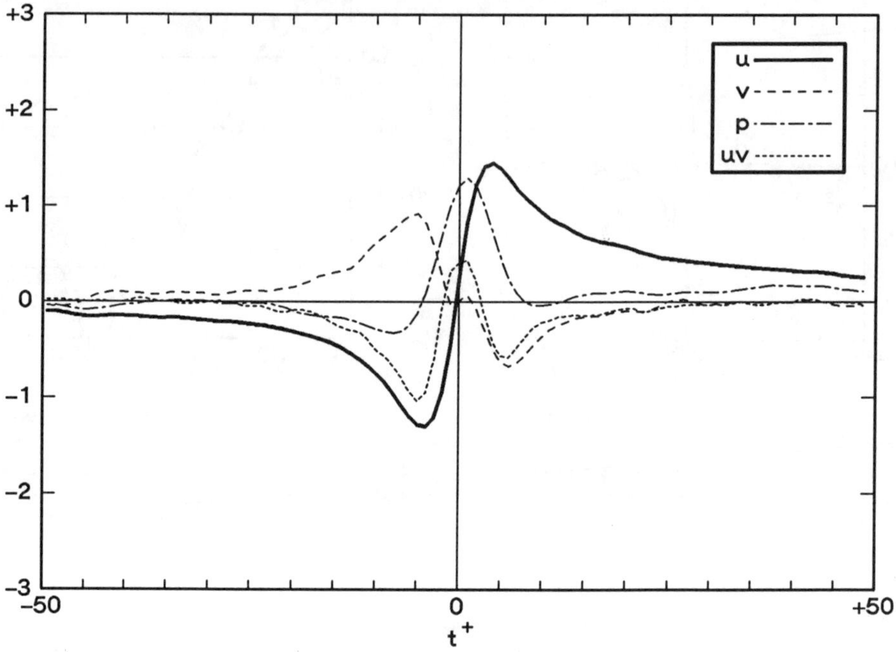

FIGURE 4a. Accelerating VITA-on-u events. $k = 1.0$, $T^+ = 11.0$.

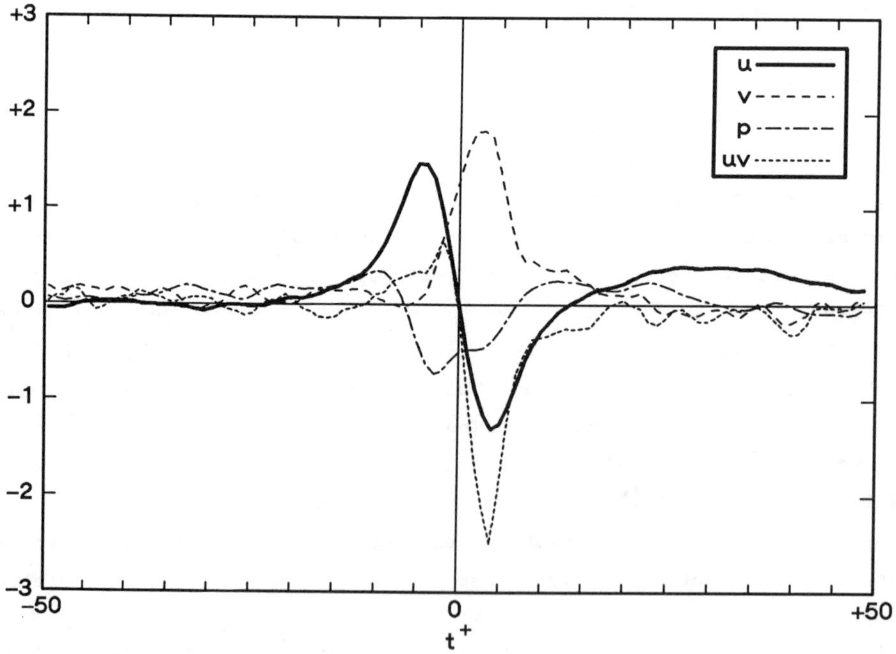

FIGURE 4b. Decelerating VITA-on-u events. $k = 1.0$, $T^+ = 11.0$.

Positive pressure peak detection, Fig. 5a, produces qualitatively the same results as the accelerating VITA events, indicating a qualitative reciprocity. The negative pressure peak events, Fig. 5b, are associated again directly with a positive dv/dt but not with the Reynolds stress, since it is practically zero throughout. Thus, the common thread between pressure peaks and the velocity field is dv/dt.

All of the uv-quadrant results, Figs. 6a-d, produce pressure signatures that are again tied to dv/dt. However, the amplitude of the corresponding pressure peaks is much smaller than those found in the VITA events even though the dv/dts in the present case are about an order of magnitude larger for the same pressure amplitude. The conclusion to be drawn so far is that <u>isolated</u> uv peaks are not intimitely related to pressure peaks.

Ensemble averages based on accelerating VITA-on-u and pressure peak events, predictably show a signal of zero amplitude for the spanwise component w. However, conditional ensemble averages with $w>0$ (or $w<0$) at the center of the detected event, Figs. 7a and 7b, show that there is a well defined w signature whose peak coinsides with the pressure peaks. Similar results are obtained for decelerating VITA-on-u and negative pressure peak events. It therefore seems that VITA-on-u and pressure peak events are associated with non-symmetrical structures with respect to the x-y plane (x is the coordinate in the streamwise direction and y the coordinate normal to the wall).

A different perspective of the wall pressure can be obtained from the weighted probability density P(u,v) of u and v with the pressure p, i.e. p*P(u,v), shown in Fig. 8. Positive pressures are primarily associated with the fourth quadrant or sweep-type events while negative pressures are associated with both the second and third quadrants or ejections and interactions inward respectively. Positive pressures, in general, appear to be the result of wallward moving fluid being retarded by the presence of the wall. Negative pressures are primarily distinguished by the fact that u is negative and it is difficult on the basis of this information alone to come to any definite conclusions.

However, the conditional averages based on the VITA-on-u and pressure peak events are in accord with the long time p*P(u,v) distribution. For accelerating VITA-on-u events, the pressure is low during the ejection

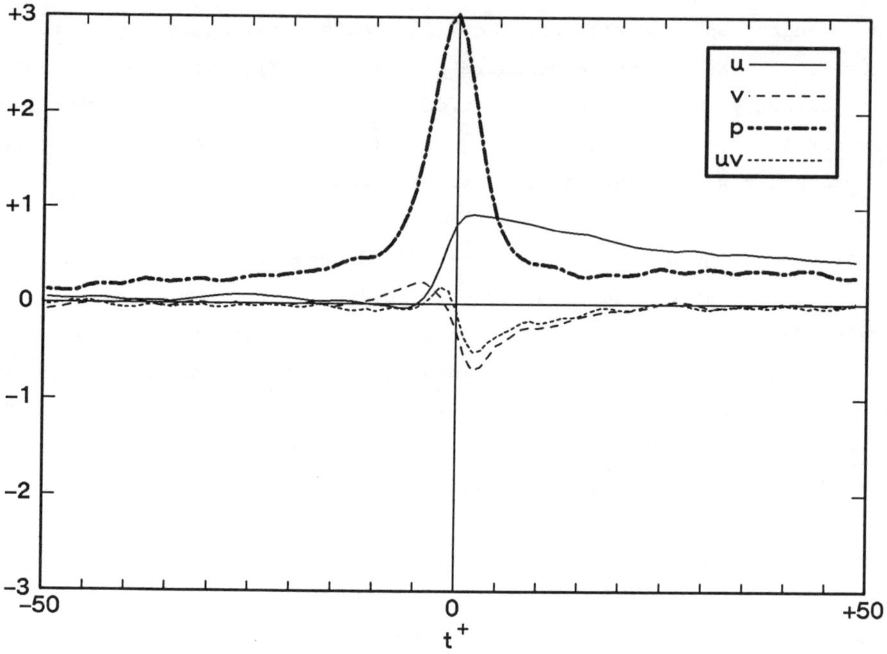

FIGURE 5a. Positive pressure peak events. k = 2.5.

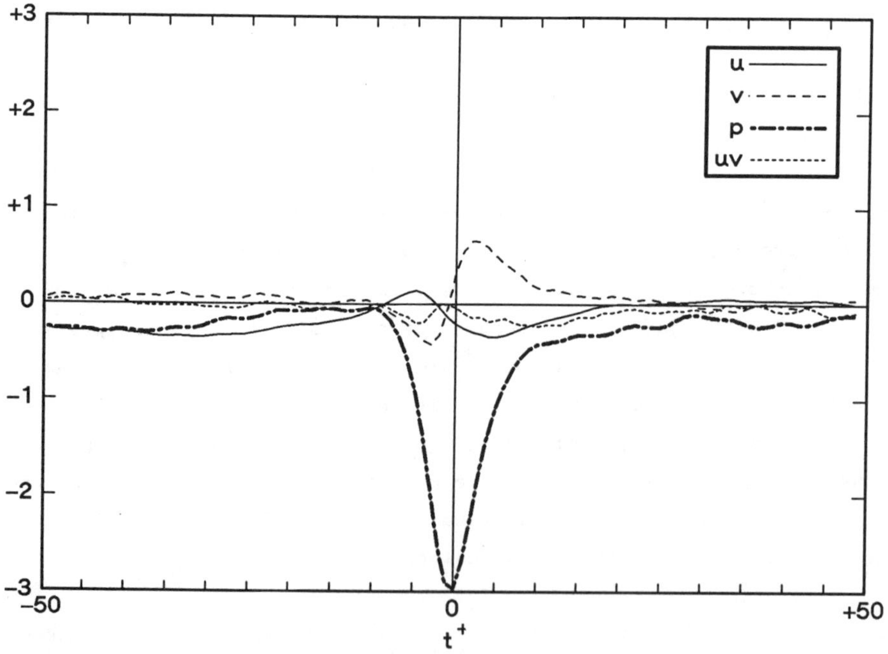

FIGURE 5b. Negative pressure peak events. k = 2.5.

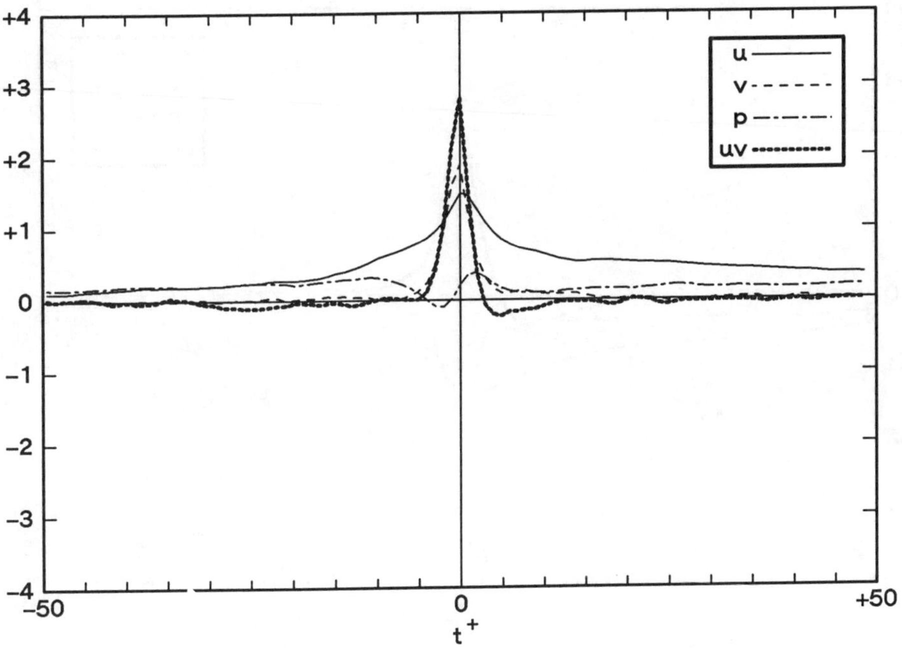

FIGURE 6. uv events with k = 2.0. a: quadrant I; b: quadrant II; c: quadrant III; d: quadrant IV.

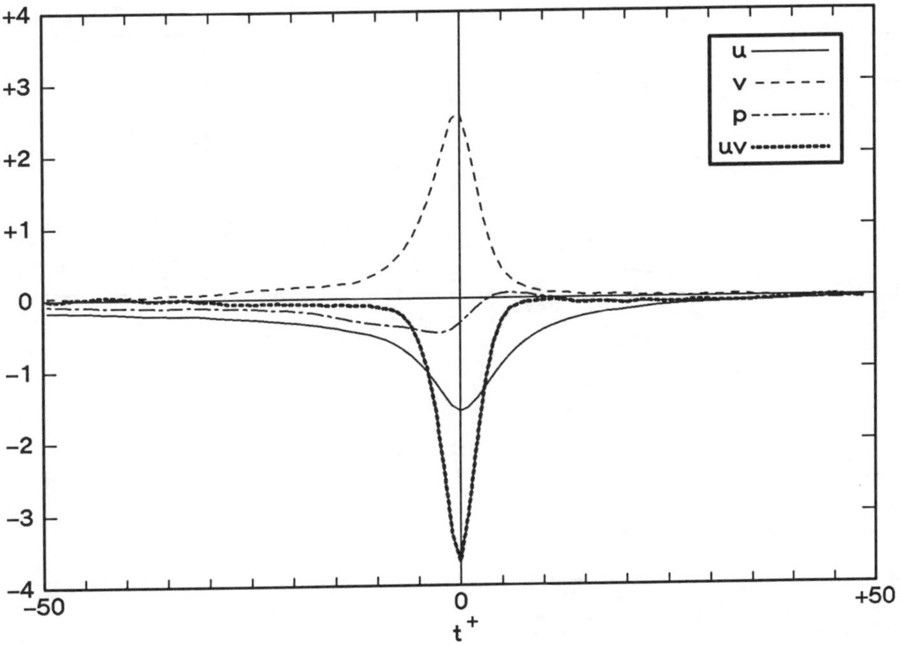

FIGURE 6b.

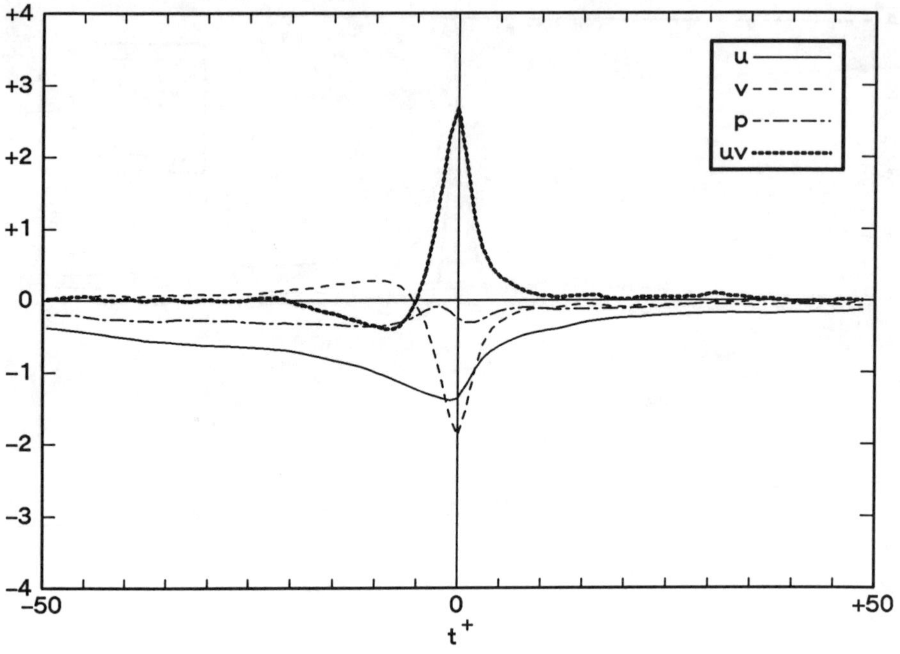

FIGURE 6c.

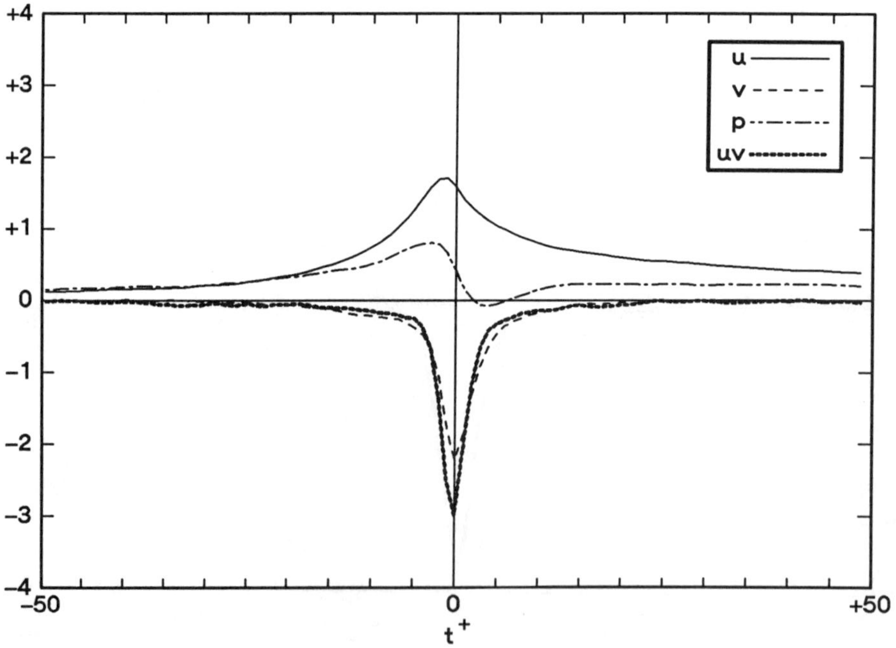

FIGURE 6d.

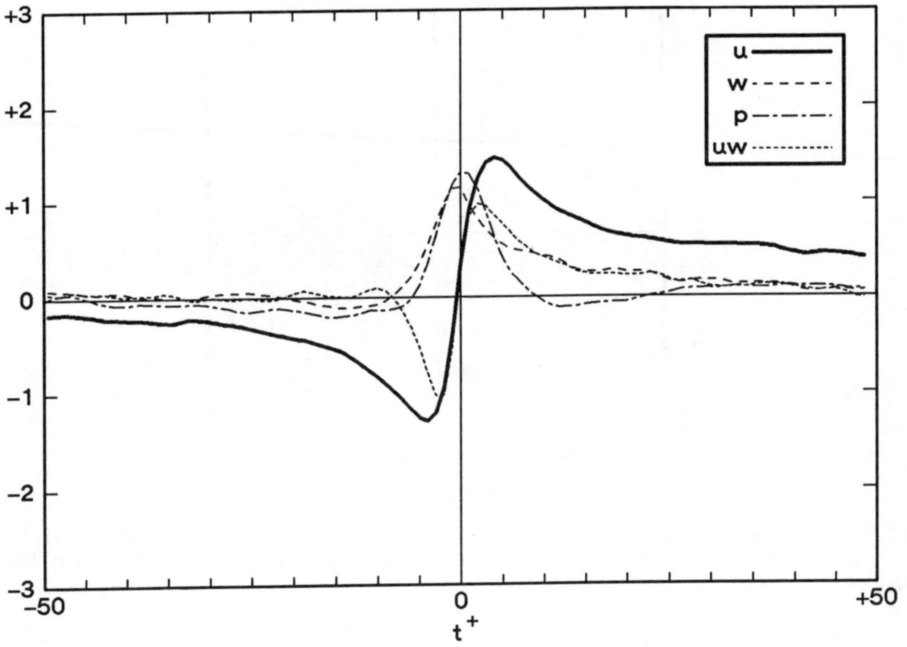

FIGURE 7a. Accelerating VITA-on-u events with w>0 at the time of detection. k = 1.0, T^+ = 11.0.

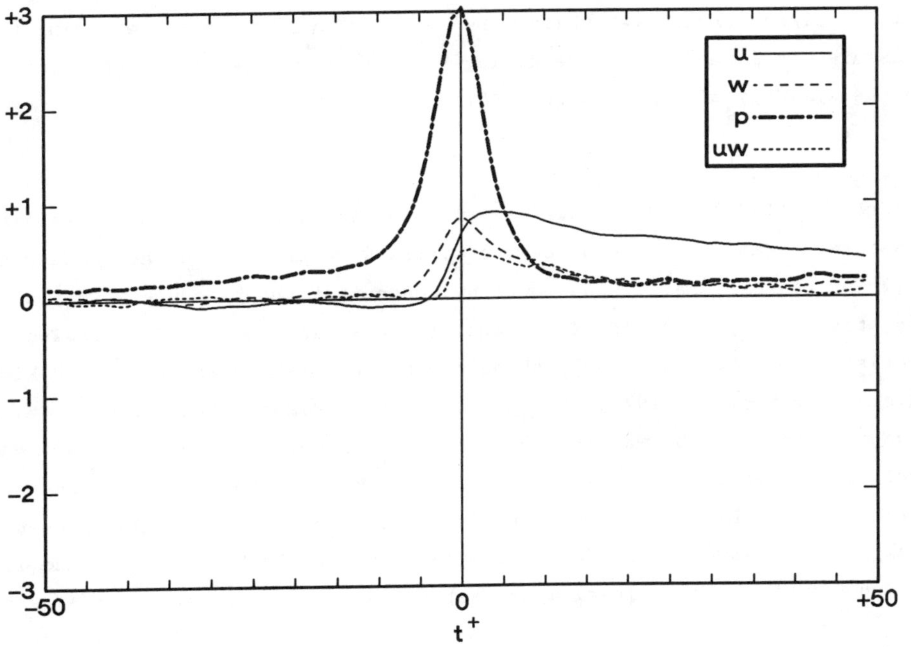

FIGURE 7b. Positive pressure peak events with w>0 at the time of detection. k = 2.5.

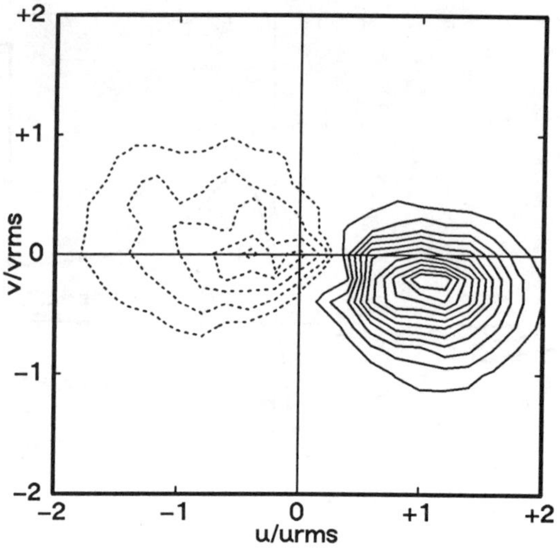

FIGURE 8. p*P(u,v), the weighted probability density of u and v with p. ------, negative pressures; ————, positive pressures. Contour level interval = .001.

part of the event and positive during the sweep part of the event. For decelerating events, again, the pressure is low during most of the ejection part. For negative pressure peak events, the u-velocity is almost always negative while the v-velocity is negative during the pressure fall and positive during the pressure rise, thus contributing to the symmetry of p*P(u,v) for u<0.

WAVE DETECTION

Emmerling (1973) and Schewe (1983) showed clearly that both isolated pressure peaks and pressure wave trains were present in their measured wall pressures. Her (1986) found evidence that the pressure wave trains are the "footprint" of flow oscillations in the buffer region as suggested by Schewe (1983) through his pressure measurements, and Kim, Kline & Reynolds (1971) through their flow visualization and hot-wire measurements in the wall region. The SPD method, described earlier, was applied to accelerating VITA-on-u events, Fig. 9, and positive pressure peak events, Fig. 10. Decelerating VITA events and negative pressure peak events gave essentially similar results. In both Figs. 9 and 10, the power threshold level was set to twice the average power at each frequency.

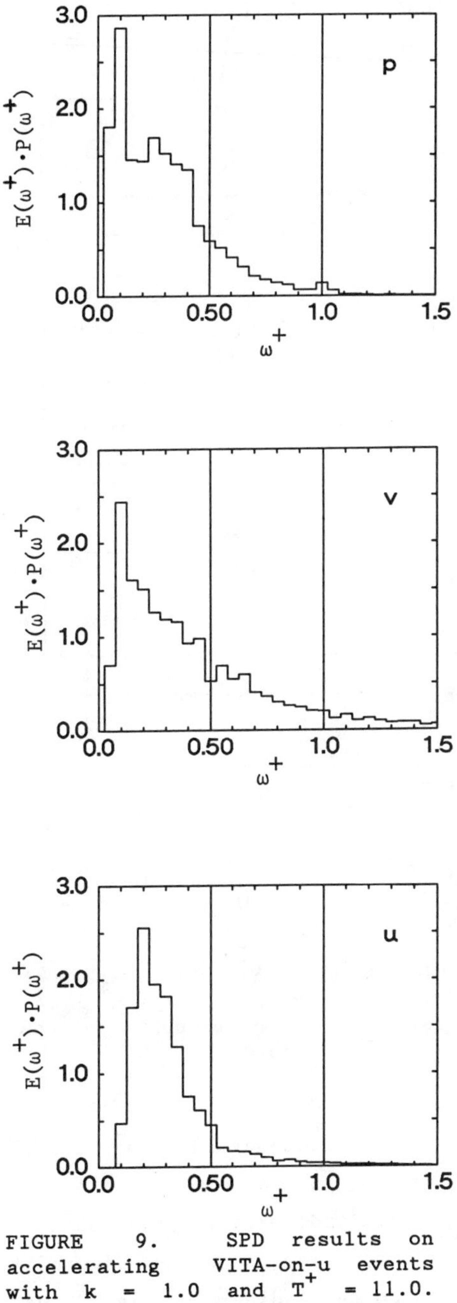

FIGURE 9. SPD results on accelerating VITA-on-u events with k = 1.0 and T^+ = 11.0. Power threshold level is two times the average power at each frequency.

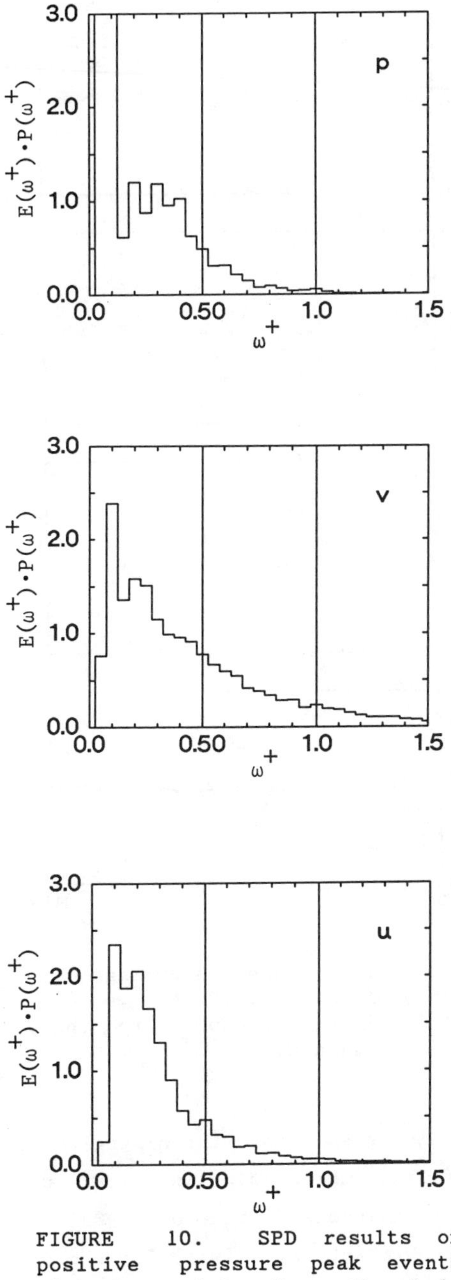

FIGURE 10. SPD results on positive pressure peak events with k = 2.5. Power threshold level is two times the average power at each frequency.

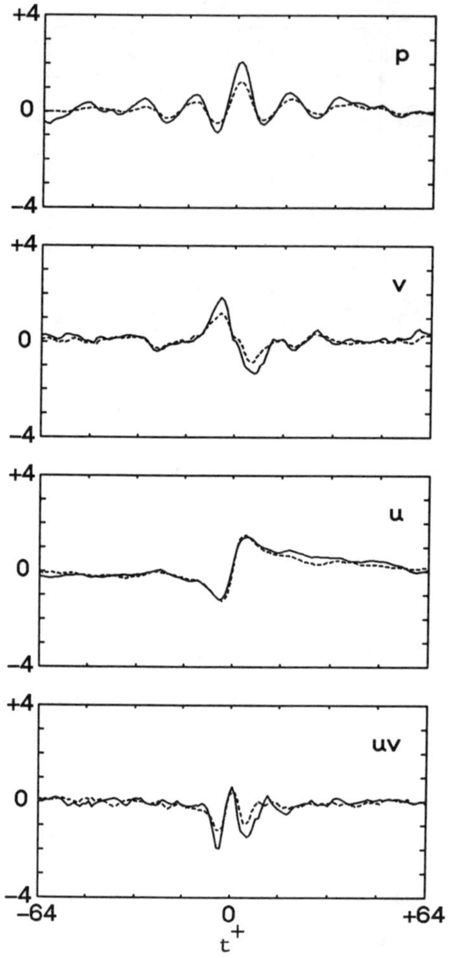

FIGURE 11. Ensemble average of events from Fig. 9 at $\omega^+ = .4$. Solid lines correspond to the aligned signals.

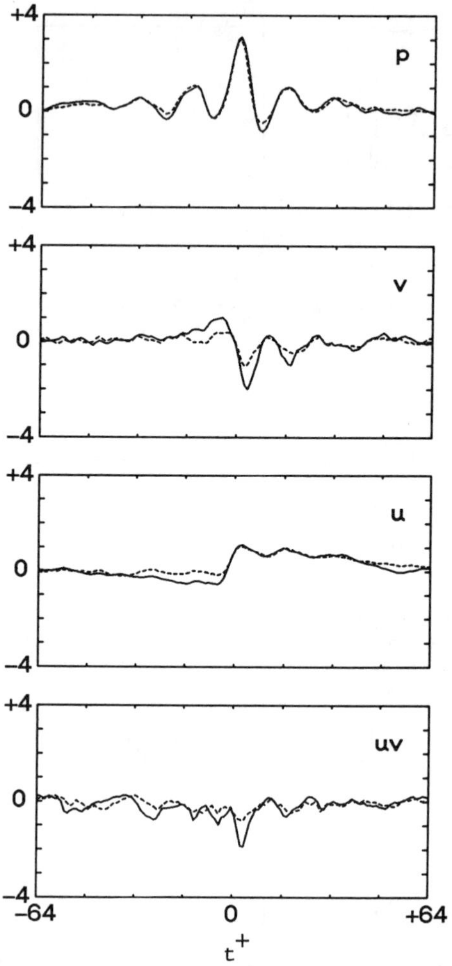

FIGURE 12. Ensemble average of events from Fig. 10 at $\omega^+ = .4$. Solid lines correspond to the aligned signals.

It is seen that the histograms peak in the frequency range $\omega^+ \approx .2-.4$ for all three signals, i.e. u, v and p, in agreement with the observations of Schewe (1983) for high amplitude pressure fluctuations and Kim et al. (1971) for the streamwise velocity component. Figure 11 shows the ensemble average of events corresponding to Fig. 9 that contributed power to a frequency of .4, and likewise in Fig. 12 for events corresponding to Fig. 10. The velocity signals have been aligned separately from the pressure signals using the basic alignment technique of Zilberman, Wygnanski & Kaplan (1977). This consists of

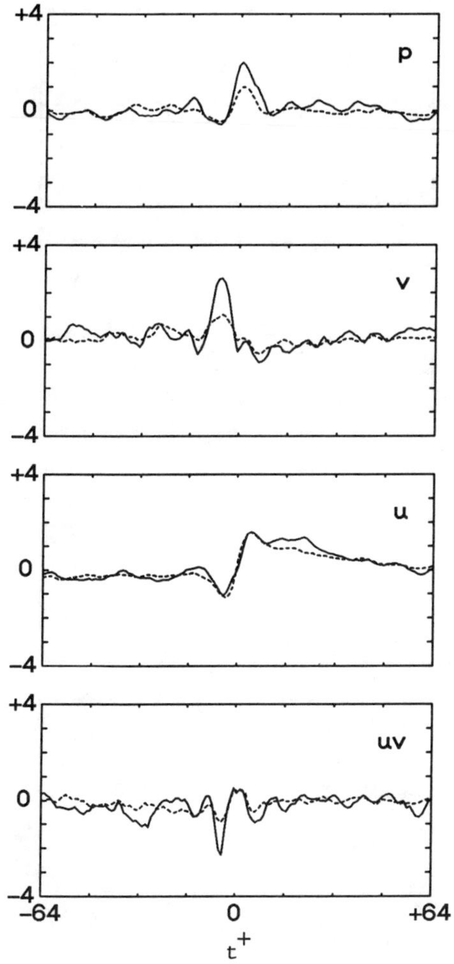

FIGURE 15. Ensemble average of events from Fig. 9 at $\omega^+ = .8$. Solid lines correspond to the aligned signals.

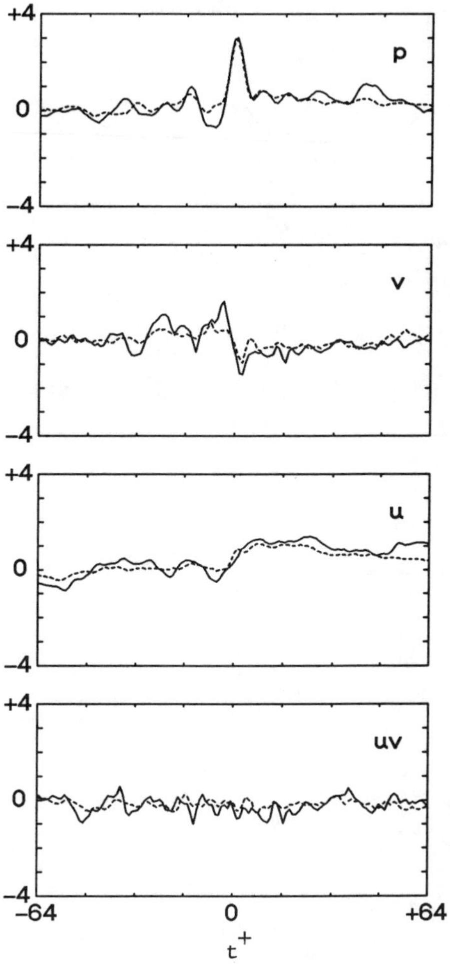

FIGURE 16. Ensemble average of events from Fig. 10 at $\omega^+ = .8$. Solid lines correspond to aligned signals.

be a manifestation of a secondary wave train or instability. At higher frequencies (Figs. 15 & 16), the pressure signal appears to be a single peak as opposed to a wave train. The latter result may be due to the poor resolution of the pressure transducer, whose diameter is approximately 2/3 of the wavelength of the pressure peak.

A comparison can be made of the above results to the computations (Navier-Stokes) of Breuer (1988) of a localized disturbance in a laminar boundary layer. The disturbance developed internal shear layers, as it

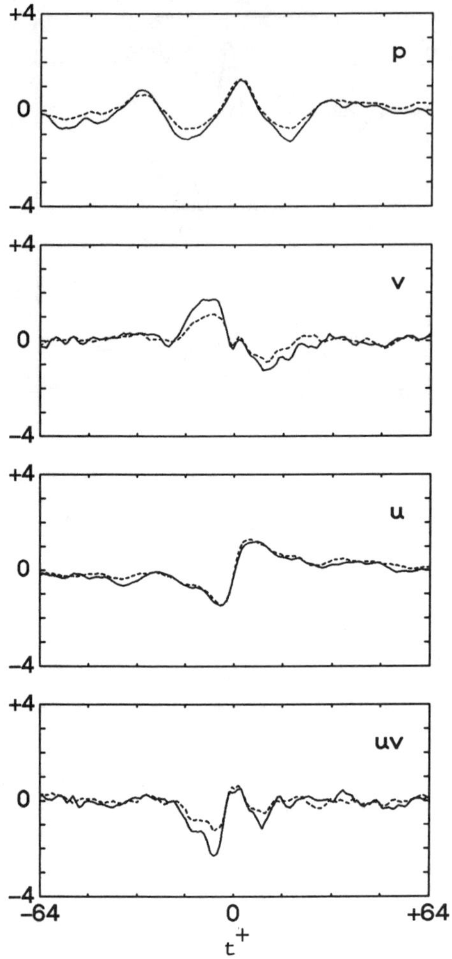

FIGURE 13. Ensemble average of events from Fig. 9 at $\omega^+ = .2$. Solid lines correspond to the aligned signals.

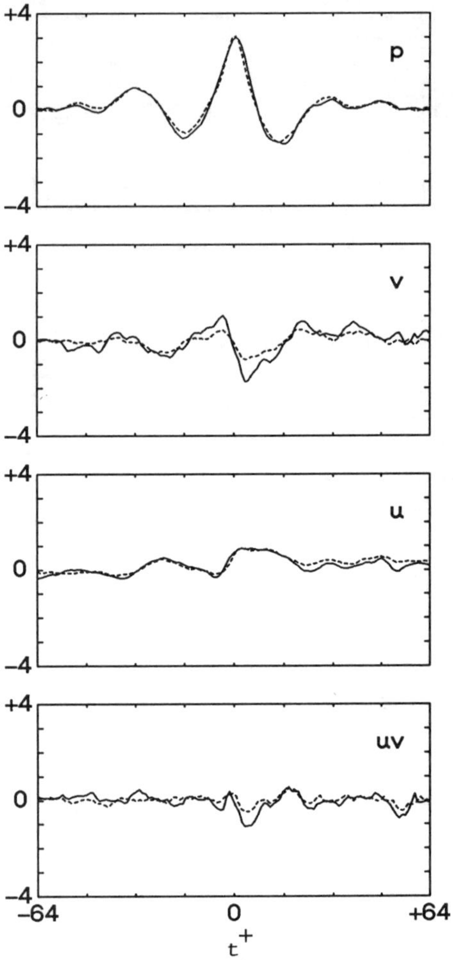

FIGURE 14. Ensemble average of events from Fig. 10 at $\omega^+ = .2$. Solid lines correspond to the aligned signals.

cross-correlating the members of the ensemble with the current ensemble average and then repeating the ensemble averaging procedure by shifting the signals according to the cross-correlation time delays. In the present version, in addition to the above, the signals were windowed as discribed earlier in the SPD technique. After about two iterations no improvement in the ensemble average is observed. At lower frequencies (Figs. 13 & 14), the results are qualitatively similar. Close observation of the v-velocity ensemble averages around the detection time (Fig. 13 (v-velocity) as well as Fig. 4a), shows a kink which may

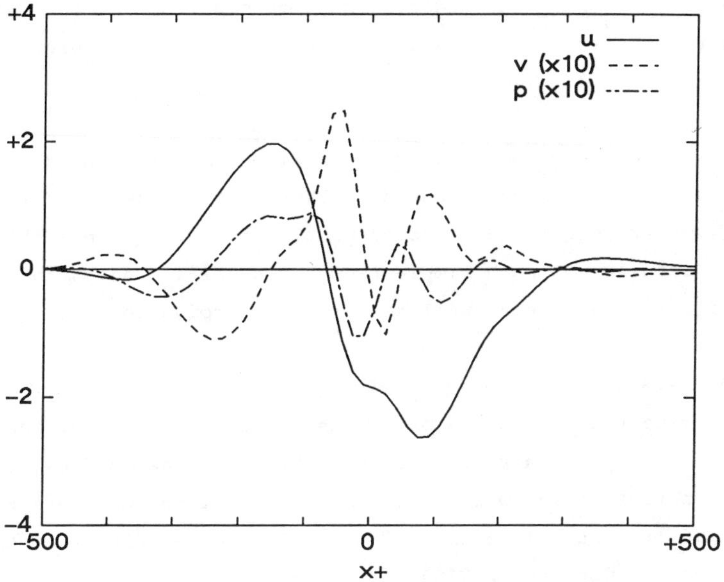

FIGURE 17. u and v velocities and wall pressure on the centerline of a localized disturbance in a laminar boundary layer. $Re_{\delta}* \approx 950$, $y^+ \approx 20$.

progressed downstream, which then became unstable. Fig. 17 shows the u and v signals at $y^+ \approx 20$ and the wall pressure p. The frequency of the wave is ≈ .42 and the wavelength ≈ 2361^*, in good agreement with the observations of Schewe (1983) and the present results.

CONCLUSIONS

Conditional sampling and spectral techniques applied to all three velocity signals and the wall pressure in a turbulent boundary layer were used to show the relation between the wall pressure and the near wall flow. The most prominent connection of the wall pressure and the flow field is through the normal component of velocity, v. Wall pressure peaks and wave trains are related to corresponding structures in the near wall flow field. The wave trains may well be the manifestation of the instability associated with the lift-up of streaks discussed by Kim et al. (1971) and simulated very convincingly by Acarlar & Smith (1987). No definite statement can be made about the formation of high amplitude pressure peaks, as opposed to waves, other than that pressure peaks may be associated with unusually rapid velocity fluctuations. The ensemble averaged v-signals show evidence of secondary wave trains most likely associated with a secondary instability.

It is apparent that in order to resolve small scale, but dynamically significant, pressure fluctuations, that smaller pressure transducers are needed.

A comparison of the present results with computations of a localized disturbance in a laminar boundary layer indicates that both flows produce similar structures. Consequently, it may be profitable to look into the details of such simple flows in order to understand the basic physics of wall pressure as well as turbulent production.

ACKNOWLEDGMENTS

The work reported in this paper was supported by the Office of Naval Research through Contract N00014-87-K-0048. One of us (KSB) was supported through the Office of Naval Research Graduate Fellowship Program and for an interim by the Air Force Office of Scientific Research through Contract F49620-83-C-0019.

REFERENCES

Acarlar, M.S. & Smith, C.R. "A Study of Hairpin Vortices in a Laminar Boundary Layer. Part 2. Hairpin Vortices Generated by Fluid Injection". J. Fluid Mech., 175, p. 43, 1987.

Blackwelder, R.F. & Kaplan, R.E. "On the Wall Structure of the Turbulent Boundary Layer", J. Fluid. Mech. 76, p. 89, 1976.

Breuer, K.S. "The Development of a Localized Disturbance in a Boundary Layer". Ph.D. Thesis, Department of Aeronautics & Astronautics, Massachusetts Institute of Technology, 1988.

Burton, T.E. "The Connection Between Intermittent Turbulent Activity Near the Wall of a Turbulent Boundary Layer With Pressure Fluctuations at the Wall". Acoustics and Vibrations Lab. MIT Rep. no. 70208-10, 1974.

Dinkelacker, A, Hessel, M., Meir, G.E.A. & Schewe, G. "Investigation of Pressure Fluctuations Beneath a Turbulent Boundary Layer by Means of an Optical Method", Phys. Fluids, 20, no. 10, Pt. II, p. S216, 1977.

Dinkelacker, A. & Langeheineken, T. "Relations Between Wall Pressure Fluctuations and Velocity Fluctuations in Turbulent Flow". IUTAM Symposium on Structure of Complex Turbulent Shear Flow, Marseille. (R. Dumas & L. Fulachier, ed.), 1982.

Emmerling, R. "The Instantaneous Structure of the Wall Pressure Under a Turbulent Boundary Layer Flow". Translation of an extended version of Mitteilungen aus dem Max-Planck-Institut fur Stromungsforschung und der Aerodynamischen Versuchsanstalt, no. 56, 1973.

Her, J-Y. "The Relation Between Wall Pressure and the Flow Field in the Wall Region of a Turbulent Boundary Layer". Ph.D. Thesis, Department of Ocean Engineering, Massachusetts Institute of Technology, 1986.

Johansson, A.V. & Alfredsson, P.H. "On the Structure of Turbulent Channel Flow". J. Fluid Mech., 122, p. 295, 1982.

Johansson, A.V., Her, J-Y. & Haritonidis, J.H. "On the Generation of High-Amplitude Wall Pressure Peaks in Turbulent Boundary Layers and Spots". J. Fluid Mech., 175, p. 119, 1987.

Kim, H.T., Kline, S.J. & Reynolds, W.C. "The Production of Turbulence Near a Smooth Wall in a Turbulent Boundary Layer". J. Fluid Mech., 50, p. 133, 1971.

Lu, S.S. & Willmarth, W.W. "Measurements of the Structure of the Reynolds Stress in a Turbulent Boundary Layer". J. Fluid Mech., 60, p. 481, 1973.

Lueptow, R.M., Breuer, K.S. & Haritonidis, J.H. "Computer-Aided Calibration of X-Probes Using a Look-Up Table". Exp. in Fluids, 6, p. 115, 1988.

Mangus, J.F. "Preliminary Measurements of Drag and Bursting Frequency in a Manipulated Turbulent Boundary Layer". M.S. Thesis, Department of Aeronautics & Astronautics, Massachusetts Institute of Technology, 1984.

Schewe, G. "On the Structure and Resolution of Wall-Pressure Fluctuations Associated With Turbulent Boundary-Layer Flow". J. Fluid Mech., 134, p. 311, 1983.

Thomas, A.S.W. & Bull, M.K. "On the Role of Wall-Pressure Fluctuations in Deterministic Motions in the Turbulent Boundary Layer". J. Fluid Mech., 128, p. 283, 1983.

Willmarth, W.W. "Pressure Fluctuations Beneath Turbulent Boundary Layers". Ann. Rev. Fluid Mech., 7, p. 13, 1975.

Zilberman, M., Wygnanski, I. & Kaplan, R.E. "Transitional Boundary Layer Spot in a Fully Turbulent Environment". Phys. Fluids, 20, no. 10, Pt.II, p. S258, 1977.

Part 5
INFORMATION DERIVABLE FROM CORRELATION MEASUREMENTS

SUMMARY LECTURE

Linking Correlations and Structure: Stochastic Estimation and Conditional Averaging

RONALD J. ADRIAN
Department of Theoretical and Applied Mechanics
University of Illinois
Urbana, Illinois 61801, USA

INTRODUCTION

The statistical description of structure in turbulent flow is a central problem in the study of turbulent motion. Various statistical quantities that have been used in this endeavor are listed in Table 1, and have been reviewed in two recent papers (Moin 1984, Williams 1987). Of the items listed the second order spatial correlation functions (and the closely related wave number spectra) provide the most elementary description, and they are the cornerstones of turbulence theory. Evenso, spatial correlation functions by themselves do not yield the degree of structural detail that is often needed in turbulence research. For example, isocorrelation contours of velocity give a measure of the size and orientation of turbulent eddies and some qualitative indication of their geometry, spherical, rod-like, sheet-like, etc., but, they cannot be interpreted too literally. They certainly cannot represent velocity fields, as they have neither the correct units, nor the correct tensor order.

TABLE 1. Methods of Analyzing the Structure of Turbulent Fields

- 2nd ORDER SPATIAL CORRELATION FUNCTIONS
- WAVE NUMBER SPECTRA
- HIGHER ORDER SPATIAL CORRELATION FUNCTIONS
- CONDITIONAL AVERAGES
- PROPER ORTHOGONAL DECOMPOSITION
- STOCHASTIC ESTIMATION
- STOCHASTIC ESTIMATION OF CONDITIONAL EDDIES

Conditional averages are an attempt to achieve a more representative picture of large scale characteristic motions than spatial correlations, and in general they accomplish this goal rather well, provided that the conditional event is properly selected. The conditional average of $\mathbf{u}' = \mathbf{u}(\mathbf{x}',t)$ given that the event $\mathbf{E}$ occurs, $\langle \mathbf{u}' | \mathbf{E} \rangle$, is a well defined vector field than can be interpreted kinematically and dynamically, and related fairly directly to the underlying random fields. In many ways, conditional averages have supplanted correlation functions in the study of organized structues (Antonia 1981, Cantwell 1981).

Lumley's proper orthogonal decomposition (Lumley 1970, 1981) expands the turbulent field in terms of orthogonal functions that are determined by the spatial correlation functions. The approach is statistically unconditional, and therefore the relationship between structures found from conditional analysis and characteristic eddies is not obvious a priori. However, the method does provide a link between spatial correlations and the vector fields described by various terms in the expansion.

Stochastic estimation is another technique that re-formulates the structural information contained in the velocity correlation functions into vector fields. As applied to $\mathbf{u}' = \mathbf{u}(\mathbf{x}',t')$, classical stochastic estimation approximates $\mathbf{u}'$ in terms of the variables contained in the data event vector $\mathbf{E}$ (Papoulis 1984). Interpretation of the resulting vector field requires conceptual development. A simple interpretation is provided by taking a modified approach in which stochastic estimation is applied to the conditional average $\langle \mathbf{u}'|\mathbf{E}\rangle$ instead of $\mathbf{u}'$ (Adrian 1977,1979). The answer is readily interpreted as a vector field that approximates the conditional average, and it can be shown that it possesses most of the properties of the conditional average, with certain important exceptions. We shall see that the linear stochastic estimate can be calculated from the two-point, second order spatial correlation function, and that it does not require conditional sampling. Stochastic estimation is appealing because it brings conditional sampling and correlation functions together in a single approach. By means of Mercer's theorem for proper orthogonal decomposition of the correlation functions, the stochastic estimate can also be related to Lumley's characteristic eddies.

In this paper we shall review the principles and applications of stochastic estimation briefly, and then present the results of various investigations as they pertain to the structure of shear flow turbulence. The results support a picture in which vortex rings (Falco 1977) and/or hairpin vortices (Theodorsen, 1952, Rogers and Moin 1987) dominate in the outer layer, whereas structures elongated in the streamwise direction and possessing vorticity in the streamwise direction dominate the wall layer (Hanratty 1988, Kim and Moin 1986). They may result from vortices moving adjacent to the wall, or the vortices may result from the interaction of the streamwise streaks with themselves, and with the wall.

STOCHASTIC ESTIMATION OF CONDITIONAL AVERAGES

Consider $\langle \mathbf{u}'|\mathbf{E}\rangle$, the condition average of $\mathbf{u}'$ given the event vector $\mathbf{E}$, consisting in general of many different random variables $\mathbf{E} = (E_1,\ldots,E_m)$. The events are arbitrary, but characteristically they take the form of components of velocity vectors at different points. We shall assume that both $\mathbf{u}'$ and the random variables defining each event component have zero mean value and that they involve N points $(\mathbf{x}_1, \mathbf{x}_2,\ldots, \mathbf{x}_N)$. $\langle \mathbf{u}'|\mathbf{E}\rangle$ can be expanded in terms of the event data, and the expansion can be truncated at the first order level to yield

$$\hat{u}_i' = L_{ij} E_j \quad , \quad i = 1,2,3 \qquad (1)$$

$$j = 1,\ldots,m \ .$$

where $L_{ij}(\mathbf{x}', \mathbf{x}_1, \ldots, \mathbf{x}_N)$ is an as-yet undetermined estimation coefficient. (Repeated subscripts are summed).

L_{ij} is found by requiring minimum mean square error

$$\langle (u_i' - L_{ij} E_j)^2 \rangle = \text{minimum} \tag{2}$$

which leads to the following set of linear algebraic equations for L_{ij}:

$$\langle E_j E_k \rangle L_{ij} = \langle u_i' E_k \rangle , \quad i = 1,2,3 \tag{3}$$

$$j,k = 1,\ldots,m .$$

For each value of 'i' (3) provides m equations for the unknowns $L_{i1},\ldots,L_{im}$. The coefficients of these linear algebraic equations are the correlations between the data $\langle E_j E_k \rangle$, and the inhomogeneous terms are determined by the correlations between the data and the variable being estimated $\langle u_i' E_k \rangle$. The formal solution for the linearly estimated conditional average is

$$\hat{u}_i' = \langle E_j E_k \rangle^{-1} \langle u_i' E_k \rangle E_j . \tag{4}$$

When $\mathbf{E}$ involves velocities at different points the correlations reduce to two-point, second order spatial correlations of velocity. Thus, the structural information contained in spatial correlations having the rank of second order tensors is re-organized into a vector field. Linear estimation is exact for processes in which all of the variables u_i', $E_1,\ldots E_m$ are joint normally distributed for every combination of $\mathbf{x}'$, $\mathbf{x}_1,\ldots \mathbf{x}_N$.

Nonlinear stochastic estimation is accomplished by expanding $\langle u_i' | \mathbf{E} \rangle$ in a Taylor series about $\mathbf{E} = \mathbf{0}$, yielding the estimate

$$\hat{u}_i' = A_{ij} E_j + B_{ijk} E_j E_k + C_{ijk\ell} E_j E_k E_\ell + \ldots , \tag{5}$$

where, again, one derives equations for **A**, **B**, **C**, by minimizing the mean square error (Tung and Adrian 1980). The resulting equations involve, in general, multi-point correlations of order two and higher. The <u>best</u> nonlinear stochastic estimate of u given the data **E** is $\langle \mathbf{u}' | \mathbf{E} \rangle$.

It can be shown (Adrian 1979, Adrian & Moin 1988) that the linear estimation equations guarantee that $\hat{\mathbf{u}}'$ has vector properties similar to $\langle \mathbf{u}' | \mathbf{E} \rangle$. In particular,

$$\nabla' \cdot \hat{\mathbf{u}}' = 0 \tag{6}$$

if the flow is incompressible. Also, the vorticity of the linear estimate is equal to the linear estimate of the vorticity

$$\nabla' \times \hat{\mathbf{u}}' = \hat{\boldsymbol{\omega}} , \tag{7}$$

and, more generally, the deformation tensor of the estimate equals the estimate of the deformation tensor. Thus, we can apply all of the usual kinematic analysis to **u'** .

Conditional Events

Stochastic estimation shares with conditional averaging the problem of selecting appropriate events. Experience in this area indicates that there is no single event that is of universal importance. A strongly coherent turbulent motion can be characterized by many different parts of its anatomy. Thus, it is important to evaluate stochastic estimates for many different events, or event values. Fortunately, the calculation in (1) is trivial numerically, thereby permitting easy evaluation of **u'** once the estimation coefficients L_{ij} have been determined. Even calculation of L_{ij} is not onerous, because it only needs to be done once. The primary effort lies in obtaining the correlation function $\langle E_j E_K \rangle$ and especially $\langle u_j' E_k \rangle$.

The choice of conditional events is best guided by theory, and in this regard the Lundgren-Monin hierarchy of probability density equations (Lundgren 1967) provides a very useful theoretical framework. The joint p.d.f. equation for the velocities at N - points $\mathbf{u}_1, \ldots, \mathbf{u}_N$ contains the conditional averages $\langle \mathbf{u}' | \mathbf{u}_1, \mathbf{u}_2, \ldots \mathbf{u}_N \rangle$ and $\langle \mathbf{u}'\mathbf{u}' | \mathbf{u}_1, \ldots \mathbf{u}_N \rangle$. Thus, it is natural to select events consisting of the velocity vector at one or more points in space. Table 2 lists the various events that have been studied. In addition to multi-point events, multi-time events and space-

TABLE 2. Events used in analyses of turbulent fields by stochastic estimation

E	Estimated Property	Reference
$\mathbf{u}(\mathbf{x},t)$	$\mathbf{u}' = \mathbf{u}(\mathbf{x},t')$	Adrian (1979), Hassan (1980), Adrian, et al. (1987)
$u(\mathbf{x},t)$	$u(\mathbf{x}',t)$	Adrian (1978,1979), Hassan (1980) Tung and Adrian (1980), Tung (1982) Chang (1985), Chin (1987), Moin, Adrian and Kim (1987)
$\mathbf{u}(\mathbf{x},t)$	$u_i(\mathbf{x}',t)u_j(\mathbf{x}',t)$	Tung (1982), Elam (1987)
$\mathbf{u}(\mathbf{x},t)$	$p(\mathbf{x}',t)$	Chang (1985)
$p_{wall}(\mathbf{x},t)$	$\mathbf{u}(\mathbf{x}',t)$	Willmarth & Wooldridge (1962)
$p(\mathbf{x},t)$	$\mathbf{u}(\mathbf{x}',t)$	Nithianandan (1980)
$\mathbf{u}(\mathbf{x}_1,t),\mathbf{u}(\mathbf{x}_2,t)$	$\mathbf{u}(\mathbf{x}',t),\boldsymbol{\omega}(\mathbf{x}',t)$	Adrian, Moin & Moser (1988)
$u(\mathbf{x},t),\mathbf{d}(\mathbf{x},t)$	$\mathbf{u}(\mathbf{x}',t),\boldsymbol{\omega}(\mathbf{x}',t)$	Adrian & Moin (1988) Ditter (1987)
$\tau_{wall}(\mathbf{x},t)$	$\mathbf{u}(\mathbf{x}',t),\boldsymbol{\omega}(\mathbf{x}',t)$	Guezennec (1987)
$\phi(\mathbf{x},t)$	$\phi(\mathbf{x}',t)$	Pescelli (1985,1987)

time events have also been used, in which case the correlations are correlations with time delay and space-time correlations, respectively. The event consisting of $\mathbf{u}(\mathbf{x},t)$ and the deformation tensor $d(\mathbf{x},t)$ ($d_{ij} = \partial u_i/\partial x_j$) used by Adrian & Moin (1988) is a '1 1/2-point' event in the sense that the determination of d requires only information from that part of the full two-point p.d.f. that is contained in small separations.

In certain experiments the choice of $\mathbf{E}$ is guided by experimental design. For example, Guezennec (1987) conditioned estimates on the wall shear stress, and Willmarth and Wooldridge (1962) conditioned on wall pressure in order to relate the events at the wall to the overlying flow fields. Nithianandan (1980) conditioned measurements of the velocity field on the turbulent pressure fluctuation in order to identify the state of the flow field associated with positive and negative pressure fluctuations. It is interesting to note that the vector fields defined intuitively by Willmarth and Wooldridge (1962) were, indeed linear estimates of $\langle \mathbf{u}' | p_{wall} \rangle$. Pescelli (1985,1987) has applied stochastic estimation to scalar voltage potential in turbulent plasma flows. Perhaps the best known event, the quadrant average event, (Willmarth and Lu, 1972), is an inequality rather that on equality.

STOCHASTIC ESTIMATES IN ELEMENTARY TURBULENT FLOWS

It is natural to ask what information can be gained from stochastic estimation. The remainder of this paper will seek to answer this question by examining the vector fields that are estimated for a set of (relatively) simple turbulent flows.

Isotropic Turbulence

By virtue of its symmetries isotropic turbulence must contain less structure than any other turbulent flow. The linear stochastic estimate of $\langle u_i' | u(\mathbf{x}) \rangle$ for isotropic turbulence is [Adrian 1979]

$$u_i' = R_{ji}(\mathbf{r})\ u_j(\mathbf{x}) /\ {}^2 \tag{8}$$

where R_{ji} is the isotropic two-point correlation tensor

$$R_{ji} = \langle u_j(\mathbf{x})\ u_i(\mathbf{x}+\mathbf{r}) \rangle, \tag{9}$$

${}^2 = \langle u_i u_i \rangle / 3$, and $\mathbf{x}' = \mathbf{x} + \mathbf{r}$. In this case the event correlation $\langle E_i E_k \rangle = \langle u_i(\mathbf{x})\ u_j(\mathbf{x}) \rangle = {}^2\delta_{ij}$ is particularly simple, and (8) follows immediately from (3), exposing the dependence of the estimate upon the spatial correlation function R_{ji} in an obvious way. It can be shown (Adrian 1979) that the vector field of the linear stochastic estimate is axisymmetric about $\mathbf{u}(\mathbf{x})$ with axisymmetric streamfunction streamfunction

$$\Psi^{(1)} = \frac{1}{2} |u(\mathbf{x})| f(r)\ r^2 \sin^2\theta \tag{10}$$

where f(r) is the longitudinal correlation function. The streamline pattern, Figure 1 reveals an axisymmetric vortex ring in which the point $\mathbf{x}$ is located at the center. A simple exponential longitudinal correlation function yields an unbounded ring that looks much like the field of a doublet, Figure 1(a). A more realistic correlation such as that of Van Atta and Chen (1968), predicts a double ring with a bounded inner ring, Figure 1(b). It is not clear whether the outer ring represents a vortex that occurs in certain physical realizations, or whether it represents the average field due to randomly oriented surrounding vortex rings.

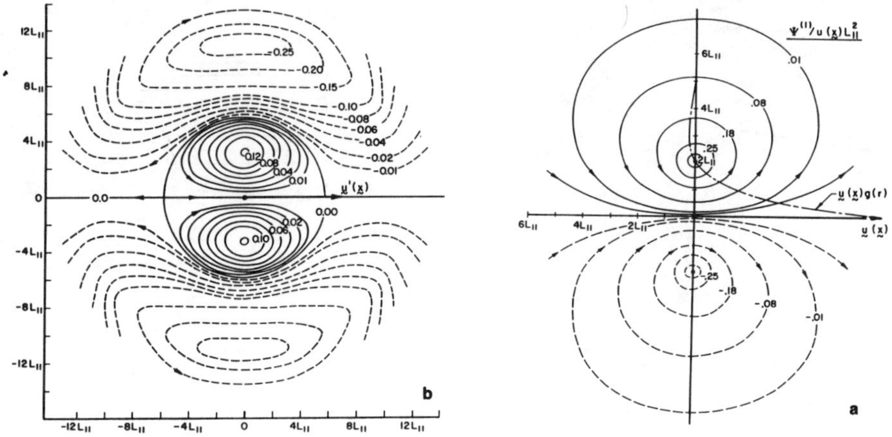

Figure 1. Streamlines of the stochastic estimate of $\langle \mathbf{u}(\mathbf{x}+\mathbf{r}) | \mathbf{u}(\mathbf{x}) \rangle$ in isotropic grid turbulence. (a) $f(r) = \exp - |r|/L_{11}$; (b) f(r) from Van Atta and Chen (1968). [Ref. 1]

Homogeneous Shear Flow

The coherent structures of a homogeneous, low Reynolds number flow with uniform mean shear $U_1 = Sx_2$ have been studied by direct numerical simulation (Rogers and Moin 1987). They find many hairpin vortex structures imbedded in the random fields of a nature very similar to that of the hairpin vortices first proposed by Theodorsen (1952). The hairpins are inclined, on average at 45° ("up-hairpins") and-135° ("down-hairpins") to the mean flow in the x_1-direction.

Linear stochastic estimates of the conditional eddy given the local kinematic state $\langle \mathbf{u}(\mathbf{x},t) | \mathbf{u}(\mathbf{x},t), \mathbf{d}(\mathbf{x},t) \rangle$ have been calculated from the direct simulation of the homogeneous shear flow (Adrian and Moin 1988). The equations for this calculation are

$$\hat{u}_i'(\mathbf{r}) = A_{ij}(\mathbf{r})\, u_j(\mathbf{x},t) + B_{ijk}(\mathbf{r})\, d_{jk}(\mathbf{x},t) \tag{11}$$

$$A_{ij}R_{j\ell}(\mathbf{0}) + B_{ijk}\, R_{\ell j,k}(\mathbf{0}) = R_{\ell i}(\mathbf{r}) \tag{12a}$$

$$A_{ij}R_{j\ell,m}(\mathbf{0}) - B_{ijk}\, R_{j\ell,km}(\mathbf{0}) = - R_{\ell i,m}(\mathbf{r}). \tag{12b}$$

The conditional event $\{\mathbf{u}(\mathbf{x},t),\ \mathbf{d}(\mathbf{x},t)\}$ was evaluated by reading the values directly from the turbulent field at a particular instant and at several points in the vicinity of a recognizable hairpin. Figure 2 presents the vortex lines of the linear estimate that was obtained when the point $\mathbf{x}$ was located under the head of the random hairpin. Clearly, an average hairpin structure emerges. The information concerning the size and shape of the hairpin is contained in the correlation function as it appears in equation (12).

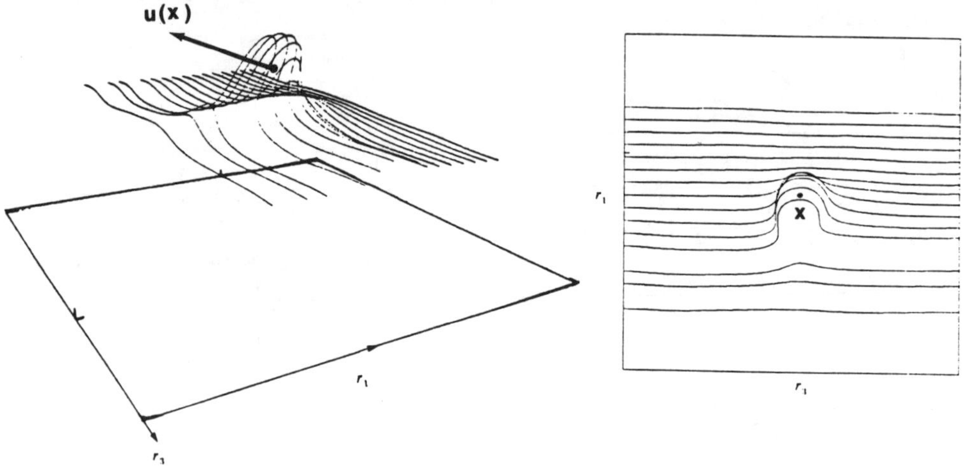

Figure 2. Vortex lines of the linear stochastic estimate in homogeneous turbulent shear flow indicate an upwards oriented hairpin vortex structure. [Ref. 4]

In order to explore the mechanisms for generation of Reynolds shear stress $\langle u_1u_2\rangle$ in turbulent shear flow Adrian and Moin (1988) examined the linearly estimated field associated with the values of $\mathbf{u}(\mathbf{x},t)$ that make the greatest contribution to $\langle u_1u_2\rangle$, in the sense that the product of u_1u_2 and the probability density function of (u_1,u_2) was a local maximum. Two maxima occur, defining a second quadrant Q2-event with velocity vector at approximately 135° to the mean flow and a fourth-quadrant Q4-event with velocity vector at approximately -45° to the mean flow. When these vectors are used as conditions in the linear estimate, with $\mathbf{d} = \mathbf{0}$, a hairpin vortex results which is very similar to that found in Figure 2. The maximum Reynolds stress occurs at a point $\mathbf{x}$ lying in the plane of the hairpin and between its legs. High production of $\langle u_1u_2\rangle$ is associated with large velocities induced at this point by the mutual vortex induction of the head and legs, as sketched in Figure 3 for the case of a Q2-event. We shall see that there are reasons to believe that this flow event occurs in turbulent boundary layers as well, and that it may be a fundamental mechanism for Reynolds stress creation that is common to many, if not all, turbulent shear flows.

A unification of the results for isotropic turbulence and anisotropic shear flow is achieved by observing that a hairpin vortex is created kinematically by the simple addition of a vortex ring to locally uniform mean shear flow, Fig. 4. The plane of the hairpin lies in the plane of the vortex ring, e.g. perpendicular to $\mathbf{u}(\mathbf{x},t)$, and the details of its pattern depend upon the dimensionless ratio $SL_{11}/\ |\mathbf{u}(\mathbf{x},t)|$ where S is the mean shear and L_{11} is the integral length scale of the correlation used to calculate the vortex ring. This simple relationship between rings and hairpins may resolve many of the conflicts that exist in the body of phenomenological observations that we have on turbulence structure. Clearly, flow visualization of the entire field will yield hairpins if the vortices lie in regions of moderate shear, $SL_{11}/|\mathbf{u}(\mathbf{x},t)| > 1$, but it will yield vortex rings if the rings move in regions of weak shear, $SL_{11}/|\mathbf{u}(\mathbf{x},t)| \ll 1$.

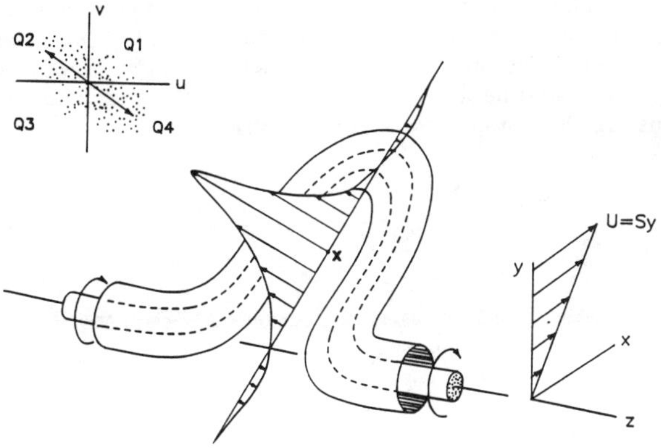

Figure 3. Creation of Reynolds shear stresses by hairpin vortices

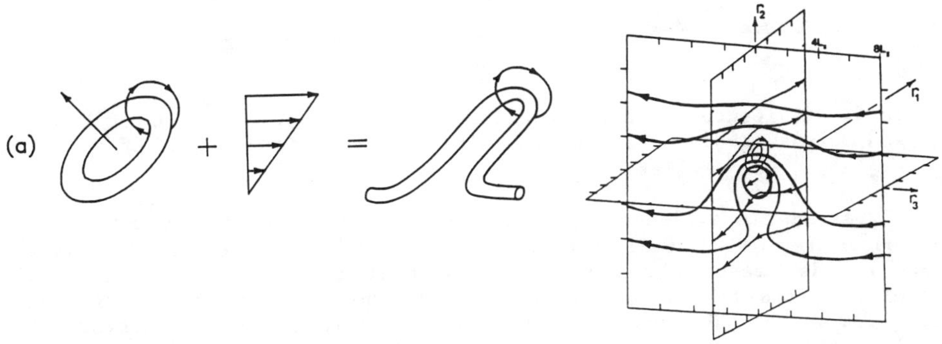

Figure 4. (a) A vortex ring plus a mean shear flow equals a hairpin vortex; (b) vortex lines of the flow obtained by adding a uniform mean shear flow to the isotropic vortex ring in Fig. 1 with $\mathbf{u}(\mathbf{x},t) = (-SL_{11}, 0,0)$. [Ref. 11]

Outer Region of Wall Bounded Shear Flow

Adrian, Moser and Moin (1987) have computed at $y^+ = 99$ in turbulent channel flow $\langle \mathbf{u}' | Q2 \rangle$, the conditionally averaged velocity field given that the (u_1, u_2) vector was equal to the second quadrant vector found by Kim and Moin (1986). Contours of constant vorticity in this field are shown in Figure 5(a). The linear stochastic estimate of $\langle \mathbf{u}' | \mathbf{u}(\mathbf{x},t) \rangle$ has been averaged over all values of $\mathbf{u}(\mathbf{x},t)$ lying in the second quadrant to obtain a comparable result, Figure 5(b). The comparison of the linear estimate with the true conditional average is remarkably close, giving some indication of the accuracy of linear estimation. Moreover, each calculation indicates the existence of a vortex ring structure centered at $y^+ = 99$, and inducing a flow at approximately 135° to the mean flow direction. These results reveal a great deal of similarity between the flow in the weakly inhomogeneous outer regions of the channel flow and the vortical motions in homogeneous shear flow.

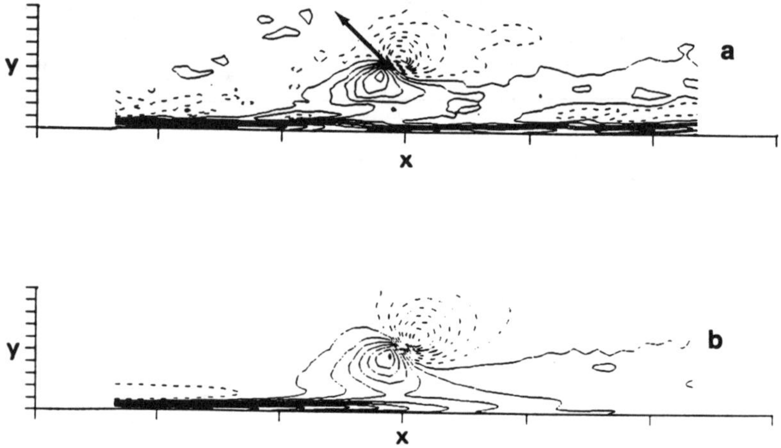

Figure 5. (a) Contours of the conditional average of $\omega_z'(\mathbf{x}')$ using Kim and Moin's (1986) Q2 event, (b) Linear stochastic estimate of $\langle \omega_z'(\mathbf{x}') \mid Q2 \rangle$. [Ref. 5]

Experimental evidence concerning the presence of vortex motions in the outer region of wall turbulence has also been obtained by Hassan (1980). He measured the two-point spatial correlation of the axial velocity u_1 and the normal velocity u_2 in the x_1-x_2 plane of turbulent pipe flow at 50,000 Reynolds number, fit this result to an axisymmetric model of the correlation tensor, and used the curve fitted correlation to compute the linear estimate of (u_1', u_2') in the axial-radial plane. The flow pattern he found for the case of a second quadrant turbulent fluctuation inclined at 135° to the mean flow, Figure 6, is consistent with the channel flow computations in Figure 5.

Ditter (1987) has performed preliminary explorations of the extent to which an isotropic vortex ring plus a uniform mean shear can emulate the flow patterns observed in inhomogeneous, anisotropic turbulence. Figure 7 was computed for the case of a vortex ring from isotropic turbulence being imbedded in a uniform shear flow with the plane of the ring at 45° to the mean flow direction, that is, the event velocity vector **u** at the center of the isotropic conditional eddy was $\mathbf{u}(\mathbf{x},t) = (-SL_{11}/\sqrt{2}, SL_{11}/\sqrt{2}, 0)$. The similarity between the streamlines in Figure 7(a) and the measurements of Hassan, Figure 6, is striking. Likewise, the vortex lines in Figure 7(b) are very similar to those found in channel flow by Kim and Moin (1986), Figure 8. On the basis of these comparisons it appears that the hairpin vortex of shear flow can be represented by a vortex ring plus mean shear, and that the representation is not very sensitive to the detailed structure of the ring.

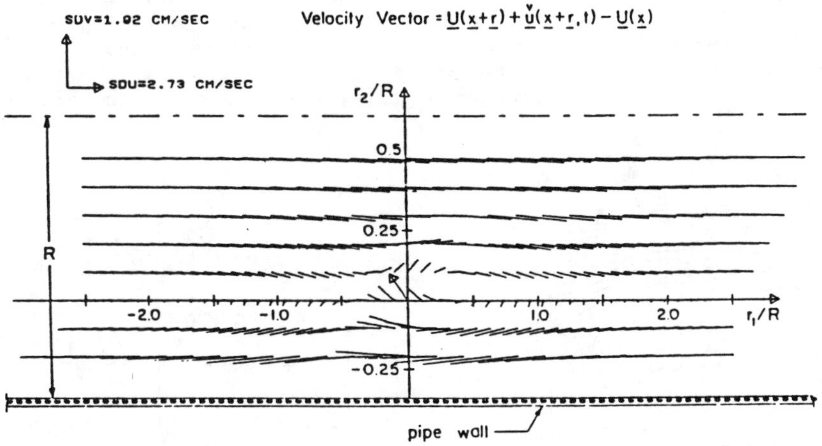

Figure 6. Linear stochastic estimate of the mean flow plus the turbulent fluctuation given a second quadrant velocity fluctuation in pipe flow at 50,000 Reynolds number. [Ref. 16]

Clearly, mean and turbulent straining fields will stretch ring vortices so that many of them will be anisotropic. It is interesting however, to explore further the conjecture that some aspects of the anisotropic structure of weak turbulent shear flow can be modelled by randomly superimposing a collection of vortex rings whose structures are isotropic, e.g. given by equation (8), but whose orientations, as specified by $\mathbf{u}(\mathbf{x},t)$ are anisotropically distributed. Let us denote the isotropic spatial correlation function by R^{iso}_{ji}. It is completely specified by a longitudinal correlation function $f(r)$. Then, from the relation (Adrian 1977)

$$\langle u_k(\mathbf{x},t)\ u_i(\mathbf{x}',t)\rangle = \langle u_k(\mathbf{x},t)\ \hat{u}_i(\mathbf{x}',t)\rangle \tag{13}$$

and equation (8), it follows that

$$R_{ki}(\mathbf{r}) = R^{iso}_{ji}(\mathbf{r})\ \langle u_j(\mathbf{x},t)\ u_k(\mathbf{x},t)\rangle \tag{14}$$

where $\langle u_j u_k \rangle$ is the Reynolds stress tensor for the anisotropic flow, as determined by the probability distribution for the random orientation of the vortex rings and their strengths. Since $R_{ij}^{(iso)}(\mathbf{0}) = \eta^2 \delta_{ij}$, equation (14) reduces to an identity for $\mathbf{r} = \mathbf{0}$. The validity of this equation as a representation of the anisotropic spatial correlation needs to be explored.

Near Wall Turbulence

Moin, Adrian and Kim (1987) have linearly estimated the structure close to the wall of the NASA Ames direct numerical simulation of channel flow using a condition on $\mathbf{u}(\mathbf{x},t)$ at a single point. A typical result, Figure 9, reveals flow patterns in the cross-stream planes that are indicative of pairs of streamwise vortices. Thus, stochastic estimation close to the wall predicts structures that are consistent with evidence

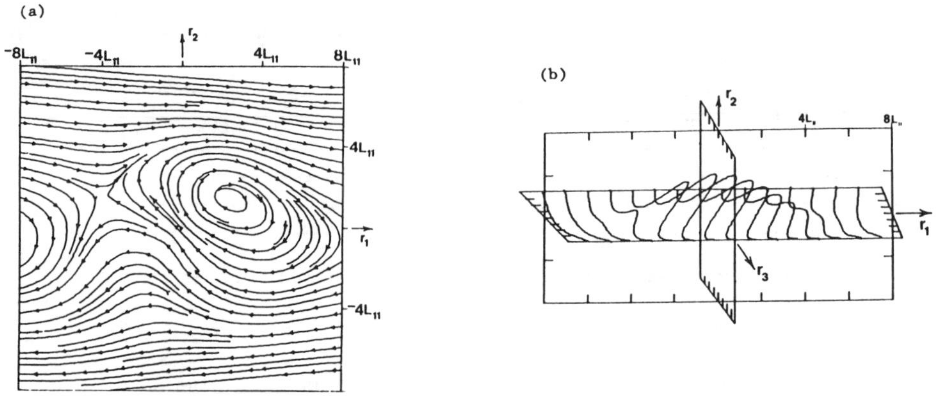

Figure 7. Superposition of the vortex ring found in isotropic turbulence with a uniform mean shear flow. The fluctuating event vector $\mathbf{u}(\mathbf{x},t)$ equals $[-SL_{11}/\sqrt{2},\ SL_{11}/\sqrt{2}\ ]$, corresponding to a turbulent fluctuation in the second quadrant. [Ref. 11]

from conditional averaging and flow visualization, and the stochastic estimate is markedly different from a vortex ring. As the point $\mathbf{x}$ at which $\mathbf{u}(\mathbf{x},t)$ is specified is moved away from the wall, the diameters of the longitudinal vortices increase in a manner consistent with Townsend's (1976) attached wall eddies. Contours of the u_2-component of the linear estimate at $y^+ = 20$ show that the wall vortices are elongated in the streamwise direction, as expected, Figure 10.

Close to the wall, the velocity vectors that make the greatest contributions to the Reynolds shear stress are second and fourth quadrant events whose angles of inclination with respect to the mean flow are very small. These angles increase from zero at the wall to +140° and −30°, respectively at the centerline of the channel (Moin, Adrian and Kim 1987). Thus, it is apparent that the structure of the stochastic estimate makes a significant transition as one moves away from the wall. Kim and Moin (1986) show a conditionally averaged flow field that manifests this behavior in the form of a hairpin vortex whose legs lay in the streamwise direction close to the wall, presumably as a consequence

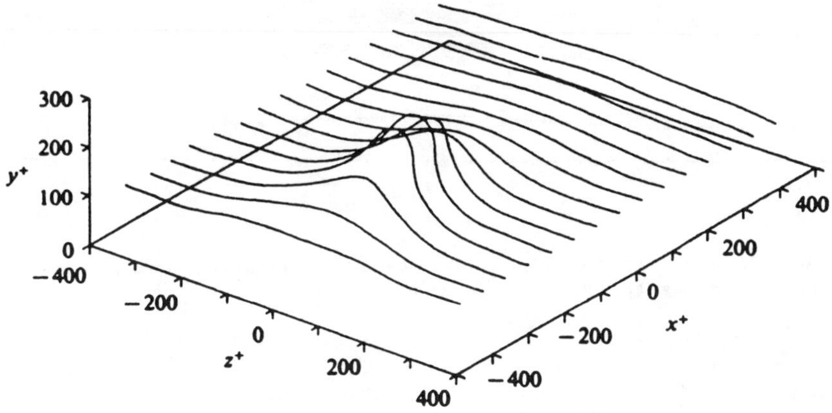

Figure 8. Conditionally averaged vorticity structure in numerically simulated turbulent channel flow. [Ref. 20]

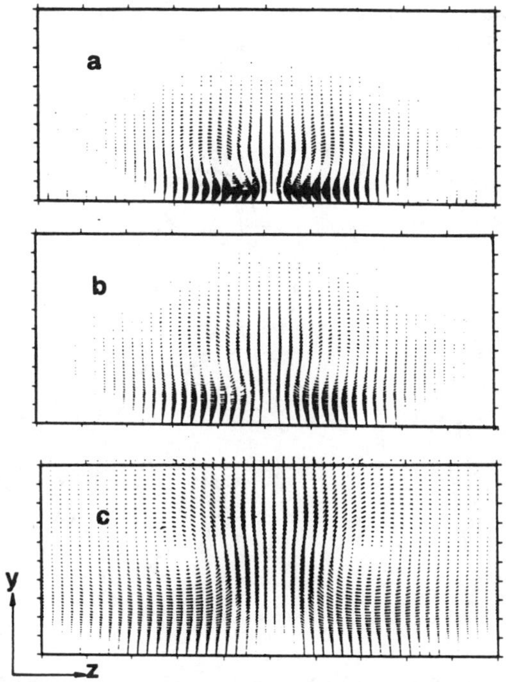

Figure 9. Linearly estimated vector fields for ejection-type events that contribute most to ⟨uv⟩. (a) $y^+ = 19.2$, (b) $y^+ = 49.6$, (c) $y^+ = 103$. [Ref. 25]

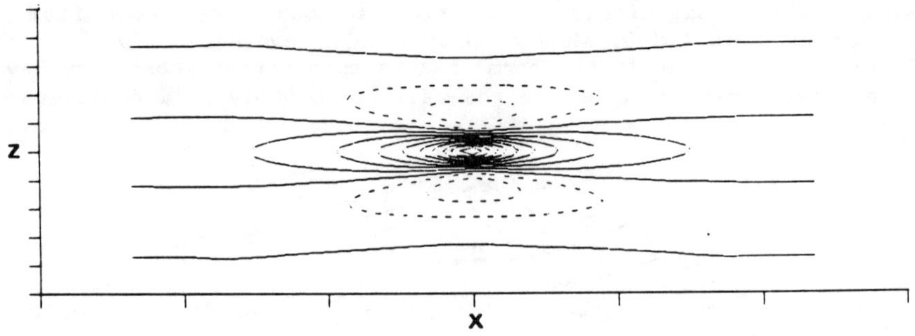

Figure 10. Contours of the linear estimate of the u_2-component of turbulence close to the wall in channel flow. [Ref. 5]

of vortex stretching. Its head lifts up from the wall in a plane that is inclined at about 45° to wall, so that it induces velocity at approximately 135°. This flow is indicated schematically in Figure 11, and Kim (1985) has studied its evolution. On the basis of present observations it is clear that the wall hairpin can be decomposed into an inclined vortex ring in the outer flow plus two longitudinal vortices along the wall. In this way, some linkage is established between the vortex ring-like motions observed most notably by Falco (1977) and the wall streaks observed by many authors (c.f. Hanratty 1988).

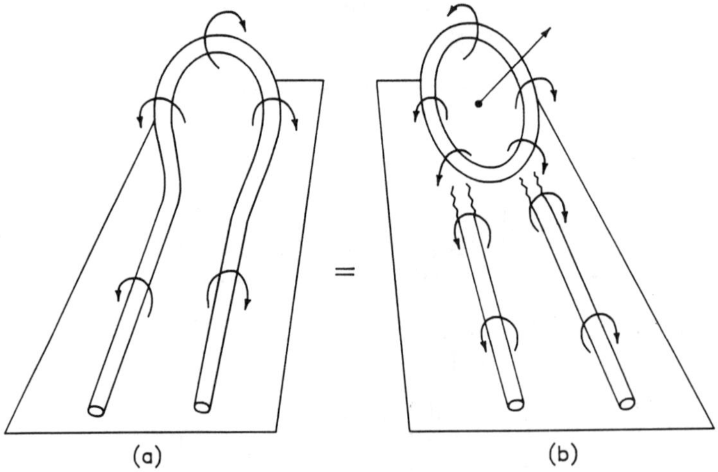

Figure 11. (a) Hairpin vortex close to the wall, (b) decomposition of a wall hairpin into vortex ring plus mean shear plus two streamwise vortices.

SUMMARY AND CONCLUSIONS

Stochastic estimation offers the interesting possibility of linking together correlations, conditional averages and coherent structures. The estimates are approximate evaluations of conditional averages, and they are evaluated using correlation functions as inputs. In one sense, stochastic estimation provides a means of finding conditional averages without conditional sampling. In another sense, it reorganizes the information contained in spatial correlation functions into vector fields that can be interpreted as coherent structures. The evidence indicates that the stochastic estimates are essentially the same as large scale coherent structures found in earlier studies.

Stochastic estimates given the velocity at one point, the velocity and deformation at one point, or the velocity at two or more points are well suited to the description of the local fluid motion. One of the strengths of stochastic estimation in terms of these variables is that it can be applied to many different flows. This leads to a relatively simple comparison and classification of coherent structures in different flows: the conditional eddy of isotropic turbulence is a vortex ring; in homogeneous turbulent shear flow it is a vortex ring plus mean shear; in inhomogeneous wall shear flow it is a vortex ring in the outer region plus viscous streamwise vortices in the wall layer, plus mean shear.

Specifying a single point event is almost certainly not sufficient to identify a coherent motion completely. Hence, linear estimates given data at one or even several points, should _not_ be interpreted as accurate representations of instantaneous flow patterns. They are averaged quantities and the variance of the flow field about the average is demonstrably large. It is better to interpret the estimates as mathematical quantities like $\mathbf{U}$ and R_{ij} that provide _some_, _but not all of the structural information_ about the instantaneous flow. As information is added to the event vector one expects the accuracy of the conditional eddy to increase. For example, Adrian, Moin and Moser (1987) observe that certain two-point events in channel flow generate patterns that evidently correspond to the collision and interaction of two vortex systems. Thus, multipoint data may be needed if one wishes to study many body interactions.

ACKNOWLEDGEMENT

This research was supported by the NASA Ames/Stanford Center for Turbulence Research and by NSF Grant ATM 86-00509.

REFERENCES

1. Adrian, R. J., "Conditional Eddies in Isotropic Turbulence," _Phys. Fluids_ Vol. 22, pp. 2065-2070, 1979.

2. Adrian, R. J., "On the Role of Conditional Averages in Turbulence Theory," _Turbulence in Liquids: Proc. of the Fourth Biennial Symposium on Turbulence in Liquids, September 1975_, ed. J. L. Zakin and G. K. Patterson, Science Press, Princeton, NJ, pp. 323-332, 1977.

3. Adrian, R. J., "Structural Information Obtained from Analysis Using Conditional Vector Events: A Potential Tool for the Study of Coherent Structures, Coherent Structure of Turbulent Boundary Layers, ed. C. R. Smith and D. E. Abbott, AFOSR Lehigh Univ. Workshop, Dept. Mech. Engr. & Mech., Bethlehem, PA, pp. 416-421, 1978.

4. Adrian, R. J., and Moin, P., "Stochastic Estimation of Organized Turbulent Structure: Homogeneous Shear Flow," to appear in J. Fluid Mech. 1988.

5. Adrian, R. J., Moin, P., and Moser, R. D., "Stochastic Estimation of Conditional Eddies in Channel Flow," Report CTR-S87, Center for Turbulence Research, NASA Ames/Stanford, 1987.

6. Adrian, R. J., Chung, M. K., Hassan, Y., Jones, B. G., Nithanandan, C. K., and Tung, A.T.C., "Experimental Study of Stochastic Estimation of Turbulent Conditional Averages," Proc. 5th Turbulent Shear Flow Symposium.

7. Antonia, R. A., "Conditional Sampling in Turbulence Measurement," Ann. Rev. Fluid Mech. Vol. 13, p. 131, 1981.

8. Cantwell, B., "Organized Motion in Turbulent Flow," Ann. Rev. Fluid Mech. Vol. 13, p. 457, 1981.

9. Chang, P. H., "Fluctuating Pressure and Velocity Fields in the Near Field of a Round Jet," Ph.D. Thesis, Univ. of Illinois, 1985. Also, Chang, P. H., Adrian, R. J. and Jones, B. G., Theor. & Appl. Mechanics Report No. 475, UILU-ENG 85-6006, Univ. of Illinois, Urbana, IL, 1985.

10. Chin, K. D., "Conditional Fluctuating Velocity Fields in the Near Field of a Round Jet," M.S. Thesis, Univ. of Illinois, Urbana, 1987.

11. Ditter, J. L., "Stochastic Estimation of Eddies Conditioned on Local Kinematics: Isotropic Turbulence," M.S. Thesis, Univ. of Illinois, Urbana, 1987.

12. Elam, K. S., "Conditional Reynolds Stresses in the Near Field of a Round Jet," M.S. Thesis, Univ. of Illinois, Urbana, 1987.

13. Falco, R. E., "Coherent Motions in the Outer Region of Turbulent Boundary Layers," Phys. Fluids, Vol. 20, S124, 1970.

14. Guezennec, Y. G., "Stochastic Estimation of Large Coherent Structures in Turbulent Boundary Layers," submitted to Phys. Fluids, 1987.

15. Hanratty, T. J., "A Conceptual Model of the Viscous Wall Region," Proc. Z. Zaric Memorial Intl. Seminar on Near Wall Turbulence, Dubrovnik, 1988.

16. Hassan, Y. A., "Experimental and Modeling Studies of Two-Point Stochastic Structure in Turbulent Pipe Flow," Ph.D. thesis, Univ. of Illinois, Urbana, 1980L. Also Hassan, Y. A., Jones, B. G. & Adrian, R. J., "Two-Point Stochastic Structure in Turbulent Pipe Flow," Nuclear Thermal-Hydraulics Report No. 2, UILU-ENG, 87-5302, Univ. of Ill., Urbana, IL, 1987.

17. Johnsen, H., Pecseli, H. L. and Trulsen, J., "Conditional Eddies in Plasma Turbulence," Phys. Fluids, Vol. 30, pp. 2239-2254, 1987.

18. Johnsen, H., Pecseli, H. L. and Trulsen, H. L., "Conditional Eddies, or Clumps, in Ion-Beam-Generated Turbulence," Phys. Rev. Lett., Vol. 55, pp. 2297-2300, 1985.

19. Kim, J., "Evolution of a Vortical Structure Associated with the Bursting Event in a Channel Flow," in Turbulent Shear Flow 5, pp. 221-232, Springer-Verlag, Berlin, 1987.

20. Kim, J. and Moin, P., "The Structure of the Vorticity Field in Turbulent Channel Flow. Part 2. Study of Ensemble-Averaged Fields," J. Fluid Mech., Vol. 162, pp. 339-363, 1986.

21. Lumley, J. L., Stochastic Tools in Turbulence, Academic, NY, 1970.

22. Lumley, J. L., "Coherent Structures in Turbulence," In Transition and Turbulence, ed. R. E. Meyer, pp. 215-241, Academic, NY, 1981.

23. Lundgren, T. S., "Distribution Functions in the Statistical Theory of Turbulence," Phys. Fluids, Vol. 10, pp. 969, 1967.

24. Moin, P., "Probing Turbulence via Large Eddy Simulation," AIAA Paper 84-0174, 1984.

25. Moin, P., Adrian, R. J. and Kim, J., "Stochastic Estimation of Organized Structures in Turbulent Channel Flow," Proc. 6th Turbulent Shear Flow Symp., Toulouse, pp. 16.9.1-16.9.8., 1987.

26. Nithianandan, C. K., "Fluctuating Velocity Pressure Field Structure in a Round Jet Turbulent Mixing Region," Ph.D. Thesis, Univ. of Illinois, Urbana, 1980. Also Nithianandan, C. K., Jones, B. G. & Adrian, R. J., "Turbulent Velocity Pressure Field Structures in an Axisymmetric Mixing Layer," Nuclear Thermal Hydraulics Rept. No. 1, UILU-ENG 87-5301, Univ. of IL, Urbana, IL, 1987.

27. Papoulis, A., Probability, Random Variables and Stochastic Theory, 2nd ed., McGraw-Hill, New York, 1984.

28. Rogers, M. M. and Moin, P., "The Structure of the Vorticity Field in Homogeneous Turbulent Flows," J. Fluid Mech, Vol. 176, p. 33, 1987.

29. Theodorsen, T. "Mechanism of Turbulence," in Proc. 2nd Midwestern Conf. on Fluid Mech., Ohio State University, Columbus, Ohio, 1952.

30. Townsend, A. A., The Structure of Turbulent Shear Flow, 2nd Ed., Cambridge Univ. Press, Cambridge, 1976.

31. Tung, A.T.C., "Properties of Conditional Eddies in Free Shear Flows," Ph.D. Thesis, Univ. of Illinois, Urbana, 1982. Also, Tung, A.T.C., Adrian, R. J. & Jones, B. G., "Conditional Velocity and Reynolds Stresses in Plane Turbulent Shear Layer," Theor. and Appl. Mechanics Report. No. 483, UILU-ENG 87-6001, Univ. of Ill., Urbana, IL, 1987.

32. Tung, T. C. and Adrian, R. J., "Higher Order Estimates of Conditional Eddies in Isotropic Turbulence," Phys. Fluids, Vol. 23, pp. 1469-1470, 1980.

33. Van Atta, C. W. and Chen, W. Y., "Correlation Measurements in Grid Turbulence Using Digital Harmonic Analysis," J. Fluid Mech., Vol. 34, pp. 497-515, 1968.

34. Williams, D. R., "Analysis of Large Data Sets Generated by Experiments," AIAA 19th Fluid Dynamics, Plasma Dynamics and Lasers Conference, Honolulu, AIAA-87-1188, 1987.

35. Willmarth, W. W. and Wooldridge, C. E., "Measurements of the Fluctuating Pressure at the Wall Beneath a Thick Turbulent Boundary Layer," J. Fluid Mech. Vol. 14, p. 187, 1962.

36. Willmarth, W. W. and Lu, S. S., "Structure of the Reynolds Stress Near the Wall, J. Fluid Mech, Vol. 55, pp. 65-92, 1972.

SUMMARY LECTURE

Observations on the Boundary Layer Based on Measured Correlations with Various Improvements

R. DUMAS
Institut de Mécanique Statistique de la Turbulence
Université d'Aix-Marseille II
Unité Mixte Université
CNRS 380033, France

ABSTRACT

A review is presented of some results concerning the structure of the turbulent boundary layer, derived from measured correlations, in particular the space-time correlation, with various improvements. The refinements were added expecting a better recognition of the structure of the flow, the conventional correlations with large band pass and long time averaging smearing the pattern of the field. Generally the measurements concern velocity components, but the temperature fluctuations were also used successfully.

The first refinement used for many years was the frequency filtering, but the physical meanings of the narrow band pass correlations may be questionable in a strong shear flow. Measurements of high-order correlations were undertaken firstly in order to check the hypothesis of Millionshchikof, but they appeared to be more broadly significant, specially the two or three-point triple products in the boundary layer. A useful improvement was the investigation of the spatial structure of the viscous sublayer using an array of hot-wires and short averaging times. Now the conditional correlations are in common use for coherent eddy recognition. But discussions remain on the choice of the detection critera to be used to correctly identify the typical structure in the flow.

1. INTRODUCTION

Space correlations and autocorrelations are convenient tools for determining the statistical properties of a turbulent field but, in addition to space separations, time shifts are necessary to search for the conservative fluctuations along the flow, and to measure their phase arrangements. In this review we shall examine in particular space-time correlations, denoted CST, which, from a dynamical point of view, represent the first statistical relationship along the stream to be taken into consideration. The CST were first introduced by Favre and measured at IMST, and special attention shall be given to the results obtained at this Institute. A review on CST has been published by Favre (1983), therefore our attention shall focus on results obtained at the IMST, pertinent to the present topic, that have not yet been widely published.

Let us recall that the balance equations of the CST were derived by Dumas et al (1976). Arzoumanian (1987) measured the several terms of this equation in the fully turbulent part of the boundary layer, except for the velocity-pressure terms which were calculated from the balance of the equation. The conclusion emphasizes that the loss of correlation, which is strong in a boundary layer in comparison to the isotropic turbulence behind a grid, is due to the high level of turbulent intensity. But the cross-correlation term representing Reynolds shear stress tends to maintain the correlation by itself. If the space separation parameter first introduced by S. Corrsin is used $(X_1/\Lambda)(u'/U)$ (where X_1 is longitudinal separation, Λ macroscale, u'/U relative intensity of turbulence), it may be observed (Dumas 1976, Favre 1983) that the decay of the maximum values of the CST along the mean streamlines is less inside the boundary layer, even close to the wall ($y/\delta = 0.03$), than in the turbulence behind a grid. Consequently it is reasonable to conjecture that the large-scale fluctuations inside the boundary layer maintain the CST as coherent structures might do.

In order to investigate the turbulent field more closely, various refinements of the CST were introduced. An early refinement was achieved through correlations of the frequency-filtered fluctuations (Favre et al 1958, 1967, Wills 1964). At IMST the triple CST products at two points and three points, which emphasize the dominant dynamic events, were measured (Dumas et al 1973, Fulachier et al 1974, Arzoumanian et al 1978). The higher moments with and without time shifts were taken into consideration previously to study the statistic distributions of the turbulent fluctuations with respect to the hypothesis of Millionshchikof (see Van Atta 1974).

A worthwhile improvement was conditional sampling introduced by Kaplan and Laufer (1968), Kovasznay and co-workers (1970) to study the intermittent zone, by Willmarth to investigate the structure of Reynolds stress near the wall (Willmarth and Lu 1972), and by many researchers to try to educe the wall structures revealed by Kline. Inside the sublayer Gupta, Laufer and Kaplan (1971) applied digital sampling to the signals from an array of hot wires, introducing variable interval time averaging (VITA).

In many experiments heat behaves as a passive contaminant and the temperature correlations are sampled from the signals provided by a rake of cold wires; at IMST the conditional correlations and probabilities were measured as a function of the temperature threshold (Dumas et al 1977, Favre 1983).

These correlations, namely the CST functions, were slightly disregarded for a few years; now, however, they are used successfully to study the wake structure (Browne et al 1986), plane jet (Antonia et al 1986), the compressible boundary layer (Spina and Smits 1986), the three-dimensional boundary layer on a moving wall (Arzoumanian et al 1984, thesis 1987) and also in works conducted by the contributors to the present session. They are frequently called upon in theoretical considerations, as we shall mention in the conclusion.

2. LONG TIME AVERAGING CORRELATIONS

2.1 Iso-correlation Contours and Fronts

The x, or x_1, coordinate is aligned with the mainstream velocity parallel to the wall. The y coordinate, or x_2, is perpendicular to the wall, and z or x_3, is the spanwise direction. The coordinates of the fixed point are x_0, y_0, z_0 and the spacial separations are in capital letters X_1, X_2, X_3. The time delay is τ. The turbulent components of velocity are u, v, w or they are indicated by the indices 1, 2, 3 respectively; mean values are written in capital letters or with average symbols.

In order to allow a general examination of the statistical structure of the turbulent field, it is useful to take into consideration space-time iso-correlation contours. Figure 1a shows a typical example of two-point iso-correlation contours in the boundary layer:

$$R_{11}(X_1, y_0, X_2, \tau) = \frac{\overline{u(0, y_0, t)\, u(X_1, y_0 + X_2, t + \tau)}}{u'(y_0)\, u'(X_1, y_0 + X_2)} \qquad (1)$$

(a prime denotes an rms quantity)

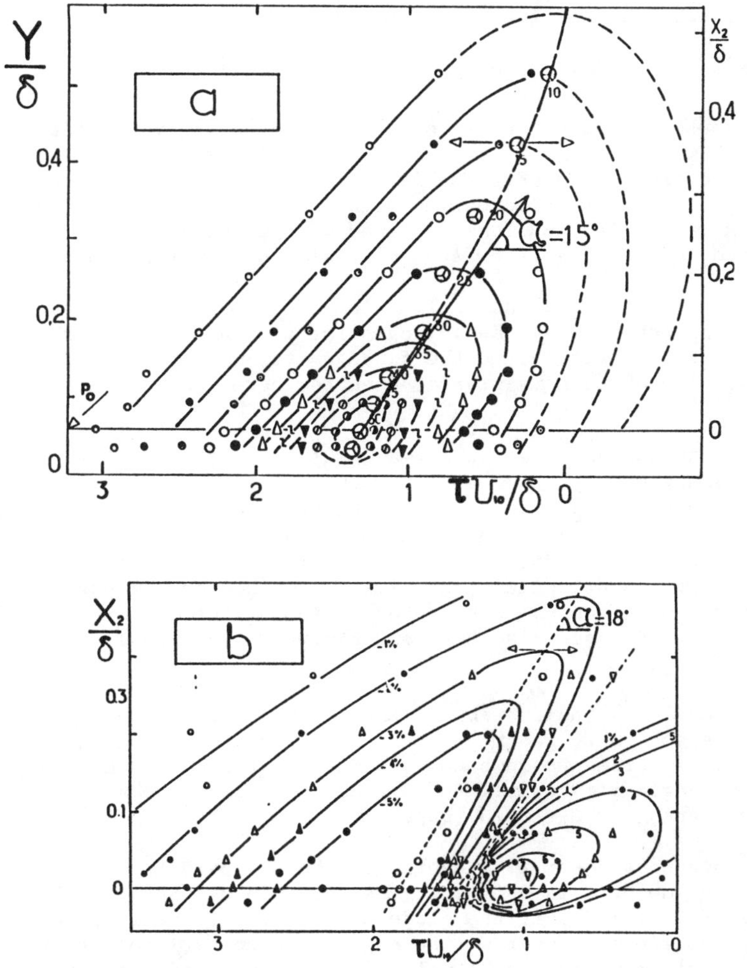

FIGURE 1. Iso-correlations a) R_{11}; b) $R_{1,11}$; $y_0 = 0.055\ \delta$, $X_1 = 1.43\ \delta$

The upstream point is located in the inertial layer at the distance $y_0 = 0.055\ \delta$ ($y_0^+ = 118$), the Reynolds number is $U_e\delta/\nu = 57{,}000$, U_e outer velocity, δ boundary layer thickness. Each contour corresponds to a constant value of the correlation R_{11} for X_2 and τ varying. The longitudinal separation X_1 ($X_1 = 1.43\ \delta$) is wide enough to provide significant results regarding large-scale fluctuations. It is generally assumed that the locus of the maximum correlation at the optimum time delay τ_m, R_{11} (X_2 = cte, $\tau = \tau_m$) is significant of the statistical inclination of the large-scale fluctuations. In our example this locus is a line with, at the origin, an inclination to the wall of $\alpha = \arctan(X_2/\tau_m U) = 15°$.

In order to find the sign of the dominant fluctuations, however, triple correlations must be measured. For example, Figure 1b shows the triple iso-correlations at the same position y_0 of the upstream point:

$$R_{1,11} = \frac{\overline{u(0,\ y_0,\ t)\ u^2(X_1,\ y_0 + X_2,\ t + \tau)}}{u'(y_0)[u^2\ (X_1,\ y_0 + X_2) - \overline{u^2(X_1,\ y_0 + X_2)}]'} \qquad (2)$$

The main contours correspond to negative fluctuating velocities, the slope of the locus of $R_{1,11}$ (X_2 = cte, τ_m) being more rectilinear than for double correlations. But smaller contours are also observed, corresponding to positive fluctuating velocities, traveling with higher convection velocities and less inclined with respect to the wall, which are related to the sweep flow events disclosed by Corino and Brodkey.

In the intermittent zone the inclination α of this locus is variable with the distance between the moving point and the wall, as shown in Figures 2a and 2b relative to the fixed upstream point at a distance $y_0 = 0.80\delta$. The angle α is greater when the normal compoment v is involved.

Figures 3a, 3b and 3c show the corresponding triple correlations. The angle α reaches 90° at the edge of the boundary layer and it follows the general decreasing trend towards the wall; but it decreases much less when the normal component v is involved. By comparing Figures 1 and 2, it appears that there is no similarity in the region where measurements are overlapping. The time τ_m depends on the position y_0 of the reference point as has been observed in earlier experiments by Favre et al (1958) (see also Head and Bandyopadhyay 1981).

Moreover, the locus inclination of the maximum correlations $R(X_2$ = cte, $\tau_m)$ is not specific to inhomogeneous shear flow, as shown in Figure 4 relative to measurements in a homogeneous shear flow by Champagne et al (1970). The angle α is nearly 60°; it may be related to a non-dimensional gradient $G = (L/u)(\partial U/\partial y)$.

In the homogeneous shear flow mentioned above, where L is the macroscale, the value of G is 0.044. In the inner layer, and at $y_0 = 0.055\ \delta$ the value of G is 1.37; if the slope tan (α) should be in inverse ratio to the distortion parameter G, the angle α in comparison with the homogeneous flow should have a value 3° instead of 15°. It appears that there is not a simple distortion effect of the large-scale eddies due to the mean shear, but a specific behaviour pattern in the boundary layer.

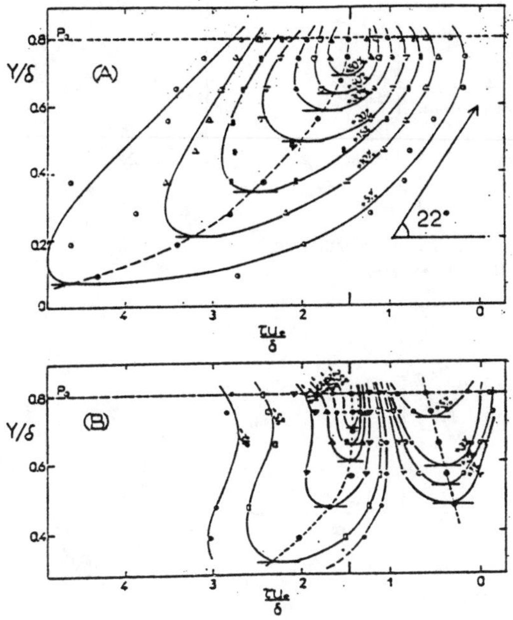

FIGURE 2. Iso-correlations.
a) R_{11}; b) R_{12}
$y_0 = 0.80\ \delta$, $X_1 = 1.43\ \delta$

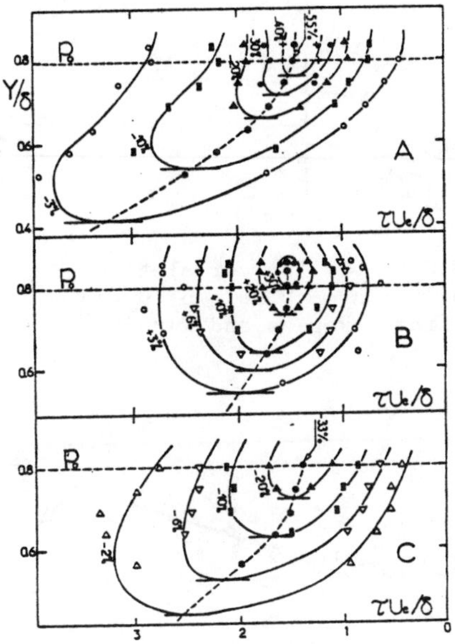

FIGURE 3. Iso-correlations.
a) $R_{1,11}$ b) $R_{1,12}$ c) $R_{1,22}$
$y_0 = 0.80\ \delta$, $X_1 = 1.43\ \delta$

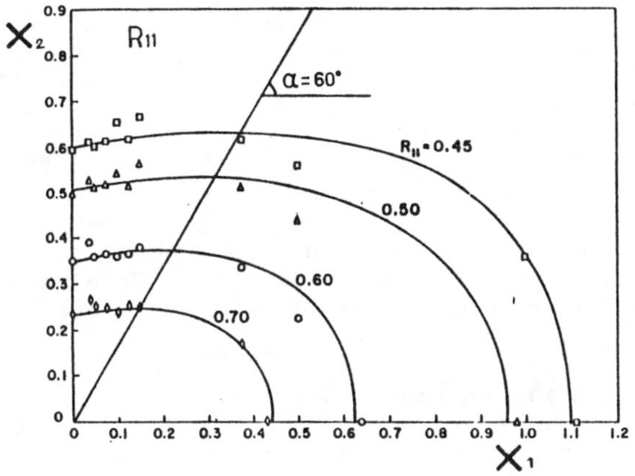

FIGURE 4. Iso-correlations R_{11} in homogeneous shear flow (from Champagne, Harris and Corrsin).

Nakagawa and Nezu (1980) have measured space-time correlations in an open-channel water flow at a low Reynolds number ($U_m h/\nu = 12{,}000$, where h is flow depth). Figures 5a, 5b and 5c show the iso-correlations R_{11} (X_1, X_2, τ = cte) at several time delays indicated respectively in the figure captions. The fixed point is at $y_0/h = 0.050$ ($y_0^+ = 38$).

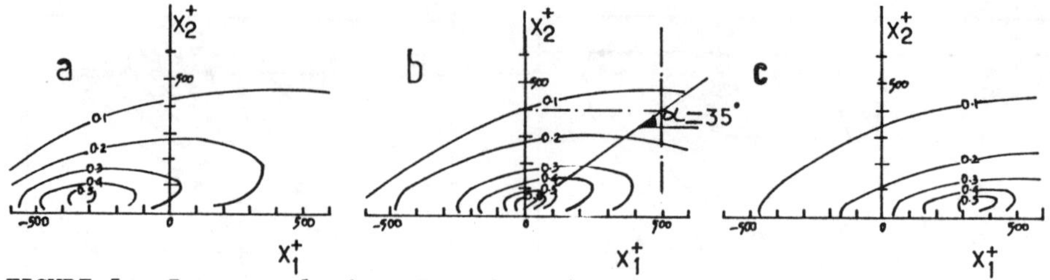

FIGURE 5. Iso-correlations R_{11} when the time τ varies.
a) $U_m\tau/h$ = -0.56 b) $U_m\tau/h$ = 0 c) $U_m\tau/h$ = +0.56
(from Nakagawa and Nezu)

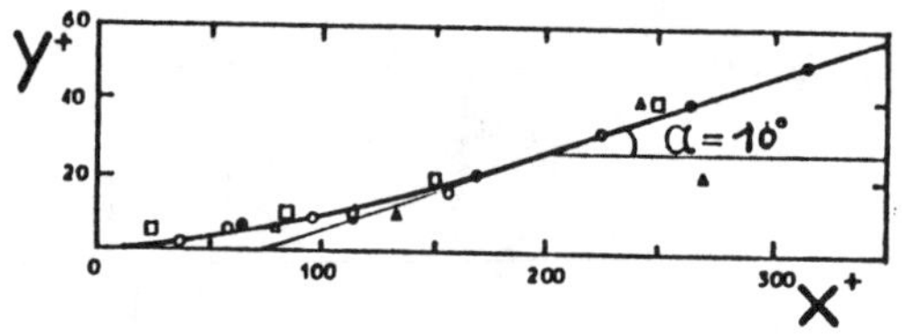

FIGURE 6. Average front.
(from Kreplin and Eckelmann)

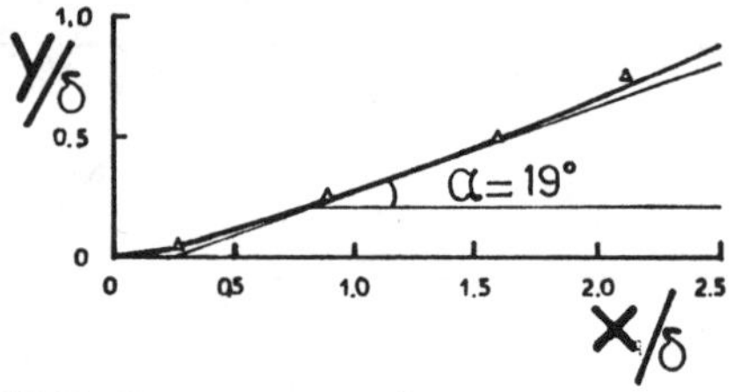

FIGURE 7. Average front.
(from Lu and Willmarth)

These figures show very clearly that the large-scale fluctuations are convected downstream without any significant deformation during this time. But the inclination angle α appears to be about 35° as is shown in 5b, which is much higher than the above value α = 15°. This discrepancy cannot be explained by the use of the Taylor hypothesis to calculate tan (α) when τ is the variable, but the correlation maximums R (X_2 = cte, X_{1m} or τ_m) are very sensitive to the tracing of the contour. Nakagawa and Nezu plot an inclination angle of 15-20°, but relative to the peak line of the contours near the centre of the maximum corelation. Usually the peak line is not very different from the line of maximum correlation at X_2 = cte.

Space-time correlations have been measured in the viscous sublayer and adjacent wall region in an oil channel flow by H. Eckelmann and co-workers at a very low Reynolds number (R_b = 7,700, where b is channel width).

The graphs by Kreplin and Eckelmann (1979) shown in Figure 6 illustrate the average front measured inside the buffer layer. The reference signal was provided by a shear stress probe at the wall. The limit angle α at the wall is 4.7° and further on 10°.

Brown and Thomas (1977) have measured CST using a hot-wire array and wall shear stress probes. They incline the hot-wires with respect to the wall in such a way that the correlations between wall shear stress and velocity reach their peak at zero time delay. The result is shown in Figure 7, where the oblique angle α is approximately 19° throughout the boundary layer.

Previously, using this technique, Lu and Willmarth (1973) found in the inner layer, the upstream point being at a distance y_0 = 16.2, an inclination α = 16°-20° on the front of the sampled stress $\langle uv\rangle/\overline{uv}$.

Lastly, conventional space-time correlations show that in spite of discrepancies, there is a front related to large-scale fluctuations and low-speed events, with an inclination that increases strongly from the wall to the outer zone, the value in the fully turbulent inner layer being about 18°. Quantitative results, however, depend on the position y_0 of the reference point and on the velocity component in question.

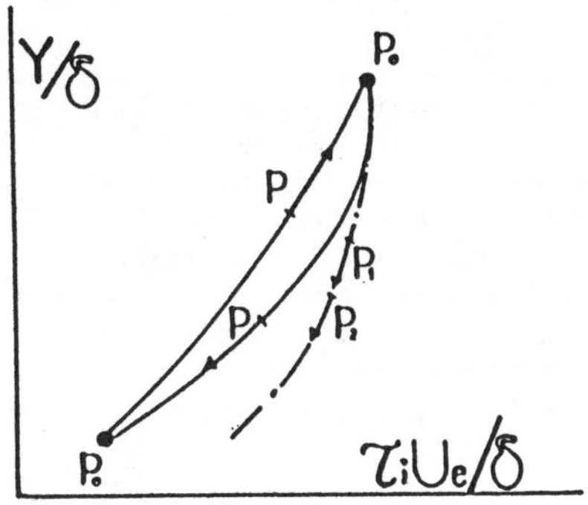

The opposite drawing shows the aspect of the variation of the time lag τ_i according to the position of the fixed point P_0 with respect to the mobile point P. The figure also illustrate the variation when points P_1 and P_2 are moving (cumulative time, see para. 3.2) starting from the outer region.

2.2. Narrow Band Correlations

In order to examine the structure of the turbulent field in a more detailed manner, the correlations of the narrow band fluctuations have been measured: $\hat{R} = \overline{(\hat{\ })(\hat{\ })}/(\hat{\ })'(\hat{\ })'$ where $(\hat{\ })$ represents filtered fluctuations.

Special attention has been given to convection velocities as a function of frequency. Let us recall that the convection velocity C_x along the mean flow ($X_2 = 0$) is defined by the relation $C_x = \tau_m/X_1$ (τ_m or τ_c being compensating time), and in reality represents phase velocity; in the perpendicular direction, velocity is defined as $C_y = \tau_i/X_2$, τ_i being the optimum time (with a space shift $X_1 \neq 0$, with a good approximation $\tau_m = \tau_c + \tau_i$); the ratio C_y is not usually interpreted as fluid motion but, as assumed in the paragraph above, as a front inclined to the wall moving at the convection velocity C_x. This is not strictly true, the visualizations showing that there are sometimes strong fluid motions through the boundary layer. Even the convection velocity C_x is a combination of fluid transport and likewise induced wavy motions, as revealed by measurements of the convection velocities of pressure fluctuations at the wall (see Willmarth 1975).

Equations representing analytical expressions of convection velocities have been derived (Favre et al 1968); the differences between these celerity values and the mean velocity values are due to the non-linear or pressure terms, as expected.

The behaviour of narrow-band CST has been studied at IMST (Favre and al 1967). The velocities C_x are more or less a function of the space separations X_1: if the separation increases, the relative effect of the large-scale fluctuations becomes greater, regardless of the frequency band. In order to obtain results with a clearer physical meaning, it is necessary to adapt the space shift X_1 to the tuning frequency f, for example $X_1 = U/2\pi f$. The relevant parameter is the number of times an eddy travels its characteristic length. Given this condition, the more conservative fluctuations have a longitudinal scale roughly equivalent to the boundary layer thickness δ (Dumas and Favre 1971). The time τ_i depends on the tuning frequency f, and decreases when the frequency increases. In other words the inclination angle α increases with the relative weight of the high-frequency fluctuations.

Another complication arises from the fact that the maximum absolute value of the narrow-band cross-correlation R_{12} is not reached at $\tau = 0$, but at an internal time τ_{i0} non-null, which is a function of the tuning frequency f and the distance from the wall y.

Figures 8a and 8b show the time τ_{i0} found by Arzoumanian (1988). In Figure 8a at the position $y_0 = 0.13\delta$, the time τ_{i0} retains negative values for the lower frequencies, and on 8b the non-dimensional wave number $k_1\delta$ is equal to 0.54 ($k_1 = 2\pi f/U$). Close to the wall the negative sign of τ_{i0} corresponds to a phase advance of the fluctuations ($\hat{v}$) with respect to the fluctuations ($\hat{u}$). In the outer part of the boundary layer it is the contrary, i.e., it appears that at very low frequencies, the anisotropic part of the v component is more sensitive to the large-scale eddies traveling in the central part of the boundary layer than to the u component.

Many measurements of narrow-band correlations have been made. Morrison, Bullock and Kronauer (1971) have investigated the sublayer in particular. Two-dimensional spectra, frequency versus wave number, were computed from Fourier transforms of only narrow-band spatial correlations, using the technique of the pioneer Wills (1964). For disturbances in the sublayer α, they found a longitudinal wave length $\lambda^+ = 135$ and a convection velocity $C_x = 8.0\ U^*$. These values are roughly in agreement with the values accepted today.

It may seem that the conventional Fourier analysis is not a suitable technique in a non-homogeneous flow where there are steep coherent fluctuations related to intermittent structures. But Brown and Thomas have applied a refined filtering technique to study the organized structures. They have examined the possibility of correlation between the slowly varying part and the high-frequency part as regards wall shear stress and velocity signals. For this purpose they measured the correlation between the low-pass filtered signal and the smoothed

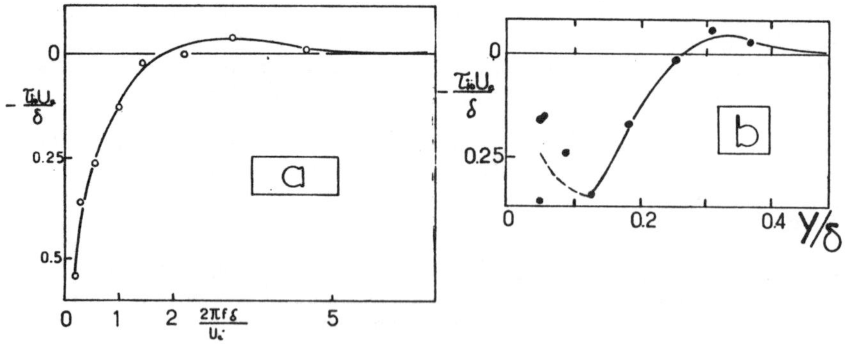

FIGURE 8. Narrow-band correlation $\hat{R}_{12}$, internal time τ_{i0}.
a) $y_0 = 0.13\ \delta$ b) $K_1\delta = 0.54$

rectified high-frequency part. Figure 9 representing the longitudinal component correlations at several distances from the wall shows a substantial coefficient was found, which may be explained only by the passage of organized structures. From the sign of the optimum time delays, the authors concluded that the sharp velocity step producing the high-frequency part occurs on the back or trailing surface of the large structure.

2.3. Spanwise Correlations

Correlations taken using spanwise space separations may prove to be quite interesting as regards the streaky structure of the velocity field close to the wall. Conventional correlations are able to reveal the spanwise scales of the turbulent field, as shown in Figures 10a and 10b by Kreplin and Eckelmann (1979) relative to the correlations R_{g11} and R_{g33} of the velocity gradients $\partial U/\partial y$ and $\partial W/\partial y$ respectively at the wall; the authors also plotted data from other experiments.

Kreplin and Eckelmann drew the conclusion that the structures at the wall are as narrow as $X_3^+ = 50$, one half the distance of the low-speed streak separation reported by Kline, and they relate this scale to the spanwise separation of the counter-rotating streamwise vortices proposed, among others, in the model of Blackwelder and Eckelmann (1979).

These measurements involve the wall itself, and it may be observed that the spanwise scale from a statistical point of view increases with the distance from the wall (Dumas et al 1976, Nakagawa and Nezu 1981). This may explain the relatively low value of X_3^+, but other explanations may be put forward. Roughly similar results were found by Hanratty et al (1977) and recently by Herzog (1986).

If the spanwise separation is only X_3, the optimum time lag is null in a two-dimensional boundary layer, but at a fixed streamwise and transverse separation the optimum time lag may vary with X_3 as can be seen in the measurements taken at IMST (Favre et al 1958) and the results concerning the velocity-pressure correlations of Willmarth (see review 1975).

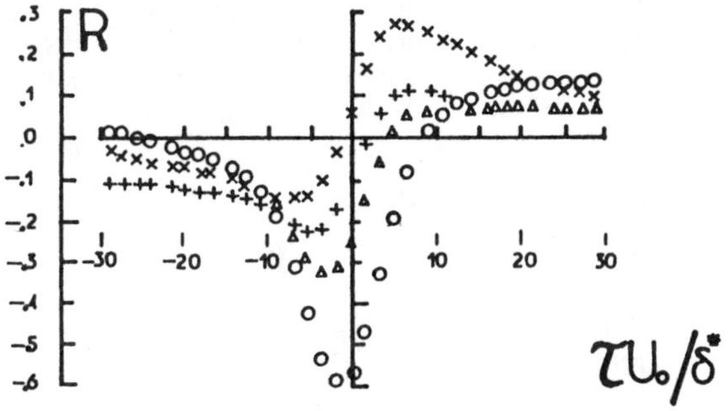

FIGURE 9. Correlation between the low-pass filtered velocity and the smoothed rectified high-frequency portion. y_0/δ : × 0.05, + 0.25, Δ 0.50, o 0.75

(from Brown and Thomas)

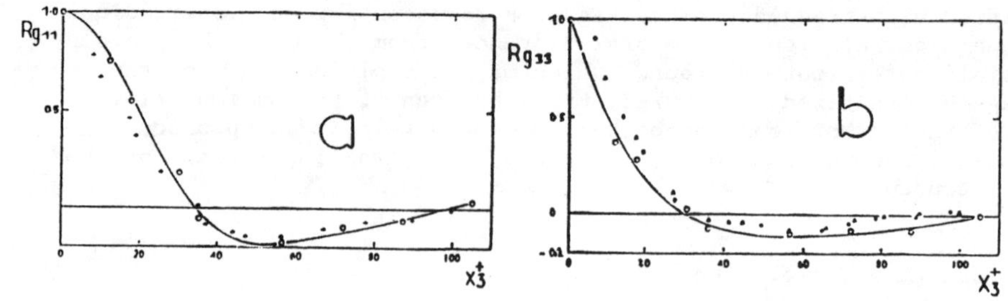

FIGURE 10. Spanwise correlations of velocity gradient at the wall. a) R_{g11} b) R_{g33} (from Kreplin and Eckelmann).

3. CONDITIONAL CORRELATIONS AND TEMPERATURE TRACER

3.1. Conditional Correlations

Conditional samplings have been introduced in order to educe the coherent structures from the whole pattern. Gupta et al (1971) first undertook space-time correlations with variable interval time averaging (VITA) to investigate the spatial structure inside the sublayer. Sampling for the spanwise correlation was defined as follows:

$$\hat{R}\ (Z,T_S) = \frac{1/N\ \Sigma_1^N\ [u(z_0,\ t) - \tilde{U}\ ]\ [u(z_0 + Z,\ t) - \tilde{U}\ ]}{\left\{ 1/N\ \Sigma_1^N\ [u(z_0,\ t) - \tilde{U}\ ]^2\ .\ 1/N\ \Sigma_1^N [u(z_0 + Z,t) - \tilde{U}\ ]^2 \right\}^{1/2}} \qquad (3)$$

where u(t) is sampled velocity, Ts = N τ variable averaging time, τ sampling rate. Various average velocities U were tried, but results were inconsequential for our purpose.

Figure 11 shows typical correlations obtained with the appropriate short time averaging. The transverse scale λ is defined as the mean value of the distance Z_{max} where the first maxima of correlations occur, or twice Z_{min} for the minima. It retains the value $\lambda+ = 100$. Nagakawa and Nezu (1981) used conditional sampling analogous to that of Willmarth and Lu (1972) and Sabot and Comte-Bellot (1976), but trying to avoid a threshold which is more or less arbitrary, they introduced a weighting function. The conditional average of a quantity q was:

$$<q> = \frac{\int_T q(t)\ [u\ (t)\ v(t)\ I(t)]_0\ dt}{\int_T [u(t)\ v(t)\ I(t)]_0\ dt}$$

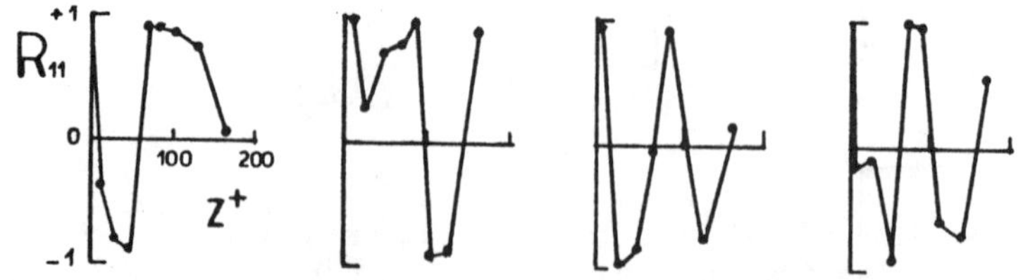

FIGURE 11. Typical short time averaging spanwise correlations R_{11} (Z). $y_0^+ = 5.4$, $T_s^+ = 1.26$ (from Gupta, Laufer and Kaplan).

I(t) is the classic detecting function in the u-v quadrant, the shear stress itself uv playing the role of the weighting function. The suffix $[\ ]_0$ denotes the position of the detecting probe. The authors do not display true conditional CST, but conditional velocity components <u> and <v>, or <uv>, time delayed with respect to the detection point. Figures 12a and 13a show the conditional velocities <u> and <v> relative to an ejection and Figures 12b and 13b are relative to the sweep events, both sequences being detected at the location $y_0^+ = 38$.

In Figures 12a and 12b the inclination α of the fronts as defined previously is roughly 20°. Figures 13a and 13b relative to the normal component v show an angle α of the same order for the ejection, but it is right for the sweeps which appear to be a motion parallel to the wall and necessarily tri-dimensional.

3.2. Temperature Tracer

It has been of interest to introduce the CST of temperature fluctuations acting as a passive scalar in order to locate the origin of the coherent motion more exactly. However, the correlation coefficient $\overline{u\theta}/u'\theta'$ reaches high values at about 70% of the fully turbulent layer, and perhaps 100% in the viscous sublayer (Fulachier 1972, and private communication). Consequently, the results found with the same averaging conditions for the temperature and the longitudinal fluctuations cannot be very different except for the small scale fluctuations. Comparisons were undertaken (Favre et al 1969) between the CST R_{11} of the component u and the CST of temperature fluctuations θ. The temperature tracer was introduced into the flow field by slowly heating the wall. The convection velocities Cx were found to be of the same order but the time delay τ_i was 40% less for the temperature than for the longitudinal component of the velocity inside the central part of the boundary layer, the fixed point being located at $y_0/\delta = 0.038$. This meant that the front determined by the conventional temperature CST was less inclined at the wall than for the component u. In these measurements, which did not include the streamwise separation X_1, the small-scale fluctuations were partly involved in the correlations.

This technique is fairly often employed, and now is used in combination with conditional sampling of the signals provided by a rake of "cold wires" (i.e. Blackwelder and Kaplan 1976, Chen and Blackwelder 1978, Subramanian et al 1982). A typical feature disclosed by the experiments is the appearance of suddenly decreasing temperature fronts extending throughout the entire boundary layer. These decreases seem to be related to internal shear layers on the back of the large-scale structures. Chen and Blackwelder defined the coincidence correlations between the short time pulses P(t) detecting the internal fronts of decreasing temperature:

$$\hat{R}\,(y_0 + \Delta y, \tau) = 1/N\ \Sigma_1^N \int_0^{\xi} P(y_0, t_n + t)\ P\,(y_0 + \Delta y, t_n + \tau + t)\ dt \qquad (5)$$

where t_n is the reference time at the detecting point, and ξ the average period between pulses.

Figure 14 shows a comparison between the non-dimensional time $\tau_i . U_e/\delta$ obtained from the temperature coincident correlations and the results found at IMST (Arzoumanian et al 1978) from the correlation of the longitudinal velocity component. The velocity correlation curve has

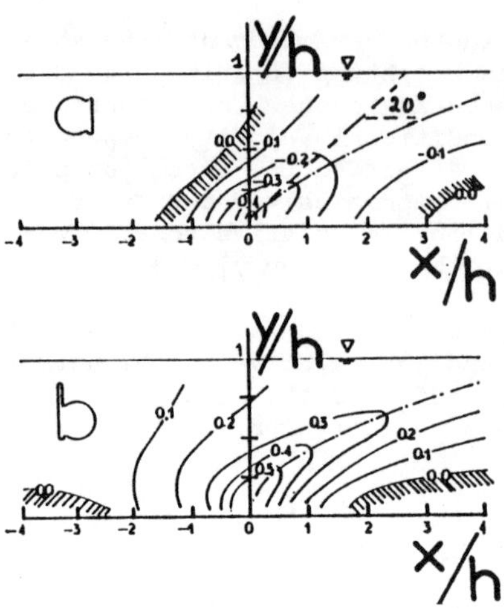

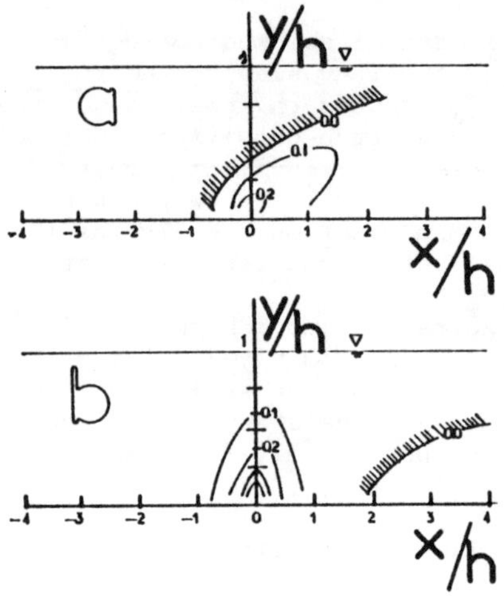

FIGURE 12. Spatial iso-pattern of <u>. a) ejection b) sweep.

FIGURE 13. Spatial iso-pattern of <v>. a) ejection b) sweep

Detecting probe $y_0^+ = 38$ (from Nakagawa and Nezu)

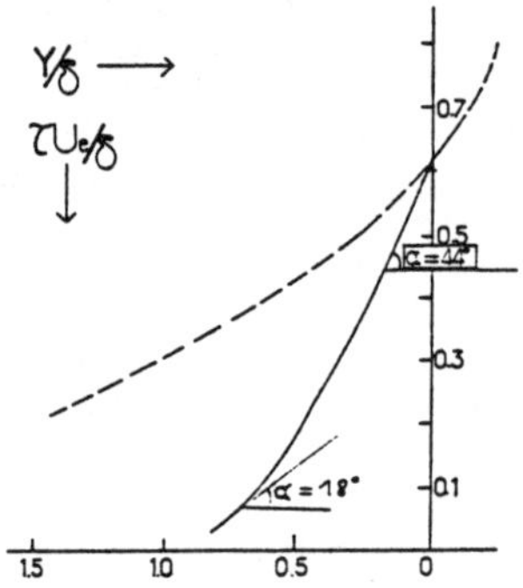

FIGURE 14. Time lag τ.
—— temperature fronts
- - - CST $y_0 = 0.80\ \delta$

been translated in a direction parallel to the time axis in order to coincide with the point $y/\delta = 0.63$, $\tau = 0.0$. The values of time lags differ strongly. This can only be partly explained by the different techniques used: in the experiments at IMST the time at each point is delayed with respect to the fixed point $y_0 = 0.80\delta$, whereas in the thermal measurements a cumulative time lag is applied at each position, calculated by summing all the individual lags between $y_0 = 0.63\delta$ and the position of interest. The slope α varies from 44° to 18° for the thermal fronts and from 26° (at $y/\delta = 0.63$) to about 11° for velocity.

These differences emphasize the fact that the large-scale fluctuations displayed by the conventional CST of the u component are not related to the above internal fronts in a simple manner. Moreover, as has been observed by Chen and Blackwelder (1978), the several results concerning temperature are found with two-dimensional (y - τ or X_1) sets of probes whereas all proposed dynamical models based on the visualizations of Kline are tri-dimensional (i.e. Offen and Kline 1975). The three-point CST of temperature fluctuations undertaken at IMST (Fulachier et al 1974 a and b, Favre 1983) showed that the large-scale eddies painted by the temperature disclose a complex pattern in the plane parallel to the wall, in agreement (Elena et al 1979) with the model proposed by Blackwelder and Kaplan (1976, 1979).

Recently, Antonia et al (1988) observed that the temperature traces reveal that in addition to the sudden temperature decrease (cooling), there exists a sudden typical increase (heating) which is also coherent throughout the boundary layer. They showed that the convection velocities C_x were higher for the temperature drops than for the temperature rises and both sequences were quite conservative along the general flow. Similar results for the convection velocities Cx have been found at IMST concerning the events of hot and cold fluctuations respectively (Dumas et al 1977, 1978).

4. CONCLUSION

Even for conventional correlations, discrepancies appear in results concerning the flow structure found either with the longitudinal velocity component, the transverse component, or the temperature tracer. These are obviously due to the turbulent eddies that are not sorted in the same manner depending on the quantities involved. From this point of view, the scales of the fluctuations which are involved in the experimental techniques are of utmost importance. It should also be recalled that the temperature fluctuations containing the variance depend on the three velocity components according to the spectral analogy proposed by Fulachier and Dumas (1976).

Moreover, Hussain (1983) argues that velocity correlations are not an acceptable approach to infer coherent structures due to the field induced by the turbulent flow in which they are embedded. This remark also applies to the fluctuating field itself. Like heat, even the tracer depends on the induced as well as the coherent movement. The vorticity correlations should be designated more exactly to educe organized vortical structure, but there are practically no measurements available.

It is my opinion that the introduction of a substantial separation with an optimum time delay emphasizes the weight of the coherent structures which are more conservative than the non-organized field. But near the wall, the fluctuations induced by the large scales traveling in the central zone are more conservative than the local coherent vortices which jitter strongly. Another observation would point to the need for tridimensional measurements. For example, at least three CST points should be established, with the detection point P_0 near the wall and the two downstream points P_1 and P_2 movable in the directions X_1 and, above all, X_2 and X_3. The time lags give the convection velocities; the geometry of the traveling structures is investigated based on the CST values for several locations of P_1 and P_2. The signs of the double and triple products provide a rough indication of the circulation.

Only theoretical considerations or numerical simulations, however, can establish a model of coherent structure. In this aim, space-time correlations drawn from a model of regular vortices at the edge of the viscous layer were computed by Hanratty (Hatziavramidis and Hanratty 1979). Experimental correlations are the inevitable validity test for these computations as Moin and Kim (1982, 1985, 1986) have shown for the large eddy simulation of a channel flow. Moreover, Moin and Kim note that the selection of the length of the computational box is initially guided by the two-point correlations measured by G. Comte-Bellot. In another way, the correlations serve to compute the coherent vortices at the wall by using the orthogonal decomposition procedure of

J.L. Lumley (Herzog 1986). Also conventional CST contribute to the linear estimations of conditional turbulent quantities as demonstrated by Adrian (Adrian et al 1987).

Obviously the correlations, as other experimental quantities, cannot provide a solution to the problem of coherent structures without the help of theory, but they remain a very useful tool for the comprehension of flow dynamics.

REFERENCES

Adrian R.J., Chung M.K., Hassan Y., Jones B.J., Nithianandan C.K., Tung A.T.C., Experimental study of stochastic estimation of turbulent condition averaging, 6th Turb. Shear Flow Conf., Proc. 611-617, Toulouse, 1987.

Antonia R.A., Chambers A.J., Britz D., Brown L.W.B., Organized structures in a turbulent plane, jet topology and contribution to momentum and heat transport, J. Fluid Mech., 172, 211-229, 1986.

Antonia R.A., Fulachier L., Krishnamoorthy L.V., Benabid T., Anselmet F., Influence of wall suction on the organized motion in a turbulent boundary layer, J. Fluid Mech., 190, 217-240, 1988.

Arzoumanian E., Corrélations spatiotemporelles en couches limites turbulentes : cas d'une mobilité de la paroi, Th. Doct.es-sc., IMST, 1987.

Arzoumanian E., Dumas R., Favre A., Corrélations doubles et triples dans une couche limite turbulente : zone externe, C.R. Acad. Sc. Paris, 286, B, 1978.

Arzoumian E., Dumas R., Fulachier L., Comportement spatiotemporel d'une couche limite soumise à une discontinuité de vitesse transversale à la paroi, C.R. Acad. Sc., 299, II, 18, 1984.

Blackwelder R.F., Eckelmann H., Streamwise vortices associated with the bursting phenomenon, J. Fluid Mech., 94, 3, 577-594, 1979.

Blackwelder R.F., Kaplan R.E., On the wall structure of the turbulent boundary layer, J. Fluid Mech., 76, 1, 89-11, 1976.

Brown G.L., Thomas A.S.W., Large structure in a turbulent boundary layer, Phys. Fluids, 20, 10, 1977.

Browne L.W.B., Antonia R.A., Bisset D.K., Coherent structure in the far field of a turbulent wake, Phys. Fluids, 29 (11), 3612, 1986.

Champagne F.H., Harris V.G., Corrsin S., Experiments on nearly homogenous turbulent shear flow, J. Fluid Mech., 41, 1, 81-139, 1970.

Chen C.P., Blackwelder R.F., Large-scale motion in a turbulent boundary layer: a study using temperature contamination, J. Fluid Mech., 89, 1, 1-31, 1978.

Dumas R., Arzoumanian E., Favre A., Correlations spatioremporelles triples dans une couche limite turbulente, C.R. Acad. Sc. Paris, 277, A, 1973.

Dumas R., Arzoumanian E., Fulachier L., Probabilités conditionnelles spatiotemporelles des fluctuations de température dans une couche limite turbulente, C.R. Acad. Sc., 284, B, 487, 1977.

Dumas R., Arzoumanian E., Fulachier L., Favre A., Conditional correlations and probabilities of temperature fluctuations in a turbulent boundary layer, Arch. Mech. Warszavia, 30, 4-5, 421-433, 1978.

Dumas R., Favre A., Some double and triple correlation measurements in a boundary layer, AFOSR Sc. Rep. no. 7, 1971.

Dumas R., Fulachier L., Arzoumanian E., Favre A., Etude de la structure de la turbulence dans une couche limite par les corrélations spatiotemporelles, Jour. de Phy., coll. C1, Sup. no. 1, 37, p. C1-181, 1976.

Elena M., Fulachier M., Dumas R., Etude expérimentale des apports et des éjections de fluide dans une couche limite turbulente, AGARD Conf. 271, 2-3, 1979.

Favre A., Turbulence: space-time statistical properties and behavior in supersonic flows, Phys. Fluids, 26, 10, 2851-2863, 1983.

Favre A., Dumas R., Verollet E., Célérité de fluctuations turbulentes de température et de vitesse dans une couche limite, Proc. 12th Int. Cong. Applied Mech., Stanford, 1968, Springer-Verlag.

Favre A., Gaviglio J., Dumas R., Couche limite turbulente: corrélations spatio-temporelles et spectres de vitesse, ONERA, Pub. no. 92, 1958.

Favre A., Gaviglio J., Dumas R., Structure of velocity space-time correlations in a boundary layer, The Phys. Fluids, sup. 1967, S138-S145.

Fulachier L., Dumas R., Spectral analogy between temperature and velocity fluctuations in a turbulent boundary layer, J. Fluid Mech., 77, 2, 257-277, 1976.

Fulachier L., Giovanangeli J.P., Dumas R., Kovasznay L.S.G., Favre A., Structure des pertubations dans une couche limite turbulente : zone interne a: C.R. Acad. Sc. Paris, 278, B, p. 683, 1974, et...: zone d'intermittence; b: p. 990.

Gupta A.K., Laufer J., Kaplan R.E., Spatial structure in the viscous sublayer, J. Fluid Mech., 50, 3, 493-512, 1971.

Hanratty J., Ghorn L.G., Hatziavramidis D.T., Turbulent fluctuations in viscous wall region for newtonian and drag reducing fluids, Phys. Fluids, 20, S112, 1977.

Hatziavramidis D.T., Hanratty T.J., The representation of the viscous wall region by a regular eddy pattern, J. Fluid Mech., 95, 4, 655-679, 1979.

Head M.P., Bandyopadhyay P., New aspects of turbulent boundary layer structure, J. Fluid Mech., 107, 297-338, 1981.

Herzog S., The large scale structure in the near-wall region of turbulent pipe flow, Ph.D. Cornell Univ. 1986.

Hussain A.K.M.F., Coherent structures - reality and myth, Phys. fluids, 26, 10, 2816-2849, 1983.

Kaplan R.E., Laufer J., The intermittently turbulent region of the boundary layer, Proc. 12th Int. Cong. Applied Mech., Stanford, 1986, Springer-Verlag.

Kovasznay L.S.G., Kibens V., Blackwelder R.F., Large-scale motion in the intermittent region of a turbulent boundary layer, J. Fluid Mech., 41, 2, 283-325, 1970.

Kreplin H.P., Eckelmann H., Propagation of perturbation in the viscous sublayer and adjacent wall region, J. Fluid Mech., 95, 2, 305-322, 1979.

Lu S.S., Wilmarth W.W., Measurements of the structure of the Reynolds stress in a turbulent boundary layer, J. Fluid Mech., 60, 3, 481-511, 1973.

Moin P., Kim J., Numerical investigation of turbulent channel flow, J. Fluid Mech., 118, 341-377, 1982.

Moin P., Kim J., The structure of the velocity field in turbulent channel flow. Part 1, Analysis of instantaneous fields and statistical correlations. Part 2, Study of ensemble-averaged field, J. Fluid Mech., P. 1, 155, 441-464, 1985; P. 2, 162, 339-363, 1986.

Morrison W.R.B., Bullock K.J., Kronauer R.E., Experimental evidence of waves in the sublayer, J. Fluid Mech., 47, 4, 639-656, 1971.

Nakagawa H., Nezu I., Structure of space-time correlations of bursting phenomena boundary in a open-channel flow, J. Fluid Mech., 104, 1-43, 1981.

Offen G.R., Kline S.J., A proposed model of the bursting process in turbulent boundary layer, J. Fluid Mech., 70, 2, 209-228, 1975.

Sabot J., Comte-Bellot G., Intermittency of coherent structures in the core region of fully developed turbulent pipe flow, J. Fluid Mech., 74, 4, 767-796, 1976.

Spina E., Smits A., Organized structures in a supersonic turbulent boundary layer, J. Fluid Mech., 182, 85-108, 1987.

Subramanian C.S., Rajagopalan S., Antonia R.A., Chambers A.J., Comparison of conditional sampling and averaging techniques in a turbulent boundary layer, J. Fluid Mech., 123, 335-362, 1982.

Van Atta C.W., Sampling techniques in turbulence boundary layer. Ann Rev. Fluid Mech., 7, 13-38, 1975.

Willmarth W.W., Pressure fluctuations beneath turbulent boundary layer, Ann. Rev. Fluid Mech., 7, 13-38, 1975.

Willmarth W.W., Lu S.S, Structure of the Reynolds stress near the wall, J. Fluid Mech., 55, 65, 1972.

Wills J.A.B., On convection velocities in turbulent shear flows, J. Fluid Mech., 20, 417-432, 1964.

Stochastic Estimation of Coherent Structures in Turbulent Boundary Layers

Y. G. GUEZENNEC and W. C. CHOI
Mechanical Engineering Department
The Ohio State University
Columbus, Ohio 43210, USA

INTRODUCTION

Although turbulent boundary layers are very common flows in engineering, the physical mechanisms responsible for the production and sustenance of turbulence are still not well understood. In the past twenty years, the identification of coherent structures first through visual studies and then by conditional sampling techniques has led to a more mechanistic description of the turbulent processes. Among the various structures identified in turbulent boundary layers are low-speed streaks (Schraub and Kline, 1965; Kline et al., 1967; Blackwelder, 1978 and others), hairpin vortices (Kim et al., 1971; Smith, 1978; Head and Bandyophadyay, 1981, Kim and Moin, 1986 and others), and the large-scale outer structures (Blackwelder and Kovasnay, 1972; Brown and Thomas, 1977; Chen and Blackwelder, 1978 and others). It has also been recognized that most of the turbulence production occurs near the wall (Klebanoff, 1954) during very localized random events (in space and time) called bursts (Kline et al., 1967; Kim et al., 1971; Blackwelder and Kaplan, 1976 and others). Due to the complexity of these unsteady three-dimensional processes, and the lack of universally accepted definition of their spatio-temporal structure, most of the evidence gathered from measurements or visual studies has been incomplete and often hard to meld into a unified view of the boundary layer.

Recent advances in computational equipment and instrumentation have triggered new approaches to improve our understanding of wall-bounded turbulent flows: first, the possibility of undertaking full Navier-Stokes simulations for low to moderate Reynolds numbers (Moin and Kim, 1982; Spalart, 1988 and others); second, the possibility of processing efficiently properly-imaged flow fields to obtain quantitative information about their instantaneous structures (Brodkey, 1987; Adrian, 1986 and others); and third, the possibility of acquiring and processing large amounts of experimental data by conventional techniques to map (at least in the statistical sense) the three-dimensional flow field (Nagib and Guezennec, 1986; Adrian et al., 1987a). It is clear that the first two approaches offer very detailed information which cannot be obtained by the latter one. Recent group efforts such as the 1987 Summer Program of the Center for Turbulence Research at NASA Ames/Stanford (see Moin, Reynolds and Kim, 1987) have focused on utilizing numerical data bases to explore new concepts and models. Other group efforts (see Brodkey and Guezennec, 1988) have concentrated on exploring new methodologies to exploit quantitative visual information. However, the interpretation of the wealth of information from instantaneous three-dimensional data bases is largely out of reach at this point and new tools (both visual and analytical) must be developed to take full advantage of the available information and formulate the conceptual mechanisms needed for engineering modeling and flow control. In particular, new techniques must be developed to take advantage of the three-dimensional nature of the new data, while processing statistically in some sense to extract common features. It is through the careful identification of these structures that mechanisms can be identified for their formation, evolution and contribution to the overall turbulence. While conditional sampling techniques introduced in the past (Blackwelder and Kaplan, 1976; Willmarth and Lu, 1972; Corke et al., 1982 and others) have had considerable success in describing local features, these structures may differ widely from instantaneous ones. Moreover, conventional ensemble averaging has been easy to implement when limited to one-dimensional measurements and a small number of "conditions", but becomes prohibitively difficult for full three-dimensional fields if a large number of conditions must be investigated.

A alternate approach has been introduced by Adrian (1974, 1979) to estimate conditional averages for arbitrary conditions. The recent availability of two- or three-dimensional data sets from experiments and Navier-Stokes direct simulations has allowed to test this approach extensively (Adrian et al., 1987a; Adrian et al., 1987b; Adrian and Moin, 1988). In the following sections, the stochastic estimation technique will be briefly described as well as its extension to space-time and two-point conditional estimates. Then, based on the experimental data of Guezennec (1985), a comparison between the stochastic estimates and the conventional ensemble averaging results (Nagib and Guezennec, 1986) will be made.

STOCHASTIC ESTIMATION

The stochastic estimation technique was first proposed by Adrian (1974) and further refined by Adrian (1979) and Tung and Adrian (1980) to include higher order estimates. In these papers, they proposed to obtain an estimate of the conditional average of the velocity vector **u** at the position **x+r** given a condition at point **x**, i.e.:

$$\hat{\mathbf{u}}(\mathbf{x;r}) = < \mathbf{u}(\mathbf{x+r}) \mid \mathbf{u}(\mathbf{x}) >$$

Normally, this quantity is evaluated by conditionally sampling the velocity at **x+r** whenever the velocity at **x** is within a given amplitude window around the specified value **u**. With the stochastic estimation technique, the conditional average is estimated by a series involving increasing powers of the velocity condition **u(x)**. An estimate for $\hat{\mathbf{u}}(\mathbf{x;r})$ is then given by:

$$u_i(\mathbf{x;r}) = A_{ij}(\mathbf{r})\, u_j(\mathbf{x}) + B_{ijk}(\mathbf{r})\, u_j(\mathbf{x})\, u_k(\mathbf{x}) + C_{ijkl}(\mathbf{r})\, u_j(\mathbf{x})\, u_k(\mathbf{x})\, u_l(\mathbf{x}) + \cdots$$

The unknown coefficients $A_{ij}(\mathbf{r})$, $B_{ijk}(\mathbf{r})$, $C_{ijkl}(\mathbf{r})$, ... are determined by a least-square procedure to minimize in the error in the mean square between the velocity at point **x+r** and the estimate:

$$e_i = < \left[\hat{u}_i(\mathbf{x;r}) - u_i(\mathbf{x+r})\right]^2 > \qquad Min.$$

The error is minimized with respect to the unknown coefficients $A_{ij}(\mathbf{r})$, $B_{ijk}(\mathbf{r})$, $C_{ijkl}(\mathbf{r})$, ... to yield:

$$\frac{\partial e_i}{\partial A_{ij}(\mathbf{r})} = \frac{\partial e_i}{\partial B_{ijk}(\mathbf{r})} = \frac{\partial e_i}{\partial C_{ijkl}(\mathbf{r})} = \cdots = 0$$

This lead to the following system of equations:

$$< u_j(\mathbf{x})u_k(\mathbf{x}) > A_{ik}(\mathbf{r}) + < u_j(\mathbf{x})u_k(\mathbf{x})u_l(\mathbf{x}) > B_{ijk}(\mathbf{r}) + < u_j(\mathbf{x})u_k(\mathbf{x})u_l(\mathbf{x})u_m(\mathbf{x}) > C_{ijkl}(\mathbf{r}) = < u_j(\mathbf{x})u_i(\mathbf{x+r}) >$$

$$< u_j(\mathbf{x})u_k(\mathbf{x})u_l(\mathbf{x}) > A_{ik}(\mathbf{r}) + < u_j(\mathbf{x})u_k(\mathbf{x})u_l(\mathbf{x})u_m(\mathbf{x}) > B_{ijk}(\mathbf{r}) + < u_j(\mathbf{x})u_k(\mathbf{x})u_l(\mathbf{x})u_m(\mathbf{x})u_n(\mathbf{x}) > C_{ijkl}(\mathbf{r}) = < u_j(\mathbf{x})u_k(\mathbf{x})u_i(\mathbf{x+r}) >$$

etc...

which can be symbolically represented as:

$$[A][C] = [B]$$

where [A] represents the matrix of moments of order up to 2^n at the reference point **x**, [C] is the unknown coefficients in the stochastic estimation expression (functions of **r**) and [C] represents the two-point, n-th order spatial correlations. In the simplest case of a linear estimate, $u_i(\mathbf{x};\mathbf{r}) = A_{ij}(\mathbf{r})\, u_j(\mathbf{x}) + O\,[u^2(\mathbf{x})]$, the determination of the coefficients $A_{ij}(\mathbf{r})$ requires the use of the **two-point, second order correlation tensor** $R_{ij}(\mathbf{r})$. The solution procedure then involves solving a small system of linear equations for the $A_{ij}(\mathbf{r})$ coefficients for each value of **r**. Then, the linear estimate of the conditional velocity field around **x** is reconstructed by choosing the condition vector **u(x)** and evaluating a simple linear form at each spatial point. Higher order estimates are obtained in a similar fashion by using higher order two-point correlations and solving for the coefficients in the expansion.

It is particularly interesting to note that the estimates of the **conditional** velocity field is obtained from **unconditional** statistics. Hence, the stochastic estimation offers a tool to quantitatively interpret correlation measurements. This tool provides a rational basis for the intuitive conclusions drawn from such measurements by Grant (1958) in turbulent wakes and Blackwelder and Kovasnay (1972) in turbulent boundary layers.

GENERALIZATION TO INCLUDE TIME DEPENDENCE

The above procedure can be generalized to estimate conditional average of **u** at **x**+**r** and $t+\tau$ given condition at **x** and t:

$$\hat{\mathbf{u}}(\mathbf{x},t;\mathbf{r},\tau) = <\mathbf{u}(\mathbf{x}+\mathbf{r};t+\tau) \mid \mathbf{u}(\mathbf{x},t)>$$

In this case, the estimate is obtained by a similar expansion of the velocity field in which the unknown coefficients are not only a function of the spatial separation **r** but also the time separation τ. The following equations outlined the solution procedure for a linear estimate for $\hat{\mathbf{u}}(\mathbf{x},t;\mathbf{r},\tau)$:

$$u_i(\mathbf{x},t;\mathbf{r},\tau) = A_{ij}(\mathbf{r},\tau)\, u_j(\mathbf{x},t) + O\,[u^2(\mathbf{x},t)]$$

The coefficients $A_{ij}(\mathbf{r},\tau)$ are once again determined by a by least-square procedure:

$$e_i = < \left[A_{ij}(\mathbf{r},\tau)u_j(\mathbf{x},t) - u_i(\mathbf{x}+\mathbf{r},t+\tau)\right]^2 > \qquad Min.$$

given by:

$$\frac{\partial e_i}{\partial A_{ij}(\mathbf{r},\tau)} = 0$$

which yields:

$$< u_j(\mathbf{x},t)u_k(\mathbf{x},t) > A_{ik}(\mathbf{r},\tau) = < u_j(\mathbf{x},t)u_i(\mathbf{x}+\mathbf{r},t+\tau) >$$

The determination of the coefficient $A_{ij}(\mathbf{r},\tau)$ requires the use of the **2-point, 2nd order space-time correlation tensor**. Extension to higher order estimates is similar and requires the determination of the two-point, n-th order space-time correlation tensor.

STOCHASTIC ESTIMATION OF TWO-POINT CONDITIONAL AVERAGES

It is clear from the previous sections that single-point conditional averages (i.e. conditional averages obtained by specifying conditions at one single point) are directly related to the correlation tensor, and hence the size of the conditional structures is commensurate with the integral scales of the flow. This point was verified experimentally and numerically by Guezennec, Piomelli and Kim (1987). However, turbulent boundary layers are characterized by a whole hierarchy of scales from approximately $l^+ = 100$ in the near wall region to several hundreds or thousands of wall units in the outer part of the layer (depending on the Reynolds number). Moreover, it is well known that important coherent motions such as the ones detected by the VITA technique are characterized by the rapid (close) interaction between second and fourth quadrant motions (Nagib and Guezennec, 1986; Kim, 1985 and others). Hence, it would be desirable to design a technique which allows to focus on structures of a particular length scale. This was partially achieved by Adrian and Moin (1988) by conditioning not only on the velocity at one point, but also on the strain rate tensor at that point. This is impractical for experimental data due to the complexity of measuring the strain rate tensor. However, a more direct procedure has recently been proposed by Adrian, Moin and Moser (1987) to generalize the stochastic estimation to multi-point conditional averages. In this fashion, the separation distance between the points at which the conditions are imposed forces a certain length scale on the estimated field.

The conditional average of $\mathbf{u}$ at point $\mathbf{r}$ given conditions at $\mathbf{x}_1$ and $\mathbf{x}_2$

$$\hat{\mathbf{u}}(\mathbf{r};\mathbf{x}_1,\mathbf{x}_2) = < \mathbf{u}(\mathbf{r}) \mid \mathbf{u}(\mathbf{x}_1),\mathbf{u}(\mathbf{x}_2) >$$

can be estimated by a linear estimate for $\hat{\mathbf{u}}(\mathbf{r};\mathbf{x}_1,\mathbf{x}_2)$ as:

$$u_i(\mathbf{r};\mathbf{x}_1,\mathbf{x}_2) = A_{ij}(\mathbf{r})\, u_j(\mathbf{x}_1) + B_{ij}(\mathbf{r})\, u_j(\mathbf{x}_2) + O\,[u^2]$$

The coefficients $A_{ij}(\mathbf{r})$ and $B_{ij}(\mathbf{r})$ are determined once again by a by least-square procedure:

$$e_i = < \left[\hat{u}_i(\mathbf{r}) - u_i(\mathbf{r})\right]^2 > \qquad Min.$$

By requiring that the error be minimum with respect to the unknown coefficients $A_{ij}(\mathbf{r})$ and $B_{ij}(\mathbf{r})$ we obtain:

$$\frac{\partial e_i}{\partial A_{ij}(\mathbf{r})} = \frac{\partial e_i}{\partial B_{ij}(\mathbf{r})} = 0$$

which yields symbolically:

$$[\mathbf{A}][\mathbf{C}] = [\mathbf{B}]$$

where [**A**] is a matrix formed by the second order one- and two-point correlation between points $\mathbf{x}_1$ and $\mathbf{x}_2$, [**C**] represent the unknowns coefficients A_{ij} and B_{ij}, and [**B**] represent the two-point spatial correlation between points $\mathbf{r}$ and $\mathbf{x}_1$ and between points $\mathbf{r}$ and $\mathbf{x}_2$. Generalization of this procedure to non-linear, time-dependent estimates of multi-point conditional averages can be obtained in a fashion similar to that described in the previous sections.

It is particularly interesting to note that if the two reference points are only displaced in the homogeneous direction(s) with respect to each other, all quantities in the matrix [**A**] and the right-hand side [**B**] can be written from the 2-point spatial correlation tensor $R_{ij}(\mathbf{r};\mathbf{x})$.

EXPERIMENTAL CONDITIONS

The experiment was performed in the M. V. Morkovin wind tunnel at the Illinois Institute of Technology. A zero pressure gradient turbulent boundary layer was developed over a flat plate suspended in a 5.8 meter long test section. All measurements were taken digitally after analog signal conditioning (amplification and filtering) through the course of this experiment. The hot-wire signals were linearized digitally and temperature compensated. The data was obtained using a fixed V-shaped array of wall-shear sensors and a movable X-wire probe traversed in space with respect to the wall probe. In order to maximize the spatial resolution of the probes,

$\Delta x^+ = 220 \quad \Delta y^+ = 55 \quad \Delta z^+ = 55$

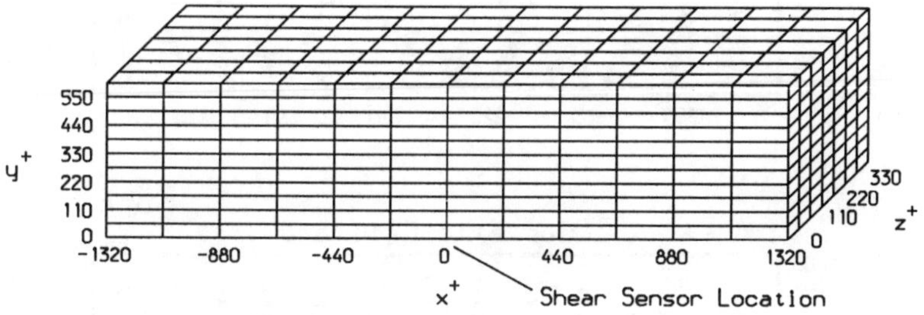

Figure 1. Experimental sampling grid.

a free-stream velocity of 11 m/s was chosen. The Reynolds number based on the momentum thickness at the measurement location was 4900. For more details about the experimental set-up and boundary layer characteristics, the reader is referred to Guezennec (1985).

Measurements consisted of four simultaneously sampled hot-wire signals, two for the fixed wall probe measuring the streamwise (τ_x) and spanwise (τ_z) wall-shear, and two for the moving X-wire measuring the streamwise (u) and normal (v) velocity components. For each position of the X-probe, a continuous record of 204800 samples per channel was acquired every two viscous time units ($T_s^+ = 2$). These measurement were repeated for a total of 1144 spatial measurement points on a 13×11×8 three-dimensional sampling grid as shown in Figure 1. The raw data was stored on magnetic tape for later processing.

RESULTS

Guezennec (1985) showed that strong fourth quadrant (Q4) and second quadrant (Q2) events can be detected by their wall shear stress signature. Moreover, they can be centered in the spanwise direction by requiring that the spanwise component of the wall shear stress be very small. Using this indirect detection technique, Nagib and Guezennec (1986) obtained conventional ensemble averages of the three-dimensional velocity field corresponding to these two types of events. Since only the streamwise and normal velocity components were measured at each point on the sampling grid, the assumption of statistical symmetry in the spanwise direction about the centerline was assumed and the spanwise velocity component was reconstructed by integrating the continuity equation from the centerline outwards. In the following sections, the same conditions on the wall shear will be used to obtain various stochastic estimates of these conditional averages and the spanwise velocity component will be obtained in the same fashion.

Figure 2 and 3 show comparisons between the conventional ensemble averaging technique and the quadratic estimate for the fourth (Q4) and second (Q2) quadrant events respectively. The figures represent vector plots of the streamwise and normal velocity perturbations on the centerline of the event ($z^+ = 0$). The small arrow in the upper right corner of each frame represent the magnitude of the friction velocity, u_τ, and will be included on all subsequent figures. For both the Q4 and the Q2 event, the two techniques agree remarquably well. The structures have similar scales and the magnitude and spatial structure of the velocity perturbations are nearly identical.

Figure 4 and 5 show comparisons between the two techniques for the streamwise evolution of the cross-flow perturbations for the Q4 and Q2 events respectively. Each vector map represent the spanwise and normal velocity perturbations associated with the two types of events. As

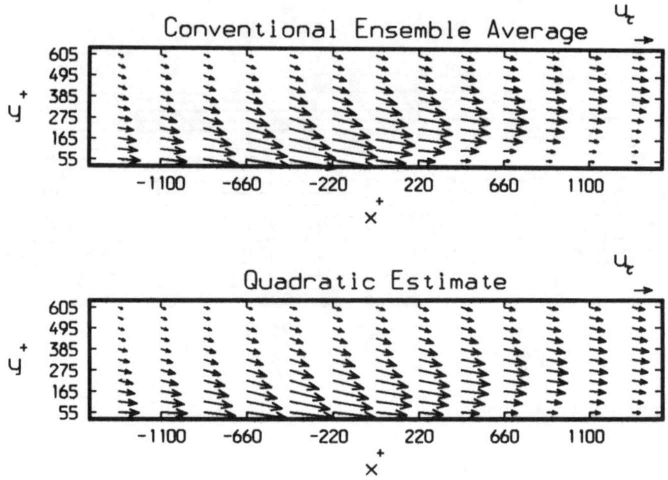

Figure 2. Comparison between conventional ensemble average (top) and quadratic estimate (bottom) of perturbation velocity vectors at $z^+ = 0$ for fourth quadrant event (Q4).

expected, the fourth quadrant motion is associated with high speed fluid moving towards the wall and the second quadrant event represent low-speed moving away from the wall. While measurements were only conducted on one side of the shear detection probe, the assumption of statistical symmetry was used to display the symmetric, counter-rotating structure on the other side. Both techniques are in excellent agreement for the Q4 event, both on the size and inclination of these structures in the streamwise direction. Some differences are observed in the Q2 case at $x^+ = -440$ and 0. This can be attributed to two factors. First, the conventional ensemble averages tend to be inherently more noisy, since the ensemble is build from a finite number of realizations (roughly 500 in this case). In the stochastic estimation case, the measured quantities are unconditional statistics which include all samples (204800) at each spatial point,

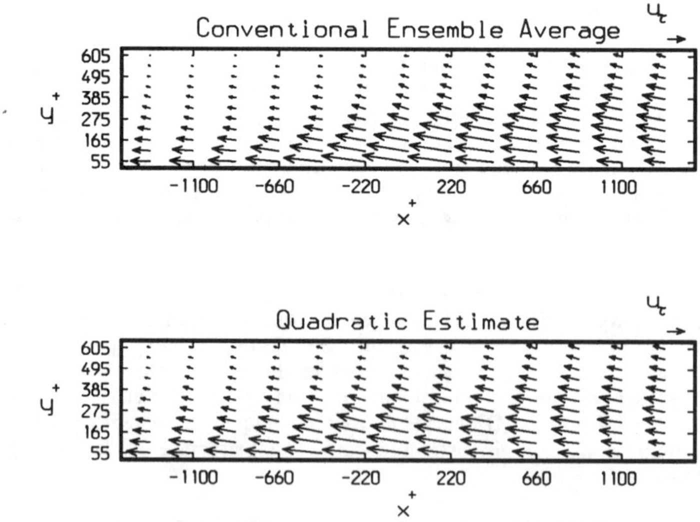

Figure 3. Comparison between conventional ensemble average (top) and quadratic estimate (bottom) of perturbation velocity vectors at $z^+ = 0$ for second quadrant event (Q2).

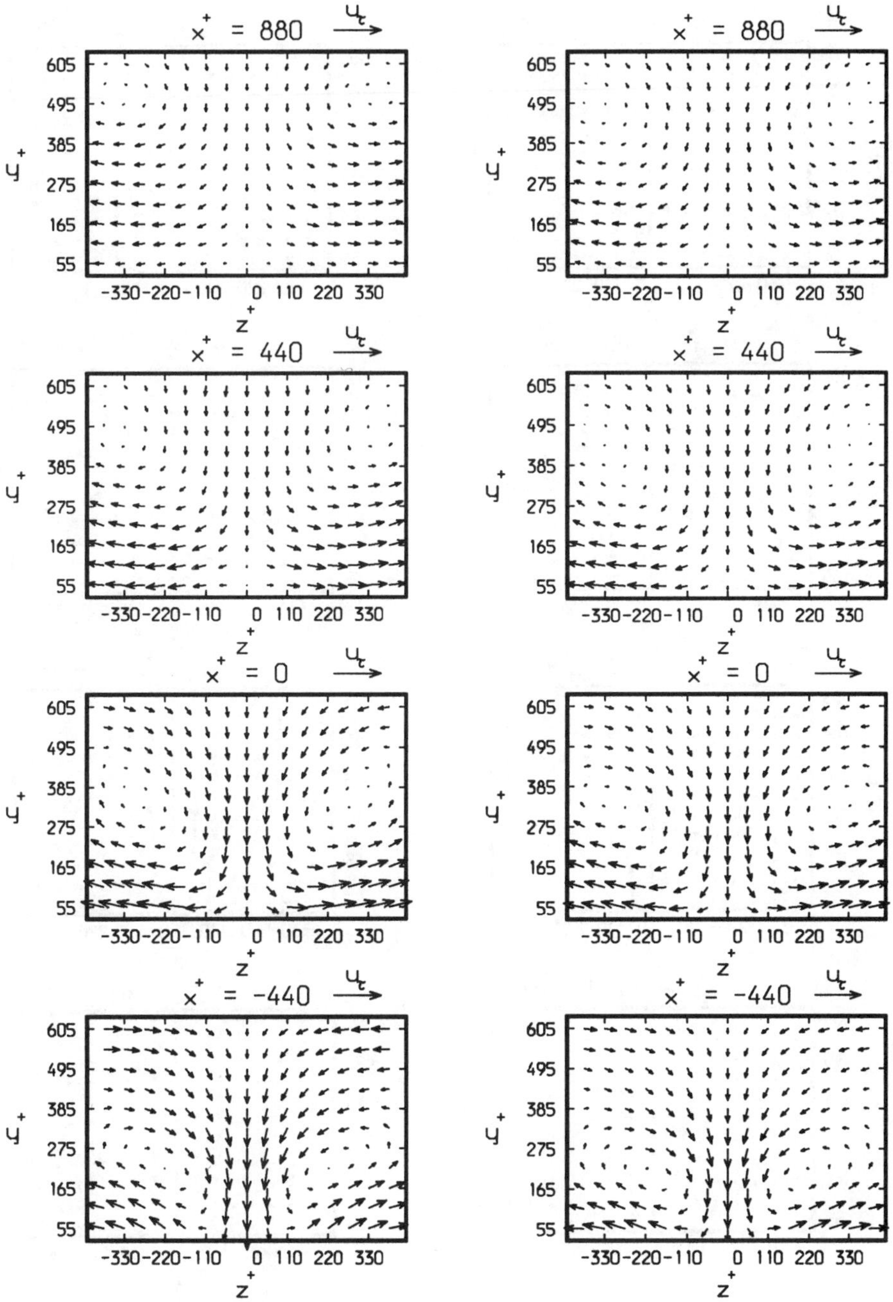

Figure 4. Comparison between conventional ensemble average (left) and quadratic estimate (right) of streamwise evolution of perturbation velocity vectors for fourth quadrant event (Q4).

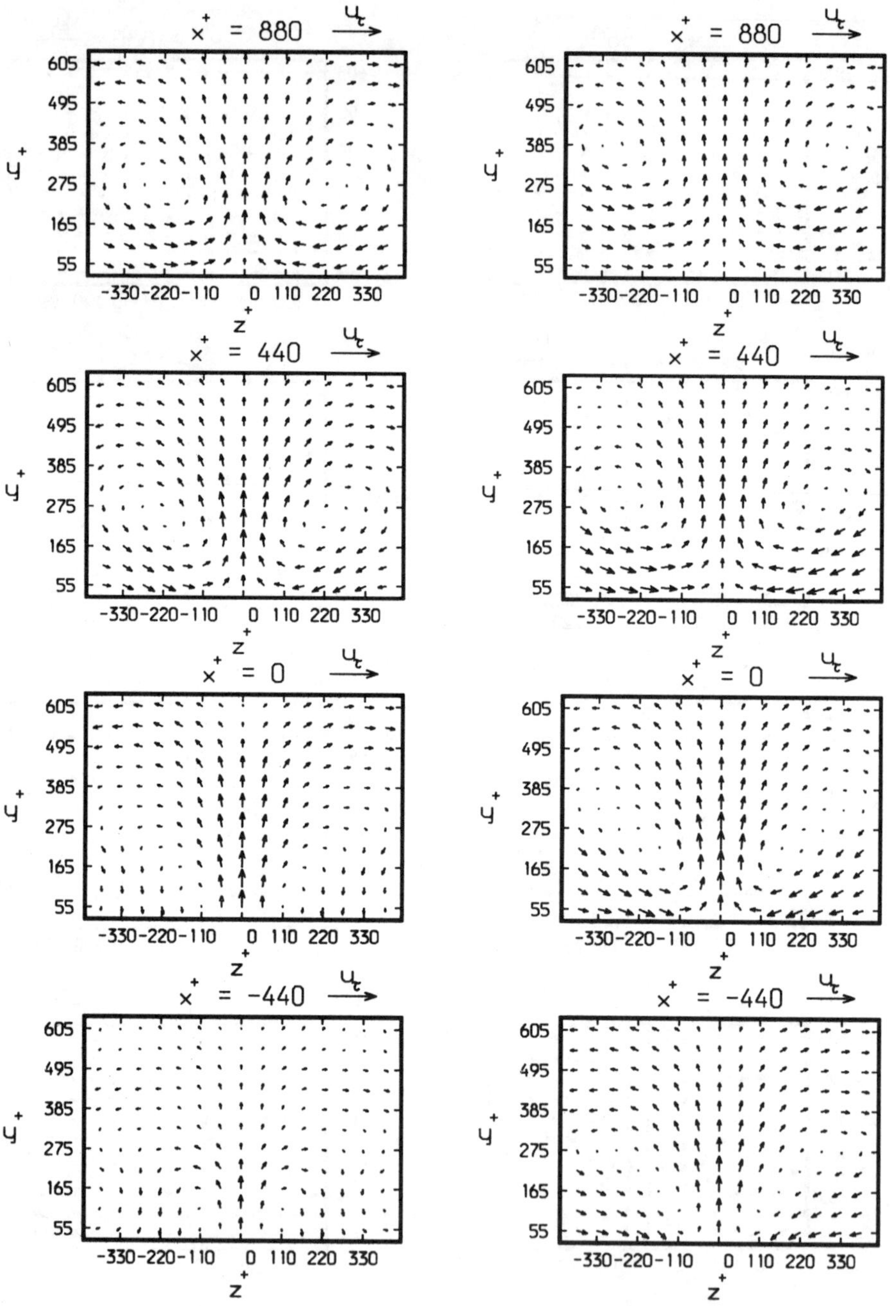

Figure 5. Comparison between conventional ensemble average (left) and quadratic estimate (right) of streamwise evolution of perturbation velocity vectors for second quadrant event (Q2).

hence ensuring a better statistical convergence. This noise problem is of course compounded by the fact that the spanwise velocity is not measured, but only integrated from the continuity equation. Second, Guezennec, Piomelli and Kim (1987) showed that the wall shear detection technique is not as good for the second quadrant events as for the fourth quadrant events. However, despite the minor differences between the two methods in the regions where the perturbation is weak, the overall structure of the flow is well captured by both.

Since the stochastic estimation technique relies on expanding the estimate of the conditional field in powers of the condition velocity vector, the question of convergence of this series must be addressed. Tung and Adrian (1980) showed that corrections from higher order terms (up to fourth order) were minor in isotropic turbulence. This was confirmed by Adrian et al. (1987) for a variety of flow situations. Figure 6 and 7 illustrate the difference between linear and quadratic estimates for the Q4 and Q2 events respectively. Each of these two figures shows the linear estimate (top), the quadratic estimate (bottom) and the contribution from the quadratic terms only. It should be noted that the coefficients of the linear terms are slightly different in the linear estimate case and the quadratic estimate case. This is due to the fact that in the quadratic estimate case, the equations for the linear terms and the quadratic terms are coupled. However, it is quite clear that the difference between the two estimates is minimal. Nonetheless, it is felt that the quadratic terms convey additional information which is important, particularly when the probability density function of the velocity (shear-stress) on which the events

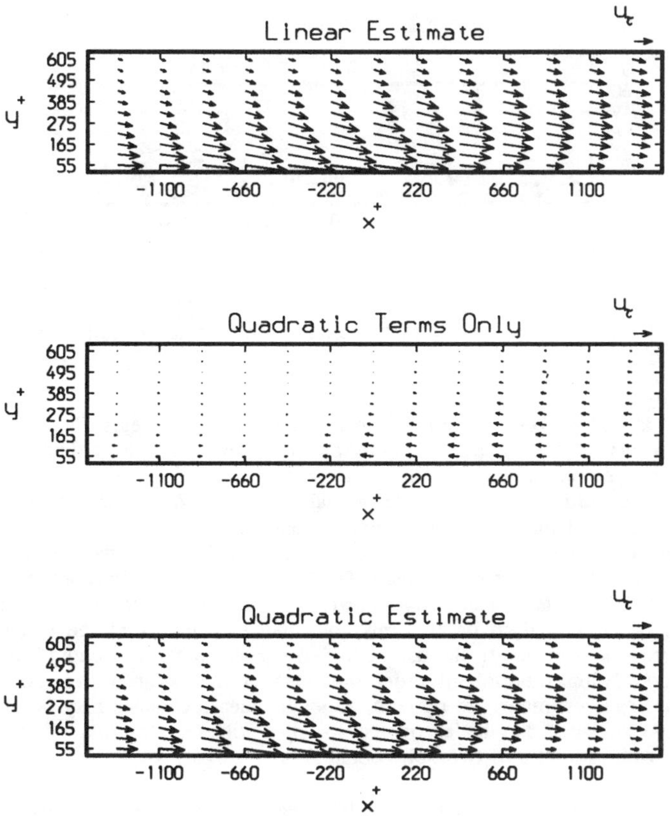

Figure 6. Linear and quadratic estimates of perturbation velocity vectors at $z^+ = 0$ for fourth quadrant event (Q4).

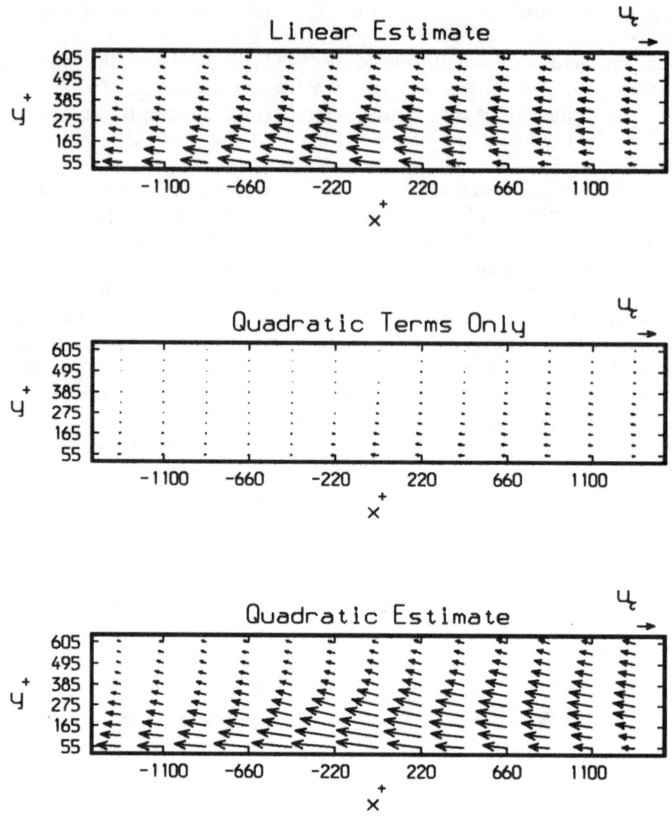

Figure 7. Linear and quadratic estimates of perturbation velocity vectors at $z^+ = 0$ for second quadrant event (Q2).

are conditioned is highly skewed. In the case of the Q4 and Q2 events, the conditions are virtually mirror images of each other, i.e. strong positive streamwise shear fluctuation for the Q4 event and strong negative streamwise shear fluctuation for the Q2 event. When evaluating the linear form, both condition yield the same spatial structure, except for a sign reversal. However, it should be noted that in the quadratic estimate case, the contribution of the quadratic terms alone has the same sign for both event, while the linear terms are of opposite sign. Hence, the quadratic (or higher order) estimate allows a spatial differentiation between "positive" and "negative" events. Results not shown here were also obtained for third-order estimates. It was found that third order terms contributed even less to the estimates, but that higher noise was observed due to the use of high-order statistical moments (fourth-order, two-point spatial correlations and eighth-order moments at the point of reference). Hence, it appears that quadratic estimates are optimal, although linear estimates may be sufficient if the skewness of the statistics can be accounted for in a different fashion. This point will be addressed later with estimates of two-point conditional averages.

Quadratic estimates of the time evolution of the conditional flow field are shown in Figure 8 and 9 for the Q4 and Q2 event respectively. These sequences of flow fields correspond to time delays of $t^+ = -30, -20, -10, 0$ and 10 around the detection time. They were obtained by applying the generalization of the stochastic estimates to include time dependence and are extracted from the two-point space-time correlation tensor. Both figures show the convection and deformation of the structures within the frame of view. The convection is the predominant

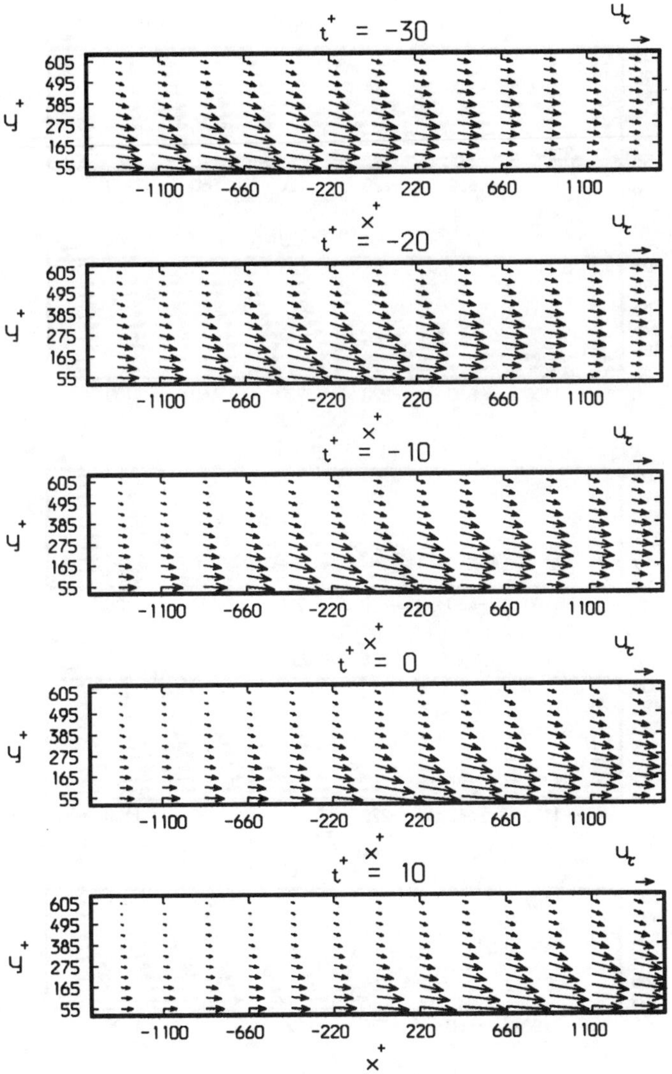

Figure 8. Quadratic estimate of time evolution of perturbation velocity vectors at $z^+ = 0$ for fourth quadrant event (Q4).

aspect of this time evolution over this time span. This fact correlates well with the long persistence of *individual* such events observed in the direct numerical simulations (see Guezennec, Piomelli and Kim (1987) for more details). Convection velocities of the various part of the structures were shown by Guezennec (1985) to vary with height above the wall (also shown by Hussain, Jeong and Kim (1987)), i.e. the structures are sheared and convected at the same time. These time sequences demonstrate the usefulness of the stochastic estimation to evaluate the local dynamics of these structures in the vicinity (in space and time) of a given condition. Once again, the use of quadratic estimates allows to differentiate the convection velocity of second and fourth quadrant motions through the use of the *third-order* statistics.

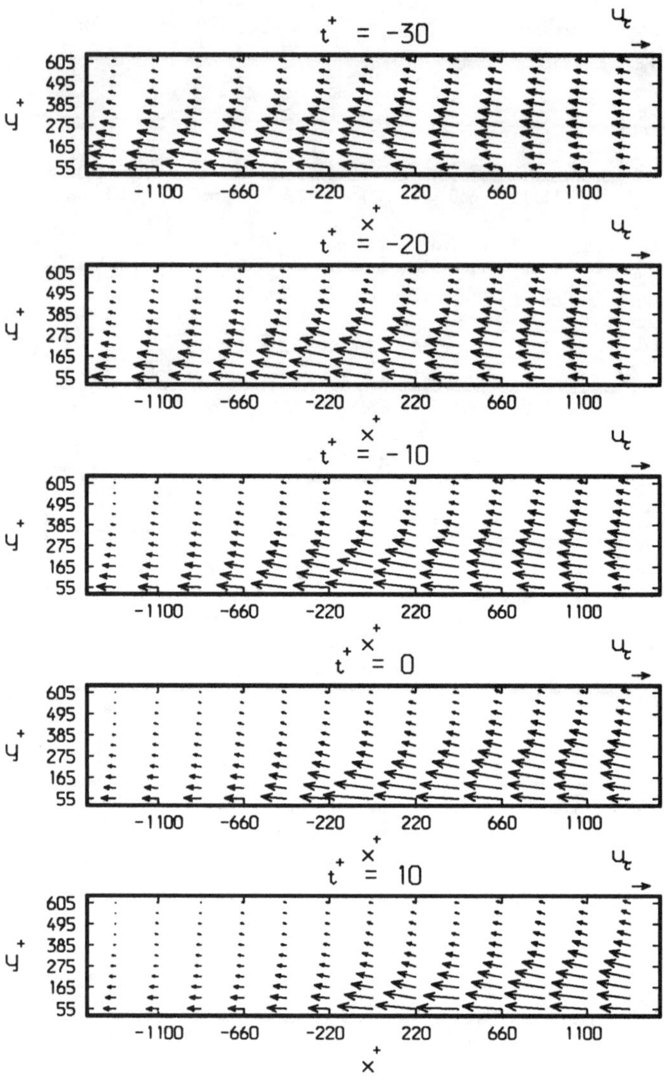

Figure 9. Quadratic estimate of time evolution of perturbation velocity vectors at $z^+ = 0$ for second quadrant event (Q2).

As pointed out earlier, the scale of the structures educted by one-point conditional averages (either by conventional means or the stochastic estimation) is linked to the integral scale of the flow. Smaller scale events such as the sharp shear layers detected by the VITA/VISA technique play an important role in the dynamics of turbulent boundary layers, but yet are not readily obtainable from one-point velocity conditions unless gradient information (such as the strain rate) is included. Two- or multi-point conditional averages circumvent this difficulty by focusing on a particular scale determined by the separation between the points at which conditions are imposed. However, such approach is usually much harder to implement by conventional ensemble averaging technique due to the difficulty of satisfying multiple conditions. However, the stochastic estimation of such quantities rely solely on unconditional statistics, and thereby eliminates that difficulty. Figure 10 and 11 show examples of estimates of two-point

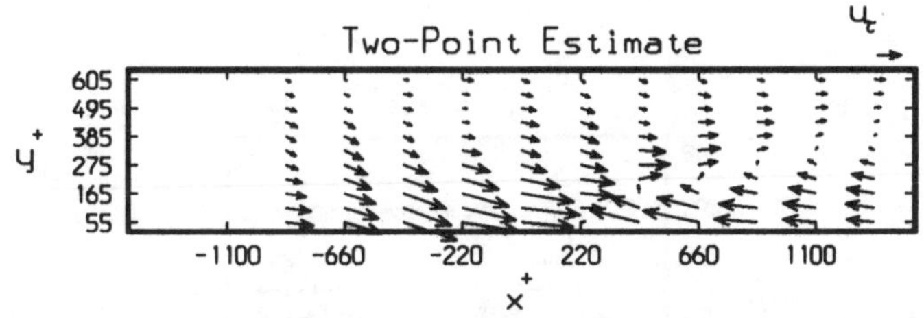

Figure 10. Linear estimate of perturbation velocity vectors at $z^+ = 0$ for "VITA" event (Q2/Q4 pair).

conditional averages. In this case, the conditions were that a fourth quadrant event occurred within a small distance ($\Delta x^+ = 440$) of a second quadrant event. This sequence has been shown by Nagib and Guezennec (1986) to represent a VITA/VISA event. The streamwise cut at $z^+ = 0$ shown in Figure 10 depicts very clearly the sharp shear layer resulting form the interaction of such events as previously observed by Kim (1985) and others. The streamwise evolution of the cross-flow velocity perturbations is shown in Figure 11. The cuts located at $x^+ = 0$ and 440 show the presence of two sets of counterrotating eddies, with a stagnation point in between, as observed by Kim (1985). The lack of direct measurements of the spanwise velocity component introduces large error through the integration of the continuity equation, in particular when focusing on sharp interactions where the grid spacing of the measurements becomes inadequate to properly represent sharp gradients. It is interesting to note that the size of the individual Q4 (Figure 2) and Q2 (Figure 3) structures is reduced in the case of their sharp interaction (Figure 10). This confirms the role of two- or multi-point conditional averages to examine specific spatial scales. Adrian, Moin and Moser (1987) have recently pointed out the close resemblance between the two-point conditional averages and the instantaneous structure matching those two conditions.

CONCLUSIONS

The stochastic estimation provides a flexible and reliable tool for the study of coherent structures in wall-bounded flows. By the proper choice of non-linear estimates, time dependent estimates and/or multi-point conditional estimates, important information can be gathered about the spatio-temporal structure of various types of events and possibly their evolution. As the measurements required for the stochastic estimation are *unconditional* in nature, the investigation of a large number of conditions is possible without any large computational expense; this aspect is of particular importance when dealing with full three-dimensional data bases.

It is felt that by measuring the spanwise velocity component, the assumption of symmetry about the centerline can be relaxed and dissymmetric conditions such as the ones noted by Guezennec et al. (1987) can be properly investigated. The generalization to multi-point conditional estimates coupled with using real (measured) time sequences of velocities as "conditions" could lead to realistic estimates of the evolution of the instantaneous velocity field, thereby providing a powerful experimental tool to gain a better understanding of the instantaneous dynamics of turbulent boundary layers.

ACKNOWLEDGMENT

The experiment and first part of this work was performed at the Fluid Dynamics Center at Illinois Institute of Technology with H. M. Nagib and supported by the Air Force Office of Scientific Research under Grant NSG-1591. The support for the latter part of the work was provided by The Ohio State University and is gratefully acknowledged.

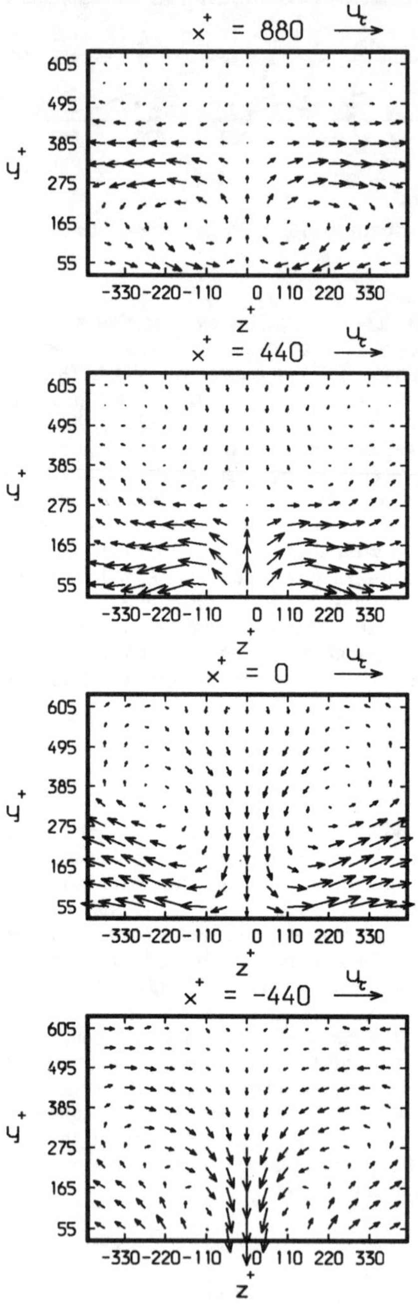

Figure 11. Linear estimate of streamwise evolution of perturbation velocity vectors for "VITA" event (Q2/Q4 pair).

REFERENCES

Adrian, R. J., On the Role of Conditional Averages in Turbulent Theory, Proceedings of the Fourth Symposium on Turbulence, University of Rolla-Missouri, 1976.

Adrian, R. J., Conditional Eddies in Isotropic Turbulence, *Phys. Fluids*, Vol. 22, pp. 2065-2070, 1979.

Adrian, R. J., Multi-Point Optical Measurements of Simultaneous Vectors in Unsteady Flows, *Int. J. Heat and Fluid Flow*, Vol. 7, No. 2, 1986.

Adrian, R. J., Chung, M. K., Hassan, Y., Jones, B. G., Nithianandan, C. K. and Tung, A. T., Experimental Study of Stochastic Estimation of Turbulent Conditional Averages, Sixth Turbulent Shear Flow Symposium, Toulouse, France, pp. 6.1.1-6.1.7, 1987.

Adrian, R. J., Moin, P. and Moser, R. D., Stochastic Estimation of Conditional Eddies, Proceedings of the 1987 Summer Program, Report CTR-S87, NASA Ames/Stanford University, 1987.

Adrian, R. J. and Moin, P., Stochastic Estimation of Organized Turbulence Structures: Homogeneous Shear Flow, to appear in *J. Fluid Mech.*, 1988.

Blackwelder, R. F. and Kaplan, R. E., On the Wall Structure of Turbulent boundary Layers, *J. Fluid Mech.*, Vol. 76, pp. 89-112, 1976.

Blackwelder, R. F. and Kovasnay, L. S. G., Time Scales and Correlations in Turbulent Boundary Layers, *Phys. Fluids*, Vol. 15, pp. 1145-1554, 1972.

Blackwelder, R. F., The Bursting Process in Turbulent Boundary Layers, Lehigh Workshop on *Coherent Structures in Turbulent Boundary Layers*, ed. C. R. Smith and D. E. Abbott, pp. 211-227, 1978.

Brodkey, R. S., Image Processing and Analysis for Turbulence Research, *Chem. Eng. Educ.*, Vol. 20, pp. 202-210, 1987.

Brodkey, R. S. and Guezennec, Y. G., Report on the NSF Imaging Workshop, The Ohio State University, 1988.

Brown, G. L. and Thomas, A. S. W., Large Structures in Turbulent Boundary Layers, *Phys. Fluids*, Vol. 20, pp. 243-252, 1977.

Chen, C. H. P. and Blackwelder, R. F., Large Scale Motion in a Turbulent Boundary Layer: a Study Using Temperature Contamination, *J. Fluid Mech.*, Vol. 89, pp. 1-31, 1978.

Corke, T. C., Nagib, H. M. and Guezennec, Y. G., A new View on Origin, Role and Manipulation of Large Scales Structures in Turbulent Boundary Layers, NASA -165861, 1982.

Grant, H. L., The Large Eddies of Turbulent Motion, *J. Fluid Mech.*, Vol. 4, pp. 149-190.

Guezennec, Y. G., Documentation of Large Coherent Structures Associated with Wall Events in Turbulent Boundary Layers, PhD Thesis, Illinois Institute of Technology, 1985.

Guezennec, Y. G., Piomelli, U. and Kim, J., Conditionally-Averaged Structures in Wall-Bounded Turbulent Flows, Proc. 1987 Summer Program of Center for Turbulence Research, Report CTR-S87, NASA Ames/Stanford University, 1987.

Head, M. R. and Bandyophadyay, P., New Aspects of Turbulent Boundary Layer Structure, *J. Fluid Mech.*, Vol. 107, pp. 297-337., 1981.

Hussain, A. K. M. F., Jeong, J. and Kim, J., Structure of Turbulent Shear Flows, Proc. 1987 Summer Program of Center for Turbulence Research, Report CTR-S87, NASA Ames/Stanford University, 1987.

Kim, J., Turbulence Structure Associated with the Bursting Event, *Phys. Fluids*, Vol. 28, pp. 52-58.

Kim, J. and Moin, P., The Structure of the Vorticity Field in Turbulent Channel Flow. Part 2. Study of the Ensemble-Averaged Fields, *J. Fluid Mech.*, Vol. 162, pp.339-363., 1986.

Kim, H. T., Kline, S. J. and Reynolds, W. C., The Production of Turbulence near a Smooth Wall in a Turbulent Boundary Layer, *J. Fluid Mech.*, Vol. 50, pp. 133-160, 1971.

Klebanoff, P. S., Characteristics of Turbulence in a Boundary Layer with Zero Pressure Gradient, NACA TN-3178, 1954.

Kline, S. J., Reynolds, W. C., Schraub, F. A. and Runstadler, P. W., The structure of Turbulent Boundary Layers, *J. Fluid Mech.*, Vol. 30, pp. 741-773, 1967.

Moin, P and Kim, J., Numerical Investigation of Turbulent Channel Flow, *J. Fluid Mech.*, Vol. 118, pp. 341-367, 1982.

Moin, P. , Reynolds, W. C. and Kim, J., *Studying Turbulence Using Numerical Simulation Databases*, Proceedings of the 1987 Summer Program, Report CTR-S87, NASA Ames/Stanford University, 1987.

Nagib, H. M. and Guezennec, Y. G., On the Structure of Turbulent Boundary Layers, Proceedings of the Tenth Symposium on Turbulence, University of Rolla-Missouri, 1986.

Schraub, F. A. and Kline, S. J., Thermosci. Div., Mech. Eng. Dept. Rep. MD-12, Stanford University, 1965.

Smith, C., Visualization of Turbulent Boundary Layer Structure Using a Moving Hydrogen-Bubble Wire Probe, Lehigh Workshop on *Coherent Structure of Turbulent Boundary Layers*, ed. C.. Smith and D. E. Abbott, pp. 48-97, 1978.

Spalart, P., Direct Simulation of a Turbulent Boundary Layer up to $Re_\theta = 1410$, *J. Fluid Mech.*, Vol. 187, pp. 61-80, 1988.

Tung, T. C. and Adrian, R. J., Higher Order Estimates of Conditional Eddies in Isotropic Turbulence, *Phys. Fluids*, Vol. 23, pp. 1469-1470, 1980.

Willmarth, W. W. and Lu, S. S., Structure of the Reynolds Stress near the Wall, *J. Fluid Mech.*, Vol. 55, pp. 65-92, 1972.

Insight into the Dynamics of Coherent Structures from a Proper Orthogonal Decomposition

WILLIAM K. GEORGE*
University of Buffalo, SUNY
Buffalo, New York 14260, USA

INTRODUCTION

The Danish cartoonist-philosopher, Piet Hein, in one of his more famous grooks, tells of the scientist who has discovered a well which is the source of truth. After carefully filling his glass from the well, he examines the liquid closely, then pronounces his conclusion: Truth is conical. A similar insight into the nature of human investigators is contained in the poem by John Godfrey Saxe in which six blind men discover an elephant. The first falls against the side and proclaims the elephant to be like a wall, and the second feeling a tusk suggests the elephant is more like a spear. The third grasps the trunk and argues that the elephant is much like a snake, while the fourth embracing a knee argues that it is really more like a tree. The fifth feeling an ear believes the elephant to be more like a fan, and the sixth seizing the tail proclaims that the elephant is clearly more like a rope.

This is a paper about coherent structures, and one of the tools we use to study them -- the proper orthogonal decomposition. Like the scientist and the blind men in the illustrations above, one's beliefs about coherent structures are largely shaped by his perspective. And like them, there is a tendency for those who study coherent structures to want the answers to be simple.

The purpose of this paper is to provide some insight into the particular perspective of the proper orthogonal decomposition. The focus will be on what the orthogonal decomposition can tell us about the mathematical representation of coherent stuctures which is necessary to capture their essential kinematics. Without such a mathematical description, coherent stuctures are likely to remain forever no more than a curiosity for the fluid mechanics community, with little possibility of influencing the manner in which turbulence models are built. The hope is that by finding a way to represent stuctures and events which is true to nature, we will open at long last the possibility of exploring the importance of coherent structures for the dynamics of turbulent flow. There will be no attempt in this paper to provide a complete history of orthogonal decomposition techniques, nor even to list the ever-increasing attempts to utilize them. Rather, the focus will be on 'What is a proper orthogonal decomposition in the context of turbulence?' and on 'What does it have to

*Presently on leave at: Fluid Mechanics Department, Technical University of Denmark, Dk 2800, Lyngby, Denmark.

offer us in the study of coherent structures?'. The hope is that this presentation will lead to a better understanding of what has already been done, and to a broader discussion of where we should go from here.

THE PROPER ORTHOGONAL DECOMPOSITION

Like most good ideas, the proper othogonal decomposition we discuss here has more than one who can lay claim to having invented it. Oceanographers and meteorologists, who have used it for the study of various geophysical phenomena, neatly by-pass the question of origin and refer to their results as Empirical Orthogonal Functions (EOF's), bristling whenever anyone suggests using another term. The mathematicians are equally adamant about referring to it as a Karhunen-Loeve expansion. The turbulence community (or at least a part of it) are happy to refer to it as Lumley's Orthogonal Decomposition. All of these choices can be defended. Certainly it is a Karhunen-Loeve expansion, and the eigenfunctions obtained are empirical. And the recognition of its applicability to turbulence (and perhaps even its generalization to multi-dimensions) is due to Lumley (1967). The viewpoint of this paper has been very strongly influenced by the latter, although some of the interpretations offered here may differ from those originally proposed. Nonetheless, for reasons of efficiency and to offend no one (or at least all equally) we shall in the remainder of this paper simply use the term Orthogonal Decomposition (or OD).

The basic idea behind the Orthogonal Decomposition (OD) is that one tries to optimally represent a random field by a set of deterministic functions which are in turn determined by the field itself. This is quite unlike the more common situation where one chooses a set of orthogonal functions (like the harmonic ones of Fourier analysis) and then seeks the coefficients necessary to represent the field. For the OD, both the functions and the coefficients arise from the statistical properties of the random field itself.

The key word for determining the orthogonal functions is "optimal". For the OD discussed here, "optimal" means that the functions are to be determined so that the most possible energy is contained in the lowest order term, then the most possible of the remaining energy in the next, and so on. This can be accomplished for a random field $u_i(x,t)$ by finding a deterministic field $\phi_i(x,t)$ for which the mean square projection of ϕ_i on u_i is maximal. If we denote this projection by α where

$$\alpha = u_i \ \phi_i = \underline{u} \cdot \underline{\phi} \tag{1}$$

then we must maximize

$$<|\alpha|^2> = <|u_i\phi_i|^2> \tag{2}$$

where < > denotes the ensemble average. From the calculus of variations, ϕ_i can be shown to be determined by the following integral equation,

$$\iiint_{\text{all space}} R_{ij}(\cdot,\cdot')\phi_j(\cdot') = <|\alpha|^2>\phi_i(\cdot) \tag{3}$$

where

$(\cdot) = (\underline{x},t)$

and

$(\cdot') = (\underline{x}',t')$

and where R_{ij} represents the cross-correlation defined by

$$R_{ij}(\cdot,\cdot') = \langle u_i(\cdot)u_j(\cdot')\rangle \tag{4}$$

Thus the problem of optimally determining the deterministic function ϕ_i has been reduced to an eigenvalue problem with the kernel given by the cross-correlation of the random vector field.

Since part of our purpose here is perspective, it is worth pausing for a moment to note that there is no longer room for judgement or subjectivity on the part of the theoretician or experimentalist. By deciding at the beginning that the functions we sought should optimally represent the field <u>in a mean square sense</u> (equation 2), the functions have been completely determined. It actually matters not whether the field represents velocity, vorticity, pressure, or temperature, the integral of equation (3) will simply have as its kernel the appropriate cross-correlation. Thus all of our subjectivity has been lumped into equation (2) and the choice of principle behind it. It might be interesting to explore the consequences of alternative choices for maximization.

The solutions to equation (2) are very strongly dependent on the kind of field being discussed. If the field is of <u>finite total energy</u>, then the solutions to equation (3) are governed by the Hilbert-Schmidt theory of linear integral equations from which it follows that:

(i) There exists a denumerable set of eigenvalues, λ_n, satisfying equation (3), and $\lambda_1 > \lambda_2 > \lambda_3$, etc.

(ii) All of the eigenvalues are real.

(iii) The eigenfunctions are orthogonal and can be normalized so that

$$\iiint\limits_{\text{Domain}} \phi_i^{(m)}(\cdot)\phi_i^{(n)}(\cdot)d(\cdot) = \delta_{mn} \tag{5}$$

(iv) The cross-correlation can be represented by them as

$$R_{ij}(\cdot,\cdot') = \sum_n \lambda_n\ \phi_i^{(n)}(\cdot)\ \phi_j^{(n)*}(\cdot') \tag{6}$$

(v) Any realization of the random field can be represented as

$$u_i(\cdot) = \sum_n a_n\ \phi_i^{(n)}(\cdot) \tag{7}$$

where the coefficients are determined for each realization of the field by

$$a_n = \iint_{\text{Domain}} u_i(\cdot)\phi_i^{(n)*}(\cdot)d(\cdot) \tag{8}$$

and where

$$\langle a_n a_m^* \rangle = \lambda_n \ \delta_{mn} \tag{9}$$

A corollary of equations (4) and (5) is that the total energy in the field can be obtained as

$$\iint_{\text{Domain}} \langle u_i u_i \rangle \ d(\cdot) = \sum_n \lambda_n \tag{10}$$

Thus, as required, the energy is recovered in an optimal manner by our eigenfunctions and eigenvalues.

Before proceeding further, it is appropriate to ask what this decomposition, however elegant and efficient, has to do with coherent structures? The answer to this question is somewhat complicated by the fact that there exists no definition of a coherent structure with any dynamical or mathematical significance. The situation was aptly described by Lumley (1982) who, drawing from a U.S. Supreme Court decision on pornography, pointed out that like pornography, coherent structures are hard to define, but you know one when you see it. Less facetiously, Lumley (1967) did suggest that the "large eddy" of a turbulent flow (coherent structures had not been invented then) be identified as the lowest order eigenfunction resulting from an orthogonal decomposition of the type outlined above.

It would appear, to this point at least, that the eigenfunctions determined by equation (3) have many of the desired properties we would associate with coherent structures. They certainly represent coherent features of the flow, and are even deterministic. Unfortunately, as will be seen later, a simple physical representation of the flow field resulting from them is not always possible. Moreover, it will also be seen that there is some evidence that a given structure is represented by different order eigenfunctions at different points in its life-cycle.

HOMOGENEOUS AND STATIONARY RANDOM FIELDS

The solution to the integral equation (3) becomes considerably more complicated if the field is not of finite extent, as is the case if it is stationary (temporally) or homogeneous (spatially) in one or more coordinates. In such cases, the eigenfunctions in these coordinates reduce to the harmonic ones and the coefficients are simply the Fourier coefficients. This would not at first sight appear to be a complication, since the eigenfunctions do not have to be determined, but are known _a priori_. In fact, because the eigenfunctions are known, the decomposition can be performed on these directions immediately without solving an integral equation. There are several equivalent ways of doing this, but probably the simplest is to Fourier transform the random field over the homogeneous and stationary variables, then solve the integral equation for this transformed field to find the optimal orthogonal functions for the remaining directions.

For example, suppose the random field to be stationary in time and homogeneous in the x_1 coordinate. Then the Fourier transform of the field is given by

$$\hat{u}_i(k_1,\omega;x_2,x_3) = \frac{1}{(2\pi)^2}\iint_{-\infty}^{\infty} u_i(x_1,x_2,x_3,t)e^{-i(k_1x+\omega t)}dx_1dt \qquad (11)$$

We now seek a deterministic function $\psi_i(x_2,x_3;k_1,\omega)$ whose mean square projection on the transformed random field is maximal. As before, we are led to an integral equation, but this time given by

$$\iint_{\text{Domain}} \Phi_{ij}(x_2,x_2',x_3,x_3';k_1,\omega)\ \psi_j(x_2,x_3;k_1,\omega)dx_2'dx_3'$$
$$= \lambda(k_1,\omega)\ \psi_i(x_2,x_3;k_1,\omega) \qquad (12)$$

Φ_{ij} is the cross-spectrum defined as the Fourier transform of the cross-correlation over the appropriate variables, i.e.

$$\Phi_{ij}(x_2,x_2',x_3,x_3';k_1,\omega) = \frac{1}{(2\pi)^2}\iint_{-\infty}^{\infty} R_{ij}\ (x_1'-x_1,x_2,x_2',x_3,x_3',t'-t)$$
$$\cdot e^{-i[k_1(x_1'-x_1)+\omega(t'-t)]}d(x_1'-x_1)d(t'-t) \qquad (13)$$

where the use of $x_1'-x_1$ and $t'-t$ reflects the fact that the field is homogeneous in the 1-direction and stationary in time.

Now, instead of the single eigenvalue problem of equation (3), we have an eigenvalue problem for each frequency/wavenumber combination. Each of the eigenvalues of equation (12) has now become an eigenspectrum, $\lambda(k_1,\omega)$. While this may present some difficulties in managing the data, they are more than compensated for by the reduced information required about the cross-correlation because of the homogeneity and stationarity. (In our example, the number of variables has been reduced from 8 to 6.)

As before, we can reconstruct the decomposed Fourier transformed field, in this case by

$$\hat{u}_i(k_1,\omega;x_2,x_3) = \sum_n a_n(k_1,\omega)\ \psi_i^{(n)}(x_2,x_3;k_1,\omega) \qquad (14)$$

where the coefficients are determined by

$$a_n(k_1,\omega) = \iint_{\text{Domain}} \hat{u}_i(k_1,\omega;x_2,x_3)\psi_i^{(n)*}(x_2,x_3;k_1,\omega)dx_2dx_3 \qquad (15)$$

and satisfy

$$\langle a_n(k_1,\omega)\ a_m^*(k_1,\omega)\rangle = \lambda_n(k_1,\omega)\ \delta_{mn} \qquad (16)$$

Equation (14) can in turn be inverse Fourier transformed to obtain the original field as

$$u_i(x_1,x_2,x_3,t) = \sum_n \left\{ \iint_{-\infty}^{\infty} a_n(k_1,\omega)\, \psi_i^{(n)}(x_2,x_3;k_1,\omega)\, e^{i(k_1x_1+\omega t)} dk_1 d\omega \right\} \tag{17}$$

Thus the original field has been completely represented in terms of the eigenfunctions and their coefficients, and as before, <u>all</u> of its statistics are intact.

The complications alluded to above then do not arise from a short-coming of the decomposition itself, since we have still optimally decomposed the random field. Nor do they arise from the reconstruction of the field, since again the original field is faithfully recovered. The problems occur when we try to reconstruct a physical picture of an individual eddy. As for the completely inhomogeneous problem above, it is straightforward to identify the spatial variation of each mode in the x_1 and t coordinates since they are simple Fourier modes. It is for the inhomogeneous directions, however, that we encounter difficulties. Because the eigenfunctions are continuous functions of wavenumber and frequency, we can determine the spatial dependence in the inhomogeneous directions only by first adding up all the Fourier modes to determine the appropriate coefficients according to the recipe of equation (17). Unfortunately, these coefficients are both complex and random so that the amount of each eigenfunction to be added varies, not only from ensemble to ensemble, but from time to time and place to place within a given ensemble. Thus without additional assumptions, there is no one way to add up the pieces to find out what a "typical" coherent structure looks like.

This inability to say unambiguously what the structures look like in the homogeneous direction(s) has led many to conclude that there is a fundamental problem with the orthogonal decomposition approach. While there may be, this is not it! The OD has identified for us a set of building blocks and is telling us that they are not always put together the same way. To see that this is not so uncommon in nature one need only consider the waves which occur on the open sea. The individual swell are made of many Fourier components, all propagating in different directions and at different rates. At the instant we observe them, they add up to form a structure which we identify as a typical wave. As we watch, the wave evolves and then disappears as the Fourier components which constitute it change. Does our inability to provide a precise recipe for the individual structure prevent us from learning about the dynamics of waves from studying the interaction of the building blocks? Of course not, and so most studies of the dynamics of the sea are based on Fourier analyis. In a similar manner, a study of the interaction of the building blocks yielded by an OD of a turbulent flow holds forth the prospect of teaching us much about the dynamics of turbulence.

We can gain considerable insight into the relation of the OD to coherent structures if we take a single eigenfunction and its coefficient from the OD, and transform it back into physical space. Choosing, for example, the lowest order term (n=0) from equation (17), the velocity field arising from it, say $u_i^{(0)}$, is given by

$$u_i^{(0)}(\underline{x},t) = \iint_{-\infty}^{\infty} a_o(k_1,\omega)\ \psi_i^{(0)}(x_2,x_3;k_1,\omega)\ e^{i(k_1x_1+\omega t)}dk_1d\omega \tag{18}$$

Now the time and spatial extent in the x_1 direction for which $u_i^{(0)}$ appears coherent will be determined by the bandwidth of the disturbances constituting it. From equation (16),

$$<|a_o(k_1,\omega)|^2> = \lambda_o(k_1,\omega) \tag{19}$$

Thus it is the eigenspectrum, λ_o, which will determine how long and over what distance the contribution of $u_i^{(0)}$ will appear coherent. If λ_o is strongly peaked in wavenumber and frequency, then the contribution will appear highly coherent, persisting for a time proportional to the inverse bandwidth in frequency and over a distance proportional to the inverse bandwidth in wavenumber. If the contribution of this mode is spread smoothly over a broad band in either wavenumber or frequency, then its coherence will be proportionately decreased.

It is the length of time for which a coherent structure exists which determines its visibility, or our ability to see it. Similarly, it is the structures spatial extent which enables us to find it. Thus visibility and spatial extent are determined for stationary and partially homogeneous flows by the spectral content of the eigenspectra.

Let us assume for the moment that the OD does in some sense reflect what is happening in a real turbulent field, and that the eigenfunctions obtained really have captured the coherent features of the flow. Then the reason for the difference in visibility of coherent structures in high and low Reynolds number turbulence becomes readily apparent. At low Reynolds number, the energy (and hence the eigenspectra) is concentrated in a narrow band of frequency and wavenumber, usually not too far removed from those characterizing the instability and transition process. As a consequence, it should be (and is) relatively easy to see and find coherent structures in these flows. As the Reynolds number is increased, however, the eigenspectra are broadened by the non-linear interactions, and the coherent structures are harder to see and find. This does not mean that they are not present, only that they are evolving more rapidly.

We can ask what the OD tells us about the probable outcome of efforts that continue to be expended to characterize coherent structures in a wide variety of stationary flows which are partially homogeneous. The answer can be best understood by returning to the wave example above. A photographer or artist can capture for us a typical wave. From it we can even determine the coefficients since we know what the eigenfunctions are already. And in turn, knowing these and the governing equations, we can come to an understanding of the evolution of the wave itself. The picture and the events, however well characterized, can only contribute to our understanding of the dynamics when we already know what the eigenfunctions are. The same would appear to be true for the elusive structures of turbulent flow as well. Their "elusivity" is a direct indication of their fluctuating composition, and would appear to doom attempts to characterize them without explicitly recognizing their nature. In view of this, efforts to find "typical" structures in such flows (even those using sophisticated shot-noise and conditional-estimation techniques) may succeed (as did the photographer above), but can contribute little to understanding of the physics in the absence of the eigenfunctions.

RANDOM FIELDS WITH PERIODICITIES

A similar situation to that described above arises when the random field is periodic in one or more dimensions. Suppose for example that the flow is axially symmetric so that the single point statistical quantities are independent of the azimuthal coordinate, θ, and the two-point statistics depend only on differences in θ, say $\theta'-\theta$. Since the random field must be periodic in θ, it can be shown that the eigenfunctions satisfying equation (3) are harmonic functions of θ, i.e.,

$e^{i\theta}$, $e^{i2\theta}$, $e^{i4\theta}$, etc.

Thus the orthogonal decomposition for axially symmetric processes reduces to the familiar Fourier series.

As for the homogeneous and stationary cases, it is most convenient to carry out the decomposition in the periodic direction first, then proceed with the remainder of the decomposition. Suppose, for example, the field is stationary in time, periodic in the azimuthal direction, θ, and inhomogeneous in the other two directions, say x and r. Then, the Fourier decomposition in the time variable yields the cross-spectrum, which must in turn be represented in terms of its azimuthal Fourier modes (Glauser 1987), i.e.,

$$A_{ij}(x,x',r,r';\omega,\alpha) = \frac{1}{2\pi}\int_0^{2\pi} \Phi_{ij}(x,x',r,r',\theta'-\theta;\omega)e^{-i\alpha(\theta'-\theta)}d(\theta'-\theta) \tag{20}$$

where Φ_{ij} is the cross-spectrum and A_{ij} is its decomposition into azimuthal modes designated by the mode number $\alpha = 0,1,2$, etc.

The orthogonal decomposition is thereby reduced to the following integral equation,

$$\iint_{\text{Domain}} A_{ij}(x,x',r,r';\omega,\alpha)\ \psi_j(x',r';\omega,\alpha)dx'dr' = \lambda(\omega,\alpha)\ \psi_i(x,r;\omega.\alpha) \tag{21}$$

The eigenspectra and eigenvectors are now functions of the frequency ω and the mode number α, and the eigenvectors are functions of the inhomogeneous variables x and r as well.

In a manner similar to the homogeneous field discussed above, the Fourier decomposed velocity field, $\hat{u}_i$, can be represented in terms of the eigenfunctions by

$$\hat{u}_i(x,r;\omega,\alpha) = \sum_n a_n(\omega,\alpha)\ \psi_i^{(n)}(x,r;\omega,\alpha) \tag{22}$$

where the coefficients for each realization of the velocity field are computed from

$$a_n(\omega,\alpha) = \iint_{\text{Domain}} \hat{u}_i(x,r;\omega,\alpha)\ \psi_i^{(n)*}(x,r;\omega,\alpha)dxdr \tag{23}$$

As before, the coefficients are related to the eigenspectra by

$$\langle a_n(\omega,\alpha) a_m^*(\omega,\alpha) \rangle = \lambda_n(\omega,\alpha)\ \delta_{mn} \tag{24}$$

The instantaneous and θ-dependent field can be recovered from equation (22) by performing the inverse transforms on $\hat{u}_i$, i.e.,

$$u_i(x,r,\theta,t) = \sum_n \left\{ \sum_{\alpha=0}^{\infty} e^{i\alpha\theta} \left[\int_{-\infty}^{\infty} e^{i\omega t}\ a_n(\omega_i{}^{(n)}(x,r;\omega,\alpha) \right] \right\} \tag{25}$$

Immediately, we see that we have the same problem we encountered above where the individual eigenfunctions arising from the decomposition in the inhomgeneous directions do not contribute in a simple way to the velocity field, since each eigenfunction is multiplied by a random coefficient which is a function of the frequency and mode number. Thus, just as for the homogeneous and stationary case, it is not possible without additional assumptions to identify a single coherent structure since its composition will continually evolve. The situation in the θ-direction is a little better than before, at least conceptually, since it is easier to visualize the roles of individual azimuthal modes than it is continuously varying ones. The zeroth mode is simply the same for all θ, while the higher order modes have a sinusoidal variation with θ, the number of oscillations corresponding to the mode number.

It is easy to see that <u>the presence of a coefficient which is dependent on both mode number and frequency allows an individual coherent structure to be transformed from one modal shape to another during its life-time.</u> If we believe that coherent structures are vortical structures which evolve in time and space, the OD would seem to have retained a feature essential for a successful representation. This situation can be contrasted with one where the coherent structure simply stays in one place and decays, which could conceivably occur only when the flow is completely inhomogeneous and non-stationary.

Given that the effect of periodicities, homogeneities, and stationarity is to make the eigenfunctions separable into harmonic and non-harmonic parts, it is difficult to see any other way for the OD to be able to describe such an evolution than by the result we have obtained. This point is to be emphasized here, because some have suggested that these random coefficients represent a deficiency of the decomposition, and have discouraged its use. Certainly, the problem is more complex than we might have wished, but it does appear that the kinematics of an evolving flow are accurately described, and probably in as simple manner as possible.

PARTIAL DECOMPOSITIONS AND EXPERIMENTS

In spite of the fact that the OD and most of our understanding of it have been around now for 22 years or so, there has been little real opportunity to find out what it tells us about coherent structures in turbulence. This is because acquiring the information on the cross-correlation or cross-spectral tensor throughout the flow field has proven beyond the vision of most experimenters. The situation has, however, improved remarkably over the past few years. First, two experimental programs have been completed which contain enough information to carry out the decomposition in three out of the four space-time dimensions necessary for a complete description (Herzog 1986, Glauser 1987). These are reviewed below. Second, the advent of full Navier-Stokes computer simulations of simple flows has opened up new possibilities which are beginning to be exploited (e.g., Moin 1984).* Unfortunately, a considerable effort has been spent (or mis-spent) trying to figure out what the eddies look like, instead of using the information we already have to study the far more important dynamical characteristics of coherent structures. Happily, this is also beginning to change (e.g.,Aubry 1987).**

Before reviewing the results of the experimental programs, it is worth reviewing precisely what information is desirable to fully exploit the OD, and what one gives up by only having some of it. As is obvious from equations (3), (11), and (21), sufficient information on the cross-correlation or cross-spectral tensor is necessary to allow the eigenvalue problem to be solved. This will yield the eigenvalues and eigenvectors. Perhaps not so obvious is that then, a complete space-time realization of the velocity field is necessary to allow the coefficients of equations (7), (14), or (23) to be computed. <u>Only when this latter element is included is there sufficient information to follow the space-time evolution of the individual structures.</u> It doesn't require much knowledge of experimental capabilities to see why the application of the OD has been rather slow in developing. It should also be easy to see why the continuing development of numerical simulation capabilities is of great interest in this context. The same can be said for the development of holographic anemometry techniques which allow an entire field to be measured simultaneously.

Even in the absence of such complete information, however, it has been possible to glean a great deal of information about the nature of these two different turbulent flows by applying the OD to experiments which provided only part of the needed information. These experiments will be discussed in the following sections. The intent here is not to present the results in detail, which has already been done elsewhere, but to focus on what even a partial decomposition is able to say about the character of structures in the flow. We begin with the experiment of Glauser in a high Reynolds number axisymmetric jet mixing layer, then consider the experiment of Herzog in a low Reynolds number pipe flow.

THE AXISYMMETRIC JET MIXING LAYER

The experiment of Glauser (1987) was carried out in the mixing layer of an axisymmetric air jet. The Reynolds number based on jet exit velocity and diameter was 10^5, and the Mach number was 0.067. The exit velocity profile was flat and the boundary layer at the lip of the jet was laminar. The experiment was performed using rakes of up to 14 hot-wire probes so that the flow could be simultaneously measured at many locations. The experiment was performed in phases of increasing complexity so that the results of simple decompostions could be studied before carrying out additional measurements. All of the two-point cross-spectra involving the streamwise and radial velocity components were

*Note added to published version: The paper presented by Moser in this meeting provides another example of this.

**Note added to published version: The paper by Aubry et al. and presented by Lumley at this symposium further illustrates this.

measured on a grid with seven positions across the layer, and 25 azimuthal positions from 0 - 180°. Thus in all, 5000 cross-spectra were measured so that with the frequency dependence, 10×10^6 cross-spectral entries were used in the eigenvalue analysis.

The first partial decomposition considered only the radial and time variation of the velocity field. Therefore, since the field was stationary in time, the cross-spectrum as a function of radius and frequency only was used to carry out the eigenvalue problem, i.e.,

$$\int_{R_1}^{R_2} \Phi_{1,1}(r,r';\omega)\psi_1^{(n)}(r';\omega)dr' = \lambda_n(\omega)\ \psi_1^{(n)}(r;\omega) \tag{26}$$

where R_1 and R_2 were chosen to span the mixing layer. Note that the dependence on x,x', and θ,θ' has been suppressed since these were all fixed. Thus the results are equivalent to summing over all the azimuthal modes for a single downstream position.

Figure (1) shows the eigenspectra for the first three modes, and shows clearly the dominant nature of the first which integrates to 40% of the energy in the field. From even the partial decomposition it is possible to reconstruct the velocity spectrum using

$$\Phi_{1,1}(r;\omega) = \sum_n \lambda_n\ |\psi_1^{(n)}(r;\omega)|^2 \tag{27}$$

Figure (2) shows a reconstruction of the one-dimensional spectrum for a position near the center of the mixing layer, and it is apparent that only three terms are required to completely recover it. The individual terms of the spectrum at each position can be integrated with respect to frequency to show what fraction of the variance at each position across the layer is contributed by each term. Figure (3) shows the results of such an integration for the first term. An even more striking illustration of the ability of the OD to capture the details of the flow is given by Figure (4) which shows how a single realization of the real part of the Fourier transform of the instantaneous signal is made up by the individual terms. Even the first term alone has captured most of the salient features, while three terms recover virtually all. In view of this, there really should be no more talk of the inability of the OD retain phase information, or of it smearing out significant events.

A reconstruction of the azimuthal variation of the field, together with the time information, would answer many questions about the behavior of coherent structures in the jet mixing layer. Unfortunately, because data were not acquired at all values of θ and r simultaneously, it was not possible to evaluate the coefficients as in equation (23). Even so, the OD applied to the azimuthal direction provides some tantalizing clues. Figures (5) - (7) show the modal constituents of two components of the velocity correlation tensor, $A_{1,1}$ and $A_{1,2}$. On the high speed (or inside of the mixing layer), both normal and Reynolds stresses show almost a complete dependence on the zeroth order mode, $\alpha=0$ (Figures 5a and 5b), while on the low speed side (or outside) the dependence has

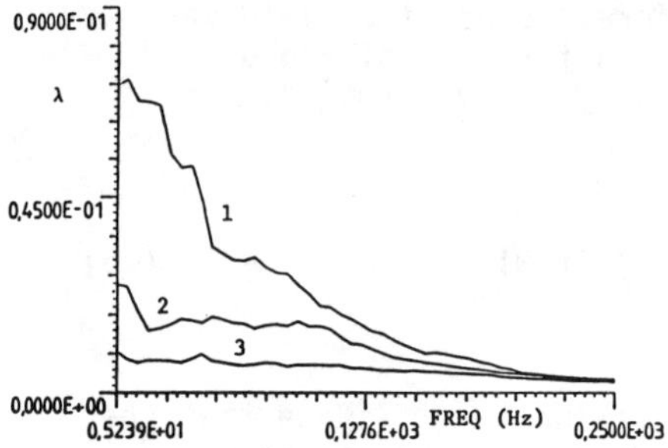

Fig. 1 Eigenspectra for Jet Mixing Layer, x/D = 3 (Glauser 1987).

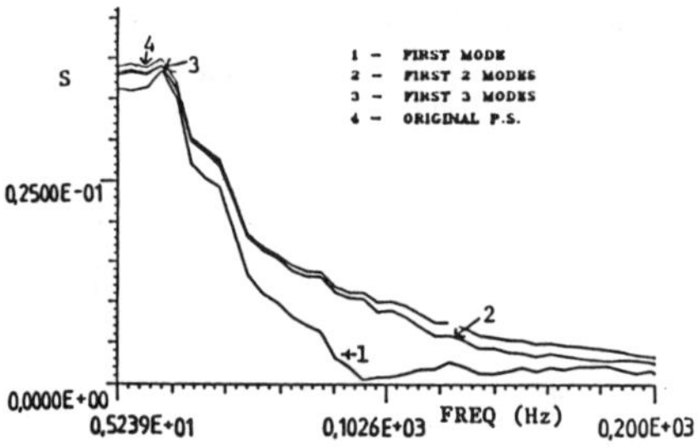

Fig. 2 Reconstruction of Power Spectrum, r/R = 0.46.

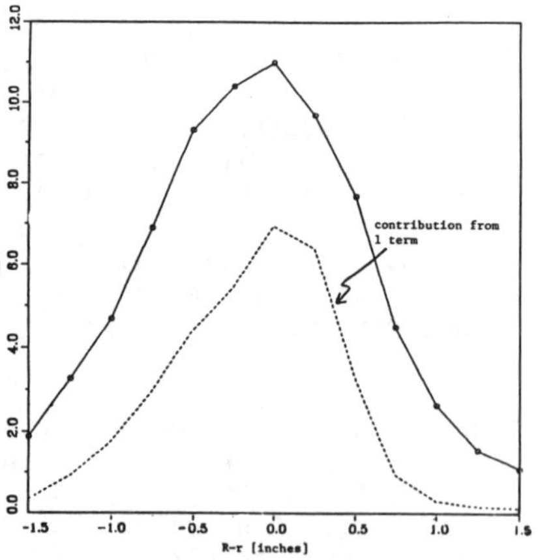

Fig. 3 Total $\langle u^2 \rangle$ and Contribution From First Term in OD.

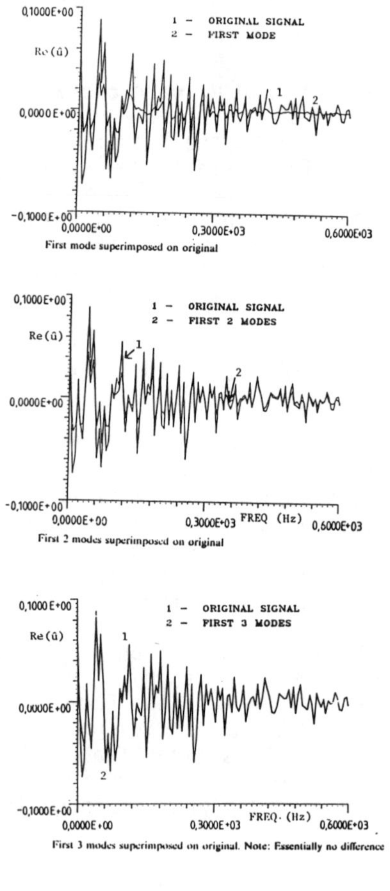

Fig. 4 Reconstruction of $Re(\hat{u}_1(\omega))$.

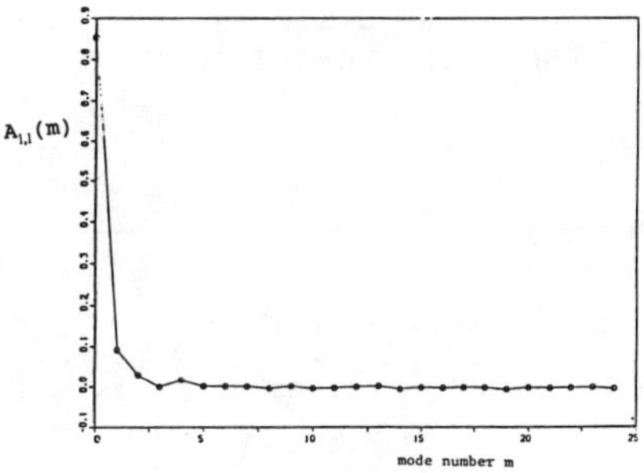

Fig. 5a $A_{1,1}$ at r/D = 0.13.

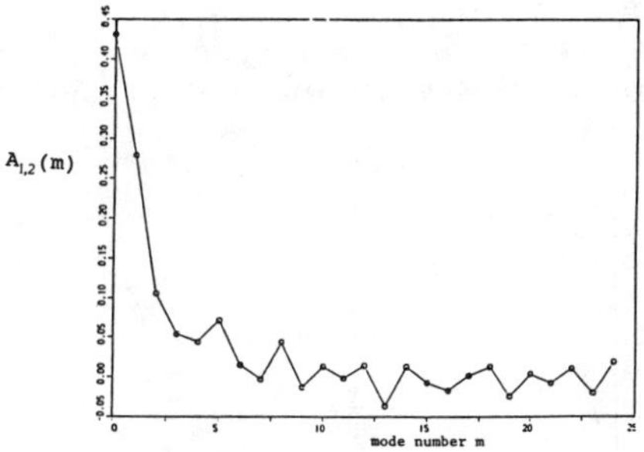

Fig. 5b $A_{1,2}$ at r/D = 0.13.

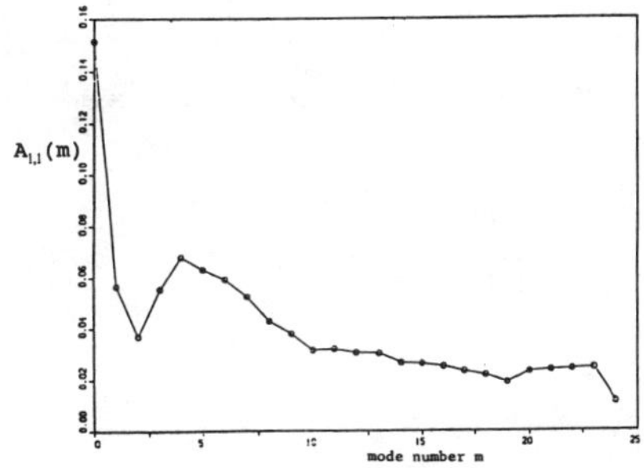

Fig. 6a $A_{1,1}$ at r/D = 0.35.

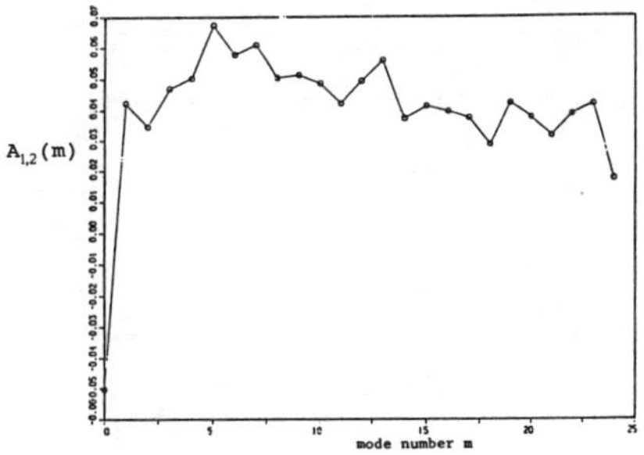

Fig. 6b $A_{1,2}$ at r/D = 0.35.

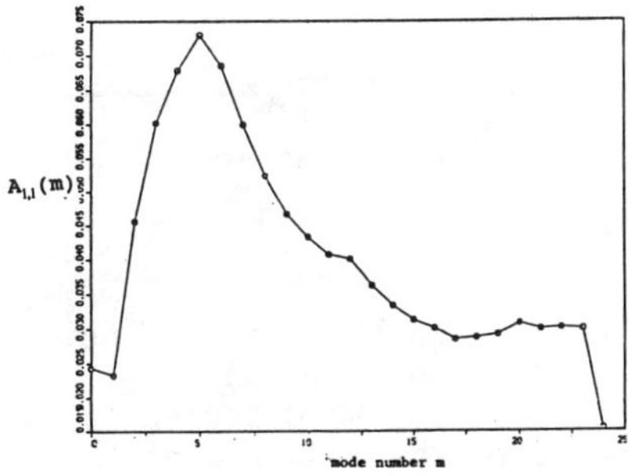

Fig. 7a $A_{1,1}$ at r/D = 0.46.

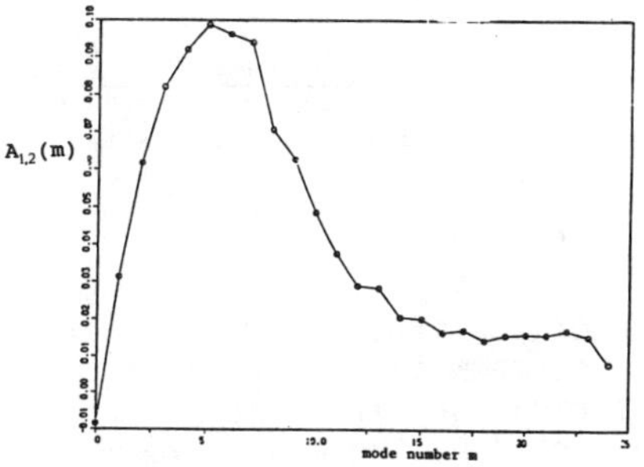

Fig. 7b $A_{1,2}$ at r/D = 0.57.

shifted almost entirely to a band of modes centered around $\alpha = 5$ or 6 (Figures 7a and 7b). In the center (Figures 6a and 6b), the normal stress component, $A_{1,1}$ has contributions from both $\alpha = 0$ and the band at $\alpha = 5 - 6$, but the Reynolds stress, $A_{1,2}$ has nearly uniform dependence on every mode but $\alpha = 0$. The first three eigenspectra for modes 1,2,3 and 5 are shown in Figure (8 a-d). Immediately apparent is the fact that the first eigenspectrum for each mode is dominant, the same being true for modes 4, 6, 7, and 8 which are not shown here. Also clear is the peaked nature of the lower order eigenspectra and the shift toward lower frequencies with increasing mode number.

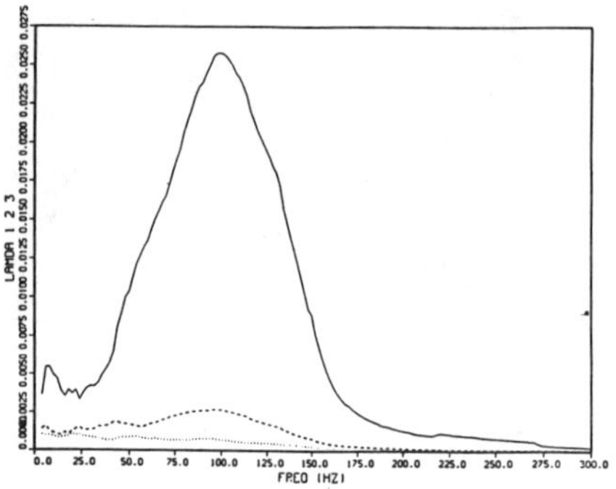

Fig. 8a First 3 Eigenspectra, Mode 0.

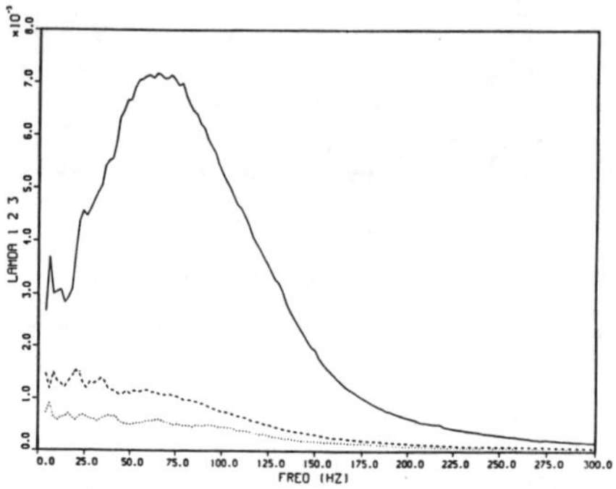

Fig. 8b First 3 Eigenspectra, Mode 1.

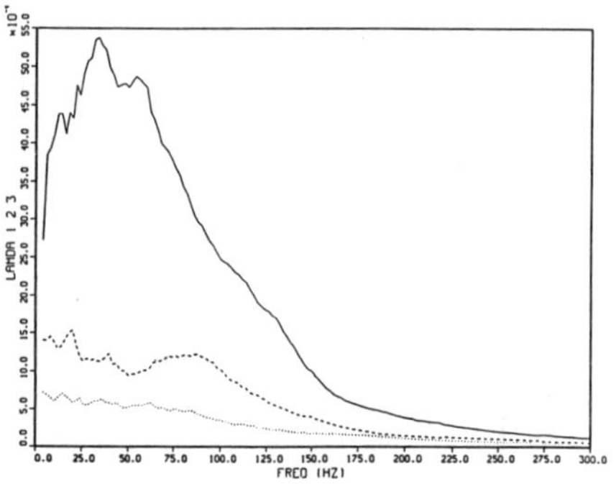

Fig. 8c First 3 Eigenspectra, Mode 2

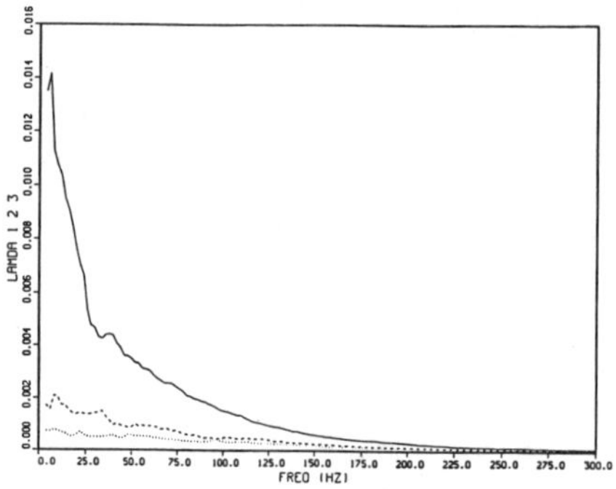

Fig. 8d First 3 Eigenspectra, Mode 5.

One of the problems with a partial decomposition which does not include the streamwise variation is that if time is included, it has contributions from the streamwise variation because the disturbances are convected by the mean stream. Thus the shift of the higher order eigenspectra to lower frequencies probably is consistent with the dominance of the higher order modes on the low speed side of mixing layer. Glauser and George (1987b) were able to use these obvervations to propose a life-cycle for the generation and decay of the coherent structures in the axisymmetric jet mixing layer. In particular, they suggested that:

"Vortex ring-like concentrations arise from an instability of the base flow, the induced velocities from vortices which have already formed providing the perturbation for those which follow. These pairs of vortices then behave like the text book examples of inviscid rings, the rear vortex ring overtaking the vortex ring ahead of it, the rearward vortex being reduced in radius and the forward one being expanded by their mutual interaction. The rearward ring is stabilized by the reduction in its vorticity (by compression), thus the predominance of the 0th mode on the high speed side (core region). The forward ring has its vorticity increased by stretching as it expands in radius. This narrowing of its core while the radius is expanding causes the vortex to become unstable, thus the predominance of the 4th-6th modes from the center of the shear layer outwards. The continued effect of the rearward vortex on the forward one and the interaction of the now highly distorted ring with itself, accelerates the instability until its vorticity is now entirely in small scale motions, in effect, an energy cascade from modes 4-6 all the way to the dissipative scales. This incoherent turbulence is swept from the outside where it has been carried, back to the center of the mixing layer as the still intact rearward vortex passes. It is this collecting of the debris, both small scale vorticity and fluid material, which has been recognized as 'pairing'. The entire process is repeated as a new rearward vortex overtakes and destabilizes the one ahead of it."

In order to evaluate whether or not this picture is correct will require application of more complete versions of the OD. The most useful and straight-forward experiment would yield information on enough points in θ and r simultaneously to enable the $a_n(\omega,\alpha)$ to be computed. Then the instantaneous field could be reconstructed in order to see the sequence by which the various modes contribute in time. Certainly there is room here also for a refined application of the shot-noise decomposition, the discussion of which has been avoided in order to avoid confusing it with the OD. Finally, and perhaps most interestingly, the decomposition should be applied to the streamwise direction so that the space-time evolution can be studied together without the problem introduced by the mean convection.

THE VISCOUS SUBLAYER

The second and last experiment to be considered here is that carried out by Herzog (1986) in the wall region of low Reynolds number turbulent pipe flow. The experiment was an extension of the earlier one of Bakewell (1966). The experiment was realized in a 12" pipe flow using glycerine as the working fluid. The Reynolds number based on pipe diameter and bulk flow velocity was 8750, and the flow was tripped at the pipe entrance to insure fully developed flow at the measurement locations. All of the measurements were taken below $x_2^+ = 40$. While the experimental techniques were quite different than those used by Glauser (1987), the overall complexity of the experiment was of comparable magnitude.

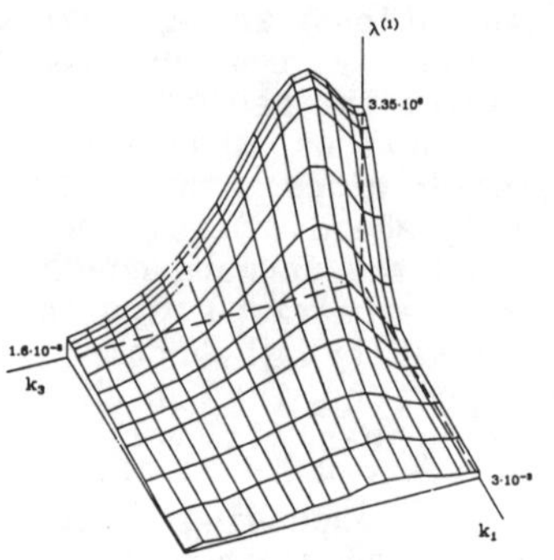

Fig. 9 Dominant Eigenspectrum for Wall Layer (Herzog 1986).

In Bakewell's original experiment only the cross-correlations involving the streamwise velocity component were measured as functions of the time, radial and spanwise coordinates. Herzog obtained all components of the cross-correlations, and included the streamwise dependence, but not the time. Therefore all correlations in the latter experiment correspond to zero time lag. Thus the measured correlations were functions of the two homogeneous coordinates, r_1 and r_3, representing the streamwise and spanwise separations, and the inhomogeneous coordinates, x_2' and x_2. (Note that it was decided to treat the spanwise direction as homogeneous instead of utilizing the azimuthal symmetry of the flow because of the small thickness of the sublayer relative to the pipe circumference.) Figure (9) shows the lowest order eigenvector (denoted by Herzog as $\lambda^{(1)}$) while Figures (10 a and b) show the cross-sections through the k_1 and k_3 axes of the first three eigenspectra. As with Bakewell's and Glauser's experiments, the first eigenspectrum is dominant at all but the very highest wavenumbers. Figures (11a through d) show how the variances of the three velocity components and the Reynolds stress are distributed among the first three terms of the OD. Particularly of interest is that the first term accounts for almost all of the streamwise velocity variance and the Reynolds stress near $x_2^+ = 20$.

The off-axis peak in the lowest order eigenspectrum (Figure 9) at $k_3^+ = 0.0035$ corresponds to a span-wise wavelength in wall units of $\lambda^+ = 286$. (Note that Herzog defines $k=1/\lambda$ omitting the usual 2π.) Herzog notes that the distance between the nodes associated with this wavelength would be half of this value which is very close to the oft-reported wall streak spacing. It is also clear from the bandwidth associated with the streamwise wavenumber variation that the streamwise extent of the coherent structure is very large indeed, perhaps as much as an order of magnitude greater than its spanwise extent. It should be noted that these observations are <u>not</u> dependent on an <u>ad hoc</u> application of the shot-noise decomposition, but depend only on the original hypothesis which led to the OD itself.

Note that it can <u>not</u> be inferred directly from the results of the OD that there are counter-rotating rolls of the type inferred originally by Bakewell (1966) and subsequently using somewhat different arguments by

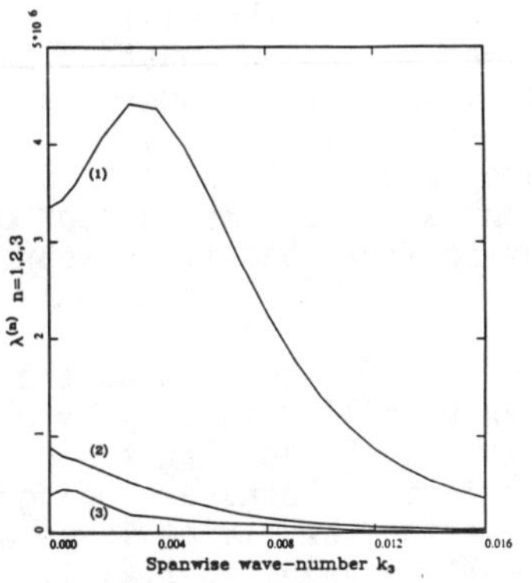

Fig. 10a First 3 Eigenspectra as Function of spanwise wavenumber, k_3.

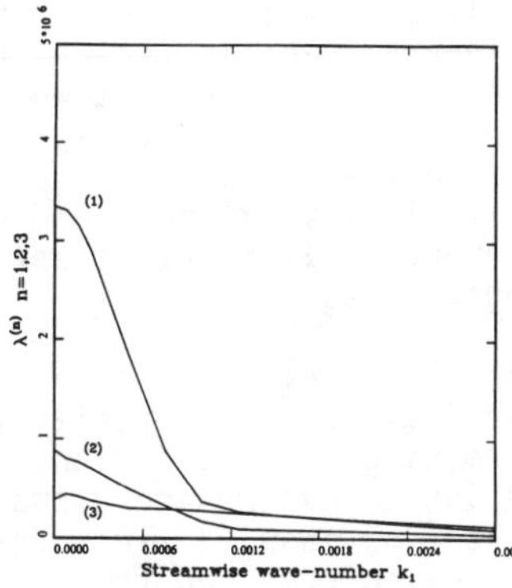

Fig. 10b First 3 Eigenspectra as Function of streamwise wavenumber, k_1.

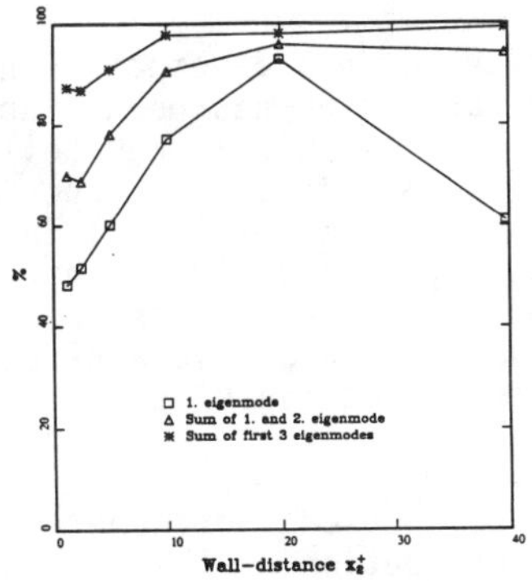

Fig. 11a Contributions of OD terms to $\langle u^2 \rangle$.

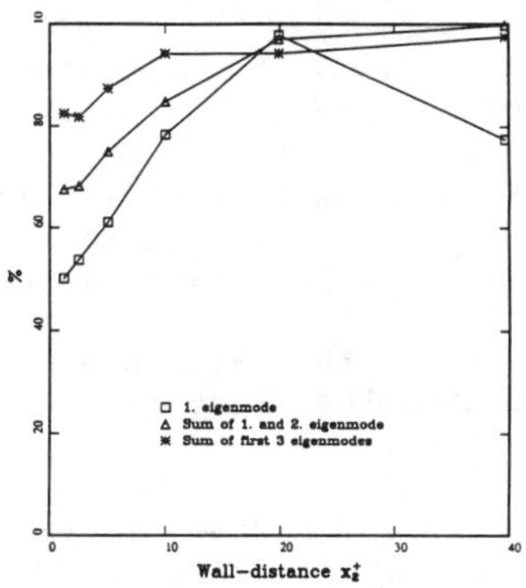

Fig. 11b Contributions of OD terms to $\langle uv \rangle$.

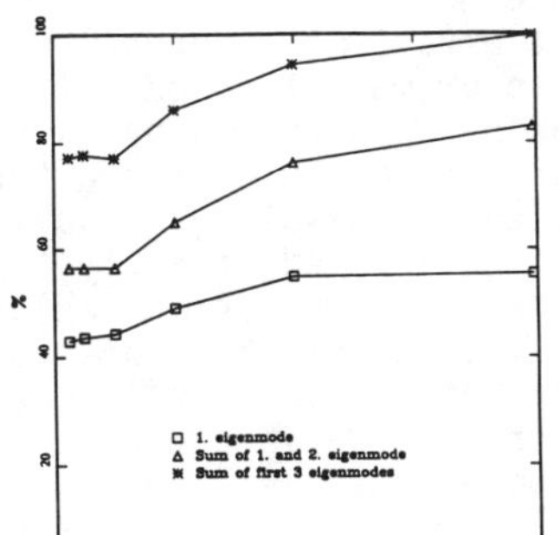

Fig. 11c Contributions of OD terms to $\langle v^2 \rangle$.

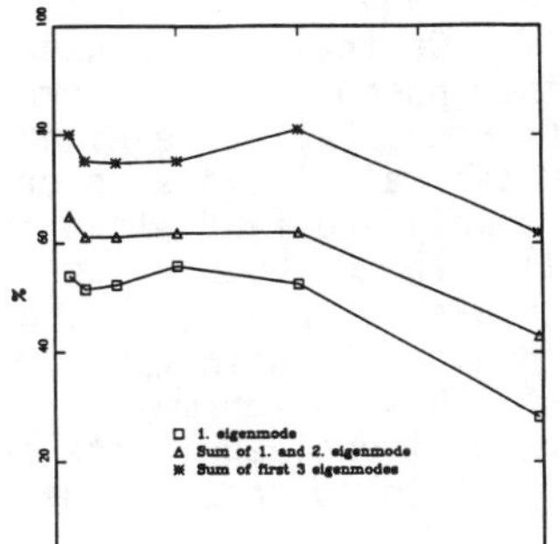

Fig. 11d Contributions of OD terms to $\langle w^2 \rangle$.

Herzog (1986). These inferences depended entirely on the introduction of other assumptions about the field and on the use of the shot-noise decomposition. Whether or not such rolls are really present depends very much on the nature of the coefficients of the reconstructed field, which in this case will depend on k_1,k_3, and also ω since the field is stationary. If, in fact, the eigenfunctions add up in constant phase as these authors suggest, then there will be counter-rotating rolls. On the other hand, when the full time-dependence of the stationary field is included with effect of the streamwise and spanwise homogeneities, the longitudinally extended and highly coherent structure may appear quite different. It might appear so different in fact that it is scarely recognized as such in a real flow.*

It should be clear that, as for the mixing layer, the part of the puzzle as to what the OD is really telling us can be filled in only by determining the coefficients for a realization of the flow field. Only by doing so can we see visually how the building blocks are being added by the flow itself. This determination, as for the mixing layer above, requires information from many points simultaneously. Certainly this would seem well within the scope of modern numerical simulations.

SUMMARY AND CONCLUSIONS

The elements of the application of proper orthogonal decomposition techniques have been reviewed. Particular attention has been placed on what can be gained from the decomposition, and on what it can tell us about the nature of coherent structures in turbulence. Two experiments have been used to illustrate the theory, one by Glauser (1987) in the axisymmetric jet mixing layer,and the other by Herzog (1986) in the viscous sublayer of a turbulent pipe flow. Both experiments show clear evidence of the existence of coherent structures, and give a strong indication that they play an important role in the dynamics of the turbulence.

There are, of course, many ways to detect the presence of coherent structures. If detection were the only contribution of the orthogonal decomposition, it would not be worth the effort. What the orthogonal decomposition provides is a mathematical description which can be used for further analysis. The results of a decomposition can be used in the governing equations to understand the interactions of the flow with itself. Or they can be used to study the effect of coherent structures on other phenomena of interest, like noise or mass transport (v. Arndt and George 1974). The eigenfunctions can be used as the basis for other dynamical calculations in order to capture the essence of the turbulence in a simplified (and often more instructive) analysis (v. Aubry 1987). Finally, the orthogonal decomposition can be applied directly to the Navier-Stokes equations to develop a hierarchy of equations which can be solved directly with appropriate closure models for the higher order terms. This latter may be the most important of all, for it holds forth the hope of engineering calculations which include in a direct manner the dynamical influences of the coherent structures.

*Note added to published version: Moser's comments in this meeting are of considerable interest here.

ACKNOWLEDGEMENTS

The author is particularly grateful to Professor Mark Glauser (of Clarkson University) with whom he has collaborated for a number of years. He would also like to express his appreciation to Dr. Siegfried Herzog (of the Pennsylvania State University) for permission to use the figures from his work, and to Dr. Hans Saustrup Kristensen (of the Danish Technical University) for calling his attention to the cartoon by Piet Hein. Finally, support of the Danish - American Fulbright Commission, the Danish Technical University, the University at Buffalo, SUNY, and the Calspan UB Research Center during this sabbatical period is gratefully acknowledged.

REFERENCES

Arndt, R.E.A. and George, W.K. (1974) Proc. Second Interagency Symp. on Univ. Res. in Transp. Noise, N.C. State Univ., Raleigh, N.C., Jun 5-7. 1984.

Aubry, N. (1987) Ph.D. Dissertation, Dept of M.E., Cornell U., Ithaca.

Bakewell, H.P (1966) Ph.D. Dissertation, Penn. State U., Univ. Pk., Pa.

Glauser, M.N. (1987) Ph.D. Dissertation, Department of Mechanical and Aerospace Engineering, University at Buffalo, SUNY, Buffalo, NY.

Glauser, M.N. and George, W.K. (1987) Proc. 6th Symp. on Turb. Shear Flows, Toulouse.

Herzog, S. (1986) Ph.D. Dissertation, Dept. of M.E., Cornell Univ., Ithaca, N.Y.

Lumley, J.L. (1967) Atm.Turb. and Radio Wave Prop. Nauka, Moscow.

Lumley, J.L (1982) Panel Discussion, APS/DFD Ann. Mtg, Rutgers Univ., Rutgers, N.J.

Moin, P. (1984) AIAA 22nd Aersp. Sci. Mtg., Reno, Nev.

Effect of Reynolds Number on the Organized Motion in a Turbulent Boundary Layer

R. A. ANTONIA, L. W. B. BROWNE, and D. K. BISSET
Department of Mechanical Engineering
University of Newcastle
N.S.W., 2308, Australia

ABSTRACT

Measurements have been made in a zero pressure gradient turbulent boundary layer in an attempt to quantify the influence of Reynolds number on features of the motion, which is organised on a scale comparable to the thickness of the layer, in the plane of main shear. These features have been obtained with an array of eight X-probes aligned in a direction normal to the wall, each probe measuring velocity fluctuations u (streamwise direction) and v (normal to the wall). Information derived from the u and v data includes conventional space-time velocity correlations, instantaneous and conditional particle paths and spanwise vorticity contours. The effect of Reynolds number on the spatial extent of isocorrelation contours is more important for u than v. No significant dependence on Reynolds number is observed for conditional particle paths but contours of conditional spanwise vorticity extend to a larger normalised distance from the wall at small Reynolds number. Contributions of the organised motion to the Reynolds stresses tend to be more important as the Reynolds number decreases.

1. INTRODUCTION

The influence of Reynolds number on the structure of a turbulent boundary layer has attracted a reasonable amount of attention. Not surprisingly, earlier work was concerned with the effect of Reynolds number on quantities such as the mean velocity and the skin friction coefficient. Coles (1962) observed that the strength of the "wake" region, as measured by the maximum departure from the log-law, asymptoted to a constant value only when the Reynolds number R_θ ($\equiv U_1\theta/\nu$, U_1 is the free stream velocity, θ is the momentum thickness, and ν is the kinematic viscosity of the fluid) was in excess of about 5000. Several investigators (e.g. Purtell et al, 1981; Subramanian and Antonia, 1981; Murlis et al, 1982; Erm et al, 1987) have established that the log-law constants are essentially unaffected by the Reynolds number. Purtell et al (1981) considered the effect of R_θ on the intensity of the longitudinal velocity fluctuation. Although distributions of this intensity across the boundary layer exhibit a fair degree of similarity when plotted using the friction velocity U_τ ($\equiv \tau_w^{1/2}$, where τ_w is the kinematic wall shear stress) and the boundary layer thickness δ, a small increase in u'/U_τ (a prime denotes an rms quantity) can be discerned in the outer layer as R_θ increases. This increase is also seen in the data of Subramanian and Antonia (1981), Murlis et al (1982) and Erm et al (1987). There is also an increase in v'/U_τ but at a much smaller rate than u'/U_τ. Murlis et al found a monotonic increase of $\overline{u^2}/\overline{v^2}$ up to $R_\theta \simeq 5000$, interpreting it as a "monotonic decrease in the general efficiency of turbulent mixing". Subramanian and Antonia (1981) found that $\overline{u^2}/\overline{v^2}$ varied little between R_θ = 1500 and R_θ = 7000 but increased at R_θ = 990. The data of Murlis et al and Erm et al indicate that $-\overline{uv}/U_\tau^2$ generally increases with R_θ, especially in the outer layer. No systematic variations can be detected in the distributions of $-\overline{uv}/U_\tau^2$ reported by Subramanian and Antonia (1981). It is difficult to comment on some of the previous discrepancies since, for a given R_θ, the quantitative agreement between hot wire data (especially X-wire data) obtained by different investigators is generally poor. Each set of investigators did however conclude that the influence of R_θ was qualitatively reflected in their data.

Flow visualisation results (Falco, 1974; Head and Bandyopadhyay, 1981) have indicated that structural flow features are affected by the Reynolds number in somewhat more spectacular fashion that has been inferred from hot wire measurements. Falco observed the effect of R_θ on typical eddies, the latter scaling on wall variables in the range 900 < R_θ < 20000. Head and Bandyopadhyay described the R_θ influence in the context of vortex pairs or hairpin vortices originating in the wall region and extending through most of the layer with an inclination of about 45° to the wall. For $R_\theta \gtrsim 2000$, Head and Bandyopadhyay describe large scale features of the motion as random arrays of highly elongated hairpin vortices whereas for $R_\theta \lesssim 800$, the large scale motion consists of vortex loops. Head and Bandyopadhyay indicated that the hairpin vortex description is consistent with the observed sudden decrease in temperature which can be followed from just outside the viscous sublayer (Chen and Blackwelder, 1978; Subramanian et al, 1982) to almost the edge of the layer. Antonia et al (1982) exploited the existence of this feature to examine the R_θ dependence of both velocity and temperature fields conditioned on the spatially coherent temperature change. The resulting conditional averages suggested a significant dependence on Reynolds number up to $R_\theta \simeq 3000$.

The previous results imply that the interaction between inner and outer regions of the boundary layer is likely to depend on the Reynolds number. This matter is important because of the bearing it may have on the appropriate scaling for the bursting frequency. Andreopoulos et al (1984) reported that statistics of u in the near wall region of a turbulent boundary layer can depend on the Reynolds number up to $R_\theta \simeq 10^4$. Such a dependence needs to be taken into account (e.g. Shah and Antonia, 1988) when distinguishing between wall variables (U_τ, ν/U_τ), outer variables (U_1, δ) or mixed variables. [In Alfredsson and Johansson's 1984 proposal, the time scale is the geometric mean of the viscous time scale (ν/U_τ^2) and the outer (δ/U_1) time scale].

A well known disadvantage of flow visualisation is the difficulty in interpretation when the Reynolds number becomes moderately large. The simultaneous use of several X-probes arranged in one or more flow cross-sections provides quantitative information as well as a "pseudo-visualisation" of the main features of the large scale motion. In principle, this approach also allows information to be collected on the small scale motion, a virtually impossible task in flow visualisation.

The main short term aims of the present work, which uses an array of X-probes in the (x,y) plane, are to extend the investigation by Antonia et al (1982), and quantify the flow features identified in the visualisation of Head and Bandyopadhyay. Longer term aims are to provide further insight into the likely relationship between bursting and the large scale motion at different Reynolds numbers and to consider the effect of R_θ on the three-dimensionality of the organised motion. Here, we first discuss the influence of R_θ on conventional velocity correlations and subsequently consider the Reynolds number variation of several topological features of the motion in the (x,y) plane. The effect on the proportion of the Reynolds stresses contributed by the organised motion is considered also.

2. EXPERIMENTAL DETAILS/CONDITIONS

The wind tunnel is of the open-return blower type; a double inlet centrifugal blower is followed by a diffuser, settling chambers, the latter including a honeycomb and a number of screens, a 6 to 1 two-dimensional contraction and a 5.4 m long working section with a height of 0.89 m and a width of 0.15 m at the exit from the contraction. The boundary layer develops on one of the vertical walls of the working section, the opposite vertical wall adjusted to achieve nominally zero pressure gradient at a particular free stream velocity U_1. The boundary layer is tripped with a 1.5 mm diameter rod, located on the wall at the contraction exit (x = 0), immediately followed by a 15 cm long strip of No. 40 sandpaper. There is provision for heating the boundary layer wall; the wall was not heated for the present experiments.

Measurements were made using an array of 8 X-probes, aligned in the y direction (normal to the wall) at the same x and z locations. The distance between adjacent X-probes is approximately constant, equal to about 5 mm. The 5 μm diameter hot wires were etched from Wollaston (Pt-10% Rh), the length to diameter ratio of each wire being of the order of 250. The wires were operated with constant temperature circuits, built in-house, at an overheat of 0.6. Output signals from these circuits were passed through buck and gain circuits and then low-pass filtered before digitising into a PDP 11/34 computer. The filter cut-off frequency was 1750 Hz for each channel. The sampling frequency (per channel) was 3000 Hz. Although slightly smaller than the Nyquist frequency of 3500 Hz, the magnitude of the sampling frequency was the maximum value attainable with the 16 channel, 12 bit data acquisition system. Digital data files of approximately 50 s duration were stored on magnetic tape and processed on a VAX 8550 computer.

The X-probe array was traversed across the boundary layer using a mechanism with a least count of 0.01 mm. The initial distance of the X-probe nearest the wall was determined using the reflection method and a theodolite with an accuracy estimated at ±0.02 mm. Velocity and yaw calibrations of the X-probes were done in the free stream using a data logger-personal computer combination. The data logger was operated at a sampling frequency of 10 Hz. Zero pressure gradient conditions in the working section were checked using wall static pressure tappings and a Baratron pressure transducer. For calibrations, a pitot static tube located in the free stream was connected to the Baratron transducer.

3. ESTABLISHMENT OF FLOW CHARACTERISTICS

Measurements were made at a distance x of 4.3 m at four values of U_1 : 3, 4.2, 10.9 and 19.7 m/s. The corresponding values of the momentum thickness Reynolds number R_θ are 1360, 2180, 6030 and 9630. Prior to taking data with the X-probe array, mean velocity profiles were measured with a pitot tube at : (i) several positions and values of x (2 to 4.5 m) and one free stream velocity (9.1 m/s); and (ii) several values of U_1 (2 to 24 m/s) at x = 4.3 m. Wall shear stress values were determined from these profiles by the Clauser chart method. They were also obtained with a Preston tube. The two approaches yielded values of the skin friction coefficient c_f (= $\tau_w/\tfrac{1}{2}U_1^2$) that agreed to within ±3.5%. The Preston tube values, shown in Table 1 for the Reynolds numbers of interest here, are in good agreement with

TABLE 1. Mean Velocity Characteristics

U_1	R_θ	c_f	c_f	δ	δ^*	θ	$\Delta\bar{U}_{max}/U_\tau$	$\Delta\bar{U}_{max}/U_\tau$
m/s		Preston Tube	Coles 1962	mm	mm	mm		Coles 1962
3	1360	0.0040	0.0040	59	10	6.8	1.5	1.8
4.2	2180	0.0034	0.0036	68	11	7.8	1.7	2.2
10.9	6030	0.0029	0.0029	66	11.6	8.3	2.8	2.8
19.7	9630	0.0027	0.0026	64	9.9	7.3	2.8	2.7

those obtained by Coles (1962) for corresponding values of R_θ. The maximum value $\Delta\bar{U}_{max}/U_\tau$ of the departure from the logarithmic profile in the outer or "wake" region was estimated using the relation $\bar{U}/U_\tau = \kappa^{-1} \ln yU_\tau/\nu + C + \Delta\bar{U}/U_\tau$, where κ is the Kármán constant (= 0.41), and C = 5.1. Values of $\Delta\bar{U}_{max}/U_\tau$ are in reasonable agreement with Coles' values for corresponding Reynolds numbers. In particular, for the two higher Reynolds numbers, $\Delta\bar{U}_{max}/U_\tau$ is very close to Coles' asymptotic value of 2.7 ($\simeq 2\Pi/\kappa$, where the parameter Π is equal to 0.55). Values of $\Delta\bar{U}_{max}/U_\tau$ and c_f deduced from mean velocity profiles at different values of x and a constant U_1, were found to be in reasonable agreement with Coles' $\Delta\bar{U}_{max}/U_\tau$ vs R_θ and c_f vs R_θ variations. These results indicate that the evolution of the boundary layer is normal, in the sense used by Coles.

Mean velocity profiles inferred from the X-probe array were in reasonable agreement with the pitot tube measurements, providing a useful indication of correct performance of the wires. Profiles of $\overline{u^2}$, $\overline{v^2}$ and $\overline{uv}$ were obtained across the boundary layer by moving the array to several locations in the flow. The resulting distributions were reasonably smooth, even in the overlap region between consecutive locations of the array, further corroborating the satisfactory performance of the 16 hot wires of the array. Profiles (not shown here) of the Reynolds normal stresses, normalised by U_τ^2 and plotted against y/δ, indicated practically no difference between the two highest Reynolds numbers. The distributions of $-\overline{uv}/U_\tau^2$ (Figure 1) at the two largest values of R_θ are in reasonable agreement with each other and with Klebanoff's (1955) distribution ($R_\theta \simeq 8000$). Reasonable agreement is also observed for the two smallest values of R_θ but these distributions fall below those at the larger Reynolds numbers in the outer layer and exhibit a larger constant stress region close to the wall. The qualitative behaviour of the data in Figure 1 is in closer agreement with that of Murlis et al (1982) or Erm et al (1987) than Subramanian and Antonia (1981).

4. SPACE-TIME CORRELATIONS OF VELOCITY FLUCTUATIONS

Before applying conditional criteria to the X-probe data, conventional (Reynolds averaged) two-point velocity correlations were first obtained. Apart from complementing the one point statistics mentioned in Section 3, these correlations can be compared with previously published data. Such a comparison provides a further check of the present measurements and, more importantly, allows the influence of Reynolds number on the correlations to be examined.

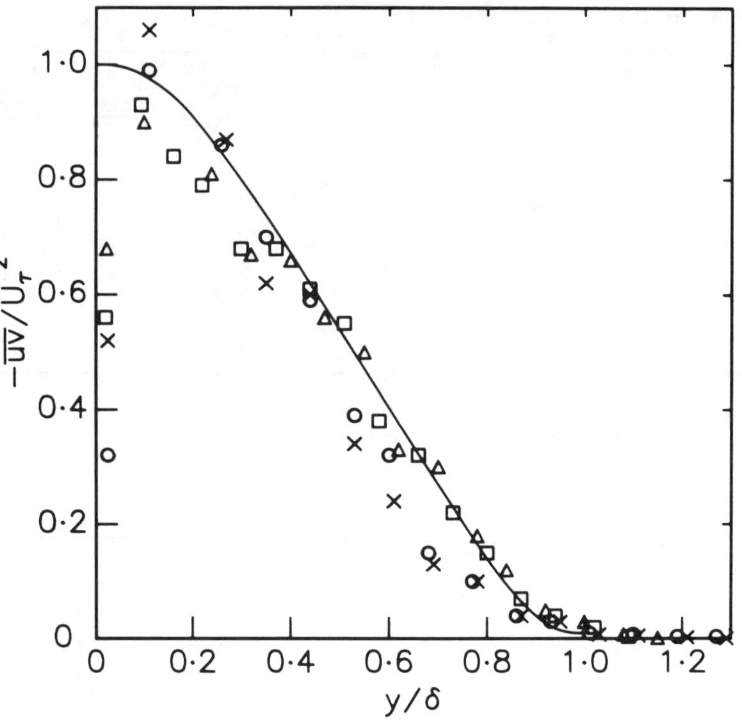

FIGURE 1. Reynolds shear stress distributions. ×, R_θ = 1360; ○, 2180; Δ, 6030; □, 9630.

Two-point correlation coefficients are defined by

$$R_{pq}(y_0,y,\tau) = \frac{\overline{p(y_0,t)\ q(y,t+\tau)}}{p'\ q'}$$

where p and q stand for velocity fluctuations (u or v, each with zero mean), the subscript zero refers to the position of the reference probe and τ is the time delay. In the present experiment, x_0 = x and z_0 = z so that R_{pq} is not expected to depend on any of these quantities if homogeneity is assumed. This assumption is expected to be good for z and reasonable for x. As noted by Kovasznay et al (1970) and Blackwelder and Kovasznay (1972), no homogeneity is expected for y, especially in the wall region, and R_{pq} depends on both y and y_0. Contours of constant R_{uu} and R_{vv} are shown in Figures 2 (R_{uu}) and 3 (R_{vv}) for the highest Reynolds number and three values of y_0/δ. The time delay τ has been normalised using U_1 and δ, as in Kovasznay et al (1970)

There is a striking difference between the R_{uu} contours in Figure 2a and the R_{vv} contours of Figure 3a for y_0/δ = 0.023, obtained when the probe nearest the wall is taken as reference. Whereas the u isocorrelations extend to relatively large values of $\tau U_1/\delta$ and y/δ and are inclined to the wall* (the

*This behaviour is similar to that inferred from the earlier R_{uu} data of Favr et al (1957) which were interpreted (Sternberg, 1967) in terms of eddies aligned downstream at an angle to the surface.

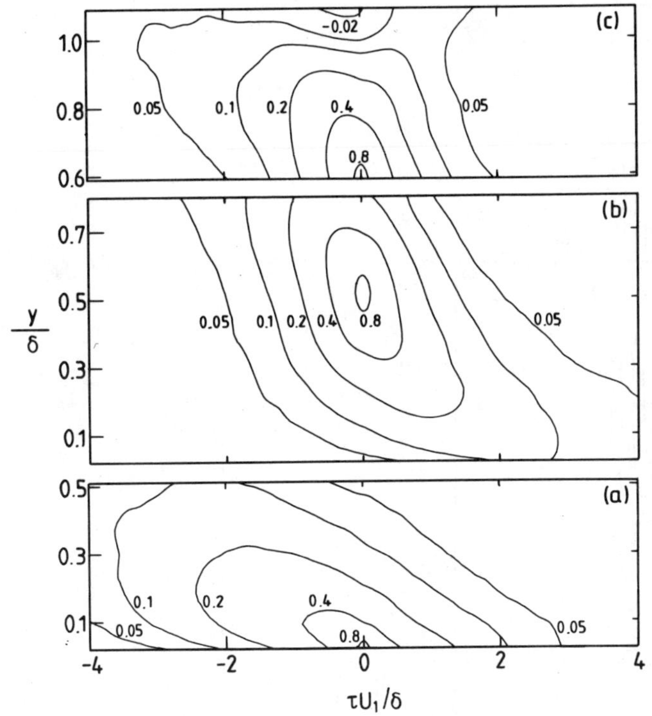

FIGURE 2. Isocorrelation contours of u at R_θ = 9630. (a) y_0/δ = 0.023; (b) 0.51; (c) 0.59.

inclination is smaller than it appears in the figure since the vertical scale is five times as large as the horizontal scale), the spatial extent of the v isocorrelations is limited, these contours appearing almost aligned in the y direction. The limited extent of R_{vv} underlines the inhibiting influence of the wall on the v fluctuation. For y_0/δ = 0.51 (the reference probe is near the centre of the array), the inclination of R_{uu} to the wall (Figure 2b) is larger for $y > y_0$ than $y < y_0$. The positive v contours (Figure 3b) are aligned in the y direction but negative contours begin to appear at large y. When the whole array is in the outer region, the positive v contours continue to extend into the irrotational flow and are preceded and followed by negative contours of small magnitude but significant extent. The vertical extent of the positive u contours is limited with a negative R_{uu} region located above the positive R_{uu} region. The contours of Figures 2c and 3c are qualitatively similar to contours of R_{uu} and R_{vv} that have been measured in our laboratory in the outer region of a self-preserving wake. The R_{vv} contours in Figure 3c are in reasonable agreement with the R_{vv} contours presented by Kovasznay et al (1970), for y_0/δ = 1. Kovasznay (1977) commented that the positive correlation of v well beyond the edge of the layer is simply the result of the motion inside the turbulent layer which "massages" the outer fluid and creates a fluctuating potential flow outside the turbulent region.

It is of some interest to compare the present contours of R_{uu} and R_{vv} with those obtained from two X-probes (one fixed, the other movable) by

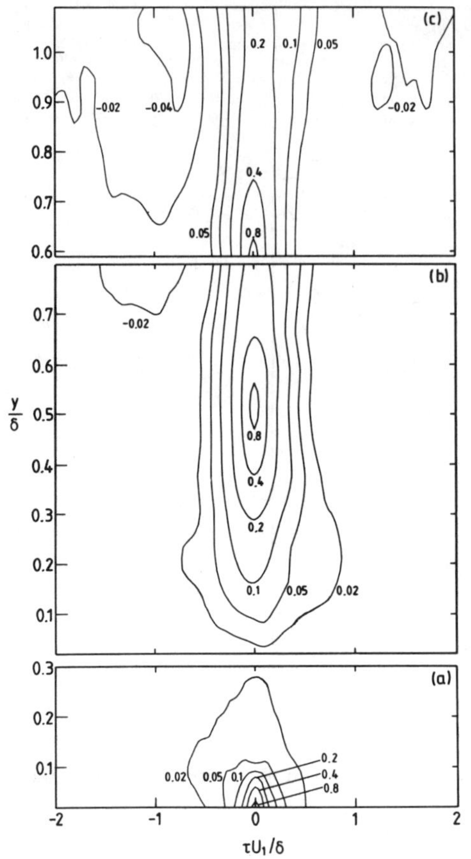

FIGURE 3. Isocorrelation contours of v at R_θ = 9630. (a) y_0/δ = 0.023; (b) 0.51; (c) 0.59.

Blackwelder and Kovasznay (1972) at comparable values of y_0/δ and R_θ. In the case of R_{uu} (Figure 4a) there is reasonable qualitative agreement. There is however less agreement in the case of R_{vv} (Figure 4b); the Blackwelder and Kovasznay contours are oriented in similar manner to the R_{uu} patterns whereas the present contours are oriented in a direction almost normal to the wall. The shape of the present R_{vv} contours at small y_0 (Figures 3a, 4b) is not inconsistent with continuity nor is it inconsistent with the notion of coherent fluid moving away from the wall. Further, the narrowness of the contours observed at small y_0 is present at large values of y_0. Blackwelder and Kovasznay argued that if the u patterns actually represent the evolution of eddies ejected near the wall, these eddies are more easily identified by their streamwise momentum defect than by their outward motion. Since u, v and w are intimately connected via the continuity equation, the R_{vv} contours should also contain some information about these eddies. The possibility that two or more scales of coherent motion co-exist, with the larger scale dominating R_{uu} and the smaller scale dominating R_{vv}, cannot be discounted, however.

The influence of the Reynolds number on a few selected values of R_{uu} and

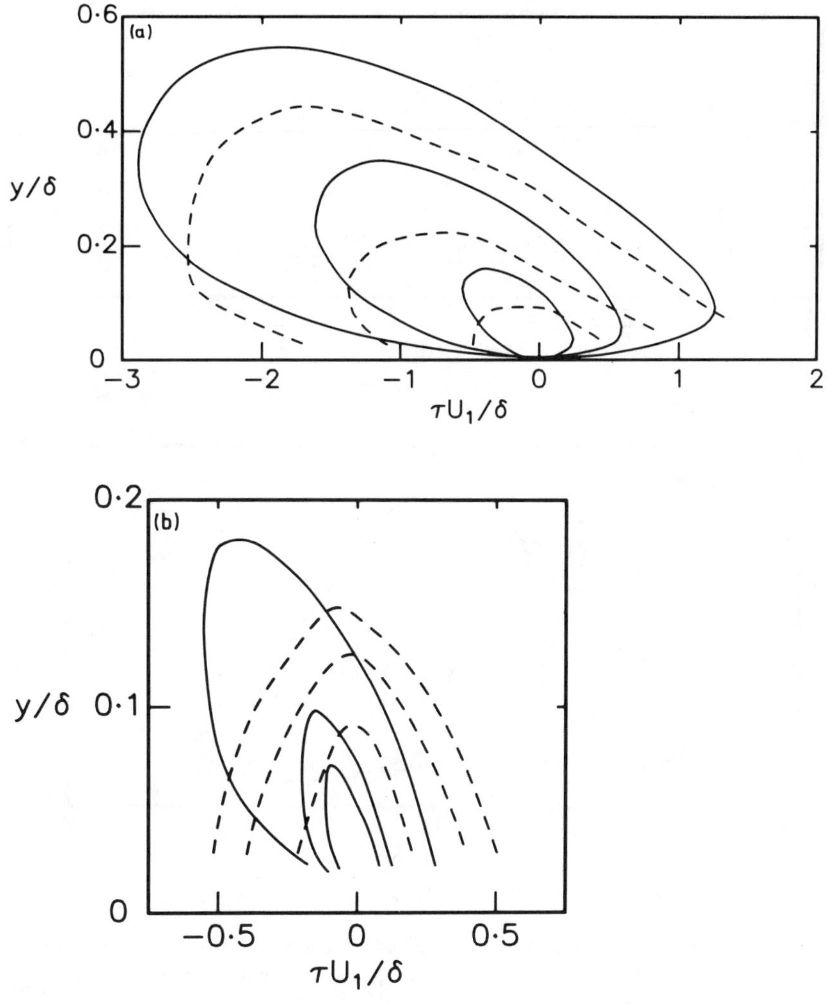

FIGURE 4. Comparison between the present isocorrelation contours (y_0/δ = 0.023, R_θ = 2180) and those of Blackwelder and Kovasznay (y_0/δ = 0.03, R_θ = 2500). (a) R_{uu} contours 0.1, 0.2, 0.4; (b) R_{vv} contours 0.05, 0.1, 0.3. ——, Blackwelder and Kovasznay; ---, present.

R_{vv} is shown in Figures 5 and 6. Except for the smallest Reynolds number, the extent of the R_{uu} contours in Figure 5a tends to increase slightly with R_θ. As y_0 increases (Figure 5b), there is an initial decrease with R_θ in the extent of R_{uu} but contours at R_θ = 6030 and R_θ = 9630 are nearly the

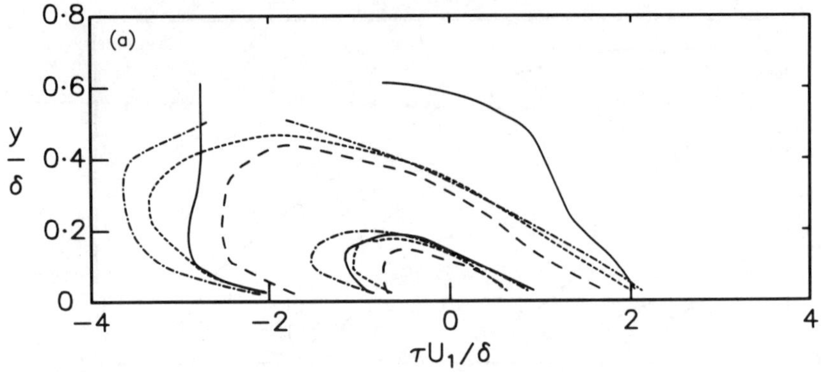

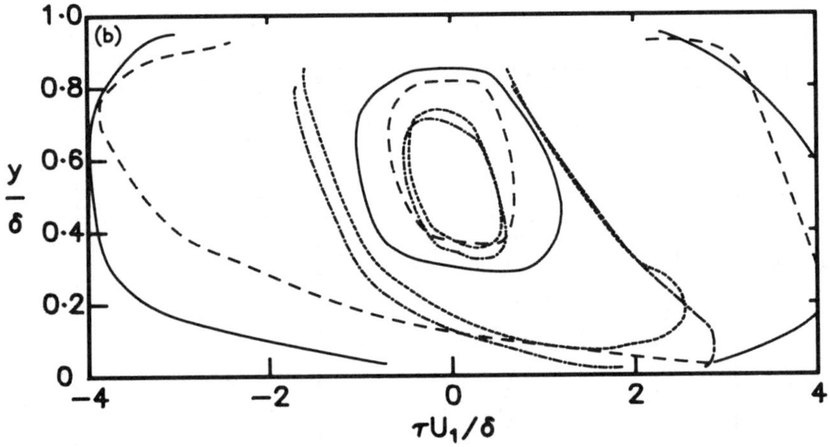

FIGURE 5. Dependence on Reynolds number of u isocorrelation contours. (a) $y_0/\delta \simeq 0.023$; (b) 0.56. At each R_θ, two contours are shown (R_{uu} = 0.1 and 0.3). ——, R_θ = 1360; — —, 2180; ---, 6030; — - —, 9630.

same. The decrease in spatial extent of the R_{vv} contours (Figure 6a) with increasing R_θ is more pronounced in the horizontal direction than in the y-direction. At $y_0/\delta \simeq 0.56$, the R_{vv} contours of Figure 6b indicate almost no Reynolds number influence.

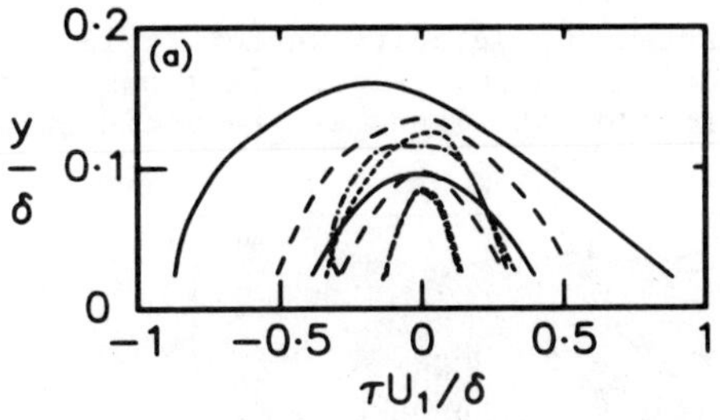

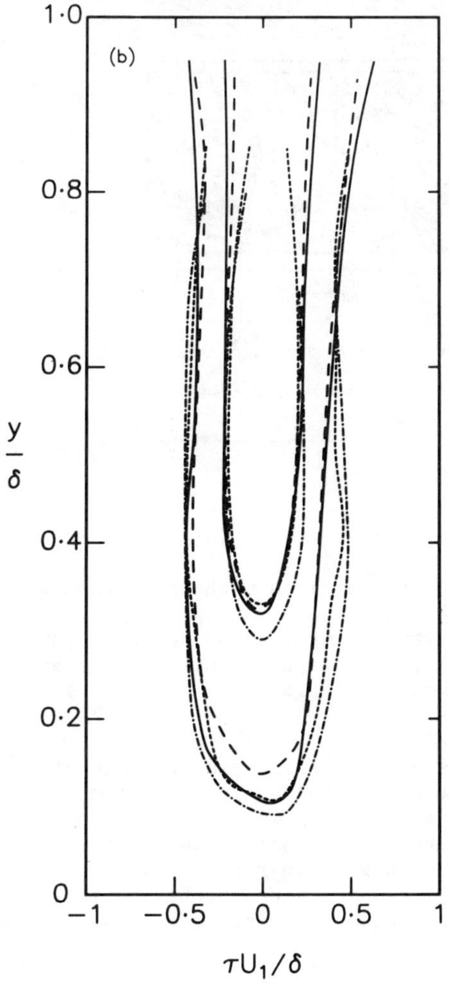

FIGURE 6. Dependence on Reynolds number of v isocorrelation contours. (a) $y_0/\delta \simeq 0.023$; (b) 0.56. At each R_θ, two contours are shown (R_{vv} = 0.05 and 0.2). ——, R_θ = 1360; — —, 2180; ---, 6030; — - —, 9630.

5. TOPOLOGICAL FEATURES

The array of X-probes allows the instantaneous velocity vector field, obtained in the plane of the array, to be displayed. Figure 7a shows velocity vectors (interpolation has been used to add data lines in the y direction)

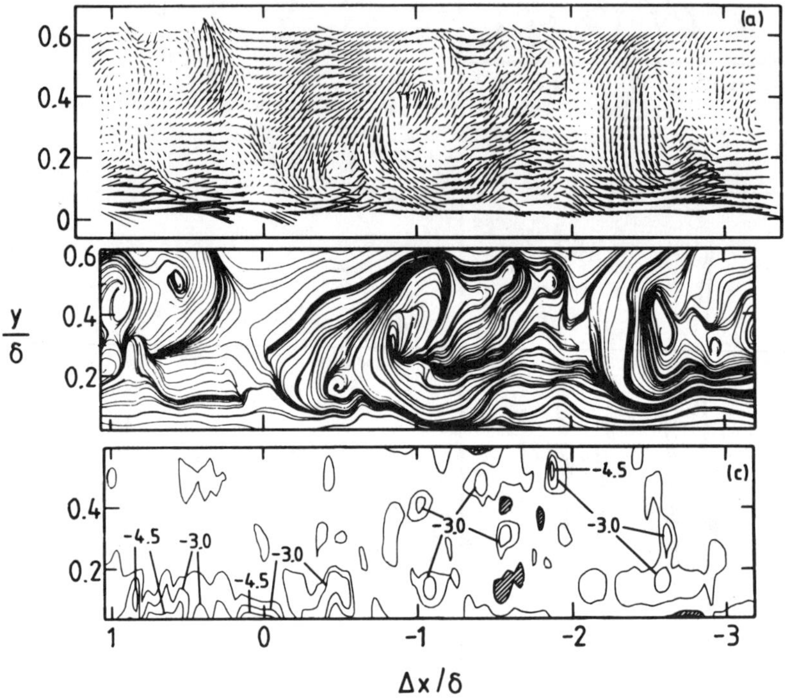

FIGURE 7. Instantaneous velocity vectors, particle paths and spanwise vorticity contours in a frame of reference which moves at $U_c = 0.8U_1$; flow is right to left. $R_\theta = 1360$. (a) velocity vectors; (b) particle paths; (c) spanwise vorticity $\omega_z\delta/U_1$. Note : cross-hatched contours have a value of +1.5. Unlabelled contours have a value of -1.5.

corresponding to $R_\theta = 1360$, in a frame of reference which travels at a convection speed U_c, equal to $0.8U_1$. The horizontal scale is a time scale, interpreted as normalised distance by multiplying with $-U_c/\delta$. One can recognise on Figure 7a the existence of relatively narrow zones at $\Delta x \simeq -2$ and 0, $0.1 \le y/\delta \le 0.61$, which are associated with relatively sharp changes in both magnitude and direction of the velocity vector (the inclination of the zones is exaggerated on the plot since the vertical scale is twice as large as the horizontal scale).

This feature can be identified in the corresponding sectional streamlines or particle paths* (Figure 7b) as lines (diverging separatrices) which are intersected by converging separatrices at saddle points. One other relatively prominent feature in Figure 7a is the appearance of foci, especially

*Details of particle path calculations are given in Bisset et al (1988).

a little downstream of the upper extremity of the diverging separatrices.

Instantaneous spanwise vorticity ω_z has been calculated using the approximation $\omega_z = -U_c^{-1}\ \partial v/\partial t - \partial u/\partial y$ on the interpolated data. Contours of constant values of $\delta\omega_z/U_1$, obtained from the data of Figure 7a, are shown in Figure 7c. Vorticity contours should strictly be independent of U_c; the above approximation includes U_c through the assumed space-time transformation but we have found that the dependence of ω_z contours on U_c is quite small for $U_c > 0.5U_1$. By contrast, it is well known that velocity vectors or streamfunctions, such as presented in Figures 7a and 7b, do depend on the choice of U_c (e.g. Coles, 1982; Hussain, 1983; Kim, 1985; Perry and Chong, 1987), but a unique choice of U_c is not possible, especially for instantaneous data such as that of Figure 7. The various values of U_c chosen by different investigators have been discussed by Antonia (1987) and Antonia and Fulachier (1988). The superposition of Figure 7b on 7c reveals that foci of fairly small spatial extent defined by the particle paths coincide reasonably well with local regions of concentrated spanwise vorticity except near the wall. Although this coincidence provides a useful corroboration of our choice for U_c, it is neither an objective nor an accurate way of determining U_c. Also, if the concentrated vorticity patches, which occur at different distances form the wall, are identified with the tips of the hairpin vortices observed by Head and Bandyopadhyay (1981), one would expect a range of convection velocities depending, inter alia, on the height distribution of these vortices.

A focus as large as the one at the centre of Figure 7b does not seem to occur very often at $R_\theta = 1360$, and occurs rarely (if at all) at higher R_θ. This focus does not have very strong or very coherent vorticity associated with it. Most foci rotate counterclockwise, but the small regions of positive vorticity are sometimes associated with clockwise rotating foci or centres (although not in Figure 7).

To quantify the effect of R_θ on features of the organised motion, the velocity data at the four Reynolds numbers were conditionally sampled and averaged. The detection focused on the features evident in Figure 7 and was based on u. The selection of u rather than v was partly suggested by the extent of the R_{uu} contours but primarily by our previous experience in detecting spatially coherent sudden changes of temperature (coolings or heatings) in the boundary layer, e.g. Subramanian et al (1982); Antonia et al (1988), Antonia and Fulachier (1988). As noted in previous work, these temperature changes are accompanied by concomitant changes in u (and v) throughout most of the layer; near the wall, however, a strong correlation has been found between u and the temperature fluctuation.

The present single point detection scheme searches for relatively sudden increases in u over a suitable time interval (details of the algorithm are given in Bisset et al, 1988; the parameter values used here were a window half length of 0.8 to 1.0 δ/U_1 and threshold $k = 0.45$). Histograms of the intervals between consecutive detections at all probe locations $y/\delta \le 0.6$ were computed. These histograms indicated that the interval T_p corresponding to the maximum of the histogram remained approximately constant ($\simeq 2.5\delta/U_1$) with Reynolds number. The histograms are very broad, however, with most intervals in the range $2\delta/U_1$ to $5\delta/U_1$. It is important to note that at $R_\theta = 1360$ and 2180, the average period between detections agreed very well with the average "bursting" period $\overline{T}$ obtained by Shah and Antonia (1988) using the u-level approach of Luchik and Tiederman (1987), although $\overline{T}$ is substantially larger than T_p, and the probability histograms are quite similar. At the two larger Reynolds numbers, the wire length and distance from the wall of the bottom X-probe are too large for a meaningful comparison with the bursting period. At lower Reynolds numbers, more than 50% of the present detections occurred at virtually the same time as the upstream end of a "burst", a spatial relationship that is consistent with the results of Kim (1985). This aspect is being investigated further.

The conditionally averaged data in Figure 8 were computed using sets of pairs of detections separated by T_p ±10% (the patch in Figure 7 is a member of one of these sets). This strategy, although not the only possible strategy, should reduce the jitter in the structures for the purpose of comparing results at different R_θ, while retaining sufficient detection pairs for conditional averaging (58 pairs at R_θ = 1360, 309 at 9630). The y location at which detections were made was just below (≃ 0.25δ) the centre of the array when the latter was nearest the wall.

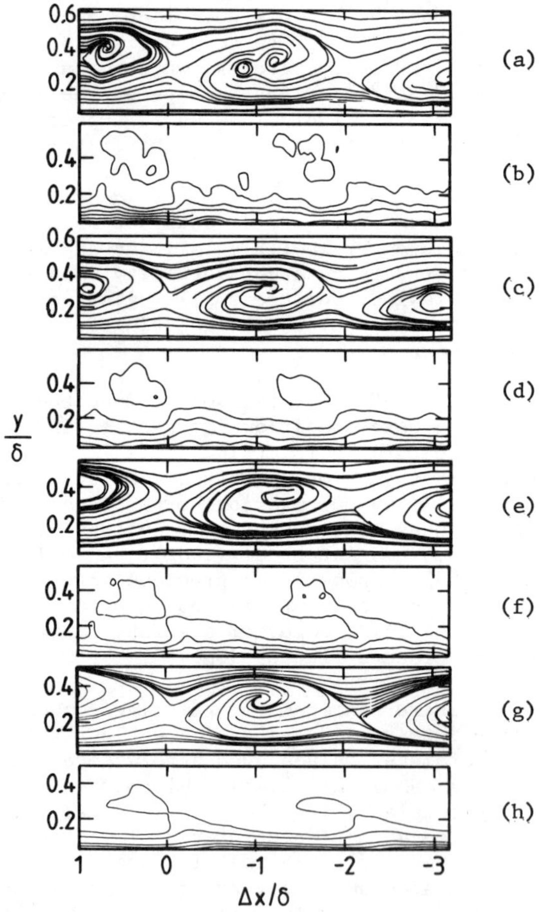

FIGURE 8. Conditional particle paths and spanwise vorticity contours in a frame of reference which moves to the left at $U_c = 0.8U_1$.
Particle paths : (a) R_θ = 1360; (c) 2180; (e) 6030; (g) 9630.
Spanwise vorticity : (b) R_θ = 1360; (d) 2180; (f) 6030; (h) 9630.
The outermost contour is $-0.7U_1/\delta$ and the contour step is $-0.5U_1/\delta$.

The conditional particle paths in Figures 8a,c,e,g clearly show the two saddles at $\Delta x/\delta = 0$ and $\Delta x/\delta \simeq -2$. The convection velocity is $U_c = 0.8U_1$. Foci now appear almost halfway between saddles, but it would be incorrect to identify these with regions of high spanwise vorticity. Conditional vorticity contours, shown (Figures 8b,d,f,h) immediately below the corresponding particle paths show a tendency for a concentration along the lower diverging separatrix and downstream of the upper one, although, as expected, the highest values occur near the wall. This behaviour is consistent with averaging a number of hairpin vortices which extend to different distances from the wall, and which are strongest and/or most likely to occur in the regions where the average vorticity $\langle\omega_z\rangle$ is concentrated. It is also consistent with Kim's (1987) observation, based on direct numerical simulations of the Navier-Stokes equations for a turbulent channel flow, that the spanwise vorticity is concentrated along the upstream edge of the inclined large scale structures.

The particle paths shown in Figures 8a,c,e,g are qualitatively similar to the velocity vector patterns obtained by Brown and Thomas (1977), in a turbulent boundary layer, the streamlines computed by Kim (1985) from data obtained from a large-eddy simulation of turbulent channel flow and the particle paths recently obtained by Antonia (1987) and Antonia and Fulachier (1988) in a turbulent boundary layer. The similarity, notwithstanding the effect of U_c and the choice of the detection location, is all the more convincing in the sense that each of the previously mentioned investigations used a different method of detection.

Figures 8b,d,f,h indicate that isovorticity contours, viz. $\langle\omega_z\rangle\delta/U_1$ = constant, tend to rise to larger values of y/δ at lower Reynolds numbers, e.g. the -0.7 contour reaches $y/\delta = 0.55$ for $R_\theta = 1360$ but only $y/\delta = 0.35$ for $R_\theta = 9630$. The comparison of vorticity contours and corresponding particle paths in Figure 8 suggests that a smaller value of U_c should have been selected at the two larger values of R_θ; a decrease of U_c/U_1 towards a constant value at large R_θ would not be unreasonable. Although not shown here, isostrain rate contours, viz. $\langle S\rangle\delta/U_1$ = constant, where $\langle S\rangle = -U_c^{-1}\,\partial\langle v\rangle/\partial t + \partial\langle u\rangle/\partial y$, stretch out in the directions of the diverging separatrices and are influenced by R_θ in the same way as the spanwise vorticity contours.

6. CONTRIBUTION FROM ORGANISED LARGE SCALE MOTION TO REYNOLDS STRESSES

Conditional averaging was used to estimate the contribution of the organised large scale motion to the Reynolds stresses $\overline{u^2}$, $\overline{v^2}$ and $\overline{uv}$, based on the triple decomposition of velocities (e.g. Hussain, 1983). The conditional average of a digitally stored quantity f_j, $j = 1,2,\ldots,N$, is defined as

$$\langle f\rangle_k = \frac{1}{n}\sum_{i=1}^{n} f_{j_i+k}$$

where the n values of j_i are a set of detection instants, and k is an offset from the detection locations. The subscript k is generally omitted unless required for clarity. Using the triple decomposition, an instantaneous quantity F is taken to be $F = \overline{F} + \tilde{f} + f_s$, where $\overline{F}$ is the conventional long-time average, $\tilde{f}$ is the organised large scale component, and f_s is a small scale (perhaps random) contribution uncorrelated with $\tilde{f}$ such that $\langle f_s\rangle \equiv 0$. The large scale components are calculated from $\tilde{f} = \langle F-\overline{F}\rangle$. Note that in general a double decomposition is used where the large scale component is taken as $\tilde{F} = \overline{F} + \tilde{f}$, e.g. the topological plots in Figure 8, but the triple decomposition is appropriate for analysing Reynolds stresses. The average value of $\langle fg\rangle$ throughout a conditionally averaged structure, where f and g may each stand for either velocity fluctuation u or v, is

$$\overline{\overline{\langle fg\rangle}} = \frac{1}{p}\sum_{k=0}^{p-1}\langle fg\rangle_k \quad ,$$

and the coherent contribution $\overline{\overline{\widetilde{fg}}}$ is defined similarly; thus

$$\overline{\overline{\langle fg\rangle}} = \overline{\overline{\widetilde{fg}}} + \overline{\overline{\langle f_s g_s\rangle}} \quad .$$

The value of p in the present case is the number of data points between the selected pairs of detections. If $\overline{\langle fg\rangle}$ is based on a set of detections that are representative of all the data from which they are derived, then $\overline{\overline{\langle fg\rangle}} \simeq \overline{fg}$.

As indicated previously, the time delays between successive detections vary over a considerable range, and this must be accounted for when determining coherent contributions. It is desirable that a considerable proportion of all possible detections be used for analysis, but without causing excessive smearing of the conditional averages. For each Reynolds number, three sub-ranges of time delays T between detection pairs were chosen, centred on $TU_1/\delta \simeq 2.1$, 2.8 and 3.8, and of width $TU_1/\delta \simeq 0.7$. Within each range, the variable number of digital u and v data corresponding to each detection pair was replaced by a number of data corresponding exactly to the centre of the chosen time delay range using linear interpolation. With this procedure, regions of fairly similar length between detections are grouped together without smearing, while regions of quite different length (and which could differ in other characteristics also) are treated separately.

For the present experiments no systematic variation of any quantity with T was found. The quantities $\overline{\langle fg\rangle}$ for each value of T generally were within 10% of the mean of the three values, especially at higher R_θ where there were several hundred detection pairs in each case. Results for the quantities $\widetilde{fg}$ were sometimes a little more scattered except at $R_\theta = 9630$. Therefore mean values were calculated from the results from the three values of T, weighted by the product of T with the number of detection pairs.

For the three lowest values of y/δ in Table 2, the ratios of $\overline{\overline{\langle fg\rangle}}/\overline{fg}$ were approximately 1.0 for $R_\theta = 1360$ and 2180, 0.95 for 6030 and 0.9 for 9630, which suggests that the selected data are reasonably representative of much of the flow. At $y/\delta \simeq 0.53$ ($R_\theta = 1360$ and 2180), the intermittency factor differs from unity and the ratios rose sharply to about 1.5 and 1.25 respectively.

The ratios $\overline{\overline{\widetilde{fg}}}/\overline{\overline{\langle fg\rangle}}$, expressed as percentages, are shown in Table 2. The values at $R_\theta = 6030$ are generally consistent with those determined by Antonia and Fulachier (1988) using a temperature-based detection scheme. It is apparent that the organised large scale motion contribues very significantly to $\overline{uv}$ and $\overline{u^2}$, but contributes only a small proportion of $\overline{v^2}$. The R_{uu} contours of Figure 2 are of a size and shape consistent with the dynamic significance of the large structures for $\overline{u^2}$, while the R_{vv} contours (Figure 3) indicate that the main contributor to $\overline{v^2}$ should be of a smaller scale.

TABLE 2. Contributions (%) From Large Scale Organised Motion to Reynolds Stresses as a Function of R_θ and y/δ.

	R_θ =	1360	2180	6030	9630
$y/\delta \simeq 0.024$	a	37.3	34.0	28.9	26.5
	b	7.7	3.1	2.4	1.7
	c	47.2	51.6	25.6	23.0
$y/\delta \simeq 0.10$	a	30.5	29.6	30.2	28.6
	b	10.6	7.6	6.2	7.1
	c	38.6	32.9	31.9	31.4
$y/\delta \simeq 0.27$	a	27.0	30.7	29.9	32.7
	b	8.8	9.4	6.3	7.2
	c	37.2	39.2	30.0	34.9
$y/\delta \simeq 0.53$	a	44.9	34.9	36.2	35.9
	b	12.2	13.2	9.9	11.2
	c	39.3	46.6	37.2	37.9

(a) $\overline{\overline{\tilde{u}^2}}/\overline{\overline{\langle u^2\rangle}}$; (b) $\overline{\overline{\tilde{v}^2}}/\overline{\overline{\langle v^2\rangle}}$; (c) $\overline{\overline{\tilde{u}\tilde{v}}}/\overline{\overline{\langle uv\rangle}}$.

Investigations are being made to determine whether the smaller vortical regions (seen in Figure 7c, for example) contribute significantly to $\overline{v^2}$ (and also to $\overline{u^2}$ and $\overline{uv}$). The trend of decreasing coherent contributions with increasing R_θ is fairly clear, especially at $y/\delta \simeq 0.024$, and would have been slightly stronger if the values had been given as $\overline{\tilde{f}\tilde{g}}/\overline{fg}$. The flow visualisation results of Head and Bandyopadhyay (1981) are consistent with this : at low R_θ the main motion is made up of the large structures which extend across the full width of the layer, while an increase in R_θ brings about an increase in smaller scale activity sufficient to obscure most of the large scale motion. The present hot-wire results indicate however that the large scale structures are still significant up to $R_\theta \sim 10^4$, and contribute one quarter to one third of the Reynolds shear stress.

7. CONCLUSION

Results obtained to date have confirmed that the influence of Reynolds number on the mean velocity and turbulence characteristics of a turbulent boundary layer is confined to momentum thickness Reynolds numbers smaller than about 5000. They have also indicated that isocorrelations of u are more affected by Reynolds number than those of v, which is consistent with the observation that $\overline{v^2}$ is less affected than $\overline{u^2}$ by the Reynolds number. For $R_\theta \gtrsim 5000$, the dependence on R_θ of the spatial extent of the correlations is negligible.

The coherent large scale motion, identifiable by an upstream feature which extends across most of the layer at an inclination to the wall, is present at all Reynolds numbers and its general form and frequency of occurrence depend only weakly on R_θ. The contribution that this motion makes to the Reynolds stresses is significant, especially as the Reynolds number decreases.

ACKNOWLEDGEMENTS

The support of the Australian Research Grants Scheme is gratefully acknowledged. We wish to thank Mr. D. A. Shah for his assistance with the experiments.

REFERENCES

Alfredsson, P. H. and Johansson, A. V. Time Scales in Turbulent Channel Flow, *Phys. Fluids*, Vol. 27, pp.1974-1981, 1984.

Andreopoulos, J., Durst, F., Zaric, Z. and Jovanovic, J. Influence of Reynolds Number on Characteristics of Turbulent Wall Boundary Layers, *Expts. in Fluids*, Vol. 2, pp.7-16, 1984.

Antonia, R. A. The Organised Motion and Its Contribution to Transport in Turbulent Shear Flows, *Proc. 2nd Int. Symposium on Transport Phenomena in Turbulent Flows, Tokyo*, pp.339-358, 1987.

Antonia, R. A. and Fulachier, L. Organised Motion in a Turbulent Boundary Layer - Influence of Wall Suction, *J. Fluid Mech.*, [submitted], 1988.

Antonia, R. A., Fulachier, L., Krishnamoorthy, L. V., Benabid, T. and Anselmet, F. Influence of Wall Suction on the Organised Motion in a Turbulent Boundary Layer, *J. Fluid Mech.*, [to appear], 1988.

Antonia, R. A., Rajagopalan, S., Subramanian, C. S. and Chambers, A. J. Reynolds-Number Dependence of the Structure of a Turbulent Boundary Layer, *J. Fluid Mech.*, Vol. 121, pp.123-140, 1982.

Bisset, D. K., Antonia, R. A. and Browne, L. W. B. Spatial Organisation of Large Structures in the Turbulent Far-Wake of a Cylinder, *J. Fluid Mech.*, [submitted], 1988.

Blackwelder, R. F. and Kovasznay, L. S. G. Time Scales and Correlations in a Turbulent Boundary Layer, *Phys. Fluids*, Vol. 15, pp.1545-1554, 1972.

Brown, G. L. and Thomas, A. S. W. Large Structure in a Turbulent Boundary Layer, *Phys. Fluids*, Vol. 20, pp.S2430S252, 1977.

Chen, C-H. P. and Blackwelder, R. F. Large Scale Motion in a Turbulent Boundary Layer : A Study Using Temperature Contamination, *J. Fluid Mech.*, Vol. 89, pp.1-31, 1978.

Coles, D. The Turbulent Boundary Layer in a Compressible Fluid, *Rand Corp. Report R-403-PR, ARC 24478*, 1962.

Coles, D. Prospects for Useful Research on Coherent Structure in Turbulent Shear Flow, in *Surveys in Fluid Mechanics*, eds. R. Narasimha and S. M. Deshpande, pp.17-33, Indian Academy of Sciences, Bangalore, 1982.

Erm, L. P., Smits, A. J. and Joubert, P. N. Low Reynolds Number Turbulent Boundary Layers on a Smooth Flat Surface in a Zero Pressure Gradient, in *Turbulent Shear Flows 5*, eds. F. Durst, B. E. Launder, J. L. Lumley, F. W. Schmidt and J. H. Whitelaw, pp.186-196, Springer, Berlin, 1987.

Falco, R. E. Some Comments on Turbulent Boundary Layer Structure Inferred from the Movements of a Passive Contaminant, *AIAA Paper No. 74-99*, presented at AIAA 12th Aerospace Sciences Meeting, Washington, 1974.

Favre, A., Gaviglio, J. and Dumas, R. Space-Time Double Correlations and Spectra in a Turbulent Boundary Layer, *J. Fluid Mech.*, Vol. 2, pp.313-342, 1957.

Head, M. R. and Bandyopadhyay, P. New Aspects of Turbulent Boundary-Layer Structure, *J. Fluid Mech.*, Vol. 107, pp.297-338, 1981.

Hussain, A. K. M. F. Coherent Structures - Reality and Myth, *Phys. Fluids*, Vol. 26, pp.303-356, 1983.

Kim, J. Turbulence Structures Associated with the Bursting Event, *Phys. Fluids*, Vol. 28, pp.52-58, 1985.

Kim, J. Investigation of Heat and Momentum Transport in Turbulent Flows via Numerical Simulation, *Proc. 2nd Int. Symposium on Transport Phenomena in Turbulent Flows, Tokyo*, pp.707-721, 1987.

KLEBANOFF, P. S. Characteristics of Turbulence in a Boundary Layer with Zero Pressure Gradient, *NACA Report 1247*, 1955.

Kovasznay, L. S. G. Coherent Structures in Turbulent Shear Flows, *Proc. 2nd Australasian Conference on Heat and Mass Transfer, Sydney*, pp.295-304, 1977.

Kovasznay, L. S. G., Kibens, V. and Blackwelder, R. F. Large-Scale Motion in the Intermittent Region of a Turbulent Boundary Layer, *J. Fluid Mech.*, Vol. 41, pp.283-325, 1970.

Luchik, T. S. and Tiederman, W. G. Timescale and Structure of Ejections and Bursts in Turbulent Channel Flows, *J. Fluid Mech.*, Vol. 174, pp.529-552, 1987.

Murlis, J., Tsai, H. M. and Bradshaw, P. The Structure of Turbulent Boundary Layers at Low Reynolds Numbers, *J. Fluid Mech.*, Vol. 122, pp.13-56, 1982.

Perry, A. E. and Chong, M. S. A Description of Eddying Motions and Flow Patterns Using Critical-Point Concepts, *Ann. Rev. Fluid Mech.*, Vol. 19, pp.125-155, 1987.

Purtell, L. P., Klebanoff, P. S. and Buckley, F. T. Turbulent Boundary Layer at Low Reynolds Number, *Phys. Fluids*, Vol. 24, pp.802-811, 1981.

Shah, D. A. and Antonia, R. A. Scaling of the "Bursting" Period in Turbulent Boundary Layer and Duct Flows, *Phys. Fluids*, [submitted], 1988.

Sternberg, J. On the Interpretation of Space-Time Correlation Measurements in Shear Flow, *Phys. Fluids Suppl.*, pp.S146-S152, 1967.

Subramanian, C. S. and Antonia, R. A. Effect of Reynolds Number on a Slightly Heated Turbulent Boundary Layer, *Int. J. Heat Mass Transfer*, Vol. 24, pp.1833-1846, 1981.

Subramanian, C. S., Rajagopalan, S., Antonia, R. A. and Chambers, A. J. Comparison of Conditional Sampling and Averaging Techniques in a Turbulent Boundary Layer, *J. Fluid Mech.*, Vol. 123, pp.335-362, 1982.

Correlation Measurements and Structure Angles in a Turbulent Boundary Layer Recovering from Convex Curvature

AMY E. ALVING and ALEXANDER J. SMITS
Department of Mechanical and Aerospace Engineering
Princeton University
Princeton, New Jersey 08544, USA

ABSTRACT

Simultaneous, multiple point turbulence measurements were made in a boundary layer upstream and downstream of a 90° convex bend. Two-point space-time correlation coefficients $R_{uu}(0,\xi,0,\tau)$ were computed for several separation distances ξ. The time delay τ_{max} corresponding to the maximum value of R_{uu} was used to define a structure angle θ, where $\theta = \tan^{-1}[\xi/(U_c\tau_{max})]$, and U_c is the best estimate for the convection velocity. For the undisturbed boundary layer, θ takes low values near the wall, increases to 30°-40° in the central portion of the layer, and increases further near the boundary layer edge. At the exit from the convex curvature, θ is largely unchanged. In the relaxing boundary layer θ appears to increase over most of the boundary layer thickness. The sensitivity of the structure angle to the probe separation distance, and the significance of the results in terms of large-scale organized motions are discussed. The instantaneous structure angle can be defined in terms of short-time correlations, and some of the difficulties associated with this concept are also discussed.

1. INTRODUCTION

This paper examines the usefulness of correlation measurements in studying the large scale structure of a turbulent boundary downstream of a short region strong convex curvature.

The flow configuration was described in detail by Alving and Smits (1986). In brief, an incompressible turbulent boundary layer was allowed to grow on a flat plate in a zero pressure gradient with a freestream velocity of 31 m/sec. At a point where the Reynolds number based on momentum thickness was approximately 6000 ($\delta_o \sim$ 23mm), the flow was turned through 90° of strong convex curvature ($\delta_o/R \sim 0.08$). Subsequently, the perturbed boundary layer relaxed on a flat plate, again in zero pressure gradient. The wind tunnel is drawn schematically in Fig. 1, and the pressure distribution is shown in Fig. 2.

Apart from the usual measurements of Reynolds stress profiles, correlation measurements were obtained from a rake of six hot-wires measuring streamwise velocity fluctuations, separated in the direction normal to the wall.

The correlation information was interpreted in terms of a large-scale "structure angle", in the same way as Spina and Smits (1987) interpreted

Test Conditions		
$p_\circ$	=	0.69×10^6 Pa (100 psi)
$T_\circ$	=	270 K
M_∞	=	2.84
Re_∞/ℓ	=	6.5×10^7/m
$\delta_\circ$	=	26 mm
δ^*	≐	6.4 mm
$\theta_\circ$	=	1.3 mm
$C_{f_\circ}$	=	0.00115

TABLE I. Boundary Layer Characteristics and Incoming Flow Conditions

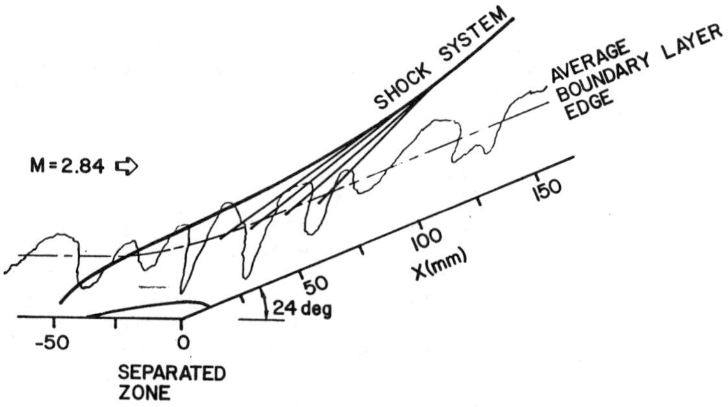

Fig. 1 **24° Compression Corner Mean Flow-Field (adapted from Settles, Vas, & Bogdonoff 1976)**

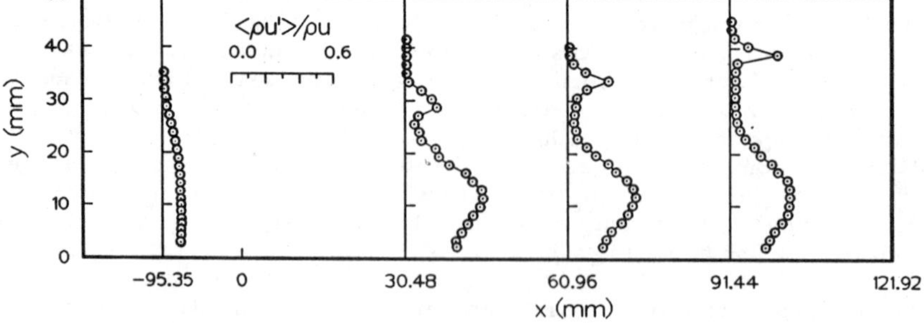

Fig. 2 **Upstream and Downstream Mass-Flux Turbulence Intensity Profiles: (*a*) Case without Blowing**

their correlation measurements in a supersonic (Mach 3) turbulent boundary layer. The concept is illustrated in Fig. 3. Consider a large-scale structure, inclined towards the wall at some characteristic angle θ, convecting downstream at U_c. If the structure is large enough to span both wires, the velocity signal resulting from this structure will be measured first by the wire farther from the wall and, at a later time, by the wire closer to the wall. The peak in the broad band cross-correlation, $R_{uu}(\tau;\xi)$ determines the time delay for which the two wires are, on the average, most similar. Along with U_c and ξ (wire separation) this time delay τ_{max} can be converted to the average structure angle, θ:

$$\theta = \tan^{-1}\left[\frac{\xi}{U_c\tau_{max}}\right]$$

Spina and Smits measured the two-point correlation functions in their supersonic turbulent boundary layer (see Fig. 4), and the structure angle was about 45° throughout the middle 70% of the layer; the angle shows a rapid decrease/increase with y near the wall/edge of the boundary layer. Clearly, θ depends on the assumed value of the convection velocity U_c. However, for $\theta = 45^o$, a 10% change in U_c results in only a 2.3^o change in θ, and θ is even less sensitive to U_c as θ decreases. Thus, the structure angle does not depend critically on the exact value of U_c.

This paper will examine the behavior of the structure angle and space-time correlations in incompressible boundary layers, upstream and downstream of strong convex wall curvature, to determine the sensitivity of θ to ξ, and to investigate the changes in θ due to a strong stabilizing extra strain rate.

1.1. The Effects of Convex Curvature

Convex curvature is fundamentally of great interest because it has a stabilizing influence on the boundary layer, as demonstrated by a simple inviscid analysis of the effect of an angular momentum gradient normal to the wall (see for example, Bradshaw, 1973). Several experimental studies have documented the suppression of turbulence intensities caused by convex curvature; see, for example, So & Mellor (1973), Gillis & Johnston (1983), and Smits, Young & Bradshaw (1979). In addition, the shear stress vanishes and the normal stresses approach an isotropic state in the outer half of the boundary layer. The "accommodation" to curvature is fairly fast; Gillis and Johnston showed that after the boundary layer turns through 30°, the Reynolds stress profiles become similar when the normal coordinate was scaled on R (radius of curvature) rather than on δ. The severe reduction of shear stress in the outer part of the mean shear layer, the near-isotropic normal stress and the scaling with R imply that curvature destroys the large-scale, anisotropic, stress-containing structures which carry the upstream "history" of the boundary layer. Gillis and Johnston proposed that, if this perturbed boundary layer were allowed to relax on a flat plate, its recovery would show a dual nature: an "outer" region containing the isotropic debris of the upstream boundary layer, and an "inner" layer containing whatever eddies survived curvature. The inner layer would grow into the outer layer with downstream distance, in a similar manner to the growth of the inner layer observed in rough-to-smooth and smooth-to-rough transitions at a rate typical of normal turbulent boundary layer growth (Smits & Wood 1985).

1.2 Overall Descriptions of the Flow Behavior

The wind-tunnel used in this experiment was designed to extend the work

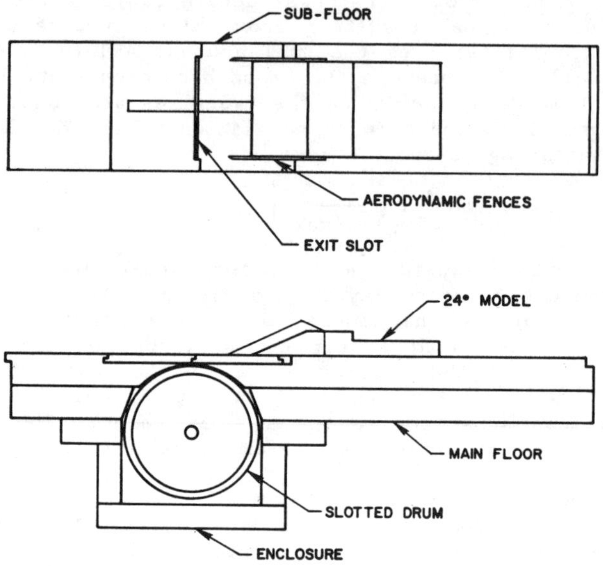

Fig. 3 Dynamic Control Apparatus—Side View and Top View

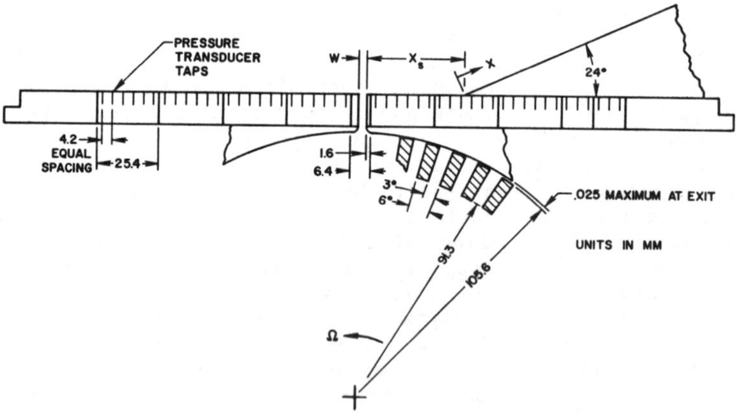

Fig. 4 Detailed Dimensions of the Dynamic Control Apparatus in the Vicinity of the Injection Slot—Cross Section

of Gillis and Johnston by adding a long flat plate recovery to the curved section, so that the far downstream relaxation process could be observed. The mean flow and Reynolds stress relaxation behavior is described in earlier papers (Alving & Smits 1986, Alving et al. 1987 and Alving 1988), and only the major results are given here. In the mean flow sense, the boundary layer relaxes in several stages. Upon first exiting from the bend, the perturbed boundary layer rapidly recovers towards its flat plate state, as indicated by the skin friction, shape factor, extent of the wall region, and wake factor. This stage lasts approximately $25\delta_o$. From these results, and by comparison with previous work, it seems that any pressure gradient history has died out by this point. Further downstream, the recovery from the effects of curvature continues but the recovery occurs at a slower rate. There is some indication that far downstream the may be a slight over-recovery of these mean flow parameters.

The Reynolds stress behavior is more dramatic, as shown in Figs. 5 and 6. The suppressed stress levels measured at the end of curvature immediately begin to regrow on the flat plate. This regrowth is initially in terms of a stress bore that moves away from the wall; however, after a distance of about $25\delta_o$ the whole boundary layer shows elevated stress levels. Interestingly, these stress levels remain high for the remainder of the recovery, up to a downstream distance s of approximately $90\delta_o$. The Reynolds stress profiles clearly demonstrate a radical reorganization of the turbulence in the relaxing boundary layer.

To understand this behavior more thoroughly, space-time correlations were measured using a six-wire probe. The aim of this paper is to study the usefulness of these cross-correlations in examining the large-scale structure in the upstream reference boundary layer, the strongly perturbed layer near the exit from curvature, and the recovering layer on the downstream flat plate.

The information available from cross-correlation measurements is not limited to the average structure angle defined above. Spina and Smits extended their method to look for instantaneous structure angles on the basis of short-time cross-correlations in regions of high turbulence activity. These instantaneous structure angles are important in understanding the physics of the flow; therefore, this paper also examines the application of correlation measurements to determining instantaneous structure angles in subsonic flows.

2. EXPERIMENTAL METHODS

The hot-wire rake used in this study was provided by S. K. Robinson of NASA/Ames; it was used by him in a high Reynolds number flow (Robinson 1986). It consists of eight normal wires (of which six were used) spaced 0.5 to 1.0mm apart in the vertical direction, resulting in a maximum span of 5.5mm. (The boundary layers of interest are 19.3, 21., 32.2, and 45.4 mm thick). For support in other flow conditions, the probe body has been reinforced with epoxy 3mm behind the tip. Initially we were concerned that in our flow blockage effects might distort the hot-wire measurements. However, spectra from these wires showed no abnormalities in the range 10Hz - 50kHz. In addition, a brief series of tests was made using a much more slender rake of two probes. In this case the probe bodies were Dantec, model 55P05, and the probe tips were located 8 mm upstream of and vertically offset from the probe support. The two- wire rake and the six-wire rake gave nearly identical results. On this basis, probe blockage was determined not to be an important factor for the six-wire rake. All results presented

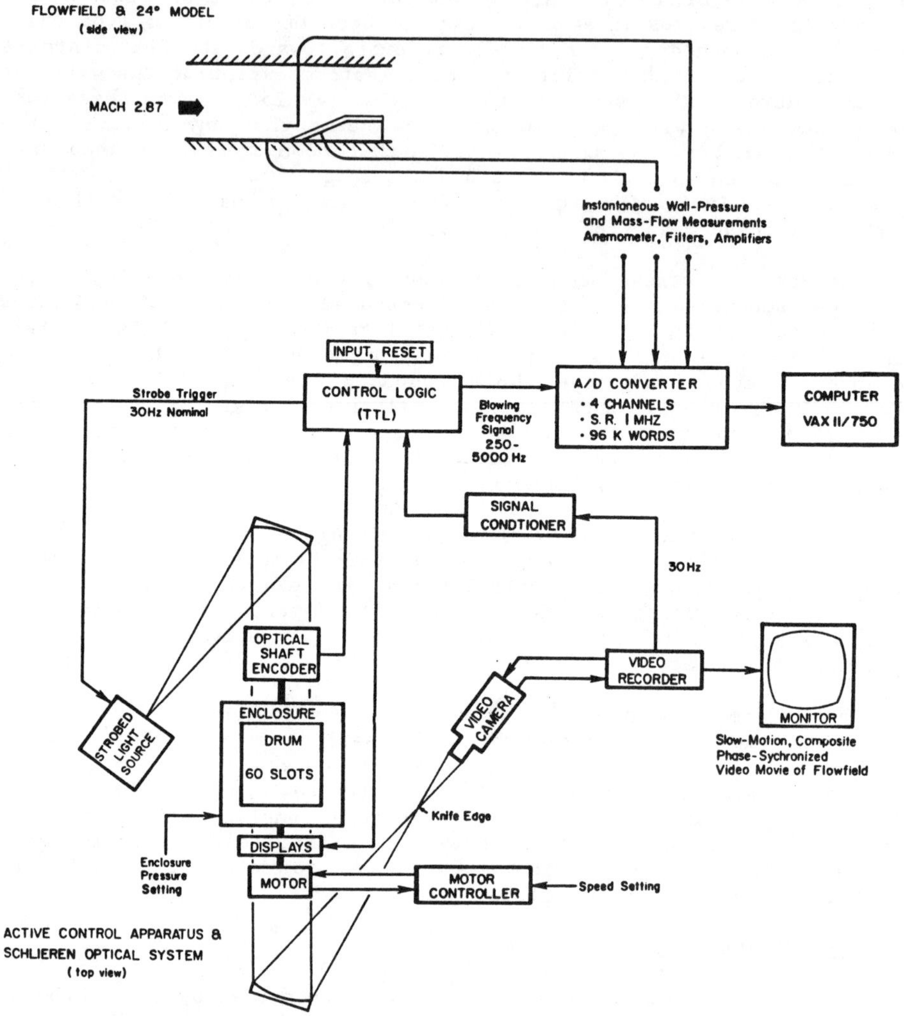

Fig. 5 Schematic of the Experimental Set-Up Including the Test Section, Data Acquisition System, Dynamic Control Apparatus, and Stroboscopic Schlieren Videography System.

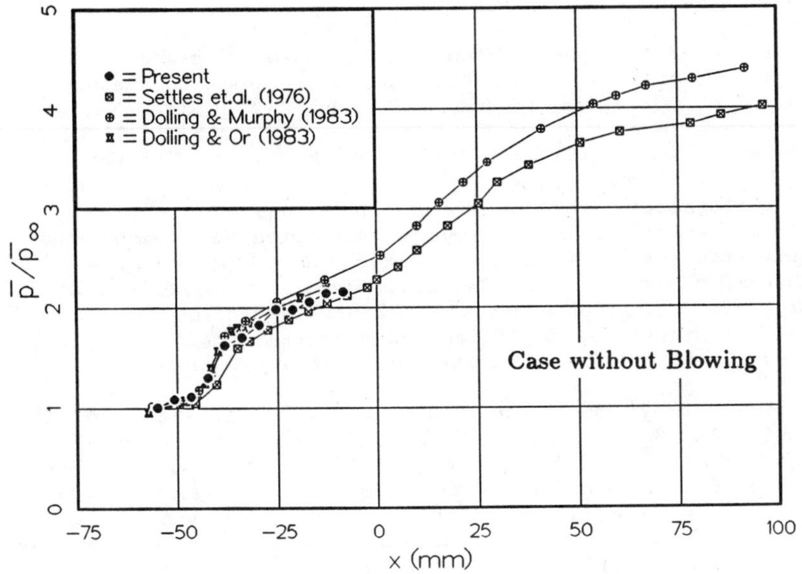

Fig. 6 Mean Wall-Pressure Distribution Compared with the Measurements of Settles, Vas & Bogdonoff (1976), Dolling & Murphy (1983), and Dolling & Or (1983)

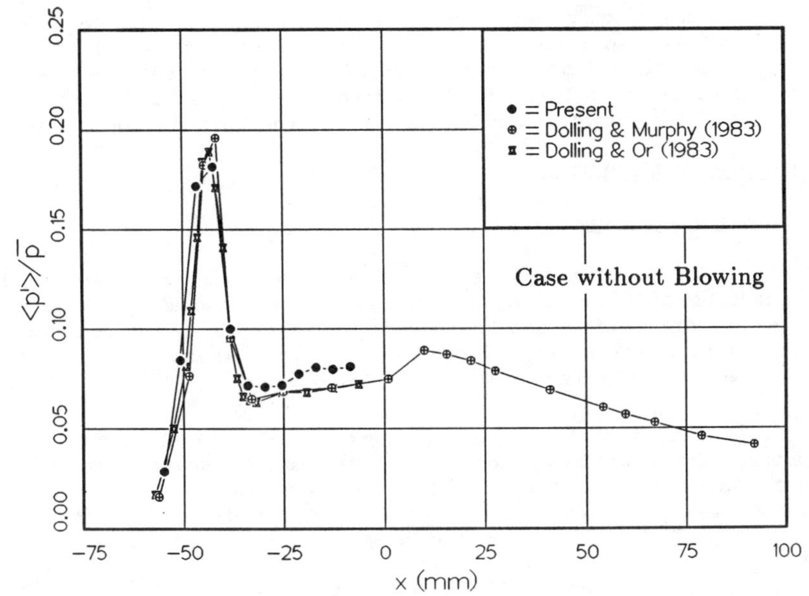

Fig. 7 Rms Wall-Pressure Distribution Compared with the Measurements of Dolling & Murphy (1983), and Dolling & Or (1983)

here were measured with this rake.

Dantec 55M10 constant temperature hot-wire anemometers with symmetrical bridges for enhanced frequency response were used to measure the fluctuating streamwise velocities on the six wires. The wires were operated at an overheat ratio of 0.7 and the frequency responses of all wires were matched at 67 kHz (> 40 U_∞/δ) as indicated by the square wave test.

Digitization of the wire signals was controlled by a VAX 11/750 minicomputer and a CAMAC (Computer Automated Measurement and Control) data acquisition system. The signals were low-pass (0-10Hz) and band-pass (10Hz - 400kHz) filtered. The low-pass signal was digitized at 200 Hz with a LeCroy 8212A A/D converter with 12 bit resolution over a $\pm$5V range. The band-pass signal was amplified and digitized at 250kHz by two synchronized LeCroy TD8210, 10 bit A/D converters with a $\pm$5V range.

Two LeCroy 8800A memory modules accompanied each of the digitizers, allowing 16,384 sequential data points per channel to be stored before a memory dump to the VAX was required. Thus, data was acquired in continuous records of 0.066 secs with a time resolution of 4μsec.; 30 such records were taken at each measuring position. The probe was positioned at several points in the boundary layer at each of four streamwise locations: upstream (unperturbed case), and δ_o = 3, 46 and 91, where s is the streamwise coordinate measured from the end of curvature and δ_o = 22.7mm, the boundary layer thickness at the beginning of the bend. At the first measuring position, the wire closest to the wall was in the wall region; the probe was then moved away from the wall in steps of less than 5mm until the outer wire was in the freestream.

The wires were calibrated in the freestream over the range of velocities they experienced in the boundary layer. The velocity was fitted to a third order polynomial in the measured voltage. This method is not as accurate as the dynamic calibration described in the earlier papers; however, it is certainly sufficient to remove the gross non-linearities in the calibration curve which could distort the correlation results.

3. BROAD-BAND RESULTS

3.1 Unperturbed boundary layer

Cross-correlations were calculated for a number of different wire combinations at several points in the upstream boundary layer. The peak in the cross-correlation and the average structure angle defined earlier are plotted versus y/δ in Fig. 7 for several wire separations, ξ. (In Fig 7(b), the convection velocity U_c was taken as 0.8 U_e).

Both plots show a strong dependence on ξ. However, for $\xi/\delta > 0.15$, the structure angle is independent of ξ within the scatter in the data. This behavior demonstrates the effect of the small-scale, isotropic eddies. For small ξ, small eddies can span both wires and bias the cross-correlation towards $\tau = 0$ ($\theta = 90^o$) on the average. As ξ increases, fewer isotropic eddies pass by both wires, and only the large-scale structures contribute to the cross-correlation. Thus, only the results from the largest wire separations are representative of the large-scale structure angle.

For a given ξ, Fig. 7(a) shows that the peak in the cross-correlation is fairly constant across most of the boundary layer. This observation supports the idea that this technique detects the same structure at different points in the boundary layer; it also implies that similar structures fill the entire shear layer. However, the peak correlation falls off near the wall; the bars in Fig. 7(a) mark the wire locations for three sets of correlations for which one wire is at $y^+ = 30$, $y/\delta = 0.02$. Clearly, the correlation in velocity fluctuations between the wall region and the outer flow is significantly less than the correlation between two points in the outer flow.

The average structure angle is replotted in Fig. 8 using the data from only the largest wire separations. The angle resolution is better than 1°, but the scatter in the data is significantly higher. Dashed lines show the best estimate of the range of θ. The structure angle is not quite constant across the layer; at $y/\delta = 0.3$, $\theta \sim 25^o$-35°, while at $y/\delta = 0.8$, $\theta \sim 35^o$-45°. For $y/\delta < 0.2$, the trend in θ and the results for smaller ξ imply that θ continues to decrease near the wall. This demonstrates the problem inherent in some previous approaches where the outer flow structure angle was measured with one sensor fixed at or near the wall; correlation calculations will show an artificially large lag time (small angle) because of the small structure angle near the wall. These are believed to be the first results to give a quantitative measurement of the local structure angle in the outer part of a subsonic boundary layer. Other workers have measured correlations between vertically spaced sensors and remarked on the peak at non-zero time shift; however, they did not interpret their results in terms of an angle, and/or they measured the correlation with respect to a sensor at or near the wall (see, for example, Chen and Blackwelder (1978), Brown and Thomas (1977), Rajagoplan and Antonia (1979) and Robinson (1986)). The present results give quantitative experimental verification of the smoke flow visualization results of Head and Bandyopadhyay (1981), who suggested that a turbulent boundary layer consists largely of hairpin vortices inclined at 40°-50° to the wall, extending throughout the boundary layer. It is not clear that the correlations presented here are explicitly due to hairpin vortices; however, they are consistent with the observation of large-scale structures spanning the entire shear layer.

These incompressible boundary layer results are similar to those of Spina and Smits taken in a flat plate, zero pressure gradient Mach 3 boundary layer (see Fig. 4). However, the correlation function itself is more shallow in this work than in the Mach 3 study. Figs. 9 and 10 show cross-correlation functions at several positions in the upstream boundary layer for the present work, and for the Mach 3 layer; the wire separations are comparable, and for emphasis the supersonic curve at $y/\delta = .41$ is shown in Fig. 10. The slope of the peak in the subsonic case is only about half that presented by Spina and Smits; in addition, the peak level is slightly lower in the incompressible case. These differences imply that individual contributions to the broad-band correlation are more similar in supersonic boundary layers, whereas in subsonic flows the variations in individual structures are more pronounced and wash out the peak. This point is considered further in the discussion of short-time correlations.

3.2 Perturbed and recovering boundary layers

The peak cross-correlation and average structure angle are plotted in Figs. 11, 12, and 13 for three streamwise locations after the end of curvature, at $s/\delta_o = 3$, 46 and 91.

At the exit from the bend, the correlation levels are significantly lower than the upstream values for comparable ξ/δ, and this effect becomes more pronounced as y decreases; this is true for all wire separations shown. The increase in correlation with y may be a reflection of the suppressed turbulence intensities in the outer half of the boundary layer; however, even well into the outer layer, the correlation level is lower than upstream.

In spite of the reduction in correlation level, the average structure angle at the end of curvature looks much the same as the upstream angle. Again, these angles measured for various ξ seem to converge for $\xi/\delta > 0.15$. Apparently, some large scale structure does survive the curvature and it is still inclined to the wall at the upstream characteristic angle. This should not be taken to imply that the structure is unaffected by curvature. As shown by Alving (1988), the boundary layer does not entrain new mass during the initial $\sim 12\delta_o$ of recovery, and, as shown in Figs. 5-6, the Reynolds stresses are strongly damped in the outer part of the layer during the initial recovery. Hence, the outer layer "structure" indicated in Fig. 11 cannot be very strong. The correlations are probably detecting the weakened remnants of the structure entering the bend, and this only because the background turbulence level is quite low.

Near the wall, the low correlation levels could indicate that the regrowing turbulence is extremely chaotic and that the concept of a "typical" structure is inappropriate. Given the relatively large size of the stress bore after only $3\delta_o$ of recovery, it is clear that some extremely energetic mixing must occur following the step change in curvature. In this case, the correlation levels could be washed out by the great variety in the individual coherent motions. Alternatively, the changes in the correlation levels could reflect a change in shape of the average structure. For example, the vortex modelling work of Perry et al. (1988) and Fernando & Smits (1988) shows that the mean velocity profile U(y) depends only on the distribution of tranverse vorticity, whereas all components of vorticity contribute to the Reynolds stresses. The reduction in mean shear caused by the step change to convex curvature may indicate that the vorticity of the large structure is realigned towards the streamwise direction; that is, the legs of the structure may be brought closer together. Similarly, after the step change to the flat plate, the new structures formed at the wall may reorient themselves again in response to the regrowth in the skin friction. As these structures grow away from the wall into the structures created within the bend, the two-point correlation measurements could be washed out by the dissimilarity of the structures contributing to the correlation.

As the perturbed boundary layer relaxes on the flat plate, the cross-correlation for a given wire separation becomes constant with y, at levels comparable to the upstream levels of similar values of ξ/δ. However, the average structure angle increases across the layer; at $s/\delta_o = 47$, θ increases by about 10°, and at $s/\delta_o = 91$ by about 20°, as compared with the upstream unperturbed layer.

The recovery of the correlation levels may indicate that the strength and/or orientation of vorticity in the structures is similar to that upstream. However, the change in average structure angle is ambiguous. It could mean that the individual structures make a steeper angle with the wall on average. This again could imply a change in structure shape and a redistribution of vorticity; this would be consistent with the elevated Reynolds stress levels measured at these locations. On the other hand, the increase in the average angle could be the result of a bi-modal probability distribution of the individual structure angles. The work of Head and Bandyopadhyay (1981) showed that the heads of the vortex structure are

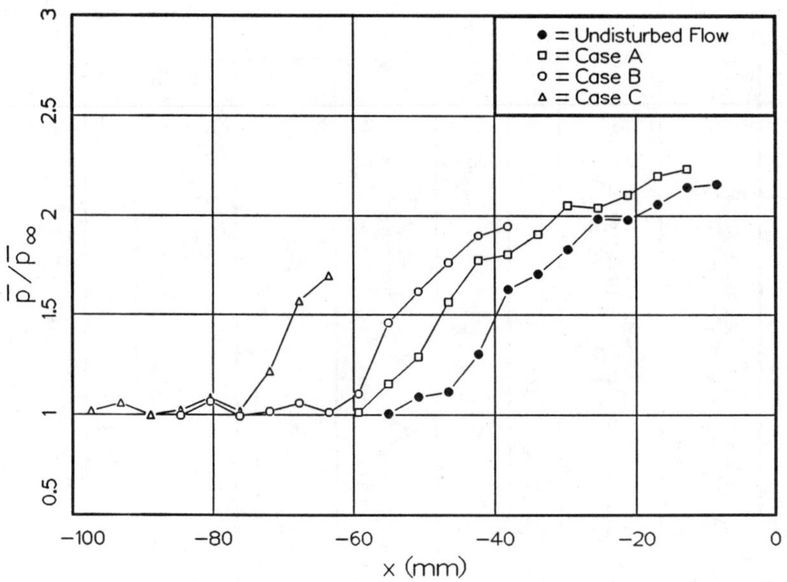

Fig. 8 Mean Wall-Pressure Distributions for Cases A, B, C, and the Case without Blowing to Illustrate the Effects of Moving the Slot

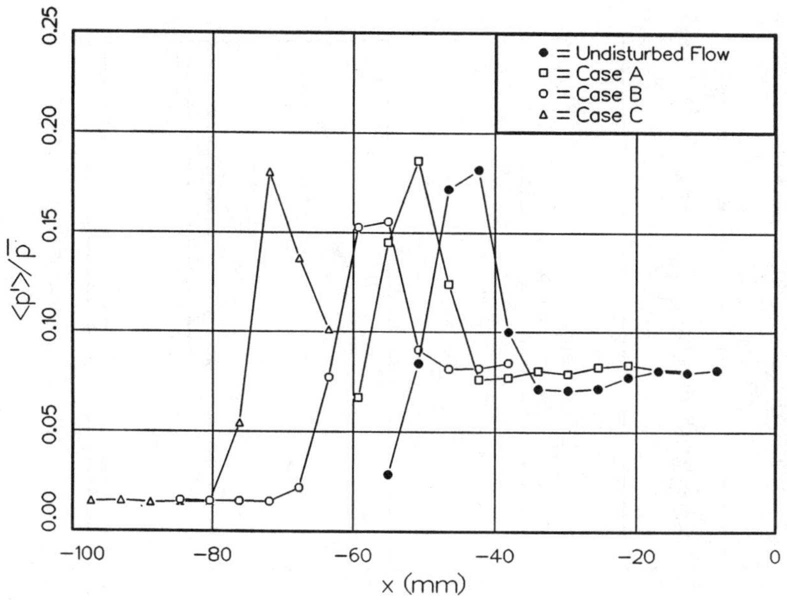

Fig. 9 Rms Wall-Pressure Distributions for Cases A, B, C, and the Case without Blowing

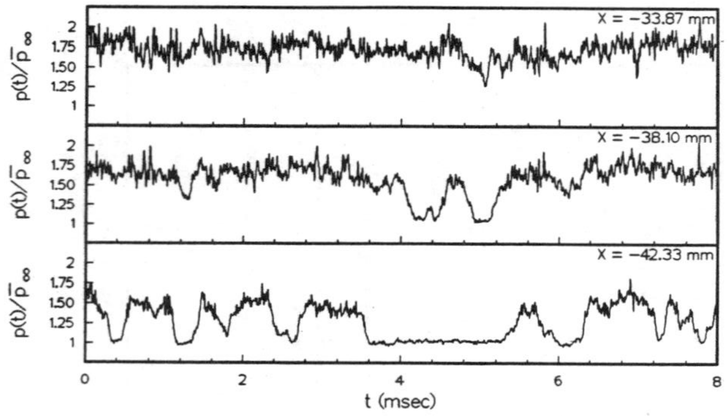

Fig. 10 Six Wall-Pressure Time Histories in the Region of the Unsteady Shock Wave (groups of three are simultaneous)

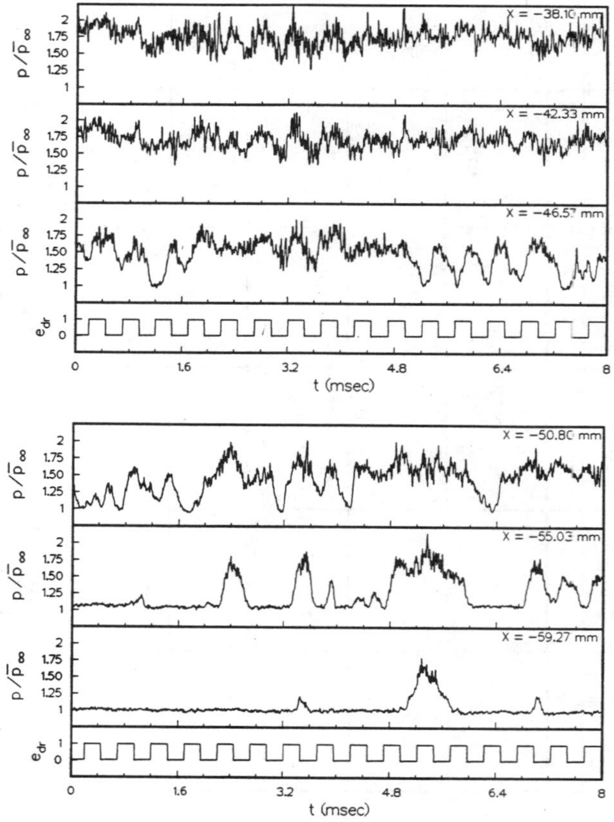

Fig. 11 Six Wall-Pressure Time Histories Upstream of the Corner for Case A (groups of three are simultaneous)

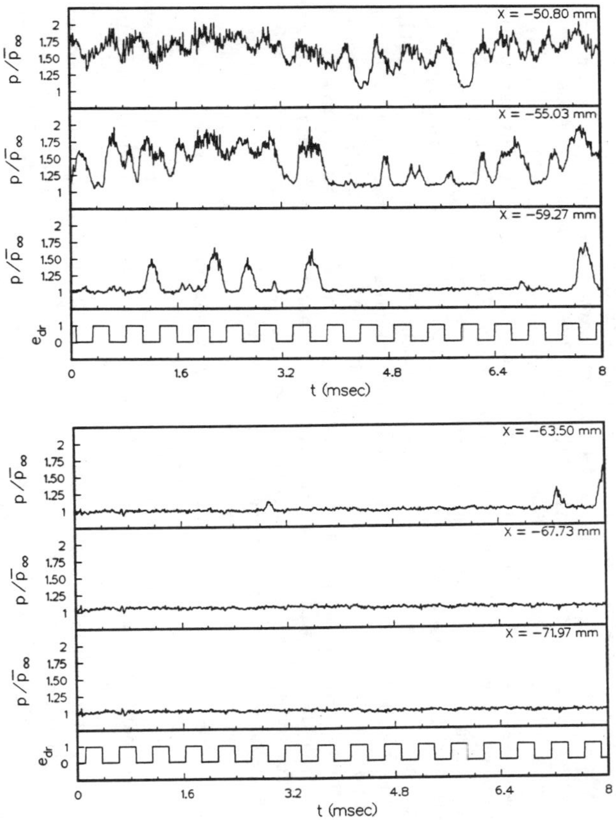

Fig. 12 Six Wall-Pressure Time Histories Upstream of the Corner for Case B (groups of three are simultaneous)

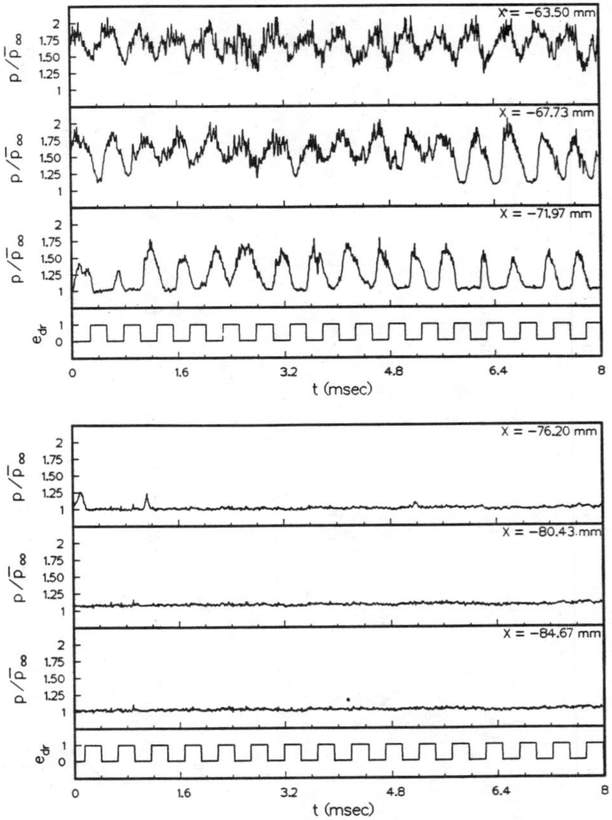

Fig. 13 Six Wall-Pressure Time Histories Upstream of the Corner for Case C (groups of three are simultaneous)

inclined at a larger angle than the rest of the structure. The increase in average structure angle measured here may represent an increase in the population of structures which do not span the entire boundary layer; their heads passing by the sensor locations could shift the average angle towards 90°. This is consistent with the idea that new structures are created near the wall and grow out into the mean shear layer as the boundary layer recovers from its stabilizing perturbation. To investigate this idea, it would be useful if we could examine correlations on a short time basis, and thereby obtain information on the individual structures.

4. SHORT-TIME CORRELATIONS

The broad-band cross-correlation technique used above was extended by Spina and Smits (1987) to look for short pieces of signal that were highly correlated; on this basis, instantaneous structure angles were defined. In brief, they found regions of high turbulence activity on each wire as indicated by the VITA technique; then they looked for high correlations over short times between neighboring VITA events to indicate the passage of an energetic structure (see Chen & Blackwelder, 1978, for a description of the VITA technique). The time delay for the peak in the short-time cross-correlation was converted to an instantaneous angle using ξ and U_c. They found this method to be quite successful in their compressible boundary layer, and the distribution of instantaneous angles agreed well with the average angle measured by the broad-band correlation. The present work discusses the extension of this technique to incompressible layers.

Interestingly, this method was found to give a fairly random distribution of instantaneous structure angles, for all combinations of VITA and cross-correlation integration times and thresholds. This problem illustrates that in this boundary layer the VITA signal obtained from velocity fluctuations is not necessarily a good starting point for the cross-correlation. Sample time histories measured simultaneously on all six wires are shown in Fig. 14(b). This figure shows one sharp rising edge on the four wires closest to the wall ($tU_e/\delta \sim 3.5$); however, before and after the rise, the signals on the four wires are not especially similar. In addition, the time-scale of the rise can be different at different positions in the boundary layer. For the sharp rise in Fig. 14(b), the largest and longest duration velocity gradient occurs on the third wire from the wall. In the case of an extreme change in time-scale, a steep rising edge can be washed out into a gradual rise; this seems to be the case for the fifth wire from the wall, $3.5 < tU_e/\delta < 4$. Besides the problems associated with rising edges, there is no clear indication that the time signals are better correlated near sudden changes in velocity than at other points in time. (Note that little significance should be placed on the high degree of similarity for the two wires closest to the wall, since they are only separated by $\xi/\delta = .02$; the effect of small ξ was discussed in Section 3.1).

Based on these considerations, we decided not to use the VITA criterion and instead we chose simply to evaluate the cross-correlation between two wire signals at a number of randomly chosen points in the time history, using an integration time that was significantly shorter than the time of the whole data series. (This is similar to the short-time correlation method used by Brown and Thomas (1977)). A high level of the peak cross-correlation (compared to the broad-band cross-correlation) was taken to indicate the passage of a large-scale structure, and the time delay for the peak correlation was used to define the instantaneous structure angle. As with the method based on VITA, the distribution of instantaneous angles appeared random, for all combinations of pattern length and correlation threshold.

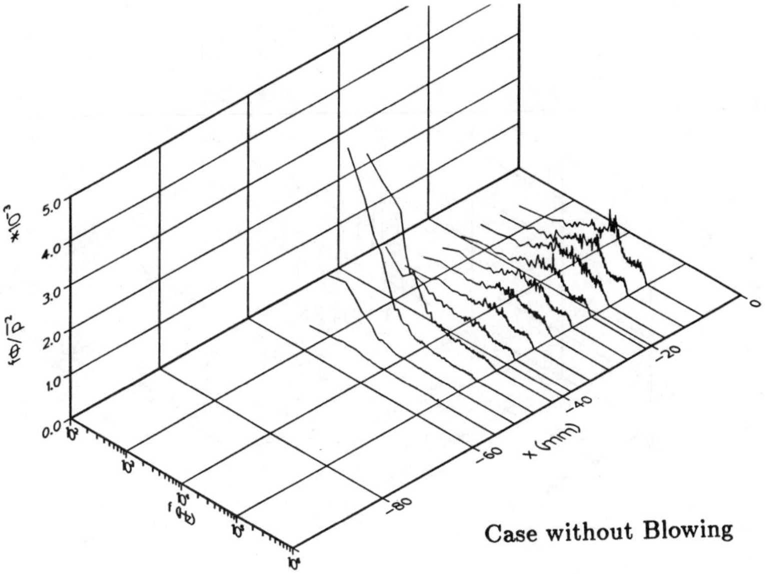

Fig. 14a Energy Spectra of the Wall-Pressure Upstream of the Corner

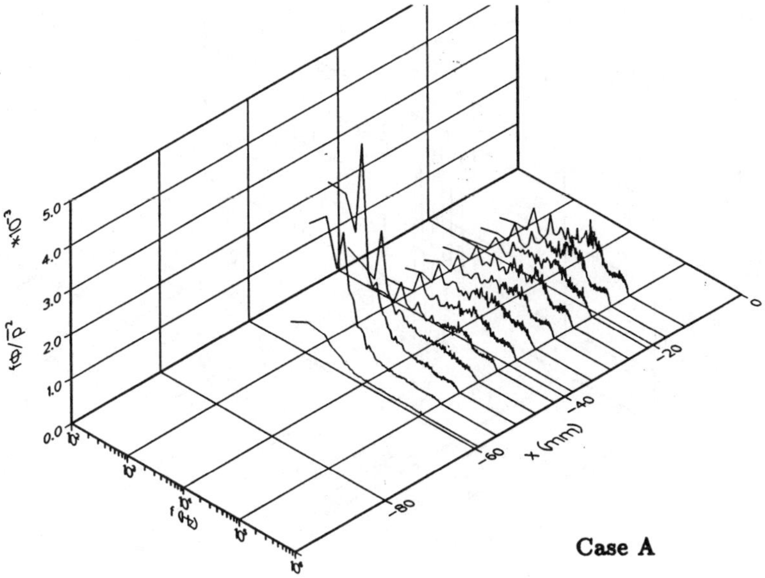

Fig. 14b Energy Spectra of the Wall-Pressure Distribution Upstream of the Corner

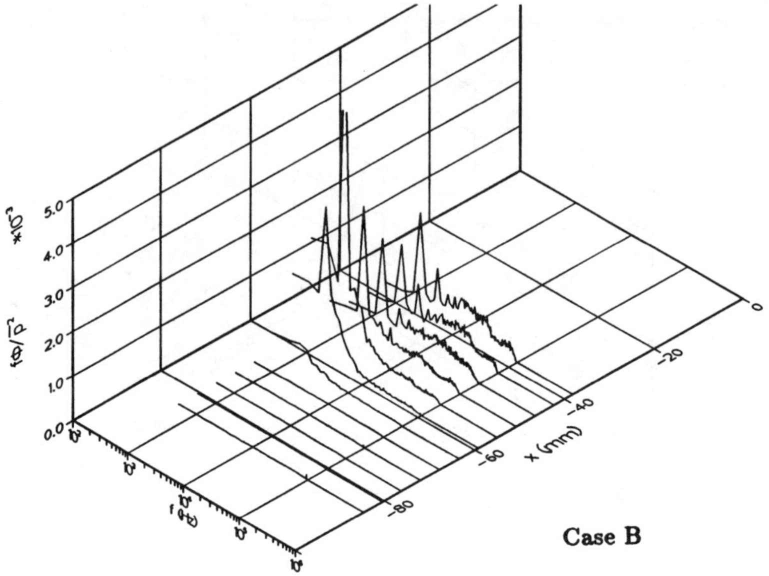

Fig. 14c Energy Spectra of the Wall-Pressure Distribution Upstream of the Corner

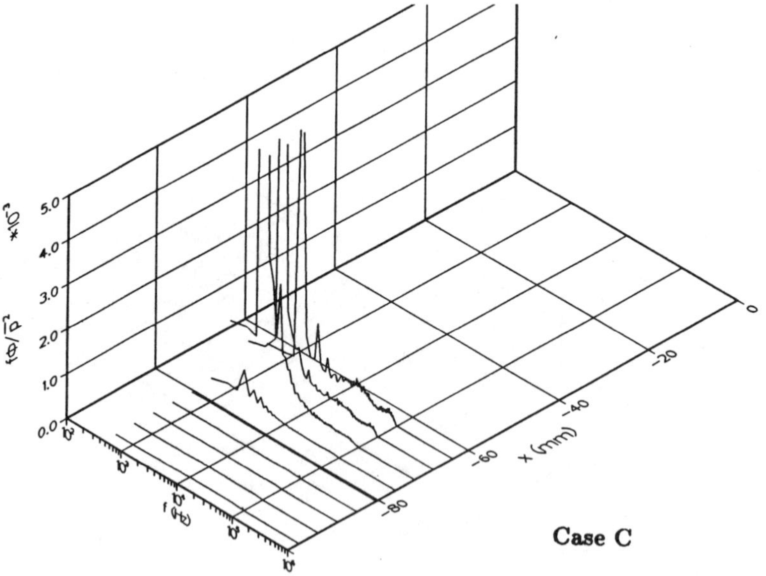

Fig. 14d Energy Spectra of the Wall-Pressure Distribution Upstream of the Corner

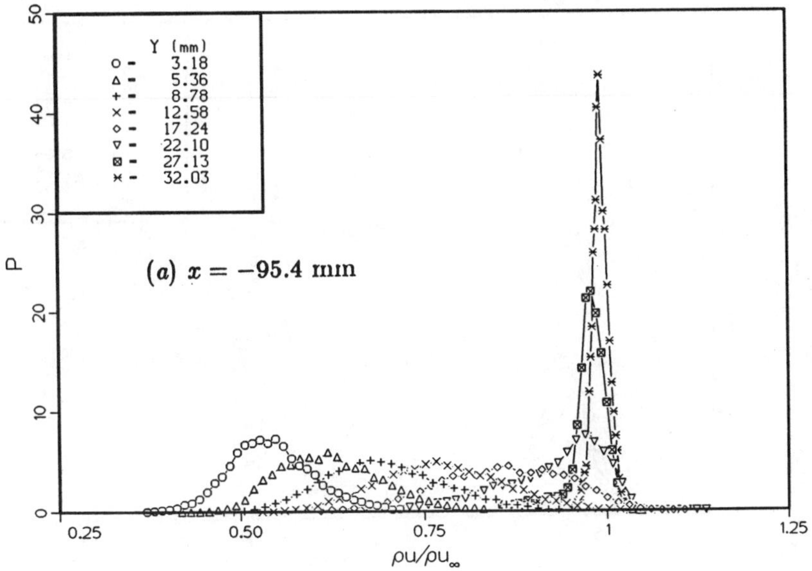

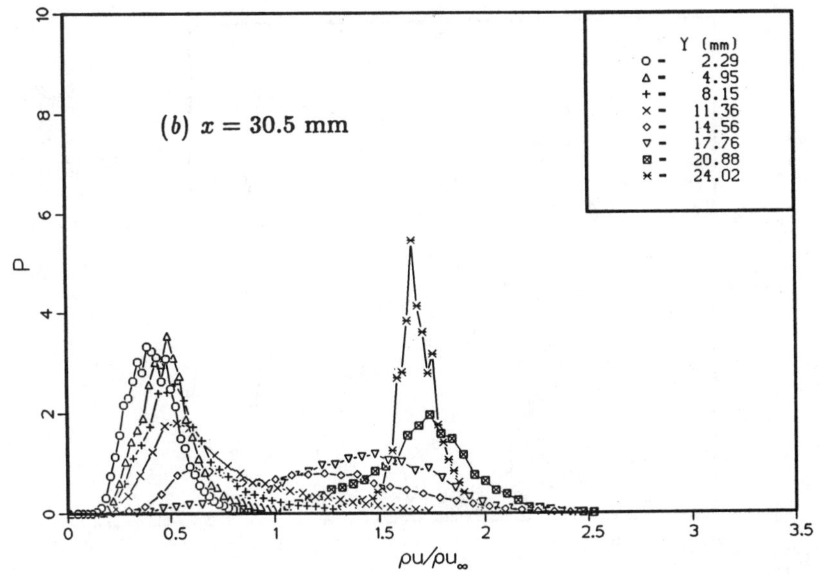

Fig. 15 Probability Density Functions of the Mass-Flux Profiles Upstream and Downstream: (*a*) $x = -95.4$ mm, (*b*) $x = 30.5$ mm, Case without Blowing

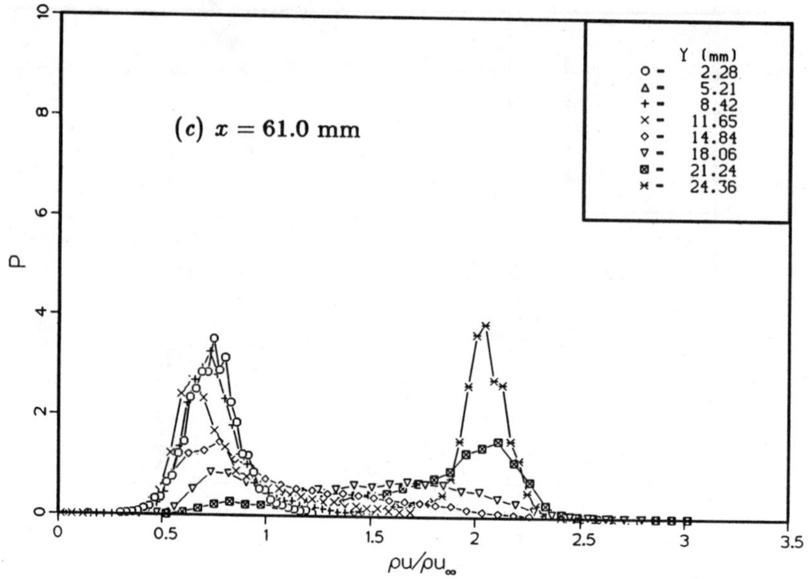

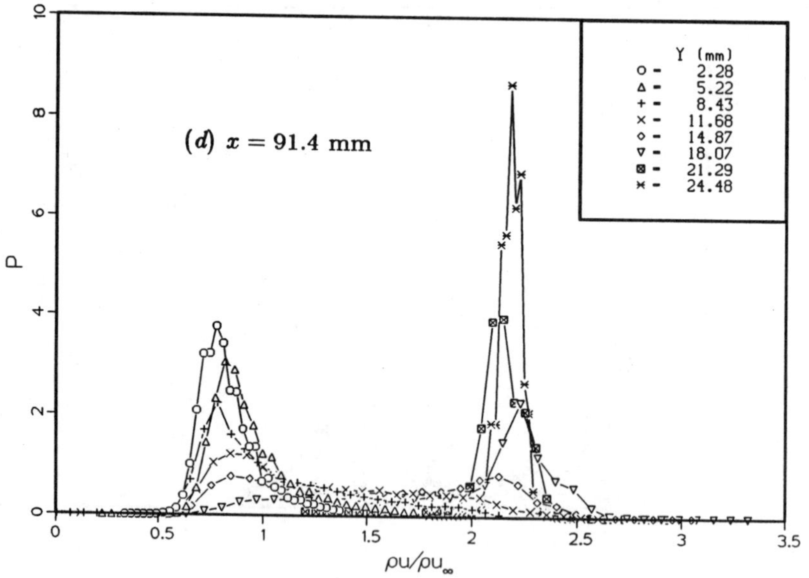

Fig. 15 Probability Density Functions of the Mass-Flux Profiles Upstream and Downstream: Case without Blowing (*c*) $x = 61.0$ mm; and (*d*) $x = 91.4$ mm

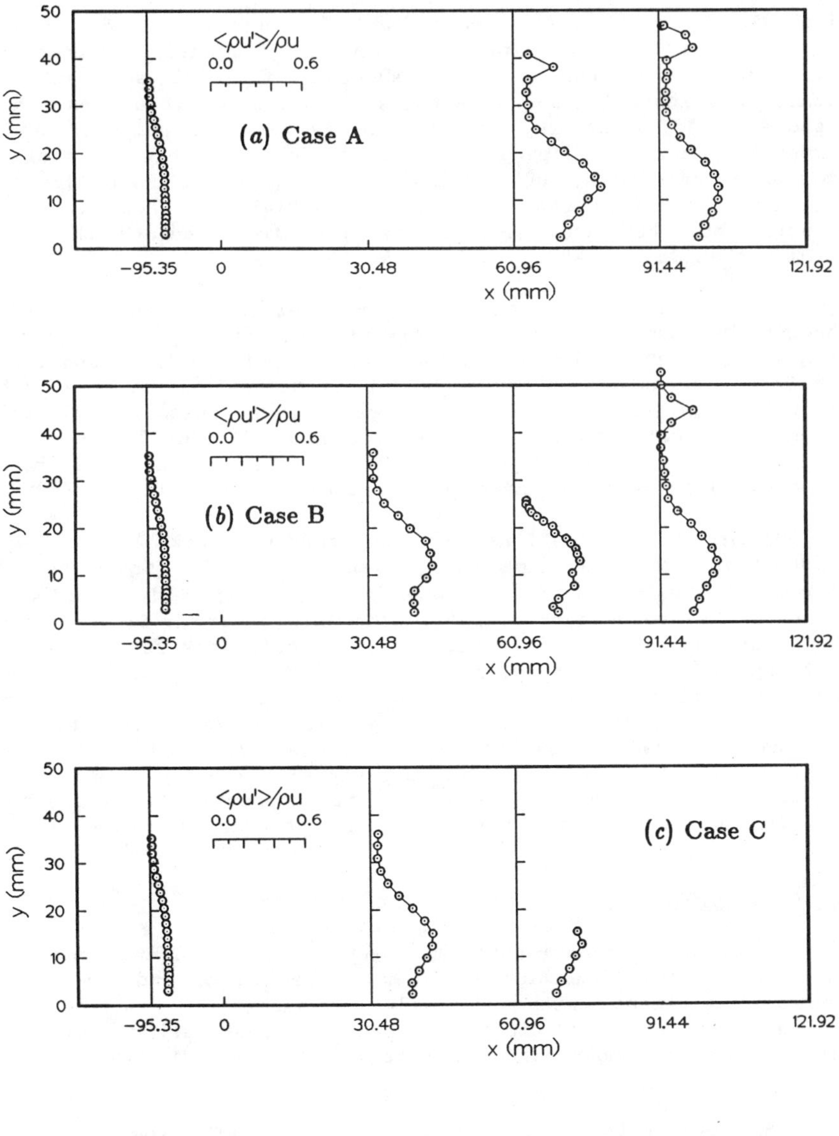

Fig. 16 **Upstream and Downstream Mass-Flux Turbulence Intensity Profiles :**
Case with Blowing

The difficulty in measuring the instantaneous structure angle implies that a velocity-velocity correlation is not ideal for detecting the passage of large-scale vortical structures. Moin and Kim's (1985) numerical results highlight the dangers of deducing information about vortical structures from velocity signals. Other workers have used other combinations of signals for correlations; for instance the temperature-tagged measurements of Chen and Blackwelder show clearly identifiable temperature edges across a large portion of the boundary layer. The success in identifying temperature edges may partially account for the success of Spina and Smits (1987) in identifying instantaneous structures. In compressible flow, the hot-wires measure $(\rho u)'$; that is, the signal is density weighted and therefore temperature marked. Fig. 14(a) shows typical time histories of $(\rho u)'$ taken from Spina and Smits, and the traces display shows sharp edges in $(\rho u)'$ at all three locations each occuring sequentially. These edges are easily detected by VITA and can then be used to calculate the instantaneous structure angle.

The current paper is not the best place to address the differences observed between subsonic and supersonic boundary layers. However, if there is a significant difference in the character of the $(\rho u)'$ and the u' signal, the Strong Reynolds Analogy suggested by Motkovin must be significantly in error. Recent experimental work, especially by Dussauge & Gaviglio (1986), strongly indicates that this is not the case. The differences may have more to do with the large Reynolds number differences between the two flows. These differences are examined more closely by Smits et al. (1988).

Given the limitations of the current work, there may be more information available from the present data. For instance, we have not looked at the effect of high-and low- pass filtering the signals; this technique has been found useful in the work of Brown and Thomas (1977), and Thomas and Bull (1983). In addition, we could look for similarities in the differences between wire signals; for three wires A, B, C, we could examine the correlations between $(u_A - u_B)$ and $(u_B - u_c)$. This technique may highlight the vortical nature of the large-scale structure. Finally, we have not extended the pattern recognition algorithms beyond the correlation level; given the complexity of the signal and the range of time-scales apparent in the time histories, more sophisticated techniques may be required to interpret the signals on an instantaneous basis.

5. CONCLUSIONS

Quantitative experimental measurements of the local structure angle in an incompressible boundary layer have been made, and the results are in general agreement with the flow visualization work of Head and Bandyopadhyay. In addition, correlating a near-wall signal with an outer flow signal is demonstrated to give an artificially low "local" structure angle.

The correlation technique used here can give some insights into the physics of the boundary layer. However, it is difficult to make unequivocal statements about the flow structure, especially in the recovering boundary layer. This problem is inherent in the use of broad-band correlations, since the correlation is not a unique function of its inputs. In addition, the available information is limited to one velocity component and one direction of separation; clearly, more information would aid in the interpretation of results, particularly simultaneous information in the spanwise direction.

Finally, the problems associated with using velocity-velocity correlations in a subsonic boundary layer make the measurement of the instantaneous structure angle quite ambiguous. It is likely that better estimates of the instantaneous structure angle could be obtained using a temperature-tagged signal.

ACKNOWLEDGEMENT

This work was supported by the U. S. Air Force Office of Scientific Research, Grant 85-0126, monitored by Dr. J. M. McMichael. The principal author was also partially supported by a Fannie and John Hertz Foundation Fellowship. The authors are indebted to Mr. E. Spina and Mr. J. Donovan for many helpful discussions on conditional sampling; to Mr. R. Smith for his work in extending the data set presented in this paper; and to Mr. S. Robinson for the use of the hot-wire rake.

REFERENCES

Alving, A. E., 1988, "Boundary Layer Relaxation from Convex Curvature," Ph.D. Thesis, Dept. Mech. and Aerosp. Engineering, Princeton Univ.

Alving, A. E. & Smits, A. J., 1986, "The Recovery of a Turbulent Boundary Layer from Longitudinal Curvature," AIAA Paper 86-0435.

Alving, A. E. & Watmuff, J. H., & Smits, A. J., 1987, "The Relaxation of a Turbulent Boundary Layer Far Downstream of a Short Region of Convex Curvature," AIAA Paper 87-0196.

Bradshaw, P., 1973, "The Effect of Streamline Curvature on Turbulent Flow.," AGARDograph No. 169.

Brown, G. L. & Thomas, A. S. W., 1977, "Large-structure in a turbulent boundary layer", Phys. Fluids, Vol. 20: 243-252.

Chen, C. P. & Blackwelder, R. F., 1978, "Large-Scale Motion in a Turbulent Boundary Layer: A Study of Using Temperature Contamination." J. Fluid. Mech., Vol. 89: 1-31.

Dussauge, J. P. & Gaviglio, J., 1987, "The Rapid Expansion of a Supersonic Turbulent Flow: Role of Bulk Dilatation", J. Fluid Mech., Vol.89, pp. 433-468.

Fernando, E. M. & Smits, A. J., 1988, "Simple Vortex Loop Model for Turbulent Boundary Layers", to be presented at First National Fluid Dynamics Conference, Cincinnatti, Ohio, July 1988.

Gillis, J. C. & Johnston, J. P., 1983, "Turbulent Boundary Layer Flow and Structure on a Convex Wall and Its Redevelopment on a Flat Wall", J. Fluid Mech., Vol. 135: 123-153.

Head, M. R. & Bandyopadhyay, P., 1981, "New Aspects of Turbulent Boundary Layer Structure", J. Fluid Mech., Vol. 107: 297-338.

Moin, P. & Kim, J., 1985, "The Structure of the Vorticity Field in Turbulent Channel Flow. Part 1. Analysis of Instantaneous Fields and Statistical Correlations", J. Fluid Mech., Vol. 155: 441-464.

Morkovin, M. V., 1962, "Effects of Compressibility on Turbulent Flows," Int. Symp. on the Mechanics of Turbulence, p. 367. C. N. R. S., Paris.

Perry, A. E., Li, J. D. and Marusic, I., 1988, "Novel Methods of Modelling Wall Turbulence", AIAA Paper 88-0219.

Rajagopalan, S. & Antonia, R. A., 1979, "Some Properties of the Large-Structure in a Fully Developed Turbulent Duct Flow," Phys. Fluids, Vol. 22:614-622.

Robinson, S. K., 1986, "Instantaneous Velocity Profile Measurements in a Turbulent Boundary Layer", Chem. Eng. Commun., Vol. 43: 347-369.

Smits, A. J., Alving, A. E., Smith, R. W., Spina, E. F. and Donovan, J.F., 1988, "A Comparison of the Turbulence Structure in Subsonic and Supersonic Boundary Layers", to be presented at the Eleventh Symp. on Turbulence, Univ. Missouri-Rolla, October 17-19.

Smits, A. J. & Wood, D. H, 1985, "The Response of Turbulent Boundary Layers to Sudden Perturbations," Ann. Rev. Fluid Mech., Vol. 17: 321-58.

So, R. M. C. & Mellor, G. L., 1973, "Experiment on Convex Curvature Effects in Turbulent Boundary Layers", J. Fluid Mech., Vol. 60: 43-62.

Spina, E. F., & Smits, A. J., 1987, "Organized Structure in a Compressible Turbulent Boundary Layer," J. Fluid Mech., Vol. 182: 85-109.

Thomas, A. S. W. & Bull, M. K., 1983, "On the Role of Wall-Pressure Fluctuations in Deterministic Motions in the Turbulent Boundary Layer," J. Fluid Mech., Vol. 128: 283-322.

Measurement of Single-Point Correlations and Root-Mean-Square Pulsations in the Wall and External Areas of the Swirled Flow in Tubes

A. A. KHALATOV
Institute of Engineering Thermophysics
Ukrainian SSR Academy of Sciences
Kiev, USSR

ABSTRACT

In this report the experiments' results of swirled flow turbulent structure in tube are given. The descriptions of experimental plant, experiment methodics are presented here and the gained results are analized. The swirled flow turbulent structure is shown to differ qualitatively from axial current. The inferences on swirled and curvilinear flows correlations are made.

1. INTRODUCTION

The flow swirling is widely used in technology for organization and intensification of various processes. It is used in gas burners, in plasmatrons for stabilization of electric arc, for intensification of heat and mass transfer in heat exchangers and rectifying columns and controlling gas expenditure through Laval nozzle and many other purposes.

There are several devices of the equipment for swirl creating at the channel inlet, namely tangential, helix, blade-tangential and axial-blade swirlers. Sneck swirlers, twisted ribbons and rotating honeycombs are used also. In certain cases partial swirling with co-axial combination of axial and swirl flows, as well as swirling of co-axial flows in different directions are used.

The averaged characteristics of swirled flow in tubes, diffusors and Laval nozzles are thoroughly studied in paper [1] . There are also given the methods of axial and rotating velocities calculation in axis-symmetrical channels under various boundary conditions (partial swirling, injection, diaphragm etc).

The measurement of root-mean-square pulsations and single point correlations in tube with flow swirled by rotating section were done by D.R.Veske and G.E.Sturov [2] . In papers [1] , [3] , [4] , [5] pulsation intensity constituents in tubes of various relative length with axial-blade, tangential-blade and tangential swirlers were studied. Turbulent pulsations and pulsation spectrums under partial swirling by tangential-slit swirler were measured in paper [6] .

Single point correlations and root-mean-square pulsations of the swirled flow in tubes with sudden widening of section after the swirler are given in papers [7],[8],[9] . Axial-blade swirlers as well as rotating honeycombs were used. Measurements of single-point correlations and pulsations in ring channels of various length were made in papers [10],[11],[12],[13] . The tangential-blade swirlers were used in papers [10],[11] whereas in papers [12],[13] the axial-blade swirlers with flow swirling in different directions were used.

The results gained in these papers testify the complicated character of swirled flows microstructure in the channels. Its main features are anisotropy and high pulsations level, sectional correlation sign change, influence of centrifugal mass forces on turbulent pulsation change character.

The experimental findings of root-mean-square pulsations and single point correlations in tube with swirling of the flow by axial-blade swirlers results are presented in the report. Wide range of swirl parameter changes are investigated. The hypothesis on connection of swirled flow in the tube and current near concave curvilinear channel wall was proposed on basis of gained results.

2. EXPERIMENTAL PURPOSE

The experiments in cylindric channel of 80 mm diameter and 12 000 mm length (l/d= 150) were done. As operating surrounding an atmosphere air was used; and Reynold's number ($Re=\bar{w}d/\nu$) was changed from $5\ 10^4$ to $1.5\ 10^5$. The flow swirling at the tube inlet was created by axial-blade swirler with central body of 43 mm diameter. The swirler blades were spofing according to the power law

$$tg\varphi = tg\varphi_H \cdot (R/r)^n \qquad (1)$$

where φ=arctg w/u- blade swirling angle on r radius; φ_H -blade swirling angle on channel radius R. w,u- velocity axial and rotating constituents. n- swirl intensity index. In these experiments φ_H value had been changing from 15 to 60 degrees, and n index from -1 to 3. Altogether 7 swirlers were studied.

Tube radius flux parameters measurement was done at the distances of 1,4,7,10,20,40,60,80,100,120 and 145 channel diameters from the swirler. Averaged velosity values, root-mean-square pulsations and swirl flow correlations were measured by thermoanemometer, complete P^* and static P pressures by three-channel probe.

3. AVERAGED CURRENT STRUCTURE

Swirled flow current in tube is characterized by two main areas. On the first ($0<\bar{X}<\bar{X}_*$) the distribution of axial and rotating velosity is determined by way of swirling, swirler construction and size. This area has from 0.5 to 3 tube diameters in length.

On the main area of the tube ($\bar{X}>\bar{X}_*$) the radial distribution of velocities and pressures doesn't depend on way of swirling and is determined only by flow swirling intensity degree. When $\bar{X} \approx \bar{X}_H$ the features conditioned by swirling disappear, and flow structure corresponds to axial stabilized current. The length of this area is calculated according to equation given in [1] and consists of up to 150 diameters.

The typical profiles of axial, rotating and summary velocities, statistic and complete pressures for one of the swirlers are given on Figs. 1 and 2. Swirled current in tube comprises a complex flow with continuous character of cross-sectional local parameters change. Such flow contains elements of more simple current types; near wall current region and outer region which contains by-axial zone of back current (or axial velocity "fall") as well as located between them stream region. Swirled flow outer region is non-viscid.

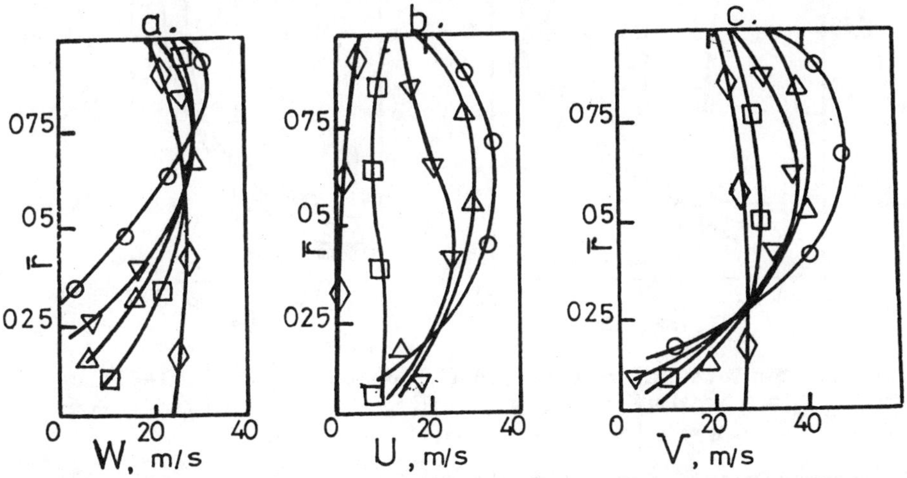

FIGURE 1. Axial (a), rotating (b), and summary (c) velocities in tube. Swirler: $\varphi_H=45°$. $n=3$. $Re=1.1\cdot10^5$. ○ – $\bar{x}=1.05$. △ – $=7.05$. ▽ – $\bar{x}=20$. □ – $\bar{x}=60$. ◇ – $\bar{x}=145$.

Under activities of centrifugical mass forces the air flow is flatten against the wall and axial and summary velocities maximal values are reached at the surface. Along the channel at the expense of friction forces and swirl lessening axial and summary velocities at the surface are reducing and increasening in the vicinity of axis. Thus, in swirled flow simultaneously exist regions with negative and positive longtitudinal velocity gradients. The interaction of these regions is displayed in gas sucktion from outlying district of the channel into by-axial one.

The profile of the rotating velocities is characterized by a curve with maximum. In near wall current region velocity abruptly increases and on a certain distance beyond its limits reaches maximum value. Afterwards rotating velocity decreases to zero on axis. Rotating velocity decreases along the channel and gradually forms "solid body" rotating profile ($u \sim r$). The circulation of rotating velocity ($\Gamma = u\cdot r$) is changing along the curve with maximum, which is located near the surface. As a result of friction forces action in outer region, the zone with constant circulation value ($u\cdot r=const$) is absent.

The distingwishing feature of swirled current is considerable radial gradient of static and complete pressure (Fig.2). Along the tube the pressure at the wall decreases, and increases while the $P^* \approx$ const region is permanently widening; thus in swirled flow simultaneously exist regions with negative and positive longtitudinal pressure gradients. Opposite to axial flows the signs of pressure and velocity gragients coincide.

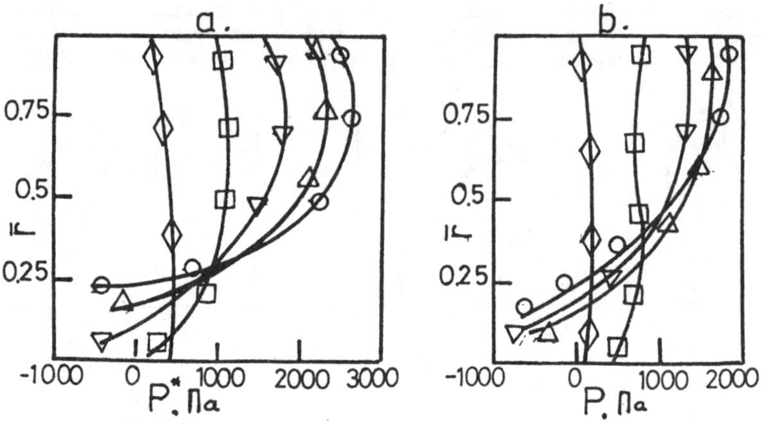

FIGURE 2. Complete (a) and static (b) pressure in swirled flow. Swirler: $\varphi_H = 45°$, $n = 3$. $Re = 1.1 \cdot 10^5$. o – $\bar{x} = 1.05$. Δ – $\bar{x} = 7.05$. ∇ – $\bar{x} = 20$. □ – $\bar{x} = 60$. ◊ – $\bar{x} = 100$.

Radial constituent of velocity in tube is several degrees less than axial and rotating ones.

The swirler (φ_H , n) parameters change doesn't change the character of velocity and pressure distribution in the main part of the tube. Typical points of profiles are preserved, and only quantative characteristics are changed as a result of swirl intensity change. As φ_H and n grows, near wall axial and summary velcčities increase, and back current region is widening. Rotating speed increases along the whole cross-section with its maximum value radius shifted to the wall.

The analysis of swirled flow structure shows that axial and rotating velocities resemblance is maintained under condition of equality of swirling parameter Φ_*, which comprises the ratio of quality flow motion moment to its axial motion quantity in channel radius scale. An one-sign connection ($tg\,\varphi_w = 1.13 \cdot \Phi_*^{0.8}$) exists between integrated swirling parameter Φ_* and swirl utmost angle tangent of channel wall ($tg\,\varphi_w$). This connection is explained by the fact that between the tube wall and back current ragion in the main part of the tube $r \cdot tg\varphi$=const is realized.

In near wall current region axial and summary velocities run in accordance with logarithmical law. The profile of axial velocity in this region is determined by equation [1]

$$\varphi_x = [5.5 + 6.13(\Phi_* - 0.07)^{0.4}] + [5.75 - 4.22(\Phi_* - 0.07)^{0.3}] \lg \eta_x \quad (2)$$

where φ_x, η_x are universal coordinates. Equation (2) was gained for $\eta_x > 30$; in paper [1] the expressions for upper limit of this equation are presented.

With the usage of profile gradual approximation of axial velocity in near wall current region power index n is determined by equation

$$n_0 / n = 1 - 0.78 \cdot (\Phi_* - 0.07) \quad (3)$$

where n_0 is power index for axial flow with Re=idem.

In near wall current region as a result of $z \cdot tg\varphi$=const condition fulfilment the flow swirl angle is changing in the range of 1-2 degrees. That's why rotating velocity circulation profile in this region is determined from equation [1] taking into consideration that $u \cdot r = tg\, \varphi_w \cdot R$.

Equation for local and integral characteristics calculation of swirled flow in tube are given in [1] . There are also given inlet cross-section diaphragm and injection through the penetrable wall experiments results on velocities and pressure distribution.

4. MEASUREMENT OF ROOT-MEAN-SQUARE PULSATIONS AND SINGLE POINT CORRELATIONS

The swirled flow turbulent characteristics measurement was made in ξ, η, ζ coordination system (Fig.3) in which axis ξ coincides with summary flow speed direction, and η axis with channel radius ($V \ll W, U$); root-mean-square pulsations $\overline{w'^2}$ were determined in 3 directions ($i = \xi, \eta, \zeta$) and single point correlations $\overline{w'_\xi w'_\eta}$, $\overline{w'_\xi \cdot w'_\zeta}$. Pulsation intensity was calculated accordingly to equation

$$\varepsilon_i = \sqrt{\overline{w_i'^2}} / V \quad (4)$$

where V is averaged value of summary velocity in point under discussion.

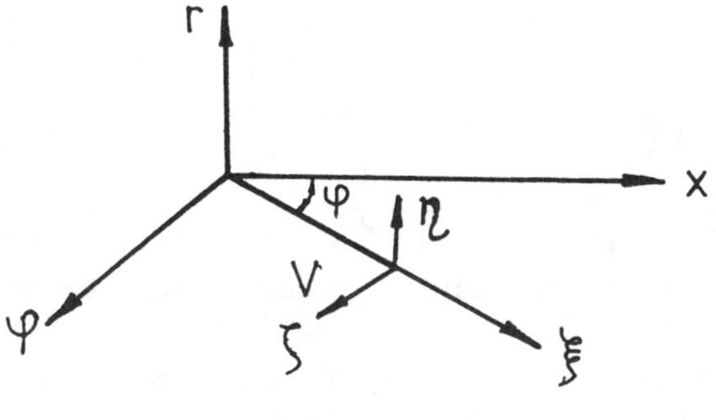

FIGURE 3. Coordinates system connected with current line.

On Fig.4 the typical intensity distribution in three directions in different tube sections is given. In region where the influence of swirl is considerable ($\bar{X} < 50$) all constituents have qualitatevely similar appearance: approximately constant section values in $\bar{\tau} = 0.5 \ldots 1.0$ region and abrupt increase in $\bar{\tau} < 0.5$ region. Under centrifugical forces activity in near wall region longtitudinal and transeversal pulsations decrease and radial pulsations increase if compared with axial stabilized flow. Along the tube the longtitudinal and transeversal pulsations increase and radial pulsations decrease as a result of swirl reduction. Along with it in by-axial region ($\bar{\tau} = 0 \ldots 0.5$) the intensity of all constituent pulsations in this area is changing weakly.

In transitional region $50<\bar{X}<100$ the further increase of ε_ξ and ε_ζ and decrease of ε_η is taking place, it leads to pulsations profile rashaping. It is condidioned by unconsiderable swirl intensity quantity as well as by turbulence diffusion from by-axial region into the near wall one . With $\bar{x}=150$ the pulsation distribution is coorespondent to axial stabilized current practically. Apparently the qualitive and quantative disaccordance of experiment and calculated data in $\bar{X}>50$ region found by different authors using model is explained not enough accurate consideration of turbulence diffusion into near wall zone.

The turbulent tangent friction tension distribution ($\tau_{\xi\eta} = -\rho\overline{w'_\xi \cdot w'_\eta}$) is given on Fig. 5a. The main feature is friction sign change (as well as correlation sign change) along tube section. Along with it the $\tau_{\xi\eta}$ zero value radius and ε_ξ and ε_ζ pulsations intensity minimum practically coincides. Along the tube length the $\tau_{\xi\eta}$ zero value radius decrease takes place, and with $\Phi_* = 0.05$ the radial friction distribution practically corresponds to linear law.

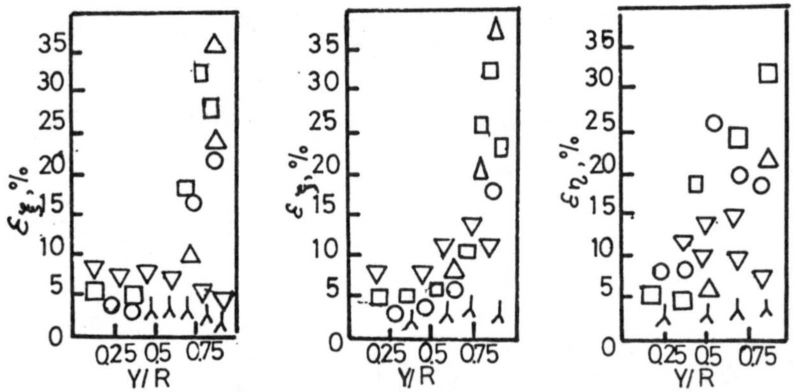

Figure 4. Pulsation intensity in swirled flow. Swirler $\varphi_H = 60°$. $n=3$. $Re=7.3 \cdot 10^4$. ○ – $\bar{X}=7.05$. △ – $\bar{X}=20$. □ – $\bar{X}=50$. ▽ – $\bar{X}=100$. ⅄ – $\bar{X}=150$

The swirled flow section turbulent friction sign change witnesses about the existance of two qualitatively different flow region. Below the τ_* radius, where $\tau_{\xi\eta}<0$ ($w'_\xi w'_\eta > 0$)the centrifugical mass forces are active and flow swirling promotes turbulence increase. In ($\bar{\tau} > \bar{\tau}_*$) near wall ragion the back influence takes place and the swirl suppresses the turbulence. Certain ε_ξ and ε_ζ pulsations intensity increase in $\bar{z}=0.7...1$ region (Fig.4) is conditioned by wall influence .

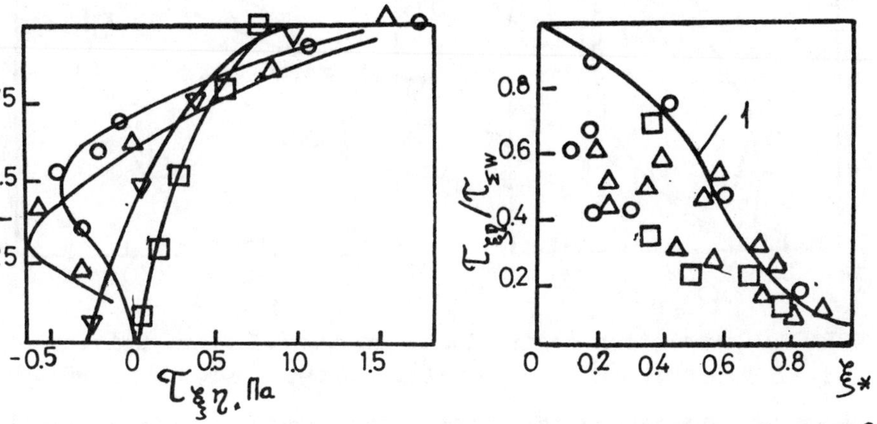

FIGURE 5. Turbulent tension in swirled flow. Swirler $\varphi_H=45°$. $n=3$. $Re=8\cdot10^4$. ○ – $\Phi_*=0.54$. △ – $\Phi_*=0.38$. ▽ – $\Phi_*=0.10$. □ – $\Phi_*=0.05$. б. □ – $\varphi_H=15°$ $n=3$. ($\Phi_*=0.03...0.36$) . ○ – $\varphi_H=45°$ $n=1$. ($\Phi_*=0.04...0.54$). △ – $\varphi_H=45°$. $n=3$. 1- flat plate .

In accordance with Reley hypothesis the correlation sign change for circular flows should occur with $\frac{d}{dz}\left(\frac{u}{z}\right)=0$ and for swirled flows with $\frac{d}{dz}\left(\rho\cdot\frac{u}{z}\cdot\frac{tg\varphi}{(1+tg^2\varphi)^{0.5}}\right)=0$ [1] . The analysis of swirled flow averaged structure shows that there is no sectional derivative sign change: on channel radius this derivative has negative value equal to $\varkappa$. Therefore r_* neutral radius should be determined by $\frac{d}{dz}\left(\rho\frac{u}{z}\cdot\frac{tg\varphi}{(1+tg^2\varphi)^{0.5}}\right)=\varkappa$ condition, where $\varkappa$ is a certain Φ_* swirl intensity function.

On Fig.5b the near wall tube region $\tau_{\xi\eta}$ change is shown. Here $\xi_*=y/\delta$, δ – is the distance from wall to radius, where $\tau_{\xi\eta}=0$. The stabilising swirl of flow in this region leads to turbulent friction reduction if compared with axial current.

The correlation coefficients were determined in accordance with the equation

$$R_{\xi\eta}=\overline{w'_\xi\cdot w'_\eta}\Big/\sqrt{\overline{w'^2_\xi}}\cdot\sqrt{\overline{w'^2_\eta}}\ \cdot\qquad R_\varkappa=\overline{w'_\xi\cdot w'_\eta}\Big/0.5\cdot\sum_i\overline{w'^2_i}\qquad (5)$$

The $R_{\xi\eta}$ and $R_\varkappa$ coefficients change along the tube section (Fig.6) is of sign change character. Near the tube surface the $R_{\xi\eta}$ and $R_\varkappa$ coefficients have small differences from $R_{\xi\eta}=0.45$ $R_\varkappa\approx0.3$ axial current. The tube section $R_{\xi\zeta}=-\overline{w'_\xi\cdot w'_\zeta}/\sqrt{\overline{w'^2_\xi}}\cdot\sqrt{\overline{w'^2_\zeta}}$ correlation coefficients preserves its positive value. Near the wall its value equals to 0.6-0.62 and then it grows up to 0.75-0.77 in channel central part and falls to 0.5 in by-axial region.

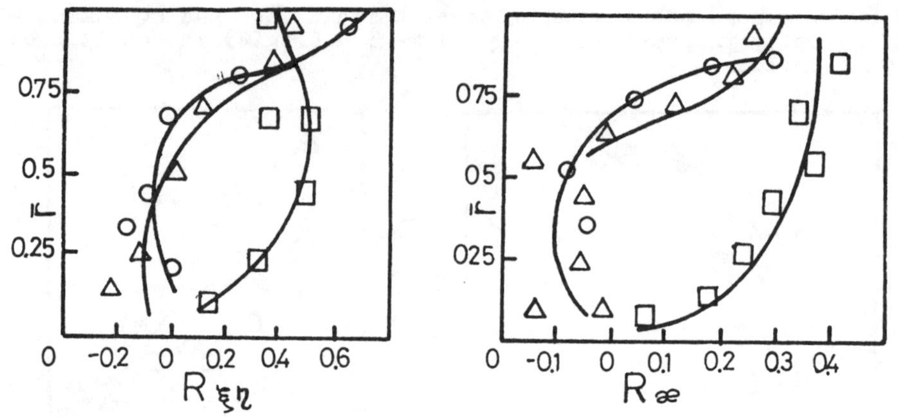

FIGURE 6. Correlation coefficients in swirled flow. Swirler $\varphi_н = 45°$. $n=3$. $Re=1.1\cdot10^5$. o- $\Phi_*=0.72$. Δ- $\Phi_*=0.54$. □- $\Phi_*=0.04$.

In technical appliances relatively short cylindric channels with length up to 15cm are used. In a case of open channel the swirled current microstructure practically doesn't differ from the current in long channel. The boundary conditions in tube outlet (diaphragm) and its surface (injection) change the turbulence characteristics distribution.

The swirled flow microstructure experiments under injection are conducted in porous tube of 30mm diameter and 1120mm length [14]. The $B_* = (\rho V)_w / \rho \overline{W}_в$ injection parameter was changing from 0.03 to 0.01. Here: $(\rho V)_w$ is mass air expenditure through the poroous wall and $\rho \overline{W}_в$ is a mass swirled flow expenditure in the inlet. The experiment were conducted under B_*=const conditions, and main results are presented on Figs.8 and 9.

The injection results in transversal and longtitudinal pulsations constituents increase in channel near the wall region, and radial one- along the whole section (Fig.8). This is conditioned by radial velocities increase in tube under injection. The $\tau_{\xi\eta}$ turbulent tensions near the wall change according to curve with maximum, as well as in axial flows. The injection increase results in surface friction reduction, absolute quantity $\tau_{\xi\eta}$ increase and friction maximum shift from the wall (Fig.9). The injection doesn't change the $R_{\xi\eta}$ and $R_{æ}$ coefficients change radial character along tube section (Fig.9). The injection doesn't change the $R_{\xi\eta}$ and $R_{æ}$ coefficients change radial character along tube section (Fig.9). Apparently it is conditioned by its equal influence on $W_\xi' \cdot W_\eta'$ correlation and root-mean-square pulsations. The $R_{\xi\eta}$ coefficients preserves its positive sign and changes ranging from 0.5 to 0.75 like in inpenetrable tube.

The swirled flow microstructure investigation under outlet section diaphragm was conducted in 80mm diameter and 1120mm length tube. As outlet diaphragm the axis-symmetrical conic confusor with $\bar{d}_к$ relative diameter from 0.5 to 1.0 was used.

Under weak flow the diaphragm doesn't influence the longtitudal pulsation distribution character (Fig.7a). Under moderate (Fig.7b) and strong (Fig.7c) swirling the diaphragm practically doesn't change ε_ξ values in near wall region ($\bar{r}$=0.75 - 1.0) and reduces it in central and by-axial zones. This is conditioned by flow averaged structure change. Under diaphragm in by-axial region the vortical cord with big positive axial velocity gradient is formed. The radius of this cord increases with $\bar{d}_K$ decrease and swirl intensity growing.

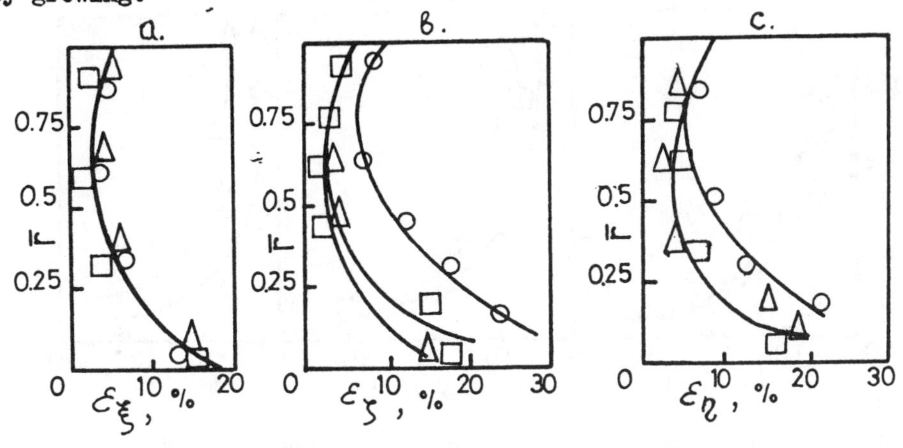

FIGURE 7. Turbulent pulsations in tube with diaphragm. $Re = 5 \cdot 10^4$. $\bar{x} = 8.7$. a: $\varphi_H = 15°$ n=3. в: $\varphi_H = 45°$ n=1. c: $\varphi_H = 45°$ n=3. o - $\bar{d}_K$ = 1.0. □ - $\bar{d}_K$ = 0.75. Δ - $\bar{d}_K$ = 0.5.

5. ON SWIRLED AND CURVILINEAR FLOWS CORRELATIONS

Swirled motion comprises a spiral the curvature radius of which under condition $V \ll W, U$ can be determined from the equation

$$P = r \cdot (1 + tg^2\varphi) / tg^2\varphi \qquad (6)$$

where r is arbitrary point radius, φ is r radius flow swirl angle. On channel surface the current curvature line radius is equal $P_w = R(1+tg^2\varphi_w)/tg^2\varphi_w$, and swirled flow current is analogical to concave surface overflow [1] .

Nevertheless the P_w spiral "geometrical" curvature radius usage doesn't take into account the flow structure. P.Bredshow showed us that current lines curvature changes the mixing way length in accordance with equation [15] :

$$\ell = \ell_0 \cdot (1 - \beta \cdot Ri) \qquad (7)$$

where ℓ, ℓ_0 is mixing way length in curvilinear and linear flows; β is constant. The Richardson's number characterizes turbulent energy production by centrifugical forces to turbulent energy production by tangential friction tensions ratio. For circular motion Richardson's number is determined by equation

$$Ri = 2U / \frac{\partial}{\partial r}(Ur) \qquad (8)$$

and for swirled motion (ρ =const)

$$Ri = 2\,\frac{u}{z^2}\cdot\frac{\partial uz}{\partial z}\Big/\left[\left(\frac{\partial w}{\partial z}\right)^2+\left(\frac{1}{z}\cdot\frac{\partial uz}{\partial z}\right)^2\right] \qquad (9)$$

In equations (8) ,(9) the hypothesis on rotating velocity circulation preservation is used.

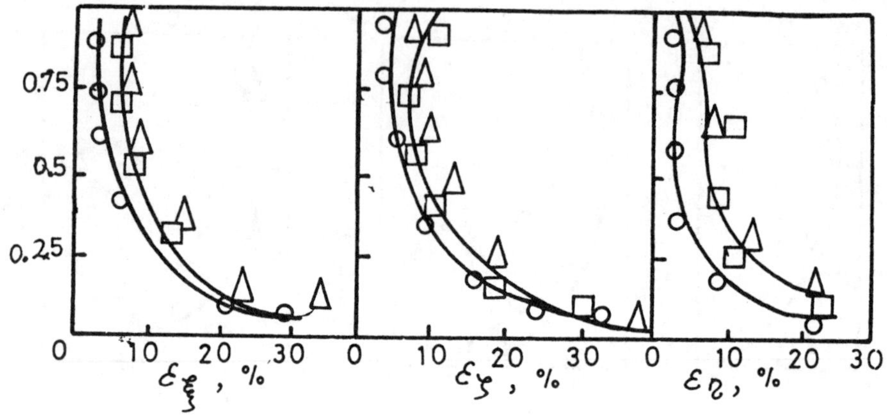

FIGURE 8. Turbulent pulsations in porous tube. Swirler $\varphi_H = 60°$. $n=3$. $\bar{x} = 8.5$. o - $B_* = 0$. Δ - $B_* = 0.0033$. □ - $B_* = 0.01$.

As tube swirled current is curvilinear, its rotating around q radius representative with V velocity, where V is summary swirled flow velocity. Then Richardson's number for such curvilinear flow in accordance with (8) would be determined by equation

$$Ri^* = 2V\Big/\frac{\partial}{\partial q}(V\cdot q) \qquad (10)$$

Let us consider that the condition of transition from swirled current to curvilinear is the equality of Richardson's number determined by equations (9) and (10).

Let us set q radius in form of equation $q = z\cdot f(tg\varphi)$ .Placing into equation (9) the expressions for axial and rotating velocity $u = V\cdot tg\varphi/\sqrt{1+tg^2\varphi}$. $w = V/\sqrt{1+tg^2\varphi}$, after equaling (9) and (10) we'll receive the equation for determining the unknown function f:

$$\frac{f\cdot tg^2\varphi}{q^2(1+tg^2\varphi)}\cdot a^2 + \frac{f^2\cdot b\cdot tg\,\varphi}{q^2\sqrt{1+tg^2\varphi}}\cdot V\cdot q = \qquad (11)$$

$$= \frac{1}{q^2}\frac{\partial q}{\partial z}\cdot a^2 + 2\,\frac{\partial q}{\partial z}\cdot\frac{c\,q + b\cdot f\cdot tg\,\varphi}{q^2\sqrt{1+tg^2\varphi}}\cdot Vq + \frac{\partial q}{\partial z}\cdot\left(c^2 + \frac{f^2}{q^2}\cdot b^2\right)$$

Here:

$$a = \frac{\partial}{\partial q}(Vq)\cdot \quad b = \frac{\partial}{\partial q}\left(f^{-1}\cdot tg\,\varphi/\sqrt{1+tg^2\varphi}\right). \quad c = \frac{\partial}{\partial q}\left(1/q\sqrt{1+tg^2\varphi}\right)$$

Equaling the coefficients under a and a^2 and in the free term we'll receive three identities

$$f \cdot tg^2\varphi/(1+tg^2\varphi) = \frac{\partial q}{\partial z} \qquad (12)$$

$$f^2 \cdot b \cdot tg\varphi = 2 \cdot \frac{\partial q}{\partial z} \cdot (c \cdot q + f \cdot b \cdot tg\varphi) \qquad (13)$$

$$\frac{\partial q}{\partial z} \cdot (c^2 + \frac{f^2}{q^2} \cdot b^2) = 0 \qquad (14)$$

As far as $\partial q/\partial z \neq 0$, from equation (14) follow the next two conditions: c=0 and b=0, which automatically suit equation (13) . These conditions under joint integrating taking into consideration that q=r · f allow to gain equation

$$z \cdot tg\varphi = const \qquad (15)$$

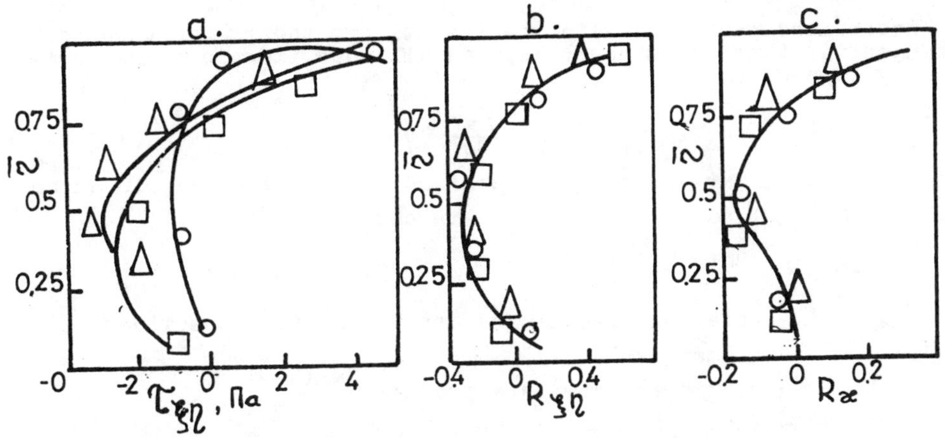

FIGURE 9. Turbulent tensions (a) and correlation coefficients (b.c). Swirler $\varphi_H=60°$. n=3 . $\bar{x}=8.5$. ○ – $B_x=0$. △ – $B_x = 0.033$. □ – $B_x = 0.010$.

The equation (15) shows that correlation of swirled flows ($Ri = Ri_*$) takes place only under condition of presence in swirled flow of the region with r·tg φ =const law. Such region higher than the back currents zone radius under $\bar{x} > \bar{x}_*$ exists and takes place under various initial swirl laws.

From equation (12) taking into consideration r · tg φ =const condition we'll receive an expression for unknown function f

$$f = tg\varphi/\sqrt{1+tg^2\varphi} \qquad (16)$$

and then an equation for q radius

$$q = z \cdot tg\varphi/\sqrt{1+tg^2\varphi} \qquad (17)$$

Thus, the swirled motion on r radius with axial w and rotating u velocity is equivalent to curvilinear concave flow which has on $q=r\cdot tg\varphi / \sqrt{1+tg^2\varphi}$ radius the $V=w\cdot\sqrt{1+tg^2\varphi}$ (tg= u/w) velocity. The equivalent concave surface curvature radius is determined by $q_w=Rtg\varphi_w / \sqrt{1+tg^2\varphi_w}$ equation and reduces along the tube ($d\,tg\varphi_w / d\,x < 0$). Thus thr swirled current in tube corresponds to concave surface overflow with the variable which is reducing in length by curvature radius. Velocity distribution in each section is determined by V swirled flow summary velocity profile according "contraction" of a radial coordinate, determined by equation (17). The same concerns the turbulence characteristics resulting from the $Ri = Ri^*$ condition realization.

The comparison of equation (6) with expression $q_w=R\cdot tg\varphi_w / \sqrt{1+tg^2\varphi_w}$ shows that spiral curvature "geometrical" radius considerably exceeds surface curvature radius gained from Richardson's number equality condition.

In accordance with aforesaid the Prandtl hypothesis for turbulent tangential tension is equivalent curvilinear flow is characterized by expression

$$\tau = \rho\cdot \ell^2\cdot\left(\frac{1}{q}\cdot\frac{\partial V\cdot q}{\partial q}\right)^2 \qquad (18)$$

where ℓ mixing way length is determined by equation (7), in which $Ri=2V/\partial Vq/\partial q$, and constant β is equal 5 [15].

6. SYMBOLS

- U, V, W - rotating, radial and axial velocity constituents
- V - summary velocity
- P, P^* - static and complete pressure
- φ - flow swirl angle
- z, φ, ξ - cylindric coordinates
- R - channel radius
- ξ, η, ζ - coordinates connected with current spiral
- Φ_* - flow swirl intensity ($\Phi_* = M/K\cdot R$)
- M, K - flow motion moment and quality
- ε_i - pulsation intensity ($i = \xi, \eta, \zeta$)
- $\overline{w'^2}$ - root-mean-square pulsations ($i = \xi, \eta, \zeta$)
- $\overline{w_i' w_j'}$ - single point correlations ($ij = \xi, \eta, \zeta, i \neq j$)
- ℓ - mixing way length
- Ri - Richardson's number
- ρ - flow density

7. REFERENCES

1. Shchukin, V.K. and Khalatov, A.A. Swirled Flows' Heat and Mass Exchange and Hydrodynamics in Axis-Symmetric Channels, pp.1-200, Mashinostroyeniye, Moscow, 1982.

2. Veske, D.R. and Sturov, G.E. Experimental Study of Turbulent Swirled Current in Cylindric Tube, News of USSR Siberia Dept. Academy of Sciences Serial, N°13, edition 3, pp.3-7, 1972

3. Glebov, G.A. and Matveev, V.B. Experimental Study of Strongly Swirled Current in Tubes. Near Wall Jet Flows, Enginnering Physics Institute of USSR Siberia Dept. Academy of Sciences, pp.81-86, Novosibirsk, 1984.

4. Wolf, L.F. Rotating Motion Fading in Turbulent Flow Measurement, Rocket Engineering and Cosmonautics, pp.214-216, 7, N°5, 1965

5. Belousov, A.N. The Study of Turbulent and Accoustic Characteristics of Swirled Air Flow in Short Vortex Chambers. Vortex Effect and its Application in Technology, Kuybishev Aviation Institute, pp.303-307, Kuybishev, 1981.

6. Volchkov, E.N., Spotar S.Y. and Terekhov V.I. Turbulent Characteristics of Boundaried Jet. Near Wall Jet Flows, Engineering Physics Institute of USSR Siberia Dept. Academy of Sciences, pp.5-13, Novosibirsk, 1984.

7. Budunov, N.F. The Study of Seoaration and Swirled Flows of Non-Pressurized Liquids in Alternate Section Channels, pp.1-19, Novosibirsk, 1973.

8. Khabib, M. and Whitelow J. The Characteristics of Boundaried Coaxial Jets With and Without Flow Swirling, Engineering Calculations Theoretical Basis, Serial D, 102, N°1, pp.163-171, 1980.

9. Ganesan, Sunder Ramon I. Murthy B. An Experimental Study of the Effect of Cold Model of an Intensity of Turbustor., Indian Journal of Technology, 13, N°9, pp.386-389, 1975.

10. Hsuan, I. Boundary Layer Along Annular Walls in a Swirling Flow. Trans. pp767-775, ASME, 80, N°4, 1958.

11. Scott, S. and Rask, D. Turbulent Viscosity in Liquid Swirled Flow in a Ring Channel, Engineering Calculations Theoretical Basis, serial D., pp.147-159, N°4, 1973.

12. Sviridenkov, A, A, and Tretyakov, V.V. The Experimental Study of Swirled Jets Turbulent Polarities Mixing on Initial Area in Ring Channel, Engineering and Physics Journal, pp.205-210, N°2, 1983.

13. Sviridenkov A.A., Tretyakov V.V. and Yagodkin V.I. On effectivity of Mixing of Swirled in Different Directions Coaxial Flows, Engineering and Physics Journal, pp.407-413, 41, N°3, 1981.

14. Shchukin, V.K. and Khalatov, A.A. Swirled Flow Microstructure in Penetrable Cylindric Channel. Vortex Effect and its Application, Kuybishev Aviation Institute, pp.267-271, Kuybishev, 1981.

15. Bradshow, P. The Analogy Between Streamline Curvature and Buyancy in Turbulent Shear Flow, Journal Fluid Mech., pp.177-191, 36 pt 1, 1969.

Part 6
INFORMATION DERIVABLE FROM HEAT TRANSFER DATA

SUMMARY LECTURE

Structure of Turbulence in Heat Transfer

E. M. KHABAKHPASHEVA
Institute of Thermophysics
Siberian Branch of the USSR Academy of Sciences
Novosibirsk, USSR

ABSTRACT

Results of experimental investigations into hydrodynamic and thermal structure of turbulent flow in the immediate proximity to the wall are presented. Experiments reveal the analogy of momentum and heat transfer mechanisms in equilibrium shear boundary layers and a decisive role of intermittent coherent structures in these processes. The level of knowledge on hydrodynamic and thermal structure of a flow under more complicated conditions is so far insufficient. Using some examples it is shown how variations in temporal and spatial scales of near-wall structures change the dimensionless heat transfer coefficients. The cases are presented when the investigation of heat transfer processes may be a useful source of information for turbulence structure studying.

1. INTRODUCTION

In recent twenty years substantial success has been achieved in the investigation of near-wall turbulence. This stimulated the new stage of turbulent heat transfer study. It was directed to detailed investigations of the role of the intermittent coherent structures in heat transfer processes.

Not only turbulent investigations enrich heat and mass transfer theory. Often data obtained in thermal investigations can yield a new information concerning turbulence and stimulate more profound study of its mechanism as well. The paper touches on both aspects of the problem.

One may point out also that hot-wire and hot-film probes long been used to study the statistical characteristics of turbulent flows, turned out to be useful to obtain information on coherent structures in a turbulent boundary layer too. Thus the arrays of hot-wire probes in [1,2] and hot-wire probe measurement complemented by visual investigations in [3-5] were utilized to study coherent structures in different flows. Using temperature as a passive contaminant sudden changes in temperature, coherent across the entire boundary layer, have been found [6-8].

Almost till the middle of our century the region of a viscous (thermal) sublayer was assigned a passive role. However at present it is well known that in a majority of cases the processes occuring in the immediate proximity to the wall (viscous and buffer layer) define the mechanism of momentum and heat transfer over the entire boundary layer. The first part of the paper is devoted to the measurements of temperature pulsations and instantaneous values of turbulent heat fluxes, their connection with the flow structure in the vicinity of the wall. It is based mostly on the original investigations of author and her colleagues.

The level of knowledge on hydrodynamic and thermal structure of a flow under complicated conditions is insufficient. Using dilute high-polymer solutions with viscoelastic properties it is shown how variations in temporal and spacial scales of near wall structures change the dimensionless heat transfer coefficients. Another examples such as heat transfer in the recirculating zones and combined forced and thermogravitational convection show the lack of turbulence structure knowledge to heat transfer predictions. The cases are also presented when the investigation of heat transfer processes may be a useful source of information for turbulence structure studying. This part of the paper based on the investigations of different authors is of surveying nature.

2. ON SIMILARITY OF VELOCITY AND TEMPERATURE FLUCTUATIONS NEAR THE WALL

It should be noted that considerably less attention was paid to the study of thermal structure of a viscous sublayer than to that of its dynamic structure. Nevertheless, in [9-25] extensive data have been obtained concerning the statistical characteristics of temperature fluctuations in flows of various fluids. Some of them contain sufficiently complete information on temperature fluctuations in the vicinity of the wall. A detailed analysis of the results of these investigations is given in a keynote report at the 8-th International Heat Transfer Conference [26] . Only some information from this survey will be given here.

One of the first results of our viscous sublayer investigations was an experimental verification of the hypothesis that the ratio of RMS fluctuations of the longitudinal velocity σ_u to the value of the mean local velocity U is constant, Figure 1-a. These data were obtained by flow visualization method. Solid particles of about micron sizes were introduced into the flow. The particles were photographed in flash light and formed tracks on the film, which consisted of sequences of points. These tracks were processed using a semiautomatic device and the kinematic characteristics were calculated by a computer [27,28] .

Temperature was measured by thermocouples with the junction sizes of about 10 microns. Figure 1-b shows RMS temperature fluctuations related to the local temperature difference. These measurements have been conducted near a thick-wall cop-

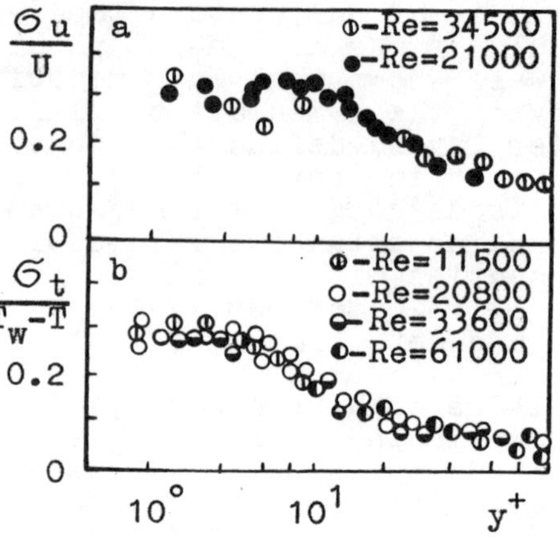

FIGURE 1. Relative intensities of velocity (a) and temperature (b) fluctuations.

per surface (for the role of surface material see [26]). As is seen, the behaviour of the longitudinal velocity and temperature fluctuations at the wall is nearly identical. But the ratios σ_u/U remain constant over the entire viscous sublayer, whereas the values σ_t/T_w-T remain constant in a somewhat thinner layer.

The distributions of skewness S and flatness factor F for temperature fluctuations and for the longitudinal velocity pulsations near the surface are analogous too. The positive values of S_u and S_t in the viscous sublayer characterize the important role of inrushes of cold liquid portions into the viscous sublayer from the bulk flow. The positive values of the flatness factors point to the considerable amplitudes of these fluctuations [20,24,26] .

Histograms of turbulent velocity and temperature fluctuations (Figure 2.) show that the most probable values of u/σ_u and t/σ_t change their signs with the increase of distance from the wall. The positive fluctuations of these values acquire the maximum probability, whereas the amplitudes of fluctuations, maximum by their absolute value, have negative signs. The latter characterize intensive ejections of decelerated (and therefore more heated) liquid from the wall to the central part of the flow.

It must be pointed out that Professor Z. Zarić was the first who studied in detail the temperature fluctuations in the proximity of the wall [9,10,12,13,22] . At the 1-st ICHMT Symposium (1968) he presented a survey concerning the investigations of the near-wall layer, including his measurement data on the statistical characteristics of velocity and temperature fluctuations. Professor Zarić suggested to consider the form of probability density distribution curves for velocity and temperature fluctuations in a viscous sublayer as

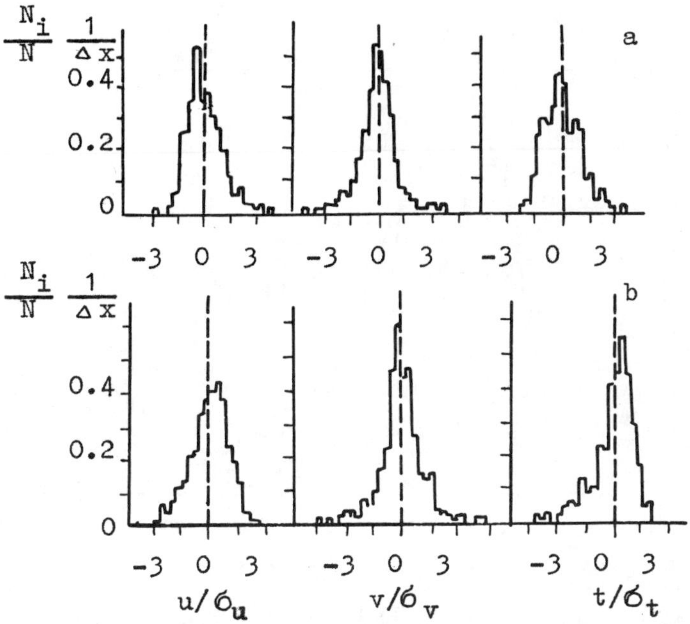

FIGURE 2. Histograms of turbulent velocity and temperature fluctuations at $y^+=3$(a) and $y^+=20$(b); Δx - interval of histogram.

a result of the interaction between the three intermittent phases of a turbulent flow: inrushes of cold liquid from the flow core, ejections of the heated liquid from the wall, and a comparatively quiescent phase. With the help of conditional sampling procedure Professor Zarić has determined the probability density distribution of velocity and temperature fluctuations for the phases of the flow. The analysis has shown that the distribution of the quiscent phase is close to the Gaussian distribution and irregularities of the complete probability distribution curve from this law are due to the superposition of the three distributions corresponding to different phases, Figure 3.

The data presented could be complemented by the results of measurements of temperature fluctuation spectra [13,16,20-22], autocorrelation coefficients [22] and their comparison with the analogous data on longitudinal velocity component fluctuations. The analysis of these data is given in [26] . It confirms the possibility to use the temperature measurements for investigation of the flow structure.

3. ANALOGY BETWEEN MOMENTUM AND HEAT TRANSFER IN NEAR-WALL REGION

Turbulent heat fluxes are determined by the correlations of velocity and temperature fluctuations. For forced convection in tubes and boundary layers the heat transfer by mean motion

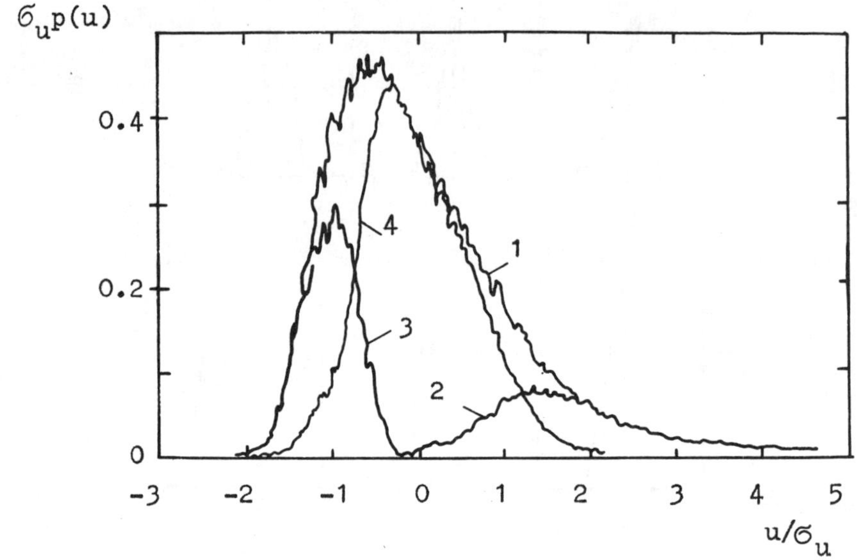

FIGURE 3-a. Separation of the intermittent phases for the velocity signal at $y^+ = 4.9$; 1 - total signal; 2 - inrush phase; 4 - quiescent phase [13] .

FIGURE 3-b. Separation of the intermittent phases for the temperature signal at $y^+ = 4.4$.

is usually well above the turbulent longitudinal heat flux. Therefore only data on vt-correlations are presented here.

There exist very few experimental measurements of correlations between temperature and velocity fluctuations within a viscous sublayer. One can mention, for example, the measurements of the correlation coefficients for temperature and longitudinal velocity fluctuations in an air flow [18] and the data on the spectral characteristics of temperature and velocity correlations at $y^+ \geq 15$ [15] . Only recently the use of stroboscopic flow visualization in our experiments [24,29,30] , laser-doppler anemometry in [31] and hot-wire anemometers with special probes in [32] has allowed to take the first steps on this way.

In our experiments measurements were made by synchronous readings of a temperature probe at a given point of the flow and the velocity field in the proximity of the probe. Data files with velocity and temperature information were recorded into the magnetic storage of a computer and processed using the algorithm allowing to determine the statistical characteristics of velocity and temperature fluctuations and their correlations, histograms of these values, as well as to analyse the structure of momentum and heat fluxes. The measurements have been made in a water flow, when the values of y^+ range within 2.7 up to 22. The authors of [31] used simultaneously a laser-doppler anemometer to measure velocity fluctuations and a hot-wire anemometer operating in the regime of a resistance thermometer. The experiments were conducted in air flows at small Reynolds numbers. In [32] a specially designed three-wire probe allowed to measure simultaneously velocity and temperature fluctuations.

Synchronous measurements of velocity and temperature fluctuations provide the possibility to obtain instantaneous values of turbulent momentum and heat fluxes and to analyse their interconnection. In our experimental investigations of the viscous sublayer structure we, as well as the other authors, made a wide use of a quadrant analysis [28,33,34,37 et.al.].

Depending on signs of fluctuations, the instantaneous values uv and vt get to one of the four quadrants on the respective planes (here $t=T_w - T$ - temperature difference pulsation):

v (axis); u (axis)	
1 $u<0$ $v>0$	2 $u>0$ $v>0$
$u<0$ $v<0$ 4	$u>0$ $v<0$ 3

v (axis); t (axis)	
1 $t<0$ $v>0$	2 $t>0$ $v>0$
$t<0$ $v<0$ 4	$t>0$ $v<0$ 3

Quadrants 1 contain instantaneous values of products uv and vt characterizing the motion of decelerated and therefore more heated liquid from the wall, quadrants 3 contain those for the inrush of cold fluid with the velocity greater than the local averaged one from a more distant regions to the wall.

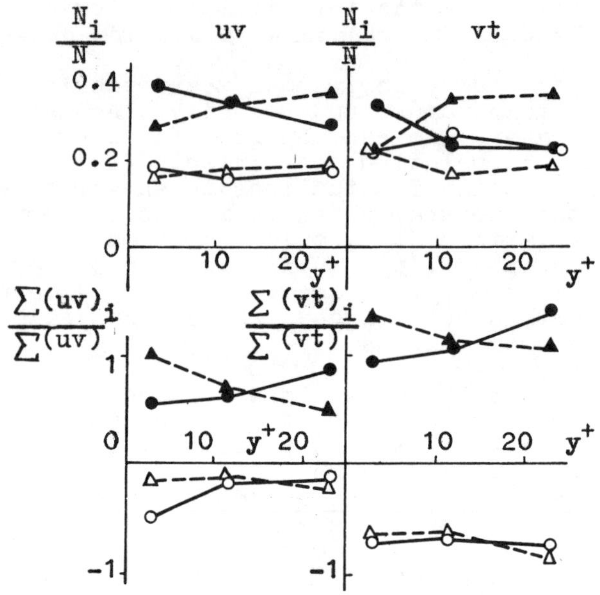

FIGURE 4. Probabilities for different combinations of u, v and t in the four quadrants, and their contributions to Reynolds stresses and heat fluxes; ●,o,▲,△ correspond to 1,2,3,4 quadrants.

Figure 4 gives probabilities for different combinations of fluctuations, as well as contributions made by the products of the fluctuations in different quadrants into the value of the turbulent shear stress and turbulent heat flux. For the velocity fluctuations in the viscous sublayer ($y^+=2.7$), the maximum probability belongs to events with $u<0$ and $v>0$ (quadrant 1), then with $u>0$ and $v<0$ (quadrant 3). At the same time, the contribution of the events in quadrant 3 to the turbulent momentum transfer is substantially higher. As regards to the instantaneous values of vt, the situation is qualitatively almost the same.

With increasing y^+, the number of instantaneous products uv in the third quadrant tends to the first one and at $y^+\approx 10$ they become equal. With inrushes of liquid, however, momentum transfer remains more intensive. At $y^+=23$ the number of events in the third quadrant is by 25 percent greater than in the first one but their contribution is lower by approximately 1.6 times. With turbulent heat transfer, one can qualitatively trace the same changes: as one moves away from the wall, the relative number of events characterizing the inrush of cold liquid into the wall region increases with a simultaneous decrease in the efficiency of inrushes with respect to ejections. The products uv with the same signs of fluctuations are usually met much more seldom and transfer momentum less efficiently. In heat transport the role of "disordered" events appears to be somewhat more essential.

Histograms of instantaneous values of $uv/\sigma_u\sigma_v$ and

$vt/\sigma_v\sigma_t$ for two values of y^+ are given in Figure 5. The left-hand part of the histograms includes events with fluctuations of different signs (quadrants 1 and 3), the right-hand part - those of the same signs (2 and 4). One can see that the spread of the instantaneous values with respect to the averaged ones (correlation coefficients $\langle uv \rangle / \sigma_u \sigma_v$ and $\langle vt \rangle / \sigma_v \sigma_t$) is very great, the maximum instantaneous values having negative signs. The data show as well, that in **the proximity of the wall the maximum instantaneous values of uv and vt** are due to inrushes of liquid into the wall region (3 quadrant), at $y^+=23$ - to ejections of decelerated liquid from the wall (quadrant 1).

All the instantaneous values of uv and vt were sorted by a computer to eight possible different combinations depending on the signs of the fluctuations u, v and t.

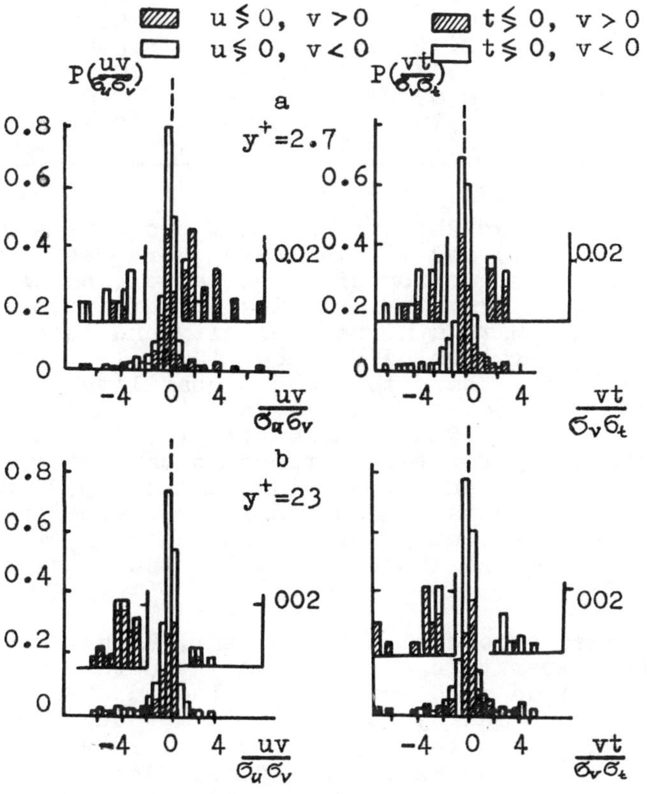

FIGURE 5. Histograms of instantaneous values for $uv/\sigma_u\sigma_v$ and $vt/\sigma_v\sigma_t$ taking into account signs of fluctuations.

(1)	$u < 0$;	$v > 0$;	$t < 0$; →	$uv < 0$;	$vt < 0$
(2)	$u < 0$;	$v > 0$;	$t > 0$; →	$uv < 0$;	$vt > 0$
(3)	$u > 0$;	$v > 0$;	$t < 0$; →	$uv > 0$;	$vt < 0$
(4)	$u > 0$;	$v > 0$;	$t > 0$; →	$uv > 0$;	$vt > 0$
(5)	$u > 0$;	$v < 0$;	$t < 0$; →	$uv < 0$;	$vt > 0$
(6)	$u > 0$;	$v < 0$;	$t > 0$; →	$uv < 0$;	$vt < 0$
(7)	$u < 0$;	$v < 0$;	$t < 0$; →	$uv > 0$;	$vt > 0$
(8)	$u < 0$;	$v < 0$;	$t > 0$; →	$uv > 0$;	$vt < 0$

Data processing has yielded probabilities of events given in the table.

TABLE 1. Probabilities of events of type (1) - (8)

Number of combination	y^+		
	2.7	11.4	23
(1)	0.24	0.20	0.17
(2)	0.12	0.12	0.12
(3)	0.07	0.05	0.06
(4)	0.10	0.11	0.12
(5)	0.12	0.07	0.09
(6)	0.17	0.27	0.28
(7)	0.07	0.08	0.07
(8)	0.12	0.10	0.10

The Table 1 shows that the maximum probability is typical for combinations (1) and (6) including "ordered" transfer mechanisms. i.e., the motion of decelerated (and therefore more heated) liquid from the wall into the flow core and inrushes of the liquid with a higher velocity and less temperature (greater temperature difference $t=T_w - T$) to the wall. It may be noted in this case that the probability of combination (1) decreases somewhat as the distance from the wall increases, and the probability of combination (6) increases. At $y^+=10÷20$ the probability for t-pulsations to be positive at inrushes ($t > 0$ at $u > 0$ and $v < 0$) is equal to 75-80 %. The correlation of the instantaneous values for momentum and heat fluxes when the liquid moves from the wall ($t < 0$ at $u < 0$ and $v > 0$), is somewhat lower and in all the studied region amounted 60-67 %.

In experiments conducted by Iritani, Kasagi and Hirata [36] the wall temperature and the flow structure were synchronously visualized with the help of liquid crystals sensitive to temperature and hydrogen bubble method. The observations have shown that"high-temperature streaks" are associated as a rule with "low-speed streaks" and "low-temperature streaks" - vise versa. The observed quantitative degree of correspondence between thermal and velocity streaks obtained in these experiments is in good agreement with the above data.

It is also interesting, whether intensive hydrodynamic ejections and inrushes are accompanied by analogous thermal events. To find it out, one may single out those events for which the instantaneous values uv and vt will simultaneously be higher than certain threshold values, i.e. $uv / \sigma_u \sigma_v \geq H$ and $vt / \sigma_v \sigma_t \geq H$. The results of such analysis are presented in Figure 6(a,b).

Figure 6-a shows that in a viscous sublayer the probabilities uv and vt combinations of type (3) and (8) become negligible already at H=1. For ejection of decelerated heated liquid from the wall (1) the limit is H=3. At H 3, one can observe in the flow only combinations of type (6) representing inrushes of cold portions of liquid to the wall, which move at velocities exceeding the mean local velocities. In approaching the flow core ($y^+ = 23$) the probability of large H-values is appreciable only for ejection processes of decelerated hot liquid from the wall. Thus powerful hydrodynamic inrushes and ejections are accompanied by analogous thermal events. There-

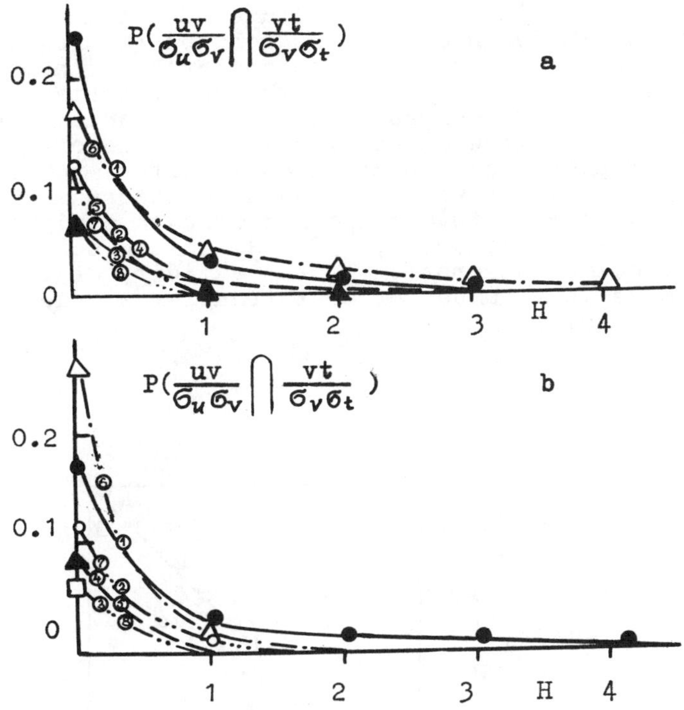

FIGURE 6. Probabilities for simultaneous values of $\frac{|uv|_i}{\sigma_u \sigma_v} \geq H$ and $\frac{|vt|_i}{\sigma_v \sigma_t} \geq H$; (i = 1,2,...8) as a function of H;
a) $y^+ = 2.7$ b) $y^+ = 23$.

fore one can conclude that turbulent heat and momentum transfer are highly correlative not only away from the surface (as in [37]) but in the proximity of it.

The above data do not cover by any means all the available experimental results. For example data of [32,36] also confirm that the kinematic structure of viscous sublayer defines the intensity of turbulent heat transfer, resulting in so called hydrodynamical analogy which is valid in channel flows and boundary layer not affected by strong perturbating factors. The below paragraphs contain some information on hydrodynamic and thermal structure of certain flows under more complicated conditions.

4. TURBULENT FLOW AND HEAT TRANSFER IN VISCOELASTIC DILUTE HIGH-POLYMER SOLUTIONS

It is well known that the addition of small amounts of certain high molecular weight polymers into a water flow changes appreciably the turbulent flow structure. The transverse velocity fluctuations in the near-wall region and the region, in which molecular and eddy viscosity are comparable, increase substantially in size. Accordingly the characteristic times of a viscous sublayer increase substantially and the sizes of near wall eddies grow as well. All these factors result in a decrease of pressure drop and heat transfer, sometimes being very substantial. It was found that the decrease in the drag coefficient usually is accompanied by even a higher reduction in heat transfer, Figure 7, [38-43]. This observation contradicts the common assumption of the validity of the Colburn analogy made in a number of heat transfer studies of viscoelastic fluids.

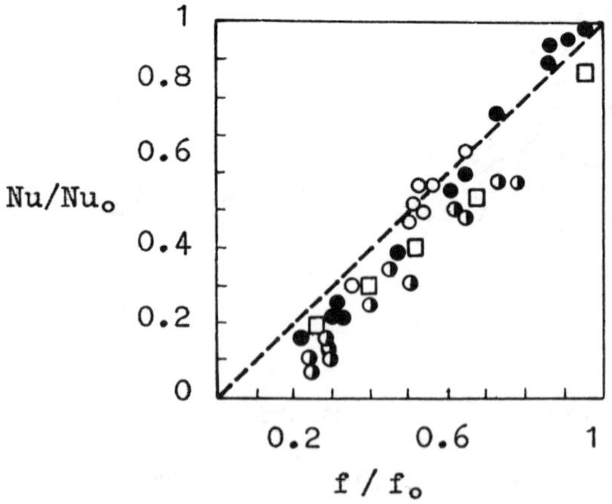

FIGURE 7. Drag and heat transfer reduction. ○ ●, ◑, □ - data of [27,39]; [41] ; [42], respectively.

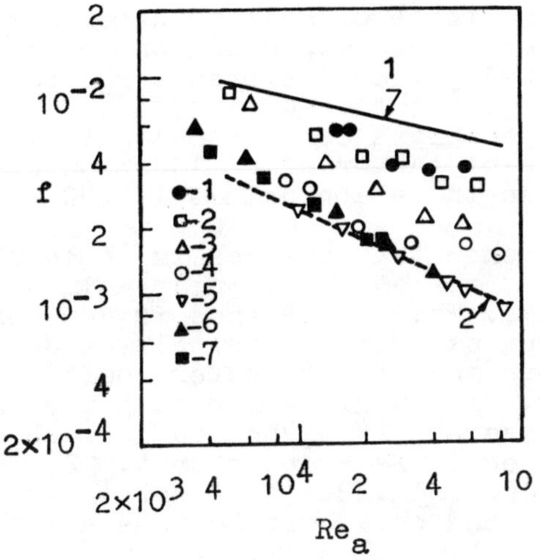

FIGURE 8. Fanning friction factor vs. Re_a [40]. 1÷7: Separan AP-273,c=5;10;20;30;100;500;1000 wppm; 1 - Newtonian; 2 - minimum drag asymptote.

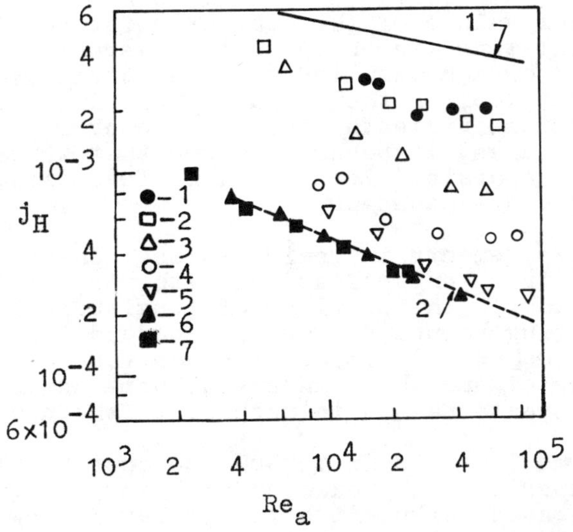

FIGURE 9. Heat transfer j_H factor vs. Re_a [40] . For legend see Figure 8. 1 - Newtonian; 2 - minimum heat transfer asymptote.

Drag and heat transfer reduction depends on many factors affecting the viscoelastic solution properties (polymer concentration, chemical properties of solvent, mechanical destruction etc.). At the same time the decreases of the shear coefficients and those of heat transfer are characterized by their asymptotic values depending on the Reynolds number alone, Figures 8,9 [40-41]. The difference between the asymptotic values of drag and heat reduction is the highest. The hydrodynamic entrance length corresponding to the asymptotic friction factors is of the order of 100 pipe diameters. The thermal entrance length is approximately 450 pipe diameters.

The above data show that changes in momentum and heat eddy diffusivities at variations in near-wall turbulence structure are not the same. This difference is not sufficiently analysed just now. In our experiments velocity and temperature profiles, friction factors and heat transfer coefficients were measured simultaneously. The results show that the turbulent Prandtl number in the near-wall region (in increased viscous and buffer layers) essentially exceeds unity [27,39]. In [42,43] this results were confirmed. But till now the simultaneous measurements of $\langle uv \rangle$ and $\langle vt \rangle$ correlations are absent.

5. TURBULENT FLOW AND HEAT TRANSFER IN THE RECIRCULATING ZONE BEYOND A BACKWARD-FACING STEP

The methods to calculate the heat and mass transfer coefficients in separated turbulent flows still remain under discussion. At the first stage attempts have been made to calculate the local values of the heat transfer coefficient in separated (recirculating) zones using the average velocity of recirculating flow and assuming that "an ordinary" turbulent boundary layer develops on the surface. Experimental values of the heat transfer coefficients exceeded, however, the calculated ones by many times. It became obvious that to calculate heat transfer in separated flows detailed investigations of the flow structure are necessary.

Flow measurements in recirculating zones revealed a high intensity of turbulent pulsations with respect to mean velocities. As an example the profiles of longitudinal velocity U and RMS fluctuating values σ_u and σ_v beyond the step are shown in Figure 10 [44]. A supposition was made [44,45] that the fluctuations of such high intensity were defined by the turbulence which is"external" with respect to the near-wall layer.

The terms included in turbulence energy balance equation have been experimentally examined in [46] for recirculating zone of separated turbulent boundary layer flow. It has been confirmed that in the backflow zone of the separated region there is no shear production as indicated by Figure 11 and advection is also insignificant. Hence the only mode by which turbulence energy can reach the backflow zone is by turbulent diffusion. So the backflow near the wall is controlled by the large-scale outer-region flow rather than by wall shear stresses. It seems however that the mechanism of this diffusion process is not clear so far.

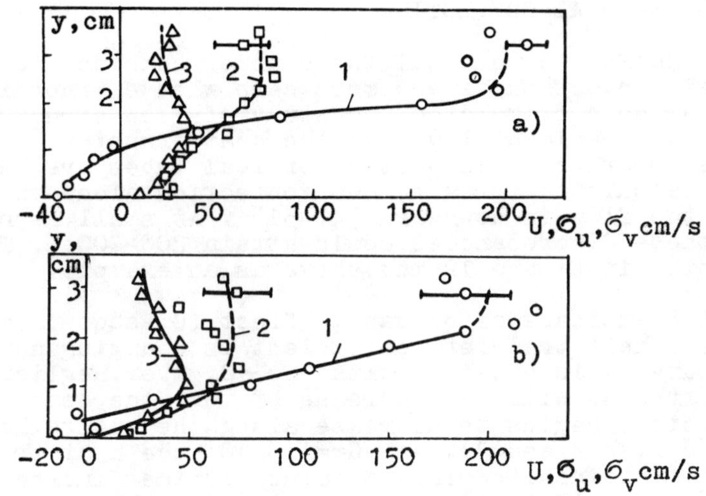

FIGURE 10. Profiles of mean longitudinal velocity U — (1) and RMS pulsations σ_u — (2) and σ_v — (3) for a) x=5.6 cm; b) x=9.8 cm (reattachment point is situated at x=10.4 cm) [44].

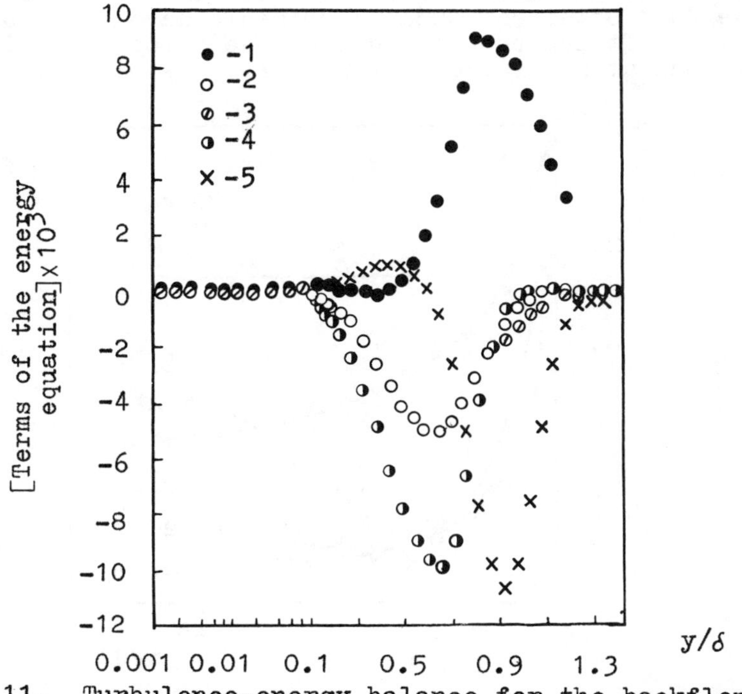

FIGURE 11. Turbulence-energy balance for the backflow ([46], Figure 13). 1 and 5 - advection terms; 2 - normal stresses production; 3 - turbulent diffusion; 4 - turbulent shear-stress production.

6. HEAT TRANSFER IN VERTICAL TUBES UNDER COMBINED FORCED AND NATURAL CONVECTION

Heat transfer in vertical tubes under combined forced and natural convection is one more case when extraordinary experimental data stimulated the study of flow structure. It has been set experimentally that the heat transfer coefficients in the upward flow in heated vertical tubes are below than those calculated for a forced convection alone and are above it if the flow is downward [47-51] . At small Reynolds numbers these discrepancies could attain 200-300 %. The data [47] presented in Figure 12 may serve as an example.

With the increase of buoyancy effect (Grashof number) dimensionless heat transfer coefficient is changing nonmonotonously. Curve 1 is obtained with $Gr \to 0$, i.e. negligible buoyancy effects. With the increase of Gr the heat transfer coefficient begins to decrease along the entire length of a tube (curves 2 and 3). At $Gr=1.5\ 10^7$ the heat transfer coefficient attains its minimum value. Further increase in a heat flux results in the increase of the Nusselt number starting from a certain section (curves 4-6). Heat transfer at large x/d may be higher than that at a forced convection of buoyancy effects (curve 6).

It became obvious that only detailed investigations of kinematic and thermal structure of this combined flow could explain the data obtained. Such experiments have been conducted in [47-49] . It has been found that the velocity profiles have maximum at a certain distance from the wall. Moving away

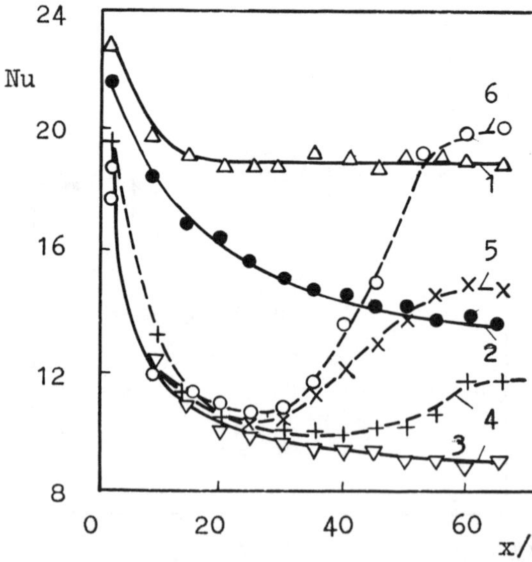

FIGURE 12. Heat transfer along the tube for different $E=Gr/Pr_4 Re^4$ numbers (Re=5100): 1. $\sqrt{E} \to 0$, 2 - $\sqrt{E}=10^{-4}$; 3 - $1.8\cdot 10^{-4}$; 4 - $2.5\cdot 10^{-4}$; 5 - $4\cdot 10^{-4}$; 6 - $7\cdot 10^{-4}$.

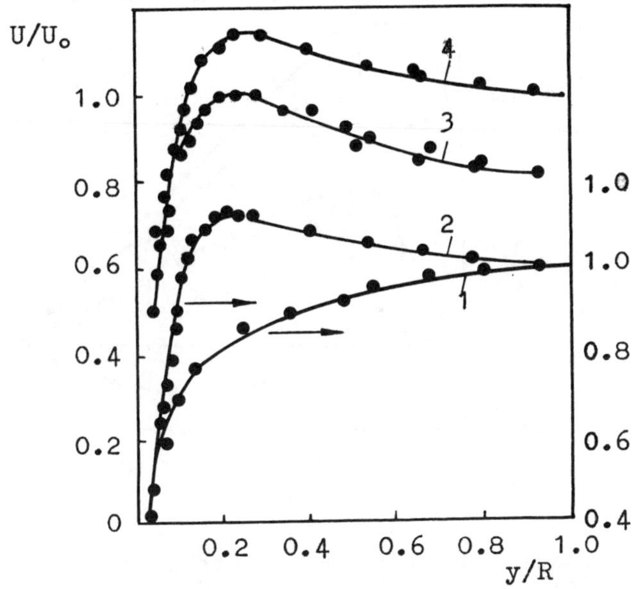

FIGURE 13. Velocity profiles at different distances from the heating onset; Re=5500; Gr=$3\cdot 10^7$; 1 - x/d=0; 2 - 29; 3 - 42; 4 - 53.

from the heating onset the value of a maximum grows first and then decreases, while the coordinate of a maximum shifts towards the tube axis, Figure 13. The intensities of velocity and temperature fluctuations decrease under the buoyancy effects and it is particularly pronounced near the wall.

Correlations of velocity and temperature pulsations, Figures 14 and 15, characterizing turbulent momentum and heat transfer, decrease with the increase of Gr number across the entire tube section. At a certain values of y/R turbulent stresses <uv> and turbulent heat flux along the axis <ut> change their signs and become negative in the flow core. If the position of U_{max} corresponds to maximum value of <uv> generation of turbulence kinetic energy by mean flow (- <uv> dU/dy) remains positive across the total cross section of a tube whereas turbulence energy generation by buoyancy effects ($-g\beta\langle ut\rangle$) in the flow core changes the sign. At further increase of buoyancy the positions of <uv>=0 and $U=U_{max}$ do not coincide and pulsation energy in the region between them is transferred to average flow. Radial turbulent heat transfer $\rho c_p\langle vt\rangle$across the entire cross section of a channel remains positive.

The kinematic energy decrease of turbulence near the wall causes a strong laminarization of the flow in this region, which manifests itself in the decrease of velocity and temperature fluctuations, heat transfer and drag reduction. Thus the necessity to understand the unusual character of experimental data on heat transfer stimulated the study of flow

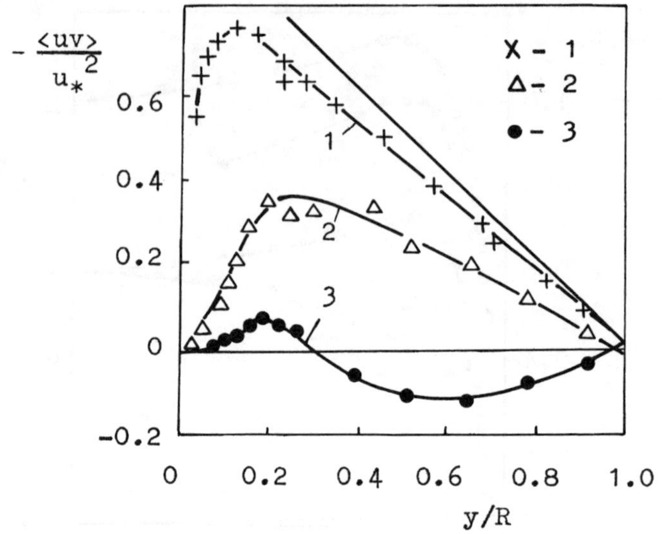

FIGURE 14. Variation of Reynolds stresses under the influence of buoyancy forces. Re=9000; 1 - Gr=0; 2 - 2 10^7; 3 - 3.4 10^7.

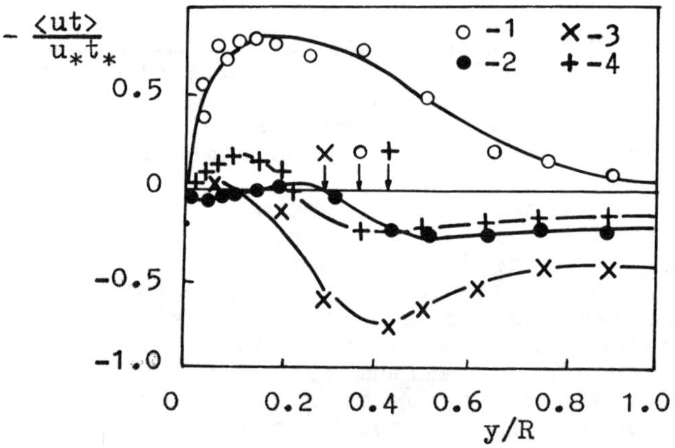

FIGURE 15. Longitudinal turbulent heat transport for Re=5100. 1 - Gr = 0; 2 - Gr=9×10^6; 3 - 1.1×10^7; 4 - 1.5×10^7.

structure under complex interaction of forced and thermogravitational convections and the development of the appropriate models to calculate such flows.

7. TRANSITION INTO TURBULENT FLOW AT THERMOGRAVITATIONAL CONVECTION

It is well known how much attention is paid to the problem of laminar-to-turbulent transition and how many difficulties face the experimentors working in this field. In the sense of stability free convective flows are more complicated than the forced ones, as they may be characterized by a wider spectrum of strongly interating hydrodynamic and thermal perturbations. At the same time using thermal methods this transition can be extended in time or space and successive stages of the flow can be studied "deliberately". Firstly unstably stratified plane horizontal layer of the liquid will be mentioned here.

It is well known that at critical value of Rayleigh number there develop a spatial-periodical flow and so called Bénard cell convection. With the increase of Rayleigh number the spatial form and flow structure gradually become complicated. In [52-53] an attempt has been made to relate the stages of flow evolution with the increase of Rayleigh number to the so-called "discrete transitions in the heat transfer curves" first discovered by Malkus [54]. Under "discrete transitions" (Figure 16) is meant a stepwise change of the factor of proportionality between measured heat flux through the liquid layer Q and temperature difference between its horizontal boundaries ΔT (or dependences $Nu \times Ra = f(Ra)$). Attempts have been made to reveal the correlations between the transition boundaries and the stages of evolution of spatial flow forms.

The most precise investigation of successive transitions has been carried out in [55-57] where the measurements of $Q(\Delta T)$ dependence were complemented by flow structure visualization.

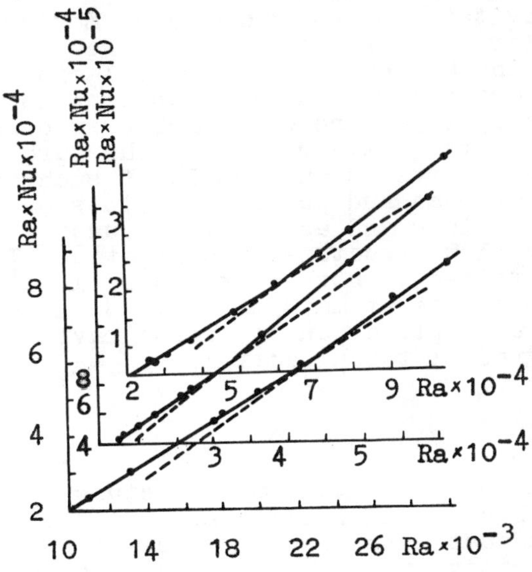

FIGURE 16. Dependence of $Nu \times Ra$ vs. Ra.

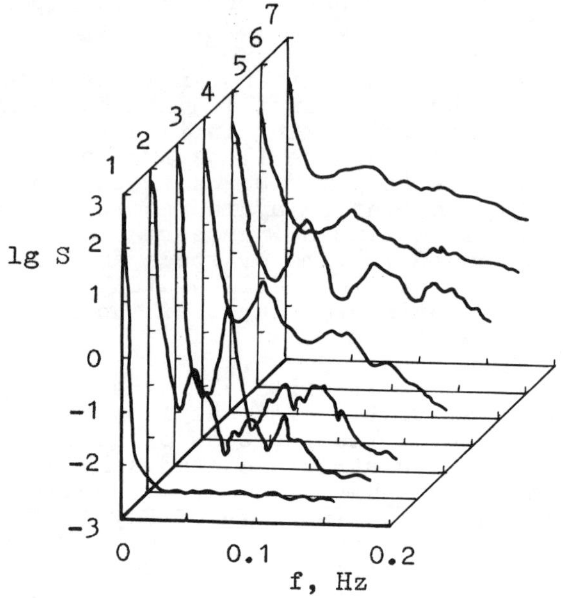

FIGURE 17. Spectral density of temperature fluctuations.

Temperature fields in the range of Rayleigh numbers from $3 \cdot 10^4$ to $3 \cdot 10^8$ have been also investigated in these experiments. The correlations of locations for discrete transitions with the boundaries of qualitative change in a frequency power spectrum of temperature fluctuations were found, Figure 17.

Thermogravitational boundary layer moving near a vertical heated plate is also of interest. Near the heating onset the flow in the boundary layer is laminar. At a certain distance from the onset the laminar flow is disturbed and along the boundary layer one can observe successive stages of transition from the laminar flow to the turbulent one. The works [56,58] may serve as an example of such investigation. The analysis of mean and pulsating values of temperatures and velocities along and across the boundary layer were complemented by flow structure visualization. The transition boundaries obtained in both these investigations may be used as reference points at mathematical modelling of the process of successive complication in thermogravitational flow structure and its gradual turbulization.

CONCLUSION

At present only the thermal structure of equilibrium boundary layer is studied in full details. The present investigations has shown that intermittent quasi-deterministic structures play a decisive role in turbulent processes including heat and mass transfer. The level of knowledge on hydrodynamic and thermal structure of a flow under more complicated conditions is so far insufficient. To attain this goal, not only a wide use of temperature as a passive makers introduced in the flow and other thermal methods of investigation but a thorough analysis of experimental data on heat and mass transfer may and should play an important role as well.

REFERENCES

1. Blackwelder, R.F., Kaplan, R.E., On the Wall Structure of the Turbulent Boundary Layer, J. Fluid Mech., vol. 76, pp. 89-113, 1976.

2. Brown, G.L., Thomas, A.S.W., Large Structure in a Turbulent Boundary Layer, Phys. of Fluids, vol. 20, pp. 243-252, 1977.

3. Falco, R.E., Combined Simultaneous Flow Visualization/ Hotwire Anemometry for the Study of Turbulent Flows, J. Fluid Eng., vol. 102, pp. 174-182, 1980.

4. Corke, T.C., Digital Image Filtering in Visualized Boundary Layers, AIAA J., vol. 22, pp. 1124-1131, 1984.

5. Bogard, D.G., Tiederman, W.G., Characteristics of Ejections in Turbulent Channel Flow, J. Fluid Mech., vol. 179, pp. 1-19, 1987.

6. Chen, C.-H.P., Blackwelder, R.F., Large-scale Motion in a Turbulent Boundary Layer: A Study Using Temperature Contamination, J. Fluid Mech., vol. 89, pp. 1-31, 1978.

7. Subramanian, C.S., Rajagopalan, S., Antonia, R.A.,Chambers, A.J., Comparison of Conditional Sampling and Averaging Techniques in a Turbulent Boundary Layer, J. Fluid Mech., vol. 123, pp. 335-362, 1982.

8. Chambers, A.J., Antonia, R.A., Browne, L.W.B., Raupach, M.R., A Multi-point Detection Method to Identify Coherent Temperature in a Turbulent Boundary Layer Over a Rough Surface, Proc. 4th Simposium on Turbulent Shear Flows, Karlsruhe, pp. 15.17-15.21, 1983.

9. Zarić, Z.,Wall Turbulence Studies, Adv. Heat Transfer, vol. 8, New York-London, pp. 285-350, 1972.

10. Zarić, Z., Statistical Analysis of Near-Wall Turbulence, Heat and Mass Transfer, vol. 9, Minsk, 1972.

11. Zhukauskas, A., Shlanchauskas, A., Heat Transfer in Turbulent Liquid Flow, "Mintis", Vilnus, pp. 1-327, 1973.

12. Zarić, Z., Statistical Analysis of Wall Turbulence Phenomena, Advances in Geophysics, vol. 18A, pp. 249-261, Academic Press, 1974.

13. Zarić, Z., Wall Turbulence Structure and Convective Heat

Transfer, Int. J. Heat Mass Transfer, vol. 18, pp. 831-842, 1975.

14. Petukhov, B.S., Polyakov, A.F., Shekhter, Yu.L., Tsipulyov, Yu.V., Statistical Characteristics of Temperature Fluctuations and Turbulent Heat Transfer in a Viscous Sublayer,- In: Wall Turbulent Flow, pt. 2, Novosibirsk, ITF SO AN SSSR, pp. 162-177, 1975.

15. Fulachier, L., Dumas, R., Spectral Analogy Between Temperature and Velocity Fluctuations in a Turbulent Boundary Layer, J. Fluid Mech., vol. 77, p. 257, 1976.

16. Perepelitsa, B.V., Statistical Characteristics of Temperature Fluctuations in a Turbulent Water Flow, Heat and Mass Transfer -Y, Proc. 5th All-Union Conference on Heat and Mass Transfer, Minsk, vol. 1, pt. 1, pp. 48-53, 1976.

17. Shlanchauskas, A.A., Vegite, N.Yu., Space Correlations of the Temperature Field in a Turbulent Boundary Layer of Different Liquids, Proceedings of the Ac Sci. Lith. SSR, ser. B, vol. 3, p. 100, 1977.

18. Elena, M., Experimental Study of Turbulence Near the Wall of a Slightly Heated Pipe, Int. J. Heat Mass Transfer, vol. 20, no. 9, pp. 935-944, 1977.

19. Ibragimov, M.Kh., Subbotin, V.I., Bobkov V.P., Sabyelev, G.N. and Taranov, G.S., The Structure of Turbulent Flow and Mechanism of Heat Transfer in Channels, pp. 1-296, Moscow, Atomizdat, 1978.

20. Petukhov, B.S., Polyakov, A.F., Tsipulyov, Yu.V., Experimental Research in Probability Characteristics of Velocity and Temperature Fluctuations in a Turbulent Air Flow in a Flat Nonsymmetrically Heated Channel, Teplofiz. Vysok. Temp., vol. 16, no. 5, pp. 987-996, 1978.

21. Kutateladze, S.S., Khabakhpasheva, E.M., Perepelitsa, B.V., Temperature Pulsations in a Viscous Sublayer, Proc. 6th Int. Heat Transfer Conf., Toronto, vol. 2, pp. 501-505, 1978.

22. Zarič, Z., Physical Evidence on Coherent Structures in the Wall Layers of Turbulent Flows, Turbulent Forced Convection in Channels and Bundles, pp. 403-440, Hemisphere, 1979.

23. Hishida, M., and Nagano, Y., Structure of Turbulent Velocity and Temperature Fluctuations in Fully Developed Pipe Flow, J. Heat Transfer, vol. 101, pp. 15-22, 1979.

24. Perepelitsa, B.V., Khabakhpasheva, E.M., Correlations Between Temperature and Velocity Fluctuations in the Wall Region of Turbulent Flow, - In: Heat and Mass Transfer-YI, vol. 1, pt.1, pp. 148-159, Minsk, 1980.

25. Perepelitsa, B.V., Khabakhpasheva, E.M. Experimental Research in Turbulent Temperature Fluctuations in the Wall Region of Turbulent Flow, - In: Mechanics of Turbulent Flows, ed. V.V.

Struminsky, Moscow, pp. 225-230, 1980.

26. Khabakhpasheva, E.M., Experimental Investigation of Turbulent Momentum and Heat Transfer in the Proximity of the Wall, Proc. 8th Int. Conf. of Heat and Mass Transfer, USA, vol. 1, pp. 79-90, 1986.

27. Kutateladze, S.S., Mironov B.P., Nakoryakov, V.Ye., Khabakhpasheva E.M., Experimental Research of Wall Turbulent Flows, pp. 1-166, Izd. "Nauka", Novosibirsk, 1975.

28. Kutateladze, S.S., Khabakhpasheva, E.M., Orlov, V.V., Perepelitsa, B.V., Mikhailova, E.S., Experimental Investigation of the Structure of Near-Wall Turbulence and Viscous Sublayer, Turbulent Shear Flows I. Selected Pap. 1st Int. Symp., Pennsylvania, State Univ., Berlin, pp. 91-99, 1977.

29. Perepelitsa, B.V., Pshenichnikov, Yu.M., Khabakhpasheva, E.M., Statistical Characteristics of Temperature Fluctuations in Unsteady Turbulent Flow, - In: Heat and Mass Transfer - YII, vol. 1, pt. II, pp. 133-137, Minsk, 1984.

30. Kutateladze, S.S., Volchkov, E.P., Mironov, B.P., Rubtsov, N.A., Khabakhpasheva, E.M., Turbulent Heat and Mass Transfer Under Interaction of Various Disturbing Factors, Proc. 7th Int. Heat Transfer Conf., München, vol. 3, pp. 269-281, 1982.

31. Polyakov, A.F., Shindin, S.A., The Joint Use of Thermal Laser Hot-wire Anemometers to Study Turbulent Heat Transfer Characteristics, - In: Convective Heat Transfer Methods and Results of Investigations, High-Temperature Institute, Moscow, pp. 169-180, 1982.

32. Hishida, M., Nagano, Y. and Tagawa, M., Transport Processes of Heat and Momentum in the Wall Region of Turbulent Pipe Flow, Proc. 8th Int Heat Transfer Conf., San Francisco, CA USA, vol. 3, pp. 925-930, 1986.

33. Khabakhpasheva, E.M., Mikhailova, E.S., Turbulence Investigation at Water Flow through a Channel With Parallel Walls, - In: Experimental Investigation into the Structure of Near-Wall Turbulence and Viscous Sublayer, pp. 33-57, Novosibirsk, 1976.

34. Willmarth, W.W., Lu, S.S., Structure of Reynolds Stress Near the Wall, J. Fluid Mech., vol. 55, pt. 1, pp. 65-92, 1972.

35. Brodkey, R.S., Wallace, J.M., Eckelmann, H., Some Properties of Truncated Turbulence Signals in Bounded Shear Flows, J. Fluid Mech., vol. 63, pt. 2, pp. 209-224, 1974.

36. Iritani, Y., Kasagi, N. and Hirata, M., Heat Transfer Mechanism and Associated Turbulence Structure in the Near-Wall Region of a Turbulent Boundary Layer, - In: Turbulent Shear Flows 4, Springer, pp. 1-397, 1985.

37. Perry, A.E., Hoffman, P.H., An Experimental Study of Turbulent Convective Heat Transfer from a Flat Plate, J. Fluid Mech., vol. 77, pt. 2, pp. 355-368, 1976.

38. Khabakhpasheva, E.M., Popov, V.I., Perepelitsa, B.V., Heat Transfer in Viscoelastic Fluids, Proc. 4th Int. Heat Transfer Conf., Paris-Versailes, vol. IY, p. Rh2, 1970.

39. Khabakhpasheva, E.M., Forced Convection With Rheological Systems, Proc. 6th Int. Heat Transfer Conf., Toronto, Canada, pt. YI, pp. 127-142, 1978.

40. Kwack, E.Y., Cho, Y.I., Hartnett, J.P., Effect of Weissenberg Number on Turbulent Heat Transfer of Aqueous Polyacrilamide Solutions, Proc. 7th Int. Heat Transfer Conf., vol. 3, pp. 63-68, 1982.

41. Kwack, E.Y., Hartnett, J.P. and Cho, Y.I., Turbulent Heat Transfer in Circular Tube Flows of Viscoelastic Fluids, Wärme-und Stoffübertragung, b. 16, s. 35-44, 1982.

42. Mizushina, T., Usui, H., Reduction of Eddy Diffusion for Momentum and Heat in Viscoelastic Fluid Flow in a Circular Tube, Physics of Fluids, 20, S 100, 1977.

43. Yoon, H.K. and Ghajar, A.J., A New Heat Eddy Diffusivity Equation for Calculation of Heat Transfer to Drag Reducing Turbulent Pipe Flows, Proc. 8th Int. Heat Transfer Conf., San Francisco, vol. 3, pp. 937-942, 1986.

44. Orlov, V.V., Ovchinnikov, V.V., Mukhina, N.V., Karsten, V.M., Some Results of Turbulence and Heat and Mass Transfer Investigation in Separated Flow Beyond a Step, - In: Structure of Forced and Thermogravitational Flows, pp. 87-97, Novosibirsk, 1983.

45. Spalding, D.B., Heat Transfer from Turbulent Separated Flows, J. Fluid Mech., vol. 27, no. 1, pp. 97-109, 1967.

46. Simpson, R.L., Chew, Y.-T., Shivaprasad, B.G., The Structure of a Separating Turbulent Boundary Layer, part 2, Higher-Order Tutrbulence Results, J. Fluid Mech., vol.113, pp. 53-73, 1981.

47. Polyakov, A.F., Shindin, S.A., Turbulent Momentum and Heat Transfer at Mixed Convection in Vertical Channels, TVT, vol. 24, no. 5, p. 1031, 1986.

48. Petukhov, B.S., Polyakov, A.F. and Martynenko, O.G., Buoyancy Effect on Heat Transfer in Forced Channel Flows, Proc. 7th Int. Heat Transfer Conf., Washington-N.Y.-L. Hemisphere Publishing Corporation, vol. 1, RK5, pp. 343-362, 1982.

49. Kitamura, K. and Inagaki, T., Turbulent Heat and Momentum Transfer of Combined Forced and Natural Convection Along a Vertical Flat Plate-aiding Flow, Int. J. Heat Mass

Transfer, vol. 30, no. 1, pp. 23-42, 1987.

50. Tanaka, H., Maruyama, S. and Hatano, S., Combined Forces and Natural Convection Heat Transfer for Upward Flow in a Uniformly Heated Vertical Pipe, Int. J. Heat Mass Transfer, vol. 30, no. 1, pp. 165-174, 1987.

51. Jackson, J.D. and Hall, W.B., Influences of Buoyancy on Heat Transfer to Fluids Flowing in Vertical Tubes Under Turbulent Conditions, - In: Turbulent Forced Convection in Channels and Bundles, ed. S. Kakac and D.B. Spalding, pp. 613-640, vol. 2, Hemisphere, Washington, DC, 1979.

52. Krishnamurti, R. On the Transition to Turbulent Convection. Pt. 1. The Transition From Two- to Three-Dimensional Flow, J. Fluid Mech., vol. 42, pt. 2, pp. 295-307, 1970.

53. Krishnamurti, R., On the Transition to Turbulent Convection. Pt. 2. The Transition to Time-Dependent Flow, J. Fluid Mech., vol. 42, pt. 2, pp. 309-320, 1970.

54. Malkus, W.V.R., Discrete Transitions in Turbulent Convection, Proc. Roy. Soc., A225, no. 1161, pp. 185-195, 1954.

55. Berdnikov, V.S., Markov, V.A., Kim, O.B., Thermogravitational Convection in Flat Horizontal and Inclined Liquid Layers Heated From Below, - In: Structure of Forced and Thermogravitational Flows, pp. 122-146, Novosibirsk, 1983.

56. Kutateladze, S.S., Berdnikov, V.S., Structure of Thermogravitational Convection in Flat Variously Oriented Layers of Liquid and on a Vertical Wall, Int. J. Heat Mass Transfer, vol. 27, no. 9, pp. 1595-1611, 1984.

57. Berdnikov, V.S., Malyshev, V.I., Markov, V.A., Experimental Studies Into Influence of Aspect Ratio of Liquid Horizontal Layer on Some Thermogravitational Convection Characteristics, - In: Transfer Processes in Forced and Free Convective Flows, pp. 50-71, Novosibirsk, 1987.

58. Berdnikov, V.S., Dyatlov, A.V., Semyonov, V.I., Thermogravitational Convection Structure in a Vertical Layer of Liquid and on a Vertical Wall at Transient Flow Regime, - In: Transfer Processes in Forced and Free Convective Flows, pp. 71-96, Novosibirsk, 1987.

Turbulent Heat Transfer Associated with Coherent Structures Near the Wall

Y. NAGANO
Department of Mechanical Engineering
Nagoya Institute of Technology
Gokiso-cho, Showa-ku, Nagoya 466, Japan

M. HISHIDA
Department of Mechanical Engineering
Chubu University
Matsumoto-cho, Kasugai 487, Japan

ABSTRACT

An objective method for the recognition and description of coherent structures is developed and applied to identify the role of coherent motions in the turbulent heat transport process. Four distinct patterns are recognized in the sequence of coherent motions near the wall. These four basic flow patterns have their own characteristic eddy structures and reflect the succession of stages that might occur in the evolution of coherent motions. Also, these recognized patterns reveal how the turbulent heat transfer is affected by the coherent eddies originating in the wall region and extending through a greater part of the flow zone.

INTRODUCTION

Visual observations of Kline and coworkers [1], and of Corino and Brodkey [2] revealed that turbulent flows near the wall are not as wholly chaotic as previously believed from a cursory survey of hot wire traces of velocity fluctuation signals; but some fairly organized motions (i.e., coherent structures) do exist. In general, turbulent heat transfer is dominated essentially by the flow near a heat-transfer surface. Accordingly, the basic mechanism of heat transfer is expected to be strongly associated with the coherent turbulent motions near the wall. However, experimental evidence which reveals the mechanistic picture of turbulent heat transport in the wall region is rare. This is mainly due to the great difficulties of simultaneous measurement of the Reynolds stress and turbulent heat flux fluctuations in close proximity to the wall.

In our series of studies of turbulent heat transfer (Hishida and Nagano [3-5]; Nagano and Hishida [6]; Hishida, Nagano and Tagawa [7]; Nagano and Tagawa [8,9]), we have developed two novel measurement techniques: one for the simultaneous measurements of velocity and temperature in nonisothermal flows (Hishida and Nagano [10]), and the other for turbulence measurements near the wall with specially devised, symmetrically bent 'V'-shaped hot wires (Hishida and Nagano [11,12]). A V-shaped hot wire works in the same manner as an inclined straight wire. The great difference is that a V-shaped wire can be supported just like a cantilever; thus, the wire may be brought closer to the wall until it touches the wall surface, whereas a conventional inclined straight wire is kept away from the wall by its supports, and a probe body significantly distorts the flow field and introduces serious errors into the measurements.

Employing these techniques, we investigated the turbulent heat transport process in wall-turbulent shear flows along with the dynamics of near-wall turbulence. The present paper describes some significant mechanisms pertaining to the turbulent transport of heat and momentum. The mechanisms we have observed are strongly associated with and dominated by the coherent turbulent structures near the wall. The structural similarity between momentum and heat transfer near the wall is demonstrated in concrete terms. The data to be presented are those obtained in an air flow in a 45.68 mm ID pipe heated to a uniform wall temperature of 100°C at a Reynolds number of 40,000.

OVERALL VIEW OF COHERENT TURBULENT HEAT TRANSFER

Description of Coherent Motions

To make clear the relation between the coherent structures and turbulent momentum and heat transfer quantitatively, it is indispensable to properly extract organized motions from superficially random time-series data of hot-wire measurement. Bogard and Tiederman [13] evaluated the capability of several representative detection-algorithms from the simultaneous measurements with a hot-wire and flow-visualization technique in a wall-bounded flow. As a result, they concluded that a uv-quadrant method had the greatest reliability with a high probability of detecting the organized motions and a low probability of false detections. Thus, in the present study, we also use the uv-quadrant analysis.

The coherent motions can be classified in four types of events according to each quadrant of the (u,v)-plane (u axial, and v normal): outward interaction, v>0 and u>0 (i=1); ejection, v>0, u<0 (i=2); inward interaction, v<0, u<0 (i=3); and sweep, v<0 and u>0 (i=4). As shown in Fig. 1, each classified motion has the deterministic mean period $\overline{T_i}$ and mean duration $\overline{\Delta T_i}$ (Nagano and Tagawa [8]). Here, the time $\overline{T_i}$ is

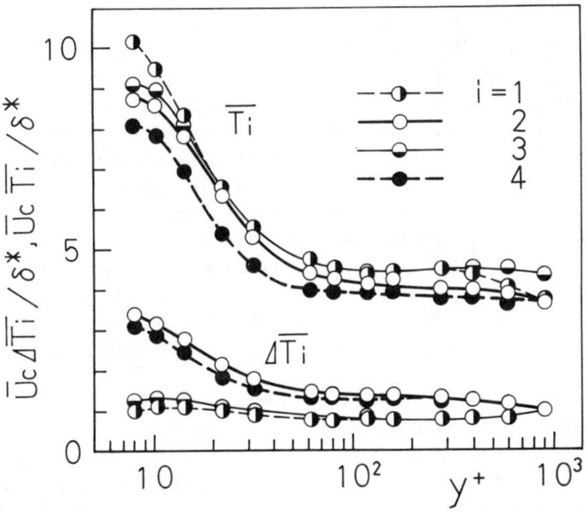

FIGURE 1. Mean period between events $\overline{T_i}$ and duration of the respective motions $\overline{\Delta T_i}$.

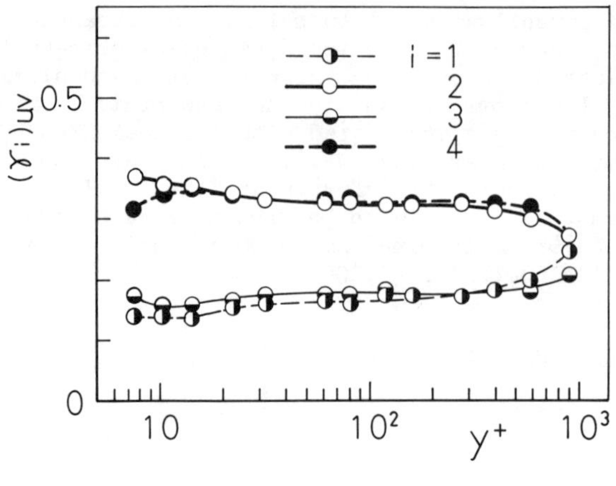

FIGURE 2. Intermittency factors of respective coherent motions.

normalized by the centerline velocity $\overline{U}_c$ and the displacement thickness δ^*. The mean periods for each motion are essentially equal at a location, indicating that each classified motion on the average occurs at the same interval. In contrast to this, the mean durations of the two interaction-type motions (i=1, 3) are about half as large as those of the ejection- (i=2) and sweep-type (i=4) motions. Thus, as shown in Fig. 2, the fractions of time occupied by the ejections and sweeps are generally larger than those of the interactions.

Instantaneous Signals of Transport Processes of Heat and Momentum

Typical instantaneous fluctuations of the velocity components u and v, temperature t, turbulent momentum transfer (i.e., Reynolds shear stress) uv, and axial and radial turbulent heat fluxes, ut and vt, recorded simultaneously (Hishida, Nagano and Tagawa [7]), are presented in Fig. 3. Figure 3(a) shows the records of the signals taken at a location near the maximum point of turbulence energy production in the buffer layer. From this figure, the highly intermittent and spiky nature of turbulent heat and momentum transfer (i.e., uv, ut and vt fluctuations) is apparent. Almost all large peaks of $uv<0$ and $vt>0$ are the consequence of inrushes (or sweeps) of high-momentum and low-enthalpy fluid from the outer region ($v<0$, $u>0$, $t<0$), and low-momentum and high-enthalpy fluid ejections from the near-wall region ($v>0$, $u<0$, $t>0$). Also observed in Fig. 3(a) are the typical u-signal patterns of the ejection- and sweep-type motions consisting of a gradual deceleration from a local maximum followed by a strong acceleration. The mean period of the coherent motions has a value of 3.2 ms at this location determined from a short-time averaged autocorrelation coefficients of u fluctuations (Hishida and Nagano [4]).

Despite increasing distance from the wall, as in the core region of the flow shown in Fig. 3(b), violent fluctuations of Reynolds shear stress and turbulent heat fluxes are still associated with ejection- and sweep-type motions. This indicates that the coherent motions originating in the near-wall region also play a very dominant role in determining the heat transport process in the outer region.

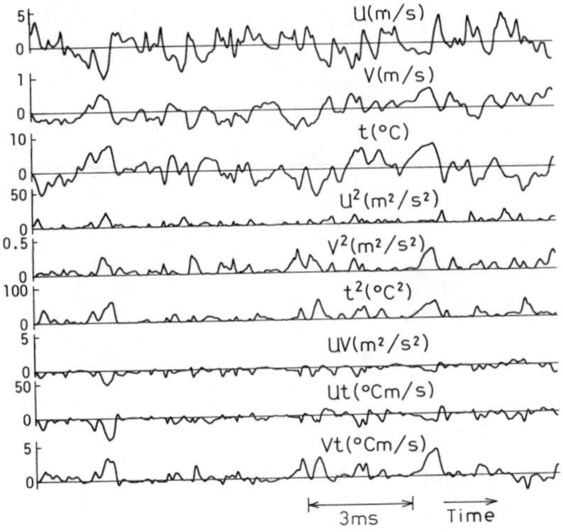

(a) $y^{+}=10.4$, $y/r_o=0.012$

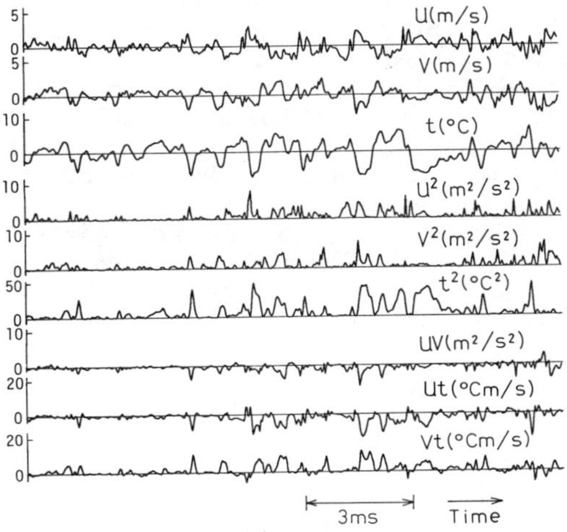

(b) $y^{+}=387.7$, $y/r_o=0.441$

FIGURE 3. Simultaneous traces of turbulent quantities related to heat and momentum transfer.

Contributions of Coherent Motions to Turbulent Heat Transfer

The measurement of a joint probability density function $P(\hat{u},\hat{v})$ shown in Fig. 4 (Nagano and Tagawa [9]) indicates that the ejections and sweeps (i=2, 4) predominate in the near-wall structure. Although the wall region occupies a small fraction of the radius (i.e., $y<0.2r_o$, where r_o is the pipe radius), it influences the entire flows. In fact, roughly 80% of the total turbulence energy production, and about 85% of the total production of both mean square temperature variance and turbulent heat flux occur within the wall region (Nagano and Hishida [6]).

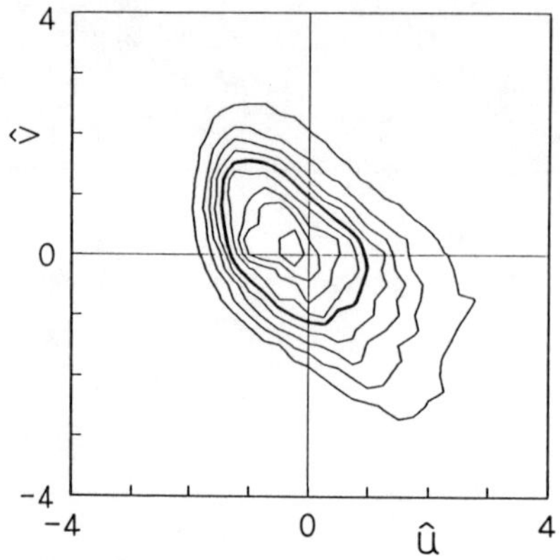

FIGURE 4. Joint probability density of u and v ($y^+=7.6$).

The v-component power spectral density, and the u-v and v-t cross-spectral densities, denoted respectively as $E_v(f')$, $E_{uv}(f')$, $E_{vt}(f')$, were measured and the spectra of production of turbulent quantities were calculated (Nagano and Hishida 1985). This spectral analysis of production process shows that the production of turbulent quantities pertaining to turbulent heat and momentum transfer is totally dominated by the coherent turbulent structures near the wall, as evidenced in Figs. 5-7. The mean period of the coherent motions, $\overline{T}_B$, has a value of $\overline{U}_c\overline{T}_B/\delta^* \cong 10$ near the wall, and $\overline{U}_c\overline{T}_B/\delta^* \cong 4$-$5$ in the fully turbulent region (see Fig. 1). As shown in Figs. 5-7, the peak contributions to the total production of the turbulence energy $\overline{q^2}$ (= $\overline{u_i u_i}$), temperature variance $\overline{t^2}$ and turbulent heat flux $\overline{vt}$ occur at about the bursting frequency $f'(=f\,\delta^*/\overline{U}_c) \cong 0.1$ (=1/10) near the wall and $f' \cong 0.2$-0.25 (=1/5-1/4) in the logarithmic region.

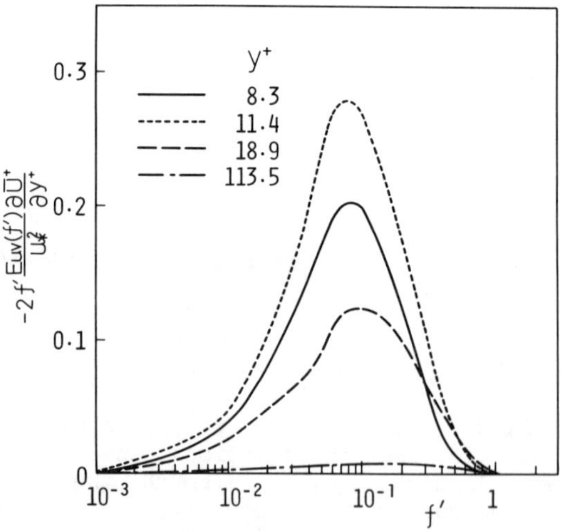

FIGURE 5. Spectra of the production $\overline{q^2}$.

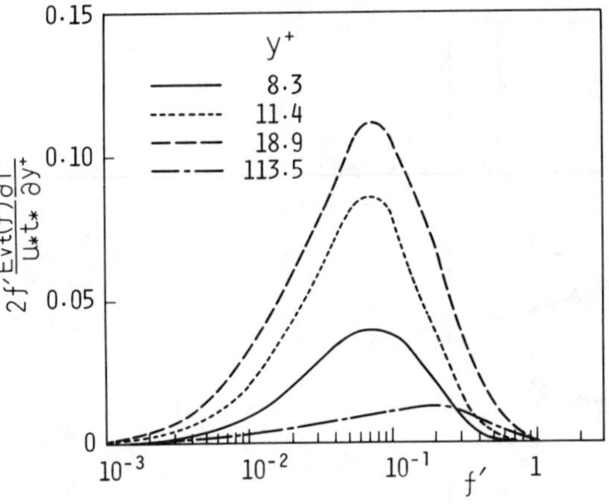

FIGURE 6. Spectra of the production $\overline{t^2}$.

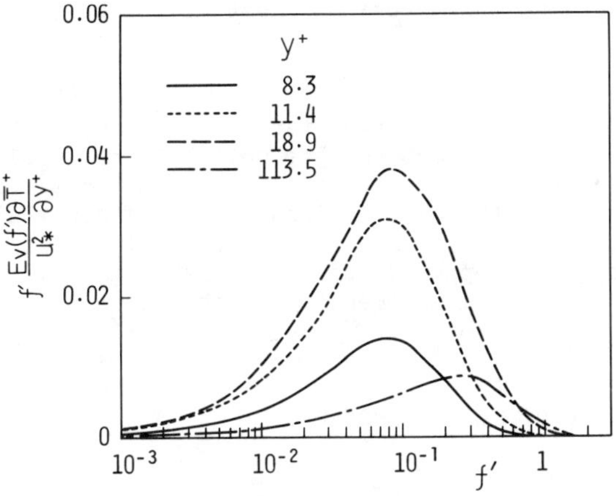

FIGURE 7. Spectra of the production $\overline{vt}$.

As described previously, the ejection- and sweep-type coherent motions are the primary mechanisms of the production of the Reynolds shear stress and turbulent heat flux. The fractional contributions to mean turbulent heat flux $\overline{vt}$ from four distinct types of motion (Hishida and Nagano [5]) (i.e., i=1:outward interaction, i=2: ejection, i=3: inward interaction, i=4: sweep) are presented in Fig. 8. The ejection-type outflow of high-enthalpy fluid from regions very near the wall is the largest contributor through a greater part of the pipe section. However, sweeps of low-enthalpy fluid from regions away from the wall are prominent as the wall is approached. This is consistent with the mechanism of Reynolds shear stress production; sweeps are the largest contributors in the immediate neighborhood of the wall, and ejections become predominant further away from the buffer layer. The contributions of both inward and outward interactions are negligible, suggesting the highly organized structure in the wall region can be described in terms of the sequence of ejection- and

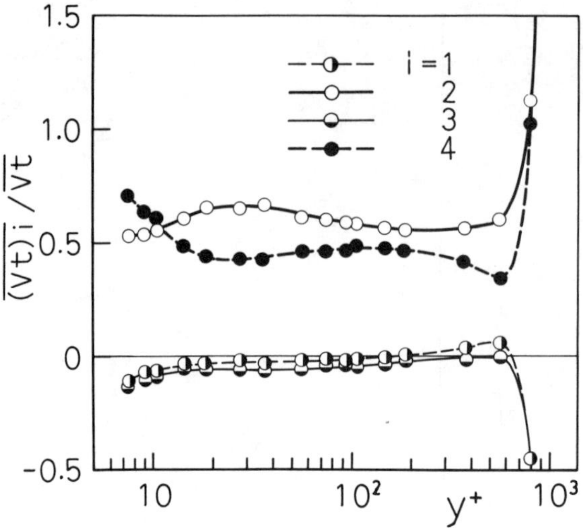

FIGURE 8. Fractional contributions to turbulent heat flux $\overline{vt}$ from different motions.

sweep-type coherent motions. In the near-centerline region, the contributions of the sweeps and two interaction-type motions are again seen to increase. This is due to complex effects of the ejections of high-enthalpy fluid originating from the wall regions in any azimuthal direction. The inward interaction (i=3), for instance, includes the ejections from the wall region of the opposite side of the wall ($v<0$, $t>0$). This also reflects trend towards homogeneity in the velocity and temperature fluctuations at the pipe centerline.

DETAILED ASPECTS OF COHERENT TURBULENT HEAT TRANSFER

Basic Flow Patterns of Coherent Structures

From the foregoing evidence, it is clear that the coherent structures near the wall afford the key to an understanding of the mechanism of turbulent heat transfer. To delineate the mechanistic picture of transfer processes, however, we must determine objectively the coherent structures consisting of organized motions from random time-series data and then describe them appropriately. The following procedure has been developed to judge whether the detected motions are the coherent motions (Nagano and Tagawa [8]).

If any organized motions exist truly in a flow, there should be a certain regularity in the trajectories of quadrant-sequences on the (u,v)-plane because of continuity of turbulent motions. In order to detect the coherent motions in a wall turbulent shear flow, an investigation of all possible trajectories on the (u,v)-plane is necessary, since various stages are observed in the evolution of coherent structures. Figure 9 shows the sample result near the wall. Numbers appended to a trajectory denote the detected frequencies of quadrant-sequence. For example, the frequency of the quadrant-sequence [4]-[1] is 717, and that of [1]-[2]

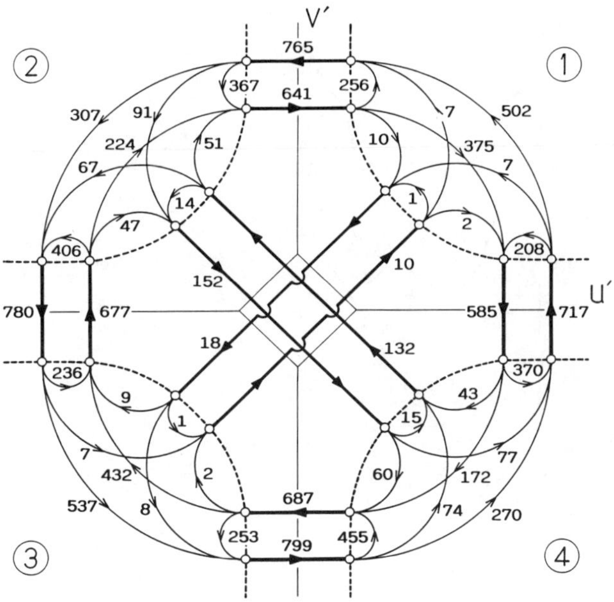

FIGURE 9. All possible trajectories on the u-v plane (y^{+}=7.6).

which follows the previous sequence [4]-[1] is 502. The frequencies of [1]-[3] and [1]-[4] which follow [4]-[1] are 7 and 208, respectively. Similarly, the frequency of [2]-[3] is 780, which is followed by 537 [3]-[4], seven [3]-[1] and 236 [3]-[2] sequences of events. If wall turbulent shear flows should consist of completely chaotic motions, a specific trajectory does not occur with such a high frequency. We have also sampled a set of trajectories moving on any four quadrants of the (u,v)-plane, e.g., [4]-[1]-[2]-[3]. The frequencies of occurrence for this case are nearly identical with those predicted from the values given in Fig. 9.

From these results, it becomes now evident that the deterministic regularity exists in the trajectory of individual coherent events and that we can detect and describe the coherent motions objectively. Four distinct patterns are eventually recognized in the sequence of coherent motions as: (i) pattern 1, [4]-[1]-[2]-[1]-[4]; (ii) pattern 2, [4]-[1]-[2]-[3]-[4]; (iii) pattern 3, [4]-[3]-[2]-[3]-[4]; (iv) pattern 4, [4]-[3]-[2]-[1]-[4].

The above finding makes it possible to recognize the most likely patterns of coherent motions, to program a computer to extract and ensemble-average sequence of events, and to delineate the characteristics of the structures. The ensemble-averaged velocity and temperature fluctuations are established by arranging temporal duration and period of each event in the mean duration and the mean period of the respective motions. Thus, the patterns recognized here are the phase averaged ones. Sample patterns (Nagano and Tagawa [8]) near the heat-transfer surface (y^{+}=7.6) are shown in Fig. 10, where $\langle Q' \rangle$ denotes the conditional average of a function $Q'(\tau)$ and of u, v and t are normalized by the respective conventional r.m.s. values of the signals. The time τ on the abscissa

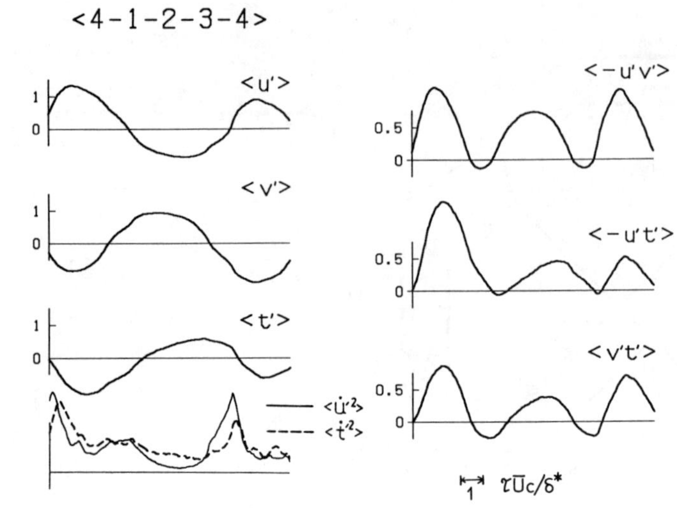

(a) Pattern 2

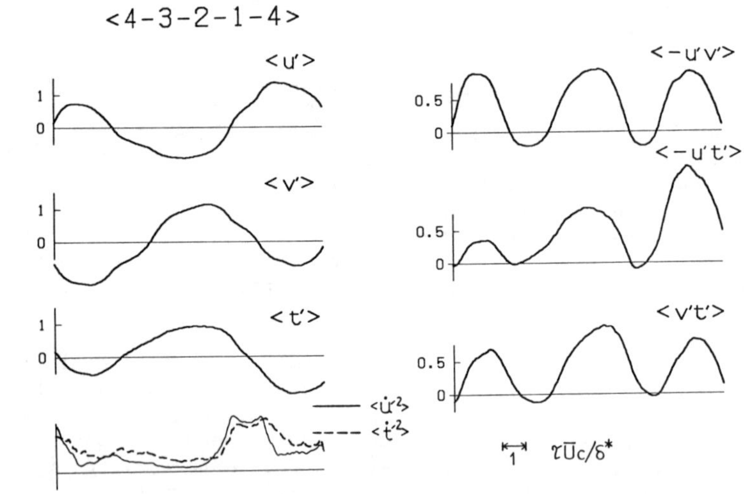

(b) Pattern 4

FIGURE 10. Basic flow-patterns of coherent motions in the near wall region (y^+=7.6).

is normalized by the centerline velocity $\overline{U}_c$ and the displacement thickness δ^*. The quantities $\langle \dot{u}'^2 \rangle$ and $\langle \dot{t}'^2 \rangle$ denote $\langle (\partial u'/\partial \tau)^2 \rangle$ and $\langle (\partial t'/\partial \tau)^2 \rangle$, and correspond to the dissipation rates of velocity and temperature fluctuations. The conditional averages of $\langle \dot{u}'^2 \rangle$ and $\langle \dot{t}'^2 \rangle$ are normalized by the respective appropriate common values so that the magnitude of the two becomes nearly the same.

The conditional averages of <u'>, <v'> and <t'> are found to show striking resemblances among one another, thus indicating that these are governed by the identical physical law. The <u'>-signal is approximately 180° out of phase with the <v'>- and <t'>-signal. The interaction-type motions are very important events providing a phase-information for recognition and classification of organized structures. It is also found that the dissipation of both velocity and temperature fluctuations happens mainly during the interaction-type motions associated with a change of flow patterns from the ejections to sweeps. The conditional averages of <u'> and <v'> of patterns 2 and 4 are bilaterally asymmetrical, with a larger <u'> amplitude of pattern 2 in the first fourth-quadrant and with a completely opposite trend in pattern 4. However, waveforms of <u'> and <v'> related to classified patterns 1 and 3 (not shown) are bilaterally symmetrical at a midpoint of each waveform. The conditional averages of <-u'v'>, <-u't'> and <v't'> prove that almost all net heat and momentum transfer occur during the sweeps and ejections.

In addition, the important finding is that heat transport becomes greatest when the pattern-4 motions dominate a flow field. The <u'>-waveform of pattern 4 depicts the typical u-signal patterns of the coherent structures consisting of a gradual deceleration from a local maximum followed by a strong acceleration. The <t'>-waveform of pattern 4 has a steep negative gradient when <u'> is under strong acceleration. This might reflect the possible existence of a large-scale temperature-front, noted by Chen and Blackwelder [14].

Coherent Eddies Near the Wall

It was confirmed from simultaneous measurements of instantaneous streamwise velocity fluctuations in the wall region that the same flow-pattern really occurred at the same time over a considerable area in the direction normal to the wall. Thus, temporal variation of velocity vectors obtained from the ensemble-averages of <u'> and <v'> can be transformed into its spatial variation by using Taylor's hypothesis. The velocity vector-diagram of each flow-pattern is established from a set of vectors, which can visualize the spatial structure of the coherent motions observed from a moving "camera".

The eddy structures of flow-patterns 2 and 4 in the wall region (Nagano and Tagawa [8]) are depicted in Fig. 11, with the convection velocity U_V of an observer equal to a mean streamwise velocity at $y^+=37.8$. Here, the relative velocity vector $\tilde{u}$ is normalized by the friction velocity u_*. The fluid motions of each flow pattern appear to be considerably different from those expected from Fig. 10. Pattern 2 has a 'small'-scale vortical motion, i.e., coherent eddy, in the immediate neighborhood of the wall. The outflowing motion from the wall seems rather weak. However, the ejection- and sweep-type motions interact violently with each other within a very small region. This is a reason why the velocity dissipation rate has its maximum in this limited region (see Fig. 10). On the other hand, pattern 4 has a 'large'-scale vortical motion centering around $y^+=70$. The fluid downstream of the coherent eddy penetrates into the region in immediate proximity to the wall and is ejected outward from the wall as if it were involving the near-wall fluid in its motion.

It is more reasonable to regard each of the four basic flow-patterns as an elemental motion which does not represent an independent motion of one another but the stage of development of the coherent structures (Nagano and Tagawa [8]). Especially pattern 4 can be considered to be the flow

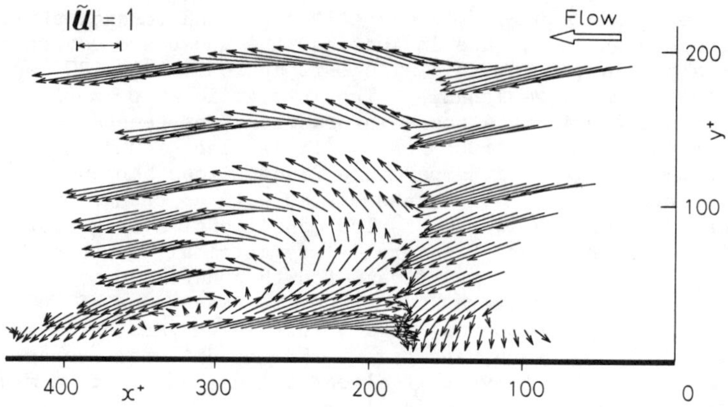

(a) Pattern 2

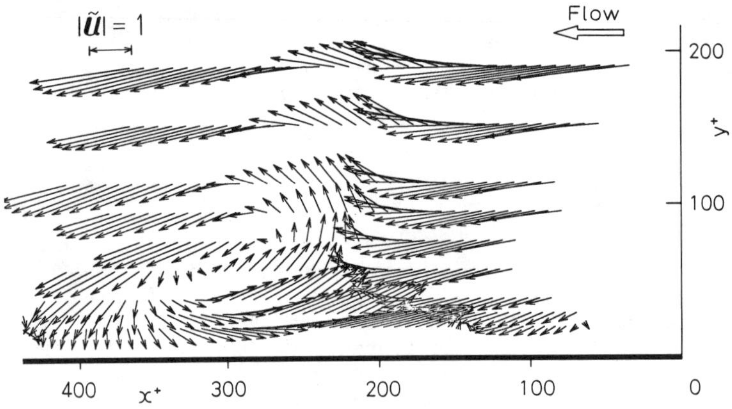

(b) Pattern 4

FIGURE 11. Eddy structures of basic flow-patterns that might be seen by an observer moving at a speed $U_V = 0.64\ \overline{U}_c$.

in the final stage of the bursting phenomena (i.e., breakup), because very active motions with vigorous ejection are observed in Fig. 11. The present scale and scope of conditionally averaged coherent eddies also agree quite well with the flow-visualization results of Nychas, Hershey and Brodkey [15], and the calculated results of LES by Kim and Moin [16].

Relationship between Coherent Motions and Heat Transfer

As the details of elemental flow-patterns for the coherent structures have been understood, we lastly investigate the relation between the coherent motions and heat transport processes. The diagrams of turbulent velocity and heat-flux vectors, normalized by the friction velocity and friction temperature, of flow-patterns 2 and 4 are demonstrated in Fig. 12 as a function of y^+. Taylor's hypothesis is again used for conversion of a

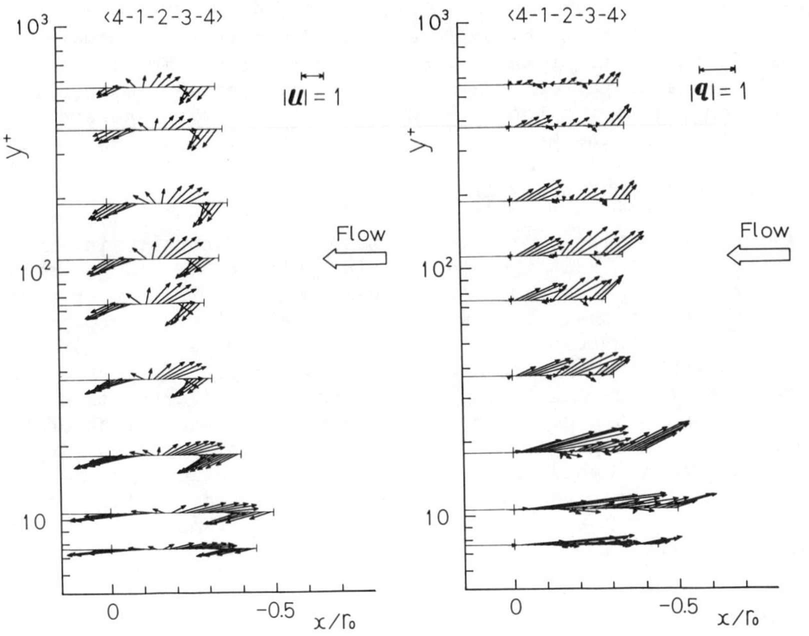

(a) Pattern 2

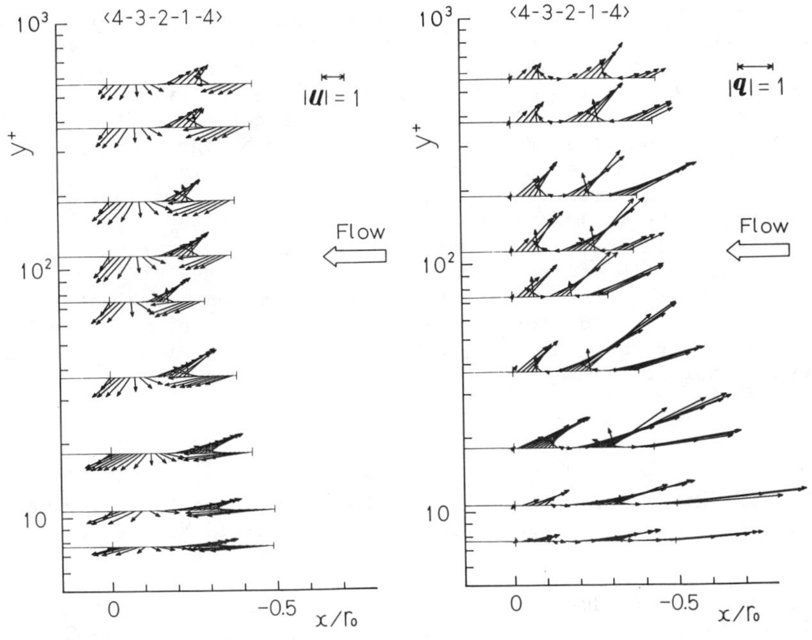

(b) Pattern 4

FIGURE 12. Velocity and turbulent heat-flux vectors for the basic flow-patterns.

temporal pattern-length to its spatial scale in the streamwise direction. The zero mark on the abscissa is the beginning point where the pattern is detected. One can see that the difference in fluid vectors between two flow patterns is much larger than that deduced from Fig. 10. For example, the ejections of pattern 2 disperse upward, whereas the ejections of pattern 4 concentrate in the outward direction. The sweeps for pattern 2 concentrate in the wallward direction with small angles to the wall, whereas those for pattern 4 occur at larger angles to the wall as if fluids were crashing against the wall.

These different characteristics in flow patterns significantly influence the state of turbulent heat transfer. In pattern 2, large axial transports <-u't'> are observed near the wall during the sweep-phase downstream of the ejections, but away from the wall (say $y^+>40$) they become smaller; thus turbulent heat-flux vectors diverge upward. In pattern 4, much turbulent heat transport occurs overall. In particular, the magnitude of <v't'> during the ejection-phase is far larger than for pattern 2. This result can be understood when viewed in the light of the eddy structures shown in Fig. 11. The radial turbulent heat-flux <v't'> is very large in pattern 4 which involves the strong ejection-type motions (i.e., fluid approaching to the wall being carried outward spontaneously), and the heated fluid near the heat-transfer surface is transported outward largely by this pattern.

CONCLUDING REMARKS

Turbulent heat-flux and Reynolds stress fluctuations in a wall turbulent shear flow have been measured by employing a combined cold wire and V-shaped hot wire anemometry technique developed by the authors, and then conditionally sampled and averaged. Instantaneous turbulent heat transfer is strongly associated with and dominated by the coherent turbulent structures, and thus the inherent intermittency of turbulent heat transport process is the consequence of the intermittent coherent motions near the wall. The detailed relationship between the coherent motions and turbulent heat transfer has been analyzed by introducing a new conditional sampling technique which possesses objectivity and reliability for the detection and description of coherent turbulent motions. The coherent structures in wall turbulence are well represented in terms of four basic well-ordered motions. These four basic flow patterns have their own characteristic coherent eddies. Among other things, the important finding is that the outflowing low-momentum wall-region fluid organizing very strong eddy motions with the incoming higher-momentum fluid from regions away from the wall is the principal contributor to the turbulent heat transfer.

REFERENCES

1. Kline, S. J., Reynolds, W. C., Schraub, F. A., and Runstadler, P. W., The Structure of Turbulent Boundary Layers, J. Fluid Mech., vol. 30, pp. 741-773, 1967.

2. Corino, E. R., and Brodkey, R. S., A Visual Investigation of the Wall Region in Turbulent Flow, J. Fluid Mech., vol. 37, pp. 1-30, 1969.

3. Hishida, M., and Nagano, Y., Structure of Turbulent Temperature and Velocity Fluctuations in the Thermal Entrance Region of a Pipe, Proc. 6th Int. Heat Transfer Conf., Toronto, vol. 2, pp. 531-536,

Hemisphere, 1978.

4. Hishida, M., and Nagano, Y., Structure of Turbulent Velocity and Temperature Fluctuations in Fully Developed Pipe Flow, Trans. ASME, J. Heat Transfer, vol. 101, pp. 15-22, 1979.

5. Hishida, M., and Nagano, Y., Structures of Turbulent Heat Transfer, Heat Transfer in High Technology and Power Engineering (eds. W.-J. Yang and Y. Mori), pp. 488-499, Hemisphere, 1987.

6. Nagano, Y., and Hishida, M., Production and Dissipation of Turbulent Velocity and Temperature Fluctuations in Fully Developed Pipe Flow, Proc. 5th Int. Symp. on Turbulent Shear Flows, Cornell University, pp. 14.19-14.24, 1985.

7. Hishida, M., Nagano, Y., and Tagawa, M., Transport Processes of Heat and Momentum in the Wall Region of Turbulent Pipe Flow, Proc. 8th Int. Heat Transfer Conf., San Francisco, vol. 3, pp. 925-930, Hemisphere, 1986.

8. Nagano, Y., and Tagawa, M., Coherent Motions and Their Role in Heat Transport Processes in a Wall Turbulent Shear Flow, J. Fluid Mech., submitted, 1987.

9. Nagano, Y., and Tagawa, M., Statistical Characteristics of Transfer Processes in a Wall Turbulent Shear Flow, Proc. 2nd Int. Symp. on Transport Phenomena in Turbulent Flows, Tokyo, pp. 481-494, 1987.

10. Hishida, M., and Nagano, Y., Simultaneous Measurements of Velocity and Temperature in Nonisothermal Flows, Trans. ASME, J. Heat Transfer, vol. 100, pp. 340-345, 1978.

11. Hishida, M., and Nagano, Y., Turbulence Measurements with Symmetrically Bent V-Shaped Hot-Wires (Part 1, Principles of Operation), Trans. ASME, J. Fluids Engineering, to appear, 1988.

12. Hishida, M., and Nagano, Y., Turbulence Measurements with Symmetrically Bent V-Shaped Hot-Wires (Part 2, Measuring Velocity Components and Turbulent Shear Stresses), Trans. ASME, J. Fluids Engineering, to appear, 1988.

13. Bogard, D. G., and Tiederman, W. G., Burst Detection with Single-Point Velocity Measurements, J. Fluid Mech., vol. 162, pp. 389-413, 1986.

14. Chen, C.-H. P., and Blackwelder, R. F., Large-Scale Motion in a Turbulent Boundary Layer: A Study Using Temperature Contamination, J. Fluid Mech., vol. 89, pp. 1-31, 1978.

15. Nychas, S. G., Hershey, H. C., and Brodkey, R. S., A Visual Study of Turbulent Shear Flow, J. Fluid Mech., vol. 61, pp. 513-540, 1973.

16. Kim, J., and Moin, P., The Structure of the Vorticity Field in Turbulent Channel Flow (Part 2, Study of Ensemble-Averaged Fields), J. Fluid Mech., vol. 162, pp. 339-363, 1986.

Effect of Turbulent Flow Structure on Temperature Field Formation in Near-Wall Region

B. V. PEREPELITSA
Institute of Thermophysics
Siberian Branch of the USSR Academy of Sciences
Novosibirsk, USSR

ABSTRACT

Experimental results of temperature field investigations in turbulent flow of water at a stepwise change of heat generation in the channel wall are presented. Measurements are performed at different distances from the beginning of heating in the range of Reynolds numbers (0.98...11.2) 10^4 and cover a close near-wall region including a viscous sublayer. The turbulence structure effect on the temperature field formation is shown. The results on mean temperature and temperature fluctuation measurements are generalized in dimensionless coordinates.

1. INTRODUCTION

The process of temperature field formation in turbulent flow at variation of temperature boundary conditions is closely related to the flow structure in the near-wall region. Turbulent velocity fluctuations result in temperature fluctuations at heat transfer. The knowledge of detailed structure of the velocity and temperature fields in turbulent flow is extremely important for better insight into complex nature of turbulent transfer processes. For this reason a great attention is paid to statistical investigations of both velocity and temperature fluctuations in the boundary flow /1-6/.

Coherent structures which have been found first as a result of visual investigations /7-9/ play an important role in the mechanism of heat and momentum transfer in the near-wall region of turbulent flow. These investigations show that the liquid streaks highly extended along the flow direction occur on a streamlined surface. These streaks in the course of their evolution are destructed generating conventional high-frequency turbulence. The destruction of streaks is of "bursting" nature and is accompanied by the ejection of retarding liquid from the wall and successive intrusion into the near-wall region of liquid with higher velocity.

To single out coherent structures and to define their characteristic scales appear to be a rather difficult experimental problem which so far remains unsolved unambiguously. The average period T between two successive "bursts" or "sweeps" is commonly used as characteristic time scale of coherent structures. Various techniques have been suggested to determine the characteristic period $\overline{T}$. Of wide use were the conventional averaging methods in which a signal was processed in selective time intervals satisfying definite conditions /10-12/. When selecting these conditions one should proceed from physical knowledge of the turbulence mechanism and the final result depends to a large extent on the experimentor skill. Various methods of the signal preliminary processing are widely used to determine the value of $\overline{T}$. Attempts have been made to relate the period $\overline{T}$ to the time scales obtained in the calculations of autocorrelative and spectral functions with the different length of assumed realizations /13-15/. The agreement between characteristic dimensionless period and the data /7-9/ obtained by the flow visualization is generally the reliability degree of the obtained results.

One of the examples of the investigation into the time scales of near-wall coherent structures is the work /16-17/ by the author in which the temperature signal is used to study the turbulence structure. The temperature measurement in the flow indicated the ejection of heated liquid from the wall and the intrusion of cold liquid portions from the external region of the flow. The mean time interval between two successive one-direction temperature signal transitions through the given threshold level is determined in the experiment. The measurements have shown that if minimum time interval T_{min} is used as characteristic time of large scale structures then the quantity $T_{min} v_*^2/\nu$ does not depend on distance to the wall and the Reynolds number in the region of viscous sublayer. The numeric value of this dimensionless quantity appeared to be 80-100, Figure 1 , that agrees well with the

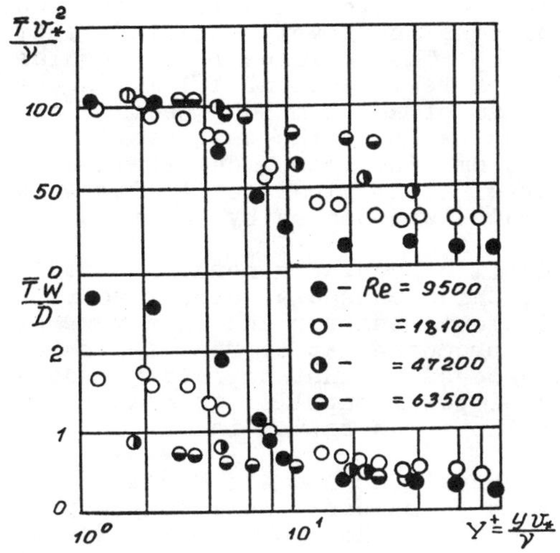

FIGURE 1. Distribution of dimensionless characteristic interval at different Reynolds numbers.

results of other researchers. Beyond viscous sublayer the value of dimensionless time scale decreases and the curves split with respect to the Reynolds numbers. In this region (measurements are performed up to $Y^+ < 100$) the data processing with using the external flow parameters w and D turns out to be more preferable.

Coherent structures should greatly affect the temperature field formation at switching on/or off the heat load. The temperature variation in the flow is defined by the ratio between intensities of turbulent and molecular heat transfer. In this case the development of temperature field in a turbulent flow depends on the Reynolds number and the distance to the wall. At present despite of many theoretical and experimental works on unsteady heat transfer /18-20/ the problem concerning the extent of the unsteadiness influence on the turbulent heat transfer mechanism remains so far unsolved.

The present paper aimed at experimental investigation of statistical characteristics of unsteady temperature field in the turbulent flow at a stepwise change of heat generation in the channel wall. Special attention was given to the effect of turbulent transfer on the formation of temperature field in a close near-wall region including that of viscous sublayer and the initial moment after switching on/or off the heat load when the value of heat flux changed most significantly. For these purposes the experimental setup and method to measure temperature in the turbulent liquid flow at unsteady heat transfer were developed /21,22/.

2. EXPERIMENTAL EQUIPMENT, MEASUREMENT PROCEDURE AND DATA PROCESSING

The experiments were conducted in a closed hydrodynamic circuit which consisted of constant level tank, test section, intake tank, refrigerator and pump. The horizontal rectangular 20 x 40 mm^2 channel was used as a test section. A 1 m long heating section was placed downstream of a 2.5 m long section of hydrodynamic stabilization. Three walls of the test section were made of plexiglass and one (40 mm wide) was made of glass fabric with 0.1 mm thick stainless steel strip glued to it. The working fluid flowed around the strip which was heated by electric current.

The heater was supplied by d.c. generator through a thyristor switch, Figure 2 , which caused a current to flow from the heater to auxiliary equivalent resistor and in a reverse order. The duration of transient processes at current switching is not more than 0.5 ms. The process of unsteady heat transfer was caused by a stepwise change of power supplied to heater that resulted in a stepwise change of heat generation in it.

The temperature fluctuations and its average values in the liquid flow were measured by a special thermocouple probe with the hot junction size about 5 μm. The probe design was chosen to affect as small as possible the temperature distribution in the near-wall region of the flow. The probe consisted of curved stainless steel tube (2 mm in dia) with a two-channel 30 mm long glass capillary fixed at the end of its horizontal section. The constantan-

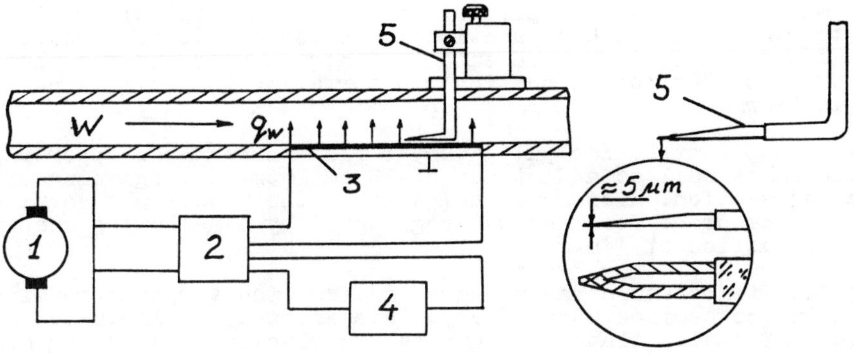

FIGURE 2. Scheme of measuring section:1 - d.c. amplifier; 2 - thyristor switch; 3 - heater; 4 - equivalent resistance; 5 - thermocouple probe.

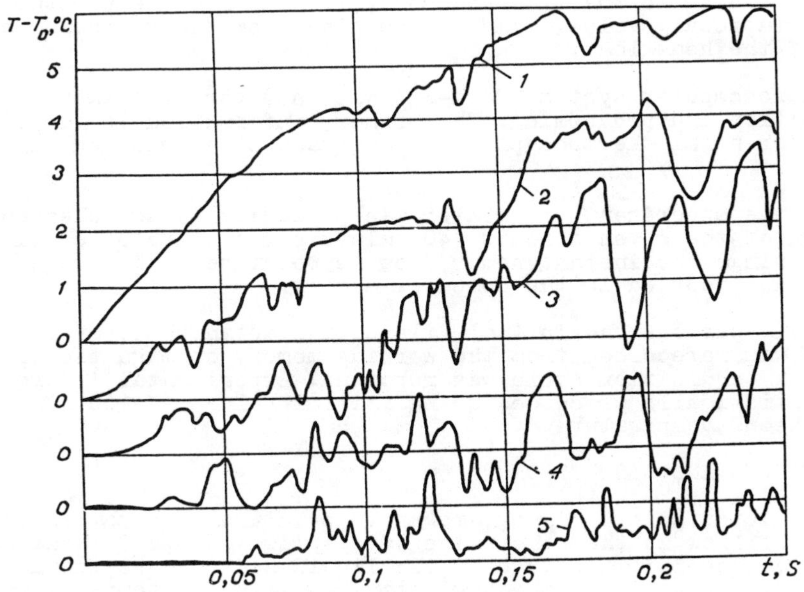

FIGURE 3. Temperature oscillograms at initial instant of time with a stepwise increase of heat load; Re=29100, 1÷5 - y=20;45;80;130;300 mkm.

nichrome thermal electrodes with the diameter of 0.03 mm were welded into the capillary. The capillary end about 10 mm long was a highly extended cone with the size at the end not more than 0.1 x 0.2 mm. The thermocouple junction was at a distance of 1-1.5 mm from the glass capillary end. After mechanical treatment of the thermocouple hot junction its cross-sectional dimension was about 5 μm. At measurements the thermocouple end was placed so that its horizontal section was normal to the wall, i.e. along an isotherm.

Thermocouple probe was put into the flow through the upper unheated channel wall at a distance of 78 cm from the beginning of the heating section. Micrometer screw moved the probe in a vertical plane. The probe movement was registered by a measuring head with the graduation of 1 mkm.

The flow temperature was measured against the temperature of cold wall in the section under study. This allowed to eliminate the effect of small inlet flow temperature fluctuations during measurements on the experimental results. The temperature difference between heated and cold walls did not exeed 10 K. The effect of thermocouple inertia manifest itself at frequencies above 1 KHz.

The triggering period of the converter was software-changed during the heating or cooling phase in order to reveal the most characteristic parts of transient processes at the minimum number of readings. At the initial stage the converter triggering period was by an order of magnitude smaller than that at the final stage. The duration of heating and cooling was chosen for each flow regime according to the time necessary for the flow temperature to reach stationary value. Time was counted from the switching on/ and off the heat load.

The microcomputer system "Mera-60" was used for experimental data acquisition and processing. The primary information is the files of differential thermocouple discrete values recorded on magnetic tape.

During the experiment the heater was periodically switched on and off and at the given moments (40 values per a phase of cooling and heating) the instantaneous flow temperature values were measured and recorded in the computer memory.

After 100 periods the initial file of instantaneous temperature values was rerecorded from the working memory on magnetic tape and data acquisition cycle was reported. Experimental information was statistically processed using initial values (N=16000) of instantaneous temperature.

3. RESULTS AND DISCUSSION

The characteristic oscillograms of temperature fluctuation in the turbulent flow at the same value of stepwise change in heat generation at the initial stage of the liquid heating are shown in Figure 3, where T is the instantaneous value of temperature in the point under study, T_o is the temperature of inlet flow.

Data are obtained at different distances to the wall. Curves 1-5 correspond to the coordinates y=20,45,80,130,300μm. The Reynolds number is defined as $Re=WD_h/\nu$, where W is the mean flow rate, D_h is the hydraulic diameter of a channel, ν is the kinematic viscosity of a liquid. The temperature of inlet liquid is assumed as determining one when calculating the physical properties of a liquid. The data presented show that there is practically a linear increase of temperature near the wall at the initial moment. The process of heat transfer at this evolution stage is basically defined by the mechanism of thermal conductivity. Turbulent mixing results in the temperature deviation towards decrease. These deviations are due to the cold liquid moles coming from the external region of the flow and continue to increase with time. Temperature increases slower with the increase of the Reynolds number, i.e. with intensity growth of turbulent mixing.

Away from the wall the temperature evolution is different. At large distances sharp deviations of temperature towards increase are due to the turbulent ejections of hot liquid from the more heated near-wall region.

The observed characteristic processes of turbulent mixing affect appropriately the evolution of averaged temperature in the flow. Figure 4 shows the curves of averaged temperature evolution in two characteristic points of the flow. Data are obtained at different Reynolds numbers and the same stepwise change of heat generation. Here the data on temperature variation in a stationary liquid are presented as well. In this case the temperature increase is due to the mechanism of thermal conductivity. As is seen from the curves, in the near-wall region during entire evolution the values of averaged temperature in the turbulent flow are below than those in a stationary liquid. These deviations increase with the growth of Re. The estimates show that pronounced deviations occur within the time comparable with the time scale between turbulent ejections $t \approx 100\, \nu/v_*^2$.

The nature of the averaged temperature evolution is different away from the wall. Prior to the thermal wave arrival to the region under study the turbulent ejections result in the fact that current averaged temperature values exceed its values in a stationary liquid. These deviations increase with the increase of the Reynolds number. After some time due to turbulent mixing the ratio between the temperatures under consideration changes to an opposite one. Figure 5 shows the difference between current values of temperature in a stationary liquid and its corresponding values in the turbulent flow. This difference is positive in the region near the wall. Sections with the value as a linear function of time can be singled out on the plots. Away from the wall the quantity under consideration takes a negative value in the initial instant of time.

The curves of relatively averaged temperature evolution in the regions of thermal stabilization (L=0.07 m) and stabilized heat transfer (L=0.78 m) are presented in Figure 6 and 7, respectively, where T_i is the averaged temperature value in the point under study at moment t, T_s - is its stationary value with the heater switched on. Data are given at switching on (a) and off

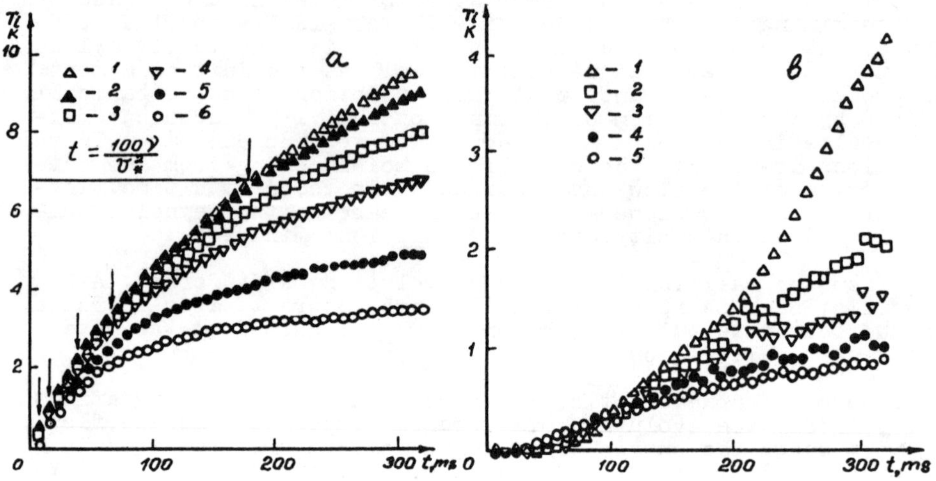

FIGURE 4. Temperature variation in stationary liquid and turbulent flow: a - y=20 mkm; 1÷6 Re=0;9800;15700;23600;37900;b - y=240 mkm; 1÷5 Re=0;14700;23100;36600;55200.

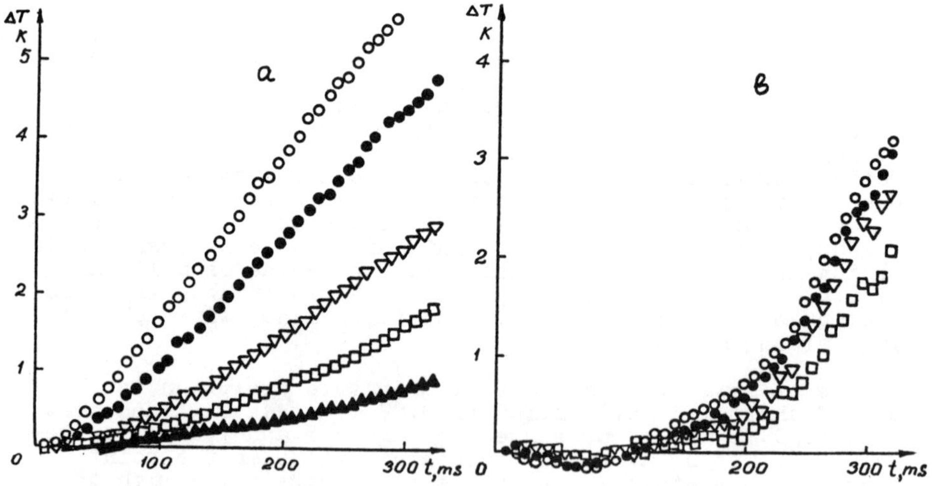

FIGURE 5. Temperature difference in stationary liquid and its values at different Reynolds numbers (for symbols see Figure 4).

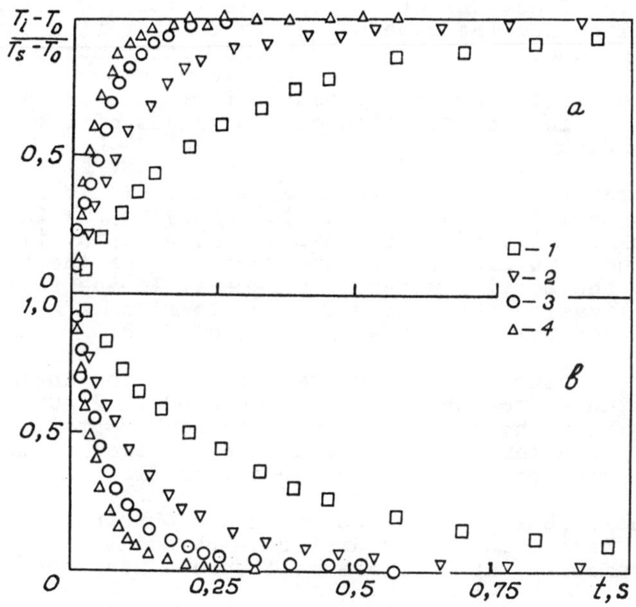

FIGURE 6. **Temperature** evolution **in** a flow in region of thermal stabilization at different Reynolds numbers y=10 mkm; a - heating **phase**; b - cooling phase; 1÷4 Re=12700; 27300; 48100; 64500.

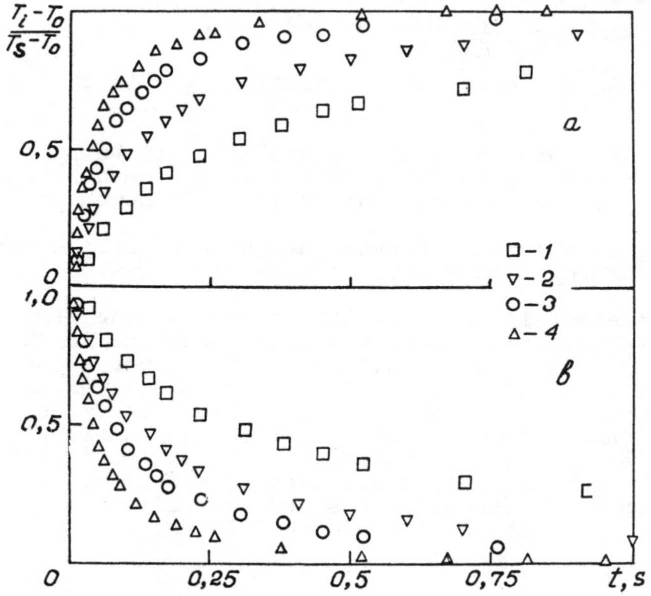

FIGURE 7. Temperature evolution in a flow in region of stabilized heat transfer at different Reynolds numbers; a - heating phase; b - cooling phase; 1÷4 Re=12800; 24400; 41800; 66900.

(b) the heat load. These results are obtained practically at the same distance to the wall ($y \approx 10\ \mu m$) and at different Reynolds numbers. As is seen from the given data, the time necessary for the flow temperature to reach a stationary value at large Re numbers amounts ten fractions of a second in total. In the region of thermal stabilization the time necessary for the flow temperature to reach a stationary value is less by several factors than that at L=0.78 m. The variationa of averaged temperature in the flow at heating and cooling are of symmetric form. The variation nature is analogous in the evolution of RMS values of temperature fluctuations, Figure 8.

If the parameter $\tau = tW/L$ is taken as a dimensionless time scale, then the time necessary to reach a stationary regime in the region of stabilized heat transfer is $t \approx L/W$. This is the time necessary for a liquid to flow the channel region from the beginning of heating to the section under measurement. In the region of thermal stabilization the time necessary to reach a stationary value exceeds the value of L/W by several fold.

The measuring results of the temperature evolution in the flow at different distances to the wall are presented in Figure 9. These data indicate that the time necessary to reach a stationary value is practically constant across the channel section. This is due to the intensive turbulent mixing across the flow.

Data in dimensionless coordinates which are obtained in the region of stabilized heat transfer at different Reynolds numbers but at the same distance to the wall are well spread near one line.

For the region of viscous sublayer all measuring results are generalized by one line. These data are presented in semilogarithmic coordinates in Figure 10. The behaviour of the dependence obtained for a heating phase can be described by the Gaussian curve

$$\Phi(x) = \frac{1}{\sigma\sqrt{2\pi}} \int_{-\infty}^{x} e^{-\frac{1}{2}\left(\frac{x-\zeta}{\sigma}\right)^2} dx$$

where $\sigma = 0.58$ is the stationary deviation, ζ - is the distribution centre.

The obtained data on the intensity evolution of turbulent pulsations in the region of stabilized heat transfer show that the time necessary to reach a stationary value is somewhat less.

The measuring results at different Reynolds numbers at the same distance to the wall are generalized by one curve.

Of special interest is the variation nature of the temperature field characteristics across the channel section. As an example the temperature profiles in universal dimensionless coordinates for different times at Re=29100 (L=0.78 m) are given in Figure 11. As is seen from the plot, these data split in different instants of time. With small times at heating the temperature profiles are below a stationary one that is related to the heating of liquid only in a narrow near-wall layer. When liquid is cooled, the profiles shift upwards. The data obtained indicate that quasistationary hypothesis is not satisfied at different times.

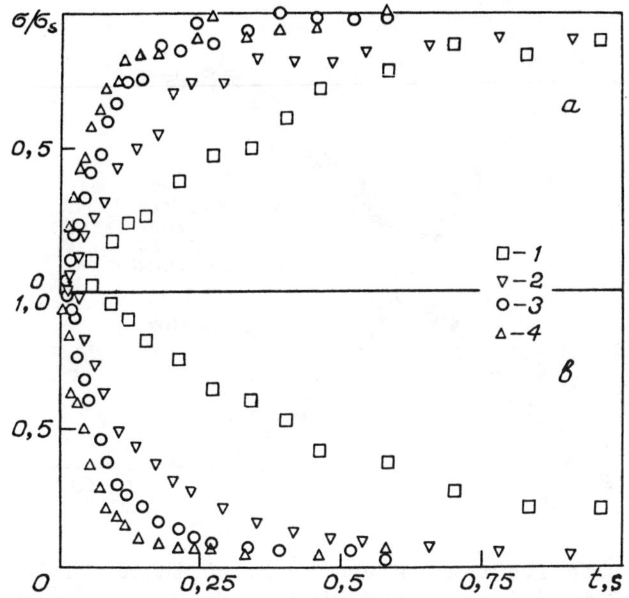

FIGURE 8. Evolution of temperature fluctuation intensity (symbols are the same as in Figure 7).

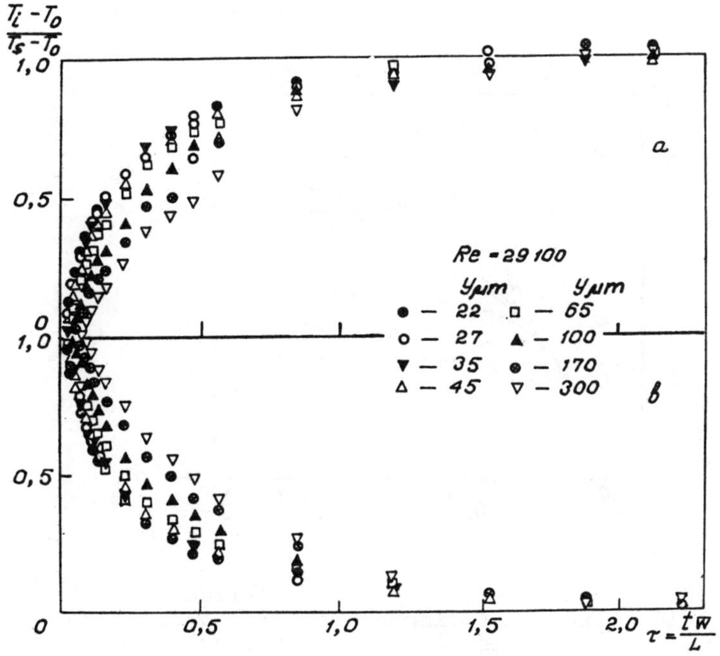

FIGURE 9. Temperature evolution in a flow at different distances to the wall L = 0.78 m; Re = 29100.

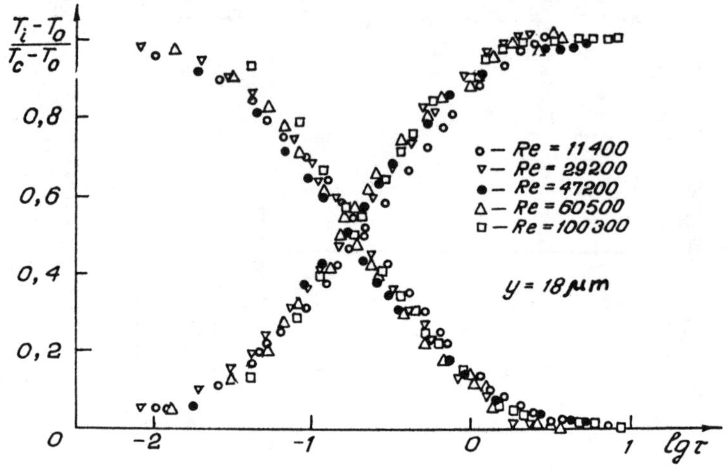

FIGURE 10. Relative temperature profiles in region of viscous sublayer at different Reynolds numbers (L = 0.78).

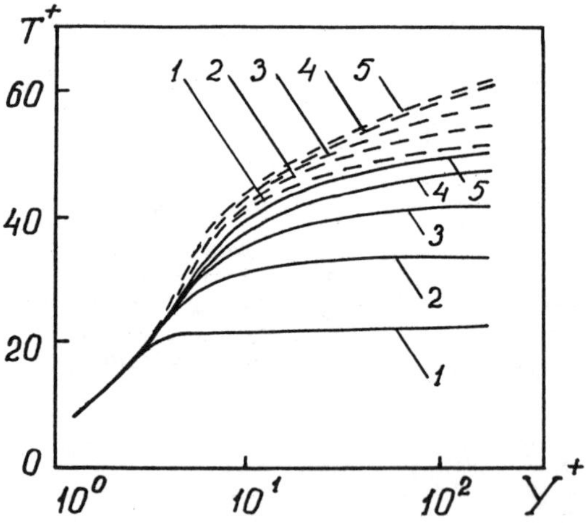

FIGURE 11. Dimensionless temperature profiles at different instants: solid lines - heating phase; 1÷5 t=0.012; 0.036; 0.12; 0.376; 1.336 s; dashed lines - cooling phase 1÷5 t= 0.008; 0.012; 0.024; 0.072; 0.408 s.

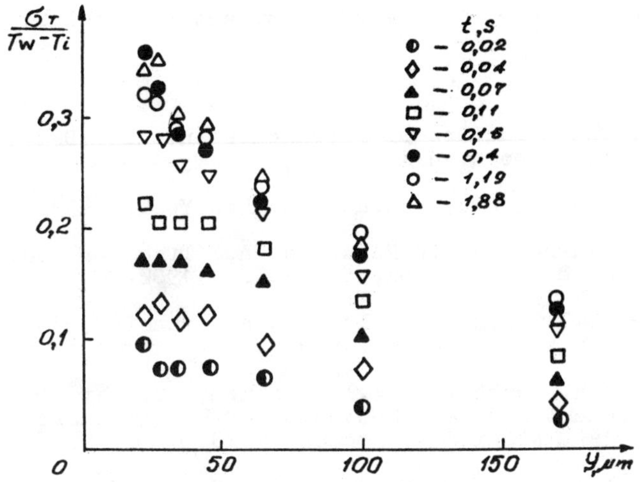

FIGURE 12. Distribution of relative temperature fluctuation intensities in different heating phases.

Figure 12 shows the intensity distribution of temperature fluctuations across the channel section at different times. The relation between RMS temperature fluctuations and local temperature difference $T_w(t) - T(t)$ is plotted along the y-axis. As is seen from the plot, this relation doesn't preserve the constant value at different times.

4. CONCLUSION

The formation of temperature field in the turbulent flow at the variation of boundary conditions is defined by the mechanisms of turbulent transfer in nonstationary fields of averaged values. The mechanism of such processes and their relation to the turbulence structure are not sufficiently well studied at present.

The results of the present measurements show that generalization of temperature curves takes place in the region of stabilized heat transfer at a sudden change of heat generation in the channel wall, if the value $\tau = tW/L$ is used as a dimensionless time scale. Such generalization of temperature curves is related to the mechanisms of turbulent transfer and the peculiarities of the turbulent flow structure. To find these relations it is necessary to analyse in detail both experimental data and our notions on the turbulence structure. Nonstationary temperature signal as well as a stationary one carries an information on the turbulence structure and its analysis can provide a new useful information about it.

It should be noted that the problem of unsteady heat transfer is conjugate i.e. the temperature evolution in the flow is related to the temperature evolution in the channel wall. However for different "working fluid-wall" combinations the evolution peculiarities related to the structure of turbulent flow should be preserved.

REFERENCES

1. Laufer, J., The structure of Turbulence in Fully Developed Pipe Flow, NACA-Report, N 1174, 1954.

2. Hinze, J.O., Turbulence, McGraw-Hill New York, 1959.

3. Structure of Turbulence in Heat and Mass Transfer, ed. Zarič, Z.P., Hemisphere, Washington, pp. 1-585, 1982.

4. Kutateladze, S.S., Fundamentals of Heat Transfer, 5-th ed., pp. 1-415, Atomizdat, Moscow, 1979.

5. Ibragimov, M.Kh., Subbotin, V.I., Bobkov, V.P. et al., Structure of Turbulent Flow and Mechanism of Heat Transfer in Channels, pp. 1-296, Atomizdat, Moscow, 1978.

6. Zhukauskas, A.A., Convective Heat Transfer in Heat Exchangers, pp. 1-472, Nauka, Moscow, 1982.

7. Kline, S.J., Reynolds, W.C., Schraub, F.A., Rundstadlier, P.W., Structure of Turbulent Boundary Layers, J. of Fluid Mechanics, vol. 30, pp. 741-773, 1967.

8. Kim, H.T., Kline, S.J., Reynolds, W.C., The Production of Turbulence Near a Smooth Wall in a Turbulent Boundary Layer, J. of Fluid Mechanics, vol. 50, pt. 1, pp. 133-160, 1971.

9. Corino, E.R., Brodkey, R.S., A Visual Investigation of the Wall Region in Turbulent Flow, J. of Fluid Mechanics, vol. 37, pp. 1-30. 1969.

10. Blackwelder, R.F., Kaplan, R.E., On the Wall Structure of the Turbulent Boundary Layer, J. of Fluid Mechanics, vol. 76, N 1, pp. 89-112, 1976.

11. Wallace, J.M., Brodkey, R.S., Eckelmann, H., Pattern Recognized Structures in Bounded Turbulent Shear Flows, J. of Fluid Mechanics, vol. 83, pt. 4, pp. 673-693, 1977.

12. Rao, K.N., Narasimha, R., Badri Narayanan, M.A., The "Bursting" Phenomenon in a Turbulent Boundary Layer, J. of Fluid Mechanics, vol. 48, pt. 2, pp. 339-352, 1971.

13. Gupta, A.K., Laufer, J., Kaplan, R.E., Spatial Structure in the Viscous Sublayer, J. of Fluid Mechanics, vol. 50, pt. 3, p. 403, 1971.

14. Strickland, J.H., Simpson, R.L., "Bursting" Frequencies Obtained From Wall Shear Stress Fluctuations in a Turbulent Boundary Layer, Phys. Fluids, vol. 18, N 3, pp. 306-308, 1975.

15. Repik, E.U., Sosedko, Yu. P., Spectral Investigation of Quasiordered Structure of the Flow in a Turbulent Boundary Layer,

Izv. AN SSSR, MZHG, N 3, pp. 10-17, 1982.

16. Perepelitsa, B.V., On Time Scale of Quasiperiodic Structures in a Turbulent Flow, In: Numeric Methods of Continuum Mechanics, vol. 13, N 3, pp. 91-98, Novosibirsk, 1982.

17. Perepelitsa, B.V., Some Properties of Large Scale Structures in a Turbulent Acqueous Flow and Dilute Polymer Solutions, In: Structure of Forced and Thermogravitational Flows, ITF SO AN SSSR, Novosibirsk, 1983.

18. Kuznetsov, Yu. N., Belousov, V.P., Numeric Solution of Problem Concerning Unsteady Heat Transfer at Turbulent Flow of Liquid in a Tube, TVT, vol. 8, N 6, pp. 1218-1227, 1970.

19. Kalinin, E.K., Dreizer, G.A., Kostuk, V.V., Berlin I.I., Methods to Calculate Conjugate Problems of Heat Transfer, pp. 1-232, Nashinostrojenie, Moscow, 1983.

20. Hiroshi Kawamura, Experimental and Analytical Study of Transient Heat Transfer for Turbulent Flow in a Circular Tube, Int. J. Heat Mass Transfer, vol. 20, N 5, pp. 443-450, 1977.

21. Perepelitsa, B.V., Pshenichnikov, Yu.M., Study of Temperature Field Development in Turbulent Flow at Unsteady Heat Transfer, Zhoornal Prikladnoi Mekhaniki i Tekhnicheskoi physiki, N 4, pp. 84-88, 1986.

Structural Study of Near-Wall Turbulence and Its Heat Transfer Mechanism

NOBUHIDE KASAGI
Department of Mechanical Engineering
The University of Tokyo
Hongo 7-3-1, Bunkyo-ku, Tokyo 113, Japan

INTRODUCTION

Since the deterministic nature of wall turbulence was first explored in the 1950's (Hama, 1957, Kline and Runstadler, 1959), a number of studies have been made to further our understandings of the physics of turbulent flows (see, e.g., Cantwell, 1981, and Coles, 1985). A series of extensive studies made by Stanford researchers, such as Kline et al. (1967) and Kim et al. (1971), revealed that the near-wall region of turbulent boundary layers exhibits recognizable streaky structures which are fairly periodic in the spanwise direction. They also documented a deterministic behavior of fluid motion, which is divided into the following three successive stages: a gradual lift-up of low-speed fluid from the near-wall region, rapid growth of oscillations, and an abrupt breakup into small scale motions, and labeled this entire process as 'bursting'. Since the near-wall region is crucial for momentum, heat and mass transfer, understanding of these structures is of great importance from both a scientific and engineering standpoint.

In the 1980's, rapid advances in electronic equipment have made it possible for experimental and numerical investigations to probe the physics of turbulence in more detail. For simple flow geometries, direct and large-eddy simulations are now possible and provide databases useful for analyzing and modeling turbulence (see, e.g., Rogallo and Moin, 1984). Flow visualization has played an equally important role in experimental and numerical investigations by providing valuable information about major flow structures for more in depth study and by allowing researchers to reduce complex flow processes to simple pictures. In particular, the techniques which employ computer-aided digital image processing are quite promising and may provide a new information on three-dimensional structures of turbulence.

This paper is a brief review of the experimental work on near-wall turbulence and heat transfer processes at the University of Tokyo during the last decade. An emphasis will be placed upon the visual studies of sublayer streaks and streamwise vortical structures. The following two flow fields are investigated: a turbulent boundary layer without a pressure gradient and a two-dimensional turbulent channel flow. An attempt is made to construct a simple deterministic turbulence model to represent the statistics and dynamics of near-wall turbulence.

CHARACTERISTICS OF THE STREAKY STRUCTURES IN THE VISCOUS WALL REGION

The streaky structures near the wall have been investigated in detail by Nakagawa and Nezu (1981), Smith and Metzler (1983) and Iritani et al.

Table 1 Basic flow conditions for a 2-D channel flow at x/H=68

U_c (m/s)	H (m)	δ^* (mm)	θ (mm)	Re_θ	Re_H	u_τ (mm/s)	ν/u_τ^2 (s)
0.0852	0.080	5.87	3.54	300	6790	4.07	0.0606

(1985). The experiment by Iritani et al. (1985) was carried out in a two-dimensional water channel facility at the University of Tokyo's Mechanical Engineering Department. The plexiglas channel has a cross section of 80mm x 800mm and is 6.5m in length. The measurements were made at a streamwise position located about 68 channel heights downstream of the channel inlet, where a fully-developed, two-dimensional turbulent flow was established. Basic flow conditions at the test section are summarized in Table 1 (for details, see Kasagi et al., 1984).

Typical examples of flow visualization by the hydrogen bubble technique are shown in Fig. 1. A single electrode wire is positioned parallel to the wall at various distances from the wall, as indicated in the figure. When the transverse wire is positioned in the viscous sublayer (denoted by A and B in Fig. 1) and in the buffer layer (C in Fig. 1), the high- and low-speed streaks are clearly seen, and appear more or less periodically in the spanwise direction. As is well known, these low-speed streaks are elongated in the streamwise direction. Even when the wire is moved into the logarithmic region and further away from the wall (D and E in Fig. 1), local regions of momentum deficit can be still recognized. However, their spanwise spacing is much larger and their streamwise length is much shorter compared to the streaky structures near the wall. The structures in a turbulent boundary layer visualized by Kline et al. (1967) and Kim et al. (1971) do not differ qualitatively from the present results; thus, there is no essential difference between the near-wall structures in external and internal flows.

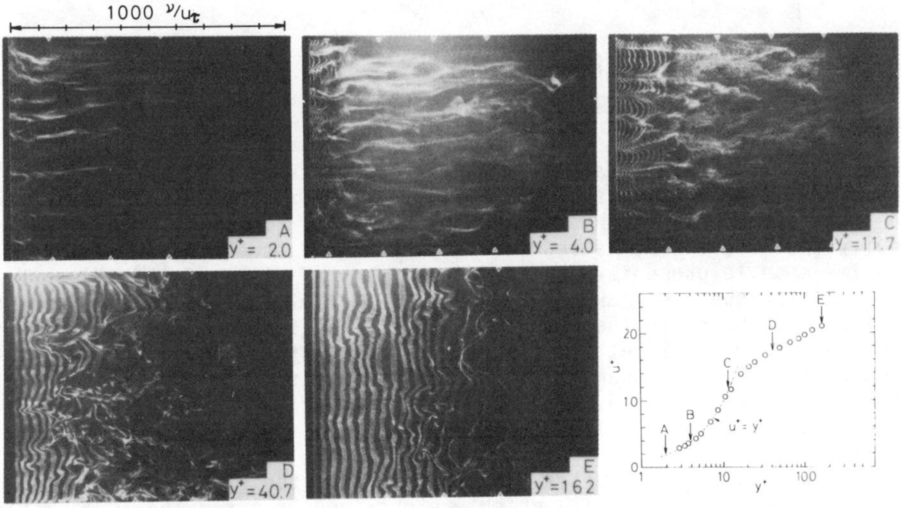

Figure 1 Flow visualization in a two-dimensional turbulent channel flow by the hydrogen-bubble technique.

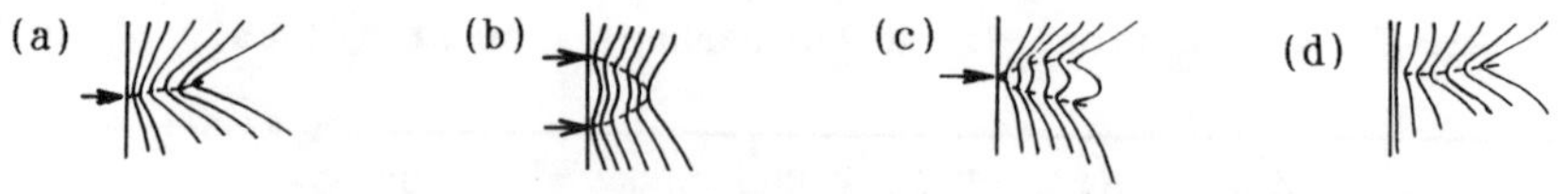

Figure 2 Definition of the low-speed streaks; patterns should persist for 0.5 second ($8.25\nu/u_\tau^2$).

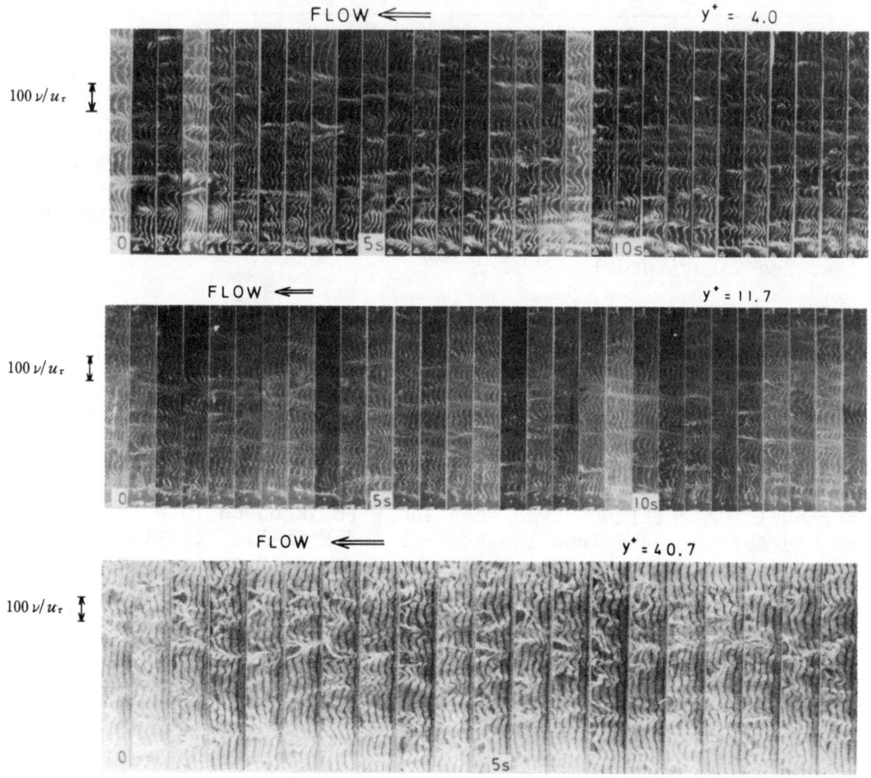

Figure 3 Persistence of the low-speed streaks.

In order to obtain quantitative data of streaky structures, the streaks are defined as follows. As shown in Fig. 2(a), a low-speed streak is defined as a local minimum in the hydrogen-bubbles' time lines in the spanwise velocity distribution downstream of the wire for a prescribed time span Δt. This time span is presently taken as Δt=0.5 second (i.e. Δt^+=8.25). Note that the low-speed streak often branches and coalesces, as shown in Figs. 2(b) and 2(c). Even in these cases, the above definition is used. When coalescence occurs, if either of two coalescing streaks continues to exist for more than the time span Δt^+, then the streak is counted to have occurred. When a low-speed streak is observed apart from the wire, as shown in Fig. 2(d), this streak is not counted.

Figure 3 shows typical examples of time traces of the streaky structures at three distances from the wall, which are obtained by arranging 16mm motion pictures at a time interval of Δt^+=8.25. The majority of low-speed streaks appear to persist over a time span much longer than that used for the

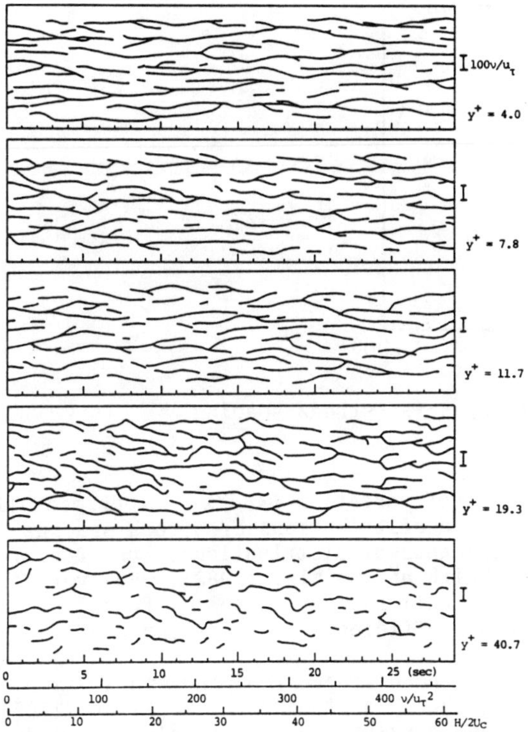

Figure 4 Time trace diagrams of the low-speed streaks.

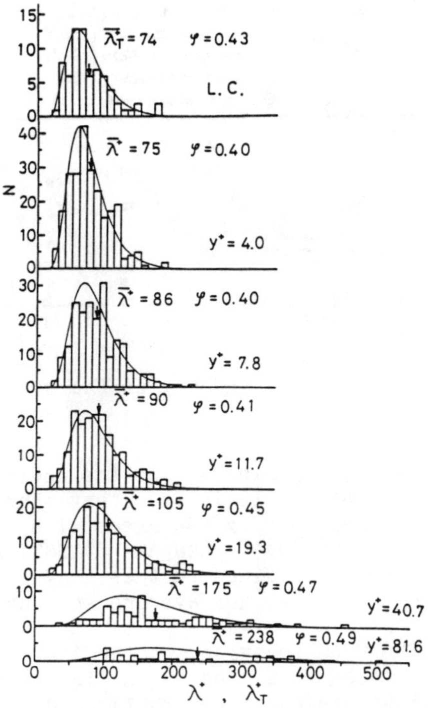

Figure 5 Probability-density histograms of the spanwise streak spacing.

definition of low-speed streaks. It is seen that they exhibit characteristic behavior, such as coalescing, branching and meandering. This is also pointed out in open channel flows (see, Utami and Ueno, 1977, and Nakagawa and Nezu, 1981) and boundary layer flows (Smith and Metzler, 1983). At $y^+=40.7$, the lifespan of the low-speed streaks is comparatively short and their meandering appears to be much enhanced.

The diagrams of the time traces of low-speed streaks over a much longer time span are shown in Fig. 4. Branching, coalescing and meandering of the low-speed streaks are clearly observed in these diagrams. The features observed in the region $y^+\leq 20$, where the branching and coalescing of the streaks occur very often, are considerably different from those observed at $y^+=40.7$. It is evident that the low-speed streaks persist for a longer time as the wall is approached; at $y^+=4.0$ it is easy to find the streaks which persist for more than 300 wall time units (ν/u_τ^2). A similar result was obtained from the direct numerical simulation database by Kim and Spalart (1987). However, the distance for which the streaks persist downstream does not agree quantitatively with the experimentally observed value, due to a difference in the definition of the low-speed streaks.

Figure 5 summarizes the probability-density histograms of the spanwise streak spacing, which are obtained from the time-trace diagrams of the low-speed streaks described above. The upper most histogram in Fig. 5 is the probability density of the thermal streaks' spacing, which will be later discussed. The curves are the lognormal distributions calculated from the

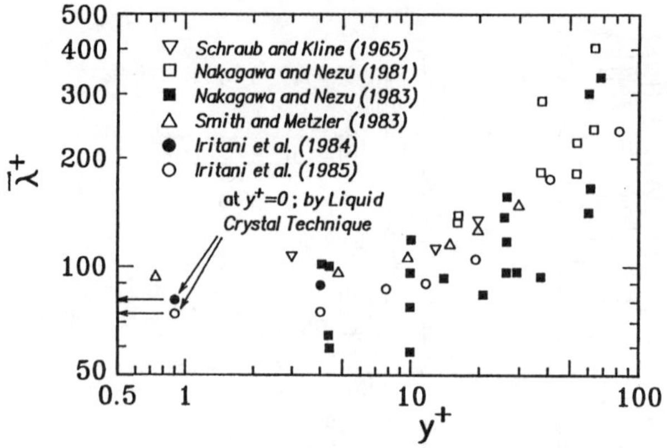

Figure 6 Dependency of the average spanwise streak spacing.

average spacing $\overline{\lambda}^+$ and the coefficient of variation $\psi=\sigma_\lambda/\overline{\lambda}^+$ experimentally obtained. It is seen that the measured distributions of streak spacing in the region of $y^+\leq 20$ agree with these lognormal distributions, as reported by Nezu and Nakagawa (1981) and Smith and Metzler (1983). However, the results obtained further away from the wall show a quite different trend with much larger average spacings and coefficients of variation.

The mean spanwise spacing increases as the distance from the wall is increased, as shown in Fig. 6, where the data of Schraub and Kline (1965), Nakagawa and Nezu (1981), and Smith and Metzler (1983) are also included. This trend suggests that the number of low-speed streaks decreases as the distance from the wall is increased, since the low-speed streaks are mostly associated with the outward fluid motion originating from the viscous wall region (i.e. ejection) as described later.

The mean passing period t_p and the mean persistence t_d of the low-speed streaks are shown in Fig. 8. The former is defined as the time span between which two successive streaks are detected at a given spanwise position as shown in Fig. 7. The latter is defined as the average lifespan of a streak, when the points of branching and coalescence are assumed to be the points of termination and/or initiation. Both time spans remain nearly constant independent of the distance from the wall in the region of $y^+\leq 20$. In the region further away from the wall, however, the period increases appreciably while the persistence decreases. The mean period of $t_p^+\simeq 55$ near the wall is consistent with the mean bursting period of $T_B^+\simeq 90$ measured by Luchik and Tiederman (1987) and also $T_B^+\simeq 60$ suggested by Kim and Spalart (1987); the quantitative difference in these values is a result of the different detection algorithms used. In the probe measurements with the VITA technique proposed by Blackwelder and Kaplan (1976), ensemble-averaged velocity signals are characterized by a rapid deceleration at the leading edge followed by a strong positive velocity gradient at the trailing edge (Bogard and Tiederman, 1987). In addition, a positive peak in the v-velocity occurs immediately after the leading edge. These velocity traces would be experienced by a probe over which the meandering streaks, such as those described above and illustrated in Fig. 7, are flowing. The mean persistence near the wall is about $t_d^+\simeq 40$ and agrees roughly with $t_d^+\simeq 26$ obtained by Bogard and Tiederman (1987); the present definition of t_d^+ tends to give a longer persistence. The value $t_d^+\simeq 480$ by Smith and Metzler (1983) is much larger; they used a comparatively long sampling period.

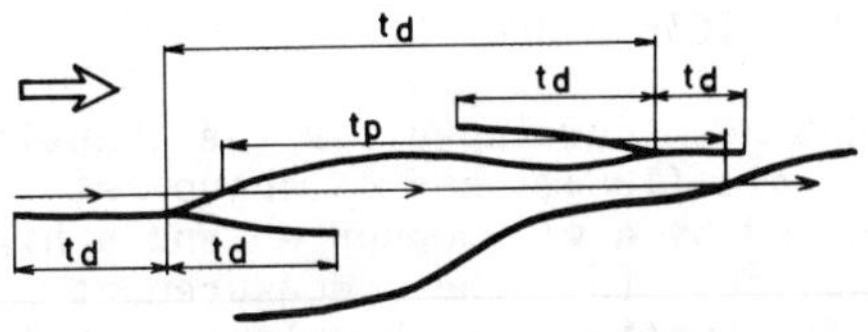

Figure 7 Definitions of the average period of passing and the average persistence of the low-speed streak.

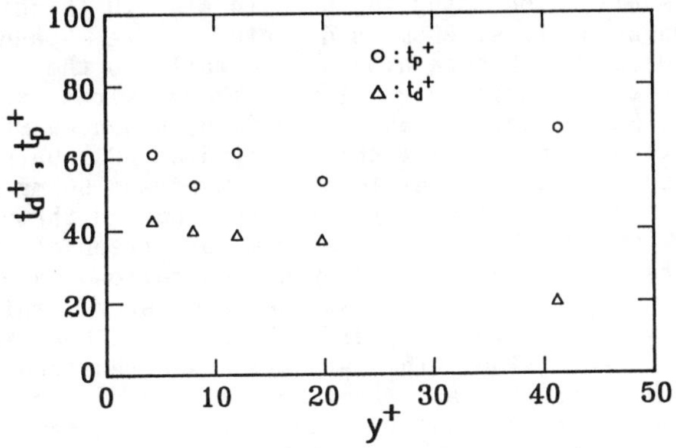

Figure 8 The average period t_p^+ and persistence t_d^+ of the low-speed streaks.

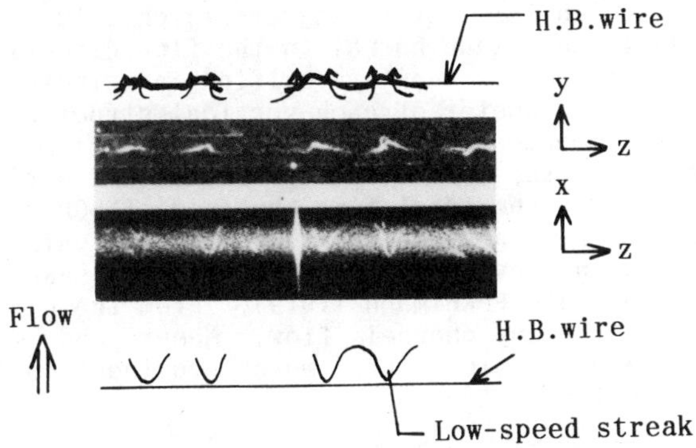

Figure 9 Stereoscopic observation of a hydrogen-bubble timeline near the wall (y^+=16.7).

The wall-normal movement of the low-speed streaks is qualitatively examined in Fig. 9, where a hydrogen-bubble timeline is observed in two planes simultaneously. As schematically indicated, each low-speed streak seems to be associated with upward swelling motions. This extremely high correspondence between the momentum deficit and the outward motion has also been directly observed by Kasagi and Hirata (1976) and Smith and Schwartz (1983). The observed fluid motions may suggest the existence of streamwise counter-rotating vortices, but recognizing that a large number of vortical motions show rotational angles much smaller than 180 degrees, Kasagi et al. (1984) and Nishino et al. (1987) referred to these coherent motions as 'pseudo-vortical structures'.

DIRECT OBSERVATION OF STREAMWISE PSEUDO-VORTICAL STRUCTURES

A digital image processing technique has been developed for the quantitative analysis of visual data by Kasagi et al. (1984) and Nishino et al. (1987). This technique, which is supported by a microcomputer and a high-performance image processor, is applied to the measurement of instantaneous, two-component velocity fluctuations in the plane normal to the streamwise direction in the 2-D channel flow described above.

The schematic of the flow visualization setup is shown in Fig. 10 (Nishino et al., 1987). Seven fine tungsten wires, 20μm in diameter, were spanned between two wing-shaped supports, which were attached normally to the test plate 150mm ($600\nu/u_\tau$) apart in the spanwise direction. These wires were electrically insulated at 2-3mm intervals and, therefore, produced many discrete hydrogen-bubble clusters when pulsed with a periodic DC charge. The period and duration of the pulses were set to be 999 msec and 60 msec, respectively. These clusters became visible only while they passed through the 15mm long lighted section, which was located 20mm downstream of the wires. Motions of the clusters were then recorded by a 16mm motion camera at a frame speed of 48 frames/second. Note that bubble clusters, which were generated simultaneously with a certain DC pulse charge, reached the lighted section at nearly the same time. This situation was realized by staggering the wires so that the wires far away from the wall were located farther upstream than those close to the wall. Quantitative data was obtained from 211 photograph sets, which were printed from the 16mm film at five frame intervals (for details, see Nishino et al., 1987).

Figure 11 shows examples of the sequential vector diagrams which illustrate the instantaneous flow fields in the plane normal to the flow direction. The existence of the vortical structures is readily verified, as indicated by arrows in these diagrams. The diameter of each vortical structure is about $40\nu/u_\tau$ and their centers are located at $y^+=25\sim70$. Aside from the strong vortical motions indicated by the arrows, weaker rotational motions can also be observed in Fig. 11. Over the total time observed ($3500\nu/u_\tau^2$), many of the vortices appear as 'solitary vortices', unlike the pairs of counter-rotating vortices proposed in previous studies, such as Bakewell and Lumley (1967) and Blackwelder and Eckelmann (1979). From the direct numerical simulation of a curved turbulent channel flow, Moser and Moin (1984) pointed out that solitary vortices are more prominent in the vicinity of the wall than vortex pairs.

It is also found that the pseudo-vortical structures have a small streamwise length scale compared to that of streaky structures. Because of the short time interval ($16.7\ \nu/u_\tau^2$) between two successive vector diagrams and the local convection velocity ($\simeq15u_\tau$) in the near-wall region, a given vortex should be present at nearly the same position over several successive vector diagrams if the vortex has as long a structure as the streaky structures. From detailed observation of the vector diagrams, however, it is revealed that very few vortices can be traced, even between two successive vector diagrams. Kim (1987) and Maruyama and Tanaka (1987) also documented that the streamwise vortical structures have a streamwise length scale much smaller than that of the sublayer streaks. Maruyama and Tanaka suggested a streamwise length scale of $300\sim400\nu/u_\tau$ from the visual data of a plane Couette flow. In a numerical simulation of a channel flow, Kim (1987) found large-scale structures in the wall region; these structures propagated downstream retaining their coherence over a streamwise distance on the order of $1000\nu/u_\tau$ with a typical velocity of $10u_\tau$ and seemed to have a length scale of $150\nu/u_\tau$. In addition, these large scale structures are

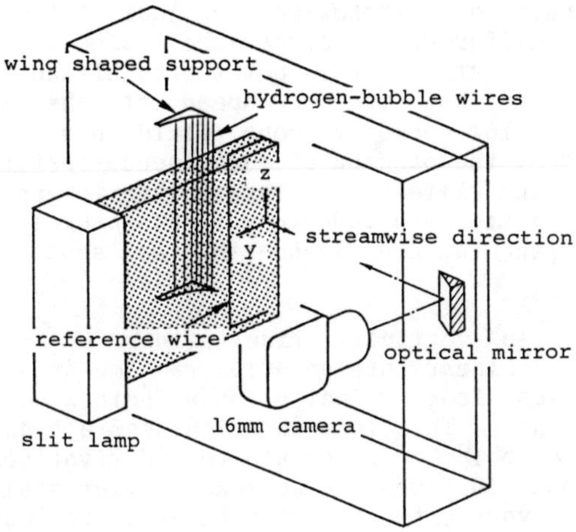

Figure 10 Schematic of the flow visualization in the plane normal to the flow direction in a 2-D turbulent channel flow.

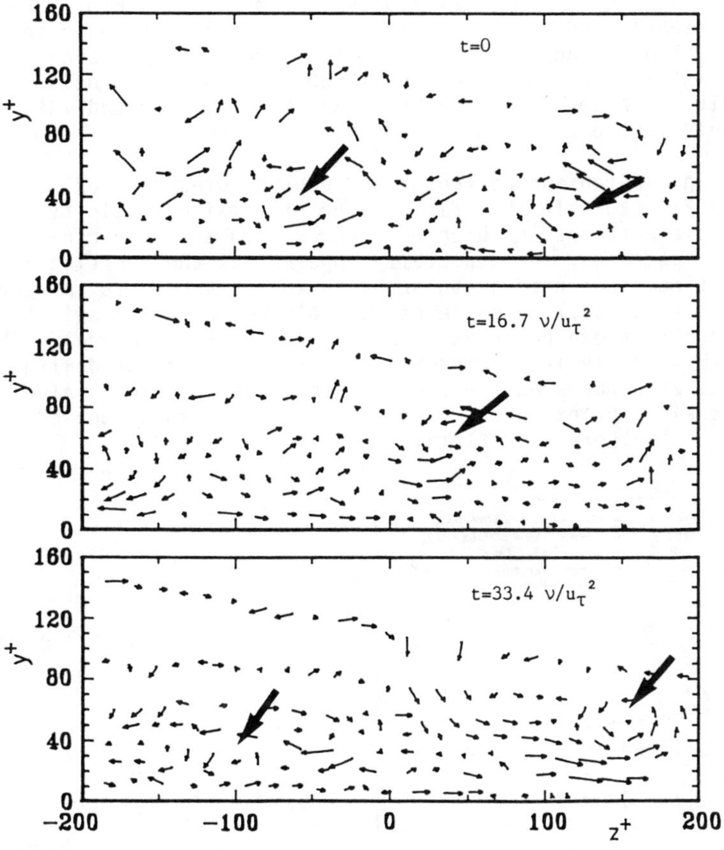

Figure 11 Sequential vector diagrams obtained by digital image processing.

associated with a single or a pair of streamwise vortical motions. Although the reason for the large difference in streamwise scales of the low-speed streak and the pseudo-vortical structure is not thoroughly understood, one plausible explanation is that the low-speed streaks are essentially footprints on the lake of low-speed viscous fluid near the wall, and these footprints result from the passing of many pseudo-vortical structures of various size, strength and lifespan. Thus, the low-speed streaks persist for a relatively long time and undergo various changes and interactions, such as coalescing, branching and meandering as shown in Fig. 4 (see also Fig. 21).

From each vector diagram, the streamwise vorticity fluctuations ω_x were calculated by Nishino et al. (1987). Linear interpolation was employed in order to obtain the spatial derivatives from velocity data points distributed randomly in the y-z plane. This interpolation used all data points located within $26\nu/u_\tau$ from each point to calculate the derivatives. Note that $26\nu/u_\tau$ corresponds to about 10 times the Kolmogorov microscale. Since this resolution is coarse, the vorticity thus calculated is considered to be associated with only the large vortical structures. Distributions of the rms streamwise vorticity fluctuation are shown in Fig. 12. Although the present rms values are smaller than the other experimental data due to the limitation of spatial resolution, they clearly show that the rms value reaches a minimum value at $y^+ \simeq 10$. The existence of this minimum has been predicted in numerical studies by Moin and Kim (1982), Moser and Moin (1984) and Kim et al. (1987), and attributed to the 'splatting effect' by Moin and Kim (1982) and Kim (1985). By using vorticity meters, Kastrinakis and Eckelmann (1983) and Balint et al. (1987) obtained the results in accordance with Moser and Moin (1984) and Kim et al. (1987), but only in the region of $y^+>10$.

An example of typical instantaneous vector and contour diagrams of v, w, and ω_x is shown in Fig. 13. In this figure, the streamwise vorticity field exhibits recognizable patterns in which ω_x changes sign in the vicinity of the wall. These patterns are especially apparent at the position where strong vortical motion, indicated by an arrow in the figure, is present. The existence of these typical patterns was pointed out by Kasagi et al. (1984), who referred to these patterns as a 'double-layer structure'. They suggested that this double-layer structure is closely associated with the streamwise vortical structures and responsible for the minimum value of $\tilde{\omega}_x^+$. Using a Rankine vortex model, Kim et al. (1987) also discussed this typical structure of the vorticity field.

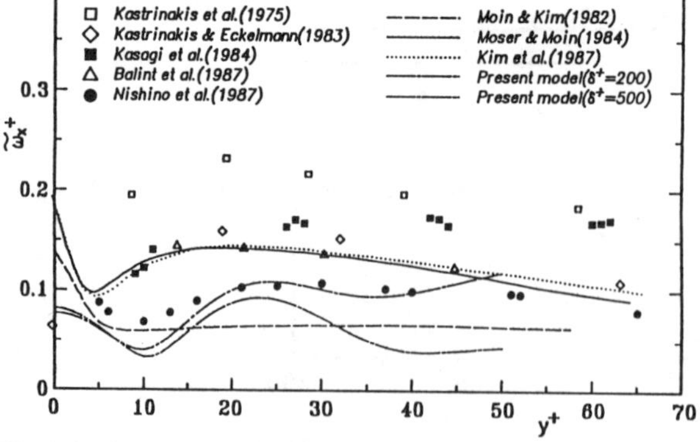

Figure 12 Distributions of the rms streamwise vorticity fluctuation.

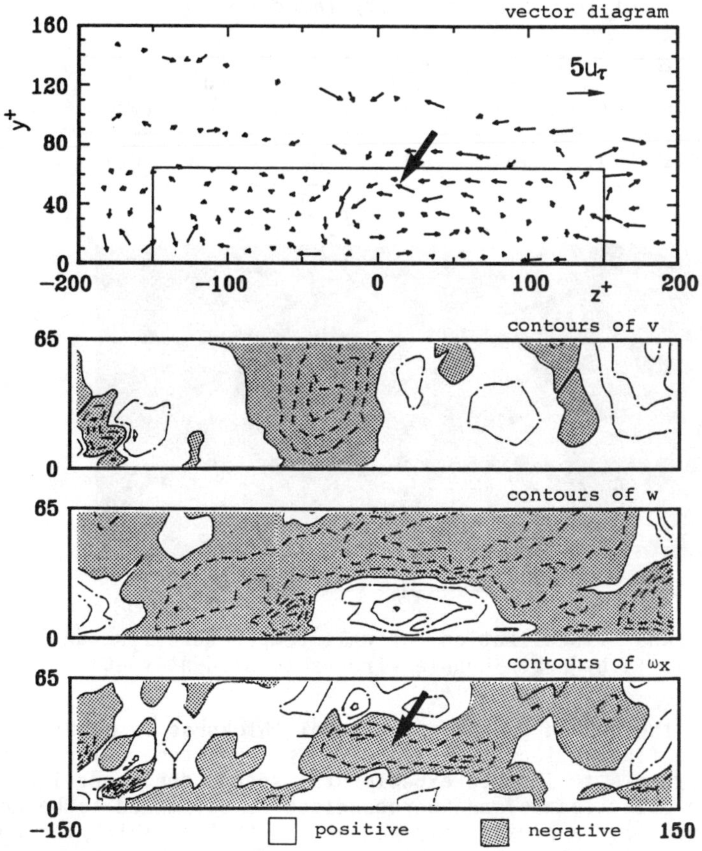

Figure 13 Instantaneous vector and contour diagrams of v, w and ω_x.

STRUCTURE OF TURBULENT SCALAR TRANSPORT ASSOCIATED WITH COHERENT STRUCTURES NEAR THE WALL

Structural studies of near-wall turbulent heat transfer have been reviewed recently by Hirata et al. (1982) and Khabakhpasheva (1986); they emphasized visual data obtained by a liquid crystal technique (for details of the technique, see Kasagi et al., 1988a). Iritani et al. (1984) combined this technique with the hydrogen bubble technique to visualize simultaneously wall temperature fluctuations and the flow structure in a turbulent boundary layer. Experiments are carried out in a low-speed water channel, and the measurements are made in a 2m-long test section downstream of a contraction nozzle. The test section is made of plexiglas and has a cross section of 0.3m x 0.3m. The heat transfer apparatus is attached to one of the test-section side walls, and a boundary layer is established throughout the test section without a pressure gradient.

The test plate specially designed for the visualization of wall temperature fluctuation is installed at the position 1600mm downstream of the boundary layer tripping. A constant heat-flux wall condition is achieved by applying an AC current supply to 30μm thick stainless steel foil sheets, which are bonded to the wall surface. A liquid crystal sheet, which is composed of polyester film of 50μm thickness, liquid crystals and black backing, is

Table 2 Basic flow conditions for a boundary layer flow

U_∞ (m/s)	δ (m)	δ^* (mm)	θ (mm)	Re_θ	$H=\delta^*/\theta$	u_τ (mm/s)	ν/u_τ^2 (s)	δ_T (mm)
0.204	0.050	6.8	4.9	990	1.4	9.35	0.0116	9.4

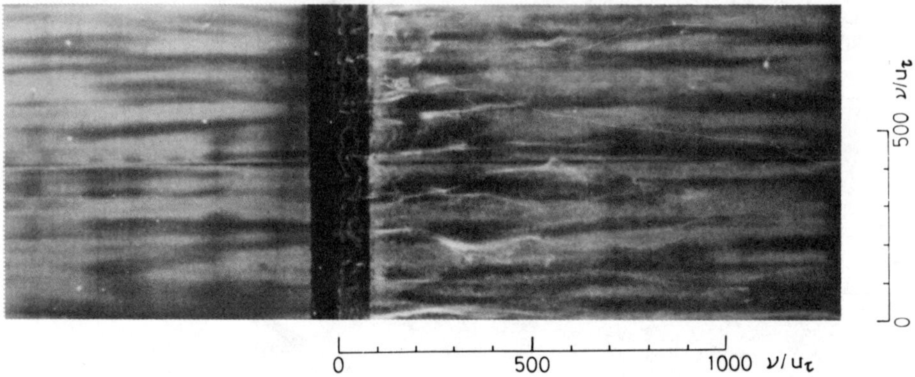

Figure 14 Simultaneous visualization of wall temperature fluctuation and flow field (a hydrogen-bubble wire as located at $y^+=6$).

attached flush to the heated surface; the total thickness is about 100μm. The wall temperature fluctuations are visualized through the color change of the liquid crystals.† In this experimental setup, the upper limit of frequency where the color change can be recognized is estimated to be about 2Hz; this frequency is approximately equal to that at which the power spectrum of the streamwise velocity fluctuation is one tenth of its maximum value. The flow conditions at the measurement position are listed in Table 2.

A typical photograph of the hydrogen-bubble/liquid crystal visualization in the turbulent boundary layer is shown in Fig. 14; here the location of transverse wire is at $y^+=6$. The wall temperature fluctuations visualized by liquid crystals are characterized by a streaky pattern, and the thermal field consists of high-temperature streaks (black in Fig. 14, but actually blue) and low-temperature streaks (white but actually orange). These thermal streaks are observed periodically in the spanwise direction and elongated in the streamwise direction, much the same as the momentum streaks discussed previously. The streamwise extent of these thermal streaks reaches $1000\sim2000\nu/u_\tau$. In the 16mm pictures, thermal streaks appear randomly in space and time and eventually vanish at some point downstream. In Fig. 14, it is qualitatively confirmed that the high-temperature streaks are mostly associated with the low-speed streaks and low-temperature streaks with the high-speed streaks.

†) Note that temperature fluctuations close to the wall (e.g., $y^+\leq3$ for Pr=7) are influenced considerably by the thermal properties of the wall, which depend upon the wall material and thickness, as discussed by Kasagi et al. (1987, 1988b). Thus, the color change of liquid crystals actually reflects this conjugate convection/conduction heat transfer.

The correspondence between the thermal and momentum streaks is shown in Fig. 15, as a function of distance from the wall. Since this data is obtained from a comparatively small number of streaks at each wire position, the measurement uncertainty is considerable. The observed correspondence is remarkable in the region $y^+ \leq 10$ and more than 70% of the thermal streaks are uniquely associated with the prescribed momentum streaks. The recent measurement by Subramanian and Antonia (1981) and Hishida et al. (1986) show that the cross correlation between the streamwise velocity and temperature fluctuations is more than 0.8 in the vicinity of the wall. This result has also been obtained from the direct simulation of a turbulent scalar field by Kim and Moin (1987). This close relationship, however, fades out in the region $y^+ \geq 30$ with the disappearance of streaky structures. Simultaneous observations of the wall thermal streaks and the flow pattern in the x-y plane have also been made with a vertical hydrogen-bubble wire by Iritani et al. (1984). With some quantitative definitions of the lift-up and sweep events, 70% of the total lift-up events are associated with the high-temperature streaks, while 72% of the sweep events are associated with the low-temperature streaks. A conceptual model of the relationship between the momentum and thermal streaks is presented in Fig. 16.

From the experimental results above, the near-wall heat transfer mechanism is considered to be governed primarily by the streaky structures in the region at $y^+ \leq 20$, or at least $y^+ \leq 10$. When the local inrush of fluid towards the wall is induced by the streamwise pseudo-vortical structures, the local wall temperature is rapidly decreased, and this results in the formation of

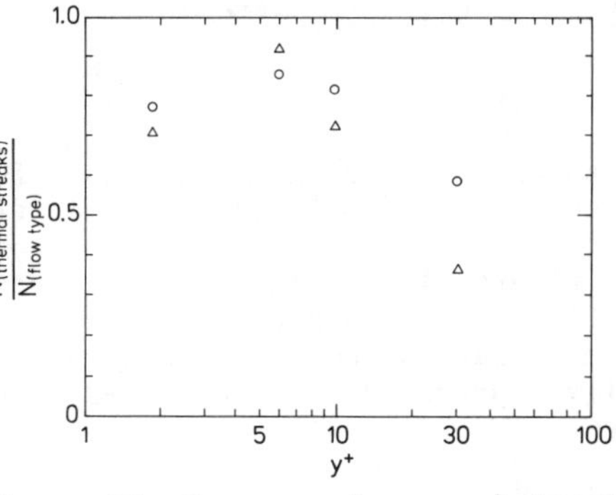

Figure 15 Correspondence between thermal and momentum streaks as a function of the wall distance.

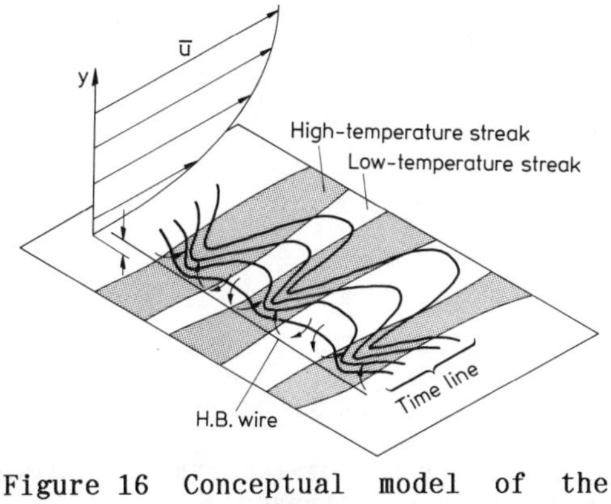

Figure 16 Conceptual model of the formation of thermal and momentum streaks.

a low-temperature high-speed streak. During the streamwise movement, the fluid is decelerated by the wall shear and, at the same time, heated by the wall, and this results in a high-temperature low-speed streak which is eventually ejected away from the wall.

The 16mm motion pictures of the visualization at $y^+=4$ are analyzed on a motion picture analyzer. The spanwise spacing of low-speed streaks λ is sampled and measured. The average spacing is $89\nu/u_\tau^2$, and the probability density distribution is again approximated by a lognormal distribution. In order to determine the spanwise spacing of the thermal streaks, high- and low-temperature streaks are distinguished by a temperature threshold, which corresponds to the border between green and orange-yellow in liquid crystals' color. The mean spacing of the thermal streaks is determined to be $\bar{\lambda}_T=81\nu/u_\tau^2$ and agrees closely with that of momentum streaks mentioned above. Similar results in a 2-D channel flow have already been included in Fig. 5, where the mean spacings near the wall are $\bar{\lambda}^+=74$ and $\bar{\lambda}_T^+=75$, respectively.

The visual breakup frequency of the low-speed streaks and the passing frequency of low-temperature streaks are counted in the motion pictures with the transverse wire located at $y^+=4$. The result for momentum streaks is shown as a straight line in Fig. 17, where those of thermal streaks are plotted with symbols representing two observers. A good agreement is obtained between the breakup frequency of the low-speed streaks and the passing frequency of the low-temperature streaks. These results again suggest that the coherent turbulent motions, which lead to the formation of momentum streaks, produce a similar pattern of near-wall temperature fluctuations (see Fig. 14). If the bursting period is defined as $T_B=1/F\bar{\lambda}$ (Kline et al., 1967), then T_B^+ is determined to be 112, which is in good agreement with the values obtained previously.

The structure of the temperature fluctuations is studied by the analysis of spatial correlations. The correlation coefficients between wall and fluid temperature fluctuations are measured by using fine thermocouple probes; the results are shown in Fig. 18. This figure also includes the correlation coefficient of spanwise velocity fluctuations measured by Blackwelder and Eckelmann (1979). The temperature correlation shows a rapid decrease in the region of $y^+\leq 10$ and gradually approaches zero but never becomes negative, in contrast to the spanwise velocity fluctuations.

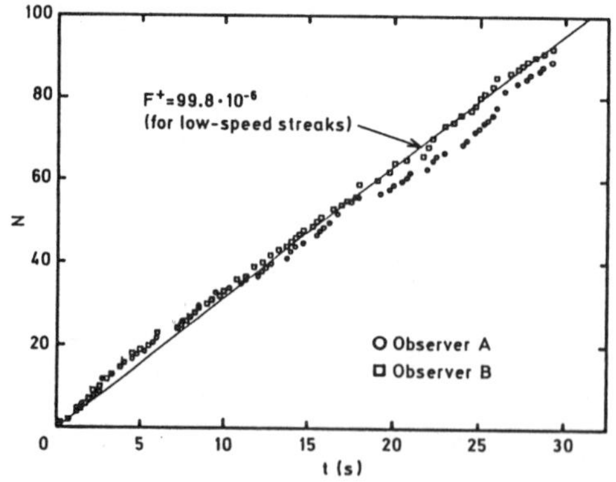

Figure 17 Comparison of the passing frequency of low-temperature streaks with the break-up frequency of low-speed streaks.

The spanwise correlation coefficients are shown in Fig. 19. At the edge of the viscous sublayer ($y^+=5.5$), there is a negative peak with the separation of $\Delta z^+ \simeq 50$, and a small positive peak with $\Delta z^+ \simeq 100$. Thus, the quasi-periodicity with a wavelength of about $\Delta z^+ \simeq 100$ is again confirmed. As the distance from the wall is increased, the separation which gives the first zero-crossing increases, and a negative correlation is hardly observed beyond $y^+=34$. The present data at $y^+=5.5$ is very similar to the correlations of the spanwise velocity gradient fluctuation at the wall but not so much to those of the streamwise velocity gradient. The streamwise correlation coefficient at $y^+=5.5$ is shown in Fig. 20, where a positive correlation can be found as far as $\Delta x^+=1000$. This length scale agrees qualitatively with the streamwise extent of thermal streaks visualized by liquid crystals (see Fig. 14). The curve for $y^+=5.5$ is again similar to that for the spanwise velocity gradient rather than the streamwise velocity gradient. All of these spatial temperature correlations are compatible with the hypothesis of streamwise pseudo-vortical structures and suggest the important role of these structures in heat transfer processes in the viscous wall region.

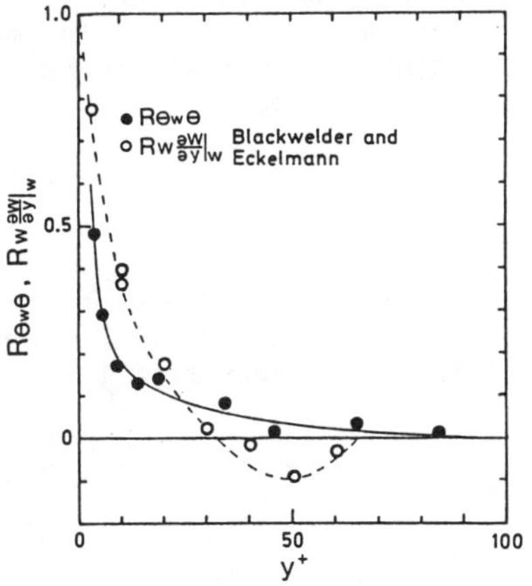

Figure 18 Cross correlations of the temperature fluctuations with the separation normal to the wall.

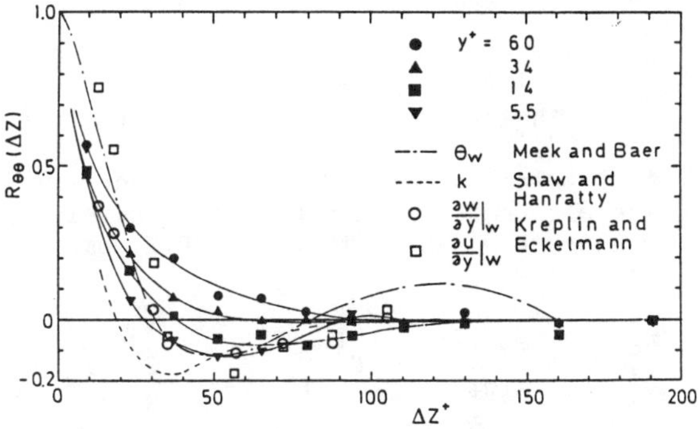

Figure 19 Spanwise correlations of the temperature fluctuations.

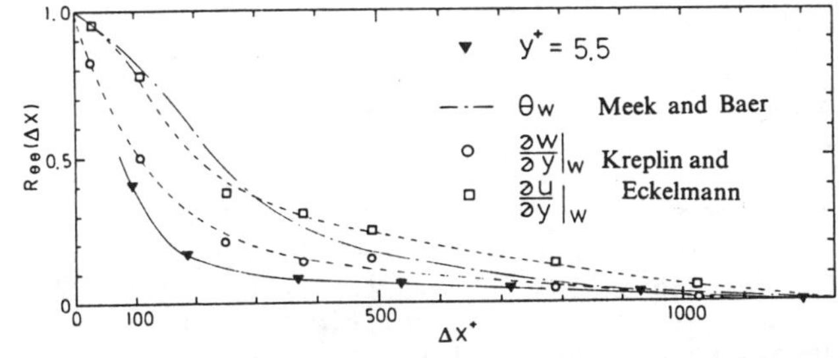

Figure 20 Streamwise correlations of the temperature fluctuations.

REPRESENTATION OF NEAR-WALL TURBULENCE BY THE TWO-COMPONENT WALL-LAYER MODEL

Based on accumulated experimental knowledge, an attempt is made to construct a simple deterministic turbulence model to represent the statistics and dynamics of near-wall turbulence. Several models have been proposed to account for deterministic properties of near-wall turbulence (see, Coles, 1978, Hatziavramidis and Hanratty, 1979, Kasagi et al., 1984, and Chapman and Kuhn, 1986). In the present work, a model is constructed on the basis of the hypothesis that turbulent motion in the inner layer can be divided into two parts, active and inactive (Bradshaw, 1967). The active part is the streamwise pseudo-vortical fluid motion (SPVM) and produces a large portion of the Reynolds stress in the vicinity of the wall, while the inactive part is a disturbance induced through pressure fluctuations caused by large scale motions (LSM) in the outer layer (Brown and Thomas, 1977, Rajagopalan and Antonia, 1979). The characteristic time and length scales associated with these two components are scaled with the appropriate parameters of the inner and outer layers, respectively.

Figure 21 shows a conceptual view of the pseudo-vortical structures and the low-speed streaks. In addition, turbulence in the inner layer contains a component due to disturbances imposed by the outer-layer, large-scale motion. Thus, each of the three velocity fluctuations is represented by the superposition of these two components in the following manner:

$$u = u_L + u_S, \quad v = v_L + v_S, \quad w = w_L + w_S \tag{1}$$

where the subscripts of L and S denote LSM and SPVM, respectively.

A regular pattern of turbulent eddies is assumed for the SPVM components.† The algebraic expressions for the velocity components produced by SPVM are constructed such that the SPVM components diminish far away from the wall as follows:

$$u_S^+ = a_1 \sin(2\pi z^+/\lambda^+) \sin(2\pi t^+/T_S^+ + \Phi_S)\, y^+ \exp(-y^+/b_1) \tag{2}$$

$$v_S^+ = a_2 \sin(2\pi z^+/\lambda^+) \sin(2\pi t^+/T_S^+)\, y^{+2} \exp(-y^{+2}/b_1^2) \tag{3}$$

$$w_S^+ = a_3 \cos(2\pi z^+/\lambda^+) \sin(2\pi t^+/T_S^+)\, [2y^+(1-y^{+2}/b_2^2)] \exp(-y^{+2}/b_2^2) \tag{4}$$

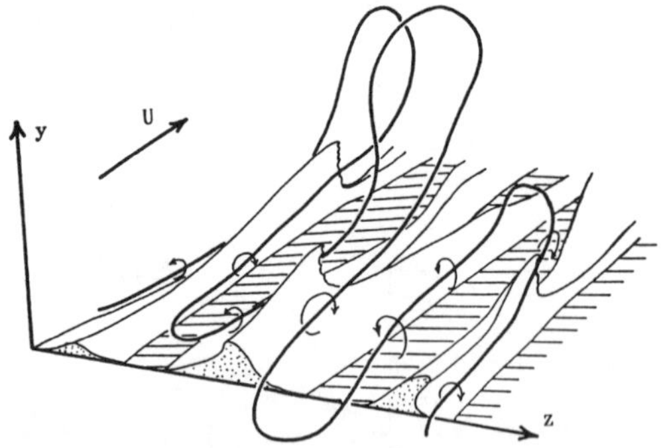

Figure 21 Conceptual model of the wall-layer structure.

†) The author does not intend to claim that counter-rotating vortices occur in pairs at neighboring positions in the instantaneous flow field. Spanwise symmetrical expressions are used to satisfy the constraint that the average flow structure has equal possibility of being positive and negative in the spanwise direction.

where $(\ )^+$ denotes the value non-dimensionalized by the wall variables. Also, in the above equations, λ^+ and T_S^+ are the streak spacing in the spanwise direction and the characteristic time scale of the streamwise pseudo-vortical structures, respectively.

The constants b_i as well as the functions a_i are determined by the following conditions and assumptions:

(a) Continuity requires that $a_2 = 2\pi/\lambda^+ a_3$.
(b) $\lambda^+ = 100$.
(c) The choice of T_S^+ is somewhat arbitrary and does not influence the turbulence statistics calculated. It is assumed to be an invariant constant $T_S^+ = 100$ (Blackwelder and Haritonidis, 1983, Luchik and Tiederman, 1987, and Kim and Spalart, 1987).
(f) The relative correlation coefficient is taken as $Ru_S v_S = -0.7$ (Hishida et al., 1980).

It is noted that a_i may be a function of λ^+, but it is presently assumed to be constant, independent of λ^+. The model, which takes into account the probability density function of λ^+, has been tested by Kasagi et al. (1986), but the turbulence statistics are found not to be influenced greatly by λ^+. Hence, presently λ^+ is assumed to be constant.

If one considers that the large scale motion induces a wave-like disturbance which has a length scale on the order of the boundary layer thickness, the velocity fluctuations induced in the inner layer can be modeled as follows:

$$u_L = f(y^+)\sin(\alpha x^+ + \beta y^+ + \gamma t^+)\sin(\varepsilon z^+) \quad (5)$$
$$v_L = g(y^+)\sin(\alpha x^+ + \beta y^+ + \gamma t^+ + \Phi_L)\sin(\varepsilon z^+) \quad (6)$$
$$w_L = h(y^+)\cos(\alpha x^+ + \beta y^+ + \gamma t^+ + \Phi_L)\cos(\varepsilon z^+) \quad (7)$$

The constants included and the functions of $f(y^+)$, $g(y^+)$ and $h(y^+)$ are determined based upon the following conditions:

(a) Continuity.
(b) Determination of α, β, γ and ε is based on the scales of the large scale motion (Cantwell, 1981). The length scales are about 2δ in the streamwise direction and about δ in the spanwise direction; thus, $\alpha = -2\pi/2\delta^+$ and $\varepsilon = 2\pi/\delta^+$. $\beta = -\alpha/\tan(\theta_L)$, where θ_L is the angle of inclination of the large scale eddies and is approximately 20°. From the time scale in the outer layer, $\gamma = 2\pi/T_L^+$. Since the convection velocity of the large scale motion in the outer layer is about $0.8u_\infty^+$, $T_L^+ = 2\delta^+/0.8u_\infty^+$. From the 1/7 power law and the empirical correlation of friction coefficient, $u_\infty^+ = 8.7\delta^{+1/7}$.
(c) The relative correlation coefficient $Ru_L v_L$ is about -0.45 in the inner region (e.g., see Laufer, 1954).

Note that the non-dimensionalized boundary layer thickness δ^+ and free stream velocity u_∞^+ must be dependent on the Reynolds number. The functional forms of f, g and h cannot be uniquely determined. Presently, f is first modeled by assuming a damping behavior of oscilatory motion near the wall, and then g an h are determined from continuity.

$$f(y^+) = a\,[1-\exp(-y^+/A)] \quad (8)$$
$$g(y^+) = -\,a\,\alpha\,\sin\Phi_L\,[y^+ + A\{-1+\exp(-y^+/A)\}] \quad (9)$$
$$h(y^+) = 1/\varepsilon\,[\alpha\cos\Phi_L\, f(y^+) + \beta g(y^+)] \quad (10)$$

Since both the amplitude a and the damping scale A are dependent on δ^+, their values are adjusted such that the model accurately predicts the mean velocity profiles. All the statistical means are defined by the following equation:

$$\bar{s}(y^+) = \lim_{t^+, x^+, z^+ \to \infty} \frac{1}{t^+ x^+ z^+} \int_0^{t^+} \int_0^{x^+} \int_0^{z^+} s \, dt^+ \, dx^+ \, dz^+ \tag{11}$$

The values of constants presently given are summarized in Table 3. The statistical quantities for δ^+=200 and 500 are discussed hereafter.

The three rms velocity fluctuations are shown in Fig. 22. Overall agreement between the model and the experiments is reasonably good, except for the region close to the wall, where the calculated v and w are smaller than the experimental values. It is known that the distributions of velocity fluctuations show Reynolds number dependence roughly beyond y^+=20, even though the mean velocity profile obeys the universal wall law much further away from the wall. In particular, the streamwise fluctuation is sensitive to the Reynolds number (i.e. δ^+) and decreases appreciably as the Reynolds number is decreased. This trend is well reproduced in the calculated results.

The Reynolds stress and the cross correlation between u and v are shown in Fig. 23. The contribution of the pseudo-vortical motion to the Reynolds stress is predominant near the wall, but that of the large scale motion becomes larger beyond $y^+ \geq 30$. The almost constant values of R_{uv}=-0.4~0.45

Table 3 Constants in the two-component wall-layer model.

a_1	b_1	a_2	b_2	a_3	λ^+	T_S^+	$\cos\Phi_S$	θ_L	$\cos\Phi_L$	δ^+	a	A
-1.25	11	0.0045	25	$\lambda^+ a_2/2\pi$	100	100	0.7	20	-0.45	200	3.4	13.0
										500	5.6	15.5

(Φ_S<0, Φ_L>0)

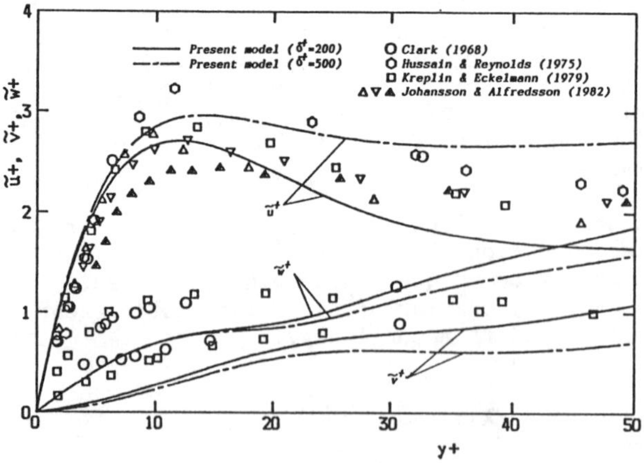

Figure 22 Three components of turbulence normal stresses near the wall.

have been measured over the entire shear layer thickness. According to the recent measurements by Hishida et al. (1980), however, the correlation coefficient has a maximum value of -0.6~0.7 near $y^+=10$ and the maximum value increases with increasing Reynolds number. These facts are also reproduced in the present calculation, by assuming $Ru_Lv_L=-0.45$ and $Ru_Sv_S=-0.7$, respectively.

The mean velocity profiles are obtained by assuming a constant stress layer (see Fig. 24). The calculated results are almost completely insensitive to the value of δ^+, although the profiles of turbulence energy is affected appreciably, as shown in Fig. 24. The rms streamwise vorticity fluctuation is included in Fig. 12. The model's prediction is qualitatively in agreement with the results obtained by the digital image processing described previously (Nishino et al., 1987). It is noted that the minimum in the rms streamwise vorticity fluctuation near the wall is reproduced by the model.

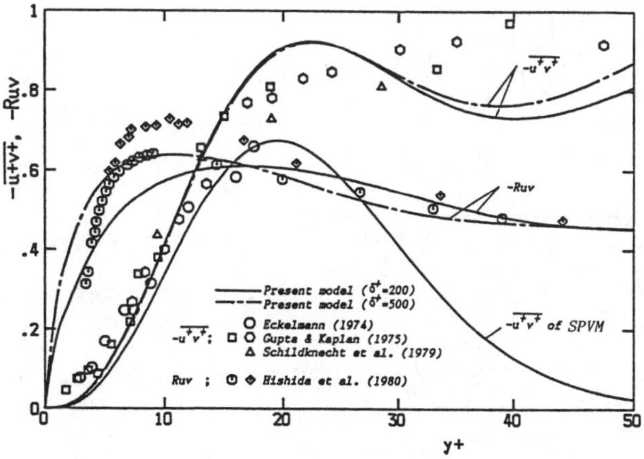

Figure 23 Distributions of the Reynolds stress and the cross correlation between u and v.

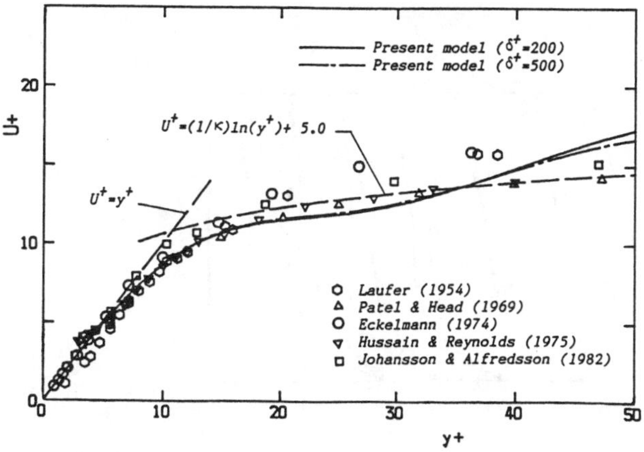

Figure 24 Mean velocity profiles.

Figures 25 and 26 show the contour diagrams of turbulence components and the time trace of the velocity profiles in the y-direction, respectively. Both of the results are qualitatively in good agreement with the recent experimental and numerical results (Blackwelder and Kaplan, 1976, Moin and Kim, 1985).

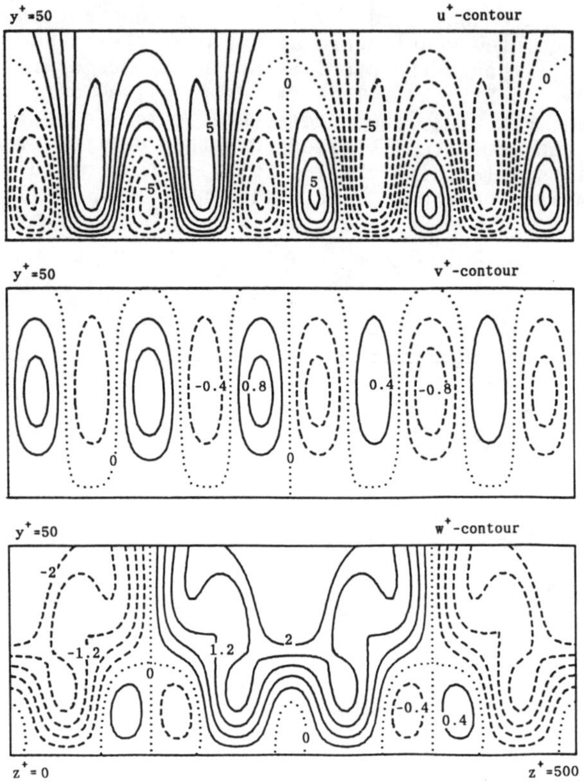

Figure 25 Contour diagrams of instantaneous velocity fluctuations in the plane of $x^+=0$ at $t^+=831.33$ ($\delta^+=500$).
(——— ; positive, ········; zero, ------- ; negative)

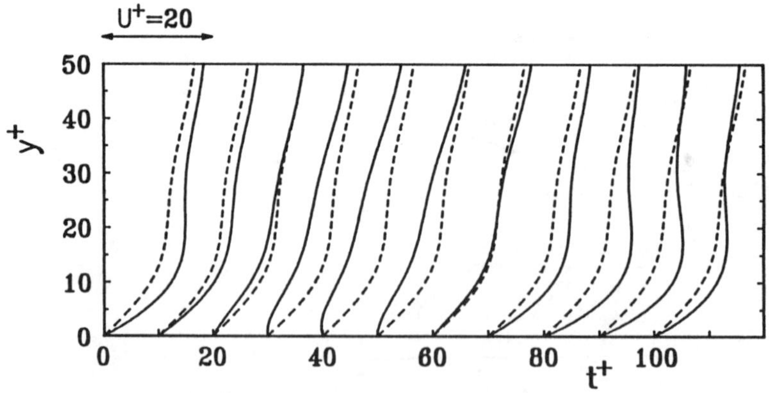

Figure 26 Time-dependent velocity profiles at $x^+=0$ and $z^+=25$ ($\delta^+=500$).

CONCLUSIONS

Detailed investigations of streaky structures are carried out for boundary layer and channel flows. The streaky structures are found clearly to exist in the region of $y^+ \leq 20$ for both flows, but they are hardly seen in the region $y^+ \geq 40$. The low-speed streaks exhibit characteristic behavior such as coalescing, branching and meandering. The average spanwise spacing of the streaks is measured quantitatively. The streamwise extent is estimated to be an order of magnitude larger than the spacing. This result is also in accordance with the considerably long persistence of the streaks.

Instantaneous vector diagrams reveal that solitary streamwise vortices are more prevalent than vortex pairs in the near wall region. Their average diameter is estimated as about $40\nu/u_\tau$, and their center is located at $y^+=25\sim70$. Many weaker pseudo-vortical motions in the wall region are also observed. In addition, sequential vector diagrams suggest that the vortical structures are not as long as the streaky structures. The instantaneous vorticity field exhibits a 'double-layer structure' associated with the pseudo-vortical motions, and this double-layer structure is responsible for the minimum of the streamwise vorticity fluctuations close to the wall.

The heat transfer mechanism of the coherent structures is also investigated experimentally. From the visual data obtained with the aid of the liquid crystal technique, it is directly confirmed that the high-temperature streaks are mostly associated with the low-speed streaks and the low-temperature streaks with the high-speed streaks. In the region of $y^+ \leq 20$, the fluctuating temperature field has a quasi-periodicity in the spanwise direction and is much elongated in the streamwise direction, much the same as the low-speed streaks. The spacing and frequency of thermal and momentum streaks are shown to agree quantitatively. Thus, the scalar field is highly correlated with the coherent motions near the wall.

From these experimental results, it is concluded that fluid motions which result from the interaction between the lake of viscous low-momentum fluid and the streamwise pseudo-vortical structures in the region roughly at $y^+ \leq 20$ are extremely deterministic and coherent. In distinct contrast, in the region $y^+ \geq 30$, where the upper part of the pseudo-vortical structures interact with large-scale turbulent eddies in the logarithmic layer, these motions appear obscure and comparatively random. The streamwise pseudo-vortical motions also play a primary role in turbulent heat transfer in the viscous-wall region.

Based on the experimental knowledge, a deterministic turbulence model is tested. This two-component, wall-layer model is found to represent the statistics and dynamic behavior of turbulence in equilibrium wall layers qualitatively in good agreement with the experimentally and theoretically established behavior.

ACKNOWLEDGEMENTS

The author is grateful to Professor M. Hirata, Dr. Y. Iritani and also Mr. K. Nishino for their cooperations and contributions in the experimental work herein described. This work was supported through the Grant-in Aid for General Scientific Research (No. 61460103) by the Japanese Ministry of Education, Science and Culture.

REFERENCES

Bogard, D. G. and Tiederman, W. G., Characteristics of Ejections in Turbulent Channel Flow, J. Fluid Mech., Vol. 179, pp. 1-19, 1987.

Bradshaw, P., 'Inactive' Motion and Pressure Fluctuations in Turbulent Boundary Layers, J. Fluid Mech., Vol. 30, pp. 241-258, 1967.

Brown, G. L. and Thomas, A. S. W., Large Structure in a Turbulent Boundary Layer, Phys. Fluids, Vol. 20, pp.s243-252, 1977.

Bakewell, H. P. and Lumley, J. L., Viscous Sublayer and Adjacent Wall Region in Turbulent Pipe Flow, Phys. Fluids, Vol. 10, pp. 1880-1889, 1967.

Balint, J. L., Vukoslavcevic, P. and Wallace, J. M., A Study of the Vortical Structure of the Turbulent Boundary Layer, in Advances in Turbulence, ed. G. Comte-Bellot and J. Mathieu, Springer-Verlag, pp. 456-464, 1987.

Blackwelder, R. F. and Kaplan, R. E., On the Wall Structure of the Turbulent Boundary Layer, J. Fluid Mech., Vol. 76, pp. 89-112, 1976.

Blackwelder, R. F. and Eckelmann, H., Streamwise Vortices Associated with the Bursting Phenomenon, J. Fluid Mech., Vol. 94, pp. 577-594, 1979.

Blackwelder, R. F. and Haritonidis, J. H., Scaling of the Bursting Frequency in Turbulent Boundary Layers, J. Fluid Mech., Vol. 132, pp. 87-103, 1983.

Cantwell, B. J., Organized Motion in Turbulent Flow, Ann. Rev. Fluid Mech., Vol. 13, pp. 457-515, 1981.

Chapman, D. R. and Kuhn, G. D., The Limiting Behavior of Turbulence Near a Wall, J. Fluid Mech., Vol. 170, pp. 265-292. 1986.

Clark, J. A., A Study of Incompressible Turbulent Boundary Layers in Channel Flow, Trans. ASME, Ser. D, Vol. 90, pp. 455-468, 1968.

Coles, D., A Model for Flow in the Viscous Sublayer, AFOSR/Lehigh Workshop on Coherent Structure of Turbulent Boundary Layers, Lehigh University, pp. 462-475, 1978.

Coles, D., The Uses of Coherent Structure, AIAA Paper 85-0506, 1985.

Eckelmann, H., The Structure of the Viscous Sublayer and the Adjacent Wall Region in a Turbulent Channel Flow, J. Fluid Mech., Vol. 65, pp. 439-459, 1974.

Gupta, A. K. and Kaplan, R. E., Statistical Characteristics of Reynolds Stress in a Turbulent Boundary Layer, Phys. Fluids, Vol. 15, pp. 981-985, 1972.

Hama, R., see Corrsin, S., Some Current Problems in Turbulent Shear Flows, Naval Hydrodyn. Publ. 515, Ch. XV, pp. 373-400, 1957.

Hirata, M., Tanaka, H., Kawamura, H., and Kasagi, N., Heat Transfer in Turbulent Flows, Proc. 7th Int. Heat Transfer Conf., Hemisphere, Washington, D.C., Vol. 1, pp. 31-57, 1982.

Hatziavramidis, D. T. and Hanratty, T. J., The Representation of the Viscous Wall Region by a Regular Eddy Pattern, J. Fluid Mech., Vol. 95, pp. 655-679, 1979.

Hishida, M., Nagano, Y. and Morimoto, Y., Structure of Pipe Flow Turbulence (1st Rep., Characteristics of Velocity Fluctuations and Reynolds Stress), Trans. JSME, Vol. 46B, pp. 1455-1466, 1980.

Hishida, M., Nagano, Y. and Tagawa, M., Transport Processes of Heat and Momentum in the Wall Region of Turbulent Pipe Flow, Proc. 8th Int. Heat Transfer Conf., Hemisphere, Washington, D. C., Vol. 3, pp. 925-930, 1986.

Hussain, A. K. M. F. and Reynolds, W. C., Measurement in Fully Developed Turbulent Channel Flow, ASME J. Fluids Eng., Vol. 97, pp. 568-580, 1975.

Iritani, Y., Kasagi, N. and Hirata, M., Heat Transfer Mechanism and Associated Turbulence Structure in the Near-Wall Region of a Turbulent Boundary Layer, Proc. 4th Symp. Turbulent Shear Flows, Karlsruhe, 17.31, 1983, (also in Turbulent Shear Flows IV, L. Bradbury et al., ed., Springer-Verlag, Berlin, pp. 223-234, 1984).

Iritani, Y., Kasagi, N. and Hirata, M., Streaky Structure in a Two-Dimensional Turbulent Channel Flow, Trans. JSME, Vol. 51B, pp. 3092-3101, 1985.

Johansson, A. V., and Alfredsson, P. H., On the Structure of Turbulent Channel Flow, J. Fluid Mech., Vol. 122, pp. 295-314, 1982.

Kasagi, N. and Hirata, M., Bursting Phenomena in Turbulent Boundary Layer on a Horizontal Flat Plate Heated From Below, Proc. ICHMT Sem. Turbulent Buoyant Convection, Dubrovnik, pp.27-38, 1976, (also in Heat Transfer and Turbulent Buoyant Convection, D. B. Spalding and N. Afgan, ed., Vol. 1, Hemisphere, pp.27-38, 1977).

Kasagi, N., Hirata, M. and Nishino, K., Streamwise Pseudo-Vortical Structures and Associated Vorticity in the Near-Wall Region of a Wall-Bounded Shear Flow, Proc. 9th Symp. Turbulence, University of Missouri-Rolla, pp. 24.1-24.10, 1984, (also in Exp. Fluids, Vol. 4, No. 6, pp. 309-318, 1986).

Kasagi, N., Nishino, K. and Hirata, M., Toward Structural Modeling of the Wall-Bounded Turbulent Shear Flow, Proc. 3rd Asian Cong. Fluid Mech., Tokyo, pp. 549-552, 1986.

Kasagi, N., Kuroda, A. and Hirata, M., Numerical Investigation of the Turbulent Heat Transfer Near the Wall by the Unsteady Streamwise Pseudo-Vortical Motion Model, Proc. 2nd ASME/JSME Joint Thermal Eng. Conf., Honolulu, Vol. 1, pp. 477-482, 1987.

Kasagi, N., Moffat, R. J. and Hirata, M., Liquid Crystals, in Handbook of Flow Visualization, W. J. Yang, ed., to be published by Hemisphere, Washington D. C., 1988a.

Kasagi, N., Kuroda, A. and Hirata, M., Numerical Investigation of Near-Wall Turbulent Heat Transfer Taking into Account the Unsteady Heat Conduction in the Solid Wall, submitted to ASME J. Heat Transfer, 1988b.

Khabakhpasheva, Y. M., Experimental Investigation of Turbulent Momentum and

Heat Transfer in the Proximity of the Wall, Proc. 8th Int. Heat Transfer Conf., Hemisphere, Washington, D. C., Vol. 1, pp. 79-90, 1986.

Kastrinakis, E. G. and Eckelmann, H., Measurement of Streamwise Vorticity Fluctuations in a Turbulent Channel Flow, J. Fluid Mech., Vol. 137, pp. 165-186, 1983.

Kastrinakis, E. G., Wallace, J. M. and Willmarth, W. W., Measurements of Small Scale Streamwise Vorticity in a Turbulent Channel Flow, Bull. Am. Phys. Soc., Vol. 20, No. 11, pp. 1422, 1975.

Kim, H. T., Kline, S. J. and Reynolds, W. C., Turbulence near a Smooth Wall in a Turbulent Boundary Layer, J. Fluid Mech., Vol. 50, pp. 133-160, 1971.

Kim, J., Turbulence Structures Associated With the Bursting Event, Phys. Fluids, Vol. 28, pp. 52-58, 1983.

Kim, J., Moin, P. and Moser, R., Turbulence Statistics in Fully Developed Channel Flow at Low Reynolds Number, J. Fluid Mech., Vol. 177, pp. 133-166, 1987.

Kim, J. and Moin, P., Transport of Passive Scalars in a Turbulent Channel Flow, Proc. 6th Symp. Turbulent Shear Flows, Toulouse, pp. 5.2.1-6, 1987.

Kim, J., Investigation of Heat and Momentum Transport in Turbulent Flows via Numerical Simulations, Proc. 2nd Int. Symp. Transport Phenomena in Turbulent Flows, Tokyo, pp. 707-721, 1987.

Kim, J. and Spalart, P. R., Scaling of the Bursting Frequency in Turbulent Boundary Layers at Low Reynolds Numbers, Phys. Fluids, Vol. 30, pp. 3326-3328, 1987.

Kline, S. J. and Runstadler, P. W., Some Preliminary Results of Visual Studies of the Flow Model of the Wall Layers of the Turbulent Boundary Layer, Trans. ASME, Ser. E, Vol. 2, pp. 166-170, 1959.

Kline, S. J., Reynolds, W. C., Schraub, F. A. and Runstadler, P. W., The Structure of Turbulent Boundary Layers, J. Fluid Mech., Vol. 30, pp. 741-773, 1967.

Kreplin, H. P. and Eckelmann, H., Behavior of the Three Fluctuating Velocity Components in the Wall Region of a Turbulent Channel Flow, Phys. Fluids, Vol. 22, pp. 1233-1239, 1979.

Laufer, J., The Structure of Turbulence in Fully Developed Pipe Flow, NACA Rep. 1174, 1954.

Luchik, T. S., and Tiederman, W. G., Timescale and Structure of Ejections and Bursts in Turbulent Channel Flows, J. Fluid Mech., Vol. 174, pp. 529-552, 1987.

Maruyama, S. and Tanaka, H., Coherent Structures in the Inner Layer of Wall Turbulence Under the Spatial Restriction, Proc. 2nd Int. Symp. Transport Phenomena in Turbulent Flows, Tokyo, pp. 61-74, 1987.

Meek and Baer, The Periodic Viscous Sublayer in Turbulent Flow, AIChE J., Vol. 16, pp. 841-848, 1970.

Moin, P. and Kim, J., Numerical Investigation of Turbulent Channel Flow, J. Fluid Mech., Vol. 118, pp. 341-377, 1982.

Moser, R. D. and Moin, P., Direct Numerical Simulation of Curved Turbulent Channel Flow, Rep. TF-20, Mech. Eng. Dept., Stanford University, 1984.

Nakagawa, H. and Nezu, I., Structure of Space-Time Correlation of Bursting Phenomena in an Open-Channel Flow, J. Fluid Mech., Vol. 104, pp. 1-43, 1981.

Nishino, K., Kasagi, N. and Hirata, M., Study of Streamwise Vortical Structures in a Two-Dimensional Turbulent Channel Flow by Digital Image Processing, Proc. 2nd Int. Symp. Transport Phenomena in Turbulent Flows, Tokyo, pp. 47-60, 1987.

Patel, V.C. and Head, M. R., Some Observation on Skin Friction and Velocity Profiles in Fully Developed Pipe and Channel Flows, J. Fluid Mech., Vol. 38, pp. 181-201, 1969.

Rajagopalan, S. and Antonia, R. A., Some Properties of the Large Structure in a Fully Developed Turbulent Duct Flow, Phys. Fluids, Vol. 22, pp. 614-622, 1979.

Rogallo, R. S. and Moin, P., Numerical Simulation of Turbulent Flows, Ann. Rev. Fluid Mech., Vol. 16, pp. 99-137, 1984.

Schildknecht, A. V., Miller, J. A., and Meier G. E. A., The Influence of Suction on the Structure of Turbulence in Fully Developed Pipe Flow, J. Fluid Mech., Vol. 90, pp. 67-107, 1979.

Schraub, F. A., and Kline, S. J., Study of the Structure of the Turbulent Boundary Layer With and Without Longitudinal Pressure Gradients, Rep. MD-12, Mech. Eng. Dept., Stanford University, 1965.

Shaw, D. A. and Hanratty, T. J., Turbulent Mass Transfer Rates to a Wall for Large Schmidt Numbers, AIChE J., Vol. 23, pp. 28-37, 1977.

Smith, C. R. and Metzler, S. P., The Characteristics of Low-Speed Streaks in the Near-Wall Region of a Turbulent Boundary Layer, J. Fluid Mech., Vol. 129, pp. 27-54, 1983.

Smith, C. R. and Schwartz, S. P., Observation of Streamwise Rotation in the Near-Wall Region of a Turbulent Boundary layer, Phys. Fluids, Vol. 26, pp. 641-652, 1983.

Subramanian, C. S. and Antonia, R. A., Effect of Reynolds Number on a Slightly Heated Turbulent Boundary Layer, Int. J. Heat Mass Transfer, Vol. 24, pp. 1833-1846, 1981.

Utami, T., and Ueno, T., Lagrangian and Eulerian Measurement of Large Scale Turbulence by Flow Visualization Technique, Proc. Int. Symp. Flow Visualization, Tokyo, pp. 145-150, 1977.

Spatial Correlations of the Temperature Fluctuation in Boundary Layers

A. ŠLANČIAUSKAS
Institute of Physical and Technical
Problems of Energetics
Academy of Sciences of the Lithuanian SSR
Kaunas, USSR

ABSTRACT

Correlations of the space temperature fluctuations were measured by microthermocouples in the boundary layers over smooth plates and under the effects of surface roughness, flow acceleration and free stream turbulence. The vortices in the boundary layers with spacings equal to the half of the boundary layer thickness were either enhanced or supressed by the rate of momentum transfer to the wall.

1. INTRODUCTION

The origin of generation of vortical flows is still in the scope of a deep interest of researchers. There appear prediction techniques based on large-scale vortices. Turbulent boundary layers are constituted of cyclic generation and disintegration of vortical structures,which merge into homogeneous and approximately steady-state flows. We know already several models simulating such flows. On the other hand turbulent boundary layers present a convenient object of study and a good deal has already been done in this field. We may well expect also simplified models of complex vortical flows here.

A most exact vortical representation of turbulent shear flows was done by A.A. Townsend, and the article by Grant /1/ is an example of a highly successful work. It has supplied a basis for numerous researchers in their analyses of correlation among the three components of velocity fluctuation. The problem received a new stimulation from Favre et. al. /2/. Their time and space analysis of velocity fluctuation fields was commonly accepted and widely applied. Such measurements were made by Kovasznay /3/ to study intermittency on the edge of the boundary layer. One of the elements of his study referring to the turbulent core is presented in Figure 1. The upper part of the Figure represent the isocorrelation curves of the longitudinal velocity fluctuations in the crosssection of boundary layer. Note a negative correlations of the data when two sensors are $r_z/\delta = 0.5$ apart.

At the same time at our Institute we studied specific features of turbulent heat transport by measuring the correlation of temperature fluctuations by two microthermocouples. We arrived at a si-

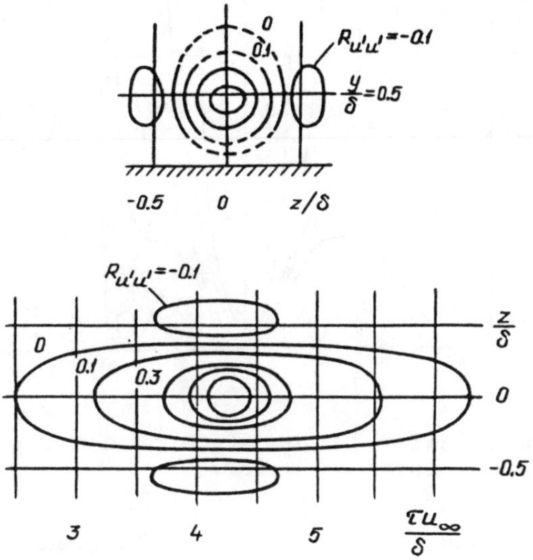

FIGURE 1. Space-time correlations of the longitudinal velocity component in the boundary layer measured by a single sensor fixed at $y/\delta = 0.5$.

milar pattern of transverse motions /4/. A negative correlations was also observed at a similar distance, but with the centres somewhat closer to the wall. A more illustrative representation was obtained by moving two thermocouples away from the wall and changing distances from each other. Thanks to the small size of the sensors, accumulation of hot and cold masses could be observed on the edge of the viscous sublayer. Some of such measurements are shown in Figure 2. They reflect the observation of Kline et. al. /5/ of flows near the wall by forming a serie of alongated vortices with a transverse pitch of $r_z u_* / \nu = 100$.

In any study of a turbulent flow and heat transfer under different boundary conditions and in different geometries one must not forget the existence of large scale vortices. Our measurements shed some light on this phenomenon, and we remained loyal to our technique of two microthermocouples in our further studies. The results of this technique are evident from the presented examples.

2. EXPERIMENTAL TECHNIQUES

The study was carried out in flows of air, water and transformer oil in closed loop channels. Free stream turbulence was 0.3 % in air, 1.5 % in water and 0.8 % in transformer oil. The 1120 mm long flat test-plate was mounted as one of the walls in a channel of 200 x 80 mm at the entrance. The plate which had a sharp edge, was preceded by a fluid-suction slot. A natural laminar-turbulent

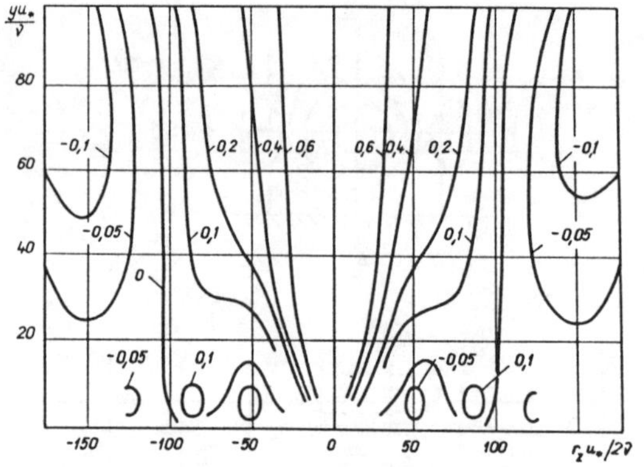

FIGURE 2. Isocorrelation curves of temperature fluctuations in a turbulent boundary layer of water.

transition of the boundary layer was influenced solely by free stream turbulence. The plate was heated at the leading edge to facilitate a simultaneous development of the thermal and hydrodynamical boundary layers. The plate was made of an insulator and covered with a DC heated stainless steel foil to maintain the heat flux $q_w(x)$ = const. The surface-fluid temperature drops were 30°C for air, 8°C for water and 15°C for transformer oil. The test plate was mounted as the bottom wall of the channel. The flow was fully developed turbulent and Re = 10^6. This was the condition of neglectful gravity forces in the tests.

For experiments in high turbulence levels a square grid was inserted in the channel upstream of the test section. For accelerating flows, profiled plates were mounted on the upper wall of the channel over the test plate.

Temperature fluctuations in air were measured by DISA thermoanemometers with resistance thermometers of a 2 μm thick and 100 diameters long wire. Needle chromel-alumel thermocouples were used to measure temperature fluctuations in transformer oil, their sensors were 4 μm in diameter. The thermocouples were fabricated by electrochemical polishing a large diameter wire. For the lower frequency range in flows of water microthermocouples of a larger diameter were used to prolong the lifetime of thermocouples for the negative effects of electrochemical and electric potentials.

In different fluids of the same Re the signal frequency is proportional to viscosity, that is to free stream velocity. If we assume the length of disturbances in a viscous sublayer to be approximately fourfold its thickness, then the maximum fluctuation frequencies are 5 kHz in air, 1.5 kHz in oil and 0.4 kHz in water. The amplitude and frequency ranges were fully met by the sensors applied. Their signals were amplified by a special amplifier with transmission frequency of 0.1-1000 Hz on the level of 3 dB. A DISA 55D70 correlometer was applied.

Nomenclature and the locations of the sensors are shown on Figure 3. The sensors were arranged parallely and could be moved simultaneously upwards away from the plate across the depth of the boundary layer. The tests were repeated with different spacings r_z of the sensors. The results were interpreted as correlation coefficients

$$R_{\vartheta_1\vartheta_2}(0,0,r_z) = \frac{\overline{\vartheta_1'(0,0,0)\ \vartheta_2'(0,0,r_z)}}{\sqrt{\overline{\vartheta_1'(0,0,0)^2}}\ \sqrt{\overline{\vartheta_2'(0,0,r_z)^2}}} \tag{1}$$

Figure 4. presents the measured correlation function. Test points for this plot were chosen from measurements at three different distances from the wall and at different spacings of the sensors. The lower plots present isocorrelation curves of test points in a mirror representation. Most of them have twice lower values of r_z.

Note that needle thermocouples thanks to their small size are more suitable for this type of study then the wire sensors of the thermoanemometers. The resistance thermometer made of a 0.8 mm long wire occupies just the length of $r_z u_*/\nu = 50$ in the viscous sublayer, plus its holder.

3. RESULTS

We first compare the results for flows over smooth and rough surfaces, which are close in their nature. The effects of free stream turbulence and of a strong acceleration with laminarisation should be treated separately.

Figure 5. presents measurements in air at $Re_{x-x_0} = 2.7 \cdot 10^6$. Here the value of Re refers to free stream velocity and to the distance from the virtual initiation of the turbulent boundary layer, which differs from the geometric edge by the value of x_0. The results in Filgure 5. were found on a longer test plate of 1.90 m.

An important initial data on the boundary layer parameters and on the test facilities are described elsewhere /6/. We just con-

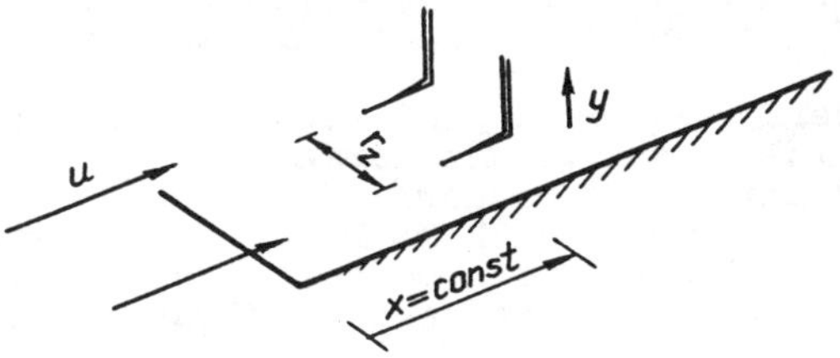

FIGURE 3. Arrangement of sensors.

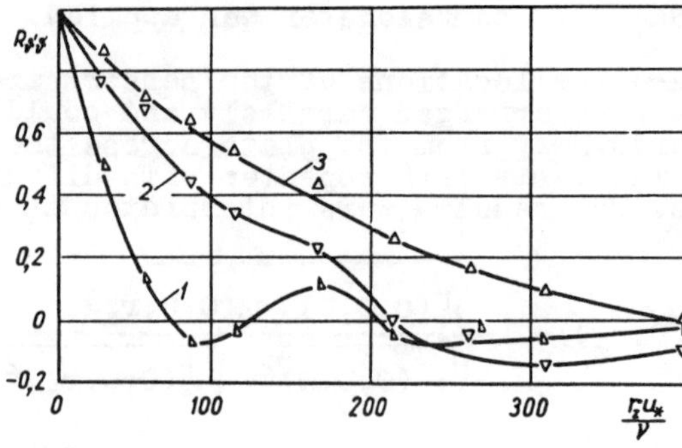

FIGURE 4. Depthwise variation of the spatial correlation coefficient in water.

sider friction resistance of a plate to establish a relation between previous data and the present results. At $Re_{x-x_o} < 2 \cdot 10^6$ pre-history of the transition and restriction of the test section begin to manifest themselves. Our results are described by

$$c_f = 0.39\ (\lg Re_{x-x_o})^{-2.6} \tag{2}$$

For higher Re the results are approximated by the equation of Ludwieg, Tillmann. For the heat transfer rate in a turbulent boundary layer on a smooth plate use

$$\frac{c_f/2}{St} = 0.93 + 12.5\ \sqrt{c_f/2}\ (Pr^{2/3} - 1). \tag{3}$$

In Figure 5 the isocorrelation curves are symmetric to $r_z = 0$ with spacings of $r_z/2$. A situation of rising and falling masses is evident for the turbulent boundary layer. The average spacings of such events equals $0.5\ \delta$, where δ is the boundary layer thickness.

Figure 6. represents isocorrelation curves of $R_{\vartheta\vartheta}$ on a rough wall. A strong shear stress at the wall causes increase of the vortical motion. A negative correlation is well pronounced, and the spacings of the opposite events are somewhat larger.

This test was performed on a plate covered with a rough foil which had pyramidal protrusions formed by rolling. The rhombic foots of the pyramids had corners of 60 and 120 deg, and 1 mm long sides. They were h = 0.22 mm high. Such surfaces are a close simulation of a homogeneous sand roughness and have no pronounced separation regions. The case in Figure 6. corresponds to $hu_*/\nu = 30$.

Measurements in transformer oil shown in Figure 7 where accumu-

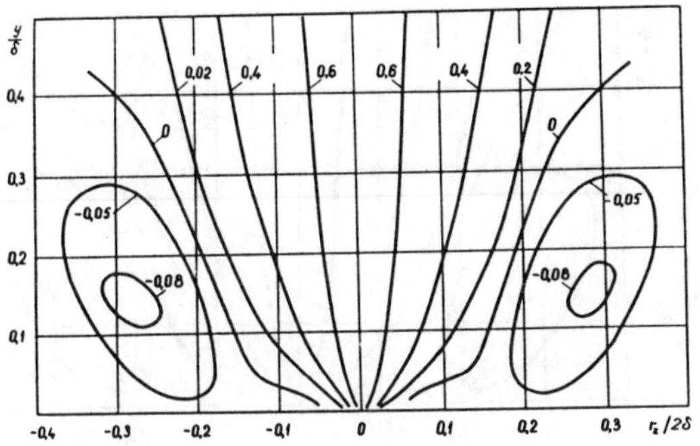

FIGURE 5. Isocorrelation curves for temperature fluctuations in air on a smooth plate.

lated at $Re_{x-x_0} = 6.1 \cdot 10^5$ and Pr = 85. Here the ability of microthermocouples to reveal intermittent hot and cold masses is particularly evident. Wire sensors of thermoanemometers were not able to demonstrate this phenomenon in air, but the large scale vortices in oil have their analogies in air. Their behaviour in oil over a rough plate is similar to the observations in air, Figure 8., namely, negative correlation is inforced with an insignificant increase of the spacing of the opposite events. The other rough plate had a heated surface foil with a 0.4 mm thick wire grid of 1.6 x 1.6 mm mesh size welded on heated surface to form a 1 mm high surface roughness. The flow conditions were Pr = 72 and $hu_*/\nu = 25$.

Isocorrelation lines near the wall are most easily observed in

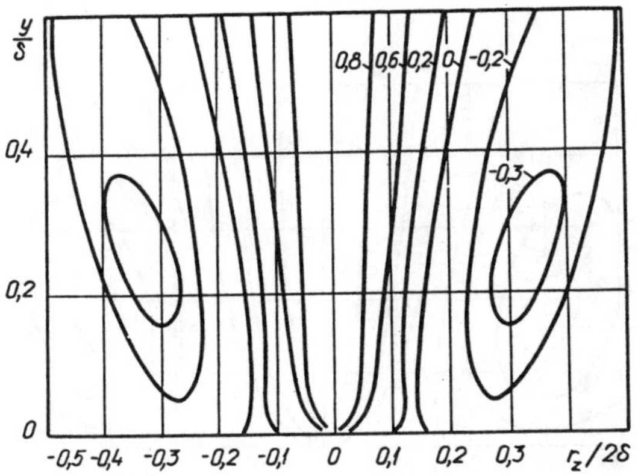

FIGURE 6. Isocorrelation lines of temperature fluctuations in a flow of air over a rough plate.

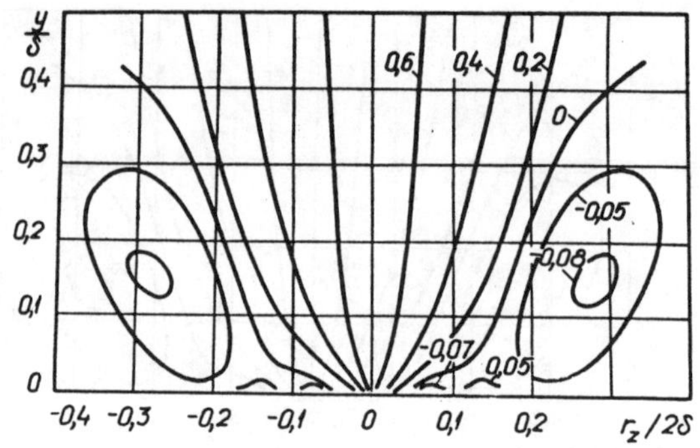

FIGURE 7. Isocorrelation curves of temperature fluctuations in transformer oil over a smooth plate.

water. Results in Figure 4. correspond to $Re_{x-x_0} = 6.9 \cdot 10^5$ and $Pr = 5.4$. At such Pr a significant thermal resistance is concentrated near the wall, and turbulent disturbances are pronounced in the fluctuations of the temperature field. Note a $r_z u_* / \nu = 70$-100 spacing of the hot and cold masses accumulated over the viscous sublayer.

The presented results are in a satisfactory agreement with the results of other authors. But a study of turbulent behaviour always includes the problem of a higher Re. New elements in the chain of vortical exchange are always to be expected. Figure 9 presents measurements in air at a fourfold velocity, that is larger than in Figure 4. The centers of the large scale vortices remain at an approximately the same transverse spacing, but are two times closer to the wall. The sensors were thinner than the viscous sublayer, but intermittent cold and hot masses were not

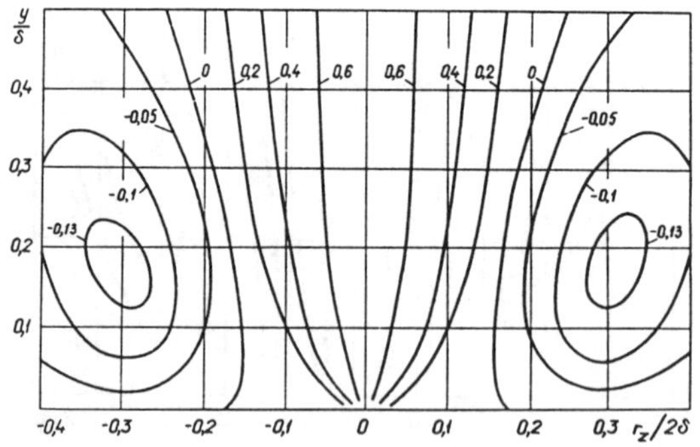

FIGURE 8. Isocorrelation curves of temperature fluctuations in transformer oil over a rough plate.

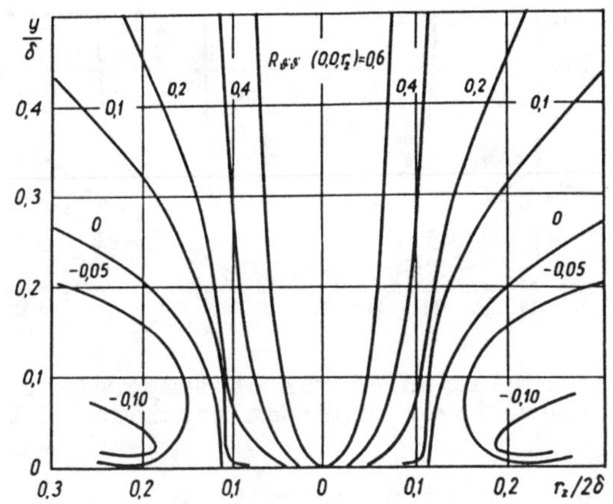

FIGURE 9. Isocorrelation curves of temperature fluctuations in water at $Re_{x-x_0} = 2.5 \cdot 10^6$.

observed over the viscous sublayer.

Figure 10 presents results for a highly turbulent flow of air. Note that the increased free stream turbulence destroys the vortical structures. The contribution of the Reynolds stresses generated at the wall is lower, and no vortices could be observed.

Figure 11 presents measurements in a strongly accelerated flow. The case corresponds to $(\nu/\rho u_\infty^3)(dp/dx) = 2 \cdot 10^{-6}$ and $Re_{\delta^{**}} = 1864$ where δ^{**} - momentum thickness. In such conditions laminarisation is already evident /6/, and from Figure 11 the temperature correlation distribution shows a strongly deformed behaviour. In the outher region of boundary layer the large scale vortices are weaker, and two sensors placed over the viscous sublayer indicate high correlation coefficients. Note a maximum correlation at a certain distance from the wall. The high correlation rate observed

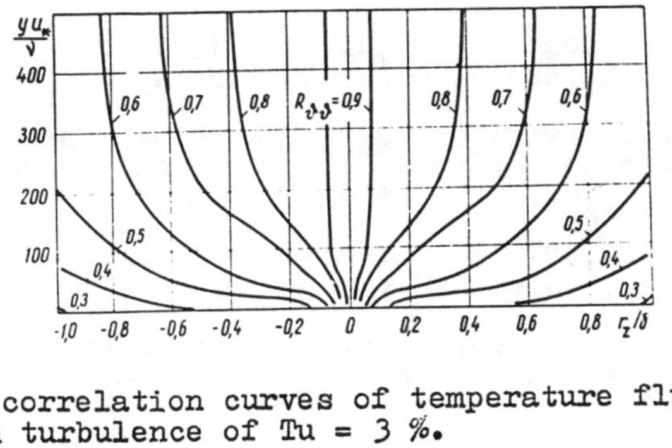

FIGURE 10. Isocorrelation curves of temperature fluctuations for free stream turbulence of Tu = 3 %.

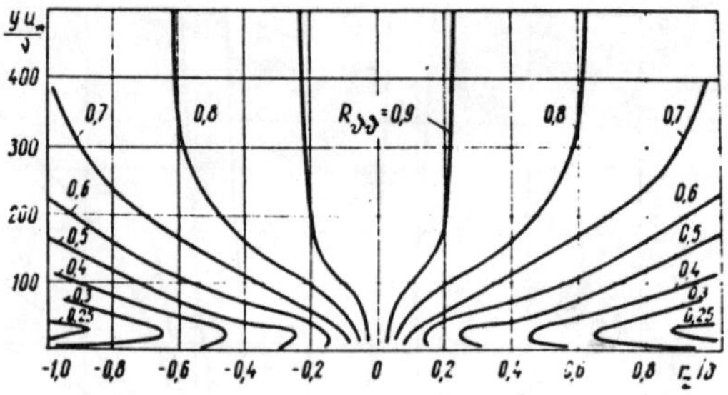

FIGURE 11. Isocorrelation curves of temperature fluctuations in a turbulent boundary layer under an accelerated flow.

here deserves to be studied separately.

CONCLUSIONS

In outer region of the boundary layer exist a vortical motions which are similar to longitudinal eddies with spacings $r_z = 0.5\delta$ and maximum of negative correlation at $y/\delta = 0.15$ from the wall. This structure is universal and independent of Re number.

The longitudinal eddies are naturally flow pattern at the high level of instability. The most important conditions for its generation is the necessity to transfer the large momentum to the wall.

The behaviour of vortical structure with Re is an interesting problem. A deformation of the elongated vortices attached to the wall might be expected.

Measurements of space correlations by two small sensors such as microthermocouples may by considered an usefull tool for determination of predominantly vortices.

REFERENCES

1. Grant, H.L., The Large Eddies of Turbulent Motion, J. Fluid Mech., Vol. 4, part 2, P. 149, 1958.

2. Favre, A.J., Gaviglio, J.J., Dumas, R., Space-Time Double Correlations and Spectra in a Turbulent Boundary Layer, J. Fluid Mech., vol. 2, part 4, p. 313, 1957.

3. Kovasznay, L.S.G., The Structure of Turbulence in Shear Flows, Agard, paper D1-14, 1971.

4. Drižius, R., Šlančiauskas, A., Temperatūrinė koreliacija turbulentiniame pasienio sluoksnyje, Mokslas ir technika, Nr5, p. 63, 1971.

5. Kline, S.J., Reynolds, W.C., Schraub, F.A., Runstadler, P.W., The Structure of Turbulent Boundary Layer, J. Fluid Mech., vol. 30, part 2, p. 741, 1967.

6. Žukauskas, A., Šlančiauskas, A., Heat Transfer in Turbulent Fluid Flows, Hemisphere, Washington, 1987.

The Near-Wall Region of the Turbulent Boundary Layer: Some Results from Heat Transfer Measurements

W. M. KAYS and R. J. MOFFAT
Department of Mechanical Engineering
Stanford University
Stanford, California 94305, USA

The objective of this paper is to present two examples of heat transfer measurements on turbulent boundary layers that might shed some light on the structure of the near-wall region. The first of these is some measurements of turbulent Prandtl number which show nothing unusual over most of the boundary layer, but then show a sharp rise in the viscous sublayer. The second example is that of a strongly accelerated flow (negative pressure gradient) where the friction coefficient tends to increase while the Stanton number decreases. It is demonstrated that this behavior can be accounted for by an increase in the apparent thickness of the viscous sublayer evidently caused by a negative pressure gradient.

TURBULENT PRANDTL NUMBER EXPERIMENTS

As a part of a comprehensive experimental study of heat transfer through the turbulent boundary layer carried out in the authors' laboratory, a considerable body of data on turbulent Prandtl number was accumulated. (Turbulent Prandtl number, Pr_t, is here defined as the eddy diffusivity for momentum transfer divided by the eddy diffusivity for heat transfer.) A description of the apparatus and a summary of the results is given in Moffat and Kays [1]. The particular results to be presented here were obtained by Blackwell [2].

The experiments involved air in equilibrium and near-equilibrium turbulent boundary layers at momentum thickness Reynolds numbers in the range 1000 to 6000. The virtual origin of the thermal boundary layers was approximately the same as that of the momentum boundary layers, so that the enthalpy thickness Reynolds numbers were the same order of magnitude as for the momentum boundary layers (and somewhat lower for the cases where pressure gradient was adverse.) The experiments covered the case of no pressure gradient, two cases of adverse pressure gradients, three cases of adverse pressure gradient with positive transpiration (blowing), and three cases of adverse pressure gradients with negative transpiration (suction).

Measurement of turbulent Prandtl number involves four quantities -- turbulent shear stress, turbulent heat flux, velocity gradient, and temperature gradient. For these experiments shear stress and velocity gradient were determined from hot- wire measurements, heat flux was measured directly at the wall and estimated at points removed from the wall, and temperature gradient was measured using a 0.001 inch thermocouple probe.

The results are presented on Figures 1,2 3, and 4, where Turbulent Prandtl Number is plotted as a function of y+. The momentum thickness and enthalpy thickness Reynolds numbers for each of the numbered test runs are given in Table 1. Fig. 1 shows the results for no pressure gradient, and two cases of adverse pressure gradient, all with no transpiration. These are all equilibrium boundary layers. Fig. 2 shows three cases with mild adverse pressure gradient, but with increasing positive transpiration (blowing). These

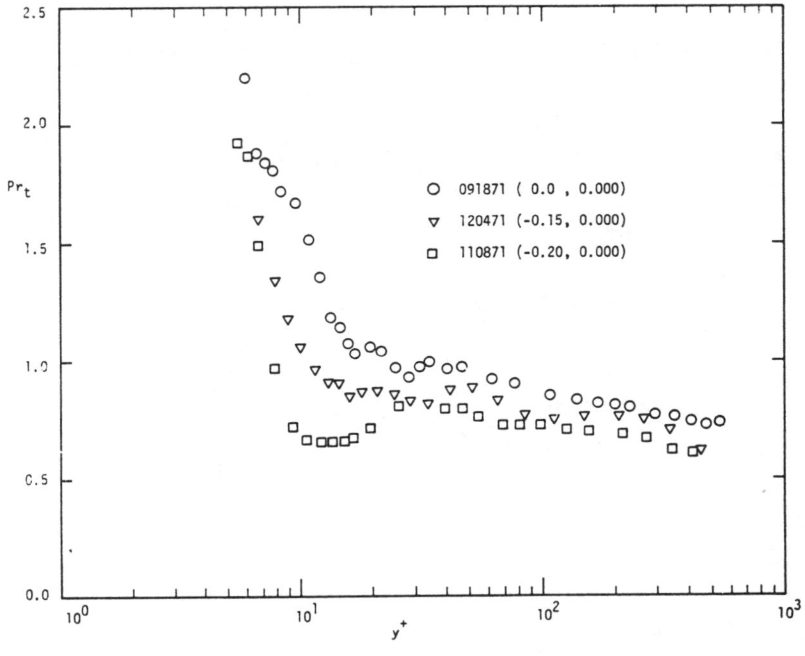

FIGURE 1. Prt - No pressure gradient, and adverse pressure gradients

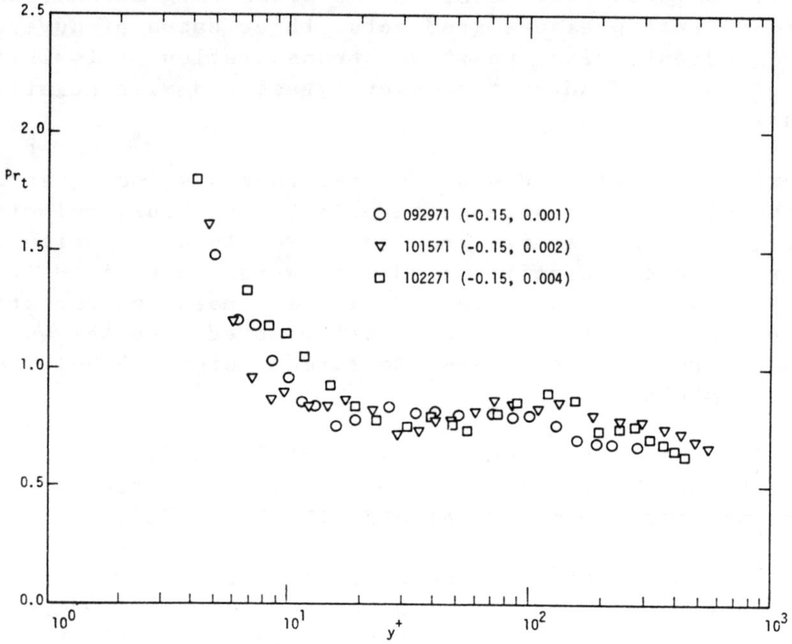

FIGURE 2. Prt - Adverse pressure gradient, with blowing

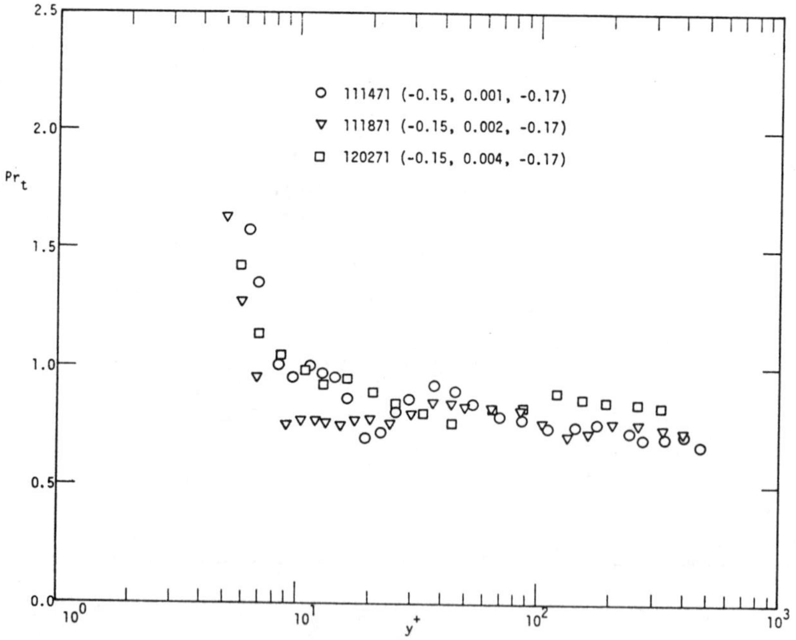

FIGURE 3. Prt - Adverse pressure gradient, with blowing, equilibrium boundary layers

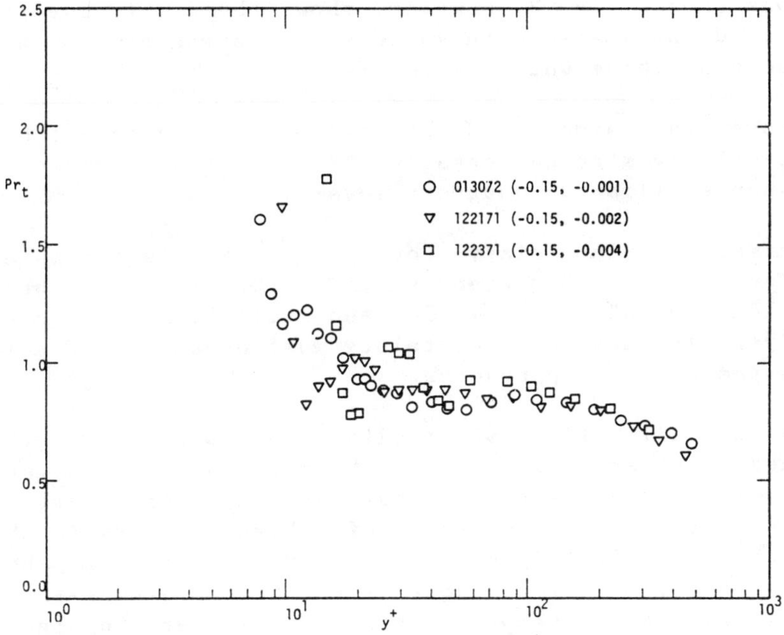

FIGURE 4. Prt - Adverse pressure gradient, with suction

TABLE 1. Reynolds Numbers for the Turbulent Prandtl Number Experiments

Run Number	Re-m	Re-h
91871	2481	2648
120471	3229	2165
110871	3659	2055
92971	4014	2669
101571	4823	3184
102271	6992	4354
111471	3740	3528
111871	4504	3060
120271	6149	3795
13072	2324	1693
122171	1747	1330
122371	789	622

are not equilibrium cases as the blowing fraction is main tained constant. Fig. 3, however, shows three mild adverse pressure gradient cases with successively stronger blowing, but all are equilibium boundary layers.

Fig. 4 shows the same mild adverse pressure gradient but with successively stronger negative transpiration (suction). These are non-equilibrium boundary layers.

The experimental uncertainty for these data is estimated to be about + or - 5 percent at y+ = 100, rising to about + or - 10 percent at y+ = 10, and still higher at lower values of y+. In fact the uncertainty must become indefinitely high as the wall is approached.

For y+ > 20 all of these results give approximately the same values for Pr_t, i.e., in the range 0.7 to 1.0, and this is quite consistent with what has long been observed. But the striking characteristic of these results is the rise to considerably higher values of Pr_t for y+ < 20, and especially for y+ < 10. Despite the higher experimental uncertainty in the near-wall region, the trend to higher Pr_t seems convincing. Earlier experiments in the author's laboratory, using quite different instrumentation, had shown the same trend, but it was apparent that the experimental uncertainty at low values of y+ was unacceptably high. In fact, the experiments reported here were undertaken to further investigate that trend.

Kim and Moin [3] carried out a direct simulation of a turbulent channel flow and (among other things) determined turbulent Prandtl number for three values of molecular Prandtl number, 0.1, 0.71, and 2. The Reynolds numbers for these calculations are rather low (3300 based on mean centerline velocity and channel half-width) and thus the value of y+ at the centerline is low (about 180), so it not clear that their results are totally comparable to the experiments reported here. However, it is interesting that they do show a rise of Pr_t as the wall is approached. But they show a maximum of Pr_t = 1.1 at about y+ = 40 for Pr = 0.71. For all three Prandtl numbers they get Pr_t = 1.1 at y+ = 0.0.

The experimental results on Fig. 1 also suggest that there may be an effect of streamwise pressure gradient on turbulent Prandtl number, with lower Pr_t obtained with an adverse pressure gradient. Some of the author's earlier experiments suggest an increase in Pr_t with a favorable pressure gradient, although the experimental uncertainty was so large that it was difficult to draw a definitive conclusion. The Kim and Moin calculations for a fully developed channel flow would involve a favorable pressure gradient, and because of the low Reynolds number the relative pressure gradient is fairly large. So it may be that the high Pr_t observed by Kim and Moin at y+ = 40 is associated with this pressure gradient.

STRONGLY ACCELERATED FLOWS

Figs. 5, 6, and 7 show two examples of highly accelerated boundary layers (large negative streamwise pressure gradient). The test system and procedures are described in Moffat and Kays [1], but the detailed results for the

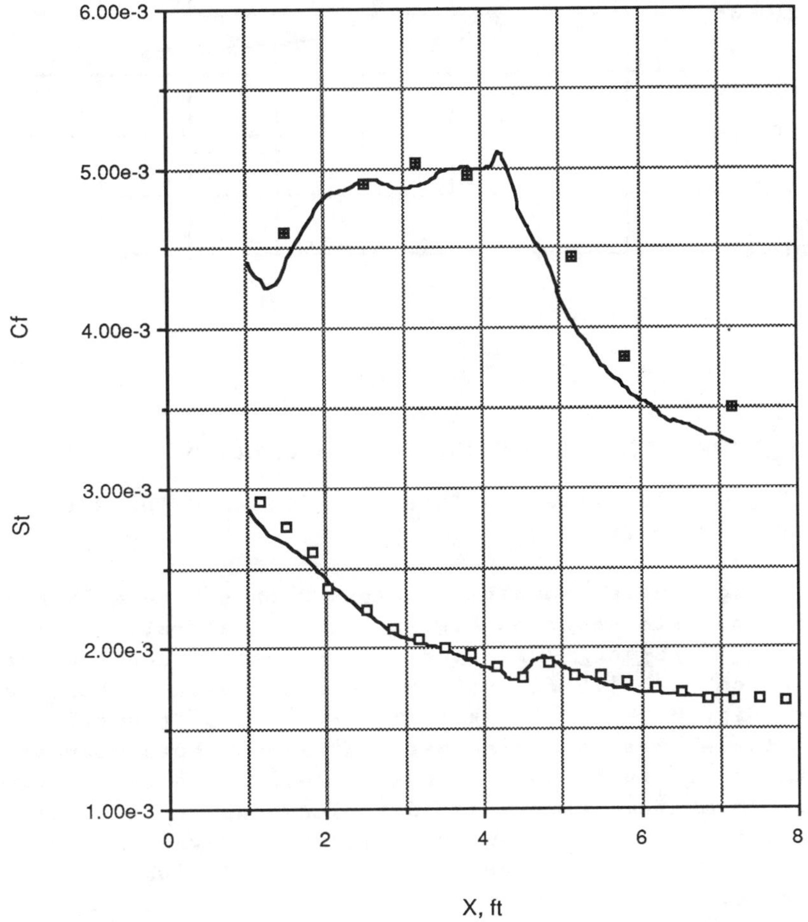

FIGURE 5. RUN 72968 (Thielbahr) K = 1.47E-06

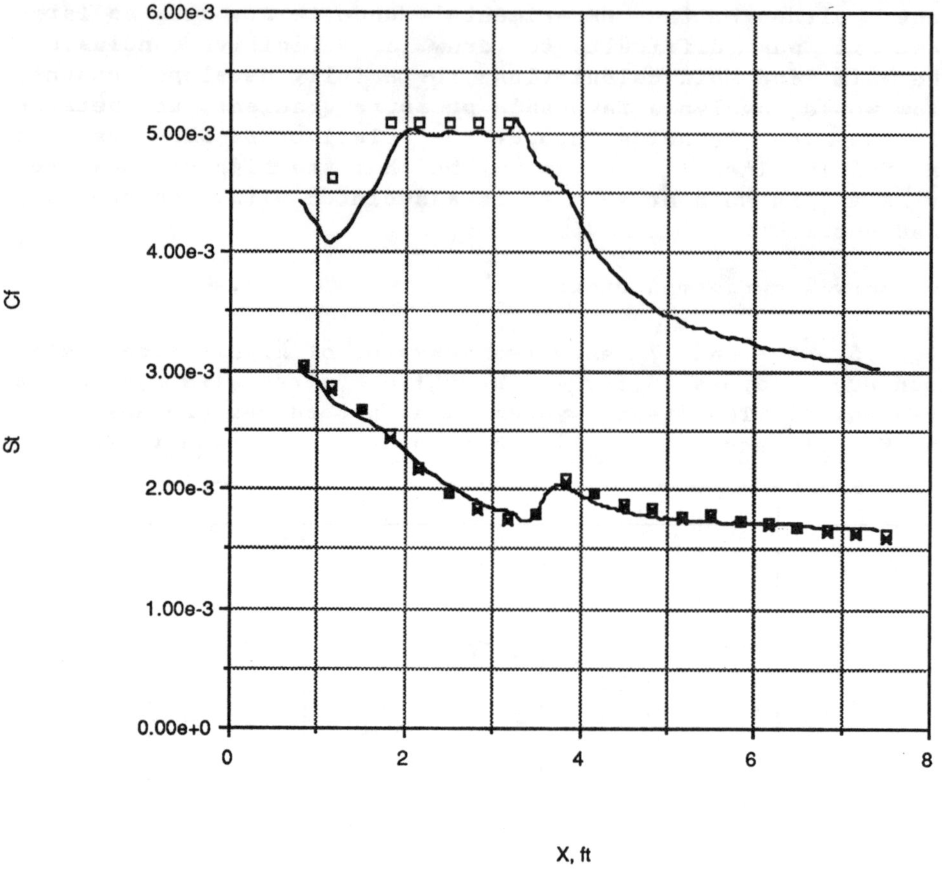

FIGURE 6. RUN 91069 (Kearney) K = 2.06E-06

data in Fig. 5 is given by Thielbahr [3], and for data in Figs. 6 and 7 by Kearney [4].

Both of these tests involve a close approach to a "sink" flow. For the data shown on Fig. 5 the acceleleration parameter, $K = (\nu / U_\infty^2) dU_\infty / dx$, was approximately constant at 1.47×10^{-6} from x= 1.8 ft until x = 4.5 ft, being otherwise 0.0. For Fig. 6 $K = 2.0 \times 10^{-6}$ for x = 1.8 ft until x = 3.3 ft. These are not "laminarized" flows; both boundary layers are true turbulent boundary layers, although the assymptotic momentum thickness Reynolds number is quite low. For Fig. 5 the momentum thickness Reynolds number at the end of the accelerated region is 747, while it is 677 for Fig. 6.

The heat transfer rate, and thus Stanton number, was measured directly from the electrical power supplied to the surface heaters. The friction coefficient was determined by a momentum balance on the boundary layer. For a sink flow this is a fairly accurate procedure because the Reynolds number is essentially constant, as is C_f.

For both of these tests the effect of the negative pressure gradient is to increase C_f but to decrease St. When the pressure gradient is removed C_f and St return to their expected no-pressure-gradient values. The magnitude of the effect we are discussing is illustrated on Fig. 7 where Stanton number for the stronger pressure gradient case (Kearney Run #91069) is plotted as a function of enthalpy thickness Reynolds number, and is compared with a correlation for no pressure gradient. It is a substantial effect.

It is no surprise that streamwise pressure gradient has a different effect upon Stanton number than upon friction

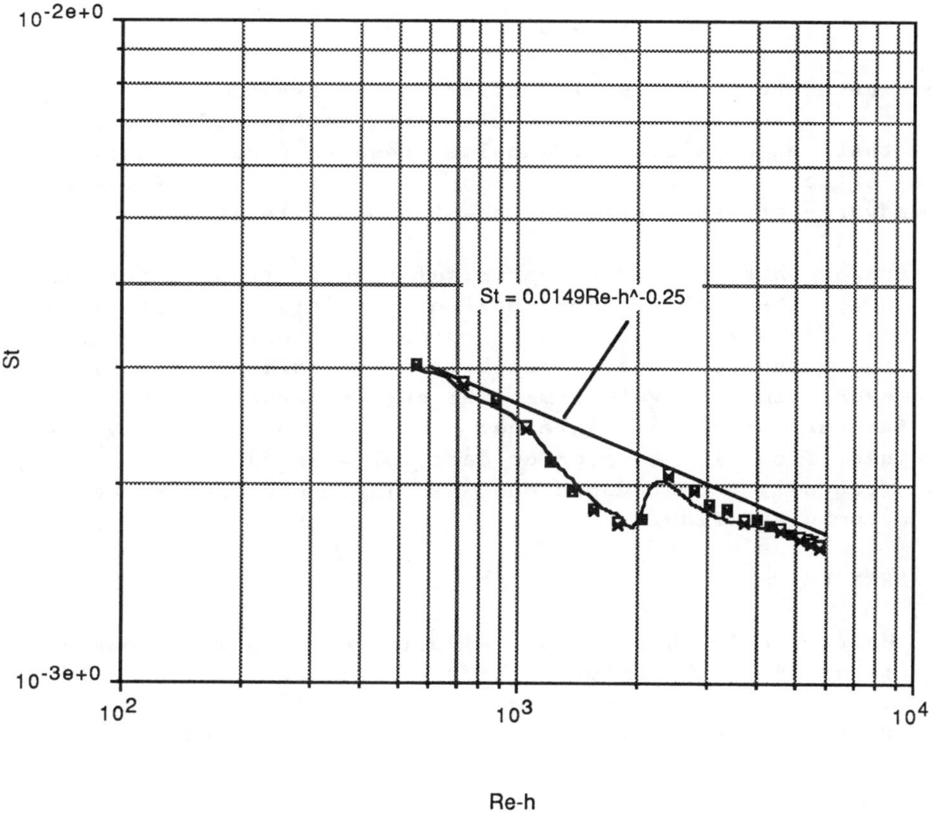

FIGURE 7. RUN 91069, St vs. Enthalpy Thickness Reynolds Number

coefficient; the momentum equation has in it an explicit pressure gradient term, while there is no comparable pressure gradient term in the energy equation. Nevertheless it is clear that any turbulence model proposed for calculating the boundary layer must not only be able to account for the behavior of the friction coefficient, but must also be capable of predicting the opposite Stanton number behavior.

The authors have found that the behavior described can be quite adequately predicted using a mixing-length model in which the near-wall region is calculated using the Van Driest damping function, but with the constant A^+ increased in a negative pressure gradient. In fact the following equation was used to perform the calculations which are indicated by solid lines on Figs. 5, 6, and 7:

$$A^+ = 25/(30p^+ + 1)$$

$$\text{where } p^+ = \mu(dP/dx)/(\rho^{1/2}\tau_o^{3/2})$$

The parameter A^+ is in effect a measure of the thickness of the viscous sublayer. The implication of this result is that a negative streamwise pressure gradient must have the effect of increasing sublayer thickness. Presumably an adverse pressure gradient has the opposite effect.

Although this is only an empirical result, it is interesting to note that A^+ becomes indefinitely large when p^+ becomes -0.033. This is close to what is observed when highly accelerated boundary layers reach laminarization; the viscous sublayer simply overwhelms the entire boundary layer. And a further point of interest is that for fully developed laminar flow in a pipe or duct, $p^+ = -0.033$ at Re = 2000, the generally accepted minimum Reynolds number for transition to a turbulent flow.

REFERENCES

1. Moffat, R.J., Kays, W.M., A Review of Turbulent-Boundary-Layer Heat Transfer Research at Stanford, 1958-1983, in Advances in Heat Transfer, Vol. 16, 1984, ed., J.P. Hartnett, T. F. Irvine, Jr., Academic Press, Inc., New York, 1984

2. Blackwell, B.F., Kays, W.M., Moffat, R.J., The Turbulent Boundary Layer on a Porous Plate: An Experimental Study of the Heat Transfer Behavior with Adverse Pressure Gradients, Report No. HMT-16, Thermosciences Division,

Department of Mechanical Engineering, Stanford University, Stanford, CA 94305, August 1972. (Also available as the PhD dissertation of B. F. Blackwell, 1972.)

3. Kim, J., Moin, P., Transport of Passive Scalars in a Turbulent Channel Flow, Proceedings of Sixth Symposium on Turbulent Shear Flows, September 7-9, 1987, Toulouse, France

4. Thielbahr, W.H., Kays, W.M., Moffat, R.J., The Turbulent Boundary Layer: Experimental Heat Transfer with Blowing, Suction, and Favorable Pressure Gradient, Report No. HMT-5,, Thermoscinces Division, Department of Mechanical Engineering, Stanford University, Stanford, CA 94305, April 1969. (Also available as the PhD Dissertation of W. H. Thielbahr, 1969.)

5. Kearney, D.W., Moffat, R.J., Kays, W.M., The Turbulent Boundary Layer: Experimental Heat Transfer With Strong Favorable Pressure Gradients and Blowing, Report No. HMT-12, Thermosciences Division, Department of Mechanical Engineering, Stanford University, Stanford, CA 94305. (Also available as the PhD Dissertation of D.W. Kearney.)

The Effects of High Free-Stream Turbulence on Heat Transfer in Turbulent Boundary Layer

PAUL K. MACIEJEWSKI and ROBERT J. MOFFAT
Thermosciences Division of Mechanical Engineering
Stanford University
Stanford, California 94305, USA

The present research is related to one of the most persistent and difficult problems in the gas turbine industry: predicting the heat transfer to turbine blades and vanes, particularly the first stages. Experience over the past ten or more years has shown that not even the most modern, most carefully applied heat transfer theory can guarantee to predict the heat transfer to the first stage blades and vanes of a newly designed gas turbine within a factor of two or three under some engine conditions. Only when the models are tuned-up using test data from a geometrically and hydrodynamically similar engine can acceptably accurate results be guaranteed. This state of affairs has persisted for years in spite of all the careful heat transfer research which has been done in industrial and academic laboratories. An enormous effort has been expended in developing predictive codes and describing the separate effects of roughness, pressure gradients, curvature, and free-stream turbulence and yet accurate prediction still cannot be made in an unfamiliar situation. In reviewing this situation, it seemed clear that something important was being overlooked and that a new method of attack was needed. When this research on the effects of high free-stream turbulence was planned, a somewhat different approach was charted. Rather than establish a particular high turbulence flow field and measure its effect on heat transfer, we set out to find a set of flow fields with very aggressive heat transfer characteristics and then learn which aspects of those flows were responsible for the high heat transfer.

The first part of the study, which is reported here, consisted of a series of heat transfer measurements using a variety of flows, each characterized by a high level of turbulence. The objective was to see how high the heat transfer could be raised by high free-stream turbulence, of whatever nature. The results of these tests were used to identify the most aggressive heat transfer environments, and the turbulence properties of those flows were later studied in detail. None of the turbulence details are presented in this report.

REVIEW OF PRIOR WORK

Recent studies in closed loop wind tunnels with grid generated turbulence performed by Simonich and Bradshaw (1978), Pedisius, et. al. (1979,1983), Hancock and Bradshaw (1983), and Blair (1983) each conclude that free stream turbulence augments heat transfer and/or skin friction on a flat plate. Simonich and Bradshaw [1] show Stanton number increases of 30% for turbulence intensity of 6% and suggest that increases in Stanton number are a function of turbulence intensity. Hancock and Bradshaw [2] show increases in skin friction of 20% for turbulence intensity of 6% and assert that both intensity and L/δ contribute to the observed increases. Pedisius et. al. [3] report Stanton number increases of 40% for turbulence levels of about 15% and correlate Stanton number by plate Reynolds number alone. In a subsequent paper Pedisius et. al.

[4] claim that 40 to 50% increase in Stanton number is an asymptotic limit approached for turbulence levels above 10% and describe the behavior in the domain below this limit in terms of intensity and L/δ. Blair [5,6] reports Stanton number increases of up to 20% for intensities up to 7% and follows the analysis of Hancock and Bradshaw [2] to support the contention that intensity and L/δ determine the increases.*

From a practical point of view, for example the point of view of a gas turbine designer who would like to account for free stream turbulence effects, the combined experience reflected in the literature might reasonably be distilled into three claims. Firstly, free stream turbulence augments heat transfer [all referenced authors]. Secondly, heat transfer augmentation due to free stream turbulence reaches an asymptotic value of about 50% at turbulence levels of greater than 10% [Pedisius et. al. (1983), not contradicted by any referenced author]. Thirdly, the set of parameters {Re, Tu , L/δ} determine Stanton number [all referenced authors].

These facts raise a final question. Should a designer feel secure knowing he has accounted for free stream turbulence using these guidelines? One can test these three claims by posing new questions. For example, can one find cases where a significant level of free stream turbulence does not augment heat transfer? Can one find cases where free stream turbulence augments heat transfer by significantly more than 50%? Can one find cases which contradict the claim that the parameters {Re, Tu , L/δ}, or a subset of them, determines Stanton number?.

The present research challenges the existing heat transfer literature, experimentally, by seeking examples counter to its claims. By doing so, the present research also challenges one of the generally held assumptions of boundary layer turbulence modeling: i.e., that the turbulent transports of momentum and heat are uniquely determined by the local turbulence field whose evolution, in turn, can be traced using equations sensitive to the parameters Re, Tu, and L/δ.

EXPERIMENT AND DISCUSSION

The intention of the research is to test the current belief set by searching for cases which contradict the following claims for constant properties turbulent boundary layer flow in air:

1) $St/Sto \geq 1$

2) $St/Sto \leq 1.5$

3) $St=St(Re)$
or $St=St(Re,Tu)$
or $St=St(Re,L/\delta)$
or $St=St(Re,Tu,L/\delta)$

The experimental basis for the tentative validity of these claims rests largely on data taken in closed loop wind tunnels with turbulence generated by grids [all referenced works]. We chose to look for cases counter to these claims in a turbulent flow field with a different character. Turbulent jets are well suited for studying the effects of very high turbulence on boundary layer heat transfer. Within turbulent jets local (relative) turbulence intensities up to 50% or more are common, and by moving from one location in the jet to another one can easily vary the local mean velocity, turbulence intensity, and

* We will use the symbol L/δ to refer to the ratio of some measure of turbulence scale to some integral measure of the boundary layer thickness. In Blair's work, the integral length scale and the dissipation length scale differ only by a scalar constant, suggesting that for our purposes one measure of scale is as good as another.

integral length scale by factors of two or more. Consequently, a single free jet apparatus can produce a wide variety of local flows. The present study was performed by placing a constant temperature heat transfer surface in the marginal flow field of a large diameter, round, turbulent free jet. The apparatus is shown schematically in Figure 1.

The jet was generated by a centrifugal fan (0.4 m wheel diameter) dumping to a 0.35 m diameter round exit plane. Outlet velocity was adjusted by setting a slide gate at the fan inlet. The fan was typically run off design, making the velocity field in the exit plane non-uniform and highly turbulent. Given the objectives of this project this was judged to be quite acceptable, and no attempt was made to "clean up" the flow. Adjusting the fan inlet gate was equivalent to changing the character of the velocity field in the fan exit plane. Because of this, each different "buildup" can be considered to be a qualitatively different flow field. This was not intended to be a study of heat transfer in a classical, self-similar free turbulent jet. It is not even a study of heat transfer in jets of the same character. It is a survey of heat transfer under highly turbulent flow conditions to test the general validity of the hypotheses (1)-(3) given above.

An aerodynamically smooth test surface 2.4 m long and 0.5 m wide was placed in different locations in the flow, with the leading edge from two to four meters downstream of the nozzle face, and the plate at different radial locations. Side walls were attached to the surface in an attempt to "freeze" the flow over the surface at the conditions found at the leading edge. Measurements were taken in the streamwise direction with a hot wire probe located 12 cm from the surface at the spanwise center of the test section. On average, the mean velocity at the leading edge of the surface was 13% higher than the mean velocity measured at the mid-length of the surface. The mean velocity at the trailing edge of the surface was on the average 5% lower than the mean velocity at the mid-length. Similarly, turbulence intensity on the average varied from 1% higher than the mid-length value at the leading edge to 8% lower at the trailing edge, and the integral scale varied from 12% lower to 9% greater. Given that the section is 2.4 m in length, streamwise gradients in all of these quantities are small. Run conditions are described in terms of the values at the mid-length.

To isolate the heated section of the test surface from side wall effects, only the central 60% of the surface was used to study wall heat transfer. This region consisted of eight 0.01 m thick aluminum plates each 0.3 m wide. The first of these is 0.25 m in length, has a sharp leading edge, and is not heated. The remaining seven are each 0.3 m in length and each has bonded to its backside a resistance "patch" heater. The backside of each plate is heavily insulated with fiberglass insulation. The plates are insulated from each other by strips of balsa. Five thermocouples are embedded in each of the plates in a cross pattern to measure plate temperature and to check for spanwise and streamwise temperature uniformity. During test runs, the power delivered to each plate is controlled individually so that at equilibrium all seven heated plates have the same average temperature to +/- 0.25 degrees C. On average, the plate-to-free stream temperature difference was between 10 and 20 degrees C. Conduction losses out the back and sides were typically about 5% of the measured input power. Radiative losses were estimated to be less than 1% of the input power.

Velocity measurements were made using a single component normal hot-wire (tungston, 5 micron diameter) operated in constant temperature mode using a DISA 55M01 Main Unit and a DISA 55M10 CTA Standard Bridge. The signal from the bridge was linearized using a TSI Model 1072 Linearizer before being sampled through an analog to digital converter. During test runs independent samples of the linearized signal where taken until the running mean value became steady to +/-1% at 20:1 odds. Reported turbulence intensities are simply the sample standard deviations divided by the sample means. Reported integral length scales were inferred by way of Taylor's hypothesis from autocorrelations performed on the linearized signal using an HP 3721A Correlator.

A single component normal hot wire produces accurate turbulence statistics only when the turbulence intensity is less than 30%. However, a simplified analysis of the response of hot wires to isotropic turbulence of increasing intensity [7] shows that the parameter estimates incur errors which have the form of a bias whose magnitude strictly increases with turbulence intensity. The free-jet flow used in these experiments was shown to be nearly isotropic, using a triple hot-wire probe, [8]. Therefore, the estimators we calculate from our data can reasonably be considered related to the unbiased estimators by a simple monotonic transformation. Since the objective of this paper is not to establish correlations for Stanton number but to find examples contradicting claims that Stanton number can be uniquely correlated on specific sets of parameters, demonstrating that Stanton number cannot uniquely be determined by a set of parameters each subjected to a monotonic transformation is sufficient.

The heat transfer surface was qualified by installing it in a closed loop wind tunnel and testing with negligible levels of free stream turbulence. Figure 2 shows that Stanton number measurements on the surface for two separate runs each match the accepted flat plate correlation [9] for normal turbulent boundary layer behavior to within +/-2.5%. From this test we conclude that the absolute level of the heat transfer coefficient can be measured on this test surface to within +/-2.5%. Figure 2 also shows the second part of the qualification runs--demonstration of repeatability. Stanton number measurements were made on two runs in the free jet, on separate days under identical experimental conditions (i.e., with the plate put into the same position in the flow with the blower at the same setting). The agreement between these two data sets again lies within +/-2.5%, establishing that the heat transfer coefficient on the surface when in the free jet is repeatable for identical buildups to within the accuracy within which we can measure it absolutely. These tests, by themselves, contradict the claim that St=St(Re): the data clearly show that Reynolds number alone cannot determine Stanton number when there is free-stream turbulence.

Ability to meet the baseline correlation for negligible levels of free stream turbulence combined with the fact that measurements in the free jet are repeatable for identical buildups establish the credibility of the present results, but our final conclusion will lead to a warning that tests with similar objectives conducted in flows other than free jets may yield different results. The major conclusions we will draw have yet to be confirmed by independent studies at enough other labs to establish their generality. The levels of heat transfer augmentation found at Stanford has been reproduced, over part of the range, by Rivir [10] who has been studying the effect of high free stream turbulence generated by a wall jet on heat transfer to a flat plate. Figure 3 shows that Rivir's work confirms the levels of augmentation observed both by the authors at Stanford and by Blair [5,6] at United Technologies Research Center. Figure 3 suggests that Rivir's wall jet may capture important features of both free jet generated and grid generated turbulence.

Figure 4 shows Stanton number data plotted against enthalpy thickness Reynolds number for 26 runs in the free jet. For these runs mean velocities varied between 0.5 and 4 m/s, turbulence intensities varied between 10% and 65%, and integral length scales varied between 4 and 17 cm. On the whole, the Stanton number data for individual runs parallel the expected response for turbulent boundary layer behavior but are 1.8 to 5 times greater than the baseline level, far higher than the previous literature would suggest. All 26 runs contradict the claim that St/Sto remains less than or equal to 1.5.

Figure 5 shows the increases in Stanton number compared to a standard turbulent boundary layer correlation [6] for several runs plotted against plate Reynolds number with individual runs labeled by their mean velocity, turbulence intensity, and integral length scale in the free stream, as measured over the center of the test section. Notice that two of the runs have nearly identical measures for mean velocity, turbulence intensity and integral length scale, yet have significantly different Stanton number ratios. In terms of the measured hydrodynamic variables the run labeled U=0.87 m/s, Tu=48%, Scale=9 cm and the run labeled U=0.92 m/s, Tu=47%, Scale=8 cm can be considered the same in mean velocity to +/-3%, the same in turbulence intensity to +/-2%, and the same in integral length scale to +/-6%. At the same time these two runs differ in Stanton number by 20% for fixed Re_x. Since we have demonstrated that heat transfer on our surface when placed in the jet is repeatable to +/- 2.5% for identical buildups, the measured 20% difference in Stanton number between these two runs is highly significant. Now compare the run labeled Tu=63% to the run labeled Tu=48%. Here Stanton number is the same despite the difference in turbulence intensities. The behavior for these pairs of runs strongly contradicts the claim that Stanton number is determined by a parameter set based on Re, U, Tu, and L.

Two other features displayed in Figure 5 are worth mentioning. The first is that the runs with augmentation greater than 2.5 (approximately) are insensitive to Reynolds number. If we consider only cases where St/Sto is less than 2.5 then correlating Stanton number data in the form of St/Sto factors out Reynolds number dependence. The highest augmentation runs were in regions of the jet where the flow wa' highly intermittent. Cases which explicitly contradict the claim St/Sto=St/Sto(Tu and/or L/δ) implicitly contradict the claim St=St(Re, Tu and/or L/δ). The second feature is that there is a large cluster of data around St/Sto=1.9 despite the fact that U, Tu , and L each vary by approximately a factor of two between the individual runs that comprise this grouping. This suggests the existence of a flow regime within which Stanton number has been significantly augmented but within which the augmentation has reached a limiting value for that regime.

Figure 6 represents each test run by a single data point in L/δ versus Tu coordinates, identifying each run by its value of St/Sto. In this plot, L/δ is represented by the ratio of integral length scale, λ, to enthalpy thickness Δ_2 and Tu is the turbulence intensity, all measured at the mid-length of the plate. The value St/Sto was calculated from the Stanton number measured at the mid-length of the plate divided by the value Sto for the same enthalpy thickness Reynolds number. To be consistent with use of the local variable Δ_2, in figure 6, Sto is here determined by a standard turbulent correlation based on enthalpy thickness Reynolds number [9]. (This removes the effects of the upstream history of the boundary layer on the local value of St). Points near Tu=24% differ in St/Sto by as much as 50% when L/δ varies. Points near L/δ=7 differ in St/Sto by as much as 40% when Tu varies. Points near the operating point near Tu=48% and L/δ=7 differ in St/Sto by 20% and points near Tu=21% and L/δ=13 differ in St/Sto differ by 30%. These measured differences in St/Sto are highly significant, and together they contradict all claims implicit in the statement that St/Sto can be represented as a function of Tu and L/δ. Based on the arguments already made, this demonstration is sufficient to contradict the general validity of the claim that Stanton number is uniquely determined by the set of parameters {Re, Tu, L/δ}.

CONCLUSIONS

On boundary layer heat transfer, studies using only grid generated free stream turbulence fail to capture some very important effects of high free stream turbulence. The present data, from free jet tests, agree with the grid-generated literature in that high turbulence augments heat transfer, but strongly contradicts that literature as to the possible levels of that effect and as to the parameters which can be used to correlate the effect. Heat transfer augmentation may be as high as a factor of 5 and cannot be uniquely determined by the most commonly measured parameters, i.e., in terms of Re, Tu and L/δ. The heat transfer designer of today, who must be concerned with what might possibly happen in highly turbulent environments more than with what happens in grid generated turbulence, has good reason to feel insecure.

At a more basic level, the study of surface heat transfer in highly turbulent flows contributes a cautionary note to our understanding of the velocity field at the wall. Efforts to model turbulent behavior begin with the assumption that the details of the turbulence field establish the transport coefficients in the momentum and energy equations, and hence control the evolution of both the velocity and temperature fields. We have demonstrated, at least, that the set of parameters consisting of $\{Re, Tu, L/\delta\}$ do not uniquely determine Stanton number. This means this set of parameters does not uniquely determine the mean temperature gradient at the wall. But if the velocity field uniquely determines the temperature field (given for a given set of Re, Tu, and L/δ, and fixed temperature boundary conditions), then evidence that the temperature field is not unique constitutes evidence that the velocity field is not unique. From this it also follows that correlations or models based either implicitly or explicitly on only these parameters (or a subset of them), cannot be valid except over limited ranges.

The fact that even the full set of commonly measured hydrodynamic parameters does not serve to uniquely specify the heat transfer in highly turbulent flow fields leaves the experimental research community with two alternative strategies for attacking this problem. The first approach is to seek a generally valid deterministic set; i.e., extending the list of locally measurable parameters (by, for example, including spectra and correlations) until we determine a set sufficient to make the heat transfer behavior unique for all possible turbulent flows. This approach assumes the existence of a general turbulent mechanism which controls behavior at the wall independent of the mechanism which generated the free stream turbulence. This approach corresponds to the more global issue of a search for a general model for turbulence, valid in all zones of a flow field. Such an approach implies that the main features of the set of full solutions to the N-S equations are dominated by the general solution and not the particular solutions (which are determined by the initial and boundary conditions). This approach may not, in fact, be fruitful. It may be that the main features of the solution are dominated by the particular solution. If this is so the present work suggests that the concept of "a boundary layer" must be refined. It may be necessary to consider the boundary layer situation as a global concept, permitting of several different solution "zones" in the Kline sense, where the zonality is defined not based on the geometry (which establishes the general solution) but on the hydrodynamic initial and boundary condition (which establish the particular solution for the turbulence field). The different zones may have to be treated by similar but not identical models. One could, for example, establish a small list of parameters sufficient to predict heat transfer within a particular zone (i.e. under a particular type of free stream turbulence) with the form of the model, or the values of its constants, specific to the type of free stream turbulence involved. Assuming that the initial and boundary conditions (i.e., the particular part of the full solution) strongly influence the physics of the full solutions to the N-S equations, then even boundary layer flows should be classifiable based on similarities in the mechanisms which generated the turbulence. It would then remain to determine which parameters form the most robust set from which to build the various "zonal" models.

ACKNOWLEDGMENT

The work described in this report was sponsored by the NASA HOST Program (Hot Section Technology) as NASA NAG 3-522. The work was started under the technical administration of Dr. John Rohde and continued under Dr. Raymond Gaugler, both of the Lewis Research Laboratories

Nomenclature

δ	generic measure of boundary layer thickness
Δ_2	enthalpy thickness
L	generic measure of gross turbulence scale
λ	integral length scale
Re_x	Reynolds number based on plate location
$Re_{\Delta 2}$	Reynolds number based on enthalpy thickness
St	Stanton number
Sto	reference Stanton number
Tu	ratio of rms velocity fluctuation to mean value
U	mean free stream convective velocity

References

1. Simonich, J.C., and Bradshaw, P., "Effect of Free-Stream Turbulence on Heat Transfer through a Turbulent Boundary Layer," ASME Journal of Heat Transfer, Vol. 100, 1978, pp. 673-677.

2. Hancock, P.E., and Bradshaw, P., "The Effect of Free-Stream Turbulence on Turbulent Boundary Layers," ASME Journal of Fluids Engineering, Vol. 105, pp. 284-289.

3. Pedisius, A.A., Kazimekas, P.-V.A., and Slanciauskas, A.A., "Heat Transfer from a Plate to a High-Turbulence Air Flow," Soviet Research, Vol. 11, No. 5, 1979, pp.125-134.

4. Pedisius, A.A., Kazimekas, P.-V.A., and Zukauskas, A.A., "Effect of Free-Stream Turbulence on Heat Transfer in the Turbulent Boundary Layer," Soviet Research, Vol. 15, No. 4, 1983, pp.1-14.

5. Blair,M.F., "Influence of Free-Stream Turbulence on Turbulent Boundary Layer Heat Transfer and Mean Profile Development, Part I- Experimental Data," ASME Journal of Heat Transfer, Vol. 105, 1983, pp. 33-40.

6. Blair,M.F., "Influence of Free-Stream Turbulence on Turbulent Boundary Layer Heat Transfer and Mean Profile Development, Part II- Analysis of Results" ASME Journal of Heat Transfer, Vol. 105, 1983, pp. 41-47.

7. Moffat, R.J., Maciejewski, P., Eaton, J.K., and Pauley, W., "Heat transfer with Very High Free-Stream Turbulence and Heat Transfer with Streamwise Vortices," Proceedings of the NASA HOST Meeting, Cleveland, Ohio, October 1987.

8. Ongoing research, data available through P.K. Maciejewski, Thermosciences Division of Mechanical Engineering, Stanford University, Stanford, California.

9. Kays, W.M., and Crawford, M.E., Convective Heat and Mass Transfer, Second Edition, McGraw-Hill, New York, 1980.

10. Rivir, R.B., Dept. of the Air Force, Aeropropulsion Laboratory, AFWAL/POTC. Wright-Patterson AFB, Ohio. Personal communication.

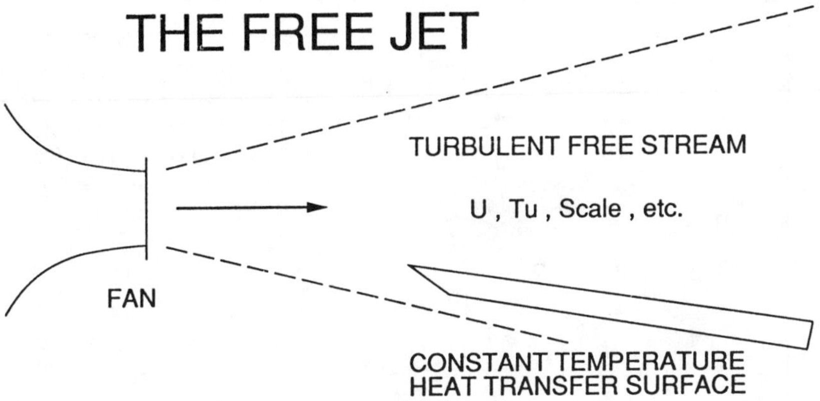

Figure 1. Schematic of the free jet test facility.

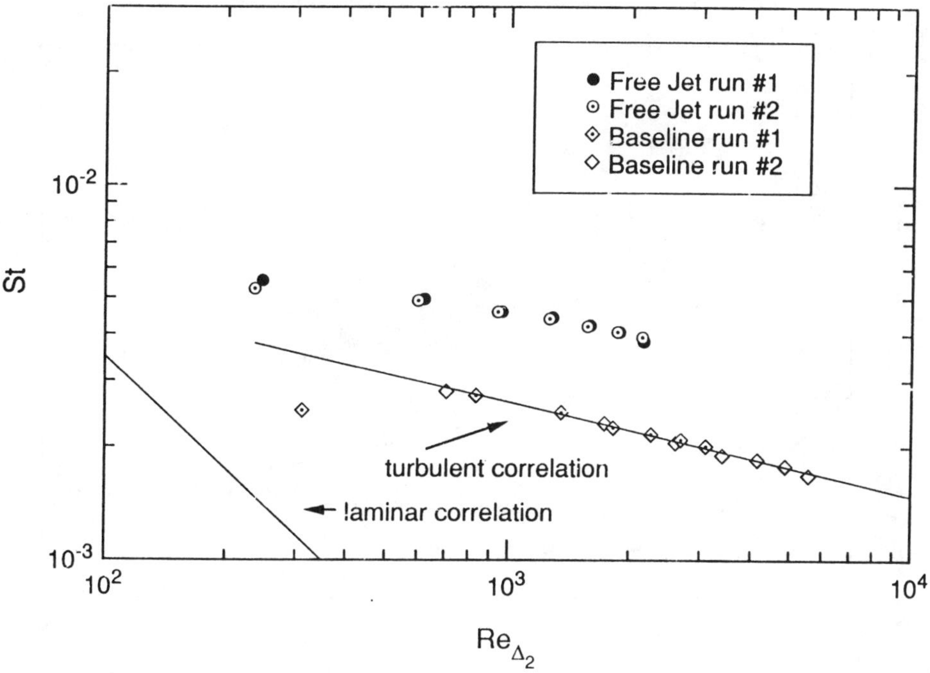

Figure 2. Baseline and repeatability tests.

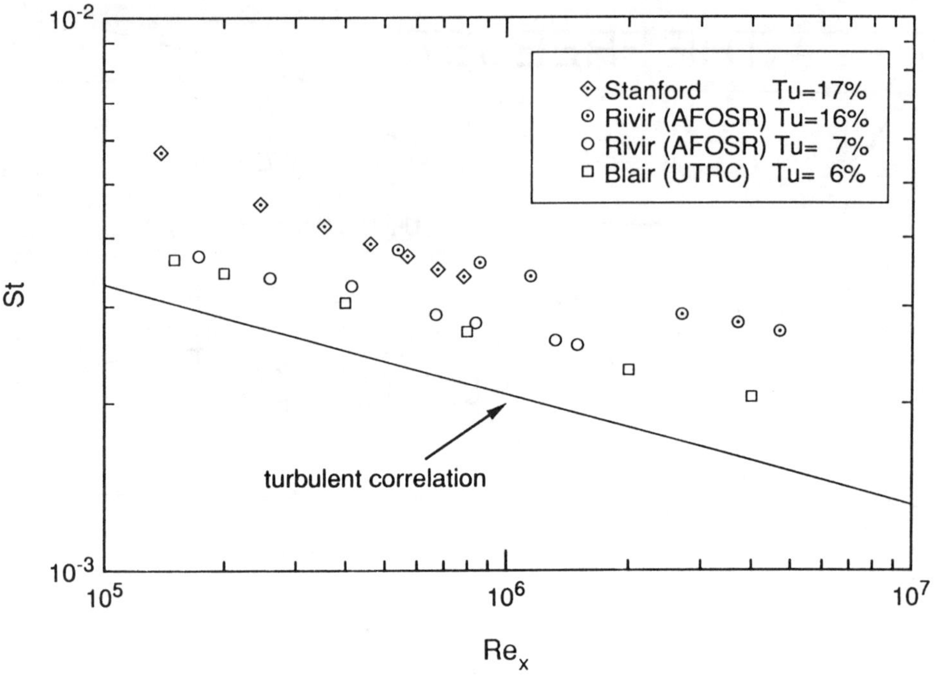

Figure 3. Cross validation with other laboratories.

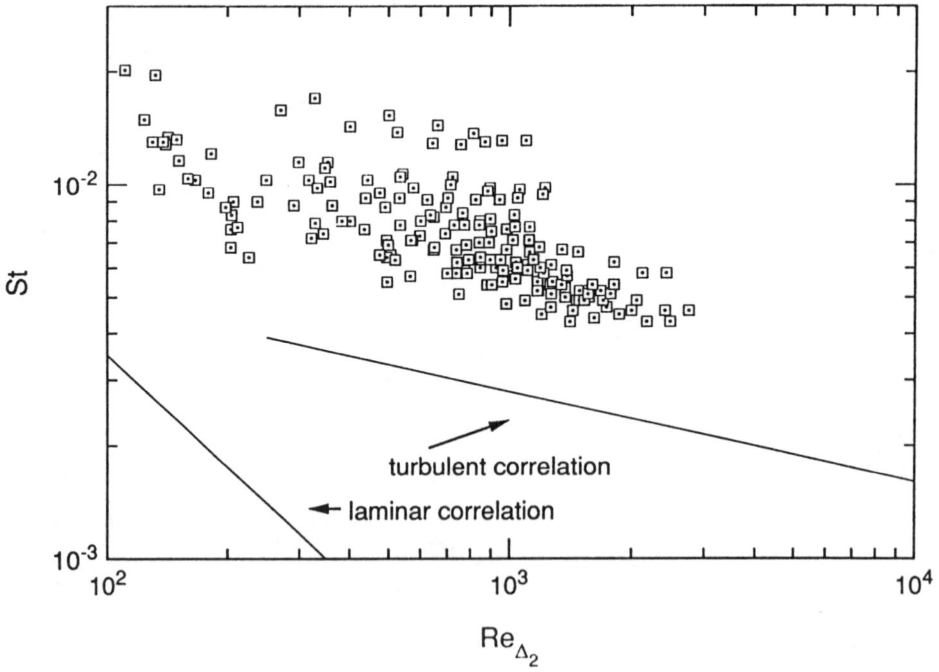

Figure 4. Stanton number data in the free jet.

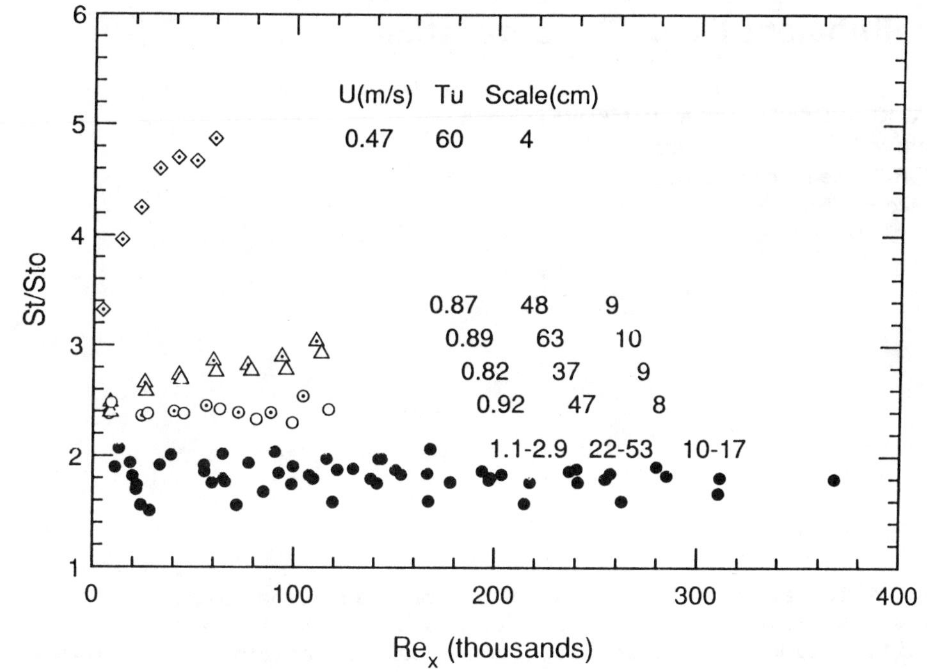

Figure 5. Hydrodynamic conditions associated with augmented heat transfer.

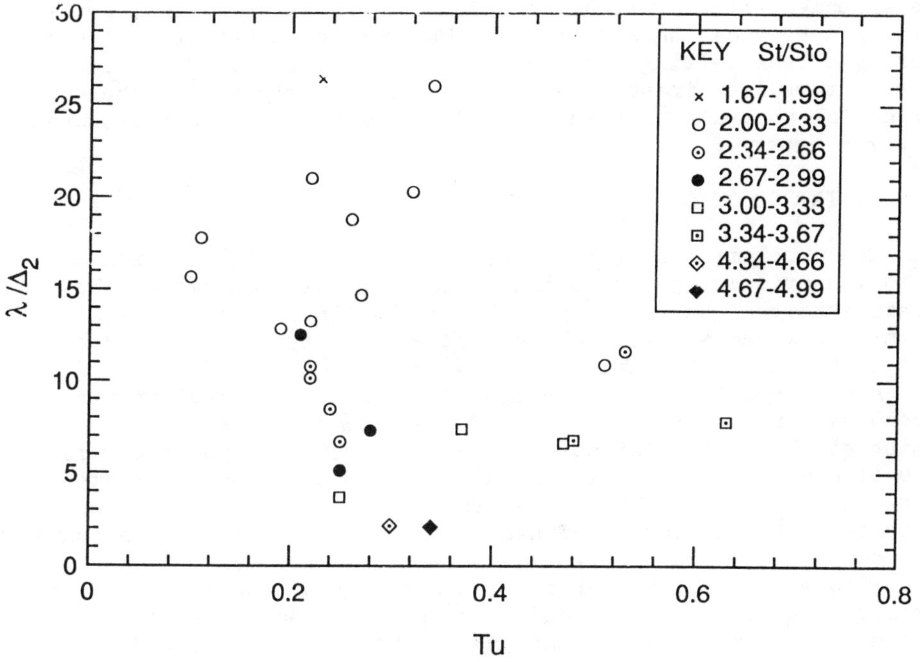

Figure 6. Effects of scale and intensity on heat transfer augmentation.

Influence of Supercritical Fluid Parameters on Turbulent Flow Characteristics

V. A. KURGHANOV and N. V. MEDVETSKAYA
Institute for High Temperatures
USSR Academy of Sciences
Moscow, USSR

ABSTRACT

This paper presents the calculation and experimental results, by which one can trace along the length of the heated part modifications of behaviour of such local characteristics as velocity profiles, shear stresses, heat fluxes, coefficients of turbulent diffusivities and others.

The calculation and experimental data have been obtained for the turbulent liquid stream of CO_2 under the supercritical pressure in a vertical heated pipe at the designed constant wall heat flux. The task setting, including dependences for the coefficients of turbulent transfer as well as the description of the experimental apparatus and the method of measurement are dwelt upon in early papers.

1. INTRODUCTION

The problem of heat transfer in the single-phase medium near to the critical point is one of the topical problems in the modern theory of heat transfer. The ample bibleography dedicated to various aspects of the above problem testify to it being dealt with in a number of critical reviews [1 - 2].

A wide practical usage explains the great interest in the investigation of the principles of heat transfer and resistance of the single-phase heat carriers of near critical point parameters of state. At the same time heavy and nonmonotonic variation of physical properties near the critical point under these conditions ranks the problem of heat transfer and resistance among the most complicated and interesting scientific problems.

Figure 1 presents characteristic change of the physical properties of water in this field. The specific heat capacity at constant pressure, Pr number and the thermal expansion coefficient have their maximums in the region of pseudocritical temperature, becoming more acute the closer pressure approaches the critical point. The coefficients of thermal conductivity and viscosity are at their minimum points. The density is dicreasing considerably, changing by an orders at the temperatures close to pseudocritical temperature t_m. Such a heavy and special change of the physical properties of heat carrier with temperature and pressure results in specific behaviour of heat transfer and resistance, leading in particular, to "deteriorated" regimes of heat transfer, which may cause excessive heating and destruction of heat exchange surfaces.

The experimental and calculation data on heat transfer for flow in channels of supercritical parameters of liquids have been analysed in the first known survey [I], in which the terms of "normal", "deteriorated" and "improved" regimes of heat transfer were introduced. Although this classification of the flow and heat transfer regimes is quite symbolic, however it reflects the specific features of heat transfer and can serve as a basis for its further thorough study. This classification is used practically in all the investigations on heat transfer for the single-phase heat carrier flow in the near critical point parameters of state.

In a number of papers, to quote but a few [2-4] , "deteriorated" regimes of heat transfer for the single-phase heat carrier flow in the near critical point parameters of state is attributed to the influence of buoyancy and thermal acceleration of the flow on the fluctuating and average characteristics of stream.

The overwhelming majority of papers, dedicated to the study of heat transfer for the flow in tubes, are experimental investigations. Measurements of heat transfer have been made for various heat carriers such as water, carbon dioxide, hydrogen, nitrogen, helium, organic heat carriers and some other liquids. Up till recently the investigations of resistance, velosity profiles (and temperatures) have brought limited and disputable results in view of extensive difficulties of methodical character.

Practically the only complex work, in which measurements of heat exchange and the coefficients of resistance were made with the method of "two pressure drops" applied, is presented in paper [5]. In the recent publications, for instance in [6], one can find the results of profile velocity measurements and the flow turbulent characteristic acquired on the basis of their processing.

The experimental data on heat transfer for the turbulent flow in tubes of the onephase heat carrier of the near critical parameters and a number of investigations dedicated to their analysis make it possible to point out the specific features of heat exchange under these conditions.

The "normal" regimes of heat exchange taking place under small heat flux are subject to the same regularities as is in the case of constant physical liquid properties. The changes in the level of heat transfer and friction resistance under these regimes are determined by the local variations of the liquid properties in the cross section of the stream.

Considerable influence on the characteristic of heat exchange and resistance of the thermal acceleration of the flow takes place under the conditions of rather a high density of heat flux on the walls of the tubes of a comparatively small diameter: "deterioration" of heat transfer in this case occurs both in vertical pipes for upward and downward flow and in horisontal pipes.

The influence of buoyancy on heat transfer in vertical pipes for upward and downward flows is substantial and varied, for the upward flow along the length of the heated part under the combination of regime parameters there may be one, two or more maximums of wall temperatures; for the downward flow in vertical pipes the effects of buoyancy inereases heat transfer.

The experimental data on local characteristic of resistance are quite limited. Most of them were obtained for the ratios $Gr_{\rho}/Re^2 \leq 0,02$; the experimental results available testify to the fact that under the conditions of heating the local coefficients of friction resistance are lower than for constant physical properties; at the same time the pressure drop component dependent on the acceleration of the stream considerably increases reaching the maximum values in the region of "deteriorated" heat transfer; the experimental data on the local values of the coefficients of resistance under the influence of buoyancy are scarce.

The numerical and theoretical study of recent years were based on the system of partial differential equations represented in the form of boundary layer equations or in approaching "narrow" channel, like in [7 - 9]. By and large they were aimed at finding the relations for turbulent transfer coefficients, which make it possible to describe the experimental data.

In the majority of cases the eddy diffusivity of momentum and heat coefficients (usually $Pr_t = 1$) were calculated by means of the equations worked out for the constant physical properties, in which the influence of variable physical properties is taken into account either by choosing the physical properties at a local temperatures, or by the introduction of the generalized variables, or by means of some corrections taking into account the influence of density fluctuation; termal acceleration and buoyancy on turbulent characteristics.

The calculations effected have shown that with the help of rather simple dependences for the eddy diffusivity of momentum coefficients (for instance, dependence [10]), it is possible to describe the behaviour of the wall temperature and resistance under the "normal" regimes of heat exchange, see, for instance, [7]. The input of corrections with the influence of thermal acceleration and buoyancy in view has made it possible to obtain a qualitative and in some cases quantitative coincidence of the predicted and experimental values of the wall temperature and local coefficients of resistance [7,9].

The experimental and numerical data obtained were presented as a distribution of the wall temperature or coeffitients of heat transfer along the length of the heated part of the tube or channel. They indicate the integral characteristics of flow and provide only indirect assessment of the turbulent characteristics behaviour under these conditions. To grasp the mechanism of heat transfer it is necessary to have data on the local parameters of flow to be able to analyse the changes in the structure and the turbulent characteristics occuring with its heating and explain the reasons of heat transfer "deterioration".

This paper presents the calculation and experimental results, by which one can trace along the length of the heated part modifications of behaviour of such local characteristics as velocity profiles, shear stresses, heat fluxes, coefficients of turbulent diffusivities and others.

The calculation and experimental data have been obtained for the turbulent liquid stream of CO_2 under the supercritical pressure in a vertical heated pipe at the designed constant wall heat flux.

The task setting, including dependences for the coefficients of turbulent transfer as well as the description of experimental apparatus and the method of measurement are dwelt upon in papers [7] and [5,6].

2. CALCULATION RESULTS

Under the conditions of small influence of buoyancy the turbulent flow and heat transfer are characterised by a number of specific features connected with considerable changes in the thermodynamic and transport properties of liquid and the influence of thermal acceleration. As a rule, the influence of these effects is most essential when the wall temperature is higher, whereas the bulk liquid temperature is lower than the value of pseudocritical temperature.

2.1. INFLUENCE OF THERMAL ACCELERATION

The "normal" regimes of heat transfer are characterized by the fact that the main alterations of properties occur in the layer immediately adjacent to the wall in the region of comparatively small intensity of turbulent exchange. At the same time the main characteristics of flow such as velocity profiles, temperature, shear stress and heat flux along the length of the heated part undergo slight changes.

Fig. 2.1. presents profiles of physical properties, velocity, temperature, shear stress and heat flow typical for the regimes with the ration of $q_c/\overline{\rho u} \ll 1$. As is shown in figure 2 - I under comparatively small values of the heat flux and temperature drops $(t_c - t_м)$ 1-3°, inherent in these regimes, the physical properties in the given cross-section undergo a comparatively insignificant change. The profiles of mass flow rate and the longitudinal velocity component are subject to practically no distortion. The profiles of shear stress and turbulent heat flux are close to the linear profiles. Alongside of this the physical properties can change essentially along the length of the heated part from sec-

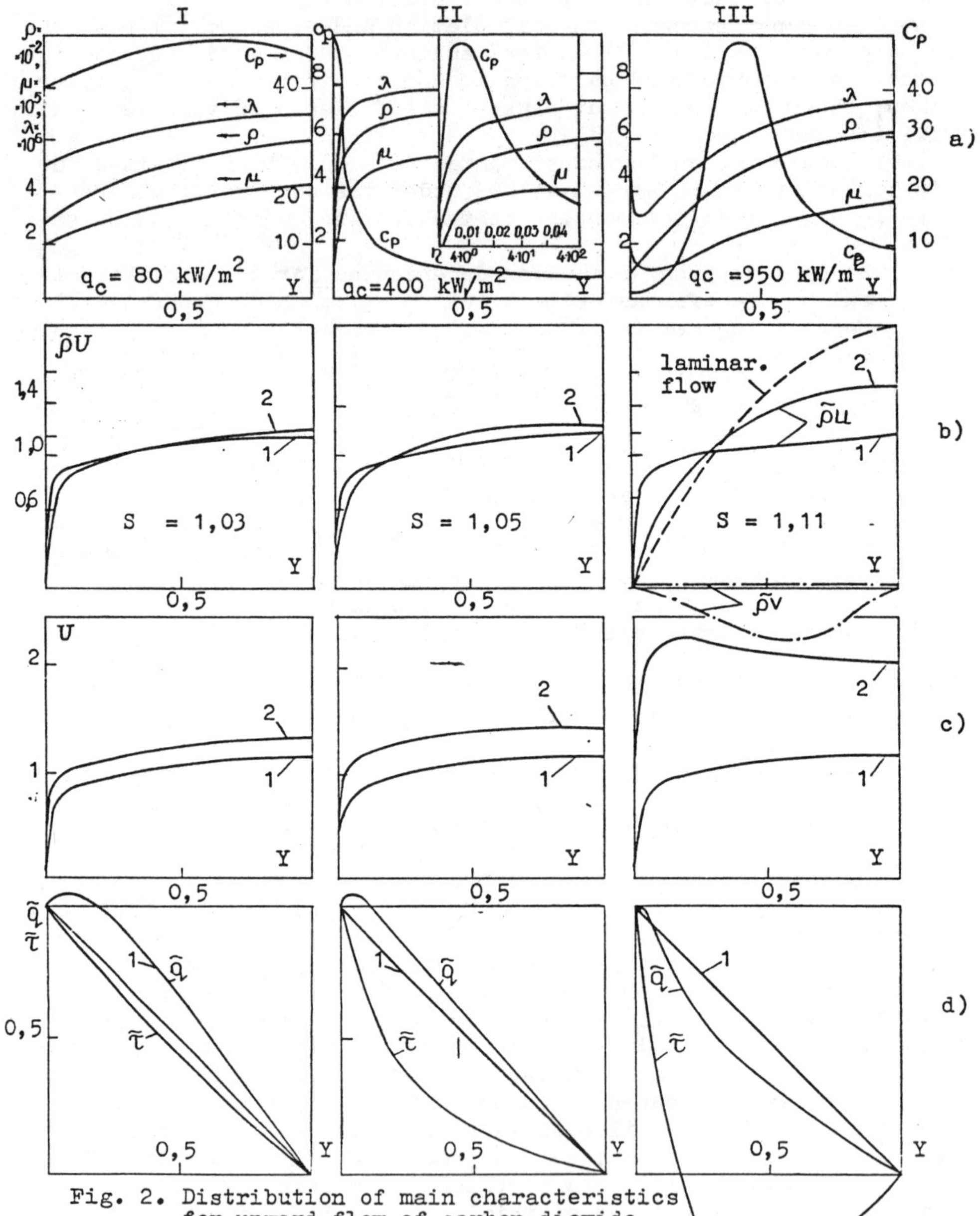

Fig. 2. Distribution of main characteristics for upward flow of carbon dioxide in heated tube; d= 8 mm, $t_{\delta x}$ = 20 C, $\overline{\rho u}$= 3200 kg/ms, q_c= I - 80 kW/m², II - 400 kW/m², III - 950 kW/m², x/d= 1 - 0, 2 - 90

tion to section, remaining practically constant in the cross-section of the pipe. Moreover under $t_c > t_m > t_{н}$ for the most part of the pipe cross section the peculiar features are the high values of heat capacity per unit weight and Prandtl numbers and rather large values of coefficients of thermal conductivity. Thus on the part of the pipe where $t_m \approx t_{н}$ the process of heat transfer is carried out under comparatively large Prandtl numbers compared to the sections up and down the flow, where as the coefficient of heat transfer somewhat increases.

It compared to the regime in figure 2-I the value of the wall heat flux (for t_{bx} and Re_{bx} being equal) is increased it brings about an increase in temperature difference between the wall and the liquid and the thermophysical liquid properties in the cross-section of the pipe change intrinsically. It is shown in figure 2-I, II, III, presenting changes of the main characteristics of the flow for three calculated regimes for the flow of carbon dioxide (P = 77,7 MPa) under equal conditions at the entry but for various values of heat flow on the wall.

With the heat flux being q_c = 400 kW /m^2 , as is shown in fig. 2.II.a, the profiles of physical properties change considerably along the radius of the pipe. However, the region of the main property changes is concentrated in the layer immediately adjacent to the wall, with the main characteristics of the flow and heat transfer changing but slightly. The coefficients of turbulent transfer along the length of the heated part somewhat increase. This increase is attributed to the dependence on local η, and local Re number accordingly, which increases at an equal distance from the start of heating with increased q_c as a result of decrease of viscosity with temperature. Turbulent Prandtl number for the most part of the pipe section is close to $Pr_t \approx 0,88$, therefore ε_q/ν is somewhat higher than ε_τ/ν. The profiles of shear stress and heat flux change slightly, although for q_c =400kW/m^2 the values τ / τ_c somewhat dicrease, which testifies to the levelling of the velocity profile. Apart from this, as is shown in figure 2-II, c, the flow along the length of the heated part is somewhat accelerated. However the degree of flow acceleration intrinsic in this regime practically does not influence the turbulent characteristics.

Figure 3 presents a comparison of calculated and experimental wall temperature distributions , coefficients of hydraulic and frictional resistance for one of the "normal" regimes. Fig. 3 shows that the results of wall temperature prediction, of the distribution hydraulic resistance and friction resistance coefficients indicate satisfactory relevance with the experimental data. Deviation does not exceed 15 per cent. Values ξ/ξ_0 along the length of the heated part are much less than 1 and are in weak dependence on X/d. The coefficient of hydraulic resistance increases along the length of the pipe almost linearly, with $x/d \sim 130$ ζ/ξ_0 is twice as much as ξ/ξ_0 . This shows

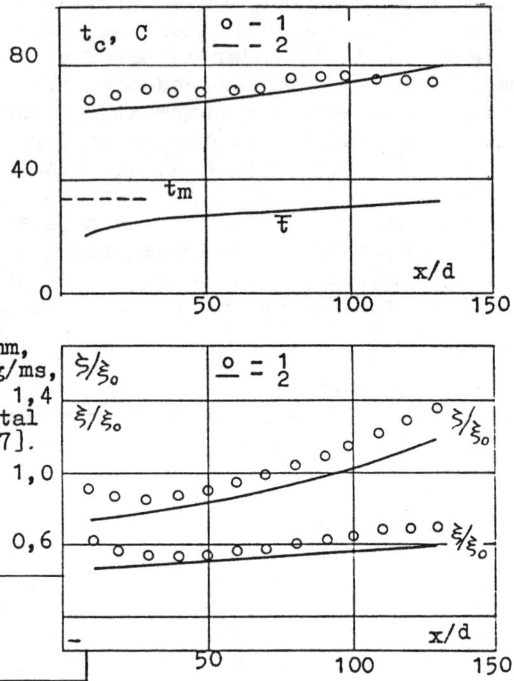

Fig. 3. Comparison of numerical results to experimental data of 5, carbon dioxide, d = 8 mm, P= 7,7 MPa, $\overline{\rho u}$= 3300 kg/ms, q_c= 400 kW/m^2, $t_{бx}$ = 19,4 C ; 1 - experimental data; 2- calculation [7].

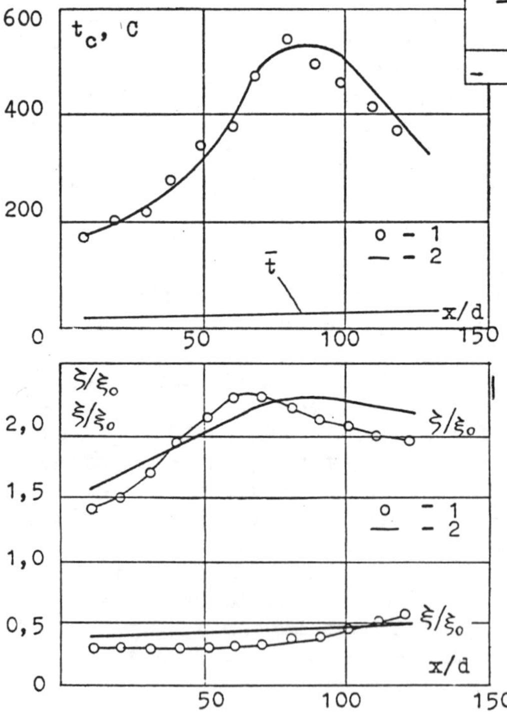

Fig. 4. Comparison of numerical results to experimental data of 5; carbon dioxide, d = 8 mm, P = 7,7 MPa, $\overline{\rho_2 u}$= 4034 kg/m^{2}s, q_c= 950 kW/ m^2, $t_{бx}$ = 19 C ; 1- experimental data , 2 - calculation [7].

that under these conditions the pressure drop component, determined by the flow acceleration, increases. However, the flow acceleration occuring in this regime, does not exert any considerable influence on the distribution of turbulent diffusivities coefficients, velocities, temperatures and other characteristics as was shown above (see Figure 2,II).

Further inerease in the heat flux (see Figure 2,III) leads to a considerable acceleration of liquid flow, although q_{cIII} grows compared to q_{cII} approximately twice as much. Decrease of density towards the wall (near t_m) leads to an acceleration of the near wall layer in relation to the flow core, forming the maximum velocity $U = u/u_{6x}$. This results in a decrease of the turbulent transfer coefficient in the near wall layer. Simultaneously the coefficient of turbulent diffusivity of heat also reduces. It is also noteworthy that in case of considerable influence of thermal acceleration the distribution of shear stress is heavily deformed compared to the linear type. In the region of velocity maximum τ passes through zero, and affer that acquires a negative value. The presence of turbulent exchange in the near wall layer with minimal values results in the greatest part of heat being accumulated in it, which leads to the overheating of the near wall layer. It is in this particular region that the wall temperature reaches its maximum and heat transfer reduces most heavily.

Under these conditions the flow momentum coefficient grows considerably. As is shown in Fugire 2,III b, for the value of heat flux q_c = 95 kW /m^2 - $S_{H} = 1,11$ ($S_{H} = 1,02$ for the isothermal flow). Besides the profile of mass flow rate is distorted approaching the laminar profile (see Figure 2,III b).

Alongside of liquid heating the pseudocritical temperature moves fast to the flow core, which happens due to comparatively large turbulent exchange in the core. The density profile in the cross-section of the pipe becomes more uniform. At the same time the values of density diminish along the length of the heated part and the flow accelerates still further. Coefficient of turbulent diffusivities of momentum remaining rather small. The region of maximum values of heat capacity together with t_m moves towards the axis of the pipe. In the near wall layer values C_p drop considerably. At the same time the coefficient of turbulent diffusivities of heat increases considerably, the values of the local heat flux also grow and the wall temperature drops.

Thus the changes of the main flow and heat transfer characteristics in herent in "deteriorated" regimes indicate that "deterioration" of heat transfer, occurring under small influence of buoyancy, evidently, is strongly determined by the axial acceleration and decrease of the turbulent exchange connected with it.

Fig. 4 presents a comparison of calculated and experimental wall temperature distributions coefficients of hydraulic resistance and frictional resistance for a "deteriorated" regime. As you can see, the coefficients of frictional resistance for this regime are over two times lower than the values of frictional resistance coefficients for constant physical properties. Besides, they are 3 - 4 times lower than the coefficients of hydraulic resistance. In indicates that the main part of pressure drop under these conditions is spent on the flow acceleration.

In case of the downward liquid flow and the domineering influence of thermal acceleration the main characteristics of the flow and heat transfer change just as is in the upward liquid flow. With the heat flux being increased under other equal parameters the acceleration of the flow is considerable: the profile of longutudinal velocity component is at the maximum; the eddy diffusivity of momentum coefficient and the eddy diffusivity of heat coefficient near the wall decrease ; the temperature profiles become more filling; the values of shear stresses and effective heat flux in the area of "deteriorated" heat transfer decrease tangibly. However in the case of carbon dioxide flow regimes some influence of buoyancy was marked. It manifested itself in the value of momentum coefficient S_{M} in downward flow being lower than in upward flow. In particular, the values of momentum coefficient S_{M} in the downward flow were lower than for a similar regime in the upward flow. The eddy diffusivity of momentum coefficient in the case of downward flow was somewhat higher, while the wall temperature was a little lower than in upward flow. Thus, in this case there is a certain influence of buoyancy on heat exchange, although it remains rather small.

2.2. INFLUENCE OF BUOYANCY

In case of the turbulent flow of heat carriers of near critical parameters the influence of buoyancy on the flow and heat transfer is most significant due to considerable temperature - dependent density change. With the ration Gr_{ρ}/Re^2 increasing, the influence of buoyancy becomes more intrinsic, which leads to absolutely opposite results in heat transfer behaviour for the upward and downward flow of liquid in vertical pipes.

Let's analyse the results obtained for the upward liquid flow in a heated vertical pipe. Figure 5-I shows the calculated profiles of physical properties, velocity, coefficients of turbulent exchange, shear stress and heat flux for the regime, in which according to predictions [3] the flow and heat exchange are influenced by acceleration and buoyancy. Besides, this Figure also presents the results of model calculation, carried out for the same boundary conditions without taking into account the influence of buoyancy in the momentum equation and the equation for the coefficient of eddy viscosity. Judging by the comparison of the calculation results the influence of buoyancy on the eddy diffusivity of momentum coefficient, velocity profiles and other characteristic is very perceptible. At the same time the calculation results, disregarding buoyancy in the calculation model, show that the flow and heat transfer are under the influence of the flow axial acce-

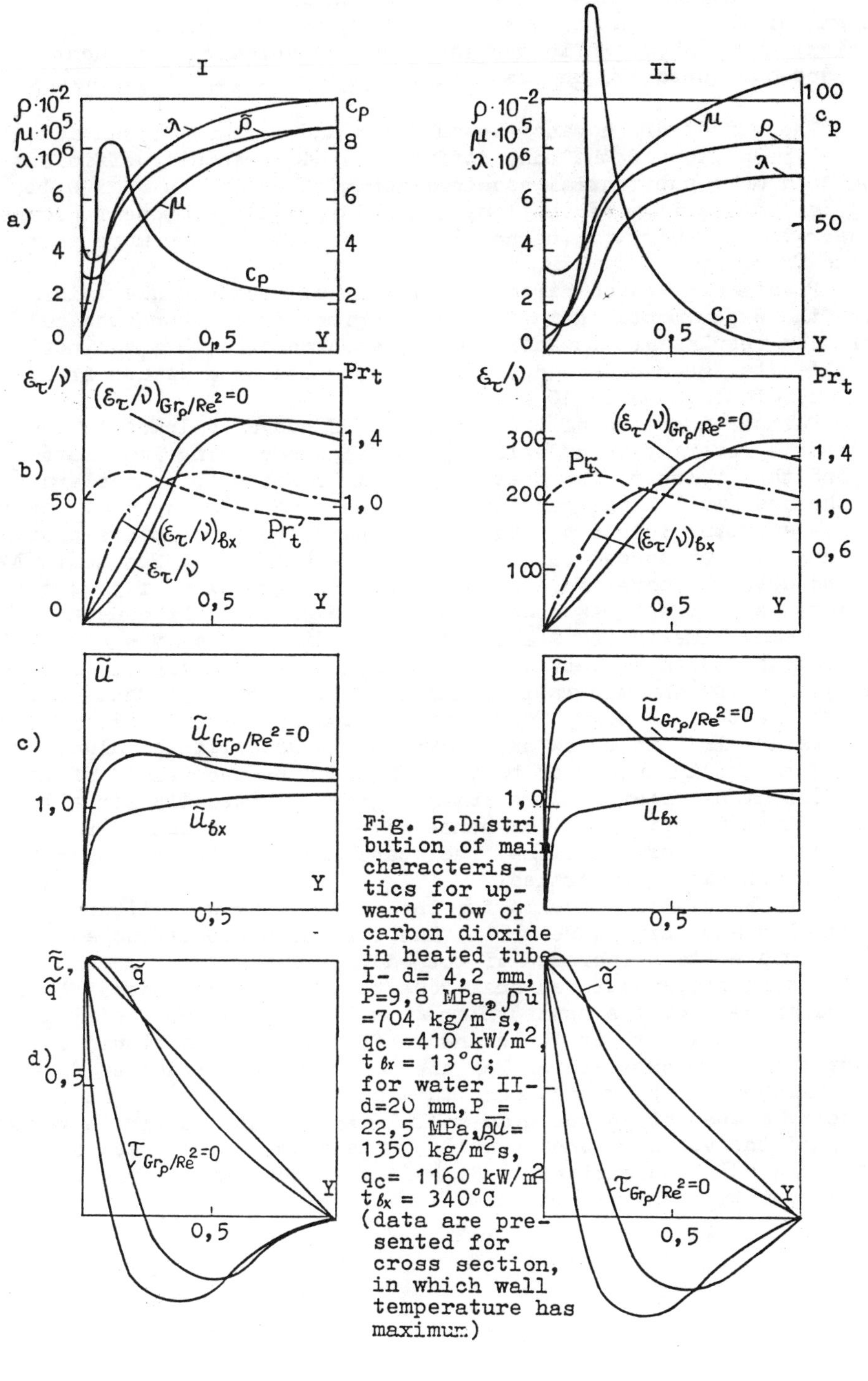

Fig. 5. Distribution of main characteristics for upward flow of carbon dioxide in heated tube I- d= 4,2 mm, P=9,8 MPa, $\overline{\rho u}$ =704 kg/m^{2}s, q_c =410 kW/m^2, t_{bx} = 13°C; for water II- d=20 mm, P = 22,5 MPa, $\overline{\rho u}$= 1350 kg/m^{2}s, q_c= 1160 kW/m^2 t_{bx} = 340°C (data are presented for cross section, in which wall temperature has maximum)

leration, occurring as a result of connected with liquid density dicrease in the near wall region.

The results of the model calculation show that along the length of the heated part the eddy diffusivity of momentum coefficient diminishes, while the longitudinal velocity component increases considerably, so the maximum is formed in section $x/d = 60$.

The effect of buoyancy leads first and foremost to an additional reduction of the eddy diffusivity of momentum coefficient and to a more substantial acceleration of flow in the near wall region. The maximum of the longitudinal velocity component occurs closer to the entry of the heating region and is of a more sharp acute form.

Figure 6 gives compared calculated wall temperature values with the experimental data for this regime. As is shown the calculation results are in satisfactory accordance with the experimental data. Mean square deviation of the measured values from the calculated ones is 10 - 12 per cent.

Further increase of the influence of buoyancy leads to considerable reformation of velocity profiles occurring immediately beyond the inlet to the heated part. As considerable acceleration of the near wall layer caused by the influence of buoyancy, occurs at the starting length of the thermal part the temperature profiles are not yet formed (do not manage to be formed). The heat flux values near the core are close to zero. The profiles of physical properties in the tubes cross section are practically constant. At the same time there is a considerable dicrease of the eddy diffusivity of momentum coefficient, caused by the influence of buoyancy, and the maximum wall temperature is reached at the starting length the thermal part. Such maximums can occur when the wall temperature is both below and above the pseudocritical value.

With the influence of buoyancy increasing the values of the coefficient of frictional resistance grow. Practically for all the regimes, in which the influence of buoyancy was considerable, the coefficient of frictional resistance was greater than for the constant physical properties.

The calculation results for the downward flow of liquid in a vertical heated pipe have shown that the influence of buoyancy results in a certain increase of the eddy diffusivity of momentum coefficient compared to an analogous regime for the upward flow. And if in case of the upward flow the wall temperature distribution reaches its maximum, in the downward flow we have monotonic change of the wall temperature. For the downward heat transfer flow the influence of buoyancy leads to a decrease of the longitudinal velocity component in the near wall layer and at a definite value Gr_ρ / Re^2 the velocity on the wall equals zero. Under these conditions the main assumptions leading to the equations of the boundary layer type are not valid.

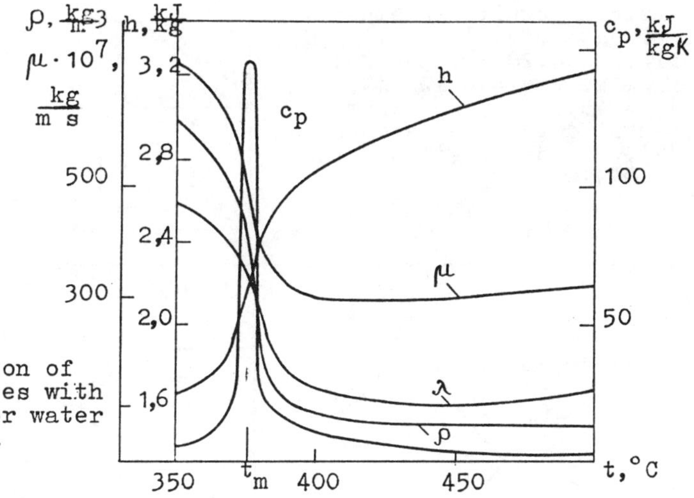

Fig.1. Variation of fluid properties with temperature for water at P= 24.5 MPa

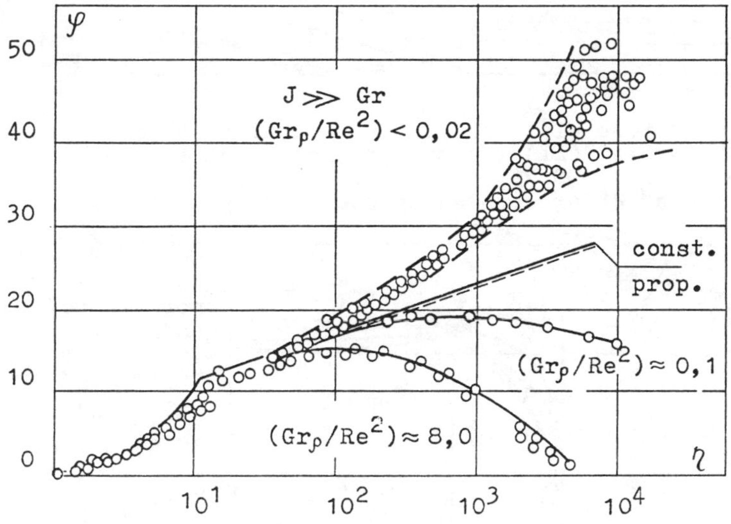

Fig. 7. Universal velosity profile of flow carbon dioxide in the pipe

Fig. 6. Comparison of numerical results to experimental data of [5]; CO_2, P=9,8 MPa, d= 4,28 mm, $\overline{\rho u}$= 704 kg/m^{2}s, q_c= 410 kW/ m^2, $t_{вх}$ = 13,3°C ; 1 - experimental data, 2 - calculation [7], 3 - calculation with ε_t/ν [10]

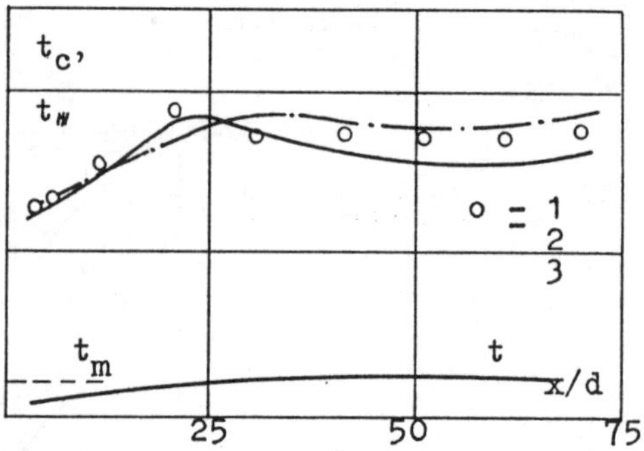

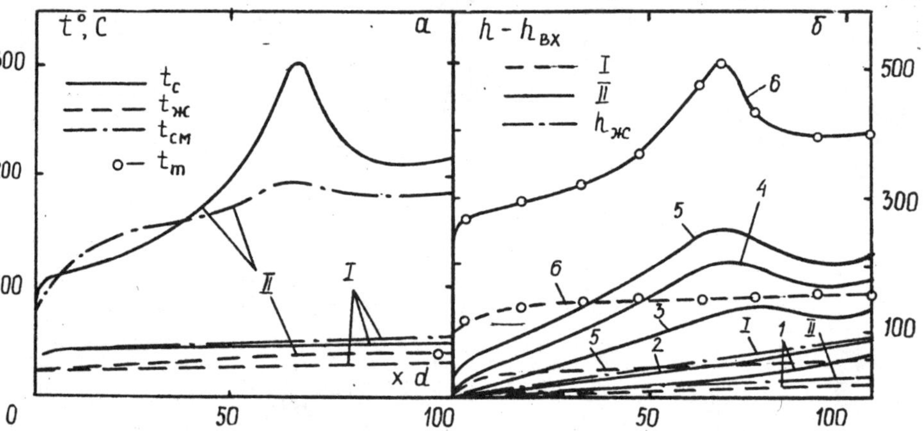

Fig. 8. Distributions of t_c and t (a), enthalpy (b) in "normal" (I) and "deteriorated"(II) regimes of heat transfer; 1- R= 0, 2 - 0,7, 3 - 0,94, 4 - 0,99, 5 - 0,997, 6 - 1,0 ; t_c - calculation by equations [12] for constant properties

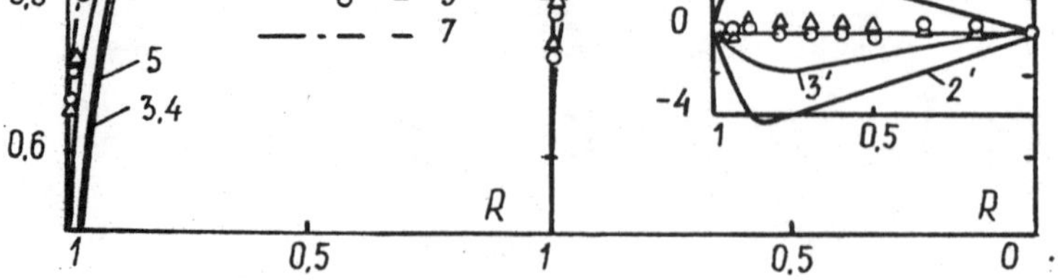

Fig.9. Profiles of $\rho u/\overline{\rho u}$, $u/u_ж$ and $\rho v/\overline{\rho u}$ in normal(points) and "deteriorated" (lines) regimes of heat transfer; 1 - x/d= 2,5; 2 - 32,5; 3 - 62,5; 4 - 77,5; 5 - 82,5; 6 - 107,5; 7 - Reichardt's profile for constant properties

2.3. VELOSITY PROFILE IN UNIVERSAL COORDINATES

The results of systematic calculations for the upward flow of the heat carrier made it possible to obtain distribution of velocity, temperature and other characteristics of flow within a wide range of determining parameters. The velocity profiles in universal coordinates are presented in Figure 7. Under small values of Gr_ρ/Re^2 in the regimes, meeting the condition $\mathcal{J} \gg Gr$ (in which $\mathcal{J}$ - is a parameter of acceleration, offered in [3]), the data for $\eta < 2 \cdot 10^3$ are well grouped along one and the same curve. With Gr_ρ/Re^2 increasing the influence of buoyancy becomes stronger, the liquid velocity in the core reduces, the minimum occurs on the tube axis, whereas the area between the wall and the axis is characterised by maximum. However all the calculation data under $\eta < 50$ coincide with the universal velocity profile for the constant physical properties.

3. EXPERIMENTAL RESULTS

Figures 8 - 12 show the experimental data on the stream structure and turbulent transfer in the "normal" and "deteriorated" regimes of heat transfer for the upward flow of carbon dioxide of supercritical parameters (P = 9 MPa, P/P_k = 1,23) with q_c=const in a heated vertical pipe of diameter d = 22,7 mm. In these regimes $\overline{\rho u}$ = 2100 kg/m^{2}s, h_{bx} = 552 kJ/kg (h_m = 640 kJ/kg) and $q_c/\overline{\rho u}$ = 0,079 and 0,205 kJ/kg.

The maximum values of the free convection parameter Gr_ρ/Re^2 are 0,02 and 0,026 respectively. A slight difference of these values indicates that free convection in itself is not the only reason for deteriorating heat transfer; in this case its role comes to amplifying the effects of the flow thermal acceleration. As is shown in Figure 8a, the wall temperature in the deteriorated regime of heat exchange passes through the maximum, around which convective heat exchange is 40 per cent lower than in formula [II] for "normal" heat transfer. Figure 8b shows the entalphy fields in these regimes. In case of normal regime of heat exchange (the dotted lines) beyond the beginning of the thermal part ($x/d > 20$) the flow is heated unifomly across the whole section, so that $dh/dx \approx dh_m/dx$, the same as in the case of stabilized heat transfer for the constant properties [12]. In the "deteriorated" regime of heat exchange the entalphy field on the given length of the pipe ($l/d \leqslant 110$) does not yet come to a stabilized condition. The part of the pipe from the inlet to maximum t_c is characterized by a higher rate of heating of the near wall flow layers (compared with the change of bulk entalphy of the flow) and slower heating of the core. Quite the reverse happens on the part beyond maximum t_c, where the "cooling" of the near wall layers is accompanied by the fast heating of the flow central part. These changes of the entalphy field lead to considerable fluctuations of heat transfer along the length of the pipe.

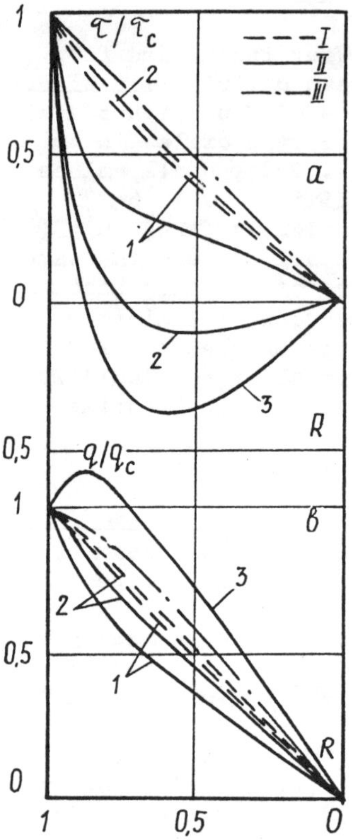

Fig. 10 •

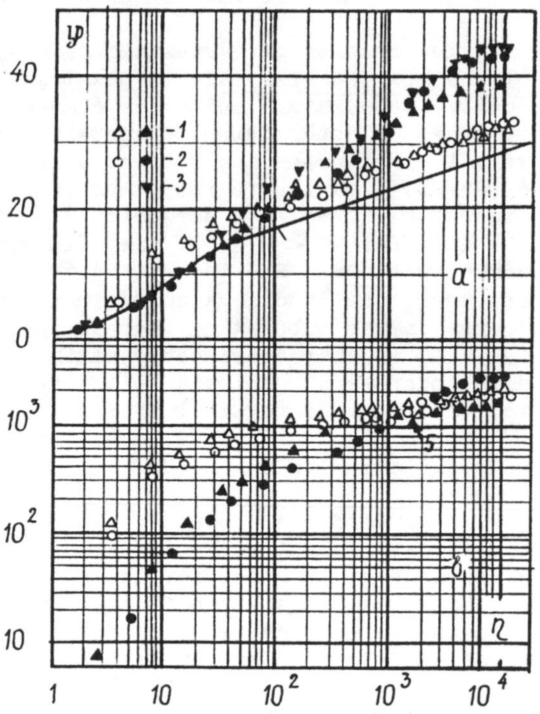

Fig. 11 .

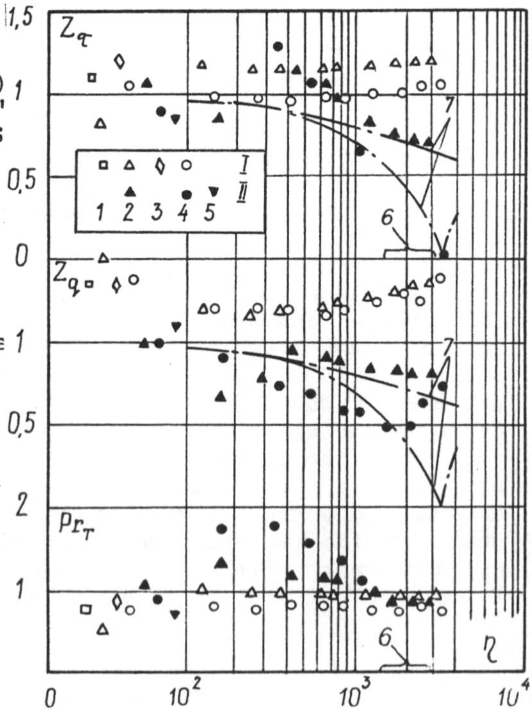

Fig. 12.

Fig.10 Distributions of τ/τ_c (a) and q/q_c (b) with R in "normal" (I) and "deteriorated" regimes; 1- x/d=32,5; 2- 67,5(I) and 62,5(II);3-92,5; III- for constant properties

Fig. 11. Velocity profiles (a) and enthalpy resistance (b) in "normal"(I) and "deteriorated (II) regimes; 1- x/d=32,5;2 - 67,5(I) and 62,5(II);3- 92,5 .

Fig.12. Turbulent diffusivities and Pr_t in "normal"(I) and "deteriorated"(II) regimes;1 - x/d=17,5;2- 32,5 ; 3 - 52,5; 4 - 67,5(I) and 62,5(II); 5 - 77,5 ; 6 - region of negative values τ/τ_c ; 7 - by equation [9] .

The data in Figure 9 indicate principal distinctions between the velocity fields in the "normal" and "deteriorated" regimes of heat transfer. In the first case velocity profiles U/U_m retain convex form with a maximum on the axis of the flow, which changes a little along the length of the pipe far from the inlet. The values of radial bulk velocity are low; profile $\rho u/\overline{\rho u}$ under the examined conditions ($h_{bx} < h_m$) is almost the same as for the isothermal flow. In the "deteriorated" heat transfer regime velocity profiles U/U_m are deformed noticeably along the length of the pipe and at a certain distance from the inlet acquire the M - shape. The part of the pipe in which the M-shaped velocity profile is formed, coincides with the region of deteriorated heat transfer, whereas the sharpest velocity maximums occur in the region of maximum t_c. Considerable changes of fields h and ρ lead to noticeable destortions of mass flow rate profiles $\rho u/\overline{\rho u}$, and to heavy radial mass transport.

Figure 10 presents distributions of shear stresses and heat fluxes along the radius of the pipe, typical for "normal" and "deteriorated" regimes. It in clear that in the "normal" regime of heat transfer (the dotted lines) distributions τ/τ_c and q/q_c remain close to the distributions common under stabilised flow and heat transfer [12]. In case of "deteriorated" heat transfer (the solid lines in Figure 10) profile τ/τ_c changes heavily in comparison with the linear law. The reason of these deformations is explained by the joint impact of thermal acceleration and buoyancy. In the region of deteriorated heat transfer ($32,5 < x/d < 62,5$) distributions τ/τ_c are located between curves I and 2 Figure 3a, i.e. they have a large area of low, near-zero values, with curve τ/τ_c passing through zero occurs in the vicinity of velocity maximum. Further on profiles τ/τ_c occurs a region of considerably large in absolute value negative shear stresses. Such profiles are typical for the region of improved heat transfer beyond maximum t_c. As is shown in Figure 10b, distributions q/q_c are also heavily deformed in the regime of "deteriorated" heat transfer: values q/q_c are lower in the region of deteriorated heat transfer (curves I and 2) than under stabilized conditions whereas immediately beyond maximum t_c they are higher.

Figure II presents analysed experimental velocity profiles and entalphies in universal coordinates. Values $v_* = \sqrt{\tau_c/\rho}$ and $\eta = y v_*/\nu$ are determined for local μ and ρ. Figure IIa indicates that in case of the "normal" regime of heat exchange the velocity profile in $\varphi(\eta)$ coordinates has a form with the logarithmic core common for gradientless flows. Considerable deviations of experimental points from Karman universal dependence [12] in the laminar and transition regions are attributed to a heavy change of viscosity in the laminar sublayer, where transition through t_m takes place. The velocity profile in the "deteriorated" regime of heat transfer for $\eta < 30$ practically coincides with universal, however deviating from it upwards in the core of the flow. Figure IIb illustrates the structure of thermal (entalphy) flow resistance in various regimes of heat transfer. Thermal resistance equal to $1/St_h$ is defined as

$$2\int_c^1 \varphi \overset{+}{h} R dR = \overline{\rho u}(h_c - h_м)/q_c = 1/St_h \qquad (1)$$

where St_h - is Stanton number. In case of "normal" heat transfer (it becomes clear, from Figure IIb), the thermal flow resistance is to a great extent determined by the laminar and transition regions of flow, and in the turbulent core value φh^+ increases rather insignificantly. However in case of "deteriorated" heat transfer (curve 2 in Figure IIb) value φh^+ in the turbulent core ($\eta > 50$) increases more than 10-fold with the largest growth occuring in the zone of velocity maximum and near zero values of τ/τ_c.

Figure 12 demonstrates the experimental values of the turbulent transfer coefficients

$$\varepsilon_\tau/\nu = \tau/(\mu|\partial u/\partial r|) - 1 \; ; \; \varepsilon_q/\nu = q \cdot c_p/(\lambda|\partial h/\partial r) - 1 \quad (2)$$

referred to values $(\varepsilon_\tau/\nu)_R$ and $Pr\, \varepsilon_\tau/\nu)_R$, difined according to Reichardt model [10]. With $\eta > 100$ local values are given for $R > 0,75$, and with $\eta < 100-200$ (for $R \geqslant 0,99$) - mean values, calculated by integrating the ratios (I) by the experimental values of U and h up to the coincidence of the calculated and experimental values of U and h for $R = 0,99$.

Figure 12, regardless of the spread of experimental points, shows that in case of "normal" heat transfer the turbulent transfer is very well described by the ratios (I), and for ε_τ and ε_q it in possible to use classical Reichardt model for $Pr_t \approx 0,9$. To create a certain t_c reserve it in possible to disregard some decrease of Pr_t in the near wall zone which becomes especially noticeable when t_m - is in the laminar sublayer. In the regime of "deteriorated" heat transfer a dicrease of the heat transfer coefficients becomes distinct in that part of the turbulent core of flow, where values τ and radial velocity gradients have minimal values. All this prompts us that first of all, the character of deteriorated heat transfer should be connected with disturbed upset generation of turbulent energy which occurs as a result of velocity fields and shear stresses being deformed due to the influence of acceleration and buoyancy. Account of the influence of profile τ/τ_c deformations by the method offered in [9], in itself helps to provide a qualitative description of the experimental data (see curves 7). Figure 12 also shows that in case of considerable influence of buoyancy description of momentum transfer on the basis of the "gradient model" in the region of velocity maximums becomes very difficult.

4. CONCLUSION

The calculation and experimental results quoted above have shown as follows:

1. The calculation results for heat transfer and resistance based on simple dependences for the turbulent diffusivities coefficient and turbulent Prandtl number are in satistactory accordance with the experimental data, obtained under the conditions of the small influence of buoyancy and thermal acceleration.

2. Under negligible influence of buoyancy on the characteristics of turbulent transfer the maximal acceleration of flow plays a certain role as it leads to the dicrease of the eddy diffusivity of momentum and heat coefficients which in its turn determines "deterioration" of heat transfer.

3. The influence of buoyancy occurs even for large Reynolds number, where the influence of thermal acceleration is the determinant feature.

4. Under the determining influence of buoyancy their effects in so significant that it causes heavy reformation of velocity profiles immediately beyond the inlet to the heated part, decrease (reduction) of the eddy diffusivity of momentum coefficient and "deterioration" of heat transfer under these conditions is caused by the influence of buoyancy. The coefficient of frictional resistance in these regimes exceeds the values of the coefficient of resistance for constant physical properties several times.

5. In the region of "deteriorated" heat transfer the profiles of shear stress and heat flux differ greatly from the linear distribution both in the case of influence of buoyancy and thermal acceleration of flow.

6. The velocity distribution, presented in universal coordinates in which the property values were determined by a local temperature, indicate that the data meeting the conditions of influence of thermal acceleration are well grouped with one another, and lie above the universal velocity profile for constant physical properties.

7. The numerical theoretical and experimental data fully coincide in that under the "normal" and "deteriorated" regimes of heat exchange there are essential distinctions (differences) in the structure of the averaged flow - velocity fields and shear stresses, occuring under the influence of additional factors of the process-thermal acceleration and buoyancy.

If in case of "normal" heat transfer the stream structure of flow and turbulent transfer remain close to the values relevant to developed flow for constant properties, whereas in the zone of "deteriorated" heat transfer M-shaped velocity profiles develop against the background of dicreased shear stresses, which in the aggregate leads to the formation in the turbulent core of the flow of large zones with small generation of turbulent energy and small turbulent exchange hindering heat transfer from the wall to the central parts of the flow.

NOMENCLATURE

c_p - specific heat at constant pressure, (kJ/kgK);
d - tube diameter, (m);
h - specific enthalpy, (KJ/kg);
h_{bx} - enthalpy of fluid at entry, (KJ/kg);
h_m - pseudocritical enthalpy (KJ/kg);
h^+ - universal enthalpy, $h^+ = (h_c - h)\ V_x/Q_c$;
P - pressure, (mPa);
P_k - critical pressure, (mPa);
q - local value of total heat flux in transverse direction, (kW/m^2);
q_t - local value of turbulent heat flux, (kW/m^2);
q_c - heat flux at the wall, (kW/m^2);
r - radial coordinate, (m);
r_t - radius of pipe, $r_t = d/2$, (m);
R - dimensionless radial coordinate, $R = r/r_t$;
S_M - coefficient of momentum, $S_M = 2\int_0^1 \rho_M \rho u^2/(\overline{\rho u})^2 RdR$
t - temperature, (^oC or K);
t_{bx} - temperature of fluid at entry, (^oC);
t - bulk temperature, (^oC);
t_c - wall temperature, (^oC);
t_m - pseudocritical temperature, (^oC);
u - X-direction component of velocity;
u - bulk velocity, (m/S);
u_{bx} - mean velocity at entry, (m/S);
ρu - mass flow rate, (kg/S);
U - dimensionless X-direction velocity, $U = u/u_{bx}$;
v - r-direction component of velocity, (m/S);
v_* - $\sqrt{\tau_c/\rho}$,
V - dimensionless r-direction velocity, $V = v/u_{bx}$;
x - axial coordinate, (m);
y - transverse coordinate measured from wall, $Y = r_t - r$;
Z_τ - turbulent diffusivities of momentum, in reference to $(\varepsilon_\tau/\nu)_R$ as defined by equation [10] $(\varepsilon_\tau/\nu)/(\varepsilon_\tau/\nu)_R$;
Z_q - turbulent diffusivities of heat, $(\varepsilon_q/\nu)/Pr(\varepsilon_q/\nu)_R$;
β - thermal expansion coefficient, $\bar{\beta} = 1/\tilde{\rho}(\rho_M - \rho_c)/(t_c - t_M)$;
$\varepsilon_\tau, \varepsilon_q$ - turbulent diffusivities of momentum and heat;
ζ - hydraulic coefficient, $\zeta = -4r_t dP/dX \cdot \rho_M/(\overline{\rho u})^2$
μ - dynamic viscosity, (kg/mS);
ξ - friction coefficient,
ρ - density, (kg/m^3);
$\tilde{\rho}$ - density at reference temperature, $t = (t_c + t_M)/2$, (kg/m^3);
ν - kinematic viscosity, $\nu = \mu/\rho$ (m^2/S);
λ - thermal conductivity, (kW/mK);
τ - local value of total shear stress, (N/m^2);
τ_c - wall shear stress, (N/m^2);

η - universal parameter, $\eta = y \cdot \sqrt{\tau_c/\rho}/\mu$;
φ - universal velosity, $\varphi = u/v_*$
J - parameter of acceleration, $J = 4\frac{Re}{\overline{Pr}}\left(\frac{q_c d}{\lambda}\bar{\beta}\frac{\rho_{ж}}{\bar{\beta}}\right)$;
Re - Reynolds number, Re $= \rho u \cdot d/\mu$;
Pr - Prandtl number, Pr $= \mu \cdot c_p/\lambda$;
$\overline{Pr} = \mu \bar{c}_p/\lambda$, $\bar{c}_p = (h_c - h_{ж})/(t_c - t_{ж})$;
Pr_t - turbulent Prandtl number, $Pr_t = \varepsilon_\tau/\varepsilon_q$;
Gr - Grashof number, $Gr = g\beta q_c d^4/(\lambda_{ж} \nu_{ж}^2)$;
$Gr_\rho = g(\rho_{ж} - \rho_c)\rho_{ж} \cdot d^3/\mu_{ж}^2$;
St_h - Stanton number, $St_h = q_c/\overline{\rho u}(h_c - h_{ж})$.

SUBSCRIPTS

ж - value evaluated at bulk temperature t ;
c - -"- -"- wall temperature t_c;
0 - value at constant physical properties;
bx - value at reference temperature t_{bx};
R - as defined by equation [10].

REFERENCES

1. Petukhov B.S., Heat Transfer in single phase medium under supercritical conditions, Teplofizika Vysokikh Temperatur, vol. 6, No. 4, pp. 732-745, 1968.
2. Hall W.B., Jackson J.D., Heat Transfer Near the Critical Point, Proc. 6-th Int. Heat Transfer Conf., Toronto, Canada, vol. 6, pp. 377-392, 1978.
3. Polyakov A.F., On the Mechanism and Boundaries of Deteriorated Heat Transfer Regimes under Supercritical Pressure of Heat Carrier, Teplofizika Vysokikh Temperatur, vol. 13, No. 6, pp. 1210-1219, 1975.
4. Protopopov V.S., Generating Dependencies for Local Heat Transfer Coefficients for Turbulent Flow of Water and Carbon Dioxide under of Supercritical Pressure in Uniformly Heated Big Pipes, Teplofizika Vysokikh Temperatur, vol. 15, No.4, pp. 815-821, 1977.
5. Petukhov B.S., Kurghanov V.A., Ankudinov V.B., Grigoriev V.S., Experimental Investigation of Resistance and Heat Transfer for Turbulent Flow of Liquid under Supercritical Pressure, Teplofizika Vysokikh Temperatur, vol. 18, No.1, pp. 100-111, 1980.
6. Kurghanov V.A., Ankudinov V.B., Kaptilniy A.G., Experimental Investigation of Velocity Fields and Temperature for Upward Flow of Carbon Dioxide under Supercritical Pressure in Vertical Heated Pipe, Teplofizika Vysokikh Temperatur, vol.24, No. 6, pp. 1104-1111, 1986.

7. Petukhov B.S., Medvetskaya N.V. Calculation of Turbulent Flow and Heat Transfer in Heated Pipes for Single-Phase Heat Carriers of Near Critical Parameters, Teplofizika Vysokikh Temperatur, vo. 17, No. 2, pp. 343-350, 1979.
8. Schurr N.M., Sastry V.S., Shapiro A.B., A Numerical Analysis of Heat Transfer to Fluids near the Thermodynamic Critical Point including the Thermal entrance region, Journal Heat Transfer, vol. 98, No. 4, pp. 609 - 615, 1976.
9. Popov V.N., Belyaev V.M., Valuyeva E.P., Calculation of Heat Transfer and Resistance under Turbulent Regime of Gelium Flow in Round Pipe under Supercritical Pressure, Teplofizika Vysokikh Temperatur, vol. 16, No. 5, pp. 1018-1027, 1978.
10. Reichard H., Schoeck P.A., in Advances in Heat and Mass Transfer, ed. Hartnett, pp. 223-251, McCraw-Hill, New York, 1961.
11. Kurghanov V.A., Ankudinov V.B., Calculation of Normal and Deteriorated Heat Transfer in Pipes for Turbulent Flow of Liquids in the near critical Point, Teploenergetika, vol. 31, No. 6, pp. 53-57, 1985.
12. Petukhov B.S., Genin L.G., Kovalev S.A., Heat Transfer in Nuclear Energy Plants, pp. 407, Atomenergoizdat, Moskva,1974.

Part 7
ANALYTICAL THEORIES OF TURBULENCE

SUMMARY LECTURE

Behavior of Coherent Structures in the Wall Region by Dynamical Systems Theory

NADINE AUBRY
Sibley School of Mechanical
and Aerospace Engineering
Cornell University
Ithaca, New York 14853, USA

PHILIP HOLMES
Department of Theoretical and Applied
Mathematics and Department of Mathematics
and Center for Applied Mathematics
Cornell University
Ithaca, New York 14853, USA

JOHN L. LUMLEY
Sibley School of Mechanical
and Aerospace Engineering
Cornell University
Ithaca, New York 14853, USA

EMILE STONE
Department of Theoretical and
Applied Mathematics
Cornell University
Ithaca, New York 14853, USA

ABSTRACT

Objective analysis of experimental measurements indicates that there are recurrent streamwise rolls present in the wall region, at least in the quadratic mean sense. Representation theorems permit optimal expansion of the instantaneous velocity field in the wall region in terms of these streamwise rolls. Without involving ourselves in the question of the source of these rolls, we ask how they will behave dynamically. Severely truncating our system, and using Galerkin projection, we obtain a closed set of non-linear ordinary differential equations with ten degrees of freedom. The methods of dynamical systems theory are applied to these equations. Loss to unresolved modes is represented by a Heisenberg parameter. We find that for large values of the Heisenberg parameter (large loss), we obtain stable streamwise rolls having the experimentally observed spacing. For smaller values of the parameter, we have traveling waves (corresponding to cross-stream drift of the rolls); we also find a heteroclinic attracting orbit giving rise to intermittency; and finally a chaotic state showing ghosts of all of the above. The behavior is quite robust, much of it being due to the symmetries present. We have examined eigenvalues and coefficients obtained from experiment, and from exact simulation, which differ in magnitude. Similar behavior is obtained in both cases; in the latter case, the heteroclinic orbits connect limit cycles instead of fixed points, corresponding to cross-stream waving of the streamwise rolls. The bifurcation diagram remains structurally similar, but somewhat distorted. The intermittent behavior appears to mimic the bursting behavior observed in the wall region, although the production of high wavenumber Reynolds stress associated with bursting is limited by the truncation of the high wavenumber components. The role of the pressure term is made clear - it triggers the intermittent jumps, which otherwise would occur at longer and longer intervals, as the system trajectory is attracted closer and closer to the heteroclinic cycle. The pressure term results in the jumps occurring at essentially random times, and the magnitude of the signal determines the average timing. This clarifies the question of whether bursting scales with wall variables or with outer variables - evidently the structure of a burst scales with wall variables, while the time between bursts should scale in a complex way with both inner and outer variables. Several applications of this approach are suggested.

Supported primarily by the U. S. Office of Naval Research under the program Select Research Opportunity IV. Prepared for presentation at the Zoran Zaric Memorial Meeting, Dubrovnik, Yugoslavia, May 16-20, 1988. An extended version with full details of the analysis will appear in *The Journal of Fluid Mechanics* under the title The Dynamics of Coherent Structures in the Wall Region of a Turbulent Boundary Layer.

INTRODUCTION.

There has been much interest in recent years regarding the application of methods, ideas and models from dynamical systems and bifurcation theory to the study of turbulent fluid flow. The first paper in which an explicit link was made between strange attractors and turbulence is that of Ruelle and Takens [1971], although, unknown to those authors, Lorenz [1963] had proposed a three dimensional set of ordinary differential equations (ODEs) as a minimal model for two dimensional Bénard convection considerably earlier.

Recent work in this area, of which accounts can be found in collections such as Rand and Young [1981], Swinney and Gollub [1981] and Campbell and Rose [1983], has concentrated on "closed" flow systems such as the Rayleigh-Bénard or Taylor-Couette problems. Even in this situation, however, rigorous results have not gone beyond the first preturbulent bifurcations from steady, laminar flows. There is nothing close to a proof of Ruelle and Takens' conjecture that the Navier-Stokes equation has a strange attractor for a realistic flow geometry.

In view of this, the present study of an open flow: the fully developed boundary layer on a flat plate, might seem hopeless. As Monin [1978] has pointed out, low dimensional models cannot hope to account for the detailed, high wavenumber, spatially chaotic aspects of fully developed turbulence. Moreover in open flow systems, natural modal decompositions such as the Fourier modes of linear theory are lacking or of limited use in the inhomogeneous directions. Thus another expansion is necessary and a coherent structure approach seems to be worth investigating for a judicious determination of relevant modes when large eddies are experimentally seen to play an important role. Since the early work of Theodorsen [1952], which suggested the importance of large scale horseshoe-shaped vortices in turbulent transport in a turbulent boundary layer, flow visualization and conditional sampling techniques have revealed the existence of organized structures in many shear flows.

Nevertheless the extraction of deterministic features from a random, fine grained turbulent flow has been a challenging problem. Zilberman et. al. [1977] write: "there are no consistent methods for identification which are independent of the techniques and the observer" and "we cannot unambiguously define the signature of an eddy without a priori knowledge of its shape and its location relative to the observation station and cannot map such an eddy because we do not have a proper criterion for pattern recognition". In contrast, Lumley [1967] proposed an unbiased technique for identifying such structures. The method consists of extracting the candidate which is the best correlated, in a statistical sense, with the background velocity field. The different structures are identified with the orthogonal eigenfunctions of the proper orthogonal or Karhunen-Loève decomposition theorem of probability theory (Loève [1955]). This is thus a systematic way to find organized motions in a given set of realizations of a random field. The method applied here is optimal in the sense that the series of eigenmodes converges more rapidly (in quadratic mean) than any other representation.

The use of these modes for a low dimension dynamical system study requires a very fast convergence of the series. The method we propose here is limited in application to certain types of flows in which large coherent structures contain a major fraction of the energy. It has been demonstrated that axisymmetric turbulent jet mixing layers (Glauser et al. [1985]) and wall regions of turbulent boundary layers (Moin [1984], Herzog [1986]) belong to this group. In the wall region of a turbulent boundary layer, large eddies, which manifest themselves as streamwise streaks in visualizations, and are thought to be characteristic organized structures, are experimentally observed to be intermittent in time and space (in the streamwise direction). They are associated with a succession of events in which their sudden breakup corresponds to peaks in the turbulent energy production. These violent motions of the streaks, called *bursting events*, are also thought to be the basic phenomenon for turbulent energy transfer between the inner and outer regions of the layer, and hence they contribute significantly to the entire turbulence process. Despite the contribution of many researchers, the description of these eddies is still vague.

The time dependent behavior of these structures has also remained unclear. Experimental results obtained from flow visualization provide a qualitative description in terms of a quasi-cyclic succession of events which consists of a gradual lift up of the longitudinal, streamwise streaks, a sudden oscillation, a breakup or *bursting event*, a downward motion of high speed fluid towards the wall, called a *sweep event,* and a violent upward motion of low velocity fluid called an *ejection event*. This intense updraft is postulated to be closely related to the high local production of turbulent energy associated with the breakup of the streaks. For more details see Kline et al. [1967]. Quantitative estimates have been given for the bursting period, i.e. the average time between two consecutive bursting events at a given spatial location (Kline et al. [1967]). Many assumptions have been made to explain the bursting phenomenon. The most common is a secondary instability of an inflexional velocity profile in the updraft between the counterrotating eddies, but nothing has really been proved.

A major purpose of this study is to understand better the patterns of the wall region structures, their dynamics, the role they play in the bursting phenomenon and their importance in the turbulence production mechanism. Specifically, we will develop a model for the wall region of the boundary layer (from $x_2{+}=0$ to $x_2{+}=40$ in wall units (Tennekes and Lumley [1972])), using the proper orthogonal decomposition of Lumley [1967, 1970, 1980] in the direction normal to the wall, in which the flow is strongly non-homogeneous. In the streamwise and spanwise directions the flow is essentially homogeneous, and Fourier modes will suffice. Used in conjunction with Galerkin projection, the proper orthogonal decomposition yields an optimal set of basis functions in the sense that the resulting truncated system of ODEs captures the maximum amount of kinetic energy among all possible truncations of the same order. The method has obvious advantages over *a priori* decompositions, based on linear normal modes, but it does not appear to have been used before due to the difficulty of computing the proper orthogonal modes. For this one requires three dimensional autocorrelation tensors averaged over many realizations of the flow in question, data only obtainable from lengthy experiments and analyses or from detailed numerical simulations. In our case complete data is only available from experimental work in a glycerine tunnel (Herzog [1986]), although Moin [1984], has derived two-dimensional orthogonal modes from large eddy simulations. However, as we shall see, knowledge of the autocorrelation tensor, and use of the Navier Stokes equations, does allow one to uniquely determine the unsteady flow, in contrast to Cantwell's [1981] expectation.

Using this modal decomposition, we derive and study minimal truncations of various orders of the Navier Stokes equations in the wall layer. The modes neglected in the truncation are accounted for by a Heisenberg model and we include a feedback effect to account for the stabilizing role of the interaction between the mean (driving) velocity profile and the fluctuating velocity field in the wall region as disturbances grow. The final models are sets of 3-6 ODEs for the complex modal amplitudes. We study these equations analytically and numerically, and interpret their behavior in the light of dynamical systems theory. This study should been seen as preliminary, and substantial modelling and mathematical issues remain to be resolved, but the initial indications are very encouraging.

THE PROPER ORTHOGONAL DECOMPOSITION

Lumley [1967] proposed a method of identification of coherent structures in a random turbulent flow. An advantage of the method is its objectivity and lack of bias. Given a realization of an inhomogeneous, energy integrable velocity field, it consists of projecting the random field on a candidate structure, and selecting the structure which maximizes the projection in quadratic mean. In other words, we are interested in the structure which is the best correlated with the random, energy-integrable field. More precisely, given an ensemble of realizations of the field, the purpose is to find the structure which is the best correlated with all the elements of the ensemble. Thus we want to maximize a statistical measure of the magnitude of the projection, which can be given by the mean square of its absolute value. The calculus of variations reduces this problem of maximization to a Fredholm integral equation of the first kind whose symmetric kernel is the autocorrelation matrix. The properties of this integral equation are given by Hilbert Schmidt theory. There is a denumerable set of eigenfunctions (structures). The eigenfunctions form a complete orthogonal set, which means that the random field can be reconstructed. The coefficients are uncorrelated and their mean square values are the eigenvalues themselves. The Kernel can be expanded in a uniformly and absolutely convergent series of the eigenfunctions and the turbulent kinetic energy is the sum of the eigenvalues. Thus every structure makes an independent contribution to the kinetic energy and Reynolds stress.

The most significant point of the decomposition is perhaps the fact that the convergence of the representation is optimally fast since the coefficients of the expansion have been maximized in a mean square sense. The mean square of the first coefficient is as large as possible, the second is the largest in the remainder of the series once the first term has been subtracted, etc.

We have described here the simplest case, that of a completely inhomogeneous, square-integrable, field. If the random field is homogeneous in one or more directions, the spectrum of the eigenvalues becomes continuous, and the eigenfunctions become Fourier modes, so that the proper orthogonal decomposition reduces to the harmonic orthogonal decomposition in those directions. See Lumley [1967, 1970, 1980] for more details.

Application of the Proper Orthogonal Decomposition to the Shear Flow of the Wall Region

The flow of interest here is three dimensional, approximately homogeneous in the streamwise direction (x_1) and spanwise direction (x_3), approximately stationary in time (t), inhomogeneous and of integrable energy in the normal direction (x_2).

We want a three dimensional decomposition which can be substituted in the Navier-Stokes equations in order to recover the phase information carried by the coefficients. We have to decide which variable we want to keep. Time is a good candidate since we are particularly interested in the temporal dynamics of the structures. Such a decomposition is possible and we do not need a separation of variables in the eigenfunctions of the type $\phi(x,\omega) = A(\omega)\psi(x)$ (as suggested by Glauser et al. [1985]) if we do not use any decomposition in time and choose the appropriate autocorrelation tensor. The idea is to measure the two velocities at the same time and determine $\langle u_i(x_1,x_2,x_3,t)\, u_j(x'_1,x'_2,x'_3,t)\rangle = R_{ij}$. Since the flow is quasistationary, R_{ij} does not depend on time and nor do the eigenvalues and eigenfunctions. The information in time is carried by the coefficients $a^{(n)}$ which are still "stochastic", but now evolve under the constraint of the equations of motion. Thus the decomposition becomes

$$(1)\quad u_i(x_1,x_2,x_3,t) = \sum_n \int a^{(n)}_{k_1k_3}(t)\, e^{2\pi i(k_1x_1+k_3x_3)}\, \phi^{(n)}_{i_{k_1k_3}}(x_2)\, dk_1 dk_3$$

$$(2)\quad \int \Phi_{ij}(x_2, x'_2)\phi^{(n)}(x'_2)dx'_2 = \lambda^{(n)}\phi_i^{(n)}(x_2),$$

and we have to solve equation (2) for each pair of wave numbers (k_1,k_3). Φ_{ij} now denotes the Fourier transform of the autocorrelation tensor in the x_1, x_3 directions.

Our second change to the decomposition is a transformation of the Fourier integral into a Fourier series, assuming that the flow is periodic in the x_1 and x_3 directions. The periods

L_1, L_3 are determined by the first non zero wave numbers chosen. Finally, each component of the velocity field can be expanded as the triple sum

$$(3) \quad u_i(x_1,x_2,x_3,t) = \frac{1}{\sqrt{L_1L_3}} \sum_{k_1k_3n} e^{2\pi i(k_1x_1+k_3x_3)} a^{(n)}_{k_1k_3}(t)\phi^{(n)}_{i_{k_1k_3}}(x_2),$$

In this case, a "structure" is defined by:

$$\frac{1}{\sqrt{L_1L_3}} \sum_{k_1k_3} e^{2\pi i(k_1x_1+k_3x_3)} a^{(n)}_{k_1k_3}\phi^{(n)}_{i_{k_1k_3}}(x_2)$$

and the entire velocity field is recovered by the sum of all the structures (over n).

EXPERIMENTAL RESULTS

The candidate flow we are investigating is the wall region (which reaches $x_2+=40$) of a pipe flow with almost pure glycerine (98%) as the working fluid (Herzog [1986]). The Reynolds number based on the centerline mean velocity and the diameter of the pipe is 8750. The corresponding Reynolds number based on the shear velocity u_τ is 531. From this data the autocorrelation tensor at zero time lag (t-t' = 0 between the two velocities), $R_{ij}(x_1-x'_1, x_2,x'_2, x_3-x'_3)_{t-t'=0}$, was obtained and the spatial eigenfunctions were extracted by numerical solution of the eigenvalue problem. The results show that approximately 60% of the total kinetic energy and Reynolds stress is contained in the first eigenmode and that the first three eigenmodes capture essentially the entire flow field as far as these statistics are concerned. This very fast convergence of the decomposition in the near wall region is in good agreement with Moin's results [1984]. From a large eddy simulation data base, Moin uses the proper orthogonal decomposition successively in one and two dimensions in the wall region (up to $x_2+= 65$). His first structure contains 60% of the total kinetic energy and 120% of the Reynolds stress (this apparent paradox occurs because the contribution of higher order structures to the Reynolds stress is negative). 90% of the kinetic energy is captured by the first three terms.

THE DYNAMICAL EQUATIONS

We decompose the velocity - or the pressure - into the mean (defined using a spatial average) and fluctuation in the usual way. We substitute this decomposition into the Navier-Stokes equations. Taking the spatial average of these equations we obtain, in the quasi stationary case, an approximate relation between the divergence of the Reynolds stress and the mean pressure and velocity.

$$(4) \quad \langle u_{i,j}u_j \rangle = -1/\rho\, P_{,i} + \nu\, U_{,jj}\delta_{i1}.$$

This is substituted in the Navier-Stokes equations, giving an equation for the fluctuating velocity. Equation (4) may be solved to give The mean veloctity U in terms of the Reynolds stress $\langle u_1u_2 \rangle$ in a channel flow in a manner which gives some feedback to the system of equations as the fluctuation varies. We will see that this feedback is necessarily stabilizing for the first structure (according to the experimental results) and increases as the Reynolds stress gets stronger. In other words, this term controls the intensity of the rolls, by reducing the mean velocity gradient as the rolls intensify, thus weakening the source of energy.

The expansion of the Fourier transform $\hat{u}_i$ of the fluctuating velocity u_i, defined by

$$(5) \qquad u_i(x_1,x_2,x_3,t) = \frac{1}{L_1L_3}\sum_{k_1k_3} e^{2\pi i(k_1x_3+k_3x_3)}\hat{u}_i(k_1,k_3,x_2,t)\,),$$

is achieved by use of the complete set of eigenfunctions $\phi^{(n)}$'s in an infinite sum:

$$(6) \qquad \hat{u}_i(k_1,k_3,x_2,t) = \sqrt{L_1L_3}\sum_{m=1}^{\infty} a^{(m)}_{k_1k_3}(t)\phi^{(m)}_{i_{k_1k_3}}(x_2)\,.$$

Since we want to truncate this sum, we use a Galerkin projection which minimizes the error due to the truncation and yields a set of ordinary differential equations for the coefficients. After taking the Fourier transform of the Navier Stokes equations $N_i=0$ and introducing the truncated expansion, we apply Galerkin projection by taking the inner product

$$(7) \qquad \left(\phi^{(n)},N\right) = \int_0^{X_2} N_i(x_2)\phi_i^{(n)*}(x_2)dx_2 = 0.$$

Finally we obtain a set of ordinary differential equations of the form:

$$(8) \qquad A\frac{da^{(n)}_{\mathbf{k}}}{dt} = Ba^{(n)}_{\mathbf{k}} + \text{N.L.}$$

where A and B are matrices. Here A is the identity matrix (since the complete set of eigenfunctions is orthogonal) and N.L. are non linear terms. The non linear terms are of two sorts: quadratic and cubic. The quadratic terms come from the non linear fluctuation-fluctuation interactions and represent energy transfer between the different eigenmodes and Fourier modes. Their signs vary. The role of the Reynolds stresses $<u_iu_j>$ on these terms should be mentioned. They vanish for all wave number pairs except for $(k_1,k_3) = (0,0)$ for which they exactly cancel the quadratic term. Therefore they prevent this mode from having any kind of quadratic interactions with other Fourier modes. Since the cubic terms are zero too, the (0,0) mode just decays by action of viscosity and does not participate in the dynamics of the system.

The cubic terms come from the mean velocity - fluctuation interaction corresponding to the Reynolds stress $<u_1u_2>$ in the mean velocity equation (the other part of this equation leads to a linear term). Since the streamwise and normal components of the first eigenfunction have opposite signs, they make a positive contribution to the turbulence production and hence provide negative cubic terms which are thus stabilizing. We remark that this is not necessarily the case for higher-order eigenfunctions.

By use of the continuity equation and the boundary conditions

$$\phi^{(n)}_{2_{\mathbf{k}}} = 0 \text{ at } x_2 = 0 \text{ (at the wall),}$$

and

$$x_2 \to \infty, \; \phi^{(n)}_{2_{\mathbf{k}}} \to 0 \text{ (decay at infinity)},$$

it can be seen by integration by parts that the pressure term would disappear if the domain of integration covered the entire flow volume. Since this is not the case (rather the domain is limited to $X_2+ = 40$), there remains the value of the pressure term at the upper edge X_2 of the integration domain which represents an external perturbation coming from the outer flow.

ENERGY TRANSFER MODEL.

The exact form of the equations obtained from the decomposition, truncated at some cut-off point (k_{1c}, k_{3c}, n_c), does not account for the energy transfer between the resolved (included) modes and the unresolved smaller scales. The influence of the missing scales will be parameterized by a simple generalization of the Heisenberg spectral model in homogeneous turbulence. Such a model is fairly crude, but we feel that its details will have little influence on the behavior of the energy-containing scales, just as the details of a sub-grid scale model have relatively little influence on the behavior of the resolved scales in a large eddy simulation. This is a sort of St. Venant's principle, admittedly unproved here, but amply demonstrated experimentally by the universal nature of the energy containing scales in turbulence in diverse media having different fine structures and dissipation mechanisms (see Tennekes & Lumley, 1972 for a fuller discussion). The only important parameter is the amount of energy absorbed.

We begin by defining a moving spatial filter which removes from the total field the unresolved modes. The details of the definition are not important - it is sufficient to conceive of the possibility of such a filter. This filter is also an averaging operator. The velocity field may now be divided into the resolved and unresolved field by using this filter. The Reynolds stress of the unresolved field may now be defined as the average using our filter operator, of the product of the unresolved velocities; this acts on the resolved field. We suppose that the deviator of this Reynolds stress is proportional to the strain rate of the resolved field. We neglect the Leonard stresses, which essentially supposes that there is more of a spectral gap than really exists. This is what is done in the Heisenberg model, without ill effect. The way in which we are treating the effect of the unresolved modes on the resolved ones is very much like what is done in large eddy simulation, and is called sub-grid scale modeling; our model would probably be called a Smagorinsky model (there are minor differences in the definition of the equivalent transport coefficient). Let us agree to designate the resolved field as $u_{i<}$ and the unresolved field as $u_{i>}$, while an average of ϕ, say, over the unresolved modes (the filtering process) can be designated as $<\phi>_>$. Thus:

(9) $$\tau_{ij>} = -2\alpha_1 \nu_T S_{ij<}$$

with $\tau_{ij>} = <u_{i>}u_{j>}>_> - <<u_{i>}u_{j>}>_>>_> - 1/3\, \delta_{ij}(<u_{k>}u_{k>}>_> - <<u_{k>}u_{k>}>_>>_>)$

and $S_{ij<} = 1/2\,(u_{i<,j} + u_{j<,i})$. Here

$<$ denotes the sum over all the modes (k_1,k_3,n) such that $k_1 \leq k_{1c}$, $k_3 \leq k_{3c}$, $n \leq n_c$ and

$>$ denotes the sum over all the modes (k_1,k_3,n) such that $k_1 > k_{1c}$ or $k_3 > k_{3c}$ or $n > n_c$, where (k_{1c},k_{3c},n_c) is the cut-off mode. The characteristic scales of the parameter ν_T are those of the higher modes. We have introduced an explicit dimensionless parameter α_1, and will exclude adjustable constants from ν_T. By observation that the energy decreases rapidly with increasing n and **k** (Figure 1), we assume that these relevant scales are given by

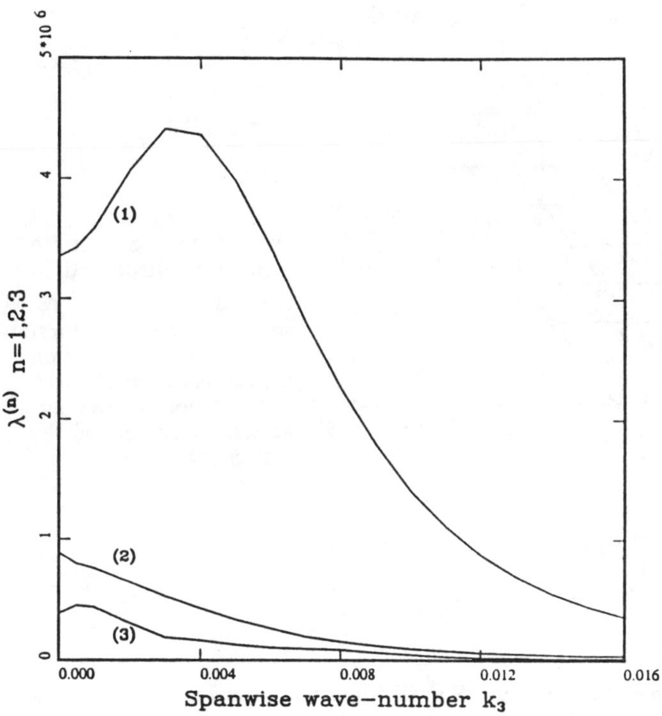

Figure 1. Convergence of the proper orthogonal decomposition in the near-wall region (x_{2+} = 40) of a pipe flow according to experimental data. Turbulent kinetic energy in the first three eigenmodes. $\lambda^{(n)}$(n=1, 2, 3) function of:the spanwise wave number (from Herzog [1986]).

characteristic scales of the <u>first</u> neglected modes. This is probably a good approximation as far as the eigenmodes are concerned since they are separated by large gaps in the spectrum and it is a reasonable assumption for the Fourier modes since the steps of our Fourier series are also large.

Finally, the parameter ν_T is taken equal to

$$(10) \qquad u_> l_> = \frac{\int_0^{X_2} \langle u_{i>} u_{i>} \rangle dx_2}{\left(X_2 \int_0^{X_2} \langle u_{i>,j} u_{i>,j} \rangle dx_2 \right)^{1/2}}$$

(where $u_>$ and $l_>$ are characteristic scales of the neglected modes). This can be expressed in terms of the eigenvalues and eigenfunctions of the first neglected modes in the three directions (see Figure 2.)

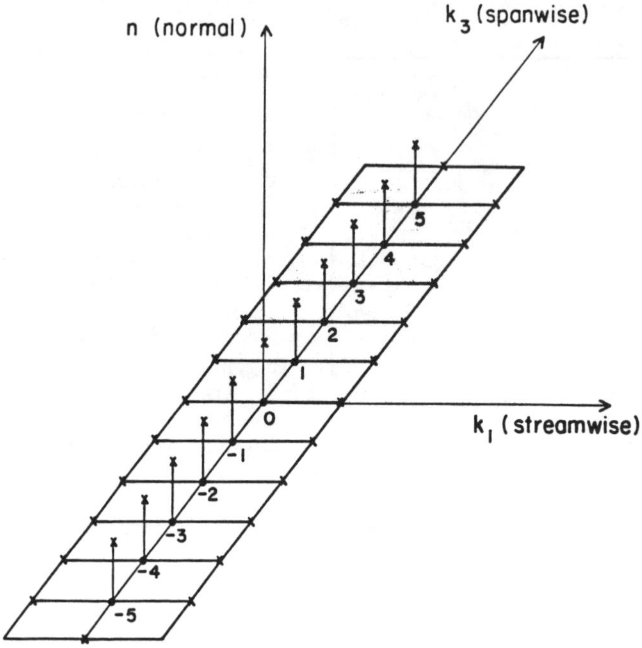

Figure 2. Inhomogeneous Heisenberg model applied to the specific truncation discussed in the text. Legend: • resolved modes x First neglected modes which are considered for the computation of the characteristic scales of the Heisenberg model.

We will refer to α_1 as a Heisenberg parameter. We will adjust α_1 upward and downward to simulate greater and smaller energy loss to the unresolved modes, corresponding to the presence of a greater or smaller intensity of smaller scale turbulence in the neighborhood of the wall. This might correspond, for example, to the environment just before or just after a bursting event, which produces a large burst of small scale turbulence, which is then diffused to the outer part of the layer.

A term $-1/3\delta_{ij}(<u_{k>}u_{k>}>_{>} - <<u_{k>}u_{k>}>_{>}>)$ appears in the equation for the resolved field. This term could be combined with the pressure term and would not have any dynamical effect if the integration domain covered the entire flow volume. In our case, it needs to be computed since, like the pressure term, it leads to a term evaluated at X_2. We assume that the deviation (on the resolved scale) in the kinetic energy of the unresolved scales is proportional to the rate of loss of energy by the resolved scales to the unresolved scales. This pseudo-pressure term gives some quadratic feed-back. The rate of loss of energy from the resolved scales to the unresolved scales is $2\alpha_1 \nu_T S_{ij<} S_{ij<}$, so that a free parameter appears also in this term. Because this approximation involves a further assumption, and to give ourselves greater flexibility, we call this parameter α_2, although in all work presented in this paper, we have set $\alpha_1 = \alpha_2$.

Thus the Heisenberg model introduces two parameters in the system of equations, one, α_1, in the linear term, the other one, α_2, in the quadratic term. The equations therefore have the following form:

(11) $$da^{(n)}_{k_1 k_3}/dt = L + (\nu + \alpha_1 \nu_T)L' + Q + \alpha_2 Q' + C$$

where L and L' represent the linear terms, Q the direct quadratic terms, Q' the quadratic pseudo-pressure term and C the cubic terms arising from the Reynolds stress.

SPECIFIC TRUNCATIONS.

This study has two conflicting requirements. On one hand we wish to keep as few modes as possible in order to obtain a low-dimensional system, permitting us to apply the techniques of dynamical system analysis. On the other hand, we would like to retain at least a qualitatively correct dynamical representation of the turbulence production phenomenon. A necessary condition for the second requirement consists of including as much of the energy and Reynolds stress as possible in our system. This is already satisfied in the inhomogeneous direction by the proper orthogonal decomposition itself, since the first structure is the most energetic. Given the energetic gap between the two first eigenfunctions (Figure 1), keeping only one structure seems quite reasonable. The choice of wave numbers is now of great importance, especially as far as the Reynolds stress is concerned (an important part of the Reynolds stress is contained in the higher modes. The best selection is probably the one for which the cut-off modes correspond to the experimental measurement cut-off. Since we retain only a few modes, the steps of our Fourier series are large. However, provided that the periodic length is larger than a few integral scales, the Fourier transform of the autocorrelation tensor is unchanged and the eigenfunctions are still the same.

We now seek a minimum truncation. The experimental results show that the ratio between the streamwise and spanwise characteristic length scales is of order ten (Figure 1). Our first approximation therefore neglects streamwise variations. We need at least three terms in the spanwise direction (see below). Thus the minimum truncation consists of one eigenmode, one streamwise wave number ($k_1 = 0$) and three (i.e. two active) spanwise wave numbers (0,k,2k). In this paper, up to six spanwise wavenumbers (0,..., 5k) will be considered. In this case k has the value 3×10^{-3} and the lengths of the periodic box are $L_1 = L_3 = 333$. Even the model having six spanwise Fourier modes is still very crude, although the truncation seems to be the one which contains an optimally large amount of the total energy among those of the same dimension. The zero cut-off mode in the streamwise direction in particular is a very rough approximation. Such a truncation causes a drop of the spanwise and normal root mean square values of the velocity which is particularly significant in the upper half of the layer. For this reason, we do not expect our velocity field reconstructed without rescaling to have more than qualitative significance.

We began with the smaller truncation, $k_1 = 0$ and (0,k,2k) in the spanwise direction, and one eigenmode. It must be noted that a certain level of complexity is necessary: such processes as vortex self-induction, vortex stretching and the like are non-linear processes, and we expect to need three active Fourier modes in order to reproduce even qualitatively these effects. Hence, we were not very surprised when our smaller truncation was somewhat lacking in dynamically interesting behavior. We then moved to five active cross-stream modes, which produced a rich dynamical behavior, which we will describe below. We are continuing to add modes, notably in the streamwise direction; this extension is non-trivial, and will be described in detail elsewhere. It is mentioned here simply to make the point that we understand the properties (notably the symmetries) of the equations responsible for the behavior observed; that the equations do not have these properties for lower order truncations, and that the properties are preserved in higher order truncations. Thus, we may expect the behavior observed at the level of truncation studied here to continue to have its counterpart in higher order truncations.

THE MODEL DIFFERENTIAL EQUATIONS.

As a preliminary approach, we study the set of equations for a truncation limited to the first eigenfunction (n=p=q=r=1), the zero streamwise wavenumber $k_1 = 0$ and up to six spanwise wavenumbers $k_3 = 0, k, ...5k$, for a suitably chosen k.

When only the zero streamwise wavenumber is considered, the equations become much simpler. Indeed, because of the symmetries of the eigenfunctions (Herzog, [1986]) in the (k_1,k_3) wavenumber plane, the first and second components are purely real and the third component is imaginary on the k_3 - axis (i.e. for $k_1 = 0$). Using these symmetries,

the equations for the complex modal coefficients $a^{(1)}_{k_1=0,k_3}$ can be readily derived. Letting

k_3 take the values jk, j = 0, 1, ..., 5 and writing $a^{(1)}_{k_1 = 0,\, k_3 = jk} = x_j + y_j$, a typical equation takes the form:

$$\text{(12)} \qquad dx_1/dt = a_1x_1$$

$$+ c_{-1,2}(x_1x_2 + y_1y_2) + c_{-2,3}(x_2x_3 + y_2y_3) + c_{-3,4}(x_3x_4 + y_3y_4) + c_{-4,5}(x_4x_5 + y_4y_5)$$

$$+ x_1(d_{11}(x_1^2 + y_1^2) + d_{12}(x_2^2 + y_2^2) + d_{13}(x_3^2 + y_3^2) + d_{14}(x_4^2 + y_4^2) + d_{15}(x_5^2 + y_5^2)$$

In these equations the zero Fourier mode $a^{(1)}_{00} = x_0+iy_0$ has been removed since its imaginary part is identically zero and its real part decays to zero under the influence of the viscosity ($dx_0/dt = a_0x_0$; $a_0 <0$). The coefficients of the quadratic and cubic terms are each the sums of two quantities: $c_{k',k-k'}$, $c_{k-k',k}$ and $d_{k,k'}$, $d_{k,-k'}$. The influence of the Heisenberg model appears in the linear and quadratic terms as follows:

$$a_k = a_k^1 + (1 + \alpha_1/Re_T)\, a_k^2$$

$$c_{k',k-k'} = c^1_{k',k-k'} + \alpha_2 c^2_{k',k-k'}$$

where α_1 and α_2 are the proportionality (Heisenberg) coefficients already mentioned. The computation of the nondimensionalized transport coefficient ν_T for this particular truncation gives the value $1/Re_T = 6.28$.

NUMERICAL SIMULATIONS AND FURTHER ANALYSIS OF THE MODEL ODES.

Numerical integrations of 3, 4, 5 and 6 mode models have been carried out, but we shall only report in detail on the 6 mode (5 active mode) simulations here.

We summarize here the behavior of the system. For $\alpha > 1.61$ a unique circle of globally attracting stable fixed points exists in the 2/4 subspace (with a window of instability between 2.0 and 2.3): the r_1,r_3,r_5 components, along with the trivial r_0 component, are zero. As α increases, the magnitude of the r_2 and r_4 components decreases until, at $\alpha \sim 2.409$, this non-trivial circle of fixed points coalesces with the origin, which for $\alpha > 2.409$ is the unique and globally attracting fixed point for the problem. For $1.37 < \alpha < 1.61$, an S^1-symmetric family of globally attracting double homoclinic cycles Γ exists, connecting pairs of saddle points which are π out of phase with respect to their second (x_2,y_2) components. The points $\mathbf{r}+$, $\mathbf{r}-$ discussed above are typical members of this family. The existence of the cycles Γ implies that, after a relatively brief and possibly chaotic transient, almost all solutions enter a tubular neighborhood of Γ and thereafter follow it more and more closely. As they approach Γ, the duration of the "laminar" phase of behavior (in which r_2 and r_4 remain non-zero and almost constant and r_1,r_3,r_5 grow exponentially in an oscillatory fashion) increases while the bursts (in which r_1,r_3 and r_5 collapse) remain short.

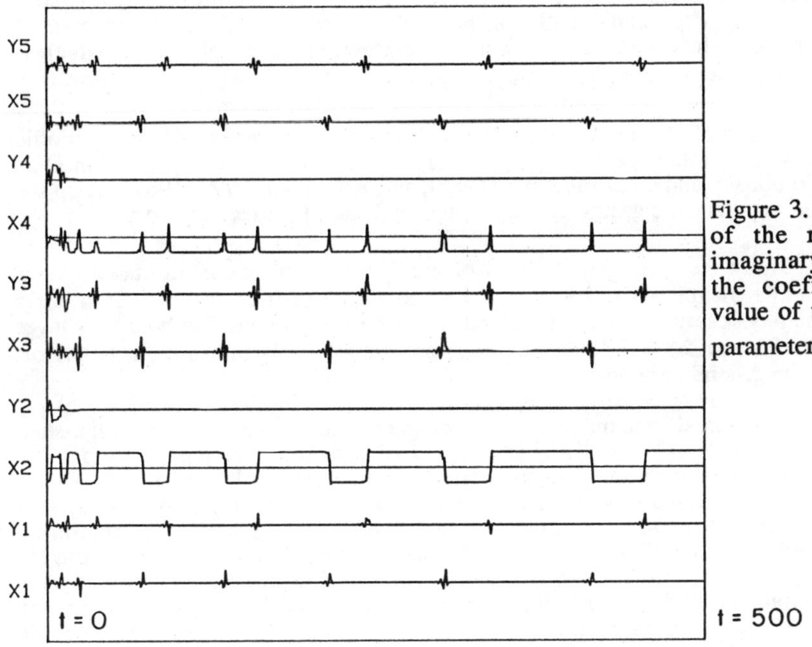

Figure 3. Time histories of the real (x_i) and imaginary (y_i) parts of the coefficients for a value of the Heisenberg parameter $\alpha = 1.4$;

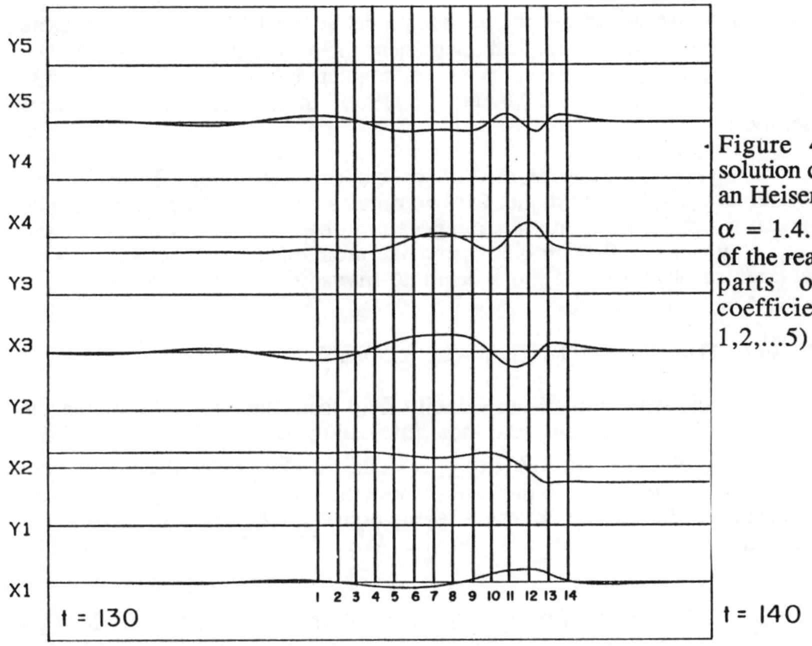

Figure 4. Intermittent solution corresponding to an Heisenberg parameter $\alpha = 1.4$. Time history of the real and imaginary parts of the model coefficients x_i, y_i ($i = 1,2,\ldots 5$)

In an ideal, unperturbed system, the laminar duration would grow without bound, but small numerical perturbations, such as truncation errors, presumably prevent this occurring in our numerical simulations. More significantly, the pressure perturbation will limit the growth of the laminar periods. Thus there is an effective maximum duration of events, which is reduced as α is decreased from the critical value $\alpha_b \sim 1.61$.

The existence of some of these solutions - notably the traveling waves and quasiperiodic motions - follow from rather general group theoretic considerations, cf papers on O(2) symmetry in Golubitsky and Guckenheimer [1986], and Armbruster *et al* [1987]. Also see § 13. We intend to pursue the group theoretic issues in detail in subsequent work.

For the present study the intermittent behavior exhibited by the six mode model for α between 1.35 and 1.61 appears to be of greatest interest, since it corresponds in a fairly clear way to the physical instability, sweep and ejection event observed in boundary layer experiments (cf §1 and see §12, below). We now describe this intermittent behavior, starting with some general remarks.

In the theory of dynamical systems three types of intermittency have been distinguished (Pomeau and Manneville [1980]). They are associated with solutions repeatedly passing close to a weakly unstable fixed point or periodic orbit. The solution spends a long "laminar" phase near the point or orbit until it reaches a critical amplitude and a brief turbulent "burst" ensues, in which it travels far and fast in phase space before returning. (These terms were appropriated by the dynamical systems community and have been used in a predominantly metaphorical fashion thus far). The laminar phase is governed by the linearized dynamics near the fixed point or periodic orbit, but the burst and return are associated with a "global reinjection mechanism", usually a homoclinic orbit or heteroclinic cycle (Tresser *et al* [1980], Silnikov [1965, 1968, 1970], Tresser [1984], Sparrow [1982], Guckenheimer & Holmes [1983, §6.5].)

While evidence of intermittency has been detected before in fluid systems and models (Pomeau & Manneville [1980], Bergé *et al* [1980], Maurer & Libchaber [1980], Dubois *et al* [1983]) there has been little evidence of type II, associated with a subcritical Hopf bifurcation, which is what we observe in the present model at $\alpha = 1.61$. We will therefore explore the connection between this dynamical phenomenon and its analogue in the turbulent boundary layer in some detail. Our model displays a "regular" intermittency, in contrast to the chaotic intermittency of Pomeau and Manneville [1980], in which event durations are distributed randomly.

As this paper was being completed we learned of work in progress on the "Kolmogorov-Sivashinsky" equation: a scalar field model for two dimensional turbulence (Nicolaenko [1986]). Remarkably similar bifurcation sequences are observed and symmetries and homoclinic cycles play roles in that problem very like their roles in the present model. Related observations of solutions of the Kuromoto-Sivashinsky equation have been published (Hyman et al. [1985, 1986]).

IMPLICATIONS FOR THE FLOW IN THE WALL REGION.

We describe here the reconstructions of the velocity field from the expansion, using the computed values of x_i and y_i. Probably the most interesting sets of solutions are those exhibiting intermittency, obtained for $1.3 \leq \alpha \leq 1.61$. In the flow field, the rapid event which follows the slowly growing oscillation and the repetition of the process reminds one of the bursting events experimentally observed (Kline et al. [1967], Corino and Brodkey [1969], etc.). For that reason we call it a "burst". We will analyse its effect on the streamwise vortices.

In Figure 4 we show an enlargement of the time histories of the modal coefficients during a burst for $\alpha = 1.4$ (cf. Figure 3). A description of the motion of the eddies during the burst is given in Figure 5 for $\alpha = 1.4$ by plotting u_2 and u_3 at the different times indicated on Figure 4. Before and after the event, two pairs of streamwise vortices are present in the periodic box, a structure very similar to that previously obtained for $1.61 \leq \alpha \leq 2.41$. Nonetheless pictures 1 and 14 are shifted in the spanwise direction by 333+/4, corresponding to the phase shift $\Delta\theta_2 = \pi$, $\Delta\theta_4 = 2\pi$. (see §10.2) The bursting event leads to variations of positions and amplitudes of the basic streamwise rolls and formation of other vortices. The oscillation, death and rebirth of vortices make the streak spacing vary, in agreement with experimental results (Kline et al. [1967]). However, since the intermittent solution is always very close to the real subspace or a rotation of it, the vortices remain symmetric and paired. Moreover it is possible to adjust the value of the viscosity parameter ($\alpha \sim 1.5$) so that the bursting period is 100 wall units as experimentally observed. It is found that, in this case, the "burst" lasts 10 wall units which is also the right order of magnitude.

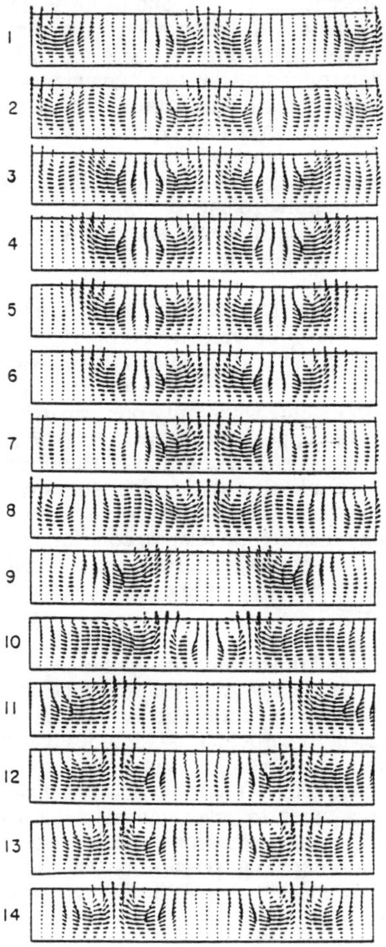

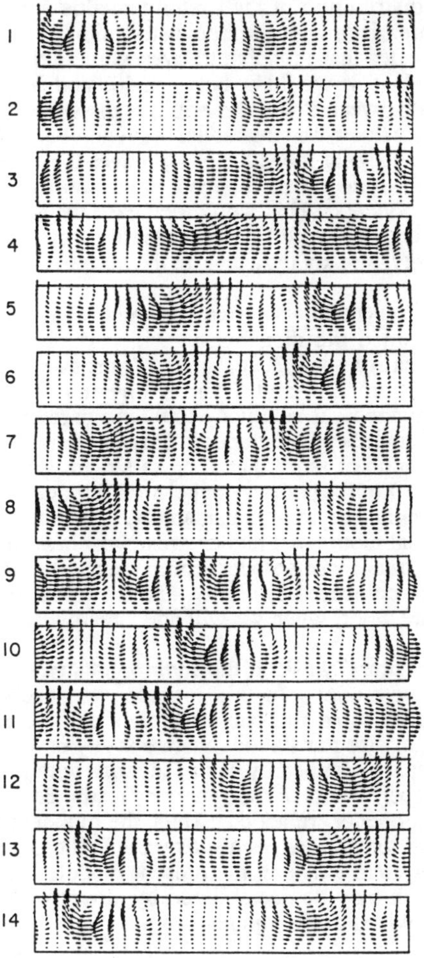

Figure 5 (left). Intermittent solution corresponding to an Heisenberg parameter $\alpha = 1.4$. Vector representation of u_2, u_3 in an x_2-x_3 plane at times indicated on figure 4.

Figure 6 (right). Chaotic solution for Heisenberg parameter $\alpha = 1.2$. Vector representation of u_2, u_3 in an x_2 - x_3 plane at uniformly spaced times (close enough to resolve the motion).

The behavior of the eddies corresponding to a chaotic solution ($1.0 \leq \alpha \leq 1.3$) is shown in Figure 6, as before, by plotting the u_2, u_3 velocity components at specific times. The behavior is less regular and isolated vortices sometimes emerge. This is consistent with the flow visualization experiments of Smith and Schwarz [1983] who observed a significant number of solitary vortices among the predominant vortex pairs. These patterns are also very rich in dynamics. We intend to study the regime further in future work.

PHYSICAL INTERPRETATION.

Due to the truncation of our model to a single eigenmode and one streamwise Fourier component, the absolute and relative energy of the various velocity components is affected. First, the choice of streamwise length scale affects the absolute values of all the component energies, but not their relative values. We chose a streamwise length scale of 333. Keeping only the Fourier mode at zero streamwise wavenumber, this is equivalent to supposing that the distribution of the first eigenvalue with wavenumber is flat to 333, and drops to zero. Comparison with the true distribution of the first eigenvalue with streamwise wavenumber indicates that this should overestimate the energy in this eigenvalue. A more reasonable length scale would be between 400 and 500, a value determined by the requirement that the area between it and the origin is the same as the true area under the distribution of the first eigenvalue with streamwise wavenumber. However, examination of the equations indicates that changing the value of L_1 leaves the equations invariant. Hence all phenomena reported here would be unchanged, with simply a decrease in energy level of all the velocity components. This would decrease the overestimated streamwise energy contained in our system but would not necessarily improve our model since it would also decrease the energy, already deficient, in the normal and cross-stream components.

The truncation also affects the relative energy in the various velocity components. We have examined the distribution with x_2^+ of the energy in the three velocity components for truncations of 3, 6, and 17 cross-stream wavenumbers, 1, 2 and 7 streamwise wavenumbers, and 1 and 3 eigenmodes. It is clear that the energy in the u_2 and u_3 components is relatively low (compared to the u_1 component) when only one streamwise wavenumber is included. This is particularly true in the upper part of the layer. While inclusion of more than 6 cross-stream wavenumbers has no effect, addition of streamwise wave numbers does make a difference. In particular addition of another streamwise wavenumber helps considerably. We must conclude that there is energy in cross-stream and normal motions of higher streamwise wavenumbers, that makes little relative contribution to the streamwise velocity component. As a consequence, our predicted rolls are somewhat weak compared to reality, and we rescaled them. This rescaling will not be necessary in future work, when we will add streamwise wavenumbers.

One naturally asks if the addition of streamwise modes will drastically change the behavior of our model. In this respect, we observe that the same symmetries, inherited from invariance of the Navier-Stokes equations to spanwise translation and reflection, and streamwise translation, operate at any order of truncation. In particular, this implies that the subspace spanned by basis functions with zero streamwise wavenumber is invariant and that the coefficients of the equations restricted to this subspace remain unchanged as streamwise components are added. Thus, the model studied in this paper, together with its dynamics, exists *unchanged*, as a *subsystem* in the larger set of differential equations, much as the 2-4 subsystem exists in the present model. This raises the interesting possibility that the 2-4 branch may be destabilized by streamwise modes as well as by the 1, 3, 5 "intersticial" spanwise Fourier modes.

Before embarking on a detailed study of models with additional Fourier components and eigenfunctions, we decided to investigate the effect of eigenfunction variations on the six mode model studied in this paper. We obtained a set of eigenfunctions computed from direct numerical simulations of a channel flow at Re = 13,800 through the courtesy of

Parviz Moin at the Stanford-Ames Center for Turbulence Research. While the flow conditions are somewhat different from those of Herzog's [1986] pipe flow, the eigenfunctions are generally similar in form, and they are defined more precisely, since the discretization is considerably finer. The resulting coefficients in the model equations have the same signs and general relative magnitudes as those of the present model, the largest differences being of the order of 20%. While this affects the detailed structure of the bifurcation diagram, the model still exhibits bursting behavior, albiet now for a lower value of α ($\approx$ 0.74) and involving heteroclinic cycles connecting small periodic orbits (limit cycles) which contain all Fourier components, not merely 2 and 4. Thus, in this model, the quasi-steady state is of weakly oscillating rolls which have a somewhat richer spanwise structure. The general features of the burst are very similar to the present model. For higher α one again observes steady rolls containing only Fourier components 2 and 4. We feel that these findings, which we will report in detail later, give much credibility to our present simple model.

We do not have present in our modal population a mechanism to represent the production of higher wavenumber energy when an intense updraft is formed, presumably as a result of a secondary instability. Thus, although our eddies are capable of exhibiting the basic bursting and ejection process, the labor is in vain - there is no sequel, no production of intense higher wavenumber turbulence. A contribution is made only to the low wavenumber part of the streamwise fluctuating velocity and the Reynolds stress. We could easily simulate the production of this high wavenumber turbulence, although we do not expect its inclusion to change the dynamics of our system qualitatively. However, we have held the value of our Heisenberg parameter constant , whereas its value should rise and fall with the level of this intense higher wavenumber turbulence in the vicinity of the wall. The transport effectiveness of the intense higher wavenumber turbulence would be expected to damp the system, suppressing the interesting dynamics until the higher levels are either blown

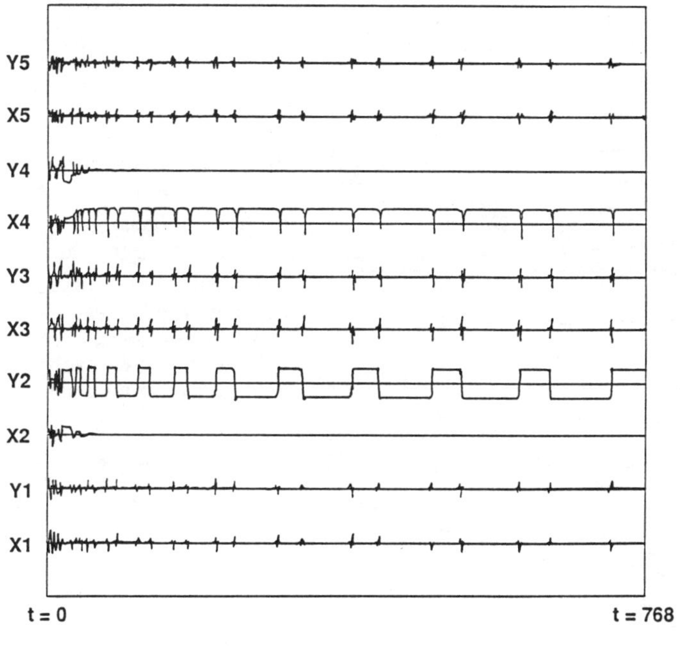

Figure 7. Pressure perturbations: solutions for α = 1.6 with pressure term.

downstream or lifted and diffused to the outer edge of the boundary layer. The production of higher values of the Heisenberg parameter could easily be parameterized by a single time constant first order equation tied to a measure of the amplitude of the coefficients. The major effect would simply be to cut off each burst more rapidly than at present. We plan to study the behavior of such a model in the future.

The occurrence of the bursts for values of the Heisenberg parameter between 1.4 and 1.6 appears to be pseudo-random. This is essentially due to round-off error in the calculations. The transitions from one solution to the other and back are extremely sensitive to the precise solution trajectory, and a minute change can make a considerable change in the time at which the next transition occurs. Initially we did not exercise the pressure term. Recall that the pressure term appeared due to the finite domain of integration. It represents the interaction of the part of the eddy that we have resolved with the part above the domain of integration, which is unresolved. The order of magnitude that we estimated for this term was small, and for that reason we at first neglected it. It has, however, an important effect, while not changing the qualitative nature of the solution. The term has the form of a random function of time, with a small amplitude. This slightly perturbs the solution trajectory constantly; away from the points **r**+ and **r**- this has little effect, but when the solution trajectory is very close to these points, the perturbation has the effect of throwing the solution away from the fixed point, so that it need not wait long to spiral outward. This results in a thorough randomization of the transition time from one solution to the other, while having little effect on the structure of the solution during a burst. While in the absence of the pressure term (and round-off error), the interburst time tends to lengthen as the solution trajectory is attracted closer and closer to the heteroclinic cycle (this effect is visible in figure 3), with the pressure term, the mean time stabilizes (see figure 7). The mean interburst times for various values of the pressure signal are shown in figure 8.

Probably the most significant finding of this work is the identification of the etiology of the bursting phenomenon. That is, the bursts appear to be produced autonomously by the wall

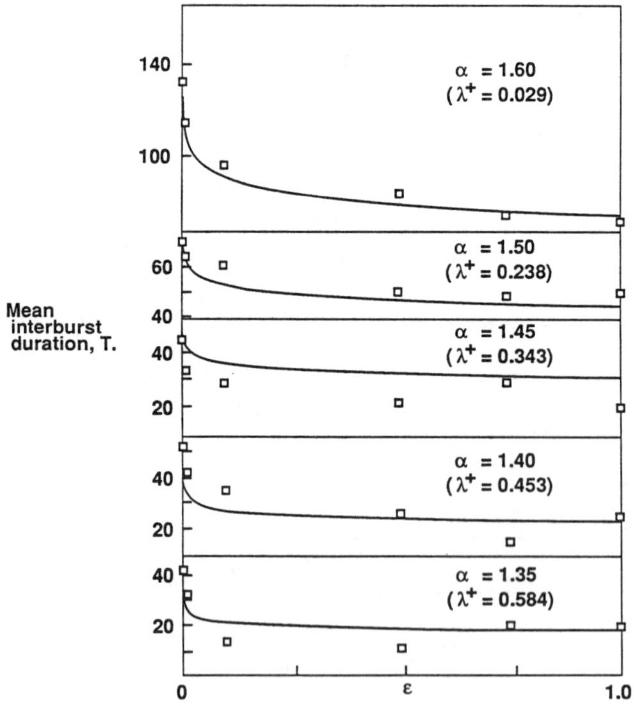

Figure 8. Mean interburst duration T as a function of eigenvalue λ_+ and pressure signal amplitude ε. Solid curves indicate prediction of Aubry *et al* (1987), open squares indicate results of numerical simulations.

region, but to be triggered by pressure signals from the outer layer. Whether the bursting period scales with inner or outer variables has been a controversy in the turbulence literature for a number of years. The matter has been obscured by the fact that the experimental evidence has been measured in boundary layers with fairly low Reynolds numbers lying in a narrow range, so that it is not really possible to distinguish between the two types of scaling. The turbulent polymer drag reduction literature is particularly instructive, however, since the sizes of the large eddies, and the bursting period, all change scale with the introduction of the polymer (see, for example, Kubo and Lumley, [1980], Lumley and Kubo [1984]). The present work indicates clearly that the wall region is capable of producing bursts autonomously, but the timing is determined by trigger signals from the outer layer. This suggests that events during a burst should scale unambiguously with wall variables. Time between bursts will have a more complex scaling, since it is dependent on the first occurrence of a *large* enough pressure signal *long* enough after a previous burst; "long enough" is determined by wall variables, but the pressure signal should scale with outer variables.

BIBLIOGRAPHY.

Aubry, N., Holmes, P., Lumley, J. L., & Stone, E. 1987. *The Dynamics of Coherent Structures in the Wall Region of a Turbulent Boundary Layer.* Sibley School of Mechanical and Aerospace Engineering Report No. FDA-86-15. Ithaca, NY: Cornell.

Armbruster, D., Guckenheimer, J. and Holmes, P. 1987. Heteroclinic cycles and modulated traveling waves in systems with O(2) symmetry. *Physica D* (to appear).

Bergé, P., Dubois, M. Manneville, P. and Pomeau, Y. 1980. Intermittency in Rayleigh-Bénard convection. *J. Phys. Lett* 41 L341.

Campbell, D. and Rose, H. 1983. *Order in Chaos.* North Holland, Amsterdam.

Cantwell, B.J. 1981. Organized motion in turbulent flow. *Ann. Rev. Fluid Mech.* 13: 457-515.

Corino, E. R., and Brodkey, R.S. 1969. A visual investigation of the wall region in turbulent flow. *J. Fluid Mech.* 37(1):1-30.

Dubois, M., Rubio, M.A. and Bergé, 1983. Experimental evidence of intermittencies associated with a subharmonic bifurcation. *Phys. Rev. Lett.* 51, 1446-1449.

Glauser, M.N., Leib, S.J. and George W.K. 1985. Coherent structure in the axisymmetric jet mixing layer. Proc. of the 5th Symp. of the Turbulent Shear Flow Conf. Cornell University. (Springer selected papers from TSF).

Golubitsky, M. & Guckenheimer, J. 1986. (eds). *Multiparameter Bifurcation Theory.* A.M.S. Contemporary Mathematics Series, No. 56. American Mathematical Society, Providence, R.I.

Guckenheimer, J. 1986. Strange Attractors in Fluids: Another View. *Ann. Rev. Fluid Mech.* 18, 15-31.

Guckenheimer, J. and Holmes, P.J. 1983. *Nonlinear Oscillations, Dynamical Systems and Bifurcations of Vector Fields.* Springer-Verlag, New York. Corrected second printing, 1986.

Herzog, S. 1986. The large scale structure in the near-wall region of turbulent pipe flow. Ph.D. thesis, Cornell University.

Hyman, J.M., Nicolaenko, B. and Zaleski, S. 1986. Order and complexity in the Kuramoto Sivashinsky model of weakly turbulent interfaces. Physica D.(to appear).

Kline, S.J., Reynolds, W.C., Schraub, F.A. and Rundstadler, P.W. 1967. The structure of turbulent boundary layers. *J. Fluid Mech.* 30(4): 741-773.

Kubo, I. and Lumley, J. L. 1980. A study to assess the potential for using long chain polymers dissolved in water to study turbulence. Annual Report, NASA-Ames Grant No. NSG-2382. Ithaca, NY: Cornell.

Loeve, M. 1955. Probability Theory. Van Nostrand, New York.

Lorenz, E.N. 1963. Deterministic non-periodic flow. *J. Atmos. Sciences* 20, 130-141.

Lumley, J.L. 1967. The structure of inhomogeneous turbulent flows. In atmospheric turbulence and radio wave propagation, A.M. Yaglom and V.I. Tatarski:, eds.: 166-178. Moscow: Nauka.

Lumley, J.L. 1970. Stochastic tools in turbulence. Academic Press, New York.

Lumley, J.L. 1981. Coherent structures in turbulence. Transition and turbulence, edited by R.E. Meyer, Academic Press, New York: 215-242.

Lumley, J. L. and Kubo, I. 1984. Turbulent drag reduction by polymer additives: a survey. In *The Influence of Polymer Additives on Velocity and Temperature Fields*. IUTAM Symposium Essen 1984. Ed. B. Gampert. pp. 3-21. Springer Berlin/Heidelberg:

Maurer, J. and Libchaber, A. 1980. Effects of the Prandtl number on the onset of turbulence in liquid helium. *J. Phys. Paris Lett.* 41, L-515-.

Moin, P. 1984. Probing turbulence via large eddy simulation. AIAA 22nd Aerospace Sciences Meeting.

Monin, A.S. 1978. On the nature of turbulence. *Soc. Phys. Usp.* 21(5), May. pp 429-442.

Nicolaenko, B. 1986. Lecture delivered at workshop on Computational Aspects of Dynamical Systems, Mathematical Sciences Institute, Cornell University, September 8-10, 1986.

Pomeau, Y. and Manneville, P. 1980. Intermittent transition to turbulence. *Comm. Math. Phys.* 74, 189-197.

Rand, D.A. and Young, L.S. 1981. (eds). 1981. *Dynamical Systems and Turbulence.* Springer Lecture Notes in Mathematics 898, Springer-Verlag, New York, Heidelberg, Berlin.

Ruelle, D. and Takens, F. 1971. On the nature of turbulence. *Comm. Math. Phys.* 82. 137-151.

Silnikov, L.P. 1965. A case of the existence of a denumerable set of periodic motions. *Soviet Math. Dokl* 6: 163-166.

Silnikov, L.P. 1968. On the generation of a periodic motion from trajectories Doubly asymptotic to an equilibrium State of Saddle type. *Math. U.S.S.R. Sbornik* 6, 427-438.

Silnikov, L.P. 1970. A contribution to the problem of the structure of an extended neighborhood of a rough equilibrium state of saddle-focus type. *Math U.S.S.R Sbornik* 10 (1), 91-102.

Smith, C.R. and Schwarz, S.P. 1983. Observation of streamwise rotation in the near-wall region of a turbulent boundary layer. Phys. Fluids 26(3) 641-652.

Sparrow, C.T. 1982. *The Lorenz Equations: Bifurcations, Chaos and Strange Attractors.* Springer-Verlag, New York, Heidelberg, Berlin.

Swinney, H.L. and Gollub, J.P. 1981. (eds) *Hydrodynamic Instabilities and the Transition to Turbulence,* Springer-Verlag, New York.

Tennekes, H. and Lumley, J.L. 1972. *A first course in turbulence.* Cambridge, MA: M.I.T. Press.

Theodorsen, T. 1952. Mechanism of turbulence. Proc. 2nd Midwestern Conf. on Fluid Mech. Ohio State U., Colombus, Ohio.

Tresser, C. 1984. About some theorems by L.P. Silnikov. *Ann de L'Inst. H. Poincané,* 40: 441-461.

Tresser, C. Coullet, P. and Arneodo, A. 1980. On the existence of hysteresis in a transition to chaos after a single bifurcation. *J. Physics (Paris)* Lett. 41, L243-246.

Zilberman, M., Wygnanski, I. , and Kaplan, R.E. 1977. Transitional boundary layer spot in a fully turbulent environment. *Phys. Fluids Supp.* 20(10): S258-S271.

Wave Interaction and Stability Problems

D. J. BENNEY
Massachusetts Institute of Technology
Cambridge, Massachusetts, USA

INTRODUCTION

In this paper two classes of stability problems are discussed and the relationship to other theoretical developments is explored.

The first of these concerns the possibility of a linear resonance between the vertical velocity and vertical vorticity modes in parallel flows. Several laminar flows have such resonances as does the mean profile for a turbulent boundary layer. Such resonances are associated with three dimensional disturbances and correspond to slightly damped modes. Each resonance has its own critical Reynolds number. As the Reynolds number increases these waves become longitudinal streaks with a well defined wave length. Nonlinear consequences are considered for such disturbances.

The second class of problems concerns an alternative theoretical approach to the three dimensional instability of parallel flows. This theory involves interactions between first harmonics and the mean flow. At high Reynolds numbers such interactions are inherently nonlinear and occur on a relatively fast time scale. Different waves interact with each other through the mean flow. The mean flow itself is modified to zeroth order. One special case that is studied in some detail concerns a comparison between the stability characteristics of a given flow with those of the same flow together with a small amplitude wave. Possible extensions of these ideas are considered.

DIRECT RESONANCE

Wave interactions have been studied in some detail especially in fluid mechanics and many of the theoretical predictions have been validated by careful experiments. One of the mechanisms of importance in such interactions concerns resonance sets of wave numbers $\underline{k}_j$ and the associated frequencies $w_j = w(\underline{k}_j)$ such that

$$\sum_{j=1}^{n} \underline{k}_j = 0\,, \quad \sum_{j=1}^{n} w_j = 0 \tag{1}$$

Energy sharing and instabilities exist between modes that satisfy the constraints of equation (1) and the time scale for the process is $t = 0(\varepsilon^{2-n})$ where ε is the amplitude parameter. Whether such resonances are possible and if so for what values of n depends on the details of the dispersion relationship. The simplest case is when $n = 3$ so that there is a resonant triad. A degenerate case of a triad arises when two wave numbers are identical and subharmonic instabilities are possible. These resonances that occur in conventional wave theory have their counterparts in shear flow instability phenomena but while in some cases they play important roles they are often overwhelmed by other more rapid and dramatic physical mechanisms.

One such mechanism is that of direct resonance where two frequencies (in general complex) and two wave numbers may coincide. This idea was proposed and explored by Gustavsson [1] for parallel flow stability problems. The possibility of such multiple eigenvalues has some interesting consequences in both the linear and nonlinear regimes. Some of these possibilities especially with respect to the initial value problem have been studied by Gustavsson [2] and with respect to nonlinear theory by Benney and Gustavsson [3].

To be specific consider the waves on a quasi-parallel flow $(\overline{u}(y), 0, 0)$. The linear perturbation equations are

$$\begin{aligned}
\frac{\partial u}{\partial x} + \frac{\partial v}{\partial y} + \frac{\partial w}{\partial z} &= 0 \\
\left(\frac{\partial u}{\partial t} + \overline{u}\frac{\partial}{\partial x}\right) u + \frac{d\overline{u}}{dy} v &= -\frac{\partial p}{\partial x} + \frac{1}{R}\Delta u \\
\left(\frac{\partial}{\partial t} + \overline{u}\frac{\partial}{\partial x}\right) v &= -\frac{\partial p}{\partial y} + \frac{1}{R}\Delta v \\
\left(\frac{\partial}{\partial t} + \overline{u}\frac{\partial}{\partial x}\right) w &= -\frac{\partial p}{\partial z} + \frac{1}{R}\Delta w
\end{aligned}$$

and these can be reduced to the two degree of freedom system

$$\begin{aligned}
\left(\frac{\partial}{\partial t} + \overline{u}\frac{\partial}{\partial x}\right) \Delta v - \frac{d^2\overline{u}}{dy^2}\frac{\partial v}{\partial x} - \frac{1}{R}\Delta\Delta v &= 0 \\
\left(\frac{\partial}{\partial t} + \overline{u}\frac{\partial}{\partial x}\right) \eta - \frac{1}{R}\Delta\eta &= -\frac{d\overline{u}}{dy}\frac{\partial v}{\partial z}
\end{aligned} \tag{3}$$

Here Δ denotes the three dimensional Laplacian, R is the Reynolds number, (u, v, w) the velocity components, p the pressure and η the y component of the vorticity perturbation. From the equations (3) it is seen that a general three dimensional disturbance has two degrees of freedom, one corresponding to the vertical velocity pertubation and the other to the vertical vorticity perturbation η. Once v and η are known the u and w perturbations are obtained from

$$\begin{aligned}
\left(\frac{\partial^2}{\partial x^2} + \frac{\partial^2}{\partial z^2}\right) u &= -\frac{\partial^2 v}{\partial x \partial y} + \frac{\partial \eta}{\partial z} \\
\left(\frac{\partial^2}{\partial x^2} + \frac{\partial^2}{\partial z^2}\right) w &= -\frac{\partial^2 v}{\partial z \partial y} - \frac{\partial \eta}{\partial z}
\end{aligned} \tag{4}$$

In general there are normal mode solutions to (3) of the form

$$\begin{aligned}
v &= A(t)\hat{v}(y)e^{i\alpha x + i\beta z} + * \\
\eta &= \left(B(t)\hat{\eta}(y) + A(t)\hat{\eta}_1(y)\right) e^{i\alpha x + i\beta z} + *
\end{aligned} \tag{5}$$

where

$$\frac{dA}{dt} = -i\alpha c A\,, \quad \frac{dB}{dt} = -i\alpha c' B \tag{6}$$

and

$$\left[\left(\frac{d^2}{dy^2} - k^2\right)^2 - i\alpha R\left\{(\overline{u} - c)\left(\frac{d^2}{dy^2} - k^2\right) - \frac{d^2\overline{u}}{dy^2}\right\}\right]\hat{v} \;=\; 0$$

$$\left[\frac{d^2}{dy^2} - k^2 - i\alpha R(\overline{u} - c')\right]\hat{\eta} \;=\; 0 \tag{7}$$

$$\left[\frac{d^2}{dy^2} - k^2 - i\alpha R(\overline{u} - c)\right]\hat{\eta}_1 \;=\; i\beta R\frac{d\overline{u}}{dy}\hat{v}$$

Here α and β are the real wave numbers, $k^2 = \alpha^2 + \beta^2$, c and c' are eigenvalues of the vertical velocity and vertical vorticity equation. We shall suppose that the flow configuration corresponds to rigid finite boundaries and/or vanishing perturbations at infinity. The first equation in (7) is the Orr Sommerfeld equation and at Reynolds numbers $R > R_c$ there will in general be instability, that is $\mathrm{Im}(\alpha c) = \alpha c_i > 0$. However it is readily shown that the verticity vorticity spectrum is always damped, $\alpha c_i' < 0$. For this reason this second degree of freedom is usually neglected in standard stability studies.

However if $c = c'$ the decomposition prescribed by equations (6) and (7) fails and there is a resonance in the linear problem. The last equations in (6) and (7) are now replaced by

$$\frac{dB}{dt} \;=\; -i\alpha c B + KA$$

$$\left[\frac{d^2}{dy^2} - k^2 - i\alpha R(\overline{u} - c)\right]\hat{\eta} \;=\; 0 \tag{8}$$

$$\left[\frac{d^2}{dy^2} - k^2 - i\alpha R(\overline{u} - c)\right]\hat{\eta}_1 \;=\; i\beta R\frac{d\overline{u}}{dy}\hat{v} + KR\hat{\eta}.$$

It follows that the constant K is determined from

$$K = -i\beta\frac{\int_{y_1}^{y_2}\frac{d\overline{u}}{dy}\hat{v}\hat{\eta}dy}{\int_{y_1}^{y_2}\hat{\eta}^2 dy} \tag{9}$$

where y_1, y_2 are the boundary values for y. A resonance of this type is called a direct resonance. It is clear that such a resonance will occur at particular values of

$$\alpha^2 + \beta^2 = a_0^2\,, \quad \alpha R = b_0 \tag{10}$$

At a given Reynolds number $R > R_0 = b_0/a_0$ it is possible to find real values of α and β for which the resonance condition is satisfied. If R_0 is such a critical resonant Reynolds number then as R increases, α decreases, while β increases so that the resonant structures tend more toward longitudinal streaks. Calculations have shown that typical values for R_0 and usually well below the critical Reynolds number R_c. It should be noted that as R increases the damping factor αc_i decreases and in this circumstance nonlinear effects would become important.

If nonlinear terms are retained in the analysis the equations (3) are replaced by

$$\begin{aligned}\left(\frac{\partial}{\partial t}+\overline{u}\frac{\partial}{\partial x}\right)\Delta v-\frac{d^2\overline{u}}{dy^2}\frac{\partial v}{\partial x}-\frac{1}{R}\Delta\Delta v&\\ +\varepsilon\left\{\left(\frac{\partial^2}{\partial x^2}+\frac{\partial^2}{\partial z^2}\right)S_2-\frac{\partial^2 S_1}{\partial x\partial y}-\frac{\partial^2 S_3}{\partial z\partial y}\right\}&=0\\ \left(\frac{\partial}{\partial t}+\overline{u}\frac{\partial}{\partial x}\right)\eta+\frac{d\overline{u}}{dy}\frac{\partial v}{\partial z}-\frac{1}{R}\Delta\eta&\\ +\varepsilon\left\{\frac{\partial S_1}{\partial z}-\frac{\partial S_3}{\partial x}\right\}&=0\end{aligned}\tag{11}$$

Here ε is the amplitude parameter and the S_j are given by

$$\begin{aligned}S_1&=\frac{\partial}{\partial x}(uu)+\frac{\partial}{\partial y}(vu)+\frac{\partial}{\partial z}(wu)\\ S_2&=\frac{\partial}{\partial x}(uv)+\frac{\partial}{\partial y}(vv)+\frac{\partial}{\partial z}(wv)\\ S_3&=\frac{\partial}{\partial x}(uw)+\frac{\partial}{\partial y}(vw)+\frac{\partial}{\partial z}(ww)\end{aligned}\tag{12}$$

To accommodate direct resonance with wave numbers $(\alpha,\pm\beta)$ Benney and Gustavsson have shown that the necessary scalings are

$$v=V\,,\quad (u,w,\eta)=\varepsilon^{-\frac{1}{2}}(U,W,N)\,,\quad T=\varepsilon^{\frac{1}{2}}t\tag{13}$$

in which case equations (11) become

$$\begin{aligned}&\left[\left\{\left(\frac{\partial}{\partial t}+\overline{u}\frac{\partial}{\partial x}\right)\Delta-\frac{d^2\overline{u}}{dy^2}\frac{\partial}{\partial x}-\frac{1}{R}\Delta^2\right\}V-\frac{\partial^3}{\partial x^2\partial y}(UU)\right.\\ -&\left.2\frac{\partial^3}{\partial x\partial z\partial y}(UW)-\frac{\partial^3}{\partial z^2\partial y}(WW)\right]\\ +&\varepsilon^{\frac{1}{2}}\left[\left(\frac{\partial^2}{\partial x^2}+\frac{\partial^2}{\partial z^2}-\frac{\partial^2}{\partial y^2}\right)\left(\frac{\partial}{\partial x}(UV)+\frac{\partial}{\partial z}(WV)\right)\right]\\ +&\varepsilon\left[\left(\frac{\partial^2}{\partial x^2}+\frac{\partial^2}{\partial z^2}\right)\frac{\partial}{\partial y}(VV)\right]=0\end{aligned}\tag{14}$$

$$\begin{aligned}&\left[\left(\frac{\partial}{\partial t}+\overline{u}\frac{\partial}{\partial x}\right)-\frac{1}{R}\Delta\right]N+\varepsilon^{\frac{1}{2}}\left[\frac{d\overline{u}}{dy}\frac{\partial V}{\partial z}\right.\\ +&\left.\frac{\partial^2}{\partial x\partial z}(UU-WW)+\left(\frac{\partial^2}{\partial z^2}-\frac{\partial^2}{\partial x^2}\right)(UW)\right]\\ +&\varepsilon\left[\frac{\partial^2}{\partial y\partial z}(UV)-\frac{\partial^2}{\partial x\partial y}(VW)\right]=0\end{aligned}$$

In this framework the Orr Sommerfeld equation is no longer the zeroth order approximation. Nevertheless a systematic perturbation can be devised and the analysis leads to a second order evolution equation of the form

$$\left(\frac{d}{dT} - M\right)\left(\frac{d}{dT} - N\right) B = PB^2B^* \tag{15}$$

where M, N and P are complex constants.

Direct resonance is a dangerous interaction mechanism. The time scale is rapid compared to the usual interaction times. However the modes are linearly damped so that nonlinearity must be sufficiently strong. In systems where linearly unstable modes exist direct resonance is unlikely to be important, but in damped systems it would appear to offer a preferential model mechanism. In this context the ideas suggested here were applied to the mean profile of a turbulent boundary layer by Jang, Benney and Gran [4]. Spanwise wave lengths for the induced mean flow were found to compare favorably with experimental values for the streak spacing.

MEAN FLOW FIRST HARMONIC THEORY

Many experiments show that during instability nonlinear three dimensional distortions of the mean flow dominate over the production of higher harmonics. Indeed it is precisely these features that have placed severe limitations on most theoretical developments. However recent work by Benney [5] and Benney and Chow [6, 7] indicates that there is an interesting alternative to conventional nonlinear stability theories.

Much of the motivation will be omitted here save to say that at high Reynolds number all three dimensional waves are resonant with the mean flow on a time scale $t = 0(\varepsilon^{-1})$. The consequence of this fact necessitates that the velocities be expanded in the forms

$$\begin{aligned} u &= u_0 + \varepsilon(u_1 e^{i\theta} + u_1^* e^{-i\theta}) \\ &\quad + \varepsilon^2(u_2 e^{2i\theta} + u_2^* e^{-2i\theta}) + \ldots \\ v &= \varepsilon(v_0 + v_1 e^{i\theta} + v_1^* e^{-i\theta}) \\ &\quad + \varepsilon^2(v_2 e^{2i\theta} + v_2^* e^{-2i\theta}) + \ldots \\ w &= \varepsilon(w_0 + w_1 e^{i\theta} + w_1^* e^{-i\theta}) \\ &\quad + \varepsilon^2(w_2 e^{2i\theta} + w_2^* e^{-2i\theta}) + \ldots \end{aligned} \tag{16}$$

where θ is assumed to be a real phase function with

$$\theta = \frac{\Theta(X,T)}{\varepsilon}, \quad X = \varepsilon x, \quad T = \varepsilon t \tag{17}$$

and the local wave number α and phase speed c are given by

$$\alpha = \frac{\partial \Theta}{\partial X}, \quad \alpha c = -\frac{\partial \Theta}{\partial T} \tag{18}$$

while for each function $f_r = f_r(y, z, X, T)$. The forms (16) are necessary in order that a weakly nonlinear thoery be valid for $T = 0(1)$ and it should be noted that there are several important differences from conventional expansions. Most noticeable of these if the fact that the transverse mean flows are of the same order as the imposed oscillations while the mean flow $u_0(y, z, X, T)$ evolves from the initial condition

$$u_0(u, z, X, 0) = \overline{u}_0(y) \tag{19}$$

It is a relatively simple matter to obtain a closed nonlinear system of equations for the functions f_0 and f_1, at least to leading order, namely

$$i\alpha u_1 + \frac{\partial v_1}{\partial y} + \frac{\partial w_1}{\partial z} = 0$$

$$\begin{aligned}
i\alpha(u_0 - c)u_1 + \frac{\partial u_0}{\partial y}v_1 + \frac{\partial v_0}{\partial z}w_1 & \\
= & -i\alpha p_1 + \frac{1}{R}\left(\frac{\partial^2}{\partial y^2} + \frac{\partial^2}{\partial z^2} - \alpha^2\right)u_1 \\
i\alpha(u_0 - c)v_1 = & -\frac{\partial p_1}{\partial y} + \frac{1}{R}\left(\frac{\partial^2}{\partial y^2} + \frac{\partial^2}{\partial z^2} - \alpha^2\right)v_1 \\
i\alpha(u_0 - c)w_1 = & -\frac{\partial p_1}{\partial z} + \frac{1}{R}\left(\frac{\partial^2}{\partial y^2} + \frac{\partial^2}{\partial z^2} - \alpha^2\right)w_1
\end{aligned} \tag{20}$$

$$\begin{aligned}
\frac{\partial u_0}{\partial X} + \frac{\partial v_0}{\partial y} + \frac{\partial w_0}{\partial z} &= 0 \\
\frac{\partial u_0}{\partial T} + \frac{\partial}{\partial X}(u_0 u_0) &+ \frac{\partial}{\partial y}(v_0 u_0) + \frac{\partial}{\partial z}(w_0 u_0) \\
&= -\frac{L}{R_0} + \frac{1}{R_0}\left(\frac{\partial^2}{\partial y^2} + \frac{\partial^2}{\partial z^2}\right)u_0 \\
\frac{\partial v_0}{\partial T} + \frac{\partial}{\partial X}(u_0 v_0) &+ \frac{\partial}{\partial y}(v_0 v_0) + \frac{\partial}{\partial z}(w_0 v_0) + 2\frac{\partial}{\partial y}(v_1 v_1^*) \\
+\frac{\partial}{\partial z}(v_1 w_1^* + v_1^* w_1) &= -\frac{\partial p_0}{\partial y} + \frac{1}{R_0}\left(\frac{\partial^2}{\partial y^2} + \frac{\partial^2}{\partial z^2}\right)v_0 \\
\frac{\partial w_0}{\partial T} + \frac{\partial}{\partial X}(u_0 w_0) + \frac{\partial}{\partial y}(v_0 w_0) &+ \frac{\partial}{\partial z}(w_0 w_0) + \frac{\partial}{\partial y}(v_1 w_1^* + v_1^* w_1) \\
+2\frac{\partial}{\partial z}(w_1 w_1^*) &= -\frac{\partial p_0}{\partial z} + \frac{1}{R_0}\left(\frac{\partial^2}{\partial y^2} + \frac{\partial^2}{\partial z^2}\right)w_0
\end{aligned} \tag{21}$$

Here an asterisk denotes a complex conjugate and $R_0 = \varepsilon R$. These are the fundamental equations for the mean flow first harmonic theory. Other modes can be included and it becomes clear that all modes interact via the mean flow on the time scale $t = 0(\varepsilon^{-1})$. No resonant conditions need be satisfied as opposed to classical conservative nonlinear wave systems. These equations also result from the direct resonance analysis at large Reynolds numbers if we identify the resonance mode with the mean flow. Such a connection is currently under study by Gustavsson and Benney.

One special solution to equations (20, 21) is

$$u_0 = \overline{u}(y) \tag{22}$$

and all other functions zero. The linear stability of this solution leads to the classical vertical velocity vertical vorticity equations and the properties of this system are now well known.

The second special solution is

$$u_0 = \overline{u}_0(y)\,, \quad v_1 = \overline{v}_1(y)\,, \quad c = \overline{c} \tag{23}$$

and all other functions zero. This corresponds to a slightly wavy flow. The linear stability of this system will be referred to as the second stability problem. The equations for the second stability problem are obtained by writing

$$\begin{aligned}
u_0 &= \overline{u}_0(y) + \varepsilon'\left[u_0(y)e^{iK(X-CT)} + *\right]\cos\beta' z \\
v_0 &= \varepsilon'\left[v_0(y)e^{iK(X-CT)} + *\right]\cos\beta' z \\
w_0 &= \varepsilon'\left[w_0(y)e^{iK(X-CT)} + *\right]\sin\beta' z \\
v_1 &= \overline{v}_1(y) + \varepsilon'\left[v_{1+}(y)e^{iK(X-CT)} + v_{1-}(y)e^{-iK(X-C^*T)}\right]\cos\beta' z \\
\eta_1 &= \varepsilon'\left[\eta_{1+}(y)e^{iK(X-CT)} + \eta_{1-}(y)e^{-iK(X-C^*T)}\right]\sin\beta' z
\end{aligned} \tag{24}$$

Substitutions into equations (20, 21) with quadratic terms in ε' neglected it is found that

$$\begin{aligned}
i\alpha&\left\{(\overline{u}_0 - \overline{c})\left(\frac{d^2}{dy^2} - \alpha^2 - \beta'^2\right)v_{1+} - \frac{d^2\overline{u}_0}{dy^2}v_{1+}\right\} \\
&+ i\alpha\left\{\left(\frac{d^2\overline{v}_1}{dy^2} - (\alpha^2 + \beta'^2)\overline{v}_1\right)u_0 - \overline{v}_1\frac{d^2u_0}{dy^2}\right\} \\
&= \frac{1}{R}\left(\frac{d^2}{dy^2} - \alpha^2 - \beta'^2\right)^2 v_{1+}
\end{aligned}$$

$$\begin{aligned}
i\alpha(\overline{u}_0 - \overline{c})\eta_{1+} - \beta'\frac{d\overline{u}_0}{dy}v_{1+} &- \beta'\left(\overline{v}_1\frac{du_0}{dy} - \frac{d\overline{v}_1}{dy}u_0\right) \\
&= \frac{1}{R}\left(\frac{d^2}{dy^2} - \alpha^2 - \beta'^2\right)\eta_{1+}
\end{aligned}$$

$$
\begin{aligned}
-i\alpha\left\{(\overline{u}_0-\overline{c})\left(\frac{d^2}{dy^2}-\alpha^2-\beta'^2\right)v^*_{1-}\right. &- \left.\frac{d^2\overline{u}_0}{dy^2}v^*_{1-}\right\} \\
-i\alpha\left\{\left(\frac{d^2\overline{v}^*_1}{dy^2}-(\alpha^2+\beta'^2)\overline{v}^*_1\right)u_0\right. &- \left.\overline{v}^*_1\frac{d^2u_0}{dy^2}\right\} \\
&= \frac{1}{R}\left(\frac{d^2}{dy^2}-\alpha^2-\beta'^2\right)^2 v^*_{1-} \\
-i\alpha(\overline{u}_0-\overline{c})\eta^*_{1-}-\beta'\frac{d\overline{u}_0}{dy}v^*_{1-} &- \beta'\left(\overline{v}^*_1\frac{du_0}{dy}-\frac{d\overline{v}^*_1}{dy}u_0\right) \\
&= \frac{1}{R}\left(\frac{d^2}{dy^2}-\alpha^2-\beta'^2\right)\eta^*_{1-} \\
iK(\overline{u}_0-C)u_0+\frac{d\overline{u}_0}{dy}v_0 &= \frac{1}{R_0}\left(\frac{d^2}{dy^2}-\beta'^2\right)u_0 \\
iK\left\{(\overline{u}_0-C)\left(\frac{d^2}{dy^2}-\beta'^2\right)v_0\right. &- \left.\frac{d^2\overline{u}_0}{dy^2}v_0\right\} \\
-2\beta'^2\frac{d}{dy}(\overline{v}^*_1v_{1+}+\overline{v}_1v^*_{1-}) &+ \frac{\beta'}{\alpha^2+\beta'^2}\left(\frac{d^2}{dy^2}+\beta'^2\right) \\
\left\{\beta'\left(\overline{v}^*_1\frac{dv_{1+}}{dy}+\overline{v}_1\frac{dv^*_{1-}}{dy}\right)\right. &- \left.i\alpha(\overline{v}^*_1\eta_{1+}-\overline{v}_1\eta^*_{1-})\right\} \\
&= \frac{1}{R_0}\left(\frac{d^2}{dy^2}-\beta'^2\right)^2 v_0
\end{aligned}
\tag{25}
$$

The equations (25) is an eighteenth order eigenvalue problem for the eigenvalue C, it being supposed that there is a given base flow $\overline{u}_0$, first eigenfunction $\overline{v}_1$, eigenvalue $\overline{c}$, spanwise wave number β', Reynolds number R, amplitude ε and packet wave number K. This problem must be solved numerically by Benney and Chow [8]. Two independent computational methods have been used to obtain explicit results for Poiseuille flow, the Blasius boundary and in a simulation of free stream disturbances outside of a boundary layer. In all cases it is found that at values of $\varepsilon = 0(10^{-2})$ three dimensional instabilities exist that have growth rates larger than those obtained from the first stability problem.

This theory provides a link between existing linear theory and recent numerical simulations of flow instabilities.

REFERENCES

[1] Gustavsson, L. H., Initial Value Problem for Boundary Layer Flows, *Physics of Fluids*, vol. 22, pp. 1602-1605, 1979.

[2] Gustavsson, L. H., Excitation of Direct Resonances in Plane Poiseuille Flow, *Studies in Applied Mathematics*, vol. 75, pp. 227-248, 1986.

[3] Benney, D. J., and Gustavsson, L. H., A New Mechanism for Linear and Nonlinear Hydrodynamic Instability, *Studies in Applied Mathematics*, vol. 64, pp. 185-209, 1981.

[4] Jang, P. S., Benney, D. J., and Gran, R. L., On the Origin of Streamwise Vortices in a Turbulent Boundary Layer, *Journal of Fluid Mechanics*, vol. 169, pp. 109 - 123, 1986.

[5] Benney, D. J., The Evolution of Disturbances in Shear Flows at High Reynolds Numbers, *Studies in Applied Mathematics*, vol. 70, pp. 1-19, 1984.

[6] Benney, D. J., and Chow, K., An Alternative Approach to Nonlinear Instabilities in Hydrodynamics, *Studies in Applied Mathematics*, vol. 73, pp. 261-267, 1985.

[7] Benney, D. J., and Chow, K., Instabilities of Waves on a Free Surface, *Studies in Applied Mathematics*, vol. 74, pp. 227-243, 1986.

On Mathematical Formulation on Near-Wall Turbulence Problem

PAVEL I. GESHEV
Institute of Thermophysics
630090, Novosibirsk, USSR

ABSTRACT

New types of equations for the velocity and temperature fluctuations in a turbulent flow in the channel and near the wall are obtained. Similarity variables are introduced and equations describing vortex structures in self-similar variables are obtained in a logarithmic layer. Corollaries of the second similarity hypothesis of near-wall turbulence (scaling hypothesis) are analyzed.

INTRODUCTION

Highly encouraging results on both average and pulsation values have been already obtained by direct numerical methods of the turbulence modelling /1,2/. Thus Karman's constant has been calculated in the turbulent flow through a channel in /1/ and appeared to be $\varkappa = 0.46 \pm 0.05$. About $4 \cdot 10^6$ calculation points have been used in the recent work /3/ devoted to direct simulation of turbulent flow in the channel and computation continued at CRAY-XMP computer for about 250 hours.

The results of the work /3/ are stimulating: comparison of calculations with a large set of experimental data showed rather good agreement.

After the whole series of direct turbulent flow computation performed of late /1-5/ there is no doubt that turbulence can be described by the Navier-Stokes equations. Further increase of the computer capacity and computation algorithm quality presumes that the way of direct simulation is likely the correct way to solve the turbulence problem.

At present it is impossible, however, to model directly developed turbulent flows with great Reynolds numbers $\sim 10^5$ and higher at which the near-wall turbulence is characterized by highly pronounced universal properties in the so-called "wall variables" and there exists a rather wide region of distances up to the wall in which logarithmic profiles of mean velocity and temperature occur and fluctuations of different values are of scale invariant nature. These properties of developed near-wall turbulence are formulated in a compact form as two similarity hypotheses /6-7/ similar to those set up by Kolmogoroff /8/ for the case of local isotropic turbulence. The variable transformation and the derivation of a new type of equations for turbulent velocity and temperature fluctuations are performed in § 4 in accordance with these hypotheses. One can suppose that direct simulating of turbulent fluctuations based on the derived equations will allow to increase the Reynolds number and the calculation accuracy in a viscous sublayer and in a buffer region.

In § 2 of the present paper the Navier-Stokes equations are transformed into two dynamic equations and four constraint equations for the components of the velocity and vorticity vectors. These equations can be used, for example, to model the accelerating turbulent flow in the channel or that arising at the pulsation pressure superposition. In § 3 the system of equations with no parameters for complex amplitudes of velocity, vorticity and temperature fluctuations is obtained in a Fourier space. The solutions of this system describe vortex structures in the logarithmic near-wall region in a frequency-wavenumber representation.

In § 5 the equations of smaller dimensional representation describing space - and time-localized vortex structures in a logarithmic layer are obtained after conventional averaging of equations from § 4. The Fourier-transformation of these equations results in equations of § 3. In § 6 general results of the whole work are discussed.

2. DERIVATION OF FLUCTUATION EQUATIONS

Consider the turbulent flow in a plane channel of the height H. Assume turbulence along the axes x and z to be statistically homogeneous. Homogeneity may be used to introduce space averaging

$$\langle f(x,y,z,t)\rangle = \frac{1}{\ell_x \ell_z}\int_0^{\ell_x}\int_0^{\ell_z} f(x,y,z,t)\,dx\,dz. \tag{1}$$

Such average values will be hereafter designated by capital letters $U(y,t)$, $W(y,t)$, $T(y,t)$,are the averaged velocities in the x- and z directions and temperature, respectively. They depend on time t and coordinate y . The fluctuation components of velocities and temperature are designated by small letters u, v, w, φ. From the averaged discontinuity equation

$$\frac{\partial V}{\partial y} = 0 \tag{2}$$

and the boundary impermeability condition on the wall $V(y=0)=0$ it follows that the averaged velocity component normal to the wall is everywhere equalto zero $V \equiv 0$.

The averaging of the Navier-Stokes equations and convective heat transfer equation yields

$$\frac{\partial U}{\partial t} + \frac{\partial}{\partial y}\langle uv\rangle + \frac{\partial}{\partial x}\langle P/\rho\rangle = \nu\frac{\partial^2 U}{\partial y^2}, \tag{3}$$

$$\frac{\partial W}{\partial t} + \frac{\partial}{\partial y}\langle wv\rangle = \nu\frac{\partial^2 W}{\partial y^2}, \tag{4}$$

$$\frac{\partial T}{\partial t} + \frac{\partial}{\partial y}\langle \varphi v\rangle = a\frac{\partial^2 T}{\partial y^2}, \tag{5}$$

where p is pressure, ρ is density, ν is the kinematic viscosity, a is the temperature conductivity of medium. Heat is assumed to be transferred only along the y - axis (e.g. due to different temperatures of the channel walls). Note that the second averaged component of the Navier-Stokes equations yields after integrating with respect to y

$$\langle v^2 \rangle + \langle p/\rho \rangle = P_w(x,t)/\rho$$

where P_w is the pressure on the channel walls which can be only the linear function of x as the value $(\partial/\partial x)\langle p/\rho\rangle$ can not be the function of x according to Eq.(3).

Substracting the averaged equations (2)-(5) from the complete Navier-Stokes and discontinuity equations we obtain the fluctuation equations

$$\frac{\partial u_j}{\partial x_j} = 0, \tag{6}$$

$$\frac{\partial u_i}{\partial t} + U_j\frac{\partial u_i}{\partial x_j} + \frac{\partial}{\partial x_j}(u_i u_j - \langle u_i u_j\rangle) + v\frac{\partial U_i}{\partial y} + \frac{1}{\rho}\frac{\partial p}{\partial x_i} = \nu \Delta u_i \tag{7}$$

$$\frac{\partial \varphi}{\partial t} + U_j\frac{\partial \varphi}{\partial x_j} + \frac{\partial}{\partial x_j}(u_j\varphi - \langle u_j \varphi\rangle) + v\frac{\partial T}{\partial y} = a\Delta\varphi, \tag{8}$$

where the vectors of mean and fluctuation velocities $U_i = (U, 0, W)$, $u_i = (u, v, w)$ are introduced. Further manipulations with these equations are as follows: on representing nonlinear terms in the Gromeco-Lamb form we obtain two equations for *rot rot* and *rot* from the assumed ones and having retained the y - component alone out of them and performed the Fourier transformation according to formula

$$\tilde{f}(\vec{\kappa}) = \frac{1}{e_x e_z}\int_0^{e_x}\int_0^{e_z} exp(-i\kappa_x x - i\kappa_z z) f(x,z)\, dx\, dz, \tag{9}$$

we obtain

$$\left(\frac{\partial}{\partial t} + i\vec{\kappa}\cdot\vec{U} - \nu\tilde{\Delta}\right)\tilde{\Delta}\tilde{v} - i\tilde{v}\frac{\partial^2(\vec{\kappa}\cdot\vec{U})}{\partial y^2} = \frac{\partial}{\partial y}(i\vec{\kappa}\cdot\vec{f}) + \kappa^2 f_y, \tag{10}$$

$$\left(\frac{\partial}{\partial t} + i\vec{\kappa}\cdot\vec{U} - \nu\tilde{\Delta}\right)\tilde{\omega}_y + i\tilde{v}\frac{\partial[\vec{\kappa}\times\vec{U}]_y}{\partial y} = i\kappa_x f_z - i\kappa_z f_x, \tag{11}$$

$$\left(\frac{\partial}{\partial t} + i\vec{\kappa}\cdot\vec{U} - a\tilde{\Delta}\right)\tilde{\varphi} + \tilde{v}\frac{\partial T}{\partial y} = -(i\vec{\kappa}\cdot\vec{q}) - \frac{\partial}{\partial y}(q_y - \langle q_y\rangle), \tag{12}$$

where the wave denotes the Fourier transforms of functions; $\vec{\kappa} = (\kappa_x, \kappa_z)$ is the wave vector; $\tilde{\Delta} = \partial^2/\partial y^2 - \kappa^2$; $\kappa^2 = \kappa_x^2 + \kappa_z^2$; $\vec{f} = [\widetilde{\vec{\omega}\times\vec{u}}]$; $\vec{q} = (\widetilde{\vec{u}\,\varphi})$.

Inhomogeneous equation (10) contains the linear Orr-Sommerfeld operator in the left-hand side /9/. This fact proves the direct connection of the suggested approach with the theory of hydrodynamic stability. A great number of results obtained in this theory may be applied in the formulation of initial conditions for the problems of direct modelling. Eq.(10) has been obtained in /7/ at studying the linear model of turbulence transfer. Equations similar to Eqs.(10)-(11) were used in /3/ only without explicit representation of mean velocity. Note that the Fourier components (9) coincide with the average value of the functions defined by formula (1) at $\vec{k} = 0$. Therefore the equality $f(\vec{k}=0)=0$ is valid for all fluctuations.

From the discontinuty equation and the second component of vorticity

$$i k_x \tilde{u} + i k_z \tilde{w} = - \frac{\partial \tilde{v}}{\partial y}, \tag{13}$$

$$i k_z \tilde{u} - i k_x \tilde{w} = \tilde{\omega}_y, \tag{14}$$

we obtain the boundary conditions on the upper and lower sides of the channel

$$\tilde{v} = 0; \quad \frac{\partial \tilde{v}}{\partial y} = 0; \quad \tilde{\omega}_y = 0; \quad \tilde{\varphi} = 0. \tag{15}$$

The boundary condition for temperature fluctuations is also chosen to be zero (isothermal walls). From Eqs.(13)-(14) we shall find

$$\tilde{u} = \frac{i k_x}{k^2} \frac{\partial \tilde{v}}{\partial y} - \frac{i k_z}{k^2} \tilde{\omega}_y, \tag{16}$$

$$\tilde{w} = \frac{i k_z}{k^2} \frac{\partial \tilde{v}}{\partial y} + \frac{i k_x}{k^2} \tilde{\omega}_y. \tag{17}$$

Substituting these expressions into formulae defining the remaining components of vorticity we obtain

$$\tilde{\omega}_x = \frac{i k_z}{k^2} \tilde{\Delta} \tilde{v} + \frac{i k_x}{k^2} \frac{\partial \tilde{\omega}_y}{\partial y}, \tag{18}$$

$$\tilde{\omega}_z = - \frac{i k_x}{k^2} \tilde{\Delta} \tilde{v} + \frac{i k_z}{k^2} \frac{\partial \tilde{\omega}_y}{\partial y}. \tag{19}$$

By direct substitution of Exps.(16)-(17) and (18)-(19) into the equations $(\nabla \cdot \vec{u}) = 0$ and $(\nabla \cdot \vec{\omega}) = 0$ we are convinced that they are satisfied identically. Thus we could express two three-dimensional fields only via two characteristic functions so that these vector fields satisfy identically the solenoidal condition, and the functions $\tilde{v}$ and $\tilde{\omega}_y$ are described by dynamic equations (10)-(11). The obtained closed system of Eqs.(3)-(5), (10)-(12), (16)-(19) may be applied further for direct modelling of turbulence in the plane channel. To calculate the right-hand sides of Eqs.(10)-(12) which contain nonlinear terms one should normally use the pseudospectral method developed in /1/.

3. EQUATIONS FOR VORTEX STRUCTURES IN A LOG-LAYER

The developed turbulence in the near-wall region is essentially the same for different geometries (plate, tube or channel). Here the statistical regime of the velocity fluctuations is defined by the parameters of dynamic velocity $v_* = \sqrt{\tau_w / \rho}$ and viscosity ν and for the temperature fluctuations by the parameters of dynamic temperature $\theta_* = q_w / \rho c_p v_*$ and thermal diffusivity a as well, where τ_w is the shear stress and q_w is the heat flux density at the wall, c_p is the heat capacity of liquid (this is the first similarity hypothesis).

Beyond the viscous sublayer the predominant motions have large scales $\sim y$ and move rather fast so here, in fact, viscosity should be excluded from the set of typical parameters. However the universal logarithmic velocity profile

$$\frac{U}{v_*} = \frac{1}{\varkappa} \ln \frac{v_* y}{\nu} + const \tag{20}$$

involves viscosity ν and hence it will define the characteristic frequencies of vortices passing by the observer.

It is possible to eliminate the apparent viscosity effect on the characteristics of large vortices by converting into the inertial frame of reference moving with the velocity U_0 which is equal to the mean velocity of liquid at a certain point y_0 satisfying inequality $H >> y_0 >> \nu/v_*$. Then in this frame of reference we move with the turbulent flow and ignore small details, i.e. vortices of size $\sim \nu/v_*$ and viscous sublayer /7/.

The vortices of intermediate sizes in the region $H >> y >> \nu/v_*$ are not sensitive to the effect of the second wall of the channel (H) and viscosity and move self-consistently forming the dynamic cascade of structures interacting and decreasing toward the wall. If the appropriate system of coordinates related to the flow itself is chosen then the vortice characteristics are defined only by the parameters v_* and θ_* (2-nd hypothesis of similarity /7/). This near-wall system of vortices is intended mainly to transfer momentum flux ρv_*^2 to the wall.

The similarity hypotheses have been applied earlier to averaged equations for correlative functions /7/. Now a question arises if it is possible to describe a universal vortex structure directly in terms of complex amplitudes with the help of these hypotheses. So the dimensions of the variable space could be reduced to unity assuming the distance y to the wall as a natural scale to make the remaining quantities dimensionless.

The derivatives of U and T in a log-layer are defined from the dimension considerations with an accuracy up to some constants $\varkappa$ and $\varkappa_T$ (Karman's constants) /10,11/

$$\frac{dU}{dy} = \frac{v_*}{\varkappa y}; \quad \frac{d^2U}{dy^2} = -\frac{v_*}{\varkappa y^2}; \quad \frac{dT}{dy} = \frac{\theta_*}{\varkappa_T y} . \tag{21}$$

Further only stationary turbulence will be considered. In this case one can also perform the Fourier transformation in time. Then the frequency ω appears Substituting Exps.(20)-(21) into Eqs.(10)-(12) and going into dimensionless variables

$$\sigma = \frac{(\omega - k_x U(y))\varkappa y}{v_*}; \quad \vec{\gamma} = (\gamma_x, \gamma_z) = y\vec{k};$$
$$\vec{V} = (V_1, V_2, V_3) = \frac{\tilde{\vec{u}}\varkappa}{v_*}; \quad \vec{\Omega} = (\Omega_1, \Omega_2, \Omega_3) = \frac{\tilde{\vec{\omega}}\varkappa y}{v_*}; \tag{22}$$
$$\Phi = \frac{\varphi \varkappa_T}{\theta_*}; \quad \vec{Q} = \frac{(\widetilde{\vec{u}\varphi})\varkappa_T \varkappa}{\theta_* v_*}; \quad \vec{F} = \frac{-[\widetilde{\vec{u}\times\vec{\omega}}]\varkappa^2 y}{v_*^2},$$

and introducing new differentiation operators

$$\hat{\Delta} = \hat{\mathcal{D}}^2 - \hat{\mathcal{D}} - \gamma^2 ,$$
$$\hat{\mathcal{D}} = (\sigma - \gamma_x)\frac{\partial}{\partial\sigma} + \gamma\frac{\partial}{\partial\gamma} , \qquad (23)$$

we shall rearrange the system of Eqs.(10)-(12), (16)-(19) to the form

$$i\sigma\hat{\Delta}V_2 - i\gamma_x V_2 = -(\hat{\mathcal{D}}-2)(i\vec{\gamma}\cdot\vec{F}) - \gamma^2 F_2 ; \qquad (24)$$
$$i\sigma\Omega_2 - i\gamma_z V_2 = i[\vec{\gamma}\times\vec{F}]_2 ; \qquad (25)$$
$$i\sigma\Phi - V_2 = (i\vec{\gamma}\cdot\vec{Q}) + \hat{\mathcal{D}}Q_2 ; \qquad (26)$$
$$\gamma^2 V_1 = i\gamma_x\hat{\mathcal{D}}V_2 - i\gamma_z\Omega_2 ; \quad \gamma^2 V_3 = i\gamma_z\hat{\mathcal{D}}V_2 + i\gamma_x\Omega_2 ; \qquad (27)$$
$$\gamma^2\Omega_1 = i\gamma_z\hat{\Delta}V_2 + i\gamma_x(\hat{\mathcal{D}}-1)\Omega_2 ; \quad \gamma^2\Omega_3 = -i\gamma_x\hat{\Delta}V_2 + i\gamma_z(\hat{\mathcal{D}}-1)\Omega_2 ;$$
$$\vec{F} = \sum_{\sigma',\gamma'}\left[\vec{\Omega}(\sigma',\vec{\gamma}')\times\vec{V}(\sigma-\sigma',\vec{\gamma}-\vec{\gamma}')\right] ; \quad Q = \sum_{\sigma',\gamma'}\Phi(\sigma',\vec{\gamma}')\vec{V}(\sigma-\sigma',\vec{\gamma}-\vec{\gamma}') \qquad (28)$$

The last expressions are obviously convolutions in $\sigma - \vec{\gamma}$ - space. When deriving the system of Eqs.(24)-(28) we neglected the molecular effects of heat and momentum transfer and therefore only the impermeability condition $V_2(\sigma = \vec{\gamma} = 0) = 0$ will remain out of the overall conditions.

All amplitudes should decrease sufficiently fast up to zero at large distances from the coordinate origin in $\sigma - \vec{\gamma}$ - space. For this reason the second boundary condition at $|\sigma| = \gamma \to \infty$ holds true: $\vec{V}, \vec{\Omega}, \Phi \to 0$.

Note the important property of the obtained system of Eqs (24)-(28), i.e. the system contains no parameters. The system solutions i.e. complex amplitudes of the velocity, vorticity and temperature fluctuations will be some universal functions of three variables $\sigma, \gamma_x, \gamma_z$.

Let us consider how to pass to new similarity variables in conventional space to elucidate the meaning of obtained equations (24)-(28).

4. EQUATION ANALYSIS IN X-T - SPACE

In this paragraph let us introduce the coordinate and function transformations more general than (22) and such that will be principle scales v^* and y as previously in the logarithmic sublayer, but at $y \to 0$ the values v_* and ν become the principle parameters in a viscous sublayer. An important point is the introduction of new dimensionless coordinate proportional to mean velocity instead of the fourth variable y . New variables are of the form

$$\eta=\frac{\varkappa U(y)}{v_*};\quad \xi=\frac{\varkappa U'}{v_*}(x-Ut);\quad \zeta=\frac{\varkappa U'}{v_*}z;\quad \tau=U't;$$

$$\vec{V}=\frac{\varkappa}{v_*}\vec{u};\quad \Phi=\frac{\varkappa_T}{\theta_*}\varphi;\quad \vec{\Omega}=[\hat{\nabla}\times\vec{V}];\quad \hat{\nabla}=\left(\frac{\partial}{\partial\xi},\hat{\mathcal{D}},\frac{\partial}{\partial\zeta}\right);\tag{29}$$

$$\mathcal{P}=\frac{\varkappa^2}{v_*^2}\left(\frac{p}{\rho}+\frac{\vec{u}^2}{2}-\left\langle\frac{\vec{u}^2}{2}\right\rangle\right);$$

where operator $\hat{\nabla}$ is expressed in terms of dimensionless operator of differentiating with respect to y

$$\hat{\mathcal{D}}=\frac{v_*}{\varkappa U'}\left(\frac{\partial}{\partial y}\right)_{x,z,t}=\frac{\partial}{\partial\eta}-\tau\frac{\partial}{\partial\xi}-\frac{y''(\eta)}{y'(\eta)}\left(\tau\frac{\partial}{\partial\tau}+\xi\frac{\partial}{\partial\xi}+\zeta\frac{\partial}{\partial\zeta}\right).\tag{30}$$

Jacobian of transformation (2) equal

$$\frac{\partial(\eta,\xi,\zeta,\tau)}{\partial(y,x,z,t)}=\frac{(U')^4}{v_*^3}.$$

Hence if $U'\neq 0$ in the entire region there exists a one-to-one correspondence between new and former coordinates. For example, Couette's plane flow satisfies such condition. In Poiseuille's flow we have $U'=0$ on the axis and the introduction of new coordinates (29) is possible only in one of the channel halves. However coordinates (29) can be used for all types of near-wall flows in the near-wall region (in a viscous sulayer and in the logarithmic region).

After transforming to new variables the discontinuity equation (6), dynamic equations (7)-(8) for velocity and temperature fluctuations as well as the Reynolds equations integrated with respect to y

$$v_*^2=\nu\frac{dU}{dy}-\langle uv\rangle;$$
$$v_*\theta_*=\alpha\frac{dT}{dy}-\langle\varphi v\rangle;\tag{31}$$

(we have additionally averaged Eqs.(3)-(5) in terms of rather large time interval T_0) are written in the form

$$(\hat{\nabla}_j V_j)=0;\tag{32}$$

$$\frac{\partial V_i}{\partial\tau}+\delta_{i1}V_2+\hat{\nabla}_i\mathcal{P}=\varepsilon_{ijk}\left(V_j\Omega_k-\langle V_j\Omega_k\rangle-\nabla_j\left(\frac{\Omega_k}{R}\right)\right);\tag{33}$$

$$\frac{\partial\Phi}{\partial\tau}+V_2\frac{d\theta}{d\eta}=-\hat{\nabla}_j\left(V_j\Phi-\langle V_j\Phi\rangle-\frac{1}{Pr\,R}\nabla_j\Phi\right);\tag{34}$$

$$\varkappa^2=1/R(\eta)-\langle V_1V_2\rangle;$$
$$\varkappa\varkappa_T=1/(Pr\,R(\eta))-\langle\Phi V_2\rangle;$$

where the local Reynolds number $R(\eta)=\frac{1}{\varkappa}\frac{dy^+}{d\eta}$ the Prandtl number $Pr=\nu/\alpha$ and dimensionless mean temperature $\theta=\varkappa_T T/\theta_*$ are introduced. The function $R(\eta)$ assumes minimum value on the wall at $y=0$: $R(0)=\varkappa^{-2}=6,25$, and should have an asymptote $R \sim e^{\eta}$ in the logarithmic sublayer so that here $\eta = \ln y_+ + const$. Operator $\mathcal{D}$ incorporates the relation $y''/y' = R'/R$ which reduces to unity at a distance from the wall. The equality $d\theta/d\eta = 1$ should be satisfied also at a sufficient distance from the wall that is related to logarithmic asymptote for the mean temperature profile ($\theta = \eta + const$). Note important properties characteristic for the system of equations (32)-(34) at a distance from the wall: 1) the value R increases rather quickly (as e^{η}) with the growth of η and the terms containing $1/R$ can be neglected; 2) the term $\langle V \Phi \rangle$ in Eq.(34) is eliminated at differentiating by virtue of the fact that it is constant in the logarithmic sublayer; 3) the term $\langle V_j \Omega_k \rangle$ also vanishes here from Eq.(33) as it is obtained from the terms $(\partial/\partial x_j)\langle u_i u_j \rangle$ in Eq.(7) and all single - point correlations including $\langle u_i u_j \rangle$ are constant in the logarithmic region (corollary of the 2-nd similarity hypothesis). Therefore in the logarithmic layer at $\eta \gg 1$ the system (32)-(34) takes the form

$$\frac{\partial V_1}{\partial \xi} + \mathcal{D} V_2 + \frac{\partial V_3}{\partial \zeta} = 0 \; ; \tag{35}$$

$$\frac{\partial V_1}{\partial \tau} + V_2 + \frac{\partial \mathcal{P}}{\partial \xi} = V_2 \Omega_3 - V_3 \Omega_2 \; ; \tag{36}$$

$$\frac{\partial V_2}{\partial \tau} + \mathcal{D}\mathcal{P} = V_3 \Omega_1 - V_1 \Omega_3 \; ; \tag{37}$$

$$\frac{\partial V_3}{\partial \tau} + \frac{\partial \mathcal{P}}{\partial \zeta} = V_1 \Omega_2 - V_2 \Omega_1 \; ; \tag{38}$$

$$\frac{\partial \phi}{\partial \tau} + V_2 = -\frac{\partial}{\partial \xi}(V_1 \Phi) - \mathcal{D}(V_2 \Phi) - \frac{\partial}{\partial \zeta}(V_3 \Phi) ; \tag{39}$$

$$\varkappa^2 = -\langle V_1 V_2 \rangle \; ; \tag{40}$$

$$\varkappa \varkappa_T = -\langle \Phi V_2 \rangle \; ; \tag{41}$$

where operator $\mathcal{D}$ is

$$\mathcal{D} = \frac{\partial}{\partial \eta} - \left(\tau \frac{\partial}{\partial \tau} + (\xi + \tau)\frac{\partial}{\partial \xi} + \zeta \frac{\partial}{\partial \zeta}\right) . \tag{42}$$

The obtained system of equations (35)-(39) can be transformed to system (24) neglecting the dependence on η ,using the Fourier transformation and eliminating pressure $\mathcal{P}$ from equations (36)-(38).

If the dependence on η is critical (as it will be shown further) then the system (35)-(39) describes apparently some continuous chaotic motion in the space of variables η, ξ, ζ, τ without any external sources of chaos and only due to the presence of nonlinear terms in the right-hand sides of equations. Here we most likely deal with a strange attractor in a dimensionless phase space of system(35)-(39).

The variable η is included into equations (35)-(39) only under the sign of a derivative whereas ξ, ζ, τ are incorporated in operator explicitly. Therefore it is natural to suppose that solutions of the system (35-39) represent homogeneous random process with respect to η and nonhomogeneous one with respect to ξ, ζ, τ. This can be proved by considering transformations to new variables of any correlative function. To find general form of correlative functions it is necessary to make use of similarity hypotheses.

5. SIMILARITY HYPOTHESES AND COROLLARIES

Let us formulate completely the similarity hypotheses:

Hypothesis 1: At rather large Reynolds numbers $Re = \frac{U_{av} H}{\nu} >> 1$ statistic regime of turbulent velocity fluctuations near the wall is defined only by parameters v_* and ν for $y << H$ but for temperature fluctuations also by parameters θ_* and α.

Hypothesis 2: At $Re >> 1$ in the region $H >> y >> \nu / v_*$ turbulent regime of velocity fluctuations is defined only by value v_* (and for temperature fluctuations also by θ_*) if the coordinate system moving with the mean liquid velocity in one of the points out of the given distance interval y is chosen.

Let us choose, for example, a two-point correlative function $\langle v' v \rangle$. According to homogeneity condition with respect to x and z variables and the second similarity hypothesis this function is presented in the logarithmic region as

$$\langle v'v \rangle = v_*^2 \Psi\left(\frac{y'-y}{\sqrt{y'y}};\ \frac{x'-x-U(\sqrt{y'y})(t'-t)}{\sqrt{y'y}};\ \frac{z'-z}{\sqrt{y'y}};\ \frac{t'-t}{\sqrt{y'y}} v_*\right), \quad (43)$$

where Ψ is the universal function of four arguments in which only velocity $U(y,\nu)$ depends on viscosity in the second argument. We have assumed geometrical average out of two distances from the wall ($\sqrt{y'y}$) for dimensionless scale and have chosen the coordinate system moving with the average arithmetic velocity $U(\sqrt{y'y}) = (U(y') + U(y))/2$. Transforming Exps. (43) to new variables (29) we obtain

$$\langle v'v \rangle = v_*^2 \Psi\left(e^{\frac{\eta'-\eta}{2}} - e^{\frac{\eta-\eta'}{2}};\ \left(\xi' - \tau'\frac{\eta'-\eta}{2}\right)e^{\frac{\eta'-\eta}{2}} - \left(\xi - \tau\frac{\eta-\eta'}{2}\right)e^{\frac{\eta-\eta'}{2}}; \quad (44)$$
$$\zeta' e^{\frac{\eta'-\eta}{2}} - \zeta e^{\frac{\eta-\eta'}{2}};\ \ \tau' e^{\frac{\eta'-\eta}{2}} - \tau e^{\frac{\eta-\eta'}{2}}\right).$$

Hence is seen that the correlative function depends only on difference ($\eta' - \eta$) and since it is valid for any correlative functions (not only two-point ones) one can conclude that there is a statistic homogeneity in terms of the variable η in the system solutions (35)-(39) in the log-layer. In terms of variables ξ, ζ, τ the correlative functions will depend on differences alone at $\eta' = \eta$. If in $x-t$ space the coordinate origin is carried to some point with the coordinates x_o, z_o, t_o

$$x' = x - x_0;\quad z' = z - z_0;\quad t' = t - t_0; \quad (45)$$

then it is equivalent to going to new variables

$$\xi' = \xi - \xi_0(\eta);\quad \zeta' = \zeta - \zeta_0(\eta);$$
$$\tau' = \tau - \tau_0(\eta);\quad \eta' = \eta; \quad (46)$$
$$\xi_0 = (x_0 - \eta v_* t_0/\varkappa)/y'(\eta);\ \zeta_0 = z_0/y'(\eta);\ \tau_0 = v_* t_0/\varkappa y'(\eta).$$

In new coordinates the operators $\hat{\nabla}, \mathcal{D}$ are of the same form as before

$$\hat{\nabla}' = \left(\frac{\partial}{\partial \xi}\,;\ \mathcal{D}\,;\ \frac{\partial}{\partial \varsigma}\right);$$
$$\mathcal{D}' = \frac{\partial}{\partial \eta'} - \left(\tau' \frac{\partial}{\partial \tau'} + (\tau' + \xi')\frac{\partial}{\partial \xi'} + \varsigma' \frac{\partial}{\partial \varsigma'}\right). \tag{47}$$

Thus invariance in terms of shifts (45) with respect to variables x, z, t which is attributed to statistic stationarity and homogeneity of all random fields manifests it-self in new variables as invariance with respect to transformations (46).

Let us make use of this invariance to perform conventional averaging as a result of which the dependence of all values on variable η should vanish in Eqs.(35)-(39). By way of example, for the Lorentz attractor let us apply the conventional average method developed in /12/.

Assume that regions with some characteristic attributes can be singled out in turbulent random field ($\vec{V}, \mathcal{P}$). For example, a high level of the product $V_1 V_2$ with due account of signs ($h + \Delta h > -V_1 V_2 > h$) can serve as such attribute that can be classified (at $h \geqslant 1$) as "bursts" giving, as known, a substantial contribution in complete Reynolds stresses /3/.

Having chosen at given y $\mathcal{N}$ of the points x_n, z_n, t_n in which formulated conditions are satisfied we can perform $\mathcal{N}$ shifts of the coordinate system in these points. In their neighbourhood Eqs.(35)-(39) will be of the same form according to transformation (45)-(47). Having added $\mathcal{N}$ equations written in the vicinity of given points at constant ξ, ς, τ and divided the sum by $\mathcal{N}$ we shall obtain conventionally averaged equations. The sign of such averaging is the bar over the functions. At such averaging the closure problem arises since after averaging the nonlinear terms we have

$$\overline{[\vec{V} \times \vec{\Omega}]} = [\overline{\vec{V}} \times \overline{\vec{\Omega}}] + \overline{[\vec{V}' \times \vec{\Omega}']},$$
$$\overline{\vec{V}\Phi} = \overline{\vec{V}}\,\overline{\Phi} + \overline{\vec{V}'\Phi'}, \tag{48}$$

where the prime designates the deviations of actual fields from conventionally averaged ones, e.g. $\vec{V}' = \vec{V} - \overline{\vec{V}}$. All the rules of Reynolds averaging are valid for the procedure of conventional averaging as well. There-fore the formula

$$\overline{\frac{\partial}{\partial \eta} f} = \frac{\partial}{\partial \eta} \overline{f} \tag{49}$$

is valid.

But can the averaged value $\overline{f}$ depend on η along which (in the region of log-layer in new variables $\eta, \xi, \varsigma, \tau$) turbulence is statistically homogeneous? Such explicit dependence on η in averaged values would mean their dependence on viscosity in normal coordinates as η is proportional to mean velocity $U(y, \nu)$.

This, however, contradicts to the second similarity hypothesis. There fore all conventionally averaged values should not depend on η and all derivatives with respect to η vanish in Eqs.(35)-(39).

In this work we shall not dwell on the enclosure problem, i.e. the problem to define the terms $[\overline{\overline{V' \times \Omega'}}]$ through $\overline{\overline{V}}$ and $\overline{\overline{\Omega}}$. We shall merely ignore these terms as the equation obtained for vortex structures are strongly nonlinear.

As a result of all transformations the system (35-39) in averaged variables will be of the same form as previously only now operator $\mathcal{D}$ no longer contains differentiation with respect to η

$$\bar{\mathcal{D}} = -\left(\tau \frac{\partial}{\partial \tau} + (\xi + \tau)\frac{\partial}{\partial \xi} + \zeta \frac{\partial}{\partial \zeta}\right). \tag{50}$$

Assuming that the near-wall turbulent process is the result of the representing point wandering in phase space along a certain chaotic attractor it is natural at large values of variables to lay down conditions which are zero with respect to all conventionally averaged functions

$$\overline{\overline{V}} = 0; \quad \overline{\overline{\Omega}} = 0; \quad \bar{\mathcal{P}} = 0; \quad \bar{\Phi} = 0 \qquad at \quad \xi, \zeta, \tau \to \pm\infty. \tag{51}$$

These conditions are related to a quick loss of "memory" and the absense of further order in turbulent random values.

The obtained conventionally averaged system of equations can be Fourier transformed and after eliminating pressure, increasing the equation order and performing all transformations previous to the derivation of the system (24)-(28) we shall obtain precisely this system of equations. However the representation of the system (35)-(39) in the variables ξ, ζ, τ is more advantageous as it contains no sums or integrals of the convolution type.

The question concerning the existence and number of soliton solutions of the system (35)-(39) remains open. A number of such solutions satisfying different conditions of averaging and describing the events of different types such as ejections or bursts (according to terminology adopted in experimental methods of conventional averaging /12/) should apparently exist.

6. DISCUSSION

The system of dynamic equations averaged with respect to homogeneous variables x and z for the flow in a plane channel has been derived in § 2. Equations for fluctuation components of velocity are reduced to two dynamic equations: inhomogeneous Orr-Sommerfeld's equation for transverse fluctuation of velocity v and equation for transverse component of vorticity ω_2. The remaining components of fluctuations for the velocity and vorticity vectors are expressed in terms v and ω_2 by constraint equations so that solenoidal conditions of these vector fields are satisfied identically. The appearance of Orr-Sommerfeld's equation, the principle equation of the theory of hydrodynamic stability, among the basic dynamic fluctuation equations is of interest from our viewpoint and requires further consideration.

Statistically stationary flow near the wall in a log-layer was considered in § 3 and a system of self-similar equations was obtained for complex amplitudes of temperature and velocity fluctuations. However further analysis in § 4-5 showed that these equations could not describe statistically stationary and homogeneous actual turbulent processes. § 5 elucidated the meaning of Eqs. (24-28) obtained in a Fourier space. They appear to describe in a Fourier space conventionally averaged vortex structures in a log-layer with statistic damping or decreasing of correlations at large times and distances in $x - t$ - space from the origin where these structures are located.

New variables which agreed with the similarity hypotheses of near-wall turbulence were introduced in § 4. The equations obtained in these variables describe actual turbulent process near the wall (in a particular case of Couette flow these equations can be used for the entire channel). In the introduced variables the region of viscous sublayer $\eta < 2$ and the so-called buffer region cover a substantial space $2 < \eta < 5.5$ whereas the log-layer region contracts essentially ($5.5 < \eta < 15 \div 20$). Therefore one can assume that the derived equations will be useful when studying the processes of turbulence generation in the near-wall region.

It is of interest to note that Karman's constants $\varkappa$ and $\varkappa_T$ defining the slope of logarithmic profiles of mean velocity and temperature can be obtained using the solutions of the complete system of equations (35-39) obtained in the form of correlations (40-41). Thus Karman's constants are not the eigenvalues of any nonlinear problem. They are determined using single-point correlations (40-41)of dimensionless velocity and temperature fields,the equations (35-39) for which contain no parameters and describe probably some chaotic attractor in an infinite-dimensional phase space.

The peculiarity of equations derived in § 4 is that they contain variables ξ, ζ and τ explicitly and the variable η only under the differentiation sign. For this reason the fluctuation described in these new coordinates should be homogeneous with respect to η and inhomogeneous with respect to the remaining variables. In § 5 this conclusion is confirmed by considering the example of transformation of two-point correlative function written according to the second similarity hypothesis in the form of function of four arguments to new coordinates.

The similarity hypotheses of near-wall turbulence are exactly formulated in § 5. Normally the second similarity hypothesis is formulated without stating that it is valid for two-point or multipoint correlations with respect to time only in a moving frame of reference "fixed" to a turbulent flow. We used substantially this important concept of passing to a moving frame of reference to eliminate the effects of convection with the mean velocity. Similarity hypotheses are discussed in /6,7,11,13-17/.

A new system of equations with no variable η was obtained after conventional averaging in § 5 from the complete system (35)-(39). The solutions of this conventionally averaged system of equations (35-39) which damp at infinity in space and time are assumed to describe vortex structures in a log-layer in self-similar variables.

It should be noted that this self-similarity is attributed to the absence of characteristic scales for fluctuations in a log-layer, i.e. hypothesis of scale invariance (2-nd hypothesis). The obtained conventionally averaged system of equations (35)-(39) can be transformed to the system (24)-(28) that accounts for the latter.

We believe that the given formulation of the second hypothesis will stimulate experimentors to represent correctly multipoint spectra and correlative functions in the similarity variables (e.g. correlative function (43)). The similarity variables should be also used when measuring vortex structures in a log-layer.

REFERENCES

1. Orszag S.A., Patera A.T. Calculation of von Karman's constant for turbulent channel flow. Phys.Rev.Lett., vol.47, no.12, pp.832-835, 1981.

2. Rozhdestvensky B.L., Simakin I.N. Modelling of turbulent flows in a plane channel. J.Comp.Math. and Math.Phys., vol.25, no.1, pp.96-121, 1985(Russian).

3. Kim J., Moin P., Moser R., Turbulent statistics in fully developed channel flow at low Reynolds number. J.Fluid Mech., vol.177, pp.133-166, 1987.

4. Leonard A., Wray A.A., A new numerical method for the simulation of three-dimensional flow in a pipe. Lect.Notes Phys., vol.170, pp.336-342, 1982.

5. Moser R.D., Moin P., Leonard A., A spectral numerical method for the Navier-Stokes equations with applications to Taylor-Couette flow. J.Comp.Phys., vol.52, pp.524-544,1983.

6. Kader B.A.,Yaglom A.M. Similarity laws for near-wall turbulent flows. Res. of Sci. and Techn. Fluid and Gas Mech., VINITI, Moscow, vol.15, 1981 (Russian).

7. Geshev P.I. Linear model of near-wall turbulence transfer. - Novosibirsk., Preprint N 73-81, Institute of Thermophysics, 1981 (Russian).

8. Kolmogoroff A.N., Local structure of turbulence in incompressible liquid at very large Reynolds numbers.- Rept. AN SSSR, vol.30, n.4, pp.299-303, 1941 (Russian).

9. Goldshtik M.A., Stern V.N., Hydrodynamic stability and turbulence. - "Nauka", Novosibirsk, 1977. - 336 p. (Russian).

10. Landau L.D., Lifshits E.M., Hydrodynamics, is.3., p.244, "Nauka", Moscow,1986(Russian).

11. Monin A.S., Yaglom A.M. Statistic hydromechanics, Part 1. - "Nauka", Moscow, 1965 (Russian).

12. Goldshtik M.A., Stern V.N. Structural turbulence in dissipative systems. - Preprint 74-81, Institute of Thermophysics, Novosibirsk, 1981(Russian).

13. Bogard D.G., Tiederman W.G. Burst detection with single-Point velocity measurements. J.Fluid Mech., vol.162, pp.389-413, 1986.

14. Perry A.E., Abbel C.J. Asymptotic Similarity of turbulence structures in smooth - and rough-walled pipes. J.Fluid Mech., vol.79, no.4, pp.785-800, 1977.

15. Bullok K.J., Cooper R.E., Abernathy F.H., Structural similarity in radial correlations and spectra of longitudinal velocity fluctuations in pipe flow. J.Fluid Mech., vol.88, no.3, pp.385-408, 1978.

16. Perry A.E., Lim K.L., Henbest S.M., An experimental study of the turbulence in smooth - and rough-wall boundary layers. J.Fluid Mech., vol.177, pp437-466, 1987.

17. Kader B.A., Yaglom A.M. Spectra of anisotropic turbulent fluctuations of velocity and temperature in near-wall turbulent flows. - Problem of turbulent flows, "Nauka", Moscow, 1987 (Russian).

The Attached Eddy Hypothesis in Wall Turbulence

A. E. PERRY, J. D. LI, S. M. HENBEST, and I. MARUŠIĆ
Department of Mechanical Engineering
University of Melbourne
Parkville, Victoria 3052, Australia

SUMMARY

In this paper a closure scheme for turbulent wall shear flow is developed which makes use of a coherent structure concept, i.e. the attached eddy hypothesis. Recent experimental verification of this hypothesis is also given here.

INTRODUCTION

The purpose of this paper is to review recent developments in the theory of wall turbulence based on the attached eddy hypothesis of Townsend (1976) as extended by Perry & Chong (1982) and Perry, Henbest and Chong (1986).

Let us briefly summarize this theory in words. Wall turbulence is considered to consist of a "forest" of randomly positioned hair-pin, horseshoe or "Λ" shaped vortices which lean in the streamwise direction (at about 45°) and have their legs extending to the boundary. These eddies maintain their orientation as they "slip" relative to the boundary while being convected downstream.

From some crude calculations of Perry & Chong (1982), it appears that the eddies change their shape as they evolve from their initial formation to ultimate death. Applying the Biot-Savart law to an isolated "Λ" shaped vortex together with its image in the wall indicates a stretching process in which the height of the eddy increases linearly with time and the distance between the legs decreases, leading to eddy death by vorticity cancellation. Such an assemblage of eddies, all of varying heights and randomly positioned over the surface, is

---

S. M. Henbest's present address is at the Aeronautical Research Laboratory, Fishermans Bend, Victoria, Australia.

as a "hierarchy". However, this is not the end of the description. A turbulent boundary layer consists of many geometrically similar "hierarchies" all having a different characteristic length scale δ and there is an associated p.d.f., $P_H(\delta)$, of hierarchy length scales, i.e. we have in effect a "hierarchy of hierarchies". For zero pressure gradient boundary layers, and for pipe and duct flow it has been assumed that the characteristic velocity scale is the wall shear velocity U_τ and is the same for all hierarchies although it will be seen later that a distribution of characteristic velocity scales can be incorporated into the model enabling the possibility of extending the theory to adverse pressure gradient flows.

These attached "coherent" eddies or "coherent substructures" are assumed to be responsible for all of the mean flow vorticity and Reynolds shear stress distributions and for most of the broadband turbulence kinetic energy. However it was shown by Perry et al (1986) that these motions alone are not sufficient for explaining the energy dissipation. It was therefore postulated that superimposed on the coherent attached eddy motions are detached incoherent statistically isotropic and statistically irrotational assemblages of eddying motions which contribute to a classical Kolmogorov inertial subrange. These motions are thought to be made up of the debris of broken up attached eddying motions.

A MATHEMATICAL DESCRIPTION OF THE ATTACHED EDDY HYPOTHESIS

Strictly speaking the analysis presented here is applicable only to flat surfaces i.e., flow in boundary layers and wide (or high aspect ratio) rectangular ducts. However the functional forms should also be applicable to circular pipes on dimensional grounds. The experimental results of Perry et al (1986) support this.

Let $\overline{U_i}$ be the mean velocity and i=1 is in the streamwise x direction, i=2 is in the spanwise y direction and i=3 is in the z direction which is normal to the wall. The mean vorticity is given by:

$$\frac{d\overline{U_1}}{dz} = \int_{\delta_1}^{\Delta_E} \xi_H\left(\frac{z}{\delta}\right) P_H(\delta)\, d\delta \tag{1}$$

where $\xi_H(z/\delta)$ is the contribution to the mean vorticity from one hierarchy and equation (1) gives the contribution to the mean vorticity at a distance z from the wall from all hierarchies whose scale ranges from δ_1 (the smallest hierarchy) to Δ_E the largest hierarchy (which is assumed to correspond with the boundary layer thickness). The smallest hierarchy δ_1 is assumed to scale with the Kline scaling i.e. $\delta_1 \cong 100\ \nu/U_\tau$ if the wall were smooth or with the scale of the roughness if the wall were rough, e.g. see Perry, Lim and Henbest (1987). δ_1 corresponds to the scale of the smallest coherent attached motion but there are probably scales smaller which belong to the detached "incoherent" motions.

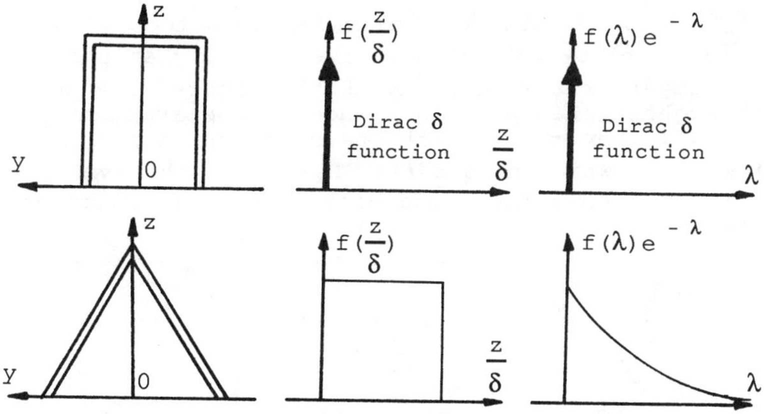

FIGURE 1. A 'Π' and 'Λ' shaped respresentative eddy, looking downstream, with their corresponding vorticity functions f.

It can be shown by dimensional analysis that:

$$\xi_H = \frac{U_\tau}{\delta} f\left(\frac{z}{\delta}\right) \tag{2}$$

where f depends on the geometry of the hierarchy. For simplicity let it be assumed that a hierarchy can be described by what Townsend would call a simple "representative eddy". Actually an assemblage of eddies would be required with eddies being at a different state of stretching. Incorporation of this aspect into the modelling is a refinement for future work. Figure 1 shows some simple representative eddy geometries together with $f(z/\delta)$. The probability density function $P_H(\delta)$ was assumed by Perry & Chong (1982) to be:

$$p_H = \frac{M}{\delta} \tag{3}$$

since this leads to the logarithmic law of the wall for $\delta_1 << z$

$<<\Delta_E$. Here M is a constant. Substituting (3) and (2) into (1) and using the logarithmic variables $\lambda = \ln(\delta/z)$; $\lambda_1 = \ln(\delta_1/z)$ and $\lambda_E = \ln(\Delta_E/z)$ we obtain:

$$\frac{dU_D^*}{d\lambda_E} = \int_{\lambda_1}^{\lambda_E} M\ f(\lambda)\ e^{-\lambda}\ d\lambda \qquad (4)$$

where U_D^* is the nondimensional velocity defect, i.e. $U_D^* = (\overline{U}_{1E} - \overline{U_1})/U_\tau$ where $\overline{U_{1E}}$ is the velocity at the edge of the boundary layer or the velocity at the centre of a duct. (Note that throughout this paper a change in the argument of a function will be written with the same function symbol i.e. $f(z/\delta)$ will be written as $f(\lambda)$. Strictly, the symbol f should be altered but this may cause confusion). Integrating equation (4) gives:

$$U_D^* = \int_0^{\lambda_E}\int_{\lambda_1}^{\lambda_E} M\ f(\lambda)\ e^{-\lambda}\ d\lambda d\lambda_E \qquad (5)$$

For $\lambda_E >> o$ and $\lambda_1 << o$, it is found that (5) will always lead to the logarithmic law of the wall and so the constant M is set by the constraint,

$$\frac{dU_D^*}{d\lambda_E} = \frac{1}{\kappa} \quad \text{for } \lambda_E >> 0 \text{ and } \lambda_1 << 0 \qquad (6)$$

(Here κ is the Kármán constant =0.41)

Perry et al (1986) and more recently Perry (1987) have extended equation (4) so that not only does it produce a logarithmic law of the wall but also a realistic Coles wake function (Coles 1956). This also opens up the possibility to include adverse pressure gradient boundary layers in the model description. A weighting function $q_V(\delta/\Delta_E)$ for the vorticity field is introduced and so:

$$\frac{dU_D^*}{d\lambda_E} = \int_{-\infty}^{+\infty} M\,f(\lambda)\ e^{-\lambda}\ q_v(\lambda-\lambda_E)\ d\lambda \qquad (7)$$

The function q_V has incorporated in it a "box-car" function which switches on at λ_1 and off at λ_E thus changing the limits of integration to give a convolution integral equation. If $dU_D^*/d\lambda_E$ is known and $f(\lambda)$ is known by assuming an eddy geometry then $q_V(\lambda-\lambda_E)$ can be found by a deconvolution process using Fourier Transforms. Figure 2 shows $q_V(\lambda-\lambda_E)$ for a zero pressure gradient boundary layer assuming Π shaped eddies as given in figure 1.

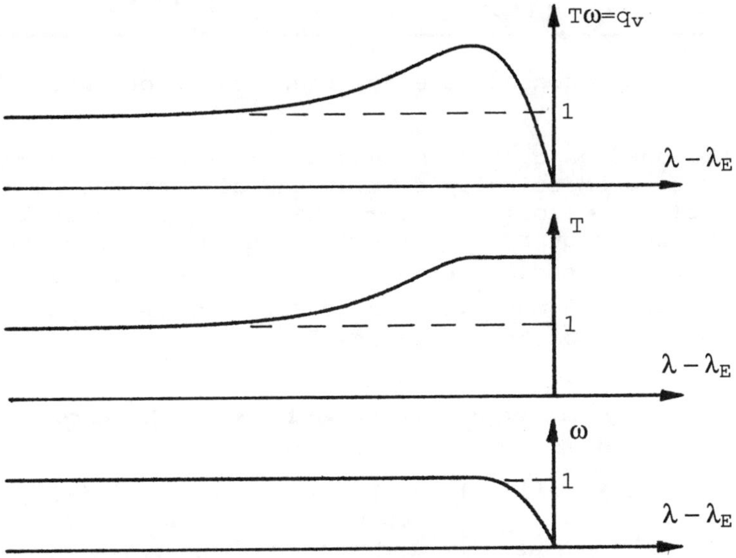

FIGURE 2. Decomposition of weighting function q_V into T and ω. After Perry et al (1988).

Similarly it can be shown (Perry 1987) that the contribution to the Reynolds normal and shear stresses from the attached eddies is given by:

$$\frac{\overline{u_i u_j}}{U_\tau^2} = \int_{-\infty}^{+\infty} M \; I_{ij}(\lambda) \; q_s(\lambda-\lambda_E) \; d\lambda \tag{8}$$

where u_i is the fluctuating component and $I_{ij}(z/\delta)$ are the Townsend (1976) "eddy intensity" functions and q_S is the weighting function for the stress field. These functions are shown sketched in figure 3 and the most important feature is their property as $(z/\delta) \to 0$. This has a dominant effect on the functional form of the broadband turbulence intensity distributions and Townsend (1976) derived the properties shown using inviscid boundary conditions at the wall. The weighting function q_s was assumed by Perry et al (1986) and Perry (1987)

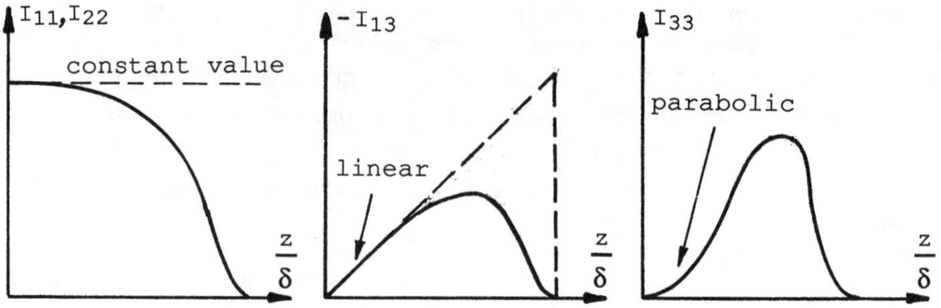

FIGURE 3. Representative eddy intensity functions.

to be the same as q_v. This would be true if all representative eddies were geometrically similar and all had the same characteristic velocity scale. Then $q_v = q_s = q$ would reflect a weighting to account for variations in eddy population density (i.e. number of eddies per area δ^2) with hierarchy scale. It would represent a modification to $P_H(\delta)$ in the region beyond the logarithmic law of the wall zone.

A Preliminary Attempt to Formulate a Closure Hypothesis

Perry, Li and Marušić (1988) have done some initial ground work in the development of a closure scheme. Again for simplicity a Π shaped eddy was assumed and the eddy intensity function $I_{13}(z/\delta)$ was assumed to be a triangle as shown dotted in figure (3). In order to successfully describe adverse pressure gradients boundary layers it was necessary to write:

$$q_v(\lambda-\lambda_E) = T(\lambda-\lambda_E)\ \omega(\lambda-\lambda_E) \qquad (9)$$

where $\omega(\lambda-\lambda_E)$ reflects eddy population density within a hierarchy scale and $T(\lambda-\lambda_E)$ reflects a change in the characteristic velocity scale with hierarchy scale. The weighting function q_s was written as:

$$q_s(\lambda-\lambda_E) = S(\lambda-\lambda_E)\ T^2(\lambda-\lambda_E)\,\omega(\lambda-\lambda_E) \qquad (10)$$

Note how we must include a T^2 in (10) and the quantity $S(\lambda-\lambda_E)$ was included to account for changes in eddy shape. This was restricted to changes in aspect ratio of the Π eddies which is assumed to leave $f(\lambda)$ unchanged but appropriate adjustments to $T\omega$ are made. For the moment it is hoped that the shape change can be accounted for by the inclusion of S in (10) alone. Actually the inclusion of $S(\lambda-\lambda_E)$ was found to be necessary to ensure a negative gradient in Reynolds stress with wall distance for zero pressure gradient boundary layers. The form of S may be different for each of the Reynolds stress components but this will not be pursued here. For $\lambda-\lambda_E$ large and negative, i.e. in the law of the wall region S, T and ω are all unity. The weighting q_V can be found by a deconvolution process by assuming the Coles law of the wall and law of the wake formulation i.e.:

$$U_D^* = -\frac{1}{\kappa}\ln\eta + \frac{\Pi}{\kappa}w(1) - \frac{\Pi}{\kappa}w(\eta) \qquad (11)$$

where $\eta = z/\Delta_E$, Π is the Coles wake factor, and $w(\eta)$ is the Coles wake function. Once $q_V = T\omega$ is known, there is a problem of separating the T variations from the ω variations. Figure 2 illustrates how this was done and what assumptions were made to do it. Firstly the rise in q_V with $(\lambda-\lambda_E)$ was attributed to T which is assumed to reach a plateau. The fall in q_V is attributed entirely to ω which falls monotonically like an intermittency factor. (ω was assumed to be a universal function of z/Δ_E and is independent of Π). The quantity S was also assumed to be universal and independent of Π and quite satisfactory results can be obtained for adverse pressure gradient layers by fixing S and ω from zero pressure gradient results. Figure 4 shows results of Reynolds shear stress profiles compared with experiment for various values of Π and agreement is quite encouraging. The dashed curves in figure 4 correspond to total shear stress derived from a momentum balance, see Perry et al (1988) for details. What the model implies is that:

$$U_D^* = \phi(\eta, \Pi)$$

$$\frac{\overline{u_i u_j}}{U_\tau^2} = g(\eta, \Pi) \qquad (12)$$

$$\frac{\varepsilon}{U_\tau \Delta_E} = \psi(\eta, \Pi)$$

where ϕ, g and ψ are universal, i.e. there is a one to one correspondence between the mean flow velocity defect law and the Reynolds shear stress profiles, implying a universal distribution for eddy viscosity ε as given in (12). In fact as $\eta \to 0$, $\varepsilon/\Delta_E U_\tau = \kappa\eta$ is the classical result for the logarithmic law of the wall region. We know from the Stanford Conference (1969) that this one to one correspondence is not strictly correct and that S should also depend on the skin friction coefficient, Reynolds number and other quantities. These are refinements which might possibly be incorporated later. If S is allowed to be a function of Π, the data fits the model over a wider ranges of Π and the form of the results given by (12) are still applicable.

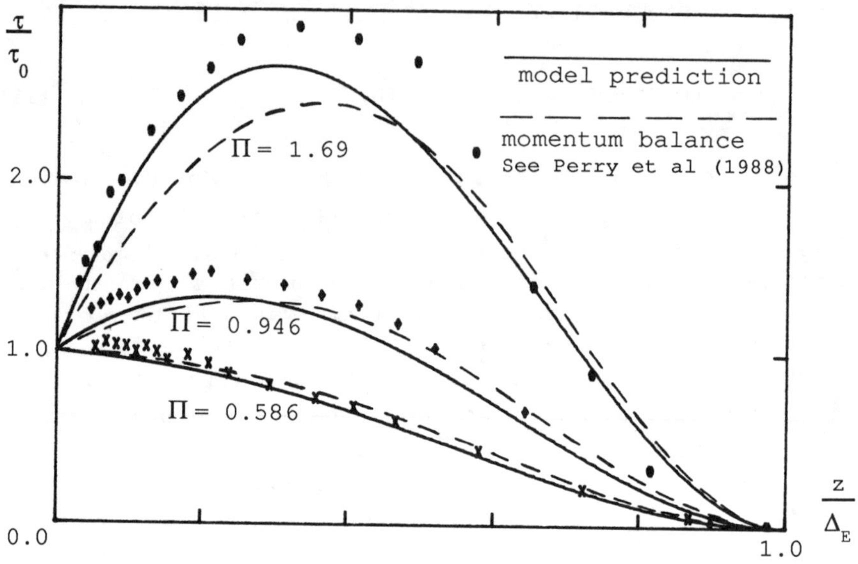

FIGURE 4. Comparison of shear stress predictions with experimental data of East et al (1979) [✗ -flow no. 3, ♦ -flow no. 4, ● -flow no.5]

SPECTRA AND NEAR WALL SCALING LAWS

Here attention will be confined to the fully turbulent region which for the purpose of this paper is defined as $z_+ \geq 100$; $z/\Delta_E \leq 0.1$ where $z_+ = zU_\tau/\nu$ and ν is the kinematic viscosity. Let $\gamma_{ij}(k_1 z, z/\delta)$ be the one dimensional power spectral density from one hierarchy of eddies and k_1 is the streamwise wavenumber. It should be noted here that the non-dimensional wavenumber argument denotes the unit interval over which power spectral density is being measured. The function γ_{ij} can be found by a fast Fourier Transform analysis of the velocity signatures generated by the eddy in isolation with its wall image, using the Biot-Savart law. Perry et al (1986) carried out such a calculation for a "Λ" shaped representative eddy. Let $\psi_{ij} = k_1 z \gamma_{ij}(k_1 z, z/\delta)$ be the pre-multiplied spectral function. Pre-multiplied spectra have the advantage that when plotted semi-logarithmically, the area under the curve is the energy contribution. Using the logarithmic variables λ, λ_E and $\alpha = \ln(k_1\delta)$, the total power spectral density from all hiearchies is given by,

(a)

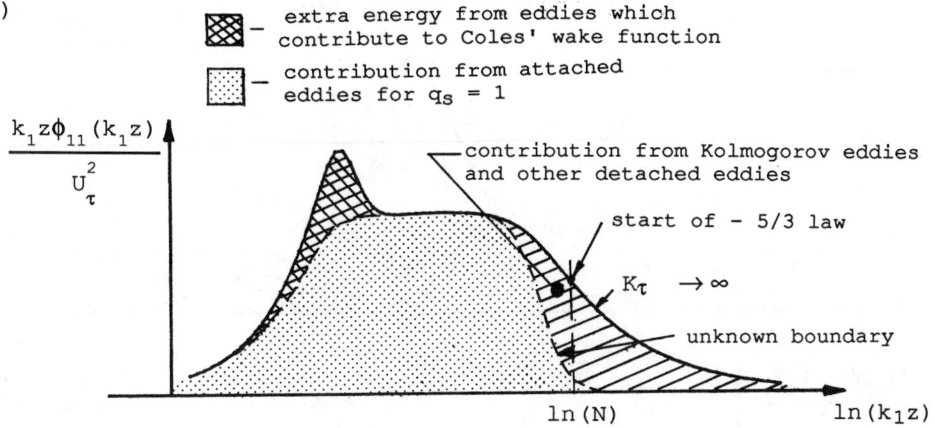

(b)

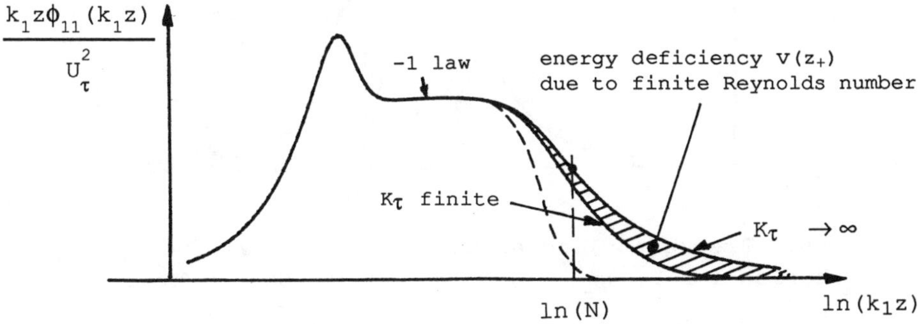

FIGURE 5. Pre-multipied streamwise spectra. (a) Energy contribution for $K_\tau \rightarrow \infty$. (b) finite Reynolds number case.

$$\frac{k_1 z \phi_{ij}(k_1 z, \frac{z}{\Delta_E})}{U_\tau^2} = \int_{-\infty}^{+\infty} M \left. \frac{\psi_{ij}(\alpha,\lambda)}{U_\tau^2} \right|_{k_1 z = \text{const.}} q_s(\lambda - \lambda_E)\ d\lambda \tag{13}$$

which is analogous to equation (8). The function ϕ_{ij} is normalised such that:

$$\int_0^\infty \frac{\phi_{ij}(k_1 z, \frac{z}{\Delta_E})}{U_\tau^2}\ dk_1 z = \frac{\overline{u_i u_j}}{U_\tau^2}(\frac{z}{\Delta_E}) \tag{14}$$

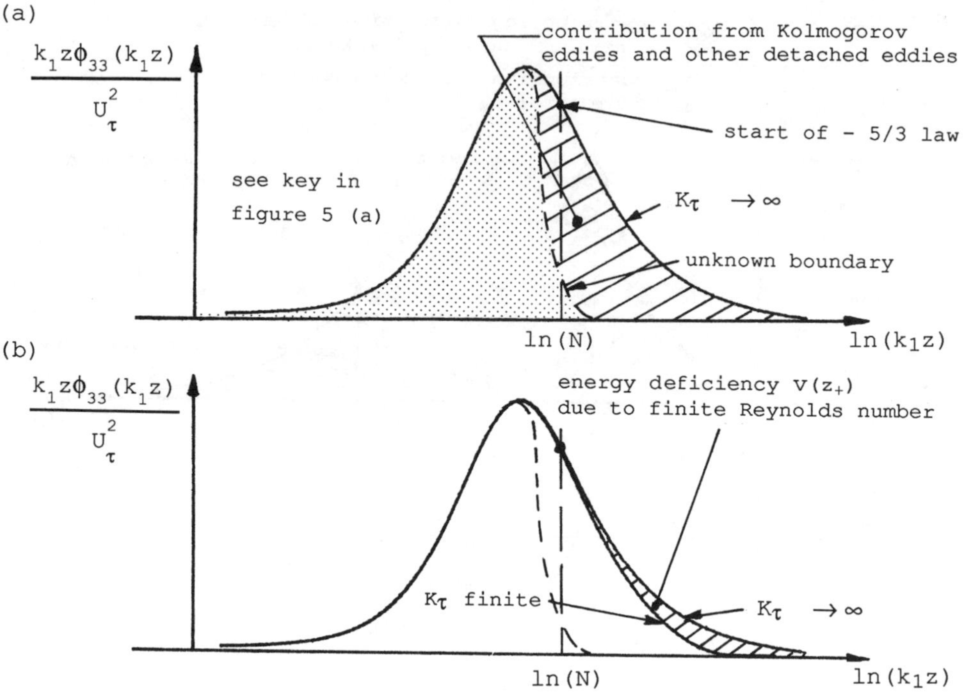

FIGURE 6. Pre-multiplied normal spectra. (a) energy contribution for $K_\tau \rightarrow \infty$. (b) finite Reynolds number case.

As a matter of interest, the spectral function ψ_{ij} are directly related to the Townsend eddy intensity function, i.e.,

$$I_{ij}(\lambda) = \int_{-\infty}^{+\infty} \frac{\psi_{ij}(\alpha,\lambda)}{U_\tau^2}\, d\alpha \qquad (15)$$

Figure 5 shows the pre-multiplied spectra ϕ_{11} for infinite Reynolds number and the contribution from the various eddy structures are shown by shaded areas. The extra energy coming from the detached motions has been added and it can be seen that its contribution is by no means negligible. The ϕ_{22} spectra are expected to have similar properties. Figure 6 shows the corresponding ϕ_{33} plots and there are significant differences from the ϕ_{11}.

The spectral plots shown in this paper are approximate curve fits to real laboratory data of Perry et al (1986, 1987) using the scaling laws as formulated by Perry et al (1986) from the attached eddy hypothesis and the classical theory of local isotropy. The low wavenumber regions of these curve-fitted spectra are applicable only to zero pressure gradient boundary layers, whereas the the formulation for the high wavenumbers should be applicable also to pipe and duct flow. These scaling laws are as follows. For ϕ_{11} and ϕ_{22} at low wavenumber, there is outer flow scaling (i.e. $\phi_{11}(k_1\Delta_E)/U_\tau^2 = g_{11}(k_1\Delta_E)$), at intermediate wavenumber there is inner flow scaling (i.e. $\phi_{11}(k_1 z)/U_\tau^2 = h_{11}(k_1 z)$) and for high wavenumber there is the Kolmogorov scaling such that $\phi_{11}(k_1\eta)/\upsilon^2 = K_{11}(k_1\eta)$. `Here η and υ are Kolmogorov length and velocity scales respectively. As was first shown by Perry & Abell (1977), a region of overlap between g_{11} and h_{11} leads to a -1 power law and a region of overlap between h_{11} and K_{11} leads to a -5/3 power law. For ϕ_{33} we have only inner flow scaling and Kolmogorov scaling. The curve-fitted functions used which incorporate these scaling laws are as follows:

$$\frac{\phi_{11}(k_1\Delta_E)}{U_\tau^2} = \frac{A'}{G_1\,G_2} \tag{16}$$

where $$G_1 = \left\{1+\left[\frac{1}{2}\,\{1+\tanh(\theta(k_1\Delta_E-q))\}\,\alpha k_1\Delta_E\right]^2\right\}^{\frac{1}{2}}$$

$$G_2 = \left\{1+\left[\beta\,\frac{U_\tau^2}{\upsilon^2}\,A'\,\frac{1}{(k_1\eta)\;K_{11}(k_1\eta)}\right]^m\right\}^{\frac{1}{m}}$$

where A' = 2.06, α =2.0; β =0.500, θ =1.8; q=0.9; m=4.0

$$\frac{\phi_{33}(k_1 z)}{U_\tau^2} = \frac{A'}{\left\{1+\left[\beta\,\frac{U_\tau^2}{\upsilon^2}\,(\frac{z}{\eta})\,\frac{1}{K_{33}(k_1\eta)}\right]^m\right\}^{\frac{1}{m}}} \tag{17}$$

where A'=0.9; m=0.8; and β =0.903 and the functional forms have no physical significance except the asymptotic forms such as the -5/3 and -1 power law regions. Although the functions are arbitrarily chosen, they are designed to give the reader a fairly accurate indictation of what the spectra look like. *The precise value of the various constants are subject to alteration without notice.* The Kolmogorov length scale and velocity scale are respectively:

$$\eta = \left(\frac{\nu^3\kappa z}{U_\tau^3}\right)^{\frac{1}{4}} \quad \text{and} \quad \upsilon = \left(\frac{\nu U_\tau^3}{\kappa z}\right)^{\frac{1}{4}}$$

and this comes from the assumption that in the turbulent wall region dissipation of turbulent energy is in balance with turbulent kinetic energy production. $K_{11}(k_1\eta)$ and $K_{33}(k_1\eta)$ have been computed using the Kovasznay (1948) spectral formula, which has been found to agree very well with the spectral formulation of Pao (1965). These formulae are :

$$K_{11}(k_1\eta) = 0.500(k_1\eta)^{-\frac{5}{3}} - 6.04(k_1\eta)^{-\frac{1}{3}} - 0.587(k_1\eta) - 0.306(k_1\eta)^2 + 7.03 \tag{18}$$

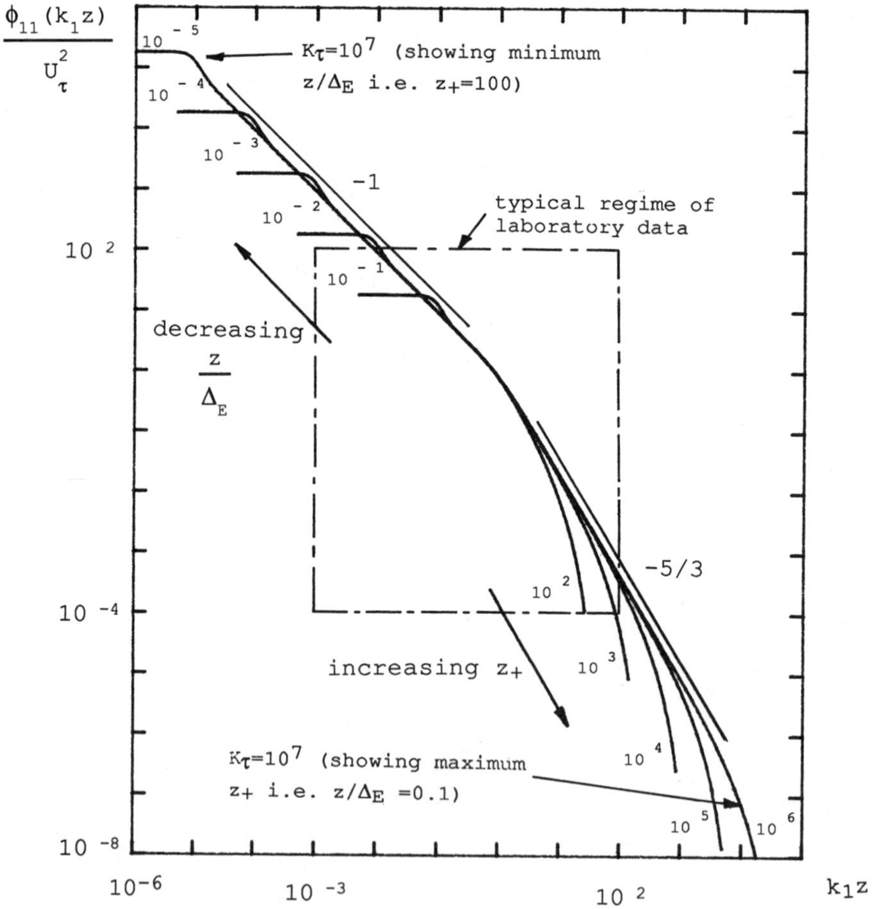

FIGURE 7. u_1 spectra for the turbulent wall region generated using equation (16).

$$K_{33}(k_1\eta) = 0.669(k_1\eta)^{-\frac{5}{3}} - 4.02(k_1\eta)^{-\frac{1}{3}}$$

$$-0.153(k_1\eta)^{2} + 3.51 \tag{19}$$

The asymptotic laws for ϕ_{11} are:

$$\frac{\phi_{11}(k_1 z)}{U_\tau^2} = A_1 (k_1 z)^{-1} \tag{20}$$

and $$\frac{\phi_{11}(k_1 z)}{U_\tau^2} = K_0\, \kappa^{-\frac{2}{3}} (k_1 z)^{-\frac{5}{3}} \tag{21}$$

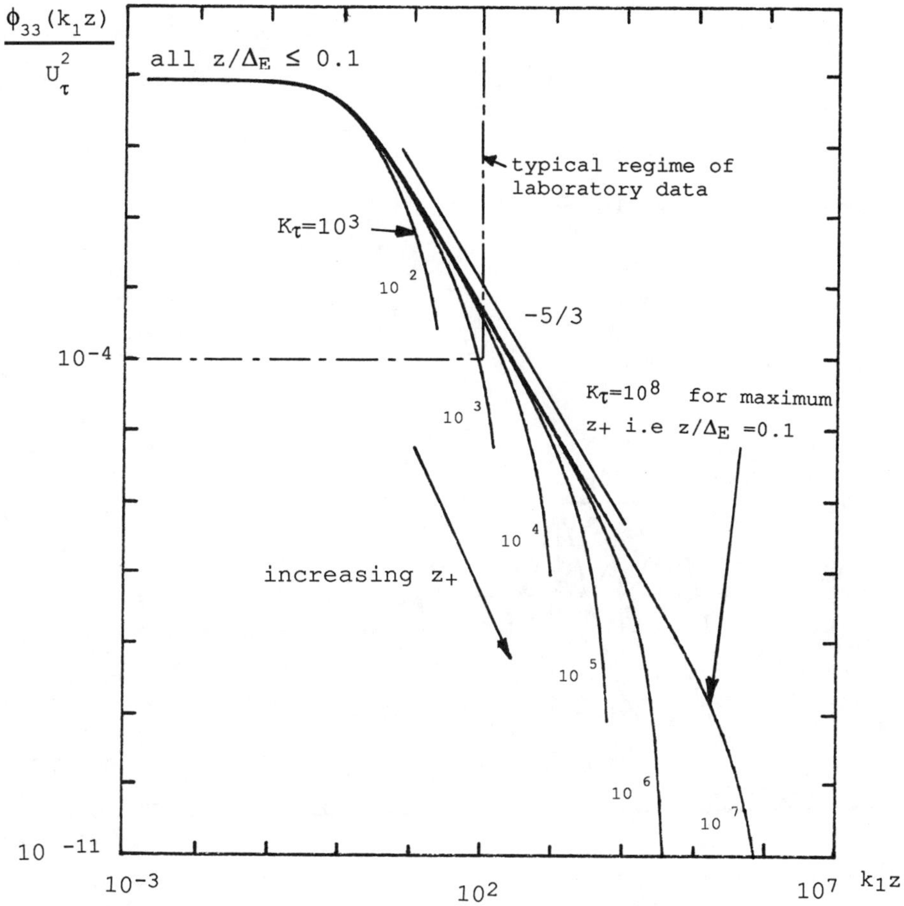

FIGURE 8. u_3 spectra for the turbulent wall region generated using equation (17)

where A_1=1.03 and K_0 is the Kolmogorov constant =0.5. The Kovasznay formula was modified to give this constant. One of the annoying things about attempts to prove the asymptotic laws is the insufficient Reynolds number range which can be achieved in the laboratory. The best form of Reynolds number to use is the Kármán number $K_\tau = \Delta_E U_\tau / \nu$ and figure 7 shows that reasonable lengths of -1 and -5/3 power law are obtained only for $K_\tau \rightarrow 10^6$ which is approaching meteorological data. The spectra show the minimum value of z/Δ_E and maximum value of z_+ obtainable while still remaining in the turbulent wall region for the value of K_τ shown. Figure 8 shows the corresponding plots for ϕ_{33}.

It can be shown that by integrating up the spectral contributions, the laws for the broadband turbulent intensities (or normal stresses) are:

$$\frac{\overline{u_1^2}}{U_\tau^2} = B_1 - A_1 \ln\left(\frac{z}{\Delta_E}\right) - V(z_+) \tag{22}$$

$$\frac{\overline{u_2^2}}{U_\tau^2} = B_2 - A_2 \ln\left(\frac{z}{\Delta_E}\right) - V(z_+) \tag{23}$$

$$\frac{\overline{u_3^2}}{U_\tau^2} = A_3 - V(z_+) \tag{24}$$

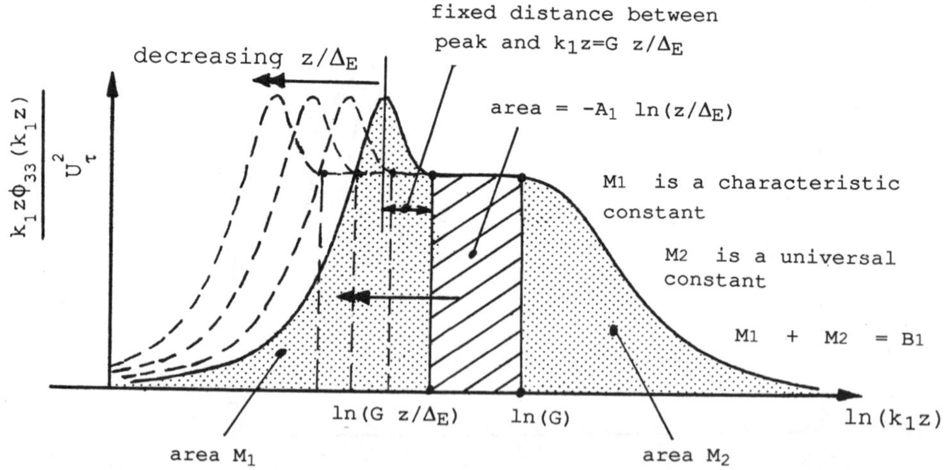

FIGURE 9. Graphic representation of the terms in equation (22) for $K_\tau \rightarrow \infty$. G is a universal constant appropriately chosen so that $k_1 z$=G and $k_1 z$=G(z/Δ_E) fall in the -1 region.

Figure 9 shows the various contributions to the spectra and one can see why B_1 and B_2 are characteristic constants (i.e. they depend on the large scale geometry of the flow, i.e. they depend on whether we are in a pipe, duct, zero pressure gradient or adverse pressure gradient boundary layer flow) while all the other constants are universal. $V(z_+)$ represents the deficiency in energy due to the size of the smallest eddies being limited by viscosity, as shown in figure 5(b). In the past, e.g. Perry et al (1986) and (1987), $V(z_+)$ was calculated assuming that all energy was cut off at $k_1\eta = M = 0.085$ and for $k_1\eta < M$ we have the asymptotic -5/3 law. This leads to the form:

$$V(z_+) = C(z_+)^{-\frac{1}{2}} \tag{25}$$

for the $\overline{u_1^2}$ component but

$$V(z_+) = \frac{4}{3} C(z_+)^{-\frac{1}{2}} \tag{26}$$

for the $\overline{u_2^2}$ and $\overline{u_3^2}$ components where C = 6.08. However as pointed out by Spalart (1986), the correction for the viscous eddies should be isotropic and the factor 4/3 should not appear, yet it is known that the - 5/3 Kolmogorov region for ϕ_{22} and ϕ_{33} must include the factor 4/3. This paradox was resolved by realizing that the isotropic condition

$$\phi_{33}(k_1) = \frac{1}{2} \left\{ \phi_{11}(k_1) - k_1 \frac{d\phi_{11}(k_1)}{dk_1} \right\} \tag{27}$$

implies that a sudden cut off of $\phi_{11}(k_1)$ results in a delta function for $d\phi_{11}(k_1)/dk_1$. When this is included in the formulation the factor 4/3 in (26) becomes unity, giving an isotropic correction. The authors have attempted to compute $V(z_+)$ by more sophisticated methods. If one is to consider $\phi_{11}(k_1 z)$ and to assume that for $k_1 z > N(=2.5)$ we have the Kovasznay spectrum and for $k_1 z < N$ we have the infinite Reynolds number spectra, then this leads to:

$$V(z_+) = 5.58\, z_+^{-\frac{1}{2}} - 22.4\, z_+^{-1} + 22.0\, z_+^{-\frac{5}{4}} - 5.62\, z_+^{-2} + 1.27\, z_+^{-\frac{11}{4}} \tag{28}$$

This formula has now been used to test the broadband turbulent intensity data of Perry et al (1987) and is shown in figures 10

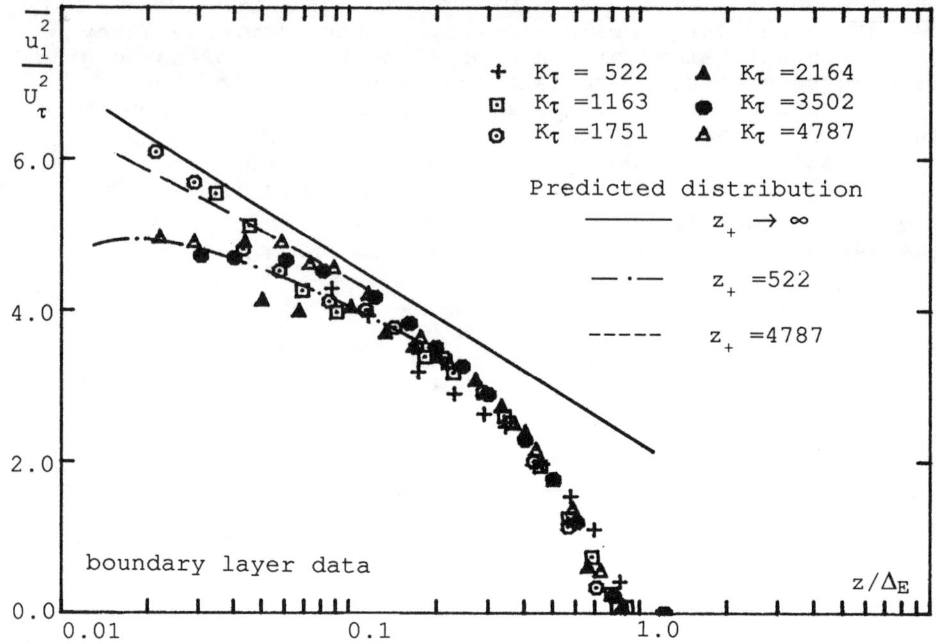

FIGURE 10. Streamwise turbulence intensities. Prediction calculated using (22) and (28). Data from Lim (1985) or Perry et al. (1987)

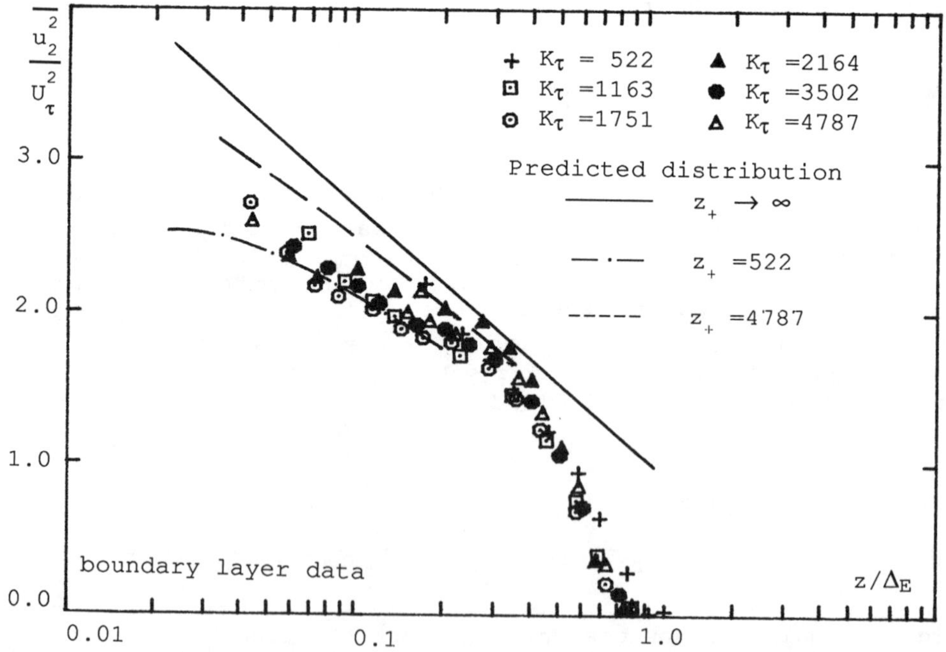

FIGURE 11. Spanwise turbulent intensities. Prediction calculated using (23) and (28). Data from Lim (1985) or Perry et al (1987).

and 11 for $\overline{u_1^2}$ and $\overline{u_2^2}$. The values of B_1 and B_2 have been set to $B_1 = 2.25$ and $B_2 = 1.02$. The most crucial test however is for $\overline{u_3^2}$ and this is the most difficult. Figure 12 shows the data from many workers for $\overline{u_3^2}/U_\tau^2$ at $z/\Delta_E = 0.1$ for various values of Kármán number K_τ and comparison is made with equation (24) with $V(z_+)$ given by the Perry et al (1986, 1987) formula (26) for different values of A_3. See Perry et al (1987) for details of the data sources. The results are very dissapointing and must reflect poor hot wire anemometry practice since it can be shown by the Buckingham Π-theorem alone that $\overline{u_3^2}/U_\tau^2$ at $z/\Delta_E = 0.1$ versus K_τ should collapse on a universal curve at least for pipe and duct flow since in general

$$\frac{\overline{u_3^2}}{U_\tau^2} = \phi\left(\frac{z}{\Delta_E}, \frac{\Delta_E U_\tau}{\nu}\right) \qquad (29)$$

and this should also be applicable for zero pressure gradient boundary layers where streamwise development is extremely slow. Many problems exist with hot wire anemometry and one of the main difficulties is that spatial resolution errors have a large fractional effect on $\overline{u_3^2}$. These are not so important for the $\overline{u_1^2}$ and $\overline{u_2^2}$ data. Also errors are caused by cross contamination of signals from the three velocity components and velocity vectors exceeding allowable cone angles for X-wires. The authors have gathered a new set of experimental data and also have carefully tested past data. The data had to satisfy the following: 1.) If spatial resolution corrections were necessary then we require accompanying spectra so as to apply the Wyngaard (1968)

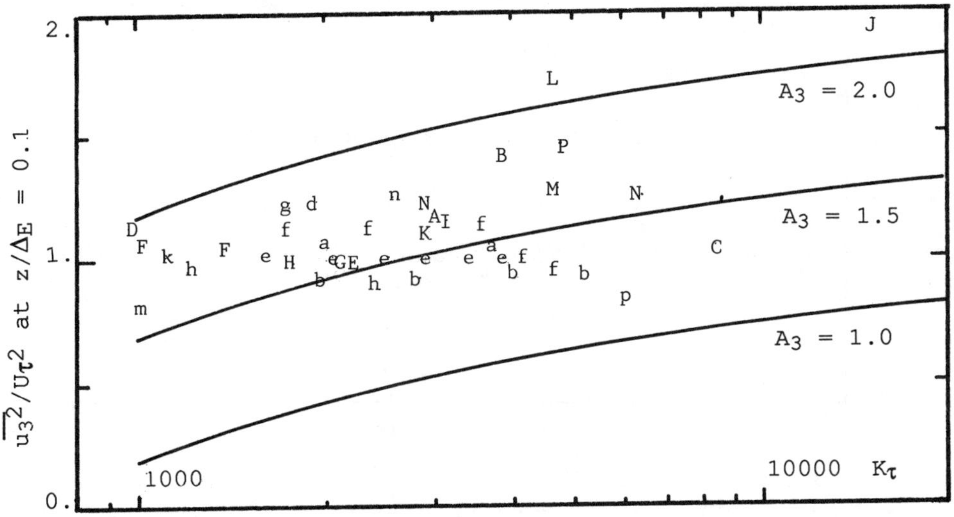

FIGURE 12. Normal turbulent intensity at $z/\Delta_E = 0.1$ versus K_τ. For details of data sources, see Perry et al (1987)

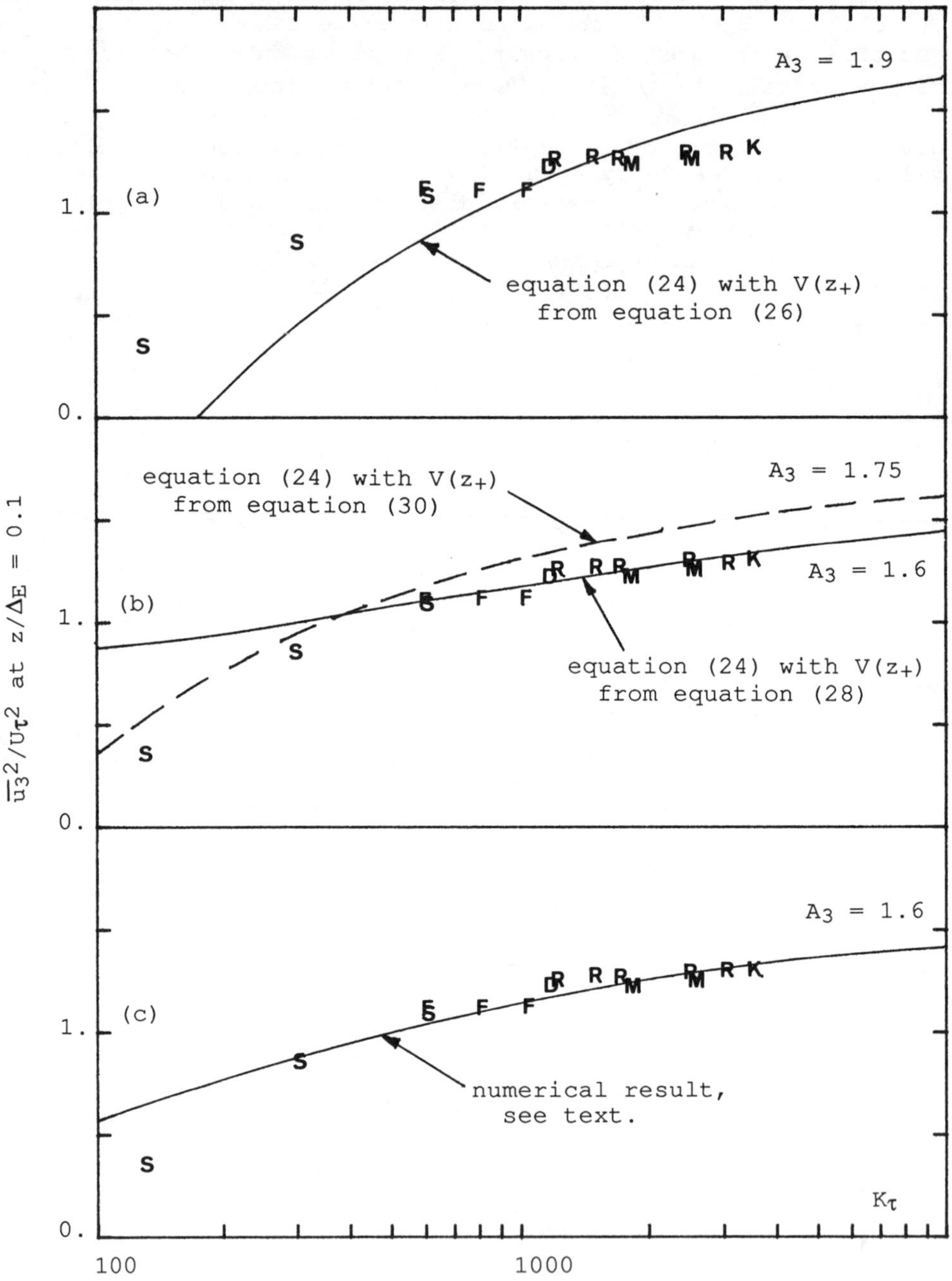

FIGURE 13. Normal turbulence intensity at z/Δ_E=0.1 versus K_τ. **S** - Spalart (1986) super-computer, **D** - 'D' type rough wall, **F**- Erm et al (1986) low Reynolds number, **K** - 'k' type rough wall, 'flying' wire result, spatial resolution effect for **D**, **F** and **K** is small, **R** - present smooth wall results, spatial resolution effect has been corrected, **M** - Ligrani et al (1987), subminiature hot-wire. Theoretical curves shown.

correction, 2.) If cone angle problems were seen to be a difficulty e.g. over rough surfaces, then only use flying hot wire data (see Perry et al 1987), 3.) If the X-wire data showed values for the Reynolds shear stress inconsistent with Clauser chart or Preston tube values then reject the data.
Figure 13 (a) shows the data comparedwith (24) with $V(z_+)$ given by the Perry et al (1986) formula (26). Figure 13 (b) shows the data compared with Spalart's modification of the Perry et al formula. This modified form is

$$V(z_+) = 4.37\, z_+^{-\frac{1}{2}} \tag{30}$$

for all three components. This curve is shown as a broken line. Figure 13 (b) also shows the data compared with (24) with $V(z_+)$ given by (28). This is shown as a full curve. Figure 13 (c) shows the data compared with the numerical values found for $V(z_+)$ from the hatched area shown in figure 5 (b) generated from equation (16) and (18) by assuming that $V(z_+)$ is isotropic.

The two lowest Reynolds number results of Spalart should strictly not be included since $z/\Delta_E = 0.1$ finishes up in the buffer zone. Also calculated values for $V(z_+)$ shown in figure 13 (b) become inacurate for K_τ less than 500 and the numerical results in 13 (c) are more belivable. The agreement of data with calculation is encouraging.

CONCLUSIONS AND DISCUSSION

The attached eddy hypothesis is showing considerable promise in leading to a useful closure formulation for the prediction of turbulent boundary layer development with streamwise pressure gradients. This approach is still new and some what crude and speculative but at least the concept of coherent structures is starting to head towards predictive schemes for engineering purposes. Furthermore, a study of the near wall region has provided very useful data for verifying whether or not the proposed similarity laws are valid and the results are very encouraging. Unfortunately we still do not have any reliable laws for the broadband turbulence intensities for flow beyond the turbulent wall region since we cannot make the assumption that production and dissipation are in balance. Work on this is continuing and many of the details of the results outlined here are under preparation for a more complete presentation of the work.

ACKNOWLEDGEMENT

The authors wish to thank the financal assistance of the Australian Research Grants Scheme.

REFERENCES

AFOSR-IFP-Stanford Conference 1968, Proceedings-Computation of Turbulent Boundary Layers. Vol. 1 (ed. Kline, S.J., Sovran, G., Morkovin, M.V. and Cockrell, D.J.) 1969

Coles, D. E. 1956 Law of The Wake in Turbulent Boundary Layer *J. Fluid Mech.* **1**, 191-226.

East, L. F., Sawyer, W. G. and Nash, C. R. 1979 An Investigation of The Structure of Equilibrium Turbulent Boundary Layers. R.A.E *Tech. Report* 79040.

Erm, L. P. Smits, A. J. & Joubert, P. N. 1986 Low Reynolds Number Turbulent Boundary Layer on a Smooth Surface in a Zero Pressure Gradient. *Turbulent Shear Flow* (ed. F. Durst, B. E. Launder, J. L. Lumley, F. W. Schmidt & J. H. Whitelaw), Vol. 5. Springer-Verlag.

Kovasznay, L. S. G. , 1948 Spectrum of Locally Isotropic Turbulence. *J. Aero. Science* . **15** 745-53

Ligrani, P. M. , Westphal, R. V. & Lemos, F. R. , 1987 Development of Subminiature Multi-Sensor Hot-wire Results NASA *Technical Memorandum.*

Lim, K. L. 1985, Wall Shear Flows Ph.D Thesis, University of Melbourne.

Pao, Y. H., Structure of Turbulent Velocity and Scalar Fields at Large Wavenumbers. *Phys. Fluids,* **8** 1063-1108

Perry, A. E., Abell, C. J. 1977 Asymptotic Similarity of Turbulence Structure in Smooth and Rough -Walled Pipes *J. Fluid Mech.* **79** 785-799

Perry, A. E. , Chong, M. S. 1982 On The Mechanism of Wall Turbulence. *J. Fluid Mech.* **119** 173-217

Perry, A. E., Henbest, S. M. & Chong, M. S. 1986 A Theortical and Experimental Study of Wall Turbulence. *J. Fluid Mech.* **165** 163-199

Perry, A. E. 1987 Turbulence Modelling Using Coherent Structures in Wakes, Plane Mixing Layers and Wall Turbulence. *Perspectives in Turbulence Studies* (ed. Meier, H. U. and Bradshaw, P.) Springer-Verlag 115-153.

Perry, A. E., Lim, K. L. & Henbest, S. M. 1987 An Experimental Study of The Turbulence Structure in Smooth and Rough-Wall Boundary Layers. *J. Fluid Mech.* **177** 437-466

Perry, A. E., Li, J. D. & Marušić, I. 1988 Novel Methods of Modelling Wall Turbulence. AIAA - 88 - 0219

Spalart, P. R. 1986 Direct simulation of a Turbulent Boundary Layer Up to R_θ=1410. NASA *Tech. Memo.* 89407.

Townsend, A. A. 1976 *The structure of Turbulent Shear Flow.* Cambridge University Press. pp.150 - 156

Wyngaard, J. C. 1968 Measurement of Small-Scale Turbulence Structure With Hot-Wires *J. Sci. Instrum.,* Ser. 2, **1** 1105-1108

The Generalized Lagrangian Mean Equation: Turbulent Channel Flow and the Aetiology of Streamwise Vortices

W. R. C. PHILLIPS
Sibley School of Mechanical and Aerospace Engineering
Cornell University
Upson Hall
Ithaca, New York 14853, USA

ABSTRACT

The formation of streamwise vortices in the wall region of turbulent channel flow is investigated using the Generalized Lagrangian mean equations of Andrews & McIntyre; these are an exact and very general Lagrangian-mean description of the back effect of oscillatory disturbances upon the mean state. The problem is solved numerically over the full half-channel, and vortex pairs are observed to form and rise gradually from the wall. The results are preliminary, but are in accord with observations made using flow visualization.

1. INTRODUCTION

Of the coherent structures present in a turbulent boundary layer, perhaps the simplest occur in the wall region as streaks. First observed by Hama (see Corrsin 1957), streaks have a mean spacing of approximately 100 wall units and end by being lifted away from the wall (Kline et al. 1967). As they are lifted, the streaks start to oscillate with increasing amplitude, a phenomenon known as breakdown. A large scale motion then emerges from the outer flow, approaches the wall and stabilizes the entire region (Corino & Brodkey 1969); this process is called the sweep. Together, this quasi-deterministic, randomly located sequence of these and other coherent structures, is collectively called the bursting process and is believed to play a dominant role in the production and maintenance of a mean turbulent flow (Cantwell 1981).

The relationship between streamwise streaks, bursts and sweeps is not altogether clear, nor is the connection between the wall-region flow and the inclined vortex loops or hairpins observed further out in the layer (Head & Bandyopadhyay 1981). It is plausible that the streaks are a visualization resulting from counter-rotating streamwise vortices. But if streamwise vortices are present it is not clear by what mechanism they arise, so the aim of the present work is to identify a plausible physical mechanism.

Several mechanisms for the generation of streamwise vortices have been proposed: a Taylor-Gortler type instability (Brown & Thomas 1977, Coles 1977); a non-linear interaction of a two- and a three-dimensional wave (Benney 1961, Benney & Lin 1962), and direct-resonance Jang et al. (1986).

Streamwise vortices are also observed to form near the surface of lakes and oceans. Here known as Langmuir circulations, they appear to develop via the Craik-Leibovich instability mechanism: background, initially randomly oriented vorticity, is oriented by a fluctuating field and amplified in the streamwise direction by an imposed weak shear, probably driven by wind (Craik 1977; Leibovich 1977; see Leibovich 1983 for a survey). The theory deals with an averaged set of equations in which only the rectified effects of the fluctuating field appear in the form of a Stokes drift (later defined). Such equations follow precisely from the exact wave/mean flow equations of Andrews & McIntyre (1978), the Generalized Lagrangian mean equations.

In the present work we employ the Generalized Lagrangian mean (GLM) equations to study the aetiology of streamwise vortices in the wall region of turbulent channel flow. We do not, however, restrict our analysis solely to the wall region but account for the whole half-channel. We solve the GLM equations modelling the flow numerically, and find that streamwise vortices indeed form. Furthermore we observe that as the vortices grow they rise from the wall. We consider our results to date preliminary, yet they appear to capture qualitative features of streaks made using flow visualization. A detailed version of this work is given in Phillips (1988).

2. THE GENERALIZED LAGRANGIAN MEAN EQUATIONS

The Generalized Lagrangian Mean equations of Andrews & McIntyre (1978) are an exact and very general Lagrangian-mean description of the back effect of oscillatory disturbances upon the mean state. The Lagrangian-mean velocity described by these equations cannot be defined as the 'mean following a single fluid particle', but as the velocity field describing trajectories about which the fluctuating particle motions have zero mean, when *any* averaging process suitable to the particular problem (e.g. time, spatial or ensemble averaging) is applied.

This GLM description is a hybrid Eulerian-Lagrangian description of wave, mean-flow interactions, since it describes the generalised Lagrangian mean flow by means of equations in Eulerian form, with position $\mathbf{x}$ and time t as independent variables. In the present work we shall restrict our discussion to fluids of constant density.

Consider two sets of trajectories with Lagrangian velocities

$$d\mathbf{x}/dt = \bar{\mathbf{u}}^{L}(\mathbf{x},t) \tag{1}$$

and

$$d\mathbf{X}/dt = \mathbf{u}\,(\mathbf{X},t)$$

respectively, where $\mathbf{u}$ is the actual instantaneous velocity of a fluid particle and $\bar{\mathbf{u}}^{L}$ is the corresponding quantity for a reference fluid motion that takes place with the Lagrangian-mean velocity. An oscillatory displacement from the reference path is then assumed to be described by the vector field $\xi(\mathbf{x},t)$, so that the two sets of trajectories are related by

$$\mathbf{x} \rightarrow \mathbf{X}(\mathbf{x},t) \equiv \mathbf{x} + \xi(\mathbf{x},t). \tag{2}$$

Let an overbar refer to any suitable averaging operation that may be contemplated in an Eulerian framework (such as time, space, or ensemble averaging). Then the GLM operator

$\overline{(\;)}^{L}$ applied to a scalar or tensor field (ϕ say) of any rank, is defined in terms of $\overline{(\;)}$ by

$$\overline{\phi(\mathbf{x},t)}^{L} = \overline{\phi(\mathbf{X}(\mathbf{x},t),t)}\ .$$

Andrews & McIntyre also introduce a vector $\mathbf{p}$, which they call the wave pseudomomentum per unit mass, defined to be

$$p_i = -\overline{\xi_{j,i}\,u_j} - \overline{\xi_{j,i}\,\xi_k}\;\bar{u}_{j,k} \tag{3}$$

Furthermore since the total motion $\mathbf{u}(\mathbf{x},t)$ conserves mass, then

$\bar{D}^L J + J\, \nabla \bullet \bar{\mathbf{u}}^L = 0$,

where, correct to $0(\varepsilon^2)$

$$J = 1 - \frac{1}{2}\overline{(\xi_j\xi_k)}_{,jk} \; .$$

The Jacobian is assumed to be non-zero for all t and **x**, so the transformation induced by (2) is invertible.

The Eulerian and Lagrangian-mean velocities are related by the Stokes drift $\bar{\mathbf{u}}^S$, as

$$\bar{\mathbf{u}}^L = \bar{\mathbf{u}} + \bar{\mathbf{u}}^S$$

where, for small displacements ξ from the mean trajectories, measured by a parameter ε,

$$\bar{u}_i{}^S = \overline{\xi_j u_{i,'j}} + \frac{1}{2}\,\overline{\xi_j\xi_k}\,\bar{u}_{i,jk} + O(\varepsilon^3)$$

For present purposes it is convenient to write the GLM equation as

$$\frac{\partial \mathbf{q}}{\partial t} + \mathbf{q} \bullet \nabla \mathbf{q} = \mathbf{p} \times \text{curl}\, \mathbf{q} - \nabla\left(\frac{1}{2}\mathbf{q} \bullet \mathbf{q} + \mathbf{q} \bullet \mathbf{p} + \Pi\right) + \bar{\mathbf{F}}^L \tag{4}$$

with $\mathbf{q} = \bar{\mathbf{u}} + \bar{\mathbf{u}}^S - \mathbf{p}$

and model (Leibovich 1980) $\bar{\mathbf{F}}^L = \nu\nabla^2\bar{\mathbf{u}}$ + higher order terms. Moreover for the flow we wish to consider (Phillips 1988),

$$\nabla \bullet (\mathbf{q} + \mathbf{p}) = 0\,. \tag{5}$$

3. SPECIFIED FUNCTIONS

We plan to apply (4) to fully developed turbulent channel flow which we shall assume is initially structureless and parallel; our aim is to observe the evolution of structure upon that flow. To proceed, therefore, we require the Eulerian mean velocity field of the flow and, to specify (4), the pseudomomentum, **p**. Consider then a channel of height $2h^*$ whose walls lie in the x^*-0-y^* plane and with streamwise flow in the x^*-direction, as shown in figure 1. If we assume homogeneity in the x^* and y^* directions and adopt a spatial average over x^*, then all mean quantities will be independent of x^*. Our mean velocity field is then simply the fully developed velocity profile for turbulent channel flow, $U^*(z^*)$, which we shall describe in § 3.1. The form of **p** will be discussed in § 3.2.

3.1 The mean velocity

Many attempts have been made to provide a simple description of the mean velocity profile in turbulent channel flow, or in a turbulent boundary layer; most are piece-wise continuous and not continuously differentiable. Those that are continuously differentiable often do not capture the precise asymptotics of the layer as $z = z^*/h^* \to 0$, such as the profile by Van Driest. We therefore employ the profiles derived by Phillips (1987) for $z \to 0$ and Ratnanather & Phillips (1988) for $z \to 1$, which take care to ensure that the asymptotics are correct. For channel flow $Re = U_\tau{}^* h^*/\nu$ and $U_\tau{}^* = [\nu (dU^*/dz^*)\,|_{z^*=0}]^{1/2}$.

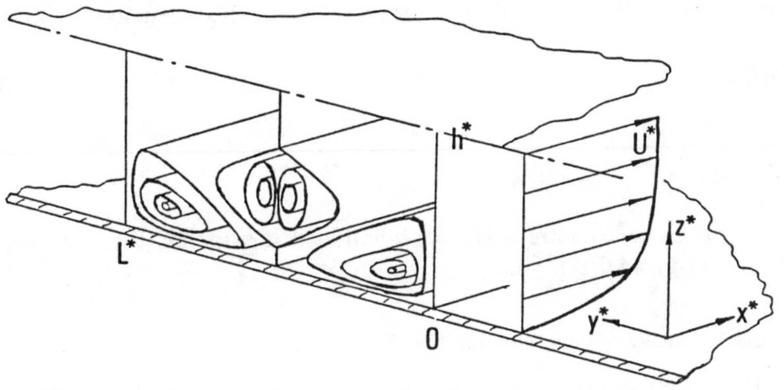

FIGURE 1. Schematic of the half channel showing vortex structure.

3.2 Pseudomomentum

Pseudomomentum is not easy to describe physically, although, as we shall see, it may be expressed in terms of space-time correlations, as may the Stokes drift (Lumley 1986). Very loosely, therefore, both the Stokes drift and the pseudomomentum may be thought of as derivable from gradients of Reynolds stress.
By employing techniques similar to those of Taylor (1921) and Kampé de Fériet (1939), we find that

$$-p_i = \frac{\partial}{\partial r_i}\Big[\int (R_{11} + R_{22} + R_{33})d\tau + \frac{d\bar{u}}{dz}\int (t-\tau)R_{13}d\tau\Big]$$

$$+2\,\frac{d\bar{u}}{dz}\frac{\partial}{\partial r_i}\Big[\int (t-\tau)R_{13}d\tau + \frac{d\bar{u}}{dz}\iint (t-\tau)R_{33}dt\,d\tau\Big]. \qquad (6)$$

where $R_{ij} = \overline{u_i(x,t)u_j(x+r,t+\tau)}$ is the space-time correlation evaluated at optimal delay, or nearly, where $\mathbf{r} = (\tau U,y,z)$.

Observe that any time dependence $\mathbf{p}$ exhibits is probably short in comparison with the time scales of the R_{ij}'s. Provided that our calculations are over a time scale comparable or longer than those of the R_{ij}'s, therefore, an assumption that $\mathbf{p}$ is independent of time is a plausible first guess, which we shall make. The same is true for $\mathbf{u}^s$.

From (6) we see by homogeneity that p_y is identically zero, but that it is not clear whether p_x is dominant and that p_z satisfies $p_z/p_x << 1$. For the sake of the present work we shall assume $p_z = 0$. Owing to homogeneity in the x and y directions then, we may write (5) as

$$\nabla \cdot q = 0 \qquad (7)$$

Equation (6) is precise, but not immediately tractable, even with Favre's (1957, 1965) measurements of R_{ij}. We therefore pursue another approach to deduce the essence of the function $p(z)$.

Consider equation (11) at t = 0 when the flow is assumed structureless, then u=v=w=0 and

$$-\frac{d^2}{dz^2}(u^s - \mathsf{p}) = Re + U'' \quad , \tag{8}$$

$$\text{so } \mathsf{p}' = U' + u'^s - Re\,(1-z) \tag{9}$$

where the prime denotes differentiation with respect to z.

For our present purposes we require a simple function that satisfies (9); one simple possibility is $u^{s'} = Re\,(1-z)$, so that p = U.

4. THE GENERALIZED LAGRANGIAN MEAN IN CHANNEL FLOW

With the mean flow, Stokes drift and pseudomomentum assumed to be parallel, as discussed above, then

$\bar{\mathbf{u}} = (U^*(z^*), 0, 0)$,

$\bar{\mathbf{u}}^s = (u^{*s}(z^*), 0, 0)$,

$\mathsf{p} = (\mathsf{p}^*(z^*), 0, 0)$,

where the normal (z*) coordinate dependence of U*, u^{*S} and p* for the case of fully developed flow will be assumed known.

We may then write

$\mathbf{q}(y^*, z^*, t^*) = (U^* + u^* + u^{*S} - \mathsf{p}^*, v^*, w^*)$

where u* is the streamwise perturbation velocity. Consonant with x*-independence of the flow and (7), we may introduce a streamfunction ψ^* to account for the perturbation velocity components v* and w* in the (y*, z*) plane. Dimensionless variables for the perturbed system are adopted as follows; the asterisk identifies dimensioned quantities. First, writing

$\beta^* = u^* + u^{*S} - \mathsf{p}^*$,

then

$\beta^* = U_\tau^*\beta$;

$(y^*, z^*) = h^*(y,z), \quad t^* = th^{*2}/\nu$;

and $(v^*, w^*) = \nu(v,w)/h^*$, $\psi = \nu\psi^*$,

with $v = \partial\psi/\partial z, \; w = -\,\partial\psi/\partial y$.

When substituted into (4), and after cross-differentiating the v and w equations, this scaling leads to the following dimensionless equations:

$$\frac{\partial\zeta}{\partial t} - \nabla^2\zeta = J(\psi, \zeta) - Re^2\frac{d\mathsf{p}}{dz}\bullet\frac{\partial\beta}{\partial y} \tag{10}$$

$$\frac{\partial \beta}{\partial t} - \nabla^2\beta = J(\psi, \beta) + Re + \frac{d^2U}{dz^2} + \frac{dU}{dz} \bullet \frac{\partial \psi}{\partial y} \tag{11}$$

$$\zeta = - \nabla^2\psi \ , \tag{12}$$

where ∇^2 is the Laplacian and J is the Jacobian with respect to (y, z), as

$$J(f,g) = \frac{\partial f}{\partial y} \bullet \frac{\partial g}{\partial z} - \frac{\partial f}{\partial z} \bullet \frac{\partial g}{\partial y} \quad ;$$

If we make the further assumption that all Eulerian velocities are symmetric on the channel centre line then the following boundary conditions may be applied:

$$\beta = \zeta = \psi = 0 \text{ on } z = 0;$$

$$d\beta/dz = \zeta = \psi = 0 \text{ on } z = 1; \tag{13}$$
$$\beta(0,z) = \beta(L,z),\ \zeta(0,z) = \zeta(L,z),\ \psi(0,z) = \psi(L,z) \ .$$

Where we have assumed periodicity of the flow (with period L) in the y-direction.

From (10) and (11) we observe that any instabilities are driven by the action of w on the gradient of the mean flow, thereby extracting energy from it, and the action of the lateral gradient of u, acting upon the gradient of the pseudomomentum.

5. THE MOST UNSTABLE MODE

Everything is now specified in our problem save the wavelength of the most unstable mode L. Experiments and earlier studies of instabilities in the wall region suggest that $50 < L < 120$; by necessity, however, these studies were confined solely to the wall region. The flow domain in the present work covers the whole half-channel, and we should like to proceed with a linear stability analysis of it to determine the most unstable mode.

Following Chandrasekhar (1961), we suppose that the perturbations ζ, u and ψ have the forms:

$$\zeta = Z(z) \exp(iky + \sigma t)$$

$$u = B(z) \exp(iky + \sigma t)$$

$$\psi = S(z) \exp(iky + \sigma t)$$

where k is the wavenumber of the disturbance and σ is a constant which can be complex. Then, after subtracting (9) from (11), equations (10), (11) and (12) become,

$$\sigma Z = (D^2 - k^2)Z - ip'Re^2kB \tag{14}$$

$$\sigma B = (D^2 - k^2)B + iU'kS \tag{15}$$

$$Z = - (D^2 - k^2)S, \tag{16}$$

where $D = d/dz$. Equations (14), (15) and (16) must satisfy the boundary conditions:

$B = Z = S = 0$ on $z = 0$

and

$dB/dz = Z = S = 0$ on $z = 1$.

Now if we assume that the principle of exchange of stabilities applies, σ is real and the marginal states are characterised by $\sigma = 0$. So, after elimination of Z from (14), we have

$$(D^2 - k^2)(D^2 - k^2) S = - ik\rho' Re^2 B \tag{17}$$

$$(D^2 - k^2)B = - ikU'S. \tag{18}$$

An analytical solution to (17) and (18) is not readily transparent, but a numerical solution is possible and in progress. From the term $-k\rho' Re^2$ however, it is evident that $k = f(Re)$; this follows because although U is only weakly dependent on Re, the turbulence intensities, and in all probability ρ, are dependent on Re. For k to be independent, or almost independent, of Re, ρ must go as Re^{-2}, and that is unlikely. For our present, work we shall assume that LRe = 100 wall units.

6. NUMERICAL METHOD

The governing equations were solved by finite difference methods, over a uniform grid with all terms discretized to second-order accuracy. The dependent variable ζ is solved on a nodal grid with the Jacobian $J(\psi,\zeta)$ discretized using the scheme developed by Arakawa (1966), which exactly conserves the finite difference analogues of vorticity, mean-square vorticity and kinetic energy of motion in the y-z plane. The dependent variable β is solved on a staggered grid and the Jacobian for $J(\psi,\beta)$ is discretized using the scheme developed by Lilly (1965), which conserves the finite difference analogues of x-momentum, and the kinetic energy associated with the x-velocity component of $\mathbf{u}$. All diffusion terms are approximated using standard central differences.

To avoid the time step restriction on the stability associated with an explicit treatment of diffusion terms, a time splitting method (Orszag & Patera 1983) is adopted. The partial advances accounting for the nondiffusive terms are carried out by a standard Bashford & Adams (1883) scheme, while advancing diffusive terms is done by the alternating direction implicit method of Peaceman & Rachford (1955). Periodic boundary conditions are treated by appropriate partitioning of the coefficient matrix.

The Poisson equation relating the streamfunction to the secondary vorticity field is solved using the direct Poisson solver developed at NCAR by Swarztrauber & Sweet (1975), which provides the streamfunction field to machine accuracy.

The computer code was validated by solving a set of test problems which permit exact solutions, such as simple problems of diffusion, advection by a known velocity field, and linear and nonlinear Bénard convection with simple boundary conditions. The accuracy and convergence performance of the code were considered acceptable.

As mentioned above our domain of interest spans the complete half channel, but experiments suggest that the streamwise rolls occur in regions less than 100 wall units from the wall; to amplify the wall region, therefore, the independent variable in the z-direction is written as

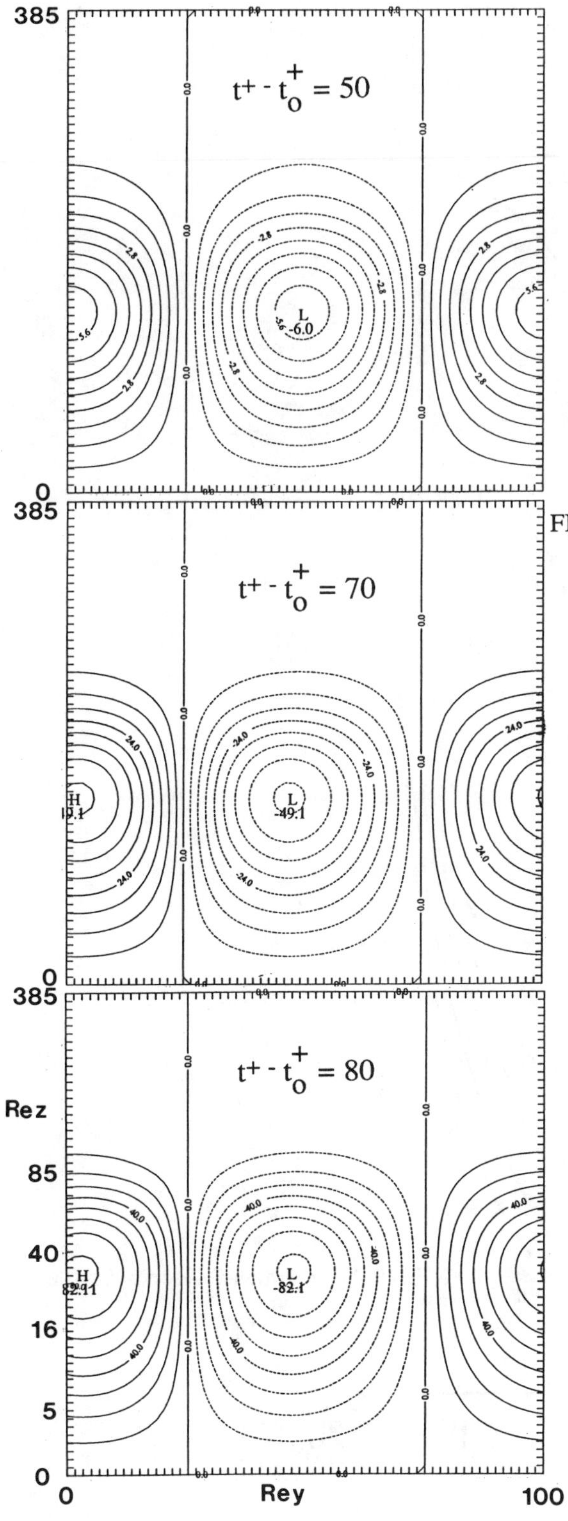

FIGURE 2. Instantaneous streamlines of the evolving flow field.

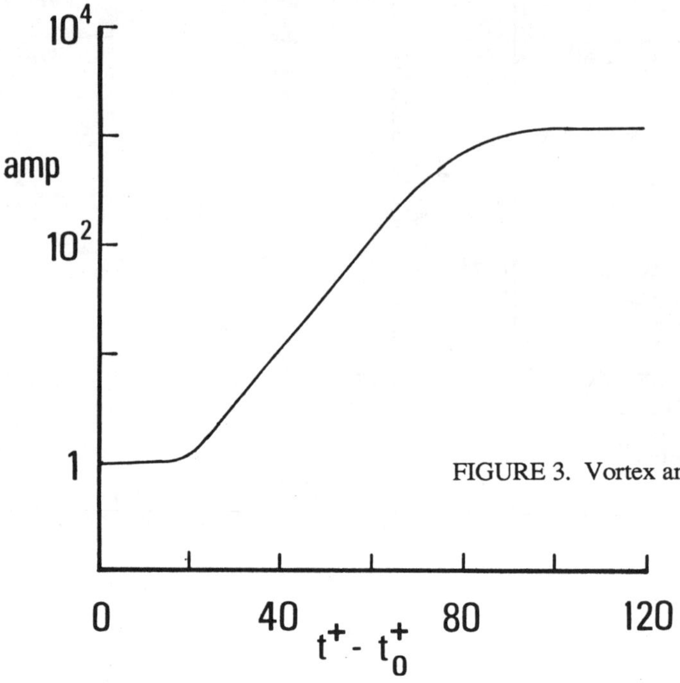

FIGURE 3. Vortex amplification with time.

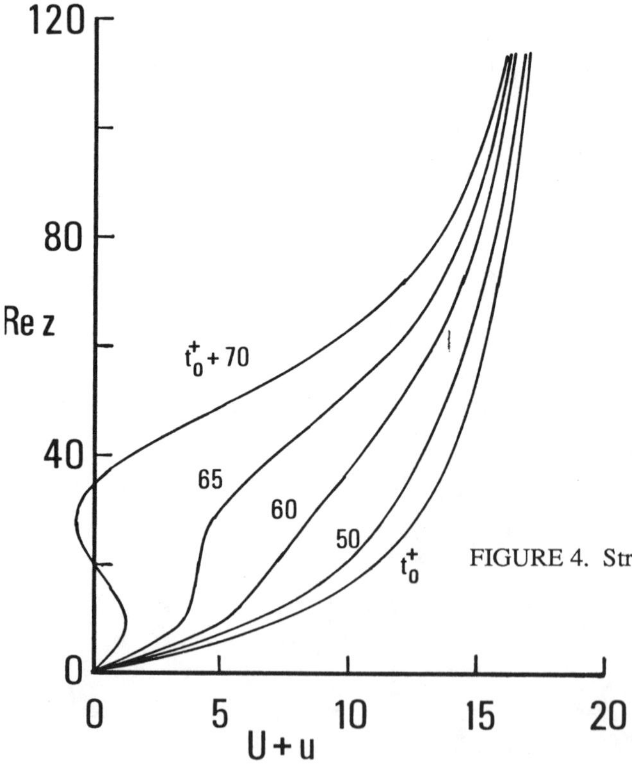

FIGURE 4. Streamwise velocity profile.

$z = C(e^{n\lambda}-1)$ (n, C constants).

Since the program require that $\Delta y = \Delta\lambda$ we choose C so that the solution domain is a square box; calculations were carried out on a 60x60 grid, with a time step $\Delta t = 0.0001$.

7. RESULTS AND DISCUSSION

Realistic values for Re in turbulent channel flow are circa 400; in the present work we use Re = 385, the value of Kreplin & Eckelmann's (1979) experiments. We set L = 100/Re and tickle the flow with a v, w flow field that satisfies the boundary conditions (11), and for which the maximum value of v, $w/U(1) < 0.05$. Instantaneous streamlines of the evolving flow field are shown in figure 2.

Streamwise vortices with a vertical extent of about 90 wall units and with axes near $z^+ = 25$ quickly form. They appear fairly docile, remaining roughly constant in size and lateral location with time; all they experience is a slight distortion and a gradual motion away from the surface, as seen in figure 2.

Now it has long been realized that the presence of a vortex pair in a boundary layer may lead to an inflexion point in the streamwise velocity profile, and that is so here, but it is evident that the resulting instability is initially weak. For some time, till t^+_o say, (where t^+_o is typically > 100), the vortices undergo only weak amplification (see figure 3) and the velocity profile barely deviates from the mean (figure 4). Eventually, however, the vortices do intensify and their influence upon the mean flow increases. It appears that the strength of the instability also increases because the vortices undergo a sudden exponential amplification in strength which has a dramatic effect upon the flow, culminating in an instantaneous flow reversal. The flow reversal is short lived because, perhaps as a result of it, the vortices move slightly upward, amplification diminishes, and the streamwise velocity profile relaxes to close to its mean. Meanwhile beneath the first pair of vortices a new pair form.

It is evident, therefore, that a warped or fluctuating transverse vorticity field can give rise to streamwise vortices in a turbulent channel flow. Moreover, we might infer that flow visualization of this flow field may produce some of the features associated with streaks.

ACKNOWLEDGEMENT

This work was initiated by S. Leibovich and J.L.Lumley and has benefited from discussions with them. The work was supported in part by the U.S. Office of Naval Research under the program Select Research Opportunity IV, grant no. N00014-85-K-0172

REFERENCES

Andrews, D.G.and McIntyre, M.E. An exact theory of non-linear waves on a Lagrangian-mean flow, *J. Fluid Mech.*, **89**, 609-646, 1978.

Arakawa, A. Computational design for long-term numerical integration of the equations of fluid motion: two dimensional incompressible flow. *J. Comp. Phys.* **1**, 119-143, 1966.

Bashforth, F. and Adams, J.C. *An attempt to test the theories of capillary action by comparing the theoretical and measured forms of drops of fluid.* Cambridge Univ. Press, 1883.

Benney, D.J. A non-linear theory for oscillations in parallel flow. *J. Fluid Mech.*, **10**, 209-236, 1961.

Benney, D.J. and Lin, C.C. On the secondary motion induced by oscillations in a shear flow. *Phys. Fluids*, **3**, 656-657, 1962.

Brown, G.L. and Thomas, A.S.W. Large structures in a turbulent boundary layer. *Phys. Fluids Suppl.* **20**, 243-252, 1977.

Cantwell, B.J. Organized motion in turbulent flow. *Ann. Rev. Fluid Mech.*, **13**, 457-515, 1981.

Chandrasekhar, S., *Hydrodynamic and Hydromagnetic stability* Dover, 1961.

Coles, D. In *Coherent structures of turbulent boundary layers* (ed C.R. Smith & D.E. Abbott), p 462, Lehigh Univ. 1979.

Corino, E.R. and Brodkey, R.S. A visual investigation of the wall region in turbulent flow. *J. Fluid Mech.*, **37**, 1-30, 1969.

Corrsin, S. Symposium on Naval Hydrodynamics, Publn 515, NAS-NRC, p 373. 1957.

Craik, A.D.D. The generation of Langmuir circulation by an instability mechanism, *J. Fluid Mech.*, **81**, 209-223, 1977.

Favre, A.J. Review of space time correlations in turbulent fluids. *J. Applied Mech.*, pp 17, 1965.

Favre, A.J., Gaviglio, J.J. and Dumas, R. Space-time double correlations and spectra in a turbulent boundary layer. *J. Fluid Mech.* **2**, 313-342, 1957.

Head, M.R. and Bandyopadhyay, P. New aspects of turbulent boundary layer structure. *J. Fluid Mech.*, **107**, 297-338, 1981.

Jang, P.S., Benney, D.J. and Gran, R.L. On the origin of streamwise vortices in a turbulent boundary layer. *J. Fluid Mech.*, **169**, 109-123, 1986.

Kampe de Feriet, *J. Ann Soc. Sci. Bruxelles*, Ser I, **59**, 145, 1939.

Kline, S.J., Reynolds, W.C., Schraub, F.A., and Runstadler, P.W. The structure of turbulent boundary layers. *J. Fluid Mech.*, **30**, 741-773, 1967.

Kreplin, H.P. and Eckelmann, H. Behaviour of the three fluctuating velocity components in the wall region of a turbulent channel flow. *Phys. Fluids*, **22**, 1233-1239, 1979.

Leibovich, S. On the evolution of the system of wind drift currents and Langmuir circulations in the ocean. *J. Fluid Mech .*, **79**, 715-743, 1977.

Leibovich, S. On wave-current interaction theories of Langmuir circulations. *J. Fluid Mech.*, **99**, 715-724. 1980.

Leibovich, S. The form and dynamics of Langmuir circulations, *Ann Rev. Fluid Mech.*, **15**, 391-427, 1983.

Lilly, D.K. On the computational stability of numerical solutions of time dependent geophysical fluid dynamics problems. *Monthly Weather Rev.* **93**, 11-26, 1965.

Lumley, J.L. Stokes drift in a turbulent boundary layer, Cornell Univ. FDA 86-03, 1986.

Orszag, S.A. and Patera, A. Secondary instability of wall bounded shear flows, *J. Fluid Mech.*, **128**, 347-385, 1983.

Peaceman, D.W. and Rachford, H.H. The numerical solution of parabolic and elliptic differential equations. *J. Soc. Indust. Appl. Math.* **3**, 28-41, 1955.

Phillips, W.R.C. The wall region of a turbulent boundary layer. *Phys. Fluids*, **30**, 2354-2361, 1987.

Phillips, W.R.C. Streaks. Cornell University. FDA-88-06, 1988.

Ratnanather, J.T. and Phillips, W.R.C. The outer region of a turbulent boundary layer. Cornell Univ. FDA-88-04, 1988.

Swarztrauber, P. and Sweet, R. Efficient fortran subprograms for the solution of elliptic partial differential equations. NCAR TN/IA-109, 1975.

Taylor, G.I. *Proc. Lond. Math.* Soc. **20**, 196, 1921.

Allowance for Many Length Scales in the Description of Turbulent Mixing

V. A. SOSINOVICH
Heat and Mass Transfer Institute of the Byelorussian Academy of Sciences
220728, Minsk, USSR

INTRODUCTION

When studying the turbulent mixing of scalar and vector fields it is easy, at a qualitative level, to distinguish two aspects of the process: dynamics of length scales and dynamics of the fluctuation intensity of these fields. Consider, for example, mixing of a scalar field in isotropic turbulent flow. If in a gas or liquid grid flow another grid brings about the scalar fluctuation field, then the evolution of this scalar field is a good example of turbulent mixing. The scalar field near the second grid has mainly only two values: zero and unity, which is in accordance with the condition of full unmixedness at the molecular level. The typical length scale of unmixed gas volumes is specified by introducing this scalar field into the flow. Different values of this field appear downstream, and at the same time there occurs the whole spectrum of length scales. It is clear that these two aspects of turbulent mixing permanently interact. The evolution of the length scale spectrum creates a steadily varying condition for the effect of molecular diffusion, thereby influencing the rate of a scalar field fluctuation spectrum. The different parts of the length-scale spectrum affect the dissipation rate of the fluctuation intensity in a different way. From physical consideration it is seen that small length scales that immediately affect the dissipation rate, influence the structure of the value spectrum more strongly than the large ones. In turn, the distribution of fluctuation directly specifying the turbulent mixing rate affects the length-scale spectrum. This effect is different for different fluctuation values and for different parts of the length-scale spectrum.

This complex dynamical system must in some way be allowed for in developing the theoretical model for turbulent mixing. To describe the turbulent mixing process, use is usually made of the one-point probability density function (pdf) of scalar field fluctuation, $f_t(c)$ [1]. However, the presence of the spectrum of length scales is, as a rule, ignored. The state of the scalar field in this relation is taken into account in terms of some average length scale or average dissipation rate of the fluctuation intensity. It is likely that the disregard of the many-scale character of turbulent mixing is responsible for the absence of a quite satisfactory closed system of equations for the process (though there are several closed models for the pdf, which are used in practial calculations [1-7]).

This work is aimed at demonstrating the feasibility of deriving an expression for the pdf by employing the notion of the many-scale character of turbulent mixing. The expression for $f_t(c)$ has been obtained as a result of the following procedure. First, a closed equation for the pdf is derived which, according to the closure assumptions, is valid for a scalar field with only one length scale, λ. The solution of this equation, $f_t(c,\lambda)$, is averaged over the spectrum of length scales, $R_t^c(\lambda)$. This yields a certain expression for the pdf $f_t(c)$, which can be regarded valid for a turbulent scalar field with a developed spectrum of length scales.

DERIVATION OF A CLOSED EQUATION FOR THE PROBABILITY DENSITY FUNCTION IN THE CASE OF ONE LENGTH SCALE

Derivation of a closed equation for the pdf is based on the equation for the amplitude $\hat{c}(t)$ of a one-length-scale scalar field of fluctuations. The function $\hat{c}(t)$ is defined from the equality

$$c(\bar{x},t) = \hat{c}(t) \sin\left(\frac{x}{\lambda}\right) \sin\left(\frac{y}{\lambda}\right) \sin\left(\frac{z}{\lambda}\right) \quad (1)$$

where λ is the fixed length scale. The equation for $c(t)$ can be written as:

$$\frac{d\hat{c}(t)}{dt} = - \beta\, \hat{c}(t) + \phi(\hat{c},t) \quad (2)$$

Here

$$\beta = \frac{3D}{\lambda^2} \quad (3)$$

where D is the diffusion coefficient. The first term on the RHS of Eq. (2) describes the action of molecular diffusion on the amplitude $\hat{c}(t)$. The second term $\phi(\hat{c},t)$ simulates the supply of a fresh mixture to the flow. Generally speaking, the term "fresh mixture" means the scalar field nonmixed at the molecular level. Formula (1) describes the mixture slightly mixed up to the molecular level. The term "one-length-scale field" will be understood to mean the scalar field (1) with the maximum amplitude $\hat{c}(t) = a$.

Write

$$\phi(\hat{c},t) = q(\hat{c})\, z(t) \quad (4)$$

Here z(t) is the random function, whose statistical characteristics will be chosen from physical considerations; $q(\hat{c})$ is a function which must restrict the supply of the fluctuation intensity to within the limits $|\hat{c}(t)| < a$, where a is the maximum concentration of the field $c(\bar{x},t)$.

The function z(t) has to be taken so that the last term in Eq. (2) would simulate the supply of a fresh mixture. Let z(t) have the form of a sum of positive and negative impulses:

$$z(t) = \sum_{i=1}^{n} \xi_i\, \delta(t - t_i) \quad (5)$$

where $\xi_i(t)$ are the statistically independent amplitudes with the probability density function $P(\xi)$. The form of $P(\xi)$ will be discussed below; the t_i are the random points equally distributed in the interval (0,T) so that the number of this point, n, is distributed in accordance with the Poisson law with $\nu = n/T$ as a parameter of this distribution; $\delta(t - t_i)$ is the Dirac function. This form of the random force simulates an impulse with the amplitude $\xi_i\ \delta(t - t_i)$ within an infinitely small time interval. This force, with a proper choice of parameters, must produce instantaneous changes of the amplitude $\hat{c}(t)$ up to a maximum value. The choice and the interpretation of the parameters of the random function z(t) will be made in deriving the equation for the pdf.

Define the function

$$\phi_t(c) = \delta[\hat{c}(t) - c]$$

Using Eq. (2) we can get, for function $\phi_t(c)$, the probabilistic equation

$$\frac{\partial\phi_t(c)}{\partial t} = \beta \frac{\partial}{\partial c} [c\ \phi_t(c)] - z(t) \frac{\partial}{\partial c} [q(c)\ \phi_t(c)] \tag{6}$$

Averaging Eq. (6) over the implemented sets of the function z(t) and taking into consideration that

$$\langle\phi_t(c)\rangle = f_t(c,\lambda)$$

where $f_t\ (c,\lambda)$ is the pdf for the amplitude of a one-length-scale scalar field, we find [8]

$$\frac{\partial f_t(c,\lambda)}{\partial t} = \beta \frac{\partial}{\partial c} [c\ f_t(c,\lambda)] + \langle\dot{\theta}_t[\frac{\delta}{i\delta\ z(\tau)}]\ \phi_t(c)\rangle \tag{7}$$

Here $\theta_t[v(\tau)] = \ell n\ \Phi[v(\tau)]$, where $\Phi[v(\tau)]$ is the characteristic functional of the random field z(t); $\dot{\theta}_t \equiv d/dt(\theta_t)$; and $\delta/\delta z(\tau)$ is the operator of variational differentiation. In the case of the δ-correlation random process, the following equality is valid

$$\theta_t[v(\tau)] = \theta_t[v(t)]$$

Variational differentiation of Eq. (6) gives

$$\frac{\delta\phi_t(c)}{\delta z(t)} = - \frac{\partial}{\partial c} [q(c)\ \phi_t(c)] \tag{8}$$

Using equality (8), we can transform the last term of Eq. (7) to the form

$$I \equiv \langle\dot{\theta}_t[\frac{\delta}{i\delta z(\tau)}]\ \phi_t(c)\rangle = \dot{\theta}_t[i \frac{\partial}{\partial c} q(c)]\ f_t(c,\lambda) \tag{9}$$

In the case of the Poisson δ-time-correlated process, the expression for the functional $\theta[v(\tau)]$ has the form [8]:

$$\theta_t[v(\tau)] = \int_0^t d\tau\{ \int_{-\infty}^{\infty} d\xi\ P(\xi)\ \exp[i\xi v(\tau)] - 1\} \tag{10}$$

With regard to (10) we can write equality (9) as

$$I = \nu\{ \int_{-\infty}^{\infty} d\xi \ P(\xi) \exp[-\xi \frac{\partial}{\partial c} q(c)] - 1\} f_t(c,\lambda) \tag{11}$$

The function q(c) must be chosen so that the supply of a fresh mixture should be realized only within the attainable concentration interval $|\hat{c}(t)| < a$. To satisfy the condition, chose the form

$$q(c) = \begin{cases} a - c , & \text{if } z(t) > 0 \\ a + c , & \text{if } z(t) < 0 \end{cases} \tag{12}$$

A detailed account of dynamics of a single supply of a fresh mixture using Eq. (2) allows the following conclusion to be made about the form of the probability distribution of the amplitude ξ_i:

$$P(\xi_i) = \frac{1}{2} \lim_{B\to\infty}[\delta(\xi - B) + \delta(\xi + B)] \tag{13}$$

Taking (12) and (13) into account, expression (11) for I can be written in the form

$$I = \frac{\nu}{2} \lim_{B\to\infty}\{e^B f_t[a - e^B(a-c),\lambda] + e^B f_t[-a + e^B(a+c),\lambda] - 2f_t(c,\lambda)\} \tag{14}$$

Substituting this result into (7) and passing to $B \to \infty$, we obtain a closed equation for the one-length-scale pdf $f_t(c,\lambda)$:

$$\frac{\partial f_t(c,\lambda)}{\partial t} = \beta \frac{\partial}{\partial c} [cf_t(c,\lambda)] + \frac{\nu}{2} [\delta(c - a) + \delta(c + a) - 2f_t(c,\lambda)] \tag{15}$$

This equation has all necessary properties of the equations for the pdf. Specifically, the normalization condition is fulfilled and the zero value of $f_t(c,\lambda)$ is preserved beyond the interval [a,-a], if these properties are guaranteed at t = 0. We can show that the structure of the last term in (15) ensures a decrease of the value of the function $f_t(c,\lambda)$ near the point c = 0 and an increase of this function near the points $c = \pm a$. This is in accordance with the physical meaning of this term. The behavior of this term is such that if the first term is absent in the RHS of Eq. (15) and provides mixing, then the stationary solution of Eq. (15) has the form of a sum of the δ-functions:

$$f_t(c,\lambda) = \frac{1}{2} [\delta(c - a) + \delta(c + a)] \tag{16}$$

This result is quite natural and necessary.

It is easy to get the following equation for the rms fluctuations $\langle c^2(t,\lambda)\rangle$ by multiplying (15) by c^2 and integrating it with respect to c. We have

$$\frac{d\langle c^2(t,\lambda)\rangle}{dt} = -2\beta \langle c^2(t,\lambda)\rangle + \nu[a^2 - \langle c^2(t,\lambda)\rangle] \tag{17}$$

We can see from (17) that the parameter ν is the intensity of supplying a fresh mixture, and the efficiency of supply decreases as far as the fluctuational scalar field approaches its limiting condition when $\langle c^2(t,\lambda)\rangle = a^2$.

Note that a similar equation for the pdf without the last term was derived in [10]. The parameter λ means a microscale of a turbulent scalar field. To derive a closed equation for the pdf, the authors [10] employed the least-squares method as well as the additional assumption on the linear dependance of the function responsible for unresolution of the equation for the pdf on the argument c. It is evident that with such an approach it is difficult to indicate the degree of validity of the equation. It is seen from the derivation of Eq. (15) that it can be regarded valid only for the case of a one-length-scale scalar field.

Using the Laplace and Fourier transformations, it is possible to get a solution to Eq. (15) [11]:

$$f_t(c,\lambda) = \theta\left(a - |c|\right)\, \theta\left(|c| - ae^{-\beta t}\right) \frac{\nu}{2a\beta} \left|\frac{c}{a}\right|^{\frac{\nu}{\beta}-1} + e^{(\beta-\nu)t} f_o\left(ce^{\beta t},\lambda\right) \quad (18)$$

Here $\theta(x)$ is the Heaviside function

$$\theta(x) = \begin{cases} 0\,, & \text{if} \quad x < 0 \\ 1\,, & \text{if} \quad x > 0 \end{cases} \quad (19)$$

Note that (18) leads to the normalization condition for $f_t(c,\lambda)$ only if this function has this property at the initial time moment. This property is guaranteed by the form of general solution (18). When the supply of a fresh mixture is absent ($\nu = 0$) for $f_t(c,\lambda)$, the known solution is obtained [1, 10] and describes the process of turbulent mixing from any initial state to a fully mixed state

$$f_t(c,\lambda) = e^{\beta t} f_o(ce^{\beta t},\lambda) \quad (20)$$

The solution of (15) in the stationary limit when $t \to \infty$ assumes the form:

$$f_\infty(c,\lambda) = \frac{\nu}{2a\beta} \left|\frac{c}{a}\right|^{\frac{\nu}{\beta}-1} \quad (21)$$

Using equality (18) it is easy to get the formula for evolution of rms value fluctuations of the scalar field:

$$\langle c^2(t,\lambda)\rangle = \langle c^2(0)\rangle\, e^{-(2\beta+\nu)t} + \frac{a^2\nu}{\beta+\nu}\left[1 - e^{-(2\beta+\nu)t}\right] \quad (22)$$

The cited characteristics of Eq. (15) show that this equation is self-consistent and can in principle serve as a modified turbulent mixing model. However, the general solution (18) still has some essential shortcomings. Figure 1 demonstrates the evolution of the function $f_t(c,\lambda)$ at several different values of ν/β if function (16) is taken as the initial function. From Figure 1 it is seen that the process of turbulent mixing

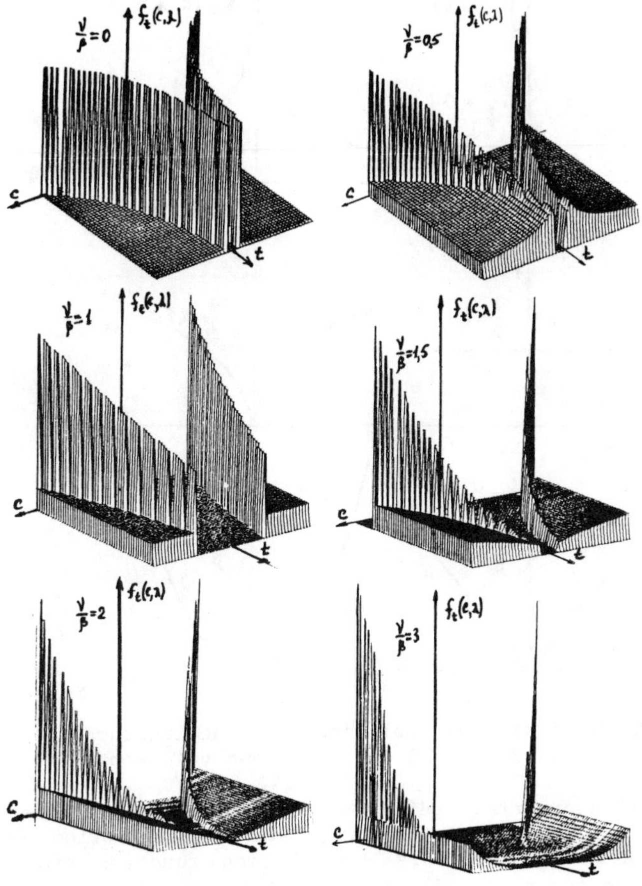

FIGURE 1. Evolution of the function $f_t(c,\lambda)$ in accordance with a one-length-scale mixing model for different values of ν/β (arrows stand for the Dirac function).

written by Eq. (15) is not real. With no supply of fresh mixture ($\nu = 0$), the form of the function is unchanged and only two fluctuation values are present. If $\nu \neq 0$, the form of the function is also nonplausible. From the stationary solution (21) it is easy to see in Figure 2 that the pdf defined by (15) for various values of ν/β is not bi-modal; this contradicts experiment [1]. The last property is probably the most serious shortcoming of the one-length-scale mixing model (15), which cannot be used for realistic simulation of the turbulent mixing process.

DERIVATION OF THE EXPRESSION FOR THE PROBABILITY DENSITY FUNCTION IN THE CASE OF A MANY-LENGTH-SCALE SCALAR FIELD

From the aforesaid concern with the properties of Eq. (15) and from the derivation of this equation, it is clear that Eq. (15) is correct for calculation of the pdf of a one-length-scale scalar field. The real turbulent flow is characterized by a full spectrum of length scales. Hence,

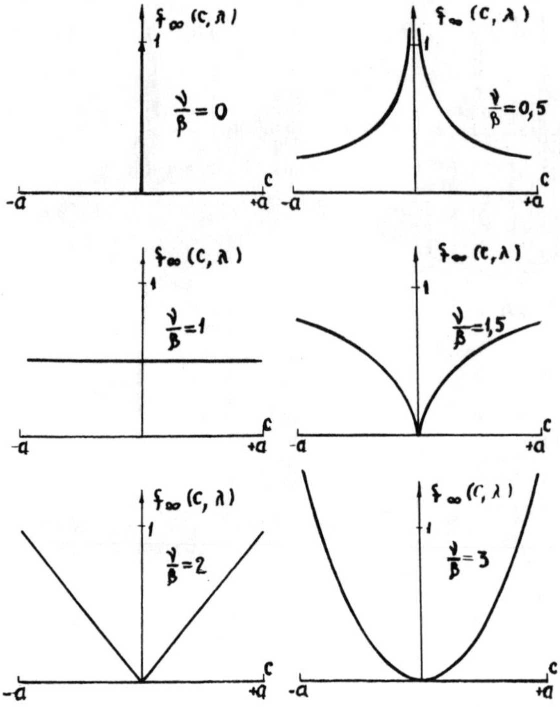

FIGURE 2. Stationary solution of a one-length-scale mixing model for different values of ν/β.

calculation of an appropriate pdf requires that we average the solution of Eq. (15) with respect to the probability distribution of length scales. Designate the necessary distribution function trough $R_t^c(\lambda)$. Then the expression for $f_t(c)$ will take the following form:

$$f_t(c) = \int_\lambda f_t(c,\lambda)\, R_t^c(\lambda)\, d\lambda \tag{23}$$

The integration extends over all λ, and $R_t^c(\lambda) \neq 0$. Using expression (18) for $f_t(c,\lambda)$ we get

$$f_t(c) = \int_\lambda e^{(\beta-\nu)t}\, f_o\left(c\, e^{\beta t}, \lambda\right) d\lambda \, +$$

$$+ \,\nu\theta\left(\left|c\right| - a\right) \int_\lambda \theta\left(\left|c\right| - ae^{-\beta t}\right) \frac{\lambda}{6aD^2} \left|\frac{c}{a}\right|^{\frac{\nu\lambda^2}{3D}-1} R_t^c(\lambda)\, d\lambda \tag{24}$$

Consider a specific example of calculation of $f_t(c)$ by formula (24). For this purpose it is necessary to select an explicit expression for $f_o(c,\lambda)$ and $R_t^c(\lambda)$. Let the initial pdf, $f_o(c,x)$, describe a binary, completely nonmixed at the molecular level, scalar field of concentration in form (16). Substituting (16) into (24) and evaluating the integral gives the following expression for $f_t(c)$ in terms of the function $R_t^c(\lambda)$:

$$f_t(c) = \frac{e^{-\nu t}\ \sqrt{3Dt}\ \ R_t^c\left(\sqrt{\dfrac{3Dt}{\ell n\ |a/c|}}\right)}{2|c|\ \left|\ell n\ |a/c|\right|^{3/2}} +$$

$$+ \frac{\nu\theta(|c| - a)}{6Dc} \int_\lambda \theta\left[\sqrt{\frac{3Dt}{\ell n\ |a/c|}} - \lambda\right] \lambda^2 \left|\frac{c}{a}\right|^{\frac{\nu\lambda^2}{3D}} R_t^c(\lambda)\ d\lambda \qquad (25)$$

Calculation of $f_t(c)$ by formula (25) requires knowledge of the form of the function $R_t^c(\lambda)$. The general problem of finding the function $R_t^c(\lambda)$ will be discussed below. Here we shall consider an example that illustrates the new type of results that may be obtained from the new method of calculating the pdf. In this example calculation we assign the following form of $R_t^c(\lambda)$ in Eq. (25), allowing for the distribution of the fluctuation intensity over different length scales:

$$R_t^c(\lambda) = \frac{2\lambda}{L^2}\, e^{-\lambda^2/L^2}, \qquad 0 < \lambda < \infty \qquad (26)$$

where L is the parameter for the mean length scale of the scalar field. Equation (26) describes a rather narrow band of the fluctuation intensity. The similar expression for a velocity field corresponds to the existence of the vortices of only one length scale L in the flow. Nevertheless, even such a narrow-band distribution is enough to demonstrate the properties of averaging. Substituting (26) into (25) gives

$$f_t(c) = \frac{e^{-\nu t}\ 3Dt\ \exp\left[-\dfrac{3Dt}{L^2\ \ell n\ |a/c|}\right]}{2\ |c|\ L^2\ \ell n\ |a/c|} +$$

$$+ \frac{\nu\theta(|c| - a)}{3D\ |c|\ L^2} \int_0^{\sqrt{\frac{3Dt}{\ell n\ |a/c|}}} \lambda^3\ \exp\left[-\frac{\nu}{3D}\ \ell n\ \left|\frac{a}{c}\right|\ \lambda^2 - \frac{\lambda^2}{L^2}\right] d\lambda \qquad (27)$$

Calculating the last integral in (27), we have

$$f_t(c) = \frac{3Dt\ \exp\left[-\nu t - \dfrac{3Dt}{L^2\ \ell n\ |a/c|}\right]}{2\ |c|\ L^2\ |a/c|} +$$

$$+ \frac{\nu\theta(|c|\ a)}{6D|c|\ L^2\left[\frac{\nu}{3D}\ \ell n|a/c| + L^{-2}\right]}\left\{1 - \left(1 + \nu t + \frac{3Dt}{L^2\ \ell n|a/c|}\right) \exp\left(-\nu t - \frac{3Dt}{L^2\ \ell n|a/c|}\right)\right\}$$

$$(28)$$

This expression for $f_t(c)$ describes turbulent mixing of a supply of fresh mixture in a many-length-scale turbulent scalar field. It is seen from

Figure 3 that the many-length-scale model simulates the process of mixing in a more realistic fashion as against the one-length-scale model. With no supply of fresh mixture, ($\nu = 0$), Eq. (28) describes turbulent mixing from the completely unmixed condition up to the molecular level, demonstrating the presence of diverse concentrations during the evolution time. With the supply of a fresh mixture, ($\nu \neq 0$), the mixing evolves according to (28) more naturally than by (18). Comparison of Figures 1 and 3 supports this observation.

In the stationary limit (Figure 4), when $t \to \infty$, in accordance with the many-length-scale model the pdf becomes a two-mode form, reflecting the possibility of the simultaneous presence of a fresh mixture and mixtures up to the molecular level in any relations, which is in accordance with experimental data [1]. This possibility is absent in the one-length-scale model.

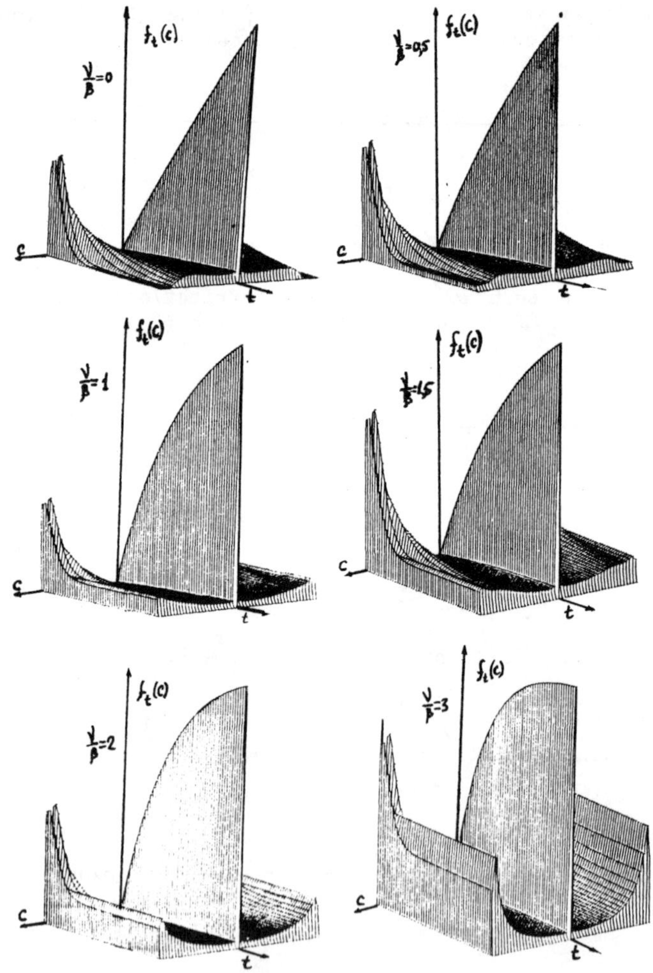

FIGURE 3. Evolution of pdf $f_t(c)$ in accordance with the many-length-scale model for turbulent mixing at different values of ν/β.

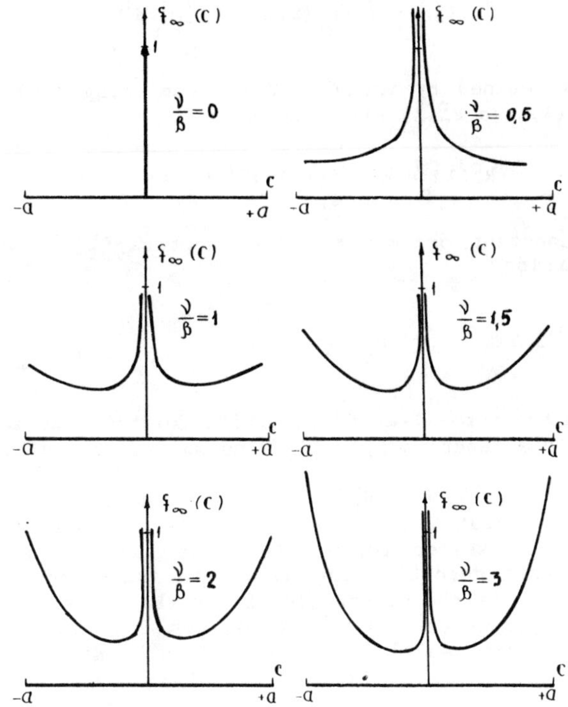

FIGURE 4. Stationary solution to the many-length-scale model for turbulent mixing at different values of ν/β.

Thus the suggested model for turbulent mixing gives a qualitatively good description of this process.

With the many-length-scale approach being used for calculation of the pdf it is necessary to relate the spectral function $R_t^c(\lambda)$ to some function $P_t^c(\lambda)$, which can be calculated from the closed evolution equation. In [6] the closed system is derived to calculate the function $P_t^c(\lambda)$ which describes a distribution of rms fluctuations of a turbulent scalar field over different length scales. This function is defined so that the following equality is valid:

$$\langle c^2(t)\rangle = \int_0^\infty P_t^c(\lambda)\, d\lambda \tag{29}$$

An attempt will be made to relate the function $P_t^c(\lambda)$ and the function $R_t^c(\lambda)$ which is needed for calculation of the pdf in accordance with formula (25). For this goal we define the value of $\langle c^2(t)\rangle$ by the formula

$$\langle c^2(t)\rangle = \int_{-a}^{a} c^2\, f_t(c)\, dc \tag{30}$$

Using expression (23) for $f_t(c)$ we get:

$$<c^2 \, |t|> = \int_\lambda <c^2(t,\lambda)> R_t^c(\lambda) \, d\lambda \qquad (31)$$

where $<c^2(t,\lambda)>$ is defined by formula (21). Comparing (29) and (30) we may assume that $R_t^c(\lambda)$ and $P_t^c(\lambda)$ are related by

$$R_t^c(\lambda) = \delta <c^2(t,\lambda)>^{-1} R_t^c(\lambda) \qquad (32)$$

where δ is some constant parameter. The function $R_t^c(\lambda)$ must obey the normalization condition

$$\int_\lambda R_t^c(\lambda(\, d\lambda = 1 \qquad (33)$$

which results from the normalization condition for the function $f_t(c)$ and from the fact that the function $f_t(c,\lambda)$ is normalized to unity.

When expression (32) for $R_t^c(\lambda)$ is used, condition (33) can be satisfied, if the integration interval is limited by some value $\lambda = \lambda_o$. Using the explicit expression for $<c^2(t,\lambda)>$ in form (21), it is quite easy to show that this condition can always be fulfilled, and, in this case $\lambda_o < \lambda_m$, where λ_m is the microscale of a turbulent scalar field. It is clear then that parameter δ slightly differs from unity because the integrand in (31) is very small at $\lambda < \lambda_o$. Then condition (33) written in the form

$$\delta \int_{\lambda_o}^{\infty} <c^2(t,\lambda)>^{-1} P_t^c(\lambda) \, d\lambda = 1 \qquad (34)$$

together with condition (31) rewritten with regard to equality (32)

$$\delta \int_{\lambda_o}^{\infty} P_t(\lambda) \, d\lambda = <c^2(t)> \qquad (35)$$

is used for calculation of the parameters λ_o and δ. The relationship between $R_t^c(\lambda)$ and $P_t^c(\lambda)$ can be written as

$$R_t^c(\lambda) = \theta(\lambda - \lambda_o) \, \delta <c^2(t,\lambda)>^{-1} P_t^c(\lambda) \qquad (36)$$

For calculation of the function $P_t^c(\lambda)$ use can be made of the closed system of the equations for $P_t^c(\lambda)$ [6].

CONCLUSIONS

An attempt has been made to demonstrate the contribution of the many-length-scale character of a turbulent scalar field to describe the process of turbulent mixing. It is shown that the many-length-scale model is capable of simulating this process more adequately than the corresponding one-length-scale model. Specifically, the derived expression (28) reflects a good experimental two-mode form of the pdf.

From expression (25) it is seen that the structure of the pdf is defined in all details of the spectral function $R_t^c(\lambda)$, but not only by its integral characteristic. This result seems very natural.

Comparison of the results calculated from the many-length-scale model and the one-length-scale model reveal an essential difference in describing turbulent mixing. This difference is explained by the fact that the many-length-scale model correctly takes into consideration the influence of the small-length-scale part of a spectrum on evolution of the pdf. In spite of their relatively small contribution to the integral intensity of fluctuations, the role of small length scales in the evolution of the pdf is very important, because the molecular diffusion mechanism and generation of a mixed mixture occurs on just these length scales. The theories based on use of the average characteristics of a length-scale spectrum are not capable of simulating small length scales and, consequently, they cannot describe turbulent mixing adequately.

REFERENCES

1. O'Brien, E.E. In: Topics in Applied Physics, Vol. 44, P. Libby and F. Williams, Eds., pp. 185-218. Springer-Verlag, Berlin, 1980.

2. Sabelnikov, V.A. Model equation for the velocity and concentration probability distribution in turbulent mixing and diffusion combustion of gases. Fiz. Gor. VZRYVA, No. 2, pp. 37-46, 1983.

3. Kuznetsov, V.R., Lebedev, A.B., Sekundow, A.N., and Smirnowa, I.P. Investigation of quasi-homogeneous turbulent diffusion combustion with the use of the equation for the function of the concentration probability density distribution. Izv. Acad. Nauk SSSR, Mech. Zhidk. Gasa, No. 4, pp. 3-11, 1980.

4. Frost, V.A. Model of turbulent diffusion combustion. Energetika i Transport, No. 6, pp. 108-116, 1973.

5. Kolovandin, B.A., Sosinovich, V.A., and Tsyganov, V.A. System of the equations used for calculating a mean chemical reaction rate in the isotropic turbulent flow. In: Proceedings of Workshop on the Gas Flame Structure, Prof. V.K. Baev, Ed., pp. 161-177. Novosibirsk, July 12-15, 1983.

6. Sosinovich, V.A., and Tsyganov, V.A. Description of turbulent mixing of reagents using the equation for the probability density of length scales. J. Eng. Phys., Vol. 16, No. 2, pp. 219-225, 1984.

7. Sirignano, W.A. Molecular mixing in a turbulent flow: some fundamental considerations. AIAA 22nd Aerospace Sciences Meeting, Reno, Nevada, Paper No. 84-0375, January 1984.

8. Klatskin, V.I. Stochastic Equations and Waves in Random Inhomogeneous Media. Nauka, Moscow, 1980.

9. Sosinovich, V.A. On theoretical description of turbulent mixing of scalar fields. Preprint No. 20. Minsk, HMTI, 1986.

10. Dopazo, C., and O'Brien, E.E. Statistical treatment of nonisothermal chemical reactions in turbulence. Combust. Sci. Techn., Vol. 13, pp. 99-122, 1976.

11. Sosinovich, V.A. In: Mathematical Models, Analytical and Numerical Methods in the Transfer Theory. Proceedings Intern. School-Seminar, October, 1986, pp. 125-134. Minsk HMTI, 1986.

Part 8
PRACTICAL PREDICTIONS BY TWO–EQUATION MODELS AND OTHER FAST METHODS

SUMMARY LECTURE

Practical Predictions by Single-Point Closure Methods—Twenty Years of Experience

KEMAL HANJALIĆ
Mašinski fakultet
Sarajevo, Yugoslavia

ABSTRACT

This paper attempts to present a retrospective over two-decades of development of the one-point closure models and to highlight major achievements, with specific emphasis on near-wall turbulence. It also recalls some of the pertinent issues that are still open and pose challenges to modelers such as:

- Certain weaknesses, which have been obvious since the very beginning of the development - and are still present - are today perhaps more intriguing than ever

- Some lucid ideas, that have appeared from time to time in the course of development and use, although often substantiated by seemingly very convincing physical arguments, have not, to the surprise of most researchers, brought about the expected improvements

- Some still existing differences in the views of various groups of modelers

The lecture will serve as an introduction that is complemented and elaborated by the papers contributed by the other invited authors in the session.

INTRODUCTION

Ever since the first successes almost two decades ago, in reproducing computationally the experimental data on field properties in a wide range of turbulent flows with a single set of modelled equations and empirical coefficients, one-point closure models rapidly gained in popularity. These methods quickly emerged - in conjunction with suitable numerical methods - as probably the most widely employed predictive tools in Fluid Mechanics, Heat and Mass Transfer. Numerous researchers and users all over the world contributed to their testing, development and refinement.

Many scientists have disputed their merits, some question the physical soundness and the mathematical exacteness of their foundation. However, the popularity of one-point closure models as a simple, but sufficiently reliable engineering tool has not yet faded regardless of known shortcomings and failures. One-point closures, so far, show no signs of decline in use in spite of thrilling advances of direct numerical simulation founded on the impressive capabilities of supercomputers.

Turbulence modelers know that what they do is just an approximation - often intuitive, with varying degree of sophistication, sometimes based on hard experimental evidence, but still an approximation. However, the approximations have always been a blend of art and science. They draw as much as possible on scientific evidence and rigorous analysis, but when this resource becomes exhausted, modeling becomes an art. And, like any other art, it requires a careful balance of ingredients: colors, sounds, movements. Here what is required is a careful balance of levels of approximation of all the important physical processes represented by various terms in the governing equations.

It is perhaps this lack of mathematical rigour and a lack of convincing experimental verification that have discouraged some orthodox scientists from giving much trust to the whole affair. However, nobody can disavow the enormous impact that modeling and numerical computation have made upon the development, optimisation and improvement of many engineering technologies and equipments. Equally important is the incentive that the modeling has given to further analysis and experimental investigations of turbulence and transport phenomena.

In spite of a wide recognition of general achievements in mathematical modelling of turbulence over the past twenty years, there are still some physical effects the modeling of which has not yet resulted in a comparable feeling of satisfaction among the turbulence research community. Streamline curvature and extra strain rate, unsteadiness and periodicity, viscosity and wall-proximity, three-dimensionality, flow separation and reversal, buoyancy and rotation etc. are, perhaps, the the most prominent examples. Most of these effects are often present in many turbulent flows. Yet their modeling still poses many difficulties not only because of an insufficient understanding of the physics, but also because of an apparent need to employ mathematical formulations and numerical schemes of greater complexity than had been hitherto used for modeling simpler, "well behaved" flows. Various attempts to incorporate some, or all of the above mentioned effects and to devise a general turbulence model have produced only partial successes, and usually, at the expense of physical clarity and computing economy. Such a discouraging outcome has caused some of the proposed concepts to become obsolete, but the necessity for proper accounting of these effects prompts the search for better models to be continued. Frequent interludes have helped the incubation and maturing of new ideas, and we can certainly expect further advances. Some recent developments in experimental techniques, in particular the LD Anemometry, produce more and more new experimental evidence on some turbulence interactions that have not been possible to measure accurately hitherto. Also, new advances in direct numerical simulation of turbulence offer a new possibility of obtaining information and of verifying the hypotheses on some turbulence properties that are still impossible to encompass by the presently available experimental techniques.

And, while efforts at improving and refining the models are continuing, in particular at a few reputed schools, (and even under the auspices of some international organizations such as the IAHR Group on Refined Modeling of Flows), numerous engineers all over the world have succumbed to the appeal of an atractive blend of simplicity and high predictive ability of popular models, such as k-ε, to employ them for the computation and predictions of turbulent flows and associated transport phenomena in very complex geometries. Some of these results indeed look very impressive with predicted details of velocity field, turbulence properties and other variables that were inconceivable only few years ago. However, Lumley s warning, issued a decade ago (1978), that the application of a technique in situations in which data do not exist "must be regarded as a dangerous practice since the limitations of the technique are not known with any precision", is still actual as ever. Although we know now much more than a decade ago, active researchers feel that a dose of caution is still often advisable and not all of the computed results are to be fully trusted. A simple turbulence model may be incapable of capturing certain interactions when applied to what Reynolds (1976) calls an "exotic" flow geometry - simply because such interactions may have been unimportant in previously considered flows. Inadequate specification of boundary conditions and/or numerical defects that the user may be unaware of may yield very misleading results and their subsequent abuse. A minimum of both, the modeling and computing expertise is a prerequisite for trustworthy results.

However, in spite of all these cautionary notes, often the prediction results could serve engineering purposes simply because the uncertainties in modeling other phenomena involved may be much higher than those implied in the model of flow hydrodynamics. Or, as Bradshaw, wrote recently (1987) in his review of secondary flows "the best modern methods allow almost all flows to be calculated to higher accuracy then the best-informed guess, which means that the methods are genuinely useful even if they cannot replace experiments."

A few selected examples, shown in Figure 1. give an impression of current achievements. These results originated in the ambients where sufficient experience has been gained both in modeling and numerical computation. But could they all be regarded as fully trustworthy?

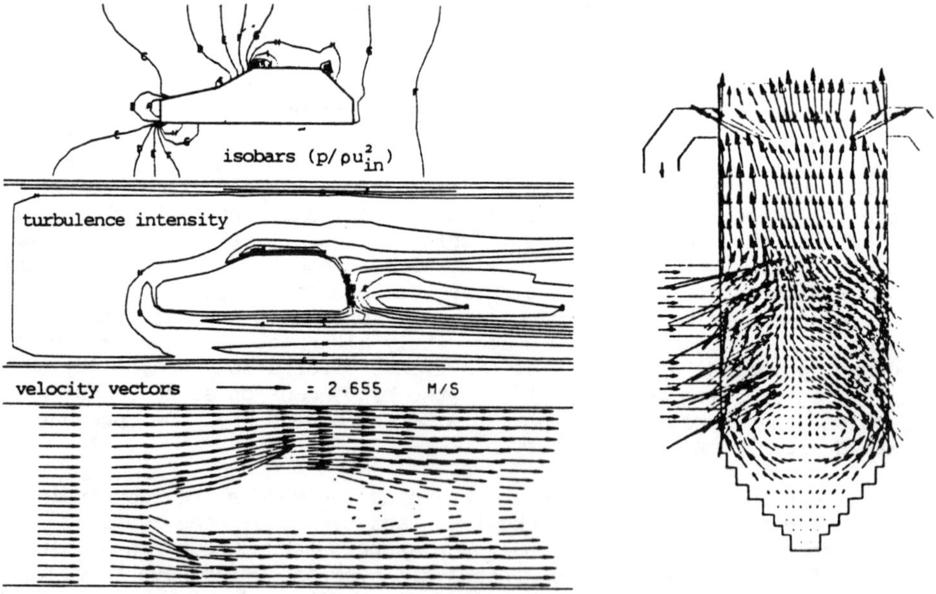

FIGURE 1. Examples of complex flows computed by k-ε model: a-flow around an automobile (Perić M., priv. commun. 1985), b- 3-D calculation of flow field in a boiler furnace (Gorner and Zinser, 1986)

A FEW HISTORICAL RECOLLECTIONS

The topic of turbulence modeling and computation has been very popular since its very appearance and a number of authors have taken up the challenge in the past to write review articles (e.g. Bradshaw 1972, Mellor and Herring 1973, Reynolds 1976, Lumley 1978, Rodi 1982, Launder, Reynolds and Rodi 1984, Lakshminarayana 1986, Nallasamy 1987, and others). To produce yet another survey (although a personal view may always throw a new light on the problem) is not very gratifying nowadays for the simple reason that the gradient of progress in refining turbulence models is now considerably slower than during the first decade of development. I took up the task, with a rather limited ambition to give only a brief overview (hopefully complemented by the papers of the other invited speakers in the session) of recent advances and some of the sustaining problems of modeling turbulence in the near-wall region, for two reasons. Firstly, to pay a tribute to the late Zoran Zarić, whose experimental research of near-wall turbulence structure contributed to our present knowledge and indirectly helped our advances in modeling and, secondly, because the present Conference happens to occur just two decades after the - we may say historic - 1968 AFOSR-IFP Stanford Conference on Computation of Turbulent Boundary Layers, which in the words of J. Lumley (1978) "was the birth of the technique which has become known as second order modeling.." and, as such, calls for a summary of twenty years of development and progress.

Of course, the 1968 Stanford meeting was not the beginning, but only the turning-point, which legitimized the superiority of the field approach to integral methods, and stimulated its further development. As is well known, it all started much earlier,with the works of Kolmogorov(1942), Chou (1945,1947), Rotta (1951) and Davidov (1958-61), whose ideas materialized in practical computation only after the advent of high-speed computers. Already in the year of the Stanford meeting frantic research was going on in parallel at a few places in the world. At the Los Alamos Laboratory Harlow and Nakayama, (1968), Hirt (1969) and Daly (1970) were working on the development of a "generalised transport theory of anisotropic turbulence" employing the partial differential equations for turbulent stresses and energy decay rate following earlier suggestions of Chou. Donaldson and Rosenbaum (1968) pursued invariant modeling to close the Re-stress equations by employing a precribed lenghth scale. At the Imperial College the successful development of the Patankar-Spalding numerical code for solving the partial differential equations in 1968 prompted a rapid development of turbulence models. Ng, Rodi and Spalding (1969) pursued the k-(kL) type of model, while Hanjalić, Jones and Launder (1970) preferred to investigate the two- and three equation models with energy dissipation rate as the scale parameter, which even at that time showed some practical advantages - that became generally recognized some years later. A number of parabolic wall- and free flows were solved with a single set of equations and empirical coefficients, that have changed little ever since, (Hanjalić, 1970). The k-ε model was already extended to be applicable to low-Re-number flows (Jones, 1971) .

However, although nothing substantial has really changed from those original formulations of two-equation models, the intensive research efforts over the past two decades have expanded the frontiers of our knowledge and resulted in numerous proposals for models upgradings and improvements. A rich experience on the extent and limitations of predictive abilities of these and higher order models has been accumulated by enlarged turbulence modeling community. This was best proved at another Stanford meeting, the 1980-81 AFOSR-HTTM Conference on Complex Turbulent Flows, where about 40 research teams demonstrated the high potential and wide scope of current modeling techniques and computational methods.

These fragmental reminiscences of the early history of one-point closure models has been recalled here partly to mark the twentieth aniversary of the method developments, and partly to remind us of the early steps and lucid ideas of the researchers of the pre-computer era who did not have the satisfaction that contemporary modelers have of verifying their thoughts, while they are still fresh.

CURRENT "FAST" MODELS

Three types of turbulence models could be identified as "fast" engineering methods (despite the need to solve partial differential equations):

- two-equation eddy viscosity models (TEM)

- differential Re-stress equation model (RSM)-(second-moment closure)

- hybrid models that hierarchically fall in between these two and take some advantage from each: -algebraic stress models, (ASM), and incomplete or partial stress models (PRSM)

Although the foundations of two-equation models are well known, in a search for model refinements it is helpful to reiterate the basic assumptions:

- turbulent stresses are expressed as directly proportional to the mean rate of strain

- the proportionality coefficient in the stress-strain relationship - the eddy viscosity- is expressed in terms of two turbulence parameters that can be grouped as to represent a characteristic turbulence scale

- the two turbulent parameters are obtained from modeled pde's, which are based upon (or, at least, have some roots in) the exact transport equations for these quantities, which take into account the time memory and the spatial awareness of the local turbulence state

Two-equation models in their rudimentary forms have major advantage in their simplicity and practical usability. A computational advantage (but also a physical deficiency!) is a tensorially invariant eddy viscosity which allows the model to be incorporated into any existing Navier-Stokes computer code.

The differential Re-stress model is regarded as the natural and most logical level of modeling within the framework of the Reynolds averaging approach since it provides the "superfluous" turbulent momentum fluxes from the solution of full transport equations, which are derivable from the Navier-Stokes equations. In addition to the solution of equations for each Re-stress component, it requires a length-scale-supplying equation for which the great majority of models employs the dissipation rate equation. In such a form the RS model still represents the most comprehensive description of turbulent flows that can be employed for practical computations with the present generation of computers.

In addition to a more faithful decription of the turbulence dynamics, second-moment closure models are legitimized by a stricter compliance with the general modeling principles on which some modelers particularly insist. These principles can be grouped into two classes: the first class can be regarded as a mathematical formalism and it implies the dimensional coherence of all terms in equations, the tensorial-order consistency, coordinate-frame indifferece, as well as full satisfaction of realizability conditions, limiting properties of two-dimensional turbulence, etc. The second class of principles are of physical nature and they are more difficult to quantify. They can be defined as principles of physical coherence which postulate e.g. that the turbulence correlations be primarily modeled in terms of turbulence parameters that are known to govern directly the descibed interactions, instead of indirectly influential mean flow-, or some external parameters. A useful principle of this kind is that of receding influence of higher order moments upon the mean flow properties.

As is well known, a more general compliance with these and other rules necessarily burdens models with more complex mathematical formalism (Mjölsness, 1979, proposed no less than eight rules, including the principle of super-realizability!), which for many practical flow situations may not produce any qualitative improvements of predictions. This is the reason why there is no general consent with respect to the strict obedience of the modeling principles. Their disobedience will cause discomfort for some modelers, while the others, more technologically oriented, tend to sacrifice mathematical puritanism for the sake of practicality. The question arises, however, whether a compliance with a wider set of constraints ensures a higher degree of model universality. This seems to be generally the case and with further advances of computers we may expect more modeling principles to be implemented in models of all levels.

Hybrid methods should be regarded as a compromise of today. In most cases they will be used by knowledgable workers who know enough about the affair to be able to estimate to what extent they can prune the full transport equations to achieve computational advantages, but still gain benefits beyond those offered by the simple two-equation models.

Some weaknesses and shortcomings that have been known from the early days of models and have persisted as still unresolved will be listed briefly. For two-equation models these are:

- linear (Newtonian) stress-strain relationship through eddy viscosity model

- scalar character of eddy viscosity (insensitivity to the orientation of the turbulence structure and its transporting and mixing mechanisms)

- scalar character of turbulence scales -insensitivity to eddy anisotropy

- limitations to define only one time- or length scale of turbulence for characterizing all turbulence interactions

- failure to account for all physical processes governing the behaviour of ε by virtue of the simplistic form of the basic equation

- inadequate incorporation of viscosity damping effects on turbulence structure (low Re-number models)

- inability to mimic the effects of pressure reflection and eddy-flattening and squeezing mechanisms due to the proximity of solid- or interphase surface

- frequently inadequate treatment of boundary conditions, in particular at the solid wall

Differential Re-stress models overcome the first two deficiencies, but all the others remain to a greater or lesser extent. In addition, this class of models involves some further uncertainties such as:

- deficiencies in modeling various terms in u u differential equation (in particular the pressure redistribution and turbulent diffusion).

The modeling of various turbulence interactions, represented by different terms in transport equations, requires a specification of characteristic turbulence time-, length- and velocity scales. Of the two turbulence properties used to define chacteristic scales, the kinetic energy of turbulent fluctuations $k=\overline{u_i u_i}/2$ has been employed without exception, since it has proved to be the best defined and most readily obtainable turbulence parameter. As the second turbulence property the most popular has been the isotropic part of the rate of energy dissipation $\varepsilon = \nu(\partial u_i/\partial x_j)^2$ which in combination with k produces the turbulence time and length scale. Other variables have also been persistently used by other researchers, such as a length scale L and the product (kL) , or $\tilde{w}^2$- the mean square of the fluctuating vorticity ($w_k = \mathcal{E}_{ijk}\partial u_i/\partial x_j$), the square root of which could be interpereted as the characteristic frequency of the large-eddy structure, or as the specific dissipation rate per unit of kinetic energy. For each of the variables an equation has been derived empirically (though an origin can be traced to the Navier-Stokes equations) which have the same conservation form with source and transport terms. The choice is more a matter of taste since all variables and their equations are derivable one from another differing only in the way the diffusion terms are modeled. The grater popularity of the dissipation equation can be attributed to the practical advantage of its simpler form as compared with other equations.

The ε-equation has a form of the simplest transport equations, with terms representing the basic processes that govern its dynamics. Despite its striking simplicity the basic equation for the turbulence energy dissipation has proved to possess a surprising level of universality and thus its usableness as the length-scale-supplying equation. Extensive tests have revealed, however, a number of shortcomings which come to the surface particularly in the flows with a pronounced inequilibrium of basic turbulence mechanisms.

Estimate of performances

The most comprehensive survey of the performance of various models can be found in the Proceedings of the 1980-81 AFOSR-HTTM Stanford Conference on Complex Turbulent Flows (Kline et al.). In the last two years several up-to-date reviews have appeared covering more specific aspects: Lakshminarayana (1986) surveyed practices of modeling the curvature, rotation and three-dimensionality, Cousteix (1986) reviewed modeling and computation of three-dimensional- and unsteady boundary layers, Bradshaw (1987) gave an interesting account of current models ability to predict secondary flows of various origins, while Nallasamy (1987) gave a review of the application of turbulence models to the prediction of internal flows. A general report on EUROMECH 180 Colloquium on turbulence modeling for incompressible flows by Rodi (1986) gives an excellent review of the state-of-the-art as seen by the Colloquium participants.

Although the comments and conclusions in these reviews reflect personal convictions and some of them may not hold generally, most of them are shared overwhelmingly by the majority of active members of turbulence modeling community since they derive from the experience of a vast number of current users. It would be beyond the scope of the present paper's objective and an unnecessary repetition to summarize conclusions, but some may be recalled as a reminder for the forthcoming discussion (limited to incompressible flows).

- Two-equation models in their basic form yield satisfactory predictions of two dimensional thin shear flows, (attached boundary layers and two-dimensional flows in conduits, with arbitrary pressure gradients, wall suction and blowing, free flows), as well as some recircualating flows which are dominated by pressure gradients, and even of some flows with streamline curvature and body force when these effects are weak.

- Two-dimensional flows with separation and stronger curvature, rotation, buoyancy and other specific effects can be predicted satisfactorily with two-equation models if adequate modifications are made. The simplest - though not always effective - way is to make the coefficient in the eddy viscosity, C_μ a function of P/ε ratio as well as of "extra" effects (preferably derived from an ASM model and expressed in forms of the nondimensional similarity parameters: Richardson-, Rossby-, Rayleigh numbers). A still sounder approach is that which implies the inclusion of additional terms into the dissipation equation, preferably based on the exact terms in the original equations.

- Better predictions of two-dimensional flows with significant rotation and curvature are obtained with AS models that are derived from parent differential stress equations, provided that flows are not far from the equilibrium, i.e. if the stress transport is of minor importance. The AS models by their origin are supposed to capture better flow physiscs e.g. the streamline curvatures and body forces on Re stresses.

- The full Re-stress model yields superior predictions of two-dimensional nonequilibrium flows (sudden and strong streamwise variations of boundary and external conditions, in particular those imposing the abrupt changes of strain rate). It also performs better in unsteady and periodic flows and everywhere where the response of turbulence field exhibits a lag and a consequent hysteresis as compared with the variation of mean flow properties.

- An increased interest in the application of turbulence models to the calculation of three-dimensional flows over tha past few years reveals that none of the models in their present forms can fully capture the significant changes in turbulence structure imposed by even mild three-dimensionalities (Bradshaw,1987). Physical understanding of the effects of strong skew-induced vorticity, embedded vortices and even stress-induced vortices, upon the turbulence structure is regarded as still insufficient (partly due to the lack of experimental data) as to suggest plausible model extensions (e.g. the influence on presssure redistributing process), although some recent predictions with standard models show an interesting outcome (Liandrat et al.,1987), e.g. Fig.2.

Computationally, of course, two-equation models offer frequently decisive advantages over the RS models. Almost the same could be claimed for the AS models, provided simpler geometries are considered. However, they are often shown to pose numerical instabilities or other kinds of computational inconveniences and the question arises whether the direct use of full or pruned differential Re-stress models would be a sounder approach in general, provided adequate computing facilities are available. Also for irregular flow domains, where the use of non-orthogonal coordinate systems is required, and for three-dimensional flows, the AS expressions become cumbersome and their advantages over the RS models diminish.

These are more general statements with which some meticulous modelers would not agree since there is evidence of many yet unresolved paradoxes which conflict with the above conclusions, to mention only the unsatisfactory predictions of the rate of spread of round jets, or the inability to predict the weakly swirling jet (contrary to the case of the strongly swirling jet which can sometimes be satisfactorily predicted by the standard $k-\varepsilon$ model) etc.

Controversies arise sometimes from uncertainties and inconsistencies in the definition of inflow turbulence quantities, in the treatment of boundary conditions (whether wall functions or a straight-to-wall integration is applied), as well as from insufficient accuracy of the applied numerical schemes; all these may obscure the real performances of the turbulence models and hence lead to misjudgements.

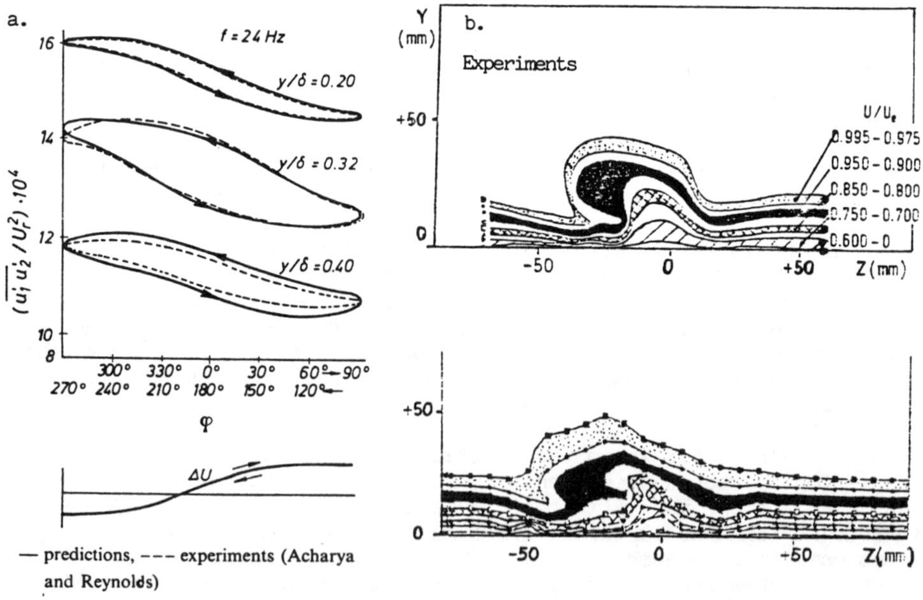

FIGURE 2. Some illustrations of Re-stress model predictions: a-hysteresis of shear stress at different flow depths in a pulsating boundary layer (Hanjalić and Stošić, 1983), b- ISO-U contours in a turbulent boundary layer with imbedded longitudinal vortices (Liandrat et al., 1987)

DIRECTIONS TOWARDS THE IMPROVEMENT OF CURRENT MODELS

The Dissipation Rate Equation

It is remarkable that the ε-equation has persistently remained in use in its very primitive form as it first appeared for high Re number flows (Davidov 1961, Hanjalić 1970):

$$\frac{D\varepsilon}{Dt} = (C_{\varepsilon 1}P + C_{\varepsilon 3}G - C_{\varepsilon 2}\varepsilon)\frac{\varepsilon}{k} + \frac{\partial}{\partial x_k}\left(C_\varepsilon \frac{k}{\varepsilon}\overline{u_k u_l}\frac{\partial \varepsilon}{\partial x_l}\right)$$

Although regarded from the very beginning as "naively simple" (Launder 1987) for representing the dynamics of the turbulence energy decay rate and for serving as a source of the characteristic length scale, it exhibited a surprising degree of universality. However, numerous testings over a period of time have revealed also its limitations and shortcomings (usually identifiable in inadequate predictions of turbulence energy) urging a search for a more general form. So far, however, none of the numerous proposals for its refinements and generalizations have been able to withstand the scrutiny.

The first attempts at modeling the terms in the ε-equation tended to draw the information from its exact form, but the outcome has in the past relied mainly on intuition. The source of ε, associated with the promotion of energy transfer through the spectrum due to self-stretching mechanism of vortex filament has been modeled in terms of turbulence energy production P (scaled by turbulence time scale $\tau = k/\varepsilon$). This is a radical over-simplification which imposes a number of constraints. Firstly, it presumes that the process is equally affected by both the linear and angular mean flow deformation. Secondly, it does not provide room for modeling separately the effect of the mean flow vorticity. Furthermore, it implies that the same time scale controls the process of production and viscous destruction of ε. A more general model is sought which would remove all of the mentioned and other limitations as well as allow the effect of turbulence anisotropy and other parameters to play adequate roles. In fact, a model of production of ε in terms of turbulence (rather than mean flow) parameters should be regarded as physically most justified and one which would satisfy the principle of physical coherence. As is well known, Lumley and Khajeh Nouri (1974) proposed to model the source of ε solely in terms of turbulence anisotropy second invariant $A_2 = a_{ij}a_{ij}$. Although well-argued, the idea did not prove to be very helpful when tested. However, like some other ideas, which rest on sound physical reasoning, it deserves to be reincarnated and retested. A possible general source term can be recast in a functional form

$$\pi = f\left[\overline{u_i u_j}\frac{\partial U_i}{\partial x_j}\frac{\varepsilon}{k},\ k\left(\frac{\partial U_i}{\partial x_j}\right)^2,\ k\left(\varepsilon_{ijk}\frac{\partial U_i}{\partial x_j}\right)^2,\ A_2\frac{\varepsilon^2}{k},\ \cdots\ \beta g_i \overline{u_i\theta}\frac{\varepsilon}{k},\ \beta g_i k\frac{\partial T}{\partial x_i},\ \frac{(\overline{u_i\theta\varepsilon})^2}{k^3},\ \ldots\right]$$

In addition to more flexibility in modeling various effects, we have now two time scales at our disposal: the turbulence scale $\tau_t = k/\varepsilon$ and the time scale of the mean flow deformation $\tau_m = (\partial U_i/\partial x_j)^{-1}$

Several previous proposals in the above sense have yielded improvements in predictions of several classes of flows, (incuding 2-D separated- and 3-D curved channel flows), but have eventually been found to worsen others, to mention Pope (1978), Hanjalić and Launder (1980), Leschziner and Rodi (1981). Recent experiments at UMIST, Launder$_2$ (1987), with the inclusion of the additional source term in form $0.6A_2\varepsilon/k$ and a decrease of the value of $C_{\varepsilon 1}$ from standard 1.4 to 1.1 have resulted in the diminishing of anomalous predictions of the axisymmetric jet.

Another possibility lies in the revision of the model of the diffusion term. The neglect of transport terms in the exact equation for the correlation $Q_k = \nu u_k (\partial u_i / \partial x_j)^2$ yields the expression

$$Q_k = -C \frac{k}{\varepsilon} \left(\overline{u_k u_j} \frac{\partial \varepsilon}{\partial x_j} + C_q \varepsilon \frac{\partial \overline{u_j u_k}}{\partial x_j} + Q_j \frac{\partial U_k}{\partial x_j} \right).$$

Only the first term has been retained in the standard model, a simplification justified for wall boundary layers on the basis of an order-of-magnitude estimate but for more complex flows these arguments do not hold. No serious testing of the above more general formulation are known to me, although some numerical experiments with other additional diffusion-type terms, formulated in an ad hoc manner, e.g.

$$\frac{\varepsilon}{k} \frac{\partial}{\partial x_j} \left(\frac{\mu_t}{\varrho} \frac{\partial k}{\partial x_j} \right); \quad \frac{\partial}{\partial x_j} \left(k \frac{\partial k}{\partial x_j} \right); \quad \left(\frac{\partial k}{\partial x_j} \right)^2$$

to improve the predictions of nonequilibrium wall boundary layers did not yield the desired effects. Some attempts in a different direction - to ensure the experimentally detected constancy of the length-scale gradient in the near wall region in strongly nonequilibrium flows and close to the reattachment points in separating flows - through the dependence of C_ε on the ratio P/ε - produced only modest improvements (Launder 1982).

New chances lie in revisiting some other discarded or not fully explored ideas of the past. Following the direct-interaction (DI) approximation of Kraichnan, Yoshizawa (1985) derived arguments for mixed- or cross-diffusion terms in both the k- and ε-equation, to be followed by Nisizima with practical computations. Another route has been followed by Aupoix et al. who derived the ε-equation for the homogeneous turbulence following the MIS ("Methode Integrale Spectrale") approach, which assumes the evolution of an a-priori defined energy spectrum shape.

These and other experience suggest that more research along these lines may still result in a more general and universal form of dissipation equation.

Reynolds-Stress Transport Equation

The transport equation for the Reynolds stresses can be expressed in symbolic form as

$$C_{ij} - D_{ij} = P_{ij} - \varepsilon_{ij} + \phi_{ij}$$

transport = production-dissipation+pressure strain

The modeling of pressure-straining process remains still the most discomforting and most challenging task. A combination of Rotta's (1951) simple model of non-linear return-to-isotropy process ϕ_{ij1} and the equally simple "isotropization-of-production" model (IP) of the mean-strain pressure-redistribution process ϕ_{ij2} (Naot et al. 1970), with the relationship between the empirical constants $(1-C_2)/C_1=0.23$ gained much popularity because of its simplicity. It was reported to perform even better in predicting some more complex flows in comparison with the marginally more elaborate "quasi-isotropic" (QI) model of the rapid term ϕ_{ijk}, (Launder et al, 1975), though this advantage is not decisive. Both models have been extended to accommodate pressure-reflection and turbulence damping due to the proximity of a solid wall. This success has lead to a similar approach in modeling the pressure interactions in scalar transport equations.

However, noted failures to predict satisfactorily some seemingly simple homogeneous turbulent flows - selected for the 1980-81 AFOSR-HTTM Stanford Conference on Complex Turbulent Flows -, cast a shadow on these models. A recent revival of interest in second-moment closure led Fu, Launder and Tselepidakis (1987) to reanalyse the current practice of modelling the pressure-strain term and to conclude that besides the earlier detected inability to predict homogeneous flows with a large mean shear, neither the IP nor the QI models satisfy the exact limit of the two-dimensional turbulence . Because this is a state to which the turbulence field tends asymptotically as the wall is approached, the authors concluded that the above mentioned deficiency may be the cause of relatively poor predictions of the normal-to-the-wall component of turbulent stress tensor by most models. They assumed first that the coefficient of the return-to-isotropy term ϕ_{ij1} would be dependent on the "two-dimensionality" or "flatness" parameter $A=1-9/8(A_2-A_3)$, which vanishes in two-dimensional turbulence (where A_2 and A_3 are second and third invariants of stress anisotropy a_{ij} respectively), and reintroduced the non-linear (second order) terms as proposed earlier by Lumley and Khajeh Nouri (1974). By following the same line Fu et al. expanded the QI model of the mean-strain term ϕ_{ij2} by including higher order terms in Re-stress, while satisfying all required symmetry conditions. An attempt with quadratic terms did not prove fully satisfactory, but cubic terms yielded at least qualitatively the desired improvements. The apparently impractically cumbersome expressions are reduced - after the application of symmetry conditions - to surprisingly compact forms. The new model requires further testing before it could be fully judged, but the approach sounds very promising.

A general trend towards the inclusion of higher order terms into the modeling expressions serves to bring back earlier, frequently discussed dilemmas on modeling triple velocity moments representing the turbulent stress transport. The invariant form suggested by Hanjalić and Launder (1972),although found to model better the experimental data in asymmetric channel (Cormack et al.,1978) and, recently, in a channel with a backward facing step (Amano and Goel, 1986), has often been discarded in favor of a more simple single-term representation because it gives rise to a large number of component terms in non-Cartesian coordinates. However, a consistent extension of modeling towards the inclusion of higher order terms calls also for a more adequate treatment of the turbulent transport terms. The exact transport equation for the triple moment may be a useful source of modeling hints. Some tests with the inclusion of some of the usually neglected terms in this equation (mean-strain generation-, pressure interactions and viscous destruction terms - the latter two modeled similarly to the analogous terms in the second-moment equation) by Dekeyser and Launder (1983), gave somewhat better agreement with the measurements of triple correlations in an asymmetric heated jet.

Rotta (1984) believes that improvements to existing second-order models could also be accomplished by considering the empirical coefficients as functions of suitable flow parameters. As discussed earlier, this has already been practiced by many modelers to account for viscosity effects, extra strain, curvature, rotation and buoyancy, but Rotta recommended some other quantities such as gradients of length scale, kinetic energy and Re stresses and intermittency factor.

The Wall-Proximity and Viscosity Effects

The influences of viscosity and wall proximity upon the turbulent motion are very different by nature. Yet, these effects have in the past been frequently modeled jointly, because they both manifest themselves in turbulence damping. However, as is well known, if the turbulence Re-number is small enough, viscosity affects all turbulent interactions, causes a departure from local isotropy (upon which some useful modeling principles rest), and promotes a consequent influence of the mean strain field upon the fine scale turbulence. In contrast, a solid wall or a phase interface "flattens" the turbulent structure, and imposes a selective damping primarily of the normal-to-the-wall fluctuations, causing the turbulence to approach a two-dimensional state. Furthermore, the wall reflects the pressure pulsations, affecting further the stress redistribution process in the region which extends well into the turbulent flow, while the influence of viscosity upon the Re-stresses at high Re-number flows remains restricted to the viscous sublayer, and even here does not seem to be very strong.

The simple two-equation models cannot differentiate these effects. Yet the demand for accurate computations of a large number of practically important flows, bounded by walls and/or interfaces, has prompted the appearance of many proposals for treating the near-wall region. The practice in the past fifteen years has meandered between the computationally more economical wall-function bridging and a more elaborate straight-to-the-wall integration. The latter approach has relied in most cases on the introduction of a number of functions in terms of turbulence Reynolds number and wall distance by means of which their effects on the various terms in the modeled transport equations are accounted for. A review by Patel et al., (1985) has shown that the two-equation models have invariably been accommodated by functions which were made to depend purely upon viscosity. Such an ill-founded practice obscures the separate effects of viscosity and wall proximity and results in poor predictions of flow regions where these effects take different relative magnitudes than in a simple wall flows. A more consistent approach requires each of the effects to be modeled separately, with coefficients and functions tuned on the basis of experimental data obtained in flows where each effect can be isolated. An illustrative example is the transition from the initial- to the final period of decay of isotropic turbulence which enables a coefficient in the dissipation equation to be determined purely on the basis of the Re-number effect. Another example is the wall damping of the stress components in the near-wall region outside the viscosity affected zone. An illustration of a possibility to account separately for these two effects has been shown in Fig. 3 where the f_μ function (used commonly to damp the eddy viscosity in the sublayer region) has been decomposed into two parts, i.e. $f_\mu = f'_\mu f_W$ The expression has been derived from the Low-Re-number stress model of Hanjalić and Launder (1976). As shown in the figure, f'_μ represents the effect of the low turbulence-Re-number, while f_W represents the wall damping effect, i. e. the ratio of $\overline{u_2^2}/k$ expressed in terms of shear stress to kinetic energy ratio. The resulting expression shows a very similar behaviour as the "experimental" function of Patel et al. (1985) generated from a collection of measured data.

Of course, because of high anisotropy of turbulence in the near-wall region, the Reynolds stress model (or at least ASM), which allows computation of each stress component, has much better prospects for simulating wall proximity effects. Hanjalić and Launder (1976)

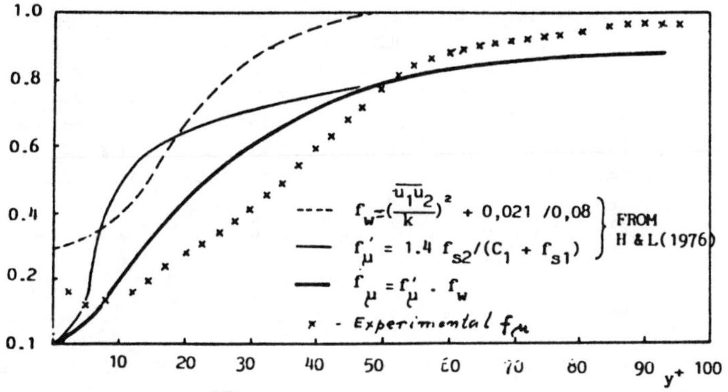

FIGURE 3. Variation of $\overline{u_2^2}/k$, f_μ and f_W functions in near-wall region

differentiated to some extent the influences of the two mentined agencies in their low-Re-number stress model, but departed from some of the adopted principles after they switched to a simpler three-equation model that was subsequently tested to yield satisfactory predictions of several classes of two-dimensional parabolic flows. The performance of this and other full-stress low-Re- number models (e.g. Prud'homme and Elghobashi 1983) were never thoroughly explored in more complex flows because of prohibitive computational demands thus a majority of computors has preferred to employ wall functions so far.

There has been a marked trend recently to abandon wall functions and to revert again to a more reliable integration straigth to the wall even by employing locally only very simple mixing-length-type models (in spite of the well known fact that this simple model neglects the turbulent transport by secondary motion, noticed to persist even in the very close vicinity of the walls). The PSL (Parabolic Sub Layer) approach of Iacovides and Launder (1983) offers a possibility for employing higher order turbulence models in the near-wall region with a substantial reduction of computation time, but it may lead to serious errors if applied to flows where normal-to-the-flow pressure gradients are large (as reported e.g by Choi, Iacovides and Launder,(1987),in the corners of the square-sectioned curved channel).

Recently Launder (1987) announced some fresh developments aimed at replacing the " outmoded" approaches of the 70's. By noting that the above mentioned experimental f_μ function of Patel et al., multiplied by 0.35, follows closely the variation of $\overline{u_2^2}/k$, through the viscous sublayer, Launder concluded that the shear stress damping is "largely independent of viscosity". This implies that a proper modelling of the pressure-correlation in the vicinity of a rigid wall would almost eliminate a need to modify the model for viscous damping. A way to do this is to employ the mentioned two-dimensionality ("flatness") factor A, which Launder employed to modify also the individual componets of the dissipation tensor as to account for the strong anisotropy, i.e.

$$\varepsilon_{ij} = \frac{2}{3}\delta_{ij}\ A^{1/2} + (\ 1 - A^{1/2})\ \frac{\overline{u_i u_j}}{k}\,\varepsilon$$

The above remedy may not be quite sufficient for the zone very close to the wall, since a kind of viscous damping function may still be needed in order to satisfy the limiting wall values of all terms in the transport equations. Certainly, contrary to earlier models, these functions should die out fast as the turbulence Re number increases, in the same way as does the function f_ε used by Hanjalic and Launder to accommodate a changeover from the inertial- to the viscous regime of the isotropic turbulence decay. Launder and Reynolds (1982) deduced from the polynomial expansion that the limiting ratios of individual dissipation-to-stress components at the wall are related as $\varepsilon_{ij}/\overline{u_iu_j}=\varepsilon/k$ if i and j take values 1 or 3, while $\varepsilon_{22}/\overline{u_2^2}=4\varepsilon/k$ and $\varepsilon_{12}/\overline{u_1u_2}=2\varepsilon/k$ .Hence a general expression which satisfies these limiting constraints should contain an invariant corrective factor, i.e.:

$$\varepsilon_{ij} = \eta \frac{\overline{u_iu_j}}{k}\varepsilon \quad \text{where} \quad \eta = 1 + \delta_{in}\delta_{in} + \delta_{jn}\delta_{jn} + \delta_{in}\delta_{jn}\delta_{ij}$$

where the index n denotes the coordinate normal to the wall (summation convention does not apply to n).

Likewise, the components of viscous diffusion satisfy the following limiting conditions:

$D_{ij}/\overline{u_iu_j} = D_k / k$ if i and j take values 1 or 3, while

$D_{12}/\overline{u_1u_2} = 3 D_k / k$ and $D_{22}/ \overline{u_2^2} = 6 D_k / k$

Hence, a general relationship can be written as

$$D_{ij} = \xi \frac{\overline{u_iu_j}}{k} D_K \quad \text{where} \quad \xi = 1 + 2 (\delta_{in}\delta_{in} + \delta_{jn}\delta_{jn}) + \delta_{ij}\delta_{in}\delta_{jn}$$

It follows that for i=j=1 or 3 the viscous diffusion balances the dissipation rate, and the pressure strain terms must vanish (an exact requirement) - a condition which Rotta's return-to-isotropy model does not satisfy, since a_{ij} reach constant values at the wall. For i=j=2 or either i or j equals 2 the viscous diffusion does not balance the dissipation at the wall. The balance is probably closed by pressure diffusion. The earlier mentioned extension of Rotta s model of ϕ_{ij1} by making the coefficient C_1 dependent on the "flatness" factor A satisfies this constraint. Another way to compensate for this while retaining the traditional form of ϕ_{ij1} is to introduce a correcting invariant parameter θ and a viscous damping function into the return-to-isotropy term.A model consistent with the above expressions and satisfying the limiting wall conditions may read as:

$$\phi_{ij1} = - (C_1 - C_{1w}) [1 + f_\theta (\theta - 1)] a_{ij} \varepsilon$$

where

$$\theta = \delta_{in}\delta_{in} + \delta_{jn}\delta_{jn}$$

Here C_{1w} is the coefficient in the ϕ_{ij1} term (wall correction for ϕ_{ij1w}, e.g. Launder et al. (1975) or Gibson and Launder (1978), and f_θ is a viscosity function like fs that takes a value of 1 at the condition of vanishing turbulence, but dies out fast as the turbulence Re number increases.

With the introduction of the above suggested corrective invariant parameters, which ensure limiting wall-behaviour of all variables and the equation terms, the stress-transport equation can be truncated with a choice of transport term treatment (e.g. Launder 1981), to yield consistent algebraic expressions for the stress component that are valid throughout the sublayer (Hanjalić, 1985, unpublished)

Turbulence Scales

Although the turbulence scales are primarily defined to characterize the turbulent mixing through the definition of eddy viscosity at a point in the flow, they are also utilized to model all "superfluous" terms in the transport equations for turbulence parameters. Since these terms represent physically different turbulence interactions that are known to occur at different rates, the employment of a single time and length scale is one of the basic approximations pertinent to all single-scale models, the abandonment of which represents an unexploited source of refinements of the model.

The multiple-scale concept, concieved by Schiestel (1974,1983) and subsequently elaborated on somewhat different physical arguments by Hanjalić et al. (1979), offers an intriguing possibility to construct a model which employs two (or more) independently calculated turbulence time scale with which to characterize the dynamics of different turbulence interactions. The idea rests on the division of turbulence spectra into two parts which respond at different rates and in different ways to changes in the environment. As such the concept may be regarded physically as an intermediate level of modeling between the single-point and two-point closure schemes, but computationally only marginally more laborious than the single-point models. The spectrum is divided at the wave number above which no significant mean-strain production occurs i.e. the low wave number "production" region and the higher wave number "energy transfer and dissipation" region. Separate transport equations are to be modeled and solved for the Re-stress components (or for the turbulence energy in two-equation models) for each part of the wave number spactra. The transport equations are solved also for the energy transfer rate out of the production range and through the inertial subrange, the latter being assumed to equal the dissipation rate, ε.

The computational results of a number of "notorious" non-equilibrium homogeneous-, free- and wall thin shear flows showed significantly closer agreement with experiments in comparison with single-scale-model predictions. This encouraging outcome was not pursued much further perhaps because of the intimidating task of retesting a vast number of flows and tuning a number of new coefficients that inevitably appear as a consequence of the increased model diversity. But the idea certainly appeals and we may expect some further developments along these lines.

Tensorial Length Scale in Eddy-Viscosity Models. Another interesting idea, proposed many years ago, (Monin, 1950, in Monin and Yaglom, 1971, Vol.1.), but not exploited much for obvious reasons, is the stress-strain relationship that employs a length scale in the form of a symmetric tensor of the second rank, l_{ij},(the scale tensor) namely

$$\overline{u_i u_j} = 2/3\, k\, \delta_{ij} - k^{1/2}\,(\, l_{ik}S_{kj} + l_{jk}S_{ki}\,)$$

This approach produces a nonisotropic eddy viscosity, regarded by some researchers as unavoidable for more accurate computation of three dimensional flows (see. e.g. Bradshaw 1987, Cousteix 1986, Lakshminarayana 1986) for which the use of any model of order higher than the two-equation model is still formidably impractical. The obvious drawback is a need to define the elipsoid of the length scales l_{ij}. A way to incorporate this idea into the currently popular models might be to keep the scalar length $l=k^{3/2}/\varepsilon$ as the spherical average of l_{ij} and introduce the deviatoric part of it, γ_{ij} which should account for the length scale directional orientation, i.e. $l_{ij}=l\gamma_{ij}$. The deviatoric can be specified by an algebraic relationship like that proposed by Naot et al. (1972) (see Wolfshtein et al.,1975):

$$\gamma_{ij} = (1-\Lambda)\,\delta_{ij}/3 + \Lambda\overline{u_i u_j}/(2k)$$

where Λ is a coefficient that could be evaluated on the basis of ratios of macro-scales of the quasi-isotropic correlation functions (based on plane channel flow data Naot suggested 0.3 to 0.4). Close to a solid wall the expression will produce $l_{22} < l_{33} < l_{11}$, as expected. This approach requires, however, a mode to express the ratio $\overline{u_i u_j}/k$, that calls for some kind of ASM model.(The idea of a tensorial scale may, however, be easily incorporated into the full RSE or ASM models.)

Nonlinear stress-strain relationship. The recognized similarities between the mean turbulent flow of a Newtonian fluid and the laminar flow of viscoelastic fluids has frequently served in the past as an inspiration to extend the validity of simple models by the asymptotic expansion of the basic stress-strain relationship and inclusion of higher order terms (e.g. Lumley 1970). Following that route in a consistent manner, satisfying most of the mathematical constraints, Speziale (1987) extended the standard k-ε (and k-L) model by adding two more terms into the stress-strain expression, which are quadratic in the mean velocity gradients - analogous to the visco-elastic terms in the second order model of a Rivlin-Ericsen fluid. Such a non-linear eddy viscosity model incorporates a trace of memory of the mean flow deformation into the expression for the Reynolds stresses:

$$\overline{u_i u_j} = 2/3\,k\,\delta_{ij} - k^{1/2} l\, S_{ij} + C_D\, l^2 (S_{im}S_{mj} - S_{mn}S_{mn}\delta_{ij}/3) + (\mathring{S}_{ij} - \mathring{S}_{mm}\delta_{ij}/3)$$

where $\mathring{S}_{ij}$ is the Oldroyd frame-indifferent derivative of S_{ij}, and C_D is an empirical coefficient to which a value of 1.68 was asigned.

In the words of Speziale, his model "is a special case of a much more complex nonlinear eddy viscosity model derived recently by Yoshizawa (1984) using Kraichnan's DIA formalism". Numerical tests of the model produced acceptable predictions of normal stresses in a fully developed plane channel (though not very satisfactory in the close-to-wall region), and of the secondary flow induced by a difference of normal Re-stresses in non-circular ducts (neither phenomena can be predicted with standard two-equation models). A substantial improvement of the prediction of the reattachment length of the recirculating zone behind a backward facing step was also obtained. This line of research has recently been pursued by Nisizama and Yoshizawa, (1988).

Although this nonlinear model extends the range of validity of the standard models, while maintaining most of their popular features, (satisfying at the same time most mathematical constraints), like its linear antecedant, it violates the principle of physical coherence, since it expresses turbulence properties (here Re-stresses) as directly proportional to the mean flow properties (mean rate of strain). As such the model can not remedy some noted failures of the eddy viscosity model such as the non-coincidence of the positions of zero shear stress and mean rate of strain in asymmetric flows (a fact which may sound trivial from a practical point of view, but which is a sensitive indicator of a model' s ability to mimic turbulent transport), not to mention local regions close to separation and reattachment points, corners and other singularities.

CONCLUDING REMARKS

In spite of all the research efforts and substantial progress in developing more general simulation models, two-equation models still remain the most widely used "fast" engineering methods for computing complex turbulent flows. In comparison with second order Re-stress models they demand more modest computational effort - an advantage that becomes progressively important in complex flows (three-dimensionality, irregular flow boundaries, multi-phase flows, chemical reactions etc.). Various remedies, designed to account for "extra effects", may improve predictions of complex flows, if employed knowledgeably.

There is, however, an evident disparity between the efforts on modelling refinements on the one hand and the use of models for calculation of complex flows on the other: very few industrial users take advantage of reported improvements - they seem to be generally satisfied with the standard models with simple wall functions. Is it a question of distrust or insufficient understanding of the whole affair that leads to a blind use of whatever the simplest models offer?

Second-order closure models are still burdened with a high degree of uncertainty in modeling higher order correlations due to a lack of sufficient knowledge of their physics. However, most model creators believe that the second order closure models will be more in use in the future and one can expect increasing efforts directed towards their refinement and generalization. Some new developments, discussed earlier sound very promising and a breakthrough may already be in the offing.

The successes in the development of modeling and computation techniques over the past twenty years have attracted many users among both the scientific and industrial communities. Some users even believe that they have at their disposal a tool powerful enough to calculate any complex flow. But, to quote Rotta s warning (1984) , " ..It should never be forgotten that the second order closure principles means a drastic simplificatioon in the description of the turbulent motion. Very complex cases like the flow field arround an arbitrarily shaped body... will hardly be accessible by such methods. A really universal turbulence model is a dream and will be a dream possibly for ever."

But do we really need a turbulence model for ever in the light of prospects for direct numerical simulation and supercomputers?

REFERENCES

Amano, R.S., Goel,P., Triple-velocity products in a channel with a backward-facing step, AIAA J., vol. 24, No 6, p. 1040, 1986

Aupoix, B., Cousteix,J., Liandrat, J., MIS: A way to derive the dissipation equation, Turbulent Shear Flows 5, 1987

Bradshaw, P., The understanding and predictions of turbulent flow, The Aeronautical J., vol. 76, No 739, p. 403, 1972

Bradshaw, P., Turbulent secondary flows, Ann. Rev. Fluid Mech., vol. 19, p. 53, 1987

Choi, Y.D., Iacovides,H., Launder, B.E., Numerical computation of turbulent flow in a square-sectioned 180-degree Bend, UMIST TFD/87/4,1987

Chou, P.Y., On velocity correlations and the solution of the equations of turbulent fluctuations, Quart. Appl. Math.,3, p.31, 1945

Cousteix, J., Three-dimensional and unsteady boundary-layers computations, Ann. Rev. Fluid Mech., vol. 18, p. 173, 1986

Daly, J., Harlow, F.H. Transport theory of turbulence, Los. Alamos Sci. Lab. DC 11304, 1970

Davidov, B.I., On the statistical dynamics,of an incompressible turbulent fluid", Dokl. AN SSSR, 136, p .47, 1961, (also 127,ps.768 & 980)

Dekeyser, I., Launder, B.E., A comparison of triple temperature-velocity correlations in asymmetric heated jet with alternative closure methods, Turb. Shear Flows 4, 1983

Fu,S., Launder, B.E., Tselepidakis,D.P., Accommodating the effects of high strain rates in modelling the pressure-strain correlation, UMIST TFD/87/5, 1987

Gorner, K., Zinser, W., Mathematsche Modellbildung der Kohlenstaubver- brennungs in Grossfeuerungsanlagen, 2-TECFLAM Seminar, Stutgart, 1986

Hanjalić, K., Ph.D. thesis, Imp. Coll of Sci and Techn.,Univ. of London, 1970

Hanjalić, K., Jones, W.P., Launder, B.E., Some notes on energy-dissipation model of turbulence, Int. Memor., Imp. Coll. of Sci. and Techn., London, 1970

Hanjalić, K., Launder, B.E., A Reynolds stress model of turbulence and its application to thin shear flows, J. Fluid Mech., vol.52, pt.4, p.609, 1972

Hanjalić, K., Launder, B.E., Contribution towards a Reynolds stress closure for low-Re-number turbulence, J. Fluid Mech., vol.74, pt.4, p. 593, 1976

Hanjalić, K., Launder, B.E., Sensitizing the dissipation equation to irrotational strains, J. Fluids Eng., vol. 102/1, p. 34, 1979

Hanjalić, K., Launder, B.E., Schiestel, R., Multiple time-scale oncept in turbulent transport, Turbulent Shear Flows-2, (ed. Bradbury, L.J.S. et al.) Springer, p. 36, 1980

Hanjalić, K., Stošić N., Hysteresis of turbulent stresses in wall flows subjected to periodic disturbances, turb. Shear Flows 4 (Bradbury, L.J.S. et al.) Springer, p.287, 1984

Harlow, F.H., Nakayama, P.I., Transport of turbulence energy decay rate, Los Alamos Sci. Lab.,Rept. LA-3854, 1968

Hirt, C. W., Generalized turbulent transport equations, Los Alamos Sci. Lab. 1969

Iacovides, H., Launder, B.E., PSL - An economical approach to the numerical analysis of near-wall-eliptic flow, UMIST TFD/83/3, 1983

Jones, W.P., Ph.D. Thesis, Imp. Coll of Sci. and Techn., Univ. of London, 1971

Kline, S.J. et al.(eds.), Proc.AFOSR-HTTM Stanford Conf."Comp. of Complex Turb.Flows", 1982

Lakshminarayana, B., Turbulence modeling for complex shear flows, AIAA J. vol.24 No.12, p.1900, 1986

Launder, B. E., Second moment closure: methodology and practice, UMIST TFD/82/4, 1982

Launder, B.E., An introduction to single-point closure methodology, UMIST TFD/87/7, 1987

Launder, B.E., Tselepidakis, D.P., Younis, B.A., A second- moment closure study of rotating channel flow,J. Fluid Mech., vol. 183, p.63, 1987

Launder, B.E., Reece, G.J., Rodi, W., Progress in the development of a Re-stress turbulence closure, J. Fluid Mech., vol.68 p. 537, 1975

Launder, B.E., Reynolds, W.C., Rodi, W., Turbulence models and their application, Editionms, Eyrolles, Paris, 1984

Liandrat,J., Aupoix,B., Cousteix, J., Calculation of longitudinal vortices imbedded in a turbulen boundary layer, Turb. Shear Flows 5, Springer, p.253, 1987

Lumley, J.L., Towards a turbulent constitutive relation, J.Fluid Mech. vol.41, p. 413, 1970

Lumley,J.L., Computational modeling of turbulent flows, Adv. in Appl. Mech., vol.18, 1978

Lumley, J.L., Khajeh Nouri, B., Computational modeling of turbulent transport, Adv. Geophys., vol. 18, p. 169, 1974

Mellor,G.L., Herring, H.J., A survey of the mean turbulent field closure models, AIAA J. vol.11 No. 5, p. 590, 1973

Mjolssness R.C., Systematic modelling rules for Reynolds stress closures in free shear layers, The Joint ASME-CSME Appl.Mech., Fluids Eng. and Bioengin. Conf. Niagara Falls, 1979

Nallasamy, M., Turbulence models and their applications to the prediction of internal flows: a review, Comp. & Fluids, vol. 15, No 2, p. 151, 1987

Naot, D., Shavit,A.,Wolfshtein,M.,Interactions between components of the turbulent velocity correlation tensor due to pressuere fluctuations, Izrael J.of Techn.,vol.8, No3, p.259,1970

Nisizima, S., Yoshizawa, A., A numerical study of turbulent square-duct flows using an anisotropic k- model, (To be published - Priv. commun.1988)

Patel, V.C., Rodi, W., Scheuerer, G., Turbulence models for near-wall and low-Re number flows: a review, AIAA J., vol.23, p. 1308, 1985

Reynolds,W.C., Computation of turbulent flows, Ann.Review of Fluid Mech., vol.8, p.183,1976

Rodi, W., Examples of turbulence models for Incompressible flows, AIAA J.,vol.20,No 7, 1982

Rodi, W., Turbulence modelling for incompressible flows-A report on the Euromech 180 Colloquium, PhysicoChem. Hydr. vol.7, No 5/6, p. 207, 1986

Rotta, J.C., Statistische Theorie Nichthomogener Turbulenz, Z. Phys. 129, p.547, 1951

Rotta, J.C., On my experience with second order closure models of turbulent flow, DFVLR Bericht No. IB 222-84 A 04 (presented at 180 Euromech Colloquium, Karsruhe, 1984)

Schiestel, R., Multiple-scale concepts in turbulence modelling, J. de Mech. Theor. et Applique, vol.2. No.3, p. 417 and No.4 p. 601, 1983

Speziale,C.G., On nonlinear k-l and k- models of turbulence, J.Fluid Mech., vol.178, 1987

Wolfshtein, M., Naot,D., Lin,A., Models of turbulence, in Topics in Transport phenomena, (ed. C. Gutfinger), Hemisphere, 1975

Yoshizawa, A., A statistically derived system of equations for turbulent shear flows, Phys. of Fluids, vol. 28, No. 1, p. 59, 1988

Modeling of Unsteady Boundary Layers

J. COUSTEIX and R. HOUDEVILLE
ONERA/CERT
Aerothermodynamics Department
2, Avenue Edouard Belin
31055 Toulouse Cedex, France

1. INTRODUCTION

Many industrial problems in aerodynamics, hydrodynamics, biomechanics, oceanography lead to an important development of studies on unsteady flows. In addition, the response of a flow to a forced unsteadiness is a basic element of understanding of flows and of turbulence.

In this paper, the emphasis will be placed on the behaviour of turbulent boundary layers submitted to the influence of an unsteady external flow.

The first part of this paper is devoted to a description of the physical aspects of the flow. This includes the evolution of the velocity in the boundary layer, the evolution of the skin-friction and the behaviour of turbulence.

The second part is devoted to applications of calculation methods of unsteady turbulent boundary layers. Several turbulence models have been used and are compared to available experimental data.

2. RESPONSE OF A FLAT PLATE BOUNDARY LAYER TO AN OSCILLATING EXTERNAL FLOW

In order to separate the unsteady effects, the mean value of any quantity is defined in a statistical way as an ensemble average denoted by $\langle f \rangle$. The turbulent fluctuation of the quantity f is defined as the difference between the instantaneous value f and its average $\langle f \rangle$:

$$f = \langle f \rangle + f' \tag{1}$$

In the case of a periodic flow, $\langle f \rangle$ is also defined as a phase average. $\langle f \rangle$ is a periodic function of time.

As far as calculation methods are concerned this double decomposition is introduced in the Navier-Stokes equations and the ensemble average of equations is taken. For calculating the evolution of the average flow, turbulence models are needed. These models are used to represent the effect of turbulence on the average flow through the action of Reynolds stresses $-\rho \langle u'v' \rangle$. This problem is discussed in paragraph 3.

For analyzing experimental data or for comparing calculation results to experiments it is interesting to decompose the average velocity by using a harmonic analysis :

$$\langle U \rangle = \overline{U} + \Delta U \cos(\omega t + \phi_u) + \ldots \qquad (2)$$

$\overline{U}$ is the time mean velocity, ΔU is the amplitude and ϕ_u is the phase angle.

In the same way global characteristics are decomposed as :

$$\delta_1 = \overline{\delta_1} + \Delta\delta_1 \cos(\omega t + \phi_{\delta_1}) + \ldots \qquad (3)$$

2.1. Time mean component of velocity. All the existing boundary layer data demonstrate that the behaviour of the time mean velocity is not affected by the unsteadiness of the flow. This means that the time mean velocity profiles are the same as those which would be obtained in a steady boundary layer submitted to the time mean external velocity distribution. This is true even in the presence of an adverse pressure gradient. This is certainly an interesting property for practical applications but this does not mean that the unsteady effects are negligible because very large instantaneous departures from the time mean state can exist. In addition, there can be very large non linear effects through the interaction between the boundary layer and the external flow. Even at low frequency the characteristics of the whole flow can be different from the characteristics of the equivalent steady flow : it is well known that hysteresis effects are observed on the evolution of the flow around a helicopter blade for example.

Experimental data discussed below have been obtained in a wind tunnel of Eiffel type (Cousteix-Houdeville, 1983 ; Javelle, 1982). The boundary layer develops on a flat plate whose length is 800 mm, the width is 220 mm and the thickness is 15 mm. The leading edge is elliptic. A wire of diameter 0.80 mm trips transition at a distance 70 mm from the leading edge. The unsteadiness of the flow is produced by a rotating valve placed in the diffuser. Figure 1 shows the variation of the external flow.

Figure 2 shows the evolution of global characteristics of the boundary layer which are compared with data obtained in steady flow in the same wind tunnel for nearly the same time mean external velocity. The unsteady effects result in a small ondulation of the global thicknesses.

2.2. Periodic variation of velocity. A first global impression of the response of the boundary layer to an oscillating flow is given in figure 3 where the evolution of the amplitude and of the phase shift of the displacement thickness are represented as a function of the reduced frequency parameter. The experimental conditions are given in table 1. The variable X is the distance measured from the virtual origin Xo of the time mean boundary layer deduced from a classical flat plate law. X_{BA} is the abscissa of the flat plate leading edge.

The striking result is the quasi-periodic oscillation of amplitude and phase shift as a function of the reduced frequency. The damping of this oscillation is small. Expressed in reduced frequency $\omega X/\overline{U}e$ the periodicity of this oscillation is about 5. The same is observed on the boundary layer profiles of amplitude and phase shift of velocity (figure 4).

To analyze more deeply this phenomenon, the boundary layer thickness δ has been determined (Houdeville et al., 1987). The value of δ has been obtained from measurements of the shear stress $-\rho \langle u'v' \rangle$ and δ is determined as the distance from the wall where $-\rho \langle u'v' \rangle$ is equal to 5.10^{-3} times its maximum value in the boundary layer. It is interesting to observe that the amplitude of δ oscillates as a function of the reduced frequency but the phase angle of δ has a monotonic evolution (figure 5). At low frequency, the limiting values of amplitude and phase angle are given by a quasi-steady analysis. If it is assumed that the steady boundary layer follows the laws :

$$\frac{\delta_1}{X} = \frac{0.048}{R_x^{1/5}} \qquad \frac{\delta}{X} = \frac{0.38}{R_x^{1/5}} \tag{4}$$

it is found that at very low frequency we have :

$$\frac{\Delta\delta_1}{\bar{\delta}_1} \bigg/ \frac{\Delta Ue}{\overline{Ue}} = \frac{\Delta\delta}{\bar{\delta}} \bigg/ \frac{\Delta Ue}{\overline{Ue}} = 0.2 \tag{5}$$

$$\phi_{\delta_1} - \phi_{Ue} = \phi_{\delta} - \phi_{Ue} = 180°$$

It is difficult to say if these laws are verified by the experimental data. The main difficulty is that the quasi-steady limit corresponds to the virtual origin of the boundary layer and the transition tripping makes the analysis not clear in this region.

When the frequency is large enough, the phase angle of the boundary layer thickness varies linearly with $\omega X/\overline{Ue}$; this means that a constant convection velocity can be determined : its value is about 0.8 $\overline{Ue}$. This explains the response of the boundary layer to the oscillation of the flow . The average velocity in the outer part of the boundary layer is directly influenced by the behaviour of turbulence. Now the evolution of turbulence results from an equilibrium between several processes among which convection plays an important role. In particular, the convection of turbulence controls the outer part of the boundary layer and therefore the evolution of δ. The combination of the convection velocity and of the forced pulsation of the flow leads to a reduced frequency $\omega X/\overline{Ue} = 5.5$. This value is very close to the periodicity observed on the amplitude and phase angle of displacement thickness and on the evolution of velocity profiles.

The general feature of the response of the boundary layer can be analyzed from the integrated form of the momentum equation. When the external velocity is purely periodic : $Ue = \overline{Ue} + A\cos\omega t$ ($\overline{Ue}$ = cst   A = cst), this equation can be written as :

$$\frac{\partial\delta_1}{\partial t} + Ue\frac{\partial\theta}{\partial x} = \frac{Cf}{2}Ue - \frac{\delta_1}{Ue}\frac{\partial Ue}{\partial t} \tag{6}$$

where δ_1 and θ are the average values of the displacement and momentum thicknesses. If it is assumed that the shape factor $H = \delta_1/\theta$ is constant , this equation clearly shows a convection phenomenon with a velocity Ue/H. The right hand side of the momentum equation plays a role of a source term.

The equation (6) can be solved analytically by using the following hypotheses.It is assumed that H = cst and that the skin friction follows a steady flat plate law $Cf = 0.0172\ R_\theta^{-1/5}$ ($R_\theta = Ue\,\theta/\nu$). These hypotheses are very crude but they enable us to obtain an analytical solution which clearly shows the nature of this solution. Furthermore it is assumed that the amplitude of variation of any quantity is small. We have :

$$Ue = \overline{Ue} + A_{Ue}\, e^{i\omega t}$$

$$\delta_1 = \overline{\delta_1} + A_{\delta_1}\, e^{i\omega t}$$

where A_{Ue} and A_{δ_1} are complex. The solution of equation (6) is :

$$\frac{A_{\delta_1}/\overline{\delta_1}}{A_{Ue}/\overline{Ue}} = -1 - \frac{5}{6}\, i\, \frac{(1 - e^{-iX^*})}{X^*} \qquad (7)$$

with $X^* = \omega X \overline{H}/\overline{Ue}$.

The solution (7) shows a quasi-periodicity with a wave length $\frac{\omega L}{\overline{Ue}} = \frac{2\pi}{\overline{H}} \sim 4.5$. The first maximum in amplitude of the displacement thickness occurs around $\frac{\omega X}{\overline{Ue}} = 3$. These values are in reasonable agreement with experiments.

By considering the dynamics of vorticity Reynolds - Carr, 1985 gave another point of view . From their analysis they conclude that the response of the boundary layer is a result of the reinforcement of surface generated vorticity by the leading edge vorticity. Indeed it is another manner to interpret equation 6 which tells that the variation of displacement thickness created at the leading edge is convected and superimposed on the variation of displacement thickness forced by the unsteadiness of the flow all along the flat plate. It should be noticed that the influence of the leading edge is lost when the distance from it increases. Indeed the asymptotic solution is $\delta_1\, Ue = cst$.

Another piece of information about this problem is provided by laminar flow calculations of an oscillating flat plate boundary layer. The same qualitative behaviour of the displacement thickness is observed but the damping of the oscillation is stronger than in the turbulent case (figure 6).

The study of the structure of the oscillating flat plate boundary layer has been completed by turbulence measurements. In particular, detailed measurements of the Reynolds stresses $\rho \langle u'^2 \rangle$, $\rho \langle v'^2 \rangle$, $-\rho \langle u'v' \rangle$ have been performed.

Figure 7 shows the evolution of the phase angle of $\langle u'v' \rangle$ at various locations in the boundary layer. In the outer region ($y/\delta = 0.6$ and $y/\delta = 0.8$) the phase angle varies linearly with the reduced frequency. A constant convection velocity can be defined at $y/\delta = 0.8$; its value is $0.73\, \overline{Ue}$. Near the wall ($y/\delta = 0.2$ and $y/\delta = 0.4$) the convection velocity is much larger . This means that the outer region is dominated by the convection phenomenon. The combination of the convection velocity with the forced pulsation of the flow leads to a spatial periodicity in the evolution of the boundary layer characterictics. With a convection velocity of 0.73 Ue the periodicity expressed in reduced frequency is $\frac{\omega X}{\overline{Ue}} = 2\pi \times 0.73 = 4.6$

Near the wall the phase angle of $\langle u'v' \rangle$ does not vary very much with the distance from the leading edge.

In fact the structure of the boundary near the wall is influenced by the pressure gradient $(-\frac{1}{\rho}\frac{\partial P}{\partial x})$, the turbulent diffusion and the molecular diffusion (very close to the wall). If the convection terms are neglected in the momentum equation we have :

$$\frac{\partial}{\partial y}\left(\mu\frac{\partial U}{\partial y} - \rho\langle u'v'\rangle\right) = \frac{1}{\rho}\frac{\partial P}{\partial x} \tag{8}$$

If the external velocity has the form $Ue = \overline{Ue} + \Delta Ue \sin \omega t$ where $\overline{Ue}$ and ΔUe are constant, the phase angle for the forcing term $\frac{\partial P}{\partial x}$ in equation (8) does not depend on X. Then the phase angles of U and $-\rho\langle u'v'\rangle$ do not depend on X.

2.3. Skin friction coefficient. In steady flow, the skin friction coefficient is often determined by using the technique of Clauser plot which consists of fitting the experimental velocity profile to the law of the wall. The same technique has been used in our experiments in unsteady flow. It is assumed that the law of the wall remains valid at every instant in the cycle. The following law has been used :

$$\langle U^+\rangle = \frac{1}{\chi}\ln\langle y^+\rangle + C \tag{9}$$

with $\langle U^+\rangle = \frac{\langle U\rangle}{\langle U_\tau\rangle}$; $\langle y^+\rangle = \frac{y\langle U_\tau\rangle}{\nu}$

$C = 5.25$; $\chi = 0.41$

In figure 8, profiles measured at a given location but at different instants during a cycle are plotted in semi-logarithmic coordinates. At each instant the value of $\langle U_\tau\rangle$ which gives the best fit is used to calculate $\langle y^+\rangle$ and $\langle U^+\rangle$. A good agreement is observed between the experimental velocity profiles and the law of the wall over a rather wide range of $\langle y^+\rangle$. Following the classical use, $\langle U_\tau\rangle$ represents the ensemble average friction velocity from which the wall shear stress is deduced $\langle\tau_w\rangle = \rho\langle U_\tau\rangle^2$. The characteristics (amplitude and phase shift) of the wall shear stress so determined are given in figure 9.

It has been shown by Simpson, 1977, that the existence of a logarithmic region between the law of the wall and the law of the wake implies that the velocity phase shift remains constant in such a region. In fact, this conclusion is a direct consequence of the assumption of the existence of a general universal law of the form :

$$\langle U^+\rangle = f(\langle y^+\rangle) \tag{10}$$

In this law, time does not appear explicitely.

A small perturbation expansion of this law (10) shows that the velocity phase shift is constant.

Experimentally this conclusion is not verified in our experiments. In fact, the velocity phase shift is nearly constant in the region where the logarithmic law is valid but the phase angle varies between the wall and this region ($0 < \langle y^+\rangle < 40$). This means that the scaling parameter $\langle U_\tau\rangle$ used in the logarithmic law of the wall is not equal to the friction velocity. This conclusion is in agreement with direct skin friction measurements performed by using hot film gauges.

The direct skin friction measurements give nearly the same evaluation of skin friction as the Clauser plot technique only for small reduced frequencies ($\omega X/\overline{U}e < 5$). For higher frequencies the two evaluations are very different. In particular, we note that the Clauser technique leads to a wall shear stress phase angle which tends towards zero as the frequency increases whereas the direct measurements indicate an increase.

At a reduced frequency of 12 the measured phase angle is 42°. At high frequency, as far as the Stokes solution represents the results the phase angle is 45° and the relative amplitude of wall shear stress can be calculated as : (Binder - Kueny, 1981) :

$$\frac{\Delta\tau w}{\overline{\tau} w} / \frac{\Delta Ue}{\overline{U}e} = \sqrt{2}\ \frac{\overline{U}e}{\overline{U}_\tau}\ \frac{1}{\ell_s^+} \qquad (11)$$

where ℓ_s^+ is the Stokes length expressed in wall units :

$$\ell_s^+ = \left(\frac{2\nu}{\omega}\right)^{1/2} \frac{\overline{U}_\tau}{\nu} \qquad (12)$$

In our experimental case f = 62 Hz $\omega X/\overline{U}e = 17.8$ $\ell_s^+ = 13.8$

we obtain $\dfrac{\Delta\tau w}{\overline{\tau} w} / \dfrac{\Delta Ue}{\overline{U}e} = 2.56$. This value compares rather well with the measurements.

In fact the study of the boundary layer structure in oscillating flow leads to use two reduced frequencies (Menendez - Ramaprian, 1983) : a) the classical reduced frequency $\omega X/\overline{U}e$ and b) the parameter $\omega\nu/\overline{U}_\tau^2$ (or ℓs^+).

At low frequency the concept of logarithmic overlap between the inner and the outer layer remains valid and the calculation of the wall shear stress is posed in the same terms as for steady flow .

The regime of high frequency is governed by ℓ_s^+ (Binder - Kueny, 1981). The unsteady effects are confined in a thin layer near the wall ; outside of this layer, the amplitude and the phase angle of velocity are constant ($\Delta U = \Delta Ue$; $\phi_U = \phi_{Ue}$). Near the wall the velocity follows the Stokes solution.

In the intermediate regime, the situation is more complicated, and results are still uncertain. Figure 10 collects experimental data on the phase angle of the wall shear stress obtained in an oscillating flat plate boundary layer and in an oscillating channel flow. By using the parameter $\omega\nu/\overline{U}_\tau^2$ we obtain results that are well grouped, whereas a wide spread is obtained when $\omega X/\overline{U}e$ is used. However it is uncertain if $\omega\nu/\overline{U}_\tau^2$ is the unique governing parameter.

3. CALCULATION METHODS OF UNSTEADY BOUNDARY LAYERS

The basic equations to solve are the continuity equation and the momentum equation :

$$\frac{\partial U}{\partial x} + \frac{\partial V}{\partial y} = 0 \qquad (13a)$$

$$\frac{\partial U}{\partial t} + U\frac{\partial U}{\partial x} + V\frac{\partial U}{\partial y} = -\frac{1}{\rho}\frac{\partial P}{\partial x} + \frac{\partial}{\partial y}\left(\nu\frac{\partial U}{\partial y} - \langle u'v'\rangle\right) \tag{13b}$$

$$\frac{\partial P}{\partial y} = 0 \tag{13c}$$

To simplify the writing, the quantities P, U, V denote ensemble averages.

In these equations, the pressure distribution is a prescribed datum. The pressure gradient is calculated form the external velocity :

$$-\frac{1}{\rho}\frac{\partial P}{\partial x} = \frac{\partial Ue}{\partial t} + Ue\frac{\partial Ue}{\partial x} \tag{14}$$

The equations (13) are completed with a turbulence model. Several models have been tested. They are described below.

3.1. Turbulence models. In the applications which will be discussed, two-layer models have been employed. Near the wall, a mixing length scheme is used. The turbulent shear stress is expressed as :

$$-\langle u'v'\rangle = F^2 \ell^2 \left(\frac{\partial U}{\partial y}\right)^2 \tag{15}$$

where ℓ is the mixing length :

$$\frac{\ell}{\delta} = 0.085 \tanh\left(\frac{\chi}{0.085}\frac{y}{\delta}\right) \quad ; \quad \chi = 0.41 \tag{16}$$

and F is the wall damping function :

$$F = 1 - \exp\left(-\frac{\ell}{26\chi\mu}(\tau\rho)^{1/2}\right) \tag{17}$$

When this model is used throughout the boundary layer, the thickness δ is defined as the point where $U/Ue = 0.99$. When the mixing length is used only near the wall, its expression is taken as $\ell = \chi y$.

In the outer layer several models have been tested. The simplest is the mixing length scheme described above. The other tested models are based on transport equations. In this case the mixing length model is used near the wall for $y^+ < 50 - 100$.

Two transport equation models have been used :

Model 1 $k - \varepsilon$ transport equation model.

The turbulent shear stress is given by :

$$-\langle u'v'\rangle = \nu_t \frac{\partial U}{\partial y} \tag{18}$$

with $\nu_t = f\mu\, C\mu \frac{k^2}{\varepsilon}$

According to Launder-Sharma, 1974, the function $f\mu$ is :

$$f\mu = \exp -\frac{3.4}{(1 + R_t/50)^2} \quad ; \quad R_t = \frac{k^2}{\nu\varepsilon} \tag{19}$$

The turbulent kinetic energy and its dissipation rate are calculated from transport equations.

$$\frac{Dk}{Dt} = -\langle u'v'\rangle \frac{\partial U}{\partial y} - \varepsilon + \frac{\partial}{\partial y}\left(\left(\nu + \frac{\nu_t}{\sigma k}\right)\frac{\partial k}{\partial y}\right) \tag{20a}$$

$$\frac{D\varepsilon}{Dt} = -C_{\varepsilon_1}\frac{\varepsilon}{k}\langle u'v'\rangle\frac{\partial U}{\partial y} - C_{\varepsilon_2}\frac{\varepsilon^2}{k} + \frac{\partial}{\partial y}\left(\left(\nu + \frac{\nu_t}{\sigma_\varepsilon}\right)\frac{\partial \varepsilon}{\partial y}\right) \tag{20b}$$

The set of constants is given by :

$C_{\varepsilon_1} = 1.44$ $C_{\varepsilon_2} = 1.92$ $C\mu = 0.09$ $\sigma_k = 1$ $\sigma_\varepsilon = 1.3$

Model 2 $k - \varepsilon - \langle u'v'\rangle$ transport equation model.

The eddy viscosity model is replaced by a transport equation for calculating the evolution of the turbulent shear stress : (Hanjalic - Launder, 1976) (21):

$$\frac{D}{Dt}\langle u'v'\rangle = -f_{s_1} C_{\tau_1} k \frac{\partial U}{\partial y} - (C_{\tau_2} + f_{s_2})\frac{\varepsilon}{k}\langle u'v'\rangle + \frac{\partial}{\partial y}\left(\left(\nu + \frac{\nu_t}{\sigma_\tau}\right)\frac{\partial}{\partial y}\langle u'v'\rangle\right)$$

$C_{\tau_1} = 0.135$ $C_{\tau_2} = 1.5$ $\sigma_\tau = 0.9$

$$f_{s_1} = \exp\left(\frac{-1.86}{0.5 + R_t/30}\right) \quad ; \quad f_{s_2} = \frac{1}{1 + R_t/10} \quad ; \quad R_t = \frac{k^2}{\nu\varepsilon}$$

3.2. Integral method. For engineering calculations, fast methods are necessary. In steady flow integral methods are a valuable tool. These methods have been extended to unsteady boundary layers as follows. The basic equations are the global continuity and momentum equations.

$$\frac{\partial \delta}{\partial x} - \frac{ve}{Ue} = \frac{1}{Ue}\frac{\partial}{\partial x}(Ue(\delta - \delta_1)) \tag{22a}$$

$$\frac{Cf}{2} = \frac{d\theta}{dX} + \theta\frac{H+2}{Ue}\frac{\partial Ue}{\partial x} + \frac{1}{Ue^2}\frac{\partial}{\partial t}(Ue\,\delta_1) \tag{22b}$$

In these equations all quantities are ensemble averages.

The closure relationships needed to solve the set of global equations are deduced from self similarity solutions. These solutions are constructed with a mixing length scheme.

The closure relationships are :

$$\frac{\delta_1}{\delta} = \left(\frac{Cf}{2}\right)^{1/2} F_1(G) \quad ; \quad G = \frac{H-1}{H\,(Cf/2)^{1/2}} \tag{23a}$$

$$\left(\frac{Cf}{2}\right)^{-\frac{1}{2}} = \frac{1}{\chi}\ln\frac{Ue\,\delta_1}{\nu} + D^*(G) \quad ; \quad \chi = 0.41 \tag{23b}$$

$$\frac{\partial \delta}{\partial x} - \frac{ve}{Ue} = \left(\frac{Cf}{2}\right)^{1/2} P(G) - \frac{1}{Ue}\frac{\partial \delta}{\partial t} \tag{23c}$$

The functions F_1, D^*, P are obtained from the analysis of self similarity solutions. They are represented by analytical formulae :

$F_1 = 0.613\,G - (3.6 + 76.86\,(1/G - 0.154)^2)/G$
$D^* = 2G - 4.25\,G^{1/2} + 2.12$
$P = 0.074\,G - 1.0957/G$

It should be noticed that the skin-friction law is obtained by assuming that there exists a universal law of wall. It has been shown that this hypothesis is not verified when the reduced frequency is high. This implies that the integral method has a limited domain of application.

4. APPLICATIONS OF CALCULATION METHODS

The calculation methods presented above have been applied to several available experiments. In this paper, three applications are presented. The first one consists of the calculation of an oscillating flat plate boundary layer. The second case is the study performed at Stanford University (Jayaraman et al., 1982) on the influence of an oscillating pressure gradient. The third case is the influence of an oscillation on a boundary layer with a mean adverse pressure gradient.

4.1. Oscillating flat plate boundary layer. Results obtained with the integral method and the three equation model are presented (Javelle, 1982 ; Cousteix - Houdeville, 1983). Figure 11 shows the evolution of the time-mean evolution of the displacement thickness, momentum thickness and skin friction coefficient for the experiment whose external velocity distribution is shown in figure 1.

Both methods lead to a good agreement for time-mean integral thicknesses, but the transport equation model gives a too low time-mean skin friction coefficient. It must be noted that this problem is not connected with the unsteadiness of the flow but is purely a steady problem. In fact, the time-mean flow is very close to a steady flow.

The evolutions of the amplitude and phase shift of the displacement thickness are well predicted by both methods for reduced frequency lower than 5 (figure 12). Above this value, the integral method gives a too strong damping of amplitude and phase shift. The results are significantly improved with the transport equation model. It is also noted that both methods correctly represent the evolutions of the quasi-periodicity in the behaviour of the displacement thickness.

A major difficulty is the prediction of the skin friction (figure 13). The amplitudes of skin friction calculated by both methods seem too large but the measurements of wall shear amplitude are perhaps not very accurate ; it is believed that the wall shear stress phase angles are more reliable and, undoubtedly, the results obtained with the integral method are wrong for reduced frequencies larger than 5. The transport equation model improves the results but the phase angle is still too low. In the case of the transport equation model the near wall treatment needs to be refined.

More detailed results are given in figure 14 where are plotted time-mean velocity profiles, amplitudes and phase shifts of average velocities. The calculated results are obtained with the three transport equation model. The outer part of the boundary layer is rather well represented. In particular, the rapid variation of amplitude and phase shift profiles measured between $X = 0.505$ m and $X = 0.545$ m is qualitatively reproduced by the model.

4.2. Calculation of Jayaraman et al. experiments. The experimental study of Jayaraman et al., 1982, is such that the flow is steady in the uspstream part of the test section. In this way the initial conditions are well known. Downstream of a certain abscissa, the external velocity decreases linearly with a slope which varies periodically in time. The desired free-stream velocity distribution is shown in figure 15.

All the calculations have been started at the theoretical beginning of the free-stream velocity oscillation (X = 200 cm on figure 15). At this station the experimental boundary layer data have been taken as initial conditions. The free-stream velocity distribution used in calculations is also the experimental result and not the ideal distribution shown in figure 15. Figure 16 shows the comparison between experiments and calculations of the amplitude and phase angle of the displacement thickness. Two cases have been selected ; the frequencies are 0.5 Hz and 2 Hz (in the experiments the selected configurations correspond to the low amplitude cases). The calculations have been performed by using the integral method and the three equation model. The agreement of both methods with experimental data is good. It is interesting to note that the calculations indicate a quasi-periodicity in the evolution of the amplitude and phase angle of the displacement thickness as in the case of the oscillating flat plate boundary layer.

More detailed results are presented in figure 17 which shows comparisons between experiments and calculations of the velocity profiles. The two equation model and the three equation model give nearly the same results. The overall agreement is not too bad but it should be noticed that the profiles of amplitude and phase angle vary rapidly with the abscissa. This means that a slight displacement of the comparison station can lead to a significant change in the shape of these profiles.

4.3. <u>Oscillating boundary layer with a mean adverse pressure gradient</u>.
The pressure distribution is shown in figure 18 (Houdeville - Cousteix, 1979).

Figure 19 stresses the influence of the turbulence model. This figure shows profiles of time-mean velocity, amplitude and phase angle of average velocity at a given location (X = 0.34 m) where the adverse pressure gradient is not very severe. For each calculation the initial station is X = 0.048 m and the initial conditions have been prescribed from experimental data.

The poorest result is obtained with a mixing length scheme : the overshoot of the amplitude is too small, the negative phase shifts in the outer part of the boundary layer are not well predicted and the near wall behaviour of the phase shift is not correct. The results obtained with the two and the three equation models are significantly better. In particular the magnitude of the overshoot in the amplitude profile is well predicted but some discrepancies are noted in the details of the amplitude and phase shift profiles.

5. CONCLUSIONS

The experimental results have shown that the time-mean flow in the boundary layer has nearly the same characteristics as a steady flow submitted to the same time-mean external velocity distribution. This does not mean that unsteady effects are small. The outer part of the boundary layer is dominated by convection effects whose combination with the forced oscillation of the flow leads to a quasi-periodicity in the integral characteristics of the boundary layer. Near the wall the behaviour of the flow is dominated by diffusion effects and by the periodic variation of the pressure gradient. In the near wall region strong unsteady effects are observed. In particular at high frequency the law of the wall does not apply for the average velocity profiles ($\omega X/\overline{U}e > 5$). It follows that the logarithmic law of the wall is not valid for determining the skin friction and can not be used in calculations as a basis for the near wall treatment.

In the range of low reduced frequencies ($\omega X/\overline{U}e < 5$) a simple integral method leads to acceptable results. For pratical applications this is important because a number of engineering configurations are concerned with this range of reduced frequencies. Even at relatively low reduced frequencies a classic mixing length model gives rather poor results. Better results are obtained from transport equation models (two or three equation models). However a major difficulty arises in the near wall region. As said above, the logarithmic law of the wall does not hold at high frequencies. Therefore, the near wall region must be calculated but the available techniques are far from being well established.

REFERENCES

Binder, G., Kueny, J. L., Measurements of the periodic velocity oscillations near the wall in unsteady turbulent channel flow, eds R. Michel, J. Cousteix, R. Houdeville. Springer Verlag, (1981).

Binder, G., Tardu, S., Kueny, J. L. and Blackwelder, R. F. Large amplitude periodic oscillations in bounded shear flows with zero and adverse pressure gradients. Symp. Turb. Shear Flows, 5th. Ithaca, N.Y. (1985).

Cousteix, J., Houdeville, R. Effects of unsteadiness on turbulent boundary layers. VKI Lecture series on Turbulent shear layers (1983).

Eldem, N. Couche limite et sillage d'un profil d'aile muni d'un volet battant. Thèse de l'ENSAE (1987).

Hanjalic, K., Launder, B. E. Contribution towards a Reynolds stress closure for low Reynolds number turbulence. J.F.M., vol.74, part 4 (1976).

Houdeville, R., Cousteix, J. Couches limites turbulentes en écoulement pulsé avec gradient de pression moyen défavorable. La Recherche Aérospatiale n° 1 (1979).

Houdeville, R., Eldem, N., Belanger, J. Couches limites et sillages instationnaires. Rapport Technique de Synthèse OA n° 65/2259 AYD, (Internal Report) (1987).

Javelle, J. Réponse d'une couche limite turbulente à une pulsation de l'écoulement général. Influence de la fréquence réduite. Thèse de l'ENSAE (1982).

Jayaraman, R., Parikh, P., Reynolds, W. C. An experimental study of the dynamics of an unsteady turbulent boundary layer. Technical Report n° TF18 Stanford University (1982);

Launder, B. E., Sharma, B. I. Application of the energy dissipation model of turbulence to the calculation of flow near a spinning disc. Letters in Heat and Mass transfer, vol. 1 (1974).

Menendez, A. N., Ramaprian, B. R. Study of unsteady turbulent boundary layers, I.I.H.R., rep. n° 270, Iowa Inst. Hydraul. Res., Iowa City (1983).

Reynolds, W. C., Carr, L. W. Review of unsteady, driven, separated flows. A.I.A.A. Paper 85-0527 (1985).

Simpson, R. L. Features of unsteady turbulent boundary layers as revealed from experiments. AGARD CP n° 227 (1977).

F(Hz)	38	62
$\overline{U}_e (ms^{-1})$	22	16,9
$\Delta U_e/\overline{U}_e$	0,15 à 0,11	0,10 à 0,5
R_X	3,2 à 9,0 10^5	2,4 à 6,9 10^5
$\frac{\omega X}{\overline{U}_e}$	2,3 à 7,0	4,9 à 14,2
$X_o - X_{BA}$ (m)	0,020	0,020

Table 1 - Flat plate experiment

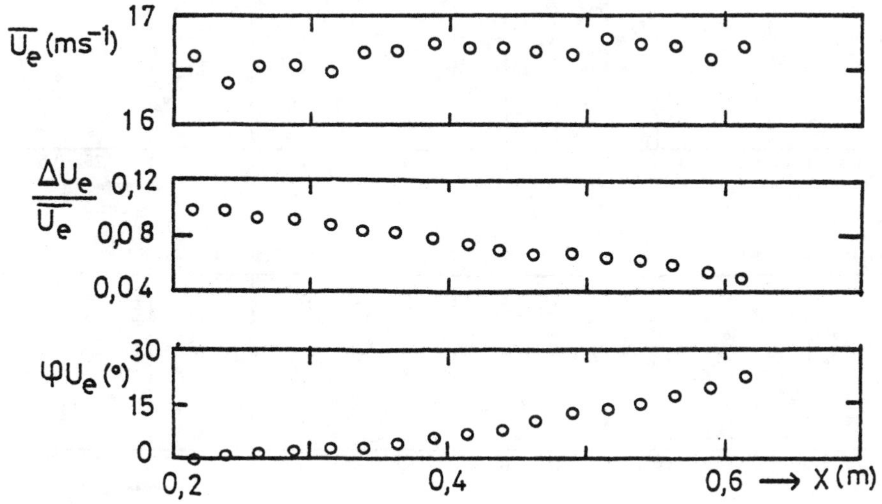

Fig. 1 - Flat plate experiment. Harmonic analysis of the external velcocity. f = 62 Hz

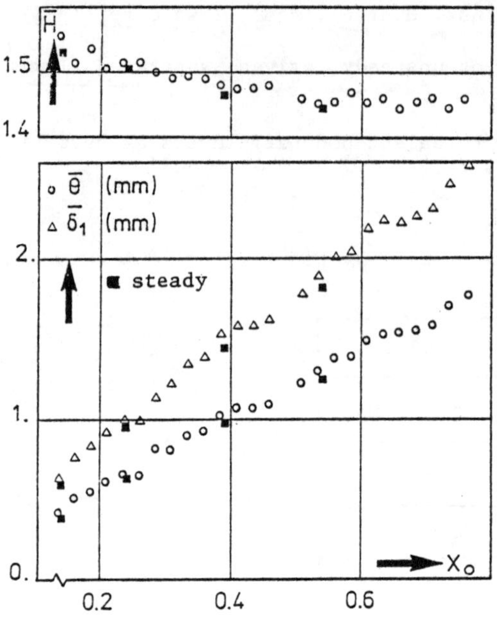

Fig. 2 - Flat plate experiment
F = 62 Hz. Time mean values of integral thicknesses and shape factor

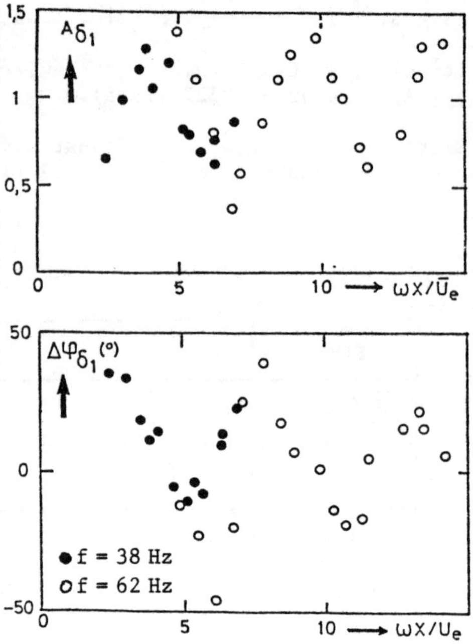

Fig. 3 - Flat plate experiment. Harmonic analysis of the displacement thickness. $A_{\delta_1} = \dfrac{\Delta\delta_1/\overline{\delta_1}}{\Delta Ue/\overline{U}e}$

$\Delta\phi_{\delta_1} = \phi_{\delta_1} - \phi_{Ue} + 180°$

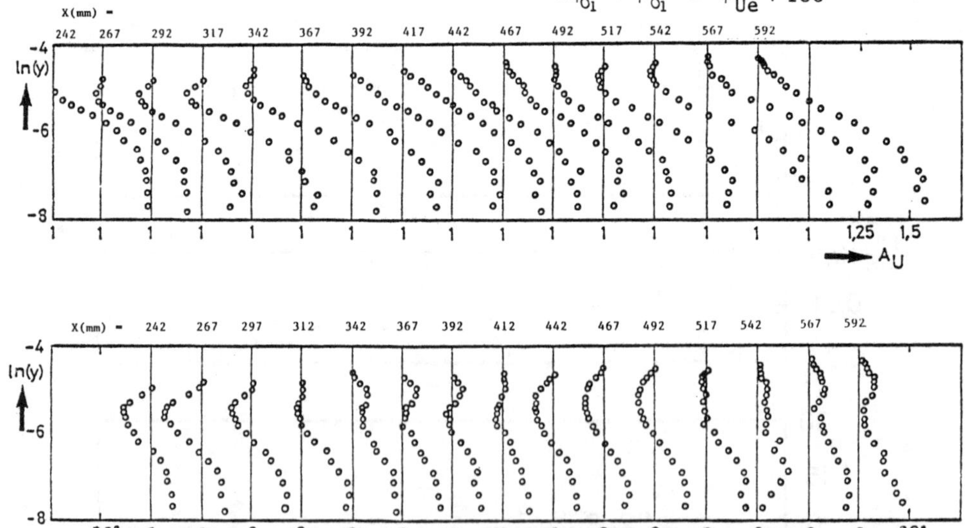

Fig. 4 - Flat plate experiment. F = 62 Hz
Amplitude and phase angle of the boundary layer velocity

$AU = \dfrac{\Delta U/\overline{U}}{\Delta Ue/\overline{U}e}$ $\Delta\phi_U = \phi_U - \phi_{Ue}$ y : distance to the wall (mm)

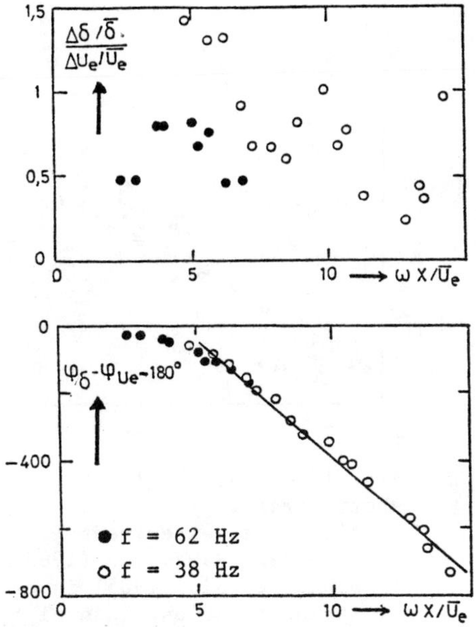

Fig. 5 - Flat plate experiment
Amplitude and phase of the boundary layer thickness :
at $y = \delta$: $\langle u'v' \rangle = 5\ 10^{-3}\ \langle u'v' \rangle_{max}$

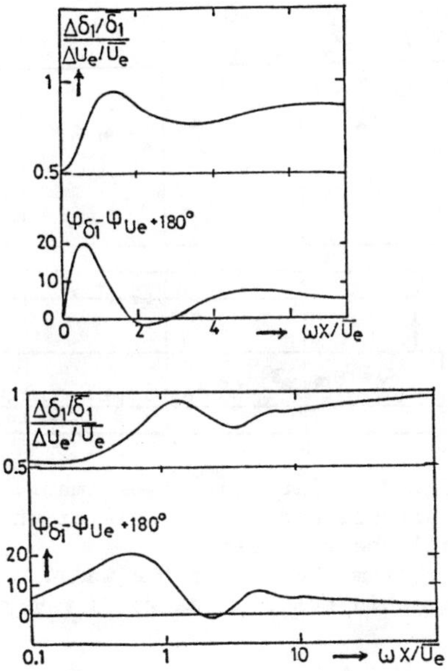

Fig. 6 - Oscillating laminar boundary layer on a flat plate.
— Numerical solution of small perturbation equations

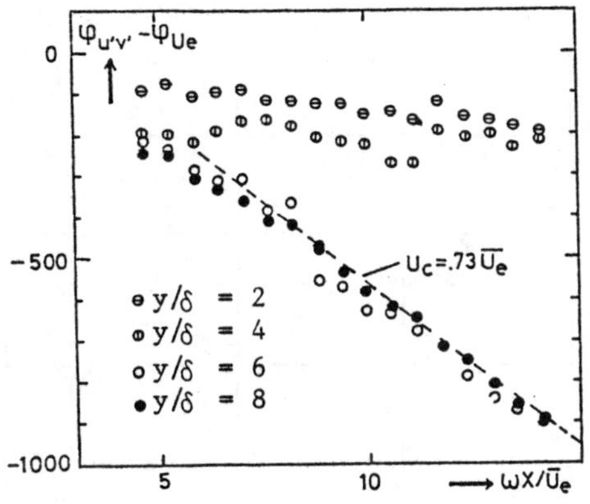

Fig. 7 - Flat plate experiment
f = 62 Hz
Phase of <u'v'> at y/δ = cst

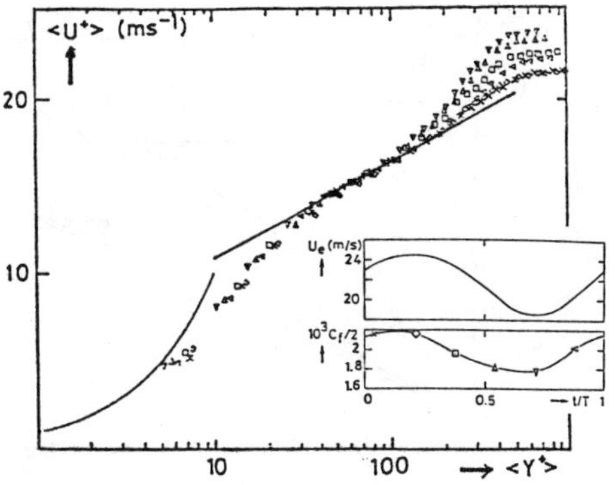

Fig. 8 - Flat plate experiment
Xo = 0.34 f = 38 Hz. Instantaneous velocity profiles in semi-log coordinates.

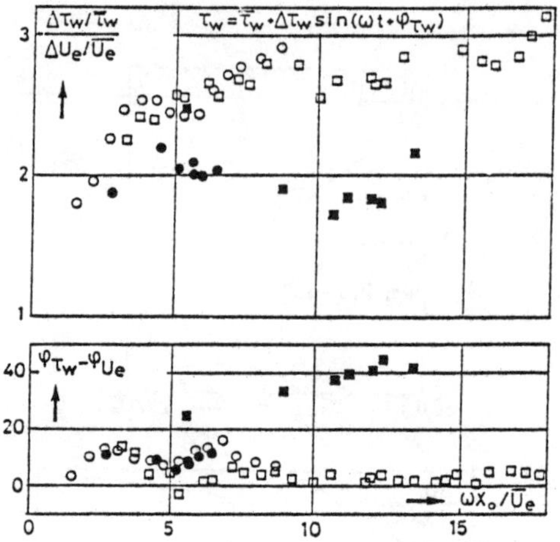

Fig. 9 - Flat plate experiment. Amplitude and phase angle of the wall shear stress.
○ □ from the law of the wall.
● ■ skin friction gauge. f = 38-62 Hz

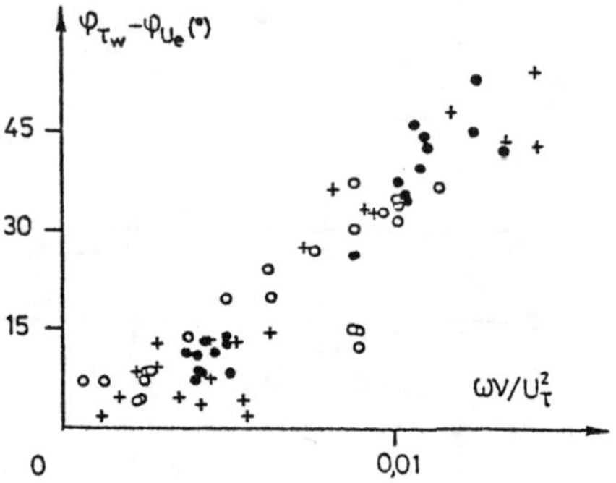

Fig. 10 - Phase angle of the wall shear stress.
● flat plate experiment
○ Menendez - Ramaprian (1983)
+ Binder et al. (1985) in oscillating turbulent channel flow

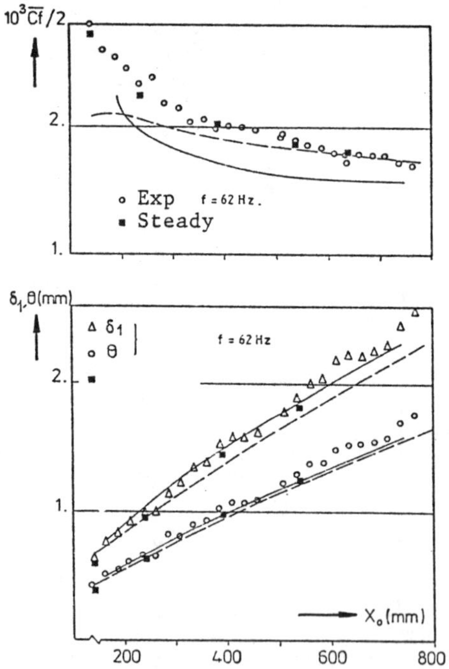

Fig. 11 - Flat plate experiment. Mean values of displacement and momentum thicknesses and skin friction coefficient
—— three equation model
···· integral method

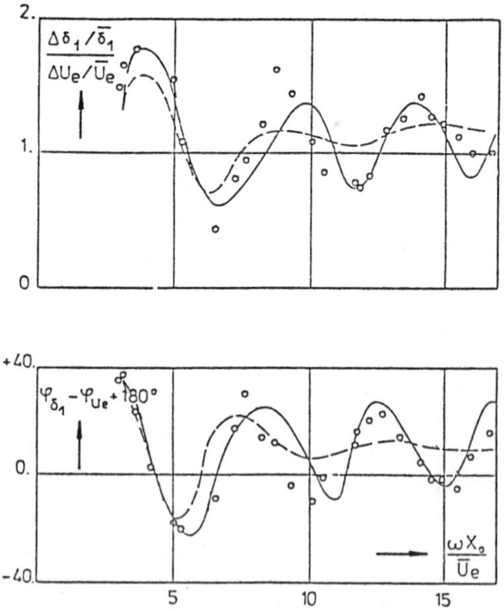

Fig. 12 - Flat plate experiment. Amplitude and phase angle of the displacement thickness
○ Experiment f = 62 Hz
—— three-equation model
.... integral method

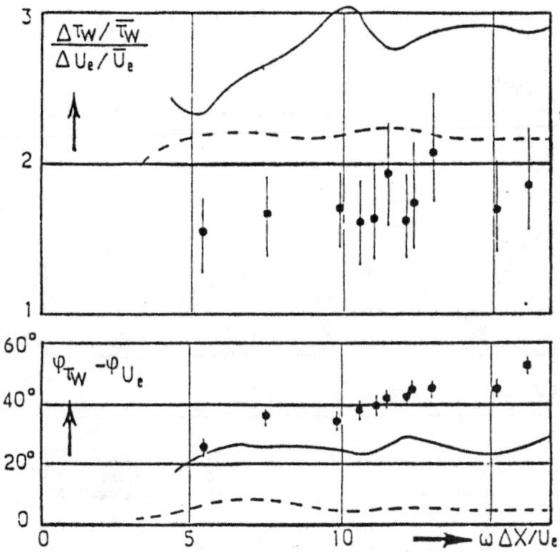

Fig. 13 - Flat plate experiment
f = 62 Hz. Amplitude and phase angle of the wall shear stress.
⧫ measurement. —— three-equation model. .... integral method.

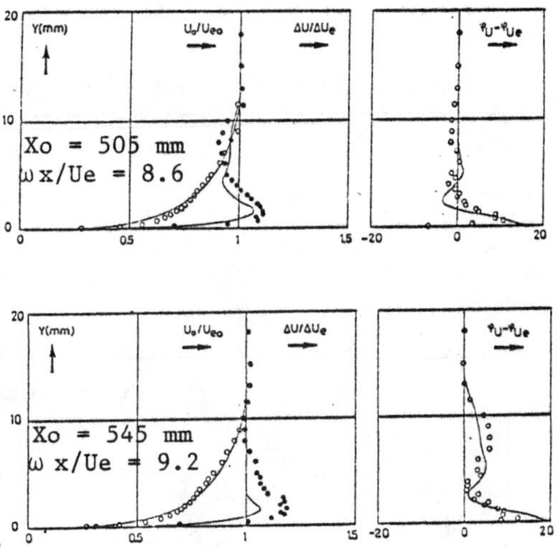

Fig. 14 - Flat plate experiment.
f = 62 Hz. Profiles of time mean velocity, amplitude and phase shift
—— three-equation model

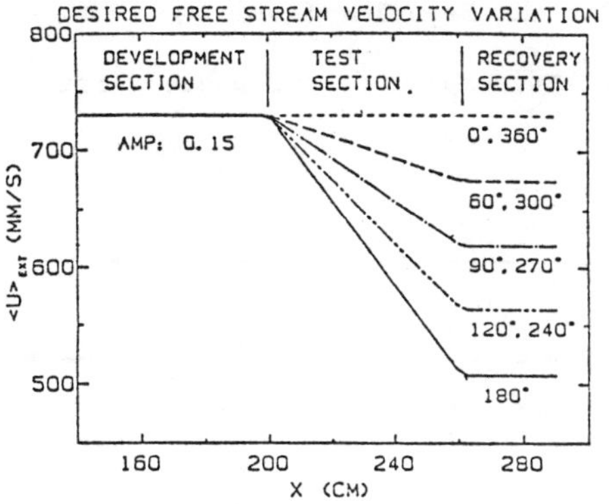

Fig. 15 - Experiment of Jayaraman Parikh - Reynolds (1982)

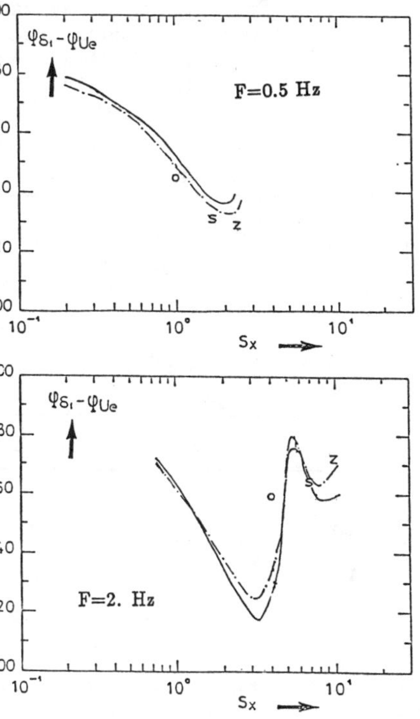

Fig. 16a-16b - Experiment of Jayaraman Parikh - Reynolds(1982).Small amplitude case

—·— Integral method
—— Three-transport equation model

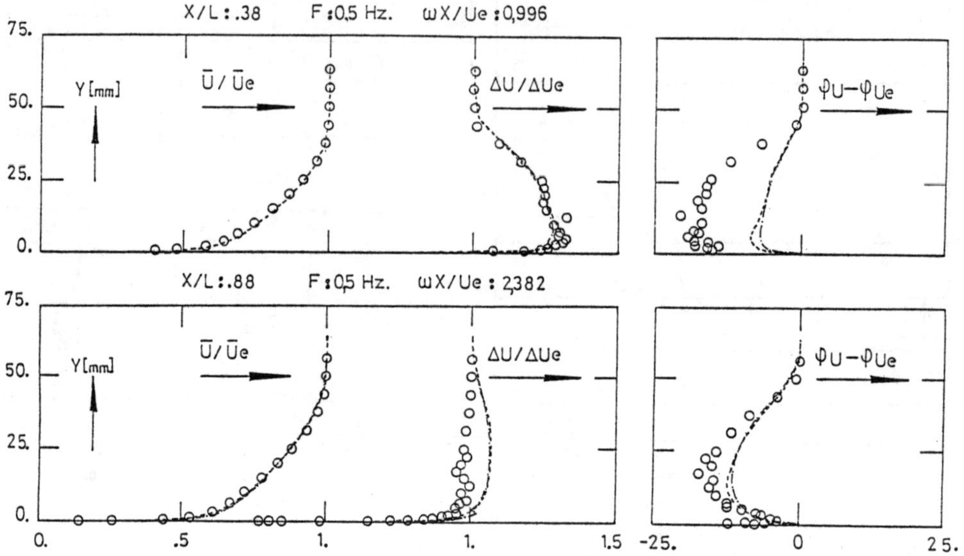

Fig. 17a - Experiment of Jayaraman, Parikh, Reynolds (1982)
Small amplitude case
Profiles of time mean velocity, amplitude and phase shift
.... k, ε , <u'v'> —.— k, ε

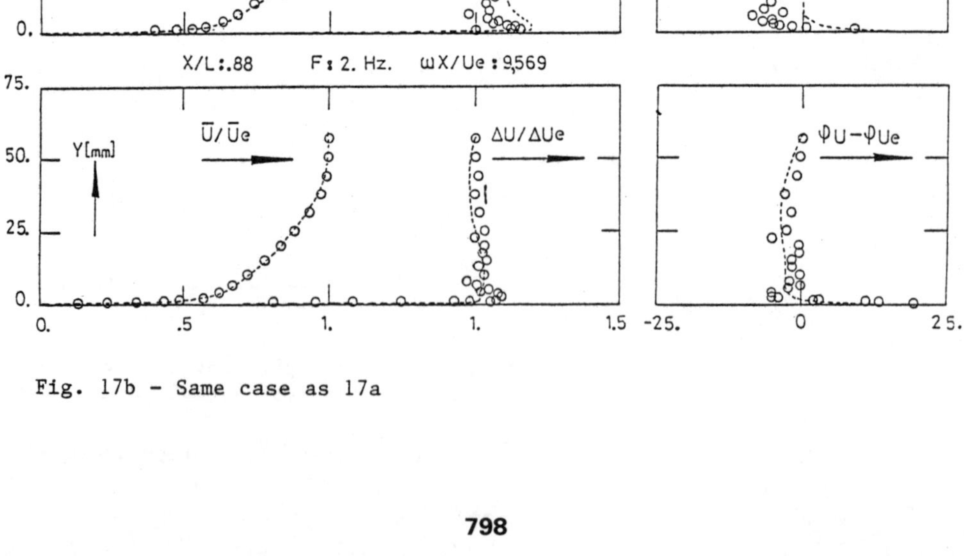

Fig. 17b - Same case as 17a

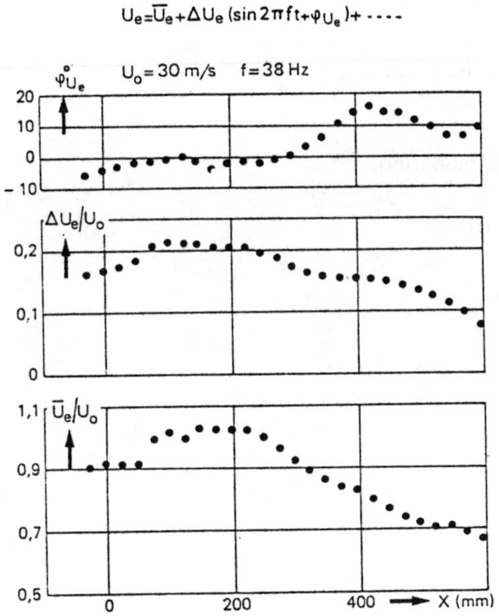

Fig. 18 - Experiment with adverse pressure gradient
(Houdeville - Cousteix, 1979)
Harmonic analysis of the external velocity

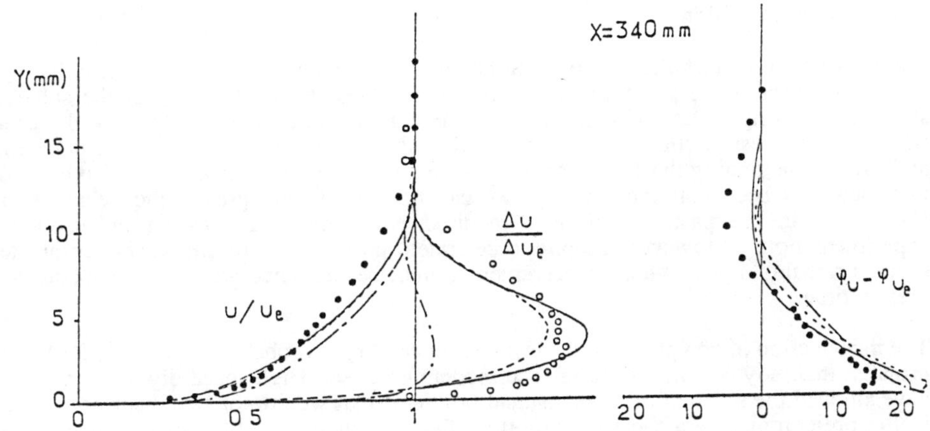

Fig. 19 - Experiment with adverse pressure gradient

—·— mixing length model
······· k - ε
——— k - ε - <u'v'>

Zonal Modeling of Turbulent Flows—Philosophy and Accomplishments

J. H. FERZIGER, S. J. KLINE, R. K. AVVA, S. N. BORDALO, and K.-L. TZUOO
Thermosciences Division
Department of Mechanical Engineering
Stanford University
Stanford, California 94305, USA

I. TURBULENCE COMPUTATION TODAY

A. General Introduction

The importance of turbulence modeling has been documented so often that it is hardly necessary to make the point again. However, the hundred year history of the subject shows a repeating pattern of suggestion of a new model, enthusiasm as the model meets with initial success, partial disillusionment when the model fails for more difficult cases, and, finally, beginning the cycle again. Lest this view appear cynical, let it also be said that there has been real achievement; we can now accurately predict the behavior of flows that, in the past, would have required either outright guessing or elaborate experimentation. However, despite accelerated progress in recent years, complete success remains a goal whose achievement cannot be predicted with confidence at the present time.

The introduction of computers in the last quarter century is probably responsible for more progress than any advance in turbulence modeling per se. High speed digital computers have allowed the solution of mathematical problems that would not have been attempted in the pre-computer era thereby permitting the introduction of turbulence models that could not have received serious attention earlier. We expect that computers will continue to increase in speed, and the solution of more complex problems will therefore be attempted. Large eddy and direct simulations of turbulent flows should find more application and their role in turbulence research will continue to increase. They may begin to be used as engineering tools but only on a limited basis. However, since most engineering applications require only the mean flow characteristics, turbulence models that provide closure for the Reynolds averaged Navier-Stokes (RANS) equations will continue to be the mainstay of flow prediction for high technology applications. These models are commonly referred to as one-point closure models.

In the past, many people have developed computer programs designed to treat a wide range of flows; these might be called 'universal codes'. Such general-purpose codes contain a grid generation procedure, a solution method, and a turbulence model; in some cases, more than one option is offered for some of the elements. In the opinion of the authors, such codes are, like many other multi-purpose tools, fairly good for some applications but rarely as good as a well-designed single-purpose tool for a given application. Since designers usually have only a single application (or a few related applications) to deal with, single-purpose codes should be preferred; they are both more accurate and faster. Zonal models, examples of which will be described in latter sections, are intended for use in single-purpose codes; they are not intended to be universal.

B. Universal Models

Many researchers now active in turbulence modeling research are involved in the search for a 'universal' turbulence model, i.e., a single model with a fixed set of equations and a fixed set of constant coefficients, capable of predicting all flows with the accuracy users require in feasible CPU times on widely available computers.

The attraction of a universal model is obvious. With such a model, it might be possible to predict a wide variety of flows with a single computer code. (We assume that the numerical techniques pose no difficulty.) In a sense, turbulence modeling would be a closed field if a universal model could be found.

Turbulence models have improved tremendously over the years. A very popular model for advanced applications today is the two-equation k-ε model. It does a wide variety of flows with moderate accuracy, and is useful for estimating the behavior of turbulent flows. For applications not requiring high accuracy, it may adequately serve as a 'universal' model. More advanced models, such as the Reynolds stress model, are widely believed to be more general than two-equation models but their record of success thus far does not appear to justify this opinion.

C. Numerical Accuracy

It is essential that accurate solutions to the equations be obtained if turbulence models of any kind are to be thoroughly tested. Everyone subscribes to the need for numerical accuracy but, too often, it is only lip service. Authors often claim that their solutions are grid-independent but careful scrutiny sometimes reveals that this is not the case.

It is reasonable to assume that the combination of partial differential equations and initial and boundary conditions constitute a problem with a unique solution; this is difficult to prove but little harm is done by accepting it. If this is the case, any numerical method which is mathematically consistent will provide the correct result if a sufficiently fine grid is used. It therefore makes little sense to compare solutions of the same problem obtained with different numerical methods; the differences are entirely due to numerical errors and one is essentially comparing two sets of errors. Unfortunately, many authors continue to do so. In comparing methods, the important issue is which method produces a solution of given accuracy at lowest cost.

Clearly, numerical errors need to be considerably smaller than errors due to turbulence modeling if turbulence models are to be evaluated fairly. For most turbulent flows, the solutions are relatively smooth and it is not difficult to estimate the errors. In our opinion (one which is now shared by a number of journal editors) all authors ought to be required to report estimates of numerical uncertainty in precisely the way that journals require experimenters to report the uncertainties in their measurements.

The choice of numerical method, especially the differencing method, is, at least in part, a matter of taste. However, in the authors' opinion there is no longer any reason to use the first-order accurate upwind method that dominated the field of computational fluid dynamics for so long. The arguments used to justify it are of dubious validity. Its major attractions for most users were the convergence of iterative methods and the smoothness of the solutions it produced. The former has been shown to depend on the choice of iterative method; there are methods whose rate of convergence is nearly independent of the difference approximation. With respect to the latter, it is worth noting that smoothness and accuracy are two different things; a pretty solution may not be a correct one. Good solutions can only come from a combination of accurate methods and sufficiently fine grids. The authors prefer second order central differences; other second and higher order methods produce equally good results at reasonable cost.

II. ZONAL MODELING AS AN ALTERNATIVE

A. Shortcomings of the Universal Approach

We noted in Section I.B that a majority of turbulence researchers are involved in the search for a single turbulence model that will predict all turbulent flows or, at least, all turbulent flows of some large class, adequately. It is important to point out that what is regarded as adequate is application dependent. A satisfactory model for environmental flows, which do not need to be simulated very accurately, should be easier to construct than models for high technology flows that demand the highest accuracy. In this paper, we shall concentrate on flows of the latter class.

The desirability of a universal model is not in question. The question of concern here is "Will a universal model adequate for the engineering applications be found in the near future?". The search for a universal model has indeed improved the quality of modeling to the point at which qualitative results for a range of interesting flows can be expected. However, high technology applications require further improvements in existing turbulence models. It is not clear that the required improvements can be achieved in the near future.

There is another serious difficulty with the search for universal models. By the very nature of the enterprise, failure of a model to perform adequately for even a single flow disqualifies it from any claim to true universality. The consequences of this rather innocuous statement are both profound and serious. The universal approach to turbulence modeling demands that any time a new model is proposed, it must be tested for a wide range of flows before it can be accepted as 'universal'. Even if the computer programs needed for such testing are available, the manpower and the computer time required to complete the calculations are both large. Even if a model is successful at early stages, it is always possible that some new flow will be found for which the model does not perform well. As stated earlier, this has been the historical outcome of the search for universal models. The 1980-81 AFOSR-HTTM-Stanford Conference on Complex Turbulent Flows demonstrated this unfortunate trend of the universal approach.

It is for these reasons that the authors, although they sincerely wish that a universal model could be found, believe that another approach is likely to serve better in the near future. It is these arguments that led to the concept of zonal modeling.

B. The Concept of Zonal Modeling

Although a universal model is desirable, the preceding discussion indicates that no such model exists at present and that the development of such a model in the near future is highly uncertain. In this situation, it is best to return to fundamentals and ask why the development of universal models has proved so difficult. In the opinion of the authors, the problem lies in the very nature of turbulence. Turbulence is probably not a single phenomenon, but rather a collection of related complex phenomena. For instance, the inner part of a boundary layer is different from a mixing layer and both are considerably influenced by the presence of 'extra' strains such as rotation. If such is the case, it follows that each type of flow might need a different model. Many authors have tried this approach, which is the extreme opposite of universal modeling. However, it is likely to be successful only so long as the flows used to determine the model cover the range of flows to which it is applied. Whenever the range of flows is extended, the model needs to be retuned, a process which is slow and expensive. Moreover, since the number of possible flows is very large, use of a different model for each flow is not practical.

It may be more practical to view a complex flow field as a composite of several structurally different regions or zones. For example, there may be mixing-layer-like regions and boundary-layer-like regions, among others, in a single flow. This leads to the idea that the model ought to depend on the physical nature of the region or zone rather than on the overall flow field, i.e., the model should be tied to the local characteristics of the flow. This can be accomplished by letting the model vary as a function of relevant flow parameters and is the basic idea behind zonal modeling. This concept will be explored further in subsequent sections.

Before going on, it is important to note that the zonal modeling approach permits a great deal more freedom in choosing the constants than does the universal approach. This is often cited as a disadvantage. Indeed, zonal modeling can degenerate into a kind of freestyle curve fitting if care is not exercised. For this reason, it is important that the model for each type of zone be fixed once and for all unless extremely strong reasons for modification arise. When this is done, zonal modeling offers a step by step approach to computing various flows based on firm 'building blocks' as elaborated in the next section.

III. MODELING STRATEGY FOR COMPLEX FLOWS

Having decided to pursue the idea of zonal modeling, we next have to find a means of putting it into practice for complex flows. If one begins with the most difficult flows, failure is almost certain. On the other hand, too cautious an approach will delay achievement of the goal. So, a sensible strategy is to begin with a tentative catalog of the kinds of zones that one is likely to encounter and to develop models for some of them. Then, one can attempt simulation of moderately complex flows, ones which contain a limited number of zones of the types found in the more complex flows to be considered later.

At this stage one encounters the need to patch the models for zones together in a smooth manner. Thus, at separation, the physical nature of the flow changes from that of a boundary layer to that of a mixing layer. If zonal models for boundary layer and mixing layer are available, some means of joining them is required. We call these joints between the zones readjustments or bridges. To make the readjustments as simple as possible, the models for all zones should have the same structural form, *e.g.*, they could be the same base model with different values for the model coefficients.

In this chapter, we will discuss in a somewhat general way the issues of cataloging the zones, the choice of a base model, the form of the bridges, and a strategy for developing zonal programs for complex flows.

A. Zone Catalog

An exhaustive list of all possible zones is impossible to construct *a priori*. Kline (1981) listed twenty zones; even though this list is not complete, it does include many frequently found zones and some rare ones. Some zones in this list may be combined into a single zone while others need to be expanded into several zones; in fact, it will be shown below that the definition of a zone may depend on the choice of base model. The significance of Kline's catalog is that the number of zones is actually not very large.

In view of the remarks just made, it is pointless to try to give a complete list of zones. Rather we shall suggest a classification scheme and leave the details of zone definitions to be determined later. Probably the simplest classification scheme has a tree structure similar to the one given by Ferziger *et al.* (1981). At each level, there is a binary or ternary split:

1. Homogeneous or inhomogeneous
2. Incompressible or compressible
   a. If compressible: subsonic, transonic, or supersonic
3. Wall bounded or free shear
4. Simple strain or 'extra' strain
   a. If 'extra' strain, type of strain

The 'extra' strains referred to in the last line include: rotation, curvature, lateral stretching, buoyancy and/or stratification, transpiration, heat transfer, phase change, and chemical reaction among others.

B. The Choice of Base Model and Bridges

General Considerations: The ideal base model would, of course, be a 'universal' model. If such a model existed, there would be nothing more to do. Given that such a model does not exist, the best choice is a model that is as close to this ideal as possible. At present, if the codes are to be fast and the model is to be capable of treating a wide variety of flows, one-point closure models are the only sensible alternative.

Among the one-point closure models, the mixing length model is attractive for its accuracy in treating boundary layers and a range of free shear flows. Its principal drawback is difficulty in dealing with separated and/or reattaching flows. This eliminated it from serious consideration.

One-equation or turbulence kinetic energy models have not proven very useful and are rarely used. They require prescription of length scales, something which is very difficult to do.

That leaves only two-equation and Reynolds stress models in contention. Both are capable of doing a wide variety of flows with reasonable accuracy. However, despite the widespread expectation that Reynolds stress models ought to be more accurate, the evidence to date favors neither model. At this point, one ought to choose the simpler model and, since there is wider experience with two-equations models, the latter was chosen. It is important to note that this choice is not unique; the strategy outlined below is more important than the particulars connected with the k-ε model.

The k-ε Model and its origins: It is important to consider the base model and its origins in more detail. In the k-ε model, the Reynolds stress is represented via a generalized Boussinesq-type eddy viscosity concept:

$$-\overline{u_i'u_j'} = \nu_t \left(\frac{\partial U_i}{\partial x_j} + \frac{\partial U_j}{\partial x_i} \right) - \frac{2}{3} k \delta_{ij} \tag{1}$$

ν_t, the eddy viscosity is given by

$$\nu_t = C_\mu \frac{k^2}{\epsilon} \tag{2}$$

where k is the turbulent kinetic energy and ε is rate of dissipation of the turbulent kinetic energy; these quantities are obtained by solving the following partial differential equations:

$$\frac{\partial (U_i k)}{\partial x_i} = \frac{\partial}{\partial x_i}\left(\frac{\nu_t}{\sigma_k}\frac{\partial k}{\partial x_i}\right) + I\!P - \epsilon \tag{3}$$

$$\frac{\partial (U_i \epsilon)}{\partial x_i} = \frac{\partial}{\partial x_i}\left(\frac{\nu_t}{\sigma_\epsilon}\frac{\partial \epsilon}{\partial x_i}\right) + C_{\epsilon_1}\frac{I\!P\epsilon}{k} - C_{\epsilon_2}\frac{\epsilon^2}{k} \tag{4}$$

In these equations, $I\!P$ is the rate of production of turbulent kinetic energy while the other symbols are the model coefficients; these coefficients are five in number and are treated as constants in the standard model. The methodology used to determine their values is discussed below.

The coefficient C_μ defines the level of the eddy viscosity and its value is obtained by noting that, in two-dimensional equilibrium shear layers, the ratio of the Reynolds shear stress to the turbulent kinetic energy is approximately constant and:

$$C_\mu = \left(\frac{-\overline{u'v'}}{k}\right)^2 \tag{5}$$

The value of the coefficient $C\varepsilon_2$ is chosen to match the decay of isotropic turbulence while $C\varepsilon_1$ is adjusted to match the von Karman constant or the slope of the logarithmic region in the turbulent boundary layer. The values for the two turbulent Prandtl numbers σ_k and σ_ε were selected by requiring the model to correctly predict the spreading rates of a set of free shear layers. When the values of the five coefficients found in this way <u>are held constant</u> throughout the flow field, the model is called "the standard K-ε model." In the zonal modeling described in this paper, we retain the same form of differential equations (K-ε), but tune the value of the coefficients for each physical type of flow, *i.e.* zone. We do not alter the values of the coefficients for any kind of zone once they have been fixed.

<u>Bridges or readjustments</u>: When zonal models are used in the sense just described, there will be regions where the flow changes from one type to another. The model must represent this by adjusting the coefficients from those appropriate to the first zone to those for the second. To avoid abrupt changes in the results, the adjustments should be smooth and as simple as possible. In the present work, they are either based on linear variation of the coefficients through the readjustment region or on the solution of a first order linear differential equation (which is equivalent to an exponential blending of the models).

Zones join in at least two distinct ways. There are lateral boundaries in which the flow is essentially parallel to the boundary. In the cases of this type so far encountered, it was found that no bridge is necessary; alternatively, simple smoothing could be used. In the other type of boundary, the fluid literally flows from one zone into another (separation and reattachment are examples of this type of readjustment). In this case, there needs to be a small bridge region in which the readjustment takes place; in flows simulated to date, it has been found that the results are not very sensitive to the length of the bridge region or to the type of smoothing used.

C. Overall Strategy for Zonal Model Development

If not employed carefully, the zonal strategy may become an arbitrary recipe for fitting a turbulence model to experimental data. For this reason (and others), it is essential that a systematic procedure be adopted and that the model be tested on far more cases than are used in constructing the model.

The following scheme has been used successfully in a number of cases and is recommended. It consists of six steps:

1. The flow is computed with the base model. In the present instance, the standard k-ε model or appropriate zonal models constructed and verified earlier are used. It is essential that the numerical errors be kept small.

2. The results are compared with experimental data. Any discrepancies are noted. An analysis is made to decide whether these discrepancies could be due to sources other than the turbulence model *e.g.*, numerical and/or experimental error.

3. If the discrepancy appears to be due to inaccurate modeling, a physical explanation is sought. Usually, this means that some physical effect (*e.g.*, curvature or rotation) has not been accounted for. Non-dimensional parameters that measure the importance of this effect are introduced.

4. A zonal model whose coefficients are functions of the non-dimensional parameters selected in Step 3 is constructed.

5. The flow is simulated with the candidate model.

6. Steps 2-5 are repeated until satisfactory results are obtained.

Although the flows treated to date are not very complex, this method has been successful in every instance, and there are no indications that this will not continue to be the case.

The above procedure can be applied to quite complex flows. Since codes based on zonal models are single-purpose, something is known about the physics of the flows to be predicted prior to writing the code. Consequently, the zones usually can be identified, at least tentatively, from the beginning. If this is not the case, one can calculate the flow using the base model; then the results can be used to suggest what zones occur in the flow.

IV. ACCOMPLISHMENTS

The approach outlined above has been applied to a number of flows. The overall strategy is to consider relatively simple flows first; these are building blocks for the more complex flows dealt with later. In this chapter, we shall give an overview of what has been done and a few selected results; details of the models and further results can be found in the references.

A. Homogeneous Flows

The simplest of all turbulent flows are the homogeneous flows so it is natural to apply zonal modeling to them first. As these are, by definition, flows in which there is no spatial variation of the statistical quantities, one might think that the zonal approach does not apply to them. They do develop in time, however, and the zonal idea can be applied in the time domain.

All of the flows considered begin as isotropic turbulence. Strain is applied to them and they begin to develop a kind of structure that depends on the type of strain. It is assumed that each flow develops to an asymptotic final state in which the turbulence energy changes with time but the anisotropy of the Reynolds stress tensor remains fixed. Since none of the experimental flows actually reach an asymptotic state (there is no proof that one exists), the entire flow is a bridge between the initial isotropic state and the final state.

Tzuoo *et al.* (1986) developed models for the asymptotic states and used linear first order differential equations as the bridges. Results for turbulence undergoing shear and plane strain are shown in Figs. 1 and 2, respectively. These results provided an initial indication that the zonal modeling concept is sound.

B. Free Shear Flows

The next flows considered were the free shear flows; see Tzuoo *et al.* (1986). These include the plane mixing layer, jet, and wake and the axisymmetric jet and wake. Initial trials showed that the standard k-ε model predicts the plane jet without co-flow and the near field part of the plane wake quite well but is somewhat inaccurate for the far field of the plane wake, co-flowing jet and all the axisymmetric flows. These results are generally known.

In the search for a feature that differentiates flows that are well-predicted from those that are not, it was found that the parameter

$$\beta = \frac{U_\infty}{\Delta U_m} \tag{6}$$

which measures the relative strength of the inertia and shear in the flow could be used to distinguish them. U_∞ stands for the free stream speed while ΔU_m is velocity difference at the centerline of the free shear layer. Note that β is a global parameter; no local parameter could be found that successfully collapsed the data. For the plane wake, β is small in the near field and large in the far field so that the standard k-ε model can be adopted as the 'near-field' (small β) model. The quotes around the term near-field are used to indicate that the distinction based on β does not correspond to the near- and far-fields in all flows. It is interesting that the parameter β is equivalent to the Sabin-Abramovich parameter which correlates mixing layers of arbitrary velocity ratios.

For the 'far-field', the problem was traced to the length scales and diffusion in the outer part of the shear layer. For this reason, the far-field or large-β zonal model was obtained by modifying the values of the model coefficients $C_{\varepsilon 1}$ and σ_ε.

Since the two limiting models correspond to small and large values of b respectively, it is natural to base the bridge on the value of this parameter. Tzuoo *et al.* chose an exponential function for the bridge and obtained the results shown in Figs. 3, 4, and 5 for the plane wake, mixing layer and jet, respectively.

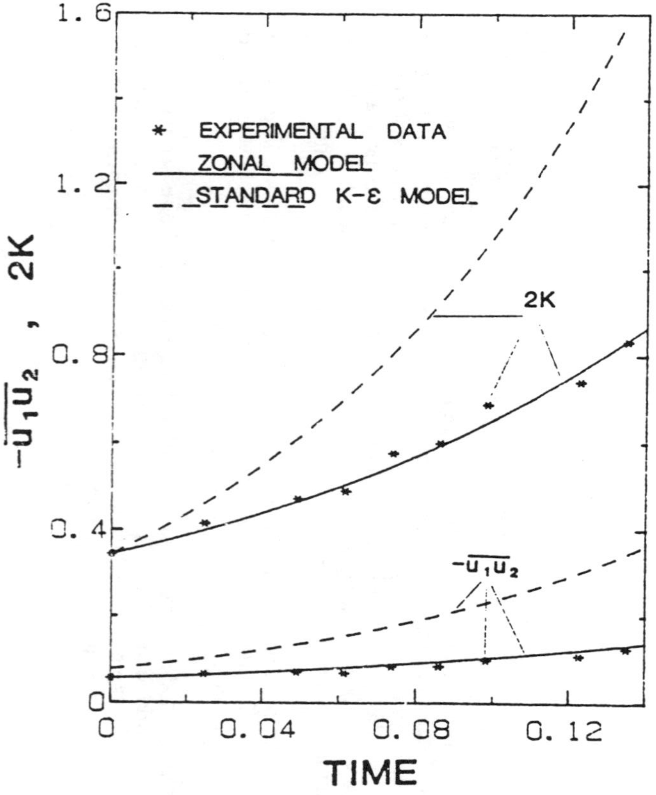

Fig. 1 Homogeneous shear flow, Case 376B in 1980-81 Stanford Conference; from Tzuoo *et al.* (1986)

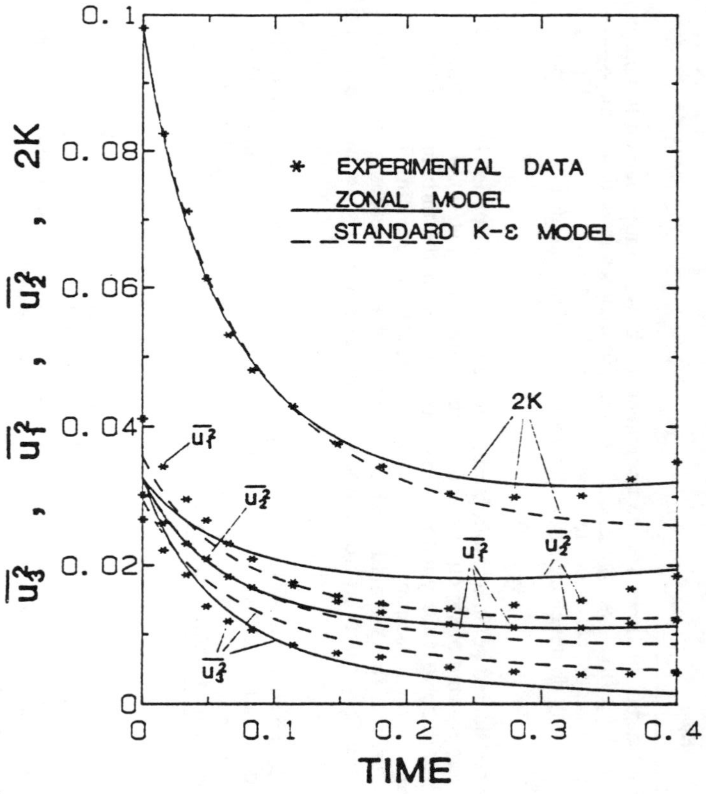

Fig. 2 Homogeneous turbulence undergoing plane strain; Case 374B in 1980-81 Stanford Conference; from Tzuoo *et al.* (1986)

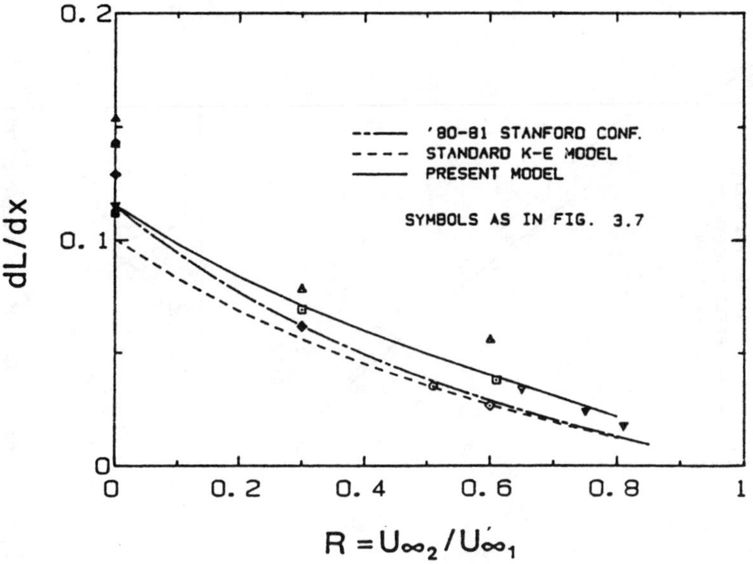

Fig. 3 Spreading rate of the plane mixing layer; from Tzuoo *et al.* (1986)

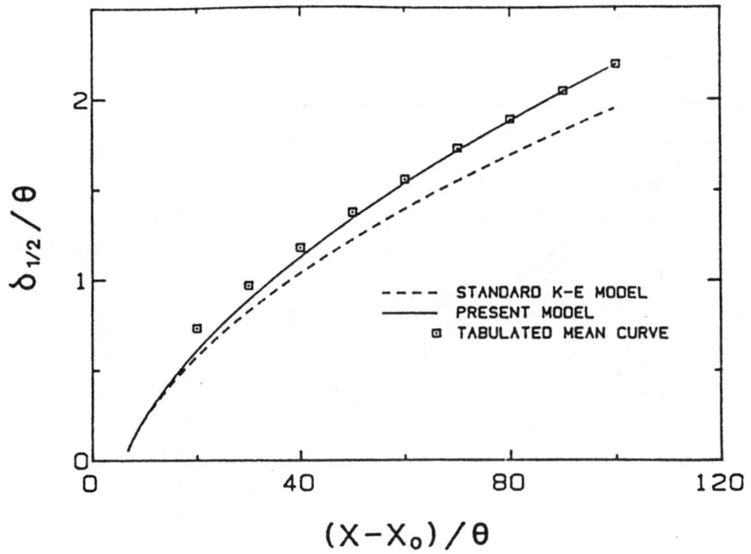

Fig. 4 Spreading rate of the plane co-flowing jet; from Tzuoo *et al.* (1986)

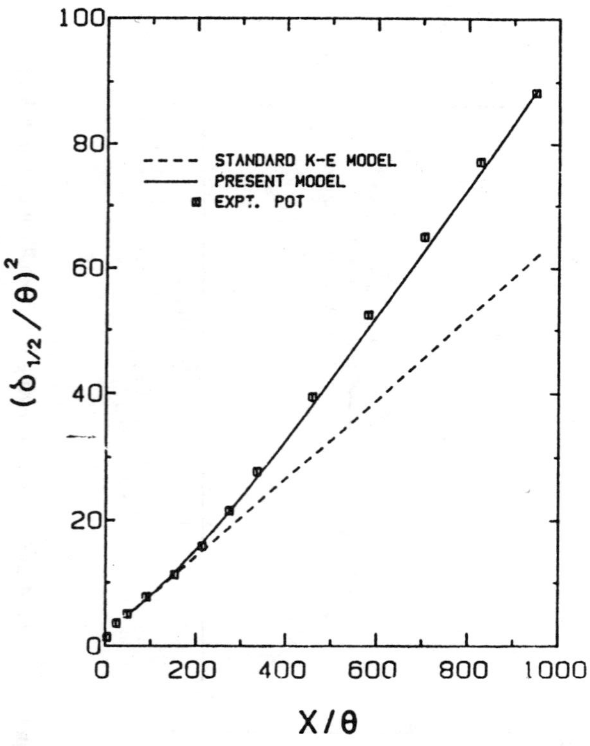

Fig. 5 Spreading rate of the plane wake; from Tzuoo *et al.* (1986)

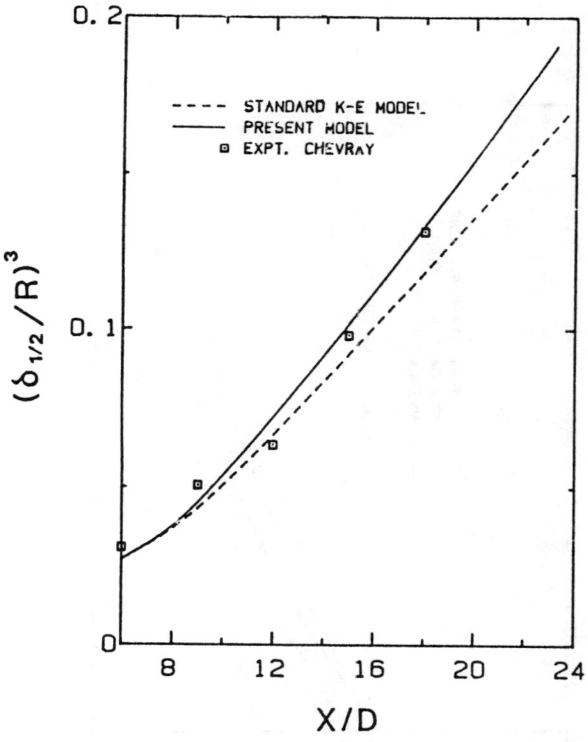

Fig. 6 Spreading rate of the axisymmetric wake; from Tzuoo *et al.* (1986)

The standard k-ε model predictions of the axisymmetric free shear layers are less than adequate. Tzuoo *et al.* postulated that the standard model does not adequately represent the effect of vortex stretching on turbulence. Vortex stretching is a prominent feature of the near-field of axisymmetric flows, but is insignificant in the far-field; it was taken into account by treating the coefficient C_μ as a function of a non-dimensional vortex stretching parameter S. The model reduces to the plane free shear flow model in the far-field. Typical results for axisymmetric flows are shown in Figs. 6 and 7.

To illustrate the kind of relationship between model coefficients and flow parameters that might be used in constructing zonal models, the model of Tzuoo *et al.* is given below.

$$C_{\epsilon 1} = 1.44 - 0.4\left(1 - e^{-0.18\beta}\right) \quad (7)$$

$$\sigma_k = 1.0 - 0.35\left(1 - e^{-0.18\beta}\right) \quad (8)$$

$$\begin{aligned} C_\mu &= 0.05 + 0.04\left(1 - e^{-0.1(S-11.3)}\right) \quad \text{for } S \geq 11.3 \\ &= 0.05 \qquad\qquad\qquad\qquad\qquad\qquad < 11.3 \end{aligned} \quad (9)$$

$$S = \frac{R/\delta}{d\delta/dx} \left. \begin{array}{l} R = \text{radius of ring vortex} \\ \delta = 1/2 \text{ thickness of shear layer} \end{array} \right.$$

The results presented in this section highlight two salient features of zonal modeling: (i) delineation of governing parameters and (ii) the accuracy achievable by zonal models. These two features will be described briefly below.

The methodology necessitates a search for the parameters that govern the physics of each flow zone. Identification of the governing parameters not only provide the basis for constructing zonal models, but may increase understanding of the underlying physics. For instance, use of the parameter β quantifies previously known differences between the near-field (strong shear) and far-field (weak shear) regions. Similarly, the parameter S quantifies the effect of vortex stretching in axisymmetric shear layers. Equally importantly, these parameters made it possible to find a 'unified' model that can predict a wide class of free shear flows accurately.

In order to provide an overview of the superior accuracy obtained with the zonal model, the results of Tzuoo *et al.* are summarized in Fig. 8. For all cases analyzed, the maximum difference between the data and the zonal model prediction is less than 10%, which we believe to be the level of uncertainty in the data. On the other hand, the standard model predictions differ from the data by more than 50% for about a fourth of the cases.

C. Boundary Layers

In modeling turbulent boundary layers, it is common to make a distinction between the inner and outer layers. The former contains the viscous, buffer, and logarithmic regions while the latter is comprised of part of the logarithmic region and the wake; the two layers overlap in the logarithmic region. The inner layer is treated via the use of wall-functions, a common approach.

In the standard k-ε model, the coefficients $C\varepsilon_1$ and C_μ were chosen to correctly simulate the logarithmic region in the boundary layer without extra strains; for this reason, the logarithmic region is predicted well in the case of zero pressure gradient. The wake is reasonably well predicted, but close inspection reveals that the growth of this region is

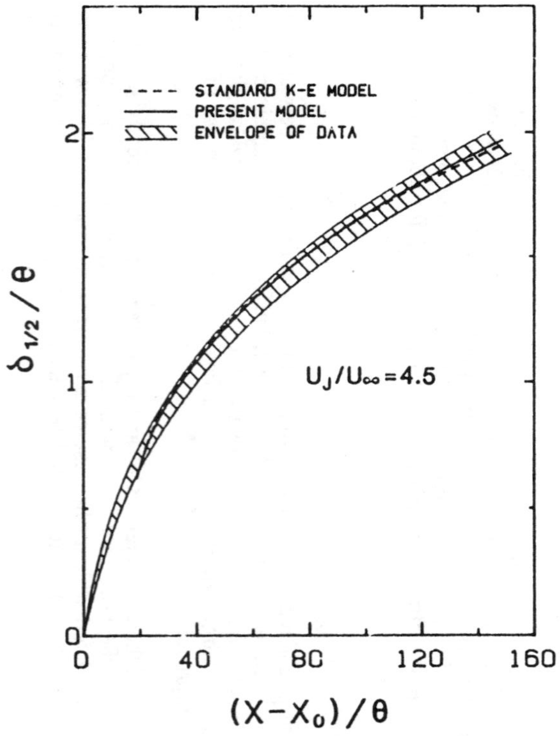

Fig. 7 Spreading rate of the axisymmetric co-flowing jet; from Tzuoo *et al.* (1986)

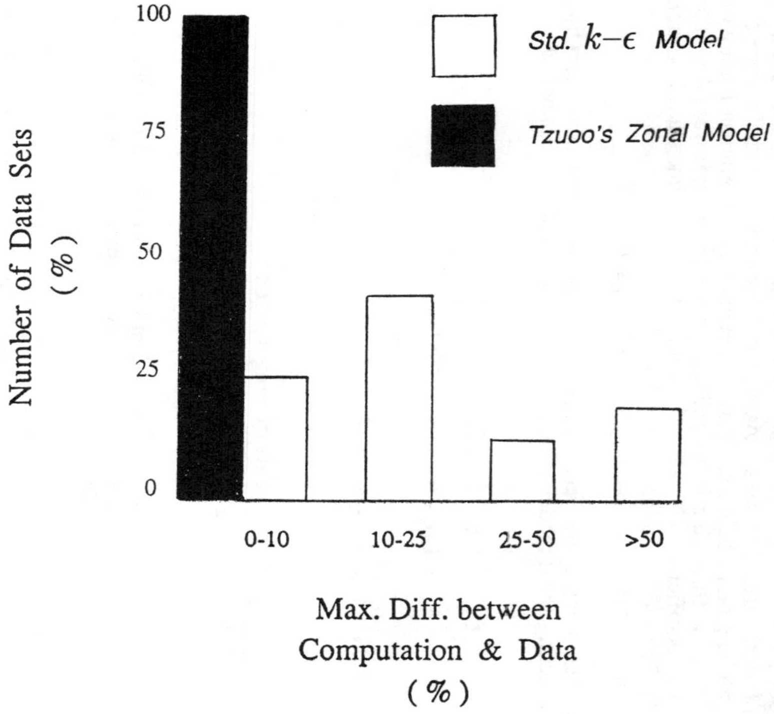

Fig. 8 Comparison of the performace of zonal and standard k–ε models in predicting free shear flows - summary of the computations of Tzuoo *et al.* (1986)

somewhat overpredicted. This problem was traced to σ_ε. As explained earlier, the standard value of σ_ε was determined by tuning the model to a set of plane free shear flows. Bordalo et al. (1988) showed that the standard value of σ_ε is not appropriate for the wake regions of boundary layers and proposed that a different value be adopted for this zone. Improved results were obtained.

Bordalo et al. also investigated flows with pressure gradient and curvature, cases that have proved difficult for invariant k-ε models. The flexibility provided by the zonal approach makes the construction of models easier. Bordalo et al. were able to improve the accuracy of the k-ε model by allowing the model coefficients $C_{\varepsilon 1}$ and C_μ to be functions of flow parameters that represent the effects of these 'extra' strains. Results for a typical pressure gradient case are given in Fig. 9. This zonal model has not yet been applied to separated flows, but an attempt to do so will be made in the near future.

D. Separating and Reattaching Flows

The motivation behind the development of zonal models and most research in turbulence modeling in general, is the need to be able to predict complex flows including fully separated and reattaching flows. The flows considered in the preceding subsections can be predicted reasonably well by mixing length models. It is the difficulty of extending mixing length models to complex flows that led to the search for more advanced models including zonal models. So, it is imperative to apply the zonal modeling concept to complex flows.

As the simplest, and best studied, complex flow, the turbulent flow over a backward-facing step is the obvious first case to consider. Avva *et al.* (1988) have done so. Before reviewing that work, it is worth mentioning that, in the course of Avva's work it was discovered that approximately half the discrepancy between the results reported at the 1980-81 AFOSR-HTTM-Stanford Conference on Complex Turbulent Flows and experimental data is numerical error; only half is actually due to the inadequacy of the turbulence models. This again underscores the need for careful numerical work.

As prescribed in Section III.C, the flow was first simulated using both the standard k-ε model and a zonal model in which the models for the free shear layer and the adverse pressure gradient boundary layer described above were employed in the appropriate regions. The results indicated that the reattachment length is underpredicted by about 15% and most of the other discrepancies result from this error. Closer inspection revealed that the probable cause for this discrepancy is the lack of sensitivity of the standard model to streamline curvature.

The reattaching shear layer has stabilizing curvature in the reattachment region. Various experimental studies on curved shear layers have demonstrated that stabilizing curvature suppresses turbulence production and the Reynolds shear stress. Consequently, shear layers with stabilizing curvature grow more slowly than their straight counterparts. Various studies have shown that the effect of extra strain due to curvature is not adequately accounted for by the standard model; the effect of this 'extra' strain can be measured by the non-dimensional curvature Richardson number. Avva *et al.* developed a modification of the standard model which accounts for the effect of streamline curvature. In this model, which is based on the studies by Leschziner and Rodi (1981) and by Rodi and Scheuerer (1983), the coefficient C_μ is treated as a function of the local curvature Richardson number. This model is applied to a number of backward-facing step flows covering a wide range of expansion ratios (downstream channel height divided by the upstream channel height) and separating boundary layer thicknesses.

We shall give a few selected results. The reattachment length as a function of expansion ratio is given in Fig. 10. The zonal model predicts the reattachment length within the experimental uncertainty for all the cases. Fig. 11 gives the pressure distribution in one

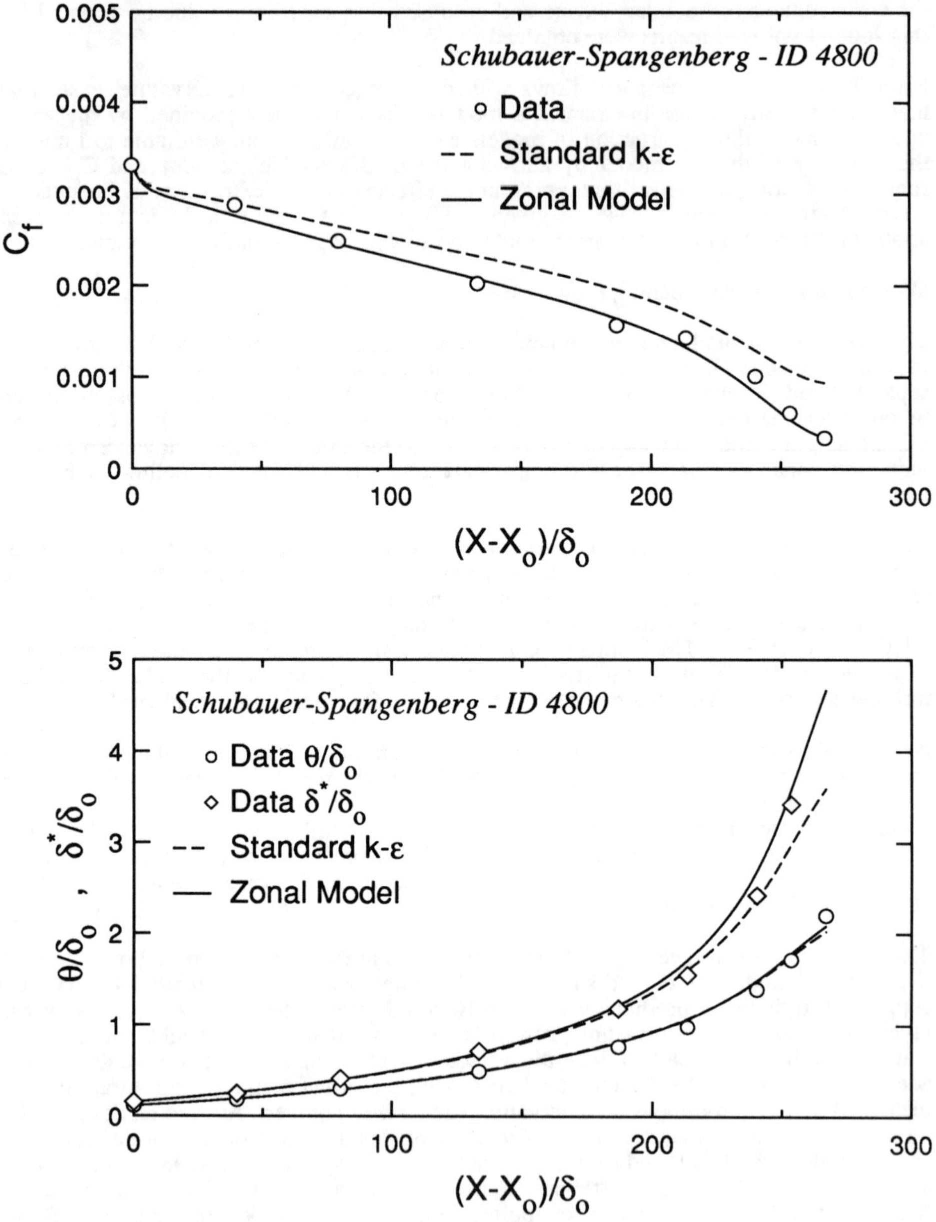

Fig. 9 Skin friction and boundary layer integral parameters for an adverse pressure gradient boundary layer; from Bordalo *et al.* (1988)

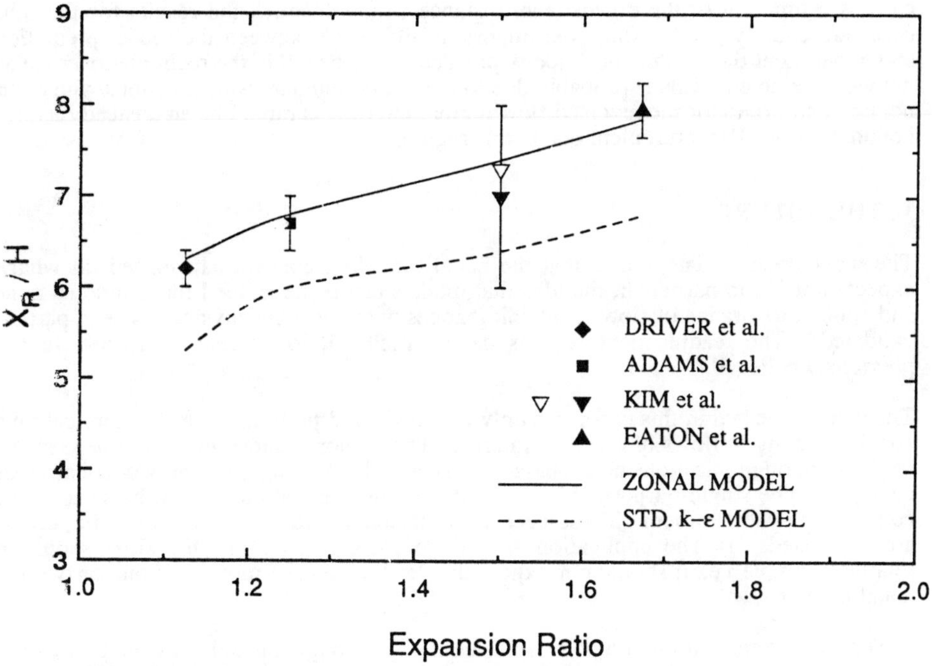

Fig. 10 Reattachment length as a function of the expansion ratio of the backward-facing step; from Avva *et al.* (1988) (The open symbol for the case of Kim *et al.* denotes the estimate from static pressure data.)

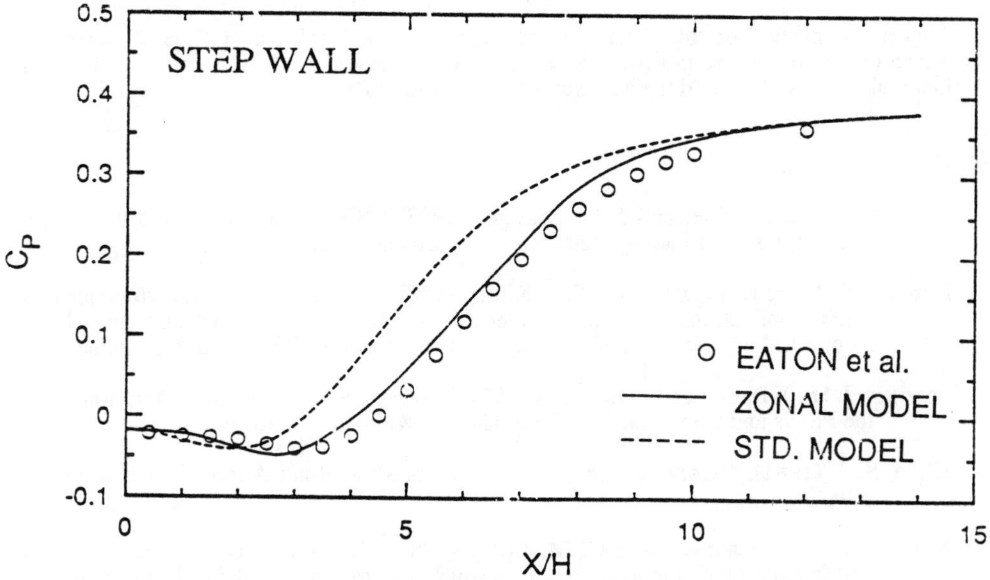

Fig. 11 Pressure coefficient in the backward facing step flow as a function of downstream distance: the case of Eaton and Johnston; from Avva *et al.* (1988)

case as a function of the downstream distance in the channel; the results for the other cases are equally good. The most important difference between the model predictions and experiment data is that the velocity profiles near the wall in the recirculation zone are not well predicted. This is probably due to the use of standard wall functions which may not be appropriate for the reversed flow region; the flow is more like an unsteady laminar boundary layer than a turbulent one in this region.

V. THE FUTURE

The work done to date shows that the zonal modeling concept can indeed do what is expected of it. In particular, the idea that models can be developed for individual zones and applied to a range of flows containing zones with the same physics has been partially validated. The readjustment regions have turned out to be less of a problem than previously believed.

The work reviewed in this paper can only be considered preliminary to the application of zonal modeling to flows of practical interest. The authors believe that the time is ripe to begin work on applications codes based on the zonal modeling concept. As zonal model codes are to be single-purpose, this means that a few applications should be selected and codes developed for them. These need to be selected with care. The following criteria are suggested. i) The application should be important and ii) the flow should be relatively simple i.e., it should not require the development of more than one or two new zonal models.

We end with a reiteration of a point made earlier. Although the authors would very much like to see a universal model developed, it appears to be unlikely that this will be accomplished in the foreseeable future. In our view, zonal modeling is a viable alternative for the development of fast and accurate codes in the near future.

ACKNOWLEDGEMENTS

This major portion of this work was sponsored by the Air Force Office of Scientific Research. Important support was also provided by the Allison Gas Turbine Division of General Motors Corp. and the Ministry of Education of Brazil.

REFERENCES

Avva, R. K., S. J. Kline, and J. H. Ferziger (1988), "Computation of Turbulent Flow over a Backward-Facing Step - Zonal Approach," AIAA-88-0611.

Bordalo, S. N., J. H. Ferziger and S. J. Kline (1988), "Zonal Models of Turbulence and Their Application to Boundary Layer Flows," Rept. TF-34, Thermosciences Div., Dept. of Mech. Eng., Stanford University, Stanford, California (to be published).

Ferziger, J. H., J. Bardina and G. Allen (1981), "Overview of Taxonomy: Morphology of the Flows and Computational Methods," in Kline et al (1981)

Kline, S. J. (1981), "Universal or Zonal Modeling - the Road Ahead," in Kline et al (1981)

Kline, S. J., B. J. Cantwell and G. M. Lilley (1981), *1980-81 AFOSR-HTTM-Stanford Conference on Complex Turbulent Flows*, Thermosciences Div., Dept. of Mech. Engr., Stanford Univ., Stanford, California.

Leschziner, M. A. and W. Rodi (1981), "Calculation of Annular and Twin Parallel Jets using Various Discretization Schemes and Turbulence Model Variations," J. Fluids Engr., Vol. 103, pp. 352-360.

Rodi, W. (1980), *Turbulence Models and their Application in Hydraulics*, IAHR Publ., Delft, The Netherlands.

Rodi, W. and G. Scheuerer (1983), "Calculation of Curved Shear Layers with Two-Equation Turbulence Models," Physics of Fluids, Vol. 26, pp. 1422-36.

Tzuoo, K-L., J. H. Ferziger and S. J. Kline (1986), "Zonal Models of Turbulence and Their Application to Free Shear Flows, Rept. TF-27, Thermosciences Div., Dept. of Mech. Engr., Stanford Univ., Stanford, California.

Contribution to the Second-Moment Modeling of Sublayer Turbulent Transport

B. E. LAUNDER and D. P. TSELEPIDAKIS
Mechanical Engineering Department
UMIST
Manchester M60 1QD, UK

ABSTRACT

The paper examines the question of developing a full second-moment closure within the viscosity affected sublayer near a wall. The effects of different processes are considered in turn. For the influential pressure-strain process, new forms are adopted that have emerged from research at UMIST on free shear flows. Our results indicate the importance of the strongly non-isotropic nature of the dissipation process and the role of pressure-driven transport of both Reynolds stress and the turbulence energy dissipation rate. The question of whether the turbulence Reynolds number must be retained as a parameter in the pressure-containing correlations remains unresolved though, if it does, its influence will be less important than in earlier closure proposals.

INTRODUCTION

The present paper is concerned with modelling turbulent transport processes in the immediate vicinity of a rigid surface where direct effects of viscosity are important. We refer to this sublayer as the "low-Reynolds-number region" and to models which attempt, in some way, to account for the processes taking place there as "low-Reynolds-number" models.

The earliest remembered model of this type is Van Driest's [1] adaptation of the mixing-length hypothesis which - more than thirty years on - is still extensively used. It is particularly effective in situations such as flow near spinning discs or in flow around bends where the near-wall velocity vector undergoes strong skewing within the low-Reynolds-number region and the conventional logarithmic laws fail [2,3]. In many other situations, however, involving severe streamwise pressure gradients, separated flows or strong body forces, the mixing-length hypothesis provides an inadequate basis for accounting for heat and momentum transport rates across this near-wall sublayer. While one-equation models that introduce as a dependent variable a turbulent velocity scale (the square root of the turbulence energy, k) achieve moderate success in separated flows, most workers now look to two-equation eddy-viscosity models to try and account for situations where the near-wall sublayer is far from local equilibrium [4-6]. However, even at this level, the use of an isotropic eddy viscosity to link the stress and strain fields intrinsically limits the types of flow phenomena that can be successfully tackled: diffusion-controlled transition or laminarization caused by flow acceleration on arbitrary, rotating surfaces are associated with the selective amplification or damping of certain of the Reynolds-stress components relative to the others. Changes in the (scalar) turbulent kinetic energy cannot adequately characterize such subtleties.

It is thus to full second-moment closure that one must turn in seeking to account for such very complex but industrially very important flows. There have been numerous proposals for models of this type beginning with Donaldson's [7] early paper on transition. Often, however, the low-Reynolds-number aspects were either uninfluential in the limited comparisons made [8] or seemed to be the principal source of disagreement with experiment [9].

A common feature of virtually all previous models has been the idea of adding viscous "damping" in the low-Reynolds-number region. In Van Driest's model [1] the inclusion of his damping function changes the limiting turbulent viscosity variation near the wall from y^2 to y^4; without such damping the turbulent viscosity exceeds the molecular viscosity at a distance from the wall only 30% of that which experiment indicates. The argument of the damping function in Van Driest's case was y^+, the dimensionless wall distance. In two-equation models, which usually provide transport equations for the turbulence energy and its dissipation rate ϵ, the near-wall parameter is usually taken as the "turbulent Reynolds number" $R_t \equiv k^2/\nu\epsilon$. Again, a functional dependence on R_t is introduced to reduce the level of turbulent viscosity as the wall is approached.

Patel et al [6], in a review of ten variants of the low-Reynolds-number $k-\epsilon$ model, pointed out that in fact the terms dependent on the turbulent Reynolds number exerted an influence on the model at distances from the wall far greater than that at which molecular viscosity would normally be thought to be significant. The "far-reaching" viscous influence in the model was, according to Patel et al, trying to compensate for the fact that the wall damped fluctuations normal to the wall relative to horizontal ones by mechanisms not directly affected by viscosity. This suggestion can in fact be tested as follows. The stress-strain relation for the $k-\epsilon$ eddy-viscosity model in low-Reynolds-number regions may be written:

$$-\overline{uv} = 0.09 f_\mu (k^2/\epsilon) \frac{\partial U}{\partial y} \tag{1}$$

where f_μ is the "viscous" damping function that approaches the value unity at large values of R_t. Patel et al estimated the distribution of f_μ from measured values of all the other quantities in equation (1). Now, in local equilibrium, the (high Reynolds number) second-moment closure of Gibson and Launder [10] itself leads to a simple eddy-viscosity-like relationship for $\overline{uv}$:

$$-\overline{uv} = 0.263 \frac{\overline{v^2}}{k} \frac{k^2}{\epsilon} \frac{\partial U}{\partial y} \tag{2}$$

Thus, one should be able to distinguish the extent to which f_μ was accounting for genuine viscous effects rather than just the damping of $\overline{v^2}/k$ by plotting $\overline{v^2}/k$ and $0.09f_\mu/0.263$ versus y^+. The result of this comparison is shown in Figure 1. What is evident is that, within experimental uncertainty, the two curves vary in the same way all the way to the wall itself! In other words, the rapid reduction in $\overline{uv}$ as the wall is approached is due to preferential damping of $\overline{v^2}$ rather than to direct viscous effects.

Now, the near-wall sublayer is not, of course, a region of local equilibrium, but the above demonstration strongly suggests that the key to predicting the near-wall behaviour with a low-Reynolds-number second-moment closure lies in achieving the correct diminution of $\overline{v^2}/k$ as the wall is approached. Our discoveries so far fall short of providing a turbulence model for general application in low-Reynolds-number regions. Nevertheless, some interesting results and experiences have emerged. Our aim has been to try and respect more closely than have previous studies certain exact limiting forms that real turbulence must approach at the wall itself. Moreover, in consonance with Figure 1, we have eschewed the use of the turbulent Reynolds number as a parameter to modify processes that display no clear evidence of depending on viscosity. In particular we have permitted no such influence in the model of the pressure-strain process, a step which sets our work apart from virtually all previous modelling efforts. It may yet be the case that such a Reynolds-number dependence will have to be introduced into our model before fully satisfactory behaviour is produced but, in that event, it will be as a final measure rather than a first expedient and the influence of the term will be relatively small.

2. TOWARDS A SECOND-MOMENT CLOSURE

2.1 The Exact $\overline{u_i u_j}$ and ϵ Equations and the Limiting Near-wall Behaviour

Exact equations for the transport of the kinematic Reynolds stress and the turbulence energy dissipation rate may readily be obtained by taking appropriate moments of the Navier-Stokes equations (c.f. ref. [7,9,11]). These may be written for a stationary turbulent flow as:

$$U_k \frac{\partial \overline{u_i u_j}}{\partial x_k} = \underbrace{- \left\{ \overline{u_j u_k} \frac{\partial U_i}{\partial x_k} + \overline{u_i u_k} \frac{\partial U_j}{\partial x_k} \right\}}_{P_{ij}} \underbrace{- 2\nu \overline{\frac{\partial u_i}{\partial x_k} \frac{\partial u_j}{\partial x_k}}}_{\epsilon_{ij}}$$

$$- \frac{1}{\rho} \left[\overline{u_j \frac{\partial p}{\partial x_i}} + \overline{u_i \frac{\partial p}{\partial x_j}} \right] - \frac{\partial}{\partial x_k} \left[\overline{u_i u_j u_k} - \nu \frac{\partial \overline{u_i u_j}}{\partial x_k} \right] \tag{3}$$

$$U_k \frac{\partial \epsilon}{\partial x_k} = - \Big[\underbrace{\epsilon_{ik}}_{(a)} + \underbrace{2\nu \overline{\frac{\partial u_\ell}{\partial x_i} \frac{\partial u_\ell}{\partial x_k}}}_{(b)} \Big] \frac{\partial U_i}{\partial x_k} \underbrace{- 2\nu \overline{u_k \frac{\partial u_i}{\partial x_\ell}} \frac{\partial^2 U_i}{\partial x_k \partial x_\ell}}_{(c)}$$

$$\underbrace{- 2\nu \overline{\frac{\partial u_i}{\partial x_k} \frac{\partial u_i}{\partial x_\ell} \frac{\partial u_k}{\partial x_\ell}}}_{(d)} \underbrace{- 2\nu^2 \overline{\left[\frac{\partial^2 u_i}{\partial x_k \partial x_\ell} \right]^2}}_{(e)} \tag{4}$$

$$- \frac{\partial}{\partial x_k} \Big\{ \underbrace{\nu \overline{u_k \frac{\partial u_i}{\partial x_\ell} \frac{\partial u_i}{\partial x_\ell}}}_{(f)} + \underbrace{2 \frac{\nu}{\rho} \overline{\frac{\partial u_k}{\partial x_\ell} \frac{\partial p}{\partial x_\ell}}}_{(g)} - \underbrace{\nu \frac{\partial \epsilon}{\partial x_k}}_{(h)} \Big\}$$

where formally $\epsilon \equiv \nu \overline{(\partial u_i/\partial x_k)^2}$.

The relative near-wall variation of the Reynolds stresses may be inferred from a Taylor-series expansion about the wall which we suppose to lie in the plane $x_2 = 0$. Since, by continuity, $\partial u_i/\partial x_i = 0$ and, at the wall, $\partial u_1/\partial x_1$ and $\partial u_3/\partial x_3$ both vanish, it follows that $\partial u_2/\partial x_2$ vanishes identically at $x_2 = 0$. However, $\partial u_1/\partial x_2$ and $\partial u_3/\partial x_2$ are ordinarily not zero there; consequently we may write

$$\begin{aligned} u_1 &= a_1 x_2 + a_2 x_2^2 + \ldots\ldots \\ u_2 &= \qquad\quad\; b_2 x_2^2 + \ldots\ldots \\ u_3 &= c_1 x_2^2 + c_2 x_2^2 + \ldots\ldots \end{aligned} \tag{5}$$

where the coefficients of x_2 are functions of time with zero mean value. Equation (5) leads immediately to

$$\overline{u_1^2} = \overline{a_1^2}\, x_2^2 + \ldots\ldots$$

$$\overline{u_3^2} = \overline{c_1^2}\, x_2^2 + \ldots\ldots$$

$$\overline{u_2^2} = \overline{b_2^2}\, x_2^4 + \ldots\ldots \tag{6}$$

$$k = \tfrac{1}{2}(\overline{a_1^2} + \overline{c_1^2})x_2^2 + \ldots\ldots$$

$$\overline{u_1u_2} = \overline{a_1b_2}\, x_2^3 + \ldots\ldots$$

where k is the turbulent kinetic energy. As will be seen later, one of the principal problems in devising a model for the $\overline{u_iu_j}$ equation is that of finding a form that makes $\overline{u_2^2}/k$ vary as x_2^2 near the wall, as indicated by (6).

Analogous results may be obtained for the energy dissipation rate and the ϵ_{ij} components. We note in particular that

$$\epsilon\Big|_o = \nu(\overline{a_1^2} + \overline{c_1^2}) = 2\nu\left[\frac{\partial k^{\frac{1}{2}}}{\partial x_2}\right]^2\Big|_o \tag{7}$$

Equation (7) provides the correct, albeit seldom used, wall boundary condition for the ϵ equation (Launder and Shima [12]). It also suggests that the modified dissipation rate

$$\tilde{\epsilon} \equiv \epsilon - 2\nu(\partial k^{\frac{1}{2}}/\partial x_2)^2 \tag{8}$$

which vanishes at the wall might well be useful in turbulence modelling, Hanjalic and Launder [11]. The components of ϵ_{ij} at the wall are easily shown [13] to be related to ϵ and the Reynolds stresses by:

$$\frac{\epsilon_{11}}{\overline{u_1^2}} = \frac{\epsilon_{33}}{\overline{u_3^2}} = \frac{1}{4}\frac{\epsilon_{22}}{\overline{u_2^2}} = \frac{1}{2}\frac{\epsilon_{12}}{\overline{u_1u_2}} = \frac{\epsilon}{k} \tag{9}$$

Thus, the supposition frequently made that, at the wall:

$$\epsilon_{ij} = \frac{\overline{u_iu_j}}{k}\,\epsilon \tag{10}$$

would underestimate the dissipation rate of $\overline{u_2^2}$ by a factor of 4 (if the other components in the equation were correct).

With these preliminary results available we turn to the unknown processes in equations (3) and (4).

2.2 Closing the $\overline{u_iu_j}$ Equation

2.2.1 The pressure–strain interactions

It is evident from equation (3) that the fluctuating pressure interaction $(\overline{u_i\ \partial p/\partial x_j + u_j\ \partial p/\partial x_i})$ vanishes at the wall in all components. It is usual to re-group this term into two parts:

$$-\frac{1}{\rho}\left[\overline{u_i \frac{\partial p}{\partial x_j}} + \overline{u_j \frac{\partial p}{\partial x_i}}\right] = \underbrace{\overline{\frac{p}{\rho}\left[\frac{\partial u_i}{\partial x_j} + \frac{\partial u_j}{\partial x_i}\right]}}_{\varphi_{ij}} - \frac{1}{\rho}\frac{\partial}{\partial x_k}(\overline{pu_j}\,\delta_{ik} + \overline{pu_i}\,\delta_{jk}) \tag{11}$$

the first of these being interpretable as the redistributive pressure-strain correlation φ_{ij} and the second as pressure-driven diffusion. Now, for the normal stresses, both the redistributive and diffusive parts of (11) vanish at the wall. This is not evidently so for the shear stress (for example, $\overline{p\partial u_1/\partial x_2}$ is not obviously zero at $x_2 = 0$), but there seem no drawbacks from making both terms on the right side of (11) vanish there since their sum should evidently do so.

While, in second-moment closures, it is indeed quite usual to make the pressure-strain process vanish at the wall, the common means of bringing this about is by incorporating in the model some functional dependence on turbulent Reynolds number that makes the term go to zero with R_t [8,11,12]. However, the Poisson equation for pressure contains no viscous term and, in any event, φ_{ij} should also vanish at an interface (for example, at a horizontal water/air surface) in the normal stress component perpendicular to this interface where the turbulent Reynolds number R_t remains high. Lumley [14] has pointed out that such a vanishing can be achieved by noting that the difference between the second and third invariants of the anisotropic Reynolds stress takes a constant value in two-component turbulence irrespective of how the intensities of fluctuations in the plane normal to that in which u_i vanishes are distributed. Thus, with the definitions

$$a_{ij} \equiv (\overline{u_i u_j} - \frac{1}{3}\delta_{ij}\,\overline{u_k u_k})/k; \quad A_2 \equiv a_{ij}\,a_{ij}; \quad A_3 \equiv a_{ij}\,a_{jk}\,a_{ki}$$

we readily find that $(A_2 - A_3) = \frac{8}{9}$ in the two-component limit. Hence, the "flatness" parameter $A \equiv \{1 - \frac{9}{8}(A_2 - A_3)\}$ ranges from unity in isotropic turbulence to zero at a rigid wall or interface. This parameter thus offers a way of constructing model forms that vanish at the wall [14] without recourse to the turbulent Reynolds number.

The pressure-strain process is traditionally modelled in two parts, one arising from purely turbulent interactions (φ_{ij1}) and the other from mean-strain influences on the fluctuating pressure field (φ_{ij2}). The forms adopted in the present study may be written

$$\varphi_{ij1} = c_1\,(AA_2)^{\frac{1}{2}}\left|a_{ij} + 1.1\left[a_{ik}\,a_{kj} - \frac{1}{3}\delta_{ij}\,A_2\right]\right|\epsilon - a_{ij}\,\epsilon \tag{12}$$

$$\begin{aligned}\varphi_{ij2} = {} & -0.6(P_{ij} - \frac{1}{3}\delta_{ij}\,P_{kk}) + 0.3\epsilon a_{ij}(P_{kk}/\epsilon) - 0.2(R_{im}P_{mj} + R_{mj}P_{im}) \\ & - 0.4k\,R_{ii}R_{kj}S_{ik} - r[A_2(P_{ij} - D_{ij}) + 3a_{ii}a_{kj}(P_{ik} - D_{ik})]\end{aligned} \tag{13}$$

These representations were evolved by the authors in a preliminary exploration of distorted homogeneous plane distortions and shear flows [15] and have subsequently been found to bring significant improvements to the prediction of the far wake [16], the round jet [17] and swirling jets [18] over that previously obtained with the Gibson-Launder model [10]. The process φ_{ij1} is strongly non-linear and, due to the presence of A in its coefficient, vanishes at a wall.

Equation (13) is an elaboration of the so-called quasi-isotropic model [19] and was obtained by following a similar route to that proposed by Shih and Lumley [20]. The quantity φ_{ij2} was first written as

$$\varphi_{ij2} = (a_{\ell j}^{mi} + a_{\ell i}^{mj})\frac{\partial U_\ell}{\partial x_m} \tag{15}$$

and the fourth-rank tensor $a_{\ell j}^{mi}$ represented by the most general form containing up to cubic products in the anisotropic stress a_{ij}. The tensor contains twenty unknown coefficients, all except of two of which may be determined by requiring that the tensor (like the exact form) be traceless, that $a_{kk}^{mi} = 2\overline{u_m u_i}$ and that φ_{222} should vanish if $\overline{u_2 u_2}$ equalled zero [14]. In fact, equation (13) contains only one freely assignable coefficient (r), not two. The second coefficient multiplied such a huge array of terms as to make the model extremely unattractive for computing inhomogeneous flows and for this reason its value has been taken as zero.† As in the free-shear-flow explorations noted above, r has been taken as 0.6 in the present study.

A further contribution has normally been included in the model of φ_{ij} associated with pressure reflections from the wall [10,19,21]. In the past this process has been seen as important in depressing the level of velocity fluctuations perpendicular to the wall relative to those parallel to it. However, the versions of φ_{ij1} and φ_{ij2} employed in [10] and [19] do not satisfy the two-component limit. One might suppose that it was this failure, to a large extent, that necessitated the introduction of a near-wall correction – or (at the least) required a spuriously large contribution from this source. In the present study our initial computations are made without a wall "echo" term; thereafter the effect of including the wall-reflection model of [10] is investigated.

2.2.2 The dissipation components ϵ_{ij}

Equation (9) has indicated that the components of ϵ_{ij} exhibit a curious non-isotropic pattern at the wall itself. The following form, proposed in [13], captures this limiting behaviour:

$$\epsilon_{ij} = \frac{\epsilon}{k} (\overline{u_i u_j} + \overline{u_i u_k}\, n_k n_j + \overline{u_j u_k}\, n_k n_i + \delta_{ij}\, \overline{u_k u_\ell}\, n_k n_\ell) \quad (16)$$

Here n_j denotes a unit vector normal to a wall. In arriving at the limiting form, use has been made of the fact that $\overline{u_2^2}/k = 0$ at the wall. It is assumed that there is a gradual changeover from this limit to local isotropy as the turbulent Reynolds number becomes indefinitely large. Thus we write

$$\epsilon_{ij} = f_\epsilon\, \epsilon_{ij}^* + (1 - f_\epsilon) \frac{2}{3} \delta_{ij}\, \epsilon \quad (17)$$

where the standard form of f_ϵ is exp – $R_t/20$ [22] (though the behaviour from using other forms will be shown) and

$$\epsilon_{ij}^* = \frac{\epsilon}{k} (\overline{u_i u_j} + \overline{u_i u_k}\, n_k n_j + \overline{u_j u_k}\, n_k n_i + \delta_{ij}\, \overline{u_k u_\ell}\, n_k n_\ell) \Big/ \left[1 + \frac{5}{2} \frac{\overline{u_p u_q}\, n_p n_q}{k}\right] \quad (18)$$

The right side of (18) is the same as (17) except for the appearance in the denominator of the factor $(1 + \frac{5}{2}\, \overline{u_2^2}/k)$. This addition is necessary because, except at the wall itself, $\overline{u_2^2}/k$ is not zero, and so without it equation (17) would not contract properly [22].

2.2.3 Diffusion, d_{ij}

While some very elaborate proposals for handling stress diffusion have been made, [14,25,26], most second-moment computations have used rather simple models of d_{ij} on the grounds that this process had little effect on the predicted behaviour of the stress field, still less on the mean flow. The writers and their colleagues [27,28] have thus usually adopted the Daly-Harlow proposal

---

† Fu [18] has made a few computations for homogeneous shear flows with the complete form and has achieved some improvement for the two cases considered.

$$d_{ij} = \frac{\partial}{\partial x_k} \left[c_s \overline{u_k u_\ell} \frac{k}{\epsilon} \frac{\partial}{\partial x_\ell} \overline{u_i u_j} \right] \tag{19}$$

an approximation that is implicitly taken to account for both pressure contributions and velocity–driven transport. Other recent contributions have employed even simpler formulations, e.g. [29]. While equation (19) has been retained in the present study, a few very recent runs have been made including broadly Lumley's [30] isotropic model for pressure transport

$$-\frac{\overline{pu_i}}{\rho} = c_\varphi \overline{u_j u_j u_i} \tag{20}$$

According to equation (20), the process effectively opposes and makes more isotropic the transport arising from velocity fluctuations, which seems in character with what we know of the role of pressure fluctuations in other correlations. The coefficient c_φ equals 0.2 in isotropic turbulence. In our computations we have used a smaller value 0.15 because, as noted above, equation (19) for $\overline{u_i u_j u_k}$ is thought at least partially to be accounting for both velocity and pressure diffusion. The process of pressure transport is one we now believe to be of substantial importance in shaping the sublayer turbulence structure. Indeed, the above model of pressure transport can mean that, in the $\overline{u_2^2}$ budget, the total contribution of pressure interactions is to reduce the level of $\overline{u_2^2}$ in the sublayer even though the x_2 velocity fluctuations are less than in other directions. Although such an eventuality seems intuitively unlikely, we note that it is indeed what the direct numerical simulations of Moser and Moin [31] indicate.

2.3 Closing the ϵ Equation

In our work reported hereunder we have, except where otherwise noted, adopted the form of the dissipation rate equation proposed by Kebede et al [22] (which in turn evolved from versions developed in [23], [24] and [11]):

$$\frac{D\epsilon}{Dt} = c_{\epsilon 1} \frac{\epsilon}{k} P_{kk} - c_{\epsilon 2} \frac{\epsilon \tilde{\epsilon}}{k} + c_{\epsilon 3} 2\nu \frac{k}{\epsilon} \overline{u_\ell u_k} \frac{\partial^2 U_i}{\partial x_\ell \partial x_j} \frac{\partial^2 U_i}{\partial x_k \partial x_j} + \frac{\partial}{\partial x_k} \left[\nu \frac{\partial \epsilon}{\partial x_k} + c_\epsilon \frac{k}{\epsilon} \overline{u_k u_\ell} \frac{\partial \epsilon}{\partial x_\ell} \right] \tag{21}$$

At high Reynolds numbers the exact dissipation equation (equation (4)) is of no direct help in establishing a suitable closed form for the ϵ equation because it brings into prominence fine–scale interactions (whereas the dissipation rate is believed to be controlled by larger–scale processes whose actions create the dynamically passive dissipative scales). In the low–Reynolds–number limit, however, the energy–containing and dissipative scales are one and the same thing, so the equation can throw some light on the asymptotic behaviour as the wall is approached.

If turbulence is locally isotropic, the terms in the exact ϵ equation (equation (4)) containing the mean strain vanish; moreover, the terms also vanish at the wall itself. Nevertheless, over at least a portion of the viscosity–affected sublayer, the terms seem likely to be important. We proposed above a form for ϵ_{ik} (equation (17)) and this is simply retained in treating element (a) from equation (4). Note that, since $\partial U_1/\partial x_2$ will usually be the only significant strain component in the sublayer, the asymptotic form is ϵP_{kk}. The benefit of this exact result is diminished by the fact that term (b) – which is similar in appearance – does not have an exactly determinable limit. Its largest elements appear to be:

$$- 2\nu \left[\overline{\frac{\partial u_1}{\partial x_1} \frac{\partial u_1}{\partial x_2}} + \overline{\frac{\partial u_3}{\partial x_1} \frac{\partial u_3}{\partial x_2}}\right] \frac{\partial U_1}{\partial x_2}$$

Continuity suggests that the first of these will scale with

$$- 2\nu \overline{\frac{\partial u_2}{\partial x_2} \frac{\partial u_1}{\partial x_2}} \frac{\partial U_1}{\partial x_2}$$

which also tends to ϵP_{kk} at the wall; but now neither the magnitude of the coefficient nor its sign are known. From these considerations we may conclude that the coefficient $c_{\epsilon 1}$ in equation (21) will vary within the low-Reynolds-number region, though not necessarily monotonically. For the moment, following [22], we have simply neglected term (b) in the modelled ϵ equation while the variation in $c_{\epsilon 1}$ from to 1 to 0.72 (the usual empirical value adopted at high Reynolds number) is achieved via the same function f_ϵ used in equation (17).

The third term in equation (21) arises from the application of the generalized gradient diffusion hypothesis to the corresponding term in equation (4). As in [22], the value 0.3 is retained for $c_{\epsilon 3}$.

An interesting, controversial feature of the direct numerical simulations of near-wall turbulence is that the peak value of ϵ occurs at the wall [31]. A term-by-term examination of the elements in the exact ϵ equation indicates that only the sink term (e) and the diffusion processes (g and h) can be non-zero there. Thus, in the immediate vicinity of the surface, there must be an energy inflow by diffusion that just balances the dissipative loss. Viscous diffusion, however, proceeds <u>down</u> the gradient of the diffused quantity - so one would ordinarily expect that the slope of ϵ at the wall would be <u>positive</u> (i.e. that the maximum value of ϵ was <u>not</u> at the wall). The explanation of the paradox (if the numerical simulations prove to be accurate) must be that the pressure diffusion need not vanish at $x_2 = 0$ (at the wall the term comprises

$- \frac{2\nu}{\rho} \left[\overline{\frac{\partial u_1}{\partial x_2} \frac{\partial^2 p}{\partial x_1 \partial x_2}} + \overline{\frac{\partial u_3}{\partial x_2} \frac{\partial^2 p}{\partial x_3 \partial x_2}}\right]$) and that it gives rise to an up-gradient transfer. The following simple approximation has been proposed [32]:

$$- \frac{2\nu}{\rho} \frac{\partial}{\partial x_k} \left[\overline{\frac{\partial p}{\partial x_i} \frac{\partial u_k}{\partial x_i}}\right] = 2c_{\epsilon 4} \frac{\nu \partial}{\partial x_k} \left[\frac{\tilde{\epsilon}}{k} \frac{\partial k}{\partial x_k}\right] \tag{22}$$

which meets at least some of the basic requirements. Some of our runs have included this term.

For ease of reference, the table gives a complete statement of the initial values of the empirical coefficients and the function of R_t adopted in this study. The effects of departing from these values are shown in the figures; revised coefficients are listed in the captions.

c_1	r	c_1'	c_2'	c_s	c_φ	c_ϵ	$c_{\epsilon 1}$	$c_{\epsilon 2}$	$c_{\epsilon 3}$	$c_{\epsilon 4}$	f_ϵ
3.05	0.68	0	0	0.22	0	0.18	0.72	1.90	0.3	0	$\exp - R_t/20$

3. APPLICATION AND ASSESSMENT OF THE MODEL

The test case considered has been that of fully-developed flow between parallel planes principally at a Reynolds number of 5.0 x 10^4, i.e. a sufficiently high value for there to be no significant variation of shear stress across the viscosity affected sublayer. The equations were solved employing a version of the parabolic solver PASSABLE [33] in which the turbulence model described above had been introduced. A total of 90 nodes was used to cover the domain between the wall and the flow symmetry plane with about half of these covering the low-Reynolds-number region. The main focus of the computations has been on the predicted variation of $\overline{v^2}/k$ in the vicinity of the wall. This dimensionless characteristic is mainly affected by the form of the $\overline{u_i u_j}$ equations, so the considerable weaknesses of the dissipation rate equation have little influence on the profile. The behaviour presented in Figure 2 shows the effect of various changes in the model.

The first computations were attempted employing equation (10) as the limiting dissipation rate as $R_t \to 0$. No converged solution to the system of equations could then be obtained, however. Although this problem could probably have been overcome, no serious attempt was made in this direction. It was recognized that, while equation (12) for φ_{ij_1} was compatible with the exact two-component limit, it did not enforce it. It thus seemed essential to ensure that this process vanished at the wall independently of $\overline{u_2^2}/k$ vanishing. Now, the quantity $\tilde{\epsilon}$ introduced above ($\equiv \epsilon - 2\nu(\partial k^{\frac{1}{2}}/\partial x_2^2)^2$) does vanish at the wall and we have thus simply replaced ϵ in equation (12) by $\tilde{\epsilon}$. This step removed the problem of instability but gave the unexpected result shown in Figure 2(a): as the wall was approached, $\overline{u_2^2}/k$ actually increased rather than decreased. The next step was to replace equation (12) – which, though widely used, was known to be wrong – by equation (18), which gave the exact limiting form at the wall. This substitution in fact brings about a great change in the profile of $\overline{u_2^2}$ as shown in Figure 2(a): there is now a smooth decrease in $\overline{u_2^2}/k$ from $x_2^+ \simeq 20$ to the wall. This region, we note, is that where $\partial k/\partial x_2$ is large and thus where $\tilde{\epsilon}$ progressively differs from ϵ. All subsequent computations have retained the two replacements introduced in Figure 2(a).

Despite the improvement discussed above, the difference between the measured and experimental behaviour was still great, the computed level of $\overline{u_2^2}/k$ being several times too large across most of the low-Reynolds sublayer. The question to be addressed was thus how to depress $\overline{u_2^2}$ over the region $x_2^+ < 60$. The obvious first step was to modify the function f_ϵ. The results of employing two variants of the original f_ϵ function are shown in Figure 2(b).

$$f_\epsilon = \exp - R_t/60 \tag{23}$$

$$f_\epsilon = \begin{cases} (1.0 - R_t/60) & \text{for } R_t < 60 \\ 1.0 & \text{for } R_t \geqslant 60 \end{cases} \tag{24}$$

The former essentially stretches viscous effects on dissipation to values of R_t three times larger than the original, while the latter inserts a linear changeover in place of the original exponential form. The alternative versions do bring minor improvements in the variation of $\overline{u_2^2}/k$ but they are hardly worth mentioning compared with the difference that remains from the experimental data. For subsequent runs we have retained the linear function as this was again more stable.

At this point the influence of wall-reflection agencies on φ_{ij} was explored. Basically, the model of ref.[10] was adopted except that the two empirical coefficients appearing in it were made roughly 50% smaller; if that adjustment had not been made, a far too large wall-reflection effect would have been produced (Ref.[10] had adopted a much simpler representation of φ_{ij_1} and φ_{ij_2} that gives $\overline{u_2^2} = \overline{u_3^2}$ in a simple shear without wall-reflection terms). A further variant explored was the replacement of $(k^{3/2}/\tilde{\epsilon}x_2)$ by $(k\overline{u_2^2}^{\frac{1}{2}}/0.64\tilde{\epsilon}x_2)$ as the wall-proximity scaling factor. The latter was felt to be better on physical grounds (for example, it allows the free surface jet to be tackled in precisely the same manner as a near-wall flow) as well as giving better numerical stability. The outcome of these calculations appears in Figure 2(c). The addition of wall-reflection terms evidently drops the level of $\overline{u_2^2}/k$ for values of $x_2^+ > 10$, but not in the manner exhibited by the data, the terms reducing $\overline{u_2^2}/k$ by a nearly uniform amount. There is, in fact, an insignificant difference between the alternative choices of wall-scaling factor and for x_2^+ below 10 the process as a whole has a negligible effect.

The final explorations made so far concern the inclusion of pressure-diffusion effects, here modelled by equation (20). The inclusion of the term gives rise to a good deal of numerical instability due, at least in part, to the overly simple expression that emerges for $\overline{u_i u_i u_j}$ (i.e. equation (19). From Figure 2(d), this step has significantly modified the profile of $\overline{u_2^2}/k$ near the wall, depressing considerably the predicted levels for $2 < x_2^+ < 15$. There remains, of course, a large discrepancy with the experimental data but, with a more realistic model for $\overline{u_i u_i u_j}$, it is possible that much better accord may be achieved. For example, the exact equation for $\overline{u_i u_j u_k}$ contains mean strain generation terms (see [23]) whose effects are implicitly ignored. These terms appear to make a substantial contribution in the buffer region.

It has been remarked above that changes in the ϵ transport equation have little effect on the profile of $\overline{u_2^2}/k$. They do, of course, affect the profile of $\overline{u_2^2}$ itself and the mean velocity distribution. Figure 3 shows changes in the near-wall ϵ profile brought about by including the simple proposal for pressure-induced transport of ϵ, equation (22). The inclusion raises the wall value of ϵ while reducing the level further from the wall. The net effect is to raise the mean level of the Reynolds stress within the sublayer causing a marked reduction in the additive constant in the conventional log-law representation of the mean velocity, Figure 4. Even with the inclusion of this up-gradient diffusion process, however, the computed level of ϵ shows a sharp decrease between $y^+ = 10$ and the wall, in contrast to the full simulations of Moser and Moin [31] and Spalart [34] and the LDA measurements of Johansson and Karlsson [35] (the last two for the flat-plate boundary layer). A possible remedy would appear to be to make $c_{\epsilon 2}$ diminish as Re_t became small. Such is the non-linearity of the ϵ equation, however, that a small reduction produces the *opposite* effect to that desired, while a larger reduction renders the equation unacceptably unstable from a numerical standpoint. At the time of writing, the most effective way we have found of raising the wall value of ϵ is by ***reducing*** the source term coefficient $c_{\epsilon 3}$. Again the effect is opposite from what intuition would suggest. Evidently there is a good deal still to be learnt before a satisfactory equation for ϵ can be devised for the near-wall sublayer.

CONCLUSIONS

While the research reported is still some way away from producing a satisfactory model of the low-Reynolds-number region, a number of important results have emerged:

(i) The use of a strongly non-isotropic dissipation (ϵ_{ij}) satisfying the exact limits and the employment of $k/\tilde{\epsilon}$ (rather than k/ϵ) as a turbulent time scale are important factors in making $\overline{u_2^2}/k$ tend to zero as the wall is approached. On their own, however, their effect does not extend sufficiently far from the wall.

(ii) While some "wall-reflection" process still seems desirable, with the new models of φ_{ij1} and φ_{ij2} adopted in this study the contribution required from such a correction is much less than needed in with earlier models.

(iii) It appears likely that pressure transport, producing effectively an up-gradient diffusion of $\overline{u_2^2}$ in parts of the near-wall sublayer, is influential in depressing the levels of $\overline{u_2^2}/k$. This conclusion points to the need to adopt more realistic models of $\overline{u_i u_j u_k}$ within the sublayer than hitherto.

(iv) In the dissipation rate equation, creating a maximum of ϵ at the wall (a feature reported in the exact simulations), requires the explicit inclusion of a model for pressure transport producing a strong up-gradient transfer. Modifications are also needed to the currently adopted ϵ source/sink terms. However, the equation for ϵ is so non-linear that evolving a suitable form may prove difficult.

On the motivating question for this research of whether the turbulent Reynolds number can be eliminated from the pressure-containing processes, it is still too early to say. Modifications to the diffusion processes implied by points (iii) and (iv) could conceivably bring the predicted distribution of $\overline{u_2^2}$ into line with the observed behaviour. Of course, the amount of change needed is large and it may well be that conformity with data will in fact demand a measure of Reynolds-number damping. If that proves to be the case, however, the degree of control exercised by such damping will be considerably less than in earlier models. Getting the correct relative contributions of the many complex processes in low-Reynolds-number turbulence is, we believe, vital to tackling the many industrially important problems associated with either incompletely turbulent flows or with strongly augmented turbulence levels as in impinging and separated flows.

ACKNOWLEDGEMENTS

We are pleased to acknowledge the support of this research by the SERC under grant GR/D82852. The manuscript has been prepared to camera-ready standards against a very tight deadline by Mrs. L.J. Ball.

Authors' names appear alphabetically.

REFERENCES

1. Van Driest, E.R. J. Aero. Sci., vol. 23, p. 1007, 1956.

2. Koosinlin, M.L., Launder, B.E. and Sharma, B.I. ASME J. Heat Transfer, vol. 96, p. 204, 1974.

3. Azzola, J., Humphrey, J.A.C., Iacovides, H. and Launder, B.E. ASME J. Fluids Engrg., vol. 108, p. 214, 1986.

4. Jones, W.P. and Launder, B.E. Int. J. Heat Mass Trans., vol. 15, p. 301, 1972.

5. Tanaka, H., Kawamura, H., Tateno, A. and Hatamiya, A. ASME J. Heat Trans., vol. 104, p. 363, 1982.

6. Patel, V.C., Rodi, W. and Scheuerer, G. AIAA J., vol. 23, p. 1320, 1985.

7. Donaldson, C. du P. AIAA J., vol. 7, p. 271, 1969.

8. Prud'homme, M. and Elghobashi, S. Proc. 4th Turbulent Shear Flows Symp., Karlsruhe, Paper 1.7, 1983.

9. Daly, B.J. and Harlow, F.H. Phys. Fluids, vol. 13, p. 2364, 1970.

10. Gibson, M.M. and Launder, B.E. J. Fluid Mech., vol. 86, p. 491, 1978.

11. Hanjalic, K. and Launder, B.E. J. Fluid Mech., vol. 74, p. 593, 1976.

12. Launder, B.E. and Shima, N. "A Second-Moment Closure for the Near-Wall Sublayer: Development and Application," Rep. TFD/87/12, Mech. Eng. Dept.,

UMIST, Manchester, 1987. (Submitted to AIAA J.).

13. Launder, B.E. and Reynolds, W.C. Phys. Fluids, vol. 26, p. 1157, 1983.

14. Lumley, J.L. Advances in Applied Mechanics, vol. 18, p. 123, 1978.

15. Fu, S. Launder, B.E. and Tselepidakis, D.P. "Accommodating the Effects of High Strain Rates in Modelling the Pressure-Strain Correlation," Rep. TFD/87/5, Mech. Eng. Dept., UMIST, Manchester, 1987.

16. Nemouchi, Z. PhD Thesis, Faculty of Technology, Univ. Manchester, 1988.

17. Ince, N. Personal communication, 1988.

18. Fu, S. PhD Thesis, Faculty of Technology, Univ. Manchester, 1988.

19. Launder, B.E., Reece, G.J. and Rodi, W. J. Fluid Mech., vol. 67, p. 569, 1975.

20. Shih, T-H and Lumley, J.L. "Modelling of Pressure Correlation Terms in Reynolds Stress and Scalar Flux Equations," Rep. FDA-85-3, Sibley School of Mech. and Aerospace Eng., Cornell University, 1985.

21. Shih, T-H and Lumley, J.L. Physics of Fluids, vol. 29, 1986.

22. Kebede, W., Launder, B.E. and Younis, B.A. Proc. 5th Turbulent Shear Flows Symp., Cornell, p. 16.23, 1985.

23. Hanjalic, K. and Launder, B.E. J. Fluid Mech., vol. 52, ,p. 609, 1972.

24. Davidov, B.I. Dokl. Akad. Nauk. SSSR, vol. 136, p. 47, 1961.

25. Ettestad, D. and Lumley, J.L. Turbulent Shear Flows 4, pp. 87-101, Springer Verlag, Heidelberg, 1983.

26. Dekeyser, I. and Launder, B.E. Turbulent Shear Flows 4, pp. 102-117, Springer Verlag, Heidelberg, 1984.

27. Launder, B.E., Tselepidakis, D.P. and Younis, B.A. J. Fluid Mech., vol. 183, p. 63, 1987.

28. Fu, S., Launder, B.E. and Leschziner, M.A. Proc. 6th Turbulent Shear Flows Symp., Toulouse, Paper 17.6, 1987.

29. Jones, W.P. and Manners, A. Proc. 6th Turbulent Shear Flows Symp., Toulouse, Paper 17.7, 1987.

30. Lumley, J.L. "Prediction Methods for Turbulent Flow - Introduction," Lecture Series 76, Von Karman Inst., Rhode-St-Genese, Belgium, 1975.

31. Moser, R.D. and Moin, P. "Direct Numerical Simulation of Curved Channel Flow," NASA Rep. TM 85974, 1984.

32. Launder, B.E. "Low-Reynolds-Number Turbulence Near Walls," Rep. TFD/86/4, Mech. Eng. Dept., UMIST, Manchester, 1986.

33. Leschziner, M.A. "A User's Guide to the PASSABLE Computer Program," Mech. Eng. Dept. Report, UMIST, Manchester, 1981.

34. Spalart, P. "Direct Simulation of a Turbulent Boundary Layer up to R_θ = 1410," J. Fluid Mech., vol. 187, p. 61, 1988.

35. Johansson, T.G. and Karlsson, R.I. "The Energy Budget in the Near-Wall Region of a Turbulent Boundary Layer," Proc. 4th Int. Symp. Applic. of LDA to Fluid Mechanics, Lisbon, Portugal, 1988.

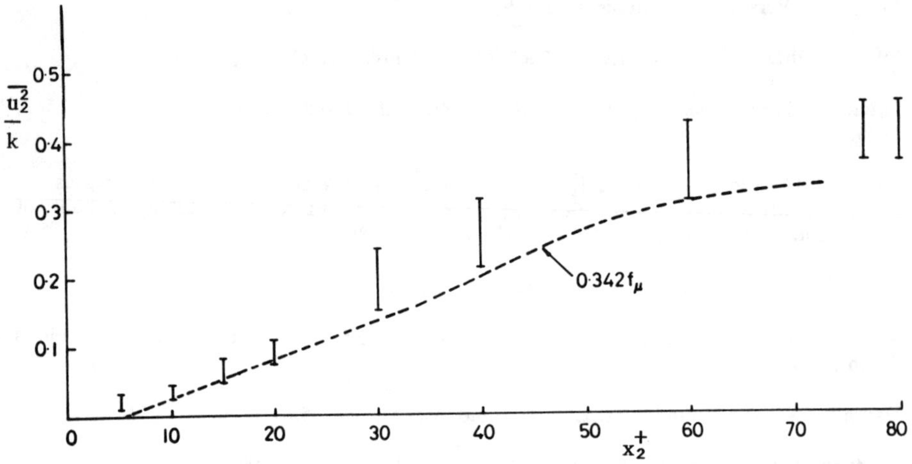

FIGURE 1. Variation of $\overline{u_2^2}/k$ across near-wall sublayer implied by G-L model in local equilibrium [10]

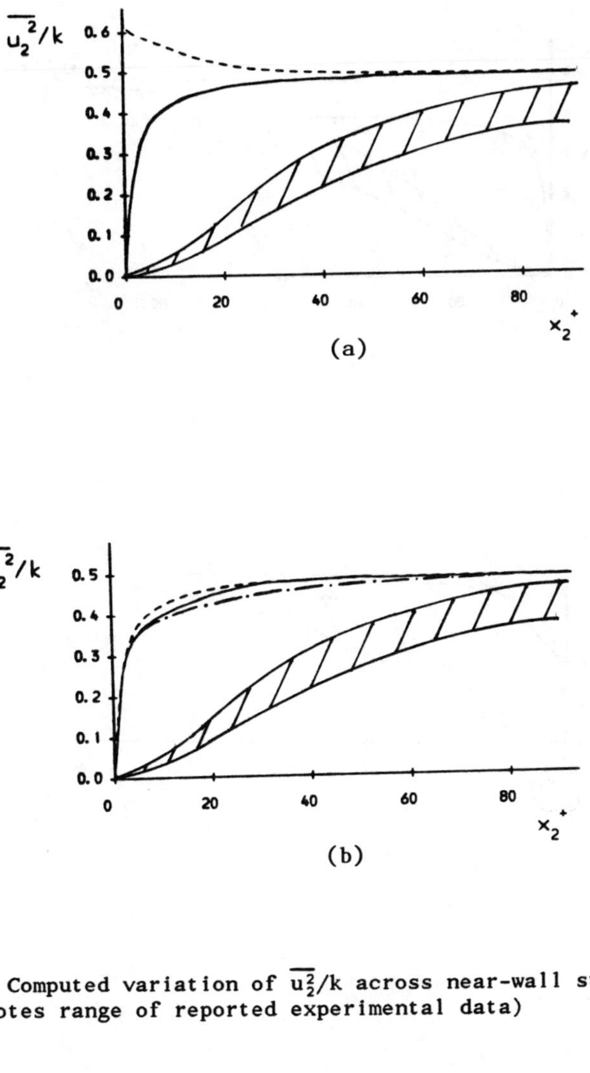

FIGURE 2. Computed variation of $\overline{u_2^2}/k$ across near-wall sublayer (Shaded region denotes range of reported experimental data)

(a) Effects of ϵ_{ij}

- - - - use of $\tilde{\epsilon}$ in place of ϵ in eq.(12)
——— use of ϵ in place of ϵ in eq.(12) plus eq.(18) in place of eq.(10)

(b) Effects of R_t in f_ϵ

- - - - solid line from Fig. 2(a) (f_ϵ = exp - $R_t/20$)
——— f_ϵ given by eq.(24); — · — f_ϵ given by eq.(23)

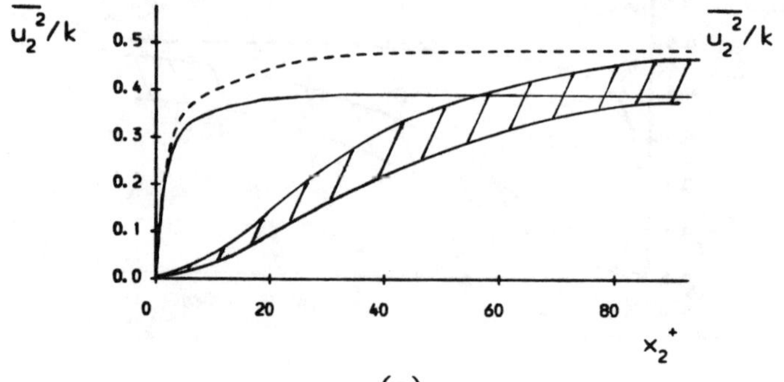

(c)

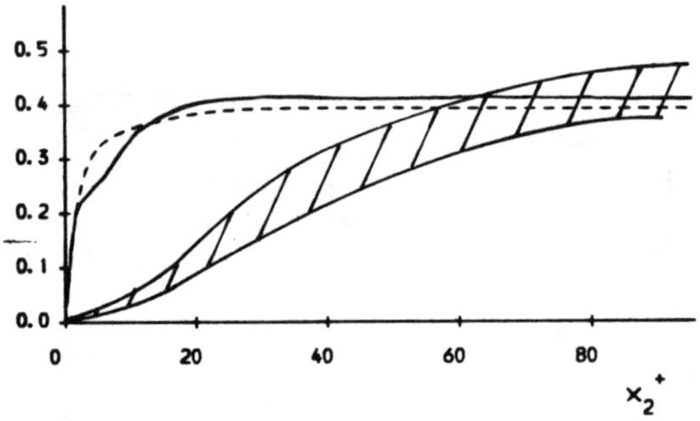

(d)

FIGURE 2 cont.

(c) Wall reflection effects

- - - - solid line from Fig. 2(b), $c_1' = c_2' = 0$
——— wall reflection model [10] included, $c_1' = 0.05$; $c_2' = 0.15$

(d) Pressure diffusion effects

- - - - solid line from Fig. 2(c), $c_\varphi = 0$
——— pressure-diffusion included, $c_\varphi = 0.15$

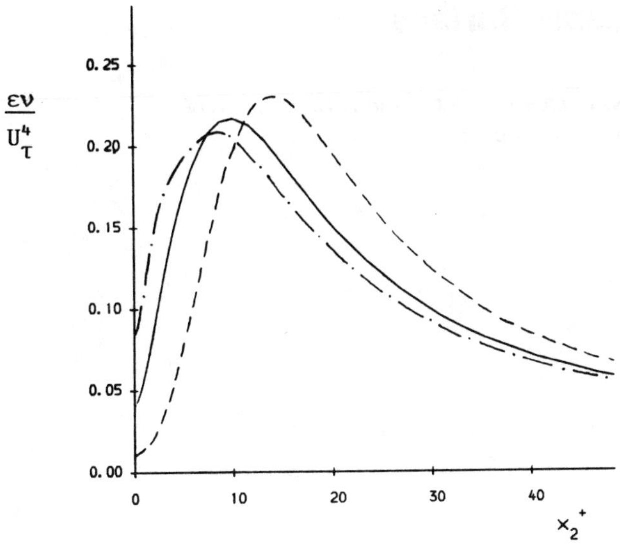

FIGURE 3. Computed near-wall ϵ profile

- - - - $c_{\epsilon 4} = 0$; ——— $c_{\epsilon 4} = 1.0$; —·— $c_{\epsilon 4} = 1.5$

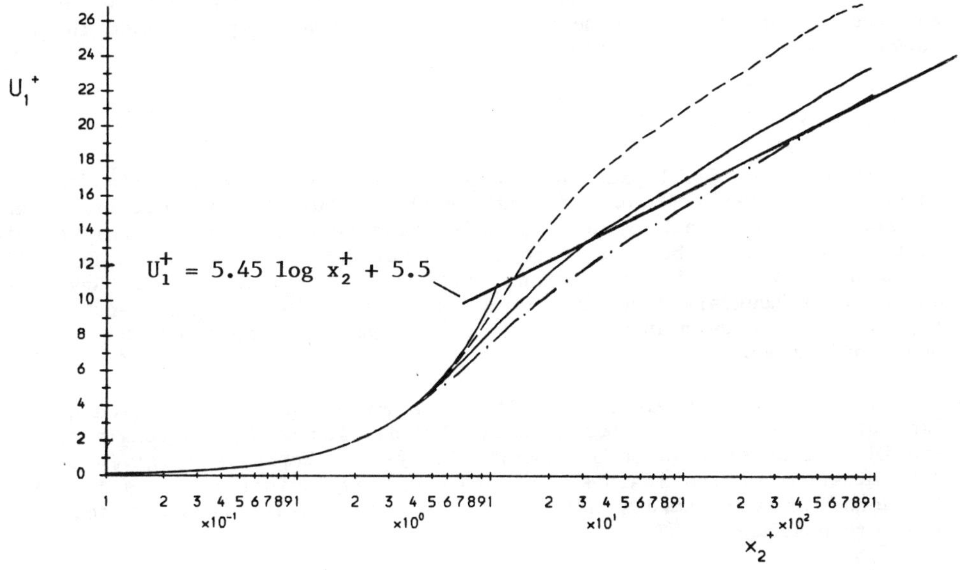

FIGURE 4. Mean velocity profile in wall coordinates
Key as Figure 3

Two-Parametric Models of a Boundary Layer on a Permeable Surface

A. I. LEONTIEV, V. P. ZABOLOTSKY, and P. S. ROGANOV
Moscow Higher Technical School
Moscow, USSR

ABSTRACT

This report summarizes the results of diagnostics of the turbulence fine structure in two-dimensional boundary layers developed under complex hydrodynamic and thermal boundary conditions, such as the presence of longitudinal pressure gradients and injection through a porous surface. Investigations were made using multichannel complex 55-M "DISA" probes. Several interesting hypotheses concerning turbulent transport mechanisms and the effect of disturbances on the Reynolds analogy are made in this work.

INTRODUCTION

At present, development of two-parameter and more complicated turbulence models that reproduce accurately the turbulent transport processes is slowed by the lack of a deep understanding concerning the nature of turbulence. This becomes particularly evident when considering wall flows developing under conditions that cause breakdown of the Reynolds analogy. A boundary layer developing in a region of strongly accelerating flow on a permeable surface is a typical example of the above-mentioned flows.

This paper contains the results of a purposeful investigation of the fine structure of the turbulent wall flow developing in a strongly accelerating region of flow on a permeable surface. Emphasis is given to the determination of the reasons that provoke flow laminarization and to the influence of the named conditions on the turbulent transport and heat loss processes.

The experimental investigation was carried out on a subsonic open-return wind tunnel. The schematic diagram and basic dimensions are given in Refs. [1,2]. The investigations were performed by a "hot-wire" technique, using a DISA 55-M hot-wire thermoanemometer with a specially designed three-filament probe with a filament thickness of 2.5 μm. The temperature fluctuation spectra and covariation measurements were corrected for the amplitude-frequency distortions due to thermal lag of the thermoanemometer probe filament [3].

DYNAMIC SCHEME OF THE FLOW

In Figures 1 and 2, representing velocity profiles, a considerable profile deformation in the strongly accelerating region of the flow is observable. In terms of the boundary layer thickness, the velocity profile appears more filled, which explains the substantial growth of the surface friction coefficient, C_f (Figure 3). These phenomena are accompanied by a considerable increase in the viscous sublayer thickness compared to the case of nongradient flow past a plate (Figure 4). Moreover, a region described by "law-of-trace" terms is not observed in the outer region of the "law-of-the-wall" coordinates velocity profile, indicating the cessation of the process of entrainment into the turbulent nucleus of the boundary layer. The mentioned phenomena are the main manifestations of the influence of the considerable negative pressure gradient on the wall turbulence structure. Let us try to analyze the reason for these phenomena, taking into account the peculiarities of the mathematical turbulence model used. The above-mentioned wall flow "laminarization" effects are explained, in our opinion, mainly by the increasing turbulence anisotropy caused by the negative pressure gradient that leads to the increasing role of normal stresses in the production of turbulence energy:

$$P = \overline{u'v'}\,\frac{\partial U}{\partial y} - \left(\overline{u'^2} - \overline{v'^2}\right)\frac{\partial U}{\partial x}$$

Hence it follows that the production of turbulence energy in the outer region of the boundary layer ceases, and the "production" component of the balance equation of the turbulent energy transfer transforms from the source into the "sink" (Figure 5). Therefore, two-parameter models using the "k" equation in modelling the flows in the accelerating region should consider both tangential and normal stress contributions to the production of turbulence energy. The importance of this assumption is confirmed by the analysis of the correlation of both components in the process of turbulence energy generation under the condition of air injection through a porous surface. The transverse flow produced by the injection turbulizes the wall region and reduces the velocity fluctuation anisotropy, i.e., the $(u'^2 - v'^2)$ difference; thus, normal stresses cease to play the "laminarizing" role in the process of turbulence energy production. Moreover, air injection causes a decrease in viscous sublayer thickness (Figure 4). It is reasonable to make the subsequent analysis of the turbulence structure in the strongly accelerating region on a permeable surface both for dynamic and thermal parameters.

THERMAL PARAMETERS OF THE FLOW

The temperature profile of Figure 6 appears to be substantially less sensitive to the longitudinal pressure gradient than the mean velocity profile. The reduction of the heat loss coefficient in the strongly accelerating region of the flow is the most promising one, from the point of view of developing a reliable two-parameter turbulence model. Probably, the reduction of St-number on a permeable surface is caused by the thickening of the viscous sublayer that serves as the main thermal resistance. The destruction of the viscous sublayer on a permeable surface would rather explain the relative increase of St-numbers. If the accepted assumption is correct, then porous injection through a permeable surface exerts double action on the process of heat loss: intensifying it in the strongly accelerating region of the flow, and at the same time

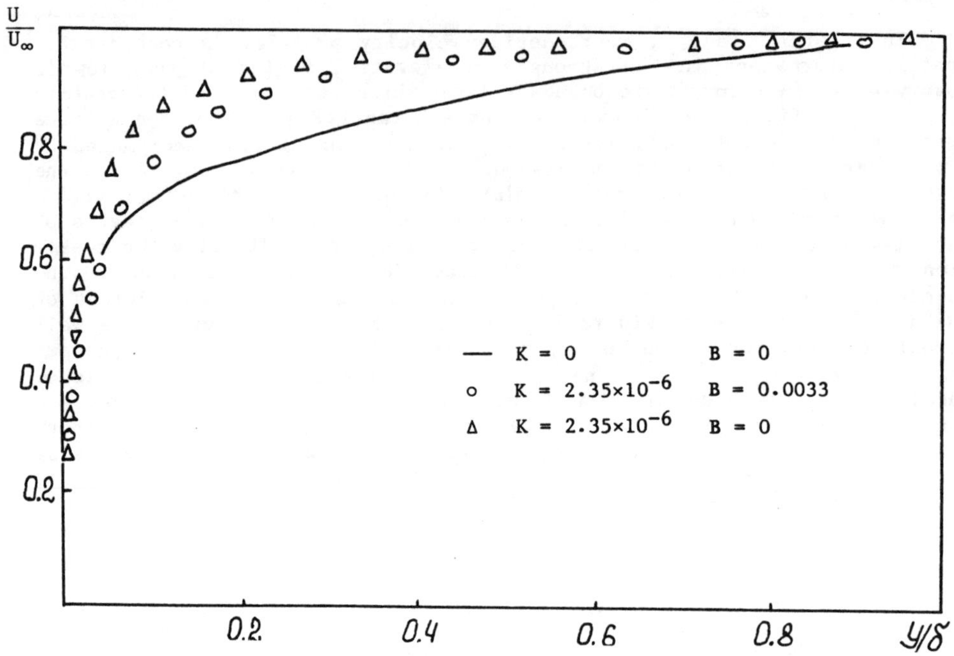

Figure 1. Velocity distribution in the flows investigated.

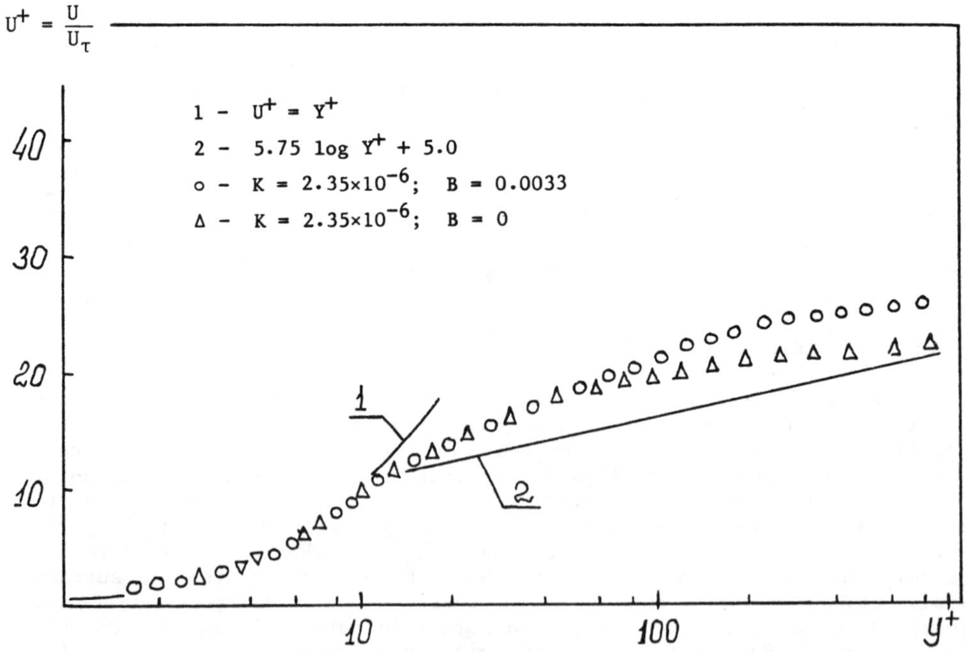

Figure 2. Velocity profiles in the law-of-the-wall coordinates.

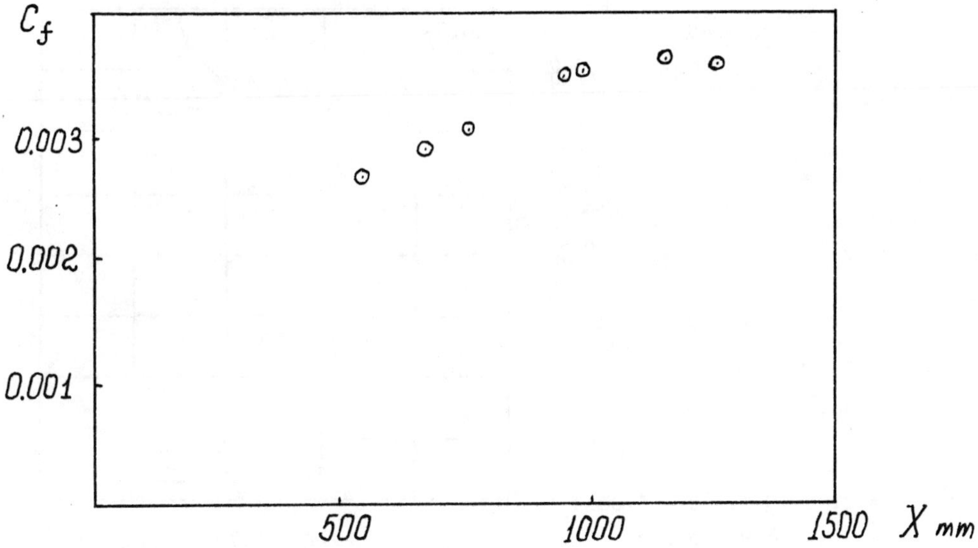

Figure 3. Friction coefficient C_f distribution.

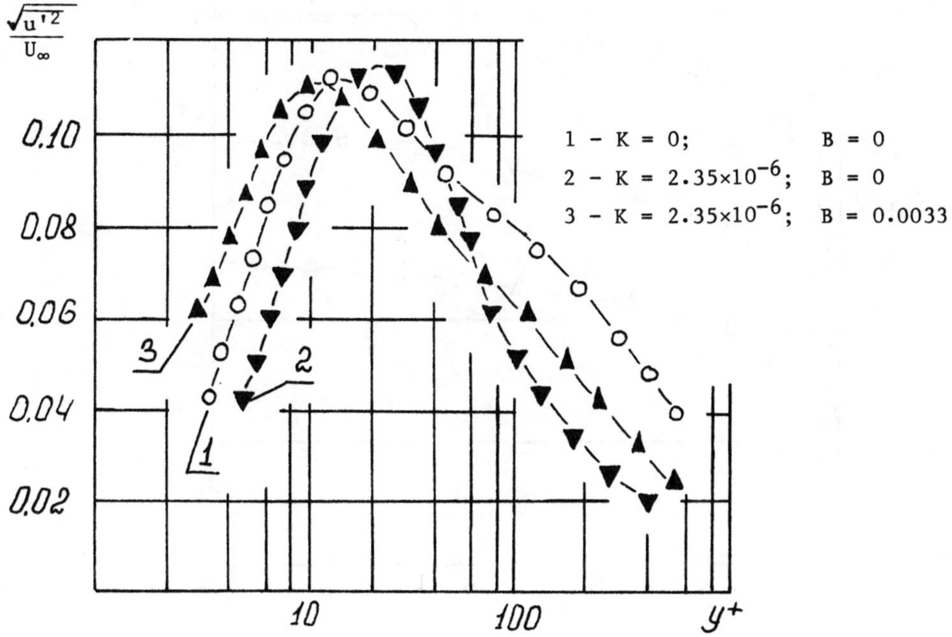

Figure 4. Velocity fluctuations distribution in the wall region.

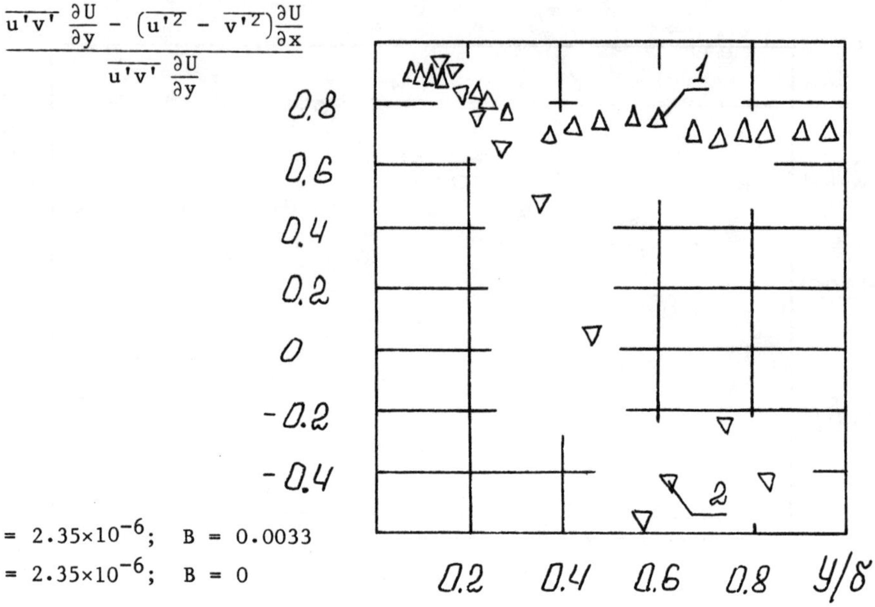

1 - K = 2.35×10^{-6}; B = 0.0033

2 - K = 2.35×10^{-6}; B = 0

Figure 5. The role of normal components in the turbulence energy production.

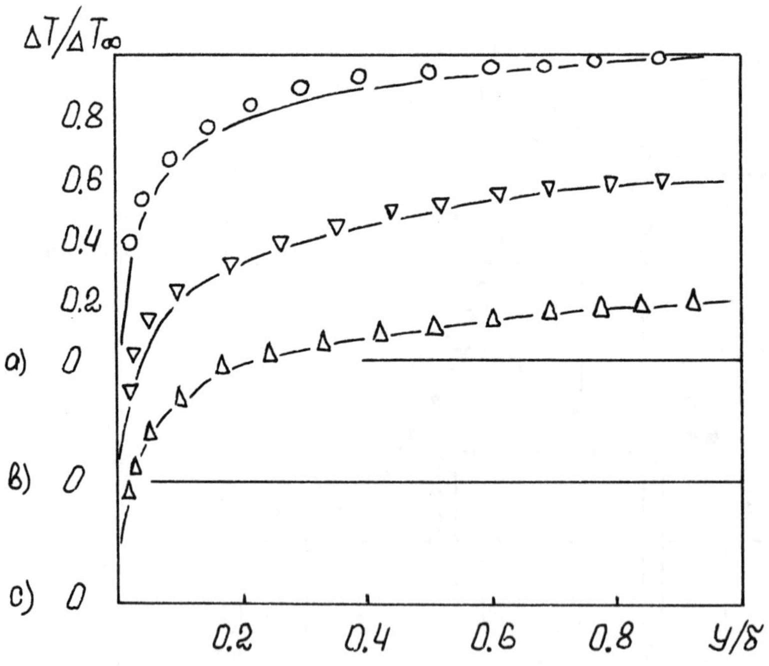

a) K = 0; B = 0

b) K = 2.35×10^{-6}; B = 0

c) K = 2.35×10^{-6}; B = 0.0033

Figure 6. Temperature distribution in the boundary layer.

reducing St-numbers in the case of nongradient flow past a plate. The above-mentioned peculiarities of the heat loss processes in the strongly accelerating regions of the flow on permeable surfaces can be explained by the presence of two opposite effects of the air-injected and mainstream flows, attributable to the two-dimensional nature of the flow in the accelerated boundary layer. The interaction of these flows leads to the appearance of the velocity fluctuations, v', which are not correlated due to different generation mechanisms with both the longitudinal velocity fluctuations, u', and temperature fluctuations, Θ'. The validity of the assumption is proved by the analysis of distribution of the correlation coefficients, R_{uv} and $R_{v\theta}$ (Figure 7), testifying that interaction between the wall and the outer regions of the boundary layer is decreased. Probably, the decrease of the turbulent heat flux level (Figure 8) caused by less correlated v' and Θ' fluctuations is the main reason for the St-number reduction on permeable surfaces under the conditions of non-gradient flow past a plate.

The analysis of the obtained heat loss coefficients shows that under given flow conditions the "standard" heat transfer law is not followed. Moreover, it is very important that the combined effect of acceleration and air injection on the heat loss results does not obey the superposition principle.

The allowance for the analogy of the heat and momentum transfer processes leads to the necessity of introducing the notion of Pr_t-number. A substantial dependence of Pr_t-distribution on the boundary conditions makes the application of these methods for calculation of the heat exchange in the turbulent wall flows doubtful. Formerly, the transport equation for turbulent heat flux has been used to complete the energy equation. The balance equation for turbulent heat flux transfer in the case of a steady-state two-dimensional boundary layer can be written as follows:

$$U \frac{\partial \overline{v'\Theta'}}{\partial x} + V \frac{\partial \overline{v'\Theta'}}{\partial y} = -\left(\overline{u'v'} \frac{\partial T}{\partial x} + \overline{v'^2} \frac{\partial T}{\partial y}\right) - \left(\overline{u'\Theta'} \frac{\partial V}{\partial x} + \overline{v'\Theta'} \frac{\partial V}{\partial y}\right) -$$

$$- \frac{1}{\rho} \overline{p' \frac{\partial \Theta'}{\partial y}} - \left(\frac{\partial \overline{u'v'\Theta'}}{\partial x} + \frac{\partial \overline{v'^2\Theta'}}{\partial y} + \frac{1}{\rho} \frac{\partial \overline{p'\Theta'}}{\partial y}\right) + \left(a\ \overline{v'\nabla^2\Theta'} + \nu\ \overline{\Theta'\nabla^2 v'}\right) \qquad (1)$$

The pressure fluctuation terms were determined as the difference for the corresponding balance equation. Expressed in dimensionless form by means of δ_t, U_∞, and ΔT_∞-scales, the components of Eq. (1) are shown in Figure 9. It is not difficult to see that under the given conditions the generation terms and the correlation of pressure and temperature fluctuations are the most substantial. Most complicated still is the choice of the approximating correlation for the last term. The most interesting approach is that suggested by Launder and Samsraweera [4]. Multiplying the Poisson equation by the mean square magnitude of the longitudinal temperature fluctuations derivative, the following relation is obtained:

$$\phi_{ic} = \frac{1}{\rho} \overline{p' \frac{\partial \Theta'}{\partial x_i}} = \frac{1}{4\pi} \iint \Bigg(\underbrace{\overline{\frac{\partial^2 u'_e u'_m}{\partial x_e \partial x_m} \frac{\partial \Theta'}{\partial x_i}}}_{\phi_{ic1}} + 2 \underbrace{\overline{\frac{\partial u'_e}{\partial x_m} \frac{\partial u'_m \partial \Theta'}{\partial x_e \partial x_i}}}_{\phi_{ic2}} + \underbrace{a \frac{g_i}{C_i} \frac{\partial \Theta'}{\partial x_e} \frac{\partial \Theta'}{\partial x_e}}_{\phi_{ic3}} \Bigg) \frac{d\ vol}{2} \qquad (2)$$

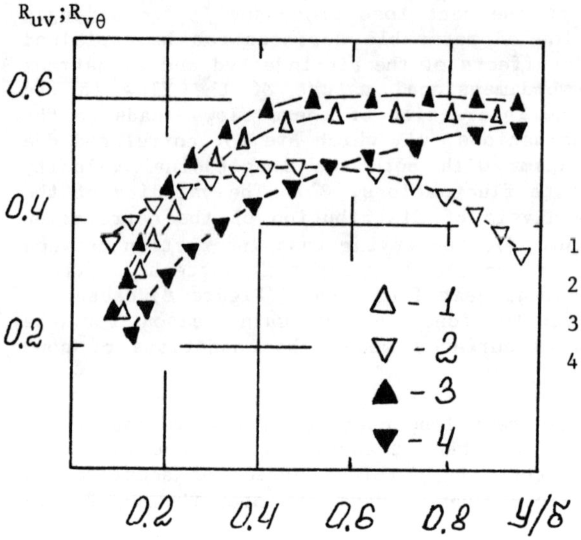

1 - $K = 2.35 \times 10^{-6}$; $B = 0.0033$

2 - $K = 2.35 \times 10^{-6}$; $B = 0$

3 - $K = 2.35 \times 10^{-6}$; $B = 0.0033$

4 - $K = 2.35 \times 10^{-6}$; $B = 0$

R_{uv} = closed points

$R_{v\theta}$ = open points

Figure 7. Correlation coefficients distribution.

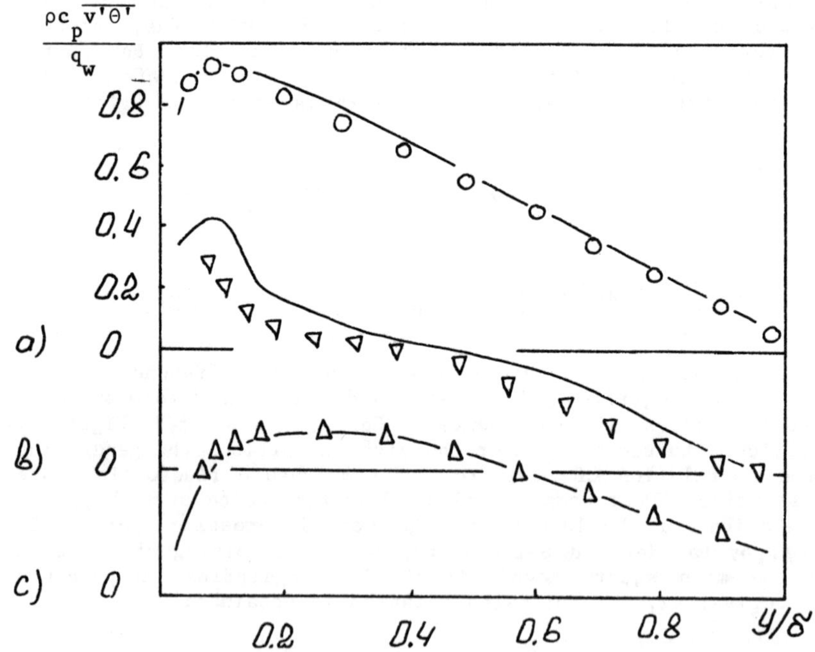

a) - $K = 0$ $B = 0$

b) - $K = 2.35 \times 10^{-6}$; $B = 0$

c) - $K = 2.35 \times 10^{-6}$; $B = 0.0033$

Figure 8. Turbulent heat flux distribution.

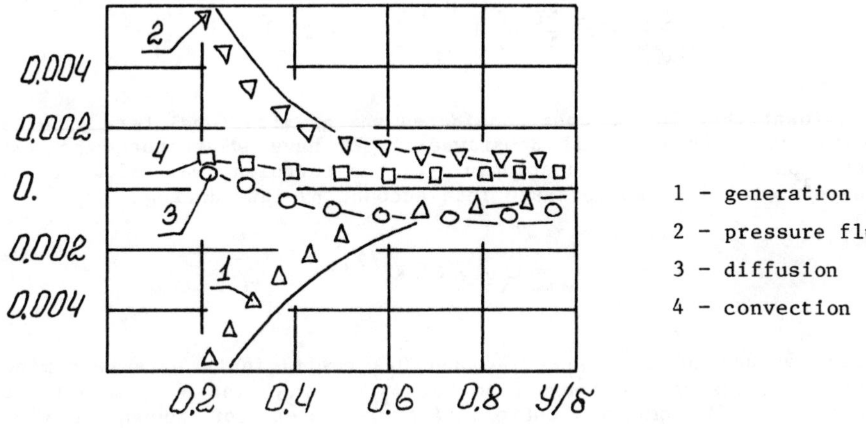

Figure 9. Turbulent heat flow balance.

Three terms in the RHS of Eq. (2) determine the turbulent, mean-deformation, and gravitational components of the correlation sought. Following Monin [5], the quantity ϕ_{ic1} is usually represented in the form

$$\phi_{ic1} = - C_{i1} \frac{\varepsilon}{k} \overline{u_i'\Theta'}$$

Launder [6] has suggested that the mean-deformation component ϕ_{ic2} should be determined in the "quasi-isotropic" form

$$\phi_{ic2} = 0.8\, \overline{u_m'\Theta'} \frac{\partial U_i}{\partial x_m} - 0.2\, \overline{u_m'\Theta'} \frac{\partial U_m}{\partial x_i}$$

Neglecting terms of the second order of smallness, the following relation can be obtained:

$$\phi_{ic2} = C_{i2} \overline{u'\Theta'} \frac{\partial U}{\partial y}$$

This introduces a new unknown quantity--the longitudinal heat flow U--into the equation system considered. The u'Θ'-quantity can be determined from the equation for the longitudinal heat flux on the assumption that diffusion and convection are neglected:

$$0 = - \overline{u'v'} \frac{\partial T}{\partial x} + \overline{v'\Theta'} \frac{\partial V}{\partial y} + C_{i1} \frac{\varepsilon}{k} \overline{u'\Theta'} + C_{i2} \overline{v'\Theta'} \frac{\partial U}{\partial y}$$

Here it follows

$$\overline{u'\Theta'} = \frac{1}{C_{i2}} \frac{k}{\varepsilon} \left(\overline{u'v'} \frac{\partial T}{\partial y} + (1 - C_{i2}) \overline{v'\Theta'} \frac{\partial U}{\partial x}\right.$$

The relation obtained makes it possible to write the following:

$$\phi_{ic2} = \frac{C_{i2}}{C_{i1}} \frac{k}{\varepsilon} \overline{u'v'} \frac{\partial T}{\partial y} \frac{\partial V}{\partial y} + \frac{C_{i2}}{C_{i1}} \frac{k}{\varepsilon} \left(1 - C_{i2}\right) \overline{v'\Theta'} \left(\frac{\partial U}{\partial y}\right)^2$$

It is evident that in the case considered the gravitational term ϕ_{ic3} can be neglected. Launder and Samsraweera [4] have shown that the wall influence on the fluctuation structure of the turbulent wall flows in the process of modelling can be taken into account by introducing

$$\phi_w = C_{i3} \overline{v'\Theta'} K^{1/2}/y$$

Numerical values of the C_{i1}, C_{i2}, and C_{i3} coefficients were determined neglecting diffusion and convection for ϕ_{ic1}, ϕ_{ic2}, and ϕ_{ic3} quantities, respectively. A computer optimization was used for comparing $v'\Theta'$-component distributions. The diffusion term in Eq. (1) was approximated by means of a standard gradient representation

$$\left(\frac{\partial \overline{u'v'\Theta'}}{\partial x} + \frac{\partial \overline{u'^2\Theta'}}{\partial y} + \frac{1}{\rho} \frac{\partial \overline{p'\Theta'}}{\partial y}\right) \approx \frac{\partial}{\partial y} \left(\frac{\mu_t}{C_{i4}} \frac{\partial \overline{v'\Theta'}}{\partial y}\right)$$

Finally, the model equation for turbulent heat flux transfer takes the form:

$$\rho U \frac{\partial \overline{v'\Theta'}}{\partial x} + \rho V \frac{\partial \overline{v'\Theta'}}{\partial y} = - \rho\left(\overline{u'v'} \frac{\partial T}{\partial x} + 1.6\, \overline{u'v'} \frac{\partial T}{\partial y}\right) + \frac{\partial}{\partial y} \left[\left(\frac{\mu}{\rho_2} + \mu_t\right) \frac{\partial \overline{v'\Theta'}}{\partial y}\right]$$

$$- 3.8 \frac{\overline{v'\Theta'}}{k} \varepsilon - 0.3\, \rho\overline{v'\Theta'} \frac{K^{1/2}}{y} + 0.065\, \rho\overline{u'v'} \frac{k}{\varepsilon} \frac{\partial T}{\partial y} \frac{\partial U}{\partial y} - 0.035\, \rho\overline{v'\Theta'} \frac{k}{\varepsilon} \left(\frac{\partial U}{\partial y}\right)^2 \tag{3}$$

The authors of the present paper have investigated and calculated different patterns of the wall flows to prove the reliability of the proposed approximations and models developed for the heat-exchange calculations, where Eq. (1) is used to complete the energy equation. The results are presented in Ref. [7]. The two-parameter "k-ε" model, described in Ref. [8], was taken as the turbulence model for hydrodynamical calculations. The accuracy of the model equation (3) is proved by a good agreement between the experimental data and the results of calculations of the components of Eq. (1), shown in Figure 9. The coincidence of St, Pr_t, and turbulent heat flux distributions for different patterns of wall flows confirms that the applied approximations are physically reasonable (Figures 10 and 11). Thus, it is shown that the heat-exchange calculation method suggested in the present paper can be recommended for modelling the heat-exchange processes in strongly accelerating flows on permeable surfaces in the range of variations of acceleration parameter $K = 0$ to 2.35×10^{-6} and injection parameter $B = 0$ to 0.004, respectively.

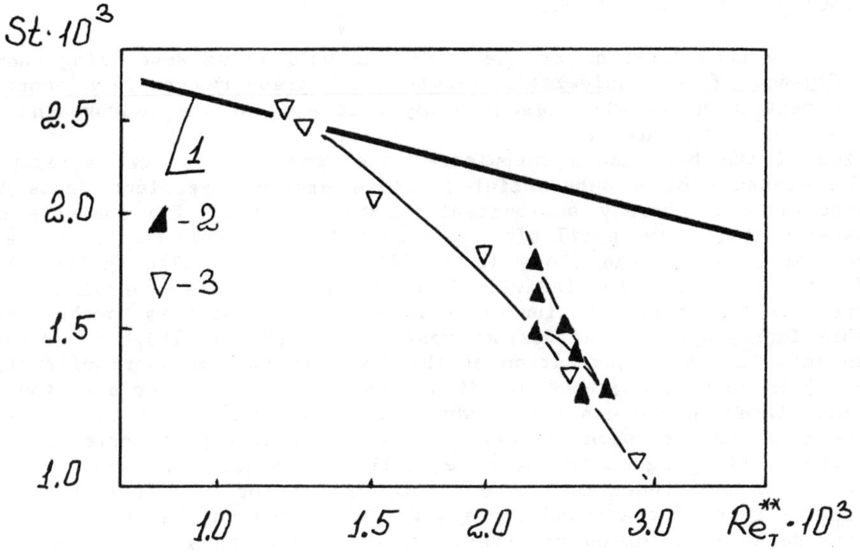

1 – St = 0.0144 $Re_t^{-0.25}$

2 – K = 2.35×10^{-6}; B = 0

3 – K = 2.35×10^{-6}; B = 0.0033

Figure 10. St-numbers distribution.

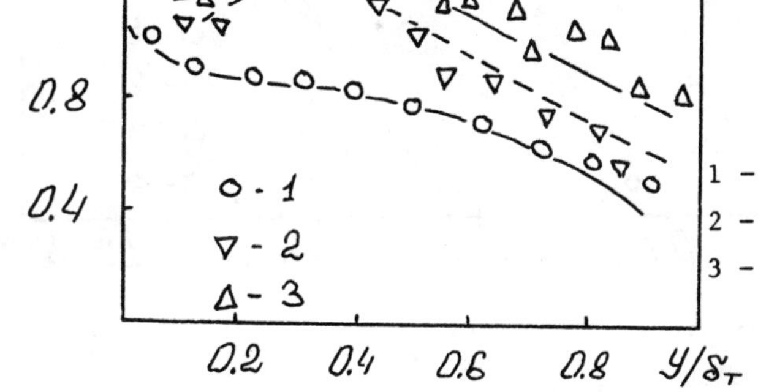

1 – K = 0; B = 0

2 – K = 2.35×10^{-6}; B = 0

3 – K = 2.35×10^{-6}; B = 0.0033

Figure 11. Pr_t-number distribution.

MODELLING OF THE TURBULENT TRANSPORT PROCESSES IN FLOWS WITH LONGITUDINAL PRESSURE GRADIENTS

The consideration of the turbulent wall flows developing under the influence of a considerable longitudinal pressure gradient constitutes the next stage of the present study that aims at the improvement of the employed turbulence models. Thus, for instance, until now the application of the best known turbulence "k-ε" models for flows developing in the presence of a substantial positive pressure gradient leads to the appearance of barely surmountable obstacles [9]. The analysis of the experimental data available is directed to reveal the characteristic peculiarities of the flows in a diffuser region. The analysis of the distribution of the individual components of the turbulence-energy-transfer transport equation given in Ref. [10] testifies to the considerable increase of the diffusion transfer role (Figure 12). Unfortunately, in Ref. [2] the contribution of the fluctuational pressure diffusion has not been determined; however, if we consider the total role of the turbulent diffusion and unbalance, which have been determined on the basis of the difference between all definite terms, it is easy to note that in the diffuser flow region the local equilibrium is not realized. The turbulence energy production is not compensated for by dissipation in the greater part of the boundary layer. This leads to the appearance of the considerable diffusion transfer. First of all, this fact makes possible application of the turbulent viscosity hypothesis doubtful; this causes the inadequacy of the two-parameter turbulence "k-ε" models in the process of modelling turbulent transfer in a diffuser flow region.

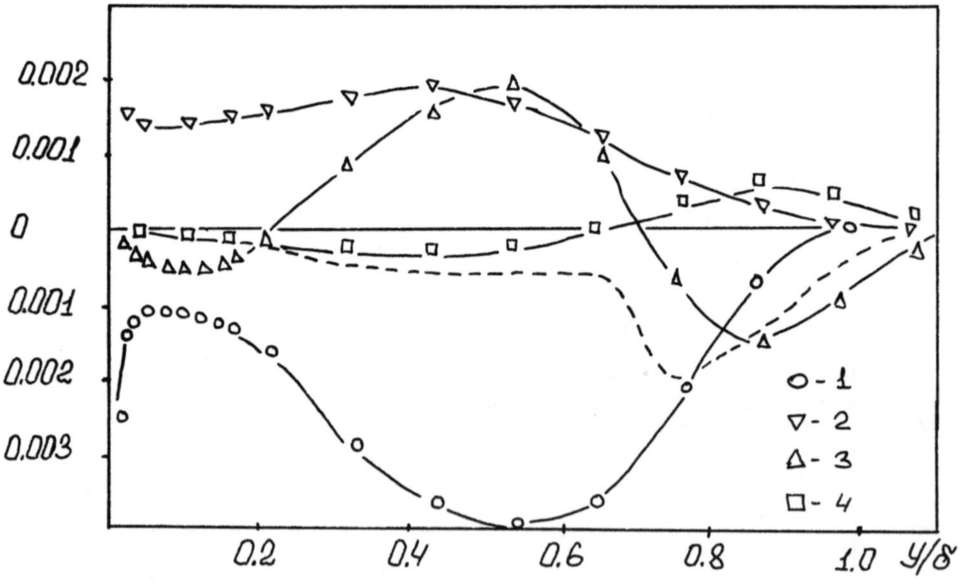

1 - generation; 2 - dissipation; 3 - diffusion;

4 - convection; ---- disbalance.

Figure 12. Turbulent energy balance.

In Ref. [9] employment of more generalized algebraic stress models in comparison with the "k-ε" model for tangent stresses was suggested. This can be obtained from the corresponding transport equation assuming the Rodi hypothesis [11], which considers paired convection and diffusion processes of the turbulence energy and tangent stresses proportional to each other. One obtains

$$\frac{\overline{uv}}{k}(P - \varepsilon) = P_{\overline{uv}} + R_{\overline{uv}} + \varepsilon_{\overline{uv}} \tag{4}$$

where $P_{\overline{uv}}$, $R_{\overline{uv}}$, $\varepsilon_{\overline{uv}}$ are generation, redistribution, and dissipation of the tangent stresses, respectively. Using the most well-known representation of the corresponding quantities it is possible for flows of the boundary layer type to obtain (see, for example, [9,12,13]):

$$P_{\overline{uv}} = -\left\{\overline{u'_j u'_k}\frac{\partial U_i}{\partial x_j} + \overline{u'_j u'_i}\frac{\partial U_x}{\partial x_j}\right\} = -v^2\frac{\partial U}{\partial y} \tag{5}$$

$$R_{\overline{uv}} = R_{uv_1} + R_{uv_2} + R_{uv_{1w}} + R_{uv_{2w}} \tag{6}$$

$$\varepsilon_{\overline{uv}} = 0 \quad \text{at} \quad Re \gg 1 \tag{7}$$

For corresponding terms of expression (6), the following approximations are suggested [12, 14-18]:

$$R_{uv_1} = -C1\frac{\varepsilon}{k}\left(\overline{u_i u_k} - \frac{2}{3}\delta_{ik}K\right) = -C1\frac{\varepsilon}{k}\overline{uv} \tag{8}$$

$$R_{uv_2} = -C2\left(P_{ik} - \frac{2}{3}\delta_{ik}P\right) = C2\, v^2\frac{\partial U}{\partial y} \tag{9}$$

$$R_{uv_{1w}} = C1W\frac{\varepsilon}{k}\left[\overline{u'^2_n}\delta_{ik} = \frac{3}{2}\left(\overline{u'_n u'_i}\delta_{nk} + \overline{u'_n u'_k}\delta_{ni}\right)\right]F_w = -\frac{3}{2}C1W\frac{\varepsilon}{k}\overline{uv}\,F_w \tag{10}$$

$$R_{uv_{2w}} = C2W\left[R_{nn_2}\delta_{ik} - \frac{3}{2}\left(R_{ni_2}\delta_{nk} + R_{nk_2}\delta_{ni}\right)\right]F_w = -\frac{3}{2}C2W\, 2\overline{v^2}\frac{\partial U}{\partial y}F_w \tag{11}$$

These terms are given the following meaning: the first two terms should account for the interaction of the fluctuational components between each other and the interaction of the definite friction stress with the fluctuational velocity components, respectively: the last two characterize the damping wall influence on the two first layers. The given correlations for $R_{\overline{uv}}$ (Eqs. 8-11) are the most widely used [9,12,13]. However, some attempts are made to use other correlations. The majority of authors developing the Rodi hypothesis obtain from Eq. (4) the algebraic expression for tangent stresses, whose concrete form depends on the representation type of the R_{uv}-term. Using the given correlations (4-11), it is possible to obtain:

$$\overline{uv} = \frac{\frac{k}{\varepsilon}\overline{v^2}\frac{\partial U}{\partial y}\left(C2 - 1 - \frac{3}{2}C2W\,C2F_w\right)}{\frac{P}{\varepsilon} - 1 + C1 + \frac{3}{2}C1W\,F_w} \tag{12}$$

To determine Reynolds normal stresses appearing in the $P_{\overline{uv}}$-term that characterizes tangent stresses "generation, the Rodi hypothesis is also used in the following form:

$$\frac{\overline{v^2}}{k}(P - \varepsilon) = P_{v^2} + R_{v^2} + \varepsilon_{v^2} \tag{13}$$

Using representations identical to the above-mentioned, we can write:

$$P_{v^2} = -\left\{\overline{u'_j u'_k}\frac{\partial U_i}{\partial x_j} + \overline{u'_j u'_i}\frac{\partial U_k}{\partial x_j}\right\} = 0 \tag{14}$$

$$R_{v^2} = R_{v^21} + R_{v^22} + R_{v^21w} + R_{v^22w} \tag{15}$$

$$\varepsilon_{v^2} = \frac{2}{3}\varepsilon \tag{16}$$

$$R_{v^21} = -C1\frac{\varepsilon}{k}\left(\overline{u'_i u'_k} - \frac{2}{3}\delta_{ik}k\right) = -C1\frac{\varepsilon}{k}\left(\overline{v^2} - \frac{2}{3}k\right) \tag{17}$$

$$R_{v^22} = -C2\left(P_{ik} - \frac{2}{3}\delta_{ik}P\right) = \frac{2}{3}C2P \tag{18}$$

$$R_{v^21w} = C1W\frac{\varepsilon}{k}\left[\overline{u'^2_n}\delta_{ik} - \frac{3}{2}\left(\overline{u'_n u'_i}\delta_{nk} + \overline{u'_n u'_k}\delta_{ni}\right)\right]F_w = 2C1W\frac{\varepsilon}{k}\overline{v^2}F_w \tag{19}$$

$$R_{v^22w} = C2W\left[R_{nn2}\delta_{ik} - \frac{3}{2}\left(R_{ni2}\delta_{nk} + R_{nk2}\delta_{ni}\right)\right]F_w = \frac{4}{3}C2W\,C2P\,F_w \tag{20}$$

Using the given representations it is easy to obtain from Eq. (10):

$$\overline{v^2} = \frac{\frac{2}{3}k\left(C1 - 1 + \frac{P}{\varepsilon}C2(1 + 2C2W\,F_w)\right)}{\frac{P}{\varepsilon} - 1 + C1 + 2C1W\,F_w} \tag{21}$$

In correlations of the (12, 21)-type, two different representations are used to describe wall functions. Thus in Ref. [13] the following expression is suggested:

$$F_w = \frac{L}{Y} = \frac{k^{3/2}}{\varepsilon Y} \tag{22}$$

The authors of Ref. [9] use another expression,

$$F_w = \frac{L}{Y} = \frac{C_\nu^{0.75}\,k^{3/2}}{\varepsilon\,\kappa\,Y} \tag{23}$$

There is a description of attempts to use quadratic functions of the following type:

$$F_w = \left(\frac{L}{Y}\right)^2 \tag{24}$$

The available descriptions of the attempts to use different representations for R_{uv}, R_{v^2}, and F_w are usually given with respect to absolute different conditions, which makes it difficult to conduct a comparative analysis between them. Moreover, different two-parameter turbulence models with different solution schemes are used rather frequently. The mentioned reasons make the comparative analysis of the employed suggestions practically impossible.

In the present paper another approach is employed: the improved concrete equations for R_{uv} and $R_{\overline{v^2}}$, which account for recommended coefficients and different F_w-functions, are performed instead of using the Rodi hypothesis in the form of Eqs. (4, 13) that has been confirmed repeatedly. dU/dy, k, ε, v^2, and UV distributions used in the present paper were determined experimentally in the flow with positive pressure gradient [19] and considerable negative pressure gradient [20]. As an example of non-gradient flow past a plate, data for the first cross section obtained in Ref. [20] are used that can be considered as practically non-gradient flow. Calculation results for V magnitude according to Eq. (21) for three considered flow patterns are shown in Figures 13-15 as a ratio V_e/V_c, where e and c subscripts correspond to experimental and calculated values, respectively. For initial coefficient values the values suggested in Ref. [13] have been taken: C1 = 1.8; C2 = 0.6; C1W = 0.52; and C2W = 0.28. The following variants of coefficient values are approved:

Variant 1 corresponds to C1 = 1.8; C2 = C1W = C2W = 0, i.e., it is suggested that the usual term is described only by a term that describes the "tendency to isotropy," i.e., $R_{\overline{v^2}} = R_{v^2 1}$ according to Eq. (17).

Variant 2 corresponds to C1 = 1.8; C2 = 0.6; C1W = C2W = 0, i.e., it is suggested that the exchange term represents the sum without taking into account the wall influence; $R_{\overline{v^2}} = R_{v^2 1} + R_{v^2 2}$.

Variant 3 corresponds to C1 = 1.8; C2 = 0.6; C1W = 0.52; C2W = 0, i.e., in addition to the previous one, the wall influence on the first term is taken into account; $R_{\overline{v^2}} = R_{v^2 1} + R_{v^2 2} + R_{v^2 1w}$.

Variant 4 uses the coefficient values suggested in Ref. [13].

Variant 5 is the most contributory variant from the point of view of accounting for the wall influence, C1 = 1.8; C2 = 0.6; C1W = 0.35; C2W = 0.01.

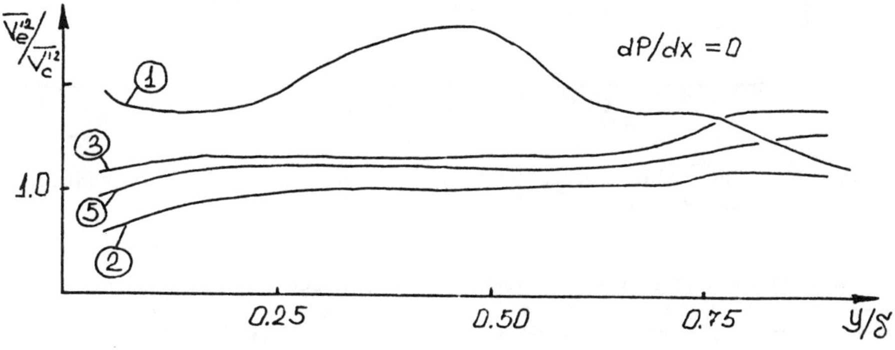

Figure 13. The relationship between the normal velocity fluctuations: dP/dx = 0.

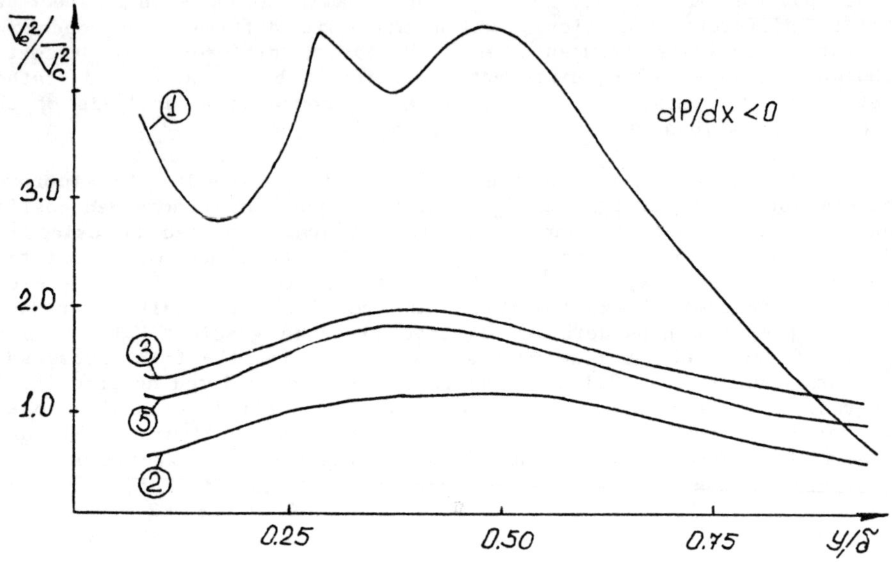

Figure 14. The relationship between the normal velocity fluctuations: dP/dx < 0.

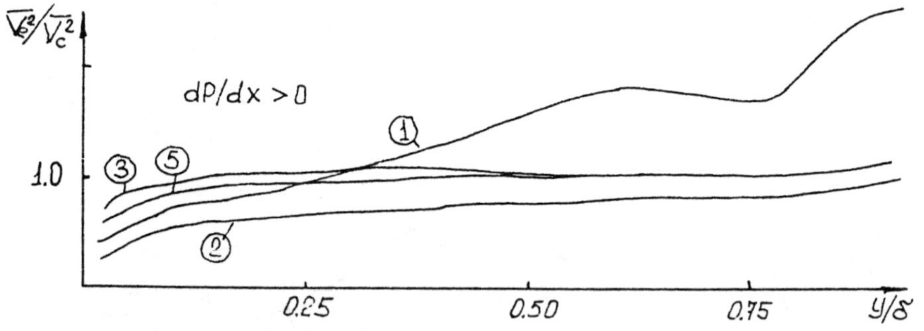

Figure 15. The relationship between the normal velocity fluctuations: dP/dx > 0.

Analysis of the results obtained for three different flow patterns shows that for normal stress models it is acceptable to use for the usual term the relation of the (15)-type but without taking into account the influence on the second term, because the employment of variant 4 has shown that for flows developing in the strongly accelerating region this variant is not suitable as far as the distribution of the calculated $\overline{v^2}$-values accept physically unreasonably magnitudes. Further, the described procedure was used to determine the uv-quantity according to Eq. (12), but while determining uv the $\overline{v^2}$-distribution calculated according to variant 5 was used. Thus it became possible to analyze the influence of the applied coefficients directly on the uv-quantity (Figures 16-18). It is easy to note that, as in the previous case, it is necessary to reduce the weight of the coefficient that accompanies the term accounting for the wall influence on the second term.

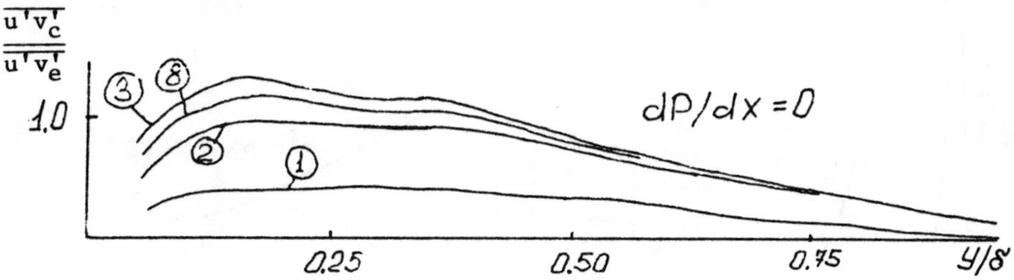

Figure 16. Relationship between the Reynolds stress; dP/dx = 0.

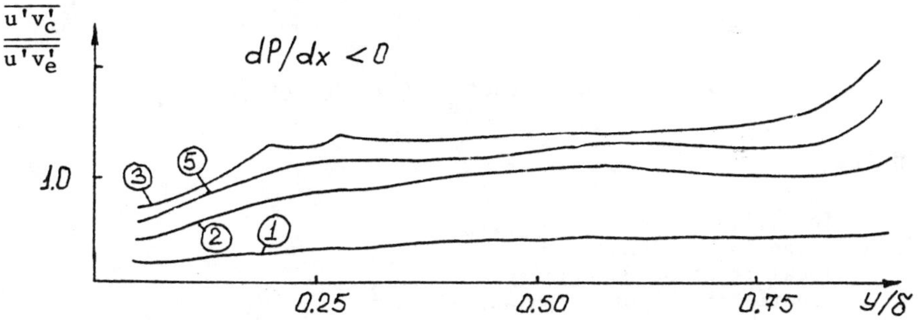

Figure 17. Relationship between the Reynolds stress; dP/dx < 0.

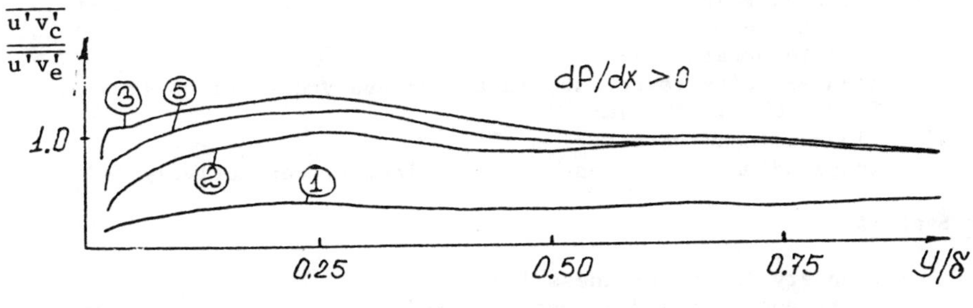

Figure 18. Relationship between the Reynolds stress; dP/dx > 0.

Analysis of attempts to use different relations for the wall functions shows that the most reliable results are obtained using relations (23). The wall functions determined by relations (23) and (22) are presented in Figure 19. Thus we consider it possible to assume the joint employment of the algebraic relations of (12)- and (21)-types together with the two-parameter turbulence model for the longitudinal pressure gradient flow modelling.

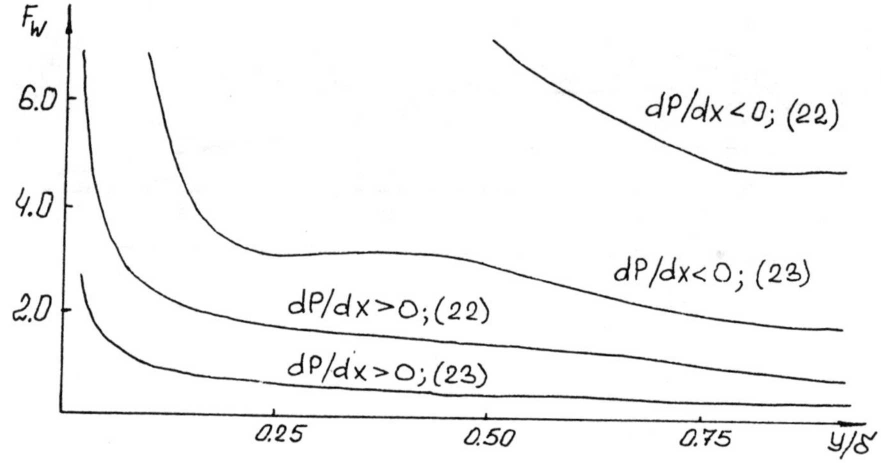

Figure 19. Distribution of F_w function.

NOMENCLATURE

B	injection parameter, $\rho V_w / \rho U_\infty$
C1,C2,CiW	dimensionless constants
C_f	local friction coefficient
k	kinetic energy of turbulence [m^2c^{-2}]
K	acceleration parameter
p'	pressure fluctuations
Pr	Prandtl number
Pr_t	turbulent Prandtl number
$R_{\overline{uv}}, R_{\overline{v\Theta}}$	correlation coefficients
Re	Reynolds number
St	Stanton number
T	mean temperature [K]
U, V	mean velocity components in the x- and y-directions [$m\text{-}s^{-1}$]
U_τ	friction velocity [$m\text{-}s^{-1}$]
u',v',w'	fluctuational velocity [$m\text{-}s^{-1}$]
X, Y	longitudinal and transverse coordinates, respectively [m]

Greek Symbols

δ	boundary layer thickness [m]
ε	turbulent energy dissipation [m-s]
Θ'	temperature fluctuation [K]

κ_{ij} Kroneker symbol
ν viscosity [m-s]
ρ density [$kg\text{-}m^{-3}$]

Subscripts

∞ in free stream
w at the wall

REFERENCES

1. Leontiev, A.I., Shishov, E.V., Yugov, V.P., et al., An experimental wind tunnel facility for the investigation of the turbulent boundary layer's structure, Trudy MLTI No. 195, pp. 5-14, 1975.
2. Zabolotsky, V.P., Roganov, P.S., and Shishov, E.V., An experimental facility and the results of investigation of the properties of an accelerated boundary layer with suction, Trudy MLTI No. 138, pp. 72-81, 1981.
3. Shishov, E.V., Roganov, P.S., Klorikian, P.V., and Zabolotsky, V.P., Correction of constant current anemometer readings in high-frequency temperature fluctuation measurements, DANTEC INFORMATION, No. 5, pp. 21-23, 1987.
4. Launder, B., and Samsraweera, D., Application of a second-moment turbulence closure to heat and mass transport in thin shear flows, Int. J. Heat Mass Transfer, Vol. 22, pp. 1631-1644, 1979.
5. Monin, A.S., The properties of symmetry in the atmospheric turbulence, Izv. Akad. Nauk SSSR, Ser. Fiz. Atm. Okeana, Vol. 1, No. 1, pp. 45-54, 1965.
6. Launder, B., Scalar property transport by turbulence, Report HTS/73/26, Dept. Mech. Eng., Imperial College, London, 1973.
7. Orlando, P., Moffat, R.J., and Kays, W.M., Turbulent transport of heat and momentum in a boundary layer subject to deceleration, suction and variable wall temperature, Report HMT-17, Dept. Mech. Engrg., Stanford University, 1974.
8. Leontiev, A.I., Shishov, E.V., Afanasiev, V.N., and Zabolotsky, V.P., The study of the fluctuational structure of a thermal turbulent boundary layer under the conditions of flow laminarization, in Heat and Mass Transfer - 6, Vol. 1, Part 2, pp. 136-146, Minsk, 1980.
9. Shishov, E.V., and Platov, S.A., Modified algebraic "k-ε" model for retarded turbulent boundary layer prediction, Izv. VUZov, Aviats. Tekh., No. 2, pp. 72-76, 1987.
10. Shishov, E.V., Roganov, P.S., Belov, V.M., and Afanasiev, V.N., Turbulence kinetic energy balance in retarded turbulent boundary layer, Izv. VUZov, Aviats. Tekh., No. 3, pp. 108-112, 1978.
11. Rodi, W., A new algebraic relation for calculating the Reynolds stresses, Z. Angew. Math. and Mech., Vol. 56, pp. 219-221, 1976.
12. Kollman, W., Prediction Methods for Turbulent Flows, Hemisphere Publishing Corporation, 1980.
13. Tahry, S.E., and Gosman, A.D., The two- and three-dimensional dispersal of a passive scalar in a turbulent boundary layer, Int. J. Heat Mass Transfer, Vol. 24, pp. 35-46, 1981.
14. Rotta, J.C., Statistische theorienicht homogener turbulenz, Zeitsch. für Physik, Vol. 131, No. 1, pp. 51-77, 1951.
15. Naot, D., Shavit, A., and Wolfstein, M., Two-point correlation model and the redistribution of Reynolds stress, Phys. of Fluids, Vol. 16, pp. 743-783, 1973.

16. So, R.M.C., The effects of extended pressure-strain models of the calculations of complex turbulent flows, Therm. Sci., Proc. 16th Southeast. Seminar, Miami, 1982, 19-21 April, 1983, Vol. I, pp. 153-170.
17. Bradshaw, P., Effects of streamline curvature on turbulent flow, AGARDograph No. 169, pp. 1-37, 1972.
18. Ljuboja, M., and Rodi, W., Calculation of turbulent wall jets with an algebraic Reynolds stress model, Trans. ASME, J. Fluids Eng., Vol. 102, No. 3, 1980.
19. Leontiev, A.I., Shishov, E.V., Afanasiev, V.N., and Belov, V.M., Mean and fluctuating parameters of temperature turbulent boundary layer and heat transfer in adverse pressure gradient flow, in Heat and Mass Transfer - 5, Vol. 1, Part 1, pp. 77-86, Minsk, 1976.
20. Roganov, P.S., Zabolotsky, V.P., Shiskov, E.V., and Leontiev, A.I., Some aspects of turbulent heat transfer in accelerated flows on permeable surfaces, Int. J. Heat Mass Transfer, Vol. 27, No. 8, pp. 1251-1259, 1984.

Heat and Mass Transfer and Turbulence Structure in the Near-Wall Region—Predictions by Single-Point Closure Methods and Some Experimental Data

K. SUZUKI
Department of Mechanical Engineering
Kyoto University
Kyoto 606, Japan

1. INTRODUCTION

Applicability of numerical prediction of heat and mass transfer employing one-point closure models of turbulence will be discussed in the following, in relation with the turbulence structure in near wall region. Turbulent heat and mass transfer is a transport phenomenon occurring in a turbulent flow field so that it is strongly affected by the characteristics of the related flow field. Thus, in some cases, its theoretical attack can be made rather easily, i.e. with minor additional effort, if a means to solve the momentum transfer in the flow has been established. Typical example of this category of heat transfer is the one in which the Reynolds analogy between momentum and heat transfers holds. However, in many cases of practical applications, the heat and mass transfer problem to be solved is not so simple. In such cases, even after the velocity field has been solved, additional difficulty still lies in solving the related thermal or concentration field. In this paper, special attention will first be paid to the numerical prediction of three different types of heat transfer problems to be categorized into a group of such cases. In the second, application of turbulence models developed for single phase flow to the two-phase flow heat transfer will be discussed. Lastly, general discussion on the applicability of single-point closure models of turbulence to heat and mass transfer problems will be given.

2. TURBULENT HEAT TRANSFER IN A PARTIALLY RECIRCULATING FLOW

Here will be discussed the turbulent heat transfer in a special type of partially recirculating flows, i.e. an axisymmetric confined jet. The axisymmetric confined jet is a coaxial jet flowing in a circular tube. The central jet is issued from a nozzle located on the tube axis into the circumferential one flowing in a tube. This particular flow was chosen as a test flow from the reason that the size of the recirculating flow region and the intensity of the flow is easily changeable only by changing the flow rate ratio between the central jet and the circumferential flow. The flow pattern to be formed in the tube becomes similar when the value of a similarity parameter called Craya-Curtet number is kept the same for different sizes of tube and different magnitudes of flow velocity. Recirculating flow appears near the wall in the region at some distance from the nozzle when the Craya-Curtet number is below its critical value, i.e. when Ct ≦ 0.75 for an axisymmetric confined jet. The Craya-Curtet number, Ct, is defined as follows.

$$Ct = \frac{U_k}{\sqrt{U_d^2 - 0.5\ U_k^2}}, \tag{1}$$

where

$$U_k = \frac{1}{A_0}\int_{A_0} U\ dA,\quad U_d^2 = \frac{1}{A_0}\int_{A_0} (U^2 - 0.5\ U_{f0}^2)\ dA.$$

Here, U denotes the local velocity in the inlet plane to the test tube or at the exit from the central nozzle, U_{f0} the circumferential velocity at the same location, A_0 the cross-sectional area of the tube.

Heat transfer from the tube wall was measured[4]. Examples of the streamwise distributions of local Nusselt number[1], are shown in Figure 1. The Nusselt number, Nu, appearing in the ordinate of the figure is normalized with the maximum value of Nusselt number, Nu_{max}, and the abscissa shows the streamwise distance from the wall, x, normalized with the tube inner diameter, D. The arrows inserted in the figure show the flow reattachment points determined from the flow visuzlization[1]. One of the features of the distribution of Nu found in the figure is that the position of maximum Nusselt number, x_m, does not necessarily coincide with the flow reattachment point. This is more clearly demonstrated in Figure 2, in which all of the flow separation point, the maximum Nusselt number and the flow reattachment point are plotted against the Craya-Curtet number[1].In case of laminar heat transfer, the flow reattachment point should coincide with the maximum Nusselt number position. This is because the development of thermal layer is suppressed there due to the flow directed towards the wall, therefore, the temperature gradient on the wall is expected to become largest there. This idea was sometimes applied also to turbulent flows so that the maximum heat transfer position was regarded to be the averaged flow reattachment point[2].

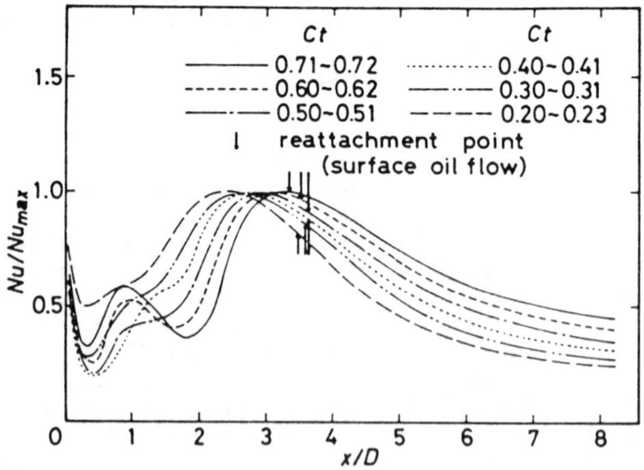

Figure 1 Streamwise distribution of local Nusselt number[1]

Therefore, the observed spatial separation between the flow reattachment point and the maximum heat transfer position is interesting. The following discussion will be given paying attention to this point.

Numerical computation was also made for the heat transfer in the flow system under the present discussion[3]. Figure 3 shows some numerical heat transfer results obtained employing the k-ε two equation model of turbulence, together with the experimental counterparts. Here, k is the turbulence kinetic energy and ε is its viscous dissipation rate. Wall functions were used in place of the wall boundary conditions for the solved variables. In the first region of small x/D, the thickness of the thermal layer remains thin compared to the grid spacing used. Therefore, steep decrease of Nu in that region cannot be calculated well. However, the thermal boundary layer develops quickly related to the deceleration of circumferential flow and the accuracy of the calculation is quickly improved downstream. After a position around x = 1.5D, the calculated Nusselt number shows fair agreement with the experimental data. For a case of Ct = 0.30, the calculated size and position of recirculating flow region is shown in the figure. Taking this

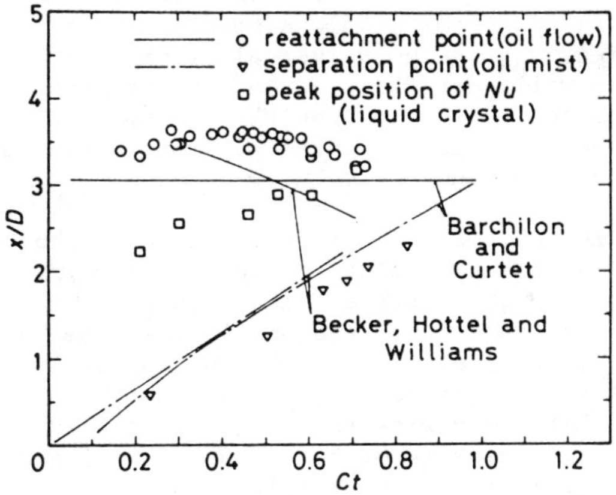

Figure 2 Maximum Nusselt number and flow reattachment points[1]

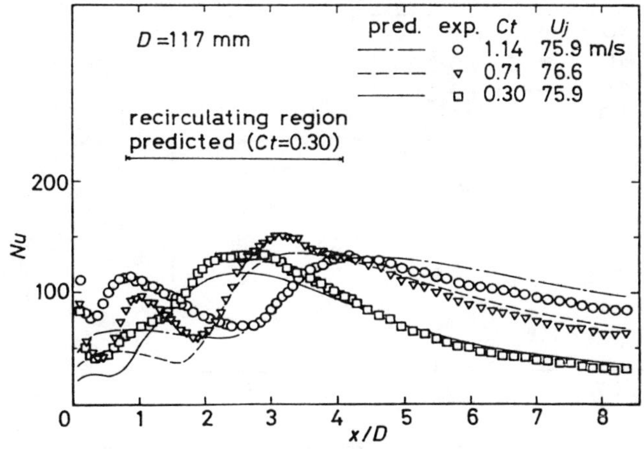

Figure 3 Calculated Nusselt number distribution[3]

into consideration, it is demonstrated that the maximum Nusselt number is calculated to be within the recirculating flow region as found in experiments.

Figure 4 shows the streamwise distributions of the calculated Nusselt number again but together with the streamwise distributions of the calculated and measured intensity of near wall turbulence[4]. The intensity of near wall turbulence plotted in the figure is the ones measured and evaluated at the location of the grid point where the wall function for the enthalpy was applied in the calculation. In the figure, it is found that the shape of Nusselt number distribution is similar with the shape of the distribution of near wall turbulence intensity. This similarity resulted from the inclusion of turbulence kinetic energy, k, in the following enthalpy wall function[47] used in the computation.

$$\rho C_D^{\frac{1}{4}} k_p^{\frac{1}{2}} \frac{h_w - h_p}{q_w} = \frac{Pr_t}{\kappa} \ln(EY_p+) + A(Pr, Pr_t) \quad (2)$$

where ρ is the fluid density, k_p and h_p are respectively the turbulence kinetic energy and the fluid enthalpy at the point P where the enthalpy wall function was applied, y_p^+ is the universal distance of the point P from the wall, q_w is the wall heat flux, Pr and Pr_t are respectively the fluid Prandtl number and the turbulent Prandtl number, $A(Pr, Pr_t)$ is a function of Pr and Pr_t relating to the enthalpy drop in the sublayer and in the buffer layer, and C_D, κ and E are the constants.

A conjecture was given in reference 4 for the physical background of the spatial separation between the two points discussed above, that the intensive near wall turbulence activates and thins the viscous and conductive sublayer. If this is true, it is reasonable that the Nusselt number becomes largest within the recirculating flow region. The intensity of near wall turbulence was confirmed to be largest within the recirculating flow region by later experiments[5,6]. Therefore, from the view point of scalar transport, inclusion of k in equation (2) has more significance than simply to take it as the results of replacing the friction velocity in the logarithmic law of the wall by the local square

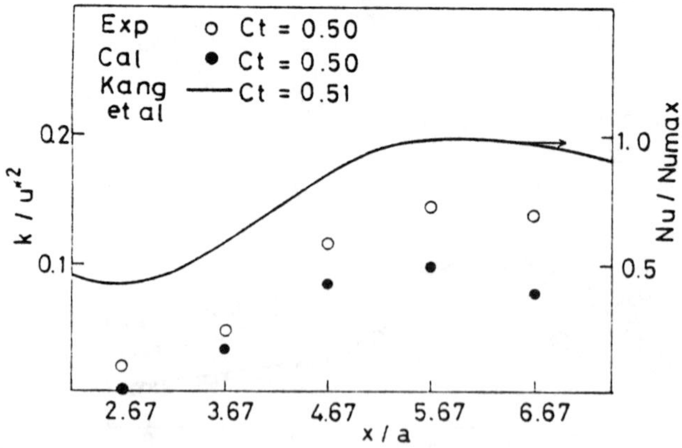

Figure 4 Distributions of Nusselt number and near wall turbulence intensity[6]

root value of k. One point worthwhile to be noticed here is that the enthalpy wall function must have been applied at the second grid point from the wall while other wall functions were applied at the first grid point. The value of k at the first grid point was found to give lower value of Nusselt number. This is because the near wall turbulence in recirculating flow region does not attain structure in quasi-equilibrium usually observed in the wall turbulent flow of boundary layer type. This was discussed in detail in reference 3. In this sense, relationship between the near wall turbulence structure and the turbulent heat transport has not been clarified yet. Especially, how the near wall turbulence activates the viscous and conductive sublayer must be studied more carefully and the results of such studies must be incorporated in the modelling of turbulent heat transfer.

Incidentally, the measured turbulence quantities in recirculating flow region were used for the test of a few turbulence models[6]. The result indicates that the k-ε two-equation model of turbulence works better than the full Reynolds stress model[6]. One of the deficiency of the former model is that each component of Reynolds stresses cannot be calculated. However, combining the k-ε model with simple algebraic Reynolds stress models[32,33] makes the calculation of them possible. The normal and shear stresses thus calculated were found to agree better with the measured ones than the directly calculated values with the full Reynolds stress model do[6]. This indicates that further improvement of the full Reynolds stress model is necessary.

3. LIQUID METAL TURBULENT HEAT TRANSFER

Liquid metal has high thermal conductivity due to large mobility of free electrons so that it is a low Prandtl number fluid. This means that the turbulent Peclet number is low even if the turbulence Reynolds number is large enough. Therefore, in liquid metal flow, fluid lumps are likely to dissipate their carrying heat to the surroundings during their irregular motion even in the region to be regarded highly turbulent from the sense of momentum transport. Thus, the turbulent heat transfer is less effective than the turbulent momentum transfer. In many previous studies on liquid metal turbulent heat transfer, this has been taken into account by setting the value of turbulent Prandtl number larger than unity and by expressing it as a function of fluid Prandtl number or of flow Peclet number. However, turbulent Prandtl number is not any physical quantity and its value can differ from one flow to another, and from one position to another even in one type of flow. Therefore, introduction of turbulent Prandtl number is not convenient to attack various types of liquid metal turbulent heat transfer. Another approach[7] to this problem will be discussed in the following.

To overcome the defficiency lying in the introduction of turbulent Prandtl number, energy equation was solved together with the equation for the turbulent heat flux, $\overline{v\theta}$. The governing equation for $v\theta$ includes unknown quantities so that it must be modelled. Denoting the radial coordinate by r, the fluid temperature by T, the intensities of radial and peripheral velocity fluctuations, respectively, by $\overline{v^2}$ and $\overline{w^2}$, and introducing two constants C and C_q, the modelled governing equation of turbulent heat flux, $\overline{v\theta}$, may be written with the cylindrical coordinate system for a hydrodynamically and thermally, fully developed state in a tube as follows.

$$\frac{1}{r}\frac{d}{dr}\{C_q \frac{k}{\varepsilon} r\overline{v^2} \frac{d\overline{v\theta}}{dr}\} - C_q \frac{k\overline{w^2}}{\varepsilon r^2}\overline{v\theta} - \overline{v^2}\frac{dT}{dr}$$

$$-3.2\frac{\varepsilon}{k}\overline{v\theta} - C(1+\frac{1}{Pr})\frac{\varepsilon}{k}\overline{v\theta} = 0 \quad (3)$$

The last term of the above equation expresses the direct dissipation of the turbulent heat flux due to high thermal conductivity. The form of this term was modified a little by de Lemos and Sesonske[8] to introduce the variation of the ratio between the length scales of velocity and temperature fluctuations with a change in turbulent Reynolds number. However, this modification does not seem to affect the final results seriously. Thus, the following discussion will be given based on the studies[9,10] made employing the original form of the term. The turbulence quantities included in equation (3) were calculated with three types of full Reynolds stress models[11-13]. These models are sometimes distinguished as models I, II and III in the figures that follow. However, the difference in the used turbulence model does not affect the final results seriously so that no further comment will be added in the following discussion on what type of full Reynolds stress model was specifically used in each case.

Computation was made for flows in circular tube and in concentric annuli. Examples of the calculated results are compared in Figure 5 through 8 together with the experimental data, some empirical relationships and numerically calculated values[14-18]. Figures 5 and 6 show the plot of Nusselt number against the flow Peclet number and the next two

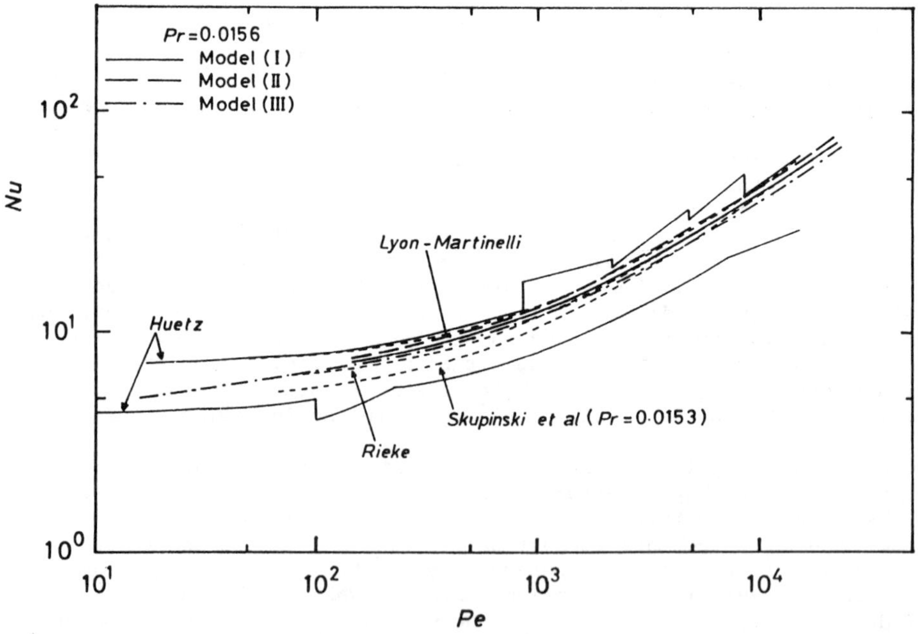

Figure 5 Nusselt number for liquid metal heat transfer in a circular tube[9]

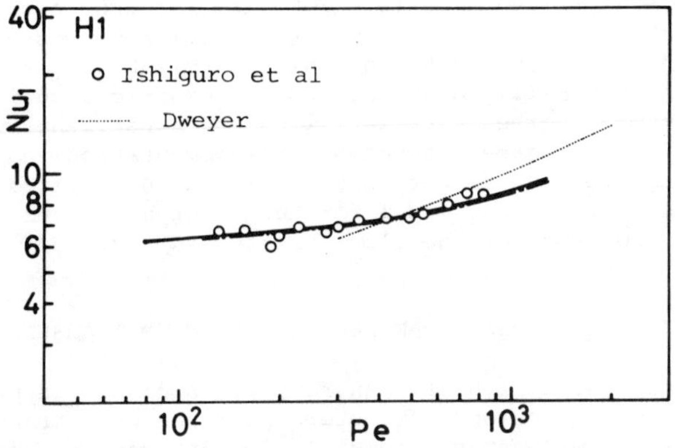

Figure 6 Nusselt number for liquid metal heat transfer in concentric annuli[10]

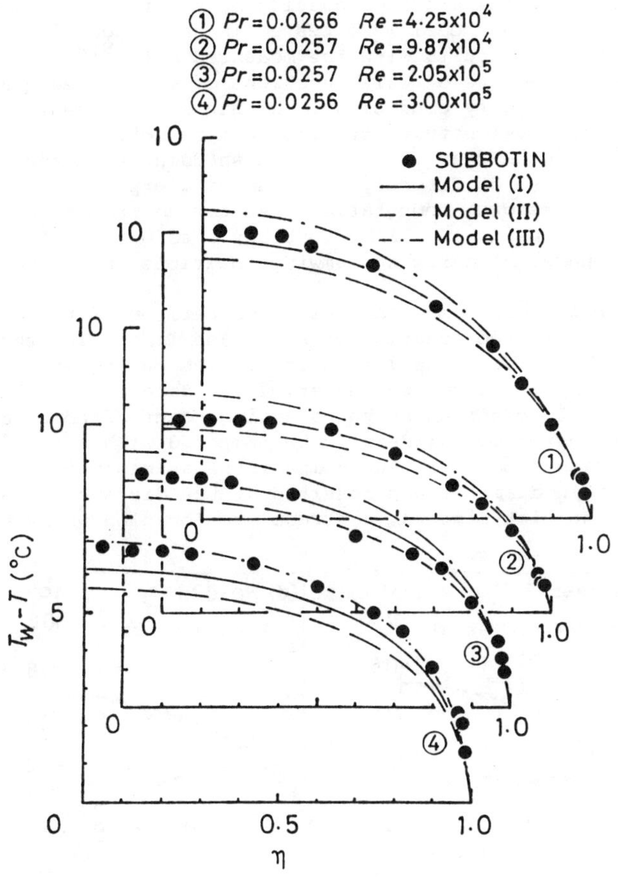

Figure 7 Temperature profiles in a circular tube (Mercury)[9]

figures compare the calculated cross-sectional profiles of temperature with the available experimental data. For circular tube flow, calculated Nusselt number lies within the upper and lower bounds of experimental data judged by Huetz to have higher reliability and very close to the Lyon-Martinelli equation. For concentric annuli, the calculated Nusselt number agrees very well with the experiments made by Ishiguro et al[17]. Good agreement between the computation and experiments is also found for the temperature distribution compared in Figures 7 and 8. Therefore, the above model for turbulent heat flux works well, at least, in the two types of the flows treated so far.

4. DISSIMILARITY BETWEEN HEAT TRANSFER AND MOMENTUM TRANSFER

In order to enhance the wall heat transfer, it is common in practical application to insert turbulence promoters near the heat transfer surface. In relation with this, studeis[19-27] were made for both a flat plate turbulent boundary layer and a channel flow disturbed by a single cylinder or by an array of cylinders. The following discussion will be given for a flat plate turbulent boundary layer disturbed by a cylinder. Mixing length hypothesis, k-ε two-equation model, three equation model[28] and full Reynolds stress model were tested in references 22 and 29 comparing the calculated results with the measured data[19,21]. These models were applied by making use of wall functions of the solved quantities for the wall side boundary conditions. Within some distance from the cylinder, periodical velocity fluctuation exists noticeably. Such fluctuation is due to the Karman like vortices shedding from the cylinder so that it has nothing to do directly with the single-point closure model of turbulence. Therefore, computation was started from a location after the position where the Karaman·like vortices died out and the calculated results were compared at a few streamwise locations in the downstream.

Figure 9 through 11 show some examples of comparison of the results calculated employing k-ε two equation model (dash line) and three equation model (dash and dot line) with the corresponding experimental data. Short dash line shows the experimental data used in the calculation as the initial condition at the starting point of computation and circles show the experimental data to be compared with the calculated profiles. The situation where the comparison was made was that a cylinder of 8 mm in diameter was inserted at the distance of 33.5 mm (H = 33.5mm) from the flat plate at a streamwise location where the bound-

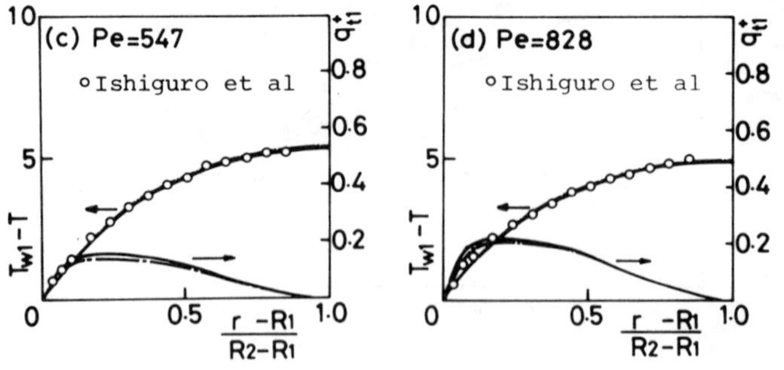

Figure 8 Temperature profiles in concentric annuli (Sodium)[10]

ary layer thickness was about 26mm when no cylinder was inserted.

The calculated results with the k-ε two equation model and the three equation model agree rather well with the experimental data. The solid line drawn in Figure 10 shows the results obtained with the mixing length hypothesis. The mixing length hypothesis does not show good agreement in the profile of Reynolds shear stress. Moreover, these studies indicate that the mixing length hypothesis does not give correct tendency of the streamwise variation of skin-friction coefficient. The results obtained with full Reynolds stress model are not plotted in the figures. But, it was found to give some over-shoot of Reynolds normal stresses. While experimentally observed relaxation process was that the profiles of all the turbulent quantities became monotonically more flattened towards the downstream, their cross-sectional profiles calculated with the full Reynolds stress model became more peaky in the first region before the start of relaxation like the one actually observed in the experiments. Therefore, the model was also found rather worse in applicability than the k-ε two-equation model and the three-equation model.

From the view point of heat transfer enhancement, the first region after the cylinder is more interesting especially when the space between the cylinder and the flat plate is small. Although any computation has not been made in such a region, detailed experimental studies have been made from various directions, including some[23,24,27] from a structural view point. A short discussion will be given on the structure of near wall turbulence in that region in a hope that the discussion will suggest what should be considered in the modelling of turbulent heat transfer in complicated flow situations to be encountered in the enhancement of heat transfer.

In Figure 12 is plotted the value of j-factor, i.e. the ratio between the local Stanton number and the skin-friction coefficient[25]. It is seen in the figure that the ratio has high value in the region close to the cylinder at two smaller value of H, or at smaller distance between the cylinder from the flat plate. As indicated from this high value of

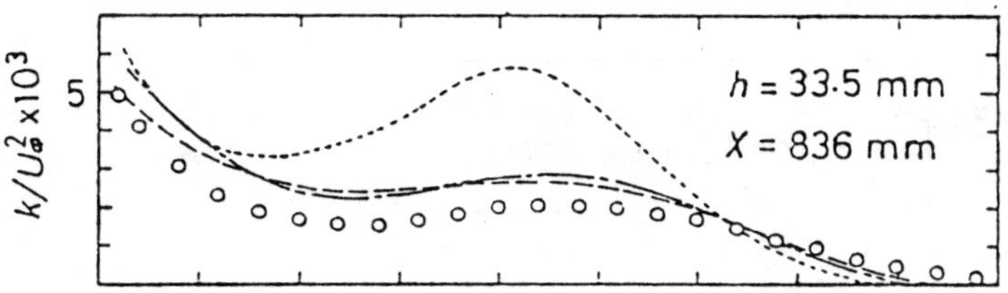

Figure 9 Lateral distribution of k[22]

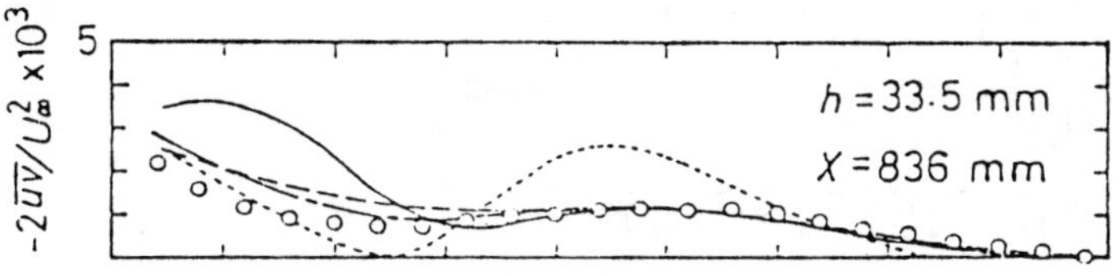

Figure 10 Lateral distribution of $\overline{uv}$[22]

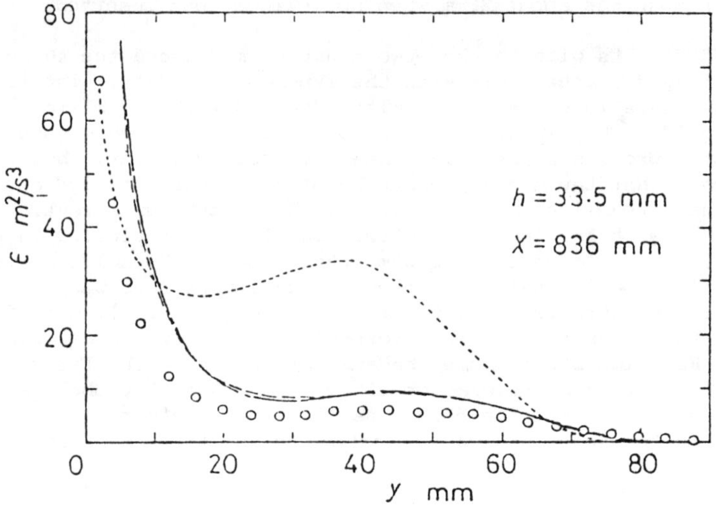

Figure 11 Lateral distribution of ε^{22}

the ratio, the value of Stanton number becomes higher because of the effect of the disturbance given by the inserted cylinder but the value of skin-friction coefficient becomes lower in that region. This means that similarity does not hold between the mechanisms of heat transfer and momentum transfer. How this dissimilarity is caused was discusssed in references 24 and 27 paying attention to the observed characteristics of bursting phenomenon.

The bursting phenomenon was studied by making use of a quadrant analysis, which is a little revised from the one firstly used by Lu and Willmarth[30] and was used in reference 24. The revised point is that temperature fluctuation was taken into consideration in order to examine the heat transfer process from the structural view point. In order to classify the instantaneous fluid motion in the u-v-θ space, an indicator

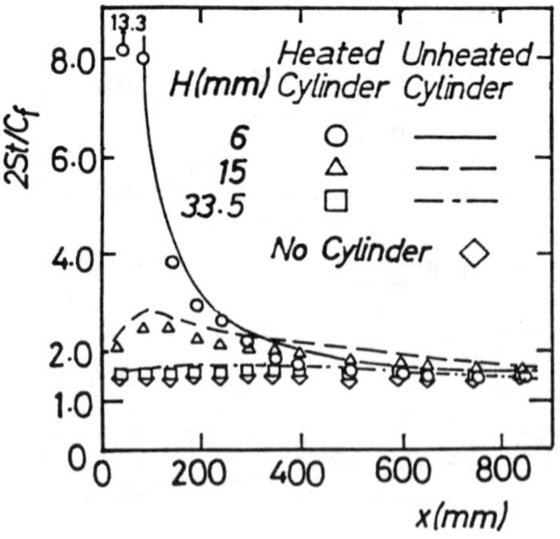

Figure 12 Streamwise distribution of $(2St/C_f)^{25}$

I_i was defined as

$$I_i = \begin{cases} 1 & \text{(when the signs of u, v and } \theta \text{ correspond to the ith quadrant of} \\ & \quad u\text{-}v\text{-}\theta \text{ space and } |uv| < H'u'v') \\ 0 & \text{(otherwise)} \end{cases}$$

where H' is the hole size parameter to descriminate weak fluid motions from those with large amplitude. Instantaneous fluid motions are classified into each of eight quadrants and identified with the symbols as follows in the figures that follow:

1st quadrant ($u > 0$, $v > 0$, $\theta < 0$); cold outward interaction (●)
2nd quadrant ($u < 0$, $v > 0$, $\theta < 0$); cold ejection-like motion (▼)
3rd quadrant ($u > 0$, $v < 0$, $\theta < 0$); cold wallward interaction (■)
4th quadrant ($u < 0$, $v < 0$, $\theta < 0$); cold sweep-like motion (▲)
5th quadrant ($u > 0$, $v > 0$, $\theta > 0$); hot outward interaction (○)
6th quadrant ($u < 0$, $v > 0$, $\theta > 0$); hot ejection-like motion (▽)
7th quadrant ($u > 0$, $v < 0$, $\theta > 0$); hot wallward interaction (□)
8th quadrant ($u < 0$, $v < 0$, $\theta > 0$); hot sweep-like motion (△)

Fluctuations classified in the even numbered quadrants contribute positively to the Reynolds shear stress but those in the odd numbered quadrants negatively to the Reynolds shear stress. On the other hand, fluctuations classified in the 3rd quadrants through 6th quadrants contribute positively to the turbulent heat flux and those in the 1st, 2nd, 7th and 8th quadrants negatively to the turbulent heat flux.

Figure 13 (a) and (b) show the lateral distributions from each quadrants, respectively, to the Reynolds shear stress and to the turbulent heat flux for a normal, undisturbed flat plate turbulent boundary layer. Figure 14 (a) and (b) show the counterparts obtained in the boundary layer disturbed by a cylinder, which was located at H = 15mm (the distance between the cylinder and the flat palte is 15mm). The plotted data in Figure 14 were measured with the hole size of H' = 0.0 and at a streamwise location of x = 37mm, where noticeable increase of heat transfer and decrease of skin friction have been noticed[25]. These figures indicate that, in the central part of the wake of cylinder, noticeable intensification was found in the contributions from hot

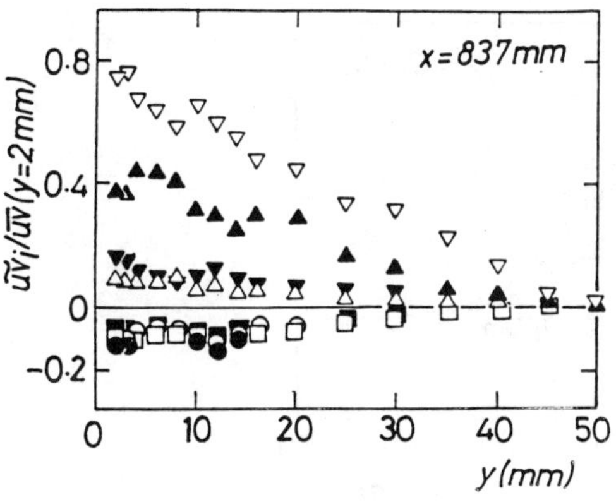

Figure 13 (a) Contribution from each quadrant to $\overline{uv}$ (Undisturbed case)[27]

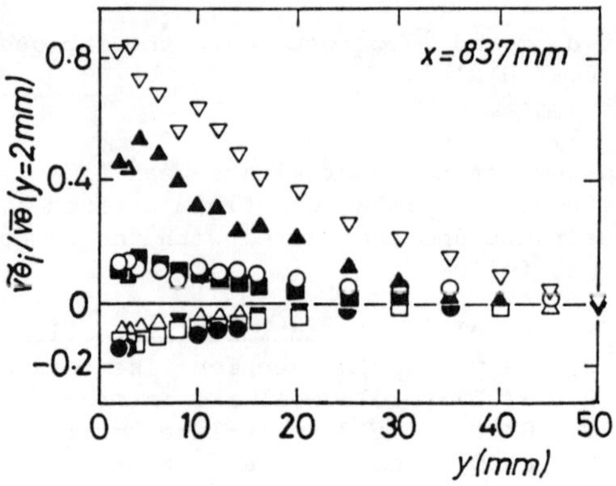

Figure 13 (b) Contribution from each quadrant to $\overline{v\theta}$ (Undisturbed case)[27]

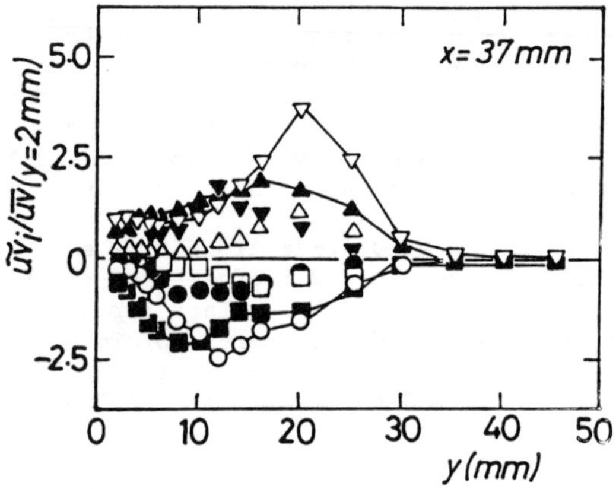

Figure 14 (a) Contribution from each quadrant to $\overline{uv}$ (Disturbed case)[27]

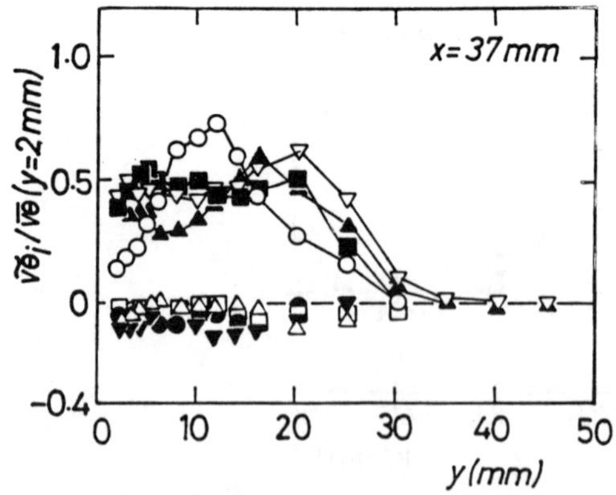

Figure 14 (b) Contribution from each quadrant to $\overline{v\theta}$ (Disturbed case)[27]

ejection-like motion, cold sweep-like motion, cold wallward interaction and hot outward interaction. This intensification affects the structure of fluid motions in the near wall region so that similar intensification is found to the positions of smaller y value. Cold wallward interaction and hot outward interaction contribute negatively to the momentum transfer but they contribute positively to the heat transfer. Therefore, the intensification of cold wallward interaction and hot outward interaction are the main cause of dissimilarity between the momentum transfer and the heat transfer found in the present type of turbulent flow.

The discussion given in the above from structural view point indicates that, in order to model the heat transfer and momentum transfer properly in turbulent flows disturbed by heat transfer promoters, to pay attention to the contributions from each qudrant to Reynolds shear stress and heat flux is important. Following the study by Nakagawa and Nezu[31], it is suggested that higher order single-point correlations among u, v, and θ are needed in roder to express relative magnitude of contribution from each quadrant to Reynolds shear stress and turbulent heat flux. To do this, much more experimental information is required. Therefore, experiments in complicated turbulent flows both from statistical and structural view points are required to be done.

5. HEAT AND MASS TRANSFER IN TWO-PHASE FLOW

Two-phase flow heat transfer (sometimes simultaneous heat and mass transfer) is encountered in many practical applications. It differs from single phase flow heat transfer in several points. In case of gas-liquid two-phase flow, interface between the both phases is easily distorted so that various flow patterns can appear depending on the flow rates of both phases. Numerical computation of two-phase flow heat transfer has been made only in limited types of flow pattern, i.e. the mist flow[34,35] and the annular flow[36-38]. In both types of flow patterns, gas flow is always turbulent so that the computation of such flows uses a single-point closure turbulence model. The following discussion will be given only for the annular two-phase flow heat transfer in a vertical tube[39].

Annular two-phase flow is defined as the flow pattern in which liquid flows forming a thin film clinging to the inner surface of the tube wall. At the interface of the two phases, evaporation occurs when the fluid is heated from the wall and condensation occurs when the fluid is cooled. These phase changes cause a gas flow normal to the interface. The effect of this normal velocity component on the gas phase flow turbulence structure can be studied in a single phase flow with fluid injection or suction from the wall. At streamwise positions far downstream from the starting point of injection, flow attains a self-preserving situation. The k-ε two-equation model[40] taking low Reynolds number turbulence effect into account seems to work well in the prediction of heat transfer and pressure drop in such flows[41]. Figure 15 shows an example of the calculated streamwise pressure gradient in a self-preserving flow region. Plotted pressure gradient is normalized with tube diameter D as a representative length scale and with a local bulk mean fluid velocity U_m as a representative velocity scale. The abscissa denotes the injection or suction rate $m = V_w/U_m$ where V_w is the normal velocity at the tube wall. Solid line shows an empirical formula given by Mizushina et al[42] ignoring the dependence of the pressure gradient on

Reynolds number. In fact, their experimental data scatter in the region bounded by dotted lines, probably showing the dependence on Reynolds number. Therefore, the calculated values can be concluded to be in a reasonable agreement with their experimental data.

Injection sometimes causes laminarization of turbulent flow[43,44], at least, in the region close to the starting point of injection. This laminarization can qualitatively be predicted again with the k-ε two-equation model[41]. However, for quantitative comparison, detailed experiments are required to be made in the future.

Large deformability of the interface causes an inception of interfacial waves and the generated waves affect the momentum, heat and mass transfer in annular two-phase flow. This occurs through the combined effects of the change in the gas flow turbulence structure near the interface and the enhanced momentum, heat and mass transfer inside the liquid

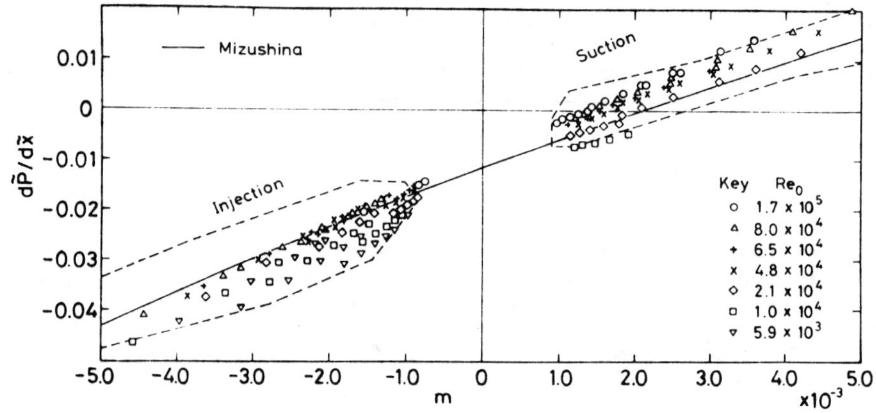

Figure 15 Dependency of pressure gradient on injection or suction rate[41]

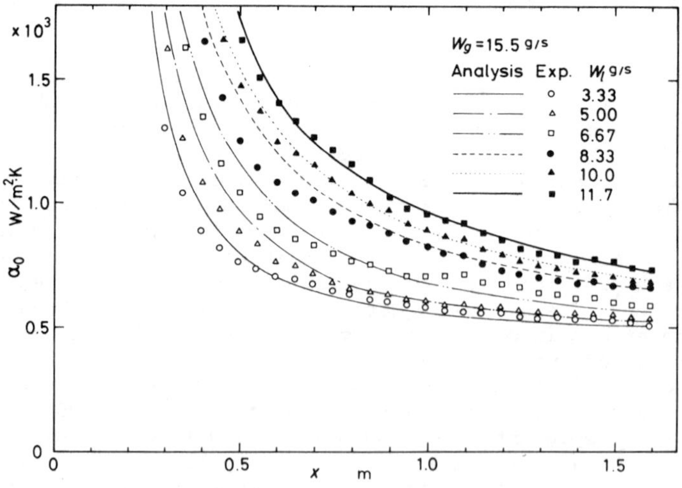

Figure 16 Streamwise change of local heat transfer coefficient[37]

film, both due to the intermittent arrival of waves. Direct numerical computation of liquid film flow is also in a primitive stage so that the effect of interfacial waves must be treated with a model[36-38]. Therefore, the discussion of two-phase flow heat transfer is somehow related to the discussion of near wall turbulence structure and to its modelling.

Figure 16 shows typical examples of the calculated heat transfer coefficient for the developing two-component (air-water) two-phase annular flow heated from the wall in a vertical tube of 26.4mm inner diameter[37]. W_g and W_l shown in the figure indicate, respectively, the total flow rates of air and water. Heat transfer coefficient was defined with the difference between the local tube wall temperature and the inlet fluid temperature. Liquid film was assumed to be in laminar state while momentum and heat transfer inside the film was considered to be affected by the interfacial waves. Enhancement of momentum and heat transfer inside the film was expressed by an introduction of the following simple model for apparent viscosity.

$$\mu^* = C\rho_l \Lambda U' \tag{4}$$

where ρ_l is the density of liquid, Λ is the length scale of waves (averaged height of waves), u' is the velocity scale of imaginary fluid motion inside the wave and C is the constant. To keep the discussion simple, van Driest model was used to calculate the gas phase flow eddy viscosity and the turbulent Prandtl and Schmidt numbers were introduced to evaluate the eddy thermal diffusivity and eddy mass diffusivity of water vapour in gas flow. Interfacial waves were also assumed to affect the momentum, heat and mass transfer in the gas flow region close to the interface. The effect was also expressed by making use of the apparent viscosity, effective only in a thin region close to the interface, which was given by replacing ρ_l in equation (4) by ρ_g, the gas density. The same model was also used in the prediction of forced-convection filmwise condensation heat transfer[38]. Good agreement between the numerically computed results and the experimental data was found in these studies. Combining this and the previous discussion on a sigle phase flow with injection or suction, it is indicated that the single-point closure model of turbulence is also useful in the calculation of two-phase flow heat transfer, at least, for an annular flow regime.

6. DISCUSSION AND CONCLUDING REMARKS

From the discussion on the heat transfer in a partially recirculating turbulent flow, the intensity of near wall turbulence was found to be important. It may be worthwhile to see how the intensity of near wall turbulence is related to the local heat transfer characteristics in the case discussed in chapter 4, i.e. the turbulent boundary layer disturbed by a cylinder.

Figure 17 shows again an example of the streamwise distribution of local Nusselt number for that case in comparison with the distribution of the intensity of streamwise velocity fluctuation measured at points distant from the flat plate by 4mm (y = 4mm)[25]. The quantities with the suffix 0 denotes the values obtained when no cylinder was inserted. In

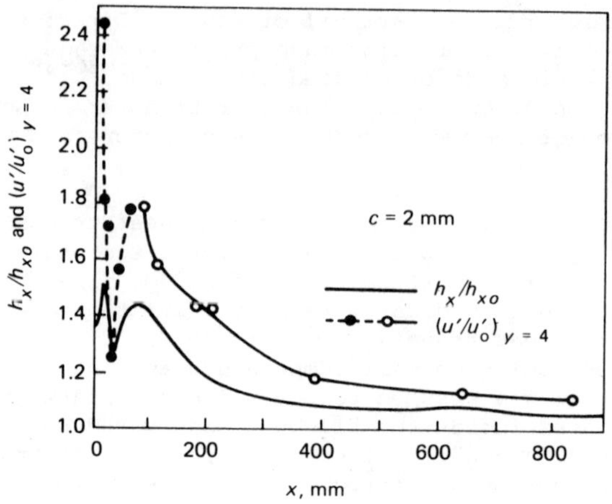

Figure 17 Local heat transfer coefficient and near wall turbulence[25]

this case, the cylinder position was much closer to the wall than in the case discussed before so that the wall heat transfer was more effectively enhanced. Since the cylinder was detached from the wall even by a small clearance, formation of a steady recirculating flow region could not be attained. However, flow becomes intermittently reversed near the wall at some streamwise locations rather close to the inserted cylinder. Because of this, accuracy of hot wire measurement at those positions was not as high as at other downstream positions. Therefore, quantitative discussion should be made with some reservation. However, similarity in the shape of distribution is again clearly observed between the local Nusselt number and the intensity of near wall turbulence.

For a turbulent boundary layer disturbed by a cylinder, dissimilarity exists between the momentum transfer and heat transfer. In order to predict heat transfer in such cases, discrimination of the contribtion from interactive fluid motions from the contributions from sweep and ejection to the turbulent heat transport was suggested to be important in Chapter 4. However, the point just discussed suggests that, even in such cases, paying attention to the near wall turbulence intensity gives a good chance in succeeding in the prediction of wall heat transfer. It is certainly better that all the components of Reynolds stresses and the heat flux can be computed with high accuracy. But, from the view point of heat transfer calculation, prediction of turbulence kinetic energy with a certain level of accuracy as well as the averaged velocity and temperature seems to be important. From this view point, further development of k-ε turbulence model into a form with wider applicability[45,46] may be one of valuable efforts to be made in the future. In relation with this, reexamination had better be made to the modelled equation of the viscous dissipation of turbulence kinetic energy[47]. As discussed in Chapter 5, the k-ε two equation model is also promising in the prediction of two-phase flow heat transfer.

REFERENCES

1. Kang, Y., Nishino, J., Suzuki, K. and Sato, T., in Flow Visualization II, ed. W. Merzkirch, pp.77-81, Hemisphere, Washington, D.C., 1982.

2. Krall, K. M. and Sparrow, E. M., Turbulent Heat Transfer in the Separated, Reattached and Development Regions of a Circular Tube, J. Heat Transfer, Vol. 88, No. , pp.131-136, 1966.

3. Kang, Y. and Suzuki, K., Numerical Study of Wall Heat Transfer in the Recirculating Flow Region of a Confined Jet, Heat Transfer-Japanese Research, Vol. 11, No. 1, pp.44-69, 1982.

4. Kang, Y., Suzuki, K. and Sato, T., Convective Heat Transfer in an Axi-symmetric Confined Jet, in Studies in Heat Transfer, ed. T. F. Irvine et al, pp.111-125, Hemisphere, Washington, D. C., 1979.

5. Suzuki, K., Ida, S. and Sato, T., Turbulence Measurements Related to Heat Transfer in An Axisymmetric Confined Jet with Laser Doppler Anemometer, Proc. 4th Symp. Turbulent Shear Flows, pp. 18.1-18.6,1983.

6. Suzuki, K., Suga, K., Oshikawa, Y. and Lee, C. G., LDV Measurement of Turbulence and Test of Turbulence Models in a Recirculating Flow, Proc. 6th Symp. Turbulent Shear Flows, pp.15-4-1 - 15-4-6, 1987.

7. Suzuki, K., An approach to Liquid Metal Turbulent Heat Transfer in a Circular Tube Solving v Equation with Local Equilibrium assumption, Letters in Heat and Mass Transfer, Vol. 9, No. 4, pp.245-254, 1982.

8. de Lemos, M. J. S. and Sesonske, A., Turbulence Modeling in Combined Convection in Mercury Pipe Flow, Int. J. Heat and Mass Transfer, Vol 28, No. 6, pp.1067-1088, 1985.

9. Suzuki, K. and Tohkaku, A., Liquid Metal Turbulent Heat Transfer in A Circular Tube, in Heat Transfer Science and Technology, ed. Bu-Xuan Wang, pp.261-268, Hemisphere, Washington, D. C., 1987.

10. Suzuki, K., Tohkaku, A. and Sato, T., Liquid Metal Turbulent Heat Transfer in Concentric Annuli, in Heat Transfer 1986, ed. C. L. Tien, V. P. Carey and J. K. Ferrell, Vol.3, pp.969-974, Hemisphere, Washington, D. C., 1986.

11. Launder, B. E., Reece, G. J. and Rodi, W., Progress in The Develop ment of Reynolds-stress Turbulence Closure, J. Fluid Mechanics, Vol.68, Part 3, pp.537-566, 1975.

12. Pope, S. B. and Whitelaw, J. H., The Calculation of Near-wake Flows, J. Fluid Mechanics, Vol. 73, Part 1, pp.9-32, 1976.

13. Prud'homme, M. and Elgobashi, S., Prediction of Wall-bounded Turbulent Flows with An Improved Version of A Reynolds-stress Model, Proc.4th Symp. Turbulent Shear Flows, pp.1.7.-1.12, 1983.

14. Huetz, J., Eddy diffusivities in Liquid Metals, in Progress in Heat and Mass Transfer, Vol. 7, ed. O.E. Dwyer, pp.3-23, Pergamon,

Oxford, 1973.

15. Jischa, M. and Rieke, H. B., Modelling Assumptions for Turbulent Heat Transfer, in Heat Transfer 1982, ed. U. Grigull et al., Vol. 3, pp.267-262, Hemisphere, Washington, D. C., 1982.

16. Subbotin, B. I., Ibragimov, M. X. and Nomofilov, E. B., Measurement of Temperature Field in Turbulent Mercury Flows in A Tube, Teploenergetika, Vol. 6, pp.70-74, 1963. (in Russian)

17. Ishiguro, R., Sugiyama, K., Sakashita, H. and Asakura, K., Sodium Flow Heat Transfer in Annuli, Proc. 17th National Heat Transfer Symp. Japan, p.100-102, 1980. (in Japanese)

18. Dwyer, O. E. and Tu, P. S., Unilateral Heat Transfer to Liquid Met als Flowing in Annuli, Nuclear Sci. Engng, Vol. 15, pp.58-68, 1962.

19. Marumo, E., Suzuki, K. and Sato, T., A turbulent Boundary Layer Dis turbed by A Cylinder, J. Fluid Mechanics, Vol. 87, Par 1, pp.121-141, 1978.

20. Marumo, E., Suzuki, K., Sasaki, T. and Sato, T., A Turbulent Bound ary Layer Disturbed by A Cylinder (1st Report: Measurements of One Point Second Order Correlation), Trans. JSME, Vol. 46, No. 407, pp.1211-1219, 1980. (in Japanese)

21. Marumo, E., Suzuki, K., Sasaki, T., Kinuta, H. and Sato, T., A Turbulent Boundary Layer Disturbed by A Cylinder (2nd Report: Produc tion rates and Dissipation Rates of Turbulence Kinetic Energy),Trans.JSME, Vol. 46, No. 407, pp.1220-1228, 1980,(in Japanese)

22. Suzuki, K., Marumo, E., Tanaka, J. and Sato, T., A Turbulent Bound ary Layer Disturbed by A Cylinder (3rd Report:Numerical computation), Trans. JSME, Vol. 46, No. 408, pp. 1474-1454, 1980. (in Japanese)

23. Kawaguchi, Y., Yano, T. and Suzuki, K., An Experimental Study in A Turbulent Boundary Layer Disturbed by A Cylinder, Proc. 8th Biennial Symp. Turbulence, ed. X. B. Reed, Jr., G. K. Patterson and J. L. Zakin, pp.26-36, 1983.

24. Kawaguchi, Y., Matsumori, Y. and Suzuki, K., Structural Study of Momentum and Heat Transport in the Wall Region of A Disturbed Tur bulent Boundary Layer, Proc. 9th Biennial Symp. Turbulence, ed. X. B. Reed, Jr., G. K. Patterson and J. L. Zakin, pp.28.1-28.10, 1985.

25. Marumo, E., Suzuki, K. and Sato, T., Turbulent Heat Transfer in A Flat Plate Boundary Layer Disturbed by A Cylinder, Int. J. Heat and Fluid Flow, Vol. 6, No.4, pp.241-248, 1985.

26. Kawaguchi, Y., Suzuki, K. and Sato, T., Heat Transfer Promotion with A Cylinder Array Located near The Wall, Int. J. Heat and Fluid Flow, Vol. 6, No. 4, pp.249-255, 1985.

27. Suzuki, H. and Suzuki, K., Dissimilarity between Heat and Momentum Transfer in A Turbulent Boundary Layer Disturbed by A Cylinder, Int.

J. Heat and Mass Transfer, to be published.

28. Hanjalic, K. and Launder, B. E., A Reynolds Stress Model of Turbulence and Its Application to Shin Shear Flows, J. Fluid Mechanics, Vol. 52, Part 4, pp.609-638, 1972.

29. Yamada, K. Kawaguchi, Y., Suzuki, K. and Sato, T., A Turbulent Boundary Layer Disturbed by A Cylinder - Test of Applicability of Full Renolds Stress Model, Preprint JSME, No. 814-5, pp.44-47, 1981. (in Japanese)

30. Lu, S. S. and Willmarth, W. W., Measurements of the structure of the Reynolds stress in a turbulent boundary layer, J. Fluid Mechanics, Vol. 60, Part 3, pp.481-511, 1973.

31. Nakagawa, H. and Nezu, I., Prediction of the contributions to the Reynolds stress from bursting events in open-channel flows, J. Fluid Mechanics, Vol. 80. Part 1, pp.99-128, 1977.

32. Launder, B. E. and Ying, W. M., Prediction of Flow and Heat Transfer in Ducts of Square Cross Section, Proc. Instn. Mech. Engrs., 187, pp.455-461, 1973.

33. Rodi, W., Turbulence Models and Their Application in Hydraulics, pp.41-43, IAHR, 1979.

34. Elgobashi, S., Abou-Arab T., Rizk, M. and Mostaffa, A., A Two-equation Turbulence Model for Two-phase Jets, Proc. 4th Symp. Turbulent Shear Flows, ed. L. J. S. Bradbury, F. Durst, B.E. Launder, F. W. Schmidt and J. H. Whitelaw, pp.12.9-12.14, 1983.

35. Suzuki, K., Nakabe, K. and Yamanaka, T., Heat Transfer Accompanied by Evaporation to A Recirculating Turbulent Dispersed Flow, PCH PhysicoChemical Hydrodynamics, Vol. 6, No. 3, pp.311-327, 1985.

36. Suzuki, K,, Hagiwara, Y. and Sato, T., Heat Transfer and Flow Characteristics of Two-phase Two-component Annular Flow,Int. J. Heat and Mass Transfer, Vol. 26, No. 4, pp.597-605, 1983.

37. Hagiwara, Y., Miwada, T. and Suzuki, K., Heat Transfer and Wave Structure in The Developing Region of Two-component Two-phase An nular Flow, PCH PhysicoChemical Hydrodynamics, Vol. 6, No. 1/2, pp.141-156, 1985.

38. Suzuki, K., Hagiwara, Y. and Izumi, H., Numerical Study of Forced-Convection Filmwise Condensation in A Vertical Tube Accounting for An Interfacial Wave Effect, Proc. 1987 ASME.JSME Thermal Engineering Joint Conference, Vol. 5, pp.55-61, 1987.

39. Suzuki, K., Heat Transfer and Hydrodynamics of Two-phase Annular Flow, in Handbook for Heat and Mass Transfer Operations,ed. N.P. Cheremisinoff, pp.681-717, Gulf Publishing, Houston, 1986, or in Encyclopedia of Fluid Mechanics, Vol. 3, ed. N. P. Cheremisinoff, pp.1356-1391, Gulf Publishing, Houston, 1986.

40. Jones, W. P. and Lauder, B. E., The Prediction of Laminarization with A Two-equation Model of Turbulence, Int. J. Heat and Mass

Transfer, Vol.15, No. , pp.301-314, 1972.

41. Hagiwara, Y., Suzuki, K. and Kawai, M., Numerical Analysis of Tur bulent Flow in a Porous Tube with Injection or Suction, Proc. 3rd Aisan Congress of Fluid Mechanics, pp.378-381, 1986.

42. Mizushina, T., Ogino, F. and Takeshita, S., Eddy Diffusivity in A Turbulent Pipe flow with Uniform Fluid Injection and Suction through The wall, J. Chemical Engng, Vol. 8, No. 3, pp.217-223, 1975.

43. Suzuki, K., Baba, T., Rahaman, A.K.A. and Hagiwara, Y., Experimental Studies on Laminarization of Turbulent Duct Flow by Fluid Injection, Proc. 2nd Int. Symp. Transport Phenomena in Turbulent Flows, pp.591-604, 1987.

44. Pennell, W. T., Eckert, E. R. G. and Sparrow, E. M., Laminarization of Turbulent Pipe flow by Fluid Injection, J. Fluid Mechanics, Vol.52, Part 3, pp.451-464, 1972.

45. Nishijima, K., Numerical Study of Turbulent Channel Flows Using and Improved Anisotropic k- Model, Trans. JSME B, Vol. 53, No. 495, pp.3166-3172, 1987.

46. Yoshizawa, A., Statistical Modeling of A Transport Equation for The Kinetic Energy Dissipation Rate, Phys. Fluids, 30-3, pp.628-631, 1987.

47. Launder, B. E. and Spalding, D. B., The Numerical Computation of Turbulent Flows, Computer Methods in Applied Mechanics and Engineering, Vol. 3, pp269-289, 1974.

Part 9
INSTRUMENTATION METHODS IN WALL TURBULENCE

Eulerian Detection of Turbulent Bursts

WILLIAM G. TIEDERMAN
School of Mechanical Engineering
Purdue University
West Lafayette, Indiana 47907, USA

INTRODUCTION

The term "burst" was used by Kim et al. (1971) to define the total process in which a streak of low momentum fluid lifts away from the wall, oscillates and then suddenly breaks up with the ejection of a sizable quantity of low momentum fluid into the outer portion of a turbulent wall layer. This definition was based upon flow visualization of burst events marked by hydrogen bubbles from a wire normal to the wall. The definition was more detailed but generally consistent with terminology used in previous flow visualization studies by Kline et al. (1967) and Corino and Brodkey (1969).

Efforts to detect individual burst events using velocity characteristics associated with low speed fluid ejecting from the wall were initiated by Lu and Willmarth (1973). They introduced two techniques. In their u-level technique, a detection occurred when the streamwise velocity decreased below the mean by an amount Lu', where L is a threshold constant and u' is the root mean square of the streamwise velocity component, U. The second method required the uv product to be less than $H\,(u'v')$ where H is a threshold and v' is the root mean square of the normal velocity component, V. In addition a quadrant two event was required. That is, the fluctuation in the streamwise velocity component, u, was required to be negative while the fluctuation in the normal velocity component, v, was positive. Lu and Willmarth discovered that a significant number of quadrant two events occurred when H was as high as 10 and that the number of these events increased monotonically as H decreased. Failure to find an unambiguous threshold for detecting burst events led Lu and Willmarth to discuss the characteristics of "relatively large bursts" where $H \approx 4$ to 4.5.

Blackwelder and Kaplan (1976) were successful in finding a universal characteristic of the streamwise velocity signal in the near wall region of turbulent flows using a variable interval time averaging (VITA) technique for the variance. Even though the shape of the conditionally averaged signal was universal, the number of events was again a monotonic function of the threshold parameter.

Offen and Kline (1975) made a one-to-one comparison of the events detected with the u-level, quadrant and VITA techniques and the burst events detected from dye slot flow visualization. Their comparison indicated no significant correlation between the flow visualization detection of bursts and the various probe detections. However Offen and Kline reported that "three times as many vortices are marked by the bubbles as by dye" at the probe position while "reasonably good correspondence" between the events marked by the hydrogen bubble wire normal to the wall and by the dye slot was obtained further downstream. Tiederman et al. (1977) also found that the number of burst events marked by a dye slot first increased and then decreased with downstream distance. The two observations suggested that Offen and Kline's probe position was too close to the dye slot and that further comparisons were needed to establish the relationship, if any, between the flow visualization burst and the velocity signals in the near wall region.

In addition Offen and Kline (1975) observed that a "given burst may go on and off several times so that dye is seen over the probe position intermittently." This observation led to a further refinement in the definition of the burst event as seen by a stationary observer. The shorter durations of dye contact with the probe were called "ejections" and the burst event was redefined to include one or more of these ejections.

By 1980 a wider variety of methods had been used to detect bursts and the methods were yielding a wide range of results for the average burst period in all the standard turbulent flows. Bogard and Tiederman (1986) showed that the average burst period from four flow visualization studies and nine probe detection studies in fully developed pipe and channel flows varied from about 2 to 25 h/U_0 at a momentum thickness Reynolds number of 1 000. Here U_0 is the center-line velocity and h is the channel height or pipe diameter. Similarly results from five flow visualization studies and eleven probe studies in zero pressure gradient boundary layers yielded an average burst period that varied from about 2 to 10 δ/U_e at a momentum thickness Reynolds number of 4 000. Here δ is the boundary layer thickness and U_e is the free stream velocity. Clearly, another one-to-one comparison was needed in an effort to resolve these differences.

One purpose of this paper is to review the successful one-to-one comparisons between probe detections and flow visualization detections by Bogard and Tiederman (1986) that began to resolve these discrepancies. The extensions to Bogard's main concepts by Barlow and Johnston (1985) and Luchik and Tiederman (1987) as well as more recent refinements also will be presented. As will be seen, the ability to make accurate time-resolved measurements in the near wall region is required to extend the present rather low Reynolds number results to high Reynolds numbers.

SIMULTANEOUS VISUAL AND PROBE DETECTIONS

Prior to making one-to-one comparisons between dye-slot flow visualization and hot-film signals, Bogard and Tiederman (1983), calibrated the dye slot technique with hydrogen bubble flow visualization. Hydrogen bubble flow visualization from a wire normal to the wall is the preferred method for detecting bursts because all bursts immediately downstream of the normal wire will be marked. Unfortunately the electric field required to generate the bubbles interferes with hot-film sensors and this interference prevents the simultaneous use of hydrogen bubbles and hot-film probes.

All of Bogard's experiments were conducted in the fully developed region of a two-dimensional water channel. The rectangular test section had a width of 575 mm and height of 60 mm. A 0.125 mm wide dye slot was located in the center of the bottom wall approximately 62 channel heights downstream of the inlet. A 25 μm diameter, platinum wire placed across the channel height normal to the bottom and top wall was used to generate hydrogen bubbles. The wire was insulated at regular intervals so that bubble "squares" could be generated by pulsing the wire with an electric current.

Bursts were identified by the detection of ejections from side views of both the hydrogen bubbles and the dye. The criteria used to define an ejection were: (1) the marked element of fluid must originate from below $y^+ = 15$; (2) an upward movement of at least $\Delta y^+ = 20$ must occur with a streamwise distance of $\Delta x^+ = 350$. The superscript "+" indicates that the variable has been normalized with the wall shear velocity and the kinematic viscosity. Ejections that were part of an overall burst pattern were classified as being multiple ejections from a single burst. A quiescent period where there was no ejection activity was required between bursts.

In this fully developed channel flow the rate of ejections per unit area must be constant. Therefore a necessary condition for the detection of all ejections marked by dye from a wall slot is either a maximum or a plateau in the histogram of the number of ejections per unit time as a function streamwise distance from the slot. When dye is seeped into the sublayer at flow rates on the order of one-tenth the sublayer flowrate, the dye immediately downstream of the slot collects into the streaky structure ($y^+ < 2$) and there are no ejections. Ejections begin to be marked only after the streaks migrate downstream and lift away from the wall to positions on the order of $y^+ = 10$. The number of ejections then increase with distance. Further downstream the number of streaks marked by the dye slot decrease and consequently the number of ejections decrease. For this experiment, the number of ejections per unit time were constant for $550 \leq x^+ \leq 1050$ where $x^+ = 0$ is the slot location. As a result, the hydrogen bubble wire was placed at $x^+ = 750$ where it would mark all of the ejections and bursts in the plateau region of the histogram of dye marked ejections.

A plane of light with a width of $\Delta z^+ = 23$ was aligned with the bubble wire and used to illuminate fluorescent dye from the slot. The hydrogen bubbles were illuminated with an orange filtered flood light that did not cause the dye to fluoresce. In this way, simultaneous views of both dye and bubble marked ejections were obtained in the thin plane centered on the hydrogen bubble wire. Analysis of motion pictures showed that 87% of the ejections detected from the hydrogen bubble wire were also detected by the dye. This comparison led Bogard and Tiederman (1983) to conclude that essentially all ejections were marked in the region $550 \leq x^+ \leq 1050$ and that dye marked ejections and bursts in this region could be used as a standard for evaluating detections based on velocity characteristics.

Velocity measurements in the channel were made using a TSI model 1249-10W X-probe that has two cylindrical hot-film sensors. Each sensor had a diameter of 25 μm and length of 0.5 mm. The spacing between the cylinders was 0.5 mm. For the conditions of the simultaneous velocity and dye slot flow visualization experiment (where the Reynolds number based on mass average velocity and channel height was 8 700) the length and spacing normalized with inner variables was 4. Two DISA CTA standard 55 M10 bridges and DISA 55 D10 linearizers were used to operate the probe. The linearized output was digitized at a rate of 70 Hz which corresponded to a dimensionless time between data samples of $t^+ = 0.9$.

The simultaneous velocity and visualization experiments were conducted with the X probe at $y^+ = 15$ above the bottom wall and $x^+ = 855$ downstream of the dye slot. This placed the probe in the center of the region where the dye slot marked essentially all the ejections. The probe was also centered in the narrow light plane used to illuminate the fluorescent dye. In this case a high speed video system was used to record side views of the dyed ejections as well as a counter that indicated when the hot-film interface was digitizing and storing velocity signals.

Criteria similar to those used by Bogard and Tiederman (1983) were used to identify ejections. The criteria were modified because the X-probe effectively destroyed the coherence of the dyed events after they passed the probe location. Hence the requirement for vertical movement was relaxed so that events originating between 100 and 350 wall units upstream of the probe needed a vertical displacement that was equal to or exceeded the minimum slope originally required. During the 200 s of synchronized recording, 271 dye marked events intersected the X probe and 164 of these events were identified as ejections.

Seven different velocity detection algorithms were tested by Bogard and Tiederman. These included methods used by Blackwelder and Eckelmann (1977), Simpson (1976), Wallace et al. (1977), and Comte-Bellot et al. (1979) in four separate experiments conducted in the oil channel at the Max-Planck-Insitüt für Strömungsforschung. Since the Reynolds number for those experiments was similar to that of Bogard and Tiederman (1986), it is important to note that each method gave nearly the same average time between detections in both channels. However, in each channel this average time between detections varied from about 2 to 16 h/U_o depending upon the method and none agreed with the average time between bursts deduced from the dye slot flow visualization in the water channel.

For the one-to-one comparisons only three methods will be discussed. They are the second quadrant uv method of Lu and Willmarth (1973), the u-level method of Lu and Willmarth (1973) and the VITA method with a slope criteria used by Johansson and Alfredsson (1982). Conditional probabilities were used to evaluate the methods. The probability for the velocity detection corresponding to a visual detection of an ejection, P(E), was the ratio of the number of visual ejections detected by the velocity method to the total number of visual detections. The probability of a false detection, P(F), was the ratio of the number of velocity detections that did not correspond to a visual detection to the total number of events given by the velocity method. A perfect one-to-one correlation would yield $P(E) = 1.0$ and $P(F) = 0.0$.

As the adjustable thresholds in the velocity methods were increased, P(F) decreased indicating that the methods are based on velocity characteristics associated with the ejections. For thresholds giving approximately 164 ejections: the quadrant technique yielded P(E) = 0.65 and P(F) = 0.14; the u-level method yielded P(E) = 0.63 and P(F) = 0.23; and the VITA with slope gave P(E) = 0.61 and P(F) = 0.35. Throughout the range of thresholds tested, P(F) for the quadrant technique was always lowest with P(F) for the u-level technique second lowest. On this basis the quadrant method was preferred and investigated further. Using an unsynchronized velocity record, it was established that the random probability for a valid detection with the quadrant two method was about 35% independent of threshold. For $H \geq 1$, the percentage of valid quadrant two detections exceeded 35% significantly.

Recall that visually detected ejections had been grouped into burst detections based upon the overall patterns. Using simultaneously recorded plan and side views of the bursting streak structure, Bogard and Tiederman (1986) showed that histograms of the distribution of time between ejections from single streaks differed significantly from the distribution of time between ejections from different streaks. For an 8 000 Reynolds number water channel flow, the maximum time between ejections from a single streak was about 1.0 s with the mode of the distribution occurring at 0.6 s. Ejections from different streaks had a minimum interval of 0.6 s and a maximum interval above 4 s. While the distributions did overlap, they were primarily distinct. Hence the probability distribution for the time between quadrant two ejections (T_E) was compared to the exponential distribution for T_E greater than or equal to the duration of the ejections (when it is not possible to detect a second ejection). A threshold of $H = 1.07$ was used where there was a reasonably high probability of detecting visual ejections and a low probability for a false detection. The data deviated from the exponential distribution when the time interval was 0.8 s which corresponded to the center of the small region of overlap for the two distributions determined from flow visualization.

Quadrant two detections whose center-to-center time interval was less than or equal to 0.8 s were then grouped together to form burst detections and the average time between bursts was deduced from the velocity record. This process was repeated for thresholds varying from 0.25 to 2.0 and it was discovered that the average time between bursts ($\overline{T}_B$) was essentially constant and equal to the flow visualization value for $0.25 \leq H \leq 1.25$. This was the first report of a velocity detection scheme that would yield the same value for T_B over a range of thresholds.

SINGLE COMPONENT BURST DETECTION

Luchik and Tiederman (1987) extended the key concept of grouping ejection detections into burst detection for the VITA and u-level methods that require only streamwise velocity measurements. The simultaneous flow visualization and two-component velocity measurements of Bogard and Tiederman (1986) were used to establish appropriate levels for the thresholds. In this process it was noted that the u-level technique frequently yielded multiple velocity detections per visual detection. By modifying the technique so that the detector was turned on at one level and turned off at a second lower level, multiple velocity detections were eliminated. The results which gave the best agreement with the visualization record were to turn the detector on when

$$u < -L u'$$ (1)

and to turn the detector off when

$$u \geq -0.25 L u' .$$ (2)

This detection algorithm was named the modified u-level method.

A modification of Bogard's approach for estimating the time period that separates ejection detections from the same burst from ejections from different bursts was introduced by Barlow and Johnston (1985). This method which worked well for quadrant 2 and u-level detections is based on plots of the cumulative probability that the time $T > T_E$. However this grouping method did not work for the VITA and modified u-level methods when the time between ejections was based on the time interval from the center of one ejection to the center of the next ejection. Bogard had used this center-to-center period for the quadrant two method because the uv signal is highly intermittent and the detector function is "on" for only relatively short durations no matter what threshold is used. The relative duration of an ejection detection is significantly longer for the VITA and modified u-level techniques. Luchik and Tiederman (1987) established a method for grouping ejections from the VITA and modified u-level techniques by matching a Poisson distribution to the distribution of ejection events with small separation times.

All four velocity detection methods (quadrant, u-level, modified u-level and VITA) were applied to two component laser velocimeter data taken at $y^+ = 30$ in four fully developed, water channel flows where the Reynolds number varied from 9 400 to 17 800. In these experiments the channel was 25 mm high and 250 mm wide. The flow was seeded so that the data rate was 2 kHz and the data records were sampled at the viscous period, $t^+ = 1.0$. At the highest Reynolds number the spanwise length of the beam crossings was 50 viscous units. In addition to the velocity detection of ejections, dye slot flow visualization was used. However, in this smaller channel simultaneous records of flow visualization and velocity were not attempted.

The most important result was that all four velocity methods as well as the flow visualization yielded the same average time between bursts in each of the four flows. The key element was grouping the ejection detections into burst detections. The results also indicated that inner variable scaling of the average time between bursts yields a constant dimensionless number of about 90 in this low Reynolds number range of channel flows.

HIGHER REYNOLDS NUMBER FLOWS

Two additional water channel flow experiments have been conducted at Purdue in an effort to determine the burst rate at higher Reynolds numbers. There are primarily two difficulties associated with these type of higher Reynolds number experiments in laboratory scale equipment. One is associated with the spanwise length of the velocity sensor and the other is the requirement for a higher sampling rate. The two are coupled for laser velocimeter data where small spanwise sampling volumes also increase the scattering particle concentration necessary to achieve high data rates. There is some point where the particle concentration will begin to either affect the flow or make the fluid opaque.

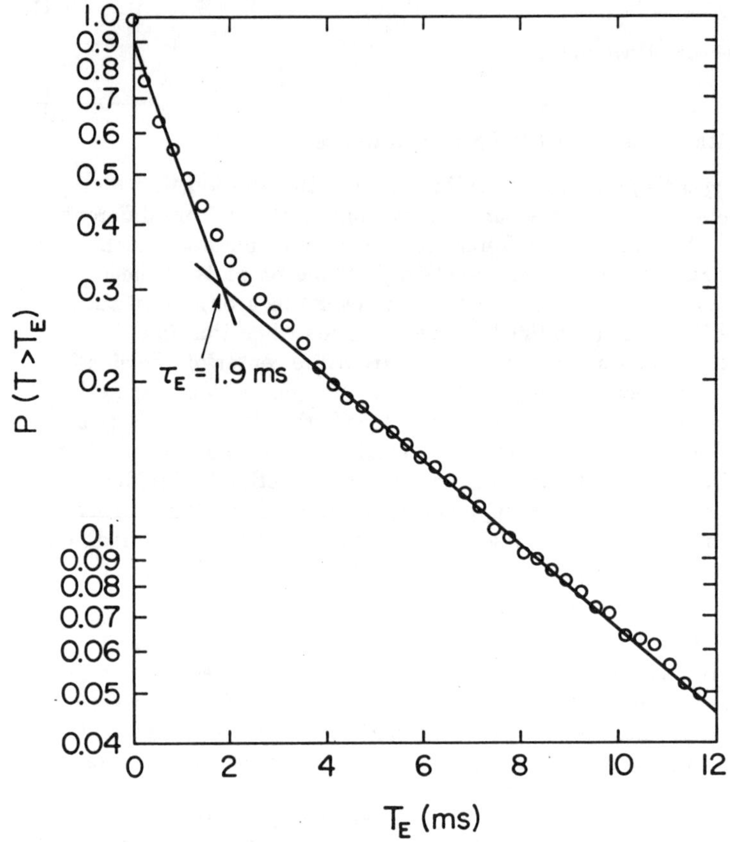

FIGURE 1. Probability distributions for the time between ejections from the revised, modified u-level method, Re = 50 000.

The issue of spanwise sensor length in the near wall region of turbulent flows originally was discussed by Willmarth and Bogar (1977). The sharp, instantaneous transverse shear layers that occur in this region, even when the turbulent statistics are homogeneous in this direction, cause averaging problems for hot-wire sensors. More recently Johansson and Alfredsson (1983), Blackwelder and Haritonidis (1983), Willmarth and Sharma (1984) and Ligrani and Bradshaw (1987) have studied the effect of spanwise length for hot-wire sensors. The conclusion is that many of the statistical properties of the streamwise velocity signal are nearly independent of the sensor length when the length is less than about 20 viscous units.

The issue is different for a laser velocimeter when there is only one scattering particle in the probe volume at any time. As demonstrated by Luchik and Tiederman (1985) there is no spatial averaging in the mean square fluctuation from this type of laser velocimeter measurements when the length is aligned with the direction of homogeneity. However, the data does come from different spanwise positions and so some requirement for a short length remains for time-resolved data. It is not clear what length is short enough but off-axis

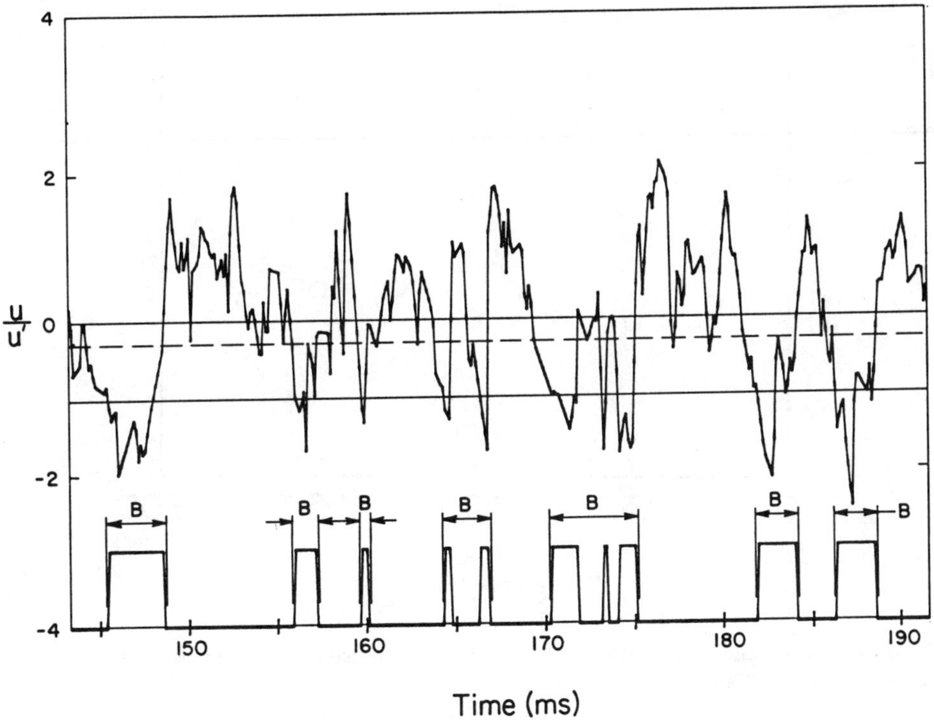

FIGURE 2. Illustration of modified u-level detections and grouping into burst events, Re = 50 000.

collection was used in both of the higher Reynolds number flows to make the spanwise length as short as practical.

One experiment was conducted in the 60 mm by 575 mm channel at a Reynolds number of 39 600. A four beam, two color laser velocimeter was used to measure the streamwise and normal velocity components. The flow was seeded with fat particles from whole milk. The sampling rate of 980 Hz was almost equal to the viscous frequency of 1.0 kHz. The probe volume was located at $y^+ = 30$ and spanwise length of the probe volume was reduced to 18 viscous units by collecting the scattered light 6 degrees off the optical axis.

The second experiment was conducted in the 25 mm by 250 mm channel at a Reynolds number of 50 000. A single component laser velocimeter was used to measure the streamwise velocity at $y^+ = 30$. The flow was uniformly seeded with 2.02 μm diameter, white latex spheres. With 45 degree off-axis collection of scattered light, the spanwise length of the probe volume was 21 viscous units and the data rate was 8 kHz. This data rate was nearly equal to the viscous frequency of 8.8 kHz.

In both cases, the average time between bursts was determined using a revised, modified u-level technique. The latest improvement was achieved by changing

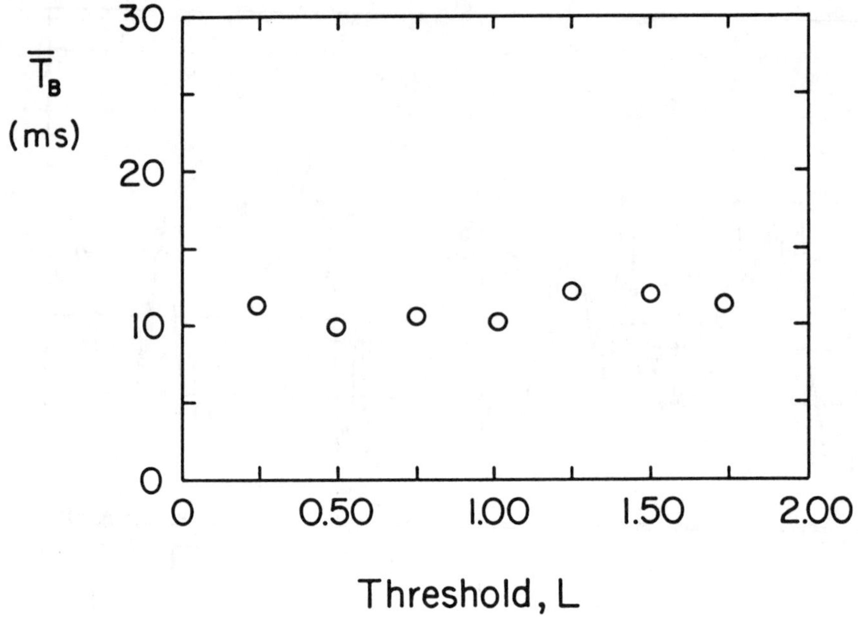

FIGURE 3. Average time between bursts for the revised, modified u-level method, Re = 50 000.

the method for measuring the time between ejections. Instead of using the "center-to-center" time, the time between the trailing edge of one ejection and the leading edge of the next ejection was used. This revision makes it possible to use the cumulative probability method of Barlow and Johnston (1985) to determine the grouping time. This method is less ambiguous than the original Poisson method for determining the grouping time.

As described by Luchik and Tiederman (1987), the data processing begins by constructing a temporal record of the detections given by a threshold of L=1.0. This record is analyzed to determine the probability that T is greater than T_E (the time between ejections) and the results are plotted in semi-log coordinates as shown in Figure 1 for the flow where the Reynolds number was 50 000. The result for $T_E > 4$ ms is the distribution given by ejections from different bursts while the result for $T_E < 1$ ms is the distribution given by ejections from the same burst. The intersection of the straight line extrapolation of these two exponential distributions yields the time interval used to group ejections. In this example, ejections where the time interval between the trailing edge of one and the leading edge of the next is less than 1.9 ms are grouped into the same burst detection. If the time interval is larger than or equal to 1.9 ms the ejections are placed in different bursts. The entire time record for the ejection detections is scanned and a new record of burst detections is created. This process is illustrated along with a short portion of the velocity record in Figure 2. The average time between bursts is then calculated by counting the burst detections and dividing the time for the total record by the number of bursts. The entire process is repeated when the threshold is changed and the threshold

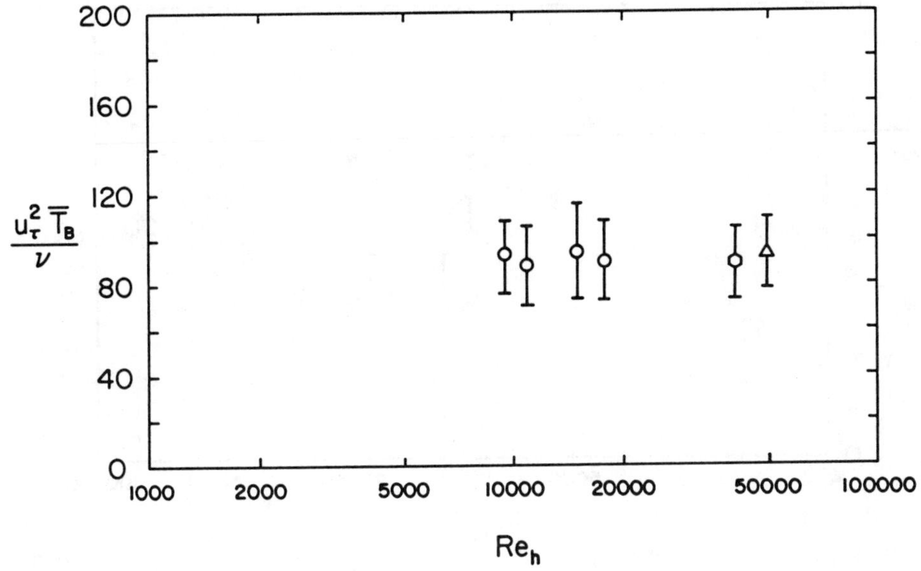

FIGURE 4. Inner variable scaling of the average time between bursts in fully developed channel flows.

is varied in all cases to ensure that there is a threshold independent range. Threshold independence is illustrated in Figure 3 where L was varied from 0.25 to 1.75 and the average time between burst is essentially constant at 10.5 ms for $0.25 \leq L \leq 1.75$.

Obviously there is some uncertainty in the value of the grouping time and this uncertainty propagates into the estimate for the average time between bursts. For the current method the uncertainty for the average time between bursts is on the order of ±15%.

Figures 4, 5 and 6 show how the present results along with the previous results of Luchik and Tiederman (1987) scale with Reynolds number. From Figure 4 it is clear that making $\overline{T}_B$ non-dimensional with the inner variables of wall shear velocity and kinematic viscosity yields a constant value of about 90. Using the friction factor and center-line velocity correlations suggested by Dean (1978) this result is plotted on Figures 5 and 6 where $\overline{T}_B$ has been normalized with the outer variables of center-line velocity and channel half-width and the mixed scaling suggested by Alfredsson and Johansson (1984). For Reynolds numbers between 9 000 and 50 000, $\overline{T}_B$ is best scaled with inner variables consistent with the numerical simulation results of Kim and Spalart (1987) and the VITA results of Blackwelder and Haritonidis (1983).

RELATIONSHIP BETWEEN TECHNIQUES

From side view flow visualization it is apparent that the interface between low speed fluid ejecting away from the wall and the surrounding fluid is often irregular and contorted. Hence it is not surprising that Eulerian velocity

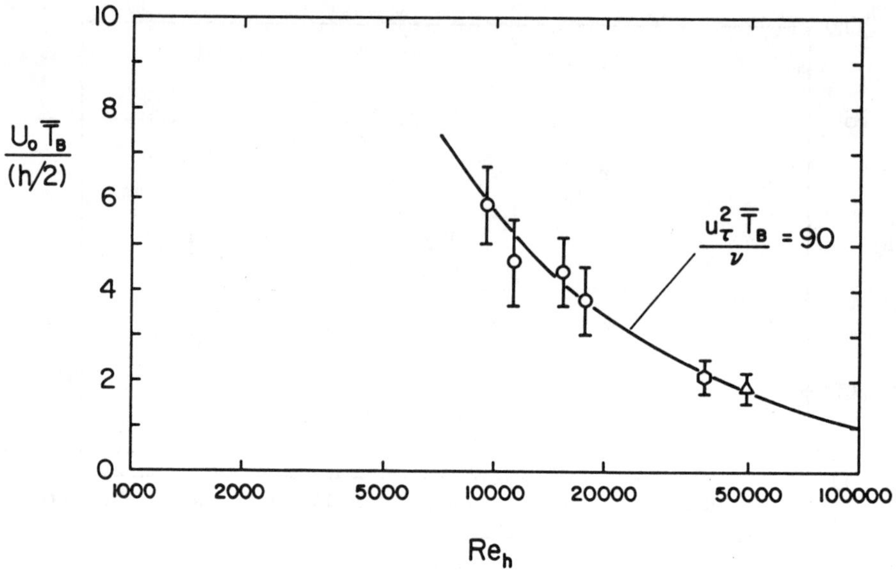

FIGURE 5. Outer variable scaling of the average time between bursts in fully developed channel flows.

detection algorithms designed to detect the ejecting low-speed fluid could give multiple detections as this irregular interface propagates past the probe position. In this sense the velocity detection techniques are ejection detectors.

By using different aspects of the dye contact with the probe to center conditional averages of the simultaneous hot-wire signals, Bogard and Tiederman (1987) were able to relate various signal characteristics with the visually detected events. For example the positive slope VITA signal of the streamwise velocity component corresponds to the trailing edge of visually detected ejections while a negative slope VITA signal occurs at the leading edge. The strongest positive slope VITA signals occur at the trailing edge of the last ejection in a burst while the strongest negative slope VITA signal occurs at the leading edge of the first ejection in a burst. Conditional averages centered on velocity characteristics show that the largest velocity normal to and away from the wall is very closely phase aligned with the large negative spike in the uv signal. In addition, the characteristics of all ensemble averages of the velocity signals are highly dependent upon the condition used to center the averages. Phase scrambling produces significant loss of coherence for those portions of the ejection away from the phase alignment or centering condition. As a result, Bogard and Tiederman (1987) concluded that various phase alignments are needed to deduce the most information about the structure of an ejection. However the complete structure of an ejection can be deduced only from conditional averages that are aligned in space as well as time.

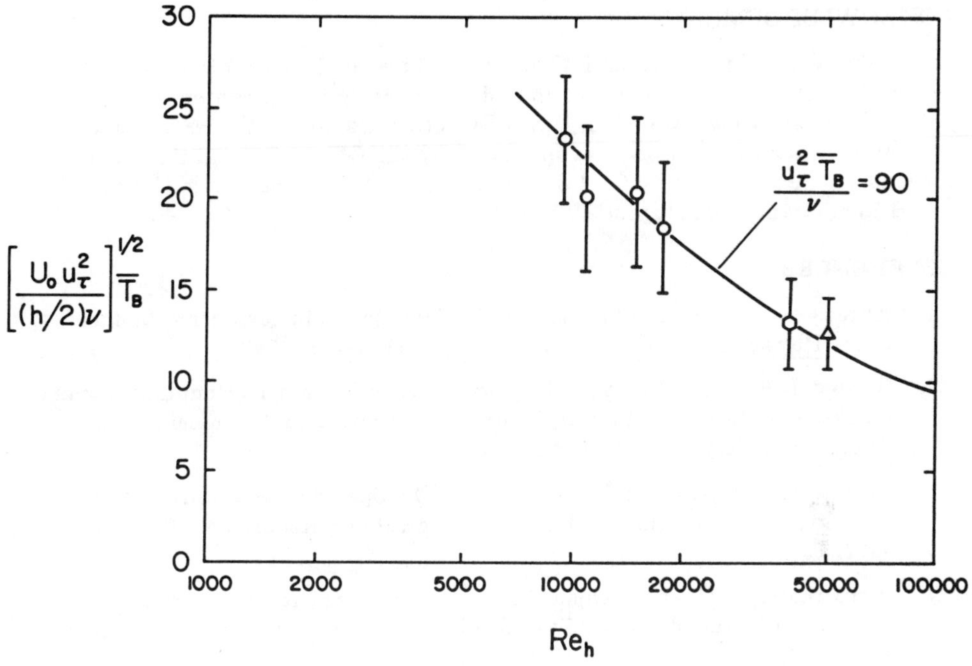

FIGURE 6. Mixed variable scaling of the average time between bursts in fully developed channel flows.

SUMMARY

Most burst detection schemes that use Eulerian velocity information from a single point identify only a portion of the total burst event, usually some portion of an ejection. From an Eulerian viewpoint, each burst may contain several ejections. Consequently the main concept in properly identifying the burst event is to group ejection detections from the same burst into a single detection. This grouping can be done reliably from semi-log plots of the probability distribution for the time interval between the trailing edge of one ejection and the leading edge of the next ejection. The subsequent average time between bursts has a range of threshold independence. Moreover, the quadrant two, u-level, modified u-level and VITA techniques all yield the same average time between bursts which also is the same as the flow visualization result when the latter is available.

Some characteristics of the burst and ejection events can be deduced from conditionally averaged velocity signals phase aligned with that portion of the event. These include characteristics of the leading and trailing edges and the negative uv spike. Away from the alignment point, features are obscured by a rapid loss of coherence due to the three dimensional nature of the flow and the bursts. Spatial and phase alignment of the velocity signals with the three-dimensional burst event are necessary before a complete description can be deduced from conditionally averaged signals.

ACKNOWLEDGEMENTS

In addition to David Bogard and Thomas Luchik several other former and current Purdue students have contributed significantly to the content of this paper. They are Amer Abdallah, David Bustetter, David T. Walker and Jack B. White. Much of this work was sponsored by the Office of Naval Research through contract number N00014-83K-183. Support also was provided by a David Ross Grant from Purdue University.

REFERENCES

1. Alfredsson, P.H. and Johansson, A.V., "Timescales in turbulent channel flow," Physics of Fluids, vol. 27, no. 8, pp. 1974-1981, 1984.

2. Barlow, R.S. and Johnston, J.P., "Structure of Turbulent Boundary Layers on Concave Surfaces," Stanford University, Department of Mechanical Engineering, Report MD-47, August 1985.

3. Blackwelder, R.F. and Eckelmann, H., "The Spanwise Structure of the Bursting Phenomenon," MPI Strömungsforschung Report no. 121, August 1977.

4. Blackwelder, R.F. and Kaplan, R.E., "On the bursting phenomenon near the wall in bounded shear flows," J. Fluid Mechanics, vol. 76, pt. 1, pp. 89-112, 1976.

5. Blackwelder, R.F. and Haritonidis, J.H., "Scaling of the bursting frequency in turbulent boundary layers," J. Fluid Mechanics, vol. 132, pp. 87-103, 1983.

6. Bogard, D.G. and Tiederman, W.G., "Investigation of flow visualization techniques for detecting turbulent bursts," Symposium on Turbulence, 1981, University of Missouri, Rolla, pp. 289-302, 1983.

7. Bogard, D.G. and Tiederman, W.G., "Burst detection with single point velocity measurements," J. Fluid Mechanics, vol. 162, pp. 389-413, 1986.

8. Bogard, D.G. and Tiederman, W.G., "Characteristics of ejections in turbulent channel flow," J. Fluid Mechanics, vol. 179, pp. 1-19, 1987.

9. Comte-Bellot, G., Sabot, J. and Saleh, I., "Detection of Intermittent Events Maintaining Reynolds Stress," Proc. Dynamic Flow Conf., Marseille-Baltimore 1978, pp. 213-229, 1979.

10. Corino, E.R. and Brodkey, R.S., "A visual study of turbulent shear flow," J. Fluid Mechanics, vol. 37, pt. 1, pp. 1-30, 1969.

11. Dean, R.B., "Reynolds Number Dependence of Skin Friction and Other Bulk Flow Variables in Two-Dimensional Rectangular Duct Flows," J. Fluids Engineering, vol. 100, no. 2, pp. 215-223, 1978.

12. Johansson, A.V. and Alfredsson, P.H., "On the structure of turbulent channel flow," J. Fluid Mechanics, vol. 122, pp. 295-314, 1982.

13. Johansson, A.V. and Alfredsson, P.H., "Effects of imperfect spatial resolution on measurements of wall bounded turbulent shear flows," J. Fluid Mechanics, vol. 137, pp. 409-421, 1983.

14. Kim, H.T., Kline, S.J. and Reynolds, W.C., "The production of turbulence near a smooth wall in a turbulent boundary layer," J. Fluid Mechanics, vol. 50, pt. 1, pp. 133-160, 1971.

15. Kim, J. and Spalart, P.R., "Scaling of the bursting frequency in turbulent boundary layers at low Reynolds numbers," Physics of Fluids, vol. 30, no. 11, pp. 3326-3328, 1987.

16. Kline, S.J., Reynolds, W.C., Schraub, F.A. and Runstadler, P.W., "The structure of turbulent boundary layers," J. Fluid Mechanics, vol. 30, pt. 4, pp. 741-773, 1967.

17. Ligrani, P.M. and Bradshaw, P., "Spatial resolution and measurement of turbulence in the viscous sublayer using sub-miniature hot-wire probes," Experiments in Fluids, vol. 5, no. 6, pp. 407-417, 1987.

18. Lu, S.S. and Willmarth, W.W., "Measurements of the structure of the Reynolds stress in a turbulent boundary layer," J. Fluid Mechanics, vol. 60, pt. 3, pp. 481-511, 1973.

19. Luchik, T.S. and Tiederman, W.G., "Effect of spanwise probe volume length on laser velocimeter measurements in wall bounded turbulent flows," Experiments in Fluids, vol. 3, no. 6, pp. 339-341, 1985.

20. Luchik, T.S. and Tiederman, W.G., "Timescale and structure of ejections and bursts in turbulent channel flows," J. Fluid Mechanics, vol. 174, pp. 529-552, 1987.

21. Offen, G.R. and Kline, S.J., "A Comparison and Analysis of Detection Methods for the Measurements of Production in a Boundary Layer," Proc. 3rd Biennial Symposium of Turbulence in Liquids, University of Missouri-Rolla, pp. 289-320, 1975.

22. Simpson, R.L., "An Investigation of the Spatial Structure of the Viscous Sublayer," MPI Strömungsforschung Report 188, August 1976.

23. Tiederman, W.G., Smith, A.J. and Oldaker, D.K., "Structure of the Viscous Sublayer in Drag-Reducing Channel Flows," Proc. 4th Biennial Symposium on Turbulence in Liquids, University of Missouri-Rolla, pp. 312-322, 1977.

24. Wallace, J.M., Brodkey, R.S. and Eckelmann, H., "Pattern-recognized structures in bounded turbulent shear flows," J. Fluid Mechanics, vol. 83, pt. 4, pp. 673-693, 1977.

25. Willmarth, W.W. and Bogar, T.J., "Survey and new measurements of turbulent structures near the wall," Physics of Fluids, vol. 20, no. 10, pt. 2, pp. S9-S21, 1977.

26. Willmarth, W.W. and Sharma, L.K., "Study of turbulent structure with hot wires smaller than the viscous length," J. Fluid Mechanics, vol. 142, pp. 121-149, 1984.

The Art of Detection of Turbulence Structures

P. HENRIK ALFREDSSON and ARNE V. JOHANSSON
Department of Mechanics
The Royal Institute of Technology
S-100 44 Stockholm, Sweden

In wall bounded shear flows a number of different coherent structures have been identified, however, the relevance to the dynamics of the flow for some of these structures is still in doubt. In the present paper we demonstrate that there is a clear connection between near-wall internal shear-layer structures detected with the VITA-technique and the long time statistics of the flow which implies that these structures are of dynamic significance.

INTRODUCTION

Turbulence was long considered to be associated with a completely chaotic and random fluid motion. However, about two decades ago flow visualization studies indicated that a large portion of the turbulence production in boundary layer flows was associated with a certain sequence of events taking place in the near wall region. The incipient event was found to be a lift-up of low-speed fluid from the viscous sublayer into the buffer region, thus forming an internal shear layer which becomes unstable and results in a chaotic break-up. Since then coherent structures have been identified in other turbulent flow fields, such as wakes, jets etc., and turbulence research has been directed from the statistical approach towards experimental and theoretical investigations of these structures.

In wall-bounded flows more than five different structures have been reported in the literature since the first discoveries of the so called bursting events. However, the dynamical significance of these structures is still largely unknown. If the coherent structure is an important part of the flow field the turbulent energy associated with it must be high. The structure will therefore strongly color the long-time statistics of the turbulent field. Also long-time information on the time scales of turbulence, i.e. spectra, micro and macro scales should be connected to the time scales of the structures. Scale information is of utmost importance for the understanding of turbulent flows. However no consensus has yet been reached on what time scale is governing e.g. turbulence production. This depends to some extent on the fact that no agreement has been reached on which are the dynamically most important structures. Therefore it may be valuable to establish the scaling behavior of more objective quantities such as spectra or micro scales, in parallel with investigations of the structural features of the flow.

One may believe that numerical simulation data bases should be able to map the velocity field in such a detail that any uncertainty on what structures are of importance should be avoided. However, this has not yet happened and for determining scaling behavior we must rely on large Reynolds number ranges which for the foreseeable future still have to be obtained experimentally.

The present paper deals with the relation between long-time statistics such as rms, skewness and flatness and the characteristics of coherent structures and the relation between the Taylor micro-scale and the typical time scale of so called VITA-event.

BACKGROUND

When studying turbulence producing processes the near wall region and especially the buffer region is of special interest as the turbulence production is largest in this region. When conditional sampling was introduced the idea was to use some feature of one or a few flow variables which is a unique attribute of the event under study. The question then arises what particular feature of the flow field should be used to extract the coherent motion from the random parts of the flow. One is therefore striving to find an intermittent triggering signal which is closely associated with the event. The intermittency implies that the signal should have a high flatness factor. In some experimental situations flow visualization could be used to determine a certain type of motion (see e.g. Bogard & Tiedermann, 1986 who used dye injection in the viscous sublayer and registered its lift-up from the wall), however this approach is always inherently subjective. Also when large amounts of data are needed or for studies at high Reynolds number visualization is usually not practical. Possible candidates to use for detection purposes are measured signals of one or several velocity components, the wall shear stress or the wall pressure. Especially velocity signals measured in the buffer region have been used extensively in numerous investigations during the last decades.

Data from numerical simulations and experiments for channel flow at a low Reynolds number has e.g. been published by Kim, Moin & Moser (1987). The rms-value of the streamwise velocity fluctuations (u) is shown to reach a maximum of almost $3u_\tau$ at $y^+=15$ and is above $2u_\tau$ within the buffer region. The intensity of the normal velocity fluctuations (v) is below $1u_\tau$ in the buffer region and reach a maximum of about $1u_\tau$ at $y^+=50$. In the buffer region the skewness of both u and v is close to zero, whereas the flatness reach fairly high levels (around 6 at $y^+=15$) for v, while it is below 3 for u. Closer to the wall the flatness becomes higher indicating an intermittent behavior. The correlation between u and v is about -0.4 in the buffer region with skewness and flatness values of uv that are close to what one would expect if u and v were Gaussian signals with the measured correlation coefficient (see Lu & Willmarth, 1973).* The overall picture is that the long time statistics of u, v and uv in the buffer region do not differ greatly from what one would expect if u and v were Gaussian. Hence, based on these long time averages there is no feature that *a priori* can be said to be suitable for detecting coherent structures in the buffer region. However both the normal velocity and especially the uv product fulfill the requirement of high flatness factors.

The spectra of u and v are broad banded and do not show any sign that deterministic features are important. However, the time-derivative of the u-velocity is skewed towards positive values (around 0.8, see e.g. Johansson & Alfredsson, 1983), i.e. a fixed probe measuring u is subjected to large and rapid variations from low to high velocity. This is probably associated with the lift up process which create sharp shear layers at the upstream side of the lifted streak.

The shear layers that are reflected in the u-velocity could therefore be detected by using this feature. This was first done by Blackwelder & Kaplan (1976) who developed the now well known VITA-technique. It uses a kind of band pass filtering to sort out large changes in the velocity of a certain time scale by calculating the short time variance, which was shown to be a highly intermittent signal. The method contains two parameters, a threshold and an averaging time. For a given threshold one can by changing the averaging time detect shear layers of various scales, and by doing so obtain a distribution of the passage time of shear layers, which gives the width of the shear layer if their propagation velocity is known. The main finding of Blackwelder & Kaplan was that by this method several more events where the velocity was going from low to high than from high to low were picked up, in accordance with what you would expect from the skewness of du/dt. A later refinement of the method is to distinguish between these different types of events.

---

* This value of the correlation coefficient is directly determinable from the mean flow equation and the measured rms-values of the fluctuating velocity components.

COHERENT STRUCTURES AND LONG-TIME STATISTICS

If the coherent structure is an important part of the flow field its characteristics must show up in the long time statistics of the flow field. For instance if the fluctuating streamwise velocity field is divided into a deterministic and a random part $u=u_d+u_r$ which are assumed to be uncorrelated, i.e. $\overline{u_d u_r}=0$ where overbar denotes that the long time average, the rms, skewness and flatness factors of u may be written

$$u_{rms}=(u_d)_{rms}(1+\gamma)^{1/2}$$

$$S(u)=\frac{S(u_d)}{(1+\gamma)^{3/2}}$$

$$F(u)=\frac{F(u_d)+6\gamma+3\gamma^2}{(1+\gamma)^2}$$

where $\gamma\ (=\overline{u_r^2}\,/\,\overline{u_d^2}\ )$ is the ratio between the energy content in the random and deterministic parts. This clearly shows that if the energy content of the deterministic structures is high this results in a strong coloring of the long-time statistics.

Figure 1 shows conditional averages obtained with the VITA technique at three different y^+-positions. Close to the wall ($y^+=2.8$) the conditional average show a positive peak which corresponds to a large positive skewness and a high flatness in the long time average, because of the large deviation from the mean value. At $y^+=12$ the conditional average is symmetric around the mean value, i.e. the skewness is close to zero and the flatness is small. Further from the wall ($y^+=220$) the conditional average is shifted to the negative side and the skewness also has a negative value and the flatness is higher than at $y^+=13$. In the VITA-technique a threshold is used which ensures that only energetic events are detected and it has been shown that these events have a significant influence on the long time statistics. Therefore the scaling of VITA-events may also have bearing on scaling of frequency resolved quantities such as spectra or e.g. micro scales.

DERIVATION OF THE TAYLOR MICRO-SCALE IN THE OUTER REGION

It is possible from rather general arguments to derive the Taylor-scale in the outer region of channel (or pipe) flows as was shown by Lawn (1971). If the contribution from viscous stresses is neglected the Reynolds stress distribution in the outer region can be written as

$$-\overline{uv}=u_\tau^2\left(1-\frac{y}{b}\right) \qquad (1)$$

where y is the coordinate normal to the wall and b is the half channel width.

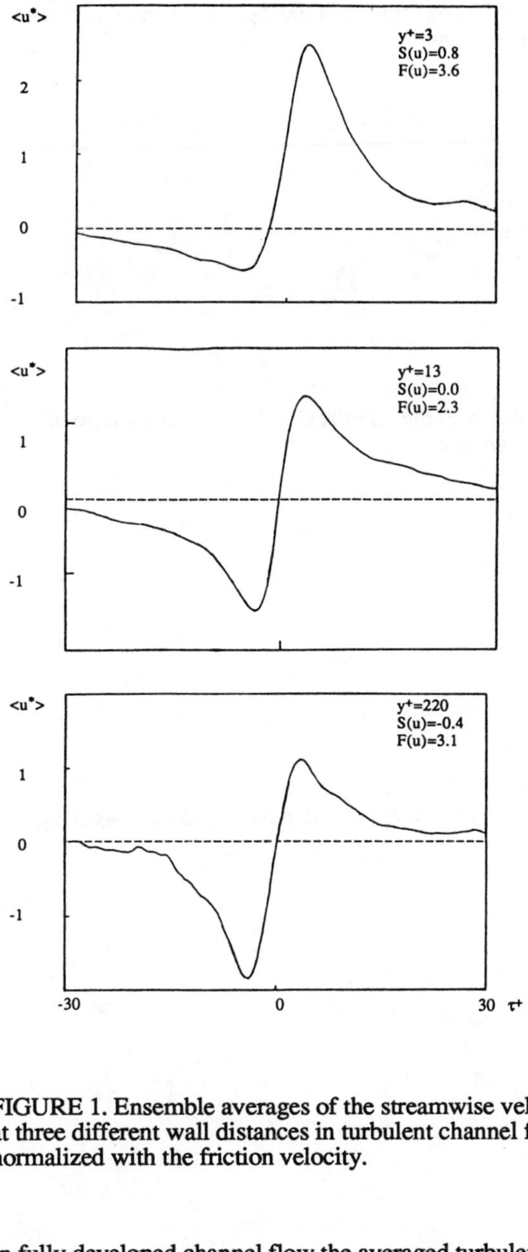

FIGURE 1. Ensemble averages of the streamwise velocity detected with the VITA-technique at three different wall distances in turbulent channel flow at Re=13800. The velocity is normalized with the friction velocity.

In fully developed channel flow the averaged turbulent kinetic energy $\overline{q}\ (=\frac{1}{2}\overline{(u^2+v^2+w^2)}\ )$ is constant in the downstream direction so there is a balance between production, dissipation (ε) and transport, i.e.

$$-\overline{uv}\frac{dU}{dy} = \varepsilon + \frac{d}{dy}\overline{v(\rho^{-1}p + q)} \qquad (2)$$

where p is the fluctuating pressure. In eq. 2 viscous transport has been neglected. In the outer region the mean velocity profile can be written

$$\frac{U_{CL}-U}{u_\tau} = f(\eta) \tag{3}$$

where $\eta = y/b$. Combining eq. 1, 2 and 3 gives

$$-(1-\eta)f' = \frac{\varepsilon b}{u_\tau^3} + \frac{d}{d\eta}\overline{v(\rho^{-1}p + q)}/u_\tau^3 \tag{4}$$

If we now assume Reynolds number similarity in the outer region for the distribution of mean velocity and the fluctuations we can write eq. 4 as

$$\varepsilon = \frac{u_\tau^3}{b} g(\eta)$$

In the outer region we may approximate the dissipation as

$$\varepsilon = \nu \overline{\frac{\partial u_i}{\partial x_j}\frac{\partial u_j}{\partial x_i}}$$

with an error of the order of the square of the ratio between the micro- and macro-scales. Hence we get

$$\overline{\left(\frac{\partial u}{\partial x}\right)^2} \sim \frac{\varepsilon}{\nu} \sim \frac{u_\tau^3}{b\nu}$$

The Taylor micro (length) scale (λ_x) is defined as

$$\lambda_x = \sqrt{2\,\overline{u^2} / \overline{\left(\frac{\partial u}{\partial x}\right)^2}}$$

which together with eq. 4 yields

$$\lambda_x \sim \left(u_\tau^2 / \frac{u_\tau^3}{b\nu}\right)^{1/2} = \sqrt{b l_*}$$

where l_* is the viscous length scale, i.e. the Taylor micro scale in the outer region is proportional to the geometric mean of the outer and viscous length scales. In experiments one usually measures the micro time scale, which in this region can be estimated from the length scale and the propagation velocity which at a given η is a given fraction of U_{CL}. This finally gives the following expression for the Taylor micro time scale

$$\lambda_t \sim \sqrt{t_o t_* \sqrt{u_\tau / U_{CL}}} \sim \sqrt{t_o t_*}\ Re^{-1/16}$$

i.e. it is essentially proportional to the geometric mean of the inner ($t_* = \nu / u_\tau^2$) and outer ($t_o = b/U_{CL}$) time scales. This time scale will in the following be denoted as the mixed time scale, which was introduced by Alfredsson & Johansson (1984).

EXPERIMENTAL RESULTS FOR THE TAYLOR MICRO SCALE

We have collected data from the literature (see table 1) from several studies which have reported the Taylor micro scale in the outer region. Included are also our own data taken both at KTH in a water tunnel, in the Göttingen oil channel and in the MIT low turbulence wind tunnel (the wind tunnel data are from a turbulent boundary layer). The data span a Reynolds number range of almost a factor of 100. In figure 2 the data are plotted normalized with the inner, the mixed and the outer time scales. For high Reynolds number the data clearly show that the Taylor micro scale is proportional to the mixed time scale in accordance with the theory. Furthermore it is of order one, in this normalization. For low Re there is a slight deviation and viscous scaling seems to be more appropriate. This may be explained by the fact that the theory assumes Reynolds number similarity which is valid only for high Re.

TABLE 1. Data on Taylor micro scale from the literature and from the present data. The data from MPI was taken in cooperation with H. Eckelmann and the data from MIT in cooperation with J. Haritonidis. L^+ denotes the probe length.

		Re	L^+	η	symbol
Comte-Bellot	channel	126 000	10	0.5	□
		266 000	21	0.5	
		514 000	35	0.5	
Laufer	pipe	50 000	10 (?)	0.5	▷
		500 000	20 (?)	0.5	
Lawn	pipe	46 000	14	0.65	x
		73 000	21	0.65	
		112 000	30	0.65	
Schildknecht *et al.*	pipe	20 000		0.6	+
Data from KTH	channel	10 250	10	0.5	o
		13 800	10	0.65	
		14 300	4	0.38	
		39 500	22	0.5	
		54 000	14	0.5	
		123 000	31	0.5	
Data from MPI	channel	5 600	2	0.5	•
		7 700	2	0.5	
Data from MIT	bound. layer	97 000	7	0.5	◇
		385 000	28	0.5	

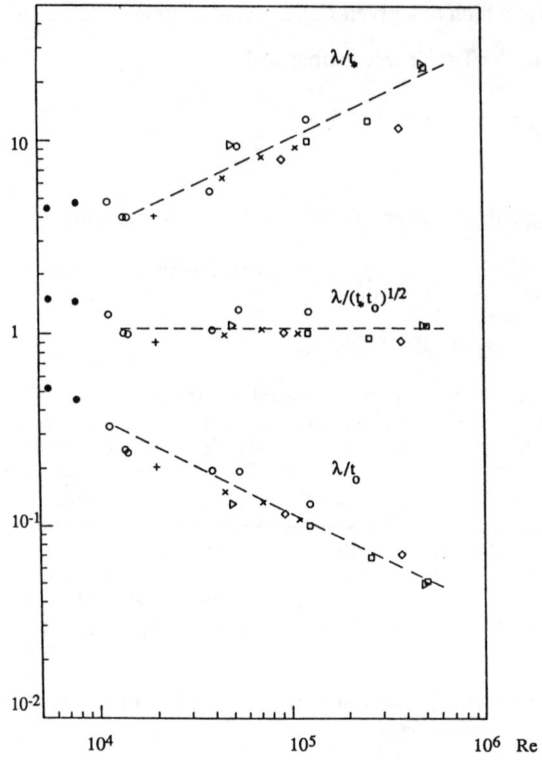

FIGURE 2. Taylor micro time scale in the outer region of channel and pipe flow, normalized with the inner, the mixed and the outer time scales. Symbols are given in table 1.

This difference in scaling indicates that experiments or simulations at low Reynolds numbers may suffer from low Reynolds-number effects, and therefore be unsuitable to use in studies of scaling behavior.

The above derivation was made for the outer parts of the flow, and it may be interesting to see the behavior near the wall. In this region the data in the literature is much more scarce, and figure 3 contains only our own data and those of Schildknecht, Miller & Meier (1979). The mixed scaling gives the best result also close to the wall, however the micro scale increases close to the wall reflecting either a slower propagation velocity of the structures contributing to the Taylor scale or that these structures become larger in the vicinity of the wall.

In channel flow it has been found that both the duration and frequency of occurrence of VITA-events scales with the mixed time scale (Alfredsson & Johansson, 1984). VITA events are shear layers passing the probe, which give large contributions to $\partial u/\partial t$. The most probable duration is of the order of Taylor time scale. VITA-events are characterized by the streamwise signal passing from negative to positive values, i.e. it has a zero-crossing. The micro-time scale is also connected to zero-crossings through Rice's theorem for a Gaussian signal (see e.g. Sreenivasan, Prabhu & Narasimha, 1983). One may hypothesize that the similarity in the scaling of the VITA-events and the Taylor micro time scale is more than a coincidence.

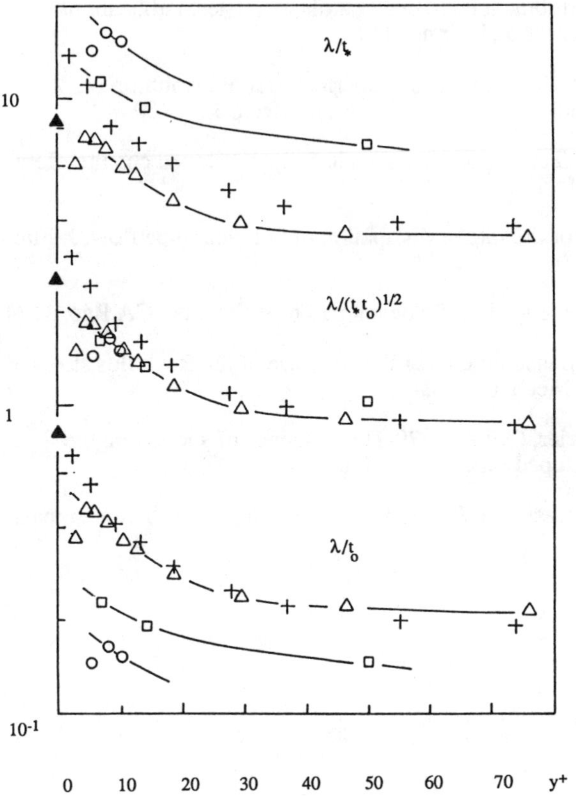

FIGURE 3. The Taylor micro scale in the near wall region. ▲: Re=10250 (from wall shear stress signal KTH), Δ: Re=13800,□: Re=54000, o: Re=123000 (channel flow KTH), +: Re=20000 (pipe flow, Schildknecht *et al)*

CONCLUSIONS

The Taylor micro time scale has been shown to scale with the mixed time scale in the outer region of channel and pipe flow over a Reynolds number range of almost a factor 100. This scaling was found experimentally to be valid also close to the wall. It is hypothesized that this is connected to the mixed scaling of VITA-events earlier found by us.

REFERENCES

Alfredsson, P.H. & Johansson, A.V. 1984 Time scales in turbulent channel flow. Phys. Fluids 27, 1974.

Blackwelder, R.F. & Kaplan, R.E. 1976 On the wall structure of the turbulent boundary layer. J. Fluid Mech. 76, 89.

Bogard, D.G. & Tiedermann, W.G. 1986 Burst detection with single-point velocity measurements. J. Fluid Mech. 162, 389.

Comte-Bellot, G. 1965 Ecoulement turbulent entre deux parois paralleles. Publications Scientifiques et Techniques du Ministere de l'Air no. 419.

Johansson, A.V. & Alfredsson, P.H. 1983 Effects of imperfect spatial resolution on measurements of wall-bounded turbulent shear flows. J. Fluid Mech. 137, 409.

Kim, J., Moin, P. & Moser, R. 1987 Turbulence statistics in fully developed channel flow at low Reynolds number. J. Fluid Mech. 177, 133.

Lawn, C.J. 1971 The determination of the rate of dissipation in turbulent pipe flow. J. Fluid Mech. 48, 477.

Laufer, J. 1954 The structure of turbulence in fully developed pipe flow. NACA Rep. 1174.

Lu, S.S. & Willmarth, W.W. 1973 Measurements of the structure of the Reynolds stress in a turbulent boundary layer. J. Fluid Mech. 60, 481.

Schildknecht, M., Miller, J.A. & Meier, G.E.A. 1979 The influence of suction on the structure of turbulence in fully developed pipe flow, J. Fluid Mech., 90, 67.

Sreenivasan, K.R., Prabhu, A. & Narasimha, R. 1983 Zero-crossings in turbulent signals. J. Fluid Mech. 137, 251.

Electrochemical Method for Studying Local Structure of Turbulent Flows

O. N. KASHINSKY
Institute of Thermophysics
Siberian Branch of the USSR Academy of Sciences
Novosibirsk, USSR

SUMMARY

The electrochemical method for studying hydrodynamic characteristics of turbulent liquid flows is a diffusional analog of the hot-wire anemometer. The technique for measurement of wall shear stress, flow velocity, and fluctuational quantities using this method is described. Examples of application of the electrochemical method for studying flows under various hydrodynamic conditions are presented.

1. INTRODUCTION

The electrochemical method for measuring local characteristics of liquid flows has recently found wide application in the study of single-phase and two-phase flows. Although the range of application of the method is narrower than, for example, for hot-wire or laser Doppler techniques, there are some peculiarities that make its use for some cases more expedient.

The main advantages of the method are:

- the ability to measure in the immediate vicinity of the wall;
- the ability to make probes of very small sizes and nonconventional shape;
- the extreme simplicity of the primary electronics which facilitates, for example, the application of the multichannel version of the method.

The electrochemical method is based on measuring the rate of ion diffusion to the surface of a polarized electrode inserted into the flow. It is necessary to use an electrolyte of a special composition as the working liquid, which creates definite difficulties in the range of applicability. Nevertheless, the electrochemical method appears to be the most suitable in the study of complicated liquid flows in laboratory setups.

2. FUNDAMENTALS OF THE ELECTROCHEMICAL METHOD

2.1 Wall-Shear-Stress Measurements

The electrochemical method was first used for the determination of wall shear stress by Mitchell and Hanratty [1]. The method consists of the following (Figure 1). Two electrodes, i.e., a small cathode and an anode of much larger area, are inserted into the flow of an electrolyte of a special composition. The cathode is a wall-shear-stress probe, a piece of platinum wire or film embedded flush with the pipe wall. When voltage is applied between the electrodes, a rapid electrochemical reaction occurs resulting in cathode polarization. The concentration of active ions at the cathode becomes zero. Active ions diffuse to the cathode due to the concentration gradient and electric current appears in the circuit. Two electrochemical systems are usually used in hydrodynamic studies [2]. The first is a solution of potassium ferri- and ferrocyanide and the excess of sodium hydroxide in distilled water. The excess of sodium hydroxide produces high electrical conductivity of the solution and prevents ion migration due to the electric field. The second system consists of a solution of iodine and potassium iodide in water with the excess of iodide being taken as both the reacting and background substance. Under these conditions electric current is defined only by the diffusion of active ions (ferricyanide or tri-iodide, respectively) to the cathode.

Solution of the concentration boundary layer equation for a cathode with a linear velocity profile near the probe,

$$u = \frac{\tau_w}{\mu} y \tag{1}$$

where u is liquid velocity, y the distance from the wall, μ the viscosity of the solution, and τ_w the wall shear stress, yields the relationship between probe current I and the wall shear stress, which in dimensionless form is

$$\frac{K\ell}{D} = 0.807 \left(\frac{\tau_w \ell^2}{\mu D}\right)^{1/3} \tag{2}$$

where ℓ is the streamwise probe size, D the diffusion coefficient, $K = I/FSC_o$ is the mass transfer coefficient, where S is the area of the electrode and C_o the concentration of active ions. The Faraday constant F

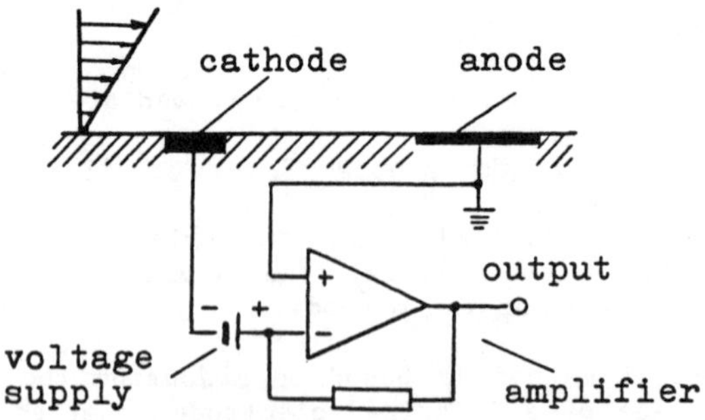

FIGURE 1. Principle of electrochemical wall-shear-stress measurements.

converts an electrical quantity, i.e., probe current, into a diffusional mass-transfer coefficient. The solution of (2) with respect to τ_w is of the form

$$\tau_w = \frac{1.87\ \mu I^3}{F^3 C_o^3 D^2 \ell^2 h^3} = AI^3 \tag{3}$$

where h is the transverse size of the probe. The coefficient A may either be calculated from the formula or determined from the calibration. When the composition and temperature of the solution are unchanged this coefficient is constant, so τ_w is proportional to the cube of the current.

The electrochemical wall-shear-stress probe differs basically from a hot-film probe. Heat transfer from the hot-film probe leads to heating of the substrate, which results in increasing the effective size of the probe compared to its actual size. Thus the use of theoretical calibration relationships for a hot-film probe is impossible. In the case of the electrochemical method, the wall is practically an ideal electrical insulator. Therefore the effective size of the probe is equal to the geochemical size of its sensitive element, and it is possible to use the calculated value of coefficient A in measurements.

The construction of the wall-shear-stress probe is shown in Figure 2.a. The sensitive element has a cross section ranging from $0.1 \times 1\ mm^2$ to $0.02 \times 0.3\ mm^2$. The use of microprobes has some peculiarities but nevertheless permits reliable measurements. If one uses the double electrochemical wall-shear-stress probe (Figure 2.b), which can also be made of very small size, it is possible to measure both value and direction of the wall shear stress, which is important when studying separated flows. This probe was first used by Karabelas and Hanratty [3].

When both electrodes operate independently (the voltage is applied only to one of them), their currents are the same if the sizes of the electrodes are equal and they are under the same hydrodynamic conditions. When the electrodes are polarized simultaneously, the diffusion wake from the upstream electrode shields the downstream one. As a result the reading of the second electrode decreases. When the flow changes direction, the smaller reading corresponds to the upstream electrode. The comparison

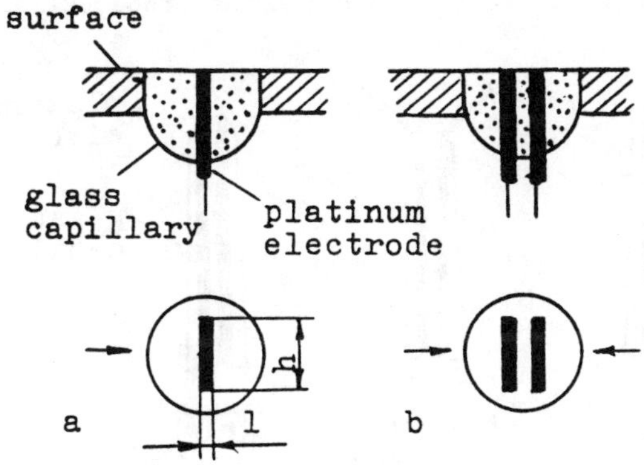

FIGURE 2. Wall-shear-stress probes.

of electrode currents permits us, therefore, to determine the flow direction, and the reading of the probe, which is the upstream one at a given moment, indicates the instantaneous τ_w value.

It is worth noting that when τ_w is close to zero, the use of relations (2) and (3) to determine τ_w is incorrect because in this case the probe current is determined by the quadratic term in velocity expansion, which is proportional to the local longitudinal pressure gradient. Nevertheless, the double electrochemical probe can be effectively used for quantitative measurements of τ_w in the regime of incipient (transient) separation when the wall shear stress fluctuates with high amplitude about zero value. In this case almost always in a given moment τ_w has a high positive or negative value. Instants when τ_w crosses zero give small contributions to the result, because both those instants are short and τ_w is small.

2.2 Liquid Velocity Measurements

If a metal electrode placed in the flow operates in the limiting diffusion current regime, its current will depend on the velocity of liquid near the electrode. This principle is the basis for the measurements of liquid velocity by the electrochemical method. In the case of a hot-wire system, the probe is a conductor and it requires two electrical leads. In the electrochemical method the probe is an electrode; therefore, only one electric lead is necessary. This permits the use of velocity probes of nonconventional shape, very small size, and point-like type. The simplest and very convenient one is the "blunt-nose" type of probe (Figure 3.a). It is a thin platinum wire welded into a glass capillary of conical shape. The tip of the probe is treated to render it streamlined. The size of the working tip of the probe is defined primarily by the requirement of mechanical rigidity and lack of vibrations in the flow. At low liquid velocity it is possible to use probes with a diameter of glass insulation ranging from 30 to 40 μm. The theoretical calibration relation for this probe is based on the solution of a mass transfer problem from an electrode placed in the forward critical point [4], which yields the result that the current is proportional to the square root of velocity. Direct application of the theoretical relation is difficult because the condition of the small ratio between electrode and cover diameters is not satisfied. Besides it is difficult to provide the correct geometrical shape of the probe nose if its dimensions are small. So it is usually more convenient to perform the probe calibration in the

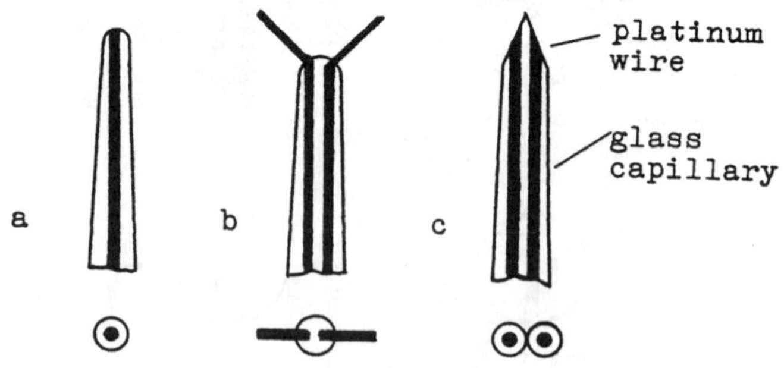

FIGURE 3. Velocity probes.

known hydrodynamic conditions. Practical calibration of the above probe is of the type of the "King's law",

$$I = a + bu^{1/2} \tag{4}$$

where a, b are constants for a given probe. Different types of electrochemical velocity probes are described by Mizushina [5] and Hogenes and Hanratty [6].

Probes with several electrodes are used for velocity vector measurements. Examples of double probes ae shown in Figure 3 (b,c). The V-shaped one is an analog of the X-type hot-wire probe. It consists of two pieces of platinum wire welded into a double glass capillary of conical shape. The free ends of the wires of 5 to 10 diameters in length are bent at an angle of ±45° with respect to probe axis. In spite of the fact that the electrodes are fixed only at one end, the probe possesses sufficient mechanical strength and is not deformed by the flow. The specific character of the electrochemical method (absence of effects similar to heat losses to prongs) allows use of probes with low wire length-to-diameter ratio. Even with a ratio of 5 to 7 the probe exhibits high angular sensitivity (up to 1 percent per degree of rotation). If the electrode diameter is 20 μm, the total size of the probe is 0.1 to 0.15 mm which is comparable to most minute X-shaped hot-wire probes. The wedge-type probe (Figure 3.c) has slightly larger dimensions, but exhibits higher mechanical strength, and is less susceptible to contaminations.

2.3 Fluctuation Quantity Measurements

When measuring unsteady wall shear stress and velocity, the frequency response of probes should be taken into account. The probe current is determined by the transfer of active ions having low mobility, so the Schmidt number of the system is high, more than 1000. Therefore the frequency response for the electrochemical probe is much poorer than for the hot-wire one. The solution of the unsteady-state concentration boundary layer equation involving time-dependent wall shear stress and velocity permits calculation of the frequency responses of the probes, defined as follows:

$$S_I = S_\tau \, |H_\tau|^2 \, , \quad S_I = S_u \, |H_u|^2 \tag{5}$$

where S_I, S_τ, S_u are spectral densities of the current, wall stress, and velocity, respectively. Approximations of H_τ and H_u are of the form [7,8],

$$|H_\tau| = \begin{cases} (1 + 0.056\,\omega_\tau^2 + 0.00126\,\omega_\tau^4)^{-1/2} & \text{when } \omega_\tau \leq 6 \\ \dfrac{3.7}{\omega_\tau}\,(1 - 1.5\,\omega_\tau^{-1/2} + 1.1\,\omega_\tau^{-1})^{1/2} & \text{when } \omega_\tau > 6 \end{cases} \tag{6}$$

$$|H_u| = \left(1 + \frac{\omega_u^2}{15\,Sc^{2/3}}\right)^{1/4} \cdot \left(\frac{1 + 2.32\times10^{-3}\,\omega_u^2}{1 + 4.45\times10^{-2}\,\omega_u^2 + 1.49\times10^{-4}\,\omega_u^4}\right)^{3/4} \tag{7}$$

where dimensionless frequencies

$$\omega_\tau = \left(\frac{\mu^2 \ell^2}{\tau_w^2\, d}\right)^{1/3} , \quad \omega_u = \frac{\omega\, d}{k_1 u}\, Sc^{1/3} , \tag{8}$$

d is the probe diameter, Sc the Schmidt number, ω the angular frequency, and the coefficient k_1 depends on the shape of the probe nose.

Figure 4 shows experimental frequency responses of the velocity "blunt-nose" probe obtained by oscillating the probe (60 μm in dia.) with the proper frequency in a jet of liquid emerging from a nozzle; f is the cycle frequency and H_0 is the value of H at 25 Hz. It can be seen that the bandwidth in which the probe signal is not damped is approximately 1 kHz for the liquid velocity of 1 m/s. This bandwidth depends linearly on mean velocity, as can be seen from (8). Therefore it is possible to use a quasi-steady-state approximation for practically all measurements. In some cases, for example when measuring the high frequency part of the turbulent fluctuation spectrum, it is necessary to take into account the attenuation of frequency response at high frequencies. In order to improve the frequency response of the probe it is necessary to reduce its dimensions. In practice, velocity probes with a diameter of 40 to 50 μm and wall-shear-stress probes with the streamline size of 20-30 μm operate always under quasi-steady-state conditions. In this case it is possible to process the probe signals by the same procedures as for hot-film probes.

2.4 Measuring Equipment

One of the main advantages of the electrochemical method is the simplicity of primary electronics. The electric circuit including an electrochemical probe determines the regime of operation of the probe (that of limiting diffusion current) and amplifies its signal. The electrochemical method is similar in principle to a constant-temperature hot wire. But in order to maintain constant concentration of active ions on the probe surface no additional electronic equipment is needed. Zero concentration of active ions is provided by rapid electrochemical reaction on the probe surface. As far as the hot wire is concerned, the probe is connected to the feedback loop of an amplifier. Therefore electronics

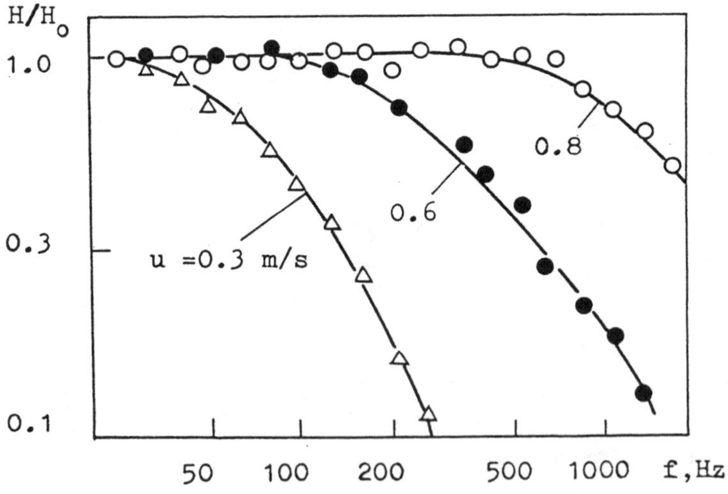

FIGURE 4. Frequency response of the velocity probes.

used in the electrochemical method are similar to those used in the constant-current hot-wire technique, but are far simpler and more stable.

The current of the electrochemical probe is usually from 1 to 20 μA, the lower value corresponding to the smallest probes described above. The frequency of the signal does not exceed several kilohertz. A typical circuit of the amplifier is shown in Figure 5. For convenience the bias voltage of the probe is fed to the noninverting input of the first operational amplifier. The second operational amplifier is used to compensate for the feed voltage and is connected as a differential amplifier. The transference coefficient is adjusted by resistor R1. If the gain has a wide range, different feedback resistors are connected as R1 by the switch. The circuit shown is very simple and stable in operation. Definite difficulties arise when eliminating electrical inductance because the probe serves as a high-resistance current source.

The simplicity of producing the probes and primary electronic equipment is an important advantage of the electrochemical method. It permits one to use a multichannel version of the method, and to perform simultaneous measurements of flow hydrodynamics at many points. In this case the configuration of the rake of probes and measuring electronics is unchanged and one channel of the amplifier corresponds to one definite probe. When either wall shear stress or velocity varies within wide limits, then the probe current alters no more than 3- to 5-fold due to nonlinearity of the calibration curves. Therefore it is possible to use a fixed resistor R1 in the amplifier circuit, which leads to the maximum simplification of measuring electronics.

An example of multichannel measurements of the turbulent flow using a rake of many electrochemical probes is presented by Hogenes and Hanratty [6].

3. EXAMPLES OF APPLICATION OF THE ELECTROCHEMICAL METHOD

3.1 Flow Between Two Close-Spaced Walls

One of the fields in which the application of the electrochemical method is effective is the flow in small-dimension channels. Results of a study of a jet flow spreading in a gap between two parallel walls are presented. At low Reynolds number based on the gap spacing, the creeping flow of this

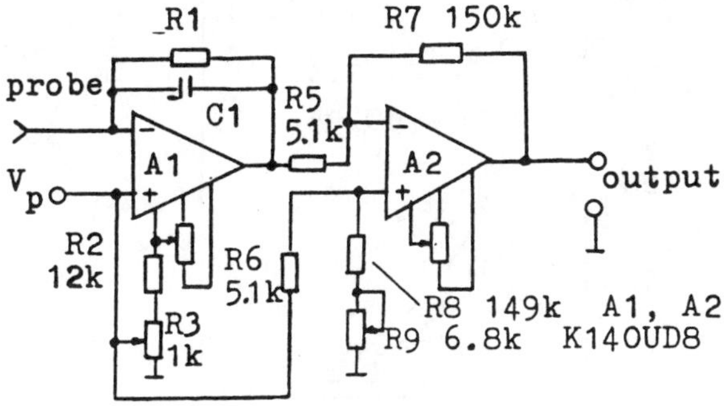

FIGURE 5. Probe current amplifier.

type simulates the two-dimensional potential flow over bodies. In real cases the condition of low Reynolds numbers is usually not satisfied; thus the inertia effects become important. The flow in a narrow gap has two characteristic linear scales, the first one being equal to the spacing between plates and the second being determined, for example, by the size of the body. The development of the flow in the direction perpendicular to the plates is impeded, whereas in the plane parallel to the walls it is not restricted, so the emergence of structures with scales much larger than the spacing between the plates is possible.

The test section was a flat channel made by two plexiglass plates separated by an insert of 1-2 mm. The liquid flowed from a rectangular nozzle and spread as a jet between the plates. Twelve electrochemical wall-shear-stress probes were inserted along the axis of one of the plates at a distance of 20 mm from each other. Figure 6 shows the characteristic wall-shear-stress distribution along the axis of the jet, the coordinate x being recorded from the nozzle. Reynolds number is based on mean velocity in the nozzle and the length scale equal to plate spacing. At small x, the first sharp increase of τ_w related to turbulization of the jet flow takes place, then τ_w attains its maximum value, afterwards decreasing rapidly. The distance at which the turbulent jet is spread increases with the nozzle width and decreases with decreasing of the spacing. This is quite natural, because a wider nozzle results in higher initial momentum of the jet and smaller spacing leads to higher energy dissipation due to wall friction.

Figure 7 shows simultaneous records of several signals of electrochemical wall-shear-stress probes placed at different distances from the nozzle in a transitional flow, Re = 4760. The two oscillograms in Figure 7 are two parts of the same record. In the system described there are two types of perturbation development. At Reynolds numbers close to critical, the appearance of short turbulent spots is typical (Figure 7.a). This picture permits one to follow the movement of spots along the jet. In addition, continuous perturbation of long duration (from 3 to 10 s) can arise. In the same geometry and the same regime, the coexistence of spots, continuous perturbations, and unperturbed liquid is possible. The

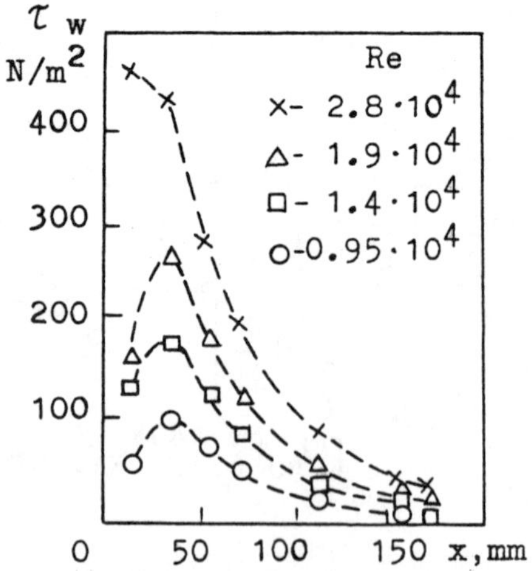

FIGURE 6. Wall-shear-stress distribution along the jet.

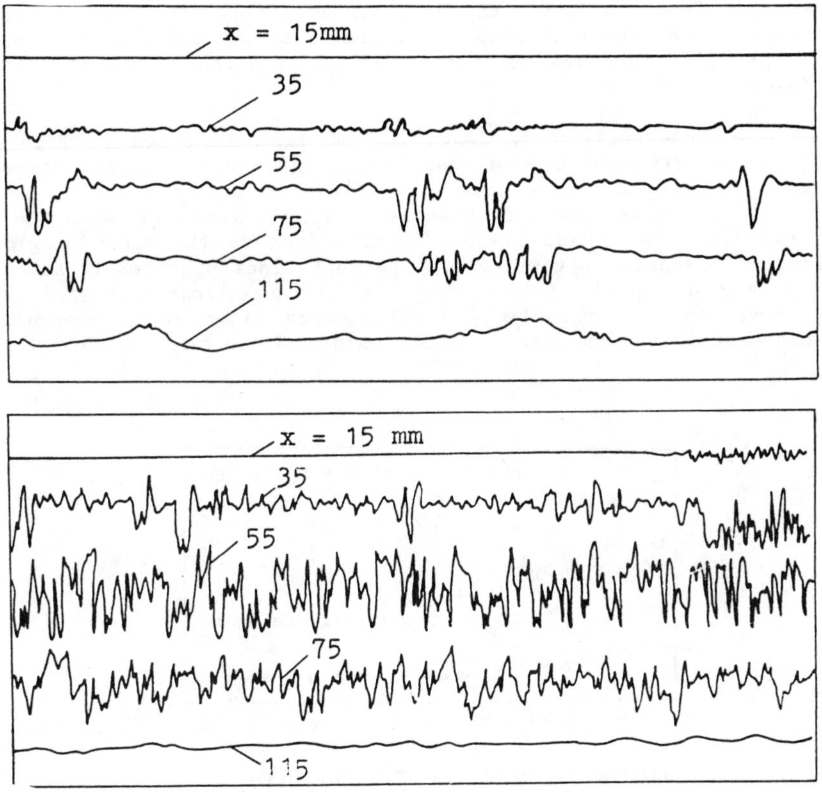

FIGURE 7. Simultaneous records of signals of wall-shear-stress probes in a transitional flow.

sequence in time may be arbitrary but usually spots appear immediately before continuous perturbations, which are followed by unperturbed flow. With the increase of Reynolds number, the latter becomes shorter, and, finally, developed turbulent flow is formed. It should be noted that the spots' attenuation along the flow is weaker than that of continuous perturbations. The transitional flow has a definite peculiarity owing to considerable friction acting on the flow by the walls.

The electrochemical method is the most suitable one for studying the flow in small channels because the wall-shear-stress probes do not introduce any hydrodynamic perturbations in the flow, and the probes are very small.

3.2 Flow in a Tank with Zero Mean Velocity

Vibration mixing devices have found an extensive application in technology. Intensive velocity fluctuations with mean velocity close to zero takes place in those systems. The liquid motion which occurs due to excitation by an oscillating grid was taken as the model of such devices. The turbulence intensity was taken to be characterized by the mass transfer rate from the liquid to the solid particle. The electrochemical velocity probe of spherical shape is affected by all velocity components; nevertheless, its readings provide useful information on the flow hydrodynamics.

Mass transfer coefficient vs depth of the tank is shown in Figure 8, where z is the distance from an oscillating grid and s is the stroke of the grid. The liquid volume under investigation can be divided into two regions. In the first one mass transfer strongly decreases with the distance from the grid. Mass transfer coefficient in the second region is practically constant. High-frequency perturbations produced by an oscillating grid are rapidly damped with the distance from the grid, their energy converted into relatively low-frequency flows which have practically constant intensity over the whole volume of the tank. Figure 9 shows

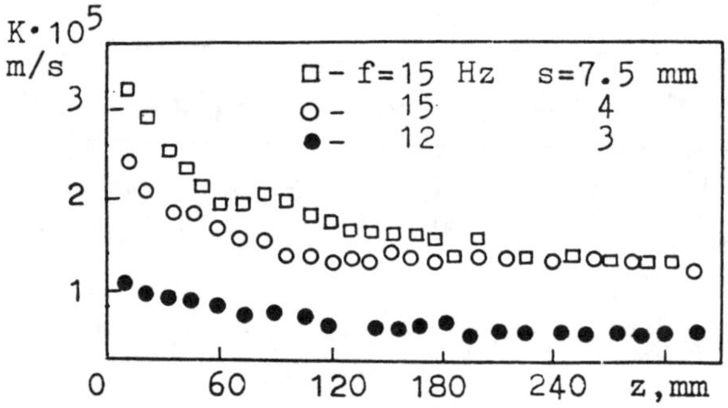

FIGURE 8. Mass transfer variation over the depth of the tank.

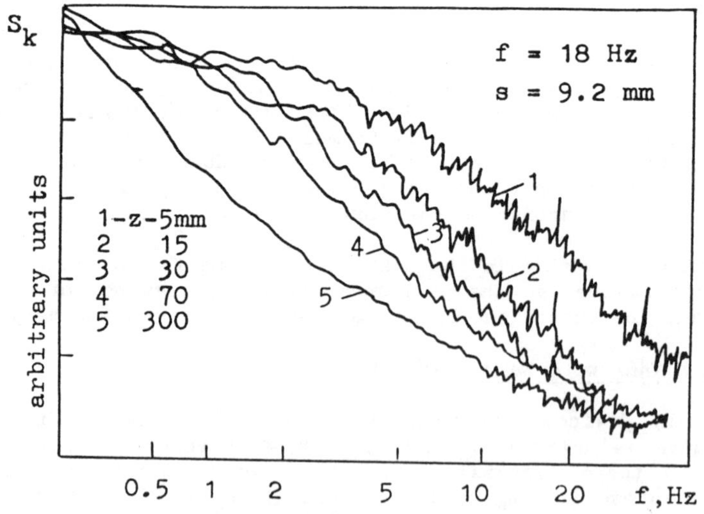

FIGURE 9. Spectral densities of mass transfer fluctuations at various distances from the grid.

the deformation of mass transfer spectra with the distance from the grid. Near it one can find higher spectral density on the oscillating frequency and its harmonics. Then spectral density acquires a characteristic fall-through, showing the decay of developed turbulence. When the intensity of mixing increases, the relative power of harmonics with respect to the main spectrum decreases.

3.3 Two-Phase Gas-Liquid Flow

Gas-liquid two-phase flow in a pipe is also one of the complicated cases of the wall turbulent flow in which the processes of turbulent mixing are superimposed due to additional perturbations produced by the relative motion of the gas phase. First the electrochemical method was used to study the hydrodynamics of gas-liquid flows by Kutateladze et al. [9]. The application of the method allowed them to perform accurate wall-shear-stress measurements in various flow regimes. The behavior of turbulent characteristics of liquid in a two-phase flow can provide an insight into single-phase flow structure.

A peculiarity of an upward bubbly flow in a vertical pipe is the higher gas-phase concentration near the walls (wall peaking), which leads to a strong increase in wall shear stress compared to a single-phase flow, even at very low void fractions. A strong asymmetry in void distribution near the wall and consequently in wall shear stress and liquid velocity is characteristic for a bubble flow. Typical profiles of void fraction α and liquid velocity U_1 in the asymmetric flow region are presented in Figure 10. U_1 is the superficial velocity of the liquid. The higher void fraction near the wall leads to more filled velocity profiles at this point. This picture is stable in time but can change over the length of the tube, and the void fraction profile can rotate from one cross section to another.

In a downward bubbly flow the gas phase migrates to the axis of the tube, and a layer of bubble-free liquid exists near the walls. Nevertheless, in a wall region of such flow at Reynolds numbers exceeding the critical one, strong damping of turbulent shear stress and velocity fluctuations takes place, as is seen in Figure 11, where spectral densities of τ_w and U_1 are shown. y is the distance from the wall and β the volumetric gas flow-rate ratio. The gas phase in the central part of the tube seems to

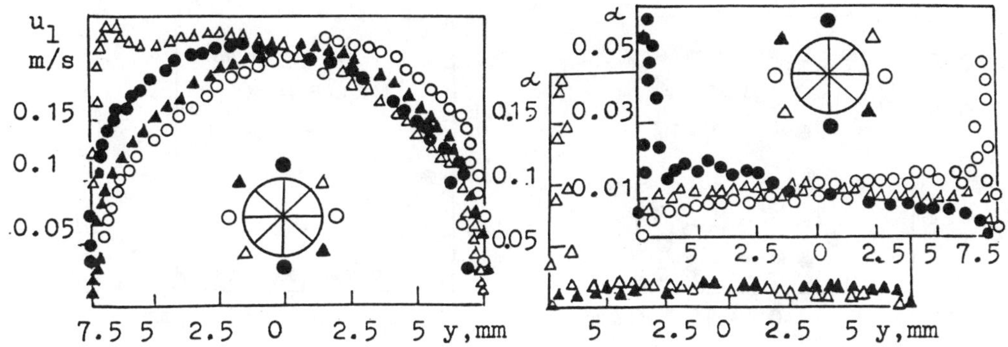

FIGURE 10. Local void fraction and velocity in upward bubbly flow.

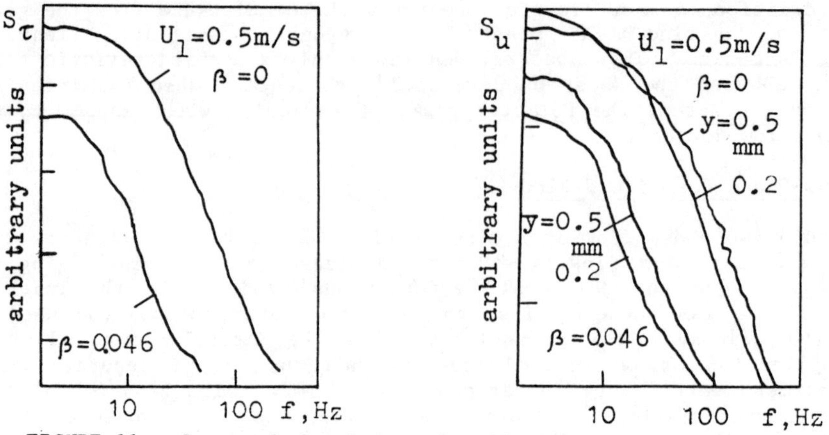

FIGURE 11. Spectral densities of wall shear stress and liquid velocity fluctuations in a downward flow.

prevent momentum transfer between the wall and the central parts of the tube. This counteracts the exciting action of gas bubbles due to their relative motion. It is interesting to note that in this case a strong decrease in the low-frequency part of the spectral density takes place.

Finally, Figure 12 shows results of multichannel measurements of wall shear stress in an upward bubbly flow. R_{11} is the normalized space correlation of wall-shear-stress fluctuations, and x and z are the corresponding axial and spanwise coordinates. Measurements were performed with two rakes of 30 electrochemical wall-shear-stress probes inserted in the tube in longitudinal and circumferential directions. It can be seen that the presence of the gas phase leads to significant decrease of the scale of longitudinal correlation with respect to a single-phase flow. This is probably related to the action of gas bubbles moving near the wall. They introduce strong perturbation in a wall region of the flow and seem to shorten longitudinal vortices responsible for the momentum exchange between the wall and the bulk of the flow. Numerous effects connected with the deformation of turbulent structure in a two-phase flow require more detailed investigation.

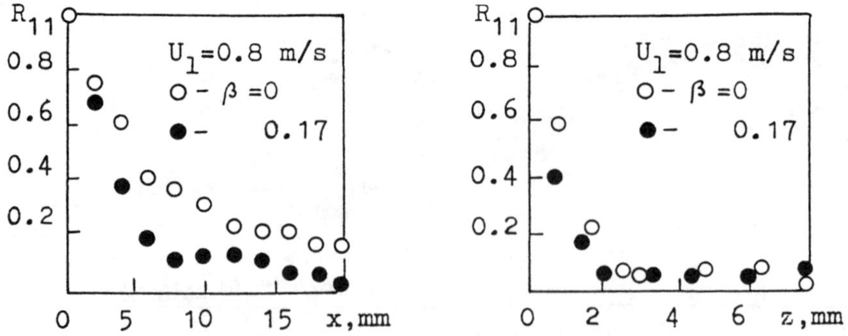

FIGURE 12. Space correlations of wall-shear-stress fluctuations in an upward bubbly flow.

4. CONCLUSIONS

Examples presented show that the electrochemical method can be effectively used to study the turbulent structure of single-phase and two-phase flows. The electrochemical method is similar to the constant-temperature hot wire in principle but has some advantages and shortcomings in comparison.

The advantage of the method is first of all the small size of probes that can be considered to a good approximation as point-like ones. The construction singularities permit reduction of hydrodynamic disturbances introduced into the flow. In addition, there is a lack of effects, such as heat losses to prongs or substrate. This permits one to equate effective and actual electrode surface, and gives the possibility of minimizing the probe size. It also permits use of an absolute version of the method, using calculated calibrations and frequency responses of the probes. The simplicity of the probes' structure and the electronic equipment allows use of the method for simultaneous measurement of the flow at many points which is extremely important for studying complicated turbulent flows that have no sufficient stability in time.

The main disadvantage of the method is the necessity to use a special electrolyte as the test liquid, resulting in great limitations in temperature range (no more than 60-80°C). Additionally, the poorer frequency response of the probe may require electronic or numerical corrections of its readings.

Therefore, the main range of applicability of the electrochemical method is for laboratory study of complex turbulent characteristics of flows under various conditions. The electrochemical method seems to be at a stage of ascent in the number of studies with its applications growing steadily. Further development of the method will, of course, open new possibilities for its use and reveal new advantages.

REFERENCES

1. Mitchell, J. E., and Hanratty, T. J. A study of turbulence at a wall using an electrochemical wall shear stress meter. J. Fluid Mech., Vol. 26, No. 1, pp. 199-221, 1966.

2. Hanratty, T. J., and Campbell, J. A. Measurements of wall shear stress. In: Fluid Mechanics Measurements, R. J. Goldstein, Ed., pp. 559-615. Springer-Verlag, Berlin a. o., 1983.

3. Karabelas, A. J., and Hanratty, T. J. Determination of the direction of surface velocity gradients in three-dimensional boundary layers. J. Fluid Mech., Vol. 34, No. 1, pp. 159-162, 1968.

4. Katsuda, H., and Yamada, J. Limiting diffusion currents in hydrodynamic voltametry. Electroanal. Chem. and Interfacial Electrochem., Vol. 30, No. 2, pp. 261-270, 1971.

5. Mizushina, T. The electrochemical method in transport phenomena. In: Advances in Heat Transfer, Vol. 7, J. P. Hartnett and T. F. Irvine, Eds., pp. 87-161. Academic Press, New York, 1971.

6. Hogenes, J. H. A., and Hanratty, T. J. The use of multiple wall probes to identify coherent patterns in the viscous wall region. J. Fluid Mech., Vol. 124, pp. 363-390, 1982.

7. Geshev, P. I., and Chernykh, A. I. On the theory of electrochemical method for velocity measurements. Izvestia Sibirskogo Otdelenia AN SSSR, ser. tekh. nauk. No. 3, pp. 23-25, 1982.

8. Nakoryakov, V. Ye., Burdukov, A. P., Kashinsky, O. N., and Geshev, P. I. Electrodiffusional Method for Studying Local Structure of Turbulent Flows. Inst. Teplofiziki, Novosibirsk, 1986. 247pp.

9. Kutateladze, S. S., Burdukov, A. P., Nakoryakov, V. Ye., and Kuzmin, V. A. Application of electrochemical method in the hydrodynamics of two-phase media. In: Teplo- i Massoperenos, Vol. 2, pp. 367-375. Nauka i Tekhnika, Minsk, 1968.

The Turbulence Burst Detection Algorithm of Z. Zarić

R. E. FALCO and C. P. GENDRICH
Turbulence Structure Laboratory
Department of Mechanical Engineering
Michigan State University
East Lansing, Michigan 48824, USA

ABSTRACT

In 1984 Zoran Zaric' introduced the first author of this paper to an algorithm which uses the rate of change of the streamwise component of the turbulence energy to detect details of the bursting process near walls. This quantity, which can be shown to be equivalent to $u'(\partial u'/\partial t)$, is more intermittent than the velocity fluctuation or velocity gradient fluctuation signal in the wall region. The underlying assumption behind the use of the detector is that the passage of a bursting event is accompanied by an increase in the time rate of change of the turbulence energy, and the end of an event is marked by a reduction in the change of turbulence energy. Through use of the sign of u', the nature of the event, i.e. whether it is an ejection or a sweep, is determined. Zaric's original scheme determined the start of an event using $u'\partial u'/\partial t$ and the length of the event by examining the magnitude and sign of $u'\partial u'/\partial t$, u' and $\partial u'/\partial t$. The technique does depend upon threshold levels. They are taken as proportional to the conditional root mean square of the three quantities $u'\partial u'/\partial t$, $\partial u'/\partial t$ and u' as found for all sweeps and ejections using a quadrant analysis. A single constant of proportionality of order one was used to generate all three thresholds. A simplified version is introduced which uses a threshold that is a constant fraction of the overall rms of $u'\partial u'/\partial t$. The quality of the techniques have been examined by applying them to the signal data obtained from an experiment in which a four wire probe was used simultaneously with sublayer flow visualization and laser sheet flow visualization in streamwise-normal planes. Thus, along with each picture of the flow in the two orthogonal planes, we had u', v', $u'v'$ and derivative information. The detection techniques (which only use u' information) are shown to quite accurately detect the start and end of individual sweeps, ejections and quiet periods (low $u'v'$) as seen in both the hot-wire data and the flow visualization.

1. INTRODUCTION

Development of an accurate turbulence detection technique would accelerate our understanding of turbulent boundary layers significantly.

It would enable us to determine the correct scaling laws for the bursting process, which is important from both an engineer's viewpoint as well as that of researchers in turbulence. Furthermore, a reliable burst detector would enable the evaluation of turbulence control experiments in the laboratory as well as in-situ.

1.1. Detection of turbulence in boundary layers

There is ample evidence from numerous experimental studies of coherent structures in wall turbulence that most of the turbulence energy and Reynolds stress production in the wall region results from two dominant features of the flow -- wallward penetration of high speed fluid from regions farther from the wall, called sweeps or inrushes, and ejections of low speed fluid from the wall layers. For a detailed discussion of the nomenclature see Kline (1978). Grass (1971) has pointed out the important contribution of sweeps to the production of turbulence, and later Brodkey et al (1974) and Willmarth and Lu (1972) found that sweeps contributed an equal amount to the Reynolds stress produced. Thus, a turbulence detection algorithm must sense both sweeps and ejections.

Fig. 1 is taken from a visual study by Falco (1987) in which flow in a flat plate turbulent boundary layer is made visible by seeding with oil-fog. The flow is from right to left. The upper part of the picture shows the plan view illuminated in flood-light. A thin sheet of laser light is introduced along the line A-A and permits a side view of the flow to be filmed simultaneously, as shown in the lower part of the picture. Regions with higher than average smoke concentration (light areas) represent regions of converging flow, and as a consequence of continuity in the incompressible fluid, they are also regions of movement away from the wall (i.e. liftups or ejections). This can be confirmed by observing that when there is a high concentration of smoke in the plan view which happens to be along the streamwise position of the laser sheet, then we see from the side view that the smoke has moved away from the wall. In experiments with simultaneous probe measurements, described later, we examined this data at $x^+ = 250$; thus a buildup of smoke in that short distance was indicative of dynamic outward movement.

In regions where there is a lower than average concentration in the plan views (dark areas), smoke has been moved away. This occurs because high speed fluid is moving towards the wall. Falco (1980a) has used simultaneous hot wire anemometry and flow visualization to confirm that this is the case. Thus, from Fig. 1 it could be observed that in the bracketed region a sweep is occurring. The effects extend very close to the wall, leaving a footprint in the plan view in the form of a "pocket" (Falco 1979, 1980a). As can be seen in Fig. 1 there is ejection from the downstream end of the pocket as well as from the lateral boundaries.

Direct association between outer region coherent motions moving towards the wall and ejections were made by Corino and Brodkey (1969). The relationship between outer region coherent motions and pockets was described by Falco (1978). Their association with the streaky structure

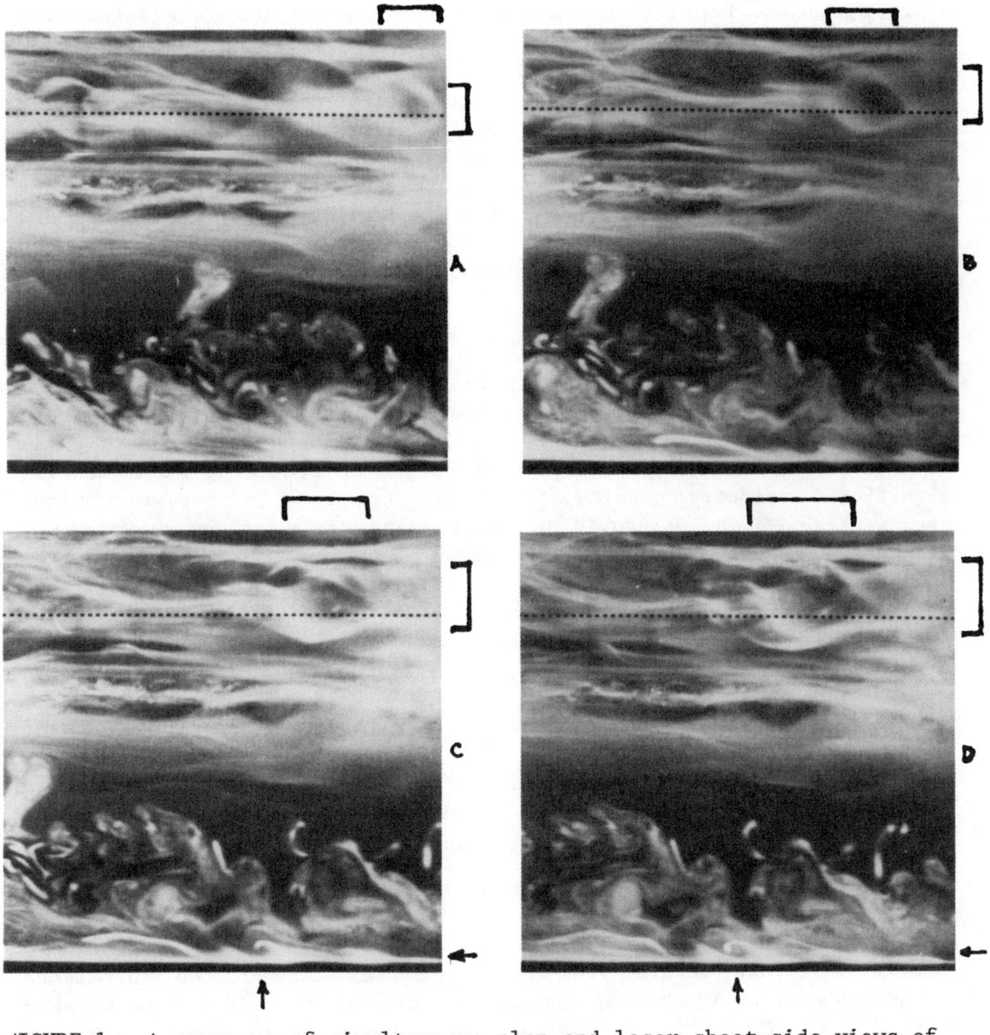

FIGURE 1. A sequence of simultaneous plan and laser sheet side views of a turbulent boundary layer. The smoke marked wall region is shown in the top of each frame illuminated by flood light. A thin sheet of laser light is introduced along the dashed line giving the side view. The bracketed region shows the formation and evolution of a sweep that forms a pocket. The high concentrations of smoke are regions of ejection, as can be seen in the laser sheet side view. Flow is from right to left.

(Kline, et al 1967) has been described by Oldaker and Tiederman (1977) in a channel flow and Falco (1979) in a boundary layer. The frequency of occurrence of pocket structures is about 25 when non-dimensionalized by wall variables (u_τ and ν) at R_θ = 1000 (Falco 1983). Falco (1980a) measured the contribution of pockets to the Reynolds stress, and found that they had ensemble averaged peak values of ten times the long time averaged Reynolds stress; initially behaving as a sweep event and later an ejection event. Pockets emphasize the close correspondence that can occur between sweeps and ejections.

There appear to be other times when smoke gathers and is ejected away from the wall. These are often associated with hairpins that form over the long streaks. Again, using movies of the type from which Fig. 1 has been obtained, we have been able to verify that when there is an increase in smoke concentration in the plan view there is liftup or ejection, and when there is a lowering of concentration there is a sweep. Thus, we can use the visual information as a reference to detect the turbulence production events. Offen and Kline (1974) and Bogard and Tiederman (1986) have also relied upon flow visualization to provide the calibration database against which to determine the quality of turbulence detectors.

Unfortunately, visual techniques, while good for determining statistically significant quantitative characterizations of length and time scales, and providing evolutionary information of the flow, are inherently unsuitable for the determination of the velocity fields. Hot-wire anemometry is a reliable technique in providing statistically significant velocity results. However, it has a number of obvious disadvantages such as providing measurements only at one point in a highly three-dimensional field. Except for very thick low Reynolds number laboratory flows, only the u-wire parallel to the wall has adequate spatial resolution to resolve all the details of the flow. Therefore it is difficult to obtain important information on the normal velocity component and Reynolds stress with sufficient accuracy.

However, the signal from a u-wire parallel to the wall contains additional important information. For example we can obtain $\partial u'/\partial t$ or any combination of the two signals, such as the $u'(\partial u'/\partial t)$, where u' is the axial velocity fluctuation $u' = u-U$. The latter is significant as it represents the rate of temporal change of the major component of the turbulent energy, since:

$$u'(\partial u'/\partial t) = 1/2\ \partial(u'^2)/\partial t$$

and is therefore related to Reynolds stress through the turbulence energy equation. The three signals that are shown in Fig. 2 were taken from Klewicki (unpublished). They were measured in the viscous sublayer of a turbulent boundary layer at $y^+ = 4.5$. R_θ was 1000 and $\delta_{99} = 203$mm. Unlike either u' or $\partial u'/\partial t$, the $u'\partial u'/\partial t$ signal is highly intermittent. The correspondence seen in the figure with the $u'v'$ signal from the same probe indicates that $u'\partial u'/\partial t$ is very strongly correlated with $u'v'$, which itself is strongly correlated with coherent structures in the wall region.

Zaric' (1982) applied his detection technique to the study of the signatures of the bursting process, and Zaric', Falco and Blackwelder (1984) made preliminary identification of the output of the detector with visual and multi-probe measurements. An important result was that the frequency of occurrence of detected events was much higher than that previously found by techniques such as VITA (Blackwelder and Kaplan 1976). Furthermore, it contained information necessary to determine the length of the events. Zaric's technique shared the simplicity of VITA in that it used a single u-wire. The investigation presented in this paper

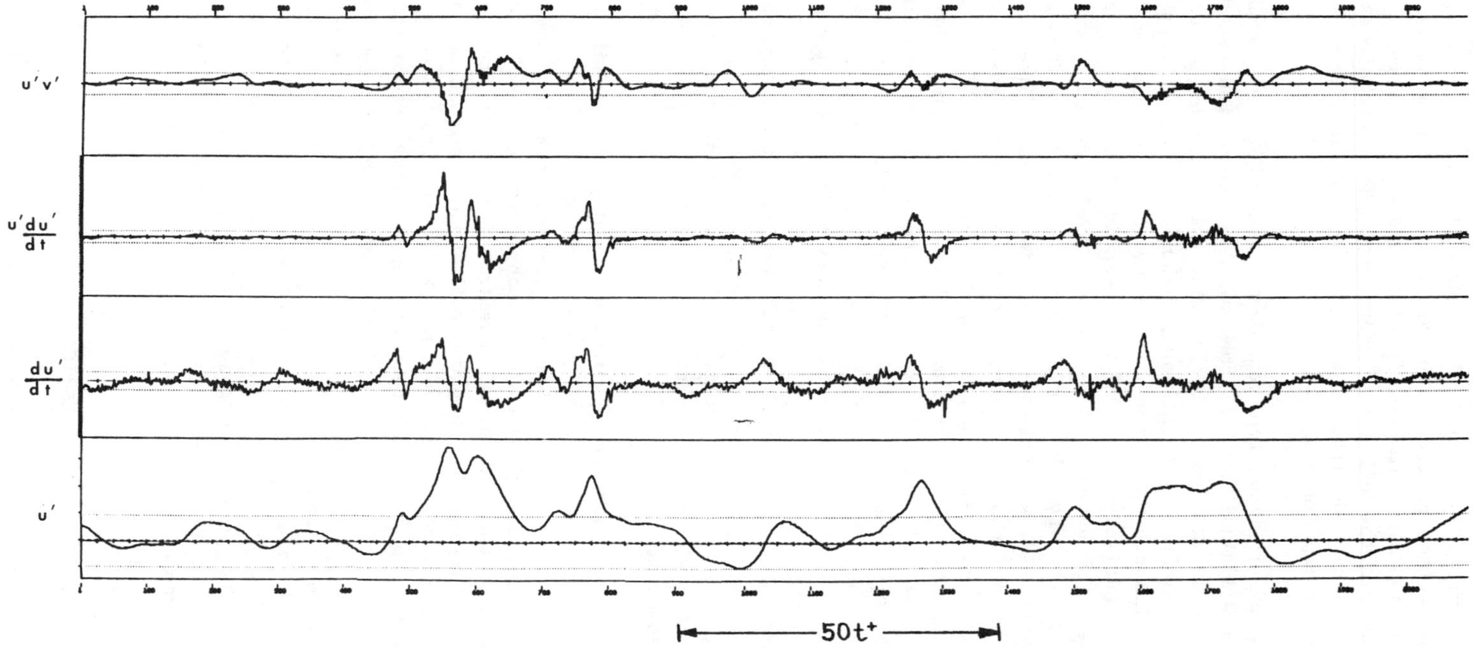

FIGURE 2. Fluctuating signals u'v', u', ∂u'/∂t, and u'∂u'/∂t at y^+ = 4.5 in a turbulent boundary layer. The intermittency of the u'∂u'/∂t signal is considerably greater than for u' or ∂u'/∂t. Fluctuations in u'∂u'/∂t correspond closely to those in the u'v' signal. The spaced dots correspond to a variation of 1 rms of the respective signals.

first indicates the quality of the detection technique used by Zaric' (1975, 1982), and Zaric', Falco and Blackwelder (1984), then examines and qualifies a simplified version of the detection technique which only uses the information contained in the signal of a u-wire parallel to the wall, and in particular the $u'\partial u'/\partial t$ information. Both techniques are compared against data obtained using the simultaneous flow visualization/hot-wire experiments.

Perhaps the most popular detection techniques are VITA, the quadrant breakdown of Lu and Willmarth (1973), the pattern recognition technique of Wallace, Brodkey and Eckelmann (1977), and the u-level of Lu and Willmarth (1973) as modified by Luchik and Tiederman (1987). It has been shown by Lu and Willmarth that the time between both ejections and sweeps varies exponentially with the threshold for the quadrant breakdown technique. Alfredsson and Johansson (1982) showed that VITA also varies exponentially with the threshold value. They also showed the very strong dependence of averaging time on detections. Essentially, neither VITA nor the quadrant breakdown technique offer a good way to determine the bursting frequency. Bogard and Tiederman (1986) developed a grouping technique in which ejections separated by a time interval shorter than some specified amount is used to group the ejections into a single burst. Using this idea Bogard and Tiederman (1986) and Luchik and Tiederman (1987) were able to argue that VITA, the quadrant breakdown in quadrant 2, and a modified u-level technique gave the same number of grouped ejections. However, the grouping was determined by statistical means, and comparison with individual events was not shown. Bogard and Tiederman did determine that on a one-to-one basis none of the techniques detected all of the ejections. In this test the quadrant technique (in quadrant 2) did the best, especially when grouping was applied and H set equal to 1.

It is highly desirable to show that a turbulence detection technique can detect individual occurrences of high Reynolds stress on a one-to-one basis, and then to build the statistics of interest from this base. In spite of the apparent success of the above techniques, it is clear that they are not detecting the individual events on a one-to-one basis. Luchik and Tiederman, for example, point out that the quadrant technique is accurate only if the sample size is long (> 200 bursts). Zaric' was particularly interested in seeing whether use of the time rate of change of turbulence energy as a detector would enable accurate one-to-one signal and visual identification. Since all aspects of turbulence production, i.e. the ejections (Kline et al 1967), the sweeps (Grass 1971), and the other interactions (Brodkey et al 1974), will involve a change in the turbulence energy, his detector could in principle detect all of the events. The primary purpose of this paper is to show the one-to-one correspondence of a technique based on descrimination using $u'\partial u'/\partial t$ for a low Reynolds number turbulent boundary layer.

2. THE DETECTION TECHNIQUE OF ZORAN ZARIC'

The technique in its different stages of development is described by

Zaric' (1982) as applied to the analysis of a turbulent flow of air in a channel at different distances from the wall. As indicated above, it uses more of the available information contained in a u-wire time series than simply u'. Fig. 3 shows a flow chart of the detection algorithm. There are three parameters to which detection thresholds are applied:

$Z_1 = u'\partial u'/\partial t, \quad Z_2 = \partial u'/\partial t, \quad Z_3 = u'$

It is hypothesized that the passage of an event involving a coherent structure over the wire has to be marked by an appreciable rate of change of turbulence energy. Thus, the obvious criterion for the recognition of an event arriving at the wire is a high value of the parameter Z_1. The nature of the event is recognized by the sign of the parameter Z_3, i.e., for $Z_3 = u' > 0$ it has to be a sweep, and for $Z_3 < 0$ it has to be an ejection. It is of considerable interest to be able to detect the length of the event or the time increment during which the event is sensed by the wire and during which the Reynolds stress produced by the event is still appreciable. For this purpose, the value and the sign of the parameters $Z_2 = \partial u'/\partial t$ and $Z_3 = u'$ are used; if the passage of the event results in a quiescent period, it could be assumed that the values of these parameters should be low and of opposite sign. It is also possible that a sweep could be followed by an ejection, or vice versa, in which case the value of Z_2 should be high and its sign equal to the sign of Z_3.

As in all turbulence structure recognition techniques, the choice of threshold levels is critical for the success of the technique. The threshold levels in TERA are based on the use of long time averaged statistical quantities obtained from analysis of the signal in question. It has to be recognized however, that the intensities of the two possible events, sweeps and ejections, might be different and also that they might be dependent on the distance from the wall. Therefore statistical analysis is performed separately for sweeps ($u' > 0$) and for ejections ($u' < 0$).

When Z_2 and Z_3 are tested in combination with one another, the thresholds take the form

$T_1(Z_2)^{S,E} = \pm \mathrm{rms}(\partial u'/\partial t)$

$T_1(Z_3)^{S,E} = \pm \mathrm{rms}(u'\partial u'/\partial t)/\mathrm{rms}(\partial u'/\partial t)$

When each parameter is tested separately, threshold levels (TR) of the first two criteria, Z_1 and Z_2 are taken as proportional to the conditional root mean square (rms) values of sweep (S) or ejection (E) events

$TR(Z_1)^{S,E} = \pm C\ (\overline{u'\partial u'/\partial t^2})^{1/2}$

$TR(Z_2)^{S,E} = \pm C\ (\overline{\partial u'/\partial t^2})^{1/2}$

For the parameter Z_3 it is taken as

$TR(Z_3)^{S,E} = \pm C\{(\overline{u'\partial u'/\partial t^2})/(\overline{\partial u'/\partial t^2})\}^{1/2}$

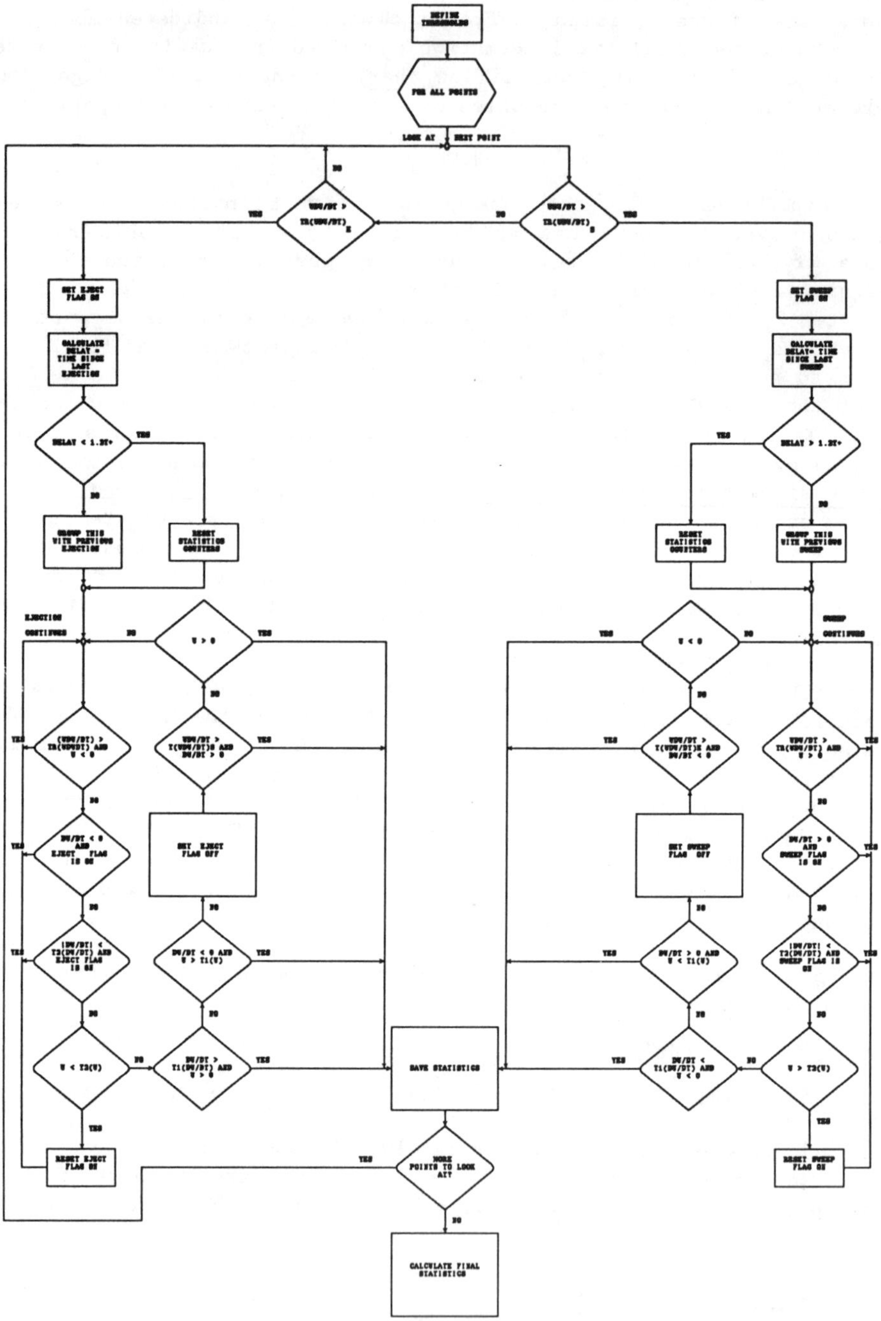

FIGURE 3. Flow chart of the turbulence detection algorithm of Zoran Zaric'.

Thus, there is only one arbitrary constant, C. Preliminary results suggested that in the range C = 0.6 to C = 1.0, the results of the analysis are only slightly sensitive to the value of C. However, the sensitivity of the results to the value of C should be tested more comprehensively.

In comparison with the other structural recognition techniques, the technique has several unique features:

a) It uses signals from single u-wires
b) It is able to recognize the nature of events in question
c) It determines the length of events
d) Its threshold levels are determined by statistics of the signal in question
e) Only one constant of proportionality is required to generate thresholding values.

2.2. Verification by comparison with a simultaneous flow visualization/hot-wire experiment

Using the combined simultaneous flow visualization/hot-wire anemometry technique developed by Falco (1977) (also see Falco 1980b), we could verify whether or not the detections accurately corresponded to ejections, sweeps or quiet periods. The flow visualization enabled us to determine the arrival of structures at the probe, and the hot wire data enabled us to examine the instantaneous values of both velocity components, the Reynolds stress, and even the spanwise vorticity component. The data was obtained in the MSU Turbulence Structure Laboratory 7.3m boundary layer flow visualization wind tunnel. Table 1 indicates the boundary layer parameters.

In the experiment a 4-wire probe was situated at $y^+ \cong 15$ in a flat plate turbulent boundary layer. The momentum thickness Reynolds number was about 680. Signals from the probe enabled u and v as well as du/dy to be obtained. The voltages were digitized at 500 Hz and recorded on disk. Two wall slits were used. One was far upstream and resulted in smoke filling most of the boundary layer. A second slit introduced marker approximately $250x^+$ upstream of the probe. The marker was a fog of 0.5 to 5.0μm oil droplets. A laser sheet was used to illuminate a streamwise slice, and it was filmed from above in floodlight using a high-speed camera framing at 225 frames per second. Output of a digital clock driven by the computer was also filmed using a split-screen technique. This enabled synchronization of the hot-wire signal with the visual information.

TABLE 1. Boundary Layer Parameters

Parameter	$y^+ = 4.5$	$y^+ = 15$
U_∞	0.61 m/s	0.91 m/s
U^*	0.028 m/s	0.044 m/s
ν	0.00001496 m^2/s	0.00001591 m^2/s
R_θ	1008	679
δ_{99}	203 mm	102 mm
wire length	1 mm	1 mm

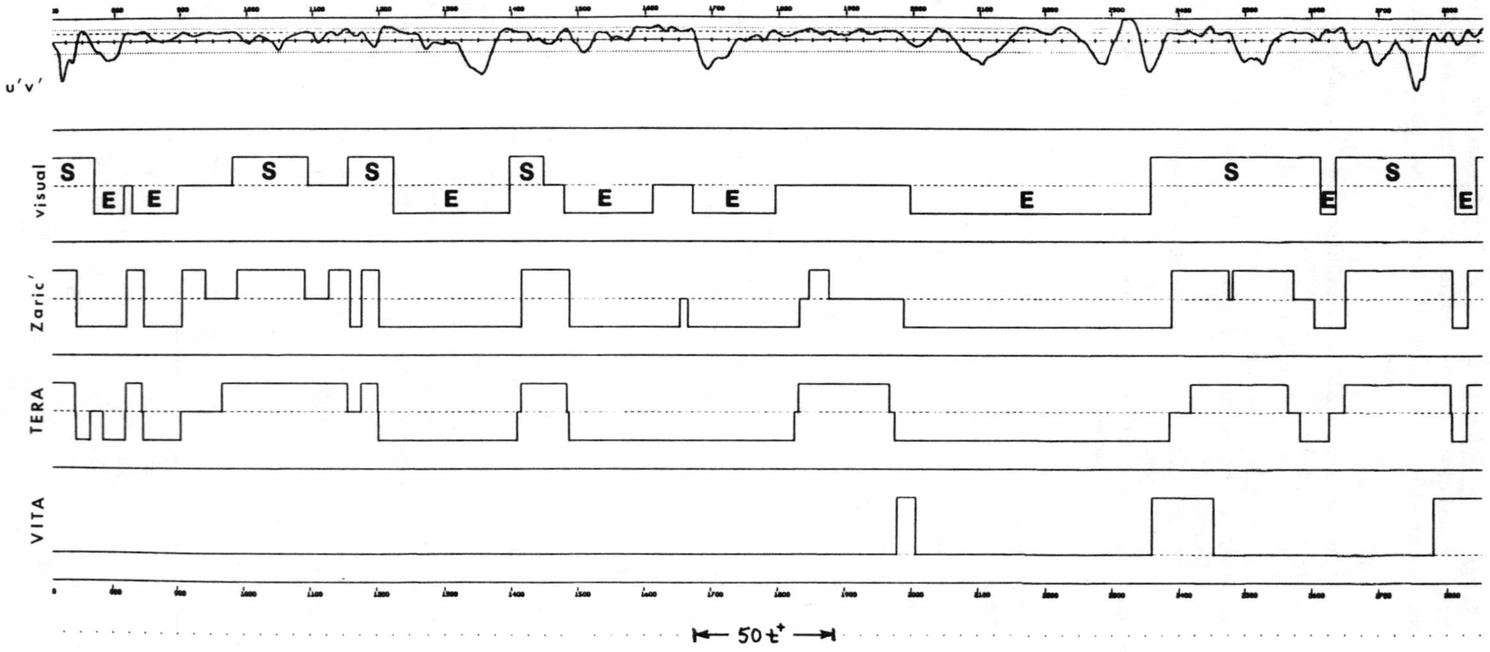

FIGURE 4. Correspondence of detection algorithms and visual identification of sweeps and ejections with the instantaneous Reynolds stress at $y^+ = 15$ in a turbulent boundary layer. A value of the intermittency functions of +1 corresponds to detection of a sweep, zero to detection of a quiet period, and -1 to detection of an ejection. For VITA a value of +1 corresponds to detection of an ejection.

Zaric's technique is applied only to the u-component of the signal. It detects sweep and ejection events arriving at the hot-wire and forms an intermittency function having values: I = +1 for sweeps; I = -1 for ejections and I = 0 for quiescent periods. Results of the technique have been compared with the u'v' signals from the 4-wire probe, and also with the events seen on film. The results are illustrated in Fig. 4. In this figure, the instantaneous Reynolds stress, the visual detections of the ejections (indicated by E), sweeps (indicated by S) and quiet periods, the Zaric' detector function, a simplified detector function which we call TERA (described below), and the VITA technique with a window optimized for maximum number of detections (see Johansson and Alfredsson 1982) and threshold equal to 1 rms(u') are presented.

It should first be noted that the visual indications of sweeps, ejections and quiet periods (see above explanation associated with Fig. 1), result in an intermittency function which closely corresponds to the sweeps, ejections and quiet periods in the u'v' signal. Examination of Zaric's detector function shows that it detects these visually-identified events with high accuracy. Both the passage of the front and the back of the events are detected. Examination of Fig. 4 indicates that the termination of the sweep event as indicated by Zaric's technique corresponds closely to the visual detection of the end of the "pocket". Pockets have been discussed in detail by Falco (1979, 1980a, 1982). They were shown to correspond to the important sweep events of the production process. The ejection events found in the visual identification are also quite accurately detected by Zaric's technique (I = -1). In this case the visually detected boundaries of high concentration gradients corresponded well with Zaric's technique.

Ratios of Total times. Table 2 shows the ratios of the total times the detector corresponded to the visual identification, and to the instantaneous Reynolds stress, where a hole size of 0.5*rms(u')*rms(v') was used.

3. TERA -- A SIMPLIFIED DETECTION ALGORITHM BASED ON THE TURBULENCE ENERGY AND THE SIGN OF U'

More careful examination of the time series of $u'\partial u'/\partial t$ revealed that the correspondence between u'v' and $u'\partial u'/\partial t$ is very high. We removed conditional tests so that the simplified detector no longer depends on the values of either u' or $\partial u'/\partial t$ for determining the end of an event. Furthermore, we use only the long time rms of $u'\partial u'/\partial t$ to determine the threshold. This may be modified by a constant (C), if desired, as a result of one-to-one comparison with visual data. Fig. 5 shows the resulting flow chart.

TERA uses the following procedure to detect events: A threshold (TR) for the absolute value of $u'\partial u'/\partial t$ is calculated -

$$TR = C*rms(u'\partial u'/\partial t)$$

An event is located by determining that $|u'\partial u'/\partial t| > TR$. At this time we start a running integral of $|u'\partial u'/\partial t|$, integrating from the

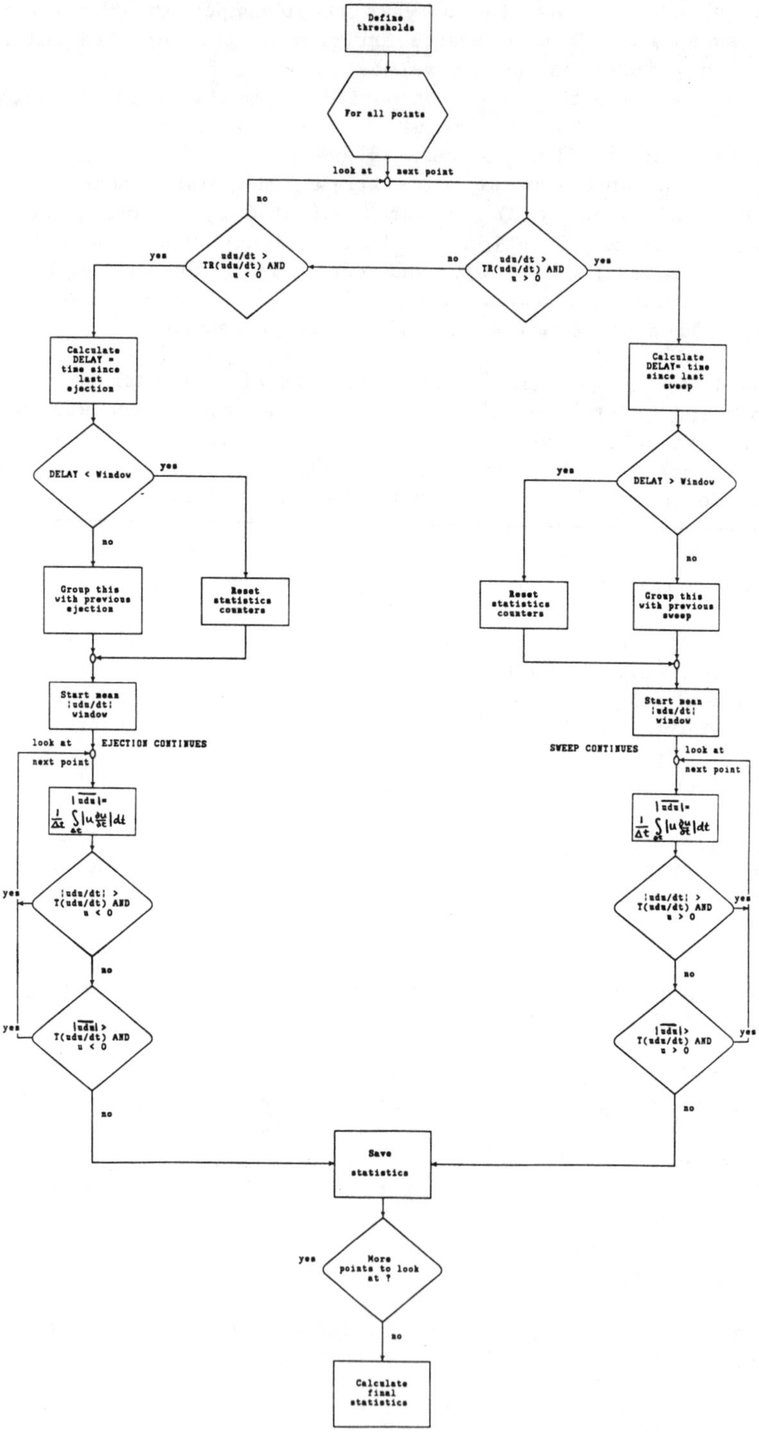

FIGURE 5. Flow chart of TERA, a simplified version of Zaric' detection algorithm.

TABLE 2. Quality of the Detector of Z. Zaric'

$\frac{\text{Total time of detected quiet periods on film}}{\text{Total time of detected quiet periods using TERA}}$	= 0.93
$\frac{\text{Total time of detected sweeps on film}}{\text{Total time of detected sweeps using TERA}}$	= 1.12
$\frac{\text{Total time of detected ejections on film}}{\text{Total time of detected ejections using TERA}}$	= 0.96
$\frac{\text{Total time of detected quiet periods from UV signals}}{\text{Total time of detected quiet periods using TERA}}$	= 0.84
$\frac{\text{Total time of detected sweeps from UV signals}}{\text{Total time of detected sweeps using TERA}}$	= 1.04
$\frac{\text{Total time of detected ejections from UV signals}}{\text{Total time of detected ejections using TERA}}$	= 1.00
$\frac{\text{Total time of detected quiet periods from UV signals}}{\text{Total time of detected quiet periods on film}}$	= 0.91
$\frac{\text{Total time of detected sweeps from UV signals}}{\text{Total time of detected sweeps on film}}$	= 0.93
$\frac{\text{Total time of detected ejections from UV signals}}{\text{Total time of detected ejections on film}}$	= 1.05

beginning of the event to a predetermined window length. At each new point we check the value of the integral and $|u'\partial u'/\partial t|$. If either the integral or the value of $|u'\partial u'/\partial t|$ is greater than the threshold, we decide that the event continues through this point in time. The detection now only depends on the magnitude of the turbulence energy averaged over a window, thus it is called TERA for Turbulence Energy Recognition Algorithm. It seems fitting to employ this acronym here, as it was arrived at during the collaboration between Zaric' and Falco in 1984, and was intended to be used with some new simplified version of Zaric's technique.

We first examine the effect of varying window size and threshold on TERA. Fig. 6 shows the correlations of TERA with visual identifications as a function of window size and threshold. For both the detection of sweeps (Fig. 6a) and of ejections (Fig. 6b) the contours indicate that there is little dependence on window size for windows greater than $5t^+$, and that there is a weak dependence on threshold in the range C = 0.1 to 0.4. It is immediately clear that TERA must be used with window sizes greater than $5t^+$, since it is otherwise too sensitive to short-term changes in $u'\partial u'/\partial t$. Since most identified coherent motions are greater than $1t^+$ in scale (approximately $12\ell^+$), this smoothing results in a grouping of events separated by as much as $60\ell^+$, which improves the correlation with visual identifications. In part, of course, the correlation is lower because the noise generated in the differentiation process may confuse the detector. Fig. 7 shows results averaged over windows from $7.2t^+$ to $24t^+$.

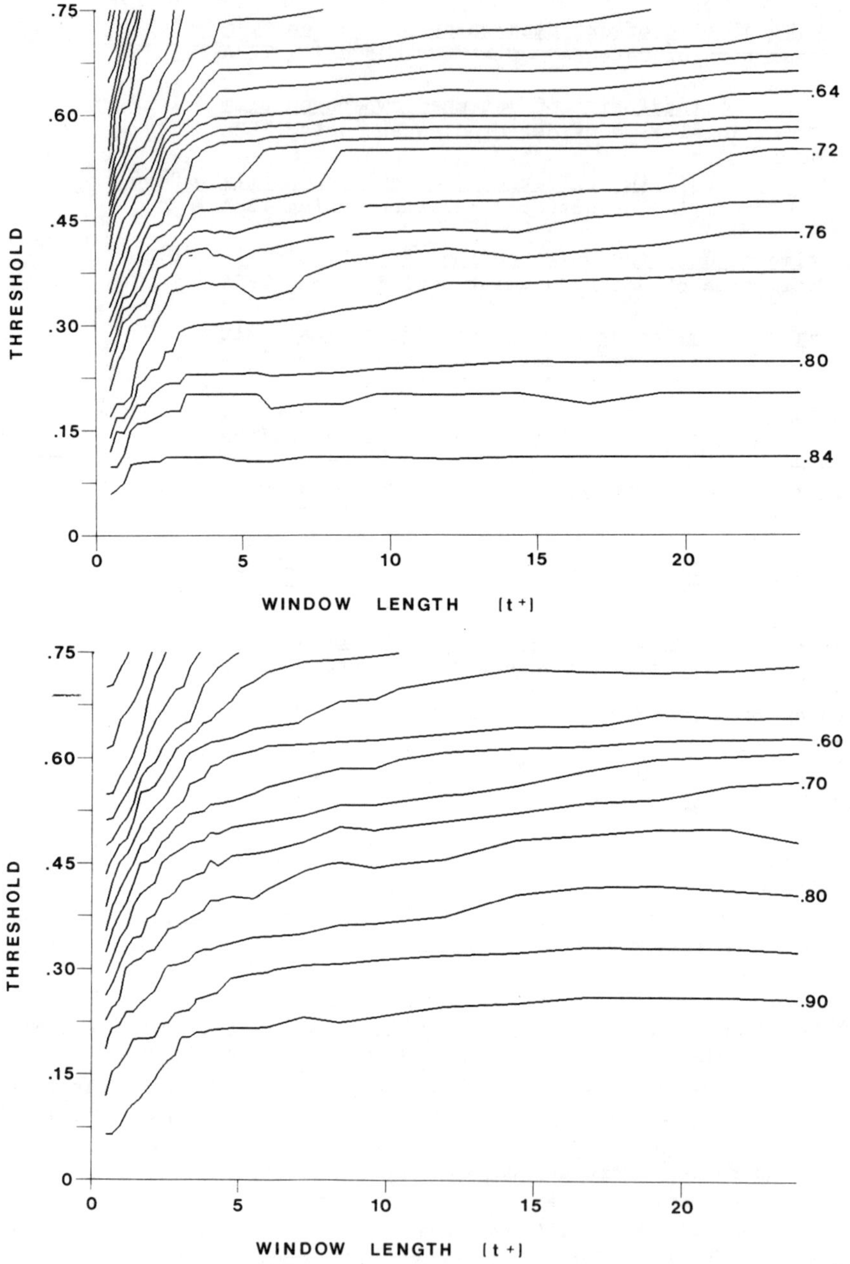

FIGURE 6. Contour plots of the magnitude of the correlation between TERA detections and visual identifications of the bursting process, as a function of the threshold and window length. a) sweeps; b) ejections. Note that for window sizes above $t^+ = 5$ the effect of window size rapidly diminishes.

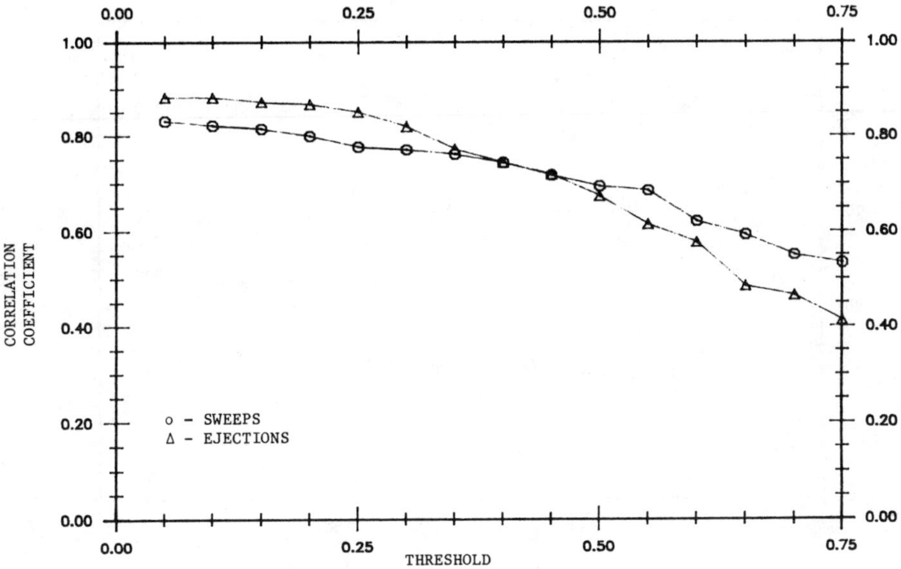

FIGURE 7. The correlation coefficient of TERA and visual identifications averaged over all window sizes between $t^+ = 7.2$ and 24, as a function of the detector threshold.

3.1. Evaluation of bursting parameters at $y^+ = 15$

The average time between ejections and sweeps and the average duration of ejections and sweeps have been determined as a function of threshold averaged over window sizes between $7.2t^+$ to $24t^+$. Fig. 8 shows the average time between ejections and the average time between sweeps. It can be seen that for C between 0.1 and 0.4 the average times between sweeps and ejections are independent of threshold. From Fig. 7 we see that this is also the range of highest average correlation of TERA with the visual detections. Fig. 9 shows the mean event duration for both sweeps and ejections. This length also varies slowly in a somewhat narrower region (C = 0.05-0.3).

3.2. Optimization of the threshold

Although TERA is a threshold-dependent turbulence detector, we have seen that average statistics are weakly dependent on the thresholding constant, C, over a range of C from 0.1 to 0.4. This encourages us to attempt to refine the detector further by determining what specific value of the threshold in this range results in the most accurate one-to-one association between TERA and the visual identifications. From Fig. 4 it can be seen that small differences exist between the visual detections and Zaric's detector indentification, as well as between the visual and TERA detections. The same order of difference was found to occur for TERA and the visual identifications when different thresholds

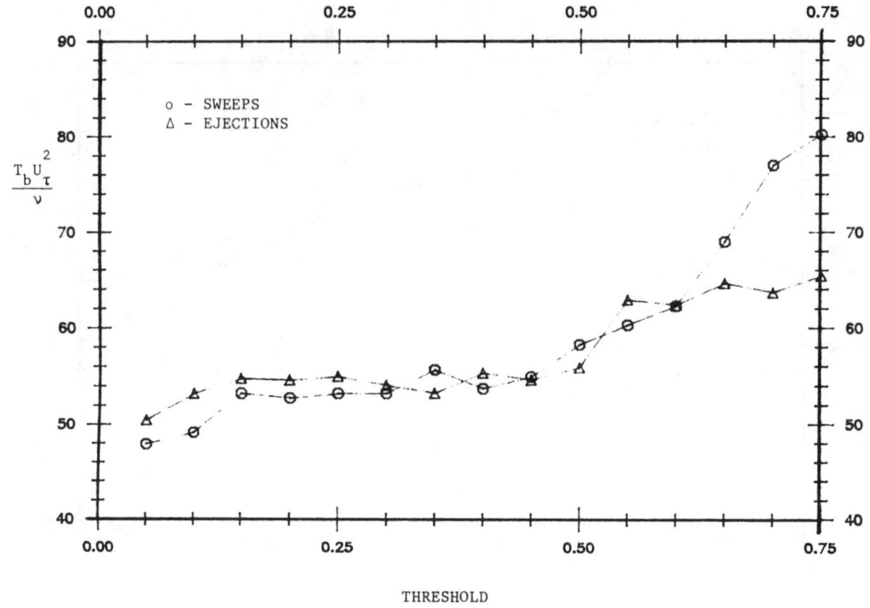

FIGURE 8. The time between the occurrence of sweeps and the time between the occurrence of ejections found using TERA, as a function of the detector threshold. As in figure 7 the results have been averaged over all window sizes between $t^+ = 7.2$ and 24.

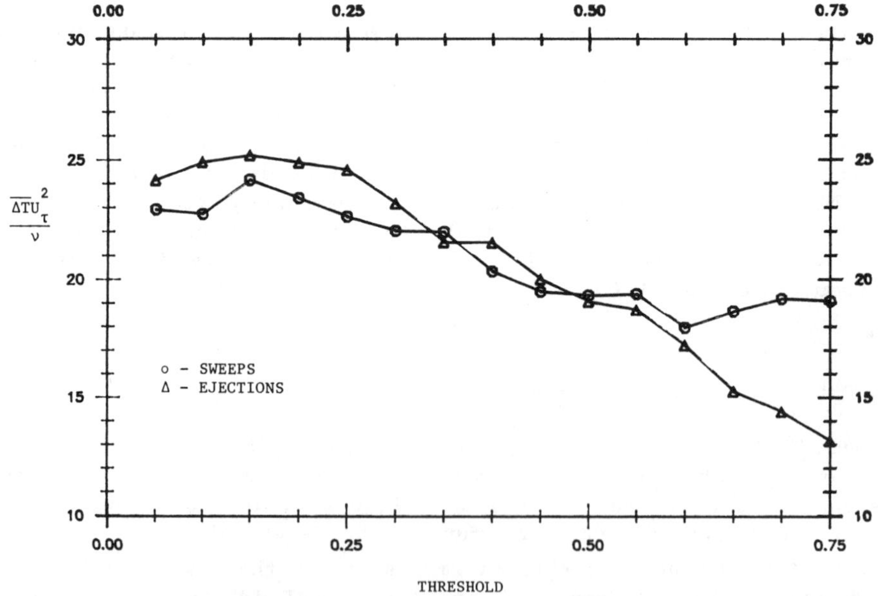

FIGURE 9. The duration of sweeps and the duration of ejections found using TERA, as a function of the detector threshold. As in figure 7 the results have been averaged over all window sizes between $t^+ = 7.2$ and 24.

where used. Thus, we sought determination of the optimum threshold for one-to-one identification. This was accomplished by looking at plots of the visual identification of sweeps and ejections and TERA for a range of thresholds and window sizes. For windows above $7t^+$, window size did not change any of the detection statistics for a given threshold.

Fig. 10 shows a sample of the data used. We also examined the detector kernel $u'\partial u'/\partial t$, as well as ω_z, u', v', and $u'v'$ for trends. We were not able to visually correlate trends in any of these signals with changes in the accuracy of the one-to-one identification. However, this examination has suggested that a threshold of $0.25*\mathrm{rms}(u'\partial u'/\partial t)$ gives the best one-to-one correspondence at $R_\theta = 680$.

For this threshold, the average time between sweeps non-dimensionalized by wall units is 34 t^+ and the time between ejections is 34.5 t^+. It should be remembered that the window size of 6 t^+ imposes a grouping on these events. For comparison, a visual counting of the occurrence of those sweeps which result from pockets resulted in $t^+ = 37$. The VITA technique results in $t^+ \cong 250$ when conditioned to detect only ejections (see Blackwelder and Haritonidis 1983). Luchik and Tiederman (1987) used a specific grouping algorithm on data from a channel flow to group ejections into bursts. Three different techniques, VITA, a modified u-level, and the $u'v'$ quadrant technique, all yielded an average time between bursts of 90 t^+.

4. DISCUSSION

Recently Bogard and Tiederman (1986) and Luchik and Tiederman (1987) have shown that by grouping ejections into bursts, the threshold dependence of the quadrant technique, VITA and the modified u-level technique all become weaker. As noted above, all three can be made to give similar results for the time between bursts. The idea of grouping is appealing. Studies (for example, Offen and Kline 1975) have shown that bursts come in groups, but it is also true that there are combinations of ejections and sweeps that can be considered to form a burst (Falco 1987, Chu and Falco 1988). When one examines visual information, whether it is side view dye lifting or simultaneous plan and side view marker movements, we are, in fact, examining changes in concentration gradients. The gradients have a history effect (i.e. once regions of high or low concentration are formed, they remain so, except for diffusion, until moved by some new motion) which can leave the appearance of the formation of a group of ejections. On the other hand simultaneous visual and hot-wire experiments and recent computer simulations have shown that often a number of ejections and a sweep are related (Corino and Brodkey (1969), Offen and Kline (1974), Chu and Falco 1988). Thus, the physical process of bursting does involve a grouping of ejections <u>and</u> sweeps. Therefore, an understanding of the relationship and phasing would further enable a detector to be honed to detect the bursting process.

Zaric' (1982), Zaric' et al (1984) realized that the phenomena of

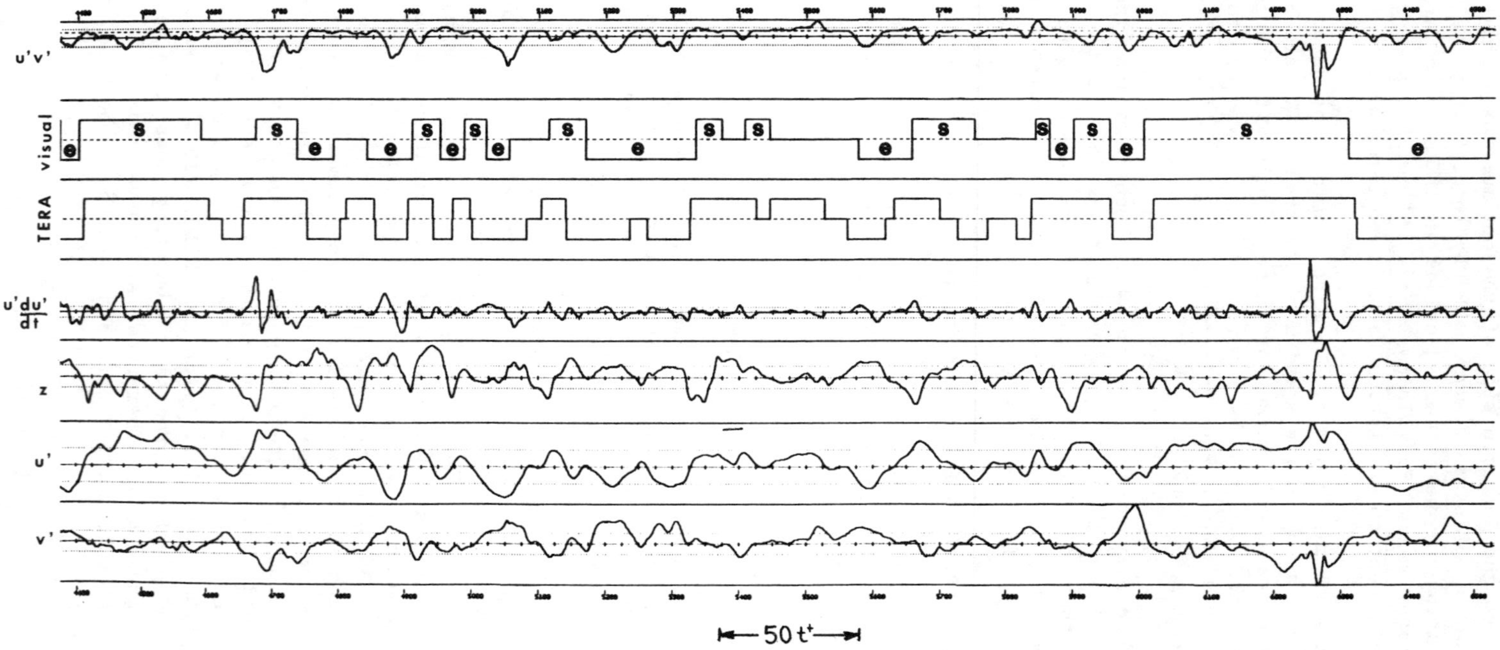

FIGURE 10. Turbulence signals at $y^+ = 15$ along with visual identifications and TERA detections. $u'v'$, ω_z, $u'\partial u'/\partial t$, u' and v' are shown. The threshold for TERA is .25 and the window size is $6t^+$. A value of the intermittency function of +1 corresponds to detection of a sweep, zero to detection of a quiet period, and -1 to detection of an ejection.

grouping can be explicitly acknowledged by using u', $\partial u'/\partial t$ and $u'\partial u'/\partial t$ in combination to determine the end of an event, because each variable responds differently to the physical processes in the burst. The simplified detector, TERA, takes a more global view by smoothing over the changes in energy level that occur in the bursting process. Since the number of detections does not change in the range of window size from 7 t^+ to 24 t^+, grouping based on this characteristic alone appears adequate. For comparison, the grouping technique of Bogard and Tiederman (1986), which was established from the probability distribution of ejection statistics, resulted in a grouping of all ejections separated by less than approximately $t^+ = 50$.

5. CONCLUSIONS

The turbulence detection technique of Zoran Zaric' which utilizes u', $\partial u'/\partial t$, and $u'\partial u'/\partial t$, along with a threshold and minimum cutoff parameter was shown to be capable of highly accurate one-to-one detection of ejections, sweeps and quiet periods. It furthermore has the inherent capability of determining the length and nature of these events.

A simplified version of the turbulence detection technique of Zaric', called TERA, which uses a threshold that is a constant fraction of the $rms(u'\partial u'/\partial t)$ has been shown to:

a) Be capable of one-to-one identifications of ejections, sweeps and quiet periods with as high a degree of accuracy as determined by comparison with visual identifications.

b) Have a weak dependency on threshold for thresholds between 0.1 and $0.4*rms(u'\partial u'/\partial t)$.

c) Be essentially independent of the interval used to calculate a moving average of $|u'\partial u'/\partial t|$ for intervals greater than 5 t^+.

6. ACKNOWLEDGEMENTS

The idea for a study of the detailed one-to-one correspondence of the detection technique with simultaneous visual and hot-wire data originated from a visit by Zoran Zaric' to Michigan State University during November 1984. This paper is dedicated to the memory of the excitement he generated by his keen desire to uncover the facts. The authors want to acknowledge the support of the Air Force Office of Scientific Research under Grant AFOSR 87-0047, monitored by Dr. Jim McMichael. The data at $y^+ = 4.5$ was kindly supplied by Joe Klewicki, to whom we owe thanks for help in numerous ways. We also want to thank Dr. Fakhri Hamady for making the visual identifications.

7. REFERENCES

Blackwelder, R. F. and Haritonidis, J. H. 1983: The bursting frequency in turbulent boundary layers. J. Fluid Mech. 132, 87.

Blackwelder, R. F. and Kaplan, R. E. 1976: On the bursting phenomena near the wall in bounded turbulent shear flows. J. Fluid Mech. 76, 89.

Bogard D. G. and Tiederman W. G. 1986: Burst detection with single-point velocity measurements. J. Fluid Mech. 162, 389.

Brodkey, R. S. Wallace, J. M. and Eckelmann, H. 1974: Some properties of truncated turbulence signals in bounded shear flows. J. Fluid Mech. 63, 209.

Chu C. C. and Falco, R. E. 1988: Vortex ring/viscous wall layer interaction model of the turbulence production process near walls. Exp. in Fluids 6, 305.

Corino E. R. and Brodkey, R. S. 1969: A visual study of turbulent shear flow. J. Fluid Mech. 37, 1.

Falco, R. E. 1977: Coherent motions in the outer part of turbulent boundary layers. Phy Fluids 20, pp S124-S132.

Falco, R. E. 1978: The role of outer flow coherent motions in the production of turbulence near a wall. In: Coherent Structure of Turbulent Boundary Layers. (ed. Smith, C. R. and Abbott, D. E.). pp. 448-461. AFOSR/Lehigh.

Falco, R. E. 1979: Structural aspects of turbulence in boundary layer flows. Proc. 6th Biennal Symposium on Turbulence, Rolla.

Falco, R.E. 1980a: The production of turbulence near a wall. AIAA Paper No. 80-1356.

Falco, R. E. 1980b: Combined simultaneous flow visualization/hot-wire anemometry for the study of turbulent flows. J. of Fluids Engr. 102, pp. 174-183.

Falco, R.E. 1982: A synthesis and model of wall region turbulence structure. In: Structure of Turbulence in Heat and Mass Transfer. (ed. Zaric', Z.). pp. 124-135. Hemisphere Press.

Falco, R. E. 1983: New results, a review and synthesis of the mechanism of turbulence production in boundary layers and its modification. AIAA Paper 83-0377.

Falco, R. E. 1987: Visualization of inner/outer layer interactions producing turbulence near walls. Flow Visualization IV, ed. C. Veret, Hemisphere, 327.

Grass, A. J. 1971: Structural features of turbulent flow over smooth and rough boundaries. J. Fluid Mech., 50, 233.

Kline, S. J. Reynolds, W. C., Schraub, F. A. and Runstadler, P. W. 1967: The structure of turbulent boundary layers. J. Fluid Mech.

Kline, S. J. 1978: Coherent Structures in Turbulent Boundary Layers. In: Coherent Structure of Turbulent Boundary Layers. (ed. Smith, C. R. and Abbott, D. E.). pp. 448-461. AFOSR/Lehigh.

Lovett, J. A. 1982: The flow fields responsible for the generation of turbulence near the wall in turbulent shear flows. MS Thesis. Michigan State Univ.

Lu, S. S. and Willmarth, W. W. 1973: Measurement of structure of the Reynolds stress in a turbulent boundary layer. J. Fluid Mech., 60, 481.

Luchik, T. S. and Tiederman, W. G. 1987: Timescale and structure of ejections and bursts in turbulent channel flows. J. Fluid Mech. 174, 529.

Oldaker, Tiederman, W. G. 1977: Spatial structure of the viscous sublayer in drag reducing channel flows. Phy Fluids 20, pp S124-S132.

Wallace, J. M. Brodkey, R. S. and Eckelmann, H. 1977: Pattern recognized structures in bounded turbulent shear flows. J. Fluid Mech. 142, 121.

Willmarth, W. W. and Lu, S. S. 1972: Structure of the Reynolds stress near the wall. J. Fluid Mech. 55, 65.

Zaric', Z. 1975: Wall turbulence structure and convective heat transfer. Int. J. Heat and Mass Transfer. 18, pp. 831-842.

Zaric', Z. 1982: Conditionally averaged patterns of coherent events in a wall bounded turbulent flow. In: Structure of Turbulence in Heat and Mass Transfer. (ed. Zaric', Z.). pp. 7-28. Hemisphere Press.

Zaric', Z., Falco, R. E. and Blackwelder, R. F. 1984: Detection of coherent structures in visual and multiple hot-wire data in boundary layers. Turbulence and Chaotic Phenomena in Fluids. Elsevier, 439.

The Transport of Enstrophy in a Turbulent Boundary Layer

J.-L. BALINT, P. VUKOSLAVČEVIĆ,*
and J. M. WALLACE
The University of Maryland
College Park, Maryland 20742, USA

1. INTRODUCTION

Research carried out over the last thirty years has revealed that ordered, coherent, vortical structures in the wall region of bounded turbulent shear flows are largely responsible for the generation of Reynolds stress. In order to successfully predict these turbulent flows, this descriptive knowledge of Reynolds stress generation must be translated into a mathematical model which captures the dynamics of the process. Before this can take place, however, a complete physical model must be developed which quantitatively characterizes the various parts of the process.

To obtain such quantitative information about the vortical processes generating Reynolds stress requires an ability to accurately measure the vorticity vector field. Such measurements have proven extremely difficult because of the requirements of spacial resolution and accuracy in estimating velocity differences. Thus vorticity measurements have long been an elusive goal of experimental research. Recently, relatively successful laboratory efforts have been made to obtain one or more of the vorticity components, and the entire vector field for some low Reynolds number flows has been obtained by direct numerical simulation.

We have developed a miniature, nine-sensor, hot-wire probe which simultaneously measures the three components of the velocity vector and all nine components of the velocity gradient fields. With this probe, therefore, the vorticity vector and the strain rate tensor can also be determined. We have used this probe to examine the kinematics and dynamics of Reynolds stress generation in a turbulent boundary layer.

In this paper we will describe measurements of the values of the terms in the time averaged transport equations for instantaneous, mean, and fluctuating enstrophy (mean square vorticity). The balance of the terms in these equations is analogous to the balance of terms in the transport equations for kinetic energy which has been known since the classical experiments in the fifties of Klebanoff (1954), Laufer (1954), and Townsend (1950). The values of the terms in the enstrophy balance have not previously been known because of the measurement difficulties described above; they are presented here for the first time. Knowledge

---

Veljko Vlahović University, 81000 Titograd, Yugoslavia.

of this balance sheds light on the vorticity dynamics which govern heat, mass and momomentum transport in the boundary layer which was of great interest to Zoran Zarić for whom this seminar is a memorial.

Additionally, some of the most important terms in these equations have also been conditionally evaluated for time periods in the data set when the Reynolds stress signal in the second quadrant of the hodograph plane is above a preset threshold. This criterion has been shown by Bogard and Tiedermann (1986) to correspond well with visually observed "ejections" of low momentum fluid away from the wall. This type of examination of the enstrophy transport during such occurances reveals some distinctive features of the vorticity dynamics related to the "ejection" process.

2. DESCRIPTION OF FLOW AND INSTRUMENTATION

2.1 The Turbulent Boundary Layer

The data were obtained in a nominally zero pressure gradient turbulent boundary layer which developed over 8 m after being tripped with a 5 mm diameter wire. The velocity and vorticity vector components were determined at locations in the buffer layer, the logarithmic region and the velocity defect region. The principal characteristics of the boundary layer investigated are:

U_∞(m/s)	δ(m)	θ(m)	Re_θ	δ^*	H	Re_x	u_τ/U_∞
2.39	0.1397	0.0136	2100	0.0175	1.28	1.3×10^6	0.042

where δ is the boundary layer thickness, θ is the momentum thickness, Re_θ is the Reynolds number based on θ, δ^* is the displacement thickness, $H=\delta^*/\theta$ is the shape factor, Re_x is the Reynolds number based on the streamwise measurement location x, and u_τ is the friction velocity.

The velocity and vorticity statistics, up to the 4th moment of the fluctations, have previously been compared by Balint, Vukoslavĉević, and Wallace (1987) with published experimental and large eddy simulation results; the agreement was satisfactory. The most important features of the vorticity field found in that study are that: (1) most of the vorticity fluctuation activity is concentrated, as expected, near the wall ($y/\delta<0.1$), and (2) in the buffer layer the measurements indicate a transfer, via rotation, of spanwise into streamwise vorticity, consistent with the hairpin vortex model of the wall layer suggested by various authors and summarized by Wallace (1982). These results have encouraged us to use the same set of data for further analysis of the enstrophy transport balance.

2.2 The Nine-Sensor Probe

The design and construction of the probe was initiated by Wassman and Wallace (1979) and greatly improved over the past few years by Balint et al. (1987). The probe arrangement, sketched and photographed in Figure 1, consists of three arrays of 2.5 μm diameter hot-wire sensors oriented at 45^0 to the mean flow. Each wire is about 0.85 mm long giving a

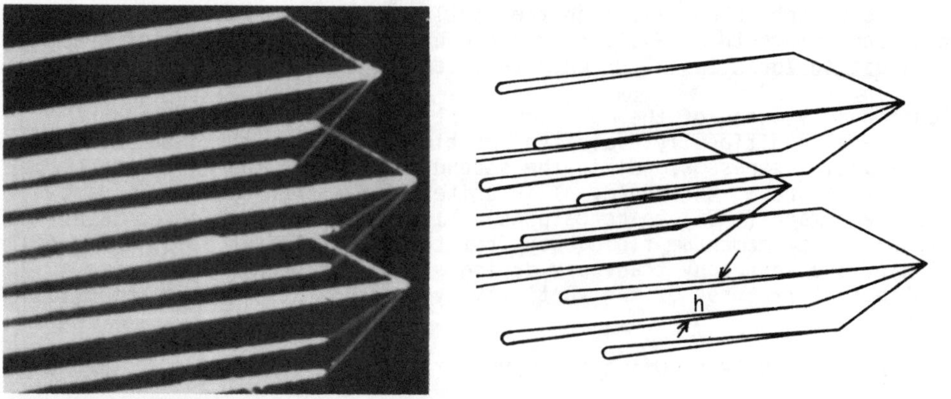

Fig. 1(a) Side view photograph and sketch of the probe.

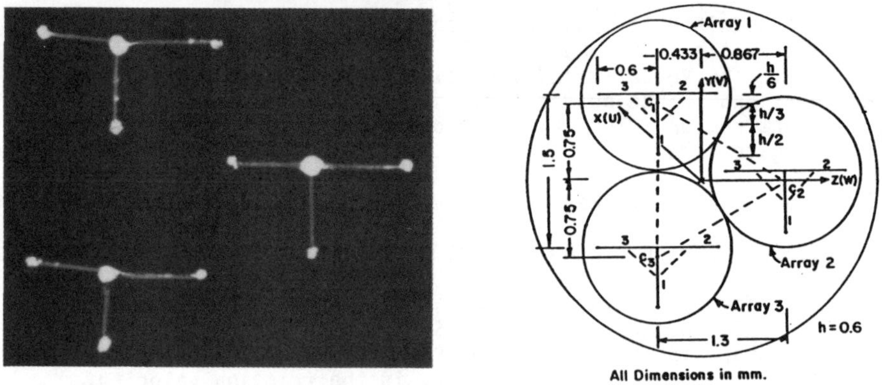

Fig. 1(b) End view photograph and sketch of the probe.

length to diameter ratio of 340. The distance between the supporting prongs is about 0.6 mm, and the distance between array centers is about 1.3-1.5 mm, thus giving an effective sensing diameter of about 2.5 mm. These dimensions give a spatial resolution of about 9.5 viscous lengths or 9.7 Kolmogorov scales at y^+=14 in the boundary layer described above.

2.3 Probe Operation

The nonlinear algebraic form taken by the expressions for the squares of the effective cooling velocities acting at the centers of each of the wires of each array in terms of the three velocity components at the centroid of the arrays, expanded in a Taylor series to first order, is given in equation (1). The array number is indicated by the subscript i and the wire number by the subscript j.

$$U_{e_{ij}}^2 = [K_{ij1}(U_0 + C_{ij1}\frac{\partial U}{\partial z} + C_{ij2}\frac{\partial U}{\partial y}) + K_{ij2}(v_0 + C_{ij3}\frac{\partial v}{\partial z} + C_{ij4}\frac{\partial v}{\partial y})]^2 + [K_{ij3}(w_0 + C_{ij5}\frac{\partial w}{\partial z} + C_{ij6}\frac{\partial w}{\partial y})]^2 \,. \quad (1)$$

The Kijks are calibration coefficients and the Cijks are constant coefficients for a given array geometry which are positive or negative fractions of h (see Figure 1). U_0, v_0 and w_0 are the unknown velocity components at the centroid of the probe, $\partial U/\partial y$, $\partial U/\partial z$, $\partial v/\partial y$, $\partial v/\partial z$, $\partial w/\partial y$ and $\partial w/\partial z$ are the velocity gradients in the plane transverse to the axis of the probe, and h is the average of the projected wire lengths on this plane. Equation (1) expresses the cooling of the vertical wires of each array. The expressions for the horizontal wires require simply exchanging the variables v and w in the equation. The effective cooling velocities are related to the measured bridge voltages through calibration relations of the King's Law form:

$$E_{ij}^2 = A_{ij} + B_{ij} U_{e_{ij}}^n \, . \tag{2}$$

The calibration coefficients were found by pitching and yawing the probe in the potential flow core of the tunnel at mean speeds corresponding to boundary layer conditions. Details of this proceedure are described in Balint (1986). With the calibration coefficients determined, the nine unknowns must be found by the solution of these nine equations. This is accomplished by the Newton-Raphson iterative procedure which is quite convergent, except for instances where the angle of attack of the flow to the probe, given by the ratios v/U and w/U, is greater than a value predetermined from the probe geometry.

One of the three streamwise gradients can be obtained directly from the continuity equation, and the other two are estimated using Taylor's hypothesis, $\partial/\partial x = -1/\overline{U}\ \partial/\partial t$. Comparison, by Piomelli (1988), of rms and correlation values of streamwise gradients obtained using Taylor's hypothesis with those obtained by spacial differencing of data from Piomelli, et al.'s (1988) large eddy simulation of the turbulent channel flow indicates the adequacy of the hypothesis for the large scales in this shear flow, at least for $y^+>30$.

The anemometer circuits used in this work are based on the design of Kaplan; their performance characteristics have been analysed by Weidman and Browand (1975). The frequency response of these circuits was flat up to about 4000 Hz, well above the largest frequencies with any significant energy in this relatively low Reynolds number flow. The signals were sampled at the Nyquist digitizing frequency of 500 Hz for 3 mn.

3. ENSTROPHY TRANSPORT EQUATIONS

Enstrophy is a scalar measure of the magnitude of vortical activity. We obtain transport equations for enstrophy by beginning with the equation of motion in vorticity form,

$$\frac{\partial \Omega_i}{\partial t} + U_j \frac{\partial \Omega_i}{\partial x_j} = \Omega_j \frac{\partial U_i}{\partial x_j} + \nu \frac{\partial^2 \Omega_i}{\partial x_j \partial x_j} \, , \tag{3}$$

which is the curl of the Navier-Stokes equation. U_i and Ω_i are the ith components of instantaneous velocity and vorticity respectively. Repeating indices imply summation over the values of the index. A transport equation for instantaneous enstrophy is obtained by multiplying equation (3), term by term, by Ω_i yielding

$$\frac{\partial(\frac{1}{2}\Omega_i\Omega_i)}{\partial t} + U_j \frac{\partial(\frac{1}{2}\Omega_i\Omega_i)}{\partial x_j} = \Omega_i\Omega_j \frac{\partial U_i}{\partial x_j} + \nu\Omega_i \frac{\partial^2\Omega_i}{\partial x_j \partial x_j} \quad . \tag{4}$$

We may divide the last term in this equation into diffusive and dissipative parts and take the mean to obtain, for steady flow,

$$\underset{(1)}{\overline{U_j \frac{\partial}{\partial x_j}(\frac{1}{2}\Omega_i\Omega_i)}} = \underset{(2)}{\overline{\Omega_i\Omega_j \frac{\partial U_i}{\partial x_j}}} + \nu \underset{(3)}{\overline{\frac{\partial^2(\frac{1}{2}\Omega_i\Omega_i)}{\partial x_j \partial x_j}}} - \nu\underset{(4)}{\overline{(\frac{\partial\Omega_i}{\partial x_j}\frac{\partial\Omega_i}{\partial x_j})}} \tag{5}$$

which expresses the averaged instantaneous enstrophy transport. The terms in this equation may be interpreted as:

- (5-1) the average advection of instantaneous enstrophy,
- (5-2) the average rotation and stretching/compression of instantaneous enstrophy by the velocity gradient field,
- (5-3) the average viscous diffusion of instantaneous enstrophy, and
- (5-4) the average viscous dissipation of instantaneous enstrophy.

We can likewise obtain an averaged transport equation for the mean enstrophy by following the same procedure but beginning with the transport equation for mean vorticity which is multiplied by $\overline{\Omega}_i$; this results in

$$\underset{(1)}{\overline{U}_j \frac{\partial}{\partial x_j}(\frac{1}{2}\overline{\Omega}_i\overline{\Omega}_i)} + \underset{(2)}{\overline{\Omega}_i \overline{u_j \frac{\partial\omega_i}{\partial x_j}}} = \underset{(3)}{\overline{\Omega}_i\overline{\Omega}_j \frac{\partial\overline{U}_i}{\partial x_j}} + \underset{(4)}{\overline{\Omega}_i \overline{\omega_j \frac{\partial u_i}{\partial x_j}}}$$

$$+ \nu \underset{(5)}{\frac{\partial^2(\frac{1}{2}\overline{\Omega}_i\overline{\Omega}_i)}{\partial x_j \partial x_j}} - \nu \underset{(6)}{(\frac{\partial\overline{\Omega}_i}{\partial x_j})(\frac{\partial\overline{\Omega}_i}{\partial x_j})} \quad . \tag{6}$$

Equations (5) and (6) are different because, in formulating equation (5), no Reynolds decomposition of the variables into mean and flucutuating parts has taken place; prior to formulating equation (6) this decomposition occured. Terms (6-1) and (6-2) in equation (6) are the two mean parts of the advection term (5-1) in equation (5). Terms (6-3) and (6-4) are the two mean parts of the rotation stretching/compression term (5-2). Term (6-5) is the mean part of the viscous diffusion term (5-3) and term (6-6) is the mean part of the viscous dissipation term (5-4).

We can further obtain an averaged transport equation for the fluctuating enstrophy by decomposing the velocity and vorticity fields of equation (5) into mean and fluctuating parts

$$U_j = \overline{U}_j + u_j \tag{7}$$

$$\Omega_i = \bar{\Omega}_i + \omega_i , \tag{8}$$

and subtracting equation (6) from equation (5) yielding

$$\underset{(1)}{\bar{U}_j \frac{\partial}{\partial x_j} (\tfrac{1}{2}\overline{\omega_i \omega_i})} + \underset{(2)}{\overline{u_j \frac{\partial}{\partial x_j} (\tfrac{1}{2}\omega_i \omega_i)}} + \underset{(3)}{\overline{u_j \omega_i} \frac{\partial \bar{\Omega}_i}{\partial x_j}} =$$

$$\underset{(4)}{\overline{\omega_i \omega_j} \frac{\partial \bar{U}_i}{\partial x_j}} + \underset{(5)}{\overline{\omega_i \omega_j \frac{\partial u_i}{\partial x_j}}} + \underset{(6)}{\overline{\bar{\Omega}_j \omega_i \frac{\partial u_i}{\partial x_j}}}$$

$$+ \underset{(7)}{\nu \frac{\partial^2 (\frac{1}{2} \overline{\omega_i \omega_i})}{\partial x_j \partial x_j}} - \underset{(8)}{\nu \overline{(\frac{\partial \omega_i}{\partial x_j})(\frac{\partial \omega_i}{\partial x_j})}} . \tag{9}$$

The first three terms, (9-1), (9-2), and (9-3), in this equation are the fluctuating parts of the advection term (5-1) in equation (5). The next three terms, (9-4), (9-5), and (9-6), are the fluctuating parts of the rotating and stretching/compression term (5-2). Term (9-7) is the fluctuating part of the viscous diffusion term (5-3), while term (9-8) is the fluctuating part of the viscous dissipation term (5-4).

Equations (6) and (9) can alternatively be expressed using the mean and fluctuating strain rates,

$$\bar{S}_{ij} = \frac{1}{2} (\frac{\partial \bar{U}_i}{\partial x_j} + \frac{\partial \bar{U}_j}{\partial x_i}) \tag{10}$$

and

$$s_{ij} = \frac{1}{2} (\frac{\partial u_i}{\partial x_j} + \frac{\partial u_j}{\partial x_i}) , \tag{11}$$

then appearing as

$$\underset{(1)}{\bar{U}_j \frac{\partial}{\partial x_j} (\tfrac{1}{2} \bar{\Omega}_i \bar{\Omega}_i)} = - \underset{(2)}{\frac{\partial}{\partial x_j}(\bar{\Omega}_i \overline{\omega_i u_j})} + \underset{(3)}{\overline{u_j \omega_i} \frac{\partial \bar{\Omega}_i}{\partial x_j}} + \underset{(4)}{\bar{\Omega}_i \bar{\Omega}_j \bar{S}_{ij}} + \underset{(5)}{\bar{\Omega}_i \overline{\omega_j s_{ij}}}$$

$$+ \underset{(6)}{\nu \frac{\partial^2 (\frac{1}{2} \bar{\Omega}_i \bar{\Omega}_i)}{\partial x_j \partial x_j}} - \underset{(7)}{\nu (\frac{\partial \bar{\Omega}_i}{\partial x_j} \frac{\partial \bar{\Omega}_i}{\partial x_j})} \tag{12}$$

and

$$\overline{U}_j \frac{\partial}{\partial x_j} (\frac{1}{2}\overline{\omega_i \omega_i}) = - \overline{u_j \omega_i} \frac{\partial \overline{\Omega}_i}{\partial x_j} - \frac{1}{2} \frac{\partial}{\partial x_j} (\overline{u_j \omega_i \omega_i}) + \overline{\omega_i \omega_j s_{ij}}$$

(1) (2) (3) (4)

$$+ \overline{\omega_i \omega_j}\, \overline{S}_{ij} + \overline{\Omega}_j \overline{\omega_i s_{ij}} + \nu \frac{\partial^2}{\partial x_j \partial x_j} (\frac{1}{2} \overline{\omega_i \omega_i})$$

(5) (6) (7)

$$- \nu \overline{(\frac{\partial \omega_i}{\partial x_j} \frac{\partial \omega_i}{\partial x_j})} \, . \qquad (13)$$

(8)

Individual terms in equation (12) are interpreted by Tennekes and Lumley (1972) as:

(12-1) advection of mean enstrophy,

(12-2) transport of mean enstrophy by turbulent vorticity-velocity interactions (analogous to the turbulent term of the mean kinetic energy equation),

(12-3) gradient production of turbulent enstrophy (analogous to the turbulence production term in the energy equation and occuring in equation (13) as term (13-2) with opposite sign),

(12-4) stretching or shrinking of the mean vorticity by the mean strainrate,

(12-5) amplification or attenuation of the mean enstrophy by the stretching of fluctuating vorticity components by fluctuating strainrates,

(12-6) viscous diffusion of mean enstrophy, and

(12-7) viscous dissipation of mean enstrophy.

They interpret the terms in equation (13) for the transport of fluctuating enstrophy as:

(13-1) advection of fluctuating enstrophy,

(13-2) gradient production of fluctuating enstrophy exchanging vorticity between fluctuating and mean enstrophy (analogous to kinetic energy production),

(13-3) transport of fluctuating enstrophy by velocity fluctuations (analogous to the transport of turbulent kinetic energy by the fluctuating velocity field),

(13-4) production of fluctuating enstrophy through stretching of fluctuating vorticity by the fluctuating strainrate,

(13-5) production/diminution of fluctuating vorticity by stretching or shrinking of vorticity fluctuations by the mean strainrate,

(13-6) mixed production which occurs in equation (12) as term (12-5) with the same sign,

(13-7) viscous diffusion of fluctuating enstrophy, and

(13-8) viscous dissipation of fluctuating enstrophy.

It is easy to show the equality of the terms in equations (12) and (13) to those in equations (8) and (9). For example,

$$- \frac{\partial}{\partial x_j} (\bar{\Omega}_i \overline{\omega_i u_j}) = - \bar{\Omega}_i \overline{u_j \frac{\partial \omega_i}{\partial x_j}} - \overline{\omega_i u_j} \frac{\partial \bar{\Omega}_i}{\partial x_j} \tag{14}$$

shows that the divergence on the right-hand side of equation (12) can be considered the sum of advection plus gradient production. Likewise, the divergence on the right-hand-side of equation (13) can be alternatively considered an advection term alone as seen in equation (15). Equations (16)-(20) are further identities needed to establish the equality of equations (12) and (13) with (8) and (9).

$$- \frac{1}{2} \frac{\partial}{\partial x_j} (\overline{u_j \omega_i \omega_i}) = - u_j \frac{\partial}{\partial x_j} (\frac{1}{2} \omega_i \omega_i), \tag{15}$$

$$\bar{\Omega}_i \bar{\Omega}_j \bar{S}_{ij} = \bar{\Omega}_i \bar{\Omega}_j \frac{\partial \bar{U}_i}{\partial x_j}, \tag{16}$$

$$\overline{\omega_i \omega_j} \, \bar{S}_{ij} = \overline{\omega_i \omega_j} \frac{\partial \bar{U}_i}{\partial x_j}, \tag{17}$$

$$\overline{\omega_i \omega_j s_{ij}} = \overline{\omega_i \omega_j \frac{\partial u_i}{\partial x_j}}, \tag{18}$$

$$\bar{\Omega}_i \overline{\omega_j s_{ij}} = \bar{\Omega}_i \overline{\omega_j \frac{\partial u_i}{\partial xj}}, \tag{19}$$

and

$$\bar{\Omega}_j \overline{\omega_i s_{ij}} = \bar{\Omega}_j \overline{\omega_i \frac{\partial u_i}{\partial x_j}}. \tag{20}$$

For the case of the steady, mean, zero pressure gradient boundary layer, equations (12) and (13) reduce to

$$0 = -\frac{\partial}{\partial x_2}(\bar{\Omega}_3\overline{\omega_3 u_2}) + \overline{u_2\omega_3}\frac{\partial\bar{\Omega}_3}{\partial x_2} + 0$$

(1) (2) (3) (4)

$$+ \bar{\Omega}_3(\overline{\omega_1 s_{13}} + \overline{\omega_2 s_{23}} + \overline{\omega_3 s_{33}})$$

(5)

$$+ \nu\frac{\partial^2(\frac{1}{2}\bar{\Omega}_3\bar{\Omega}_3)}{\partial x_2\,\partial x_2} - \nu\left(\frac{\partial\bar{\Omega}_3}{\partial x_2}\frac{\partial\bar{\Omega}_3}{\partial x_2}\right) \qquad (21)$$

(6) (7)

and

$$0 = -\overline{u_2\omega_3}\frac{\partial\bar{\Omega}_3}{\partial x_2} - \frac{1}{2}\frac{\partial}{\partial x_2}(\overline{u_2\omega^2})$$

(1) (2) (3)

$$+ (\overline{\omega_1^2 s_{11}} + \overline{\omega_2^2 s_{22}} + \overline{\omega_3^2 s_{33}} + \overline{2\omega_1\omega_2 s_{12}} + \overline{2\omega_1\omega_3 s_{13}} + \overline{2\omega_2\omega_3 s_{23}})$$

(4)

$$+ \overline{\omega_1\omega_2}\frac{\partial\bar{U}_1}{\partial x_2} + \bar{\Omega}_3(\overline{\omega_1 s_{13}} + \overline{\omega_2 s_{23}} + \overline{\omega_3 s_{33}})$$

(5) (6)

$$+ \nu\frac{\partial^2}{\partial x_2\partial x_2}\left(\frac{1}{2}\bar{\omega}^2\right) - \dot{\epsilon}\,, \qquad (22)$$

(7) (8)

where

$$\omega^2 = (\omega_1^2 + \omega_2^2 + \omega_3^2) \qquad (23)$$

and

$$\dot{\epsilon} = \nu\left(\overline{\frac{\partial\omega_1}{\partial x_1}\frac{\partial\omega_1}{\partial x_1}} + 2\,\overline{\frac{\partial\omega_1}{\partial x_2}\frac{\partial\omega_2}{\partial x_1}} + \overline{\frac{\partial\omega_2}{\partial x_2}\frac{\partial\omega_2}{\partial x_2}} + 2\,\overline{\frac{\partial\omega_1}{\partial x_3}\frac{\partial\omega_3}{\partial x_1}}\right.$$

$$\left. + \overline{\frac{\partial\omega_3}{\partial x_3}\frac{\partial\omega_3}{\partial x_3}} + 2\,\overline{\frac{\partial\omega_2}{\partial x_3}\frac{\partial\omega_3}{\partial x_2}}\right) \qquad (24)$$

4. CONDITIONAL ANALYSIS

Where possible, the data obtained in this experiment were conditionally analysed using second quadrant detection with a threshold defined as

$$H = - \langle uv \rangle_2 / u_{rms} v_{rms} . \tag{25}$$

Bogard and Tiedermann (1984) conclude that choosing H=1.07 gives the highest coincidence of visually detected low momentum fluid "ejections", at $y^+=15$, with second quadrant anemometer signal detections. The probability of detecting in the signal the same event as the visual detection was 0.69, with 0.15 for false detection. As seen in equation (25), the second quadrant detection method [introduced by Wallace, et al. (1972) and extended by Lu and Willmarth (1973)] finds contributions to the Reynolds stress of low momentum fluid with motion away from the wall. We chose the same threshold, H=1.07, to conditionally analyse the data in our experiment at $y^+\simeq14$, and also used this same value of H for all the y^+ locations in the boundary layer. Of course, there is no certainty that we continued to detect motions that would correspond to visually observed "ejections" for locations above $y^+=15$.

5. EXPERIMENTAL RESULTS

Before considering the average transport of instantaneous, mean, and fluctuating enstrophy, we should first look at the distribution of the fluctuating enstrophy itself, for locations across the boundary layer. This is shown in Figure 2 where two features are evident: (1) the enstrophy level is high near the wall, decreasing rapidly near the outer part of the boundary layer, and (2) the conditionally averaged enstrophy is signifigantly higher than the unconditioned values in and just above the buffer layer. When the individual components contributing to the mean fluctuating enstrophy are considered, as shown in the lower three sections of the figure, it is seen that the streamwise and spanwise components of the conditional enstrophy are smaller than the unconditional enstrophy for $y^+<25$.

In Figure 3, the distribution of the values of the terms in equation (21) are shown. This is the average transport of mean enstrophy by the various means discussed above. Two features stand out: (1) all of these terms are nearly zero for positions above the buffer layer, and (2) the dominant terms are viscous diffusion and direct viscous dissipation of mean enstrophy. The dashed curve shown in the figure represents the residual unbalance when the values of all the terms are summed at each position. It is evident that, near the wall ($y^+<15$), this residual error is of the order of all the terms there except the diffusion and dissipation terms.

The terms of equation (22), representing the average transport of fluctuating enstrophy, are shown in Figure 4. Here, most of the terms continue to have signifigant values well into the logarithmic region, in contrast to the terms in the mean equation. The dominant mechanism producing fluctuating enstrophy is that of stretching by the mean strainrate and by the turbulent strainrate. These proceses are counteracted by the viscous dissipation of fluctuating enstrophy. The viscous dissipation, given by term (22-8), also includes all the residual measurement error. It had to be determined as the residual unbalance of the equation, because we are unable to directly measure gradients of instantaneous vorticity with the nine-sensor probe.

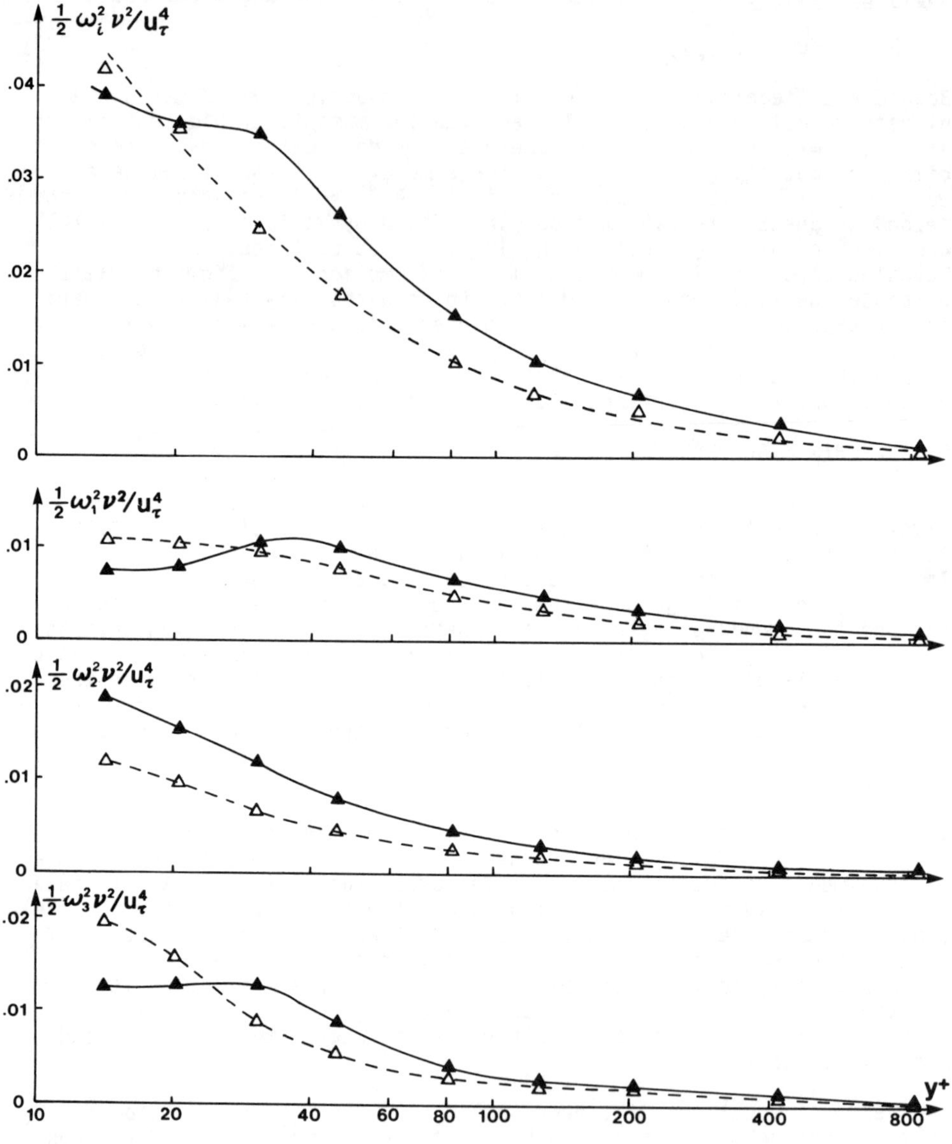

Fig. 2 Normalized enstrophy distribution (- - - - unconditioned ——— conditioned).

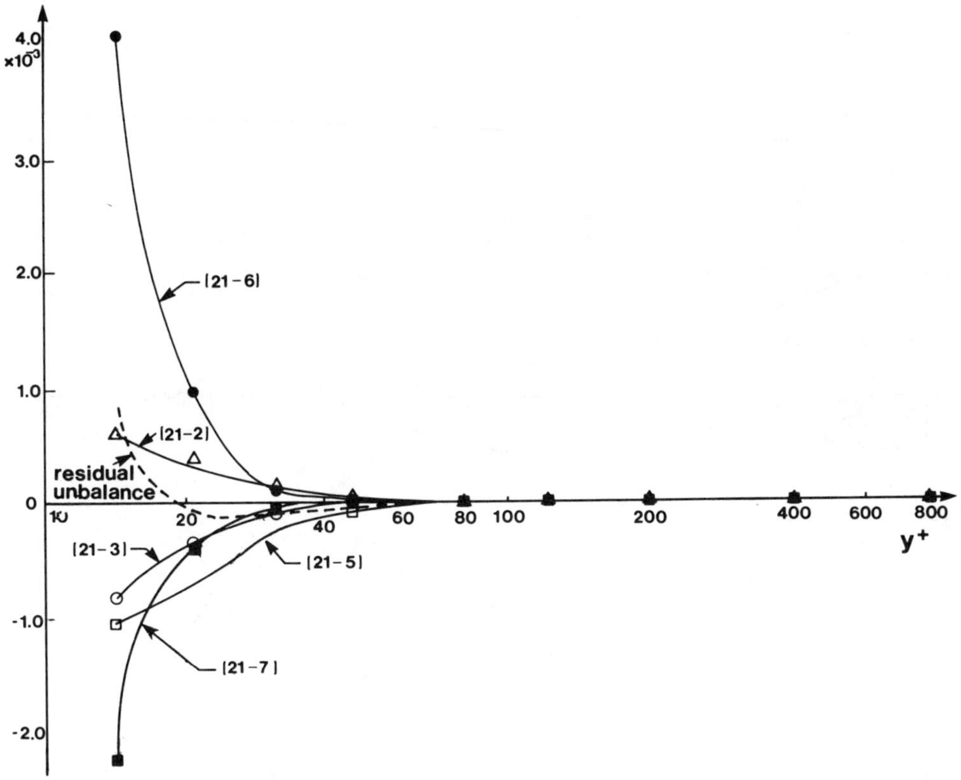

Fig. 3 Distribution of the terms in equation (21) expressing the average transport of mean enstrophy. Terms shown in Figures 3-10 are nondimensionalized by u_τ^6/ν^3.

The values of the terms in equation (22) which could be conditionally analysed, as described in Section 4., are shown in Figure 5. In the buffer layer, the mean gradient production term (22-2), conditionally averaged, becomes negative, in contrast to its unconditional average positive value throughout the boundary layer. Term (22-4), giving the production by stretching of the vorticity fluctuations by the fluctuating strainrate, is almost twice the unconditioned value for $y^+<40$. By contrast, term (22-5), expressing production by mean strainrate stretching of the vorticity fluctuations, is signifigantly diminished for the conditional case, only having about 60% of the unconditioned values for locations close to the wall. Finally, term (22-6), the so-called combined production term, changes sign at the same y^+ location as term (22-2), becoming positive, in contrast to its unconditioned always negative value. It is quite clear from these results that, for positions close to the wall, the transport of fluctuating enstrophy is radically effected when the averaging is done only over time intervals when "ejection" events take place.

Although the decomposition of the flow field into mean and fluctuating parts is necessary for predictive calculations, averaging this decomposed field often obscures and confuses some of the physical meaning of

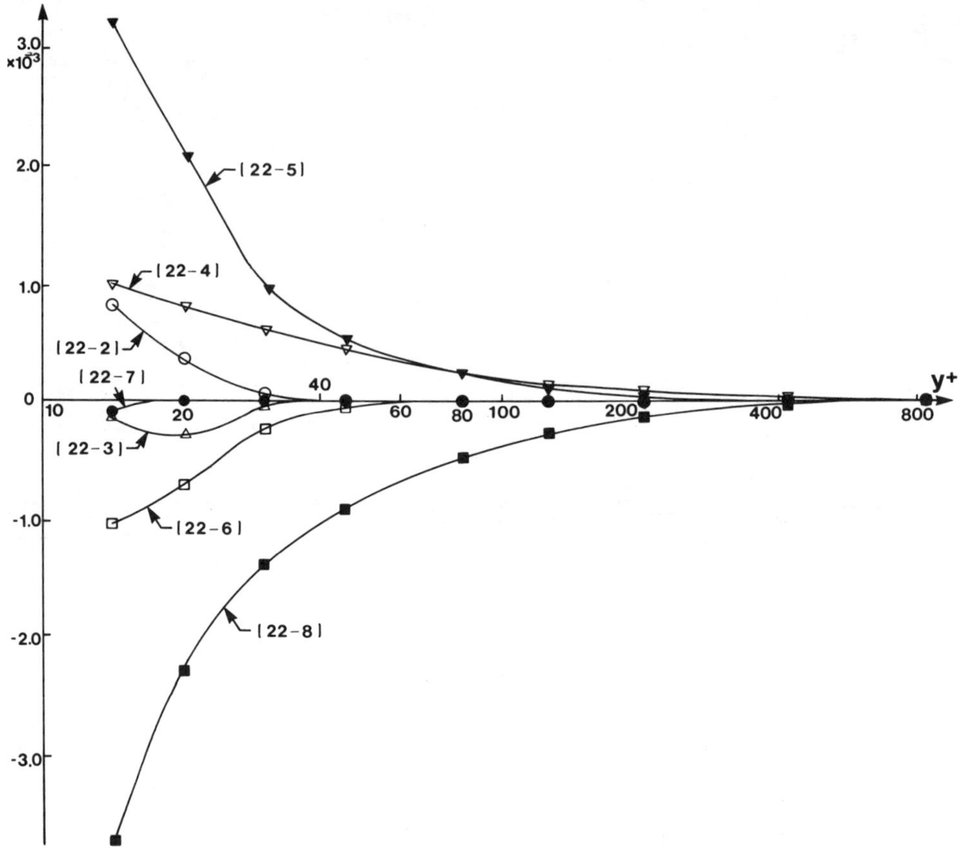

Fig. 4 Distribution of the terms in equation (22) expressing the average transport of fluctuating enstrophy.

the terms in these equations. To interpret the physical processes involved in the vorticity dynamics, it is useful to look at the average transport of instantaneous enstrophy given in equation (5). The distribution across the boundary layer of the values of the four terms in this equation, advection, rotation and stretching/compression, viscous diffusion, and viscous dissipation, are shown in Figure 6. We can see that: (1) enstrophy diffuses into a control volume near the wall at a much greater rate than further out in the boundary layer, (2) it is being created by rotation and stretching of the instantaneous vortex lines at a greater rate there too, (3) only a relatively small portion is due to advection by the instantaneous velocity field, and (4) all of these positive contributions to the rate of change of enstrophy in a control volume at a given location, are balanced by the viscous dissipation of enstrophy, which is very high near the boundary.

When term (5-1) in equation (5) is decomposed into its mean and fluctuating parts, there are five separate parts coming from equations (6)

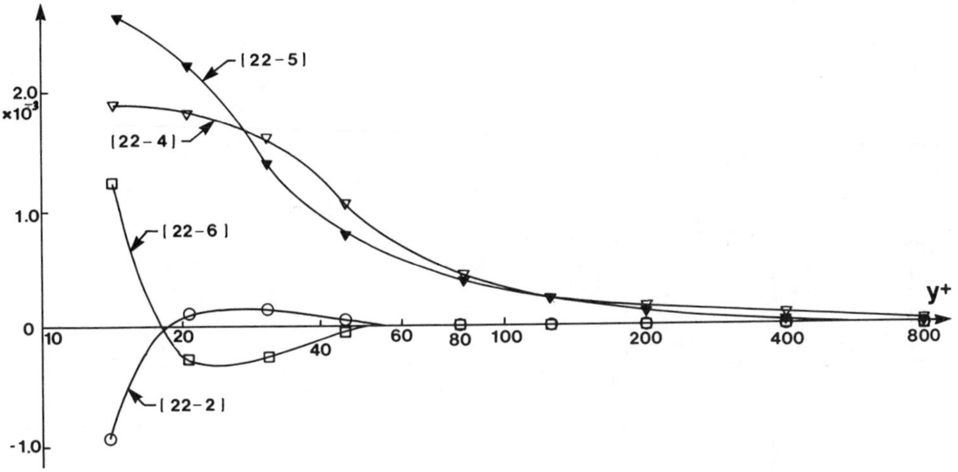

Fig. 5 Distribution of the conditional averages of some terms in equation (22) expressing the average transport of fluctuating enstrophy during "ejections".

and (9) (terms 6-1, 6-2, 9-1, 9-2, and 9-3) as described above. Term 9-1 is negligible for the flow case considered here. Thus Figure 7, shows the contributions to the averaged instantaneous advected enstrophy by these terms from equations (6) and (9), the dominant contribution coming from what we previously described as gradient production.

Likewise, the rotation and stretching/compression term (5-2) in equation (5) is made up of five terms (6-3, 6-4, 9-4, 9-5, and 9-6) from the mean and fluctuating equations. These contributions to the rotation and stretching/compression are shown in Figure 8 (where term 6-3 is zero for this flow). The dominant contribution is the rotation and stretching of the fluctuating vorticity field by the mean velocity gradient field, a somewhat surprising conclusion. The rotation and stretching of the fluctuating vorticity field by the fluctuating velocity gradient field is considerably smaller, as we have previously noted. The mixed gradient production term (6-4 or 9-6) has the average effect of diminishing the enstrophy in the control volume for $y^+ \lesssim 60$.

The average viscous diffusion of instantaneous enstrophy is almost entirely due to diffusion of mean enstrophy as seen Figure 9. The diffusion of fluctuating enstrophy is nearly negligible over the entire region of the boundary layer where measurements could be made.

For $y^+ < 30$, the average viscous dissipation of instantaneous enstrophy is due almost entirely to the dissipation from the fluctuating field. For positions closer to the wall, however, the direct dissipation from the mean vorticity field plays an increasingly important role as seen in Figure 10.

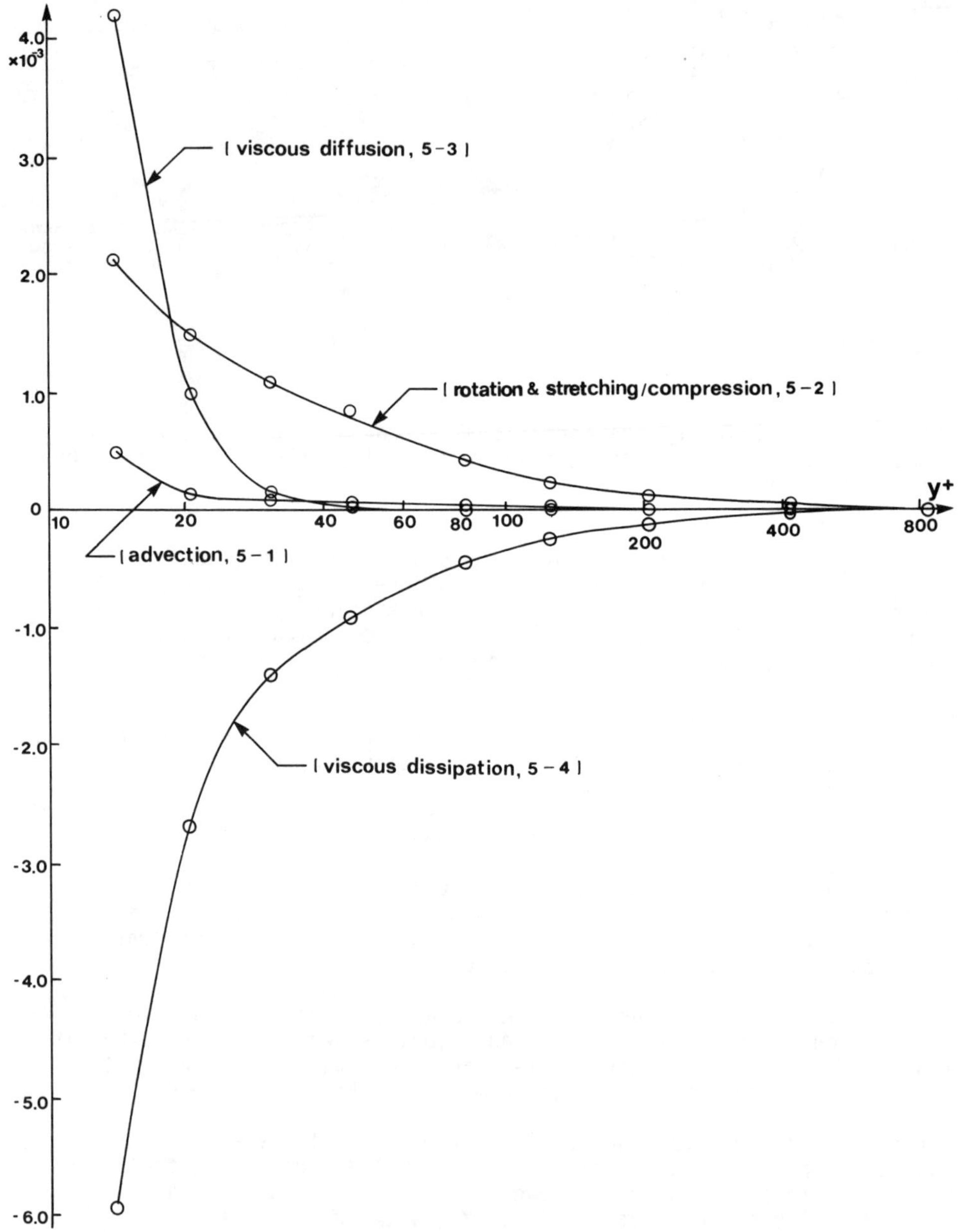

Fig. 6 Distribution of the terms in equation (5) expressing the average transport of instantaneous enstrophy.

Fig. 7 Contributions to the average instantaneous advected enstrophy from the mean and fluctuating parts.

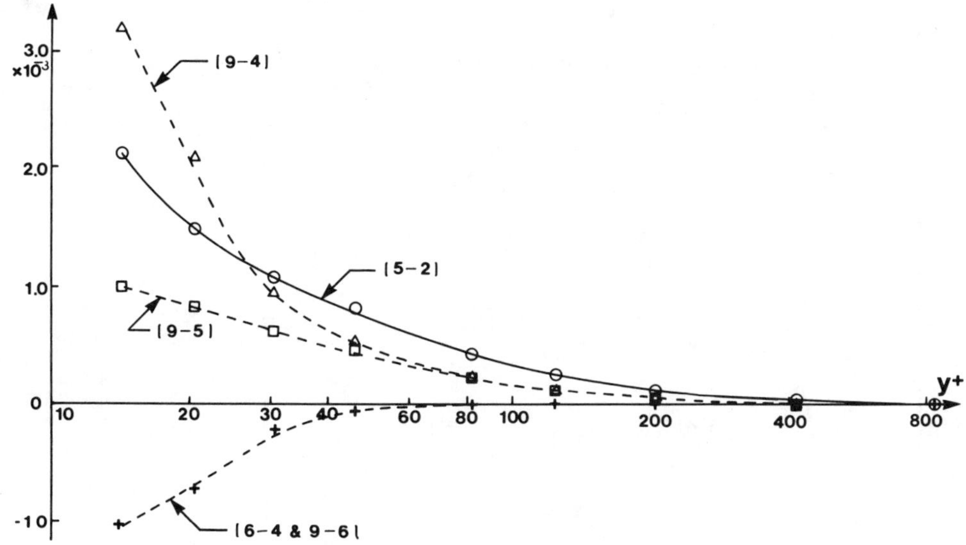

Fig. 8 Contributions to the average production/dimunition of instantaneous enstrophy by rotation and stretching/compression from the mean and fluctuating parts.

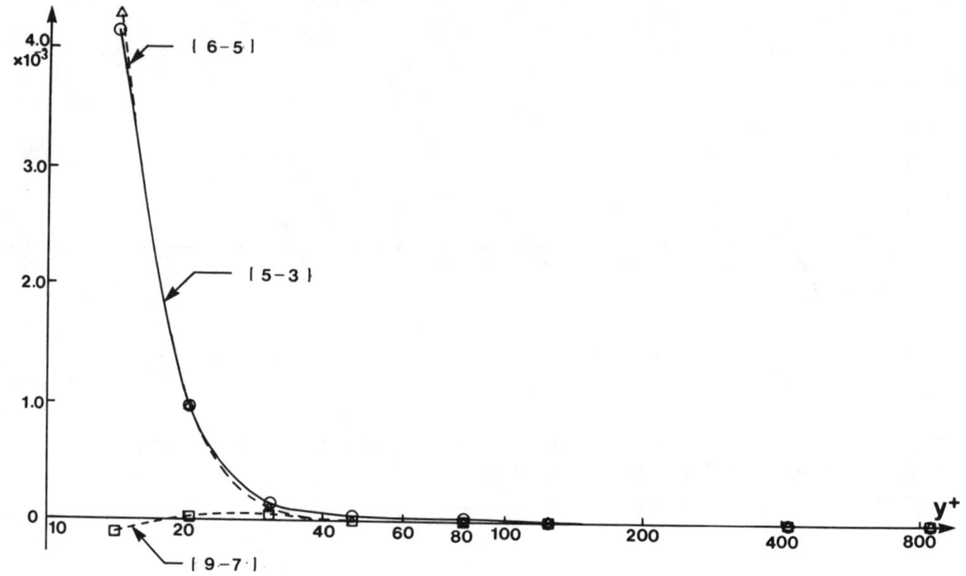

Fig. 9 Contributions to the average instantaneous diffusion of enstrophy from the mean and fluctuating parts.

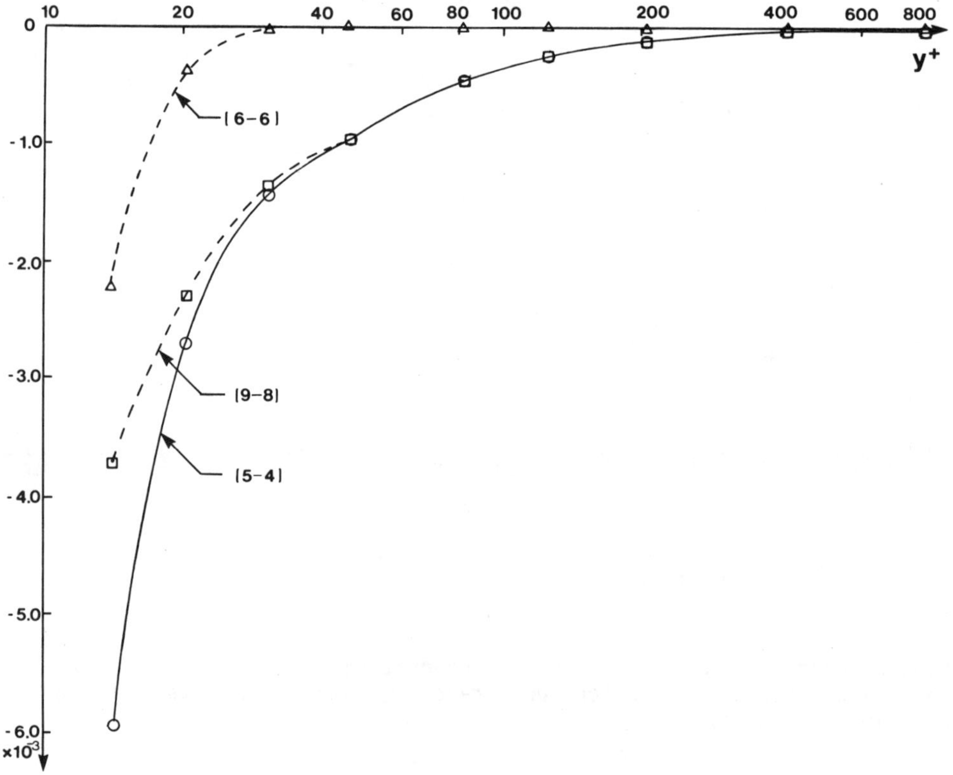

Fig. 10 Contribution to the average instantaneous dissipation of enstrophy from the mean and fluctuating parts.

6. CONCLUSIONS

(1) Conditionally averaged enstrophy is about 40% greater than the unconditioned data for $y^+<20$.

(2) Evaluation of the terms in the averaged transport equation for mean enstrophy shows the viscous diffusion and direct dissipation to be the dominant processes for $y^+<25$. All the terms in this equation have small values above the buffer layer.

(3) Evaluation of the terms in the averaged transport equation for fluctuating enstrophy shows the rotation and stretching term (by mean velocity gradients) and the viscous dissipation term to dominate. The values of these terms plus the rotation and stretching by the fluctuating velocity gradients are significant well into the logarithmic region.

(4) Conditionally averaging some of these terms, based on a criterion related to second quadrant Reynolds stress generation by "ejections", shows the transport of enstrophy to be dramatically different from the unconditioned averaged values of these same terms.

(5) The averaged transport equation for the instantaneous transport of enstrophy reveals more of the physics of the vorticity dynamics than when the equation is decomposed into mean and fluctuating parts.

(6) The average viscous diffusion, rotation and stretching/compression, and advection are balanced by viscous dissipation over the entire boundary layer. Near the wall, viscous diffussion and dissipation are the dominant processes. Above the buffer layer, rotation and stretching/compression replace viscous diffusion as the positive source of enstrophy. The average advection of instantaneous enstrophy is only non-negligible quite near the wall.

7. ACKNOWLEDGEMENT

The authors are grateful for the support of NSF Grant No. MSM-8610369 in carrying out this work.

8. REFERENCES

Balint, J.-L., Vukoslavĉević, P. and Wallace, J.M., Advances in Turbulence, ed. by G. Comte-Bellot and J. Mathieu, Springer Verlag, Berlin, 456, 1987.

Balint, J.-L., Thèse Docteur D'Etat ès Sciences, Univ. of Lyon, France (1986).

Bogard, D.G. and Tiederman, W.G., J. Fl. Mech. 162, 389, 1986.

Klebanoff, P.S., NACA TN 3178, 1954.

Laufer, J., NACA Tech. Rep. 1174, 1954.

Lu, S.S. and Willmarth, W.W., J. Fl. Mech. 60, 481, 1973.

Piomelli, U., Unpublished, 1988.

Piomelli, U., Moin, P. and Ferziger, J.H., Phys. Fluids 31, 1884, 1988.

Tennekes, H. and Lumley, J.L., A First Course in Turbulence, The MIT Press, Cambridge, 1972.

Townsend, A.A., Proc. Cambridge Phil. Soc. 47, 375, 1950.

Wallace, J.M., Developments in Theoretical and Applied Mechanics 11, ed. by T.J. Chung and G.R. Karr, University of Alabama in Huntsville, 509, 1982.

Wallace, J.M., Eckelmann, H. and Brodkey, R.S., J. Fl. Mech. 54, 39, 1972.

Wassman, W.W. amd Wallace, J.M., Bull. Am. Phys. Soc. 24, 1142, 1979.

Weidman, P.D. and Browand, F.K., J. Phys. E, 8, 553, 1975.

Index